Lecture Notes in Artificial Intelligence   9852

Subseries of Lecture Notes in Computer Science

LNAI Series Editors

Randy Goebel
  University of Alberta, Edmonton, Canada
Yuzuru Tanaka
  Hokkaido University, Sapporo, Japan
Wolfgang Wahlster
  DFKI and Saarland University, Saarbrücken, Germany

LNAI Founding Series Editor

Joerg Siekmann
  DFKI and Saarland University, Saarbrücken, Germany

Editors
Paolo Frasconi
Università degli Studi di Firenze
Florence
Italy

Giuseppe Manco
National Research Council (ICAR-CNR)
Rende
Italy

Niels Landwehr
Computer Science
University of Potsdam
Potsdam
Germany

Jilles Vreeken
MPI for Informatics
Saarland University
Saarbrucken, Saarland
Germany

ISSN 0302-9743   ISSN 1611-3349   (electronic)
Lecture Notes in Artificial Intelligence
ISBN 978-3-319-46226-4   ISBN 978-3-319-46227-1   (eBook)
DOI 10.1007/978-3-319-46227-1

Library of Congress Control Number: 2016950748

LNCS Sublibrary: SL7 – Artificial Intelligence

Printed on acid-free paper

This Springer imprint is published by Springer Nature
The registered company is Springer International Publishing AG
The registered company address is: Gewerbestrasse 11, 6330 Cham, Switzerland

Preface

These are the proceedings of the 15th European Conference on Machine Learning and Principles and Practice of Knowledge Discovery in Databases (ECML PKDD 2016), held in Riva del Garda, Italy, during September 19–23, 2016. This event is the premier European Machine Learning and Data Mining conference and builds upon a very successful series of 26 ECML and 19 PKDD conferences, which have been jointly organized for the past 15 years.

The response to our call for paper was very good. We received 353 papers for the main conference track, of which 100 were accepted, yielding an acceptance rate of about 28 %.

Traditionally, ECML PKDD provides an extensive technical program that consists of several focused tracks:

- the conference track, featuring regular conference papers, published in these proceedings;
- the journal track, featuring papers that satisfy the quality criteria of journal papers and at the same time lend themselves to conference talks (these papers are published separately in the journals *Machine Learning* and *Knowledge Discovery and Data Mining*);
- the industrial track, aiming to bring together participants from academia, industry, government, and NGOs (non-governmental organizations) in a venue that highlights practical and real-world studies of machine learning, knowledge discovery, and data mining.
- the demo track, presenting innovative prototype implementations or mature systems that use machine learning techniques and knowledge discovery processes in a real setting;
- the nectar track, offering conference attendees a compact overview of recent scientific advances at the frontier of machine learning and data mining with other disciplines, as published in related conferences and journals.

Moreover, the conference program included 3 discovery challenges, 13 workshops, and 10 tutorial presentations. The discovery challenges were organized by Elio Masciari and Alessandro Moschitti. Fabrizio Costa, Matthijs van Leeuwen, and Albrecht Zimmermann had the responsibility of selecting workshop and tutorial proposals. The PhD Forum, where junior PhD students exchange ideas, experiences, and get advise from senior researchers, was organized by Leman Akoglu and Tijl De Bie.

The program included six plenary keynotes by invited speakers Susan Athey (Stanford Graduate School of Business), Zoubin Ghahramani (University of Cambridge and Alan Turing Institute), Thore Graepel (Google DeepMind and University College London), Ravi Kumar (Google), Rasmus Pagh (IT University of Copenhagen), and Alex "Sandy" Pentland (MIT).

Putting together the program of this conference would have been impossible without the help of a large and supportive team. Our thirty Area Chairs nominated reviewers, moderated the discussion among them to find a consensus over each paper, and made a final accept/reject decision. A total of 315 reviewers (listed in this book) helped to select papers. Two best student papers were selected by Toon Calders and Hendrik Blockeel. The associated awards were sponsored by Springer and the journals *Machine Learning* and *Data Mining and Knowledge Discovery*.

For the fourth time, the conference used a double submission model: next to the regular conference tracks, papers submitted to the Springer journals Machine Learning (MACH) and Data Mining and Knowledge Discovery (DAMI) were considered for presentation at the conference. These papers were submitted to the ECML PKDD 2016 special issue of the respective journals, and underwent the normal editorial process of these journals. Those papers accepted for one of these journals were assigned a presentation slot at the ECML PKDD 2016 conference. A total of 120 original manuscripts were submitted to the journal track during this year. Some of these papers are still being refereed. Of the fully refereed papers, 8 were accepted in DAMI and 10 in MACH, together with 10 papers from last year's call, which were also scheduled for presentation at this conference.

There were two major innovations at this year's conference. First, we decided to have a full day of plenary presentation on September 21st, while the usual four parallel session tracks were run on September 20th and 22nd. These plenary oral presentations were selected by the Program and Journal Track Co-chairs from the pool of all accepted papers according to criteria such as: (1) novelty and significance of the results and their expected impact; (2) breadth of interest for both machine learners and data miners. It is our belief that this will strengthen the synergy between the ML and the DM sub-communities, allowing papers of general interest for both to be presented to the whole audience.

The second major difference is the adoption of the practices of Reproducible Research (RR). Authors were encouraged to adhere to such practices by making available data and software tools for reproducing the results reported in their papers. In total, 29 papers with accompanying software and/or data are flagged as RR-papers on the conference website http://ecmlpkdd2016.org/, which provides links to such additional material (links are also available within the paper bodies in these proceedings).

Part I and Part II of the proceedings of the ECML PKDD 2016 conference contain the full papers of the contributions presented in the scientific track and the abstracts of the scientific plenary talks. Part III of the proceedings of the ECML PKDD 2016 conference contains the full papers of the contributions presented in the industrial track, short papers describing the demonstrations, the nectar papers, and the abstracts of the industrial plenary talks. First of all, we would like to express our gratitude to the general chairs of the conference, Fosca Giannotti and Andrea Passerini, as well as to all members of the Organizing Committee, for managing this event in a very competent and professional way. In particular, we thank the demo, workshop and tutorial, industrial, and nectar track chairs. Special thanks go to the proceedings chairs, Marco Lippi and Stefano Ferilli, for the hard work of putting these proceedings together. We thank the PhD Forum organizers, the Discovery Challenge organizers, and all the people involved in the conference, who worked hard for its success. We would like to thank Microsoft for allowing us to use their CMT software for conference management. Last but not least, we would like to sincerely thank the authors for submitting their work to the conference and the reviewers and area chairs for their tremendous effort in guaranteeing the quality of the reviewing process, thereby improving the quality of these proceedings.

September 2016

Paolo Frasconi
Niels Landwehr
Giuseppe Manco
Jilles Vreeken

Organization

ECML PKDD 2016 Organization

General Chairs

Andrea Passerini   University of Trento, Italy
Fosca Giannotti   National Research Council (ISTI-CNR), Italy

Program Chairs

Paolo Frasconi   University of Florence, Italy
Niels Landwehr   University of Potsdam, Germany
Giuseppe Manco   National Research Council (ICAR-CNR), Italy
Jilles Vreeken   Cluster of Excellence MMCI, Saarland University & Max
    Planck Institute for Informatics, Germany

Journal Track Chairs

Thomas Gärtner   University of Nottingham, UK
Mirco Nanni   National Research Council (ISTI-CNR), Italy
Andrea Passerini   University of Trento, Italy
Céline Robardet   National Institute of Applied Science in Lyon, France

Industrial Track Chairs

Björn Bringmann   Deloitte GmbH, Germany
Gemma Garriga   Inria, France
Volker Tresp   Siemens AG & Ludwig Maximilian University of Munich,
    Germany

Local Organization Chairs

Simone Marinai   University of Florence, Italy
Gianluca Corrado   University of Trento, Italy
Katya Tentori   University of Trento, Italy

Workshop and Tutorial Chairs

Matthijs van Leeuwen	Leiden University, Netherlands
Fabrizio Costa	University of Freiburg, Germany
Albrecht Zimmermann	University of Caen, France

Awards Committee Chairs

Toon Calders	Free University of Bruxelles, Belgium
Hendrik Blockeel	University of Leuven, Belgium

Nectar Track Chairs

Bettina Berendt	University of Leuven, Belgium
Pauli Miettinen	Max Planck Institute for Informatics, Germany

Demo Chairs

Nikolaj Tatti	Aalto University School of Science, Finland
Élisa Fromont	Jean Monnet University, France

Discovery Challenge Chairs

Elio Masciari	National Research Council (ICAR-CNR), Italy
Alessandro Moschitti	Qatar Computing Research Institute, HBKU, Qatar & University of Trento, Italy

Sponsorship Chairs

Michelangelo Ceci	University of Bari, Italy
Chedy Raïssi	Inria, France

Publicity and Social Media Chairs

Olana Missura	University of Bonn, Germany
Nicola Barbieri	Yahoo!, UK
Gianluca Corrado	University of Trento, Italy

PhD Forum Chairs

Leman Akoglu	Stony Brook University, USA
Tijl De Bie	Ghent University, Belgium

Proceedings Chairs

Marco Lippi	University of Bologna, Italy
Stefano Ferilli	University of Bari, Italy

Web Chair

Daniele Baracchi	University of Florence, Italy

Area Chairs

Hendrik Blockeel	KU Leuven, Belgium
Francesco Bonchi	ISI Foundation Turin, Italy
Karsten Borgwardt	ETH Zurich, Switzerland
Toon Calders	University of Antwerp, Belgium
Aaron Courville	University of Montreal, Canada
Ian Davidson	University of California at Davis, USA
Tijl De Bie	Ghent University, Belgium
Luc De Raedt	KU Leuven, Belgium
Carlotta Domeniconi	George Mason University, USA
Peter Flach	University of Bristol, UK
Claudio Gentile	Università dell'Insubria, Italy
Mohammad Ghavamzadeh	Inria Lille, France
Aristides Gionis	Aalto University, Finland
Bart Goethals	University of Antwerp, Belgium
Geoff Holmes	University of Waikato, New Zealand
Andreas Hotho	University of Wurzburg, Germany
Eyke Hüllermeier	University of Paderborn, Germany
Manfred Jaeger	Aalborg University, Denmark
George Karypis	University of Minnesota, USA
Samuel Kaski	Aalto University, Finland
Kristian Kersting	University of Dortmund
Donato Malerba	University of Bari, Italy
Pauli Miettinen	Max-Planck Institute for Informatics, Germany
Dino Pedreschi	University of Pisa, Italy
Bernhard Pfahringer	University of Waikato, New Zealand
Guido Sanguinetti	University of Edinburgh, UK
Arno Siebes	University of Utrecht, Netherlands
Guy Van den Broeck	University of California at Los Angeles, USA
Marco Wiering	University of Groningen, Netherlands
Stefan Wrobel	University of Bonn, Germany

Reviewers

Prashanth A.
Mohammad Al Hasan
Carlos Alzate
Aijun An
Aris Anagnostopoulos
Fabrizio Angiulli
Annalisa Appice
Ira Assent
Martin Atzmueller
Antonio Bahamonde
Jose Balcázar
Nicolas Ballas
Nicola Barbieri
Christian Bauckhage
Roberto Bayardo
Martin Becker
Srikanta Bedathur
Jessa Bekker
Vaishak Belle
András Benczúr
Michael Berthold
Albert Bifet
Konstantinos Blekas
Paul Blomstedt
Dean Bodenham
Mario Boley
Gianluca Bontempi
Henrik Bostrom
Jean-François Boulicaut
Marc Boulle
Pavel Brazdil
Ulf Brefeld
Robert Busa-Fekete
Rui Camacho
Longbing Cao
Francisco Casacuberta
Michelangelo Ceci
Peggy Cellier
Loic Cerf
Tania Cerquitelli
Edward Chang
Thierry Charnois
Duen Horng Chau

Keke Chen
Ling Chen
Silvia Chiusano
Arthur Choi
Frans Coenen
Fabrizio Costa
Vítor Santos Costa
Bruno Crémilleux
Botond Cseke
Boris Cule
Tomaz Curk
James Cussens
Claudia d'Amato
Maria Damiani
Jesse Davis
Martine De Cock
Colin de la Higuera
Juan del Coz
Anne Denton
Christian Desrosiers
Nicola Di Mauro
Tom Diethe
Ying Ding
Stephan Doerfel
Frank Dondelinger
Anton Dries
Madalina Drugan
Wouter Duivesteijn
Robert Durrant
Ines Dutra
Sašo Džeroski
Tapio Elomaa
Dora Erdos
Floriana Esposito
Nicola Fanizzi
Elaine Faria
Fabio Fassetti
Ad Feelders
Stefano Ferilli
Carlos Ferreira
Cesar Ferri
Maurizio Filippone
Asja Fischer

Eibe Frank
Élisa Fromont
Fabio Fumarola
Johannes Fürnkranz
Victor Gabillon
Esther Galbrun
Patrick Gallinari
Joao Gama
Byron Gao
Paolo Garza
Eric Gaussier
Ricard Gavalda
Rainer Gemulla
Konstantinos Georgatzis
Pierre Geurts
Aris Gkoulalas-Divanis
Dorota Glowacka
Mehmet Gonen
Michael Granitzer
Caglar Gulcehre
Francesco Gullo
Stephan Gunnemann
Maria Halkidi
Jiawei Han
Xiao He
Denis Helic
Jose Hernandez-Orallo
Thanh Lam Hoang
Frank Hoeppner
Jaakko Hollmen
Arjen Hommersom
Tamas Horvath
Yuanhua Huang
Van-Anh Huynh-Thu
Dino Ienco
Bhattacharya Indrajit
Frederik Janssen
Nathalie Japkowicz
Szymon Jaroszewicz
Alipio Jorge
Giuseppe Jurman
Robert Jeschke
Hachem Kadri

Bo Kang
U Kang
Andreas Karwath
Hisashi Kashima
Ioannis Katakis
Yoshinobu Kawahara
Mehdi Kaytoue
John Keane
Latifur Khan
Arto Klami
Levente Kocsis
Yun Sing Koh
Alek Kolcz
Irena Koprinska
Frederic Koriche
Walter Kosters
Lars Kotthoff
Meelis Kull
Nicolas Lachiche
Helge Langseth
Thomas Lansdall-Welfare
Pedro Larranaga
Silvio Lattanzi
Niklas Lavesson
Nada Lavrač
Sangkyun Lee
Florian Lemmerich
Jiuyong Li
Juanzi Li
Limin Li
Jefrey Lijffijt
Felipe Llinares López
Daniel Hernandez Lobato
Corrado Loglisci
Peter Lucas
Elio Masciari
Andres Masegosa
Wannes Meert
Ernestina Menasalvas
Rosa Meo
Mehdi Mirza
Karthika Mohan
Anna Monreale
Guido Montufar
Joao Moreira
Katharina Morik

Mohamed Nadif
Ndapa Nakashole
Amedeo Napoli
Sriraam Natarajan
Benjamin Nguyen
Thomas Nielsen
Xia Ning
Kjetil Nørvåg
Eirini Ntoutsi
Andreas Nürnberger
Francesco Orsini
George Paliouras
Apostolos Papadopoulos
Evangelos Papalexakis
Panagiotis Papapetrou
Ioannis Partalas
Nikos Pelekis
Jing Peng
Ruggero Pensa
Francois Petitjean
Nico Piatkowksi
Andrea Pietracaprina
Gianvito Pio
Marc Plantevit
Pascal Poncelet
Philippe Preux
Kai Puolamaki
Buyue Qian
Chedy Raïssi
Jan Ramon
Huzefa Rangwala
Zbigniew Rás
Chotirat Ratanamahatana
Jan Rauch
Steffen Rendle
Chiara Renso
Achim Rettinger
Fabrizio Riguzzi
Matteo Riondato
Pedro Rodrigues
Juan Rodriguez
Simon Rogers
Damian Roqueiro
Fabrice Rossi
Juho Rousu
Celine Rouveirol

Stefan Rueping
Salvatore Ruggieri
Yvan Saeys
Alan Said
Lorenza Saitta
Ansaf Salleb-Aouissi
Scott Sanner
Claudio Sartori
Lars Schmidt-Thieme
Christoph Schommer
Matthias Schubert
Giovanni Semeraro
Sohan Seth
Vinay Setty
Junming Shao
Sameer Singh
Andrzej Skowron
Kevin Small
Marta Soare
Yangqiu Song
Mauro Sozio
Myra Spiliopoulou
Papadimitriou Spiros
Eirini Spyropoulou
Jerzy Stefanowski
Daria Stepanova
Florian Stimberg
Gerd Stumme
Mahito Sugiyama
Einoshin Suzuki
Panagiotis Symeonidis
Sandor Szedmak
Andrea Tagarelli
Domenico Talia
Letizia Tanca
Nikolaj Tatti
Maguelonne Teisseire
Aika Terada
Georgios Theocharous
Kai Ming Ting
Ljupco Todorovski
Hannu Toivonen
Luis Torgo
Roberto Trasarti
Daniel Trejo-Banos
Panagiotis Tsaparas

Vincent Tseng
Grigorios Tsoumakas
Karl Tuyls
Niall Twomey
Theodoros Tzouramanis
Antti Ukkonen
Jan Van Haaren
Matthijs van Leeuwen
Maarten van Someren
Iraklis Varlamis
Michalis Vazirgiannis
Julien Velcin
Shankar Vembu
Celine Vens
Deepak Venugopal

Vassilios Verykios
Herna Viktor
Fabio Vitale
Christel Vrain
Willem Waegeman
Jianyong Wang
Ding Wei
Cheng Weiwei
Zheng Wen
Joerg Wicker
Makoto Yamada
Jeffrey Yu
Philip Yu
Bianca Zadrozny
Marco Zaffalon

Gerson Zaverucha
Demetris Zeinalipour
Filip Železný
Bernard Zenko
Junping Zhang
Min-Ling Zhang
Nan Zhang
Shichao Zhang
Ying Zhao
Mingjun Zhong
Djamel Zighed
Arthur Zimek
Albrecht Zimmermann
Indre Zliobaite
Blaž Zupan

Sponsors

Gold Sponsors

Google   http://research.google.com
IBM   http://www.ibm.com

Silver Sponsors

Deloitte   http://www.deloitte.com
Siemens   http://www.siemens.com
Unicredit   http://www.unicreditgroup.eu
Zalando   http://www.zalando.com

Award Sponsors

Deloitte   http://www.deloitte.com
DMKD   http://link.springer.com/journal/10618
MLJ   http://link.springer.com/journal/10994

Badge Lanyard

Knime   http://www.knime.org

Additional Supporters

Springer   http://www.springer.com

Institutional Supporters

ICAR-CNR   http://www.icar.cnr.it
DISI-UNITN   http://www.disi.unitn.it
ISTI-CNR   http://www.isti.cnr.it
COGNET   http://www.cognet.5g-ppp.eu

Organizing Institutions

UNITN   http://www.unitn.it
UNIFI   http://www.unifi.it
ISTI-CNR   http://www.isti.cnr.it
ICAR-CNR   http://www.icar.cnr.it

Abstracts of Invited Talks

Causal Inference and Machine Learning: Estimating and Evaluating Policies

Susan Athey

Stanford Graduate School of Business

Abstract. In many contexts, a decision-making can choose to assign one of a number of "treatments" to individuals. The treatments may be drugs, offers, advertisements, algorithms, or government programs. One setting for evaluating such treatments involves randomized controlled trials, for example A/B testing platforms or clinical trials. In such settings, we show how to optimize supervised machine learning methods for the problem of estimating heterogeneous treatment effects, while preserving a key desiderata of randomized trials, which is providing valid confidence intervals for estimates. We also discuss approaches for estimating optimal policies and online learning. In environments with observational (non-experimental) data, different methods are required to separate correlation from causality. We show how supervised machine learning methods can be adapted to this problem.

Bio. Susan Athey is The Economics of Technology Professor at Stanford Graduate School of Business. She received her bachelor's degree from Duke University and her Ph.D. from Stanford, and she holds an honorary doctorate from Duke University. She previously taught at the economics departments at MIT, Stanford and Harvard. In 2007, Professor Athey received the John Bates Clark Medal, awarded by the American Economic Association to "that American economist under the age of forty who is adjudged to have made the most significant contribution to economic thought and knowledge." She was elected to the National Academy of Science in 2012 and to the American Academy of Arts and Sciences in 2008. Professor Athey's research focuses on the economics of the internet, online advertising, the news media, marketplace design, virtual currencies and the intersection of computer science, machine learning and economics. She advises governments and businesses on marketplace design and platform economics, notably serving since 2007 as a long-term consultant to Microsoft Corporation in a variety of roles, including consulting chief economist.

Automating Machine Learning

Zoubin Ghahramani

University of Cambridge and Alan Turing Institute

Abstract. I will describe the "Automatic Statistician"[1], a project which aims to automate the exploratory analysis and modelling of data. Our approach starts by defining a large space of related probabilistic models via a grammar over models, and then uses Bayesian marginal likelihood computations to search over this space for one or a few good models of the data. The aim is to find models which have both good predictive performance, and are somewhat interpretable. The Automatic Statistician generates a natural language summary of the analysis, producing a 10–15 page report with plots and tables describing the analysis. I will also link this to recent work we have been doing in the area of Probabilistic Programming (including an new system in Julia) to automate inference, and on the rational allocation of computational resources (and our entry in the AutoML conference).

Bio. Zoubin Ghahramani FRS is Professor of Information Engineering at the University of Cambridge, where he leads the Machine Learning Group, and the Cambridge Liaison Director of the Alan Turing Institute, the UK's national institute for Data Science. He studied computer science and cognitive science at the University of Pennsylvania, obtained his PhD from MIT in 1995, and was a postdoctoral fellow at the University of Toronto. His academic career includes concurrent appointments as one of the founding members of the Gatsby Computational Neuroscience Unit in London, and as a faculty member of CMU's Machine Learning Department for over 10 years. His current research interests include statistical machine learning, Bayesian nonparametrics, scalable inference, probabilistic programming, and building an automatic statistician. He has published over 250 research papers, and has held a number of leadership roles as programme and general chair of the leading international conferences in machine learning including: AISTATS (2005), ICML (2007, 2011), and NIPS (2013, 2014). In 2015 he was elected a Fellow of the Royal Society.

---

[1] http://www.automaticstatistician.com/.

AlphaGo - Mastering the Game of Go with Deep Neural Networks and Tree Search

Thore Graepel

Google DeepMind and University College London

Abstract. The game of Go has long been viewed as the most challenging of classic games for artificial intelligence owing to its enormous search space and the difficulty of evaluating board positions and moves. Here we introduce a new approach to computer Go that uses 'value networks' to evaluate board positions and 'policy networks' to select moves. These deep neural networks are trained by a novel combination of supervised learning from human expert games, and reinforcement learning from games of self-play. Using this search algorithm, our program AlphaGo achieved a 99.8 % winning rate against other Go programs and beat the human European Go champion Fan Hui by 5 games to 0, a feat thought to be at least a decade away by Go and AI experts alike. Finally, in a dramatic and widely publicised match, AlphaGo defeated Lee Sedol, the top player of the past decade, 4 games to 1. In this talk, I will explain how AlphaGo works, describe our process of evaluation and improvement, and discuss what we can learn about computational intuition and creativity from the way AlphaGo plays.

Bio. Thore Graepel is a research group lead at Google DeepMind and holds a part-time position as Chair of Machine Learning at University College London. He studied physics at the University of Hamburg, Imperial College London, and Technical University of Berlin, where he also obtained his PhD in machine learning in 2001. He spent time as a postdoctoral researcher at ETH Zurich and Royal Holloway College, University of London, before joining Microsoft Research in Cambridge in 2003, where he co-founded the Online Services and Advertising group. Major applications of Thore's work include Xbox Live's TrueSkill system for ranking and matchmaking, the AdPredictor framework for click-through rate prediction in Bing, and the Matchbox recommender system which inspired the recommendation engine of Xbox Live Marketplace. More recently, Thore's work on the predictability of private attributes from digital records of human behaviour has been the subject of intense discussion among privacy experts and the general public. Thore's current research interests include probabilistic graphical models and inference, reinforcement learning, games, and multi-agent systems. He has published over one hundred peer-reviewed papers, is a named co-inventor on dozens of patents, serves on the editorial boards of JMLR and MLJ, and is a founding editor of the book series Machine Learning & Pattern Recognition at Chapman & Hall/CRC. At DeepMind, Thore has returned to his original passion of understanding and creating intelligence, and recently contributed to creating AlphaGo, the first computer program to defeat a human professional player in the full-sized game of Go, a feat previously thought to be at least a decade away.

Sequences, Choices, and Their Dynamics

Ravi Kumar

Google

Abstract. Sequences arise in many online and offline settings: urls to visit, songs to listen to, videos to watch, restaurants to dine at, and so on. User-generated sequences are tightly related to mechanisms of choice, where a user must select one from a finite set of alternatives. In this talk, we will discuss a class of problems arising from studying such sequences and the role discrete choice theory plays in these problems. We will present modeling and algorithmic approaches to some of these problems and illustrate them in the context of large-scale data analysis.

Bio. Ravi Kumar has been a senior staff research scientist at Google since 2012. Prior to this, he was a research staff member at the IBM Almaden Research Center and a principal research scientist at Yahoo! Research. His research interests include Web search and data mining, algorithms for massive data, and the theory of computation.

Dimensionality Reduction with Certainty

Rasmus Pagh

IT University of Copenhagen

Abstract. Tool such as Johnson-Lindenstrauss dimensionality reduction and 1-bit minwise hashing have been successfully used to transform problems involving very high-dimensional real vectors into lower-dimensional equivalents, at the cost of introducing a random distortion of distances/similarities among vectors. While this can alleviate the computational cost associated with high dimensionality, the effect on the outcome of the computation (compared to working on the original vectors) can be hard to analyze and interpret. For example, the behavior of a basic kNN classifier is easy to describe and interpret, but if the algorithm is run on dimension-reduced vectors with distorted distances it is much less transparent what is happening. The talk starts with an introduction to randomized (data-independent) dimensionality reduction methods and gives some example applications in machine learning. Based on recent work in the theoretical computer science community we describe tools for dimension reduction that give stronger guarantees on approximation, replacing probabilistic bounds on distance/similarity with bounds that hold with certainty. For example, we describe a "distance sensitive Bloom filter": a succinct representation of high-dimensional boolean vectors that can identify vectors within distance r with certainty, while far vectors are only thought to be close with a small "false positive" probability. We also discuss work towards a deterministic alternative to random feature maps (i.e., dimension-reduced vectors from a high-dimensional feature space), and settings in which a pair of dimension-reducing mappings outperform single-mapping methods. While there are limits to what performance can be achieved with certainty, such techniques may be part of the toolbox for designing transparent and scalable machine learning and knowledge discovery methods.

Bio. Rasmus Pagh graduated from Aarhus University in 2002, and is now a full professor at the IT University of Copenhagen. His work is centered around efficient algorithms for big data, with an emphasis on randomized techniques. His publications span theoretical computer science, databases, information retrieval, knowledge discovery, and parallel computing. His most well-known work is the cuckoo hashing algorithm (2001), which has led to new developments in several fields. In 2014 he received the best paper award at the WWW Conference for a paper with Pham and Mitzenmacher on similarity estimation, and started a 5-year research project funded by the European Research Council on scalable similarity search.

Social Learning

Alex "Sandy" Pentland

MIT

Abstract. Human decisions are heavily influenced by social interaction, so that predicting or influencing individual behavior requires modeling these interaction effects. In addition the distributed learning strategies exhibited by human communities suggest methods of improving both machine learning and human-machine systems. Several practical examples will be described.

Bio. Professor Alex "Sandy" Pentland directs the MIT Connection Science and Human Dynamics labs and previously helped create and direct the MIT Media Lab and the Media Lab Asia in India. He is one of the most-cited scientists in the world, and Forbes recently declared him one of the "7 most powerful data scientists in the world" along with Google founders and the Chief Technical Officer of the United States. He has received numerous awards and prizes such as the McKinsey Award from Harvard Business Review, the 40th Anniversary of the Internet from DARPA, and the Brandeis Award for work in privacy.

He is a founding member of advisory boards for Google, AT&T, Nissan, and the UN Secretary General, a serial entrepreneur who has co-founded more than a dozen companies including social enterprises such as the Data Transparency Lab, the Harvard-ODI-MIT DataPop Alliance and the Institute for Data Driven Design. He is a member of the U.S. National Academy of Engineering and leader within the World Economic Forum.

Contents – Part II

AUC-Maximized Deep Convolutional Neural Fields for Protein Sequence Labeling

Sheng Wang[1,2], Siqi Sun[1], and Jinbo Xu[1(✉)]

[1] Toyota Technological Institute at Chicago, Chicago, IL 60615, USA
{wangsheng,siqi.sun,j3xu}@ttic.edu
[2] University of Chicago, Chicago, IL 60615, USA

Abstract. Deep Convolutional Neural Networks (DCNN) has shown excellent performance in a variety of machine learning tasks. This paper presents Deep Convolutional Neural Fields (DeepCNF), an integration of DCNN with Conditional Random Field (CRF), for sequence labeling with an imbalanced label distribution. The widely-used training methods, such as maximum-likelihood and maximum labelwise accuracy, do not work well on imbalanced data. To handle this, we present a new training algorithm called maximum-AUC for DeepCNF. That is, we train DeepCNF by directly maximizing the empirical Area Under the ROC Curve (AUC), which is an unbiased measurement for imbalanced data. To fulfill this, we formulate AUC in a pairwise ranking framework, approximate it by a polynomial function and then apply a gradient-based procedure to optimize it. Our experimental results confirm that maximum-AUC greatly outperforms the other two training methods on 8-state secondary structure prediction and disorder prediction since their label distributions are highly imbalanced and also has similar performance as the other two training methods on solvent accessibility prediction, which has three equally-distributed labels. Furthermore, our experimental results show that our AUC-trained DeepCNF models greatly outperform existing popular predictors of these three tasks. The data and software related to this paper are available at https://github.com/realbigws/DeepCNF_AUC.

1 Introduction

Deep Convolutional Neural Networks (DCNN), originated by Yann LeCun at 1998 [30] for document recognition, is being widely used in a plethora of machine learning (ML) tasks ranging from speech recognition [22], to computer vision [27], and to computational biology [9]. DCNN is good at capturing medium- and/or long-range structured information in a hierarchical manner. To handle structured data, [5] has integrated DCNN with fully connected Conditional Random Fields (CRF) for semantic image segmentation. Here we present Deep Convolutional Neural Fields (DeepCNF), which is an integration of DCNN and linear-chain

---

Electronic supplementary material The online version of this chapter (doi:10. 1007/978-3-319-46227-1_1) contains supplementary material, which is available to authorized users.

© Springer International Publishing AG 2016
P. Frasconi et al. (Eds.): ECML PKDD 2016, Part II, LNAI 9852, pp. 1–16, 2016.
DOI: 10.1007/978-3-319-46227-1_1

CRF, to address the task of sequence labeling and apply it to three important biology problems: solvent accessibility prediction (ACC), disorder prediction (DISO), and 8-state secondary structure prediction (SS8) [24,34].

A protein sequence can be viewed as a string of amino acids (also called residues in the protein context) and we want to predict a label for each residue. In this paper we consider three types of labels: solvent accessibility, disorder state and 8-state secondary structure. These three structure properties are very important to the understanding of protein structure and function. The solvent accessibility is important for protein folding [10], the order/disorder state plays an important role in many biological processes [37], and protein secondary structure(SS) relates to local backbone conformation of a protein sequence [38]. The label distribution in these problems varies from almost uniform to highly imbalanced. For example, only ∼6 % of residues are shown to be disordered [19]. Some SS labels, such as 3–10 helix, beta-bridge, and pi-helix are extremely rare [46]. The widely-used training methods, such as maximum-likelihood [29] and maximum labelwise accuracy [16], perform well on data with balanced labels but not on highly-imbalanced data [8].

This paper presents a new maximum-AUC method to train DeepCNF for imbalanced sequence data. Specifically, we train DeepCNF by maximizing Area Under the ROC Curve (AUC), which is a good measure for class-imbalanced data [7]. Taking disorder prediction as an example, random guess can obtain ∼94 % per-residue accuracy, but its AUC is only ∼0.5. AUC is insensitive to changes in class distribution because the ROC curve specifies the relationship between false positive (FP) rate and true positive (TP) rate, which are independent of class distribution [7]. However, it is very challenging to directly optimize AUC. A few algorithms have been developed to maximize AUC on unstructured data [21,23,36], but to the best of our knowledge, there is no such an algorithm for imbalanced structured data (e.g., sequence data addressed here). To train DeepCNF by maximum-AUC, we formulate the AUC function in a ranking framework, approximate it by a polynomial Chebyshev function [3] and then use L-BFGS [31] to optimize it.

Our experimental results show that when the label distribution is almost uniform, there is no big difference between the three training methods. Otherwise, maximum-AUC results in better AUC and Mcc than the other two methods. Tested on several publicly available benchmark data, our AUC-trained DeepCNF model obtains the best performance on all the three protein sequence labeling tasks. In particular, at a similar specificity level, our method obtains better precision and sensitivity for those labels with a much smaller occurring frequency.

Contributions. 1. A novel training algorithm that directly maximizes the empirical AUC to learn DeepCNF model from imbalanced structured data. 2. Studying three training methods, i.e. maximum-likelihood, maximum labelwise accuracy, and maximum-AUC, for DeepCNF and testing them on three real-world protein sequence labeling problems, in which the label distribution varies from almost uniform to highly imbalanced. 3. Achieving the state-of-the-art performance on three important protein sequence labeling problems. 4. All benchmarks are public available, and the code is available online at https://github.com/realbigws/DeepCNF_AUC. A web server is also implemented and available at http://raptorx.uchicago.edu/StructurePropertyPred/predict/ [43].

1.1 Notations

Let L denote the sequence length, $[L]$ denote the set $\{1, 2, \ldots, L\}$. For a finite set S, let $|S|$ denote its cardinality. Let $X = (X_1, X_2, \ldots, X_L), y = (y_1, y_2, \ldots, y_L)$ denote the input features and labels respectively for position $i, i \in [L]$. Denote Σ as the set of all possible labels, i.e., $y_i \in \Sigma, \forall i \in [L]$.

2 Related Work

Class imbalance issue is a long-standing notorious problem. Early works have addressed this issue through data-level methods, which change the empirical distribution of the training data to create a new balanced dataset [20]. These methods include (a) under-sampling the majority class; (b) over-sampling the minority class; or (c) combining both under-sampling and over-sampling [4, 13, 32].

As AUC is an unbiased measurement for class-imbalanced data, a variety of approaches have been proposed to directly optimize the AUC value. In particular, (a) Cortes et al. [7] optimized AUC by RankBoost algorithm; (b) Ferri et al. [15] trained a decision tree by using AUC as splitting criteria; (c) Herschtal and Raskutti [21] trained a neural network by optimizing AUC; and (d) Joachims [23] proposed a generalized Support Vector Machines (SVM) that optimizes AUC.

However, all these approaches could only be applied on *unstructured models*. Recently, Rosenfeld et al. [40] have proposed a learning algorithm for structured models with AUC loss. However, there are three fundamental differences of our method with theirs: (a) our method targets at a sequence labelling problem (of course a structured model) with an imbalance label assignment, while their model is proposed for a ranking problem. Specifically, sequence labeling requires the prediction of the label (might not necessarily be binary) at each position, while the focus of structured ranking is on prediction of binary vectors $(y_1; \ldots; y_n)$ where it is hard (or unnecessary) to exactly predict which y_i have the value 1. Instead the goal of structured ranking is to rank the items $1, \ldots, n$ such that elements with $y_i = 1$ are ranked high [40]; (b) our method is based on CRF, while they used structured SVM; and (c) we also studied deep learning extension of our method, while they did not. In summary, to the best of our knowledge, our work is the first sequence labelling study that aims to optimize the AUC value directly under a deep learning framework.

3 Method

3.1 DeepCNF Architecture

As shown in Fig. 1, DeepCNF has two modules: (i) the Conditional Random Fields (CRF) module consisting of the top layer and the label layer, and (ii) the

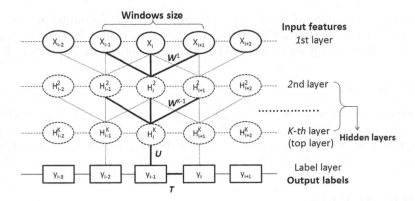

Fig. 1. Illustration of a DeepCNF. Here i is the position index and X_i the associated input features, H^k represents the k-th hidden layer, and y is the output label. All the layers from the first to the K-th (i.e., top layer) form a DCNN with parameter $W^k, k \in [K]$, where K is number of hidden layers. The K-th layer and the label layer form a CRF, in which the parameter U specifies the relationship between the output of the K-th layer and the label layer and T is the parameter for adjacent label correlation. Windows size is set to 3 only for illustration.

deep convolutional neural network (DCNN) module covering the input to the top layer. When only one hidden layer is used, DeepCNF becomes Conditional Neural Fields (CNF), a probabilistic graphical model described in [39].

Given $X = (X_1, \ldots, X_L)$ and $y = (y_1, \ldots, y_L)$, DeepCNF calculates the conditional probability of y on the input X with parameter θ as follows,

$$P_\theta(y|X) = \frac{1}{Z(X)} \exp \left(\sum_{i \in L} (f_\theta(y, X, i) + g_\theta(y, X, i)) \right), \tag{1}$$

where $f_\theta(y, X, i)$ is the binary potential function specifying correlation among adjacent labels at position i, $g_\theta(y, X, i)$ is the unary potential function modeling relationship between y_i and input features for position i, and $Z(X)$ is the partition function. Formally, $f_\theta(\cdot)$ and $g_\theta(\cdot)$ are defined as follows:

$$f_\theta(y, X, i) = f_\theta(y_{i-1}, y_i, X, i) = \sum_{a,b} T_{a,b} \delta(y_{i-1} = a) \delta(y_i = b)$$

$$g_\theta(y, X, i) = g_\theta(y_i, X, i) \quad = \sum_{a,h} U_{a,h} A_{a,h}(X, i, W) \delta(y_i = a),$$

where a and b represent two specific labels for prediction, $\delta(\cdot)$ is an indicator function, $A_{a,h}(X, i, W)$ is a deep neural network function for the h-th neuron at position i of the top layer for label a, and W, U and T are the model parameters to be trained. Specifically, W is the parameter for the neural network, U is the parameter connecting the top layer to the label layer, and T is for label correlation. The two potential functions can be merged into a single binary potential function $f_\theta(y, X, i) = f_\theta(y_{i-1}, y_i, X, i) = \sum_{a,b,h} T_{a,b,h} A_{a,b,h}(X, i, W) \delta(y_{i-1} = a) \delta(y_i = b)$. Note that these deep neural network functions for different labels could be shared to $A_h(X, i, W)$. To control model complexity and avoid over-fitting, we add a L_2-norm penalty term as the regularization factor.

Figure 1 shows two adjacent layers of DCNN. Let M_k be the number of neurons for a single position at the k-th layer. Let $X_i(h)$ be the h-th feature at the input layer for residue i and $H_i^k(h)$ denote the output value of the h-th neuron of position i at layer k. When $k = 1$, H^k is actually the input feature X. Otherwise, H^k is a matrix of dimension $L \times M_k$. Let $2N_k + 1$ be the window size at the k-th layer. Mathematically, $H_i^k(h)$ is defined as follows:

$$H_i^k(h) = X_i(h), \qquad\qquad\qquad\qquad\qquad\qquad \text{if } k = 1$$

$$H_i^{k+1}(h) = \pi\Big(\sum_{n=-N_k}^{N_k} \sum_{h'=1}^{M_k} (H_{i+n}^k(h) * W_n^k(h, h')) \Big) \qquad \text{if } k < K$$

$$A_h(X, i, W) = H_i^k(h) \qquad\qquad\qquad\qquad\qquad \text{if } k = K.$$

Meanwhile, $\pi(\cdot)$ is the activation function, either the sigmoid (i.e. $1/(1 + \exp(-x))$) or the tanh (i.e. $(1 - \exp(-2x))/(1 + \exp(-2x))$) function. $W_n^k(-N_k \le n \le N_k)$ is a 2D weight matrix for the connections between the neurons of position $i + n$ at layer k and the neurons of position i at layer $k + 1$. $W_n^k(h, h')$ is shared by all the positions in the same layer, so it is position-independent. Here h' and h index two neurons at the k-th and $(k + 1)$-th layers, respectively. See Appendix about how to calculate the gradient of DCNN by back propagation.

3.2   Objective Functions

Let T be the number of training sequences and L_t denote the length of sequence t. We study three different training methods: maximum-likelihood, maximum labelwise accuracy, and proposed maximum-AUC.

Maximum-Likelihood. The log-likelihood is a widely-used objective function for training CRF [29]. Mathematically, the log-likelihood is defined as follows:

$$LL = \sum_{t \in [T]} \log P_\theta(y^t | X^t),$$

where $P_\theta(y | X)$ is defined in Eq. (1).

Maximum Labelwise Accuracy. Gross et al. [16] proposed an objective function that could directly maximize the labelwise accuracy defined as

$$LabelwiseAccuracy = \sum_{t \in [T]} \sum_{i \in [L_t]} \delta\Big(P_\theta(y_i^{(\tau)}) > \max_{y_i \neq y_i} P_\theta(y_i) \Big),$$

where $y_i^{(\tau)}$ denotes the real label at position i, $P_\theta(y_i^{(\tau)})$ is the predicted probability of the real label at position i. It could be represented by the marginal probability

$$P_\theta(y_i^{(\tau)}|X^t) = \frac{1}{Z(X)} \sum_{y_{1:L^t}} \delta(y_i = (\tau)) \exp(F_{1:L^t}(y, X^t, \theta)),$$

where $F_{l_1:l_2}(y, X, \theta) = \sum_{i=l_1}^{l_2} f_\theta(y, X, i)$.

To obtain a smooth approximation to this objective function, [16] replaces the indicator function with a sigmoid function $Q_\lambda(x) = 1/(1 + \exp(-\lambda x))$ where the parameter λ is set to 15 by default. Then it becomes the following form:

$$Labelwise Accuracy \approx \sum_{t \in [T]} \sum_{i \in [L_t]} Q_\lambda(P_\theta(y_i^{(\tau)}|X^t) - P_\theta(\tilde{y}_i^{(\tau)}|X^t)),$$

where $\tilde{y}_i^{(\tau)}$ denote the label other than $y_i^{(\tau)}$ that has the maximum posterior probability at position i.

Maximum-AUC. The AUC of a predictor function P_θ on label τ is defined as:

$$AUC(P_\theta, \tau) = P\Big(P_\theta(y_i^\tau) > P_\theta(y_j^\tau) | i \in D^\tau, j \in D^{!\tau}\Big), \tag{2}$$

where $P(\cdot)$ is the probability over all pairs of positive and negative examples, D^τ is a set of positive examples with true label τ, and $D^{!\tau}$ is a set of negative examples with true label not being τ. Note that the union of D^τ and $D^{!\tau}$ contains all the training sequence positions, i.e., $D^\tau = \cup_{t=1}^{T} \cup_{i=1}^{L_t} \delta_{i,t}^\tau$ where $\delta_{i,t}^\tau$ is an indicator function. If the true label of the i-th position from sequence t equals to τ, then $\delta_{i,t}^\tau$ is equal to 1; otherwise 0. Again, $P_\theta(y_i^\tau)$ could be represented by the marginal probability $P_\theta(y_i^\tau|X^t)$ from the training sequence t. Since it is hard to calculate the derivatives of Eq. (2), we use the following Wilcoxon-Mann-Whitney statistic [18], which is an unbiased estimator of $AUC(P_\theta, \tau)$:

$$AUC^{WMW}(P_\theta, \tau) = \frac{\sum_{i \in D^\tau} \sum_{j \in D^{!\tau}} \delta\Big(P_\theta(y_i^\tau|X) > P_\theta(y_j^\tau|X)\Big)}{|D^\tau||D^{!\tau}|}. \tag{3}$$

Finally, by summing over labels, the overall AUC objective function is $\sum_\tau AUC^{WMW}(P_\theta, \tau)$.

For a large dataset, the computational cost of AUC by Eq. (3) is high. Recently, Calders and Jaroszewicz [3] proposed a polynomial approximation of AUC which can be computed in linear time. The key idea is to approximate the indicator function $\delta(x > 0)$, where x represents $P_\theta(y_i^\tau|X) - P_\theta(y_j^\tau|X)$ by a polynomial Chebyshev approximation. That is, we approximate $\delta(x > 0)$ by $\sum_{\mu \in [d]} c_\mu x^\mu$ where d is the degree and c_μ the coefficient of the polynomial [3]. Let $n_1 = |D^\tau|$ and $n_0 = |D^{!\tau}|$. Using the polynomial Chebyshev approximation, we can approximate Eq. (3) as follows:

$$AUC^{WMW}(P_\theta, \tau) \approx \frac{1}{n_0 n_1} \sum_{\mu \in [d]} \sum_{l \in [\mu]} \mathcal{Y}_{\mu l} s(P_\theta^l, D^\tau) v(P_\theta^{\mu-l}, D^{!\tau})$$

where $\mathcal{Y}_{\mu l} = c_\mu \binom{\mu}{l} (-1)^{\mu-l}$, $s(P^l, D^\tau) = \sum_{i \in D^\tau} P(y_i^\tau)^l$ and $v(P^l, D^{!\tau}) = \sum_{j \in D^{!\tau}} P(y_j^\tau)^l$. Note that we have $s(P^l, D^\tau) = \sum_{t \in [T]} \sum_{i \in [L_t]} \delta_{i,t}^\tau P(y_i^\tau)^l$ and a similar structure for $v(P^l, D^{!\tau})$.

4 Results

In this section presents our experimental results of the AUC-trained DeepCNF models on three protein sequencing problems, which are summarize as follows:

ACC. We used DSSP [26] to calculate the absolute accessible surface area for each residue in a protein and then normalize it by the maximum solvent accessibility to obtain the relative solvent accessibility (RSA) [6]. Solvent accessibility of one residue is classified into 3 labels: buried (B) for RSA from 0 to 10), intermediate (I) for RSA from 10 to 40 and exposed (E) for RSA from 40 to 100. The ratio of these three labels is around 1:1:1 [33].

DISO. Following the definition in [35], we label a residue as disordered (label 1) if it is in a segment of more than three residues missing atomic coordinates in the X-ray structure. Otherwise it is labeled as ordered (label 0). The distribution of these two labels (ordered vs. disordered) is 94:6 [45].

SS8. The 8-state protein secondary structure is calculated by DSSP [26]. In particular, DSSP assigns 3 types for helix (G for 310 helix, H for alpha-helix, and I for pi-helix), 2 types for strand (E for beta-strand and B for beta-bridge), and 3 types for coil (T for beta-turn, S for high curvature loop, and L for irregular) [44]. The distribution of these 8 labels (H,E,L,T,S,G,B,I) is 34:21:20:11:9:4:1:0 [43].

4.1 Dataset

To use a set of non-redundant protein sequences for training and test, we pick one representative sequence from each protein superfamily defined in CATH [42] or SCOP [1]. The test proteins are in different superfamilies than the training proteins, so we can reduce the bias incurred by the sequence profile similarity between the training and test proteins. The publicly available JPRED [11] dataset (http://www.compbio.dundee.ac.uk/jpred4/about.shtml) satisfies such a condition, which has 1338 training and 149 test proteins, respectively, each belonging to a different superfamily. We train the DeepCNF model using the JPRED training set and conduct 7-fold cross validation to determine the model hyper-parameters for each training method.

We also evaluate the predictive performance of our DeepCNF models on the CASP10 [28] and CASP11 [25] test targets (merged to a single CASP dataset) and the recent CAMEO [17] hard test targets. To remove redundancy, we filter the CASP and CAMEO datasets by removing those targets sharing >25% sequence identity with the JPRED training set. This result in 126 CASP and 147 CAMEO test targets, respectively. See Appendix for their test results.

4.2   Evaluation Criteria

We use Qx to measure the accuracy of sequence labeling where x is the number of different labels for a prediction task. Qx is defined as the percentage of residues for which the predicted labels are correct. In particular, we use Q3 accuracy for ACC prediction, Q8 accuracy for SS8 prediction and Q2 accuracy for disorder prediction.

From TP (true positives), TN (true negatives), FP (false positives) and FN (false negatives), we may also calculate sensitivity (sens), specificity (spec), precision (prec) and Matthews correlation coefficient (Mcc) as $\frac{TP}{TP+FN}$, $\frac{TN}{TN+FP}$, $\frac{TP}{TP+FP}$ and $\frac{TP \times TN - FP \times FN}{\sqrt{(TP+FP)(TN+FP)(TP+FN)(TN+FN)}}$, respectively. We also use AUC as a measure. Mcc and AUC are generally regarded

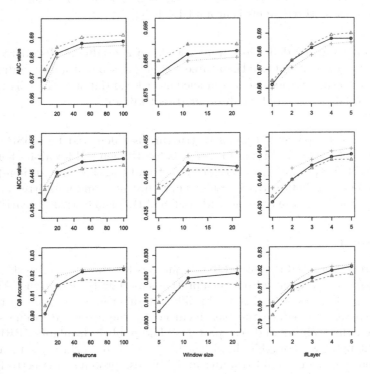

Fig. 2. Q3 accuracy, mean Mcc and AUC of solvent accessibility (ACC) prediction with respect to the DCNN architecture: (left) the number of neurons, (middle) window size, and (right) the number of hidden layers. Training methods: maximum likelihood (black), maximum labelwise accuracy (red) and maximum AUC (green). (Color figure online)

as balanced measures which can be used on class-imbalanced data. Mcc ranges from -1 to $+1$, with $+1$ representing a perfect prediction, 0 random prediction and -1 total disagreement between prediction and ground truth. AUC has a minimum value 0 and the best value 1.0. When there are more 2 different labels in a labeling problem, we may also use mean Mcc (denoted as $\bar{M}cc$) and mean AUC (denoted as $A\bar{U}C$), which are averaged over all the different labels.

4.3   Performance Comparison on Objective Functions

The architecture of the DCNN in DeepCNF model is mainly determined by the following 3 factors (see Fig. 1): (i) the number of hidden layers; (ii) the number of different neurons at each layer; and (iii) the window size at each layer. We compared three different methods for training the DeepCNF model: maximum likelihood, maximum labelwise accuracy, and maximum AUC for the prediction of three-label solvent accessibility (ACC), two-label order/disorder (DISO), and eight-label secondary structure element (SS8), respectively.

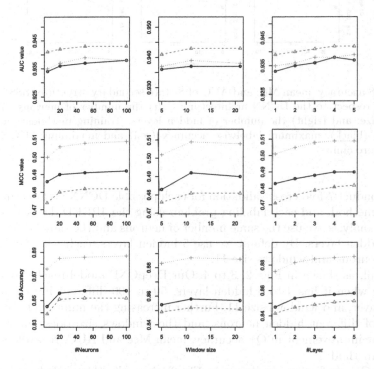

Fig. 3. Q2 accuracy, mean Mcc and AUC of disorder (DISO) prediction with respect to the DCNN architecture: (left) the number of neurons, (middle) window size, and (right) the number of hidden layers. Training methods: maximum likelihood (black), maximum labelwise accuracy (red) and maximum AUC (green). (Color figure online)

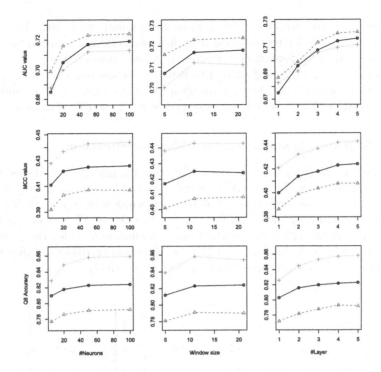

Fig. 4. Q8 accuracy, mean Mcc and AUC of 8-state secondary structure (SS8) prediction with respect to the DCNN architecture: (left) the number of neurons, (middle) window size, and (right) the number of hidden layers. Training methods: maximum likelihood (black), maximum labelwise accuracy (red) and maximum AUC (green). (Color figure online)

We conduct 7-fold cross-validation for each possible DCNN architecture, each training method, and each labeling problem using the JPRED dataset. To simplify the analysis, we use the same number of neurons and the same windows size for all hidden layers. By default we use 5 hidden layers, each with 50 different hidden neurons and windows size 11.

Overall, as shown in Figs. 2, 3 to 4, Our DeepCNF model reaches peak performance when it has 4 to 5 hidden layers, 50 to 100 different hidden neurons at each layer, and windows size 11. Further increasing the number of layers, the number of different hidden neurons, and the windows size does not result in significant improvement in Qx accuracy, mean Mcc and AUC, regardless of the training method.

For ACC prediction, as shown in Fig. 2, since the three labels are equally distributed, no matter what training methods are used, the best Q3 accuracy, the best mean Mcc and the best mean AUC are 0.69, 0.45, 0.82, respectively; For DISO prediction, since the two labels are highly imbalanced, as shown in Fig. 3, although all three training methods have similar Q2 accuracy 0.94, maximum-AUC obtains mean Mcc and AUC at 0.51 and 0.89, respectively, greatly outperforming the other two; For SS8 prediction, as shown in Fig. 4, since there are three rare labels (i.e., G for 3–10 helix, B for beta-bridge, and I for pi-helix), maximum-AUC has the overall mean Mcc at 0.44 and mean AUC at 0.86, respectively, much better than maximum labelwise accuracy, which has mean Mcc at 0.41 and mean AUC less than 0.8, respectively.

4.4 Performance Comparison with State-of-the-art

Programs to Compare. Since our method is *ab initio*, we do not compare it with consensus-based or template-based methods. Instead, we compare our method with the following ab initio predictors: (i) for ACC prediction, we compare to SPINE-X [14] and ACCpro5-ab [34]. SPINE-X uses neural networks (NN) while ACCpro5-ab uses bidirectional recurrent neural network (RNN); (ii) for DISO prediction, we compare to DNdisorder [12] and DisoPred3-ab [24]. DNdisorder uses deep belief network (DBN) while DisoPred3-ab uses support vector machine (SVM) and NN for prediction; (iii) for SS8 prediction, we compare our method with SSpro5-ab [34] and RaptorX-SS8 [46]. SSpro5-ab is based on RNN while RaptorX-SS8 uses conditional neural field (CNF) [39]. We cannot evaluate Zhous method [48] since it is not publicly available.

Overall Evaluation. Here we only compare our AUC-trained DeepCNF model (trained by the JPRED data) to the other state-of-the-art methods on the CASP and CAMEO datasets. As shown in Tables 1, 2 to 3, our AUC-trained DeepCNF model outperforms thPlease refer to appendix for a more detailed review for those problems and existing state-of-the art algorithms.e other predictors on all the three sequence labeling problems, in terms of the Qx accuracy, Mcc and AUC. When the label distribution is highly imbalanced, our method greatly exceeds the others in terms of Mcc and AUC. Specifically, for DISO prediction on the CASP data, our method achieves 0.55 Mcc and 0.89 AUC, respectively, greatly outperforming DNdisorder (0.37 Mcc and 0.81 AUC) and DisoPred3_ab (0.47 Mcc and 0.84 AUC). For SS8 prediction on the CAMEO data, our method obtains 0.42 Mcc and 0.83 AUC, respectively, much better than SSpro5_ab (0.37 Mcc and 0.78 AUC) and RaptorX-SS8 (0.38 Mcc and 0.79 AUC).

Sensitivity, Specificity, and Precision. Tables 4 and 5 list the sensitivity, specificity, and precision on each label obtained by our method and the other competing methods evaluated on the merged CASP and CAMEO data. Overall, at a high specificity level, our method obtains compatible or better precision and sensitivity for each label, especially for those rare labels such as G, I, B, S, T for SS8, and disorder state for DISO. Taking SS8 prediction as an example, for pi-helix (I), our method has sensitivity and precision 0.18 and 0.33 respectively, while the second best method obtains 0.03 and 0.12, respectively. For beta-bridge (B), our method obtains sensitivity and precision 0.13 and 0.42, respectively, while the second best method obtains 0.07 and 0.34, respectively (Table 6).

Table 1. Performance of solvent accessibility (ACC) prediction on the CASP and CAMEO data. Sens, spec, prec, Mcc and AUC are averaged on the 3 labels. The best values are shown in bold.

Method	CASP						CAMEO					
	Q3	Sens	Spec	Prec	Mcc	AUC	Q3	Sens	Spec	Prec	Mcc	AUC
OurMethod	**0.66**	**0.65**	**0.82**	**0.64**	**0.47**	**0.82**	**0.67**	**0.62**	**0.81**	**0.62**	**0.43**	**0.80**
SPINE-X	0.58	0.59	0.80	0.59	0.42	0.78	0.57	0.58	0.78	0.57	0.39	0.75
ACCpro5_ab	0.58	0.58	0.81	0.57	0.41	0.76	0.57	0.55	0.79	0.55	0.36	0.73

Table 2. Performance of order/disorder (DISO) prediction on the CASP and CAMEO data.

Method	CASP						CAMEO					
	Q2	Sens	Spec	Prec	Mcc	AUC	Q2	Sens	Spec	Prec	Mcc	AUC
OurMethod	**0.94**	**0.74**	**0.74**	**0.75**	**0.55**	**0.89**	**0.94**	**0.73**	**0.73**	**0.74**	**0.49**	**0.88**
DisoPred3_ab	0.94	0.67	0.67	0.72	0.47	0.84	0.94	0.71	0.71	0.71	0.42	0.83
DNdisorder	0.94	0.73	0.73	0.70	0.37	0.81	0.94	0.72	0.72	0.68	0.36	0.79

Table 3. Performance of 8-state secondary structure (SS8) prediction on the CASP and CAMEO data.

Method	CASP						CAMEO					
	Q8	Sens	Spec	Prec	Mcc	AUC	Q8	Sens	Spec	Prec	Mcc	AUC
OurMethod	**0.72**	**0.48**	**0.96**	**0.56**	**0.44**	**0.85**	**0.72**	**0.45**	**0.95**	**0.54**	**0.42**	**0.83**
RaptorX-SS8	0.65	0.42	0.95	0.50	0.41	0.81	0.66	0.40	0.94	0.48	0.38	0.79
SSpro5_ab	0.64	0.41	0.95	0.48	0.40	0.79	0.64	0.38	0.94	0.46	0.37	0.78

Table 4. Sensitivity, specificity, and precision of each solvent accessibility (ACC) label, tested on the combined CASP and CAMEO data.

ACC Label	Sensitivity			Specificity			Precision		
	Our	SpX*	Acc5**	Our	SpX	Acc5	Our	SpX	Acc5
B	**0.77**	0.74	0.75	**0.82**	0.81	0.80	**0.67**	0.63	0.62
M	**0.45**	0.36	0.34	**0.80**	0.78	0.79	**0.54**	0.48	0.46
E	**0.71**	0.67	0.63	**0.82**	0.79	0.80	**0.67**	0.62	0.61

       * SPINEX, ** ACCpro5_ab

Table 5. Sensitivity, specificity, and precision of each disorder label on the combined CASP and CAMEO data.

DISO Label	Sensitivity			Specificity			Precision		
	Our	Diso*	DN**	Our	Diso	DN	Our	DISO	DN
0	**0.96**	0.96	0.89	0.51	0.41	**0.55**	**0.95**	0.94	0.93
1	0.51	0.41	**0.55**	**0.96**	0.96	0.89	**0.54**	0.51	0.47

* DisoPred3_ab; ** DNdisorder

Table 6. Sensitivity, specificity, and precision of each 8-state secondary structure label on the combined CASP and CAMEO data.

SS8 Label	Sensitivity			Specificity			Precision		
	Our	Rapt*	SSp5**	Our	Rapt	SSp5	Our	Rapt	SSp5
H	**0.91**	0.89	0.90	0.92	**0.93**	0.93	**0.85**	0.84	0.84
G	**0.28**	0.21	0.19	**0.99**	0.98	0.97	**0.47**	0.43	0.41
I	**0.18**	0.03	0.02	**0.99**	0.98	0.98	**0.33**	0.12	0.06
E	**0.84**	0.78	0.77	**0.94**	0.91	0.89	**0.73**	0.72	0.69
B	**0.13**	0.05	0.07	**0.99**	0.99	0.99	**0.42**	0.33	0.34
T	**0.56**	0.49	0.51	**0.95**	0.93	0.93	**0.56**	0.50	0.49
S	**0.29**	0.21	0.18	**0.97**	0.96	0.97	**0.51**	0.43	0.45
L	0.61	0.62	**0.63**	0.86	0.86	**0.87**	**0.58**	0.58	0.54

* RaptorX-SS8; ** SSpro5_ab

5 Discussions

We have presented a novel training algorithm that directly maximizes the empirical AUC to learn DeepCNF model (DCNN+CRF) from imbalanced structured data. We also studied the behavior of three training methods: maximum-likelihood, maximum labelwise accuracy, and maximum-AUC, on three real-world protein sequence labeling problems, in which the label distribution varies from equally distributed to highly imbalanced. Evaluated by AUC and Mcc, our maximum-AUC training method achieves the state-of-the-art performance in predicting solvent accessibility, disordered regions, and 8-state secondary structure.

Instead of using a linear-chain CRF, we may model a protein by Markov Random Fields (MRF) to capture long-range residue interactions [47]. As suggested in [41], the predicted residue-residue contact information could further contribute to disorder prediction under the MRF model. In addition to the three protein sequence labeling problems tested in this work, our maximum-AUC training algorithm could be applied to many sequence labeling problems with imbalanced label distributions [20]. For example, in post-translation modification (PTM) site prediction, the phosphorylation and methylation sites occur much less frequently than normal residues [2].

Acknowledgments. The authors are grateful to the computing power provided by the UChicago Beagle and RCC allocations. The authors are also grateful to the National Institutes of Health [R01GM0897532 to J.X.] and National Science Foundation [DBI-0960390 to J.X.].

References

1. Andreeva, A., Howorth, D., Chothia, C., Kulesha, E., Murzin, A.G.: Scop2 prototype: a new approach to protein structure mining. Nucleic Acids Res. **42**(D1), D310–D314 (2014)
2. Blom, N., Sicheritz-Pontén, T., Gupta, R., Gammeltoft, S., Brunak, S.: Prediction of post-translational glycosylation and phosphorylation of proteins from the amino acid sequence. Proteomics **4**(6), 1633–1649 (2004)
3. Calders, T., Jaroszewicz, S.: Efficient AUC optimization for classification. In: Kok, J.N., Koronacki, J., Lopez de Mantaras, R., Matwin, S., Mladenič, D., Skowron, A. (eds.) PKDD 2007. LNCS(LNAI), vol. 4702, pp. 42–53. Springer, Heidelberg (2007). doi:10.1007/978-3-540-74976-9_8
4. Chawla, N.V., Bowyer, K.W., Hall, L.O., Kegelmeyer, W.P.: Smote: synthetic minority over-sampling technique. J. Artif. Intell. Res. **16**, 321–357 (2002)
5. Chen, L.-C., Papandreou, G., Kokkinos, I., Murphy, K., Yuille, A.L., Semantic image segmentation with deep convolutional nets, fully connected CRFs. arXiv preprint arXiv: 1412.7062 (2014)
6. Chothia, C.: The nature of the accessible and buried surfaces in proteins. J. Mol. Biol. **105**(1), 1–12 (1976)
7. Cortes, C., Mohri, M.: Auc optimization vs. error rate minimization. Adv. Neural Inf. Process. Syst. **16**(16), 313–320 (2004)
8. De Lannoy, G., François, D., Delbeke, J., Verleysen, M.: Weighted conditional random fields for supervised interpatient heartbeat classification. IEEE Trans. Biomed. Eng. **59**(1), 241–247 (2012)
9. Di Lena, P., Nagata, K., Baldi, P.: Deep architectures for protein contact map prediction. Bioinformatics **28**(19), 2449–2457 (2012)
10. Dill, K.A.: Dominant forces in protein folding. Biochemistry **29**(31), 7133–7155 (1990)
11. Drozdetskiy, A., Cole, C., Procter, J., Barton, G.J.: Jpred4: a protein secondary structure prediction server. Nucleic Acids Res., gkv332 (2015)
12. Eickholt, J., Cheng, J.: Dndisorder: predicting protein disorder using boosting and deep networks. BMC Bioinf. **14**(1), 88 (2013)
13. Estabrooks, A., Jo, T., Japkowicz, N.: A multiple resampling method for learning from imbalanced data sets. Comput. Intell. **20**(1), 18–36 (2004)
14. Faraggi, E., Xue, B., Zhou, Y.: Improving the prediction accuracy of residue solvent accessibility and real-value backbone torsion angles of proteins by guided-learning through a two-layer neural network. Proteins: Struct., Funct., Bioinf. **74**(4), 847–856 (2009)
15. Ferri, C., Flach, P., Hernández-Orallo, J.: Learning decision trees using the area under the ROC curve. In: ICML, vol. 2, pp. 139–146 (2002)
16. Gross, S.S., Russakovsky, O., Do, C.B., Batzoglou, S.: Training conditional random fields for maximum labelwise accuracy. In: Advances in Neural Information Processing Systems, pp. 529–536 (2006)

17. Haas, J., Roth, S., Arnold, K., Kiefer, F., Schmidt, T., Bordoli, L., Schwede, T.: The protein model portala comprehensive resource for protein structure and model information. Database **2013**, bat031 (2013)
18. Hanley, J.A., McNeil, B.J.: The meaning and use of the area under a receiver operating characteristic (ROC) curve. Radiology **143**(1), 29–36 (1982)
19. He, B., Wang, K., Liu, Y., Xue, B., Uversky, V.N., Dunker, A.K.: Predicting intrinsic disorder in proteins: an overview. Cell Res. **19**(8), 929–949 (2009)
20. He, H., Garcia, E., et al.: Learning from imbalanced data. IEEE Trans. Knowl. Data Eng. **21**(9), 1263–1284 (2009)
21. Herschtal, A., Raskutti, B.: Optimising area under the ROC curve using gradient descent. In: Proceedings of the Twenty-first International Conference on Machine Learning, p. 49. ACM (2004)
22. Hinton, G., Deng, L., Dong, Y., Dahl, G.E., Mohamed, A., Jaitly, N., Senior, A., Vanhoucke, V., Nguyen, P., Sainath, T.N., et al.: Deep neural networks for acoustic modeling in speech recognition: the shared views of four research groups. IEEE Sig. Process. Mag. **29**(6), 82–97 (2012)
23. Joachims, T.: A support vector method for multivariate performance measures. In: Proceedings of the 22nd International Conference on Machine Learning, pp. 377–384. ACM (2005)
24. Jones, D.T., Cozzetto, D.: Disopred3: precise disordered region predictions with annotated protein-binding activity. Bioinformatics **31**(6), 857–863 (2015)
25. Joo, K., Joung, I., Lee, S.Y., Kim, J.Y., Cheng, Q., Manavalan, B., Joung, J.Y., Heo, S., Lee, J., Nam, M., et al.: Template based protein structure modeling by global optimization in CASP11. Proteins: Struct., Funct., Bioinform. (2015)
26. Kabsch, W., Sander, C.: Dictionary of protein secondary structure: pattern recognition of hydrogen-bonded and geometrical features. Biopolymers **22**(12), 2577–2637 (1983)
27. Krizhevsky, A., Sutskever, I., Hinton, G.E.: Imagenet classification with deep convolutional neural networks. In: Advances in Neural Information Processing Systems, pp. 1097–1105 (2012)
28. Kryshtafovych, A., Barbato, A., Fidelis, K., Monastyrskyy, B., Schwede, T., Tramontano, A.: Assessment of the assessment: evaluation of the model quality estimates in CASP10. Proteins: Struct., Funct., Bioinform. **82**(S2), 112–126 (2014)
29. Lafferty, J., McCallum, A., Pereira, F.C.: Conditional random fields: probabilistic models for segmenting and labeling sequence data (2001)
30. LeCun, Y., Bottou, L., Bengio, Y., Haffner, P.: Gradient-based learning applied to document recognition. Proc. IEEE **86**(11), 2278–2324 (1998)
31. Liu, D.C., Nocedal, J.: On the limited memory BFGS method for large scale optimization. Math. Program. **45**(1–3), 503–528 (1989)
32. Liu, X.-Y., Jianxin, W., Zhou, Z.-H.: Exploratory undersampling for class-imbalance learning. IEEE Trans. Syst. Man, Cybern. Part B (Cybern.) **39**(2), 539–550 (2009)
33. Ma, J., Wang, S.: Acconpred: Predicting solvent accessibility and contact number simultaneously by a multitask learning framework under the conditional neural fields model. BioMed Res. Int. **2015** (2015)
34. Magnan, C.N., Baldi, P.: SSpro/ACCpro 5: almost perfect prediction of protein secondary structure and relative solvent accessibility using profiles, machine learning and structural similarity. Bioinformatics **30**(18), 2592–2597 (2014)
35. Monastyrskyy, B., Fidelis, K., Moult, J., Tramontano, A., Kryshtafovych, A.: Evaluation of disorder predictions in CASP9. Proteins: Struct., Funct., Bioinform. **79**(S10), 107–118 (2011)

36. Narasimhan, H., Agarwal, S.: A structural SVM based approach for optimizing partial AUC. In: Proceedings of the 30th International Conference on Machine Learning, pp. 516–524 (2013)
37. Oldfield, C.J., Dunker, A.K.: Intrinsically disordered proteins and intrinsically disordered protein regions. Ann. Rev. Biochem. **83**, 553–584 (2014)
38. Pauling, L., Corey, R.B., Branson, H.R.: The structure of proteins: two hydrogen-bonded helical configurations of the polypeptide chain. Proc. Nat. Acad. Sci. **37**(4), 205–211 (1951)
39. Peng, J., Bo, L., Xu, J.: Conditional neural fields. In: Advances in Neural Information Processing Systems, pp. 1419–1427 (2009)
40. Rosenfeld, N., Meshi, O., Globerson, A., Tarlow, D.: Learning structured models with the AUC loss and its generalizations. In: Proceedings of the Seventeenth International Conference on Artificial Intelligence and Statistics, pp. 841–849 (2014)
41. Schlessinger, A., Punta, M., Rost, B.: Natively unstructured regions in proteins identified from contact predictions. Bioinformatics **23**(18), 2376–2384 (2007)
42. Sillitoe, I., Lewis, T.E., Cuff, A., Das, S., Ashford, P., Dawson, N.L., Furnham, N., Laskowski, R.A., Lee, D., Lees, J.G., et al.: Cath: comprehensive structural and functional annotations for genome sequences. Nucleic Acids Res. **43**(D1), D376–D381 (2015)
43. Wang, S., Li, W., Liu, S., Jinbo, X.: Raptorx-property: a web server for protein structure property prediction. Nucleic Acids Res., gkw306 (2016)
44. Wang, S., Peng, J., Ma, J., Jinbo, X.: Protein secondary structure prediction using deep convolutional neural fields. Sci. Rep. **6** (2016)
45. Wang, S., Weng, S., Ma, J., Tang, Q.: Deepcnf-d: predicting protein order/disorder regions by weighted deep convolutional neural fields. Int. J. Mol. Sci. **16**(8), 17315–17330 (2015)
46. Wang, Z., Zhao, F., Peng, J., Jinbo, X.: Protein 8-class secondary structure prediction using conditional neural fields. Proteomics **11**(19), 3786–3792 (2011)
47. Jinbo, X., Wang, S., Ma, J.: Protein Homology Detection Through Alignment of Markov Random Fields: Using MRFalign. SpringerBriefs in Computer Science. Springer, Heidelberg (2015)
48. Zhou, J., Troyanskaya, O.G.: Deep supervised, convolutional generative stochastic network for protein secondary structure prediction. arXiv preprint arXiv:1403.1347 (2014)

A Novel Incremental Covariance-Guided One-Class Support Vector Machine

Takoua Kefi[1(✉)], Riadh Ksantini[1,2], Mohamed Bécha Kaâniche[1], and Adel Bouhoula[1]

[1] Sécurité Numérique, Higher School of Communication of Tunis, Tunis, Tunisia
{takoua.kefi,medbecha.kaaniche,adel.bouhoula}@supcom.tn
[2] University of Windsor, 401, Sunset Avenue, Windsor, ON, Canada
ksantini@uwindsor.ca

Abstract. Covariance-guided One-Class Support Vector Machine (COSVM) is a very competitive kernel classifier, as it emphasizes the low variance projectional directions of the training data, which results in high accuracy. However, COSVM training involves solving a constrained convex optimization problem, which requires large memory and enormous amount of training time, especially for large scale datasets. Moreover, it has difficulties in classifying sequentially obtained data. For these reasons, this paper introduces an incremental COSVM method by controlling the possible changes of support vectors after the addition of new data points. The control procedure is based on the relationship between the Karush-Kuhn-Tuker conditions of COSVM and the distribution of the training set. Comparative experiments have been carried out to show the effectiveness of our proposed method, both in terms of execution time and classification accuracy. Incremental COSVM results in better classification performance when compared to canonical COSVM and contemporary incremental one-class classifiers.

Keywords: One-class classification · Incremental learning · Support Vector Machine · Covariance

1 Introduction

One-Class Classification is considered as one of the most challenging areas of machine learning. It has gained a lot of attention and it can be found in many practical applications such as medical analysis [1], face recognition [2], authorship verification [3].

To solve one-class classification problems, several methods have been proposed and different concrete models have been constructed. However, the key limitation of the existing categories of one-class classification methods is that none of them consider the full scale of information available. In boundary-based methods, like the One-Class Support Vector Machine (OSVM) [4] or Support Vector Data Description (SVDD) [5], only boundary data points are considered to build the model, and the overall class is not completely considered.

© Springer International Publishing AG 2016
P. Frasconi et al. (Eds.): ECML PKDD 2016, Part II, LNAI 9852, pp. 17–32, 2016.
DOI: 10.1007/978-3-319-46227-1_2

Besides, unlike multi-class classification problems, the low variance directions of the target class distribution are crucial for one-class classification. In [6], it has been shown that projecting the data in the high variance directions (like PCA) will result in higher error (bias), while retaining the low variance directions will lower the total error. As a solution, Naimul Mefraz Khan et al. proposed in [7] to put more emphasis on the low variance directions while keeping the basic formulation of OSVM untouched, so that we still have a convex optimization problem with a unique global solution, that can be reached easily using numerical methods. Covariance Guided One-Class Support Vector Machine (COSVM) is a powerful kernel method for one-class classification, inspired from the Support Vector Machine (SVM), where the covariance matrix is incorporated into the dual optimization problem of OSVM. The covariance matrix is estimated in the kernel space. Concerning its classification performance, success of COSVM has been shown when compared to SVDD and OSVM. However, there are still some difficulties associated with COSVM application in real case problems, where data are sequentially obtained and learning has to be done from the first data. Besides, COSVM requires large memory and enormous amount of training time, especially for large dataset.

Implementations for the existing One-Class Classification methods assume that all the data are provided in advance, and learning process is carried out in the same step. Hence, these techniques are referred to as batch learning. Because of this limitation, batch techniques show a serious performance degradation in real-word applications when data are not available from the very beginning. For such situation, a new learning strategy is required. Opposed to batch learning, incremental learning is more effective when dealing with non-stationary or very large amount of data. Thus, it finds its application in a great variety of situations such as visual tracking [8], software project estimation [9], brain computer interfacing [10].

It has been defined in [11] with 4 criteria:

1. it should be able to learn additional information from new data
2. it should not require access to the original data
3. it should preserve previously acquired knowledge and use it to update an existing classifier.
4. it should be able accommodate new outliers and target samples.

Several learning algorithms have been studied and modified to incremental procedures, able to learn through time. Cauwenberghs and Poggio [12] proposed an online learning algorithm of Support Vector Machine (SVM). Their algorithm changes the coefficient of original Support Vectors (SV), and retains the Karuch-Kuhn-Tucker (KKT) conditions on all previously training data as a new sample acquired. Their approach have been extended by Laskov et al. [13] to OSVM. However, the performance evaluation was only based on multi-class SVM. From their side, Manuel Davy et al. introduced in [14] an online SVM for abnormal events detection. They proposed a strategy to perform abnormality detection over various signals by extracting relevant features from the considered signal and detecting novelty, using an incremental procedure. Incremental SVDD proposed in [15] is also based on the control of the variation of the KKT conditions as new samples are added. An other approach to improve the classification performance is introduced in [16]. Incremental Weighted One-Class Support Vector Machine (WOCSVM) is an extension of incremental OSVM. The proposed algorithm aims to assign weights to each object of the training set, then it controls its influence on the shape of the decision boundary.

All the proposed Incremental One-Class SVM inherit the problem of classic SVM method which uses only boundary points to build a model, regardless of the spread of the remaining data. Also, none of them emphasizes the low variance direction, which results in performance degradation. Therefore, in this paper we try to solve mainly this problem by using an incremental COSVM (iCOSVM) approach. In fact, iCOSVM has the advantage of incrementally emphasizing the low variance direction to improve classification performance, which is not the case for classical incremental one-class models. Our preposition aims to take advantages from the accuracy of COSVM procedure and we prove that it is a good candidate for learning in non-stationary environments.

The rest of the paper is organized as follows. Section 2 reviews the canonical COSVM method since it is the basis of our proposed method. In Sect. 3, we present in details the mathematical derivation of iCOSVM and we describe the incremental algorithm. Section 4 presents our experimental studies and comparison with canonical COSVM and other incremental one-class classifiers. Finally, Sect. 5 contains some concluding remarks and perspectives.

2   The COSVM Method

Mathematically, OSVM tries to find the hyperplane that separates the training data from the origin with maximum margin. It can be modeled by the following dual problem, formulated using Lagrange multipliers.

$$\min_{\alpha} \frac{1}{2}\alpha^T \mathbf{K}\alpha + b\left(1 - \sum_{i=1}^{N}\alpha_i\right). \tag{1}$$

$$s.t. \quad 0 \leq \alpha_i \leq \frac{1}{\nu N} = C, \quad \sum_{i=1}^{N}\alpha_i = 1.$$

Here, $\nu \in (0,1]$ is a key parameter that controls the fraction of outliers and that of support vectors, C is the penalty weight punishing the misclassified training examples, $\mathbf{K}(x_i, x_j) = \langle \Phi(x_i), \Phi(x_j)\rangle, \forall i, j \in \{1, 2, \ldots, N\}$ is the kernel matrix for the training data, and α are the Lagrange multipliers to be determined.

The covariance matrix is then plugged in the dual problem and a parameter $\eta \in [0, 1]$ is introduced to control the contribution of the kernel matrix $\mathbf{K}$ and the covariance matrix to the objective function. The COSVM optimization problem can be written as follows:

$$\min_{\alpha} W(\alpha, b) = \frac{1}{2}\alpha^T (\eta \mathbf{K} + (1-\eta)\Delta)\,\alpha - b\left(1 - \sum_{i=1}^{N}\alpha_i\right). \qquad (2)$$

$$s.t. \quad 0 \le \alpha_i \le C, \quad \sum_{i=1}^{N}\alpha_i = 1,$$

where $\Delta = \mathbf{K}(I - 1_N)\mathbf{K}^T$. The control parameter η can take values from 0 to 1.

3   The Incremental COSVM Method

The key of our method is to construct a solution recursively, by adding one point at a time [12], and retain the Karush-Kuhn-Tucker Conditions on all previously acquired data.

3.1   Karush-Kuhn-Tucker Conditions

Both the kernel matrix $\mathbf{K}$ and the covariance matrix Δ are positive definite [17]. Therefore, the proposed method still results in a convex optimization problem. Thus, the solution to this optimization problem will have one global optimum solution and can be solved efficiently using a mathematical method. Karush-Kuhn-Tucker (KKT) conditions [18] are among the most important theoretical optimization methods.

First, let's note

$$\Gamma = (\eta \mathbf{K} + (1-\eta)\Delta).$$

The slopes g_i of the cost function W in equation (2) are expressed using the KKT conditions as:

$$g_i = \frac{\partial W}{\partial \alpha} = \sum_{j}\Gamma_{i,j}\alpha_j - b \begin{cases} \ge 0; & \alpha_i = 0 \\ = 0; & 0 < \alpha_i < C \\ \le 0; & \alpha_i = C \end{cases} \qquad (3)$$

$$\frac{\partial W}{\partial b} = 1 - \sum \alpha = 0. \qquad (4)$$

According to the KKT conditions above, the target training data can be divided into three categories, shown in Figs. 1, 2 and 3:

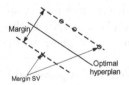

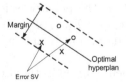

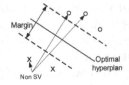

Fig. 1. Subset $\mathcal{S}$. $g_i = 0$ and $0 < \alpha_i < C$

Fig. 2. Subset $\mathcal{E}$. $g_i < 0$ and $\alpha_i = C$

Fig. 3. Subset $\mathcal{O}$. $g_i > 0$ and $\alpha_i = 0$

1. Margin or unbounded Support Vectors are training points $\mathcal{S} = \{i/0 < \alpha_i < C\}$,
2. Error or bounded Support Vectors $\mathcal{E} = \{i/\alpha_i = C\}$,
3. Non Support Vectors $\mathcal{O} = \{i/\alpha_i = 0\}$.

The KKT conditions have to be maintained for all trained data before a new data x_c is added and preserved after the new data is trained. Hence, the change of Lagrange multipliers $\Delta\alpha$ is determined to hold the KKT.

3.2 Adiabatic Increments

To maintain the equilibrium of the KKT conditions expressed in Eqs. (3) and (4), we express them differentially:

$$\Delta g_i = \Gamma_{i,c}\alpha_c + \sum_j \Gamma_{i,j}\alpha_j - \Delta b, \tag{5}$$

$$\Delta\alpha_c + \sum_{j \in \mathcal{S}} \alpha_j = 0. \tag{6}$$

The two equations above can be written as:

$$\begin{bmatrix} \Delta g_c \\ \Delta g_s \\ \Delta g_r \\ 0 \end{bmatrix} = \begin{bmatrix} 1 & \Gamma_{c,s} \\ 1 & \Gamma_{s,s} \\ 1 & \Gamma_{r,s} \\ 0 & 1 \end{bmatrix} \begin{bmatrix} -\Delta b \\ \Delta\alpha_s \end{bmatrix} + \Delta\alpha_c \begin{bmatrix} \Gamma_{c,c} \\ \Gamma_{s,c} \\ \Gamma_{r,c} \\ 1 \end{bmatrix}. \tag{7}$$

Since $\Delta g_i = 0$ when $i \in \mathcal{S}$ (it remains zero), lines 2 and 4 of the system (7) can be re-written as:

$$\begin{bmatrix} 0 \\ 0 \end{bmatrix} = \begin{bmatrix} 0 & 1 \\ 1 & \Gamma_{s,s} \end{bmatrix} \begin{bmatrix} -\Delta b \\ \Delta\alpha_s \end{bmatrix} + \Delta\alpha_c \begin{bmatrix} 1 \\ \Gamma_{s,c} \end{bmatrix}. \tag{8}$$

Thus, we can express the dependence of $\Delta\alpha_i$, $i \in \mathcal{S}$ and $\Delta g_i = 0$, $i \notin \mathcal{S}$ on $\Delta\alpha_c$ as the following:

$$\begin{bmatrix} -\Delta b \\ \Delta\alpha_s \end{bmatrix} = -\mathbf{R} \begin{bmatrix} 1 \\ \Gamma_{s,c} \end{bmatrix} \Delta\alpha_c, \tag{9}$$

with

$$\mathbf{R} = \begin{bmatrix} 0 & 1 \\ 1 & \Gamma_{s,s} \end{bmatrix}^{-1}.$$

Here, $\Gamma_{s,s}$ is the kernel matrix whose entries are support vectors, and $\Gamma_{s,c}$ is a vector of kernels between the margin support vectors and the new candidate vector x_c.

The Eq. (9) gives the following:

$$\begin{bmatrix} -\Delta b \\ \Delta\alpha_s \end{bmatrix} = \beta\Delta\alpha_c,$$

where

$$\beta = -\mathbf{R} \begin{bmatrix} 1 \\ \Gamma_{s,c} \end{bmatrix}. \tag{10}$$

In equilibrium,

$$\begin{cases} \Delta b = -\beta_b \Delta \alpha_c, \\ \Delta \alpha_j = \beta_j \Delta \alpha_c, j \in \mathcal{S} \end{cases} \tag{11}$$

and $\beta_j = 0$ for all j outside the subset $\mathcal{S}$.

Substituting Eq. (11) into lines 1 and 3 of the system (7) leads to the desired relation between Δg_i and $\Delta \alpha_c$:

$$\Delta g_i = \gamma_i \, \Delta \alpha_c, i \in \{1...n\} \cup \{c\} \tag{12}$$

where we define

$$\begin{cases} \gamma_i = \Gamma_{i,c} + \sum_{j \in \mathcal{S}} \Gamma_{i,j} \beta_j, \ i \notin \mathcal{S} \\ \gamma_i = 0, \qquad\qquad\qquad i \in \mathcal{S} \end{cases} \tag{13}$$

3.3   Vectors Entering and Leaving a Subset

During the incremental procedure, a new example x_c can be added to the previous training set, and depending on the value of the calculated parameters g_c and α_c, the x_c is recognized as a support vector, an error vector or a data vector. If x_c is classified as a support vector, the set $\mathcal{S}$, as well as the classification boundaries and margins should be updated. Since the Margin Support Vectors are our first concern in a classification process, it is worth to focus on the changes in the subset $\mathcal{S}$. Besides, we can see from the Eqs. (10), (11), (12) and (13) of the previous section, that only $\boldsymbol{R}$ matrix needs to be computed to obtain all updated parameters. Let us consider a vector x_k entering to the subset $\mathcal{S}$. Using the Woodbury formula [19], $\boldsymbol{R}$ expands as:

$$\widetilde{\mathbf{R}} = \begin{bmatrix} R & 0 \\ 0 & 0 \end{bmatrix} + \frac{1}{\gamma_c} \begin{bmatrix} \beta \\ 1 \end{bmatrix} \begin{bmatrix} \beta \\ 1 \end{bmatrix}^T. \tag{14}$$

When x_k leaves $\mathcal{S}$, and using the same formula, $\boldsymbol{R}$ contracts as:

$$\widetilde{\mathbf{R}} = \mathbf{R}_{\overline{k},\overline{k}} - \mathbf{R}_{k,k} \mathbf{R}_{\overline{k},k} \mathbf{R}_{k,\overline{k}}. \tag{15}$$

3.4   The Impact of the Tradeoff Parameter η

The contribution of our kernel matrix $\mathbf{K}$ and the covariance matrix Δ is controlled using the parameter η. Figures 4, 5 and 6 present three different cases showing the impact of the covariance matrix on the direction of the separating hyperplane in the kernel space optimality. In Fig. 4, the optimal decision hyperplane is on the same direction as the high variance direction. Hence, the low variance direction will not improve the separating direction. That is why the

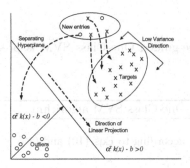

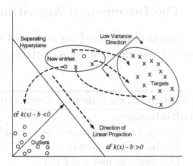

Fig. 4. Case 1: schematic depiction of the decision hyperplane for iCOSVM when the optimal control parameter value is $\eta = 1$. The optimal linear projection is along the direction of high variance.

Fig. 5. Case 2: Schematic depiction of the decision hyperplane for iCOSVM when the optimal control parameter value is $\eta = 0$. The optimal linear projection is along the direction of low variance.

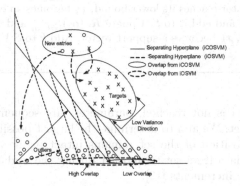

Fig. 6. General Case: schematic depiction of the decision hyperplane for iCOSVM when the optimal parameter value lies in between 0 and 1 ($0 < \eta < 1$). The linear projection direction for iOSVM (depicted by dotted arrows) results in higher overlap between the example target and hypothetical outlier data (circled by dotted boundary) than the iCOSVM projection direction (depicted by solid arrows and the overlap circled by solid boundary).

value of η should be set to 1 in order to eliminate the covariance matrix term. On the other hand, in Fig. 5, the directions of the optimal decision hyperplane and the low variance are parallel. Therefore the incremental OSVM (iOSVM) term (kernel matrix) is ignored by setting η to 0. However, in real world cases, it is very rare that the optimal decision hyperplane has the same direction as the low or high variance. For this reason, the value of η needs to be tuned so that we have less overlap between the linear projections of the target data and the outlier data. As Fig. 6 shows, by using optimal η value, iCOSVM can reduce the huge overlap caused by iOSVM projection.

3.5   The Incremental Algorithm

Our implementation of incremental Covariance-guided One-Class SVM is presented as pseudo-code in Algorithm 1.

Algorithm 1. Incremental Covariance-guided One-Class SVM algorithm

1. **Initialization**
2.   Compute R, and use it to compute β and γ according to Eqs. (10) and (13)
3.   Set α_c and $\Delta\alpha_c$ to 0
4.   Compute g_c using Eq. (3)
5. **While** $g_c < 0$ and $\alpha_c < C$ **do**
6.   if $g_c = 0$ then x_c is a margin support vector. Add c to $\mathcal{S}$ and equilibrium is reached. Set $\alpha_c = \Delta\alpha_c$ and update $(\alpha_i)_{i=1\ldots n}$. Update R, and b.
7.   if $g_c < 0$ then x_c is an error support vector. Add c to $\mathcal{E}$ and equilibrium is reached. Set $\alpha_c = C$, update $(\alpha_i)_{i=1\ldots n}$ and b.
8.   if a support vector reaches its upper bound, x_k becomes a non-support vector. Remove k from $\mathcal{S}$ and add it to $\mathcal{O}$. Update R, $(\alpha_i)_{i=1\ldots n}$ and b.
9.   if a support vector reaches its lower bound, x_k becomes an error support vector. Remove k from $\mathcal{S}$ and add it to $\mathcal{E}$. Update R, $(\alpha_i)_{i=1\ldots n}$ and b.
10.   if g_i becomes 0, x_k becomes a support vector. Add k to $\mathcal{S}$ Update R, $(\alpha_i)_{i=1\ldots n}$ and b.

If the equilibrium is not reached, parameters are sequentially moved until the equilibrium is met. We aim to determine the largest possible increment $\Delta\alpha_c$ so that the decomposition of the set remains intact, while accounting for the movement of some data from set to another during the update process. This is the idea of adiabatic increments [12].

4   Experimental Results

In this section, we present detailed experimental analysis and results for our proposed method, performed on artificially synthesized dataset and real world datasets. We have evaluated the performance of our method with two different experiment sets. In the first one, we compared the accuracy and time results with non-incremental COSVM, to tease out the advantage of our incremental model over batch learning model. In the second experiment set, we compare the iCOSVM performance against the performance of contemporary incremental one-class classifiers, to show the advantage of incrementally projecting data in low variance directions. For the implementation, we used Tax's data description toolbox [20] in Matlab. First, we provided an analysis of the effect of tuning the key control parameter η. This analysis will lead us to decide how to optimize the value of η for a particular dataset.

4.1 Optimising the Value of η

Cross validation can not be used to optimize the value of η. Therefore, a stopping criterion is considered to find the optimum value. We use a pre-defined lowest fraction of outliers allowed (f_{OL}) as a stopping criterion. For new datasets, we set η to 1, and we decrease its value, while observing the fraction of outliers. When it hits (f_{OL}), we stop and use the current value of η for the considered dataset. We have to mention that there is no conflict between (f_{OL}) and the OSVM parameter ν, and they can be set independently to fit the purpose of the dataset to be trained on. There is no strict conditions on how to choose the value of ν, it can be set to any value from 0 to 1 [21]. For our additional parameter (f_{OL}), it is set to any value between 0 to ν.

Table 1. Description of datasets.

Dataset name	Number of targets	Number of outliers	Number of features
Biomedical	67	127	5
Heart disease	160	137	13
Liver (diseased)	145	200	6
Liver (healthy)	200	145	6
Diabetes (absent)	268	500	8
Diabetes (present)	500	268	8
Arrythmia-1	237	183	278
Arrythmia-2	36	384	278
Chromosome-1	392	751	30
Chromosome-2	447	696	30
Chromosome-3	492	651	30
Chromosome-4	536	598	30

4.2 Datasets Used

We have used both artificially generated datasets and real world datasets in our experiments to tease out the effectiveness of our proposed method in different scenarios. For the experiments on artificially generated data, we have created a number of 2D two class data drawn from two different set of distributions: (1) Gaussian distribution with different covariance matrices. (2) Banana-shaped distribution with different variances. For each distribution, two datasets were generated, the first one with low overlap, and the second with high overlap. Each class of each dataset was used as a target class and outliers in turns, such that we evaluate the performance on 8 datasets (2 distributions × 2 classes × 2 overlaps). Figure 7 presents the plots of the generated datasets.

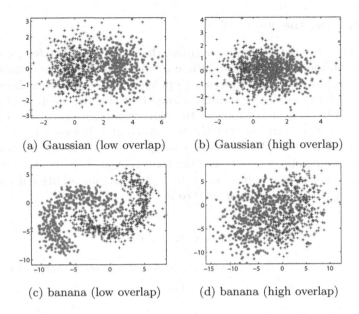

(a) Gaussian (low overlap)   (b) Gaussian (high overlap)

(c) banana (low overlap)   (d) banana (high overlap)

Fig. 7. Artificial datasets used for comparison. The two shapes denote two different classes generated from a pre-defined distribution. Each class was used as target and outlier in turns.

For the real world case, we focused on medical datasets as this domain is one of the key fields where one class classification is applied [1]. A detailed description of the used datasets can be found in Table 1. These datasets are collected from the UCI machine learning repository [22] and picked carefully, so that we have a variety of sizes and dimensions, and we can, then, test the robustness of our iCOSVM. As these datasets are originally two-class or multi-class, we used one of them as a target class and the other ones are kept outliers.

4.3   Experimental Protocol

To make sure that our results are not coincidental or overoptimistic, we used a cross-validation process [23]. The considered dataset was randomly split into 10 subsets of equal size. To build a model, one of the 10 subsets was removed, and the rest was used as the training data. The previously removed subset was added to the outliers and this whole set was used for testing. Finally the 10 accuracy estimates are averaged to provide the accuracy over all the models of a dataset. This guarantees that the achieved results were not a coincidence. Moreover, to measure the performance of one class classifier, the Receiver Operating Charac-teristic (ROC) curves [24] are usually used. The ROC curve presents a powerful measurement of the performance of studied classifier. It does not depend on the number of training or testing data points neither on the number of outliers, it only depends on rates of correct and incorrect target detection. To evaluate the methods, we have also used the Area Under Curve (AUC) [25] produced by the ROC curves, and we presented them in the results.

4.4 Classifiers

The iCOSVM was evaluated withe the comparison of its performance against the following classifiers' performance:

- COSVM: Since our incremental approach is built upon COSVM, this classifier has been described in details in Sect. 2.
- iOSVM: This method tries to find, recursively, the maximum margin hyperplane that separates targets from outliers.
- iSVDD: This method gives the sphere boundary description of the target data points with minimum volume.

The incremental classifiers, iOSVM, iSVDD and iCOSVM were implemented with the help of DDtools [20]. For the implementation of COSVM, the SVM-KM toolbox was used [26]. The radial basis kernel was used for kernelization. This kernel is calculated as $\mathcal{K}(x_i, x_j) = e^{-\|x_i - x_j\|^2/\sigma}$. It is proved to be robust and flexible [27]. Here, σ represents the positive "width" parameter. For η value optimization, the value of σ was set to 1. But, when comparing with other methods σ is optimized first. The parameter ν for COSVM, iOSVM and iCOSVM, also called *fraction of rejection* in the case of iSVDD was set to 0.2.

While optimizing η, the lowest threshold for the fraction of outliers (f_{OL}) was set to 0.1 (see Sect. 4.1). However, it is too difficult, and even not possible to define optimal values for the parameters f_{OL} and ν in real cases, where data points are unknown in the beginning of the classification process. Therefore, we have set both of the two parameters to 0.2.

4.5 Results and Discussion

To test the effectiveness of our proposed algorithm, we started by comparing iCOSVM with canonical COSVM on artificially generated datasets.

As we can see in Table 2, iCOSVM provides better results in terms of AUC values, on all datasets, by averaging over 10 different models. Figure 8 shows the average training time per model for artificial datasets of different sizes. The training speed of our algorithm is faster than the COSVM, mainly on large data sets, and presents insignificant variation as the size of the dataset increases. It has been shown in a number of recent studies [28] that incremental learning algorithms outperform batch learning algorithms in both speed and accuracy, because they provide cleaner solution.

In fact, the complexity for solving the convex optimization problem of COSVM is $O(N^3)$, where N is the number of training data points. Whereas, a key to efficiency of the iCOSVM algorithm lies in identifying performance bottlenecks associated with inverting matrices to solve the convex optimization problem. These operations were eliminated thanks to the introduction of

Table 2. Average AUC of COSVM and iCOSVM for the 8 artificial datasets. Each dataset has 1000 data points(best method in **bold**).

Dataset	COSVM	iCOSVM
Gauss. (low overlap)-1	98.33	**98.45**
Gauss. (low overlap)-2	98.28	**98.52**
Gauss. (high overlap)-1	81.47	**84.19**
Gauss. (high overlap)-2	87.14	**87.74**
Banana (low overlap)-1	98.46	**98.88**
Banana (low overlap)-2	98.33	**99.26**
Banana (high overlap)-1	85.73	**86.43**
Banana (high overlap)-2	84.88	**84.97**

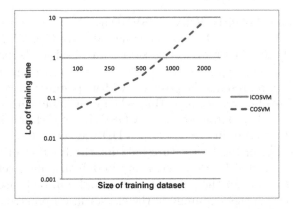

Fig. 8. Log of training times (per model) in *seconds* for COSVM and iCOSVM for the experiments on the artificial datasets of different sizes.

the Woodbery formula for the re-computation of the gradient, β and γ. This involves matrix-vector multiplications and recursive updates of the matrix $\mathbf{R}$, whose dimension is equal to the support vectors number N_s. The running time needed for an update of the matrix $\mathbf{R}$ is quadratic in the number of support vectors, which is much better than explicit inversion. Thus, in incremental learning, the complexity is $O(N_s^2)$, where $N_s \leq N$.

Tables 3 and 4 contain the average AUC for the incremental classifiers on the artificial and real datasets, respectively. As we can see, iCOSVM provides better results on all datasets. Specially in case of the biomedical and chromosome datasets, iCOSVM performs significantly better when compared to other methods. It is not surprising that iSVDD gives almost the worst accuracy values, as SVM and its derivatives are constructed to give the better separation [29].

We notice that iCOSVM outperforms the other classifiers as η values are in the neighborhood of 0.7, which puts more emphasize on the kernel matrix and fine-tune the contribution of the covariance matrix.

Table 3. Average AUC of each method for the 12 artificial datasets (best method in **bold**, second best *emphasized*).

Experiment	iSVDD	iOSVM	iCOSVM
Gauss. (low overlap)-1	90.7	*95.0*	**95.2**
Gauss. (low overlap)-2	92.2	*95.2*	**96.0**
Gauss. (high overlap)-1	71.7	*73.8*	**75.2**
Gauss. (high overlap)-2	69.8	*73.3*	**76.3**
Banana (low overlap)-1	95.2	*97.5*	**97.8**
Banana (low overlap)-2	92.2	*97.3*	**97.5**
Banana (high overlap)-1	74.8	*83.2*	**83.7**
Banana (high overlap)-2	74.0	*81.0*	**82.8**

Table 4. Average AUC of each method for the 12 real-world datasets (best method in **bold**, second best *emphasized*).

Experiment	iSVDD	iOSVM	iCOSVM
Biomedical	28.4	*77.7*	**82.9**
heart disease	49.4	*60.9*	**61.9**
Liver (diseased)	54.8	*69.1*	**69.6**
Liver (healthy)	52.5	*67.3*	**68.7**
Diabetes (present)	95.2	*97.5*	**97.8**
Diabetes (normal)	92.2	*97.3*	**97.5**
arrhythmia-1	74.8	*83.2*	**83.7**
arrhythmia-2	74.0	*81.0*	**82.8**
Chromosome-1	48.0	*65.2*	**78.2**
Chromosome-2	48.2	*63.2*	**73.4**
Chromosome-3	*47.4*	46.6	**55.5**
Chromosome-4	47.9	*58.6*	**70.8**

Since the process of computing the covariance matrix is done as a pre-processing and re-used during all training phase, in terms of training complexity, iCOSVM does not have additional overhead on top of the original iOSVM. Table 5 shows the average training times per model for both the artificial and the real-world datasets. As we expect, iCOSVM performs almost as fast as iOSVM, while providing better classification accuracy.

Also, we present some individual graphical results for the dataset models by plotting the actual ROC curves for a real world dataset. Figure 9 shows the ROC curves of the three incremental classifiers for four models of the chromosome dataset. The rule-of-thumb to judge the performance of a classifier from a ROC

Table 5. Average training times (per model) in *seconds* for iOSVM and iCOSVM for the experiments on the artificial and real-world datasets. Average training times (per model) in *seconds* for iOSVM and iCOSVM for the experiments on the artificial and real-world datasets.

Experiment	iOSVM	iCOSVM
Artificial datasets	0.0047	0.0046
Real-world datasets	0.0044	0.0043

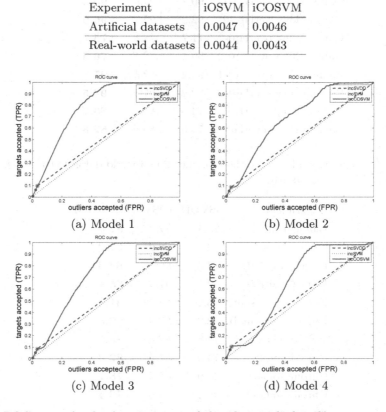

(a) Model 1   (b) Model 2

(c) Model 3   (d) Model 4

Fig. 9. ROC curves for the three incremental classifiers applied on Chromosome dataset

curve is "The best classification has the largest area under curve". We can clearly see from the Fig. 9 that iCOSVM indeed leads to better ROC curves.

5   Conclusion

In this paper, we have proposed an incremental Covariance-guided One-Class Support Vector Machine (iCOSVM) classification approach. iCOSVM improves upon the incremental One-Class Support Vector Machine method by the incorporation of the covariance matrix into the optimization problem. The new introduced term emphasized the projection in the directions of low variances of the training datasets. The contribution of both Kernel and covariance matrices are controlled via a parameter that was tuned efficiently for optimum performance. iCOSVM takes advantages from the high accuracy of the canonical

Covariance-guided One-Class Support Vector Machine (COSVM). We have presented detailed experiments on several artificial and real-world datasets, where we compared our method against contemporary batch and incremental learning methods. Results have shown the superiority of the method. Future works will consist in validating these results on strong applications such as face recognition, anomaly detection, etc.

References

1. Gardner, A.B., Krieger, A.M., Vachtsevanos, G., Litt, B.: One-class novelty detection for seizure analysis from intracranial EEG. J. Mach. Learn. Res. **7**, 1025–1044 (2006)
2. Zeng, Z., Fu, Y., Roisman, G.I., Wen, Z., Hu, Y., Huang, T.S.: Spontaneous emotional facial expression detection. J. Multimedia **1**, 1–8 (2006)
3. Koppel, M., Schler, J.: Authorship verification as a one-class classification problem. In: 21st International Conference on Machine Learning, pp. 62–68. ACM, New York (2004)
4. Schölkopf, B., Platt, J., Shawe-Taylor, J., Smola, A., Williamson, R.: Estimating the support of a high-dimensional distribution. Neural Comput. **13**, 1443–1471 (2001)
5. Tax, D.M.J.: One-class classification. Concept-learning in the absence of counter-examples. Ph.D. Thesis, Technichal University of Delft, June 2001
6. Tax, D.M.J., Müller, K.R.: Feature extraction for one-class classification. In: Artificial Neural Networks and Neural Information Processing, pp. 342–349. IEEE Press (2003)
7. Khan, N.M., Ksantini, R., Ahmad, I.S., Guan, L.: Covariance-guided one-class support vector machine. Pattern Recognit. **47**, 2165–2177 (2014)
8. Ross, D.A., Lim, J., Lin, R.S., Yang, M.H.: Incremental learning for robust visual tracking. Int. J. Comput. Vis. **77**, 125–141 (2007)
9. Giraud-Carrier, C.: A note on the utility of incremental learning. AI Commun. **13**, 215–223 (2000)
10. Blankertz, B., Dornhege, G., Schäfer, C., Krepki, R., Kohlmorgen, J., Müller, K.R., Kunzmann, V., Losch, F., Curio, G.: Boosting bit rates and error detection for the classification of fast-paced motor commands based on single-trial EEG analysis. IEEE Trans. Neural Syst. Rehabil. Eng. **11**, 127–131 (2003)
11. Polikar, R., Upda, L., Atish, S., Upda, S., Honavar, V.: Learn++: an incremental learning algorithm for supervised neural networks. IEEE Syst. Man Cybern. **31**, 497–508 (2002)
12. Cauwenberghs, G., Poggio, T.: Incremental and decremental support vector machine learning. In: Neural Information Processing Systems, pp. 409–415. NIPS (2000)
13. Laskov, P., Gehl, C., Krüger, S.: Incremental support vector learning: analysis, implementation and applications. J. Mach. Learn. Res. **7**, 1909–1936 (2006)
14. Davy, M., Desorby, F., Gretton, A., Doncarli, C.: An online support vector machine for abnormal events detection. Sig. Process. **86**, 2009–2025 (2005)
15. Hua, X., Ding, S.: Incremental learning algorithm for support vector data description. J. Softw. **6**, 1166–1173 (2011)
16. Krawczyk, B., Woźniak, M.: Incremental weighted one-class classier for mining stationary data streams. J. Comput. Sci. **9**, 19–25 (2015)

17. Horn, R.A., Charles, R.: Matrix Analysis. Cambridge University Press (1990)
18. Bazaraa, M.S., Sherali, H.D., Shetty, C.M.: The Fritz John and Karush-Kuhn-Tucker optimality conditions. In: Nonlinear Programming: Theory and Algorithms. Wiley, Singapore (2006)
19. Hager, W.W.: Updating the inverse of a matrix. Soc. Ind. Appl. Math. **31**, 221–239 (1989)
20. Tax, D.M.J.: DDtools, the Data Description Toolbox for Matlab, version 2.1.2 (2015)
21. Parra, L., Deco, L., Miesbach, S.: Statistical independence and novelty detection with information preserving nonlinear maps. Neural Comput. **8**, 260–269 (1996)
22. Lichman, M.: Machine Learning Repository. University of California, Iverine, School of Information and computer Sciences (2013)
23. Alippi, C., Roveri, M.: Virtual k-fold cross validation: an effective method for accuracy assessment. In: International Joint Conference on Neural Networks, pp. 1–6. IEEE Conference Publications (2010)
24. Fawcett, T.: An introduction to ROC analysis. Pattern Recogn. Lett. **27**, 861–874 (2006)
25. Hanley, J., McNeil, B.J.: A method of comparing the areas under receiver operating characteristic curves derived from the same cases. Radiology **148**, 839–843 (1983)
26. Canu, S., Grandvalet, Y., Guigue, V., Rakotomamonjy, A.: SVM and Kernel Methods Matlab Toolbox. Perception Systmes et Information, INSA de Rouen, Rouen, France (2005)
27. Cristianini, N., Shawe-Taylor, J.: An Introduction to Support Vector Machines and Other Kernel-Based Learning Methods. Cambridge University Press, Cambridge (2000)
28. Lorenzo, C.Y., Guan, B.R., Lu, L., Liang, Y.: Incremental and decremental affinity propagation for semisupervised clustering in multispectral images. IEEE Trans. Geosci. Remote Sens. **51**, 1666–1679 (2013)
29. Wei, H., Li, J.: Credit scoring based on eigencredits and SVDD. In: Applied Informatics and Communication, pp. 32–40 (2011)

Learning Efficiently in Semantic Based Regularization

Michelangelo Diligenti[1(✉)], Marco Gori[1], and Vincenzo Scoca[2]

[1] Dipartimento di Ingegneria dell'Informazione e Scienza Matematiche, Siena, Italy
{diligmic,marco}@diism.unisi.it
[2] IMT School for Advanced Studies, Lucca, Italy
vincenzo.scoca@imtlucca.it

Abstract. Semantic Based Regularization (SBR) is a general framework to integrate semi-supervised learning with the application specific background knowledge, which is assumed to be expressed as a collection of first-order logic (FOL) clauses. While SBR has been proved to be a useful tool in many applications, the underlying learning task often requires to solve an optimization problem that has been empirically observed to be challenging. Heuristics and experience to achieve good results are therefore the key to success in the application of SBR. The main contribution of this paper is to study why and when training in SBR is easy. In particular, this paper shows that exists a large class of prior knowledge that can be expressed as convex constraints, which can be exploited during training in a very efficient and effective way. This class of constraints provides a natural way to break the complexity of learning by building a training plan that uses the convex constraints as an effective initialization step for the final full optimization problem. Whereas previous published results on SBR have employed Kernel Machines to approximate the underlying unknown predicates, this paper employs Neural Networks for the first time, showing the flexibility of the framework. The experimental results show the effectiveness of the training plan on categorization of real world images.

Keywords: Statistical Relational Learning · First Order Logic · Convex optimization

1 Introduction

Semantic Based Regularization [5] is a Statistical Relational Learning (SRL) framework, which integrates the ability to learn from examples and data distributions, like in traditional semi-supervised learning, with the inference process typical of high level background knowledge typically used in logic inference. Prior knowledge in SBR is expressed via a set of FOL clauses expressing relationships among the tasks, or relationships among the patterns, or providing a partial definition of the mapping between the input and the output. The main advantage of SBR over other Statistical Relational Learning approaches like Markov Logic

© Springer International Publishing AG 2016
P. Frasconi et al. (Eds.): ECML PKDD 2016, Part II, LNAI 9852, pp. 33–46, 2016.
DOI: 10.1007/978-3-319-46227-1_3

Networks (MLNs) [14] or Probabilistic Soft Logic (PSL) [3] is in the tighter integration of logic and the processing of feature-based continuous sensorial input that is available in many real world applications. Indeed, while a MLN can capture a logistic regression model [7,8], it requires to deal with a large number of weights and groundings. More complex correlations between features and classes can not be captured as the resulting models would be too large to be tractable.

Deep Neural Networks [16] have been shown to be relatively successful in performing feature selection and inference over pattern constituents in their hidden layers. However, this process is opaque and it is not generally clear which is the amount of training data required to correctly instantiate the process during training. SBR provides a way to integrate (deep or shallow) learning with any explicit knowledge about the task at hand, making the learning process more controlled, easier to understand and requiring less labeled data. In the applications where this knowledge is available, it seems natural to exploit the knowledge to force the learning machine to develop more targeted intermediate pattern representations.

Unfortunately, learning is typically hard for all SRL approaches with high generality: the integration of learning with logic inference transforms the intractability of the latter (in a general setting) into the complexity of the numerical optimization problem that needs to be solved during learning. This issues also applies to SBR, for which getting good solutions often requires heuristics and experience. Some attempts at breaking the complexity of learning by subdividing the learning process into small and easier sequential tasks have been hinted by Bengio et al. [2] and later studied in a more systematic way by Yang et al. [19], and Friesen et al. [9]. This paper studies under which conditions training in SBR becomes easy. In particular, it will be shown that it exists a large class of knowledge that can be expressed as a set of convex constraints in SBR. These constraints can be exploited during training in a very efficient and effective way. This class of constraints provides a natural way to break the complexity of learning by building a training plan that uses the convex constraints as an effective initialization step for the final full optimization problem. The experimental results show the effectiveness of this training plan. Another contribution of this paper is to employ Neural Networks for the first time in the context of SBR, showing the generality and flexibility of the framework. Experimental results on image classification are presented to validate the approach.

The paper is organized as follows: Sect. 2 provides an introduction to the SBR learning framework and Sect. 3 shows how to build real-valued constraints from a FOL knowledge base. The experimental results are show in Sect. 4. Finally, some conclusions are drawn in Sect. 5.

2   Semantic Based Regularization

Consider a multi-task learning problem, where a set of T functions must be estimated (*query or unknown functions*) Let $\boldsymbol{f} = \{f_1, \ldots, f_T\}$ indicate the vector of functions.

A set of H functional constraints in the form $1 - \Phi_h(\boldsymbol{f}) = 0, 0 \leq \Phi_h(\boldsymbol{f}) \leq 1, h = 1, \ldots, H$ are provided to describe how the query functions should behave. These functionals can express a property of a single function or correlate multiple functions, so that learning can be helped by exploiting these correlations.

The j-th function is associated to a set $\boldsymbol{\mathcal{X}}_j^\circ$, which is a sample of the patterns input to the function, Each pattern in this set is represented via a vector of features. We assume that this set of patterns is partially labeled, so that the desired function output is also provided for some patterns in the sample. Multiple functions can share the same sample of patterns (e.g. $\boldsymbol{\mathcal{X}}_j^\circ = \boldsymbol{\mathcal{X}}_i^\circ \ i \neq j$). Some functions may express relations across multiple patterns, and the pattern representations associated to these functions can be generally expressed as the combination of the patterns from a set of finite domains: $\boldsymbol{\mathcal{X}}_j^\circ = \boldsymbol{\mathcal{X}}_{j1} \times \boldsymbol{\mathcal{X}}_{j2} \times \ldots$.

Let $f_k(\boldsymbol{\mathcal{X}}_k^\circ)$ indicate the vector of values obtained by applying the function f_k to the set of patterns $\boldsymbol{\mathcal{X}}_k^\circ$ and $\boldsymbol{f}(\boldsymbol{\mathcal{X}}) = f_1(\boldsymbol{\mathcal{X}}_1^\circ) \cup f_2(\boldsymbol{\mathcal{X}}_2^\circ) \cup \ldots$ collects the groundings for all functions.

Constraint satisfaction can be enforced by penalizing their violation on the sample of data:

$$C_e[\boldsymbol{f}(\boldsymbol{\mathcal{X}})] = \sum_{k=1}^{T} \|f_k\|^2 + \sum_{h=1}^{H} \lambda_h \Big(1 - \Phi_h\big(\boldsymbol{f}(\boldsymbol{\mathcal{X}})\big) \Big), \tag{1}$$

where the first term is a regularization term penalizing non-smooth solutions and λ_h is the weight for the h-th constraint.

The weights are optimized via gradient descent using a back-propagation schema, where the derivative of the cost function with respect to the j-th weight of the i-th function w_{ij} is:

$$\frac{\partial C_e}{\partial w_{ij}} = \sum_k \frac{\partial C_e}{\partial \Phi_k} \cdot \frac{\partial \Phi_k}{\partial w_{ij}} = \sum_k \frac{\partial C_e}{\partial \Phi_k} \cdot \left(\sum_{t_{\Phi_k}} \frac{\partial \Phi_k}{\partial t_{\Phi_k}} \cdot \frac{\partial t_{\Phi_k}}{\partial f_i} \cdot \frac{\partial f_i}{\partial w_{ij}} \right). \tag{2}$$

2.1 Collective Classification

Collective classification (CC) [17] is the task of performing inference over a set of instances that are connected among each other via a set of relationships. Collective classification in SBR [6] enforces that the classification output is consistent with the FOL knowledge used during training.

In particular, let $f_k(\boldsymbol{\mathcal{X}}_k')$ indicate the vector of values obtained by evaluating the function f_k over the data points of the test set $\boldsymbol{\mathcal{X}}_k'$. The set of vectors will be compactly referred to as: $\boldsymbol{f}(\boldsymbol{\mathcal{X}}') = f_1(\boldsymbol{\mathcal{X}}_1') \cup \ldots \cup f_T(\boldsymbol{\mathcal{X}}_T')$. If no neural network has been trained for f_k (no examples or no feature representations were available during training), $f_k(\boldsymbol{\mathcal{X}}_k')$ is assumed to be just filled with default values equal to 0.5.

Collective classification searches for the values $\bar{\boldsymbol{f}}(\boldsymbol{\mathcal{X}}_k') = \bar{f}_1(\boldsymbol{\mathcal{X}}_1') \cup \ldots \cup \bar{f}_T(\boldsymbol{\mathcal{X}}_T')$ respecting the FOL formulas on the test data, while being close to the prior values established by the kernel machines over the test data:

$$C_{cc}[\bar{\boldsymbol{f}}(\mathcal{X}'), \boldsymbol{f}(\mathcal{X}')] = \frac{1}{2}\sum_{k=1}^{T}|\bar{f}_k(\mathcal{X}'_k) - f_k(\mathcal{X}'_k)|^2 + \sum_{h}\left(1 - \Phi_h(\bar{\boldsymbol{f}}(\mathcal{X}'))\right)$$

Optimization can be performed via gradient descent by computing the derivative with respect to the function values.

2.2 Logic and Constraints

This section will show how to convert any First Order Logic (FOL) knowledge into a set of constraints Φ_h that can be integrated into learning using Eq. 2.

Our approach is a variation of fuzzy generalizations of First Order Logic (FOL), which have been first proposed by Novak [13]. Fuzzy FOL can transform any FOL knowledge base into a real valued constraint.

T-norm and Residuum. A *t-norm fuzzy logic* [11,20] is defined by its t-norm $t(a_1, a_2)$ that models the logical AND.

Given a variable $\bar{a}$ with continuous generalization a in $[0,1]$, its negation $\neg\bar{a}$ corresponds to $1 - a$. Once the t-norm functions corresponding to the $\wedge$ and $\neg$ are defined, they can be composed to generalize any logic proposition. Different t-norm fuzzy logics have been proposed in the literature. For example, given two Boolean values $\bar{a}_1, \bar{a}_2$ and their continuous generalizations a_1, a_2 in $[0,1]$, the *product t-norm* is defined as: $(\bar{a}_1 \wedge \bar{a}_2) \rightarrow t(a_1, a_2) = a_1 \cdot a_2$. The Lukasiewicz t-norm is instead defined as

$$(\bar{a}_1 \wedge \bar{a}_2) \quad \rightarrow \quad t(a_1, a_2) = \max(0, a_1 + a_2 - 1).$$

Any t-norm features a binary operator called *residuum*, which is used to generalize implications when dealing with continuous variables [11]. For example, the Lukasiewicz t-norm has a residuum defined as:

$$(\bar{a}_1 \Rightarrow \bar{a}_2) \quad \longrightarrow \quad t(a_1, a_2) = \begin{cases} 1 & a_1 \leq a_2 \\ 1 - a_1 + a_2 & a_1 > a_2 \end{cases}$$

Quantifiers. With no loss of generality, we focus our attention on FOL formulas in the Prenex Normal Form, having all the quantifiers at the beginning of the expression. The quantifier-free part of the expression is an assertion in fuzzy propositional logic once all the quantified variables are grounded. Hence, a t-norm fuzzy logic can be used to convert it into a continuous function. Let's consider a FOL formula with variables $x_1, x_2, \ldots$ assuming values in the finite sets $\mathcal{X}_1, \mathcal{X}_2, \ldots$. $\mathcal{P} = \{p_1, p_2, \ldots\}$ is the vector of predicates, where the j-th n-ary predicate is grounded from $\mathcal{X}_j^\circ = \mathcal{X}_{j1} \times \mathcal{X}_{j2} \times \ldots$. Let $p_j(\mathcal{X}_j^\circ)$ indicate the set of possible groundings for the j-th predicate, and $\mathcal{P}(\mathcal{X})$ indicate all possible grounded predicates, such that $\mathcal{P}(\mathcal{X}) = p_1(\mathcal{X}_1^\circ) \cup p_2(\mathcal{X}_2^\circ) \cup \ldots$.

If the atoms $\mathcal{P}(\mathcal{X})$ are generalized to assume real values in $[0,1]$, the degree of truth of a formula containing an expression E with a universally quantified variable x_i is the average of the t-norm generalization $t_E(\cdot)$, when grounding x_i over $\mathcal{X}_i$ (see Diligenti et al. [5] for more details):

$$\forall x_i \ E(\mathcal{P}(\mathcal{X})) \quad \longrightarrow \quad \Phi_\forall(\mathcal{P}(\mathcal{X})) = \frac{1}{|\mathcal{X}_i|} \sum_{x_i \in \mathcal{X}_i} t_E(\mathcal{P}(\mathcal{X}))$$

For the existential quantifier, the truth degree is instead defined as the *maximum* of the t-norm expression over the domain of the quantified variable:

$$\exists x_i \ E(\mathcal{P}(\mathcal{X})) \quad \longrightarrow \quad \Phi_\exists(\mathcal{P}(\mathcal{X})) = \max_{x_i \in \mathcal{X}_i} t_E(\mathcal{P}(\mathcal{X}))$$

When multiple universally or existentially quantified variables are present, the conversion is recursively performed from the outer to the inner variables. Please note that the fuzzy formula expression is continuous and differentiable with respect to the fuzzy value of a predicate, and it can therefore easily be integrated into learning.

2.3 Building Constraints from Logic

Let us assume to have available a knowledge base KB, consisting of a set of FOL formulas and a finite set of groundings of the variables. We assume that some of the predicates are unknown: the SBR learning process aims at finding a good approximation of each unknown predicate, so that the estimate predicates will satisfy the FOL formulas for the sample of the inputs. In particular, the function $f_j(\cdot)$ will be learned as approximation of the j-th unknown predicate. The variables in the KB that are input to any f_j are replaced with the feature-based representation of the object grounded by the variables, and we will indicate as x_i the representation of the object grounded by x_i. The groundings $\mathcal{X}_i$ of the i-th variable are therefore replaced by the set $\mathbf{\mathcal{X}}_i$, indicating the set of feature-based representations of the groundings. One constraint $1 - \Phi_i(\cdot) = 0$ for each formula F_i in the knowledge base is built by taking the fuzzy FOL generalization of the formula $\Phi_i(\cdot)$, where the unknown predicates are replaced by the learned functions, and the variables input to the learned functions are replaced by their duals iterating over the feature-based representations of the groundings. Previous literature on Semantic Based Regularization [5,6] has focused on Kernel Machines to implement the functions $f_j(\cdot)$. However, the SBR framework does not pose any restriction on the machine learning machinery used to approximate the unknown functions. In particular, Neural Networks are used for the first time in the experimental section of this paper.

3 Constraints and Local Minima

The constraint resulting from a FOL formula can be hard to optimize during learning. Let's consider universally quantified FOL formulas in DNF form:

$$\forall x_1 \ldots \forall x_n \overbrace{\left(n_{11} P_1(x_1) \wedge \ldots \wedge n_{1n} P_n(x_n) \right)}^{minterm\ 1} \vee \ldots \vee \overbrace{\left(n_{k1} P_1(x_1) \wedge \ldots \wedge n_{kn} P_n(x_n) \right)}^{minterm\ k}$$

where n_{ij} determines whether the j-th variable in the i-th minterm is negated or not. The following expression for each grounding can be obtained by applying a double negation and using the DeMorgan rule:

$$\neg\Big(\neg\big(n_{11}P_1(x_1) \wedge \ldots \wedge n_{1n}P_n(x_n)\big) \wedge \ldots \wedge \neg\big(n_{k1}P_1(x_1) \wedge \ldots \wedge n_{kn}P_n(x_n)\big)\Big)$$

For any given grounding, the resulting propositional expression can be converted using the product t-norm and replacing the atoms with the unknown function approximations, yields the constraint:

$$1 - \Phi(f(\mathcal{X})) = \frac{1}{\prod\limits_{i=1}^{n} |\mathcal{X}_i|} \sum_{x_1} \cdots \sum_{x_{n-1}} \prod_{r=1}^{k} \left(1 - \prod_{i \in A_r^p} f_i(x_i) \prod_{j \in A_r^n}(1 - f_j(x_j))\right) = 0$$

where A_r^p and A_1^n are the set of non-negated and negated atoms in the r-th minterm. It is clear that a null contribution to the summation for a given grounding is obtained as solution of a polynomial equation, where the r-th solution of the polynomial equation corresponds to the assignment satisfying the r-th minterm, that is:

$$\prod_{i \in A_r^p} f_i(x_i) \prod_{j \in A_r^n}(1 - f_j(x_j)) = 1.$$

Since all minterms are by construction different and the polynomial equation is continuous and assuming values greater or equal to zero as guaranteed by any t-norm, the resulting expression has as many local minima as the number of true configurations in the truth table for the grounded propositional formula, which is in turn equal to the number of minterms of the initial DNF.

This shows that there is a duality between the number of possible assignments of the atoms satisfying the FOL formula for a given grounding of the variables, and the number of local minima in the expression generalizing the formula to a continuous domain. The intractability of unrestricted FOL inference is therefore translated into a SBR cost function that is plagued by many local minima.

3.1   Convexity of the Constraints

While optimization remains generally intractable, however using t-norm residua to translate logic implications, significantly increases the portion of constraints that can be efficiently exploited in learning.

T-norm residua are consistent with modus ponens at the extremes of the variable range. However, they soften the conditions under which the formula is verified. Indeed, any t-norm residuum returns a 1 value whenever the head holds a value larger than the body. This specifies an interval for the admissible solution. On the other hand, the t-norm translation of the implication via modus ponens has only 3 singular points where it is fully satisfied. An interesting result of the application of the t-norm residuum to SBR theory is that a much larger set of formula correspond to a convex constraint with respect to what it would happen using a modus ponens based translation.

In particular, let's consider the class of FOL formula that are universally quantified, for which the propositional clause resulting from the evaluation of the predicates for any grounding is a definite clauses (e.g. having conjunctive body of positive atoms and a single literal head). One generic formula in this class has the following structure:

$$\forall x_1 \ldots \forall x_v \ P_1(x_{i(1)}) \wedge \ldots \wedge P_n(x_{i(n)}) \Rightarrow P_{n+1}(x_{i(n+1)})$$

where $i(j)$ is the index of the variable used by the j-th predicate. Let $x = \{x_1, \ldots, x_v\}$ and replacing the predicates with the predicates with the functions f to be learned, the constraint can be written as:

$$1 - \Phi(f(\mathcal{X})) = \frac{1}{\prod\limits_{j=1}^{v} |\mathcal{X}_j|} \sum_{x_1 \in \mathcal{X}_1} \ldots \sum_{x_v \in \mathcal{X}_v} (1 - t(f, x)) = 0 \qquad (3)$$

where $t(\cdot)$ is the t-norm representation of the definite clause.

Theorem 1. *The function $1 - \Phi(\cdot)$ translating a generic FOL formula with any number of nested universal quantifiers, conjunctive body and a single head is convex with respect to the function values if using the Lukasiewicz t-norm.*

Proof. Equation 3 shows the general form of the constraint. A positive summation of convex functions is convex, then we only need to prove that each single contribution $1 - t(f, x)$ to the summation is convex with respect to the $f_i(x)$ values.

The translation of a conjunction of n variables $A_1 \wedge \ldots \wedge A_n$ using the Lukasiewicz t-norm is equal to: $\max(0, \sum_{j=1}^{n} A_j - n + 1)$. Therefore the head and body of the clause are translated as:

$$\underbrace{P_1(x_{i(1)}) \wedge \ldots \wedge P_n(x_{i(n)})}_{\max(0, \sum\limits_{j=1}^{n} f_j(x_{i(j)}) - n + 1)} \Rightarrow \underbrace{P_{n+1}(x_{i(n+1)})}_{f_{n+1}(x_{i(n+1)})}$$

The residuum definition for the Lukasiewicz tnorm is:

$$A_1 \Rightarrow A_2 \quad \longrightarrow \quad \begin{cases} 1 & A_1 - A_2 < 0 \\ 1 - A_1 + A_2 & else \end{cases}$$

Let us call $h(f, x) = \max(0, \sum_{j=1}^{n} f_j(x_{i(j)}) - n + 1) - f_{n+1}(x_{i(n+1)})$, Therefore,

$$1 - t(f, x) = g(h(f, x)) =$$

$$= 1 - \begin{cases} 1 & h(f, x) < 0 \\ 1 - h(f, x) & else \end{cases}$$

$$= \begin{cases} 0 & h(f, x) < 0 \\ h(f, x) & else \end{cases}$$

$g(\cdot)$ is convex and non-decreasing in $h(\cdot)$, while $h(\cdot)$ is convex. Therefore, the combination $g(h(\cdot))$ is convex as well.

Let's now see some special constraints that fall in this class.

Constraint and supervised data. Let $\mathcal{X}_k^+$ be the sets of positive for the k-th unknown predicate p_k. The following logic formula expresses the fact that p_k is constrained on the values assumed over the supervised data, as it should get a 1 value on a positive example:

$$\forall x \; P_k(x) \Rightarrow p_k(x)$$

where $x \in \mathcal{X}_k$ and the predicate $P_k(x)$ is an evidence function holding true iff x is a positive example for the query predicate p_k, respectively (e.g. $x \in \mathcal{X}_k^+$). Using the Lukasiewicz t-norm and replacing p_k with its approximation f_k, this corresponds to the following constraint:

$$1 - \Phi\big(f_k(\mathcal{X}_k^+)\big) = \tfrac{1}{|\mathcal{X}_k^+|} \sum_{x \in \mathcal{X}_k^+} \max\big(0, 1 - f_k(x)\big) = 0$$

This an example showing how training using the hinge loss ($\max(0, 1 - f_k(x))$) emerges when expressing the fitting of the supervised data via a definite clause. As predicted by Theorem 1, this corresponds to a convex cost function to optimize when using a linear model (like when using an SVM to implement f_k [4]).

Manifold Regularization. Let's consider the formula expressing a manifold based on some relation R:

$$\forall x \forall y \; R(x, y) \Rightarrow (P_k(x) \Leftrightarrow P_k(y))$$

which is equivalent to the conjunction of the following two FOL formulas:

$$\forall x \forall y \; R(x, y) \wedge P_k(x) \Rightarrow P_k(y)$$
$$\forall x \forall y \; R(x, y) \wedge P_k(y) \Rightarrow P_k(x)$$

According to Theorem 1, the resulting constraint for these formulas must yield a convex constraint. Indeed, the constraint is:

$$
\begin{aligned}
1 - \Phi(f_k(\mathcal{X}_k)) &= 1 - \tfrac{1}{|\mathcal{X}_k|^2}\bigg(|\mathcal{X}_k|^2 - |\mathcal{R}| + \sum_{(x,y)\in\mathcal{R}} \max\Big(0, -1 \\
&\quad + \begin{cases} 1 - f_k(x) + f_k(y) & f_k(x) > f_k(y) \\ 1 & else \end{cases} \\
&\quad + \begin{cases} 1 - f_k(y) + f_k(x) & f_k(y) > f_k(x) \\ 1 & else \end{cases}\Big)\bigg) \\
&= \tfrac{|\mathcal{R}|}{|\mathcal{X}_k|^2} - \tfrac{1}{|\mathcal{X}_k|^2} \sum_{(x,y)\in\mathcal{R}} \max(0, 1 - |f_k(x) - f_k(y)|) \\
&= \tfrac{1}{|\mathcal{X}_k|^2} \sum_{(x,y)\in\mathcal{R}} |f_k(x) - f_k(y)| = 0
\end{aligned}
$$

which is the L1 variation of the classical manifold regularization constraint [1].

3.2 Teaching Plans

The results shown in the previous section suggest a natural heuristic to deal with harder SBR problems:

– solve the optimization problem introducing only the convex constraints: this means to optimize via gradient descent until the gradient vanishes (e.g. its module falls below some threshold) and the learning process has found a good approximation of the best solution for the convex problem.
– Introduce the other constraints into the previous problem and run the training process until convergence.

The second step can be further subdivided into multiple stages by forcing first the formulas with a lower number of possible valid (e.g. satisfying the formula) assignments to the atoms to be learned. As explained in the previous sections, these formulas introduce a lower number of local minima into the cost function. This heuristic is similar to what done in constraint satisfaction programming [15], where the variables with the smallest number of admissible values remaining in its domain are selected first during the search process over the possible assignments [10].

4   Experimental Results

The experimental analysis has been carried out on an animal identification benchmark proposed by P. Winston [18], which was initially designed to show the ability of logic programming to determine the class of an animal from some initial clues regarding its features. Unlike in the original challenge, we do not input to the test phase a sufficient set of clues to perform classification, but only the raw images, leaving to the learning framework the duty to develop the intermediate clues over which to perform inference.

The dataset is composed of 5605 images, taken from the *ImageNet*[1] database, equally divided in 7 classes, each one representing one animal category: albatross, cheetah, giraffe, ostrich, penguin, tiger and zebra. The feature vector used to represent each image is composed of bag-of-feature and color histogram descriptors. In particular, for each image SIFT descriptors [12] have been extracted, and then later clustering them into 600 *visual words*. A vector containing the normalized count of each visual word for the given image is provided as representation. We also added a 12-dimension normalized color histogram for each channel in the RGB color space to the feature representation (Fig. 1).

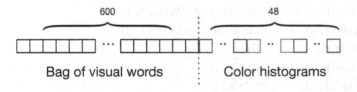

Fig. 1. The feature vector representation for each image in the Winston benchmark.

---

[1] http://www.image-net.org.

Table 1. The KB used for training the SBR model. The rules are divided into groups: only the first "definite" group is formed by definite clauses that were originally proposed by Winston to classify the animals. The "excl" rule states the fact that one and only one class should be assigned to each image. The "inter" rules add another intermediate classification level that can be exploited to perform classification over the final classes.

Type	Rule
definite	HAIR(x) $\Rightarrow$ MAMMAL(x)
definite	MILK(x) $\Rightarrow$ MAMMAL(x)
definite	FEATHER(x) $\Rightarrow$ BIRD(x)
definite	LAYEGGS(x) $\Rightarrow$ BIRD(x)
definite	MAMMAL(x) $\wedge$ MEAT(x) $\Rightarrow$ CARNIVORE(x)
definite	MAMMAL(x) $\wedge$ POINTEDTEETH(x) $\wedge$ CLAWS(x) $\wedge$ FORWARDEYES(x) $\Rightarrow$ CARNIVORE(x)
definite	MAMMAL(x) $\wedge$ HOOFS(x) $\Rightarrow$ UNGULATE(x)
definite	MAMMAL(x) $\wedge$ CUD(x) $\Rightarrow$ UNGULATE(x)
definite	CARNIVORE(x) $\wedge$ TAWNY(x) $\wedge$ DARKSPOTS(x) $\Rightarrow$ CHEETAH(x)
definite	CARNIVORE(x) $\wedge$ TAWNY(x) $\wedge$ BLACKSTRIPES(x) $\Rightarrow$ TIGER(x)
definite	UNGULATE(x) $\wedge$ LONGLEGS(x) $\wedge$ LONGNECK(x) $\wedge$ TAWNY(x) $\wedge$ DARKSPOTS(x) $\Rightarrow$ GIRAFFE(x)
definite	UNGULATE(x) $\wedge$ WHITE(x) $\wedge$ BLACKSTRIPES(x) $\Rightarrow$ ZEBRA(x)
definite	BIRD(x) $\wedge$ LONGLEGS(x) $\wedge$ LONGNECK(x) $\wedge$ BLACK(x) $\Rightarrow$ OSTRICH(x)
definite	BIRD(x) $\wedge$ SWIM(x) $\wedge$ BLACKWHITE(x) $\Rightarrow$ PENGUIN(x)
definite	BIRD(x) $\wedge$ GOODFLIER(x) $\Rightarrow$ ALBATROSS(x)
excl	CHEETAH(x) $\oplus$ TIGER(x) $\oplus$ GIRAFFE(x) $\oplus$ ZEBRA(x) $\oplus$ OSTRICH(x) $\oplus$ PENGUIN(x) $\oplus$ ALBATROSS(x)
inter	MAMMAL(x) $\oplus$ BIRD(x)
inter	HAIR(x) $\oplus$ FEATHER(x)
inter	(DARKSPOTS(x)) $\Rightarrow$ $\neg$ BLACKSTRIPES(x)
inter	(BLACKSTRIPES(x)) $\Rightarrow$ $\neg$ DARKSPOTS(x)
inter	TAWNY(x) $\Rightarrow$ $\neg$ BLACK(x)) $\wedge \neg$ WHITE(x)
inter	BLACK(x) $\Rightarrow$ $\neg$ TAWNY(x)) $\wedge \neg$ WHITE(x)
inter	WHITE(x) $\Rightarrow$ $\neg$ BLACK(x)) $\wedge \neg$ TAWNY(x)
inter	BLACK(x) $\Rightarrow$ $\neg$ WHITE(x)
inter	BLACK(x) $\Rightarrow$ $\neg$ TAWNY(x)
inter	WHITE(x)) $\Rightarrow$ $\neg$ BLACK(x)
inter	WHITE(x) $\Rightarrow$ $\neg$ TAWNY(x)
inter	TAWNY(x) $\Rightarrow$ $\neg$ WHITE(x)
inter	TAWNY(x)) $\Rightarrow$ $\neg$ BLACK(x)

The images have been split into two initial sets: the first one is composed of 2100 images utilized for building the visual vocabulary, while the second set is composed of 3505 images used in the learning process. The experimental analysis has been carried out using by randomly sampling from the overall set of the supervisions the labels to keep as training, validation and test set, randomly sampling $50, 25, 25\%$ of the supervisions, respectively.

Knowledge base. The knowledge domain is expressed in terms of FOL rules. Table 1 shows the full set of rules used in this task. A total of 33 predicates are available in the KB, but only 7 of them are considered in evaluating the results, while the other ones are intermediate predicates helping to determine the final classes during the inference process.

The rules in KB can be subdivided into subsets:

- the original set of the rules as provided in the original problem definition by Winston. These rules are definite clauses resulting into a convex constraint and they are marked as *definite* in the table;
- the *excl* rule states that each pattern should belong to one and only one final class, this rule does not translate into a convex constraint;
- the *inter* rules show how it is possible to inject any amount of additional knowledge into the classification problem. These rules do not translate into a convex constraint.

4.1   Results

The first set of experiments tests the performance of SBR in a transductive context, where all the images are available at training time, but only the training

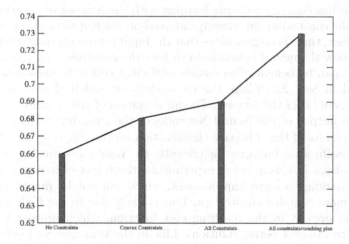

Fig. 2. Experimental results obtained in a transductive setting using standard Neural Networks with no constraints, SBR with and without using different set of rules and learning schemas.

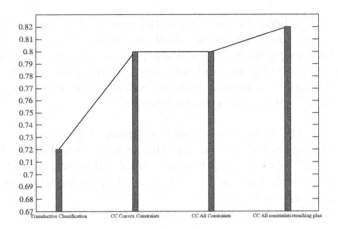

Fig. 3. Experimental results obtained by performing SBR collective classification and using different set of rules and learning schemas.

labels are made available during training. One Neural Network with one hidden layer using a sigmoidal activation function on the output layer and a rectified linear activation in the hidden layer is trained for each of the 33 predicates in the KB. Figure 2 reports a summary of the results, evaluated over the training data for the 7 final predicates corresponding to the final classes in the Winston benchmark. Using the convex constraints coming from the definite clauses provides an improvement of 2 points of F1. A small additional improvement can be obtained by adding all the available constraints at the beginning of training. A larger improvement can be obtained by using a training plan where the convex constraints are added first, and the remaining constraints are added when the cost function has converged during learning with the first set of constraints.

Even if the constraints are already enforced on the test data given the transductive context, there is no guarantee that the input representations are powerful enough to allow the neural networks to respect the constraints on the test data. Therefore, it can be beneficial to further perform a collective classification step as described in Sect. 2.1, where the constraints are enforced over the output label assignments over the test set. The initialization of the assignments is done by using the output of the Neural Networks of the transductive step. Figure 3 reports the results of the collective classification on the test set. It is clear that collective classification improves significantly the results obtained by transductive classification. In this round of experiments, the subset of convex constraints is already providing a large improvement, which can not be moved higher by adding the more complex constraints. This is likely due to the high number of local minima present in the resulting cost function, which prevent the training process to discover better solutions. Like in the transductive learning case, breaking the learning complexity into stages seems to be very useful. Indeed, using the training plan described in Sect. 3.2 delivers a significant boost of the classification performance.

5   Conclusions

Semantic Base Regularization seamlessly integrates First Order Logic into multi-task learning allowing to tackle complex learning problems even when supervised data is scarce. This is possible by leveraging unsupervised data and any domain knowledge available on the field. However, the integration sometimes requires to solve a challenging optimization problem during the learning process. This paper shows a large class of FOL knowledge that can be integrated into learning, while keeping the resulting optimization problem easy. By leveraging this class of clauses, the paper shows how to improve the trained solution in a more general case, by breaking the complexity of learning into multiple stages, which are initialized using a solution built over the "easy" clauses. The experimental results on image classification show the effectiveness of the framework and of the proposed training heuristic. While providing extensive prior knowledge is cumbersome in large and complex experimental setups, we still think that the integration of prior knowledge and learning will be a required step to achieve real human-level capabilities in vision and language understanding. As future work we plan to extend the experimental evaluation to other larger image datasets.

References

1. Belkin, M., Niyogi, P., Sindhwani, V.: Manifold regularization: a geometric framework for learning from labeled and unlabeled examples. J. Mach. Learn. Res. **7**, 2434 (2006)
2. Bengio, Y.: Curriculum learning. In: Proceedings of the 26th Annual International Conference on Machine Learning (ICML), pp. 41–48 (2009)
3. Broecheler, M., Mihalkova, L., Getoor, L.: Probabilistic similarity logic. In: Proceedings of the Twenty-Sixth Conference on Uncertainty in Artificial Intelligence (UAI), pp. 73–82 (2010)
4. Cortes, C., Vapnik, V.: Support-vector networks. Mach. Learn. **20**(3), 273–297 (1995)
5. Diligenti, M., Gori, M., Maggini, M., Rigutini, L.: Bridging logic and kernel machines. Mach. Learn. **86**(1), 57–88 (2012)
6. Diligenti, M., Gori, M., Saccà, C.: Semantic-based regularization for learning and inference. Artif. Intell. (2015)
7. Domingos, P., Sumner, M.: The alchemy tutorial (2010). http://alchemy.cs.washington.edu/tutorial/tutorial.pdf
8. Domingos, P., Richardson, M.: Markov logic: a unifying framework for statistical relational learning. In: ICML-2004 Workshop on Statistical Relational Learning, pp. 49–54 (2004)
9. Friesen, A.L., Domingos, P.: Recursive decomposition for nonconvex optimization. In: Proceedings of the 24th International Joint Conference on Artificial Intelligence (2015)
10. Golomb, S.W., Baumert, L.D.: Backtrack programming. J. ACM (JACM) **12**(4), 516–524 (1965)
11. Hajek, P.: The Metamathematics of Fuzzy Logic. Kluwer, Dordrecht (1998)

12. Lowe, D.G.: Object recognition from local scale-invariant features. In: The Proceedings of the Seventh IEEE International Conference on Computer Vision, vol. 2, pp. 1150–1157. IEEE (1999)
13. Novák, V.: First-order fuzzy logic. Studia Logica **46**(1), 87–109 (1987)
14. Richardson, M., Domingos, P.: Markov logic networks. Mach. Learn. **62**(1–2), 107–136 (2006)
15. Rossi, F., Van Beek, P., Walsh, T.: Handbook of Constraint Programming. Elsevier (2006)
16. Schmidhuber, J.: Deep learning in neural networks: an overview. Neural Netw. **61**, 85–117 (2015)
17. Sen, P., Namata, G., Bilgic, M., Getoor, L., Galligher, B., Eliassi-Rad, T.: Collective classification in network data. AI Mag. **29**(3), 93 (2008)
18. Winston, P.H., Horn, B.K.: LISP. Addison Wesley Pub., Reading (1986)
19. Yang, P., Tang, K., Yao, X.: A novel divide and conquer based approach for large-scale optimization problems. arXiv preprint (2016). arXiv:1603.03518
20. Zadeh, L.A.: Fuzzy sets. Inf. Control **8**, 338–353 (1965)

Persistent Roles in Online Social Networks

Matt Revelle$^{(\boxtimes)}$, Carlotta Domeniconi, and Aditya Johri

George Mason University, Fairfax, VA 22030, USA
{revelle,carlotta}@cs.gmu.edu, ajohri3@gmu.edu

Abstract. Users in online social networks often have very different structural positions which may be attributed to a latent factor: roles. In this paper, we analyze dynamic networks from two datasets (Facebook and Scratch) to find roles which define users' structural positions. Each dynamic network is partitioned into snapshots and we independently find roles for each network snapshot. We present our role discovery methodology and investigate how roles differ between snapshots and datasets. Six persistent roles are found and we investigate user role membership, transitions between roles, and interaction preferences.

1 Introduction

Online networks rely extensively on user contributions and participation for their vibrancy. This requires that users perform certain activities and take on specific roles within the network. In this paper we take a distinct approach to identify latent role behaviors which persist over time by examining interaction patterns and structural positions of users. Our approach provides a novel way of understanding latent mechanisms that underlie the structure and processes of dynamic networks.

Role discovery has been applied to many networks [22] and incorporated into static network models [27]. Despite the prevalence of role discovery methods and applications, no experiments have been presented that show the existence of persistent roles derived directly from data. While network-specific roles are useful for many purposes, identifying a set of roles which commonly occur in online social networks enables new methods for comparative analysis which emphasize relationships between roles.

In this paper, we present a methodology for discovering and tracing persistent roles over time. We discover roles for 26 network snapshots of online social networks from two datasets (Facebook and Scratch). These roles are found to persist both within and between the network snapshots from both datasets. We then conduct a summary analysis to demonstrate how roles may help interpret network structure by considering role membership, transitions between roles, and interaction preferences.

In our experiments, we discover six roles from the networks and show these roles are both distinct from one another and occur in every network from both datasets. These roles are: *popular, friendly, explorer, reciprocated, community member* and *active-community member*. While the discovered roles are common

© Springer International Publishing AG 2016
P. Frasconi et al. (Eds.): ECML PKDD 2016, Part II, LNAI 9852, pp. 47–62, 2016.
DOI: 10.1007/978-3-319-46227-1_4

to both datasets and persist over time, we find the relationships between roles may differ. These findings suggest common roles shared among social interaction networks are useful for modeling and comparing networks.

2 Related Work

2.1 Role Discovery

An overview of role discovery approaches is provided in [21] which discusses graph-based, feature-based, and hybrid definitions of roles and methods for their discovery from graph and node-attribute data. They show that feature-based roles are more flexible and capable of capturing more complex roles. A framework for feature-based role discovery is introduced and discusses classes of approaches for role feature construction and role assignment.

The use of non-negative matrix factorization (NMF) for discovering node roles was introduced in [9]. In that paper, the authors use a method [10] to generate features which aggregate various per-node structural attributes. This node-attribute matrix is then decomposed using NMF and the resulting basis vectors correspond to node roles in the network. Later work adds additional constraints to NMF which can be used to specify expectations of sparsity or diversity of the roles [7]. The work in our paper differs as we discover persistent roles across datasets and time using independent decompositions of network snapshots.

Other work [5,30] uses role-labeled nodes to identify the roles of unlabeled nodes. However, the roles in their work are not defined in terms of structural positions in the network but rather functional occupations in an organization (e.g., roles held in technology companies: research & development, executives, and human resources). That is, the roles are defined in terms of domain knowledge and non-structural node features. The authors then introduce a classifier for these functional roles which incorporates information derived from the network structure.

Aside from identifying patterns of structural positions, roles have also been used in the context of information cascades to identify groups of nodes which have similar influence and blockage attributes [6].

A feature-based approach for automatic detection of user roles in online forums is presented in [4]. Their method uses principal component analysis (PCA) and agglomerative clustering of feature profile data to find roles; where each cluster corresponds to a role. Another feature-based approach using a mixture model of roles is presented in [27]. Nodes are first clustered using node features derived from the network structure and then a qualitative assignment of nodes to roles follows.

Role discovery has been used to assist in creating compact representations of networks. In [25] a method is introduced for generating multi-resolution maps of networks by constructing a hierarchy of roles defined with regular equivalence. The different levels of the hierarchy are used for different resolution maps.

2.2   Network Models with Roles

In [22] the role membership for a series of network snapshots are found and analyzed and the roles are used to construct a transition model of role memberships. Every node in every snapshot is represented as a mixed membership of roles. This mixed membership may change over time and a transition model captures the likelihood of transitioning between roles. Their method assumes roles are stationary and uses the same set of basis vectors (roles) for every snapshot rather than directly estimate roles from each snapshot. The authors suggest roles may generalize over time and across datasets, but do not provide support for this statement. To our knowledge, this paper is the first to present evidence of common, persistent roles derived directly from data.

Some models which incorporate roles do not distinguish between node features derived from network structure and those external to the graph. In [28], a probabilistic model which incorporates node features as dependent on latent factors (roles) is introduced. While these features could be derived from network structure as described in [21], the experiments performed in [28] only include external features such as document terms and voting counts. The network topology is ignored.

Communities provide extra structural information which can benefit role discovery. In both [8,23], communities are simultaneously detected with roles. Roles are used as latent factors of which node attributes are dependent.

Finally, [14,29] add roles to topic models where authors may take a role when generating a document and the topic of the document is dependent on the author's role.

3   Discovering Persistent Roles

We aim to find roles which best characterize the nodes in a network. The network datasets we consider in this paper are dynamic networks which include timestamped, directed *interactions* between node pairs. Each interaction represents a single action such as one user messaging another. As our primary goal is to identify persistent roles over time, we will partition the dynamic network $\mathcal{D} = (\mathcal{N}, \mathcal{E})$ into snapshots, $\mathcal{S}_t$ for each timestep t. The original edges $\mathcal{E}$ are timestamped, directed interactions between node pairs and only edges occurring at timestep t, $\mathcal{E}_t$, are included in snapshot $\mathcal{S}_t = (\mathcal{N}_t, \mathcal{E}_t)$. The edges in $\mathcal{E}_t$ are converted from individual interactions to directed, weighted edges, where the edge weight is the total number of directed interactions occuring between the nodes in $\mathcal{S}_t$. Nodes $\mathcal{N}$ are derived from the edges $\mathcal{E}$ and all nodes present at timestep t, $\mathcal{N}_t$, participate in at least one edge in $\mathcal{E}_t$.

3.1   Temporal Network Snapshots

The snapshots we construct are non-overlapping and each snapshot $\mathcal{S}_t$ spans the same length of time, known as the observation window Ω. The structure of network snapshots are defined by the activity which occurred within the observation window, thus there is no accumulation of inactive edges. The observation window Ω is calculated so that most time deltas δt_{ij} between interactions of any two nodes i and j are smaller than Ω. Specifically, we find the average time deltas $\langle \delta t_{ij} \rangle$ for each interacting node pair. The 90th percentile of all average time deltas is then used as Ω. We assume most connected pairs do not continually disconnect and reconnect and thus choosing an Ω which preserves most edges is appropriate. This methodology is described with more detail in [15, 19].

3.2　Role Feature Selection

From the network snapshots we find D structural and behavioral features ($D = 12$ for our experiments) for all $n \in \mathcal{N}_t$ nodes and construct a matrix of node attributes $\mathbf{X}_t \in \mathbb{R}^{D \times N_t}$. The complete list of features used is shown in Table 1. Most of the features listed in Table 1 have common definitions, a few do not. The *new activity count* is computed for each node as the difference of the set of nodes reached from outgoing edges at the current snapshot $\mathcal{S}_t$ and the set of nodes reached from outgoing edges at the previous snapshot $\mathcal{S}_{t-1}$. Similarly, *social strategy* is a ratio of the count of new outgoing edges (outgoing edges at snapshot $\mathcal{S}_t$) that did not exist at the previous snapshot over the total number of outgoing edges for the given node at snapshot $\mathcal{S}_t$, $\frac{\text{num. of new outgoing edges}}{\text{num. of all outgoing edges}}$. Users with a higher social strategy value tend to prefer making new connections (*social explorer*, or simply *explorer*) rather than preserve older connections (*social keeper*) [15].

Table 1. Node features

	Name	Description
1	*In-degree*	Count of incoming edges
2	*Out-degree*	Count of outgoing edges
3	*Weighted in-degree*	Count of incoming interactions
4	*Weighted out-degree*	Count of outgoing interactions
5	*Reciprocity*	Ratio of reciprocated edges over all outgoing edges
6	*New activity count*	Count of new outgoing edges
7	*Social strategy*	Ratio of new outgoing edges over all outgoing edges [15]
8	*Betweenness centrality*	Number of all shortest paths which pass through the node
9	*PageRank*	PageRank measure of centrality [17]
10	*Weighted PageRank*	Weighted variant of PageRank
11	*Transitivity*	Probability any two neighbor nodes are connected (local clustering coefficient) [26]
12	*Weighted transitivity*	Weighted variant of transitivity [2]

These features were selected to enable the representation of the unique structural and behavioral patterns which may exist in online social networks which include individual, timestamped interactions. For example, while in-degree (count of incoming edges) captures popularity, the weighted in-degree (count of incoming interactions, e.g., in Facebook, number of incoming wall comments) captures the overall level of incoming activity for the target node. Features such as *transitivity* encode information about a node's neighborhood while *betweenness centrality* and *PageRank* capture global information about the node's position in the network. The *reciprocity*, *new activity count*, and *social strategy* pertain to interaction behaviors.

3.3 Role Discovery and Membership

To find roles, a decomposition of a node-attribute matrix is performed and the resulting basis vectors are the discovered roles. We use non-negative matrix factorization (NMF) [13] for this task. The role vectors contain values corresponding to each feature which can be used to characterize the role — features with higher values are more characteristic of the role. For example, a role with a large in-degree might be labeled as popular.

NMF decomposes a matrix $\mathbf{X} \in \mathbb{R}^{D \times N}$ into a basis matrix $\mathbf{U} \in \mathbb{R}^{D \times L}$ and a coefficient matrix $\mathbf{V} \in \mathbb{R}^{L \times N}$, where L is the factorization rank of the decomposition $\mathbf{X} \approx \mathbf{UV}$. Each of the L columns of the basis matrix $\mathbf{U}$ are the basis vectors or factors (roles) and the N columns of the coefficient matrix $\mathbf{V}$ are the coefficient (weight) vectors which explain how each observation $\mathbf{x}_i$ is represented as a mixture of roles.

NMF is independently run on the matrix of node attributes for each snapshot $\mathbf{X}_t$ with the same parameters. We use the standard Euclidean update equation and Frobenius cost function. We use non-negative double singular value decomposition (NNDSVD) [3] to initialize NMF. This helps NMF converge faster and introduces a bias for sparse factors (roles). We do not expect roles will have non-zero values for all features as we assume roles are a parts-based representation [12] of node attributes. Each role is characterized by a subset of all available features.

3.4 Model Selection

A critical parameter of NMF is the factorization rank L. The common methods for selecting the rank value include: MDL [20], AIC [1], and error curves [16]. We initially tried to use MDL but found model size dominated the description length and resulted in the selection of low-performing models.

Recent existing work on role discovery with NMF [9,22] used MDL and we attempted to use the same MDL function definition. Unfortunately, it appears the function does not appropriately balance between the model size and error for our datasets. We found that in all cases, the model with the lowest MDL had the smallest rank possible (for NMF with NNDSVD), $L = 2$.

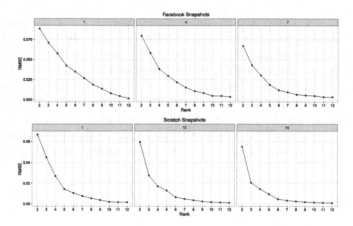

Fig. 1. Error curves for the first, mid, and final network snapshots in Facebook (top) and Scratch (bottom).

We inspected the error curves, shown in Fig. 1, and found that $L = 2$ results in a relatively large error. These curves were computed by calculating the root-mean-square error (RMSE) between the actual data $\mathbf{X}$ and corresponding NMF approximation $\mathbf{UV}$. Instead of MDL, we elected to use the knee of the error curve to estimate the rank. As shown in Fig. 1, networks across both datasets had a similar error curve. Ranks $L = 5$ and $L = 6$ correspond to the knee point for most of the curves, and therefore are appropriate choices. Rank $L = 6$ is used for the factorization of all networks in our experiments.

3.5   Tracking Roles

Given T snapshots and node-attribute matrices for each snapshot $\mathbf{X}_t$, $t = 0 \ldots T-1$, NMF is used to perform the approximate decomposition $\mathbf{X}_t \approx \mathbf{U}_t \mathbf{V}_t$. Recall the basis matrix $\mathbf{U}_t$ corresponds to role features and the coefficient matrix $\mathbf{V}_t$ corresponds to role membership weights for each user. We hypothesize that roles may persist over time and need to verify whether the same roles do occur in consecutive basis matrices; i.e., do roles from $\mathbf{U}_t$ appear in $\mathbf{U}_{t+1}$.

This role tracking is performed by measuring the similarity of every pair of role vectors between consecutive snapshots $\{\mathbf{u}_t^i \times \mathbf{u}_{t+1}^j \mid i,j \in 1 \ldots L\}$. We use cosine similarity to evaluate the pairs and ensure that each role in snapshot t maps to only one role in snapshot $t + 1$ (the mapping is injective). We use a threshold value (0.75) to determine whether a pair matches. That is, if $\text{sim}(\mathbf{u}_t^i, \mathbf{u}_{t+1}^j) > 0.75$ then the pair of role vectors match. In practice, we find most matching pairs in our data have a cosine similarity greater than 0.9.

4   Data

We use two datasets of timestamped, directed interactions to construct dynamic networks and 26 network snapshots. The first dataset is a collection of Facebook wall posts [24] available from KONECT[1]. In Facebook, users may post on each other's wall and these posts are typically comments, photos, and web links. Each of these posts is recorded as an interaction with a source user (the post author), a destination user (the owner of the wall), and a timestamp.

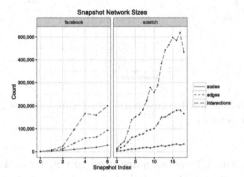

Fig. 2. Number of nodes, edges, and interactions over time in the Facebook and Scratch networks.

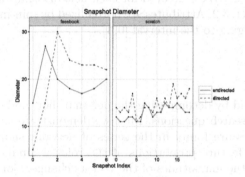

Fig. 3. Network diameter over time in the Facebook and Scratch networks.

The second dataset is a collection of Scratch project comments [18] extracted from a general Scratch dataset available from the MIT Media Lab website[2]. Scratch is an online social network and web application for writing and sharing software programs. Programming education is the primary objective of Scratch and many users are children and young adults. Scratch users write and share projects; comments may be made on each other's projects. Similar to Facebook walls, project comments in Scratch serve the purpose of public communication between users.

---

[1] http://konect.uni-koblenz.de/networks/facebook-wosn-wall.
[2] https://llk.media.mit.edu/scratch-data.

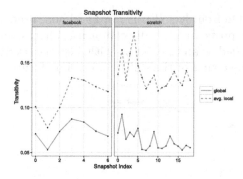

Fig. 4. Global and average local transitivity (clustering coefficient) over time in the Facebook and Scratch networks.

In both datasets, the interactions are used to construct a dynamic network and then network snapshots. The snapshots are constructed using the methodology discussed in Sect. 3.1. Figures 2, 3 and 4 show how the size and clustering of the snapshots from both datasets vary over time. Note that both the Facebook and Scratch interaction networks are growing over time.

A node-attribute matrix is created for each network snapshot using the features described in Sect. 3.2. Attributes are normalized by min-max normalization with all values belonging to the interval $[0, 1]$.

5   Results

We use the roles found by decomposing the per-snapshot, node-attribute matrix $\mathbf{X}_t$ to answer our research questions. First we demonstrate that a common set of six persistent roles are found in the series of network snapshots from both datasets. While the feature proportions of the roles is similar across datasets and over snapshots, the magnitudes of the vectors change. Correspondingly, the magnitudes of the coefficient vectors (role membership weights) differ between snapshots.

We resolve this issue by averaging the basis vectors (roles) across all snapshots and then using non-negative least squares (NNLS) [11] to find the optimal coefficient matrix for the data, given the averaged basis matrix. This normalizes the role memberships between snapshots and these membership values are used in the rest of the analysis. Note that since the original basis vectors for all network snapshots had high cosine similarity, the averaged basis vectors also have a high cosine similarity with every original basis vector.

5.1   Persistent Roles

We use the methodology discussed in Sect. 3.3 to find roles in each network snapshot from both datasets. Then we follow the methodology described in

Sect. 3.5 to determine whether the discovered roles occur in all snapshots from each dataset. We find six roles in both datasets which persist over time and perform a pairwise comparison of the sets of roles from each dataset. There is a one-to-one correspondence (bijection) of the two sets of six roles, using the same cosine similarity test as was used for testing the persistence of roles across consecutive snapshots. That is, the same set of six roles persist over time in both datasets. We note that several roles are dominated by a single feature which is not shared with any other role, this suggests a parts-based factorization of node attributes.

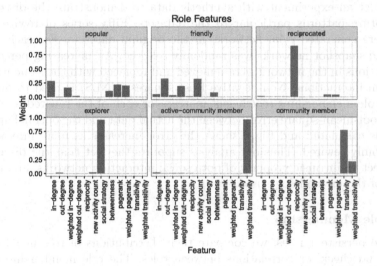

Fig. 5. Features for all roles, computed as average of role basis vectors from all network snapshots.

Figure 5 shows the discovered roles and their feature weights. The role names were selected according to the distinguishing features of the roles and we describe them here. The *popular* role is defined by the in-degree and centrality features while the *friendly* role has larger proportions in out-degree, weighted out-degree, and the number of new outgoing edges.

The *reciprocated* role is dominated by the reciprocity feature and captures the proportion of a node's outgoing edges which are reciprocated by the receiver node. A node with perfect reciprocity would have a high membership weight in this role. The *explorer* role is dominated by the social strategy feature which indicates whether a node prefers to interact with new nodes rather than maintain existing relationships. We have observed that many nodes start as explorers when they first join the network.

The final two roles, *active-community member* and *community member*, capture the clustering of nodes. Active-community member is dominated by weighted transitivity which is similar to standard transitivity (local clustering coefficient) but accounts for the strength of the edge when calculating the coefficient. As we defined edge weight as the number of directed interactions between a pair of source and destination nodes, a node with a high weighted transitivity coefficient is involved in an active community. In contrast, a node with a high unweighted transitivity coefficient simply participates in a densely-connected community and we cannot say anything about the activity of the community without further information.

5.2   Evidence of Role Dependence on Network Structure

We conduct an experiment with synthetic data to demonstrate the discovered roles capture patterns particular to the datasets. Fifty series of rewired networks were generated from networks in the original datasets. For each series, one of the snapshot networks was randomly selected. An increasing percentage of interactions in the network were removed and replaced with the same number of random interactions. Non-negative least squares (NNLS) is used to find the optimal role memberships (coefficient matrix) for each of the rewired networks.

The root-mean-square error between the actual data and the optimal approximation is calculated and Fig. 6 shows the error increases as more interactions are randomly rewired. Thus our analysis supports the fact that the discovered roles reflect an intrinsic property of both social interaction networks, and not an artifact of the methodology used.

5.3   Role Membership

Using the persistent roles, we compare their distributions of role membership weights and check for correlations between roles. The role membership corre-

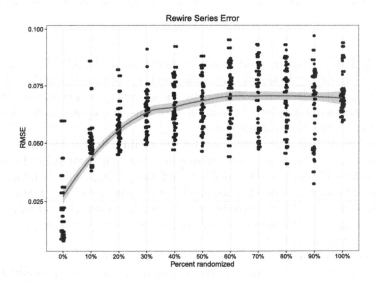

Fig. 6. Errors plotted for 50 series of randomly rewired networks.

lations (Spearman's coefficients) were calculated for every snapshot network, however due to space constraints only the results for the final snapshot from Scratch is shown in Fig. 7.

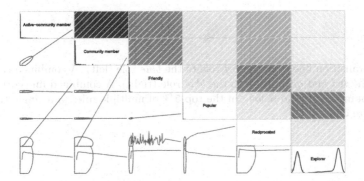

Fig. 7. Role correlations for the final snapshot from the Scratch dataset. The upper panels are colored to correspond to positive (blue) and negative (red) correlation. Darker shaded panels indicate larger correlation. The diagonal panels show the distribution of role membership weights. The lower panels show a confidence ellipse and smoothed line of the correlation. (Color figure online)

The role membership correlations tend to be similar between all network snapshots in each dataset with one notable exception. Several correlations in early Facebook snapshots (popular and friendly, community member and friendly) shifted from having a negative correlation to a positive correlation. This change in Facebook may be due to the growth and sudden increase of activity after the first few snapshots.

5.4   Role Transitions

Nodes may be members of multiple roles and their role memberships may change over time. We visualize these transitions in Fig. 9 for both the Facebook and Scratch datasets by identifying the top-5 % nodes of each role for each network snapshot and draw a line between the roles of subsequent snapshots if nodes transition from one role to the other between those two snapshots. We select the nodes with the highest role membership weights as we expect them to be exemplary representatives of the roles. The height of the bars corresponds to the number of nodes with the role. A line is drawn between two roles if at least 10 users transitioned between the roles. The transition lines are sized according to the logarithm of the number of transitioning users. Since a user may share multiple roles, some transition lines merge and show users with multiple roles in common transitioning to a role in the next timestep. Figure 8 helps explain how to interpret the transition lines.

As shown in [22], role membership of nodes may change over time and understanding these transitions allows us to construct predictive models. In this work,

Fig. 8. A transition line from the *red* role to the *blue* role (left). A combined transition line from the *red* and *green* roles to the *blue* role (right). A combined line corresponds to transitioning users who belong in the top-5 % of multiple roles in a single timestep. (Color figure online)

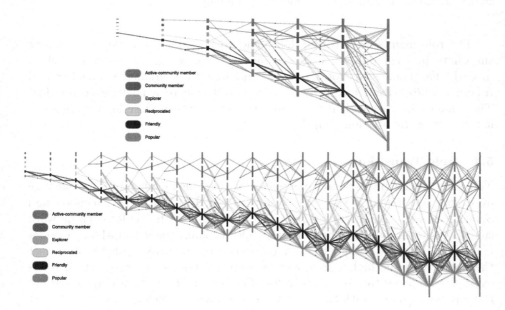

Fig. 9. The role transitions for the top-5 % users in each role over all snapshots for Facebook (top) and Scratch (bottom).

since a set of common roles has been identified, we can also perform comparative analysis of role transitions between the two datasets.

In both datasets, we see there are many transitions between *popular* and *friendly* roles as well as both community member roles. This is unsurprising as membership correlation is high for both pairs of roles. Further we note that neither *popular* nor *friendly* nodes ever transition to the *explorer* role. In contrast, users do transition from *explorer* to *popular* and *friendly*. This suggests that the most-popular users are less inclined to form new connections at the same rate as the top-5 % *explorer* users.

There are also differences in the role transitions between the two datasets. In Facebook, we observe some community member nodes transition to the *explorer* role but this does not occur in Scratch. We hypothesize this may be attributed to the different uses of the social networks. While Facebook is a general social network, Scratch is used for teaching programming by schools and it is common for students in those classes to primarily only interact with other classmates.

5.5   Role Affinity

In this section we determine whether the persistent roles affect user link preferences. As the networks used in this study are directed, we consider both how roles impact the selection of nodes to interact with (outgoing) as well as how roles affect the attractiveness of some nodes (incoming). All nodes are assigned their primary role (the role with highest membership weight) for the role affinity analysis.

In Fig. 10, we have colored nodes according to role and highlighted a subgraph for demonstration purposes. A standard force-directed layout algorithm was used to position the nodes. Note that while nodes with a higher in-degree tend to be either popular (magenta) or friendly (black), the friendly nodes have more outgoing interactions (larger outgoing edges). While friendly and popular roles reside in the core of the subgraph, explorer (green) and reciprocated (yellow) nodes appear on the periphery.

We augment the network visualization with Fig. 11 to present the exact counts of edges between roles. We note the lack of incoming edges to explorer nodes; evidence of this is also visible in the network of Fig. 10.

6   Conclusion

User roles have become a critical component for improving our understanding of user interactions in online social networks. Persistent roles, shared between multiple datasets, enable a new comparative analysis method based on relationships between roles.

In this paper, we present a methodology for identifying persistent roles across time and datasets. Using this methodology, we find the same six user roles which capture distinct structural positions in 26 network snapshots from two online social networks. To our knowledge, this paper is the first to present evidence

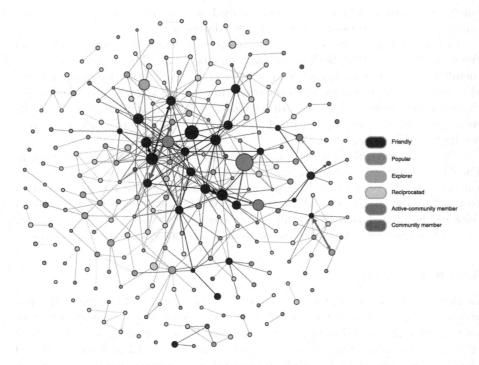

Fig. 10. A subgraph from a Facebook snapshot network. Nodes are colored by their primary role and sized according to their in-degree. Edges are sized according to the number of interactions they represent. (Color figure online)

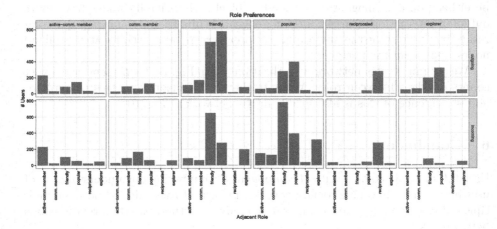

Fig. 11. The number of users with a primary role linked to/from other user roles. The column labels refer to the source node roles (for outgoing edges) and destination node roles (for incoming edges). The roles on the x-axis refer to the adjacent nodes.

of persistent roles independently derived from multiple datasets. Beyond the discovery of persistent roles, we provide an analysis of the roles and show there are differences in role membership and interaction across the snapshots.

The findings presented in this paper will be leveraged in our future work to develop probabilistic models for the prediction of role membership and node attributes. We will also investigate the composition and evolution of communities viewed as interactions of roles.

Acknowledgement. We appreciate the Lifelong Kindergarten group at MIT for publicly sharing the Scratch datasets. This work is partly based upon research supported by U.S. National Science Foundation (NSF) Awards DUE-1444277 and EEC-1408674. Any opinions, recommendations, findings, or conclusions expressed in this material are those of the authors and do not necessarily reflect the views of NSF.

References

1. Akaike, H.: Information theory and an extension of the maximum likelihood principle. In: Parzen, E., Tanabe, K., Kitagawa, G. (eds.) Selected Papers of Hirotugu Akaike. Springer Series in Statistics, pp. 199–213. Springer, New York (1998)
2. Barrat, A., Barthelemy, M., Pastor-Satorras, R., Vespignani, A.: The architecture of complex weighted networks. Proc. Nat. Acad. Sci. Unit. States Am. **101**(11), 3747–3752 (2004)
3. Boutsidis, C., Gallopoulos, E.: SVD based initialization: a head start for nonnegative matrix factorization. Pattern Recogn. **41**(4), 1350–1362 (2008)
4. Chan, J., Hayes, C., Daly, E.M.: Decomposing discussion forums and boards using user roles. In: Proceedings of the 4th International Conference on Web and Social Media, vol. 10, pp. 215–218 (2010)
5. Cheng, Y., Agrawal, A., Choudhary, A., Liu, H., Zhang, T.: Social role identification via dual uncertainty minimization regularization. In: 2014 IEEE International Conference on Data Mining (ICDM), pp. 767–772. IEEE (2014)
6. Choobdar, S., Ribeiro, P., Parthasarathy, S., Silva, F.: Dynamic inference of social roles in information cascades. Data Min. Knowl. Discov. 1–26 (2014)
7. Gilpin, S., Eliassi-Rad, T., Davidson, I.: Guided learning for role discovery (GLRD): framework, algorithms, and applications. In: Proceedings of the 19th ACM SIGKDD International Conference on Knowledge Discovery and Data Mining, pp. 113–121. ACM (2013)
8. Han, Y., Tang, J.: Probabilistic community and role model for social networks. In: Proceedings of the 21th ACM SIGKDD International Conference on Knowledge Discovery and Data Mining, pp. 407–416. ACM (2015)
9. Henderson, K., Gallagher, B., Eliassi-Rad, T., Tong, H., Basu, S., Akoglu, L., Koutra, D., Faloutsos, C., Li, L.: Rolx: structural role extraction & mining in large graphs. In: Proceedings of the Eighteenth ACM SIGKDD International Conference on Knowledge Discovery and Data Mining, pp. 1231–1239. ACM (2012)
10. Henderson, K., Gallagher, B., Li, L., Akoglu, L., Eliassi-Rad, T., Tong, H., Faloutsos, C.: It's who you know: graph mining using recursive structural features. In: Proceedings of the 17th ACM SIGKDD International Conference on Knowledge Discovery and Data Mining, pp. 663–671. ACM (2011)
11. Lawson, C.L., Hanson, R.J.: Solving Least Squares Problems, vol. 161. SIAM, Philadelphia (1974)

12. Lee, D.D., Seung, H.S.: Learning the parts of objects by non-negative matrix factorization. Nature **401**(6755), 788–791 (1999)
13. Lee, D.D., Seung, H.S.: Algorithms for non-negative matrix factorization. In: Advances in Neural Information Processing Systems, pp. 556–562 (2001)
14. McCallum, A., Wang, X., Corrada-Emmanuel, A.: Topic and role discovery in social networks with experiments on enron and academic email. J. Artif. Intell. Res. **30**(1), 249–272 (2007)
15. Miritello, G., Lara, R., Cebrian, M., Moro, E.: Limited communication capacity unveils strategies for human interaction. Scientific reports 3 (2013)
16. Owen, A.B., Perry, P.O.: Bi-cross-validation of the SVD and the nonnegative matrix factorization. Ann. Appl. Stat. **3**(2), 564–594 (2009)
17. Page, L., Brin, S., Motwani, R., Winograd, T.: The pagerank citation ranking: bringing order to the web. Technical report 1999–66, Stanford InfoLab, November 1999. http://ilpubs.stanford.edu:8090/422/, previous number= SIDL-WP-1999-0120
18. Resnick, M., Maloney, J., Monroy-Hernández, A., Rusk, N., Eastmond, E., Brennan, K., Millner, A., Rosenbaum, E., Silver, J., Silverman, B., et al.: Scratch: programming for all. Commun. ACM **52**(11), 60–67 (2009)
19. Revelle, M., Domeniconi, C., Johri, A.: Evidence of temporal artifacts in social networks. In: Proceedings of the 6th International Workshop on Mining Ubiquitous and Social Environments (MUSE), pp. 35–42 (2015)
20. Rissanen, J.: Modeling by shortest data description. Automatica **14**(5), 465–471 (1978)
21. Rossi, R., Ahmed, N.K., et al.: Role discovery in networks. IEEE Trans. Knowl. Data Eng. **27**(4), 1112–1131 (2015)
22. Rossi, R.A., Gallagher, B., Neville, J., Henderson, K.: Modeling dynamic behavior in large evolving graphs. In: Proceedings of the Sixth ACM International Conference on Web Search and Data Mining, pp. 667–676. ACM (2013)
23. Ruan, Y., Parthasarathy, S.: Simultaneous detection of communities and roles from large networks. In: Proceedings of the Second Edition of the ACM Conference on Online Social Networks, pp. 203–214. ACM (2014)
24. Viswanath, B., Mislove, A., Cha, M., Gummadi, K.P.: On the evolution of user interaction in facebook. In: Proceedings of the 2nd ACM Workshop on Online Social Networks, pp. 37–42. ACM (2009)
25. Wang, J., Chang, K.C.C., Sundaram, H.: Network cartography: seeing the forest and the trees. arXiv preprint (2015). arXiv:1512.06021
26. Watts, D.J., Strogatz, S.H.: Collective dynamics of 'small-world' networks. Nature **393**(6684), 440–442 (1998)
27. White, A.J., Chan, J., Hayes, C., Murphy, B.: Mixed membership models for exploring user roles in online fora. In: Proceedings of the 6th International Conference on Web and Social Media (2012)
28. Yu, R., He, X., Liu, Y.: Glad: group anomaly detection in social media analysis. In: Proceedings of the 20th ACM SIGKDD International Conference on Knowledge Discovery and Data Mining, pp. 372–381. ACM (2014)
29. Zhao, W.X., Wang, J., He, Y., Nie, J.Y., Wen, J.R., Li, X.: Incorporating social role theory into topic models for social media content analysis. IEEE Trans. Knowl. Data Eng. **27**(4), 1032–1044 (2015)
30. Zhao, Y., Wang, G., Yu, P.S., Liu, S., Zhang, S.: Inferring social roles and statuses in social networks. In: Proceedings of the 19th ACM SIGKDD International Conference on Knowledge Discovery and Data Mining, pp. 695–703. ACM (2013)

A Split-Merge DP-means Algorithm to Avoid Local Minima

Shigeyuki Odashima$^{(\boxtimes)}$, Miwa Ueki, and Naoyuki Sawasaki

Fujitsu Laboratories Ltd., Kawasaki, Japan
{s.odashima,ueki.miwa,sawasaki.naoyuk}@jp.fujitsu.com

Abstract. We present an extension of the DP-means algorithm, a hard-clustering approximation of nonparametric Bayesian models. Although a recent work [6] reports that the DP-means can converge to a local minimum, the condition for the DP-means to converge to a local minimum is still unknown. This paper demonstrates one reason the DP-means converges to a local minimum: the DP-means cannot assign the optimal number of clusters *when many data points exist within small distances*. As a first attempt to avoid the local minimum, we propose an extension of the DP-means by the split-merge technique. The proposed algorithm splits clusters when a cluster has many data points to assign the number of clusters near to optimal. The experimental results with multiple datasets show the robustness of the proposed algorithm.

Keywords: Clustering · DP-means · Small-variance asymptotics

1 Introduction

As we enter the age of "big data", there is no doubt that there is an increasing need for clustering algorithms that summarize data autonomously and efficiently. Nonparametric models are prospective models to address this need because of their flexibility. Unlike traditional models with fixed model complexity as a parameter, nonparametric Bayesian models [17] dynamically determine the model complexity, i.e., the number of model components, in accordance with the data. The traditional nonparametric Bayesian model often needs a high computation time because the methods need sampling algorithms or variational inference for model optimization; however, the recently introduced DP-means algorithm [20] can determine model complexity with less computational cost. The DP-means uses a technique named small-variance asymptotic (SVA) for nonparametric Bayesian models and derives a hard-clustering algorithm similar to Lloyd's k-means algorithm. The DP-means automatically determines the number of clusters and cluster centroids efficiently without sampling methods or variational inference techniques.

Convergence to a local minimum is a well-known problem in hard-clustering algorithms, especially in k-means clustering. Convergence to a local minimum of the k-means occurs when the initial clusters are assigned to close data points.

© Springer International Publishing AG 2016
P. Frasconi et al. (Eds.): ECML PKDD 2016, Part II, LNAI 9852, pp. 63–78, 2016.
DOI: 10.1007/978-3-319-46227-1_5

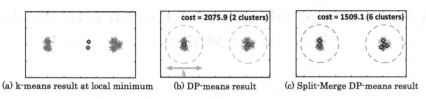

(a) k-means result at local minimum   (b) DP-means result   (c) Split-Merge DP-means result

Fig. 1. Local minima of clustering algorithms. (a) The problem of the local minimum of the k-means is well-known; the k-means converges to a local minimum when initialization of clusters is not appropriate. (b) The DP-means is believed to be robust for this type of local minimum because it assigns new clusters when the new data points are distant from existing clusters. (c) However, as shown in this paper, the DP-means has a different type of local minimum; *the DP-means cannot assign the optimal number of clusters when data points exist within small distances* (Color figure online)

For example in Fig. 1(a), if the initial two clusters are assigned to the left data points, the final solution can converge to the two clusters of the upper data points (green) and lower data points (red), although the preferred clustering solution is left data points and right data points. This convergence occurs because the k-means assigns the data points to the nearest clusters, so if the cluster initialization is inappropriate, the clustering solution converges to a local minimum. Therefore, initialization to avoid converging to a local minimum is an important step for the k-means algorithms, such as the k-means++ algorithm [4].

The DP-means also assigns the data points to the nearest clusters but generates new clusters when the distances between specific data points and existing clusters are large (Fig. 1(b)). Therefore, the DP-means is believed to be robust for convergence to local minima. However, a recent paper [6] reports that the DP-means can converge to a local minimum with fewer than the optimal number of clusters. The paper has a huge impact, but the condition for the DP-means to converge to a local minimum is still unknown.

In this paper, we present an analysis of local minima of the DP-means. As shown later, the original DP-means can converge to local minima because the DP-means cannot assign the optimal number of clusters *when the many data points exist within small distances*. For example, in Fig. 1(b), the solution with the lowest cost (i.e. the preferred solution for the DP-means) is not that with *two* clusters when the number of data points is large; a solution with lower cost can be acquired when the number of clusters is *six* (Fig. 1(c)).

To avoid these local minima, we propose an extension of the DP-means by the split-merge technique. The proposed algorithm splits clusters when the original DP-means converges to a local minimum to obtain a good solution[1] with a near-optimal number of clusters.

---

[1] In this paper, the quality of clustering is measured by the DP-means cost function value defined in Sect. 2.1. Although other clustering evaluation metrics exist, such as NMI scores or the Rand index, these metrics depend on the hyperparameter of the DP-means. Note that a similar evaluation metric is commonly used in streaming clustering [1,2,6] and robust k-means algorithms [4,5].

1.1 Related Work

DP-means and extensions. Like nonparametric Bayesian models, the DP-means has many extensions. For example, the DP-means (i.e. the small-variance asymptotic technique for nonparametric Bayesian models) has been extended to the hard-clustering version of HDP topic models [19], dependent Dirichlet process [8], Bayesian hierarchical clustering [22], nonparametric Bayesian subspace clustering [31], infinite hidden Markov models [26], and infinite support vector machines [30]. Also, the DP-means itself has been extended to efficient algorithms, such as the distributed DP-means algorithm [25], one-pass online clustering for tweet data [28], and approximate clustering with a small subset named a coreset [6]. Although the concept of the DP-means has been extended to many algorithms, the condition of the local minimum of the DP-means is still unknown. To the best of our knowledge, this paper provides the first insight into conditions when the DP-means converges to a local minimum.

Hard-clustering algorithms. The Lloyd's k-means algorithm [24] was proposed half a century ago, but it is still popular for data mining [32]. Although the original k-means algorithm is a batch clustering algorithm, the k-means algorithm has been extended to online settings [12,23] and streaming settings [1,2,27]. These algorithms are mainly based on the k-means++ algorithm [4] with analysis of the local minimum of k-means clustering. Therefore, analysis of the local minimum of the DP-means provides useful information for future efficient DP-means algorithms.

Split-merge clustering algorithms. Split-merge techniques have been used in clustering including hard-clustering [3,10,14,33]. Recently, split-merge algorithms have been extended to nonparametric Bayesian models optimized by MCMC [9,11,18,29] and by variational inference [7]. This paper is the first to apply split-merge techniques to hard-clustering methods of nonparametric Bayesian models with small-variance asymptotics.

The contributions of this paper are: (1) analysis of a condition of converging to a local minimum for the DP-means, (2) proposal of a novel DP-means algorithm with split-merge techniques to avoid converging to a local minimum, and (3) evaluation of the efficiency of the proposed algorithms with several datasets including real-world data.

2 Analysis of Existing DP-means Approaches

2.1 DP-means Clustering Problem

First, we provide a brief overview of existing DP-means algorithms. The clustering problem is selecting cluster centroids so as to minimize the distance between each data point and its closest cluster. Solving this problem exactly is NP-hard even with two clusters [15], so a local search method known as Lloyd's k-means algorithm [24] is widely used to acquire clustering solutions for a fixed number of clusters. The DP-means [20] (Algorithm 1) is a local search method to acquire clustering solutions for a variable number of clusters. Like Lloyd's k-means, the DP-means optimizes clusters by iteratively (a) assigning each data point x_i to clusters and (b) updating the centroids of each cluster by using assigned data points. However, unlike the k-means, the number of clusters optimized by the DP-means is not fixed. With hyperparameter λ to control clustering granularity, the number of clusters is automatically determined in accordance with data complexity.

In one view, clustering problems can be regarded as optimization problems that minimize objective functions between data points and extracted clusters. Similar to the k-means, the DP-means monotonically decreases the following objective function, which is called the *DP-means cost function*:

$$\text{cost}_{DP}(\mathcal{X}, \mathcal{C}) = \sum_{x \in \mathcal{X}} \min_{\mu \in \mathcal{C}} ||x - \mu||^2 + \lambda^2 k \qquad (1)$$

Here, $\mathcal{X} \in \mathbb{R}^{d \times n}$ is a set of n data points with d feature dimensions, $\mathcal{C} \in \mathbb{R}^{d \times k}$ is a set of cluster centroids, and k is the number of clusters. The first term of Eq. (1) represents the quantization error when approximating data by clusters as the k-means objective function. The second term represents penalization of the number of clusters to avoid over-fitting data with too many clusters.

The DP-means can easily be extended to online algorithms like an online extension of Lloyd's k-means [12]. Algorithm 2 shows a naive online extension of the DP-means algorithm. Instead of clustering all data at once, the online DP-means algorithm successively updates clusters as new data is loaded.

Because the batch DP-means and the online DP-means needs to perform a nearest-neighbor search for all existing clusters with each data point, the majority of computation time is consumed by this search step. Therefore, the time complexity of the batch DP-means is $\mathcal{O}(knl)$ (l is the number of iterations to convergence), and that of the online DP-means is $\mathcal{O}(kn)$. Note that because the computation time depends on the number of clusters, the computation time of the online DP-means can be larger than that of the batch DP-means.

The batch and online DP-means algorithms may assign new clusters when a new data point is loaded, so intuitively these algorithms seem to be strongly affected by the order of the data points. However, the problem is not data order: convergence to a local minimum can occur in both the existing DP-means approaches regardless of the data order, as shown in the following section.

2.2   Analysis of DP-means Algorithms

First, we provide a simplified condition to analyze DP-means clustering.

Definition 1 *(easy case for DP-means clustering). We say the data is in the "easy case" for DP-means clustering when the maximum of the squared Euclidean distance of data is lower than* λ^2.

In the easy case, the data points exist within the hypersphere whose diameter is λ (Fig. 2(a)). Note that even if the data points have multiple clusters, when

Algorithm 1. Batch DP-means	**Algorithm 2.** Online DP-means

Input: Data $\mathcal{X} = \{x_1, ..., x_n\}$, threshold λ
 Output: Centroids $\mathcal{C} = \{\mu_1, ..., \mu_k\}$
 Init. $\mathcal{C} \leftarrow \text{mean}(x_i \| x_i \in \mathcal{X}), k = 1$.
 Init cluster indicators $z_i = 1$ for all $i = 1, ..., n$
 while *not converged* **do**
    for $x_i \in \mathcal{X}$ **do**
       $c \leftarrow \arg\min_c \|x_i - \mu_c\|^2$
       if $\|x_i - \mu_c\|^2 > \lambda^2$ **then**
          $\mathcal{C} \leftarrow \mathcal{C} \cup x_i$
          set $k = k + 1$, $z_i \leftarrow k$
       else
          set $z_i \leftarrow c$
    for $\mu_c \in \mathcal{C}$ **do**
       $\mu_c \leftarrow \text{mean}(x_i \| z_i = c)$
	Input: New data x, threshold λ
	 Input: Centroids $\mathcal{C} = \{\mu_1, ..., \mu_k\}$
	 Input: The number of assigned data to each cluster
	    $w = \{w_1, ..., w_k\}$
	 Output: Updated $\mathcal{C}$ and w
	 $c \leftarrow \text{argmin}_c \|x - \mu_c\|^2$
	 if $\|x - \mu_c\|^2 > \lambda^2$ **then**
	    $\mathcal{C} \leftarrow \mathcal{C} \cup x$
	    $w \leftarrow w \cup \{1\}$
	 else
	    $\mu_c \leftarrow \frac{w_c \mu_c + x}{w_c + 1}$
	    $w_c \leftarrow w_c + 1$

the distance between clusters is sufficiently large, each individual cluster can be considered as belonging to the easy case. In this case, the solutions of the batch and online DP-means are the same, as shown in the following lemma.

Lemma 1. *In the easy case, the solutions of the batch DP-means and the online DP-means are always one cluster whose centroid is the mean of the data points regardless of data order.*

Proof. For the batch DP-means, the initial centroid is the mean of the data points regardless of data order by initialization. In this case, all data points are assigned to this centroid because the squared Euclidean distance between data points is less than λ^2. For the online DP-means, the initial centroid is the top of data, and all data points are assigned to this cluster because the squared Euclidean distance between data points is less than λ^2. In this case, the coordinates of the centroid converge to the mean of the data points regardless of the data order. □

Lemma 1 suggests the DP-means always converges to the solution with one cluster in easy cases. However, as shown in the following lemma, the solution with one cluster is not always that with the lowest DP-means cost.

Lemma 2. *In easy cases, there exists the case when the DP-means cost with two clusters is less than that with one cluster.*

Proof. Assume that $\lambda^2 = 100$ and the data consists of 1000 points on $(-1, 0)$ and 1000 points on $(1, 0)$, as shown in Fig. 2(b). This is an easy case. When the solution has one cluster, the centroid of the cluster is $(0, 0)$. In this case, the DP-means cost is $\text{cost}_{DP}(\mathcal{X}, \mathcal{C}_1) = 2000 \times 1^2 + 100 \times 1 = 2100$. However, if the solution

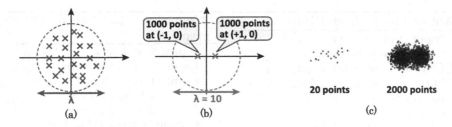

Fig. 2. Analysis of DP-means. (a) Easy case. Although DP-means always converges solution with one cluster in easy case, (b) there exists case of solution with two clusters with less DP-means cost. (c) This characteristic is natural because nonparametric Bayesian models change model complexity according to data complexity.

has two clusters on $(-1,0)$ and $(1,0)$, the DP-means cost is $\mathrm{cost}_{DP}(\mathcal{X}, \mathcal{C}_2) = 1000 \times 0^2 + 1000 \times 0^2 + 100 \times 2 = 200$. Therefore, in this case, the lower DP-means cost is acquired when the number of clusters is two.   □

Then, we can find that the DP-means can converge to a local minimum.

Theorem 1. *In easy cases, the batch DP-means and the online DP-means can converge to a local minimum with fewer than the optimal number of clusters.*

Proof. By Lemma 2, there exists a case when the number of optimal clusters is more than one. However, by Lemma 1, the batch DP-means and the online DP-means always converge to the solution with one cluster. Therefore, in this case, the DP-means can converge to a local minimum with fewer clusters than the optimal number.

This result matches a previous experimental result [6]. One reason for converging to the local minimum is that the DP-means ignores the number of data points assigned to clusters. For example, if the data consists of 10 points with $(-1,0)$ and 10 points with $(1,0)$, the cost with one cluster is 120, and the cost with two clusters is 210, so the optimal solution is one cluster. Therefore, as suggested above, the optimal solution changes as the number of data points grows. Figure 2(c) shows an intuitive interpretation of this result. On the left and the right of this figure, the data points are generated by a mixture of two Gaussian distributions centered at $(-1,0)$ and $(1,0)$, but the numbers of data points are different. When the number of data points is small (Fig. 2(c) left), the boundary between two clusters is vague. However, as the number of data points grows (Fig. 2(c) right), the boundary between two clusters becomes clear. This is the same characterization as for nonparametric Bayesian models, which is original distribution of the DP-means: the number of clusters is determined based on the complexity of data. Therefore, the DP-means should assign more clusters when the complexity of data grows with many data points.

3   Split-Merge DP-means

In this section, we discuss an extension of the DP-means named *split-merge DP-means* to avoid a local minimum with the split-merge technique. Based on the analysis in the previous section, the proposed algorithm splits clusters when the clusters contain many data points. Also, the proposed algorithm merges insufficiently split clusters.

In particular, as the first step to avoid a local minimum, we derive a split-merge DP-means with the following approximations[2]: (a) the distributions of clusters are approximated as uniform distributions, (b) the algorithm is executed with a one-pass update rule, and (c) the cluster is split in one dimension[3]. In the following section, we discuss the details of the proposed algorithm.

3.1   Condition for Splitting One Cluster into Two

Here, we provide the condition for splitting clusters to acquire the optimal DP-means cost. We assume that the data $\mathcal{X} \in \mathbb{R}^{d \times w}$ consists of w points generated by an origin-centered uniform distribution of range $\sigma = (\sigma_1, ..., \sigma_d) \in \mathbb{R}^d$. In the following, we consider two cases. The first is when the data is not split, i.e., the solution is one cluster. The second is when the data is split in dimension j, i.e., the solution is two clusters. When the data is not split, the DP-means solution is one cluster on $\mathcal{C}_1 = \{\mu_1\} = \{(0, ..., 0)\}$ with w data points. When the data is split on dimension j, the DP-means solution is two clusters on $\mathcal{C}_2 = \{\mu_{21}, \mu_{22}\} = \{(0, ..., -\sigma_j/2, ..., 0), (0, ..., \sigma_j/2, ..., 0)\}$ with $w/2$ points on each cluster because of assumption of a uniform distribution.

Below, we provide the condition when the DP-means cost with two clusters is less than the DP-means cost with one cluster. Now, we consider the expectation values of DP-means cost with one cluster $\mathrm{Exp}(\mathrm{cost}_{DP}(\mathcal{X}, \mathcal{C}_1))$ and two clusters $\mathrm{Exp}(\mathrm{cost}_{DP}(\mathcal{X}, \mathcal{C}_2))$. Because of assumption of a uniform distribution, we can compute the expectation value of the squared Euclidean distance between a data point and the cluster center in dimension l when the cluster is not split, as $\mathrm{Exp}((x_l - \mu_l)^2) = \int_{-\sigma_l/2}^{\sigma_l/2} \frac{1}{\sigma_l/2 - (-\sigma_l/2)} x^2 dx = \sigma_l^2/12$. Therefore, we have

$$\mathrm{Exp}(\mathrm{cost}_{DP}(\mathcal{X}, \mathcal{C}_1)) = w \cdot \mathrm{Exp}(\sum_{x \in \mathcal{X}} ||x - \mu_1||^2) + \lambda^2 \cdot 1$$

$$= w(\frac{\sigma_1^2}{12} + ... + \frac{\sigma_d^2}{12}) + \lambda^2 \qquad (2)$$

---

[2] Although the current form of the proposed algorithm has strong approximations, as shown in the experimental results (Sect. 4), the proposed algorithm reduces the DP-means cost in many situations. Therefore, the basic idea of the proposed method (i.e. splitting clusters with many data points to avoid local minima of DP-means) is useful for extending more exact algorithms without these approximations.

[3] Although the split operation is ideally performed in multiple dimensions, naive selection from multiple dimensions needs $\mathcal{O}(2^d)$ time. Therefore, for computational efficiency, we limit the splitting dimension to only one dimension.

When the cluster is split in dimension j, the range of distance between the data point and the cluster center is reduced to $\sigma_j/2$. Therefore, the expectation value of the squared Euclidean distance between a data point and the cluster center in dimension j is reduced to $\int_{-\sigma_j/4}^{\sigma_j/4} \frac{1}{\sigma_j/4-(-\sigma_j/4)} x^2 dx = \sigma_j^2/48$. Therefore,

$$\text{Exp}(\text{cost}_{DP}(\mathcal{X}, \mathcal{C}_2)) = \frac{w}{2} \cdot \text{Exp}(\sum_{x \in \mathcal{X}} ||x - \mu_{21}||^2) + \frac{w}{2} \cdot \text{Exp}(\sum_{x \in \mathcal{X}} ||x - \mu_{22}||^2) + 2\lambda^2$$

$$= w(\frac{\sigma_1^2}{12} + ... + \frac{\sigma_j^2}{48} + ... + \frac{\sigma_d^2}{12}) + 2\lambda^2 \tag{3}$$

Note that because the cluster range changes only dimension j, the expectation value of distance in each dimension except dimension j is the same value as that of one cluster. Then, the condition in which the solution with two clusters is better than that with one cluster is when $\text{Exp}(\text{cost}_{DP}(\mathcal{X}, \mathcal{C}_1)) > \text{Exp}(\text{cost}_{DP}(\mathcal{X}, \mathcal{C}_2))$. Therefore, by using Eqs. (2) and (3), we have the condition to split the cluster in dimension j:

$$w > 16 \left(\frac{\lambda}{\sigma_j}\right)^2 \tag{4}$$

Equation (4) means (a) clusters *with many data points* should be split and (b) clusters *with a wide range* should be split. Also, when w is fixed, this condition is first satisfied by the dimension with maximum range. Therefore, we can determine the dimension to split clusters by finding out the dimension with the maximum range of clusters. Note that the derived splitting condition is based only on the expectation values of the distance (i.e. the second moment) between clusters and data points. Therefore, this analysis can easily be extended when the cluster is approximated to other distributions, such as Gaussian distributions.

3.2   Split DP-means

In the following section, we provide an novel online DP-means algorithm by splitting clusters on the basis of the analysis. The basic idea of the proposed online DP-means algorithm is storing the range of each cluster instead of each data point. The range of clusters can easily be updated and split with online update rules. Like the online DP-means algorithm, the proposed algorithm incrementally updates clusters with new data points. However, unlike existing DP-means algorithms, the proposed algorithm splits massive clusters that satisfy Eq. (4) to avoid converging to a local minimum.

Here, we provide online update rules when adding a data point and when splitting clusters. Consider the cluster $C = (\mu, w, \sigma, p, q)$, where $\mu \in \mathbb{R}^d$ is the cluster centroid, $w \in \mathbb{R}$ is the number of data points assigned to the cluster, $\sigma \in \mathbb{R}^d$ is the range of the cluster, and $p \in \mathbb{R}^d$ and $q \in \mathbb{R}^d$ are the minimum values and the maximum values of the data points assigned to the cluster, respectively. When a new piece of data x is added to the cluster, the cluster can be updated in the following manner,

$$\mu_{new} = \frac{w_{old}\mu_{old} + x}{w_{old} + 1}, \quad w_{new} = w_{old} + 1,$$

$$\sigma_{new} = q_{new} - p_{new}, \quad p_{new} = \min(p_{old}, x), \quad q_{new} = \max(q_{old}, x) \quad (5)$$

When the splitting condition of Eq. (4) is satisfied in dimension j, the proposed algorithm splits cluster C into two clusters, $C_L = (\mu_L, w_L, \sigma_L, p_L, q_L)$ and $C_R = (\mu_R, w_R, \sigma_R, p_R, q_R)$, centered on the centroid μ_j of C in dimension j. The values of each cluster are computed in the following manner:

$$\mu_{L,m} = \begin{cases} (\mu_m + p_m)/2 & (m = j) \\ \mu_m & (m \neq j) \end{cases}, \quad \mu_{R,m} = \begin{cases} (\mu_m + q_m)/2 & (m = j) \\ \mu_m & (m \neq j) \end{cases}$$

$$w_L = w\frac{\mu_j - p_j}{\sigma_j}, \quad \sigma_L = q_L - p_L, \quad w_R = w\frac{q_j - \mu_j}{\sigma_j}, \quad \sigma_R = q_R - p_R$$

$$p_{L,m} = p_m, \quad q_{L,m} = \begin{cases} \mu_m & (m = j) \\ q_m & (m \neq j) \end{cases}, \quad p_{R,m} = \begin{cases} \mu_m & (m = j) \\ p_m & (m \neq j) \end{cases}, \quad q_{R,m} = q_m \quad (6)$$

Here, $\mu_{L,m}, \mu_{R,m}, p_{L,m}, p_{R,m}, q_{L,m}$, and $q_{R,m}$ are the values of $\mu_L, \mu_R, p_L, p_R, q_L$, and q_R in dimension m, respectively. Note that because the real data does not follow uniform distributions, splitting clusters causes approximation errors due to the assumption of uniform distribution. Therefore, reducing the number of cluster splits is desirable. Therefore, in our implementation, if a new point is *outside* the cluster and the splitting condition of Eq. (4) is satisfied, the point is regarded as a new cluster instead of splitting the cluster. Also, note that as discussed in Sect. 3.1, when selecting the dimension to split the cluster, we select the dimension with the maximum range.

Algorithm 3 shows the derived algorithm. Like the online DP-means, the majority of computation time of split DP-means is consumed by the nearest-neighbor step. Therefore, the time complexity of split DP-means is $\mathcal{O}(kn)$.

3.3 Merge DP-means

The split DP-means (Algorithm 3) uses only the local information of the data, so the solution might have much more clusters than the optimal number. Here, we discuss the condition to merge overestimated clusters.

Merging two clusters. First, we provide the condition for merging two clusters. Consider two clusters: $C_L = (\mu_L, w_L, \sigma_L, p_L, q_L)$ and $C_R = (\mu_R, w_R, \sigma_R, p_R, q_R)$. Like the discussion in Sect. 3.1, the expected value of the DP-means cost with two clusters $\mathrm{Exp}(\mathrm{cost}_{DP}(\mathcal{C}_2))$ can be computed as follows:

$$\mathrm{Exp}(\mathrm{cost}_{DP}(\mathcal{C}_2)) = w_L\left(\frac{\sigma_{L,1}^2}{12} + \dots + \frac{\sigma_{L,d}^2}{12}\right) + w_R\left(\frac{\sigma_{R,1}^2}{12} + \dots + \frac{\sigma_{R,d}^2}{12}\right) + 2\lambda^2 \quad (7)$$

Algorithm 3. Split DP-means

Input: New data x, threshold λ
Input: Clusters $\mathcal{C} = \{C_1, ..., C_k\}$, where $C_m = (\mu_m, w_m, \sigma_m, p_m, q_m)$
Output: Updated clusters $\mathcal{C}$
$\hat{C} \leftarrow \emptyset, \quad \hat{d}^2 \leftarrow \lambda^2$
for $C_m \in \mathcal{C}$ **do**
   | **if** x is outside C_m and Eq. (4) is satisfied when x is added to C_m **then**
   | | continue
   | **if** $||x - \mu_m||^2 > \hat{d}^2$ **then**
   | | $\hat{C} \leftarrow C_m, \quad \hat{d}^2 = ||x - \mu_m||^2$
if $\hat{C} = \emptyset$ **then**   // New cluster
   | $\mathcal{C} \leftarrow \mathcal{C} \cup C_{new}$, where $C_{new} = (x, 1, 0, x, x)$
else   // Update cluster
   | update $\hat{C}$ with x by Eq. (5)
   | **if** $\hat{C}$ satisfies Eq. (4) **then**   // Split cluster
   | | split $\hat{C}$ by Eq. (6)

Algorithm 4. Merge DP-means

Input: Cluster centroids $\mathcal{C} = \{\mu_1, ..., \mu_n\}$ with $w = \{w_1, ..., w_n\}$, threshold λ
Output: Cluster centroids $\mathcal{C}' = \{\mu'_1, ..., \mu'_k\}$
Compute $\Delta cost_m$ for all pairs of clusters with Eq. (10)
for all i, $\mu'_i = \mu_i$ and $z_i = i$ // Initialize cluster assignments
while $\min(\Delta cost_m) < 0$ **do**
   | $(k, l) \leftarrow \text{argmin}_{k,l}(\Delta cost_m(C_k, C_l))$
   | $z_l = k, \; C' \leftarrow C' \setminus \{\mu'_l\}$ // Merge cluster
   | $\mu'_k \leftarrow \frac{\sum_{\{i|z_i=k\}} w_i \mu_i}{\sum_{\{i|z_i=k\}} w_i}$ // Recompute centroid
   | recompute $\Delta cost_m$ for all pairs of existing clusters with Eq. (10)

When the two clusters are merged to one cluster, the merged centroid becomes $\mu_M = \frac{w_L \mu_L + w_R \mu_R}{w_L + w_R}$. In this case, the difference vectors of centroids between cluster C_L, C_R and the merged cluster C_M are $\delta_L = (\delta_{L,1}, ..., \delta_{L,d}) = (\mu_{M,1} - \mu_{L,1}, ..., \mu_{M,d} - \mu_{L,d})$ and $\delta_M = (\delta_{M,1}, ..., \delta_{M,d}) = (\mu_{R,1} - \mu_{M,1}, ..., \mu_{R,d} - \mu_{M,d})$. Under the assumption of a uniform distribution, the expectation value of the squared Euclidean distance between a data point in cluster C and the center of the merged cluster C_M in dimension l is computed as $\text{Exp}((x_l - \mu_m)^2) = \int_{\delta_l - \sigma_l/2}^{\delta_l + \sigma_l/2} \frac{1}{\delta_l + \sigma_l/2 - (\delta_l - \sigma_l/2)} x^2 dx = \delta_l^2 + \sigma_l^2/12$. Therefore, the expectation value of the DP-means with one merged cluster $\text{Exp}(cost_{DP}(\mathcal{C}_1))$ is

$$\text{Exp}(cost_{DP}(\mathcal{C}_1)) = w_L \{(\delta_{L,1}^2 + ... + \delta_{L,d}^2) + (\frac{\sigma_{L,1}^2}{12} + ... + \frac{\sigma_{L,d}^2}{12})\}$$

$$+ w_R \{(\delta_{R,1}^2 + ... + \delta_{R,d}^2) + (\frac{\sigma_{R,1}^2}{12} + ... + \frac{\sigma_{R,d}^2}{12})\} + \lambda^2 \quad (8)$$

The condition in which the solution with one merged cluster has lower cost than that with two clusters is when $\text{Exp}(cost_{DP}(\mathcal{C}_1)) < \text{Exp}(cost_{DP}(\mathcal{C}_2))$.

Therefore, by using Eqs. (7) and (8), we have the condition to merge the clusters:

$$w_L(\delta_{L,1}^2 + \dots + \delta_{L,d}^2) + w_R(\delta_{R,1}^2 + \dots + \delta_{R,d}^2) - \lambda^2 = w_L d_L^2 + w_R d_R^2 - \lambda^2 < 0 \quad (9)$$

Here, d_L^2 and d_R^2 are the squared Euclidean distance between the cluster centers of C_L and C_M, C_R and C_M, respectively.

Merging multiple clusters. Because the original cluster information is lost if we naively replace the clusters C_L and C_R with the merged clusters C_M, it is preferable to compute merged clusters with original clusters extracted by the split DP-means. Below, we discuss the merging condition when using multiple clusters extracted by the split DP-means.

Consider a cluster C_k that is originally contained in the cluster C_{old} with the centroid μ_{old} and then contained in the cluster C_{new} with the centroid μ_{new} by a merge operation. Like the discussion of merging two clusters, the expectation value of the squared Euclidean distance of the dimension l when C_k is contained by C_{old} is $\mathrm{Exp}((x_l - \mu_{old,l})^2) = d_{old}^2 + \sigma_l^2/12$ and that when C_k is contained by C_{new} is $\mathrm{Exp}((x_l - \mu_{new,l})^2) = d_{new}^2 + \sigma_l^2/12$, where d_{old}^2 and d_{new}^2 are the squared Euclidean distance between the centroids of C and C_{old}, and that between the centroids of C and C_{new}, respectively. Therefore, the cost improvement of merging two clusters $\Delta\mathrm{cost}_m(C_L, C_R)$ is computed as follows:

$$\Delta\mathrm{cost}_m(C_L, C_R) = \sum_{C_i \in C_L} w_i(d_{new,i}^2 - d_{old,i}^2) + \sum_{C_i \in C_R} w_i(d_{new,i}^2 - d_{old,i}^2) - \lambda^2 \quad (10)$$

If $\Delta\mathrm{cost}_m(C_L, C_R) < 0$, the DP-means cost function improves by applying the merge operation. Algorithm 4 shows the derived algorithm. Our algorithm greedily merges the clusters with the lowest $\Delta\mathrm{cost}_m(C_L, C_R)$ value to improve the total DP-means cost.

Figure 3 shows an example result of the split-merge DP-means. In this example, the data points are generated by five Gaussians (two Gaussians in the left side, three Gaussians in the right side). When the number of data points is small, the split DP-means extracts clusters in the same way as the original DP-means (Fig. 3(a)). However, when the number of data points grows, the split DP-means splits clusters even when the data points are within a circle with diameter λ as shown by the gray dotted circles in Fig. 3(b). Finally, the merge DP-means merges insufficiently split clusters (Fig. 3(c)).

4 Experiments

In this section, we validate the performance of the proposed algorithms. Although we should determine λ^2 for evaluation, to determine the "correct" value of λ^2 is impossible because the suitable granularity of clusters differs in each application. Therefore, we conducted experiments with multiple λ^2 for feasible results, i.e., so as not to generate too many clusters or too few clusters.

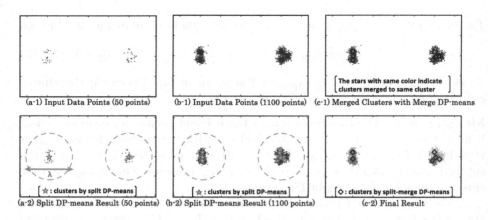

(a-1) Input Data Points (50 points)   (b-1) Input Data Points (1100 points)   (c-1) Merged Clusters with Merge DP-means

(a-2) Split DP-means Result (50 points) (b-2) Split DP-means Result (1100 points)   (c-2) Final Result

Fig. 3. Example result of split-merge DP-means for synthetic 2D data.

Datasets. We compare our algorithms with the following real data.

(1) USGS [16] contains the locations of earthquakes around the world between 1972 to 2010 mapped to 3D space with WGS 84. The value of each coordinate is normalized by the radius of the earth. USGS has $59,209$ samples with three dimensions. We use $\lambda^2 = [0.1, 0.32, 1.0, 3.2]$.

(2) MNIST [21] contains 70,000 images of handwritten digits of size 28×28 pixels. We transform these images to 10 dimensions by using randomized PCA with whitening. MNIST has $70,000$ samples with 10 dimensions. We use $\lambda^2 = [8.0 \times 10^0, 4.0 \times 10^1, 2.0 \times 10^2, 1.0 \times 10^3]$.

(3) KDD2004BIO [13] contains features extracted from native protein sequences. KDD2004BIO has 145,751 samples with 74 dimensions. We use $\lambda^2 = [3.2 \times 10^7, 1.0 \times 10^8, 3.2 \times 10^8, 1.0 \times 10^9]$.

(4) SUN SCENES 397 [34] is a widely used image database for large-scale image recognition. We use GIST features extracted from each image for evaluation. SUN SCENES 397 has $198,500$ samples with 512 dimensions. We use $\lambda^2 = [0.250, 0.354, 0.500, 0.707]$.

Note that our experimental settings include "reasonable" λ^2 parameters used in a related work [6] (USGS with $\lambda^2 = 1.0$, MNIST with $\lambda^2 = 1.0 \times 10^3$, KDD2004BIO with $\lambda^2 = 1.0 \times 10^9$), which are determined by dataset statistics. Additionally, to evaluate non-easy cases, we also conducted evaluations with smaller λ^2.

Algorithms. We compared the DP-means costs with the following algorithms:

(1) **BD**: Batch DP-means [20] (Algorithm 1).
(2) **OD**: Online DP-means (Algorithm 2).
(3) **SD** (proposed): Split DP-means (Algorithm 3).
(4) **SMD** (proposed): Split-merge DP-means (Algorithms 3 and 4)[4].

---

[4] We first applied split DP-means to the whole data in a one-pass settings and then applied merge DP-means to the result of split DP-means.

All algorithms were implemented by Python and ran in a single thread on an Intel Xeon machine with eight 2.5 GHz processors and 32 GB RAM. We measured the DP-means convergence when the change in cost was less than 0.01 or the number of iterations was over 300. We ran experiments for each data set with five different orders of data (the original order and four random permutations).

Results. Tables 1, 2, 3 and 4 show comparison results for the USGS data, MNIST data, KDD2004BIO data, and SUN SCENES 397 data, respectively. These tables show the average lowest DP-means cost with a 95 % confidence interval and the average computation time with five different data orders (±0 means the clustering results are the same in the five data orders).

As shown by the results, the proposed split-merge DP-means algorithm provided the solutions with lower cost than the existing DP-means algorithms for all datasets (including "reasonable" λ^2 parameters [6]). Also, the result shows that the solutions of the batch DP-means result are the same in five different data orders in several cases (e.g. $\lambda^2 = 1.0 \times 10^3$ with MNIST data). These cases can be interpreted as when the batch DP-means converges to a local minimum. But even in these cases, the split-merge DP-means solutions have lower DP-means cost. Therefore, the proposed algorithm avoided converging to the local minima where the original DP-means converged. Also, as expected, the solutions provided by the split-merge DP-means have lower cost than those provided by the solutions of the split DP-means algorithm. Note that the split DP-means increases the computation time by more than the batch DP-means does in many cases because the solution of split DP-means has more clusters than that of batch DP-means. For example, in the case of MNIST with $\lambda^2 = 40$, though the

Table 1. DP-means cost and runtime comparison for USGS data.

λ^2	Cost ($\times 10^2$)				Computation time (s)			
	BD	OD	SD	SMD	BD	OD	SD	SMD
0.1	4.68 ± 0.26	7.06 ± 1.13	1.54 ± 0.03	$\mathbf{1.20 \pm 0.01}$	2217.9	45.9	525.6	604.4
0.32	18.0 ± 4.31	24.8 ± 3.35	3.39 ± 0.15	$\mathbf{2.50 \pm 0.06}$	507.1	17.2	404.1	446.6
1.0	64.1 ± 3.15	80.7 ± 23.6	7.02 ± 0.40	$\mathbf{5.07 \pm 0.20}$	119.8	6.3	308.8	328.5
3.2	460 ± 0	422 ± 105	14.5 ± 0.22	$\mathbf{10.2 \pm 0.22}$	1.7	2.1	234.3	242.5

Table 2. DP-means cost and runtime comparison for MNIST data.

λ^2	Cost ($\times 10^5$)				Computation time (s)			
	BD	OD	SD	SMD	BD	OD	SD	SMD
8.0×10^0	1.31 ± 0.01	1.73 ± 0.04	1.14 ± 0.01	$\mathbf{1.12 \pm 0.00}$	58255.7	134.5	1739.0	2379.1
4.0×10^1	7.00 ± 0	6.91 ± 0.26	1.86 ± 0.07	$\mathbf{1.71 \pm 0.04}$	2.1	2.6	1006.0	1180.3
2.0×10^2	7.00 ± 0	7.00 ± 0	2.78 ± 0.13	$\mathbf{2.50 \pm 0.07}$	2.1	2.4	440.2	458.4
1.0×10^3	7.01 ± 0	7.01 ± 0	3.96 ± 0.15	$\mathbf{3.60 \pm 0.12}$	2.1	2.4	154.4	155.7

Table 3. DP-means cost and runtime comparison for KDD2004BIO data.

λ^2	Cost ($\times 10^{11}$)				Computation time (s)			
	BD	OD	SD	SMD	BD	OD	SD	SMD
3.2×10^7	1.91 ± 0.01	3.37 ± 0.13	1.57 ± 0.05	$\mathbf{1.45 \pm 0.03}$	30349.0	49.6	2340.9	2522.5
1.0×10^8	2.75 ± 0.00	4.68 ± 0.03	2.12 ± 0.12	$\mathbf{1.86 \pm 0.06}$	10810.1	27.1	1543.8	1593.9
3.2×10^8	3.19 ± 0.04	5.69 ± 0.34	2.89 ± 0.19	$\mathbf{2.45 \pm 0.09}$	7723.6	20.2	948.9	960.8
1.0×10^9	7.33 ± 0.01	9.27 ± 3.62	4.49 ± 0.75	$\mathbf{3.59 \pm 0.43}$	255.5	10.2	661.3	665.5

Table 4. DP-means cost and runtime comparison for SUN SCENES 397 data.

λ^2	Cost ($\times 10^4$)				Computation time (s)			
	BD	OD	SD	SMD	BD	OD	SD	SMD
0.250	1.15 ± 0.01	1.37 ± 0.01	$\mathbf{1.13 \pm 0.00}$	$\mathbf{1.13 \pm 0.00}$	279623.7	605.5	5084.7	5268.8
0.354	1.25 ± 0.01	1.47 ± 0.02	$\mathbf{1.17 \pm 0.00}$	$\mathbf{1.17 \pm 0.00}$	101439.4	173.3	4223.2	4346.9
0.500	1.41 ± 0.02	1.56 ± 0.04	$\mathbf{1.21 \pm 0.00}$	$\mathbf{1.21 \pm 0.00}$	12746.6	58.1	3518.2	3593.1
0.707	1.79 ± 0	1.81 ± 0.01	$\mathbf{1.24 \pm 0.01}$	$\mathbf{1.24 \pm 0.01}$	421.6	23.1	2966.5	3013.5

Table 5. Numbers of clusters for each dataset in original data order. Note that this table uses the same settings as Bachem et al. [6].

Dataset	BD	OD	SD	SMD	Optimal [6]
USGS ($\lambda^2 = 1.0$)	8	6	529	312	156
MNIST ($\lambda^2 = 1.0 \times 10^3$)	1	1	140	87	65
KDD2004BIO ($\lambda^2 = 1.0 \times 10^9$)	4	3	202	122	55

split DP-means solution had only one cluster, the split DP-means solution had average 1700 clusters.

Note that though the DP-means solutions of the proposed algorithms are worse than Bachem's recently reported result using a grid search of cluster numbers on the coresets [6] (2.50×10^{11} DP-means cost for KDD2004BIO dataset with $\lambda = 1.0 \times 10^9$), Bachem's algorithm requires an exhaustive search to determine the optimal number of clusters. In contrast, the proposed algorithms do not require an exhaustive search for the number of clusters.

Table 5 shows the number of clusters extracted by each algorithm. In this table, "optimal" is the optimal number of clusters reported by the grid-search k-means algorithm of the number of clusters [6]. As reported in this study, the batch and online DP-means tend to converge to the local minimum due to there being too few clusters (less than 1/13 of the optimal number). The proposed algorithms extract the number of clusters nearer the optimal number (the result by the split-merge DP-means is within 2.5 times the optimal number). Although the online split DP-means result tends to extract more clusters than the optimal result, refinement with the split-merge DP-means reduces overestimated clusters.

5   Conclusion

In this paper, we discussed the condition where the DP-means can converge to a local minimum and then showed an extension of the DP-means. We provided an analysis for the condition where the DP-means converges to a local minimum: though more clusters are needed when the number of data points grows, the original DP-means cannot assign the optimal number of clusters. To avoid converging to these local minima, we derived an extension of the DP-means with the split-merge technique. We empirically showed that the proposed algorithm provides solutions with lower cost values.

The limitations of the current form of our algorithm are (a) data points are approximated as a specific distribution (uniform distribution), (b) the information of detailed data points is lost due to online update rules, and (c) the split operation is performed only in one dimension. In the future, we hope to extend the proposed algorithm to an more exact one without these approximations.

References

1. Ackermann, M., Märtens, M., Raupach, C., Swierkot, K., Lammersen, C., Sohler, C.: StreamKM++: a clustering algorithm for data streams. J. Exp. Algorithms **17**, 2.4:2.1–2.4:2.30 (2012)
2. Ailon, N., Jaiswal, R., Monteleoni, C.: Streaming k-means approximation. In: NIPS (2009)
3. Appice, A., Guccione, P., Malerba, D., Ciampi, A.: Dealing with temporal and spatial correlations to classify outliers in geophysical data streams. Inf. Sci. **285**, 162–180 (2014)
4. Arthur, D., Vassilvitskii, S.: k-means++: The advantage of careful seeding. In: SODA (2007)
5. Bachem, O., Lucic, M., Hassani, S., Krause, A.: Approximate k-means++ in sublinear time. In: AAAI (2016)
6. Bachem, O., Lucic, M., Krause, A.: Coresets for nonparametric estimation - the case of DP-means. In: ICML (2015)
7. Bryant, M., Sudderth, E.: Truly nonparametric online variational inference for hierarchical Dirichlet processes. In: NIPS (2012)
8. Campbell, T., Liu, M., Kulis, B., How, J., Carin, L.: Dynamic clustering via asymptotics of the dependent Dirichlet process mixture. In: NIPS (2013)
9. Chang, J., Fisher, J.W.: Parallel sampling of DP mixture models using sub-clusters splits. In: NIPS (2013)
10. Chaudhuri, D., Chaudhuri, B., Murthy, C.: A new split-and-merge clustering technique. Pattern Recogn. Lett. **13**, 399–409 (1992)
11. Dahl, D.: An improved merge-split sampler for conjugate Dirichlet process mixture models. University of Wisconsin, Technical report (2003)
12. Dasgupta, S.: Course notes, CSE 291: Topics in unsupervised learning (2008). http://www-cse.ucsd.edu/~dasgupta/291/index.html
13. Protein Homology Dataset: KDD Cup 2004 (2004). http://www.sigkdd.org/kdd-cup-2004-particle-physics-plus-protein-homology-prediction
14. Ding, C., He, X.: Cluster merging and splitting in hierarchical clustering algorithms. In: ICDM (2003)

15. Drineas, P., Frieze, A., Kannan, R., Vempala, S., Vinay, V.: Clustering large graphs via the singular value decomposition. Mach. Learn. **56**, 9–33 (2004)
16. Global earthquakes (1.1.1972-19.3.2010): United States Geological Survey (2010). https://mldata.org/repository/data/viewslug/global-earthquakes/
17. Hjort, N., Holmes, C., Mueller, P., Walker, S. (eds.): Bayesian Nonparametrics: Principles and Practice. Cambridge University Press, Cambridge (2010)
18. Jain, S., Neal, R.: Splitting and merging components of a nonconjugate Dirichlet process mixture model. Bayesian Anal. **2**, 445–472 (2007)
19. Jiang, K., Kulis, B., Jordan, M.: Small-variance asymptotics for exponential family Dirichlet process mixture models. In: NIPS (2012)
20. Kulis, B., Jordan, M.: Revisiting k-means: new algorithms via Bayesian nonparametrics. In: ICML (2012)
21. LeCun, Y., Bottou, L., Bengio, Y., Haffner, P.: Gradient-based learning applied to document recognition. Proc. IEEE **86**, 2278–2324 (1998)
22. Lee, J., Choi, S.: Bayesian hierarchical clustering with exponential family: small-variance asymptotics and reducibility. In: AISTATS (2015)
23. Liberty, E., Sriharsha, R., Sviridenko, M.: An algorithm for online k-means clustering. In: ALENEX (2016)
24. Lloyd, S.: Least squares quantization in PCM. IEEE Trans. Inf. Theory **28**, 129–137 (1982)
25. Pan, X., Gonzalez, J., Jegelka, S., Broderick, T., Jordan, M.: Optimistic concurrency control for distributed unsupervised learning. In: NIPS (2013)
26. Roychowdhury, A., Jiang, K., Kulis, B.: Small-variance asymptotics for hidden Markov models. In: NIPS (2013)
27. Shindler, M., Wong, A.: Fast and accurate k-means for large datasets. In: NIPS (2011)
28. Shirakawa, M., Hara, T., Nishio, S.: MLJ: language-independent real-time search of tweets reported by media outlets and journalists. In: VLDB (2014)
29. Wang, C., Blei, D.: A split-merge MCMC algorithm for the hierarchical Dirichlet process (2012). arXiv:1201.1657 [stat.ML]
30. Wang, Y., Zhu, J.: Small-variance asymptotics for Dirichlet process mixture of SVMs. In: AAAI (2014)
31. Wang, Y., Zhu, J.: DP-space: Bayesian nonparametric subspace clustering with small-variance asymptotics. In: ICML (2015)
32. Wu, X., Kumar, V., Quinlan, J., Ghosh, J., Yang, Q., Motoda, H., McLachlan, G., Ng, A., Liu, B., Yu, P., Zhou, Z., Steinbach, M., Hand, D., Steinberg, D.: Top 10 algorithms in data mining. Knowl. Inf. Syst. **14**, 1–37 (2008)
33. Xiang, Q., Mao, Q., Chai, K., Chieu, H., Tsang, I., Zhao, Z.: A split-merge framework for comparing clusterings. In: ICML (2012)
34. Xiao, J., Hays, J., Ehinger, K., Oliva, A., Torralba, A.: SUN database: large-scale scene recognition from abbey to zoo. In: CVPR (2010)

Efficient Distributed Decision Trees
for Robust Regression

Tian Guo[1]([✉]), Konstantin Kutzkov[2], Mohamed Ahmed[2],
Jean-Paul Calbimonte[1], and Karl Aberer[1]

[1] École Polytechnique Fédérale de Lausanne (EPFL), Lausanne, Switzerland
{tian.guo,jean-paul.calbimonte,karl.aberer}@epfl.ch
[2] NEC Laboratories, Europe, Heidelberg, Germany
kutzkov@gmail.com, mohamed.ahmed@neclab.eu

Abstract. The availability of massive volumes of data and recent advances in data collection and processing platforms have motivated the development of distributed machine learning algorithms. In numerous real-world applications large datasets are inevitably noisy and contain outliers. These outliers can dramatically degrade the performance of standard machine learning approaches such as regression trees. To this end, we present a novel distributed regression tree approach that utilizes robust regression statistics, statistics that are more robust to outliers, for handling large and noisy data. We propose to integrate robust statistics based error criteria into the regression tree. A data summarization method is developed and used to improve the efficiency of learning regression trees in the distributed setting. We implemented the proposed approach and baselines based on Apache Spark, a popular distributed data processing platform. Extensive experiments on both synthetic and real datasets verify the effectiveness and efficiency of our approach. The data and software related to this paper are available at https://github.com/weilai0980/DRSquare_tree/tree/master/.

Keywords: Decision tree · Distributed machine learning · Robust regression · Data summarization

1 Introduction

Decision trees are at the core of several highly successful machine learning models for both regression and classification, since their introduction by Quinlan [15]. Their popularity stems from the ability to (a) select, from the set of all attributes, a subset that is most relevant for the regression and classification problem at hand; (b) identify complex, non-linear correlations between attributes; and to (c) provide highly interpretable and human-readable models [7,14,15,20]. Recently due to the increasing amount of available data and the ubiquity of distributed computation platforms and clouds, there is a rapidly growing interest in designing distributed versions of regression and classification trees [1,2,14,16,21,23],

© Springer International Publishing AG 2016
P. Frasconi et al. (Eds.): ECML PKDD 2016, Part II, LNAI 9852, pp. 79–95, 2016.
DOI: 10.1007/978-3-319-46227-1_6

for instance, the decision/regression tree in Apache Spark MLlib machine learning package[1]. Meanwhile, since many of the large datasets are from observations and measurements of physical entities and events, such data is inevitably noisy and skewed in part due to equipment malfunctions or abnormal events [9,11,22].

With this paper, we propose an efficient distributed and regression tree learning framework that is robust to noisy data with outliers. This is a significant contribution since the effect of outliers on conventional regression trees based on the mean squared error criterion is often disastrous. Noisy datasets contain outliers (e.g., grossly mis-measured target values), which deviate from the distribution followed by the bulk of the data. Ordinary (distributed) regression tree learning minimizes the squared mean error objective function and outputs the mean of the data points in the leaf nodes as predictions, which is especially problematic and sensitive to noisy data in two aspects [9,11,20]. First, during the tree growing phase (the learning phase), internal tree nodes are split so as to minimize the square-error loss function, which places much more emphasis on observations with large residuals [7,9,20]. As a result, bias on the split of a tree node due to noisy and skewed data will propagate to descendent nodes and derail the tree building process. Second, outliers drag the mean predictions away from the true values on leaf nodes, thereby leading to highly skewed predictors. Consequentially, the distributed regression tree trained on noisy data can neither identify the true patterns in data, nor provide reliable predictions [8,9,11,20,22].

Contributions. Previous methods to address robustness in the distributed regression tree fail to prevent noisy data from deviating the splits and predictions of tree nodes. In this paper, we focus on enhancing the robustness of a distributed regression tree as well as the training efficiency. Concretely, this paper makes the following contributions:

- We define the distributed robust regression tree employing robust loss functions and identify the difficulty in designing an efficient training algorithm for the distributed robust regression tree.
- We propose a novel distributed training framework for the robust regression tree, which consists an efficient data summarization method on distributed data and a tree growing approach exploiting the data summarization to evaluate robust loss functions.
- The proposed distributed robust regression tree and baselines are implemented based on Apache Spark. Extensive experiments on both synthetic and real datasets demonstrate the efficiency and effectiveness of our approach.

The organization of the paper is as follows: Sect. 2 summarizes the related work. Section 3 presents the necessary background and the problem definition. Then, Sects. 4 and 5 present proposed framework and experiment results.

2   Related Work

Robust Classification/Regression Trees. Many methods have been proposed to handle noisy data, but most of them concentrate on refining leaf nodes

---

[1] http://spark.apache.org/docs/latest/mllib-decision-tree.html.

after training or purely on the classification problem. [24] applies smoothing on the leaves of a decision tree but not inner nodes. [5] assigns a confidence score to the classifier predictions rather than improving the classification itself. [3,24] improve the classification probabilities by using regression in the leaves. Another well-known method for dealing with noisy data is fuzzy decision trees [8,13]. The fuzzy function may be domain specific and require a human expert in order to correctly define it. The other type of approaches is based on post-processing applied after a decision tree has already been built on noisy data. John [9] proposed iterative removal of instances with outlier values. [11] requires to perform back-ward path traversal for examined instances.

Our paper aims to improve the robustness of distributed regression trees by preventing the outliers from influencing the tree induction phase based on robust loss functions. Above post-processing methods can be smoothly integrated into our framework.

Distributed Classification/Regression Trees. Previous distributed regression tree algorithms do not consider the effect of data noise and outliers.

Parallel and distributed decision tree algorithms can be grouped into two main categories: task-parallelism and data-parallelism. Algorithms in the first category [4,18] divide the tree into sub-trees, which are constructed on different workers, e.g. after the first node is split, the two remaining sub-trees are constructed on separate workers. The downside of this approach is that each worker should either have a full copy of data. For large data sets, this method would lead to slowdown rather than speed-up.

In the data-parallelism approach, the training instances are divided among the different nodes of the cluster. Dividing data by features [6] requires the workers to coordinate which input data instance falls into which tree-node. This requires additional communication, which we try to avoid as we scale to very large data sets. Dividing the data by instances [16] avoids this problem. Instance-partitioning approach PLANET [14] selects splits using histograms with fixed bins constructed over the value domain of features. Such static histograms overlooks the variation of underlying data distribution as the tree grows and therefore could lead to biased splits. [2,21] put forward to construct dynamic histograms rebuilt for each layer of tree nodes and used for deliberately approximating the exact splits. [2,21] communicate the histograms re-built for each layer of tree nodes to a master worker for tree induction. [1] is a MapReduce algorithm which builds multiple random forest ensembles on distributed blocks of data and merges them into a mega-ensemble. In [10] ScalParC employs a distributed hash table to implement the splitting phase for classification problems.

In this paper, our approach falls into the instance-partition category and we build dynamic histograms to summarize the value distribution of the target variable for robust loss estimation.

3   Preliminaries and Problem Statement

In this part, we first present the regression tree employing robust loss functions. Then, we describe the robust regression tree in the distributed environment and formulate the problem of this paper.

3.1   Robust Regression Tree

In the regression problem, define a dataset $\mathcal{D} = \{(\boldsymbol{x}_i, y_i)\}$, where $\boldsymbol{x}_i \in \mathbb{N}^d$ is a vector of predictor features of a data instance and $y_i \in \mathbb{R}$ is the target variable. d is the number of features. Let $D^n \in \mathcal{D}$ denote the set of instances falling under tree node n.

Regression tree construction [15,20] proceeds by repeated greedy expansion of tree nodes layer by layer until a stopping criterion, e.g. the tree depth is met. Initially, all data instances belong to the root node of the tree. An internal tree node (e.g., D^n) is split into two children nodes respectively with data subsets $D_L(D_L \subset D^n)$ and $D_R(D_R = D^n - D_L)$ by using a predicate on a feature, so as to minimize the weighted loss criteria: $\frac{|D_L|}{|D^n|}L(D_L) + \frac{|D_R|}{|D^n|}L(D_R)$, where $L(\cdot)$ is a loss function (or error criteria) defined over a set of data instances.

This paper proposes the distributed regression tree employing robust loss functions to handle noisy datasets with outliers on the target variable (the regression tree is robust to outliers in feature space [7]). In robust regression, there are two main types of robust loss functions: accommodation and rejection [7,9,19]. Accommodation approach is to define a loss function that lessens the impact of outliers The least absolute deviation, referred to as LAD, is an accommodation method [7,19,20]. It is defined on a set of data instances D as: $L_l(D) = \frac{1}{|D|}\sum_{(\boldsymbol{x}_i y_i)\in D}|y_i - \hat{y}|$, and $\hat{y} = median_{(\boldsymbol{x}_i y_i)\in D}(\{y_i\})$, which returns the median of a set of values [20]. On the other hand, rejection approach aims to restrict the attention only to the data that seems "normal" [9]. The loss function of the rejection type is the trimmed least absolute deviation, referred to as TLAD. It is defined as $L_l(\tilde{D})$, where $\tilde{D}$ is the trimmed dataset of D derived by removing data instances with the $k\%$ largest and $k\%$ smallest target values $(0 < k < 1)$ from D and thus in TLAD $\hat{y} = median_{y_i \in \tilde{D}}(\{y_i\})$. Then, the robust regression tree in this paper is defined as:

Definition 1 (Robust Regression Tree). *In a robust regression tree, an internal tree node is split so as to minimize the weighted robust loss function $\frac{|D_L|}{|D^n|}L_l(D_L) + \frac{|D_R|}{|D^n|}L_l(D_R)$, where D_L and D_R are two (trimmed) data subsets corresponding to the children nodes. The leaf nodes take the median of target values in the leaf node as the prediction value.*

3.2   Robust Regression Tree in the Distributed Environment

In contemporary distributed computation systems [12,17], one node of the cluster is designated as the master processor and the others are the workers.

Denote the number of workers by P. The training instance set is instance-divided into P disjoint subsets stored in different workers and each worker can only access its local data subset. Let D_p be the set of data instances stored at worker p, such that $\cup_{p=1}^{P} D_p = \mathcal{D}$. For $p, q \in \{1, \ldots, P\}$, $D_p \cap D_q = \emptyset$ and $|D_p| \approx |\mathcal{D}|/P$. Denote the data instances in D_p belonging to a tree node n by D_p^n. A straightforward way to grow the robust regression tree layer by layer on the master is inefficient [2,14,17], because splitting an internal tree node requests to repeatedly access distributed data and calculate LAD (or TLAD) via expensive distributed sorting [2,17], for each trial split predicate per feature. Such a solution incurs dramatic communication and computation overheads, thereby degrading the training efficiency and scalability [2,20].

To this end, our following proposed distributed robust regression tree will exploit data summarization [2,14,21], which is able to provide compact representations of the distributed data, to enhance the training efficiency.

3.3 Problem Statement

As is presented above, it is non-trivial to design an efficient training approach for distributed robust regression tree. Therefore, the problem of this paper is formulated as:

Definition 2 (Training a Distributed Robust Regression Tree). *Given robust lost functions (LAD or TLAD) and training instance partitions $D_1, \ldots, D_p$ of a data set $\mathcal{D}$ distributed across the workers $1, \ldots, p$ of a cluster, training a robust regression tree in such a distributed setting involves two sub-problems: (1) to design an efficient data summarization method for the workers to extract sufficient information from local data and to transmit only such data summarization to the master with bounded communication cost. (2) to grow a robust regression tree on the master by estimating the robust loss function based on the data summarization.*

To keep things simple, we assume that all the features are discrete or categorical. However, all the discussion below can be easily generalized to continuous features [7]. Therefore, a split predicate on a categorical feature is a value subset. Let $\mathcal{V}_k$ represents the value set of feature k and $k \in \{1, \ldots, d\}$. For instance, given the set of data instances D^n on a tree node n and a value subset on feature k, $\mathcal{V}_k^- \subset \mathcal{V}_k$, two data subsets partitioned by $\mathcal{V}_k^-$ are $D_L = \{(\boldsymbol{x}_i, y_i) | (\boldsymbol{x}_i, y_i) \in D^n, x_{i,k} \in \mathcal{V}_k^-\}$ and $D_R = D^n - D_L$.

Often, regression tree algorithms also include a pruning phase to alleviate the problem of overfitting the training data. For the sake of simplicity, we limit our discussion to regression tree construction without pruning. However, it is relatively straightforward to modify the proposed algorithms to incorporate a variety of pruning methods [2,7].

4   Distributed Robust Regression Tree

In this part, we introduce the key contribution, the distributed robust regression tree, referred to as DR2-Tree.

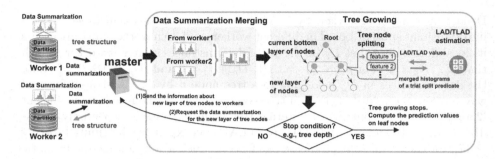

Fig. 1. Framework of the distributed robust regression tree (best viewed in colour). (Color figure online)

Overview: As is shown in Fig. 1, in DR2-Tree the master grows the regression tree layer by layer in the top-down manner. Each worker retains the split predicates of the so-far trained tree nodes for data summarization. An efficient dynamic-histogram based data summarization approach is designed for workers to communicate with the master (refer to Sect. 4.1). Then, by using such approximate descriptions of data, the master is able to efficiently evaluate robust loss functions for determining the best split of each internal tree node, thereby circumventing expensive distributed sorting for deriving LAD/TLAD (refer to Sect. 4.2). Finally, the master sends the new layer of tree nodes to each worker for the next round of node splitting.

4.1 Data Summarization on Workers

Our data summarization technique adopts the dynamic histogram, a concise and effective data structure supporting mergable operations in the distributed setting [2,21]. The one-pass nature of our proposed data summarization algorithm enables it to be adaptable to the distributed streaming learning [2] as well. Moreover, we will derive efficient robust loss function estimation algorithm based on such data summarization in the next subsection.

During the data summarization process, worker p builds a *histogram set* denoted by $\mathcal{H}_p^n = \{H_{r,v_r}^n\}$, for each tree node on the bottom layer, e.g., node n. It summarizes the target value distributions of D_p^n, the data instances belonging to tree node n in data partition D_p. H_{r,v_r}^n is a histogram describing the target value distribution of data instances having value v_r on feature r in D_p^n. H_{r,v_r}^n is a space bounded histogram of maximum β bins ($|H_{r,v_r}^n| \leq \beta$), e.g. $H_{r,v_r}^n = \{b_1, \ldots, b_\beta\}$. Let $count(H)$ (or $count(\mathcal{H})$) be the number of data instances summarized by a histogram H (or a histogram set $\mathcal{H}$). Each bin of a histogram is represented by a quad, e.g. $b_i = (l, r, c, s)$, where l and r are the minimum and maximum target values in this bin, c is the number of target values falling under this bin and s is the sum of the target values. We will see how such quad elements are used in growing the tree in the next subsection. The number of bins β in the histograms is specified through a trade-off between accuracy and computational and communication costs: a large number of bins gives a more accurate data summarization, whereas small histograms are beneficial for avoiding time, memory, and communications overloads.

Algorithm 1 presents the data summarization procedure on each worker, which updates the local data instances one by one to the corresponding histogram set. First, the tree node n_i in the bottom layer of the tree for a data instance $(x_i, y_i) \in D_p$ is found (line 1–2) and its associated $\mathcal{H}_p^{n_i}$ will be updated. For each feature value of (x_i, y_i), y_i is inserted to the corresponding histogram in $\mathcal{H}_p^{n_i}$ by either updating an existing bin having the value range covering y_i (line 3–6) or inserting a new bin $(y_i, y_i, 1, y_i)$ to the histogram (line 7–12). Second, if the size of the histogram exceeds the predefined maximum value β then the nearest bins are continuously merged until addressing the limit β (line 13–16). A temporary priority structure (e.g., $T_{v_r}^{n_i}$) is maintained for efficiently finding closest bins to merge (line 13–16). Finally, workers only send such data summarization to the master.

Complexity Analysis: In line 2–6, the binary search over bins of a histogram takes $\log \beta$ time. Then the priority structure can support in finding the nearest

---

Algorithm 1. Data summarization on a worker

Input: data partition in this worker, e.g., D_p

Output: histogram sets $\{\mathcal{H}_p^n\}$ describing the target value distribution in internal tree node n {Bins in each histogram are maintained according to the order of bin boundaries.} $\{T_{v_r}^{n_i}$: a priority queue recording the distances between neighbouring bins.}

1: **for each** data sample (x_i, y_i) in D_p **do**
2:　search the tree built so far to locate the leaf node, e.g. n_i, to which sample (x_i, y_i) belongs.
3:　**for each** feature value $x_{i,k}$ of x_i **do**
4:　　search for the bin b_{incl} such that $y_i \in [b_{incl}.l, b_{incl}.r]$ by the binary search over bins of $H_{k,x_{i,k}}^{n_i}$
5:　　**if** there exits such a bin b_{incl} for y_i **then**
6:　　　only update the bin b_{incl} by $b_{incl}.c = b_{incl}.c + 1$, $b_{incl}.s = b_{incl}.s + y_i$
7:　　**else**
8:　　　{ b_{lower} and b_{upper} are obtained during the above search process for b_{incl}.}
9:　　　$b_{lower} = \underset{b_j \in \{b_k | b_k.r \leq y_i\}}{\mathrm{argmax}} \; b_j.r$, $b_{upper} = \underset{b_j \in \{b_k | b_k.l \geq y_i\}}{\mathrm{argmin}} \; b_j.l$
10:　　　insert a new bin $(y_i, y_i, 1, y_i)$ into $H_{k,x_{i,k}}^{n_i}$ between bin b_{lower} and b_{upper}
11:　　　insert two new neighbour-bin distances $|b_{lower}.r - y_i|$ and $|b_{upper}.l - y_i|$ to the $T_{v_r}^{n_i}$
12:　　　**if** current $|H_{k,x_{i,k}}^{n_i}| >$ histogram space bound β **then**
13:　　　　for the pair of bins b_u and b_v with the minimum distance in $T_{v_r}^{n_i}$, replace the bins b_u and b_v in $H_{k,x_{i,k}}^{n_i}$ by the merged bin $(\min(b_u.l, b_v.l), \max(b_u.r, b_v.r), b_u.c + b_v.c, b_u.s + b_v.s)$
14:　　　**end if**
15:　　**end if**
16:　**end for**
17: **end for**

bins and updating bin distances in $\log \beta$ time (line 13–16). Overall, the time complexity of Algorithm 1 is $\mathcal{O}(|D_p|d\log\beta)$. Compared with the histogram building approach in [2,21], our method circumvents the sorting operation for updating individual data instances and improves the efficiency, as is demonstrated in Sect. 5. The communication complexity for transmitting data summarization of the bottom layer of nodes between the worker and master is bounded by $\mathcal{O}(\max_r(|\mathcal{V}_r|)d\beta)$ independent of the size of the data partitions. For the features with high cardinality, our data summarization can incorporate extra histograms over feature values to decorrelate the communication cost and the feature cardinality [2,21].

4.2 Tree Growing on the Master

In this part, we will first outline the tree node splitting process using the data summarization in growing the tree. Then, we present the involved two fundamental operations in detail, namely the histogram merging and LAD/TLAD estimation.

Tree Node Splitting: In order to find the best split of a tree node, we need a histogram set summarizing all the data instances falling under this node. Therefore, as is presented in Algorithm 2, a unified histogram set is built by using the histogram merging operation, which will be described in Algorithm 3. Then, it iterates over each feature to find a split predicate, i.e., a feature value subset, such that the robust lose function is minimized (Line 4–6). For a trial feature value

---

Algorithm 2. Tree node splitting
___
Input: histogram sets of tree node n from all data partitions, $\mathcal{H}_1^n, \ldots, \mathcal{H}_P^n$.
Output: the split feature and associated value set for tree node n.
1: build a unified histogram set summarizing the overall target value distribution for this tree node $\mathcal{H}^n = merge(\mathcal{H}_1^n, \ldots, \mathcal{H}_P^n)$ by using the histogram merging operation presented in Algorithm 3
2: **for each** feature $k \in \{1, \ldots, d\}$ **do**
3:    Sort the feature values in $\mathcal{V}_k$ according to the median estimations of data in the corresponding histograms [23].
4:    $\tilde{\mathcal{V}}_k$: the sorted feature values in $\mathcal{V}_k$.
5:    iterate over $\tilde{\mathcal{V}}_k$ to find a v_j and the associated feature value subsets $\mathcal{V}^- = \{v_j | j \leq i\}$ and $\mathcal{V}^+ = \mathcal{V}_k - \mathcal{V}^-$, so as to the minimize robust loss function, namely
$$\{v*, \mathcal{V}^{+*}, \mathcal{V}^{-*}\} = \underset{v_i, \mathcal{V}^+, \mathcal{V}^-}{argmin}\, \hat{L}_l(H^-)\frac{count(H^-)}{count(\mathcal{H}^n)} + \hat{L}_l(H^+)\frac{count(H^+)}{count(\mathcal{H}^n)}$$
where $H^- = merge(\{H_{v_j}^n | j \leq i\})$ and $H^+ = merge(\{H_{v_j}^n | j > i\})$ are two merged histograms respectively approximating the distributions of two data subsets defined by $\mathcal{V}^-$ and $\mathcal{V}^+$,
   $\{\,\hat{L}_l(\cdot)$ is the histogram based LAD estimation function, which is presented in Algorithm 4.$\}$
6: **end for**
7: return the feature and value subsets, which achieve the minimum robust loss.

---

subset, e.g. $\mathcal{V}^- = \{v_j | j \le i\}$ and $\mathcal{V}^+$, we need to estimate the LAD/TLAD over the data subsets defined by $\mathcal{V}^-$ and $\mathcal{V}^+$. Therefore, two temporary histograms, e.g., H^- and H^+ are built by merging the histograms in $\mathcal{H}^n$ corresponding to the feature values present in $\mathcal{V}^-$ and $\mathcal{V}^+$. $\hat{L}_l(\cdot)$ is the histogram based LAD (or TLAD) estimation function, which will be described in Algorithm 4.

Finally, when the tree reaches the stopping depth, the predictions on the leaf nodes can be exactly derived by accessing the distributed dataset. This step is only performed when the tree growing phase is finished.

Histogram Merging: Our proposed histogram merging operation is a one-pass method over the bins of histograms and creates a histogram summarizing the union of data distribution of the two histograms. As is presented in Algorithm 2, it is mainly used in two cases: (1) build a unified histogram set for each tree node on the bottom layer; (2) build temporary histograms to approximate the target value distributions of two data subsets defined by a trial feature value subset. Two histograms H_1 and H_2 are first combined in the merge-sort way. Then, bins which are closest are merged together to form a single bin. The process repeats until the histogram has β bins. Limited by the space, refer to [25] for details.

LAD/TLAD Estimation: A straightforward method to estimate LAD (or TLAD) based on a histogram is to first make a median estimate and then to sample data in each bin of the histogram to approximate individual absolute deviations [2,21]. Both the median estimation and data sampling process introduce errors into the LAD (or TLAD) estimation [20].

To this end, we propose a more efficient and precise algorithm to approximate LAD and TLAD in one-pass way. Before giving the details, we first define some notations.

Definition 3 (Quantile Bin of a Histogram). *Given a histogram $H = \{b_1, \ldots, b_\beta\}$, count$(H)$ the number of values this histogram summarizes and a quantile q over the summarized values, the quantile bin b_q addresses $\sum_{b_i < b_q} b_i.c < count(H) \cdot q$ and $\sum_{b_i \le b_q} b_i.c \ge count(H) \cdot q$*

Definition 4 (R-Partial-Sum of a Bin). *Given a bin $b = (l, r, c, s)$ of a histogram, R-Partial-Sum of bin b, $S_p(b, R)$ is defined as the sum of the R smallest values summarized in this bin.*

Recall that in the data summarization in Algorithm 1, the histogram updating process unites neighbouring bins according to the distance of bin boundaries. This allows the bins to adapt to the data distribution. Regarding the values summarized by a bin in a histogram (e.g., b), we can safely assume that they are uniformly distributed in range $[b.l, b.r]$ [2]. Therefore, we provide the lemma below, which will be used for LAD estimation, to approximate R-Partial-Sum: (Refer to [25] for the derivation.)

Lemma 1. *For a bin $b = (l, r, c, s)$ of a histogram and an integer R $(R \le b.c)$, under the assumption of the uniform distribution of values in the bin,*

R-Partial-Sum of bin b can be approximated by $S_p(b, R) \approx \hat{S}_p(b, R) =$
$\begin{cases} b.s & : R = b.c \\ R \cdot b.l + R(R-1)\delta & : otherwise \end{cases}$, where $\delta = \frac{(b.s - b.r - b.c \cdot b.l + b.l)}{(b.c-2)(b.c-1)}$.

Now we provide the following lemma for estimating the LAD/TLAD based on a histogram as:

Lemma 2. *Given a histogram $H = \{b_1, \ldots, b_\beta\}$, the LAD/TLAD over the data summarized by histogram H can be exactly computed by:*

(1) $L_l(H) = \sum\limits_{b_i > b_m} b_i.s - \sum\limits_{b_i < b_m} b_i.s + b_m.s - 2S_p(b_m, R)$, where $R = \lceil \frac{C}{2} \rceil -$
$\sum\limits_{b_i < b_m} b_i.c$, $C = count(H)$ *is the total number of data instances covered in histogram H, and an b_m is the $\frac{1}{2}$-quantile bin.*

(2) $L_l(H, \tau) = \sum\limits_{b_m < b_i < b_{\bar{q}}} b_i.s - \sum\limits_{b_q < b_i < b_m} b_i.s + S_p(b_q, R_1) - b_q.s + S_p(b_{\bar{q}}, R_2) +$
$b_m.s - 2S_p(b_m, R)$, *where b_m, b_q and $b_{\bar{q}}$ are respectively the $\frac{1}{2}$, τ and $(1 - \tau)$-quantile bins, $R = \lceil \frac{C}{2} \rceil - \sum\limits_{b_i < b_m} b_i.c$, $R_1 = C \cdot \tau - \sum\limits_{b_i < b_q} b_i.c$, $R_2 = C \cdot (1 - \tau) -$*
$\sum\limits_{b_i < \bar{q}} b_i.c$.*

Proof. Limited by the space, refer to [25] for the proof details.

Lemma 2 suggests that in estimating LAD/TLAD based on a histogram, the median estimation step is circumvented. Meanwhile, given the histogram LAD/TLAD can be estimated through replacing $S_p(\cdot)$ in Lemma 2 by $\hat{S}_p(\cdot)$ defined in Lemma 1 and exactly computing the remaining terms.

On the basis of Lemma 2, our proposed LAD/TLAD estimation algorithm is able to estimate LAD or TLAD in one-pass over the bins of the given histogram. Limited by the space limit, refer to [25] for the detailed pseudo-code of this algorithm. In our LAD/TLAD estimation, the only approximate part is $\hat{S}_p(b, R)$. Now we provide the theoretical error bound on it: (refer to [25] for the proof.)

Theorem 1. *Given a bin $b = (l, r, c, s)$ of a histogram and R $(R \leq b.c)$, if $R = 1$ or $R = b.c$, $\hat{S}_p(b, R)$ provided in Lemma 1 is the exact R-Partial-Sum. Otherwise, the approximation error of R-Partial-Sum of bin b, $S_p(b, R) - \hat{S}_p(b, R)$ is bounded within $[(R - b.c) \cdot (b.r - b.l), b.s - b.c \cdot b.l]$.*

As a result, the LAD/TLAD estimation has bounded errors as well.

5   Experimental Evaluations

In this section, we perform extensive experiments to demonstrate the efficiency and effectiveness of DR2-Tree. Due to space limits, additional results are available at [25].

5.1 Setup

Dataset: In the experiments, we use one synthetic and two real datasets. Synthetic Data: Our synthetic data generator[2] produces data instances with specified number of features. For each distinct feature value combination e.g., $(v^1, \ldots, v^d)$, where v^1 is the value of the first feature and d is the number of features, it generates several data instances having such feature values and the target values sampled from a Gaussian distribution. Such Gaussian distributions are specific *w.r.t.* feature-value combinations. Meanwhile, data instances with outliers on the target variable are injected based on a Bernoulli distribution. The probability of the Bernoulli distribution is specified through the *percentage of outliers* in the produced dataset and it is set as 0.05 initially, i.e., 5% of data instances have outlier target values. *The magnitude of outlier target values* is defined as the times of the Gaussian distribution mean. By default, the magnitude is 3, which means that the target value of an outlier data instance is sampled from a Gaussian distribution with 3 times larger mean than the mean of the corresponding feature value combination's distribution. The percentage and magnitude of outliers will be tuned later in Subsect. 5.3. Flight Dataset: It contains the scheduled and actual departure and arrival times of flights reported by certified U.S. air carriers from 1987–2008[3]. It contains data instances with abnormal values on the "arrival delay" and "departure delay" attributes, due to abnormal events, e.g., weather, security, etc. In our experiments, we use the attribute "ArrDelay" as the target variable and the categorical features as the independent variables. The cardinalities of these categorical features vary from 10 to 1032. Network Dataset: It is a dataset provided by a major European telecommunication service provider consisting of active measurements from probes within a residential ISP network. The probes measure various performance fields such as the throughput, jitter and delay between their location and chosen end-points. Furthermore, each probe and end-point are associated with various categorical and continuous features, such as the time of the measurement, the location of the endpoints and the configuration of the lines. Finally the tests cover a period of 2 days and involve 124 probes and 1314 targets. This dataset is noisy in the sense that due to network anomalies and events, the measurements could have huge outlier values (refer to the preliminary analysis of the dataset in [25]).

Baselines: ER2T is a distributed robust regression tree. SRT and DHRT are two representative distributed regression trees in the literature [2,14,21]. ER2T: It refers to the exact distributed robust regression tree. It builds the robust regression tree on the master by exactly calculating the robust loss functions in a distributed way. SRT: It refers to the distributed regression tree based on square error criteria [14] in Apache Spark machine learning tool set[4]. Prior to the tree induction, a pre-processing step is performed to obtain static and equidepth histograms for each feature and the split points are constantly selected from the

---

[2] https://github.com/weilai0980/DRSquare_tree/tree/master/dr2tree_src.
[3] http://stat-computing.org/dataexpo/2009/the-data.html.
[4] http://spark.apache.org/docs/latest/mllib-decision-tree.html.

bins of such histograms in the training phase. <u>DHRT:</u> It implements a single distributed regression tree based on [21], which employs **dynamic histograms** [2] to summarize statistics in distributed data for evaluating the square error split criterion on the master. In building histograms, it requires to sort the bins each time a data instance is added to the set already represented by the histogram [2].

In Subsect. 5.3, we will also use random forests (<u>RF</u>) and gradient boosted regression trees (<u>GBT</u>) in the distributed machine learning library Spark MLlib to compare with our robust regression tree in terms of accuracy.

Implementation: Our proposed DR2-Tree and baselines are all implemented on Apache Spark, a popular distributed data processing engine. The engine is deployed on a cluster of 23 servers, each with 16 cores (2.8GHz) and 32G of RAM.

5.2 Efficiency

In this group of experiments, we evaluate the efficiency of growing regression trees under different conditions. The training time is measured as the total time for growing a tree from the root node until the specified depth. To mitigate the effects of varying cluster conditions, all the results have been averaged over multiple runs.

We consider four parameters to tune in this set of experiments, namely the tree depth, training data size, maximum number of bins in data summarization and the number of workers. They have direct effect on the training time [2,14, 17,21]. The experiments are performed by varying one parameter while keeping the others as default values. By default, the maximum number of bins is set as 500, the number of workers is 5, the depth is 6 and the size of the training dataset is 10 million initially.

Depth: Fig. 2 presents the training time as a function of the training depth of the regression tree. DR2-Tree outperforms ER2T and DHRT by 3× and 2× faster in average. In ER2T the training time consistently takes the longest, as it computes the expensive exact median and LAD (TLAD) in the distributed setting. SRT takes 0.5 times less time than DR2-Tree. This is because SRT constantly summarizes the data using fixed bins and thus takes less time to extract statistics in bins from distributed data during the training phase.

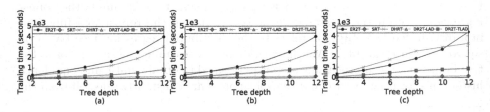

Fig. 2. Training time *w.r.t.* the depth of the tree. (a) synthetic dataset (b) flight dataset (c) network dataset. (best viewed in colour) (Color figure online)

But square-error based SRT and DHRT are less robust to noisy data than DR2-Tree, which will be shown in the next subsection.

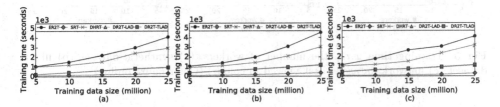

Fig. 3. Training time *w.r.t.* the size of training dataset. (a) synthetic dataset (b) flight dataset (c) network dataset. (best viewed in colour) (Color figure online)

Data Size: In Fig. 3, we present the training time as a function of the size of the training dataset, i.e., the number of data instances. Due to the one-pass nature of data summarization and tree growing processes, the training time of DR2-Tree increases linearly, highlighting the scalability. DHRT has a quickly increasing training time in part due to the quadratic computation in updating histograms. DR2-Tree takes 3× and 2× less training time in average than ER2T and DHRT.

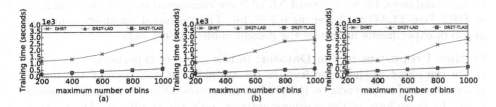

Fig. 4. Training time *w.r.t.* the maximum number of bins in data summarization. (a) synthetic dataset (b) flight dataset (c) network dataset. (best viewed in colour) (Color figure online)

Maximum Number of Bins: In Fig. 4, we investigate the effect of the maximum number of bins in data summarization on the training time. ER2T employs no data summarization. In SRT, bins are built according to the cardinality of features in data. Therefore, varying the number of bins has no effect on ER2T and SRT and in Fig. 4 only the results of DR2-Tree and DHRT are reported. The number of bins affects both the efficiency of data summarization on workers and tree growing on the master, and thus in general the training time is positively correlated with the maximum number of bins. At the highest level of bin numbers, the training time of DR2-Tree is average 4 times less than DHRT.

Number of Workers: In Fig. 5, we proceed to investigate the speedup for different numbers of workers. For large datasets, the communication between

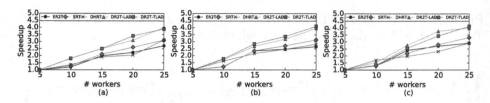

Fig. 5. Training time *w.r.t.* the number of workers. (a) synthetic dataset (b) flight dataset (c) network dataset. (best viewed in colour) (Color figure online)

workers and the master is negligible relative to the gain in the data summarization building phase. Therefore, increasing the number of workers is beneficial for speeding up the training process [2]. DR2-Tree presents 2× higher speedup than ER2T at the highest level of number of workers.

5.3 Effectiveness

In this part, we evaluate the prediction accuracy of DR2-Tree and baselines under different dataset properties and regression tree set-ups. Specifically, we aim to study the effect of the maximum number of bins, the outlier percentage and magnitude in the training dataset. The prediction accuracy is measured by normalized root mean square error (NRMSE), so as to facilitate the comparison between datasets. Lower values of NRMSE are considered better. The trim ratio in DR2-Tree-TLAD is chosen as 0.1 [7,19]. The un-tuned parameters in each group of experiments are set as the default values in Sect. 5.2.

Outlier Properties in the Dataset: In this group of experiments, we investigate the effect of the noise level of the training dataset, namely the outlier percentage and magnitude, on the prediction accuracy. Since we can only manipulate the noise level of the synthetic dataset, only the results on the synthetic dataset are reported in Fig. 6.

In Fig. 6(a), we increase the percentage of data instances with outlier target values while keeping the magnitude of the outlier values as 3× of the target value mean. It is observed that initially when the training dataset has no outliers, the accuracies of all approaches are highly close. As the percentage of outliers

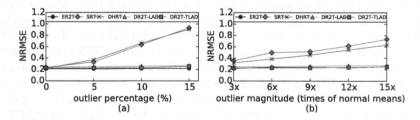

Fig. 6. Prediction accuracy *w.r.t.* the (a) outlier percentage and (b) magnitude in the training dataset (best viewed in colour). (Color figure online)

increases, the accuracy difference between the square error and robust error criterion based approaches becomes significant.

In Fig. 6(b), we study the effect of the outlier magnitude on the accuracy. In this group of experiments, 5 % data instances have outlier target values. When the magnitude of outliers increases, ER2T and DR2-Tree demonstrate stable accuracy and have 2 times less errors than SRT and DHRT at the highest level of the outlier magnitude. Compared with Fig. 6(a), we also observe that square error based approaches are more sensitive to the outlier percentage than to the outlier magnitude in the dataset.

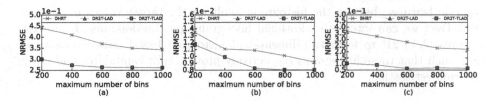

Fig. 7. Prediction accuracy $w.r.t.$ the maximum number of bins in data summarization. (a) synthetic dataset (b) flight dataset (c) network dataset. (best viewed in colour) (Color figure online)

Maximum Number of Bins: Figure 7 displays the prediction accuracy for different number of bins in the data summarization. As is presented in Sect. 4, the number of bins affects the precision of error criterion estimation in DR2-Tree. Meanwhile, it should avoid setting the number of bins too large, otherwise the training efficiency would degrade, as is shown in Fig. 4. Since only DR2-Tree and DHRT have tunable dynamic histograms, only the results of them are shown.

For the noisy synthetic and network datasets, as the number of bins increases, the master in DR2-Tree can obtain more precise data summarization based LAD/TLAD estimation and thus yields decreasing prediction errors. DR2-Tree outperforms DHRT by around 3 times. For the flight data, they present comparable accuracies.

Overall Accuracy Comparison: In this part, we perform this group of experiments by cross-validation to choose the depth in ER2T, SRT, DHRT, DR2-Tree-LAD, DR2-Tree-TLAD, depth and the number of trees in RF and the number of iterations in GBT. The results are reported in Table 1. The synthetic dataset

Table 1. Overall accuracy comparison (NRMSE).

Datasets	ER2T	SRT	DHRT	DR2-Tree-LAD	DR2-Tree-TLAD	RF	GBT
Synthetic data	0.225	0.481	0.493	0.224	0.219	0.481	0.476
Flight data	0.00882	0.00874	0.00908	0.00889	0.00890	0.00836	0.00835
Network data	0.061	0.148	0.153	0.0629	0.0581	0.145	0.181

is set to have 5 % outlier target values with magnitude 3. It shows that robust error criterion based approaches have around 50 % less error than square error based approaches, i.e., SRT, DHRT, RF and GBT. For not so noisy data, i.e. the flight data, two types of approaches have very comparable accuracy. Such results also demonstrate the wide applicability of our DR2-Tree.

6   Discussion

Our current version of DR2-Tree focuses on the robust regression with categorical features. It can be smoothly extended to handle numeric or mixed features. For numeric features, besides the histograms built on the target values in current DR2-Tree, we can integrate additional histograms on the domains of numeric features [2,14,21] to form two-dimensional histogram based data summarization, such that these histograms respectively provide split candidates and error criterion estimation.

7   Conclusion

In this paper, we propose an efficient distributed robust regression tree for handling large and noisy data. Extensive experiments reveal that: (1) Our proposed DR2-Tree is robust to datasets with various outlier percentages and magnitudes. (2) DR2-Tree exhibits comparable accuracy as the conventional distributed regression tree for relatively clean datasets with rare outliers. (3) DR2-Tree is much more efficient than exact robust regression and the dynamic histogram based regression tree [2,21].

Acknowledgements. The authors thank Mathias Niepert for his insightful input to this work. The research leading to these results has received funding from the European Union's Horizon 2020 innovation action program under grant agreement No 653449-TYPES as well as Nano-Tera.ch through the OpenSense2 project.

References

1. Basilico, J.D., et al.: COMET: a recipe for learning and using large ensembles on massive data. In: IEEE ICDM, pp. 41–50 (2011)
2. Ben-Haim, Y., Tom-Tov, E.: A streaming parallel decision tree algorithm. J. Mach. Learn. Res. **11**, 849–872 (2010)
3. Chm-les, X., et al.: Decision tree with better ranking. In: AAAI (2003)
4. Darlington, J., Guo, Y., Sutiwaraphun, J., To, H.W.: Parallel induction algorithms for data mining. In: Liu, X., Cohen, P., Berthold, M. (eds.) IDA 1997. LNCS, vol. 1280, pp. 437–445. Springer, Heidelberg (1997). doi:10.1007/BFb0052860
5. Esposito, F., Malerba, D., Semeraro, G., et al.: A comparative analysis of methods for pruning decision trees. IEEE TPAMI **19**(5), 476–491 (1997)
6. Freitas, A.A., Simon, H.L.: Mining very large databases with parallel processing. Springer Science Business Media (1998)

7. Hastie, T., Tibshirani, R., Friedman, J., et al.: The elements of statistical learning: data mining, inference and prediction. Math. Intelligencer **27**(2), 83–85 (2005)
8. Janikow, C.Z.: Fuzzy decision trees: issues and methods. IEEE Trans. Syst. Man Cybern. Part B Cybern. **28**(1), 1–14 (1998)
9. John, G.H.: Robust decision trees: removing outliers from databases. In: KDD, pp. 174–179 (1995)
10. Joshi, M.V., Karypis, G., Kumar, V.: ScalParC: A new scalable and efficient parallel classification algorithm for mining large datasets. In: 12th International Parallel Processing Symposium, pp. 573–579 (1998)
11. Katz, G., Shabtai, A., Rokach, L., et al.: ConfDTree: improving decision trees using confidence intervals. In: IEEE ICDM, pp. 339–348 (2012)
12. Lee, K.H., Lee, Y.J., Choi, H., et al.: Parallel data processing with MapReduce: a survey. ACM SIGMOD Rec. **40**(4), 11–20 (2012)
13. Olaru, C., Wehenkel, L.: A complete fuzzy decision tree technique. Fuzzy sets and systems **138**(2), 221–254 (2003)
14. Panda, B., Herbach, J.S., Basu, S., et al.: Planet: massively parallel learning of tree ensembles with mapreduce. PVLDB **2**(2), 1426–1437 (2009)
15. Quinlan, J.R.: Bagging, boosting, and C4. 5. In: AAAI/IAAI, pp. 725–730 (1996)
16. Shafer, J., Agrawal, R., Mehta, M.: SPRINT: a scalable parallel classier for data mining. In: PVLDB, pp. 544–555 (1996)
17. Shanahan, J.G., Dai, L.: Large scale distributed data science using Apache Spark. In: ACM SIGKDD, pp. 2323–2324 (2015)
18. Srivastava, A., Han, E.H., Kumar, V., et al.: Parallel formulations of decision-tree classification algorithms. In: High Performance Data Mining, Springer, US (2002)
19. Stuart, C.: Robust regression. Department of Mathematical Sciences, Durham University, vol. 169 (2011)
20. Torgo, L.: Inductive learning of tree-based regression models. Thesis (1999)
21. Tyree, S., Weinberger, K.Q., Agrawal, K., et al.: Parallel boosted regression trees for web search ranking. In: ACM WWW, pp. 387–396 (2011)
22. Wang, P., Sun, W., Yin, D., et al.: Robust tree-based causal inference for complex ad effectiveness analysis. In: ACM WSDM, pp. 67–76 (2015)
23. Ye, J., Chow, J.H., Chen, J., et al.: Stochastic gradient boosted distributed decision trees. In: ACM CIKM, pp. 2061–2064 (2009)
24. Zadrozny, B., Elkan, C.: Obtaining calibrated probability estimates from decision trees and naive Bayesian classifiers. ICML **1**, 609–616 (2001)
25. Supplementary Material. https://infoscience.epfl.ch/record/218970

Fast Hoeffding Drift Detection Method
for Evolving Data Streams

Ali Pesaranghader$^{(\boxtimes)}$ and Herna L. Viktor

Faculty of Engineering, School of Electrical Engineering and Computer Science,
University of Ottawa, Ottawa, ON K1N 6N5, Canada
{apesaran,hviktor}@uottawa.ca

Abstract. Decision makers increasingly require near-instant models to make sense of fast evolving data streams. Learning from such evolving environments is, however, a challenging task. This challenge is partially due to the fact that the distribution of data often changes over time, thus potentially leading to degradation in the overall performance. In particular, classification algorithms need to adapt their models after facing such distributional changes (also referred to as concept drifts). Usually, drift detection methods are utilized in order to accomplish this task. It follows that detecting concept drifts as soon as possible, while resulting in fewer false positives and false negatives, is a major objective of drift detectors. To this end, we introduce the Fast Hoeffding Drift Detection Method (FHDDM) which detects the drift points using a sliding window and Hoeffding's inequality. FHDDM detects a drift when a significant difference between the maximum probability of correct predictions and the most recent probability of correct predictions is observed. Experimental results confirm that FHDDM detects drifts with less detection delay, less false positive and less false negative, when compared to the state-of-the-art.

Keywords: Data stream mining · Concept drift · Hoeffding's inequality · Evolving environments

1 Introduction

Learning in evolving environments is a challenging task. This difficulty is caused, not only by the speed and volume of data arrival, but also by the changes in the distribution that may occur. Intuitively, distributional changes may cause degradation in the performance of classification models. To this end, adaptive learning algorithms utilize drift detection methods to detect such changes and then take appropriate actions [1]. Typically, classification models are updated or, alternatively, retrained when a drift has been detected. Alternatively, ensemble learning algorithms are employed in an attempt to maintain the accuracy [2–6].

It follows that, in such a setting, drift detection methods resulting in fewer false positives and less false negatives are preferred. Such detectors should also detect drifts as soon as they arrive. A drift detector with a high false positive

© Springer International Publishing AG 2016
P. Frasconi et al. (Eds.): ECML PKDD 2016, Part II, LNAI 9852, pp. 96–111, 2016.
DOI: 10.1007/978-3-319-46227-1_7

number (or rate) causes frequent retraining, leading to more resources being used [7,8]. On the other hand, a drift detector with a high false negative number causes decay in the accuracy of classification, since it does not detect drift points. These types of oversights are costly and should be avoided in many applications, e.g. in fraud detection and emergency response settings. Moreover, drift detectors should detect drifts with the least possible delay. Correct approximation of a drift point, i.e. detecting the drift with less delay, is necessary because it helps not only to make maximum usage from the data, but also aids us to realize how the drift happened. Such insights are crucial in Business Intelligence (BI) applications. Accordingly, *false positive*, *false negative* and *detection delay* are considered as evaluation measures for drift detection methods [23,24].

We introduce the Fast Hoeffding Drift Detection Method (FHDDM) based on our requirement that the accuracy of classification models should stay steady, or increase, as more instances are processed. Otherwise, the degradation in accuracy may indicate that we face concept drifts. The FHDDM algorithm, in a novel way, uses a sliding window and Hoeffding's inequality [9] to calculate and compare the maximum probability of correct predictions observed so far with the most recent probability of correct predictions for the purpose of drift detection. We will show that the FHDDM algorithm results in less detection delay, less false positive and less false negative, when compared to the state-of-the-art.

The remainder of this paper is organized as follows: We talk about related works to concept drift detection in Sect. 2. We describe the Fast Hoeffding Drift Detection Method (FHDDM) algorithm in Sect. 3. Section 4 presents an approach for evaluating drift detectors on the basis of detection delay. We conduct our experiments on synthetic and real-world datasets in Sect. 5. Finally, we conclude the paper and discuss future works in Sect. 6.

2 Related Works

Gama et al. [1] classified concept drift detectors into three general groups of: (1) *Sequential Analysis based Methods*: These methods sequentially evaluate prediction results as they become available, and they alarm for drifts when a predefined threshold is met. The Cumulative Sum (CUSUM) [10] and Geometric Moving Average (GMA) [11] are members of this group. (2) *Statistical based Methods*: These methods probe the statistical parameters such as mean and standard deviation of prediction results to detect drifts in a stream. The Drift Detection Method (DDM) [12], Early Drift Detection Method (EDDM) [13] and Exponentially Weighted Moving Average (EWMA) [14] are placed in this group. (3) *Windows based Methods*: They usually use a fixed reference window summarizing the past information and a sliding window summarizing the most recent information. A significant difference between the distributions of these windows suggests the occurrence of a drift. Statistical tests or mathematical inequalities, with the null-hypothesis saying that the distributions are equal, can be used to decide the level of difference. Kifer's [15], Nishida's [16], Bach's [17], the Adaptive Windowing (ADWIN) [18], the Hoeffding Drift Detection Methods (HDDM$_{A-test}$ and

HDDM$_{W-test}$) [19], and SeqDrift detectors [23,24] are members of this group. As discussed in [1], drift detectors in the second and third groups have shown better performances and have been frequently considered as benchmarks in the literature [7,13,16,18,19]. We will, thus, compare our FHDDM with DDM, EDDM, ADWIN, HDDM$_{A-test}$ and HDDM$_{W-test}$. We describe each one below:

DDM: Drift Detection Method – DDM, by Gama et al. [12], monitors the error-rate of the classification model to detect drifts. On the basis of PAC learning model [20], the method considers that the error-rate of a classifier decreases or stays constant as the number of instances increases. Otherwise, it suggests the occurrence of a drift. Consider p_t as the error-rate of the classifier with standard deviation of $s_t = \sqrt{(p_t(1 - p_t)/t)}$ at time t. As instances are processed, DDM updates two variables p_{min} and s_{min} when $p_t + s_t < p_{min} + s_{min}$. DDM warns for a drift when $p_t + s_t \geq p_{min} + 2 * s_{min}$, and it detects a drift when $p_t + s_t \geq p_{min} + 3 * s_{min}$. The p_{min} and s_{min} are reset in the case of drift detection.

EDDM: Early Drift Detection Method – EDDM, by Baena-Garcia et al. [13], checks the distances between wrong predictions to detect concept drifts. The algorithm is based on the observation that facing a drift is likely when the distances between errors are smaller. EDDM calculates the average distance between two recent errors, i.e. p'_t, with its standard deviation s'_t at time t. It updates two variables p'_{max} and s'_{max} when $p'_t + 2 * s'_t > p'_{max} + 2 * s'_{max}$. It warns for a drift if $(p'_t + 2 * s'_t)/(p'_{max} + 2 * s'_{max}) < \alpha$, and it detects a drift if $(p'_t + 2 * s'_t)/(p'_{max} + 2 * s'_{max}) < \beta$. They set α and β to 0.95 and 0.90 respectively. The p'_{max} and s'_{max} are reset if a drift is detected.

ADWIN: Adaptive Sliding Window – ADWIN, by Bifet et al. [18], slides the window w on the results of predictions to detect drifts. It examines two large enough sub-windows, i.e. w_0 with size n_0 and w_1 with size n_1, of w for drift detection where $w_0 \cdot w_1 = w$. A significant difference between the means of two sub-windows suggests a concept drift, i.e. $|\mu_{w_0} - \mu_{w_1}| \geq \varepsilon$ where $\varepsilon = \sqrt{\frac{1}{2m} \ln \frac{4}{\delta'}}$, m is the harmonic mean of n_0 and n_1, $\delta' = \delta/n$, δ is the confidence level and n is the size of window w. After a drift detection, elements are dropped from the tail of the window until no significant difference is seen.

HDDM$_{A-test}$ ◇ HDDM$_{W-test}$ – HDDM$_{A-test}$ and HDDM$_{W-test}$ are proposed by Frias-Blanco et al. [19]. The former compares the moving averages to detect drifts. The latter uses the EMWA forgetting scheme [14] to weight the moving averages. Then, weighted moving averages are compared to detects the drift. For both cases, Hoeffding's inequality [9] is used to set an upper bound to the level of difference between averages. The authors noted that the first and the second methods are ideal for detecting abrupt and gradual drifts, respectively.

The pros and cons of all methods will be discussed in more details in Sect. 5. However, during our preliminary experiments, we observed that the aforementioned methods may cause high numbers of false positives and false negatives. Some resulted in long detection delays, though they had short detection runtimes.

In the next section, we introduce our Fast Hoeffding Drift Detection Method (FHDDM), developed to address these shortcomings.

3   Fast Hoeffding Drift Detection Method

We present our Fast Hoeffding Drift Detection Method (FHDDM) which uses the Hoeffding's inequality [9] to detect drifts in evolving data streams. The FHDDM algorithm slides a window with a size of n on the classification results. Subsequently, it inserts a 1 into the window if the prediction result is *true*, otherwise it inserts 0. As inputs are processed, it calculates the probability of observing 1s, i.e. p_t^1, in the sliding window at time t, and also keeps the maximum probability of 1s occurring, i.e. p_{max}^1. Equation (1) shows if the value of p_1 at time t is greater than the value of p_{max}^1 then the value of p_{max}^1 will be updated.

$$if\ p_{max}^1 < p_t^1 \Rightarrow p_t^1 \to p_{max}^1 \tag{1}$$

On the basis of the probably approximately correct (PAC) learning model [20], the accuracy of classification would increase or stay steady as the number of instances increases; otherwise the possibility of facing drifts increases [12]. Thus, the value of p_{max}^1 should increase or remain steady as we process instances. In other words, the possibility of facing a concept drift increases if p_{max}^1 does not change and p_t^1 decreases over time. Eventually, as in Eq. (2), a significant difference between p_{max}^1 and p_t^1 indicates the occurrence of a drift in the stream.

$$\Delta p = p_{max}^1 - p_t^1 \geq \varepsilon_d \Rightarrow Drift := True \tag{2}$$

We use the Hoeffding's inequality to define the value of ε_d, Eq. (4). The Hoeffding's inequality has a very attractive property that it is independent of the probability distribution generating the data [9,19,21]. That is, it assigns an upper bound for the deviation between the mean of n random variables and its expected value.

Hoeffding's Inequality Theorem: Let $X_1, X_2, ..., X_n$ be n independent random variables such that $X_i \in [0,1]$, then with probability at most δ, the difference between the empirical mean $\overline{X} = \frac{1}{n}\sum_{i=1}^n X_i$ and the true mean $E[\overline{X}]$ is at least ε_H, i.e. $Pr(|\overline{X} - E[\overline{X}]| \geq \varepsilon_H) \leq \delta$, where:

$$\varepsilon_H = \sqrt{\frac{1}{2n}\ln\frac{2}{\delta}} \tag{3}$$

Corollary (FHDDM test): In a stream setting, assume p_t^1 is the probability of observing 1s in a sequence of n random entries, each in $\{0,1\}$, at time t, and p_{max}^1 is the maximum probability observed so far. Let $\Delta p = p_{max}^1 - p_t^1 \geq 0$ be the difference between those two probabilities. Then, given the desired δ, i.e. the probability of error allowed, the Hoeffding's inequality guarantees a drift has happened if $\Delta p \geq \varepsilon_d$, where:

$$\varepsilon_d = \sqrt{\frac{1}{2n}\ln\frac{1}{\delta}} \tag{4}$$

Figure 1 depicts an illustrative example of the FHDDM algorithm. In this example, n and δ are set to 10 and 0.2, respectively. Using Eq. (4), the value of ε_d will be equal to 0.28. In this example, a real drift occurs right after the 12th instance. The values of p^1 and p^1_{max} are null and zero until 10 elements are inserted into the window. We have seven 1s in the window after reading the first 10 elements, and so the p^1_{10} is equal to 0.7. The value of p^1_{max} is set to 0.7 too. The 1st element is dropped out from the window before the 11th prediction status is inserted. Since the value of prediction status is 0, the value of p^1 decreases to 0.6. The value of p^1_{max} stays the same, because it is greater than the current p^1. This progress is continued until the 18th is inserted. At this moment the difference between p^1_{max} and p^1_{18} becomes more than the value of ε_d. In this case, the FHDDM algorithm alarms for a drift.

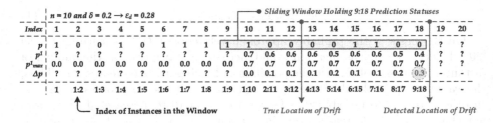

Fig. 1. Illustration of how FHDDM works

We present the pseudocode of the FHDDM approach in Algorithm 1. First, we need to instantiate an object from FHDDM and then call its DETECT function. The result of prediction, i.e. p, is sent to the DETECT function as an input in order to determine whether a drift has occurred, in line 11. The oldest element is dropped out from the sliding window if it is full; then, a new element is pushed into it, as shown in lines 12 to 15. The algorithm returns *False* in the case of having not enough elements in the window, as depicted in lines 16 and 17. Next, the values of p^1, p^1_{max}, and Δp are calculated or updated (lines 19 to 23). In the case of having $\Delta p \geq \varepsilon_d$, it resets its parameters and alarms for a drift by returning *True*.

Window-based approaches [15–18] usually compare two (sub)windows, e.g. w_1 and w_2, leading to a considerable memory usage [1]. That is, one window is used to maintain historic information (from the beginning) and the second maintains the most recent information. In contrast, FHDDM compares the current accuracy of the classifier with its best accuracy, i.e. the best experience, observed so far using one sliding window size of n. Thus, it occupies only one register, i.e. p^1_{max}, and a sliding window size of n where $n \ll |w_1|$ or $|w_2|$. Eventually, unlike [12–14], as we apply the Hoeffding's inequality, our method is independent of the probability distribution of data. The Hoeffding's inequality assumes instances are independent of each other that makes the bound independent of the probability distribution.

Algorithm 1. Pseudocode of Fast Hoeffding Drift Detection Method (FHDDM)

1: **function** INITIALIZE($windowSize, delta$)
2:     $n = windowSize$
3:     $\delta = delta$
4:     $\varepsilon_d = \sqrt{\frac{1}{2n} \ln \frac{1}{\delta}}$
5:     RESET()
6: **end function**
7: **function** RESET()
8:     $w = []$ ▷ Creating an empty sliding window.
9:     $p_{max}^1 = 0$
10: **end function**
11: **function** DETECT(p)   ▷ p is 1 for the correct predictions, 0 otherwise.
12:     **if** $w.size() = n$ **then**
13:         $w.tail.drop()$ ▷ Dropping an element from the tail.
14:     **end if**
15:     $w.push(p)$ ▷ Pushing an element into the head.
16:     **if** $w.size() < n$ **then**
17:         **return** False
18:     **else**
19:         $p^1 = w.count(1)/w.size()$ ▷ The recent probability of seeing 1s.
20:         **if** $p_{max}^1 < p^1$ **then**
21:             $p_{max}^1 = p^1$
22:         **end if**
23:         $\Delta p = p_{max}^1 - p^1$
24:         **if** $\Delta p \geq \varepsilon_d$ **then**
25:             RESET()   ▷ Resetting parameters.
26:             **return** True   ▷ Signalling for an alarm.
27:         **else**
28:             **return** False
29:         **end if**
30:     **end if**
31: **end function**

---

4   On Evaluation of Concept Drift Detectors

True Positive, False Positive and False Negative numbers are useful to evaluate the performance of concept drift detectors. Intuitively, a drift detector with the highest true positive, the lowest false positive and the lowest false negative values is preferred. Huang et al. [7] and Bifet et al. [18] used three types of tests to measure true positive, false positive, and false negative values of a drift detector. For instance, to measure the false positive, they generated a stream of bits from a stationary Bernoulli distribution. If the detector alarms for drifts, one false positive is counted for each alarm. Thus, one may use three of such tests to count true positive, false positive, and false negative numbers. However, having an approach able to count them in one test, for any stream generated by any probability distribution, is preferred. To this end, we introduce an approach to count true positive, false positive and false negative by defining the *acceptable delay length* Δ. The acceptable delay length is a threshold set to determine how far the detected drift could be from the true location of drift, for being considered as true positive. Considering the acceptable delay length Δ, we describe the true positive, false positive and false negative calculations as follows:

- *True Positive (TP)*: A drift detector truly detects a drift occurred at time t if it alarms for that at anytime in $[t - \Delta, t + \Delta]$. We call this range as the *acceptable detection interval* of true positive. Eventually, the true positive rate is defined as the number of drifts correctly identified over the total number of drifts in a stream. For evaluating reactive concept drift detectors, the acceptable detection interval is $[t, t + \Delta]$.
- *False Positive (FP)*: A drift detector falsely alarms for a drift if it detects that outside of the acceptable detection intervals. The false positive rate is defined as the number of points incorrectly considered as drifts over the total number of points which are not drifts.
- *False Negative (FN)*: A drift detector falsely overlooks a drift occurred at time t if it does not alarm for that at anytime in $[t - \Delta, t + \Delta]$. The false negative rate is defined as the number of drifts incorrectly left unidentified over the total number of drifts in a stream. For the reactive concept drift detectors, the range is $[t, t + \Delta]$.

Figure 2, as an example, illustrates how the true positive, false positive and false negative are counted. The upper stream shows the real locations of drifts, i.e. the squares with D inside, and the lower stream shows the result of detection at each location. The squares with T inside represent the drifts detected correctly (true positive), the squares with F inside represent the points incorrectly considered as drift points (false positive), and the squares with N inside indicate undiscovered drifts (false negative). The drift detector signals for a drift within the first acceptable detection interval and so the true positive number increases. Subsequently, it incorrectly alarms for a drift and the false positive number increases. Since the detector does not alarm for a drift within the second acceptable detection interval, the false negative number increases. The figure shows that the detector incorrectly alarms for a drift at the very end of the stream.

For data stream mining, the usage of resources will be high if the drift detector incorrectly alarms for drift repeatedly. Further, the error-rate or cost of classification would be high if the drift detector could not correctly detect the location of drifts. In other words, the error-rate of classification typically increases as does the false negative number [7,8,18]. Therefore, false positive and false negative are essential measures for evaluating concept drift detectors.

The *delay of detection* may be considered as a performance measure for drift detectors. Less detection delay results in losing less data for learning, it means more instances from the new distribution can be used for learning. The *detection runtime* and *detection memory usage* of drift detectors can be also used as performance measures. Intuitively, a drift detector able to correctly find drifts with less delays faster by consuming less resources is preferred.

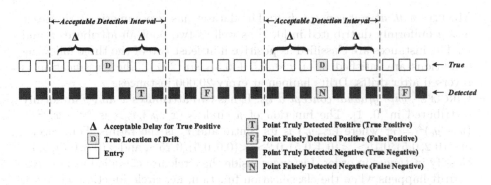

Fig. 2. Illustration of counting true positive, false positive and false negative

5   Experimental Analysis

We discuss our experimental results by comparing the performance of FHDDM against that of DDM, EDDM, ADWIN, HDDM$_{A-test}$ and HDDM$_{W-test}$. We ran the experiments on synthetic and real-world datasets often used in concept drift detection research [6,7,12–14,19]. We have considered Hoeffding Tree (HT), also known as VFDT, and Naive Bayes (NB) as our incremental classifiers; they are frequently used in the literature [6,7,12,13,16,18,19]. In all experiments, we ran the Hoeffding Tree with $\delta = 10^{-7}, \tau = 0.05$ and $n_{min} = 200$ as used in the [21]. Instances are processed prequentially, which means they are first tested and then used for training. We used MOA [22], a framework for data stream mining, to implement FHDDM in and compare it with other drift detectors. Experiments are run on Intel Core i5 @ 2.8 GHz with 16 GB of RAM running Apple OS X Yosemite.

5.1   Experiments on Synthetic Datasets

Synthetic Datasets – We generated three synthetic datasets of SINE1, MIXED and CIRCLES, as originally described in [25] and used in the literature [12,13,16], containing 100,000 instances with 2 classes. We also added 10 % noise to each dataset. In this way, we can consider how robust drift detectors are against noisy data streams by distinguishing noises from drifts. One of the advantages of synthetic datasets is being aware of the location of drifts. Therefore, we can measure the detection delay, true positive, false positive and false negative numbers (or rates). The datasets are described below:

– SINE1 · *with abrupt concept drift*: The dataset has two attributes x and y uniformly distributed in $[0, 1]$. The classification function is $y = sin(x)$. Before the first drift, instances under the curve are classified as positive and others as negative. At a drift point the classification is reversed. We put the drifts at every 20,000 instances.

– MIXED · *with abrupt concept drift*: The dataset has two numeric attributes x and y uniformly distributed in $[0, 1]$ as well as two boolean attributes v and w. The instances are classified as positive if at least two of the three following conditions are satisfied: $v, w, y < 0.5 + 0.3 * sin(2\pi x)$. The classification is reversed after drifts. Drifts happen at every 20,000 instances.

– CIRCLES · *with gradual concept drift*: It has two attributes x and y uniformly distributed in $[0, 1]$. The function of a circle $<(x_c, y_c), r_c>$ is $(x - x_c)^2 + (y - y_c)^2 = r_c^2$ where (x_c, y_c) is its centre and r_c is the radius. Four circles of $<(0.2, 0.5), 0.15>$, $<(0.4, 0.5), 0.2>$, $<(0.6, 0.5), 0.25>$, and $<(0.8, 0.5), 0.3>$ classify instances in order. Instances inside the circle are classified as positive. A drift happens when the classification function, i.e. circle function, changes. Drifts occur at every 25,000 instances.

Experiments – We ran Hoeffding Tree (HT) and Naive Bayes (NB) with each drift detector for 100 times and then averaged the detection delays, true positives, false positives, false negatives, detection runtimes (in millisecond), memory usage (in bytes) of drift detectors as well as the accuracies of classifiers. The acceptable drift detection delay length, i.e. Δ, was set to 250 for the SINE1 and MIXED datasets and to 1000 for the CIRCLES dataset. We consider a longer Δ for the CIRCLES dataset because it contains gradual concept drifts. Preliminary experiments and inspections confirmed that a longer Δ should be considered for gradual drifts, otherwise the false negative numbers would increase. We ran FHDDM with a sliding window size of 25 on the SINE1 and MIXED datasets, and with a sliding window size of 100 on the CIRCLES dataset. We considered a wider sliding window size for the CIRCLES dataset to make sure we have enough examples in the window as we are facing with the gradual drifts. Preliminary inspections helped us to adjust our window sizes for resulting in less detection delay, less false positive and less false negative. Since FHDDMs' sliding windows are small and they compare p_t^1 with the p_{max}^1, we need to set δ to a small value to make sure the ε_d is big enough. It was, therefore, set δ to 10^{-7} for our experiments. All other drift detectors were run with the default parameters as set in MOA (or as in the original papers).

Table 1(a) represents the results of experiments on the SINE1 dataset. FHDDM has the lowest false positive and false negative averages with both classifiers. HDDM$_{W-test}$ results in the lowest delay followed by FHDDM with small margins. DDM and EDDM exhibit the highest detection delay. They are also the only two drift detectors with false negative averages. EDDM and ADWIN have considerable false positive averages. As shown in Table 1 (b), we achieve the highest classification accuracies by FHDDM and HDDM$_{W-test}$ with both classifiers. It is clearly seen that ADWIN has the longest runtimes and the highest memory usages with considerable margins.

We show the results of experiments on the MIXED dataset in Table 2. FHDDM and ADWIN have the highest true positive averages without causing any false negatives. FHDDM has the smallest false positive averages while EDDM and ADWIN have the highest averages. HDDM$_{W-test}$ and FHDDM have the shortest detection delays. As represented in Table 2(b), we achieve the highest

Table 1. Results of experiments on SINE1 dataset (10 % Noise)

(a) Drift Detection Delay, True Positive, False Positive and False Negative

Classifier	Detector	Delay	TP	FP	FN
HT	FHDDM	16.58 ± 1.39	**4.0**	**0.02 ± 0.14**	**0.0**
	DDM	139.82 ± 23.87	3.42 ± 0.72	2.93 ± 1.87	0.58 ± 0.72
	EDDM	242.81 ± 16.42	0.20 ± 0.40	32.67 ± 12.97	3.80 ± 0.40
	ADWIN	21.23 ± 1.39	**4.0**	21.74 ± 5.72	**0.0**
	HDDM$_{A-test}$	32.2 ± 12.69	**4.0**	0.73 ± 1.01	**0.0**
	HDDM$_{W-test}$	**11.62 ± 1.00**	**4.0**	0.6 ± 0.72	**0.0**
NB	FHDDM	16.84 ± 1.02	**4.0**	**0.0**	**0.0**
	DDM	213.05 ± 69.57	3.39 ± 0.71	2.19 ± 1.51	0.61 ± 0.71
	EDDM	412.74 ± 73.03	1.65 ± 0.99	31.01 ± 12.15	2.35 ± 0.99
	ADWIN	22.13 ± 1.50	**4.0**	17.89 ± 5.10	**0.0**
	HDDM$_{A-test}$	59.34 ± 21.05	**4.0**	0.27 ± 0.53	**0.0**
	HDDM$_{W-test}$	**11.66 ± 1.29**	**4.0**	0.39 ± 0.69	**0.0**

(b) Drift Detection Runtime and Memory Usage with Classification Accuracy

Classifier	Detector	Runtime (ms)	Memory (bytes)	Accuracy
HT	FHDDM	35.02 ± 5.93	352	**87.07% ± 0.15**
	DDM	17.15 ± 4.47	**160**	86.16% ± 1.01
	EDDM	**11.89 ± 3.51**	**160**	84.62% ± 0.54
	ADWIN	2538.31 ± 103.53	1399.7 ± 81.43	86.77% ± 0.18
	HDDM$_{A-test}$	43.19 ± 6.35	168	86.99% ± 0.14
	HDDM$_{W-test}$	36.39 ± 5.75	**160**	87.05% ± 0.15
NB	FHDDM	35.22 ± 5.72	352	**86.08% ± 0.21**
	DDM	17.23 ± 4.69	**160**	81.70% ± 4.49
	EDDM	**13.39 ± 3.8**	**160**	83.67% ± 2.27
	ADWIN	2532.92 ± 89.88	1356.32 ± 65.79	85.99% ± 0.21
	HDDM$_{A-test}$	43.92 ± 6.16	168	85.96% ± 0.21
	HDDM$_{W-test}$	35.23 ± 6.55	**160**	**86.09% ± 0.21**

classification accuracies by FHDDM with both classifiers. EDDM and ADWIN result in the shortest and longest detection runtimes, respectively. ADWIN considerably occupies the memory.

Tables 3(a) and (b) hold the experiments results on the CIRCLES dataset. FHDDM results in the shortest detection delay, the highest false positive, the lowest false positive and the lowest false negative with Hoeffding Tree. ADWIN has the shortest detection delay and the highest true positive average with Naive Bayes and it is followed by FHDDM. EDDM and ADWIN have the highest false positive averages. In the terms of classification accuracies, we achieve the highest ones by FHDDM with either of classifiers. Like the previous experiments, EDDM has the shortest detection runtimes and ADWIN has the highest memory occupations.

We compared FHDDM with existing drift detection methods on the synthetic datasets containing abrupt and gradual concept drifts. In conclusion, FHDDM had the first or second shortest detection delay, the highest true positive average, the lowest false positive average, and the lowest false negative average. Further, the detection runtime and memory occupation was comparable to HDDM$_{A-test}$'s and HDDM$_{W-test}$'s. Importantly, FHDDM led to the highest classification accuracies with both Hoeffding Tree and Naive Bayes.

Table 2. Results of experiments on MIXED dataset (10 % Noise)

(a) Drift Detection Delay, True Positive, False Positive and False Negative

Classifier	Detector	Delay	TP	FP	FN
HT	FHDDM	16.05 ± 1.28	**4.0**	**0.24 ± 0.51**	**0.0**
	DDM	181.43 ± 24.08	2.69 ± 0.9	2.95 ± 1.92	1.31 ± 0.90
	EDDM	248.01 ± 8.95	0.07 ± 0.26	20.81 ± 7.18	3.93 ± 0.26
	ADWIN	23.85 ± 1.68	**4.0**	18.96 ± 5.27	**0.0**
	HDDM$_{A-test}$	41.12 ± 16.89	3.99 ± 0.10	1.16 ± 1.14	0.01 ± 0.10
	HDDM$_{W-test}$	**11.57 ± 5.92**	3.99 ± 0.10	3.15 ± 2.04	0.01 ± 0.10
NB	FHDDM	16.15 ± 1.43	**4.0**	**0.03 ± 0.17**	**0.0**
	DDM	178.08 ± 27.77	2.77 ± 0.93	2.26 ± 1.33	1.23 ± 0.93
	EDDM	248.13 ± 7.55	0.11 ± 0.34	20.71 ± 7.94	3.89 ± 0.34
	ADWIN	23.79 ± 1.61	**4.0**	18.24 ± 5.28	**0.0**
	HDDM$_{A-test}$	58.10 ± 23.56	3.99 ± 0.10	0.60 ± 0.76	0.01 ± 0.1
	HDDM$_{W-test}$	**11.48 ± 6.12**	3.99 ± 0.10	1.63 ± 1.16	0.01 ± 0.1

(b) Drift Detection Runtime and Memory Usage, and Classification Accuracy

Classifier	Detector	Runtime (ms)	Memory (bytes)	Accuracy
HT	FHDDM	35.98 ± 5.32	352	**83.40% ± 0.12**
	DDM	17.35 ± 4.28	**160**	81.77% ± 1.74
	EDDM	**13.09 ± 3.56**	**160**	80.58% ± 0.89
	ADWIN	2417.56 ± 89.81	1343.63 ± 69.07	83.28% ± 0.13
	HDDM$_{A-test}$	41.69 ± 6.54	168	83.30% ± 0.13
	HDDM$_{W-test}$	33.07 ± 5.69	**160**	83.27% ± 0.13
NB	FHDDM	35.71 ± 6.04	352	**83.39% ± 0.09**
	DDM	17.65 ± 4.76	**160**	80.74% ± 3.52
	EDDM	**13.38 ± 3.81**	**160**	80.53% ± 1.89
	ADWIN	2715.83 ± 99.69	1349.26 ± 58.26	83.31% ± 0.09
	HDDM$_{A-test}$	43.64 ± 6.45	168	83.27% ± 0.11
	HDDM$_{W-test}$	36.89 ± 5.87	**160**	83.37% ± 0.09

5.2   Experiments on Real-World Datasets

Real-World Datasets – We considered the AIRLINES [26], POKER HAND [27] and ELECTRICITY [28] datasets widely used in concept drift research [6,7,12, 13,18,19]. The preprocessed and normalized version of datasets are available at MOA website[1]. The datasets are described below:

- AIRLINES: This dataset was created to be used as a non-stationary data stream for evaluating learning algorithms [26]. It contains 539,383 records of flight schedules defined by 7 attributes. The task is to predict if a flight is delayed or not. Concept drift could appear as the result of changes in the flights schedules, e.g. changes in day, time, and the length of flights.
- POKER HAND: It comprises 1,000,000 instances with 11 attributes. Each instance is an example of a hand consisting of five playing cards drawn from a standard deck of 52. Each card is described by two attributes (suit and rank), for ten predictive attributes. The class predicts the poker hand. Concept drift happens as changing the card at hand, i.e. the poker hand [4].
- ELECTRICITY: It has 45,312 instances, with 8 input attributes, recorded every half an hour for a period of two years from Australian New South Wales Electricity. The classification task is to predict a rise (*Up*) or a fall (*Down*)

---

[1] http://moa.cms.waikato.ac.nz/datasets/.

Table 3. Results of experiments on CIRCLES dataset (10% Noise)

(a) Drift Detection Delay, True Positive, False Positive and False Negative

Classifier	Detector	Delay	TP	FP	FN
HT	FHDDM	**94.14 ± 28.05**	**3.0**	0.04 ± 0.20	**0.0**
	DDM	604.7 ± 118.48	2.26 ± 0.8	1.74 ± 1.58	0.74 ± 0.8
	EDDM	981.60 ± 62.64	0.11 ± 0.31	22.75 ± 10.83	2.89 ± 0.31
	ADWIN	249.45 ± 159.06	2.58 ± 0.55	17.9 ± 5.66	0.42 ± 0.55
	HDDM$_{A-test}$	123.57 ± 72.89	2.95 ± 0.22	0.48 ± 0.77	0.05 ± 0.22
	HDDM$_{W-test}$	101.84 ± 70.79	2.96 ± 0.2	0.54 ± 0.75	0.04 ± 0.20
NB	FHDDM	270.98 ± 125.98	2.83 ± 0.38	**0.20 ± 0.40**	0.17 ± 0.38
	DDM	830.15 ± 139.60	1.35 ± 0.90	2.50 ± 1.58	1.65 ± 0.90
	EDDM	949.50 ± 96.80	0.32 ± 0.51	34.84 ± 19.30	2.68 ± 0.51
	ADWIN	**225.25 ± 56.39**	**3.0**	17.84 ± 5.28	**0.0**
	HDDM$_{A-test}$	494.95 ± 155.27	2.63 ± 0.52	0.68 ± 0.68	0.37 ± 0.52
	HDDM$_{W-test}$	316.24 ± 149.4	2.68 ± 0.51	1.21 ± 1.06	0.32 ± 0.51

(b) Drift Detection Runtime and Memory Usage, and Classification Accuracy

Classifier	Detector	Runtime (ms)	Memory (bytes)	Accuracy
HT	FHDDM	64.38 ± 6.97	952	**86.66% ± 0.14**
	DDM	16.76 ± 4.39	**160**	85.84% ± 0.81
	EDDM	**11.92 ± 3.57**	**160**	84.96% ± 0.27
	ADWIN	2690.09 ± 129.28	1716.69 ± 106.23	86.04% ± 0.23
	HDDM$_{A-test}$	43.85 ± 6.83	168	86.60% ± 0.19
	HDDM$_{W-test}$	35.67 ± 5.73	**160**	86.58% ± 0.17
NB	FHDDM	64.54 ± 7.56	952	**84.31% ± 0.14**
	DDM	16.66 ± 0.90	**160**	82.73% ± 1.57
	EDDM	**13.25 ± 3.60**	**160**	83.30% ± 0.40
	ADWIN	2818.84 ± 41.55	1777.86 ± 77.53	**84.31% ± 0.12**
	HDDM$_{A-test}$	42.43 ± 6.57	168	84.24% ± 0.14
	HDDM$_{W-test}$	35.81 ± 5.67	**160**	84.27% ± 0.16

in the electricity price. The concept drift may happen because of changes in consumption habits, unexpected events and seasonality [29].

Experiments – The ground truth for drifts is not available for the real-world datasets. This implies that we do not know whether drifts occur in these datasets or where they occur [6, 7]. We, therefore, cannot measure the detection delay, true positive, false positive, and false negative numbers of drift detectors in this section. We only evaluate the number of drifts detected and the accuracy of classification. All classifiers and drift detectors were run with the default parameters. For FHDDM, we only present the results obtained by the sliding window size of 25 because we usually obtained better classification accuracies with size 25 on the real-world datasets in our preliminary experiments, though the margins of differences were small.

Tables 4, 5 and 6 summarize the results of experiments on the aforementioned datasets. The accuracy of classification has improved by using drift detectors. The classification accuracy with FHDDM is among the highest ones. It also detects less drifts compared to ADWIN, HDDM$_{A-test}$ and HDDM$_{W-test}$ while their classification accuracies are similar. As argued in [7], there are two possible cases if a drift detector detects less number of drifts compared to the other drift detectors while they all lead to similar classification accuracies: (1) That drift detector caused less false positive compared to others, or (2) Not detected

Table 4. The results of experiments on AIRLINES dataset

Classifier	Detector	Runtime (ms)	Memory (bytes)	Num. Drifts	Accuracy
HT	FHDDM	118.05 ± 12.04	352	**339**	**65.66%**
	DDM	49.75 ± 9.87	160	14	**65.29%**
	EDDM	**41.60 ± 8.37**	160	54	65.06%
	ADWIN	13439.70 ± 96.48	1879.23	**341**	**65.25%**
	HDDM$_{A-test}$	146.45 ± 12.08	168	88	64.99%
	HDDM$_{W-test}$	107.55 ± 10.67	160	652	65.02%
	No Detection	—	—	—	65.07%
NB	FHDDM	114.15 ± 7.93	352	**297**	66.44%
	DDM	48.65 ± 6.13	**160**	13	65.33%
	EDDM	**39.55 ± 7.17**	**160**	23	65.18%
	ADWIN	13034.20 ± 77.13	1896.08	**300**	66.79%
	HDDM$_{A-test}$	139.80 ± 13.34	168	72	**67.22%**
	HDDM$_{W-test}$	86.70 ± 9.63	**160**	620	65.34%
	No Detection	—	—	—	64.55%

Table 5. The results of experiments on POKER HAND dataset

Classifier	Detector	Runtime (ms)	Memory (bytes)	Num. Drifts	Accuracy
HT	FHDDM	173.85 ± 16.54	352	**1557**	76.45%
	DDM	62.00 ± 7.78	**160**	1046	72.74%
	EDDM	**56.50 ± 7.05**	**160**	4806	**77.30%**
	ADWIN	12725.25 ± 108.71	1464.42	**2373**	74.56%
	HDDM$_{A-test}$	200.10 ± 8.81	168	**2565**	76.40%
	HDDM$_{W-test}$	147.65 ± 12.84	**160**	**2211**	77.11%
	No Detection	—	—	—	76.07%
NB	FHDDM	166.60 ± 12.60	352	**1660**	76.30%
	DDM	69.05 ± 7.53	**160**	433	61.97%
	EDDM	**56.60 ± 8.43**	**160**	4863	**77.48%**
	ADWIN	12650.00 ± 248.42	1453.93	**2453**	74.60%
	HDDM$_{A-test}$	195.85 ± 13.74	168	**2615**	76.48%
	HDDM$_{W-test}$	125.55 ± 13.26	**160**	**2312**	77.11%
	No Detection	—	—	—	59.55%

Table 6. The results of experiments on ELECTRICITY dataset

Classifier	Detector	Runtime (ms)	Memory (bytes)	Num. Drifts	Accuracy
HT	FHDDM	22.70 ± 4.31	352	**77**	84.38%
	DDM	15.70 ± 4.79	**160**	169	84.41%
	EDDM	**10.40 ± 3.37**	**160**	191	84.91%
	ADWIN	738.05 ± 23.28	1468.29	**110**	83.40%
	HDDM$_{A-test}$	28.05 ± 6.67	168	210	**85.71%**
	HDDM$_{W-test}$	24.35 ± 4.66	**160**	**117**	**85.06%**
	No Detection	—	—	—	79.20%
NB	FHDDM	23.90 ± 4.17	352	**96**	82.69%
	DDM	13.25 ± 4.90	**160**	143	81.18%
	EDDM	**8.60 ± 2.24**	**160**	203	**84.83%**
	ADWIN	691.85 ± 17.82	1408.25	**128**	81.63%
	HDDM$_{A-test}$	25.40 ± 5.23	168	211	**84.92%**
	HDDM$_{W-test}$	25.45 ± 5.84	**160**	**132**	84.09%
	No Detection	—	—	—	73.36%

drifts, i.e. false negatives, were less significant drifts. The second case implicitly says having less number of drifts detected leading to lower classification accuracy suggests significant false negatives. Therefore, based on these arguments, it is more likely that FHDDM caused fewer false positives. Its detection runtime is also comparable with $\text{HDDM}_{\text{A}-\text{test}}$'s and $\text{HDDM}_{\text{W}-\text{test}}$'s. In all cases, FHDDM resulted in shorter detection runtimes, less memory occupations and higher classification accuracies compared to ADWIN.

6 Conclusion and Future Work

Adapting classification learners is essential when they are used to learn from data in evolving environments. In this paper, we introduced a new concept drift detection method, so-called FHDDM, that uses the Hoeffding's inequality. The method works based on the fact that the accuracy of a classifier should increase or stay steady as more instances arrive; otherwise it implies the existence of drift points in the stream. FHDDM slides a window with a size of n on the stream and measures the p_t^1, i.e. the probability of correct classification predictions in the most recent n instances at time t. It updates the value of p_{max}^1 that holds the maximum probability of correct predictions seen so far. A significant difference, bounded by Hoeffding's inequality, between p_t^1 and p_{max}^1 suggests a drift. In addition, we introduced an approach to count true positive, false positive and false negative of drift detectors by considering their delay of detection for evolving data streams.

We experimentally evaluated our method on the synthetic and real-world datasets. Experiments on the synthetic datasets indicated that FHDDM detects drifts with a shorter delay, leading to the highest true positive, the lowest false positive and the lowest false negative, when compared to the state-of-the-art. When considering real-world datasets, the classification accuracies of our method were consistently high.

In the future, we will investigate the performance of our FHDDM approach on imbalanced and highly noisy data streams as well as streams containing outliers. We will also consider implementing an adaptable window size, as based on the trends of prediction results. In addition, we plan to study the sensitivity of FHDDM's parameters, i.e. *size of sliding window* and *confidence level*, along with other drift detectors' and consider their performances in different domains. It would also be worthwhile to compare FHDDM with other drift detectors, as proposed in [14,24], amongst others. Finally, we intend to use our proposed method in anomaly detection and business intelligence applications.

References

1. Gama, J.A., Zliobaite, I., Bifet, A., Pecheniziky, M., Bouchachia, A.: A survey on concept drift adaptation. ACM Comput. Surv. **46**(4), 44:1–44:37 (2014)
2. Ditzler, G., Roveri, M., Alippi, C., Polikar, R.: Learning in nonstationary environments: a survey. Comput. Intell. Mag. **10**(4), 12–25 (2015)
3. Alippi, C., Boracchi, G., Roveri, M.: Just-in-time ensemble of classifiers. In: International Joint Conference on Neural Networks, pp. 1–8 (2012)
4. Olorunnimbe, M.K., Viktor, H.L., Paquet, E.: Intelligent adaptive ensembles for data stream mining: a high return on investment approach. In: Ceci, M., Loglisci, C., Manco, G., Masciari, E., Ras, Z.W. (eds.) NFMCP 2015. LNCS (LNAI), vol. 9607, pp. 61–75. Springer, Heidelberg (2016). doi:10.1007/978-3-319-39315-5_5

5. Kuncheva, L.I.: Classifier ensembles for detecting concept change in streaming data: overview and perspectives. In: 2nd Workshop SUEMA, pp. 5–9 (2008)
6. Bifet, A., Holmes, G., Pfahringer, B., Kirkby, R., Gavalda, R.: New ensemble methods for evolving data streams. In: 15th ACM SIGKDD International Conference on Knowledge Discovery and Data Mining, pp. 139–148. ACM (2009)
7. Huang, D.T.J., Koh, Y.S., Dobbie, G., Bifet, A.: Drift detection using stream volatility. In: Appice, A., Rodrigues, P.P., Santos Costa, V., Soares, C., Gama, J., Jorge, A. (eds.) ECML PKDD 2015. LNCS (LNAI), vol. 9284, pp. 417–432. Springer, Heidelberg (2015). doi:10.1007/978-3-319-23528-8_26
8. Zliobaite, I., Budka, M., Stahl, F.: Towards cost-sensitive adaptation: when is it worth updating your predictive model? Neurocomputing **150**, 240–249 (2015)
9. Hoeffding, W.: Probability inequalities for sums of bounded random variables. Am. Stat. Assoc. **58**(301), 13–30 (1963)
10. Page, E.S.: Continous inspection schemes. Biometrika **41**, 100–115 (1954)
11. Roberts, S.W.: Control chart tests based on geometric moving averages. Technometrics **42**(1), 97–101 (2000)
12. Gama, J., Medas, P., Castillo, G., Rodrigues, P.: Learning with drift detection. In: Bazzan, A.L.C., Labidi, S. (eds.) SBIA 2004. LNCS (LNAI), vol. 3171, pp. 286–295. Springer, Heidelberg (2004). doi:10.1007/978-3-540-28645-5_29
13. Baena-Garcia, M., del Campo-Avila, J., Fidalgo, R., Bifet, A., Gavalda, R., Morales-Bueno, R.: Early drift detection method. In: 4th International Workshop on Knowledge Discovery from Data Streams, vol. 6, pp. 77–86 (2006)
14. Ross, G.J., Adams, N.M., Tasoulis, D.K., Hand, D.J.: Exponentially weighted moving average charts for detecting concept drift. Pattern Recogn. Lett. **33**(2), 191–198 (2012)
15. Kifer, D., Ben-David, S., Gehrke, J.: Detecting change in data streams. In: 30th International Conference on Very Large Data Bases, vol. 30, pp. 180–191 (2004)
16. Nishida, K., Yamauchi, K.: Detecting concept drift using statistical testing. In: Corruble, V., Takeda, M., Suzuki, E. (eds.) DS 2007. LNCS (LNAI), pp. 264–269. Springer, Heidelberg (2007). doi:10.1007/978-3-540-75488-6_27
17. Bach, S.H., Maloof, M.A.: Paired learners for concept drift. In: 8th IEEE International Conference on Data Mining, ICDM 2008, pp. 23–32 (2008)
18. Bifet, A., Gavalda, R.: Learning from time-changing data with adaptive windowing. In: SIAM International Conference on Data Mining, pp. 443–448 (2007)
19. Frias-Blanco, I., del Campo-Avila, J., Ramos-Jimenez, G., Morales-Bueno, R., Ortiz-Diaz, A., Caballero-Mota, Y.: Online and non-parametric drift detection methods based on Hoeffding's bounds. IEEE Trans. Knowl. Data Eng. **27**(3), 810–823 (2015)
20. Mitchell, T.: Machine Learning. McGraw Hill (1997)
21. Domingos, P., Hulten, G.: Mining high-speed data streams. In: 6th ACM SIGKDD International Conference on Knowledge Discovery and Data Mining, pp. 71–80 (2000)
22. Bifet, A., Holmes, G., Kirkby, R., Pfahringer, B.: MOA: massive online analysis. Mach. Learn. Res. **11**, 1601–1604 (2010)
23. Sakthithasan, S., Pears, R., Koh, Y.S.: One pass concept change detection for data streams. In: Pei, J., Tseng, V.S., Cao, L., Motoda, H., Xu, G. (eds.) PAKDD 2013. LNCS (LNAI), pp. 461–472. Springer, Heidelberg (2013). doi:10.1007/978-3-642-37456-2_39
24. Pears, R., Sakthithasan, S., Koh, Y.S.: Detecting concept change in dynamic data streams. Mach. Learn. **97**(3), 259–293 (2014)

25. Kubat, M., Widmer, G.: Adapting to drift in continuous domains (Extended abstract). In: Lavrac, N., Wrobel, S. (eds.) ECML 1995. LNCS, vol. 912, pp. 307–310. Springer, Heidelberg (1995). doi:10.1007/3-540-59286-5_74
26. Ikonomovska, E.: Airline Dataset (2011). http://kt.ijs.si/elena_ikonomovska/data. html. Last Visit Happened on 15 March 2016
27. Cattral, R., Oppacher, F., Deugo, D.: Evolutionary data mining with automatic rule generalization. Recent Adv. Comput. Comput. Commun., 296–300 (202)
28. Harries, M., Wales, N.S.: Splice-2 Comparative Evaluation: Electricity pricing. Technical report, University of New South Wales, Australia (1999)
29. Zliobaite, I.: How Good is The Electricity Benchmark for Evaluating Concept Drift Adaptation. arXiv preprint (2013). arXiv:1301.3524

Differentially Private User Data Perturbation with Multi-level Privacy Controls

Yilin Shen$^{(\boxtimes)}$, Rui Chen, and Hongxia Jin

Samsung Research America, Mountain View, CA 94043, USA
{yilin.shen,rui.chen1,hongxia.jin}@samsung.com

Abstract. Service providers typically collect user data for profiling users in order to provide high-quality services, yet this brings up user privacy concerns. One hand, service providers oftentimes need to analyze multiple user data attributes that usually have different privacy concern levels. On the other hand, users often pose different trusts towards different service providers based on their reputation. However, it is unrealistic to repeatedly ask users to specify privacy levels for each data attribute towards each service provider. To solve this problem, we develop the *first* lightweight and provably framework that not only guarantees differential privacy on both *service provider* and *different data attributes* but also allows configurable *utility functions* based on service needs. Using various large-scale real-world datasets, our solution helps to significantly improve the utility up to 5 times with negligible computational overhead, especially towards numerous low reputed service providers in practice.

Keywords: Differential privacy · Multi-level privacy · Optimization

1 Introduction

The last few decades have witnessed a variety of personalized services to users, such as intelligent assistant, targeted advertising and so on, which has become key business drivers for many companies. As one can understand, such services are based on user's data and oftentimes require substantial user data in order to provide high-quality services. However, consumer fears over privacy continue to escalate due to the release of users' private data. Based on Pew Research [1], 68 % consumers think that current laws are insufficient to protect their privacy and demand tighter privacy laws; and 86 % of Internet users have taken proactive steps to remove or mask their digital footprints. Responding to increasing user privacy concerns, governments in US/EU are increasing regulations and applying/enforcing existing regulations.

More importantly, in order to provide high quality services, service providers usually profile users by analyzing multiple attributes of their private data. Recent research has showed that various attributes of data are often associated with different privacy concerns [13,24,26]. More importantly, Zhang *et al.* [26] revealed that user's perception of privacy concerns will dramatically decrease if providing them fine-grained privacy controls for different attributes of data.

© Springer International Publishing AG 2016
P. Frasconi et al. (Eds.): ECML PKDD 2016, Part II, LNAI 9852, pp. 112–128, 2016.
DOI: 10.1007/978-3-319-46227-1_8

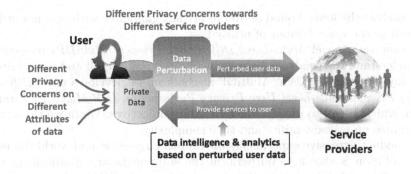

Fig. 1. Data Perturbation with Multi-Level Privacy Controls under Untrusted Server

On the other hand, while traditionally users count on service providers to protect their data privacy, recent years have witnessed a variety of privacy breaches through service providers when malicious attackers break into the cloud/server and steal user data. Target, HomeDepot, and Anaheim health insurance companies are among the largest hits. Huge number of sensitive user data is leaked through servers. Additionally, the insiders of service providers are another source of privacy threat. It would be ideal if users do not have to fully trust the service providers to protect their data; and users can impose different privacy concerns based on each service provider's reputation according to recent research [15], i.e., trust Google more than aforementioned intruded service providers.

However, it is unrealistic to repeatedly ask ordinary users to specify privacy levels for each attribute of data every time releasing to different service providers. Therefore, it is critically desirable to develop technologies that not only allow business intelligence but also preserve users' privacy needs toward both different data attributes and different service providers.

In this paper, we aim to develop the *first* lightweight and provably private framework, under *untrusted server* settings, to automate users multi-level privacy controls for releasing the aggregates of attributes associated with their private data to each service provider. As shown in Fig. 1, our adoption of *untrusted server* setting, in which user data is perturbed and anonymized on their private devices before releasing, enjoys a number of benefits as discussed in [23]. In the meanwhile, these protections should be done to still provide different reasonable utilities of perturbed data based on service needs. Our approach is developed to provide a strong and provable privacy guarantee, *differential privacy*, which is the current state-of-the-art paradigm for privacy-preserving data publishing.

Our contributions are summarized as follows:

– We formulate a novel *Multi-Level User Privacy Perturbation (MultiUPP)* problem, which aims to release perturbed aggregates on user data attributes that not only preserves both *differential privacy towards a service provider (overall privacy)* and *differential privacy on each data attribute (per-attribute privacy)*, but also optimizes a specific utility function based on service needs.

- We analyze the lower bound of overall privacy guarantee with optimal utility, as well as the lower bounds of utility loss.
- We propose a novel *Multi-Level Differential Privacy (MultiDP)* mechanism to understand the condition between utility loss and overall and per-attribute privacy preservation. Using MultiDP mechanism, we develop a novel *Differentially Private Multi-Level User Privacy Perturbation (DP-MultiUPP)* framework which allows to plug in different utility objectives. We prove theoretical guarantee on privacy, utility and time complexity.
- We conduct extensive experiments on various large-scale real-world datasets. Our solution is shown to outperform the state-of-the-art approach up to 5 times with negligible computational overhead on both PC and Android smartphones. Particularly, the utility is significantly improved toward low reputed service providers in practice.

The rest of paper is organized as follows: Sect. 2 presents notations, preliminaries and problem definition. Section 3 provides the lower bounds of utility loss and overall privacy budget. Section 4 develops the DP-MultiUPP framework via a novel Multi-Level Differential Privacy Mechanism. Experimental results and related work are presented in Sects. 5 and 6. Finally, Sect. 7 concludes the whole paper and discusses future work.

2   Preliminaries and Problem Definition

In this section, we first introduce notations and restate the definition and existing mechanism of differential privacy. Then, we define a novel *Multi-Level User Privacy Perturbation (MultiUPP)* problem definition, along with its challenges.

2.1   Notations

Let I be the public set/universe of items of size $|I| = n$. A user's raw private data is denoted as a vector $\mathbf{d^r}$ of dimension n. The ith entry in $\mathbf{d^r}$ is either 1 or 0, meaning that item i does or does not belong to user's private/raw history data. Public attribute set is defined as A of size $|A| = m$, in which each item is associated with a subset of attributes represented by a public item-attribute matrix $\mathbf{A}$ of dimension $n \times m$. The entry a_{ij} in $\mathbf{A}$ is the value that item i has for attribute j. For the attributes in A, we define their private aggregate vectors to be $\mathbf{a^r}$ such that $\mathbf{a^r} = \mathbf{A}^T \mathbf{d^r}$. The published perturbed attribute histogram is presented as a vector $\mathbf{a^P}$ (details in utility objectives of problem definition in Sect. 2.3).

This user's multi-level privacy concern on different attributes is denoted as a vector $\mathbf{t} = (t_1, \ldots, t_m)$, in which the jth entry $t_j > 0$ means the privacy budget of attribute j. This user's overall privacy concern towards the service provider is defined as $\epsilon > 0$. A smaller t_j or ϵ means a higher privacy concern (a stronger privacy guarantee) on attribute j or towards the service provider. (Note that all t_j and ϵ correspond to the privacy budget in differential privacy notion, defined in the next subsection.) For reference, we list all notations in Table 1.

Table 1. Notations

Symbol	Description
I	public item set/universe of size $|I| = n$
A	public attribute set of size $|A| = m$
$\mathbf{A}$	public item-attribute matrix $\mathbf{A} \in \{0,1\}^{n \times m}$
$\mathbf{d^r}$	user private item vector $\mathbf{d^r} \in \{0,1\}^n$
$\mathbf{a^r}$	user private attribute aggregate vector $\mathbf{a^r} \in \mathbb{R}^{*m}$
	($\mathbb{R}^*$: non-negative real numbers)
$\mathbf{a^p}$	user perturbed attribute aggregate vector $\mathbf{a^p} \in \mathbb{R}^m$
$\mathbf{t}$	user per-attribute privacy budget vector $t \in \mathbb{R}^{+m}$ ($\mathbb{R}^+$: positive real numbers)
ϵ	user overall privacy budget towards service provider

2.2 Differential Privacy

Differential privacy [9] is a recent privacy model which provides strong privacy guarantee. Informally, an algorithm $\mathcal{A}$ is differentially private if the output is insensitive to any particular record in the dataset.

Definition 1 (ϵ-Differential Privacy). *Let $\epsilon > 0$ be a small constant. A randomized function $\mathcal{A}$ is ϵ-differentially private if for all data sets D_1 and D_2 differing on at most one element, i.e., $d(D_1, D_2) = 1$, and all $\mathcal{S} \subseteq \mathsf{Range}(\mathcal{A})$,*

$$Pr[\mathcal{A}(D_1) \in \mathcal{S}] \leq \exp(\epsilon) Pr[\mathcal{A}(D_2) \in \mathcal{S}] \tag{1}$$

The probability is taken over the coin tosses of $\mathcal{A}$.

The parameter $\epsilon > 0$ is referred to as *privacy budget*, which allows us to control the level of privacy. A smaller ϵ suggests more limit posed on the influence of an individual item, which gives stronger privacy guarantee. Differential privacy enjoys the following important composition property:

Lemma 1 (Composition Property [8]**).** *If an algorithm $\mathcal{A}$ runs t randomized algorithms $\mathcal{A}_1, \mathcal{A}_2, \ldots, \mathcal{A}_t$, each of which is t_i-differentially private, and applies an arbitrary randomized algorithm ϕ to their results ($\mathcal{A}(D) = \phi(\mathcal{A}_1(D), \ldots, \mathcal{A}_t(D))$), then $\mathcal{A}$ is $\sum_i t_i$-differentially private.*

One of the most widely used mechanisms to achieve ϵ-differential privacy is Laplace mechanism [9] (Theorem 1). Laplace mechanism adds random noises to the numeric output of a query, in which the magnitude of noises follows Laplace distribution with variance $\frac{\Delta f}{\epsilon}$ where Δf represents the global sensitivity of query f (Definition 2).

Definition 2 (Global Sensitivity [9]**).** *For a query* $f : \mathcal{D} \to \mathbb{R}^k$, *the global sensitivity* Δf *of* f *is as follows:*

$$\Delta f = \max_{d(D_1, D_2)=1} \|f(D_1) - f(D_2)\|_1 \tag{2}$$

for all D_1, D_2 *differing in one element, i.e.,* $d(D_1, D_2) = 1$.

Theorem 1 (Laplace Mechanism [9]**).** *For* $f : \mathcal{D} \to \mathbb{R}^k$, *a randomized algorithm* $\mathcal{A}_f = f(D) + \mathsf{Lap}^k(\frac{\Delta f}{\epsilon})$ *is* ϵ-*differentially private.*

The Laplace distribution with parameter β, denoted $\mathsf{Lap}(\beta)$, has probability density function $\frac{1}{2\beta} \exp(-\frac{|z|}{\beta})$ and cumulative distribution function $\frac{1}{2}(1 + \mathsf{sgn}(z)(1 - \exp(-\frac{|z|}{\beta})))$.

2.3 MultiUPP Problem Definition

The goal of *Multi-Level User Privacy Perturbation (MultiUPP)* problem is to publish an accurate histogram that summaries the distribution of data attributes, which is sufficient to provide user high-quality services (e.g., personalized advertising, recommendation) in most cases [22]. In the meanwhile, MultiUPP also preserves both *overall privacy toward a specific service provider* and *different privacy needs for different data attributes*. More importantly, our *MultiUPP problem is considered as a general framework which can be coupled with different utility objectives*. Next, we specify the MultiUPP problem and its associated privacy and utility objectives respectively.

Formal Definition: Given a user's private item vector $\mathbf{d^r}$ associated with public universal item set I and a public item-attribute matrix $\mathbf{A}$; and this user's attribute-based privacy budget vector $\mathbf{t}$ as well as his overall privacy budget ϵ towards a service provider. MultiUPP outputs this user's perturbed attribute aggregates $\mathbf{a^P}$ to satisfy the following privacy and utility objectives:

Privacy Objectives of MultiUPP: We consider two privacy objectives aiming to defend against privacy leakage via public attribute information.

P1. Overall Differential Privacy Objective towards a service provider: satisfy ϵ-*differential privacy* on published histogram on all attributes with the presence or absence of an individual item in I. Each service provider is associated with an overall privacy budget ϵ based on its reputation, i.e., a smaller ϵ for a lower reputed service provider.

P2. Per-attribute Differential Privacy Objectives with Multiple Levels: satisfy t_j-*differential privacy* on published histogram on each attribute j with the presence or absence of an individual item in I. Each attribute j of data is associated with a privacy budget t_j based on each user's privacy concern on this attribute. For example, if a user considers location more private than price (attribute 1 and 2 of an item), this user will set $t_1 < t_2$.

Utility Objectives of MultiUPP: We consider publishing the histogram in which the number of bins equals to the number of attributes and the count in each bin j is perturbed summation of attribute values w.r.t. items in user's history. The published histogram is denoted as $\mathbf{a^P}$ as in Table 1. Following the convention in [25, 27], we measure the accuracy (or utility) of a perturbed histogram in terms of the following two utility loss functions between raw and perturbed attribute aggregates $\mathbf{a^r}$ and $\mathbf{a^P}$ (denoted as $\mathcal{U}$):

U1. *Expected Mean Absolute Error (MAE):* $\mathcal{U}_{MAE} = \mathsf{E}\left[\frac{1}{m}\|\mathbf{a^P} - \mathbf{a^r}\|_1\right]$

U2. *Expected Mean Square Error (MSE):* $\mathcal{U}_{MSE} = \mathsf{E}\left[\frac{1}{m}\|\mathbf{a^P} - \mathbf{a^r}\|_2^2\right]$

In addition, we also consider the following third utility regarding per-attribute utility with multi-level privacy controls, for measuring the utility loss over the best utility on the aggregate of each attribute:

U3. *Expected Mean Absolute Error Loss (MAEL):* $\mathcal{U}_{MAEL} = \mathsf{E}\left[\frac{1}{m}\sum_{j=1}^{m}\right.$ $\left.\frac{|a_j^p - a_j^r|}{BU_j}\right] - 1$, where BU_j stands for the best expected utility of attribute j. More specifically, $BU_j = \frac{\Delta f_j}{t_j}$ indicating the expectation of optimal Laplace noise $\mathsf{Lap}\left(\frac{\Delta f_j}{t_j}\right)$ for each query function $f_j : (\mathbb{Z}^+)^n \to \mathbb{R}$ [12].

Remarks: According to user study results in [24], what users most prefer is to control their different privacy concerns on limited number of relatively coarse-grained data attributes. Thus, we assume that the number of attributes m is bounded by a constant.

Challenges: (1) An item is usually associated with a number of attributes while each attribute has a different privacy concern level. How can we perturb the data to optimize the utility when satisfying all privacy guarantees? (2) When $\epsilon < \sum t_j$, the existing composition approach [8] is no longer feasible. In this case, what are the lower bounds of optimal utilities? What is the lower bound of ϵ with such optimal utilities? (3) When overall privacy budget ϵ is smaller than the above lower bound, how can we optimize the utility loss?

3   Lower Bounds

In this section, we focus on the queries $f_j : (\mathbb{Z}^+)^n \to \mathbb{R}$ in line with the aggregate (counting) of each attribute in MultiUPP problem definition. We first discuss the lower bound of overall privacy budget ϵ when optimal utilities are achieved, followed by the detailed lower bounds of the utility loss functions (optimal utilities) described in our problem.

3.1   Lower Bound of Overall Privacy Budget ϵ

We first understand the turning point when all utilities for each attribute aggregate are optimized while both overall and per-attribute privacy guarantees are satisfied. That is, we study a lower bound of ϵ on the public domain (item set/universe I), as shown in the following Theorem 2:

Theorem 2 (Lower Bound of ϵ). *For a set of queries $f_1, \ldots, f_m$ in which each $f_j : (\mathbb{Z}^+)^n \to \mathbb{R}$ is associated with its global sensitivity Δf_j and a privacy budget $t_j > 0$. If t_j-differential privacy is satisfied for each query with optimal utility, the overall privacy guarantee ϵ for all queries is lower bounded as follows:*

$$\epsilon \geq \max_{d(D_1, D_2)=1} \left\{ \sum_{j=1}^m \frac{t_j}{\Delta f_j} |f_j(D_1) - f_j(D_2)| \right\} \tag{3}$$

where $d(D_1, D_2) = 1$ stands for two neighboring datasets D_1, D_2.

Proof. According to the result by Hardt *et al.* [12], the optimal utility for an arbitrary query function $f (\mathbb{Z}^+)^n \to \mathbb{R}$ is $\Omega(\Delta f/\epsilon)$, which can be obtained by Laplace mechanism. Consider two arbitrary neighboring datasets D_1, D_2 $(d(D_1, D_2) = 1)$ and any $\mathbf{s} = (s_1, \ldots, s_m) \in Range(\mathcal{A}^N)$ when every j^{th} element is obtained by adding noise $\mathsf{Lap}(\frac{\Delta f_j}{t_j})$ to aggregate of attribute j, in which $\mathcal{A}^N$ is the naive randomized algorithm where each $\mathcal{A}_j^N$ is $\mathsf{Lap}(\frac{\Delta f_j}{t_j})$):

$$\frac{\Pr[\mathcal{A}^N(D_1) = \mathbf{s}]}{\Pr[\mathcal{A}^N(D_2) = \mathbf{s}]} = \prod_{j=1}^m \frac{\Pr[\mathcal{A}_j^N(D_1)_j = s_j]}{\Pr[\mathcal{A}_j^N(D_2)_j = s_j]} = \prod_{j=1}^m \frac{\exp(-|f_j(D_1) - s_j|\frac{t_j}{\Delta f_j})}{\exp(-|f_j(D_2) - s_j|\frac{t_j}{\Delta f_j})}$$

$$\geq \prod_{j=1}^m \exp\left(-\frac{t_j}{\Delta f_j} |f_j(D_1) - f_j(D_2)| \right) = \exp\left(\sum_{j=1}^m -\frac{t_j}{\Delta f_j} |f_j(D_1) - f_j(D_2)| \right)$$

Therefore, proof is complete.

3.2 Lower Bounds of Utility Loss

We next study the lower bounds of optimal utility loss when the privacy objectives are satisfied. These lower bounds will also be used as baseline for experimental evaluation in Sect. 5.

Theorem 3 (Lower Bounds of Utility Loss). *If t_j-differential privacy is satisfied for each query and ϵ satisfies the lower bound in (3), the lower bounds of utilities defined in Sect. 2.3 are as follows:*

$$\mathcal{U}_{MAE} \geq \frac{1}{m} \sum_{j=1}^m \frac{C_j}{t_j}; \; \mathcal{U}_{MSE} \geq \frac{2}{m} \sum_{j=1}^m \frac{C_j^2}{t_j^2}; \; \mathcal{U}_{MAEL} \geq 0$$

where $C_j = max_{1 \leq i \leq n}\{a_{ij}\}$.

Proof. As the optimal utility is obtained by Laplace mechanism in our case [12], we prove the above lower bounds based on the properties of Laplace distribution. Let X_j be the random variable following distribution $\mathsf{Lap}(\frac{\Delta f_j}{t_j})$, we have $\mathsf{E}[|X_j|] = \frac{\Delta f_j}{t_j}, \mathsf{Var}[X_j] = 2\left(\frac{\Delta f_j}{t_j}\right)^2$. Moreover, $\Delta f_j = C_j = max_{1 \leq i \leq n}\{a_{ij}\}$ based on the definition of global sensitivity.

$$\mathcal{U}_{MAE} = \mathsf{E}\left[\frac{1}{m}\|\mathbf{a^P} - \mathbf{a^r}\|_1\right] \geq \frac{1}{m}\sum_{j=1}^{m}\mathsf{E}[|X_j|] = \frac{1}{m}\sum_{j=1}^{m}\frac{C_j}{t_j}$$

$$\mathcal{U}_{MSE} = \mathsf{E}\left[\frac{1}{m}\|\mathbf{a^P} - \mathbf{a^r}\|_2^2\right] \geq \frac{1}{m}\sum_{j=1}^{m}\mathsf{Var}[X_j] = \frac{2}{m}\sum_{j=1}^{m}\frac{C_j^2}{t_j^2}$$

$$\mathcal{U}_{MAEL}(\mathbf{v}) = \mathsf{E}\left[\frac{1}{m}\sum_{j=1}^{m}\frac{|a_j^p - a_j^r|}{BU_j}\right] - 1 \geq \frac{1}{m}\sum_{j=1}^{m}\mathsf{E}\left[\frac{|X_j|}{\frac{\Delta f_j}{t_j}}\right] - 1 = 0$$

4 DP-MultiUPP Framework

In this section, we develop a novel *Differentially-Private Multi-Level User Privacy Perturbation (DP-MultiUPP)* framework to optimize the utility, especially when the condition (3) does not hold for ϵ. Specifically, we first introduce a novel differential privacy mechanism, called *Multi-Level Differential Privacy (MultiDP) Mechanism*, for trading off the utility loss and privacy guarantees. We then apply MultiDP mechanism to develop the DP-MultiUPP framework, with the provable privacy and utility guarantees and linear time complexity.

4.1 Multi-level Differential Privacy Mechanism

In this subsection, we focus on the case that ϵ is smaller than the lower bound in Theorem 2, i.e., the optimal utility cannot be achieved. In this case, we propose a novel mechanism, called *Multi-Level Differential Privacy (MultiDP) Mechanism*, to optimize the utility loss while preserving both per-attribute t_j-DP and overall ϵ-DP. In this mechanism, our goal is to find the condition for automating per-attribute privacy budgets t'_j (a reflection of the utility loss without violating per-attribute t_j-differential privacy) and overall ϵ-differential privacy guarantee.

As the determination of optimal privacy budgets t'_j is dependent on public domain, we consider the following MultiDP condition:

Definition 3 (MultiDP Condition). *For a set of queries* $f_1, \ldots, f_m$ *in which each* $f_j : (\mathbb{Z}^+)^n \rightarrow \mathbb{R}$ *is associated with its global sensitivity* Δf_j. *The set of non-negative numbers* $t'_1, \ldots, t'_m$ *satisfies MultiDP condition if the following two conditions hold:*

$$0 \leq t'_j \leq t_j, \forall 1 \leq j \leq m \tag{4}$$

$$\max_{d(D_1, D_2)=1}\left\{\sum_{j=1}^{m}\frac{t'_j}{\Delta f_j}|f_j(D_1) - f_j(D_2)|\right\} \leq \epsilon \tag{5}$$

Algorithm 1. DP-MultiUPP Algorithm

Input : user private data $\mathbf{d}^\mathbf{r}$, public item-attribute matrix $\mathbf{A}$, per-attribute
privacy budgets t_j, overall privacy budget ϵ
Output: perturbed attribute aggregates $\mathbf{a}^\mathbf{p}$

1 $\mathbf{a}^\mathbf{r} \leftarrow \mathbf{A}^T\mathbf{d}^\mathbf{r}$;
2 Solve (6) with $\mathbf{v}^T\mathbf{v}^\mathbf{r} \geq \mathbf{I}$ using [18];
3 $\mathbf{v} \leftarrow$ reciprocal of each entry in $\mathbf{v}^\mathbf{r}$;
4 **foreach** $j = 1, \ldots, m$ **do**
5 $\quad \lfloor \; a_j^p = a_j^r + \mathsf{Lap}(v_j)$;

6 **return** $\mathbf{a}^\mathbf{p}$;

Theorem 4 (MultiDP Mechanism). *Given a set of non-negative numbers*
$t_1, \ldots, t_m$, *and* $t_1', \ldots, t_m'$ *satisfying MultiDP condition in Definition 3. For a*
set of queries $f_1, \ldots, f_m$ *in which each* $f_j : (\mathbb{Z}^+)^n \to \mathbb{R}$ *is associated with its*
global sensitivity Δf_j, *a randomized algorithm* $\mathcal{A}^{MultiDP}$ *that adds independently*
generated noise $\mathsf{Lap}\left(\frac{\Delta f_j}{t_j'}\right)$ *to each query* f_j *enjoys* t_j-*differential privacy for each*
query f_j *and overall* ϵ-*differential privacy for all queries* $f_1, \ldots, f_m$.

Proof. First, it is trivial to prove that $\mathcal{A}^{MultiDP}$ achieves t_j-differential privacy
for each query since $t_j' \leq t_j$ always holds for each query j.

Next, we focus on the proof of overall differential privacy for all queries.
Let D_1, D_2 be any two neighboring datasets, i.e., $d(D_1, D_2) = 1$. For any $\mathbf{s} = (s_1, \ldots, s_m) \in Range(\mathcal{A}^{MultiDP})$,

$$\frac{\Pr[\mathcal{A}^{MultiDP}(D_1) = \mathbf{s}]}{\Pr[\mathcal{A}^{MultiDP}(D_2) = \mathbf{s}]} \geq \prod_{j=1}^{m} \exp\left(-\frac{t_j'}{\Delta f_j}|f_j(D_1) - f_j(D_2)|\right) \geq \exp(-\epsilon)$$

The first step holds due to the independent Laplace noises on each attribute
aggregate and triangle inequality; and the last step holds from the MultiDP
condition in Definition 3.

The advantage of our proposed lower bound and multi-level mechanism, over
the composition approach in [8], is that we take into account the correlation
between queries. Therefore, our approach not only provides a much better ϵ
lower bound but also helps to dramatically reduce the utility loss.

4.2 DP-MultiUPP Framework

Applying our proposed MultiDP mechanism, DP-MultiUPP framework aims to
automate per-attribute privacy budgets $t_1', \ldots, t_m'$ based on the overall privacy
levels/budgets ϵ towards the service provider.

The rest of this subsection consists of notion definition, detailed DP-
MultiUPP framework, and theoretical privacy, utility and time complexity
analysis.

Notations. We define two notations:

(1) the noise standard deviation reciprocal vector $\mathbf{v^r} = (\frac{t_1'}{\Delta f_1}, \ldots, \frac{t_m'}{\Delta f_m})$, where the jth entry is proportional to the reciprocal of standard deviation of injected Laplace noise on attribute j; and the noise standard deviation vector $\mathbf{v} = (\frac{\Delta f_1}{t_1'}, \ldots, \frac{\Delta f_m}{t_m'})$, where the jth entry is the reciprocal of corresponding j^{th} entry in $\mathbf{v^r}$, i.e., proportional to the standard deviation of injected Laplace noises. The dimension of $\mathbf{v}, \mathbf{v^r}$ is given by the number of attributes m.

(2) the global sensitivity diagonal matrix $\mathbf{GS} = diag(\Delta f_1, \ldots, \Delta f_m)$, where the jth entry is the global sensitivity of query f_j (aggregate of attribute j).

DP-MultiUPP Algorithm. The goal is to achieve optimal noise magnitude $\mathbf{v}$. To do so, we first formulate the mathematical programming as follows:

$$\begin{aligned} & minimize \ \mathcal{U}(\mathbf{v}) \\ & subject \ to \ \mathbf{Av^r} \leq \epsilon\mathbf{1}_n, \mathbf{v^r} \leq \mathbf{GS}^{-1}\mathbf{t}, \mathbf{v^r} \geq \mathbf{0} \end{aligned} \qquad (6)$$

where we optimize the utility function $\mathcal{U}$ defined in MultiUPP problem. Specifically, $\mathcal{U}$ takes noise standard deviation vector $\mathbf{v}$ as input, denoted as $\mathcal{U}(\mathbf{v})$. The three constraints imposes the MultiDP condition, which is sufficient to guarantee both t_j-differential privacy and ϵ-differential privacy as shown in MultiDP mechanism. As (6) is not convex in general with an implicit constraint $\mathbf{v}^T\mathbf{v^r} = \mathbf{I}$, we treat $\mathbf{v}, \mathbf{v^r}$ as two vector variables and add one more constraint $\mathbf{v}^T\mathbf{v^r} \geq \mathbf{I}$. The tweaked formulation has convex property. Algorithm 1 describes the DP-MultiUPP algorithm.

Formulation of Various Utilities for DP-MultiUPP Algorithm: Consider random variables $X_j \sim \mathsf{Lap}(\frac{\Delta f_j}{t_j'})$ on each attribute j. We specify utility functions $\mathcal{U}(\mathbf{v})$ for three utility objectives discussed in Sect. 2.3.

- $\mathcal{U}_{MAE}$: *Expected Mean Absolute Error.*

$$\mathcal{U}_{MAE}(\mathbf{v}) \propto \mathsf{E}\Big[\|\mathbf{a^p} - \mathbf{a^r}\|_1\Big] = \sum_{j=1}^{m} \mathsf{E}[|X_j|] = \sum_{j=1}^{m} v_j = \|\mathbf{v}\|_1$$

- $\mathcal{U}_{MSE}$: *Expected Mean Square Error.*

$$\mathcal{U}_{MSE}(\mathbf{v}) \propto \mathsf{E}\Big[\|\mathbf{a^p} - \mathbf{a^r}\|_2^2\Big] = \sum_{j=1}^{m} \mathsf{Var}[X_j] \propto \sum_{j=1}^{m} v_j^2 = \|\mathbf{v}\|_2^2$$

- $\mathcal{U}_{MAEL}$: *Expected Mean Absolute Error Loss.*

$$\mathcal{U}_{MAEL}(\mathbf{v}) \propto \mathsf{E}\Big[\sum_{j=1}^{m} \frac{|a_j^p - a_j^r|}{BU_j}\Big] - 1 \simeq \sum_{j=1}^{m} \frac{v_j}{BU_j} = \mathbf{BU^r}\mathbf{v}^T - 1$$

where $\mathbf{BU^r} = (\frac{t_1}{\Delta f_1}, \ldots, \frac{t_m}{\Delta f_m})$ stands for the reciprocal of standard deviation of injected noise with respect to each given privacy budget t_j. That is, $BU_j^r = \frac{1}{BU_j}$.

Theoretical Analysis. We provide privacy and utility analysis, as well as time complexity analysis.

Privacy analysis: *DP-MultiUPP framework enjoys t_j-differential privacy for each attribute aggregate and overall ϵ-differential privacy for all attribute aggregates.* The proof follows directly from the multi-level differential privacy mechanism proposed in Sect. 4.1.

Utility analysis: *DP-MultiUPP framework ensures the utilities upper bounded by the following:*

$$U_{MAE} \leq \frac{1}{m}\sum_{j=1}^{m}\frac{C_j}{t'_j}; \ U_{MSE} \geq \frac{2}{m}\sum_{j=1}^{m}\frac{C_j^2}{t'^2_j}; \ U_{MAEL} \geq \frac{t_j}{t'_j} - 1$$

where $t'_j = t_j\epsilon/\max_{d(D_1,D_2)=1}\left\{\sum_{j=1}^{m}\frac{t_j}{\Delta f_j}|f_j(D_1) - f_j(D_2)|\right\}$ when ϵ is smaller than lower bound in (3). This is because the equal loss of each attribute leads to feasible solution regardless of the selected utility function. In the experiment, we treat this as baseline and show that the performance of our DP-MultiPP framework is much better in practice. In addition, when overall privacy budget ϵ is larger than lower bound in Theorem 2, DP-MultiUPP automatically achieves the lower bounds of utility losses in Theorem 3.

Time complexity analysis: *DP-MultiUPP framework has $O(n)$ time complexity.* This is exactly obtained from the analysis in [18] since the number of attributes is assumed to be bounded by a constant in this paper. Also, steps 3 and 4–5 both take $O(m)$ time.

5 Experimental Evaluation

In this section, we evaluate the performance of our proposed DP-MultiUPP framework. We conduct our experiments extensively on a variety of real-world datasets. We first use different metrics to measure the performance of the utility of all perturbed attribute aggregates as well as each attribute aggregate. Then, we report the scalability of DP-MultiUPP framework on both personal computer with 1.9 GHz CPU and 8 GB RAM, and Android Phone Galaxy S5.

5.1 Datasets, Settings, Metrics and Competitors

Datasets: We use three real world datasets.

MovieLens[1]*:* a movie rating dataset collected by the GroupLens Research Project at the University of Minnesota through the website movielens.umn.edu during the 7-month period from September 19th, 1997 through April 22nd, 1998. The number of attributes is 19. We use the MovieLens-1M, with 1,000,209 ratings from 6,040 users on 3,883 movies.

---

[1] http://grouplens.org/datasets/movielens.

Yelp[2]: a business rating data provided by RecSys Challenge 2013, in which Yelp reviews, businesses and users are collected at Phoenix, AZ metropolitan area. The number of attributes is 21. We use all reviews in training dataset, with 229,907 reviews from 43,873 users on 11,537 businesses.

MSNBC[3]: an anonymous web dataset collected by the UCI Machine Learning Repository through the msnbc.com domain during a 24-hour period on September 28, 1999. We consider types of websites as their attributes and the number of attributes is 17. We use the whole dataset, with 4,698,794 reviews from 989,818 users on these 17 attributes of websites.

Settings: We consider a fixed sum of per-attribute privacy budgets, i.e., $\sum t_j = 1$, and randomly select a privacy budget t_j for each attribute to satisfy this summation. We test different overall privacy budget ϵ from 0.05 to 0.4. We run each experiment 10 times and report the average result.

We test our proposed DP-MultiUPP framework by incorporating it with different utility functions in Sect. 2.3, denoted as DP-MultiUPP (MAE), DP-MultiUPP (MSE) and DP-MultiUPP (MAEL).

Metrics: We measure the performance of our DP-MultiUPP framework on utilities of both all attribute aggregates and each attribute aggregate, referred to as *Overall Utilities* and *Per-attribute Utilities*.

Overall Utilities. We use the expected Mean Absolute Error (MAE) and the expected Mean Square Error (MSE) in Sect. 2.3.

Per-attribute Utilities. We first use expected the Mean Absolute Error Loss (MAEL) in Sect. 2.3. In addition, we also consider another metric, KL-Divergence on injected per-attribute noise variance over optimal per-attribute noise variance, to measure the difference between the variance of injected Laplace noise using the optimized t'_j and that using a given t_j. Specifically, it can be written as $D_{KL} = \sum_{j=1}^{m} \frac{(\Delta f_j/t_j)^2}{\sum_j (\Delta f_j/t_j)^2} \log \left(\frac{\frac{(\Delta f_j/t_j)^2}{\sum_j (\Delta f_j/t_j)^2}}{\frac{(\Delta f_j/t'_j)^2}{\sum_j (\Delta f_j/t'_j)^2}} \right)$.

Competitors: We consider a baseline algorithm based on the state-of-the-art composition algorithm in Lemma 1 and our proposed lower bound of ϵ in Theorem 2. In detail, this baseline algorithm first scans all items and determines if the overall privacy budget ϵ is smaller than its lower bound given by per-attribute privacy budgets t_j. In this case, the utility obtained by this baseline approach is exactly the lower bound of utility loss in Sect. 3.2. If not, we simply inject $\mathsf{Lap}(\frac{\Delta f_j}{t_j})$ noises to the aggregate of each attribute j. Otherwise, we adjust each per-attribute privacy budget t_j to $t'_j = t_j/r$ where ratio $r = \max_{d(D_1,D_2)=1} \left\{ \sum_{j=1}^{m} \frac{t_j}{\Delta f_j} |f_j(D_1) - f_j(D_2)| \right\}/\epsilon$. Then, we inject $\mathsf{Lap}(\frac{\Delta f_j}{t'_j})$ into each attribute aggregate and it is not hard to see that this also satisfies overall ϵ-differential privacy.

---

[2] https://www.kaggle.com/c/yelp-recsys-2013/data.

[3] https://archive.ics.uci.edu/ml/datasets/MSNBC.com+Anonymous+Web+Data.

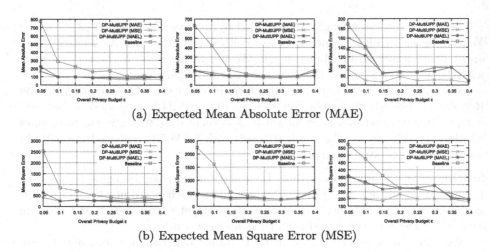

(a) Expected Mean Absolute Error (MAE)

(b) Expected Mean Square Error (MSE)

Fig. 2. Overall Utility Results (Left to Right: MovieLens, Yelp, MSNBC)

5.2 Utility Results

Overall Utility Results: Figure 2 reports the performance of DP-MultiUPP on overall utility results. As one can see, DP-MultiUPP consistently outperforms baseline algorithm regardless of its associated utility function. When ϵ is small ($\epsilon = 0.05$), DP-MultiUPP improves the performance up to 5 times out of the baseline approach. When ϵ is larger than the lower bound in Theorem 2, DP-MultiUPP continuously returns the optimal utility automatically due to its optimized utility objective.

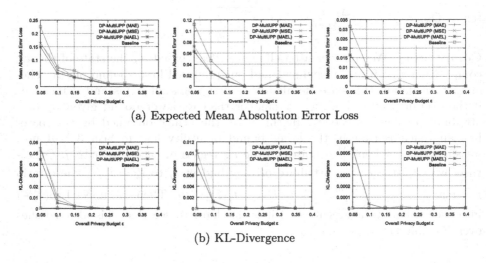

(a) Expected Mean Absolution Error Loss

(b) KL-Divergence

Fig. 3. Per-attribute Utility Results (Left to Right: MovieLens, Yelp, MSNBC)

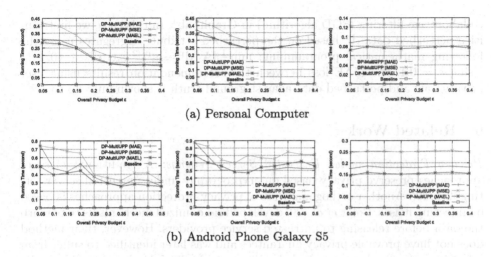

(a) Personal Computer

(b) Android Phone Galaxy S5

Fig. 4. Running Time of DP-MultiUPP (Left to Right: MovieLens, Yelp, MSNBC)

It is interesting to see that DP-MultiUPP with MSE utility function most of the time has best performance, especially in MSNBC dataset. This is because the variance of the injected noises can better capture all these utility losses. This provides us with an insight regarding how to select a better utility function.

More importantly, the smaller the overall privacy budget ϵ (w.r.t. lower reputed services) is, the bigger advantage DP-MultiUPP has over the baseline algorithm. This makes DP-MultiUPP very practically useful since users need stronger privacy guarantee especially for numerous low reputed service providers.

Per-attribute Utility Results: Figure 3 reports the performance of DP-MultiUPP on per-attribute utility results. Figure 3(a) shows DP-MultiUPP (MAEL) again improves the utility up to twice than using the baseline algorithm. As one can see in Fig. 3(b), the KL-Divergence on injected per-attribute noise variance over optimal per-attribute noise variance remains small in all datasets. This is because the optimization of (6) evenly increases privacy levels for each attribute while preserving the overall privacy level. Thus, user's preferred privacy levels for each attribute are very well maintained.

Scalability: Figure 4 reports the averaged running time of all algorithms on different datasets on both personal computer and Android Phone Galaxy S5. As one can see, our DP-MultiUPP framework takes at most 0.5 s and 1 s on PC and Android smartphone respectively and the running time almost remains invariant with different overall privacy budgets. Overall, thanks to the linear time complexity, DP-MultiUPP is very scalable on different client devices.

5.3   Case Study: Personalized Recommendation

We conduct an additional case study of personalized recommendation using perturbed data obtained by our approach on MovieLens dataset, through collaborative filtering (SGD algorithm) in GraphLab[4]. In this case, we first sanitize perturbed data $\mathbf{d^P}$ based on the perturbed attribute aggregates using the following mathematical programming: $\min \frac{1}{2}\|\mathbf{A}^T\mathbf{d^P} - \mathbf{a^P}\|^2$ s.t. $\mathbf{d^P} \in \{0,1\}^n$. Using $\epsilon = 0.1$, the MAE loss between the recommendation results using user private/raw and perturbed data against ground truth is shown only up to 8%.

6   Related Work

Privacy Protection under Untrusted Server Settings: A traditional class of approaches preserve privacy based on cryptography under untrusted server setting [2,5,19]. Another orthogonal class of privacy protection approaches is based on injecting noises. Polat *et al.* [20] developed randomized mechanisms to perturb the data before releasing to untrusted service providers. However, their method does not have provable privacy guarantees and was later identified to suffer from inference attacks. A recent work by Shen *et al.* [23] introduced a differential private data perturbation method on user's client. Although this approach has formal privacy and utility guarantee, it can only take one privacy budget and treat every type of data with the same privacy concern.

Differential Privacy: Differential privacy [7,9] has become the de facto standard for privacy preserving data analytics. Dwork *et al.* [9] established the guideline to guarantee differential privacy for individual aggregate queries by calibrating the Laplace noise to each query regarding the global sensitivity. Various works have adopted this definition for publishing histograms [25], search logs [14], mining data streams [6], and record linkage [4]. Later on, a noise mitigation mechanism was proposed by Machanavajjhala *et al.* [17].

Histogram Release via Differential Privacy: The most basic approach is to add noises of full contingency table of the whole dataset that suffers from exponential computational and space complexity. An improvement of this basic approach was proposed by Dwork *et al.* [9] to add independently generated Laplace noise to each k-way marginal table. Later on, Barak *et al.* [3] proposed the approach to add noises in the Fourier domain and improve the expected squared error by 2^k. Li *et al.* [16] proposed the matrix mechanism for counting queries. However, it still suffers from high computational complexity. In addition to these approaches, there exist many other approaches such as [10,11,21]. Unfortunately, none of these approaches provides an option for multi-level privacy concern configuration.

Multi-level Differential Privacy Preservation: The state-of-the-art method is the composition approach in [9] which preserves both per-attribute and overall differential privacy. However, it does not analyze when the achievement of all privacy guarantees is feasible, and does not provide a utility optimization mechanism when it is infeasible to achieve all privacy guarantees.

---

[4] http://select.cs.cmu.edu/code/graphlab/pmf.html.

7 Conclusion and Future Work

In this paper, we develop the first lightweight framework via differential privacy to automate multi-level privacy controls for releasing different attributes of data to service providers of different reputations. We theoretically analyze privacy, utility and time complexity. The experimental results show that our approach outperforms state-of-the-art approach up to 5 times with high scalability on both personal computer and smartphone. Particularly, our framework shows significant advantage for stronger privacy guarantee towards numerous low reputed service providers, making it very practically useful.

In the future work, we intend to extend our approach into more practical scenarios: (1) we will conduct more thorough experiments on personalized recommendation case study; (2) when the correlation among user private data attributes and the correlation among public attributes are similar, we will define a new privacy notion and mechanism to tackle the decreased privacy guarantees; (3) we will design a streaming mulit-level privacy preserving data publishing approach to tackle continuously generated user private data.

References

1. Pew research report. http://www.pewinternet.org/2013/09/05/anonymity-privacy-and-security-online-2/
2. Armknecht, F., Strufe, T.: An efficient distributed privacy-preserving recommendation system. In: Ad Hoc Networking Workshop, pp. 65–70, June 2011
3. Barak, B., Chaudhuri, K., Dwork, C., Kale, S., McSherry, F., Talwar, K.: Privacy, accuracy, and consistency too: a holistic solution to contingency table release. In: PODS, pp. 273–282. New York, NY, USA (2007)
4. Bonomi, L., Xiong, L., Lu, J.J.: LinkIT: privacy preserving record linkage and integration via transformations. In: SIGMOD, pp. 1029–1032 (2013)
5. Canny, J.: Collaborative filtering with privacy. In: IEEE Symposium on Security and Privacy, pp. 45–57 (2002)
6. Chan, T.-H.H., Li, M., Shi, E., Xu, W.: Differentially private continual monitoring of heavy hitters from distributed streams. In: Cristofaro, E., Murdoch, S.J. (eds.) PETS 2014. LNCS, vol. 8555, pp. 140–159. Springer, Heidelberg (2012). doi:10.1007/978-3-642-31680-7_8
7. Dwork, C.: Differential privacy: a survey of results. In: Jain, R., Jain, S., Stephan, F. (eds.) TAMC 2015. LNCS, vol. 9076, pp. 1–19. Springer, Heidelberg (2008). doi:10.1007/978-3-540-79228-4_1
8. Dwork, C., Lei, J.: Differential privacy and robust statistics. In: STOC, pp. 371–380. ACM, New York (2009)
9. Dwork, C., McSherry, F., Nissim, K., Smith, A.: Calibrating noise to sensitivity in private data analysis. In: Kushilevitz, E., Malkin, T. (eds.) TCC 2016. LNCS, vol. 9563, pp. 265–284. Springer, Heidelberg (2006). doi:10.1007/11681878_14
10. Gupta, A., Hardt, M., Roth, A., Ullman, J.: Privately releasing conjunctions and the statistical query barrier. In: STOC, pp. 803–812. New York, NY, USA (2011)
11. Hardt, M., Ligett, K., Mcsherry, F.: A simple and practical algorithm for differentially private data release. In: NIPS, pp. 2339–2347 (2012)

12. Hardt, M., Talwar, K.: On the geometry of differential privacy. In: STOC, pp. 705–714. ACM, New York (2010)
13. Jeckmans, A.J.P., Beye, M.R.T., Erkin, Z., Hartel, P.H., Lagendijk, R.L., Tang, Q.: Privacy in recommender systems. In: Ramzan, N., van Zwol, R., Lee, J.-S., Clüver, K., Hua, X.-S. (eds.) Social Media Retrieval. Computer Communications and Networks, pp. 263–281. Springer, London (2013)
14. Korolova, A., Kenthapadi, K., Mishra, N., Ntoulas, A.: Releasing search queries and clicks privately. In: WWW, pp. 171–180 (2009)
15. Leon, P.G., Ur, B., Wang, Y., Sleeper, M., Balebako, R., Shay, R., Bauer, L., Christodorescu, M., Cranor, L.F.: What matters to users?: factors that affect users' willingness to share information with online advertisers. In: SOUPS, pp. 7:1–7:12. ACM, New York (2013)
16. Li, C., Hay, M., Rastogi, V., Miklau, G., McGregor, A.: Optimizing linear counting queries under differential privacy. In: PODS, pp. 123–134 (2010)
17. Machanavajjhala, A., Kifer, D., Abowd, J., Gehrke, J., Vilhuber, L.: Privacy: theory meets practice on the map. In: ICDE, pp. 277–286 (2008)
18. Megiddo, N.: Linear programming in linear time when the dimension is fixed. J. ACM **31**(1), 114–127 (1984). http://doi.acm.org/10.1145/2422.322418
19. Nikolaenko, V., Ioannidis, S., Weinsberg, U., Joye, M., Taft, N., Boneh, D.: Privacy-preserving matrix factorization. In: CCS, pp. 801–812. New York (2013)
20. Polat, H., Du, W.: Privacy-preserving collaborative filtering using randomized perturbation techniques. In: ICDM, pp. 625–628 (2003)
21. Qardaji, W., Yang, W., Li, N.: PriView: practical differentially private release of marginal contingency tables. In: SIGMOD, pp. 1435–1446. New York (2014)
22. Rajaraman, A., Ullman, J.D.: Mining of Massive Datasets. Cambridge University Press, New York (2011)
23. Shen, Y., Jin, H.: Privacy-preserving personalized recommendation: an instance-based approach via differential privacy. In: ICDM, pp. 540–549 (2014)
24. Tsai, J.Y., Egelman, S., Cranor, L., Acquisti, A.: The effect of online privacy information on purchasing behavior: an experimental study. Inf. Syst. Res. **22**(2), 254–268 (2011)
25. Xu, J., Zhang, Z., Xiao, X., Yang, Y., Yu, G.: Differentially private histogram publication. In: ICDE, pp. 32–43 (2012)
26. Zhang, B., Wang, N., Jin, H.: Privacy concerns in online recommender systems: influences of control and user data input. In: SOUPS, pp. 159–173 (2014)
27. Zhang, X., Chen, R., Xu, J., Meng, X., Xie, Y.: Towards accurate histogram publication under differential privacy. In: SDM, pp. 587–595 (2014)

On Dynamic Feature Weighting for Feature Drifting Data Streams

Jean Paul Barddal[1]([⊠]), Heitor Murilo Gomes[1], Fabrício Enembreck[1],
Bernhard Pfahringer[2], and Albert Bifet[3]

[1] Graduate Program in Informatics (PPGIa),
Pontifícia Universidade Católica do Paraná, Curitiba, Brazil
{jean.barddal,hmgomes,fabricio}@ppgia.pucpr.br
[2] Department of Computer Science, University of Waikato, Hamilton, New Zealand
bernhard@cs.waikato.ac.nz
[3] Computer Science and Networks Department (INFRES), Institut Mines-Télécom,
Télécom ParisTech, Université Paris-Saclay, Paris, France
abifet@waikato.ac.nz

Abstract. The ubiquity of data streams has been encouraging the development of new incremental and adaptive learning algorithms. Data stream learners must be fast, memory-bounded, but mainly, tailored to adapt to possible changes in the data distribution, a phenomenon named concept drift. Recently, several works have shown the impact of a so far nearly neglected type of drifcccct: feature drifts. Feature drifts occur whenever a subset of features becomes, or ceases to be, relevant to the learning task. In this paper we (i) provide insights into how the relevance of features can be tracked as a stream progresses according to information theoretical Symmetrical Uncertainty; and (ii) how it can be used to boost two learning schemes: Naive Bayesian and k-Nearest Neighbor. Furthermore, we investigate the usage of these two new dynamically weighted learners as prediction models in the leaves of the Hoeffding Adaptive Tree classifier. Results show improvements in accuracy (an average of 10.69 % for k-Nearest Neighbor, 6.23 % for Naive Bayes and 4.42 % for Hoeffding Adaptive Trees) in both synthetic and real-world datasets at the expense of a bounded increase in both memory consumption and processing time.

1 Introduction

Data streams are ubiquitous, potentially unbounded and generated at a very fast pace. Examples of streaming data include, but are not limited, to: ATM transactions, readings in mobile sensor networks, social networks posts and stock trades. Motivated by these real world problems, data stream mining grew in popularity and became a very active research field with new techniques proposed every year aiming at learning from these sequences of data in an incremental, fast and memory-bounded fashion. Many of these new developments in data stream learning focus on the ephemeral characteristics of data streams, i.e. when the

© Springer International Publishing AG 2016
P. Frasconi et al. (Eds.): ECML PKDD 2016, Part II, LNAI 9852, pp. 129–144, 2016.
DOI: 10.1007/978-3-319-46227-1_9

underlying data distribution shifts with time, a phenomenon named *concept drift* [27].

More recently, studies [6,7] shed light onto a specific kind of drift which has so far practically been neglected, the so-called *feature drifts*, sometimes referred to as contextual concept drifts in seminal works [27]. In practice, a feature drift occurs whenever a subset of features of a data stream becomes, or ceases to be, relevant to the learning task. As surveyed in [7] and empirically analyzed in [6], feature drifts pose challenges that are yet to be tackled by the data stream mining community.

In this paper we propose a low complexity and memory-bounded solution to track the relevance of features in streaming data accordingly to the information theoretical Symmetrical Uncertainty. Additionally, we show how this metric can be used to enhance prediction accuracy in k-Nearest Neighbor, Naive Bayes and Hoeffding Adaptive Tree classifiers when they are applied feature drifting data streams.

2   Learning from Data Streams

As times goes by, data acquisition and storage becomes cheaper and easier. As a consequence, companies and individuals can generate and store data at an increasing rate. Some of these data are generated sequentially and are so massive that it would not be practical nor useful to store them all. For instance, the data generated by a wearable gadget may only be meaningful for a small period of time, such that the burden to store or transmit it may be unjustifiable. These abundant sources of raw data may also be unintelligible to its possessors and in these situations, data mining techniques are often employed to extract meaningful patterns from apparent chaos. Currently, a lot of effort has been directed towards mining data that is generated in a continuous stream, an area that has been commonly known as data stream mining [2,10].

Generally, data stream mining combines almost all problems of conventional batch learning (e.g. missing values, noisy data, outliers) with problems such as instability of the underlying concept and restrictive resources constraints. Specifically, data stream learners must (i) be able to process instances sequentially according to their arrival, (ii) act within limited memory space and processing time, (iii) deal with data instability (concept drifts); and (iv) be able to generalize well as instances' labels become available [17].

Ideally, algorithms for learning from data streams must include techniques for dealing with all aforementioned problems. However, not all of them must be addressed at once since it depends on the problem being tackled. For instance, a given problem setting may exhibit concept drifts but not suffer from a lack of labeled data or vice-versa.

2.1   Data Stream Classification

The most common (and widely explored) learning task in a data stream setting is undoubtedly classification. Formally, given a set of possible class labels

$Y = \{y_1, \ldots, y_c\}$ and a set of labeled training instances $X = \{(\boldsymbol{x}^1, y^1), \ldots, (\boldsymbol{x}^n, y^n)\}$, a classifier uses the training set to build a model $f : \boldsymbol{x} \to Y$ capable of predicting the class label of an unlabeled instance $\boldsymbol{x}^i$. Precisely, each instance $\boldsymbol{x}$ is a d-dimensional feature vector belonging to a feature set $\mathcal{D} = \bigcup_{i=1}^d \{D_i\}$, that is possibly categorical, ordinal, numeric or most likely mixed.

Data stream (or online) classification is a variant of the traditional batch classification, and both are concerned with the problem of predicting class labels for unlabeled instances. The main difference between the batch and the online setting remains on how data are presented to the classifier. In a batch configuration data are entirely accessible in a finite and static dataset, while streaming data are presented sequentially over time [18] while f must be updated accordingly.

2.2 Concept Drift

Due to the inherent temporal aspect of data streams, their underlying data distribution may change over time, directly influencing changes to the concept to be learned, a phenomenon often referred as concept drift.

Let Eq. 1 denote a concept C, a set of prior probabilities of the classes and class-conditional probability density function [22].

$$C = \bigcup_{y_i \in Y} \{(P[y_i], P[\boldsymbol{x}|y_i])\} \tag{1}$$

Given a stream $\mathcal{S}$, retrieved instances i_t will be generated by a concept C_t. If during every instant t_i of $\mathcal{S}$ we have $C_{t_i} = C_{t_{i-1}}$, then the concept is stable. Otherwise, if between any two timestamps t_i and $t_j = t_i + \Delta$ (with $\Delta \geq 1$) it is the case that $C_{t_i} \neq C_{t_j}$, then we have observed a concept drift [17].

3 Problem Statement

Most existing algorithms for data streams tackle the infinite length and drifting concept characteristics. However, not much attention has been given to a specific kind of drift: feature drifts. Conversely to conventional concept drifts, where changes in the data distribution are claimed to occur inside the skewing of classes in ranges of features' values, feature drifts occur whenever a subset of features becomes, or ceases to be, relevant to the concept to be learned.

Until this point, the term "relevance" was used without a proper definition. In this paper we divide features in two types: relevant and irrelevant [7]. Assuming $S_i = \mathcal{D} \setminus \{D_i\}$, a feature D_i is deemed **relevant** iff Eq. 2 holds.

$$\exists S_i' \subset S_i, \text{ such that } P[Y|D_i, S_i'] \neq P[Y|S_i'] \tag{2}$$

Otherwise, the feature D_i is said **irrelevant**. In practice, if a feature that is statistically relevant is removed from a feature set, it will reduce overall prediction power since (i) it is strongly correlated with the class; or (ii) it belongs to a subset of features that is strongly correlated with the class [29].

Changes in the relevant subset of features enforce the learning algorithm to adapt its model to ignore the irrelevant attributes and to account for the newly relevant ones [22]. Given a feature space $\mathcal{D}$ at a timestamp t, we are able to select the ground-truth relevant subset $\mathcal{D}_t^* \subseteq \mathcal{D}$ such that $\forall D_i \in \mathcal{D}_t^*$ Eq. 2 holds and $\forall D_j \in \mathcal{D} \setminus \mathcal{D}_t^*$ the same definition does not. A feature drift occurs if, at any two time instants t_i and t_j, $\mathcal{D}_{t_i}^* \neq \mathcal{D}_{t_j}^*$ holds.

Let $r(D_i, t_j) \in \{0,1\}$ denote a function which determines whether Eq. 2 holds for a feature D_i in a timestamp t_j of the stream. A positive relevance $(r(D_i, t_i) = 1)$ states that $D_i \in \mathcal{D}^*$ in a timestamp t_i. A feature drift occurs whenever the relevance of an attribute D_i changes in a timespan between t_j and t_k, as stated in Eq. 3.

$$\exists t_j \exists t_k, \ t_j < t_k, \ r(D_i, t_j) \neq r(D_i, t_k) \tag{3}$$

Changes in $r(\cdot, \cdot)$ directly affect the ground-truth decision boundary to be learned by the learning algorithm. Therefore, feature drifts can be posed as a specific type of concept drift that may occur with or without changes in the data distribution $P[\boldsymbol{x}]$ [6,7]. We emphasize that feature drifts are indeed targeted by the generic concept drift formalization, however, most existing works on concept drift detection and adaptation assume that the relevant subset of features remains the same and that drifts occur if certain values, or ranges of values, of attributes have their class distribution re-skewed.

As pointed out in [6,7], feature drifts are likely to occur in a variety of scenarios, but mainly in text stream scenarios, e.g. social media, SMS chats, online social networks (Facebook, Twitter) and e-mail spam detection systems.

As in conventional concept drifts, changes in $r(\cdot, \cdot)$ may occur during the stream. This enforces learning algorithms to detect changes in $\mathcal{D}^*$, discerning between features that became irrelevant and the ones that are now relevant and vice-versa. In order to overcome feature drifts, a learner must either (i) discard and derive an entirely new classification model that is consistent with the relevant features; or (ii) adapt its current model to relevance drifts [22].

4   Dynamic Feature Weighting

Feature weighting is broadly used in batch learning [1,11] to assign different weights to features according to their relevance to the concept to be learned and to improve prediction accuracy. As shown earlier, in opposition to static scenarios, the relevance of features may increase or decrease during a data stream, thus, techniques for tracking and quantifying the proportions of such changes are needed.

The main hypothesis behind our proposals is that features can be dynamically weighted in order to augment the importance of relevant features and diminish the importance of those which are deemed irrelevant according to observed feature drifts. In this section we show how Symmetrical Uncertainty can be swiftly computed along a sliding window based on Entropy computation. Later, we introduce how Symmetrical Uncertainty can be applied into two distinct learning schemes to boost prediction accuracy on feature drifting data streams. Finally, we detail the bounded computational overhead this proposal provides in processing time and memory usage.

4.1 Preliminaries

The relevance of a feature can be computed in diverse ways. In this section we discuss evaluation techniques for measuring the goodness of features for classification. Generally, a feature is good if it is relevant to predict the class. If one adopts correlation to measure the goodness of a feature, a feature will be deemed as relevant if its value surpasses a given threshold.

Several approaches exist to measure the correlation between two random variables. One such approach use linear correlation and another one is based on measures from information theory.

The most common formula for computing the correlation for a pair of variables (X, Y) is the linear correlation coefficient, which can be computed as follows:

$$c(X, Y) = \frac{\sum_{q \in D_i} \sum_{y_i \in Y} (q - \bar{D}_i)(y_i - \bar{Y})}{\sqrt{\sum_{q \in D_i} (q - \bar{D}_i)^2} \sqrt{\sum_{y_i \in Y} (y_i - \bar{Y})^2}} \tag{4}$$

The linear correlation coefficient is bounded in the $[-1; 1]$ interval. If X and Y are completely correlated, c takes the value of 1 or -1; and if these variables are completely uncorrelated, c is 0. Adopting linear correlation as a feature goodness measure has the benefit of eliminating completely uncorrelated features. Also, if data are linearly separable in its original representation then they will also be separable if all but one a group of linearly dependent features are removed [28]. Nevertheless, assuming linear correlations is not safe for a variety of domains. Linear correlation is likely to be unable to depict correlations which are non-linear in nature.

In our proposal, we adopt information theory approaches to compute the goodness of a feature. The first one is a measure of uncertainty of a random variable, named Entropy. The Entropy of a variable X is given by:

$$H(X) = - \sum_{x_i}^{X} P[X = x_i] \log_2 P[X = x_i] \tag{5}$$

On the other hand, the Entropy of a variable X after observing values of a variable Y (Conditional Entropy) is given by:

$$H(X|Y) = - \sum_{y_j}^{Y} P[Y = y_j] \sum_{x_i}^{X} P[X = x_i|Y = y_j] \log_2 P[X = x_i|Y = y_j] \tag{6}$$

Clearly, one of the drawbacks of picking Entropy as a goodness measure is that is it unable to work with numeric features, unless they are discretized. Since minimum (min) and maximum (max) values of features in streaming scenarios

---

Algorithm 1. Sliding window entropy. Adapted from [25].

---

    input : window size w, a data stream $\mathcal{S}$.
    output : be ready to provide the entropy h at any time.
1 Let $W \leftarrow \emptyset$ be the sliding window;
2 Let $h \leftarrow 0$ be the entropy;
3 Let $n \leftarrow 0$ be the number of instances in W;
4 Let $n_i \leftarrow 0$ be the number of instances with the y_i-th label;
5 **foreach** $(\boldsymbol{x}_i, y_i) \in \mathcal{S}$ **do**
6 | **if** $|W| = w$ **then**
7 | | Dequeue oldest element from W from the y_j-th class;
8 | └ $h \leftarrow DEC(h, n, n_j)$;
9 | $W \leftarrow W \cup \{(\boldsymbol{x}_i, y_i)\}$;
10   └ $h \leftarrow INC(h, n, n_i)$;
11 **Function** $INC(h, n, n_i)$
12 | Update $n \leftarrow n + 1$;
13 | Update $n_i \leftarrow n_i + 1$;
14   └ **return** $\frac{n-1}{n} \left(h - \log_2 \frac{n-1}{n} \right) - \frac{n_i}{n} \log_2 \frac{n_i}{n} + \frac{n_i-1}{n} \log_2 \frac{n_i-1}{n}$
15 **Function** $DEC(h, n, n_i)$
16 | Update $n \leftarrow n - 1$;
17 | Update $n_i \leftarrow n_i - 1$;
18   └ **return** $\frac{n+1}{n} \left(h + \frac{n_i+1}{n+1} \log_2 \frac{n_i+1}{n+1} - \frac{n_i}{n+1} \log_2 \frac{n_i}{n+1} \right) + \log_2 \frac{n}{n+1}$

---

are unknown a priori, we adaptively discretized features using a sliding-window version of the Partition Incremental Discretization algorithm [16] with 10 bins.

One of the advantages of Entropy is that it can be computed along sliding windows. In Algorithm 1 we present the pseudocode for Entropy computation over sliding windows. Proofs for Entropy equations (lines 14 and 18) were omitted from this paper for the sake of brevity, thus, the reader is referred to [25] for details.

Entropy is the base for computing more robust metrics. One example is Information Gain, which is the amount by which the Entropy of a variable X decreases reflecting additional information about X provided by Y, and is given by:

$$IG(X|Y) = H(X) - H(X|Y) \tag{7}$$

An important trait of Information Gain is that it is symmetrical, i.e. $IG(X|Y) = IG(Y|X)$. To prove it, one needs to verify that $H(X) - H(X|Y) = H(Y) - H(Y|X)$ and this can be derived from $H(X, Y) = H(X) + H(Y|X) = H(Y) + H(X|Y)$.

As Entropy, Information Gain is biased towards features with more values. Therefore, different metrics that compensate for this bias are preferred. In this paper we picked Symmetrical Uncertainty (SU) as a goodness measure since it atones this bias. Symmetrical Uncertainty can be computed as follows:

$$SU(X,Y) = 2\left[\frac{IG(X|Y)}{H(X)+H(Y)}\right] = 2\left[\frac{H(Y)-H(Y|X)}{H(X)+H(Y)}\right] \tag{8}$$

The range of possible values for SU is the $[0;1]$ interval, where 1 indicates that the value of a variable completely predicts the other, while 0 indicates that X and Y are completely independent.

In order to compute SU along a sliding window, one must keep track of $H(D_i)$, $H(Y)$ and $H(Y|D_i)$ entropies. Both $H(D_i)$ and $H(Y)$ can be incremented and decremented in $\mathcal{O}(1)$ accordingly to Algorithm 1, while the Conditional Entropy $H(Y|D_i)$ can be computed with separate $H(Y|D_i = q)$ entropies (see Eq. 6), also given by Algorithm 1. If we assume that $q \in D_i$ and $|D_i| = m$, then SU can be computed with low computational complexity in the $\mathcal{O}(m)$ order for a single feature and $\mathcal{O}(dm)$ for all features in a d-dimensional data stream.

Memory-wise, the cost of tracking $H(Y)$ is $\mathcal{O}(|Y|)$, while the cost for $H(D_i)$ is $\mathcal{O}(m)$, thus, the total complexity is $\mathcal{O}(md)$ for a d-dimensional stream. Finally, $H(Y|D_i = q)$ incurs a cost of $\mathcal{O}(|Y|)$, therefore the total cost is $\mathcal{O}(md \times |Y|)$, when considering all features $D_i \in \mathcal{D}$.

4.2   Applying Feature Weighting to k-Nearest Neighbor Learning

k-Nearest Neighbor (kNN) [5] is one of the most fundamental, simple and widely used classification methods, which is able to learn complex (non-linear) functions [5]. kNN is a lazy learner since it does not require building a model before actual use. It classifies unlabeled instances according to the k "closest" buffered instances. The definition of "close" means that a distance measure is used to determine how similar/dissimilar two instances are. There are several approaches to compute distances between instances, nevertheless, the most common one is the Euclidian distance, given by Eq. 9, where $\boldsymbol{x}_i$ and $\boldsymbol{x}_j$ are two arbitrary instances, and the summation occurs over all features $D_k \in \mathcal{D}$.

$$d_E(\boldsymbol{x}_i, \boldsymbol{x}_j) = \sqrt{\sum_{D_k \in \mathcal{D}} (\boldsymbol{x}_i[D_k] - \boldsymbol{x}_j[D_k])^2} \tag{9}$$

As discussed in a variety of works [3], Euclidian distances fail on representing in an effective fashion the distance between points (instances) in a high-dimensional space, since both irrelevant and redundant features have the same weight as relevant ones.

k-Nearest Neighbor with Feature Weighting (kNN-FW) is an extension to the original kNN algorithm that performs dynamic feature weighting to overcome both irrelevant features and feature drifts. kNN-FW comprises the following internal structures: an instance buffer queue and variables to track $H(D_i)$, $H(Y)$ and $H(Y|D_i = q)$. Finally, kNN-FW has two distinct steps: a training and a classification phase.

During the training step, instances i_t are retrieved from a stream $\mathcal{S}$ and enqueued in a buffer of size W. For every instance being enqueued or dequeued, the values of $H(D_i)$, $H(Y)$ and $H(Y|D_i = q)$ are updated according to Algorithm 1, thus, enabling prompt SU computation.

During the classification step, unlabeled instances $\boldsymbol{x}_t$ are classified according to the k-nearest neighbors available in buffer. In opposition to the conventional *kNN* algorithm, we modify the Euclidian distance to perform feature weighting accordingly to the discriminative power provided by Symmetrical Uncertainty, i.e. $w(D_i) = SU(D_i, Y)$.

$$d(\boldsymbol{x}_i, \boldsymbol{x}_j) = \sqrt{\sum_{D_k \in \mathcal{D}} w(D_k) \times (\boldsymbol{x}_i[D_k] - \boldsymbol{x}_j[D_k])^2} \qquad (10)$$

Due to the dynamic computation of Symmetrical Uncertainty, *kNN-FW* is expected to assign weights dynamically accordingly to their discriminative power. In feature-drifting cases, features that become, or cease to be, relevant to the learning task will be promptly detected by changes in their Symmetrical Uncertainty values, generating appropriate changes for each feature's weight.

4.3   Applying Feature Weighting to Naive Bayes

Naïve Bayes (NB) is a probabilistic classifier based on Bayes theorem that works under the naïve independence assumption between features. These predictors are easy to build, can easily be incremented, and have no complicated parameter estimation, making it useful for large datasets and data streams. Classification (labeling) of instances in this learning scheme is given by Eq. 11, that is, the class is chosen accordingly to the label y_i that maximizes the $P[y_i] \prod_{j=1}^{d} P[\boldsymbol{x}[D_j] \mid y_i]$ probability.

$$y = \underset{y_i \in Y}{\arg\max} \, P[y_i] \prod_{j=1}^{d} P[\boldsymbol{x}[D_j] \mid y_i] \qquad (11)$$

Although Naïve Bayes is commonly referred as an appropriate solution for high dimensionality problems [11], it has been shown to be prone to feature drifts [6]. Analogously to *kNN-FW*, we now propose the adoption of a dynamic weighting factor during Naive Bayes prediction. Naïve bayes with feature weight (NB-FW) also adopts Symmetrical Uncertainty as a weighting factor during classification, thus, probabilities are also weighted accordingly with $w(D_i) = SU(D_i, Y)$, thus, labeling is performed as follows:

$$y = \underset{y_i \in Y}{\arg\max} \, P[y_i] \prod_{j=1}^{d} (w(D_i) + \xi) \times P[\boldsymbol{x}[D_j] \mid y_i] \qquad (12)$$

where ξ is a small padding factor, set to 0.0001, used to avoid zero weights which would nullify the probabilities of some class values.

5   Analysis

In order to assess our proposal's performance, we built an experimentation environment encompassing both synthetic and real-world data. This analysis centers on prediction accuracy, processing time and memory usage.

5.1   Synthetic Data Stream Generators

Drifts are synthesized as the combination of two pure distributions. The probability that an instance is drawn from the prior or posterior concept inside a drift window is given by a sigmoid function. This drift framework is the default provided in the MOA framework [9] and all drift windows in our experiments have a length of 1,000 instances. All synthesized data streams contain 100,000 instances and contain 9 feature drifts. In the following, we introduce three synthetic data generators used to induce feature drifts on our experiments: AGRAWAL [4], Assets Negotiation (ASSETS) [14] and SEA-FD [6]. In the following experiments, we guarantee that feature drifts occur by changing the relevant subset of features between prior and posterior concepts.

AGRAWAL. The AGRAWAL generator [4] produces data streams with the aim of determining whether a loan should or should not be given to a bank customer. This generator is composed by the following features: salary, commission, age, education level, car make, zip-code, house value, years house is owned and loan value. There are 10 functions for mapping instances to 2 possible classes, each of which relying on different subsets of these features.

Asset Negotiation (AN). This generator was originally presented in [14], where the aim was to simulate drifting bilateral multi-agent system negotiation of assets. Assets are described by the following features: color, price, payment, amount and delivery delay. The task is to predict whether an opposing agent would be interested, or not, in an asset, making this a binary classification problem. Feature drifts are synthesized with changes on the interest of an agent by modifying the concept through time given five functions, each of which is relying on a different subset of features.

SEA-FD. Described in [6], SEA-FD extends the SEA generator [26] and synthesizes streams with $d > 2$ uniformly distributed features, where $\forall D_i \in \mathcal{D}, D_i \in [0; 10]$ and $\mathcal{D}^* = \{D_\alpha, D_\beta\}$ is randomly chosen with the guarantee that it differs from the relevant subset of features from the earlier concept. As in [26], instances are labeled using $y = 1$ if $D_\alpha + D_\beta \leq \theta$ and $y = 0$ otherwise; where θ is a user-supplied threshold. In the following experiments we chose $\theta = 7$ since it is a widely used value in many papers of the area [6,8].

5.2   Symmetrical Uncertainty Tracking in Synthetic Experiments

In order to exemplify how the dynamic weights are computed during experiments, we devote this section to present and discuss the Symmetrical Uncertainty tracking during synthetic experiments. Figure 1 presents the Symmetrical

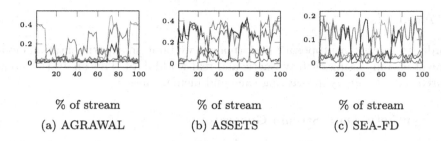

% of stream   % of stream   % of stream

(a) AGRAWAL   (b) ASSETS   (c) SEA-FD

Fig. 1. Symmetrical Uncertainty of several features during synthetic experiments.

Uncertainty of features during AGRAWAL, ASSETS and SEA-FD experiments, where each feature is represented by a curve with a different color. We highlight the fact that different features show higher SU values along the streams, thus confirming that our tracking strategy is able to depict feature drifts correctly.

5.3   Real-World Data

To complement the synthetic data, some real-world datasets were also used for the evaluation of the new algorithms. The adoption of real-world data is beneficial since they present differentiated behavior, e.g. the class distribution is often imbalanced and data are often noisy. On the other hand, it is nearly impossible to affirm whether drifts occur, making evaluation of drift detection unfeasible. We refrain from providing a detailed description of each used dataset for brevity. The used datasets are: Electricity (ELEC) [23], Kaggle's Give me Some Credit[1] (GMSC) and Spam Corpus (SPAM) [21].

5.4   Evaluated Algorithms

Besides kNN and Naive Bayes, we also report results for a Very Fast Decision Tree (VFDT) and a Hoeffding Adaptive Tree (HAT) since both perform embedded feature selection during training.

VFDT. Very Fast Decision Tree (VFDT) is an incremental decision tree learner for non-drifting data streams [13]. The tree is recursively built as instances arrive and new split nodes are generated if the information gain of the two most discriminative features differ at least by ϵ, given by the Hoeffding bound [19]. The prediction at the leaves may occur following three different strategies: majority class, Naive Bayes and Adaptive Naive Bayes. The Adaptive Naive Bayes monitors the error rate of the majority class and Naive Bayes, always employing the one that currently best fits data, as judged by their recent estimated accuracy.

---

[1] Available at: https://www.kaggle.com/c/GiveMeSomeCredit. Last access in Feb. 25th, 2016.

HAT. Hoeffding Adaptive Tree (HAT) algorithm is an extension to the VFDT to deal with drifts [20]. HAT updates its tree model over a sliding window and creates or updates decision nodes if the data distribution changes at an arbitrary split node. HAT detects data distribution changes according to the ADWIN change detector [8] provided in MOA [9]. Whenever ADWIN detects a change in a split node, the entire subtree is replaced by a new split node with the most discriminant feature if the Hoeffding bound is still met. As in VFDTs, the decision at leaf nodes may occur according to the majority class, Naive Bayes and Adaptive Naive Bayes methods.

5.5   Experimental Protocol

Accuracy is computed accordingly to the Prequential test-then-train procedure [15]. Prequential was chosen due to its way of monitoring a model's performance over time. Processing time is measured in seconds, while memory usage is given in RAM-Hours, where 1 RAM-Hour equals 1 GB of RAM dispended per hour of processing (GB-Hour). We adopted a window size $W = 1,000$ to keep track of Symmetrical Uncertainty in all experiments with the exception of Spam Corpus, where $W = 100$, due to the smaller number of instances in this dataset. An analysis of the impact of the window size W is later discussed in Sect. 5.7. All remaining parameters were set accordingly to the defaults provided in the Massive Online Analysis (MOA) framework [9].

5.6   Discussion

Table 1 presents the prequential accuracy results obtained during experiments. In all cases, the usage of our proposed feature weighting scheme was beneficial, providing an average boost of 10.69 % and 6.23 % for kNN and NB, respectively. To provide statistical significance to our claims, we performed Wilcoxon's, Friedman's and Nemenyi's tests [12]. Pairwise comparisons conducted with Wilcoxon's procedure between the original kNN and NB to their dynamically feature weighted versions with a 95 % confidence level corroborated that there is statistical difference between their accuracy rates.

Finally, with the aid of Friedman's and Nemenyi's tests, we compared all algorithms in Table 1. Results showed that, {HAT, kNN-FW, VFDT, NB-FW} $\succ$ {kNN, NB}, also with a 95 % confidence level. These results highlight that the weighting scheme is beneficial since it allows both kNN and NB to achieve comparable results with more sophisticated techniques that embed feature selection during stream learning, i.e. VFDT and HAT.

Tables 2 and 3 present processing time and memory usage obtained during the execution of experiments. With the exception of the ASSETS experiment, the adopted weighting scheme provides an computation overhead in both aspects. We claim, however, that this computational overhead is not damaging enough to prevent the usage of our weighting scheme, even in high dimensional problems, e.g. the SPAM experiment.

Table 1. Prequential accuracy (%).

Experiment	kNN	kNN-FW	NB	NB-FW	VFDT	HAT
AGRAWAL	57.74	65.64	59.18	62.68	69.98	**81.13**
ASSETS	85.02	87.87	70.51	77.11	91.57	**93.15**
SEA-FD	64.02	**84.14**	76.05	78.35	82.63	83.24
ELEC	54.31	**84.08**	57.62	73.39	79.23	83.46
GMSC	92.48	92.67	93.09	93.32	93.25	**93.37**
SPAM	80.56	83.87	66.26	75.22	79.32	**84.48**

Table 2. Processing time (s).

Experiment	kNN	kNN-FW	NB	NB-FW	VFDT	HAT
AGRAWAL	20.91	21.54	**0.45**	0.47	1.08	1.57
ASSETS	13.52	13.92	**0.28**	**0.28**	1.04	0.97
SEA-FD	104.68	107.82	**2.04**	2.08	4.85	6.15
ELEC	7.36	7.51	**0.36**	0.37	1.43	1.08
GMSC	32.35	33.64	**1.30**	1.31	7.90	15.43
SPAM	5911.54	6088.89	288.38	291.27	**253.76**	421.45

5.7 On the Impact of the Window Size W

Windowing is a common approach for both data management and dealing with drifting data. Our proposal relies on a window size parameter W that determines how much data should be considered to keep track of SU. Finding an optimal value for W is a trade-off without solution, a problem commonly referred as the stability-plasticity dilemma. While short windows reflect the current data distribution and ensures fast adaptation to drifts (plasticity), shorter ones worsen the performance of the system in stable areas. Conversely, larger windows give better performance in stable periods (stability), however, these imply a slower response to drifts.

In this section we evaluate the impact of the window size W in our proposal. We evaluated the original kNN, kNN-FW and NB-FW with different W values across the $[5; 2000]$ domain. Results for the Spam Corpus experiment were omitted since there was not enough time to run kNN-based algorithms in such high-dimensional scenarios in this amount of window sizes. In Fig. 3 we report the average accuracy obtained during experiments.

In Figs. 3a and b we present the results obtained by kNN and kNN-FW, where it is clear that finding an optimal value that achieves the best results on all datasets is not trivial. However, by comparing the results in both graphics, we highlight that regardless of the chosen W value, the adoption of the proposed weighting scheme is beneficial.

Table 3. RAM-Hours (GB-Hour).

Experiment	kNN	kNN-FW	NB	NB-FW	VFDT	HAT
AGRAWAL	7.57×10^{-7}	7.88×10^{-7}	$\mathbf{1.15 \times 10^{-9}}$	1.18×10^{-9}	5.05×10^{-8}	3.99×10^{-8}
ASSETS	3.77×10^{-7}	3.99×10^{-7}	$\mathbf{4.34 \times 10^{-10}}$	4.51×10^{-10}	7.84×10^{-8}	3.22×10^{-8}
SEA-FD	1.29×10^{-5}	1.34×10^{-5}	$\mathbf{1.40 \times 10^{-8}}$	1.45×10^{-8}	8.16×10^{-7}	2.65×10^{-7}
ELEC	2.43×10^{-7}	2.55×10^{-7}	$\mathbf{6.14 \times 10^{-10}}$	6.51×10^{-10}	3.45×10^{-8}	9.27×10^{-9}
GMSC	5.12×10^{-1}	5.43×10^{-1}	$\mathbf{2.12 \times 10^{-3}}$	2.19×10^{-3}	1.34×10^{-3}	4.21×10^{-3}
SPAM	1.20×10^{-6}	1.26×10^{-6}	$\mathbf{2.38 \times 10^{-9}}$	2.45×10^{-9}	3.88×10^{-7}	1.70×10^{-6}

Table 4. Prequential accuracy (%) for different leaf prediction strategies in HAT.

Experiment	HAT	HAT-kNN-FW	HAT-NB-FW
AGRAWAL	81.13	88.45	**91.03**
ASSETS	93.15	**95.63**	93.37
SEA-FD	83.24	81.12	**84.80**
ELEC	83.46	83.24	**83.56**
GMSC	93.37	**93.43**	93.39
SPAM	84.48	**92.62**	85.25

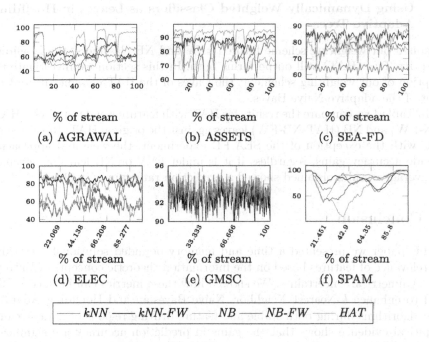

Fig. 2. Prequential accuracy (%) obtained during experiments.

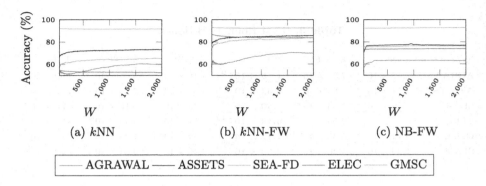

Fig. 3. Impact of W in prediction accuracy.

On the other hand, the results presented in Fig. 3c show that for NB-FW results are robust across different window sizes, although the accuracy drops very slightly for windows with $W > 1,000$. Finally, we highlight the ELEC and GMSC experiments, where the differences between the maximum and minimum accuracies were just 0.35 % and 0.97 %, respectively. This shows that the concept is relatively stable during the whole experiment, thus, the weights obtained across different window sizes are consistent.

5.8   Using Dynamically Weighted Classifiers as Leaves in Hoeffding Adaptive Trees

Although our weighting scheme favored kNN and NB classifiers, the Hoeffding Adaptive Tree (HAT) still outperforms both. In this section we investigate the adoption of our weighting scheme at the leaves of the HAT classifier in replacement of the adaptive Naive Bayes.

In Table 4 we compare the results for HAT with feature weighted KNN (HAT-kNN-FW) and NB (HAT-NB-FW) leaves against the original HAT. Results show that, with the exception of the SEA-FD experiment, the weighted approaches provide accuracy gains, regardless if it is under kNN or NB learning schemes. On average, results obtained showed a prediction rate gain of 4.42 %.

6   Conclusion

In this paper we presented a time and memory-bounded solution for tracking the relevance of features based on the information theoretic concepts of Entropy and Symmetrical Uncertainty. We showed how these metrics can be successfully used to enhance k-Nearest Neighbor, Naive Bayesian and Hoeffding Adaptive Tree algorithms during both stable and feature drifting regions of data streams. Empirical evidence shows that the gains in prediction accuracy are significant and occur in both synthetic and real-world datasets. Results point out the need for future research into feature drift detection and adaptation.

Both Entropy and Symmetrical Uncertainty are computed along a sliding window, thus allowing adaptation to feature drifts. Finding an optimal window size is a trade-off without solution, thus, future works include the adopting of change detectors (e.g. ADWIN [8] and EWMA [24]) to eliminate the need of a predefined window size, which is a drawback of the proposed method.

Finally, there is the need to investigate the usage of these adaptive metrics (Entropy and Symmetrical Uncertainty) for the task of dynamic feature selection for data streams. This would allow a generic filter method that does not depend on any specific base classifier and that would select features dynamically according to the occurrence of feature drifts.

References

1. Aggarwal, C.C.: An introduction to data classification. In: Data Classification: Algorithms and Applications, pp. 1–36 (2014). http://www.crcnetbase.com/doi/abs/10.1201/b17320-2
2. Aggarwal, C.C., Han, J., Wang, J., Yu, P.S.: A framework for clustering evolving data streams. In: Proceedings of the 29th International Conference on Very Large Data Bases, VLDB 2003, vol. 29. pp. 81–92. VLDB Endowment (2003). http://dl.acm.org/citation.cfm?id=1315451.1315460
3. Aggarwal, C.C., Hinneburg, A., Keim, D.A.: On the surprising behavior of distance metrics in high dimensional space. In: Van den Bussche, J., Vianu, V. (eds.) ICDT 2001. LNCS, vol. 1973, pp. 420–434. Springer, Heidelberg (2001)
4. Agrawal, R., Imielinski, T., Swami, A.: Database mining: a performance perspective. IEEE Trans. Knowl. Data Eng. 5(6), 914–925 (1993)
5. Aha, D., Kibler, D.: Instance-based learning algorithms. Mach. Learn. 6, 37–66 (1991)
6. Barddal, J.P., Gomes, H.M., Enembreck, F.: Analyzing the impact of feature drifts in streaming learning. In: Arik, S., Huang, T., Lai, W.K., Liu, Q. (eds.) ICONIP 2015. LNCS, pp. 21–28. Springer, Heidelberg (2015). doi:10.1007/978-3-319-26532-2_3
7. Barddal, J.P., Gomes, H.M., Enembreck, F.: A survey on feature drift adaptation. In: Proceedings of the International Conference on Tools with Artificial Intelligence. IEEE, November 2015
8. Bifet, A., Gavaldà, R.: Learning from time-changing data with adaptive windowing. In. SIAM International Conference on Data Mining (2007)
9. Bifet, A., Holmes, G., Kirkby, R., Pfahringer, B.: MOA: Massive online analysis. J. Mach. Learn. Res. 11, 1601–1604 (2010)
10. Bifet, A., Read, J., Žliobaitė, I., Pfahringer, B., Holmes, G.: Pitfalls in benchmarking data stream classification and how to avoid them. In: Blockeel, H., Kersting, K., Nijssen, S., Železný, F. (eds.) ECML PKDD 2013. LNCS (LNAI), pp. 465–479. Springer, Heidelberg (2013). doi:10.1007/978-3-642-40988-2_30
11. Chen, L., Wang, S.: Automated feature weighting in naive bayes for high-dimensional data classification. In: Proceedings of the 21st ACM International Conference on Information and Knowledge Management, CIKM 2012, pp. 1243–1252. ACM, New York (2012). http://doi.acm.org/10.1145/2396761.2398426
12. Corder, G., Foreman, D.: Nonparametric Statistics for Non-Statisticians: A Step-by-Step Approach. Wiley, Hoboken (2011). http://books.google.com.br/books?id=T3qOqdpSz6YC

13. Domingos, P., Hulten, G.: Mining high-speed data streams. In: Proceedings of the Sixth ACM SIGKDD International Conference on Knowledge Discovery and Data Mining, KDD 2000, pp. 71–80. ACM, New York (2000). http://doi.acm.org/10. 1145/347090.347107
14. Enembreck, F., Avila, B.C., Scalabrin, E.E., Barthès, J.P.A.: Learning drifting negotiations. Appl. Artif. Intell. **21**(9), 861–881 (2007). http://dblp.uni-trier.de/ db/journals/aai/aai21.html#EnembreckASB07
15. Gama, J., Rodrigues, P.: Issues in evaluation of stream learning algorithms. In: Proceedings of the 15th ACM SIGKDD International Conference on Knowledge Discovery and Data Mining, ACM SIGKDD, pp. 329–338, June 2009
16. Gama, J., Pinto, C.: Discretization from data streams: applications to histograms and data mining. In: Proceedings of the 2006 ACM Symposium on Applied Computing, SAC 2006, pp. 662–667. ACM, New York (2006). http://doi.acm.org/10. 1145/1141277.1141429
17. Gama, J., Zliobaite, I., Bifet, A., Pechenizkiy, M., Bouchachia, A.: A survey on concept drift adaptation. ACM Comput. Surv. **46**(4), 44:1–44:37 (2014). http:// doi.acm.org/10.1145/2523813
18. Gama, J.: Knowledge Discovery from Data Streams, 1st edn. Chapman & Hall/CRC, Boca Raton (2010)
19. Hoeffding, W.: Probability inequalities for sums of bounded random variables. J. Am. Stat. Assoc. **58**(301), 13–30 (1963). http://www.jstor.org/stable/2282952?
20. Hulten, G., Spencer, L., Domingos, P.: Mining time-changing data streams. In: Proceedings of the Seventh ACM SIGKDD International Conference on Knowledge Discovery and Data Mining, KDD 2001, pp. 97–106. ACM, New York (2001). http://doi.acm.org/10.1145/502512.502529
21. Katakis, I., Tsoumakas, G., Vlahavas, I.: Dynamic feature space and incremental feature selection for the classification of textual data streams. In: ECML/PKDD-2006 International Workshop on Knowledge Discovery from Data Streams, vol. 4213, p. 107. Springer (2006)
22. Nguyen, H.-L., Woon, Y.-K., Ng, W.-K., Wan, L.: Heterogeneous ensemble for feature drifts in data streams. In: Tan, P.-N., Chawla, S., Ho, C.K., Bailey, J. (eds.) PAKDD 2012. LNCS (LNAI), vol. 7302, pp. 1–12. Springer, Heidelberg (2012). doi:10.1007/978-3-642-30220-6_1
23. Rodrigues, P., Gama, J., Pedroso, J.: Hierarchical clustering of time-series data streams. IEEE Trans. Knowl. Data Eng. **20**(5), 615–627 (2008)
24. Ross, G.J., Adams, N.M., Tasoulis, D.K., Hand, D.J.: Exponentially weighted moving average charts for detecting concept drift. ArXiv e-prints (2012)
25. Sovdat, B.: Updating formulas and algorithms for computing entropy and gini index on time-changing data streams. CoRR abs/1403.6348 (2014). http://arxiv. org/abs/1403.6348
26. Street, W.N., Kim, Y.: A streaming ensemble algorithm (sea) for large-classification. In: Proceedings of the Seventh ACM SIGKDD International Conference on Knowledge Discovery and Data Mining, ACM SIGKDD, pp. 377–382, August 2001
27. Widmer, G., Kubat, M.: Learning in the presence of concept drift and hidden contexts. Mach. Learn. **23**(1), 69–101 (1996)
28. Yu, L., Liu, H.: Feature selection for high-dimensional data: a fast correlation-based filter solution. In: Proceedings of the Twentieth International Conference on Machine Learning, pp. 856–863. AAAI Press (2003)
29. Zhao, Z., Morstatter, F., Sharma, S., Alelyani, S., Anand, A., Liu, H.: Advancing feature selection research. ASU Feature Sel. Repository, 1–28 (2010)

Modeling Sequential Preferences with Dynamic User and Context Factors

Duc-Trong Le[1]([✉]), Yuan Fang[2], and Hady W. Lauw[1]

[1] School of Information Systems, Singapore Management University,
Singapore, Singapore
ductrong.le.2014@phdis.smu.edu.sg, hadywlauw@smu.edu.sg
[2] Institute for Infocomm Research, Singapore, Singapore
yfang@i2r.a-star.edu.sg

Abstract. Users express their preferences for items in diverse forms, through their liking for items, as well as through the sequence in which they consume items. The latter, referred to as "sequential preference", manifests itself in scenarios such as song or video playlists, topics one reads or writes about in social media, etc. The current approach to modeling sequential preferences relies primarily on the sequence information, i.e., which item follows another item. However, there are other important factors, due to either the user or the context, which may dynamically affect the way a sequence unfolds. In this work, we develop generative modeling of sequences, incorporating dynamic user-biased emission and context-biased transition for sequential preference. Experiments on publicly-available real-life datasets as well as synthetic data show significant improvements in accuracy at predicting the next item in a sequence.

Keywords: Sequential preference · Generative model · User-biased emission · Context-biased transition

1 Introduction

Users express their preferences in their consumption behaviors, through the products they purchase, the social media postings they like, the songs they listen to, the online videos they watch, etc. These behaviors are leaving increasingly greater traces of data that could be analyzed to model user preferences. Modeling these preferences has important applications, such as estimating consumer demand, profiling customer segments, or supporting product recommendation.

There are diverse forms of expression of preferences yielding different types of observations. Most of the previous works deal with *ordinal preference*, where the objective is to model the observed interactions between users and items [1]. In this scenario, a user's preference for an item is commonly expressed along some ordinal scale, e.g., higher rating indicating greater liking or preference.

Electronic supplementary material The online version of this chapter (doi:10.1007/978-3-319-46227-1_10) contains supplementary material, which is available to authorized users.

© Springer International Publishing AG 2016
P. Frasconi et al. (Eds.): ECML PKDD 2016, Part II, LNAI 9852, pp. 145–161, 2016.
DOI: 10.1007/978-3-319-46227-1_10

In this work, we are interested in another category, namely: *sequential prefer-ence*, where the objective is to model the sequential effect between adjacent items in a sequence. In this scenario, preference is expressed in terms of which other items may be preferred after consuming an item. For instance, a user's stream of tweets may reveal which topics tend to follow a topic, e.g., commenting on pol-itics upon reading morning news followed by more professional postings during working hours. The sequence of songs one listens to may express a preference for which genre follows another, e.g., more upbeat tempo during a workout followed by slower music while cooling down. Similarly, sequential preferences may also manifest in the books one reads, the movies one watches, etc.

Problem. Given a set of item sequences, we seek a probabilistic model for *sequential preferences*, so as to estimate the likelihood of future items in any par-ticular sequence. Each sequence (e.g., a playlist, a stream of tweets) is assumed to have been generated by a single user.

To achieve this goal, we turn to probabilistic models for *general* sequences. While there are several such models studied in the literature (see Sect. 2), here we build on the foundation of the well-accepted Hidden Markov Model (HMM) [16], which has been shown to be effective in various applications, including speech-and handwriting-recognition, etc. We review HMM in Sect. 3. Briefly, it models a number of hidden states. To generate each sequence, we move from one state to another based on *transition* probability. Each item in the sequence is sampled from the corresponding state's *emission* probability.

While HMM is fundamentally sound as a basic model for sequences, we iden-tify two significant factors, yet unexploited, which would contribute towards greater effectiveness for modeling sequential preferences. *First*, the generation of an item from a state's emission in HMM is only dependent on the state. However, as we are concerned with *user-generated* sequences, the selection of items may be affected by the user's preferences. However, due to the sparsity of information on individual users, we stop short of modeling individual emissions. Rather, we model latent groups, whereby users in the same group share similar preferences over items, i.e., emissions. *Second*, the transition to the next state in HMM is only dependent on the previous state. We posit that *context* in which a transition is about to take place also plays a role. For example, in the scenario of musical playlists, let us suppose that a particular state represents the genre of soft rock. There are different songs in this genre. If a user likes the artist of the current song, she may wish to listen to more songs by the same artist. Otherwise, she may wish to change to a different genre altogether. In this case, the artist is an observed *feature* of the context that may influence the transition dynamically.

Contributions. In this work, we make the following contributions. *First*, we develop a probabilistic model for sequences, whereby transitions from one state to another state may be *dynamically* influenced by the context features, and emissions are influenced by latent groups of users. We develop this model sys-tematically in Sect. 4, and describe how to learn the model parameters, as well as to generate item predictions in Sect. 5. *Second*, we evaluate these models com-prehensively in Sect. 6 over varied datasets. Experiments on a synthetic dataset investigate the contributions of our innovations on a dataset with known parameters. Experiments on publicly available real-life sequence datasets (song playlists from Yes.com and hashtag sequences from Twitter.com) further showcase accuracy improvements in predicting the next item in sequences.

2   Related Work

Here, we survey the literature on modeling various types of user preferences.

Ordinal Preferences. First, we look at *ordinal preferences*, which models a user's preference for an item in terms of rating or ranking. The most common framework is matrix factorization [11,17,20], where the observed user-by-item rating matrix is factorized into a number of latent factors, so as to enable prediction of missing values. Another framework is restricted Boltzmann machines [21] based on neural networks. Meanwhile, latent semantic analysis [8,9] models the association among users, items, and ratings via multinomial probabilities. These works stand orthogonally to ours, as the main interactions they seek to model are user-to-item ratings/rankings, rather than item-to-item sequences.

Sequential Preferences. Our work falls into *sequential preferences*, which models sequences of items, so as to enable prediction of future items. As mentioned in Sect. 1, our contribution is in factoring dynamic context-biased transition and user-biased emission. To make the effects of these dynamic factors clear, we build on the foundation of HMM [16], and focus our comparisons against this base platform. Aside from HMM, there could potentially be different ways to tackle this problem such as probabilistic automata [7] and recurrent neural networks [14], which are beyond the scope of this paper. Other works deal with sequences, but with different objectives. Markov decision processes [2,22,23] are concerned with how to make use of the transitions to arrive at an "optimal policy": a plan of actions to maximize some utility function. Sequential pattern mining [15] finds frequent sequential patterns, but these require exact matches of items in sequences. [4,13] model sequences in terms of Euclidean distances in metric embedding space. Aside from different objectives, these works also model explicit transitions among items, in contrast to our modeling of latent states.

Hybrid Models. Efforts to integrate ordinal and sequential preferences combine the "long-term" (items a user generally likes) and "short-term" preferences (items frequently consumed within a session). [27] models the problem as random walks in a session-based temporal graph. [26] designs a two-layer representation model for items: the first layer models interaction with previous item and the second layer models interaction with the user. [6,18] conduct joint factorization of user-by-item rating matrix and item-by-item transition matrix. It is not the focus of our current work to incorporate ordinal preferences directly, or to rely on full personalization by associating each user with an individual parameter.

Temporal Models. Aside from the notion of sequence, there are other *temporal factors* affecting recommendation. [19] assumes that users may change their

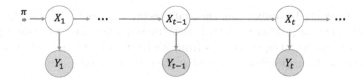

Fig. 1. A standard HMM for sequential preferences

ordinal preferences over time. [3] models the scenario where users "lose interest" over time. [10] takes into account the life stage of a consumer, e.g., products for babies of different ages, while [28] intends to model evolutions that advance "forward" in event sequences without going "backward". [25] seeks to predict not what, but rather when to recommend an item. [5] considers how changes in social relationships over time may affect a user's receptiveness or interest to change. In these and other cases, the key relationship being modeled is that between user and time, which is orthogonal to our focus in modeling item sequences.

3   Preliminaries

Towards capturing sequential preferences, our model builds upon HMM. The standard HMM assumes a series of discrete time steps $t = 1, 2, \ldots$, where an item Y_t can be observed at step t. To model the sequential effect in this series of observed items, HMM employs a Markov chain over a *latent* finite state space across the time steps. As illustrated in Fig. 1, at each time step t a latent state X_t is transitioned from the previous state X_{t-1} in a Markovian manner, i.e., $P(X_t|X_{t-1}, X_{t-2}, \ldots, X_1) \equiv P(X_t|X_{t-1})$, known as the *transition* probability.

Formally, consider an HMM with a set of observable items $\mathcal{Y}$ and a set of latent states $\mathcal{X}$. It can be fully specified by a triplet of parameters $\theta = (\pi, A, B)$, such that $\forall x, u \in \mathcal{X}, y \in \mathcal{Y}, t \in \{1, 2, \ldots\}$,

- π is the initial state distribution with $\pi_x \triangleq P(X_1 = x)$;
- A is the transition matrix with $A_{xu} = P(X_t = u|X_{t-1} = x)$;
- B is the emission matrix with $B_{xy} = P(Y_t = y|X_t = x)$.

Given a sequence of items $Y_1, \ldots, Y_t$, the optimal parameters θ^* can be learned by maximum likelihood (Eq. 1). Note that we can easily extend the likelihood function to accommodate multiple sequences, but for simplicity we only demonstrate with a single sequence throughout the technical discussion. Moreover, given θ^* and a sequence of items $Y_1, \ldots, Y_t$, the next item y^* can be predicted by maximum a posteriori probability (Eq. 2). Both learning and prediction can be efficiently solved using the forward-backward algorithm [16].

$$\theta^* = \arg\max_\theta P(Y_1, ..., Y_t; \theta) \tag{1}$$

$$y^* = \arg\max_y P(Y_{t+1} = y|Y_1, \ldots, Y_t; \theta^*) \tag{2}$$

4   Proposed Models

In a standard HMM, item emission probabilities are invariant across users, and state transition probabilities are independent of contexts at different times. However, these assumptions often deviate from real-world scenarios, in which different users and contexts may have important bearing on emissions and transitions. In this section, we model dynamic emissions and transitions respectively, and ultimately jointly, to better capture sequential preferences.

4.1   Modeling Dynamic User-Biased Emissions (SEQ-E)

It is often attractive to consider personalized preferences [18], where different user sequences may exhibit different emissions even though they share a similar transition. For instance, while two users both transit from soft rock to hard rock in their respective playlist, they might still choose songs of different artists in each genre. As another example, two users both transit from spring to summer in their apparel purchases, but still prefer different brands in each season. However, a fully personalized model catered to every individual user is often impractical due to inadequate training data for each user. We hypothesize that there exist different groups such that users across groups manifest different emission probabilities, whereas users in the same group share the same emission probabilities.

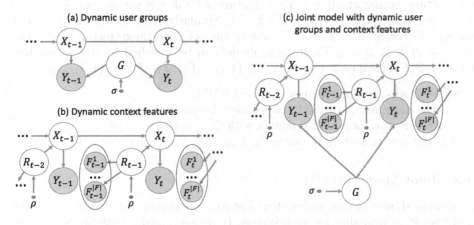

Fig. 2. Sequential models with dynamic user groups and contexts

In Fig. 2(a), we introduce a variable G_u to represent the group assignment of each user u. For simplicity, our technical formulation presents a single sequence and hence only one user. Thus, we omit the user notation u when no ambiguity arises. Assuming a set of groups $\mathcal{G}$, the new model can be formally specified by the parameters (π, σ, A, B), such that $\forall x \in \mathcal{X}, y \in \mathcal{Y}, g \in \mathcal{G}, t \in \{1, 2, \ldots\}$,

- π and A are the same as in a standard HMM;
- σ is the group distribution with $\sigma_g = P(G = g)$;
- B is the new emission tensor with $B_{gxy} = P(Y_t = y | X_t = x, G = g)$.

4.2   Modeling Dynamic Context-Biased Transitions (SEQ-T)

In standard HMM, the transition matrix is invariant over time. In real-world applications, this assumption may not hold. The transition probability may change depending on contexts that vary with time. Consider modeling a playlist of songs, where the transitions between genres are captured. The transition probabilities could be influenced by characteristics of the current song (e.g., artist, lyrics and sentiment). A fan of the current artist may break her usual pattern of genre transition and stick to genres by the same artist for the next few songs. As another example, a user purchasing apparels throughout the year may follow seasonal transitions. If satisfied with certain qualities (e.g., material and style) of past purchases, she may buy more such apparels out of season to secure discounts, breaking the usual seasonal pattern. We call such characteristics *context features*.

It is infeasible to differentiate transition probabilities by individual context features directly, which would blow up the parameter space and thus pose serious computational and data sparsity obstacles. Instead, we propose to model a single context factor that directly influences the next transition. The context factor, being latent, manifests itself through the observable context features.

As illustrated in Fig. 2(b), consider a set of context features $F = \{F^1, F^2, \ldots\}$. As feature values vary over time, let $F_t = (F_t^1, F_t^2, \ldots)$ denote the feature vector at time t. Each feature F^i takes a set of values $\mathcal{F}^i$, i.e., $F_t^i \in \mathcal{F}^i, \forall i \in \{1, ..., |F|\}, t \in \{1, 2, \ldots\}$. Similarly, let R_t denote the latent context factor at time t, and $\mathcal{R}$ denote the set of context factor levels, i.e., $R_t \in \mathcal{R}, \forall t \in \{1, 2, \ldots\}$. Finally, the model can be specified by the parameters (π, ρ, A, B, C), such that $\forall x, u \in \mathcal{X}, i \in \{1, \ldots, |F|\}, f \in \mathcal{F}_i, t \in \{1, 2, \ldots\}$,

- π and B are the same as in a standard HMM;
- ρ is the distribution of the latent context factor with $\rho_r = P(R_t = r)$;
- C is the feature probability matrix with $C_{rif} = P(F_t^i = f | R_t = r)$;
- A is the new transition tensor with $A_{rxu} = P(X_t = u | X_{t-1} = x, R_{t-1} = r)$.

4.3   Joint Model (SEQ*)

As discussed, user groups and context features can dynamically bias the emission and transition probabilities, respectively. Here, we consider both users and contexts in a joint model, as shown in Fig. 2(c). Accounting for all the parameters defined earlier, the joint model is specified by a six-tuple $\theta = (\pi, \sigma, \rho, A, B, C)$. The algorithm for learning and inference will be discussed in the next section.

5   Learning and Prediction

We now present efficient learning and prediction algorithms for the joint model. Note that the user and context-biased models are only degenerate cases of the joint model— the former assumes one context factor level (i.e., $|\mathcal{R}| = 1$) and no features (i.e., $F = \emptyset$), whereas the latter assumes one user group (i.e., $|\mathcal{G}| = 1$).

5.1 Parameter Learning

The goal of learning is to optimize the parameters $\theta = (\pi, \sigma, \rho, A, B, C)$ through maximum likelihood, given the observed items and features. Consider a sequence of $T > 1$ time steps. Let $\underline{Y} \triangleq (Y_1, \ldots, Y_T)$ as a shorthand; and similarly for $\underline{F}, \underline{X}, \underline{R}$. Subsequently, the optimal parameters can be obtained as follows.

$$\theta^* = \arg\max_\theta \log P(\underline{Y}, \underline{F}; \theta) \tag{3}$$

We demonstrate with one sequence for simpler notations. The algorithm can be trivially extended to enable multiple sequences as briefly described later.

Expectation Maximization (EM). We apply the EM algorithm to solve the above optimization problem. Each iteration consists of two steps below.

– **E-step.** Given parameters θ' from the last iteration (or random ones in the first iteration), calculate the expectation of the log likelihood function:

$$Q(\theta|\theta') = \sum_{\underline{X}, G, \underline{R}} P(\underline{X}, G, \underline{R}|\underline{Y}, \underline{F}; \theta') \log P(\underline{Y}, \underline{F}, \underline{X}, G, \underline{R}; \theta') \tag{4}$$

– **M-step.** Update the parameters $\theta = \arg\max_\theta Q(\theta|\theta')$.

Given the graphical model in Fig. 2(c), the joint probability $P(\underline{Y}, \underline{F}, \underline{X}, G, \underline{R})$ can be factorized as

$$P(G)P(X_1) \cdot \prod_{t=1}^{T} \left(P(Y_t|G, X_t)P(R_t) \prod_{i=1}^{|F|} P(F_t^i|R_t) \right) \cdot \prod_{t=1}^{T-1} P(X_{t+1}|X_t, R_t). \tag{5}$$

Maximizing the expectation $Q(\theta|\theta')$ is equivalent to maximize the following, assuming that $Y_t = y_t$ and $F_t^i = f_t^i$ are observed, $\forall t \in \{1, \ldots, T\}, i \in \{1, \ldots, |F|\}$.

$$\sum_{x \in \mathcal{X}} P(X_1 = x|\underline{Y}, \underline{F}; \theta') \log \pi_x + \sum_{g \in \mathcal{G}} P(G = g|\underline{Y}, \underline{F}; \theta') \log \sigma_g$$

$$+ \sum_{t=1}^{T} \sum_{r \in \mathcal{R}} P(R_t = r|\underline{Y}, \underline{F}; \theta') \log \rho_r$$

$$+ \sum_{t=1}^{T-1} \sum_{x \in \mathcal{X}} \sum_{u \in \mathcal{X}} \sum_{r \in \mathcal{R}} P(R_t = r, X_t = x, X_{t+1} = u|\underline{Y}, \underline{F}; \theta') \log A_{rxu}$$

$$+ \sum_{t=1}^{T} \sum_{x \in \mathcal{X}} \sum_{g \in \mathcal{G}} P(X_t = x, G = g|\underline{Y}, \underline{F}; \theta') \log B_{gxy_t}$$

$$+ \sum_{t=1}^{T} \sum_{i=1}^{|F|} \sum_{r \in \mathcal{R}} P(R_t = r|\underline{Y}, \underline{F}; \theta') \log C_{rif_t^i} \tag{6}$$

The optimization problem is further constrained by laws of probability, such that $\sum_{x \in \mathcal{X}} \pi_x = 1, \sum_{g \in \mathcal{G}} \sigma_g = 1, \sum_{r \in \mathcal{R}} \rho_r = 1, \sum_{u \in \mathcal{X}} A_{rxu} = 1, \sum_{y \in \mathcal{Y}} B_{gxy} = 1$ and $\sum_{f \in \mathcal{F}^i} C_{rif} = 1$. Applying Lagrange multipliers, we can derive the following updating rules.

$$\pi_x = \frac{P(X_1 = x|\underline{Y}, \underline{F}; \theta')}{1} = \frac{\sum_{g \in \mathcal{G}} \sum_{r \in \mathcal{R}} \gamma_{gxr}(1)}{1},$$ (7)

$$\sigma_g = \frac{P(G = g|\underline{Y}, \underline{F}; \theta')}{1} = \frac{\sum_{x \in \mathcal{X}} \sum_{r \in \mathcal{R}} \gamma_{gxr}(1)}{1},$$

$$\rho_r = \frac{\sum_{t=1}^{T} P(R_t = r|\underline{Y}, \underline{F}; \theta')}{\sum_{t=1}^{T} \sum_{k \in \mathcal{R}} P(R_t = k|\underline{Y}, \underline{F}; \theta')} = \frac{\sum_{g \in \mathcal{G}} \sum_{x \in \mathcal{X}} \sum_{t=1}^{T} \gamma_{gxr}(t)}{T},$$

$$A_{rxu} = \frac{\sum_{t=1}^{T-1} P(R_t = r, X_t = x, X_{t+1} = u|\underline{Y}, \underline{F}; \theta')}{\sum_{t=1}^{T-1} P(R_t = r, X_t = x|\underline{Y}, \underline{F}; \theta')} = \frac{\sum_{t=1}^{T-1} \sum_{g \in \mathcal{G}} \xi_{gxur}(t)}{\sum_{t=1}^{T-1} \sum_{g \in \mathcal{G}} \gamma_{gxr}(t)},$$

$$B_{gxy} = \frac{\sum_{t=1}^{T} P(X_t = x, G = g|\underline{Y}, \underline{F}; \theta')I(y_t = y)}{\sum_{t=1}^{T} P(X_t = x, G = g|\underline{Y}, \underline{F}; \theta')} = \frac{\sum_{t=1}^{T} \sum_{r \in \mathcal{R}} \gamma_{gxr}(t)I(y_t = y)}{\sum_{t=1}^{T} \sum_{r \in \mathcal{R}} \gamma_{gxr}(t)},$$

$$C_{rif} = \frac{\sum_{t=1}^{T} P(R_t = r|\underline{Y}, \underline{F}; \theta')I(f_t^i = f)}{\sum_{t=1}^{T} P(R_t = r|\underline{Y}, \underline{F}; \theta')} = \frac{\sum_{t=1}^{T} \sum_{g \in \mathcal{G}} \sum_{x \in \mathcal{X}} \gamma_{gxr}(t)I(f_t^i = f)}{\sum_{t=1}^{T} \sum_{g \in \mathcal{G}} \sum_{x \in \mathcal{X}} \gamma_{gxr}(t)},$$

where $I(\cdot)$ is an indicator function and

$$\gamma_{gxr}(t) \triangleq P(G = g, X_t = x, R_t = r|\underline{Y}, \underline{F}; \theta'),$$ (8)

$$\xi_{gxur}(t) \triangleq P(G = g, X_t = x, X_{t+1} = u, R_t = r|\underline{Y}, \underline{F}; \theta').$$ (9)

Note that, to account for multiple sequences, in each updating rule we need to respectively sum up the denominator and numerator over all the sequences.

Inference. To efficiently apply the updating rules, we must solve the inference problems for $\gamma_{gxr}(t)$ and $\xi_{gxur}(t)$ in Eqs. 8 and 9. Towards these two goals, similar to the forward-backward algorithm [16] for the standard HMM, we first need to support the efficient computation of the below probabilities.

$$\alpha_{gxr}(t) = P(Y_1, \ldots, Y_t, F_1, \ldots, F_t, X_t = x, G = g, R_t = r; \theta')$$ (10)

$$\beta_{gxr}(t) = P(Y_{t+1}, \ldots, Y_T, F_{t+1}, \ldots, F_T | X_t = x, G = g, R_t = r; \theta')$$ (11)

Letting $\theta' = (\pi', \sigma', \rho', A', B', C')$ and $C'(r, t) = \prod_{i=1}^{|F|} C'_{rif_t^i}$, both probabilities can be computed recursively, as follows.

$$\alpha_{gxr}(t) = \begin{cases} \pi'_x \sigma'_g \rho'_r C'(r, 1) B'_{gxy_1}, & t = 1 \\ \rho'_r C'(r, t) B'_{gxy_t} \sum_{u \in \mathcal{X}} \sum_{k \in \mathcal{R}} \alpha_{guk}(t-1) A'_{kux}, & \text{else} \end{cases}$$ (12)

$$\beta_{gxr}(t) = \begin{cases} B'_{gxy_T} C'(r, T), & t = T - 1 \\ \sum_{k \in \mathcal{R}} \rho'_k C'(k, t+1) \sum_{u \in \mathcal{X}} B'_{guy_{t+1}} A'_{rxu} \beta_{guk}(t+1), & \text{else} \end{cases}$$ (13)

Subsequently, $\gamma_{gxr}(t)$ and $\xi_{gxur}(t)$ can be further computed.

$$\xi_{gxur}(t) = \frac{\alpha_{gxr}(t) A'_{xur} B'_{guy_{t+1}} \sum_{k \in \mathcal{R}} \beta_{guk}(t+1) \rho'_k C'(k, t+1)}{\sum_{h \in \mathcal{G}} \sum_{v \in \mathcal{X}} \sum_{k \in \mathcal{R}} \alpha_{hvk}(T)}$$ (14)

$$\gamma_{gxr}(t) = \begin{cases} \sum_{x \in \mathcal{X}} \xi_{gxur}(t) & t = T \\ \sum_{u \in \mathcal{X}} \xi_{gxur}(t) & \text{else} \end{cases}$$ (15)

5.2 Item Prediction

Once the parameters are learnt, we can predict the next item of a user given her existing sequence of items $\{Y_1, Y_2, ..., Y_t\}$ and context features $\{F_1, F_2, ..., F_t\}$. In particular, her next item y^* can be chosen by maximum a posteriori estimation:

$$
\begin{aligned}
y^* &= \arg\max_y P(Y_{t+1} = y | Y_1, \ldots, Y_t, F_1, ..., F_t) \\
&= \arg\max_y P(Y_1, \ldots, Y_t, Y_{t+1} = y, F_1, ..., F_t) \\
&= \arg\max_y P(Y_1, \ldots, Y_t, Y_{t+1} = y, F_1, ..., F_t, F_{t+1})/P(F_{t+1}) \\
&= \arg\max_y \sum_{g \in \mathcal{G}} \sum_{x \in \mathcal{X}} \sum_{r \in \mathcal{R}} \alpha_{gxr}(t+1).
\end{aligned} \tag{16}
$$

While we do not observe features at time $t+1$, in the above we can adopt any value for F_{t+1} which does not affect the prediction. Instead of picking the best candidate item, we can rank all the candidates and suggest the top-K items.

5.3 Complexity Analysis

We conduct a complexity analysis for learning the joint model SEQ*. Consider one sequence of length T with $|\mathcal{X}|$ states, $|\mathcal{Y}|$ items, $|\mathcal{G}|$ user groups, $|\mathcal{R}|$ context factor levels, $|F|$ features and $|\mathcal{F}|$ values for each feature. For this one sequence, the complexity of one iteration of the EM is contributed by three main steps:

- *Step 1:* Calculate α, β: $O\left(T|\mathcal{G}||\mathcal{X}||\mathcal{R}|^2(|\mathcal{X}| + |F|)\right)$. Because $\rho'_r, C'(r, t)$ in Eq. 12 are independent of g, x, u, k while $\rho'_k, C'(k, t+1)$ in Eq. 13 are independent of g, x, u, r, we can further simplify this to: $O\left(T|\mathcal{R}|(|\mathcal{G}||\mathcal{X}|^2|\mathcal{R}| + |F|)\right)$.
- *Step 2:* Calculate ξ, γ using α, β: $O\left(T|\mathcal{G}||\mathcal{X}|^2|\mathcal{R}|^2|F|\right)$. As $\rho'_k C'(k, t+1)$ in Eq. 14 is independent of g, x, u, r, we reduce it to: $O\left(T|\mathcal{R}|(|\mathcal{G}||\mathcal{X}|^2|\mathcal{R}| + |F|)\right)$.
- *Step 3:* Update θ using γ, ξ: $O\left(T|\mathcal{G}||\mathcal{X}||\mathcal{R}|(|\mathcal{X}| + |F|)\right)$. As y in B_{gxy} of Eq. 7 is independent of g, x, r, we first compute the denominator, and update a normalized score to y in the B_{gxy} while computing the numerator. Likewise, i, f in C_{rif} are independent of g, x, r. Thus, we have: $O\left(T|\mathcal{R}|(|\mathcal{G}||\mathcal{X}|^2 + |F|)\right)$.

The overall complexity of SEQ* is $O\left(T|\mathcal{R}|(|\mathcal{G}||\mathcal{X}|^2|\mathcal{R}| + |F|)\right)$ for one sequence, one iteration. The complexities of lesser models are (by substitution):

- HMM with $|\mathcal{G}| = |\mathcal{R}| = 1, |F| = |\mathcal{F}| = 0$: $O\left(T|\mathcal{X}|^2\right)$
- SEQ-E with $|\mathcal{R}| = 1, |F| = |\mathcal{F}| = 0$: $O\left(T|\mathcal{G}||\mathcal{X}|^2\right)$
- SEQ-T with $|\mathcal{G}| = 1$: $O\left(T|\mathcal{R}|(|\mathcal{X}|^2|\mathcal{R}| + |F|)\right)$

The result implies that the running times of our proposed models are quadratic in the number of states and context factor levels, while linear in all the other variables. HMM is also quadratic in the number of states. Comparing to HMM with the same number of states, our joint model incurs a quadratic increase in complexity only in the number of context factor levels (which is typically small), and merely a linear increase in the number of groups and context features.

6 Experiments

The objective of experiments is to evaluate *effectiveness*. We first look into a synthetic dataset to investigate whether context-biased transition and user-biased emission could have been simulated by increasing the number of HMM's states. Next, we experiment with two real-life, publicly available datasets, to investigate whether the models result in significant improvements over the baseline.

6.1 Setup

We elaborate on the general setup here, and describe the specifics of each dataset later in the appropriate sections. Each dataset has of a set of sequences. We create random splits of 80:20 ratio of training versus testing. In this sequential preference setting, a sequence (a user) is in either training or testing, but not necessarily in both. This is different from a fully personalized ordinal preference setting (a different framework altogether), where a user would be represented in both sets.

Task. For each sequence in the testing set, given the sequence save the last item, we seek to predict the last item. Each method generates a top-K recommendation, which is evaluated against the held-out ground-truth last item.

Comparative Methods. Since we build our dynamic context and user factors upon HMM, it is the most appropriate baseline. To investigate the contribution of user-biased emission and context-based transition *separately*, we compare the two models SEQ-E and SEQ-T respectively against the baseline. To see their contributions *jointly*, we further compare SEQ* against the baseline. In addition, we include the result of the frequency-based method FREQ as a reference, which simply choose the most popular item in the training data.

Metrics. We rely on two conventional metrics for top-K recommendation. Inspired by a similar evaluation task in [24], the first metric we use is *Recall@K*.

$$Recall@K = \frac{\text{number of sequences with the ground truth item in the top } K}{\text{total number of sequences in the testing set}}$$

If we assume the ground truth item to be the only true answer, average precision can be measured similarly (dividing by K) and would show the same trend as recall. In the experiments, we primarily study top 1 % recommendation, i.e., *Recall@1 %*, but will present results for several other K's as well. Actually, it is not clear that the other items in the top-K would really be rejected by a user [24]. Instead of precision, we rely on another metric.

The second metric is Mean Reciprocal Rank or *MRR*, defined as follows.

$$MRR = \frac{1}{|S_{\text{test}}|} \times \sum_{s \in S_{\text{test}}} \frac{1}{\text{rank of target item for sequence } s}$$

We prefer a method that places the ground-truth item higher in the top-K recommendation list. Because the contribution of a very low rank is vanishingly small, we cut the list off at 200, i.e., ranks ≥ 200 contribute zero to *MRR*. Realistically, a recommendation list longer than 200 is unlikely in realistic scenarios.

For each dataset, we create five random training/testing splits. For each "fold", we run the models ten times with different random initializations (but with common seeds across comparative methods for parity). For each method, we average the *Recall@K* and *MRR* across the fifty readings. All comparisons are verified by one-sided paired-sample Student's t-test at 0.05 significance level.

6.2 Synthetic Dataset

We begin with experiments on a synthetic dataset, for two reasons. First, one advantage of a synthetic dataset is the knowledge of the actual parameters (e.g., transition and emission probabilities), which allows us to verify our model's ability to recover these parameters. Second, we seek to verify whether the effects of context-biased transition and user-biased emission could have been simulated by increasing the number of hidden states of traditional sequence model HMM.

Dataset. We define a synthetic dataset with the following configuration: 2 groups ($|\mathcal{G}| = 2$), 2 states ($|\mathcal{X}| = 2$), 2 context factor levels ($|\mathcal{R}| = 2$), 4 items ($|\mathcal{Y}| = 4$), 4 features ($|F| = 4$) each with 2 feature values (present or absent).

The complete set of synthetic parameters are specified in the supplementary material. Here, we discuss the key ideas. A six-tuple $\theta = (\pi, \sigma, \rho, A, B, C)$ is specified as follows: $\pi = [0.8, 0.2]$, $\sigma = [0.9, 0.1]$, $\rho = [0.3, 0.7]$. The transition tensor A is such that we induce self-transition to the same state for the first context factor level, and switching to the other state for the second context factor level. The emission tensor B is such that the four (state, group) combinations each tend to generate one of the four items. The feature matrix C is such that each context factor level is mainly associated with two of the four features.

We then generate 10 thousand sequences, each of length 10 ($T = 10$). For each sequence, we first draw a group according to σ. At time $t = 1$, we draw the first hidden state X_1 from π, followed by drawing the first item Y_1 from B. We also draw a context factor level from ρ and generate features via C. For time $t = 2, \ldots, 10$, we follow the same process, but each hidden state is now drawn from A according to the previous state and context factor level at time $t - 1$.

Results. We run the four comparative methods on this synthetic dataset, fixing the context factor levels and groups to 2 for the relevant methods, while varying the number of states. Figure 3(a) shows the results in terms of *Recall@1*, i.e., the ability of each method in recommending the ground truth item as the top prediction. There are several crucial observations. *First*, the proposed model SEQ* outperforms the rest, attaining recall close to 85%, while the baseline HMM hovers around 65%. SEQ* also outperforms SEQ-T and SEQ-E.

Second, as we increase the number of states, most models initially increase in performance and then converge. Evidently, increasing the number of states alone does not lift the baseline HMM to the same level of performance as SEQ*

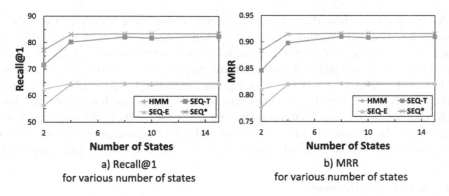

Fig. 3. Performance of comparative methods on Synthetic Data for *Recall@1* and *MRR*

or SEQ-T, indicating the effect of context-biased transition. Meanwhile, though SEQ-E and HMM are similar (due to inability to model context factor), SEQ* is slightly better than SEQ-T, indicating the contribution of user-biased emission. Figure 3(b) shows the results for *MRR*, showing similar trends and observations.

6.3 Real-Life Datasets

We now investigate the performance of the comparative methods on real-life, publicly available datasets covering two different domains: song playlists from online radio station Yes.com, and hashtag sequences from users' Twitter streams.

Playlists from Yes.com. We utilize the *yes_small* dataset[1] collected by [4]. The dataset includes about 430 thousand playlists, involving 3168 songs. Noticeably, the majority of playlits has length which is shorter than 30. To keep the playlist lengths relatively balanced, we filter out playlists with fewer than two songs and retain up to the first thirty songs in each playlist. Finally, we have 250 thousand playlists (sequences) consisting of 3168 unique songs (items).

Features. We study the effect of features on the context-biased transition model SEQ-T. Each song may have tags. There are 250 unique tags. We group tags with similar meanings (e.g., "male vocals" and "male vocalist"). As the first feature, we use a binary feature of whether the current song and the previous song shares at least one tag. For additional features, we use the most popular tags. Note that we never assume knowledge of the tags of the song to be predicted. Figure 4(a) shows the performance of SEQ-T, with two context factor levels, for various number of features. Figure 4(a) has dual vertical axes for *Recall@1 %* (left) and *MRR* (right) respectively. The trends for both metrics are similar: performance initially goes up and then stabilizes. In subsequent experiments, we use eleven features (similarity feature and ten most popular tags).

---

[1] http://www.cs.cornell.edu/~shuochen/lme/data_page.html.

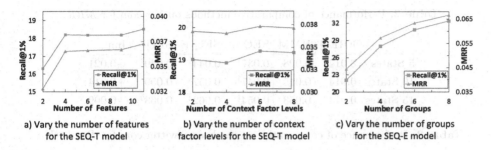

a) Vary the number of features
for the SEQ-T model

b) Vary the number of context
factor levels for the SEQ-T model

c) Vary the number of groups
for the SEQ-E model

Fig. 4. Effects of features, context factor on SEQ-T & groups on SEQ-E on Yes.com

Table 1. Performance of comparative methods on Yes.com for *Recall@K*

		FREQ	HMM	SEQ-T	SEQ-E	SEQ*	Imp.
5 States	Recall@1%	6.8	13.8	$18.4^\dagger$	$22.0^\S$	$24.1^{\dagger\S}$	+10.3
	Recall@50	9.6	19.2	$25.1^\dagger$	$29.5^\S$	$32.1^{\dagger\S}$	+13.0
	Recall@100	16.2	29.3	$37.0^\dagger$	$42.6^\S$	$46.1^{\dagger\S}$	+16.8
10 States	Recall@1%	6.8	22.3	$23.2^\dagger$	$27.8^\S$	$28.6^{\dagger\S}$	+6.3
	Recall@50	9.6	30.0	$31.1^\dagger$	$36.9^\S$	$38.1^{\dagger\S}$	+8.1
	Recall@100	16.2	43.4	$44.9^\dagger$	$52.1^\S$	$53.5^{\dagger\S}$	+10.2
15 States	Recall@1%	6.8	26.1	$26.5^\dagger$	$30.1^\S$	$30.6^{\dagger\S}$	+4.5
	Recall@50	9.6	34.7	$35.5^\dagger$	$39.4^\S$	$40.2^{\dagger\S}$	+5.5
	Recall@100	16.2	49.3	$50.8^\dagger$	$55.1^\S$	$56.3^{\dagger\S}$	+7.0

Context Factor. We then vary the number of context factor levels of SEQ-T (with eleven features). Figure 4(b) shows that for this dataset, there is not much gain from increasing the number of context factor levels beyond two. Therefore, for greater efficiency, subsequently we experiment with two context factor levels.

Latent Groups. We turn to the effect of latent groups on the user-biased emission model SEQ-E. Figure 4(c) shows the effect of increasing latent groups. More groups lead to better performance. Because of the diversity among sequences, having more groups increases the flexibility in modeling emissions while still sharing transitions. For the subsequent comparison to the baseline, we will experiment with two latent groups, as the earlier comparison has shown that the results with higher number of groups would be even higher.

Comparison to Baseline. We now compare the proposed models SEQ-T, SEQ-E, and SEQ* to the baseline HMM. Table 1 shows a comparison in terms of *Recall@K* for 5, 10, and 15 states. In addition to *Recall@1%* (corresponding to top 31), we also show results for *Recall@50* and *Recall@100*. The symbol † denotes statistical significance due to the effect of context-biased transition. In other words, the outperformance of SEQ-T over HMM, and that of SEQ* over SEQ-E, are significant. The symbol § denotes statistical significance due to the effect of user-biased emission, i.e., the outperformance of SEQ-E over HMM,

Table 2. Performance of comparative methods on Yes.com for *MRR*

	FREQ	HMM	SEQ-T	SEQ-E	SEQ*	Imp.
5 States	0.014	0.028	$0.037^{\dagger}$	$0.044^{\S}$	$0.049^{\dagger\S}$	+0.021
10 States	0.014	0.045	$0.047^{\dagger}$	$0.057^{\S}$	$0.059^{\dagger\S}$	+0.014
15 States	0.014	0.053	$0.054^{\dagger}$	$0.062^{\S}$	$0.063^{\S}$	+0.009

Table 3. Performance of comparative methods on Twitter.com for *Recall@K*

		FREQ	HMM	SEQ-T	SEQ-E	SEQ*	Imp.
5 States	Recall@1%	8.4	16.9	$17.1^{\dagger}$	$20.6^{\S}$	$21.0^{\dagger\S}$	+4.1
	Recall@50	16.1	28.3	$28.6^{\dagger}$	$33.2^{\S}$	$33.7^{\dagger\S}$	+5.4
	Recall@100	25.5	40.6	$40.9^{\dagger}$	$46.0^{\S}$	$46.5^{\dagger\S}$	+5.9
10 States	Recall@1%	8.4	21.8	$22.0^{\dagger}$	$26.5^{\S}$	$26.9^{\dagger\S}$	+5.1
	Recall@50	16.1	34.2	$34.4^{\dagger}$	$39.4^{\S}$	$39.8^{\dagger\S}$	+5.7
	Recall@100	25.5	47.2	$47.4^{\dagger}$	$52.0^{\S}$	$52.4^{\S}$	+5.2
15 States	Recall@1%	8.4	25.2	$25.3^{\dagger}$	$29.9^{\S}$	$30.0^{\dagger\S}$	+4.8
	Recall@50	16.1	38.1	$38.2^{\dagger}$	$43.1^{\S}$	$43.3^{\dagger\S}$	+5.1
	Recall@100	25.5	51.2	$51.3^{\dagger}$	$55.2^{\S}$	$55.3^{\dagger\S}$	+4.1

Table 4. Performance of comparative methods on Twitter.com for *MRR*

	FREQ	HMM	SEQ-T	SEQ-E	SEQ*	Imp.
5 States	0.019	0.045	$0.046^{\dagger}$	$0.062^{\S}$	$0.063^{\dagger\S}$	+0.0183
10 States	0.019	0.063	0.064	$0.084^{\S}$	$0.086^{\dagger\S}$	+0.0227
15 States	0.019	0.076	$0.078^{\dagger}$	$0.100^{\S}$	$0.101^{\dagger\S}$	+0.0246

and that of SEQ* over SEQ-T, are significant. Finally, our overall model SEQ* is significantly better than the baseline HMM in all cases. The absolute improvement of the former over the latter in additional percentage terms is shown in the *Imp.* column. For all models, more states generally translate to better performance, and the improvements are somewhat smaller but still significant. Table 2 shows a comparison in terms of *MRR*, where similar observations hold.

Hashtag Sequences from Twitter.com. We conduct similar experiments on the Twitter dataset[2] [12]. There are 130 thousand users. In our scenario, each sequence corresponds to the hashtags of a user. The average length of our dataset is 19. If a tweet has multiple hashtags, we retain the most popular one, so as to maintain the sequence among tweets. Similarly to the treatment of stop words and infrequent words in document modeling, we filter out hashtags that are too

---

[2] https://wiki.cites.illinois.edu/wiki/display/forward/Dataset-UDI-TwitterCrawl-Aug2012.

popular (frequency ≥ 25000) or relatively infrequent (frequency ≤ 1000). Finally, we obtain 114 thousand sequences involving 2121 unique hashtags. Similarly to Yes.com, we run the models for two levels of context factor and two latent groups, but with seven features extracted from the tweet of the current hashtag (not the one to be predicted): number of retweets, number of hashtags, time intervals to the previous one and two tweets, time interval to the next tweet, and edit distances with the previous one and two observations.

The task is essentially predicting the next hashtag in a sequence. In brief, Tables 3 and 4 support that the improvements due to context-biased transition (†) and user-biased emission (§) are mostly significant. Importantly, the overall improvements by SEQ* over the baseline HMM (*Imp.* column) are consistent and hold up across 5, 10, and 15 states for both *Recall@K* and *MRR*.

Computational efficiency is not the main focus of experiments. We comment briefly on the running times. For the Twitter dataset, the average learning time per iteration on Intel Xeon CPU X5460 3.16 GHz with 32 GB RAM for our models with 15 states, 2 groups, 2 context factor levels are 2, 3, and 6 min for SEQ-E, SEQ-T and SEQ* respectively. HMM requires less than a minute.

7   Conclusion

In this work, we develop a generative model for sequences, which models two types of dynamic factors. First, transition from one state to the next may be affected by context factor. This results in SEQ-T model, with context-biased transition. Second, we seek to incorporate how different latent user groups may have preferences for certain items. This results in SEQ-E model, with user-biased emission. Finally, we unify these two factors into a joint model SEQ*. Experiments on both synthetic and real-life datasets support the case that these dynamic factors contribute towards better performance than the baseline HMM (statistically significant) in terms of top-K recommendation for sequences.

Acknowledgments. This research is supported by the National Research Foundation, Prime Minister's Office, Singapore under its NRF Fellowship Programme (Award No. NRF-NRFF2016-07).

References

1. Adomavicius, G., Tuzhilin, A.: Toward the next generation of recommender systems: a survey of the state-of-the-art and possible extensions. IEEE Trans. Knowl. Data Eng. **17**(6), 734–749 (2005)
2. Brafman, R.I., Heckerman, D., Shani, G.: Recommendation as a stochastic sequential decision problem. In: Proceedings of the International Conference on Automated Planning and Scheduling (ICAPS), pp. 164–173 (2003)
3. Chen, J., Wang, C., Wang, J.: A personalized interest-forgetting Markov Model for recommendations. In: Proceedings of the AAAI Conference on Artificial Intelligence (AAAI), pp. 16–22 (2015)

4. Chen, S., Moore, J.L., Turnbull, D., Joachims, T.: Playlist prediction via metric embedding. In: Proceedings of the ACM SIGKDD Conference on Knowledge Discovery and Data Mining (KDD), pp. 714–722 (2012)
5. Chen, W., Hsu, W., Lee, M.L.: Modeling user's receptiveness over time for recommendation. In: Proceedings of the ACM SIGIR Conference (SIGIR), pp. 373–382 (2013)
6. Cheng, C., Yang, H., Lyu, M.R., King, I.: Where you like to go next: successive point-of-interest recommendation. In: Proceedings of the International Joint Conference on Artificial Intelligence (IJCAI) (2013)
7. Dupont, P., Denis, F., Esposito, Y.: Links between probabilistic automata and Hidden Markov Models: probability distributions, learning models and induction algorithms. Pattern Recogn. **38**(9), 1349–1371 (2005)
8. Hofmann, T.: Probabilistic latent semantic analysis. In: Proceedings of the Conference on Uncertainty in Artificial Intelligence (UAI), pp. 289–296 (1999)
9. Hofmann, T.: Latent semantic models for collaborative filtering. ACM Trans. Inf. Syst. (TOIS) **22**(1), 89–115 (2004)
10. Jiang, P., Zhu, Y., Zhang, Y., Yuan, Q.: Life-stage prediction for product recommendation in E-commerce. In: Proceedings of the ACM SIGKDD Conference on Knowledge Discovery and Data Mining (KDD), pp. 1879–1888 (2015)
11. Koren, Y., Bell, R., Volinsky, C.: Matrix factorization techniques for recommender systems. Computer **8**, 30–37 (2009)
12. Li, R., Wang, S., Deng, H., Wang, R., Chang, K.C.C.: Towards social user profiling: unified and discriminative influence model for inferring home locations. In: Proceedings of the ACM SIGKDD Conference on Knowledge Discovery and Data Mining (KDD), pp. 1023–1031 (2012)
13. Liu, X., Liu, Y., Aberer, K., Miao, C.: Personalized point-of-interest recommendation by mining users' preference transition. In: Proceedings of the ACM International Conference on Information and Knowledge Management (CIKM), pp. 733–738 (2013)
14. Mikolov, T., Karafit, M., Burget, L., Cernock, J., Khudanpur, S.: Recurrent neural network based language model. In: INTERSPEECH, vol. 2, p. 3 (2010)
15. Parameswaran, A.G., Koutrika, G., Bercovitz, B., Garcia-Molina, H.: Recsplorer: recommendation algorithms based on precedence mining. In: Proceedings of the International Conference on Management of Data (SIGMOD), pp. 87–98 (2010)
16. Rabiner, L.R., Juang, B.H.: An introduction to Hidden Markov Models. IEEE ASSP Mag. **3**(1), 4–16 (1986)
17. Rendle, S., Freudenthaler, C., Gantner, Z., Schmidt-Thieme, L.: BPR: Bayesian personalized ranking from implicit feedback. In: Proceedings of the Conference on Uncertainty in Artificial Intelligence (UAI), pp. 452–461 (2009)
18. Rendle, S., Freudenthaler, C., Schmidt-Thieme, L.: Factorizing personalized Markov chains for next-basket recommendation. In: Proceedings of the International World Wide Web Conference (WWW), pp. 811–820 (2010)
19. Sahoo, N., Singh, P.V., Mukhopadhyay, T.: A Hidden Markov Model for collaborative filtering. MIS Q. **36**(4), 1329–1356 (2012)
20. Salakhutdinov, R., Andriy, M.: Probabilistic matrix factorization. In: Proceedings of the Annual Conference on Neural Information Processing Systems (NIPS), vol. 21 (2008)
21. Salakhutdinov, R., Andriy, M., Hinton, G.: Restricted Boltzmann machines for collaborative filtering. In: Proceedings of the International Conference on Machine Learning (ICML), pp. 791–798 (2007)

22. Shani, G., Brafman, R.I., Heckerman, D.: An MDP-based recommender system. In: Proceedings of the Conference on Uncertainty in Artificial Intelligence (UAI), pp. 453–460 (2002)
23. Tavakol, M., Brefeld, U.: Factored MDPs for detecting topics of user sessions. In: Proceedings of the ACM Conference on Recommender Systems (RecSys), pp. 33–40 (2014)
24. Wang, C., Blei, D.M.: Collaborative topic modeling for recommending scientific articles. In: Proceedings of the ACM SIGKDD Conference on Knowledge Discovery and Data Mining (KDD), pp. 448–456 (2011)
25. Wang, J., Zhang, Y., Posse, C., Bhasin, A.: Is it time for a career switch? In: Proceedings of the International World Wide Web Conference (WWW), pp. 1377–1388 (2013)
26. Wang, P., Guo, J., Lan, Y., Xu, J., Wan, S., Cheng, X.: Learning hierarchical representation model for next basket recommendation. In: Proceedings of the ACM SIGIR Conference (SIGIR), pp. 403–412 (2015)
27. Xiang, L., Yuan, Q., Zhao, S., Chen, L., Zhang, X., Yang, Q., Sun, J.: Temporal recommendation on graphs via long-and short-term preference fusion. In: Proceedings of the ACM SIGKDD Conference on Knowledge Discovery and Data Mining (KDD), pp. 723–732 (2010)
28. Yang, J., McAuley, J., Leskovec, J., LePendu, P., Shah, N.: Finding progression stages in time-evolving event sequences. In: Proceedings of the International World Wide Web Conference (WWW), pp. 783–794 (2014)

Node Re-Ordering as a Means of Anomaly Detection in Time-Evolving Graphs

Lida Rashidi[1,2]([envelope]), Andrey Kan[2], James Bailey[2], Jeffrey Chan[3],
Christopher Leckie[1,2], Wei Liu[4], Sutharshan Rajasegarar[5],
and Kotagiri Ramamohanarao[2]

[1] Data61 Victoria Research Laboratory, Melbourne, Australia
lrashidi@student.unimelb.edu.au
[2] Department of Computing and Information Systems,
The University of Melbourne, Melbourne, Australia
{akan,baileyj,caleckie,kotagiri}@unimelb.edu.au
[3] Department of Computer Science and Information Technology,
RMIT University, Melbourne, Australia
jchan@rmit.edu.au
[4] Department of Engineering and IT,
University of Technology Sydney, Sydney, Australia
wei.liu@uts.edu.au
[5] School of Information Technology, Deakin University, Geelong, Australia
srajas@deakin.edu.au

Abstract. Anomaly detection is a vital task for maintaining and improving any dynamic system. In this paper, we address the problem of anomaly detection in time-evolving graphs, where graphs are a natural representation for data in many types of applications. A key challenge in this context is how to process large volumes of streaming graphs. We propose a pre-processing step before running any further analysis on the data, where we permute the rows and columns of the adjacency matrix. This pre-processing step expedites graph mining techniques such as anomaly detection, PageRank, or graph coloring. In this paper, we focus on detecting anomalies in a sequence of graphs based on rank correlations of the reordered nodes. The merits of our approach lie in its simplicity and resilience to challenges such as unsupervised input, large volumes and high velocities of data. We evaluate the scalability and accuracy of our method on real graphs, where our method facilitates graph processing while producing more deterministic orderings. We show that the proposed approach is capable of revealing anomalies in a more efficient manner based on node rankings. Furthermore, our method can produce visual representations of graphs that are useful for graph compression.

1 Introduction

Dynamic graphs are becoming ubiquitous formats for representing relational datasets such as social, collaboration, communication and computer networks.

© Springer International Publishing AG 2016
P. Frasconi et al. (Eds.): ECML PKDD 2016, Part II, LNAI 9852, pp. 162–178, 2016.
DOI: 10.1007/978-3-319-46227-1_11

One of the vital tasks for gaining an insight into the behavioral patterns of such datasets is anomaly detection. Anomaly detection in time-evolving graphs is the task of finding timestamps that correspond to an unusual event in a sequence of graphs [2]. For instance, a social network anomaly may correspond to the merging or splitting of its communities. Anomaly detection plays an important role in numerous applications, such as network intrusion detection, credit card fraud [9] and discontinuity detection in social networks [3].

However, there are many challenges associated with event detection in dynamic graphs. Networks such as Facebook or Twitter comprise billions of interacting users where the structure of the network is constantly updated. Moreover, there is often a lack of labels for normal and anomalous graph instances, which requires learning to be unsupervised. Due to these challenges, graph anomaly detection has attracted growing interest over time.

To address these challenges, many anomaly detection techniques use a pre-processing phase where they extract structural features from graph representations. These features may include node centrality [14], ego-nets [3] and eigenvalues [10]. They then apply well-known similarity measures to compare graph changes over a period of time. In this scenario, the graphs are converted into feature sets and therefore they do not pose the complexities associated with the inter-dependencies of nodes, in addition to causing a considerable decrease in the time and space requirements for the anomaly detection scheme.

However, the process of generating structure-aware features for graphs can be challenging in itself. For instance, the eigenvalues of a graph can be a suitable representation for its patterns of connectivity, but they have high storage and time requirements. A common shortcoming between these approaches is the need to perform matrix inversions, where the graphs are too sparse to be invertible. Another property of graph summarization techniques should be their interpretability. Revealing structural information such as communities, node roles or maximum independent sets can be very useful in further analysis of graphs.

To address these issues we propose an approach for detecting graph anomalies based on the ranking of the nodes. The novelty of our method lies in a scalable pre-processing scheme that produces stable results. Our matrix re-ordering approach efficiently assigns ranks to each node in the graph, where the resulting ranks can be used directly as a basis for comparing consecutive graph snapshots. Our re-ordering approach reduces the input dimension of a graph from $O(n^2)$ to $O(n)$. We can easily use a rank correlation coefficient as a similarity measure over pairs of graphs. Another advantage of our approach is its capability to produce interpretable results that identify large independent sets. The compact representation of the graphs yields faster and simpler anomaly detection schemes.

We review some of the algorithms previously introduced in the domain of graph anomaly detection in Sect. 2. We then define our notation and outline the problem statement in Sect. 3. The details of the proposed method and its properties are summarized in Sect. 4. The benchmark datasets in addition to the baseline algorithms for comparison are discussed in Sect. 5. We then show the results of anomaly detection and discuss the scalability and stability of our algorithm in Sect. 6. Finally, we conclude the paper and present future directions for research in Sect. 7.

2   Related Work

One of the most valuable tasks in data analysis is to recognize what stands out in a dataset. This type of analysis provides actionable information and improves our knowledge of the underlying data generation scheme. Various approaches have been developed for detection of such abnormalities [4], however many of these techniques disregard relational datasets where data instances demonstrate complex inter-dependencies. Due to the abundance and cross-disciplinary property of relational datasets, graph-based anomaly detection techniques have received growing attention in social networks, web graphs, road map networks and so forth [3].

We review some of the dominant techniques for the detection of anomalies. We focus on graphs that are plain where nodes and/or edges are not associated with attributes and the nodes are consistently labeled over time.

2.1   Graph-Based Anomaly Detection

Several approaches to pattern mining in graphs stem from distance based techniques, which utilize a distance measure in order to detect abnormal vs. normal structures. An example of such an approach is the k-medians algorithm [8], which employs graph edit distance as a measure of graph similarity. Other approaches take advantage of graph kernels [15], where kernel-based algorithms are applied to graphs. They compare graphs based on common sequences of nodes, or subgraphs. However, the computational complexity of these kernels can become problematic when applied to large graphs.

Other graph similarity metrics use the intuition of information flow when comparing graphs. The first step in these approaches is to compute the pairwise node affinity matrices in each graph and then determine the distance between these matrices. There are several approaches for determining node affinities in a graph, such as Pagerank and various extensions of random walks [6]. Another recent approach in this category is called Delta connectivity, which can be used for the purpose of anomaly detection. This approach calculates the graph distance by comparing node affinities [16]. It measures the differences in the immediate and second-hop neighborhoods of graphs. These approaches also suffer from the curse of dimensionality in large graphs.

Moreover, there are approaches that try to extract properties such as graph centric features before performing anomaly detection. These features can be computed from the combination of two, three or more nodes, i.e., dyads, triads and communities. They can also be extracted from the combination of all nodes in a more general manner [1]. Many anomaly detection approaches [12] have utilized graph centric features in their process of anomaly detection. Since the graph is summarized as a vector of features, the problem of graph-based anomaly detection transforms to the well-known problem of spotting outliers in an n-dimensional space. Therefore standard unsupervised anomaly detection schemes such as ellipsoidal cluster based approaches can be employed [19]. A thorough survey of such techniques can be found in [4]. It is worth noting that the extracted features cause information loss that can affect the performance of the anomaly detection scheme.

Another approach for graph mining is tensor decomposition. These techniques represent the time-evolving graphs as a tensor that can be considered as a multidimensional array, and perform tensor factorization. Tensor factorization approximates the input graph, where the reconstruction error can highlight anomalous events, subgraphs and/or vertices [20].

Although this field of research has received growing attention in recent years, the problem of scalability and interpretability of results still remains. Graph-centric features can reduce the dimensionality of the input graphs, but they may not be able to provide visually interpretable results. On the other hand, decomposition-based methods provide meaningful representations of graphs but suffer from the curse of dimensionality. The trade-off between these two issues has motivated us to find a compact representation of graphs that preserves the structural properties of networks. This can help further analysis of the data to become computationally efficient. Specifically for the task of anomaly detection, we provide experiments that demonstrate the efficiency and utility of our approach.

3 Preliminaries and Problem Statement

We start by describing the basic notation and assumptions of our anomaly detection task. A graph $G = (V, E)$ is defined as a set of nodes V and edges $E \subseteq V \times V$, where an edge $e \in E$ denotes a relationship between its corresponding nodes v_i, v_j. The degree d_i of a vertex v_i is defined as the sum of the number of its incoming (in-degree) and outgoing (out-degree) edges. A Maximum Independent Set (MIS) is the largest subset of vertices $V_{MIS} \subseteq V$ such that there is no edge between any pair of vertices in V_{MIS}.

The maximum independent set problem is closely related to common graph theoretical problems such as maximum common induced subgraphs, minimum vertex covers, graph coloring, and maximum common edge subgraphs. Finding MISs in a graph can be considered a sub-problem of indexing for shortest path and distance queries, automated labeling of maps, information coding, and signal transmission analysis [18].

Graphs are often represented by binary adjacency matrices, $A_{n \times n}$, where $n = |V|$ denotes the number of nodes. An element of the adjacency matrix $a_{ij} = 1$ if there is an edge from v_i to v_j. The simultaneous re-ordering of rows and columns of the adjacency matrix is called matrix permutation.

We formulate the problem of anomaly detection as follows: Given a sequence of graphs $\{G\}_{1...m}$, where m is the number of input graphs, we want to determine the time stamp(s), $i \in \{1...m\}$, when an event has occurred and changed the structural properties of the graph G_i. We consider the following assumptions about the input graphs:

- The vertices and edges in the graph are unweighted.
- There is no external vertex ordering.
- The input graphs are plain, i.e., no attributes are assigned to edges or vertices.
- The number of nodes remains the same throughout the graph sequence.

– The labeling of nodes between graphs is consistent.

An important issue for the design of a scalable anomaly detection scheme is the number of input features or dimensions that are required to be processed. If a graph-based anomaly detection uses a raw adjacency matrix as input, then the input dimensionality is $O(n^2)$, which is impractical for large graphs. In order to address the issue of scalability, we need to find a compact representation for each graph. We propose a pre-processing algorithm that extracts a rank feature for each node that is associated with the maximum independent sets in each graph. Therefore, instead of storing and processing an adjacency matrix of size $n \times n$, we reduce the input dimensionality and computational requirements for our anomaly detector to n.

For each graph in the sequence $\{G_1 = (V_1, E_1), G_2 = (V_2, E_2), ..., G_m = (V_m, E_m)\}$, we determine the new matrix re-ordering vector $\{V_1', V_2', ..., V_m'\}$. We then compute the rank correlation coefficient between every two consequent tuples, (V_i', V_{i+1}'). We employ the Spearman rank correlation coefficient as shown in Eq. 1 between two input rank vectors, $\overrightarrow{V}'_i, \overrightarrow{V}'_{i+1}$, where $d_i = v_i - v_{i+1}$:

$$\rho = 1 - \frac{6 \sum d_i{}^2}{n(n^2 - 1)} \tag{1}$$

The computational complexity of Eq. 1 is $O(n)$, where n is the length of the input vectors. The intuition behind our approach is to design a stable and scalable algorithm for determining the significance of each node and revealing structural information by manipulating the adjacency matrix $A_{n \times n}$. We need to find a matrix permutation that satisfies the following properties:

– *Locality*: Non-zero elements of the matrix should be in close vicinity in the ordering after the permutation.
– *Stability*: The initial ordering of the rows and columns should have no effect on the final outcome of the re-ordering.
– *Scalability*: The algorithm should have low computational complexity in order to handle large scale graphs.
– *Interpretability*: The permuted matrix should reveal structural information such as MISs about the graph.

4   Our Approach: Amplay

In order to achieve the above objectives, we propose an approach entitled Amplay (Adjacency matrix permutation based on layers). In each iteration, Amplay sorts vertices according to their total degree, and picks the vertex with the highest degree. Ties are resolved according to the ordering in the previous iteration. We then remove the vertex and its incidental edges, and recursively apply the algorithm. The outline of the re-ordering approach is given in Algorithm 1. In order to clarify the process of Amplay implementation, we have provided an example of Amplay operation in Figs. 1a and 1b.

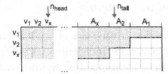

(a) A partially reordered matrix at the beginning of iteration 3 of Amplay. In this iteration, v_x will be placed at position n_{head}, and A_x will be placed before position n_{tail}. A_x are vertices that are only incidental to vertices placed before/to v_x, which results in a zero area at the bottom right corner, i.e., white squares. Elements at gray squares can contain 0 or 1.

(b) Amplay ordering for a sample graph. Each row shows an ordering at the end of each iteration. Rectangles outline sets V_i at the beginning of each iteration.

Fig. 1. Examples of Amplay algorithm operation

Algorithm 1. Amplay Permutation

Input : Graph $G = (V, E)$ and $n = |V|$
Output: Node re-ordering $V \rightarrow V'$
1 $n_{head} = 1$; $n_{tail} = n + 1$; $i = 1$; $V_i = V$; $E_i = E$; $G_i = (V_i, E_i)$;
2 **while** $n_{head} < n_{tail}$ **do**
3   Sort V_i according to the degrees of vertices resolving ties using previous ordering;
4    $v_x \in V_i \leftarrow$ a vertex with the maximum total degree;
5    $e_x \subseteq E_i \leftarrow$ edges incidental to v_x in G_i;
6    $A_x \subseteq V_i \leftarrow$ vertices incidental only to v_x in G_i;
7    $a_i = |A_x|$;
8   Place v_x in position n_{head};
9   Place A_x in position $n_{tail} - a_i, ..., n_{tail} - 1$;
10   (preserving ordering of vertices A_x from G_i);
11   $V_{i+1} = V_i \backslash v_x \cup A_x$, $E_{i+1} = E_i \backslash e_x$ ($\backslash$ denotes set difference);
12   $n_{head} = n_{head} + 1$, $n_{tail} = n_{tail} - a$, $i = i + 1$;

One of the interesting properties of Amplay is its capability to reveal MISs associated with each input graph. Figure 2 shows the permuted adjacency matrix of the Enron email dataset where the MISs are denoted as $S_1, S_2, ...$ The groupings of nodes into the MISs indicates that Amplay can be used as a heuristic to determine the MISs of a graph in various problem domains. A prominent feature of the matrices produced by the Amplay method is a front line such that all non-zero matrix elements are located above the line. Indeed, we can consider an adjacency matrix as a grid with integer coordinates. Here the first coordinate spans rows from top to bottom, the second coordinate spans columns from left to right. We define the front line as follows: $(1, n), (1, n - a_1 + 1), (2, n - a_1), (2, n - a_1 - a_2 + 1), ..., (s, s), ..., (n - a_1 + 1, 1), (n, 1)$, where $\{a_i\}$ is the sequence produced by Algorithm 1 and s is the number of iterations of the algorithm.

Lemma 1. *Every matrix element below the front line is zero.*

Proof. The front line spans intersections of vertices from sets A_x with their respective v_x. By definition, A_x are vertices that are only incidental to vertices placed before v_x or to v_x, which implies that matrix elements below and to the right from the intersections of A_x and v_x are zero.

As we explain below, the front line is important in visualization, because it allows us to grasp (1) the degree distribution of the graph, and (2) the relative size of the largest independent set revealed by Amplay. Note that the shape of the front line is defined by the sequence $\{a_i\}$, where a_i is closely related to the degree of the vertex placed at position i. As a consequence, the front line reflects the degree distribution in a graph.

    A key property of Amplay is multiple vertex sorting. Recall that at each iteration, vertices are sorted according to the total degree of the remaining graph, and ties are resolved using the ordering from the previous iteration. Such a sorting has two consequences. First, the resulting index of each vertex depends not only on the vertex degree, but also on a vertex connectivity pattern (e.g., the number of connections to high-degree nodes). This pattern is reflected in the positions of the vertex in subsequent sorting rounds. While many vertices can have the same degree, the vertices tend to differ in their connectivity patterns. As such, Amplay tends to produce a relatively deterministic ordering. This in turn results in a relatively small variance in the behavior of subsequent graph processing algorithms. Second, vertices that have a similar connectivity pattern will have similar positions during sorting across subsequent iterations, and thus have similar positions in the resulting Amplay ordering. This explains why Amplay tends to produce matrices with a smooth visual appearance.

Lemma 2. *Graph $G = (V, E)$ contains an independent set with at least $n - n_{tail}$ vertices, where n_{tail} is the value from Amplay at the moment of termination.*

Proof. At the end of each iteration of Amplay, vertices assigned to indices larger than or equal to n_{tail} are incidental only to vertices assigned to indices smaller than n_{head}. At the point of termination $n_{head} = n_{tail}$. Hence, vertices assigned to indices larger than n_{tail} are pairwise disjoint and form an independent set.

    In addition to revealing structural properties of the graph, Amplay proves to be scalable. We describe the computational complexity of this re-ordering approach in Lemma 3.

Lemma 3. *The complexity of Amplay is $O(\sum_{i=0}^{s} n_i \log n_i)$ where $n_i = |V_i|$ defined in Amplay, and $s \le |V|$ is the number of iterations.*

Proof. Each iteration of the algorithm operates on a subgraph with n_i vertices, and involves sorting (which can be performed in $O(n_i \log n_i)$ time), finding neighbors of the chosen vertex v_x (linear in n_i), and removing incidental edges (linear in n_i). As such the overall complexity of one iteration is bounded by $O(n_i \log n_i)$ and the total complexity is bounded by $O(\sum_{i=0}^{s} n_i \log n_i)$.

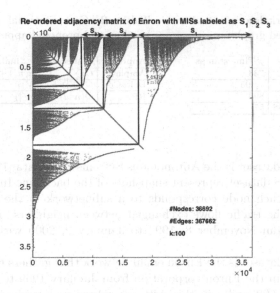

Fig. 2. The Amplay re-ordered adjacency matrix of the Enron email dataset.

It is worth mentioning that in many real-world graphs, n_i rapidly decreases, which reduces the total running time. Moreover, we can improve the scalability of Amplay further, by choosing k vertices with the largest total degrees, place them, and advance the n_{head} pointer by k at each iteration (line 4 in Algorithm 1). Furthermore, in line 6 of Algorithm 1, we can define A_x as a set of vertices incidental only to the chosen k vertices. The front line is now defined as $(k, n), (k, n - a_1 + 1), (2k, n - a_1), (2k, n - a_1 - a_2 + 1), ..., (s.k, s.k), ..., (n - a_1 + 1, k), (n, k)$, and it is easy to verify that Lemmas 1 and 2 hold. If we increase k, we can see that the prominent structural features of the graph are preserved. Moreover the computational complexity of Amplay when $k > 1$ is $O(\sum_{i=0}^{s'} n_i' \times r_i)$ where $r_i = \max(\log n_i', k)$. Using $k > 1$ is beneficial because it reduces the number of iterations s', and sequence n_i' decreases faster than n_i.

5   Evaluation Methodology

In this section, we describe each dataset used in our experiments and elaborate on the baseline algorithms for comparison.

5.1   Benchmark Datasets

For the purpose of anomaly detection, we have selected a representative sample of sparse real-world datasets. The first real dataset is the Facebook wall posts data collected from September 26th, 2006 to January 22nd, 2009 from users in the New Orleans network [22]. The number of users is 90,269, however only 60,290 exhibited activity.

Table 1. Benchmark description where * denotes undirected graphs.

Dataset	#Nodes	#Time stamps
AS *	65,535	733
Facebook *	60,290	1,495
Enron	184	893
DBLP *	1,631,698	57

Table 2. Computational complexity for baseline and proposed approaches.

Approach	Embedding + Similarity complexity
Amplay	$O(\sum_{i=0}^{s} n_i \log n_i) + O(|V|)$
DeltaCon	$O(|E|) + O(|V|)$
RP	$O(n^2 d) + O(|V|)$

The next real dataset is the Autonomous Systems (AS) data [17]. The graphs comprising the AS dataset represent snapshots of the backbone Internet routing topology, where each node corresponds to a subnetwork in the Internet. The edges represent the traffic flows exchanged between neighbors. The dataset is collected daily from November 8, 1997 to January 2, 2000 with nodes being added or deleted.

Another real dataset is the Enron email network that gathers the email communications within the Enron corporation from January 1999 to January 2003 [7]. There are 36,692 nodes in this network, where each node corresponds to an email address. We have used the nodes with a minimum activity level and reduced the graph to 184 nodes.

The final real data is the DBLP[1] dataset that consists of co-authorship information in computer science. The number of nodes is 1,631,698 and the data is gathered from 1954 to 2010. The description of these datasets is summarized in Table 1. DBLP graphs are used to test the scalability of our approach.

5.2 Baseline Algorithm

For the purpose of comparison, we have used a recent approach for computing graph similarity with applications in anomaly detection as our baseline. This algorithm is called delta connectivity (DeltaCon) [16], where the node affinity matrices for each graph are calculated using a belief propagation strategy shown in Eq. 2. This approach considers first-hop and second-hop neighborhoods for calculating the influence of the nodes on each other and has been proven to converge.

$$S = [s_{ij}] = [I + \eta^2 D - \eta A'^{-1}] \tag{2}$$

After determining the node affinity matrices, they compare the consecutive graphs by calculating the root Euclidean distance shown in Eq. 3, which varies in the range $[0, 1]$. We empirically have chosen $\eta = 0.1$ in our experiments.

$$sim(S_1, S_2) = \sqrt{\sum_{i=1}^{n} \sum_{j=1}^{j=n} (\sqrt{S_{1,ij}} - \sqrt{S_{2,ij}})^2} \tag{3}$$

The computational complexity of this algorithm is reported to be linear in the number of edges of each graph, $O(|E|)$.

---

[1] http://dblp.uni-trier.de/xml/.

Another baseline algorithm is an approach called Random Projection (RP) that has shown to be effective in determining anomalous graphs in block-structured networks [21]. The intuition behind RP comes from the Johnson and Lindenstrauss lemma [11] as presented in Lemma 4. This lemma asserts that a set of points in Euclidean space, $P^{1...n} \in \mathbb{R}^{n \times m}$, can be embedded into a d-dimensional Euclidean space, $P'^{1...n} \in R^{n \times d}$ while preserving all pairwise distances within a small factor ϵ with high probability.

Lemma 4. *Given an integer n and $\epsilon > 0$, let d be a positive integer such that $d \geq d_0 = O(\epsilon^{-2} \log n)$. For every set P of n points in $\mathbb{R}^m$, there exists $f : \mathbb{R}^m \to \mathbb{R}^d$ such that with probability $1 - n^{-\beta}$, $\beta > 0$, for all $u, v \in P$*

$$(1 - \epsilon)||u - v||^2 \leq ||f(u) - f(v)||^2 \leq (1 + \epsilon)||u - v||^2 \tag{4}$$

One of the algorithms for generating a random projection matrix that has been shown to preserve pairwise distances [11] is presented in Eq. 5:

$$r_{ij} = \sqrt{3} \begin{cases} +1 & with\, probability\, 1/6 \\ 0 & with\, probability\, 2/3 \\ -1 & with\, probability\, 1/6 \end{cases} \tag{5}$$

6 Results and Discussion

In this section, we outline our experimental setup in five sections and report the observed results. We first demonstrate the effectiveness of Amplay and rank correlation in prioritizing nodes that can contribute the most to the structural change in consecutive graphs. We then investigate the capability of our algorithm in detecting anomalous graphs based on the produced similarity score. Thereafter, we discuss the scalability of our approach empirically by changing parameter k. We provide our empirical studies regarding the stability of the Amplay algorithm on static graphs.

Experiment I: Gradual Change Detection. The effectiveness of Amplay lies in its ability to reveal maximum independent sets. The nodes that comprise each set can be considered the most influential nodes collected from every community in the graph. Figure 3 shows the gradual change in the graph structure by removing the edge $e_{3,10}$ connecting v_3 and v_{10}. $e_{3,10}$ is the connecting bridge between two of the present communities in the graph and its elimination may lead to discontinuity in the entire graph structure. As can be seen, v_3 is the node that contributes the most to the dissimilarity between G_1 and G_2.

Experiment II: Anomaly Detection. We have applied the proposed approach (with parameter $k = 1$) and the baseline algorithms on the benchmark datasets, and compared their computed similarity score between consecutive days. The implementations were run in Matlab using a machine with a 3 GHz Processor and 8 GB RAM. Due to the computational complexity of the random

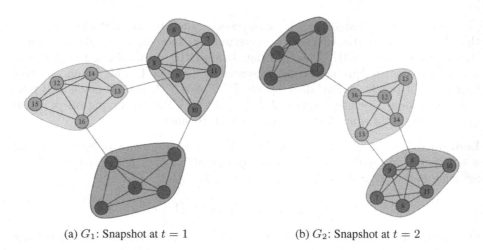

(a) G_1: Snapshot at $t = 1$ (b) G_2: Snapshot at $t = 2$

Fig. 3. Example of gradual change in the structure of the graph and the importance of each node in the overall similarity score.
Initial Node Ordering for G_1, G_2: 1, 2, 3, 4, 5, 6, 7, 8, 9, 10, 11, 12, 13, 14, 15
Amplay and Rank Correlation Node Importance: 3, 5, 13, 15, 6, 7, 1, 2, 4, 8, 9, 10, 11, 12, 14, 16
DeltaCon Node Importance: 3, 10, 14, 16, 12, 13, 2, 5, 15, 4, 6, 7, 11, 1, 9, 8

projection approach, we only use this algorithm as a baseline for comparing scalability.

Our proposed method and DeltaCon generate scores in the range $[0, 1]$. Figures 4, 5 and 6 demonstrate the graph similarity scores for the Autonomous Systems, Facebook and Enron datasets respectively. As can be seen, the trend of similarity scores is the same for DeltaCon and our proposed method.

Experiment III: Computational Scalability. The reported results for anomaly detection were achieved by setting parameter $k = 1$, where k was defined at the end of Sect. 4 as the number of vertices that are processed and removed from the graph in a single iteration. We decided to increase k and investigate the performance of our anomaly detection scheme. It is worth recalling that we are using only a subset of nodes for the purpose of anomaly detection. We consider the top l elements in the rank vectors where $l = n_{head}$ after the termination of Amplay.

Increasing parameter k leads to an exponential decrease in computation time. This observation can be explained by the sparsity of real-world graphs, i.e., the small proportion of fully-connected cliques. Since k is the number of vertices that are processed and removed from the graph within a single iteration, increasing k leads to a more rapid graph reduction. However, at some value of k, all highly connected vertices are processed within a single iteration, and the remaining graph contains only vertices with low degrees. Therefore, subsequent increases of k do not lead to a significant performance improvement. Figure 7 demonstrates the effect of parameter k on the processing time of Amplay for

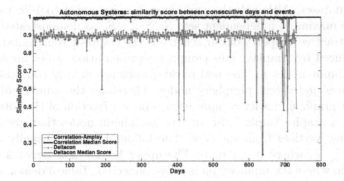

Fig. 4. Comparison of graph similarity scores based on the correlation score of the Amplay-permuted adjacency matrix and DeltaCon on the Autonomous Systems dataset.

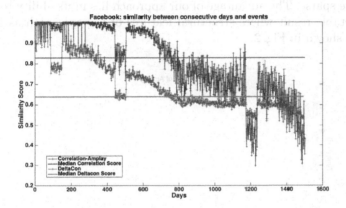

Fig. 5. Comparison of graph similarity scores based on the correlation score of the Amplay-permuted adjacency matrix and DeltaCon on the Facebook dataset.

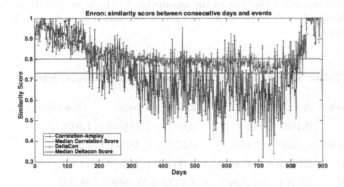

Fig. 6. Comparison of graph similarity scores based on the correlation score of the Amplay-permuted adjacency matrix and DeltaCon on the Enron dataset.

the Enron dataset. Although the parameter k is increased to 100, we can still observe the maximum independent sets $S_1, S_2, ..., S_n$ as demonstrated in Fig. 2. Another attractive property of our scheme is the compact representation of the graph produced by Amplay. This compact representation scales linearly in the number of input nodes n. The real-world graphs are mainly comprised of sets of dense cores and sparse periphery nodes. Therefore, the number of nodes to consider for graph similarity computation is only a fraction of the total number of nodes in a graph. Amplay discards the peripheral nodes that are connected to only a few vertices from the core. The influential nodes usually appear as $V_1', V_2', ..., V_{n_{head}}'$, where $n_{head} \ll n$. The upper bound of n denotes the worst case scenario where the input graph is fully-connected. Table 3 demonstrates the computation time and number of considered nodes in calculating graph similarity. The upper bounds for time complexity of the embedding approaches is demonstrated in Table 2. As can be seen, our proposed method and Delta-Con outperform random projection, and both are scalable when the adjacency matrices are sparse. The advantage of our approach lies in its ability to generate an interpretable result where structural features of a graph, such as MISs, are revealed as shown in Fig. 2.

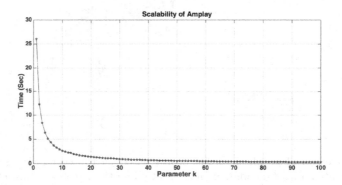

Fig. 7. Amplay computation time as the parameter k is increased in the Enron dataset where k is the number of vertices that are processed and removed from the graph in a single iteration.

Table 3. Computation time of Amplay on different datasets.

Dataset	Amplay time	DeltaCon time	#Nodes to consider
AS	0.196 ± 0.005	0.087 ± 3.616e-04	1,913
Facebook	0.0538 ± 0.003	0.072 ± 4.482e-04	1,316
Enron	0.0009 ± 0.0008	0.003 ± 8.286e-07	41
DBLP	29.707 ± 6.268e+03	1.7174 ± 0.1998	38,903

Experiment IV: Amplay Stability. We compare Amplay with other ordering methods, namely random, RCM [5], and SlashBurn [13]. Random permutation serves as a naive baseline; RCM is a classical bandwidth reduction algorithm [5]; and SlashBurn is a recent method that is shown to produce adjacency matrices with localized non-zero elements. This method is shown to be one of the best state-of-the-art methods [13].

We use a representative sample of sparse real-world graphs of different sizes for quantitative evaluation (Table 4) where all graphs were downloaded from the Stanford Large Network Collection[2]. The table shows graph names as they appear in the Collection, however in the paper we use simplified names (e.g., gnutella instead of p2p-Gnutella08).

We first load each graph as an adjacency matrix S and produce $N+1$ random permutations of the graph vertices $RND_i(S), i = 0, 1, ..., N$. We then take each random permutation as input and either leave it as it is (method Random), or apply RCM, SlashBurn or Amplay permutation -respectively, $RCM(RND_i(S))$, $SlashBurn(RND_i(S))$ and $Amplay(RND_i(S))$.

We then evaluate ordering stability by selecting one of the random permutations as a reference (e.g., $i_{ref} = 0$), and comparing the vertex ordering between each of the other permutations and the reference (e.g., compare $RND_0(S)$ with $RND_j(S)$). In this section, we use both Amplay and SlashBurn with $k = 1$. That is, we evaluate the basic forms of these algorithms, as opposed to more coarse scalable versions.

Table 4. Real-world graphs used in our stability analysis. * mark undirected graphs.

Dataset	Vertices	Edges	Dataset*	Vertices	Edges
Wiki-Vote	7115	103689	ca-HepTh*	9877	51971
p2p-Gnutella08	6301	20777	oregon1*	10670	22002
soc-epinions1	75879	508837	loc-Gowalla*	196591	1900654
Email-EuAll	265214	420045	flickr*	105936	2300660

We compare two vertex orderings using the Kendall correlation coefficient. This coefficient takes values in $[-1, 1]$, where 1 is reached in the case of equivalence of the orderings. If the two orderings are independent, one would expect the coefficient to be approximately 0. Intuitively, vertices with higher degrees tend to have a higher impact on matrix operations and visual appearance. Therefore, we also separately look at ordering stability for higher degree vertices only. Specifically, we compute the Kendall correlation while ignoring a certain proportion $(0, 80, 90, 95\%)$ of vertices with low degrees. Here 0% means that we compare orderings for all graph vertices. On the other hand, 95% means that we only consider the ordering of the top 5% of vertices with the highest degrees. We present our results in Fig. 8 and Table 5 (permutations with $k = 1$ were slow for large

---

[2] snap.stanford.edu/data.

Table 5. Stability measured with Kendall Tau at 90 % for large graphs. The table shows the means for three comparisons.

Dataset	Random	RCM	SlashBurn	Amplay
Email-Eu	< 0.01	0.02	0.11	0.46
gowalla	< 0.01	0.41	0.78	0.89
flicker	< 0.01	0.27	0.05	0.99

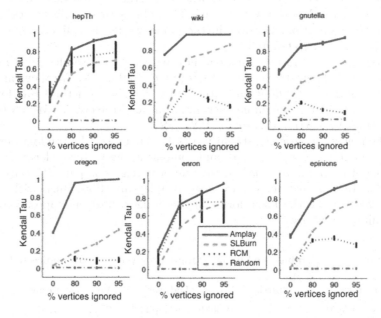

Fig. 8. Amplay stability in comparison to the rival approaches, SlashBurn [13], RCM [5] and Random ordering, as we vary the percentage of ignored low-degree vertices.

graphs, therefore we have fewer runs for large graphs). Overall, Amplay outperforms the other methods by a large margin ($p < 0.01$, Wilcoxon signed rank test). In other words, Amplay tends to be less dependent on the input ordering.

7   Conclusion and Future Work

In this paper, we presented an unsupervised approach for detecting anomalous graphs in time-evolving networks. We created a compact yet structure-aware feature set for each graph using a matrix permutation technique called Amplay. The resulting feature set included the rank of each node in a graph and this rank ordering was used by rank correlation for comparing a pair of graphs. This simple yet effective approach overcomes the issues of scalability when handling large-scale graphs. We showed the low time complexity and structure-aware property of our re-ordering approach both empirically and theoretically. Moreover, we designed experiments for the purpose of anomaly detection in four real datasets, where our approach was compared against an effective graph similarity method and proved to be successful in highlighting abnormal events. In future work, we will explore the possibilities of reducing the dimensionality of the graph even further by using a random projection approach. Since we reduce the dimensionality from $O(n^2)$ to $O(n)$, we can consider the rank vectors of each graph as a data stream. Thereafter, we will investigate a window-based approach for determining anomalous graphs given a history of past normal instances.

References

1. Akoglu, L., Vaz de Melo, P.O.S., Faloutsos, C.: Quantifying reciprocity in large weighted communication networks. In: Tan, P.-N., Chawla, S., Ho, C.K., Bailey, J. (eds.) PAKDD 2012. LNCS (LNAI), pp. 85–96. Springer, Heidelberg (2012). doi:10.1007/978-3-642-30220-6_8
2. Akoglu, L., Tong, H., Koutra, D.: Graph based anomaly detection and description: a survey. Data Mining Knowl. Discov. **29**(3), 626–688 (2014)
3. Berlingerio, M., Koutra, D., Eliassi-Rad, T., Faloutsos, C.: Netsimile: a scalable approach to size-independent network similarity. arXiv preprint (2012). arXiv:1209.2684
4. Chandola, V., Banerjee, A., Kumar, V.: Anomaly detection: a survey. ACM Comput. Surv. (CSUR) **41**(3), 15 (2009)
5. Cuthill, E., McKee, J.: Reducing the bandwidth of sparse symmetric matrices. In: Proceedings of the 1969 24th National Conference ACM, pp. 157–172. ACM (1969)
6. Del Corso, G.M., Gulli, A., Romani, F.: Fast pagerank computation via a sparse linear system. Internet Math. **2**(3), 251–273 (2005)
7. Diesner, J., Frantz, T.L., Carley, K.M.: Communication networks from the Enron email corpus it's always about the people. Enron is no different. Comput. Math. Organ. Theory **11**(3), 201–228 (2005)
8. Ferrer, M., Valveny, E., Serratosa, F., Bardají, I., Bunke, H.: Graph-based k-means clustering: a comparison of the set median versus the generalized median graph. In: Jiang, X., Petkov, N. (eds.) CAIP 2009. LNCS, vol. 5702, pp. 342–350. Springer, Heidelberg (2009). doi:10.1007/978-3-642-03767-2_42
9. Flegel, U., Vayssière, J., Bitz, G.: A state of the art survey of fraud detection technology. In: Insider Threats in Cyber Security, vol. 49, pp. 73–84. Springer, New York (2010)
10. Hirose, S., Yamanishi, K., Nakata, T., Fujimaki, R.: Network anomaly detection based on eigen equation compression. In: Proceedings of the 15th ACM SIGKDD International Conference on Knowledge Discovery and Data Mining, pp. 1185–1194. ACM (2009)
11. Johnson, W.B., Lindenstrauss, J.: Extensions of lipschitz mappings into a hilbert space. Contemp. Math. **26**(1), 189–206 (1984)
12. Kang, U., Chau, D.H., Faloutsos, C.: Mining large graphs: algorithms, inference, and discoveries. In: IEEE 27th International Conference on Data Engineering (ICDE), pp. 243–254. IEEE (2011)
13. Kang, U., Faloutsos, C.: Beyond 'caveman communities': hubs and spokes for graph compression and mining. In: IEEE 11th International Conference on Data Mining (ICDM), pp. 300–309. IEEE (2011)

14. Kang, U., Papadimitriou, S., Sun, J., Tong, H.: Centralities in large networks: algorithms and observations. In: SDM, vol. 2011, pp. 119–130. SIAM (2011)
15. Kang, U., Tong, H., Sun, J.: Fast random walk graph kernel. In: SDM, pp. 828–838. SIAM (2012)
16. Koutra, D., Vogelstein, J.T., Faloutsos, C.: Deltacon: A principled massive-graph similarity function. In: SIAM (2013)
17. Leskovec, J., Kleinberg, J., Faloutsos, C.: Graphs over time: densification laws, shrinking diameters and possible explanations. In: Proceedings of the Eleventh ACM SIGKDD International Conference on Knowledge Discovery in Data Mining, pp. 177–187. ACM (2005)
18. Liu, Y., Lu, J., Yang, H., Xiao, X., Wei, Z.: Towards maximum independent sets on massive graphs. Proc. VLDB Endowment 8(13), 2122–2133 (2015)
19. Moshtaghi, M., Leckie, C., Karunasekera, S., Bezdek, J.C., Rajasegarar, S., Palaniswami, M.: Incremental elliptical boundary estimation for anomaly detection in wireless sensor networks. In: IEEE 11th International Conference on Data Mining (ICDM), pp. 467–476. IEEE (2011)
20. Papalexakis, E.E., Faloutsos, C., Sidiropoulos, N.D.: ParCube: sparse parallelizable tensor decompositions. In: Flach, P.A., Bie, T., Cristianini, N. (eds.) ECML PKDD 2012. LNCS (LNAI), vol. 7523, pp. 521–536. Springer, Heidelberg (2012). doi:10.1007/978-3-642-33460-3_39
21. Rashidi, L., Rajasegarar, S., Leckie, C.: An embedding scheme for detecting anomalous block structured graphs. In: Cao, T., Lim, E.-P., Zhou, Z.-H., Ho, T.-B., Cheung, D., Motoda, H. (eds.) PAKDD 2015. LNCS (LNAI), vol. 9078, pp. 215–227. Springer, Heidelberg (2015). doi:10.1007/978-3-319-18032-8_17
22. Viswanath, B., Mislove, A., Cha, M., Gummadi, K.P.: On the evolution of user interaction in facebook. In: Proceedings of the 2nd ACM SIGCOMM Workshop on Social Networks (WOSN 2009), August 2009

Building Ensembles of Adaptive Nested Dichotomies with Random-Pair Selection

Tim Leathart[✉], Bernhard Pfahringer, and Eibe Frank

Department of Computer Science, University of Waikato, Hamilton, New Zealand
tml15@students.waikato.ac.nz, {bernhard,eibe}@cs.waikato.ac.nz

Abstract. A system of nested dichotomies is a method of decomposing a multi-class problem into a collection of binary problems. Such a system recursively applies binary splits to divide the set of classes into two subsets, and trains a binary classifier for each split. Although ensembles of nested dichotomies with random structure have been shown to perform well in practice, using a more sophisticated class subset selection method can be used to improve classification accuracy. We investigate an approach to this problem called random-pair selection, and evaluate its effectiveness compared to other published methods of subset selection. We show that our method outperforms other methods in many cases when forming ensembles of nested dichotomies, and is at least on par in all other cases. The software related to this paper is available at https://svn.cms.waikato.ac.nz/svn/weka/trunk/packages/internal/ensemblesOfNestedDichotomies/.

1 Introduction

Multi-class classification problems – problems with more than two classes – are commonplace in real world scenarios. Some learning methods can handle multi-class problems inherently, *e.g.*, decision tree inducers, but others may require a different approach. Even techniques such as decision tree inducers may benefit from methods that decompose a multi-class problem in some manner. Typically, a collection of binary classifiers is trained and combined in some way to produce a multi-class classification. This process is called binarization. Popular techniques for adapting binary classifiers to multi-class problems include pairwise classification [11], one-vs-all classification [15], and error correcting output codes [5]. Ensembles of nested dichotomies [8] have been shown to be an effective substitute to these methods. Depending on the base classifier used, they can outperform both pairwise classification and error-correcting output codes [8].

In a nested dichotomy, the set of classes is split into two subsets recursively until there is only one class in each subset. Nested dichotomies are represented as binary tree structures (Fig. 1). At each node of a nested dichotomy, a binary classifier is learned to classify instances as belonging to one of the two subsets of classes. A nice feature of nested dichotomies is that class probability estimates can be computed in a natural way if the binary classifier used at each node can output two-class probability estimates.

© Springer International Publishing AG 2016
P. Frasconi et al. (Eds.): ECML PKDD 2016, Part II, LNAI 9852, pp. 179–194, 2016.
DOI: 10.1007/978-3-319-46227-1_12

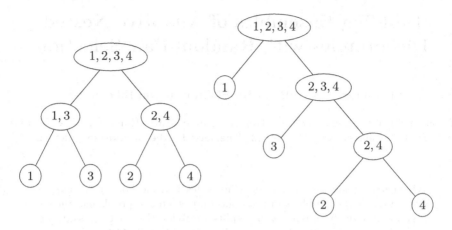

Fig. 1. Two examples of nested dichotomies for a four class problem.

The number of nested dichotomies for a c-class problem increases exponentially with the number of classes. One approach is to sample nested dichotomies at random to form an ensemble of them [8]. However, this may result in binary problems that are difficult to learn for the base classifier.

This paper is founded on the observation that some classes are generally easier to separate than others. For example, in a dataset of images of handwritten digits, the digits '5' and '6' are are much more difficult to distinguish than the digits '0' and '1'. This means that if '5' and '6' were put into opposite class subsets, the base classifier would have a more difficult task to discriminate the two subsets than if they were grouped together. Moreover, if the base classifier assigns high probability to an incorrect branch when classifying a test instance, it is unlikely that the final prediction will be correct. Therefore, we should try to group similar classes into the same class subsets whenever possible, and separate them in lower levels of the tree near the leaf nodes.

In this paper, we propose a method for semi-random class subset selection, which we call "random-pair selection", that attempts to group similar classes together for as long as possible. This means that the binary classifiers close to the root of the tree of classes can learn to distinguish higher-level features, while the ones close to the leaf nodes can focus on the more fine-grained details between similar classes. We evaluate this method against other published class subset selection strategies.

This paper is structured as follows. In Sect. 2, we give a review of other adaptations of ensembles of nested dichotomies. In Sect. 3, we describe the random-pair selection strategy and give an overview of how it works. We also cover theoretical advantages of our method over other methods, and give an analysis of how this strategy affects the space of possible nested dichotomy trees to sample from. In Sect. 4, we evaluate these methods and compare them to other class subset selection techniques.

2   Related Work

The original framework of ensembles of nested dichotomies by Frank and Kramer was proposed in 2004 [8]. In this framework, a binary tree is sampled randomly from the set of possible trees, based on the assumption that each nested dichotomy is equally likely to be useful *a priori*. By building an ensemble of nested dichotomies in this manner, Frank and Kramer achieved results that are competitive with other binarization techniques using decision trees and logistic regression as the two-class models for each node.

There have been a number of adaptations of ensembles of nested dichotomies since, mainly focusing on different class selection techniques. Dong *et al.* propose to restrict the space of nested dichotomies to only consist of structures with balanced splits [6]. Doing this regulates the depth of the trees, which can reduce the size of the training data for each binary classifier and thus has a positive effect on the runtime. It was shown empirically that this method has little effect on accuracy. Dong *et al.* also consider nested dichotomies where the number of instances per subset is approximately balanced at each split, instead of the number of classes. This also reduces the runtime, but can aversely effect the accuracy in rare cases.

The original framework of ensembles of nested dichotomies uses randomization to build an ensemble, *i.e.*, the structure of each nested dichotomy in the ensemble is randomly selected, but built from the same data. Rodriguez *et al.* explore the use of other ensemble techniques in conjunction with nested dichotomies [16]. The authors found that improvements in accuracy can be achieved by using bagging [3], AdaBoost [9] and MultiBoost [17] with random nested dichotomies as the base learner, compared to solely randomizing the structure of the nested dichotomies. The authors also experimented with different base classifiers for the nested dichotomies, and found that using ensembles of decision trees as base classifiers yielded favourable results compared to individual decision trees.

Duarte-Villaseñor *et al.* propose to split the classes more intelligently than randomly by using various clustering techniques [7]. They first compute the centroid of each class. Then, at each node of a nested dichotomy, they select the two classes with the furthest centroids as initial classes for each subset. Once the two classes have been picked, the remaining classes are assigned to one of the two subsets based on the distance of their centroids to the centroids of the initial classes. Duarte-Villaseñor *et al.* evaluate three different distance measures for determining the furthest centroids, taking into account the position of the centroids, the radius of the clusters and average distance of each instance from the centroid. They found that these class subset selection methods gave superior accuracy to the random methods previously proposed when the nested dichotomies were used for boosting.

3   Random-Pair Selection

We present a class selection strategy for choosing subsets in a nested dichotomy called random-pair selection. This has the same intention as the centroid-based methods proposed by Duarte-Villaseñor *et al.* [7]. Our method differs in that it takes a more direct approach to discovering similar classes by using the actual base classifier to decide which classes are more easily separable. Moreover, it incorporates an aspect of randomization.

3.1   The Algorithm

The process for constructing a nested dichotomy with random-pair selection is as follows:

1. Create a root node for the tree.
2. If the class set C has only one class, then create a leaf node.
3. Otherwise, split C into two subsets by the following:
   (a) Select a pair of classes $c_1, c_2 \in C$ at random, where C is the set of all classes present at the current node.
   (b) Train a binary classifier using these two classes as training data. Then, use the remaining classes as test data, and observe which of the initial classes the majority of instances of each test class are classified as.[1]
   (c) Two subsets are created, using the initial classes: $s_1 = \{c_1\}, s_2 = \{c_2\}$
   (d) The test classes $c_n \in C \setminus \{c_1, c_2\}$ are added to s_1 or s_2 based on whether c_n is more likely to be classified as c_1 or c_2.
   (e) A new binary model is trained using the full data at the node, using the new class labels s_1 and s_2 for each instance.
4. Create new nodes for both s_1 and s_2 and recurse for each child node from Step 2.

This selection algorithm is illustrated in Fig. 2. The process for making predictions when using this class selection method is identical to the process for the original ensembles of nested dichotomies. Assuming that the base classifier can produce class probability estimates, the probability of an instance belonging to a class is the product of the estimates given by the binary classifiers on the path from the root to the leaf node corresponding to the particular class.

3.2   Analysis of the Space of Nested Dichotomies

To build an ensemble of nested dichotomies, a set of nested dichotomies needs to be sampled from the space of all nested dichotomies. The size of this space grows very quickly as the number of classes increases. Frank and Kramer calculate that the number of potential nested dichotomies is $(2c - 3)!!$ for a c-class problem [8]. For a 10-class problem, this equates to $34,459,425$ distinct systems of nested

---

[1] When the dataset is large, it may be sensible to subsample the training data at each node when performing this step.

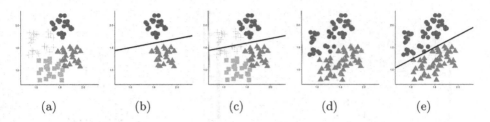

Fig. 2. Random-Pair Selection. (a) Original multi-class data. (b) Two classes are selected at random, and a binary classifier is trained on this data. (c) The binary classifier is tested on the other classes. The majority of the 'plus' class is classified as 'circle', and all of the 'square' class is classified as 'triangle'. (d) Combine the classes into subsets based on which of the original classes each new class is more likely to be classified as. (e) Learn another binary classifier, which will be used in the final nested dichotomy tree.

dichotomies. Using a class-balanced class-subset selection strategy reduces this number:

$$T(c) = \begin{cases} \frac{1}{2}\binom{c}{c/2}T(\frac{c}{2})T(\frac{c}{2}), & \text{if } c \text{ is even} \\ \binom{c}{(c+1)/2}T(\frac{c+1}{2})T(\frac{c-1}{2}), & \text{if } c \text{ is odd} \end{cases} \tag{1}$$

where $T(2) = T(1) = 1$ [6]. The number of class-balanced nested dichotomies is still very large, giving 113,400 possible nested dichotomies for a 10-class problem. The subset selection method based on clustering [7] takes this idea to the extreme, and gives only a single nested dichotomy for any given number of classes because the class subset selection is deterministic. Even though the system produced by this subset selection strategy is likely to be a useful one, it does not lend itself well to ensemble methods.

The size of the space of nested dichotomies that we sample using the random-pair selection method varies for each dataset, and is dependent on the base classifier. The upper bound for the number of possible binary problems at each node is the number of ways to select two classes at random from a c-class dataset, *i.e.*, $\binom{c}{2}$. In practice, many of these randomly chosen pairs are likely to produce the same class subsets under our method, so the number of possible class splits is likely to be lower than this value. For illustrative purposes, we empirically estimate this value for the logistic regression base learner. We enumerate and count the number of possible class splits for our splitting method at each node of a nested dichotomy for a number of datasets, and plot this number against the number of classes at the corresponding node (Fig. 3a). We also show a similar plot for the case where C4.5 is used as the base classifier (Fig. 3b). Fitting a second degree polynomial to the data for logistic regression yields

$$p(c) = 0.3812c^2 - 1.4979c + 2.9027. \tag{2}$$

Assuming we apply logistic regression, we can estimate the number of possible class splits for an arbitrary number of classes based on this expression by making a rough estimate of the distribution of classes at each node. Nested dichotomies

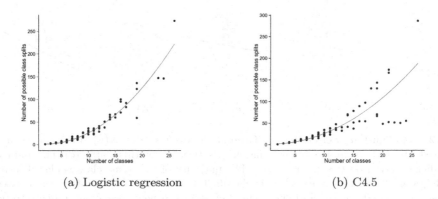

(a) Logistic regression   (b) C4.5

Fig. 3. Number of possible splits under a random-pair selection method vs number of classes for a number of UCI datasets.

Table 1. The number of possible nested dichotomies for up to 12 classes for each class subset selection technique. The first two columns are taken from [6], and the random-pair column is estimated from (3).

Number of classes	Number of nested dichotomies	Number of class-balanced nested dichotomies	Number of random-pair nested dichotomies
2	1	1	1
3	3	3	1
4	15	3	5
5	105	30	15
6	945	90	36
7	10,395	315	182
8	135,135	315	470
9	2,027,025	11,340	1,254
10	34,459,425	113,400	7,002
11	654,729,075	1,247,400	28,189
12	13,749,310,575	3,742,200	81,451

constructed with random-pair selection are not guaranteed to be balanced, so we average the class subset proportions over a large sample of nested dichotomies on different datasets to find that the two class subsets contain $\frac{1}{3}$ and $\frac{2}{3}$ respectively of the classes on average. Given this information, we can estimate the number of possible nested dichotomies with logistic regression by the recurrence relation

$$T(c) = p(c)T(\frac{c}{3})T(\frac{2c}{3}) \tag{3}$$

where $T(c) = 1$ when $c \leq 2$. Table 1 shows the number of distinct nested dichotomies that can be created for up to 12 classes for the random-pair selection

Fig. 4. Class centroids of the training component of the CIFAR-10 dataset (above). Samples from each class (below).

method, class-balanced and completely random selection when we apply this estimate.

3.3 Advantages Over Centroid Methods

Random-pair selection has two theoretical advantages compared to the centroid-based methods proposed by the authors of [7]: (a) an element of randomness makes it more suitable for ensemble learning, and (b) it adapts to the base classifier that is used.

In the centroid-based methods, each class split is deterministically chosen based on some distance metric. This means that the structure of every nested dichotomy in an ensemble will be the same. This is less important in ensemble techniques that alter the dataset or weights inside the dataset (*e.g.*, bagging or boosting). However, an additional element of randomization in ensembles is typically beneficial. When random-pair selection is employed, the two initial classes are randomly selected in all nested dichotomies, increasing the total number of nested dichotomies that can be constructed as discussed in the previous section.

Centroid-based methods assume that a smaller distance between two class centroids is indicative of class similarity. While it is true that this is often the case, sometimes the centroids can be relatively meaningless. An example is the CIFAR-10 dataset, a collection of small natural images of various categories such as cats, dogs and trucks [12]. The classes are naturally divided into two subsets – animals and vehicles. Figure 4 shows an image representation of the centroids of each class, and a sample image from the respective class below it. It is clear to see that most of these class centroids do not contain much useful information for discriminating between the classes.

This effect is clearer when evaluating a simple classifier that classifies instances according to the closest centroid of the training data. For illustrative purposes, see the confusion matrix of such a classifier when trained on the CIFAR-10 dataset (Fig. 5). It is clear to see from the confusion matrix that the centroids cannot be relied upon to produce meaningful predictions in all cases for this data.

A disadvantage of random-pair selection compared to centroid-based methods is an increase in runtime. Under our method, we need to train additional base classifiers during the class subset selection process. However, the extra base classifiers are only trained on a subset of the data at a node, *i.e.*, only two of the

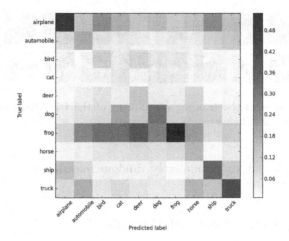

Fig. 5. Confusion matrix of a centroid classifier for the CIFAR-10 dataset. The darkness of each square corresponds with the number of instances classified as a particular class.

classes, and we can subsample this data during this step if we need to improve the runtime further.

4   Experimental Results

We present an evaluation of the random-pair selection method on 18 datasets from the UCI repository [13]. Table 2 lists and describes the datasets we used. We specifically selected datasets with at least five classes, as our method will not have a large impact on datasets with few classes. This is due to the fact that there is a relatively small number of possible nested dichotomies for small numbers of classes.

4.1   Experimental Setup

All experiments were conducted in WEKA [10], and performed with 10 times 10-fold cross validation.[2] The default settings in WEKA for the base learners and ensemble methods were used in our evaluation. We compared our class subset selection method (RPND) to nested dichotomies based on clustering (NDBC) [7], class-balanced nested dichotomies (CBND) [6], and completely random selection (ND) [8]. We did not compare against other variants of nested dichotomies such as data-balanced nested dichotomies [6], nested dichotomies based on clustering with radius [7] and nested dichotomies based on clustering with average radius [7], because they were found to either have the same or worse performance on average in [6] and [7] respectively. We used logistic regression and

---

[2] Our implementations can be found in the `ensemblesOfNestedDichotomies` package in WEKA.

Table 2. The datasets used in this evaluation

Dataset	Classes	Instances	Attributes	Dataset	Classes	Instances	Attributes
audiology	24	226	70	optdigits	10	5620	65
krkopt	18	28056	7	page-blocks	5	5473	11
LED24	10	5000	25	pendigits	10	10992	17
letter	26	20000	17	segment	7	2310	20
mfeat-factors	10	2000	217	shuttle	7	58000	10
mfeat-fourier	10	2000	77	usps	10	9298	257
mfeat-karhunen	10	2000	65	vowel	11	990	14
mfeat-morphological	10	2000	7	yeast	10	1484	9
mfeat-pixel	10	2000	241	zoo	7	101	18

C4.5 as the base learners for our experiments, as they occupy both ends of the bias-variance spectrum. In our results tables, a bullet (•) indicates a statistically significant accuracy gain, and an open circle (○) indicates a statistically significant accuracy reduction ($p = 0.05$) by using the random-pair method compared with another method. To establish significance, we used the corrected resampled paired t-test [14].

4.2 Single Nested Dichotomy

We expect that intelligent class subset selection will have a larger impact in small ensembles of nested dichotomies. This is due to the fact as ensembles grow larger, the worse performing ensemble members will not have as great an influence over the final predictions. Therefore, we first compare a single nested dichotomy using random-pair selection to a single nested dichotomy obtained with other class selection methods.

Table 3 shows the classification accuracy and standard deviations of each method when training a single nested dichotomy. When logistic regression is used as the base learner, compared to random methods (CBND and ND), we obtain a significant accuracy gain in most cases, and comparable accuracy in all others. When using C4.5 as the base learner, our method is preferable to random methods in some cases, with all other datasets showing a comparable accuracy.

In comparison to NDBC, gives similar accuracy overall, with three significantly better results, four significantly worse results, and the rest comparable over both base learners. It is to be expected that NDBC sometimes has better performance than our method when only a single nested dichotomy is built. This is because NDBC deterministically selects the class split that is likely to be the most easily separable. Our method attempts to produce an easily separable class subset selection from a pool of possible options, where each option is as likely as any other.

4.3   Ensembles of Nested Dichotomies

Ensembles of nested dichotomies typically outperform single nested dichotomies. The original method for creating an ensemble of nested dichotomies is a randomization approach, but it was later found that better performance can be obtained by bagging and boosting nested dichotomies [16]. For this reason, we consider three types of ensembles of nested dichotomies in our experiments: bagged, boosted with AdaBoost and boosted with MultiBoost (the latter two applied with resampling based on instance weights). We built ensembles of 10 nested dichotomies for these experiments.

Bagging.   Table 4 shows the results of using bagging to construct an ensemble of nested dichotomies for each method and for both base learners. When logistic

Table 3. Accuracy of a single nested dichotomy with (a) logistic regression and (b) C4.5 as the base learner.

(a)

Dataset	RPND	NDBC	CBND	ND
audiology	75.36±8.45	72.47±8.80	68.55±9.61	71.91±9.85
krkopt	33.13±0.97	33.23±0.80	28.55±1.50 •	28.70±1.56 •
LED24	72.85±2.03	72.73±2.06	67.11±4.08 •	70.26±3.28 •
letter	67.70±2.72	72.23±0.93 ○	47.98±3.08 •	53.10±4.36 •
mfeat-factors	95.04±1.99	96.62±1.19 ○	91.83±2.20 •	93.08±2.15 •
mfeat-fourier	76.37±3.22	75.17±2.76	73.17±3.34 •	74.00±3.34
mfeat-karhunen	89.83±2.32	90.83±1.75	84.96±3.75	86.53±3.06 •
mfeat-morphological	72.64±3.25	70.45±3.03 •	62.31±7.79 •	66.40±5.19 •
mfeat-pixel	71.16±9.98	88.67±2.51 ○	61.25±9.25 •	47.44±9.15 •
optdigits	92.72±2.06	92.00±1.10	87.83±3.01 •	90.95±2.60
page-blocks	96.17±0.75	95.77±0.77	95.44±0.84	95.61±0.86
pendigits	90.20±2.32	87.97±0.96 •	82.23±4.42 •	87.08±4.22
segment	94.02±2.40	88.76±1.91 •	87.36±4.16 •	89.11±3.93 •
shuttle	96.87±0.46	96.86±0.20	92.14±6.86	91.72±7.03 •
usps	87.47±1.47	87.64±1.06	84.70±2.26 •	85.83±1.97 •
vowel	81.80±4.46	80.83±4.10	47.86±8.67 •	53.08±8.98 •
yeast	58.35±3.89	59.00±3.58	56.43±4.20	55.91±3.90 •
zoo	90.41±9.15	87.55±9.32	88.88±9.34	89.00±8.65

(b)

Dataset	RPND	NDBC	CBND	ND
audiology	76.86±7.23	75.49±7.29	74.45±8.04	73.79±7.62
krkopt	70.04±2.45	69.33±0.99	64.83±1.78 •	65.13±2.19 •
LED24	72.68±2.12	72.99±1.72	72.07±2.08	72.22±2.05
letter	86.32±0.85	86.50±0.88	85.38±0.88 •	86.03±0.88
mfeat-factors	88.47±2.59	88.77±1.73	86.76±2.43	87.47±2.23
mfeat-fourier	74.46±3.09	73.97±2.90	72.63±2.97	73.03±3.29
mfeat-karhunen	82.04±2.84	82.56±2.66	80.11±3.15	80.18±3.28
mfeat-morphological	72.44±2.73	72.27±2.48	71.90±2.40	71.85±2.52
mfeat-pixel	81.83±3.23	81.36±2.79	77.13±3.61 •	79.44±3.91
optdigits	90.72±1.43	90.76±1.15	89.27±1.52 •	89.93±1.44
page-blocks	97.07±0.72	97.05±0.66	97.00±0.67	97.05±0.65
pendigits	95.92±0.70	95.81±0.62	95.60±0.67	95.79±0.68
segment	96.10±1.38	96.59±1.25	95.88±1.49	95.88±1.37
shuttle	99.97±0.02	99.98±0.02	99.97±0.02	99.97±0.03
usps	87.95±1.18	89.44±0.91 ○	86.06±1.52 •	86.48±1.37 •
vowel	79.04±4.22	76.96±4.45	76.07±4.75	75.54±4.87
yeast	57.22±3.31	57.58±3.69	56.18±3.43	56.64±3.36
zoo	91.63±8.06	88.65±8.30	90.72±7.12	90.67±8.72

Table 4. Accuracy of an ensemble of 10 bagged nested dichotomies with (a) logistic regression and (b) C4.5 as the base learner.

(a)

Dataset	RPND	NDBC	CBND	ND
audiology	81.79±7.56	81.25±7.25	80.32±7.69	82.35±7.57
krkopt	33.77±0.78	33.29±0.77 •	31.73±0.98 •	31.99±0.94 •
LED24	73.56±1.90	73.42±2.01	73.50±1.94	73.49±1.85
letter	78.65±0.94	76.16±0.96 •	73.76±1.24 •	74.51±1.27 •
mfeat-factors	98.11±1.02	97.39±1.10 •	97.72±1.09	97.94±1.01
mfeat-fourier	83.08±2.18	80.03±2.25 •	82.16±2.66	82.14±2.39
mfeat-karhunen	95.66±1.54	93.67±1.75 •	94.88±1.56	94.89±1.57
mfeat-morphological	73.71±2.79	72.33±2.87	73.19±2.94	73.55±2.45
mfeat-pixel	94.70±1.95	93.15±1.49 •	90.96±2.51 •	83.65±4.01 •
optdigits	97.15±0.68	93.56±0.93 •	96.50±0.83 •	96.83±0.68
page-blocks	96.46±0.68	96.14±0.66 •	95.92±0.72 •	96.11±0.68 •
pendigits	95.93±0.80	88.90±1.08 •	94.61±1.00 •	95.12±0.88 •
segment	95.37±1.61	89.26±1.95 •	94.03±1.96 •	94.15±1.73 •
shuttle	96.74±0.24	96.86±0.21	94.94±1.52 •	94.86±1.39 •
usps	93.83±0.69	92.02±0.91 •	93.59±0.70	93.32±0.73 •
vowel	89.76±3.04	85.72±3.49 •	77.52±4.90 •	78.30±4.61 •
yeast	58.86±3.85	59.18±3.84	58.91±3.64	58.92±3.62
zoo	94.87±6.03	91.62±8.33	93.36±7.16	93.20±7.37

(b)

Dataset	RPND	NDBC	CBND	ND
audiology	79.76±7.32	80.33±6.11	80.65±7.29	79.30±7.30
krkopt	75.70±0.95	73.93±0.90 •	74.20±1.00 •	74.82±1.00 •
LED24	73.22±1.92	73.12±1.82	73.10±1.90	73.23±1.92
letter	93.81±0.55	92.73±0.66 •	93.92±0.50	94.07±0.49
mfeat-factors	95.27±1.58	93.37±1.76 •	95.80±1.40	95.44±1.52
mfeat-fourier	81.36±2.81	78.79±2.64 •	81.30±2.83	80.94±2.76
mfeat-karhunen	92.83±1.96	90.27±2.11 •	92.98±1.42	93.13±1.67
mfeat-morphological	73.38±2.61	72.78±2.72	73.07±2.83	73.37±2.62
mfeat-pixel	92.56±1.91	87.01±2.47 •	92.24±1.82	92.65±1.79
optdigits	97.09±0.70	95.34±0.90 •	97.04±0.72	97.00±0.72
page-blocks	97.41±0.64	97.29±0.62	97.39±0.59	97.36±0.63
pendigits	98.53±0.40	97.67±0.46 •	98.68±0.35	98.64±0.38
segment	97.45±1.09	97.52±1.11	97.54±1.14	97.53±0.88
shuttle	99.98±0.02	99.97±0.02	99.98±0.02	99.98±0.02
usps	94.63±0.59	93.85±0.72 •	94.52±0.59	94.61±0.70
vowel	87.69±3.52	85.82±3.73	89.15±3.46	88.26±3.25
yeast	59.86±3.29	59.55±3.38	59.93±3.54	59.72±3.79
zoo	93.81±7.17	91.70±7.77	93.57±6.81	94.36±6.17

regression is used as a base learner, our method outperforms all other methods in many cases. When C4.5 is used as a base learner, our method compares favourably with NDBC and achieves comparable accuracy to the random methods. Our method is better in a bagging scenario than NDBC because of the first problem highlighted in Sect. 3.3, *i.e.*, using the furthest centroids to select a class split results in a deterministic class split. Evidently, with bagged datasets, this method of class subset selection is too stable to be utilized effectively. Our method, on the other hand, is sufficiently unstable to be useful in a bagged ensemble.

AdaBoost. Table 5 shows the results of using AdaBoost to build an ensemble of nested dichotomies for each method and for both base learners. When comparing with the random methods, we observe a similar result to the bagged ensembles. When using logistic regression, we see a significant improvement in accuracy in

Table 5. Accuracy of an ensemble of 10 nested dichotomies boosted with AdaBoost with (a) logistic regression and (b) C4.5 as the base learner.

(a)

Dataset	RPND	NDBC	CBND	ND
audiology	81.42±7.38	80.31±6.92	79.87± 7.49	80.78± 7.50
krkopt	32.99±1.01	32.81±0.77	28.24± 1.47 •	28.66± 1.44 •
LED24	72.41±2.16	72.93±1.99	69.17± 2.77 •	70.44± 2.72 •
letter	71.39±2.50	71.44±1.49	47.42± 3.29 •	55.16± 5.35 •
mfeat-factors	97.71±1.09	97.66±0.99	97.11± 1.25	97.52± 1.17
mfeat-fourier	81.01±2.28	79.96±2.52	80.12± 2.43	80.13± 2.64
mfeat-karhunen	94.93±1.50	94.42±1.61	93.76± 1.54 •	94.01± 1.54
mfeat-morphological	72.81±2.82	71.02±3.10	66.73± 6.80 •	69.38± 5.53
mfeat-pixel	94.15±1.81	93.87±1.59	91.16± 2.39 •	86.21± 3.48 •
nursery	92.51±0.70	92.52±0.70	92.29± 0.74	92.38± 0.69
optdigits	97.01±0.69	96.84±0.77	96.26± 0.74 •	96.37± 0.86 •
page-blocks	96.09±0.80	95.93±0.75	95.43± 0.84	95.77± 0.90
pendigits	94.94±0.93	94.83±0.77	93.86± 1.30	93.67± 1.03 •
segment	94.94±1.40	94.66±1.48	93.88± 1.93	93.82± 1.84
shuttle	96.83±0.45	96.86±0.26	96.51± 1.57	96.40± 2.18
usps	92.03±0.88	91.83±0.86	91.91± 0.91	91.66± 0.85
vowel	90.59±3.11	89.74±3.10	48.45±10.68 •	58.93±11.42 •
yeast	57.97±3.78	58.39±3.62	56.90± 4.05	56.56± 3.66
zoo	94.95±6.40	94.96±6.33	94.38± 7.44	94.77± 6.19

(b)

Dataset	RPND	NDBC	CBND	ND
audiology	83.64±7.37	83.29±6.68	82.63±6.87	82.58±7.36
krkopt	81.01±0.78	79.37±0.80 •	77.25±0.95 •	78.36±1.04 •
LED24	69.59±2.13	69.49±2.11	69.04±1.95	69.42±1.78
letter	94.58±0.49	94.37±0.48	94.30±0.49	94.60±0.55
mfeat-factors	95.75±1.36	95.31±1.48	95.49±1.38	95.62±1.37
mfeat-fourier	80.43±2.74	79.54±2.60	80.12±2.49	80.74±2.47
mfeat-karhunen	93.20±1.80	92.67±1.83	92.96±1.76	92.85±1.84
mfeat-morphological	70.48±3.10	70.45±3.19	70.13±2.84	70.50±2.45
mfeat-pixel	93.76±1.53	93.27±1.80	92.48±1.80 •	93.01±1.83
optdigits	97.31±0.72	97.23±0.70	97.25±0.68	97.20±0.70
page-blocks	97.05±0.62	97.05±0.66	97.11±0.64	97.11±0.66
pendigits	98.95±0.30	98.89±0.33	98.91±0.30	98.93±0.28
segment	98.23±0.84	98.24±0.84	98.09±0.86	98.09±0.94
shuttle	99.99±0.01	99.99±0.01	99.99±0.01	99.99±0.01
usps	94.85±0.64	94.86±0.64	94.41±0.72	94.59±0.66
vowel	91.95±2.71	90.73±3.00	91.28±2.82	91.30±2.78
yeast	57.39±3.76	57.42±4.02	56.93±3.27	57.25±4.19
zoo	95.45±6.19	95.53±6.39	95.15±6.21	95.36±6.13

many cases, and when C4.5 is used, we typically see comparable results, with a small number of significant accuracy gains. When comparing with NDBC, we see a small improvement for the vast majority of datasets, but these differences are almost never individually significant. In one instance (krkopt with C4.5 as the base learner), we achieve a significant accuracy gain using our method.

MultiBoost. Table 6 shows the results of using MultiBoost to build an ensemble of nested dichotomies for each method and for both base learners. Compared to the random methods, again we see similar results to the other ensemble methods – using logistic regression as the base learner results in many significant improvements, and using C4.5 as the base learner typically produces comparable results, with few significant improvements. In comparison to NDBC,

Table 6. Accuracy of an ensemble of 10 nested dichotomies boosted with MultiBoost with (a) logistic regression and (b) C4.5 as the base learner.

(a)

Dataset	RPND	NDBC	CBND	ND
audiology	80.55±7.80	80.05±7.20	78.90± 7.51	79.53± 7.73
krkopt	32.99±1.01	32.81±0.77	28.24± 1.47 •	28.66± 1.44 •
LED24	73.38±1.81	73.31±2.15	72.01± 2.67	72.75± 2.38
letter	77.29±1.83	75.36±1.03 •	47.42± 3.29 •	55.85± 6.25 •
mfeat-factors	97.82±1.16	97.70±1.09	97.40± 1.31	97.53± 1.17
mfeat-fourier	82.12±2.28	80.22±2.28 •	80.22± 2.35 •	80.72± 2.44
mfeat-karhunen	95.22±1.59	94.70±1.57	93.94± 1.62 •	94.17± 1.71
mfeat-morphological	73.63±2.80	72.33±2.64	67.52± 7.04 •	70.40± 5.74
mfeat-pixel	94.37±1.48	94.16±1.30	91.89± 2.71 •	86.37± 4.74 •
optdigits	97.03±0.57	96.10±0.79 •	96.26± 0.78 •	96.47± 0.83 •
page-blocks	96.39±0.69	96.10±0.72	96.01± 0.68 •	96.19± 0.74
pendigits	96.02±0.73	94.27±1.32 •	94.17± 1.05 •	94.76± 0.92 •
segment	95.56±1.40	94.11±1.92 •	94.12± 1.93 •	94.35± 1.63 •
shuttle	96.89±0.27	96.87±0.24	96.63± 1.53	96.65± 1.59
usps	93.12±0.78	92.45±0.84 •	92.62± 0.83	92.57± 0.84
vowel	89.53±3.15	87.52±3.03	48.92±11.26 •	60.91±12.38 •
yeast	58.28±4.19	58.60±3.93	57.13± 4.03	57.03± 3.88
zoo	94.97±6.49	94.65±6.79	94.46± 7.35	94.07± 7.02

(b)

Dataset	RPND	NDBC	CBND	ND
audiology	81.32±7.06	82.14±7.39	81.25±7.48	80.32±7.37
krkopt	76.22±0.80	75.05±0.84 •	73.54±1.03 •	74.58±1.14 •
LED24	72.27±2.00	71.90±1.99	71.78±1.89	71.96±1.99
letter	93.98±0.47	93.65±0.53	93.78±0.55	93.98±0.46
mfeat-factors	95.63±1.33	94.82±1.45	95.32±1.46	95.14±1.48
mfeat-fourier	80.46±2.40	79.54±2.36	80.36±2.57	80.68±3.00
mfeat-karhunen	92.88±1.95	91.82±1.91	92.16±2.03	92.64±1.81
mfeat-morphological	71.30±2.75	71.26±2.85	71.32±3.11	71.75±2.84
mfeat-pixel	93.10±1.71	91.15±1.86 •	91.75±1.67 •	92.40±1.90
optdigits	97.00±0.70	96.80±0.75	96.91±0.73	97.00±0.69
page-blocks	97.33±0.65	97.24±0.63	97.34±0.64	97.29±0.66
pendigits	98.78±0.35	98.69±0.35	98.78±0.33	98.75±0.28
segment	97.90±0.93	98.06±0.94	97.79±0.95	97.88±0.99
shuttle	99.99±0.02	99.99±0.02	99.99±0.02	99.99±0.01
usps	94.67±0.65	94.48±0.64	94.25±0.58	94.33±0.71
vowel	88.60±3.40	88.33±3.61	88.79±3.18	88.34±3.56
yeast	58.91±3.58	58.91±3.56	58.53±3.63	58.35±3.92
zoo	95.09±6.73	94.17±7.34	94.26±6.48	95.66±6.11

we see many small (although statistically insignificant) improvements across both base learners, with some significant gains in accuracy on some datasets.

4.4 Training Time

Figure 6 shows the training time in milliseconds for training a single RPND and a single NDBC, with logistic regression and C4.5 as the base learners for each of the datasets used in this evaluation. As can be seen from the plots, there is a computational cost for building an RPND over an NDBC, which is to be expected as there is an additional classifier trained and tested at each split node of the tree. The gradient of both plots is approximately one, which indicates that our method does not add additional computational complexity to the problem. The runtime is comparatively worse for logistic regression than for C4.5.

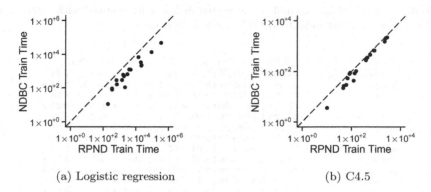

(a) Logistic regression   (b) C4.5

Fig. 6. Log-log plots of the training time for a single RPND and a single NDBC, for both base learners.

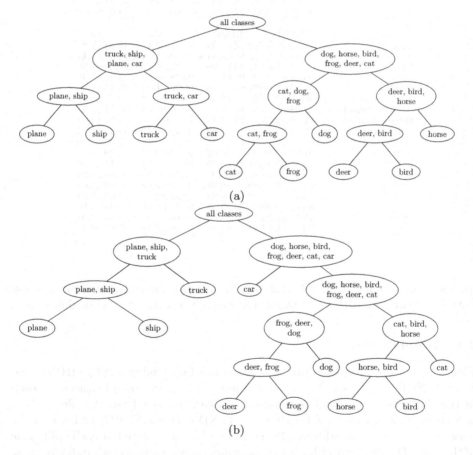

Fig. 7. Nested dichotomies trained on CIFAR-10, with (a) random-pair selection, and (b) centroid-based selection.

4.5 Case Study: CIFAR-10

To test how well our method adapts to other base learners, we trained nested dichotomies with convolutional networks as the base learners to classify the CIFAR-10 dataset [12]. Convolutional networks learn features from the data automatically, and perform well on high dimensional, highly correlated data such as images. We implemented the nested dichotomies and convolutional networks in Python using Lasagne [4], a wrapper for Theano [1,2]. The convolutional network that we used as the base learner is relatively simple; it has two convolutional layers with 32 5×5 filters each, one 3×3 maxpool layer with 2×2 stride after each convolutional layer, and one fully-connected layer of 128 units before a softmax layer.

As discussed in Sect. 3.3, the centroids for a dataset like CIFAR-10 appear to not be very descriptive, and as such, we expect NDBC with convolutional networks as the base learner to produce class splits that are not as well founded as those in RPND. We present a visualisation of the NDBC produced from the CIFAR-10 dataset, and an example of a nested dichotomy built with random-pair selection (Fig. 7). We can see that both methods produce a reasonable dichotomy structure, but there are some cases in which the random-pair method results in more intuitive splits. For example, the root node of the RPND splits the full set of classes into the two natural subsets (vehicles and animals), whereas the NDBC omits the 'car' class from the left-hand subset. Two pairs of similar classes in the animal subset – 'deer' and 'horse', and 'cat' and 'dog' – are kept together until near the leaves in the RPND, but are split up relatively early in the NDBC. Despite this, the accuracy and runtime of both methods were comparable. Of course, the quality of the nested dichotomy under random-pair selection is dependent on the initial pair of classes that is selected. If two classes that are similar to each other are selected to be the initial random pair, the tree can end up with splits that make less intuitive sense.

5 Conclusion

In this paper, we have proposed a semi-random method of class subset selection in ensembles of nested dichotomies, where the class selection is directly based on the ability of the base classifier to separate classes. Our method non-deterministically produces an easily separable class-split, which not only improves the accuracy over random methods for a single classifier, but also for ensembles of nested dichotomies. Our method also outperforms other non-random methods when nested dichotomies are used in a bagged ensemble and an ensemble boosted with MultiBoost, and otherwise gives comparable results.

In the future, it would be interesting to explore selecting several random pairs of classes at each node, and choosing the best of the pairs to create the final class subsets. This will obviously increase the runtime, but may help to produce more accurate individual classifiers and small ensembles. We also wish to explore the use of convolutional networks in nested dichotomies further.

Acknowledgements. This research was supported by the Marsden Fund Council from Government funding, administered by the Royal Society of New Zealand. The authors also thank NVIDIA for donating a K40c GPU to support this research.

References

1. Bastien, F., Lamblin, P., Pascanu, R., Bergstra, J., Goodfellow, I.J., Bergeron, A., Bouchard, N., Bengio, Y.: Theano: new features and speed improvements. In: Deep Learning and Unsupervised Feature Learning NIPS 2012 Workshop (2012)
2. Bergstra, J., Breuleux, O., Bastien, F., Lamblin, P., Pascanu, R., Desjardins, G., Turian, J., Warde-Farley, D., Bengio, Y.: Theano: a CPU and GPU math expression compiler. In: Proceedings of the Python for Scientific Computing Conference (SciPy), June 2010. Oral Presentation
3. Breiman, L.: Bagging predictors. Mach. Learn. **24**(2), 123–140 (1996)
4. Dieleman, S., Schlter, J., Raffel, C., Olson, E., Snderby, S.K., Nouri, D., Maturana, D., Thoma, M., Battenberg, E., Kelly, J., Fauw, J.D., Heilman, M., diogo149, McFee, B., Weideman, H., takacsg84, peterderivaz, Jon, instagibbs, Rasul, D.K., CongLiu, Britefury, Degrave, J.: Lasagne: First release, August 2015. http://dx.doi.org/10.5281/zenodo.27878
5. Dietterich, T.G., Bakiri, G.: Solving multiclass learning problems via error-correcting output codes. J. Artif. Intell. Res. **2**, 263–286 (1995)
6. Dong, L., Frank, E., Kramer, S.: Ensembles of balanced nested dichotomies for multi-class problems. In: Jorge, A.M., Torgo, L., Brazdil, P., Camacho, R., Gama, J. (eds.) PKDD 2005. LNCS(LNAI), vol. 3721, pp. 84–95. Springer, Heidelberg (2005). doi:10.1007/11564126_13
7. Duarte-Villaseñor, M.M., Carrasco-Ochoa, J.A., Martínez-Trinidad, J.F., Flores-Garrido, M.: Nested dichotomies based on clustering. In: Alvarez, L., Mejail, M., Gomez, L., Jacobo, J. (eds.) CIARP 2012. LNCS, pp. 162–169. Springer, Heidelberg (2012). doi:10.1007/978-3-642-33275-3_20
8. Frank, E., Kramer, S.: Ensembles of nested dichotomies for multi-class problems. In: Proceedings of the Twenty-First International Conference on Machine Learning, p. 39. ACM (2004)
9. Freund, Y., Schapire, R.E.: Game theory, on-line prediction and boosting. In: Proceedings of the Ninth Annual Conference on Computational Learning Theory, pp. 325–332. ACM (1996)
10. Hall, M., Frank, E., Holmes, G., Pfahringer, B., Reutemann, P., Witten, I.H.: The weka data mining software: an update. ACM SIGKDD Explor. Newsl. **11**(1), 10–18 (2009)
11. Hastie, T., Tibshirani, R., et al.: Classification by pairwise coupling. Ann. Stat. **26**(2), 451–471 (1998)
12. Krizhevsky, A.: Learning multiple layers of features from tiny images. Master's thesis, University of Toronto, Toronto (2009)
13. Lichman, M.: UCI machine learning repository (2013). http://archive.ics.uci.edu/ml
14. Nadeau, C., Bengio, Y.: Inference for the generalization error. Mach. Learn. **52**(3), 239–281 (2003)
15. Rifkin, R., Klautau, A.: In defense of one-vs-all classification. J. Mach. Learn. Res. **5**, 101–141 (2004)
16. Rodríguez, J.J., García-Osorio, C., Maudes, J.: Forests of nested dichotomies. Pattern Recogn. Lett. **31**(2), 125–132 (2010)
17. Webb, G.I.: MultiBoosting: a technique for combining boosting and wagging. Mach. Learn. **40**(2), 159–196 (2000)

Credible Review Detection with Limited Information Using Consistency Features

Subhabrata Mukherjee[(✉)], Sourav Dutta, and Gerhard Weikum

Max Planck Institute for Informatics, Saarbrücken, Germany
{smukherjee,sdutta,weikum}@mpi-inf.mpg.de

Abstract. Online reviews provide viewpoints on the strengths and shortcomings of products/services, influencing potential customers' purchasing decisions. However, the proliferation of *non-credible* reviews — either fake (promoting/ demoting an item), incompetent (involving irrelevant aspects), or biased — entails the problem of identifying *credible* reviews. Prior works involve classifiers harnessing rich information about items/users — which might not be readily available in several domains — that provide only limited interpretability as to why a review is deemed non-credible.

This paper presents a novel approach to address the above issues. We utilize latent topic models leveraging review texts, item ratings, and timestamps to derive *consistency features* without relying on item/user histories, unavailable for "long-tail" items/users. We develop models, for computing review credibility scores to provide interpretable evidence for non-credible reviews, that are also transferable to other domains — addressing the scarcity of labeled data. Experiments on real-world datasets demonstrate improvements over state-of-the-art baselines.

1 Introduction

Motivation: Online reviews about hotels, restaurants, consumer goods, movies, books, drugs, etc. are an invaluable resource for Internet users, providing a wealth of related information for potential customers. Unfortunately, corresponding forums such as TripAdvisor, Yelp, Amazon, and others are being increasingly game to manipulative and deceptive reviews: fake (to promote or demote some item), incompetent (rating an item based on irrelevant aspects), or biased (giving a distorted and inconsistent view of the item). For example, recent studies depict that 20 % of Yelp reviews might be fake and Yelp internally rejects 16 % of user submissions [20] as "not-recommended".

Starting with the work of [11], research efforts have been undertaken to automatically detect non-credible reviews. In parallel, industry (e.g., stakeholders such as Yelp) has developed its own standards[1] to filter out "illegitimate" reviews. Although details are not disclosed, studies suggest that these filters tend to be fairly crude [24]; for instance, exploiting user activity like the number

---

[1] officialblog.yelp.com/2009/10/why-yelp-has-a-review-filter.html.

© Springer International Publishing AG 2016
P. Frasconi et al. (Eds.): ECML PKDD 2016, Part II, LNAI 9852, pp. 195–213, 2016.
DOI: 10.1007/978-3-319-46227-1_13

of reviews posted, and treating users whose ratings show high deviation from the mean/majority ratings as suspicious. Such a policy seems to over-emphasize trusted long-term contributors and suppress outlier opinions off the mainstream. Moreover, these filters also employ several aggregated metadata, and are thus hardly viable for "long tail" items having very few reviews.

State of the Art: Existing research has cast the problem of review credibility into a binary classification task, a review is either credible or deceptive, using supervised and semi-supervised methods that largely rely on features about users and their activities as well as statistics about item ratings. Most techniques also consider spatio-temporal patterns of user activities like IP addresses or user locations (e.g., [14,15]), burstiness of posts on an item or an item group (e.g., [6]), and further correlation measures across users and items (e.g., [25]). However, the classifiers built this way are mostly geared for popular items, and the meta-information about user histories and activity correlations are not always available. For example, someone interested in opinions on a new art film or a "long-tail" bed-and-breakfast in a rarely visited town, is not helped at all by the above methods. Several existing works [21,26,27] consider the textual content of user reviews for tackling opinion spam by using word-level unigrams or bigrams as features, along with specific lexicons (e.g., LIWC [28] psycholinguistic lexicon, WordNet Affect [30]), to learn latent topic models and classifiers (e.g., [16]). Although these methods achieve high classification accuracy, they do not provide any interpretable evidence as to why a certain review is classified as non-credible.

Problem Statement: This paper focuses on detecting credible reviews *with limited information*, namely, in the absence of rich data about user histories, community-wide correlations, and for "long-tail" items. In the extreme case, we are provided with only the review texts and ratings for an item. Our goal is then to compute a *credibility score* for the reviews and to provide possibly *interpretable evidence* for explaining why certain reviews have been categorized as non-credible.

Approach: Our proposed method to this end is to learn a model based on *latent topic models* and combining them with limited metadata to provide a novel notion of *consistency features* characterizing each review. We use the LDA-based Joint Sentiment Topic model (JST) [18] to cast the user review texts into a number of informative facets — per-item, aggregating the text among all reviews for the same item, and also per-review. This allows us to identify, score, and highlight inconsistencies that may appear between a review and the community's overall characterization of an item. Additionally, we learn inconsistencies such as discrepancy between the contents of a review and its rating, and temporal "bursts" — where a number of reviews are written in a short span of time targeting an item. We propose five kinds of inconsistencies in our credibility scoring model, fed into a Support Vector Machine for classification, or for ordinal ranking.

Contribution: In summary, our contributions are summarized as:

- *Model:* We develop a novel *consistency model* for credibility analysis of reviews that works with limited information, with particular attention to "long-tail" items, and offers interpretable evidence for reviews classified as non-credible.
- *Tasks:* We investigate how credibility scores affect the overall ranking of items. To address the scarcity of labeled training data, we transfer the learned model from Yelp to Amazon to rank top-selling items based on (classified) *credible* user reviews. In the presence of proxy labels for item "goodness" (e.g., item sales rank), we develop a better ranking model for domain adaptation.
- *Experiments:* We perform extensive experiments in TripAdvisor, Yelp, and Amazon to demonstrate the viability of our method and its advantages over state-of-the-art baselines in dealing with "long-tail" items and providing interpretable evidence.

2   Related Work

Previous works on fake review/opinion spam detection focused on 2 different aspects:

Linguistic Analysis [21,26,27] – This approach exploits the distributional difference in the wordings of authentic and manually-created fake reviews using word-level features. However, such artificially created fake review datasets give away explicit features not dominant in real-world data, as confirmed by a study on Yelp filtered reviews [24], where the n-gram features performed poorly. Additionally, linguistic features such as *text sentiment* [33], *readability score* (e.g., Automated readability index (ARI), Flesch reading ease, etc.) [9], *textual coherence* [21], and rules based on *Probabilistic Context Free Grammar* (PCFG) [7] have been studied in this context.

Rating and Activity Analysis – In the absence of proper ground-truth data, prior works make simplistic assumptions, e.g., duplicates and near-duplicates are fake, and make use of *extensive* background information like brand name, item description, user history, IP addresses and location, etc. [10,11,14,17,22–24,29,32]. Thereafter, regression models trained on all these features are used to classify reviews as credible or deceptive. Some of these works also use crude or ad-hoc language features like content similarity, presence of literals, numerals, and capitalization. In contrast to these works, our approach uses limited information about users and items catering to a broad domain of applications. We harvest several consistency features from user rating and review text that give some interpretation as to why a review should be deemed non-credible.

Learning to Rank – Supervised models have also been developed to rank items from constructed item feature vectors [19]. Such techniques optimize measures like Discounted Cumulative Gain, Kendall-Tau, and Reciprocal Rank to generate item ranking similar to the training data based on the feature vectors. We use one such technique, and show its performance can be improved by removing non-credible item reviews.

3   Review Credibility Analysis

3.1   Language Model

Previous works [3, 21, 26, 27] in linguistic analysis explore distributional difference in the wordings between deceptive and authentic reviews. In general, authentic reviews tend to have more *sensorial and concrete language* than deceptive reviews, with higher usage of nouns, adjectives, prepositions, determiners, and coordinating conjunctions; whereas deceptive reviews were shown to use more verbs, adverbs, and superlatives manifested in exaggeration for imaginary writing. [26, 27] found that authentic hotel reviews are more specific about spatial configurations (small room, low ceiling, etc.) and aspects like location, amenities and cost; whereas deceptive reviews focus on aspects external to the item being reviewed (like traffic jam, children, etc.). Extreme opinions were also found to be dominant in deceptive reviews to assert stances, whereas authentic reviews have a more balanced view. Our latent facet model implicitly exploits these features to find opinion on important item facets and the overall rating distribution.

In order to explicitly capture such distributional difference in the language of credible and non-credible reviews at word-level, we use unigram and bigram language features shown to outperform other fine-grained psycholinguistic features (e.g., LIWC lexicon) and Part-of-Speech tags [27]. We also experimented with WordNet Affect to capture fine-grained emotional dimensions (like anger, hatred, and confidence), which, however, were seen not to perform well. In general, the bigram features capture context-dependent information to some extent, and together with simple unigram features performed the best, with the presence or absence of words mattering more than their frequency for credibility analysis. In our model, all the features were length normalized, retaining punctuations (like '!') and capitalization as non-credible reviews manifesting exaggeration tend to over-use the latter features (e.g., "the hotel was AWESOME !!!").

Feature vector construction: Consider a vocabulary V of unique unigrams and bigrams in the corpus (after removing stop words). For each token type $f_i \in V$ and each review d_j, we compute the presence/absence of words, w_{ij}, of type f_i occurring in d_j, thus constructing a feature vector $F^L(d_j) = \langle w_{ij} = I(w_{ij} = f_i) \, / \, length(d_j) \rangle, \forall i$, with $I(.)$ denoting an indicator function (notations used are presented in Table 1).

3.2   Facet Model

Given review snippets like "the hotel offers free wi-fi", we now aim to find the different facets present in the reviews along with their corresponding sentiment polarities by extracting the *latent* facets from the review text, without the help of any explicit facet or seed words, e.g., ideally "wi-fi" should be mapped to a latent facet cluster like "network, Internet, computer, access, ...". We also want to extract the sentiment expressed in the review about the facet. Interestingly, although "free" does not have a polarity of its own, in the above example "free" in conjunction with "wi-fi" expresses a positive sentiment of a service being offered without charge. The hope is that although "free" does not have an individual polarity, it appears in the neighborhood of words that have known polarities (from lexicons). This helps in the joint discovery of facets and sentiment labels, as "free wi-fi" and "internet without extra charge" should ideally map to the same facet cluster with similar polarities using their co-occurrence with similar words with positive polarities. In this work, we use the Joint Sentiment Topic Model approach (JST) [18] to jointly discover the latent facets along with their expressed polarities.

Consider a set of reviews $\langle D \rangle$ written by users $\langle U \rangle$ on a set of items $\langle I \rangle$, with $r_d \in R$ being the rating assigned to review $d \in D$. Each review document d consists of a sequence of words N_d denoted by $\{w_1, w_2, ... w_{N_d}\}$, and each word is drawn from a vocabulary V indexed by $1, 2, ..V$. Consider a set of facet assignments $z = \{z_1, z_2, ... z_K\}$ and sentiment label assignments $l = \{l_1, l_2, ... l_L\}$ for d, where each z_i can be from a set of K possible facets, and each label l_i is from a set of L possible sentiment labels.

JST adds a layer of sentiment in addition to the topics as in standard LDA [1]. It assumes each document d to be associated with a multinomial distribution θ_d over facets z and sentiment labels l with a symmetric Dirichlet prior α. $\theta_d(z, l)$ denotes the probability of occurrence of facet z with polarity l in document d. Topics have a multinomial distribution $\phi_{z,l}$ over words drawn from a vocabulary V with a symmetric Dirichlet prior β. $\phi_{z,l}(w)$ denotes the probability of the word w belonging to the facet z with polarity l. In the generative process, a sentiment label l is first chosen from a document-specific rating distribution π_d with a symmetric Dirichlet prior γ. Thereafter, a facet z from θ_d conditioned on l is chosen, and subsequently a word w from ϕ conditioned on z and l. Exact inference is not possible due to intractable coupling between Θ and Φ, and thus we use Collapsed Gibbs Sampling for approximate inference.

Let $n(d, z, l, w)$ denote the count of the word w occurring in document d belonging to the facet z with polarity l. The conditional distribution for the latent variable z (with components z_1 to z_K) and l (with components l_1 to l_L) is given by:

$$P(z_i = k, l_i = j | w_i = w, z_{-i}, l_{-i}, w_{-i}) \propto$$
$$\frac{n(d, k, j, .) + \alpha}{\sum_k n(d, k, j, .) + K\alpha} \times \frac{n(., k, j, w) + \beta}{\sum_w n(., k, j, w) + V\beta} \times \frac{n(d, ., j, .) + \gamma}{\sum_j n(d, ., j, .) + L\gamma} \quad (1)$$

In the above equation, the operator (.) in the count indicates marginalization, i.e., summing up the counts over all values for the corresponding position in $n(d, z, l, w)$, and the subscript $-i$ denotes the value of a variable excluding the data at the ith position.

3.3   Consistency Features

We extract the following features from the latent facet model enabling us to detect *inconsistencies* in reviews and ratings of items for credibility analysis.

1. **User Review – Facet Description:** The facet-label distribution of different items differ; i.e., for some items, certain facets (with polarity) are more important than others. For instance, the "battery life" and "ease of use" for consumer electronics are more important than "color"; for hotels, certain services are available for free (e.g., wi-fi) which may be charged elsewhere. Hence, user reviews involving less relevant facets of the item, e.g., downrating hotels for "not allowing pets", should also be detected.

Given a review $d(i)$ on an item $i \in I$ with a sequence of words $\{w\}$ and previously learned Φ, its facet label distribution $\Phi'_d(i)$ with dimension $K \times L$ is given by:

$$\phi'_{k,l} = \sum_{w:l^*=argmax_l \ \phi_{k,l}(w)} \phi_{k,l^*}(w) \tag{2}$$

For each word, w, and latent facet dimension, k, we consider the sentiment label l^* that maximizes the facet-label-word distribution $\phi_{k,l}(w)$, and is aggregated over all words. This facet-label distribution for review $\Phi'_d(i)$ (dimension $K \times L$) forms a feature vector capturing the importance of various latent dimensions and *domain-specific* facet-labels.

2. **User Review — Rating:** The user-assigned rating corresponding to the review should be consistent to her opinion expressed in the review text. For example, the user is unlikely to give an average rating to an item when she expresses a positive opinion about all the important facets of the item. The inferred rating distribution π'_d (with dimension L) of a review d consisting of a sequence of words $\{w\}$ and learned Φ is computed as:

$$\pi'_l = \sum_{w,k:\{k^*,l^*\}=argmax_{k,l} \ \phi_{k,l}(w)} \phi_{k^*,l^*}(w) \tag{3}$$

For each word, we consider the facet and label that jointly maximizes the facet-label-word distribution, and aggregate over all the words and facets. The absolute deviation (of dimension L) between the user-assigned rating π_d, and estimated rating π'_d from user text is taken as a component in the overall feature vector.

3. **User Rating:** Previous works [9, 27, 31] on opinion spam found that fake reviews tend to have overtly positive or overtly negative opinions. Therefore, we also use π'_d as a component of the overall feature vector to detect cues from such extreme ratings.

4. **Temporal Burst:** Typically observed in *group spamming*, where a number of reviews are posted in a short span of time. Consider a set of reviews $\{d_j\}$ posted at timepoints $\{t_j\}$ for a *specific* item. The burstiness of review d_i for the item is $\left(\sum_{j,j\neq i} \frac{1}{1+e^{t_i-t_j}}\right)$, with exponential decay used to weigh the temporal proximity of reviews for burst.

5. **User Review – Item Description:** In general, the description of the facets in an item review should not differ much from that of the majority. For example, if majority says the "hotel offers free wi-fi", and a user review says "internet

is charged" — this presents a possible inconsistency. For the facet model this corresponds to word clusters having the same facet label but different sentiment labels. However, experimentally we found this feature to play a weak role in the presence of other inconsistency features.

We aggregate the *per-review* facet distribution $\phi'_{k,l}$ over all the reviews $d(i)$ on the item i to obtain the facet-label distribution $\Phi''(i)$ of the item. We use the Jensen-Shannon divergence, a symmetric and smoothed version of the Kullback-Leibler divergence as a feature. This depicts how much the facet-label distribution in the given review diverges from the general opinion of other people about the item.

$$JSD(\Phi'_d(i) \parallel \Phi''(i)) = \frac{1}{2}(D(\Phi'_d(i) \parallel M) + D(\Phi''(i) \parallel M)) \qquad (4)$$

where, $M = \frac{1}{2}(\Phi'_d(i) + \Phi''(i))$, and D represents Kullback-Leibler divergence.

Feature vector construction: For each review d_j, the above *consistency features* are computed, and a facet feature vector $\langle F^T(d_j) \rangle$ of dimension $2 + K \times L + 2L$ is created.

3.4 Behavioral Model

Earlier works [10,11,17] on review spam show that user-dependent models detecting user-preferences and biases perform well in credibility analysis. However, such information is not always available, especially for newcomers, and not so active users in the community. Besides, [22,23] show that spammers tend to open multiple fake accounts to write reviews for malicious activities — using each of those accounts sparsely to avoid detection. Therefore, instead of relying on extensive user history, we use simple proxies for user activity that are easier to aggregate from the community:

1. **User Posts:** number of posts written by the user in the community.
2. **Review Length:** longer reviews tend to go off-topic (high emotional digression).
3. **User Rating Behavior:** absolute deviation of the review rating from the mean and median rating of the user to other items, as well as the first three moments of the user rating distribution — capturing whether a user has a *typical rating behavior*.
4. **Item Rating Pattern:** absolute deviation of the item rating from the mean and median rating obtained from other users captures the extent to which the user disagrees with other users about the item quality; the first three moments of the item rating distribution captures the general item rating pattern.
5. **User Friends:** number of friends of the user.
6. **User Check-in:** if the user checked-in the hotel — first hand experience of the user adds to the review credibility.
7. **Elite:** elite status of the user in the community.
8. **Review helpfulness:** number of up-votes received to capture the quality of reviews.

Note that user rating behavior and item rating pattern are also captured *implicitly* using the consistency features in the latent facet model.

Since we aim to detect credible reviews in scenarios of limited information, we split the above activity or behavioral features into two components: (a) *Activity*⁻ using features [1 − 4], obtained straightforward from the tuple $\langle userId, itemId, review, rating \rangle$ and easily available even for "long-tail" items/users; and (b) *Activity*⁺ using all the features. However the latter requires additional information (features [5 − −8]) that might not always be available, or takes long time to aggregate for new items/users.

Feature vector construction: For each review d_j by user u_k, we construct a behavioral feature vector $\langle F^B(d_j) \rangle$ using the above features.

3.5   Application Oriented Tasks

Credible Review Classification: In the first task, we *classify* reviews as *credible* or not. For each review d_j by user u_k, we construct the joint feature vector $F(d_j) = F^L(d_j) \cup F^T(d_j) \cup F^B(d_j)$, and use Support Vector Machines (SVM) [4] for classification of the reviews. SVM maps the features (using Kernels) to a high dimensional space, and constructs a hyperplane to separate the two categories. Although there can be an infinite number of such hyperplanes possible, SVM constructs the one with the largest functional margin given by the distance of the nearest point to the hyperplane on each side of it. New points are mapped to the same space and classified to a category based on which side of the hyperplane it lies. We use a linear kernel shown to perform the best for text classification tasks. We use the L_2 regularized L_2 loss SVM with dual formulation from the LibLinear package (csie.ntu.edu.tw/cjlin/liblinear) [5], and report 10-fold cross-validation classification accuracy on TripAdvisor and Yelp datasets.

Item Ranking: Due to the scarcity of ground-truth data pertaining to review credibility, a more suitable way to evaluate our model is to examine the *effect* of non-credible reviews on the relative *ranking* of items in the community. For instance, in case of popular items with large number of reviews, even if a fraction of it were non-credible, its effect would not be so severe as would be on "long-tail" items with fewer reviews.

A simple way to find the "goodness" of an item is to aggregate ratings of all reviews − using which we also obtain a ranking of items. We use our model to filter out non-credible reviews, aggregate ratings of credible reviews, and re-compute the item ranks.

Evaluation Measures − We use the *Kendall-Tau Rank Correlation Co-efficient* (τ) to find effectiveness of the rankings, against a *reference ranking* — for instance, the *sales rank* of items in Amazon. τ measures the number of concordant and discordant pairs, to find whether the ranks of two elements agree or not based on their scores, out of the total number of combinations possible. Given a set of observations $\{x, y\}$, any pair of observations (x_i, y_i) and (x_j, y_j), where $i \neq j$, are said to be *concordant* if either $x_i > x_j$ and $y_i > y_j$, or $x_i < x_j$

and $y_i < y_j$, and *discordant* otherwise. If $x_i = x_j$ or $y_i = y_j$, the ranks are tied — neither discordant, nor concordant.

We use the *Kendall-Tau-B* measure (τ_b) which allows for rank adjustment. Consider n_c, n_d, t_x, and t_y to be the number of concordant, discordant, tied pairs on x, and tied pairs on y respectively, whereby Kendall-Tau-B is given by: $\frac{n_c - n_d}{\sqrt{(n_c+n_d+t_x)(n_c+n_d+t_y)}}$.

However, this is a conservative estimate as multiple items — typically the top-selling ones in Amazon — have the same rating. Therefore, we use a second estimate (*Kendall-Tau-M* (τ_m)) considering non-zero tied ranks to be concordant. Note that, an item can have a zero-rank if all of its reviews are classified as non-credible. A high positive (or, negative) value of Kendall-Tau indicates the two series are positively (or, negatively) correlated; whereas a value close to zero indicates they are independent.

Domain Transfer from Yelp to Amazon – A typical issue in credibility analysis task is the scarcity of labeled training data. In the first task, we use labels from the Yelp Spam Filter (considered to be the industry standard) to train our model. However, such ground-truth labels are not available in Amazon. Although, in principle, we can train a model M_{Yelp} on Yelp, and use it to filter out non-credible reviews in Amazon.

Transferring the learned model from Yelp to Amazon (or other domains) entails using the learned weights of *features* in Yelp that are analogous to the ones in Amazon. However, this process encounters the following issues:

- Facet distribution of Yelp (food and restaurants) is different from that of Amazon (products such as software, and consumer electronics). Therefore, the facet-label distribution and the corresponding learned feature weights from Yelp cannot be directly used, as the latent dimensions are different.
- Additionally, specific metadata like check-in, user-friends, and elite-status are missing in Amazon.

However, the learned weights for the following features can still be directly used:

- Certain unigrams and bigrams, especially those depicting opinion, that occur in both domains.
- Behavioral features like user and item rating patterns, review count and length, and usefulness votes.
- Deviation features derived from *Amazon-specific* facet-label distribution that is obtained using the JST model on Amazon corpus:
  - Deviation (with dimension L) of the user assigned rating from that inferred from review content.
  - Distribution (with dimension L) of positive and negative sentiment as expressed in the review.
  - Divergence, as a unary feature, of the facet-label distribution in the review from the aggregated distribution over other reviews on a given item.
  - Burstiness, as a unary feature, of the review.

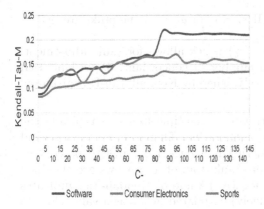

Fig. 1. Variation of Kendall-Tau-M (τ_m) on different Amazon domains with parameter C^- variation (using model M_{Yelp} trained in Yelp and tested in Amazon).

Table 1. List of variables and notations used with corresponding description.

Notation	Description
U, D, I	set of users, reviews, and items resp.
d, r_d	review text and associated rating
V, f	unigrams and bigrams vocab. &token types
w_{ij}	word of token type f_i in review d_j
$I(\cdot)$	indicator fn. for presence/absence of words
z, l	set of facets and sentiment labels resp.
K, L	cardinality of facets and sentiment labels
$\theta_d(z, l)$	multinom. prob. distr. of facet z with sentiment label l in document d
$\phi_{z,l}(w)$	multinom. prob. distr. of word w belonging to facet z with sentiment label l
Φ', Φ''	facet-label distr. of review and item resp.
α, β, γ	Dirichlet priors
π, π'	review rating distr. &inferred rating distr.
$n(\cdot)$	word count in reviews
$F^x(d_j)$	feature vec. of review d_j using lang. (x=L), consistency (x=T), and behavior (x=B)
c^+, c^-	C-SVM regularization parameters

Using the above components, that are common to both Yelp and Amazon, we *first* re-train the model M_{Yelp} from Yelp to remove the non-contributing features for Amazon.

Now, a direct transfer of the model weights from Yelp to Amazon assumes the distribution of credible to non-credible reviews, and corresponding feature importance, to be the same in both domains — which is not necessarily true. In order to boost certain features to better identify non-credible reviews in Amazon, we tune the *soft margin parameter C* in the SVM. We use *C-SVM* [2], with slack variables, that optimizes:

$$min_{\boldsymbol{w},b,\xi_i \geq 0} \frac{1}{2}\boldsymbol{w}^T\boldsymbol{w} + C^+ \sum_{y_i=+1} \xi_i + C^- \sum_{y_i=-1} \xi_i$$

$$\text{subject to } \forall\{(\boldsymbol{x_i}, y_i)\}, y_i(\boldsymbol{w}^T\boldsymbol{x_i} + b) \geq 1 - \xi_i \tag{5}$$

C^+ and C^- are regularization parameters for positive and negative class (credible and deceptive), respectively. The parameters $\{C\}$ provide a trade off as to how wide the margin can be made by moving around certain points which incurs a penalty of $\{C\xi_i\}$. A high value of C^- places a large penalty for misclassifying instances from the negative class, and therefore boosts certain features from that class. As the value of C^- increases, the model classifies more reviews as non-credible. In the worse case, all reviews of an item are deemed as non-credible, with the aggregated item rating being 0.

Table 2. Dataset statistics for review classification. (Yelp* denotes balanced dataset using random sampling.)

Dataset	Non-credible reviews	Credible reviews	Items	Users
TripAdvisor	800	800	20	-
Yelp	5169	37,500	273	24,769
Yelp*	5169	5169	151	7898

Table 3. Amazon dataset statistics for item ranking, with cumulative #items and varying #reviews.

Domain	#Users	#Reviews	#Items with reviews per-item						Total
			≤5	≤10	≤20	≤30	≤40	≤50	
Consumer electronics	94,664	121,234	14,797	16,963	18,350	18,829	19,053	19,187	19,518
Software	21,825	26,767	3,814	4,354	4,668	4,767	4,807	4,828	4,889
Sports	656	695	202	226	233	235	235	235	235

We use τ_m to find the optimal value of C^- by varying it in the interval $C^- \in \{0, 5, 10, 15, ...150\}$ using a *validation set* from Amazon as shown in Fig. 1. We observe that as C^- increases, τ_m also increases till a certain point as more and more non-credible reviews are filtered out, after which it stabilizes.

Ranking SVM – Our previous approach uses the model M_{Yelp} trained on Yelp, with the references sales ranking in Amazon being used only for evaluating the item rankings on the Kendall-Tau measure. To obtain a good item ranking based on credible reviews, a model M_{Amazon} that directly optimizes for Kendall-Tau using the reference ranking as training labels can be used. This allows the use of the entire feature space available in Amazon, including the explicit facet-label distribution and the full vocabulary. The feature space is constructed similarly to that of the Yelp dataset.

The goal of Ranking SVM [12] is to learn a ranking function which is concordant with a given ordering of items. The objective is to learn w such that $w \cdot x_i > w \cdot x_j$ for most data pairs $\{(x_i, x_j) : y_i > y_j \in R\}$. Although the problem is known to be NP-hard, it is approximated using SVM techniques with pairwise slack variables $\xi_{i,j}$. The optimization problem is equivalent to that of classifying SVM, but now operating on *pairwise difference vectors* $(x_i - x_j)$ with corresponding labels $+1/-1$ indicating which one should be ranked ahead. We use the implementation of [12] (obtained from www.cs.cornell.edu/people/tj/svm_light/svm_rank.html) that maximizes the empirical Kendall-Tau by minimizing the number of discordant pairs.

Unlike the classification task, where labels are *per-review*, the ranking task requires labels *per-item*. Consider $\langle f_{i,j,k} \rangle$ to be the feature vector for the jth review of an item i, with k indexing an element of the feature vector.

We aggregate these feature vectors element-wise over all the reviews on item i to obtain its feature vector $\langle \frac{\sum_j f_{i,j,k}}{\sum_j 1} \rangle$.

4   Experimental Setup

Parameter Initialization: The sentiment lexicon from [8] consisting of 2006 positive and 4783 negative polarity bearing words is used to initialize the review text based facet-label-word tensor Φ prior to inference. We consider the number of topics, $K = 20$ for Yelp, and $K = 50$ for Amazon with the review sentiment labels $L = \{+1, -1\}$ (corresponding to positive and negative rated reviews) initialized randomly. The symmetric Dirichlet priors are set to $\alpha = 50/K$, $\beta = 0.01$, and $\gamma = 0.1$.

Datasets and Ground-Truth: In this work, we consider the following datasets (refer to Tables 2 and 3) with available ground-truth information.
• The *TripAdvisor Dataset* [26,27] consists of 1600 reviews from TripAdvisor with positive (5 star) and negative (1 star) sentiment — comprising 20 credible and 20 non-credible reviews for *each* of 20 most popular Chicago hotels. The authors crawled the *credible* reviews from online review portals like TripAdvisor; whereas the *non-credible* ones were generated by users in Amazon Mechanical Turk. The dataset has only the review text and sentiment label (positive/negative ratings) with corresponding hotel names, with no other information on users or items.
• The *Yelp Dataset* consists of 37.5K recommended (i.e., *credible*) reviews, and 5K non-recommended (i.e., *non-credible*) reviews given by the Yelp filtering algorithm, on 273 restaurants in Chicago. For each review, we gather the following information: $\langle userId, itemId, timestamp, rating, review, metadata \rangle$. The metadata consists of some user activity information as outlined in Sect. 3.4.

The reviews marked as "not recommended" by the Yelp spam filter are considered to be the ground-truth for comparing the accuracy for credible review detection for our proposed model. The Yelp spam filter presumably relies on linguistic, behavioral, and social networking features [24].
• The *Amazon Dataset* used in [11] consists of around 149K reviews from nearly 117K users on 25K items from three domains, namely Consumer Electronics, Software, and Sports items. For each review, we gather the same information tuple as that from Yelp. However, the metadata in this dataset is not as rich as in Yelp, consisting only of helpfulness votes on the reviews.

Further, there exists no explicit ground-truth characterizing the reviews as credible or deceptive in Amazon. To this end, we re-rank the items using learning to rank, implicitly filtering out possible deceptive reviews (based on the feature vectors), and then compare the ranking to the *item sales rank* considered as the pseudo ground-truth.

Comparison Baselines: We use the following state-of-the-art baselines (given the full set of features that fit with their model) for comparison with our proposed model.

(1) Language Model Baselines: We consider the unigram and bigram language model baselines from [26,27] that have been shown to outperform other baselines using psycholinguistic features, part-of-speech tags, information gain, etc. We take the best baseline from their work which is a combination of unigrams and bigrams. Our proposed model (N-gram+Facet) enriches it by using length normalization, presence or absence of features, latent facets, etc. The recently proposed *doc-to-vec* model based on Neural Networks, overcomes the weakness of bag-of-words models by taking the context of words into account, and learns a dense vector representation for each document [13]. We train the doc-to-vec model in our dataset as a baseline model. In addition, we also consider readability (ARI) and review sentiment scores [9] under the hypothesis that writing styles would be random because of diverse customer background. ARI measures the reader's ability to comprehend a text and is measured as a function of the total number of characters, words, and sentences present, while review sentiment tries to capture the fraction of occurrences of positive/negative sentiment words to the total number of such words used.

(2) Activity and Rating Baselines: Given the tuple ⟨*userId, itemId, rating, review, metadata*⟩ from the Yelp dataset, we extract all possible activity and rating behavioral features of users as proposed in [10,11,14,17,22–24,32]. Specifically, we utilize the number of helpful feedbacks, review title length, review rating, use of brand names, percent of positive and negative sentiments, average rating, and rating deviation as features for classification. Further, based on the recent work of [29], we also use the user check-in and user elite status information as additional features for comparison.

Empirical Evaluations: Our experimental setup considers the following evaluations:

(1) Credible review classification: We study the performance of the various approaches in distinguishing a *credible* review from a *non-credible* one. Since this forms a binary classification task, we consider a balanced dataset containing equal proportion of data from each of the two classes. On the Yelp dataset, for each item we randomly sample an equal number of credible and non-credible reviews (to obtain Yelp*); while the TripAdvisor dataset is already balanced. Table 4 shows the 10-fold cross validation accuracy results for the different models on the two datasets. We observe that our proposed *consistency and behavioral features* exhibit around 15 % improvement in Yelp* for classification accuracy over the best performing baselines (refer to Table 4). Since the TripAdvisor dataset has *only* review text, the user/activity models could *not* be used there. The experiment could also not be performed on Amazon, as the ground-truth for credibility labels of reviews is absent.

(2) Item Ranking: In this task we examine the effect of non-credible reviews on the ranking of items in the community. This experiment is performed *only* on Amazon using the item *sales rank* as ground or reference ranking, as Yelp does not provide such item rankings. The sales rank provides an indication as to how

Table 4. Credible review classification accuracy with 10-fold cross validation. TripAdvisor dataset contains only review texts and no user/activity information.

Models	Features	TripAdvisor	Yelp*
Deep learning	Doc2Vec	69.56	64.84
	Doc2Vec + ARI + Sentiment	76.62	65.01
Activity & rating	Activity+Rating	-	74.68
	Activity+Rating+Elite+Check-in	-	79.43
Language	Unigram + Bigram	88.37	73.63
	Consistency	80.12	76.5
Behavioral	Activity Model$^-$	-	80.24
	Activity Model$^+$	-	86.35
Aggregated	N-gram + Consistency	**89.25**	79.72
	N-gram + Activity$^-$	-	82.84
	N-gram + Activity$^+$	-	88.44
	N-gram + Consistency + Activity$^-$	-	86.58
	N-gram + Consistency + Activity$^+$	-	**91.09**
	M_{Yelp}	-	89.87

well a product is selling on Amazon.com and highlights the item's rank in the corresponding category[2].

The baseline for the item ranking is based on the aggregated rating of all reviews on an item. The first model M_{Yelp} (C-SVM) trained on Yelp filters out the non-credible reviews, before aggregating review ratings on an item. The second model M_{Amazon} (SVM-Rank) is trained on Amazon using SVM-Rank with the reference ranking as training labels. 10-fold cross-validation results are reported on the two measures of Kendall-Tau (τ_b and τ_m) in Table 5 with respect to the reference ranking. τ_b and τ_m for SVM-Rank are the same since there are no ties. Our first model performs substantially better than the baseline, which, in turn, is outperformed by our second model.

In order to find the effectiveness of our approach in dealing with "long-tail" items, we perform an additional experiment with our best performing model i.e., M_{Amazon} (SVM-Rank). We use the model to find Kendall-Tau-M (τ_m) rank correlation (with the reference ranking) of items having less than (or equal to) $5, 10, 20, 30, 40$, and 50 reviews in different domains in Amazon (results reported in Table 6 with 10-fold cross validation). We observe that our model performs substantially well even with items having as few as *five* reviews, with the performance progressively getting better with more reviews per-item.

---

[2] www.amazon.com/gp/help/customer/display.html?nodeId=525376.

Table 5. Kendall-Tau correlation of different models across domains.

Domain	Kendall-Tau-B (τ_b)		Kendall-Tau-M (τ_m)		Kendall-Tau ($\tau_b = \tau_m$)
	Baseline	M_{Yelp} (C-SVM)	Baseline	M_{Yelp} (C-SVM)	M_{Amazon} (SVM-Rank)
CE	0.011	0.109	0.082	0.135	0.329
Software	0.007	0.184	0.088	0.216	0.426
Sports	0.021	0.155	0.102	0.170	0.325

Table 6. Variation of Kendall-Tau-M (τ_m) correlation with #reviews with M_{Amazon} (SVM-Rank).

Domain	τ_m with #reviews per-item						
	≤ 5	≤ 10	≤ 20	≤ 30	≤ 40	≤ 50	Overall
CE	0.218	0.257	0.290	0.304	0.312	0.317	0.329
Software	0.353	0.375	0.401	0.411	0.417	0.419	0.426
Sports	0.273	0.324	0.310	0.325	0.325	0.325	0.325

5   Discussions on Experimental Results

Language Model: The bigram language model performs very well (refer to Table 4) on the TripAdvisor dataset due to the setting of the task. Workers in Amazon Mechanical Turk were tasked with writing fake reviews with the guideline of knowing all the hotel amenities in its website before writing reviews. Therefore it is quite difficult for the facet model to find contradictions or mismatch in facet descriptions. Consequently, the facet model gives marginal improvement when combined with the language model.

On the other hand, the Yelp dataset is real-world, and therefore more noisy. The bigram language model and doc-to-vec hence do not perform as good as they do in the previous dataset; and neither does the facet model in isolation. However all the components put together give significant performance improvement over the ones in isolation (around 8 %).

Incorporation of writing style using ARI and sentiment measures improves performance of doc-to-vec in the TripAdvisor dataset, but not significantly in the real-world Yelp data.

Table 7 shows the top unigrams and bigrams contributing to the language feature space in the *joint model* for credibility classification — given by the feature weights of the C-SVM. We find that credible reviews contain a mix of function and content words, balanced opinions, with the highly contributing features being mostly unigrams. Whereas, non-credible reviews contain extreme opinions, less function words and more of sophisticated content words — consisting of a lot of signature bigrams — to catch the readers' attention.

Table 7. Top n-grams (by feature weights) for credibility classification.

Credible Reviews	Non-credible reviews
not, also, really, just, like, get, perfect, little, good, one, space, pretty, can, everything, come_back, still, us, right, definitely, enough, much, super, free, around, delicious, no, fresh, big, favorite, lot, selection, sure, friendly, way, dish, since, huge, etc., menu, large, easy, last, room, guests, find, location, time, probably, helpful, great, now, something, two, nice, small, better, sweet, though, loved, happy, love, anything, actually, home	dirty, mediocre, charged, customer_service, signature_lounge, view_city, nice_place, hotel_staff, good_service, never_go, overpriced, several_times, wait_staff, signature_room, outstanding, establishment, architecture_foundation, will_not, long, waste, food_great, glamour_closet, glamour, food_service, love_place, terrible, great_place, never, wonderful, atmosphere, signature, bill, will_never, good_food, management, great_food, money, worst, horrible, manager, service, rude

Behavioral Model: We find the activity based model to perform the best in isolation (refer to Table 4). Combined with language and consistency features, the joint model exhibits around 5 % improvement in performance. Additional metadata like the user elite and check-in status improves the performance of activity based baselines, which are not typically available for newcomers in the community. Our model using limited information (N-$gram$+$Consistency$+$Activity^-$) performs better than the activity baselines using fine-grained information about items (like brand description) and user history. Incorporating additional user features ($Activity^+$) further boosts its performance.

Consistency Features: In order to find the effectiveness of the facet based consistency features, we perform ablation tests (refer to Table 4). We remove the consistency model from the aggregated model, and see significant performance degradation of $3 - 4\%$ for the Yelp* dataset. In the TripAdvisor dataset the performance reduction is less compared to Yelp due to reasons outlined before.

Table 8 shows a snapshot of the non-credible reviews, with corresponding (in)consistency features in Yelp and Amazon. We see that ratings of deceptive reviews do not corroborate with the textual description, irrelevant facets influencing the rating of the target item, contradicting other users, expressing extreme opinions without explanation, depicting temporal "burst" in ratings, etc. In principle, these features can also be used to detect other anomalous phenomena like group-spamming (one of the principal indicators of which is temporal burst), which is out of scope of this work.

Ranking Task: For the ranking task in Amazon (refer to Table 5), the first model M_{Yelp} — trained on Yelp and tested on Amazon using C-SVM — performs much better than the baseline exploiting various consistency features. The second model M_{Amazon} — trained on Amazon using SVM-Rank — outperforms the

Table 8. Snapshot of non-credible reviews (reproduced verbatim) with inconsistencies.

Inconsistency features	Yelp review & [rating]	Amazon review & [rating]
user review − **rating** *(promotion/demotion)*:	never been inside James. never checked in. never visited bar. yet, one of my favorite hotels in Chicago. James has dog friendly area. my dog loves it there. [5]	Excellant product-alarm zone, technical support is almost non-existent because of this i will look to another product. this is unacceptible. [4]
user review − facet **description** *(irrelevant)*:	you will learn that they are actually EVANGELICAL CHRISTIANS working to proselytize the coffee farmers they buy from. [2]	DO NOT BUY THIS. I used turbo tax since 2003, it never let me down until now. I can't file because Turbo Tax doesn't have software updates from the IRS "because of Hurricane Katrina". [1]
user review − item **description** *(deviation from community)*:	internet is charged in a 300 dollar hotel! [3]	The book Amazon offers is a joke! All it provides is the forward which is not written by Kalanithi. I don't have any sample of HIS writing to know if it appeals. [1]
extreme user **rating:**	GREAT!!!i give 5 stars!!!Keep it up. [5]	GREAT. This camera takes pictures. [1]
temporal bursts[3]:	Dan's apartment was beautiful and a great downtown location... (3/14/2012) [5] I highly recommend working with Dan and NSRA... (3/14/2012) [5] Dan is super friendly, demonstrating that he was confident... (3/14/2012) [5] my condo listing with no activity, Dan really stepped in... (4/18/2012) [5]	

[3]these reviews have also been flagged by the Yelp Spam Filter as not-recommended (i.e., non-credible)

former exploiting the power of the entire feature space and domain-specific proxy labels unavailable to the former.

"Long-Tail" Items: Table 6 shows the gradual degradation in performance of the second model M_{Amazon} (SVM-Rank) in dealing with items with lesser number of reviews. Nevertheless, we observe it to give a substantial Kendall-Tau correlation (τ_m) with the reference ranking, with as few as *five* reviews per-item, demonstrating the effectiveness of our model in dealing with "long-tail" items.

6 Conclusions

We present a novel consistency model using limited information for detecting non-credible reviews which is shown to outperform state-of-the-art baselines. Our approach overcomes the limitation of existing works that make use of fine-grained information which are not available for "long-tail" items or newcomers in the community. Most importantly prior methods are not designed to *explain* why the detected review should be non-credible. In contrast, we make use of different *consistency features* from *latent facet model* derived from user text and ratings that can explain the assessments by our method. We develop multiple models for domain transfer and adaptation, where our model performs very well in the ranking tasks involving "long-tail" items, with as few as *five* reviews per-item.

References

1. Blei, D.M., Ng, A.Y., Jordan, M.I.: Latent dirichlet allocation. J. Mach. Learn. Res. **3**, 993–1022 (2003)
2. Chen, D.R., Wu, Q., Ying, Y., Zhou, D.X.: Support vector machine soft margin classifiers: error analysis. J. Mach. Learn. Res. **5**, 1143–1175 (2004)
3. Chen, Y.R., Chen, H.H.: Opinion spam detection in web forum:a real case study. In: WWW (2015)
4. Cortes, C., Vapnik, V.: Support-vector networks. Mach. Learn. **20**(3), 273–297 (1995)
5. Fan, R.E., Chang, K.W., Hsieh, C.J., Wang, X.R., Lin, C.J.: Liblinear: a library for large linear classification. J. Mach. Learn. Res. **9**, 1871–1874 (2008)
6. Fei, G., Mukherjee, A., Liu, B., Hsu, M., Castellanos, M., Ghosh, R.: Exploiting burstiness in reviews for review spammer detection. In: ICWSM (2013)
7. Feng, S., Banerjee, R., Choi, Y.: Syntactic stylometry for deception detection. In: ACL (2012)
8. Hu, M., Liu, B.: Mining and summarizing customer reviews. In: KDD (2004)
9. Hu, N., Bose, I., Koh, N.S., Liu, L.: Manipulation of online reviews: an analysis of ratings, readability, and sentiments. Decis. Support Syst. **52**(3), 674–684 (2012)
10. Jindal, N., Liu, B.: Analyzing and detecting review spam. In: ICDM, pp. 547–552 (2007)
11. Jindal, N., Liu, B.: Opinion spam and analysis. In: WSDM, pp. 219–230 (2008)
12. Joachims, T.: Optimizing search engines using clickthrough data. In: KDD (2002)
13. Le, Q., Mikolov, T.: Distributed representations of sentences and documents. In: ICML (2014)
14. Li, H., Chen, Z., Liu, B., Wei, X., Shao, J.: Spotting fake reviews via collective positive-unlabeled learning. In: ICDM, pp. 899–904 (2014)
15. Li, H., Chen, Z., Mukherjee, A., Liu, B., Shao, J.: Analyzing and detecting opinion spam on a large-scale dataset via temporal and spatial patterns. In: ICWSM (2015)
16. Li, J., Ott, M., Cardie, C.: Identifying manipulated offerings on review portals. In: EMNLP (2013)
17. Lim, E., Nguyen, V., Jindal, N., Liu, B., Lauw, H.W.: Detecting product review spammers using rating behaviors. In: CIKM, pp. 939–948 (2010)
18. Lin, C., He, Y.: Joint sentiment/topic model for sentiment analysis. In: CIKM (2009)
19. Liu, T.Y.: Learning to rank for information retrieval. Found. Trends Inf. Retrieval **3**(3), 225–331 (2009)
20. Luca, M., Zervas, G.: Fake it till you make it: Reputation, competition, and yelp review fraud. Technical report, Harvard Business School (2015)
21. Mihalcea, R., Strapparava, C.: The lie detector: explorations in the automatic recognition of deceptive language. In: ACL/IJCNLP (Short Papers), pp. 309–312 (2009)
22. Mukherjee, A., Kumar, A., Liu, B., Wang, J., Hsu, M., Castellanos, M., Ghosh, R.: Spotting opinion spammers using behavioral footprints. In: KDD, pp. 632–640 (2013)
23. Mukherjee, A., Liu, B., Glance, N.S.: Spotting fake reviewer groups in consumer reviews. In: WWW, pp. 191–200 (2012)
24. Mukherjee, A., Venkataraman, V., Liu, B., Glance, N.S.: What yelp fake review filter might be doing? In: ICWSM (2013)

25. Mukherjee, S., Weikum, G., Danescu-Niculescu-Mizil, C.: People on drugs: credibility of user statements in health communities. In: KDD, pp. 65–74 (2014)
26. Ott, M., Cardie, C., Hancock, J.T.: Negative deceptive opinion spam. In: NAACL (2013)
27. Ott, M., Choi, Y., Cardie, C., Hancock, J.T.: Finding deceptive opinion spam by any stretch of the imagination. In: ACL-HLT, vol. 1. pp. 309–319 (2011)
28. Pennebaker, J., Francis, M., Booth, R.: Linguistic Inquiry and Word Count: A Computerized Text Analysis Program. Psychology Press, Mahwah (2001)
29. Rahman, M., Carbunar, B., Ballesteros, J., Chau, D.H.P.: To catch a fake: curbing deceptive yelp ratings and venues. Stat. Anal. Data Min. 8(3), 147–161 (2015)
30. Strapparava, C., Valitutti, A.:WordNet-Affect: an affective extension of wordnet. In: LREC (2004)
31. Sun, H., Morales, A., Yan, X.: Synthetic review spamming and defense. In: KDD (2013)
32. Wang, G., Xie, S., Liu, B., Yu, P.S.: Review graph based online store review spammer detection. In: ICDM, pp. 1242–1247 (2011)
33. Yoo, K.H., Gretzel, U.: Comparison of deceptive and truthful travel reviews. In: ENTER (2009)

Interactive Visual Data Exploration
with Subjective Feedback

Kai Puolamäki[1]([✉]), Bo Kang[2], Jefrey Lijffijt[2], and Tijl De Bie[2]

[1] Finnish Institute of Occupational Health, Helsinki, Finland
kai.puolamaki@ttl.fi
[2] Data Science Lab, Ghent University, Ghent, Belgium
{bo.kang,jefrey.lijffijt,tijl.debie}@ugent.be

Abstract. Data visualization and iterative/interactive data mining are growing rapidly in attention, both in research as well as in industry. However, integrated methods and tools that combine advanced visualization and data mining techniques are rare, and those that exist are often specialized to a single problem or domain. In this paper, we introduce a novel generic method for interactive visual exploration of high-dimensional data. In contrast to most visualization tools, it is not based on the traditional dogma of manually zooming and rotating data. Instead, the tool initially presents the user with an 'interesting' projection of the data and then employs data randomization with constraints to allow users to flexibly and intuitively express their interests or beliefs using visual interactions that correspond to exactly defined constraints. These constraints expressed by the user are then taken into account by a projection-finding algorithm to compute a new 'interesting' projection, a process that can be iterated until the user runs out of time or finds that constraints explain everything she needs to find from the data. We present the tool by means of two case studies, one controlled study on synthetic data and another on real census data. The data and software related to this paper are available at http://www.interesting-patterns.net/forsied/interactive-visual-data-exploration-with-subjective-feedback/.

1 Introduction

Data visualization and iterative/interactive data mining are both mature, actively researched topics of great practical importance. However, while progress in both fields is abundant (see Sect. 4), methods that combine iterative data mining with visualization and interaction are rare, except for a number of tools designed for specific problem domains.

Yet, tools that combine state-of-the-art data mining with visualization and interaction are highly desirable as they would maximally exploit the strengths of both human data analysts and computer algorithms. While humans are unmatched in spotting interesting relations in low-dimensional visual representations but poor at handling high-dimensional data, computers excel in manipulating high-dimensional data but are weaker at identifying patterns that are

P. Frasconi et al. (Eds.): ECML PKDD 2016, Part II, LNAI 9852, pp. 214–229, 2016.
DOI: 10.1007/978-3-319-46227-1_14

truly relevant to the user. A symbiosis of the human data analyst and a well-designed computer system thus promises to provide the most efficient way of navigating the complex information space hidden in high-dimensional data [17].

Contributions in This Paper. In this paper we introduce a generically applicable methodology and a tool that demonstrates the proposed approach for interactive visual exploration of (high-dimensional) data. The tool iteratively cycles through three steps, as indicated in Fig. 1. Throughout these cycles, the user builds up an increasingly accurate understanding of the aspects of the data. Our tool maintains a model for this understanding—to which we refer as the *background model*.

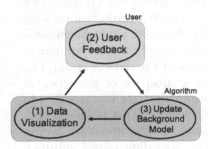

Fig. 1. This three-step cycle illustrates our tool's flow of action.

Step 1. The tool initially presents the user with an 'interesting' projection of the data, visualized as a scatter plot (Fig. 1 *step 1*). Here, interesting is formalized with respect to the initial background model; more details follow below.

Step 2. On investigating this scatter plot, the user may take note of some features of the data that contrast with, or add to, their beliefs about the data. We will refer to such features as *patterns*. In *step 2*, the user is offered the opportunity to tell the tool what patterns they have noted and assimilated.

Step 3. In *step 3*, the tool updates the background model to reflect this newly assimilated information embodied by the patterns highlighted by the user. Then the most interesting projection with respect to this updated background model can be computed, and the cycle can be reiterated until the user runs out of time or finds that patterns explain everything the user needs at the moment.

Formalizing the Background Model. A crucial challenge in realizing such a tool is the formalization of the background model. One way to do this is by specifying a randomization procedure that, when applied to the data, does not affect how plausible the user would deem it to be [7,13]. Access to such a randomized version of the data can be sufficient for determining interesting remaining structure in the data that is not yet known to the user. New patterns are then incorporated by adding corresponding *constraints* to the randomization procedure, ensuring that the patterns remain present after randomization. We will refer to this approach as the CORAND approach (for Constrained Randomization).

An Illustrative Example. As an example, consider a synthetic data set consisting of 1000 10-dimensional data vectors of which dimensions 1–4 can be clustered into five clusters, dimensions 5–6 into four clusters *involving different subsets of data points*, and of which dimensions 7–10 are Gaussian noise. All dimensions have equal variance. Figure 2 shows the scatter plots for all pairs of dimensions.

We designed this example to illustrate the two pattern types that a user can specify in the current implementation of our tool. Additionally, it shows how the tool succeeds in finding interesting projections given previously identified patterns. Thirdly, it also demonstrates how the user interactions meaningfully affect subsequent visualizations.

The first projection projects the data onto a two-dimensional subspace of the first four dimensions of the data (Fig. 3a), i.e., in a subspace of the space in which the data is clustered into 5 clusters. This is indeed sensible, as the structure within this four-dimensional subspace is arguably the strongest.

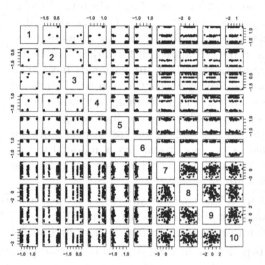

Fig. 2. Subsample of the toy data.

We then consider two possible user actions (step 2, shown in Fig. 3b). In the first possibility, the user marks all points within each cluster (cluster by cluster), indicating they have taken note of the positions of these groups of points within this particular projection. In the second possibility, the user additionally concludes that these points appear to be clustered, possibly also in other dimensions. (Details on how these two pattern types are formalized will follow.)

Both these pattern types lead to additional constraints on the randomization procedure. The effect of these constraints is identical within the two-dimensional projection of the current visualization (Fig. 3c): the projections of the randomized points onto this plane are identical to the projections of the original points onto this plane. Not visible though, is that in the second possibility the randomization is restricted also in orthogonal dimensions (possibly different ones for different clusters), to account for the additional clustering assumption.

The most interesting subsequent projection, following the user interaction, is different in the two cases (see Fig. 3d). In the first case, the remaining cluster structure within dimensions 1–4 is shown. However, in the second case this cluster structure is fully explained by the constraints, and as a result, the cluster structure in dimensions 5–6 being is shown instead.

Outline of This Paper. To use the CORAND approach, three main challenges had to be addressed, as discussed in Sect. 2: (1) defining intuitive pattern types that can be observed and specified based on a scatter plot of a two-dimensional projection of the data; (2) defining a suitable randomization scheme, that can be constrained to take account of such patterns; and (3) a way to identify the

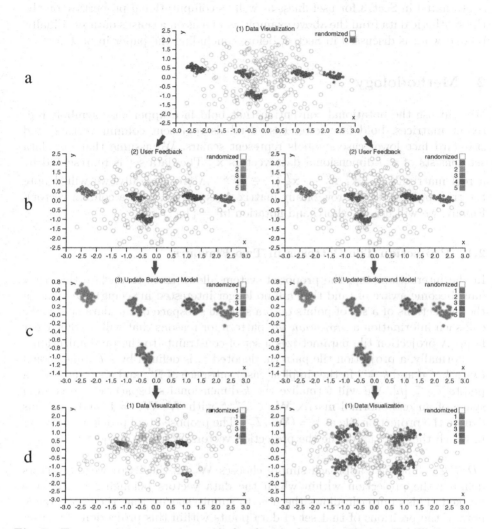

Fig. 3. Two user interaction scenarios for the toy data set. The smaller filled points represent actual data vectors, whereas the unfilled circles represent randomized data vectors. Row (a) shows the first visualization, which is the starting point for both scenarios. Row (b) shows the sets of data points marked by the user, (c) shows the randomized data and original data projected onto the same plane as (a), and (d) shows the most interesting visualization given these specified patterns. The left column shows the scenario when the user assumes nothing beyond the values of the data points in the projection in row (a), whereas the right column shows the scenario when the user assumes each of these sets of points may be clustered in other dimensions as well. (Color figure online)

most interesting projections given the background model. The resulting system is evaluated in Sect. 3 for usefulness as well as computational properties, on the the synthetic data from the above example as well as on a census dataset. Finally, related work is discussed in Sect. 4, before concluding the paper in Sect. 5.

2   Methodology

We will use the notational convention that bold face upper case symbols represent matrices, bold face lower case symbols represent column vectors, and standard face lower case symbols represent scalars. We assume that our data set consists of n d-dimensional data vectors $\mathbf{x}_i$. The data set is represented by a real matrix $\mathbf{X} = \left(\mathbf{x}_1^T \, \mathbf{x}_2^T \, \cdots \, \mathbf{x}_n^T \right)^T \in \mathbb{R}^{n \times d}$. More generally, we will denote the transpose of the ith row of any matrix $\mathbf{A}$ as $\mathbf{a}_i$ (i.e., $\mathbf{a}_i$ is a column vector). Finally, we will use the shorthand notation $[n] = \{1, \ldots, n\}$.

2.1   Projection Tile Patterns in Two Flavours

In the interaction step, the proposed system allows users to declare that they have become aware of (and thus are no longer interested in seeing) the value of the projections of a set of points onto a specific subspace of the data space. We call such information a *projection tile* pattern for reasons that will become clear later. A projection tile parametrizes a set of constraints to the randomization.

Formally, a projection tile pattern, denoted τ, is defined by a k-dimensional (with $k \leq d$ and $k = 2$ in the simplest case) subspace of $\mathbb{R}^d$, and a subset of data points $\mathcal{I}_\tau \subseteq [n]$. We will formalize the k-dimensional subspace as the column space of an orthonormal matrix $\mathbf{W}_\tau \in \mathbb{R}^{d \times k}$ with $\mathbf{W}_\tau^T \mathbf{W}_\tau = \mathbf{I}$, and can thus denote the projection tile as $\tau = (\mathbf{W}_\tau, \mathcal{I}_\tau)$. The proposed tool provides two ways in which the user can define the projection vectors $\mathbf{W}_\tau$ for a projection tile τ.

2D Tiles. The first approach simply chooses $\mathbf{W}_\tau$ as the (two) weight vectors defining the projection within which the data vectors belonging to $\mathcal{I}_\tau$ were marked. This approach allows the user to simply specify that they have taken note of the positions of that set of data points within this projection. The user makes no further assumptions – they assimilate solely what they see without drawing conclusions not supported by direct evidence, see Fig. 3b (left).

Clustering Tiles. It seems plausible, however, that when the marked points are tightly clustered, the user concludes that these points are clustered *not just within the two dimensions shown* in the scatter plot. To allow the user to express such belief, the second approach takes $\mathbf{W}_\tau$ to additionally include a basis for other dimensions along which these data points are strongly clustered, see Fig. 3b (right). This is achieved as follows.

Let $\mathbf{X}(\mathcal{I}_\tau, :)$ represent a matrix containing the rows indexed by elements from $\mathcal{I}_\tau$ from $\mathbf{X}$. Let $\mathbf{W} \in \mathbb{R}^{d \times 2}$ contain the two weight vectors onto which the data was projected for the current scatter plot. In addition to $\mathbf{W}$, we want to find any other dimensions along which these data vectors are clustered. These dimensions can be found as those along which the variance of these data points is not much larger than the variance of the projection $\mathbf{X}(\mathcal{I}_\tau, :)\mathbf{W}$.

To find these dimensions, we first project the data onto the subspace orthogonal to $\mathbf{W}$. Let us represent this subspace by a matrix with orthonormal columns, further denoted as $\mathbf{W}^\perp$. Thus, $\mathbf{W}^{\perp^T}\mathbf{W}^\perp = \mathbf{I}$ and $\mathbf{W}^T\mathbf{W}^\perp = \mathbf{0}$. Then, Principal Component Analysis (PCA) is applied to the resulting matrix $\mathbf{X}(\mathcal{I}_\tau, :)\mathbf{W}^\perp$. The principal directions corresponding to a variance smaller than a threshold are then selected and stored as columns in a matrix $\mathbf{V}$. In other words, the variance of each of the columns of $\mathbf{X}(\mathcal{I}_\tau, :)\mathbf{W}^\perp\mathbf{V}$ is below the threshold.

The matrix $\mathbf{W}_\tau$ associated to the projection tile pattern is then taken to be:

$$\mathbf{W}_\tau = \begin{pmatrix} \mathbf{W} & \mathbf{W}^\perp\mathbf{V} \end{pmatrix}.$$

The threshold on the variance used could be a tunable parameter, but was set here to twice the average of the variance of the two dimensions of $\mathbf{X}(\mathcal{I}_\tau, :)\mathbf{W}$.

2.2   The Randomization Procedure

Here we describe the approach to randomizing the data. The randomized data should represent a sample from an implicitly defined background model that represents the user's belief state about the data.

Initially, our approach assumes the user merely has an idea about the overall scale of the data. However, throughout the interactive exploration, the patterns in data described by the projection tiles will be maintained in the randomization.

Initial Randomization. The proposed randomization procedure is parametrized by n orthogonal rotation matrices $\mathbf{U}_i \in \mathbb{R}^{d \times d}$, where $i \in [n]$, and the matrices satisfy $(\mathbf{U}_i)^T = (\mathbf{U}_i)^{-1}$. We further assume that we have a bijective mapping $f : [n] \times [d] \mapsto [n] \times [d]$ that can be used to permute the indices of the data matrix. The randomization proceeds in three steps:

Random rotation of the rows. Each data vector $\mathbf{x}_i$ is rotated by multiplication with its corresponding random rotation matrix $\mathbf{U}_i$, leading to a randomised matrix $\mathbf{Y}$ with rows $\mathbf{y}_i^T$ that are defined by:

$$\forall i : \; \mathbf{y}_i = \mathbf{U}_i\mathbf{x}_i.$$

Global permutation. The matrix $\mathbf{Y}$ is further randomized by randomly permuting all its elements, leading to the matrix $\mathbf{Z}$ defined as:

$$\forall i, j : \; \mathbf{Z}_{i,j} = \mathbf{Y}_{f(i,j)}.$$

Inverse rotation of the rows. Each randomised data vector in $\mathbf{Z}$ is rotated with the inverse rotation applied in step 1, leading to the fully randomised matrix $\mathbf{X}^*$ with rows $\mathbf{x}_i^*$ defined as follows in terms of the rows $\mathbf{z}_i^T$ of $\mathbf{Z}$:

$$\forall i : \; \mathbf{x}_i^* = \mathbf{U}_i^T\mathbf{z}_i.$$

The random rotations $\mathbf{U}_i$ and the permutation f are sampled uniformly at random from all possible rotation matrices and permutations, respectively.

Intuitively, this randomization scheme preserves the scale of the data points. Indeed, the random rotations leave their lengths unchanged, and the global permutation subsequently shuffles the values of the d components of the rotated data points. Note that without the permutation step, the two rotation steps would undo each other such that $\mathbf{X}^* = \mathbf{X}$. Thus, it is the combined effect that results in a randomization of the data set.[1]

Accounting for One Projection Tile. Once the user has assimilated the information in a projection tile $\tau = (\mathbf{W}_\tau, \mathcal{I}_\tau)$, the randomization scheme should incorporate this information by ensuring that it is present also in all randomized versions of the data. This ensures that it continues to be a sample from a distribution representing the user's belief state about the data.

This is achieved by imposing the following *constraints* on the parameters defining the randomization:

Constraints on the rotation matrices. For each $i \in \mathcal{I}_\tau$, the component of $\mathbf{x}_i$ that is within the column space of $\mathbf{W}_\tau$ must be mapped onto the first k dimensions of $\mathbf{y}_i = \mathbf{U}_i \mathbf{x}_i$ by the rotation matrix $\mathbf{U}_i$. This can be achieved by ensuring that:[2]

$$\forall i \in \mathcal{I}_\tau : \ \mathbf{W}_\tau^T \mathbf{U}_i = \begin{pmatrix} \mathbf{I} & \mathbf{0} \end{pmatrix}. \tag{1}$$

Constraints on the permutation. The permutation should not affect any matrix cells with row indices $i \in \mathcal{I}_\tau$ and columns indices $j \in [k]$:

$$\forall i \in \mathcal{I}_\tau, j \in [k] : \ f(i,j) = (i,j). \tag{2}$$

Proposition 1. *Using the above constraints on the rotation matrices $\mathbf{U}_i$ and the permutation f, it holds that:*

$$\forall i \in \mathcal{I}_\tau, \mathbf{x}_i^T \mathbf{W}_\tau = \mathbf{x}_i^{*T} \mathbf{W}_\tau. \tag{3}$$

Thus, the values of the projections of the points in the projection tile remain unaltered by the constrained randomization. We omit the proof as the more general Proposition 2 is provided with proof further below.

---

[1] The random rotations may seem superfluous: the global permutation randomizes the data so dramatically that the added effect of the rotations is relatively unimportant. However, their role is to make it possible to formalize the growing understanding of the user as simple constraints on this randomization procedure, as discussed next.

[2] This explains the name *projection tile*: the information to be preserved in the randomization is concentrated in a 'tile' (i.e. the intersection of a set of rows and a set of columns) in the intermediate matrix $\mathbf{Y}$ created during the randomization procedure.

Accounting for Multiple Projection Tiles. Throughout subsequent iterations, additional projection tile patterns will be specified by the user. A set of
tiles τ_i for which $\mathcal{I}_{\tau_i} \cap \mathcal{I}_{\tau_j} = \emptyset$ if $i \neq j$ is straightforwardly combined simply by
applying the relevant constraints on the rotation matrices to the respective rows.
When the sets of data points affected by the projection tiles overlap though, the
constraints on the rotation matrices need to be combined. The aim of such a
combined constraint should be to preserve the values of the projections onto the
projection directions for *each* of the projection tiles a data vector was part of.

The combined effect of a set of tiles will thus be that the constraint on the
rotation matrix $\mathbf{U}_i$ will vary per data vector, and depends on the set of projections $\mathbf{W}_\tau$ for which $i \in \mathcal{I}_\tau$. More specifically, we propose to use the following
constraint on the rotation matrices:

Constraints on the rotation matrices. Let $\mathbf{W}_i \in \mathbb{R}^{d \times d_i}$ denote a matrix of
which the columns are an orthonormal basis for space spanned by the union
of the columns of the matrices $\mathbf{W}_\tau$ for τ with $i \in \mathcal{I}_\tau$. Thus, for any i and
$\tau : i \in \mathcal{I}_\tau$, it holds that $\mathbf{W}_\tau = \mathbf{W}_i \mathbf{v}_\tau$ for some $\mathbf{v}_\tau \in \mathbb{R}^{d_i \times \dim(\mathbf{W}_\tau)}$. Then, for
each data vector i, the rotation matrix $\mathbf{U}_i$ must satisfy:

$$\forall i \in \mathcal{I}_\tau : \ \mathbf{W}_i^T \mathbf{U}_i = \begin{pmatrix} \mathbf{I} & \mathbf{0} \end{pmatrix}. \tag{4}$$

Constraints on the permutation. Then the permutation should not affect
any matrix cells in row i and columns $[d_i]$:

$$\forall i \in [n], j \in [d_i] : \ f(i,j) = (i,j).$$

Proposition 2. *Using the above constraints on the rotation matrices* $\mathbf{U}_i$ *and
the permutation* f, *it holds that:*

$$\forall \tau, \forall i \in \mathcal{I}_\tau, \mathbf{x}_i^T \mathbf{W}_\tau = {\mathbf{x}_i^*}^T \mathbf{W}_\tau.$$

Proof. We first show that ${\mathbf{x}_i^*}^T \mathbf{W}_i = \mathbf{x}_i^T \mathbf{W}_i$:

$${\mathbf{x}_i^*}^T \mathbf{W}_i = \mathbf{z}_i^T \mathbf{U}_i^T \mathbf{W}_i = \mathbf{z}_i^T \begin{pmatrix} \mathbf{I} \\ \mathbf{0} \end{pmatrix} = \mathbf{z}_i (1 : d_i)^T = \mathbf{y}_i (1 : d_i)^T = \mathbf{y}_i^T \begin{pmatrix} \mathbf{I} \\ \mathbf{0} \end{pmatrix} = \mathbf{x}_i^T \mathbf{W}_i.$$

The result follows from the fact that $\mathbf{W}_\tau = \mathbf{W}_i \mathbf{v}_\tau$ for some $\mathbf{v}_\tau \in \mathbb{R}^{d_i \times \dim(\mathbf{W}_\tau)}$. $\square$

Technical Implementation of the Randomization Procedure. To ensure
the randomization can be carried out efficiently throughout the process, note
that the matrix $\mathbf{W}_i$ for the $i \in \mathcal{I}_\tau$ for a new projection tile τ can be updated by
computing an orthonormal basis for $\begin{pmatrix} \mathbf{W}_i & \mathbf{W} \end{pmatrix}$.[3]

Additionally, note that the tiles define an equivalence relation over the row
indices, in which i and j are equivalent if they were included in the same set
of projection tiles so far. Within each equivalence class, the matrix $\mathbf{W}_i$ will be
constant, such that it suffices to compute it only once, simply keeping track of
which points belong to which equivalence class.

---

[3] Such a basis can be found efficiently as the columns of $\mathbf{W}_i$ in addition to the columns
of an orthonormal basis of $\mathbf{W} - \mathbf{W}_i^T \mathbf{W}_i \mathbf{W}$ (the components of $\mathbf{W}$ orthogonal to $\mathbf{W}_i$),
the latter of which can be computed using the QR-decomposition.

2.3 Visualization: Finding the Most Interesting Two-Dimensional Projection

Given the data set $\mathbf{X}$ and the randomized data set $\mathbf{X}^*$, it is now possible to quantify the extent to which the empirical distribution of a projection $\mathbf{Xw}$ and $\mathbf{X}^*\mathbf{w}$ onto a weight vector $\mathbf{w}$ differ. There are various ways in which this difference can be quantified. We investigated a number of possibilities and found that the L_1-distance between the cumulative distribution functions works particularly well in practice. Thus, with $F_{\mathbf{x}}$ the empirical cumulative distribution function for the set of values in $\mathbf{x}$, the optimal projection is found by solving:

$$\max_{\mathbf{w}} \|F_{\mathbf{Xw}} - F_{\mathbf{X}^*\mathbf{w}}\|_1 .$$

The second dimension of the scatter plot can be sought by optimizing the same objective while requiring it to be orthogonal to the first dimension.

We are unaware of any special structure of this optimization problem that makes solving it particularly efficient. Yet, using the standard quasi-Newton solver in R [18][4] already yields satisfactory result. Note that runs of the method may produce different local optimum due to random initialization.

3 Experiments

We present two case studies to illustrate the framework and its utility. The case studies are completed by using a JavaScript version of our tool, made freely available along with the data used for maximum reproducibility.[5]

Table 1. Weight vectors of projections for the synthetic data.

Fig.	Axis	1	2	3	4	5	6	7	8	9	10
3a	X	0.194	0.545	−0.630	0.499	−0.119	−0.041	0.057	0.001	−0.029	0.003
	Y	−0.269	−0.754	−0.481	0.340	0.091	−0.004	0.016	−0.057	0.003	0.005
3d (left)	X	0.143	−0.118	0.005	0.981	0.001	−0.013	−0.031	−0.022	0.044	−0.031
	Y	−0.245	0.448	0.854	0.088	0.004	−0.001	0.005	0.008	−0.043	0.023
3d (right)	X	0.121	0.019	−0.232	0.017	−0.963	−0.008	0.022	0.023	0.037	0.004
	Y	−0.139	−0.067	−0.369	−0.082	0.111	−0.898	−0.083	0.086	0.005	−0.017

3.1 Synthetic Data Case Study

This section gives an extended discussion of the illustrative example from the introduction, namely the synthetic data case study. The data is described in Sect. 1 and shown in Fig. 2. The first projection shows that the projected data

---

[4] The optim optimization function with method = "BFGS" and default settings.

[5] Readers can access this tool online at: http://www.interesting-patterns.net/forsied/ interactive-visual-data-exploration-with-subjective-feedback/.

(blue dots in Fig. 3a) differs strongly from the randomized data (gray circles). The weight vectors defining the projection, shown in the 1st row of Table 1, contain large weights in dimensions 1–4. Therefore, the cluster structure seen here mainly corresponds to dimensions 1–4 of the data. A user can indicate this insight by means of a *clustering tile* for each of the clustered sets of data points (Fig. 3b, right). Encoding this into the background model, results in a randomization shown in Fig. 3c (right), where in the projection the randomized points perfectly align with data points. The new projection that differs most from this updated random background model is given by Fig. 3d (right), revealing the four clusters in dimensions 5–6 that the user was not aware of before.

If the user does not want to draw conclusions about the points being clustered in dimensions other than those shown, she can use *2D tiles* instead of *clustering tiles* (Fig. 3b, left). The updated background model then results in a randomization shown in Fig. 3c (left). In the given projection, this randomization is indistinguishable from the one with a clustering tile, but it results in a different subsequent projection. Indeed, now it shows just another view of the five clusters in dimensions 1–4 (Fig. 3d, left), as confirmed by the large weights for dimensions 1–4 in the 2nd row of Table 1.

Thus, by these simple interactions the user can choose whether she will explore more the cluster structure in dimensions 1–4 or if she already is aware of the cluster structure or does not find it interesting, in which case the system would direct her to the structure occurring in dimensions 5–6.

3.2   UCI Adult Dataset Case Study

In this case study, we demonstrate the utility of our method by exploring a real world dataset. The data is compiled from UCI Adult dataset[6]. To ensure the real time interactivity, we sub-sampled 218 data points and selected six features: "Age" $(17-90)$, "Education" $(1-16)$, "HoursPerWeek" $(1-99)$, "Ethnic Group" (White, AsianPacIlander, Black, Other), "Gender" (Female, Male), "Income" $(\geq 50k)$. Among the selected features, "Ethnic Group" is a categorical feature with five categories, "Gender" and "Income" are binary features, the rest are all numeric. To make our method applicable to this dataset, we further binarized the "Ethnic Group" feature (yielded four binary features) and the final dataset consists of 218 points and 9 features.

We assume the user uses clustering tiles throughout her exploration. Each of the patterns discovered during the exploration process thus corresponds to certain demographic clustering pattern. To illustrate how our tool helps the user rapidly gain an understanding of the data, we discuss the first three iterations of the exploration process below.

The first projection (Fig. 4a) visually consists of four clusters. The user notes that the weight vectors corresponding to the axes of the plot assign large weights to the "Ethnic Group" attributes (1st row, Table 2). As mentioned, we assume the user marks these points as part of the same clustering tile. When marking

---

[6] https://archive.ics.uci.edu/ml/datasets/Adult.

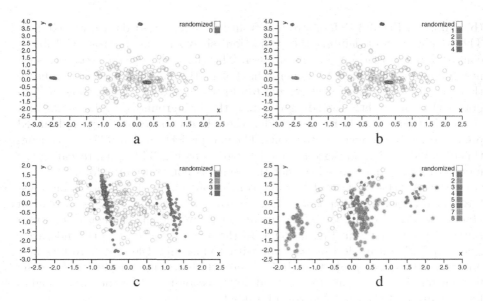

Fig. 4. Projections of UCI Adult dataset: (a) projection in the 1st iteration, (b) clusters marked by user in the 1st iteration, (c) projection in the 2nd iteration, and (d) projection in the 3rd iteration.

Table 2. Weight vectors of projections for the UCI Adult dataset.

Fig.	Axis	Age	Edu.	h/w	EG_AsPl	EG_Bl.	EG_Oth.	EG_Whi.	Gender	Income
4a	X	−0.039	−0.001	0.001	**0.312**	**−0.530**	**−0.193**	**0.763**	0.017	0.008
	Y	0.004	−0.004	−0.002	**0.816**	**−0.141**	**0.465**	**−0.313**	−0.011	0.002
4c	X	0.081	−0.028	−0.022	−0.259	−0.233	−0.104	−0.380	**−0.846**	−0.001
	Y	−0.590	0.541	0.143	−0.233	−0.380	−0.026	−0.293	0.232	0.000
4d	X	0.119	−0.149	0.047	0.102	0.191	0.104	**−0.556**	0.0581	**−0.769**
	Y	**−0.382**	**−0.626**	**−0.406**	0.346	0.317	−0.0287	0.111	−0.248	0.059

the clusters (Fig. 4b), the tool informs the user of the mean vectors of the points within each clustering tile. The 1st row of Table 3 shows that each cluster completely represents one out of four ethnic groups, which corroborates the user's understanding.

Taking the user's feedback into consideration, a new projection is generated by the tool. The new scatter plot (Fig. 4c) shows two large clusters, each consisting of some points from the previous four-cluster structure (points from these four clusters are colored differently). Thus, the new scatter plot elucidates structure not shown in the previous one. Indeed, the weight vectors (2nd row of Table 2) show that the clusters are separated mainly according to the "Gender" attribute. After marking the two clusters separately, the mean vector of each cluster (2nd row of Table 3) again confirms this: the cluster on the left represents male group, and the female group is on the right.

Table 3. Mean vectors of user marked clusters for the UCI Adult data set.

Fig.	Cluster	Age	Edu.	h/w	EG_AsPl	EG_Bl.	EG_Oth.	EG_Whi.	Gender	Income
4b	top left	35.0	8.67	34.7	0.00	0.00	**1.00**	0.00	0.667	0.333
	bott. left	37.2	9.43	40.3	0.00	**1.00**	0.00	0.00	0.286	0.071
	top right	35.6	1.3	51.1	**1.00**	0.00	0.00	0.00	0.750	0.250
	bott. right	38.4	10.2	41.6	0.00	0.00	0.00	**1.00**	0.762	0.275
4c	left	39.0	10.2	43.3	0.0377	0.0252	0.0126	0.925	**1.00**	0.321
	right	36.0	9.95	37.9	0.0339	0.169	0.0169	0.780	**0.00**	0.102
4d	left	**42.5**	**11.6**	**46.3**	0.00	0.00	0.00	**1.00**	1.00	**1.00**

The projection in the third iteration (Fig. 4d) consists of three clusters, separated only along the X-axis. Interestingly, the corresponding weight vector (3rd row of Table 2) has strongly negative weights for the attributes "Income" and "Ethnic Group - White". This indicates the left cluster mainly represents the people with high income and whose ethnic group is also "White". As this cluster has relatively low Y-value, according to the weight vector, they are also generally older and more highly educated. These observations are corroborated by the cluster mean (3rd row of Table 3).

This case study shows that the proposed tool facilitates human data exploration iteratively presenting an information projection considering what the user has already learned about the data.

3.3 Performance on Synthetic Data

Ideally interactive data exploration tools should work in close to real time. This section contains an empirical analysis of an (unoptimized) R implementation of our tool, as a function of the size, dimensionality, and complexity of the data. Note that limits on screen resolution as well as on human visual perception render it useless to display more than of the order of a few hundred data vectors, such that larger data sets can be down sampled without noticeably affecting the data exploration experience.

We evaluated the scalability on synthetic data with $d \in \{16, 32, 64, 128\}$ dimensions and $n \in \{64, 128, 256, 512\}$ data points scattered around $k \in \{2, 4, 8, 16\}$ randomly drawn cluster centroids (Table 4). The randomization is done here with the initial background model. The most costly part in randomization is the multiplication of orthogonal matrices. Indeed, the running time of the randomization scales roughly as nd^{2-3}. The results suggests the running time of the optimization is roughly proportional to the size of the data matrix nd and the complexity of data k has here only a minimal effect in the running time of the optimization.

Furthermore, in 69 % of the cases, the L_1 on the first axis is within 1 % of the best L_1 norm out of ten restarts. The optimization algorithm is thus quite stable, and in practical applications it may well be be sufficient to run the optimization algorithm only once. These results have been obtained with unoptimized and single-threaded R implementation on a 2.3 GHz Intel Xeon

Table 4. Median wall clock running times for randomization ("rand.") and for optimization ("optim.") over ten iterations of the optimization algorithm that finds the two-dimensional projection using the L_1 loss function for datasets of n data items and d dimensions with k clusters and its randomized version. We also show the number of iterations in which the L_1 norm first component ended up within 1 % of the result with the largest L_1 norm out of 10 tries; 10 means that the L_1 of the first component was within 1 % for all ten optimization runs.

n	d	rand. (s)	$k \in \{2, 4, 8, 16\}$	
			optim. (s)	in top 1 % out of 10
64	16	0.1	$\{0.4, 0.7, 0.5, 0.8\}$	$\{10, 2, 9, 5\}$
64	32	0.2	$\{1.2, 1.4, 1.5, 1.4\}$	$\{9, 1, 10, 9\}$
64	64	1.1	$\{3.1, 3.6, 3.4, 3.9\}$	$\{9, 1, 9, 5\}$
64	128	4.8	$\{9.1, 10.2, 10.4, 10.2\}$	$\{4, 2, 8, 8\}$
128	16	0.1	$\{0.8, 0.9, 1.3, 1.0\}$	$\{10, 2, 2, 8\}$
128	32	0.4	$\{1.6, 2.2, 2.4, 2.7\}$	$\{2, 10, 8, 10\}$
128	64	1.7	$\{5.6, 5.5, 5.9, 6.9\}$	$\{7, 10, 7, 10\}$
128	128	10.5	$\{11.5, 16.3, 18.2, 18.3\}$	$\{7, 7, 6, 5\}$
256	16	0.2	$\{1.2, 1.3, 1.5, 2.4\}$	$\{10, 4, 10, 9\}$
256	32	0.7	$\{3.6, 3.8, 3.6, 4.3\}$	$\{7, 8, 1, 9\}$
256	64	3.8	$\{8.8, 9.0, 9.8, 12.8\}$	$\{3, 9, 7, 9\}$
256	128	21.7	$\{24.5, 29.3, 28.0, 34.1\}$	$\{8, 9, 5, 5\}$
512	16	0.4	$\{2.8, 2.4, 3.1, 3.3\}$	$\{10, 9, 9, 10\}$
512	32	1.5	$\{5.2, 5.1, 6.8, 7.9\}$	$\{8, 8, 8, 10\}$
512	64	7.7	$\{15.6, 14.8, 17.1, 17.6\}$	$\{10, 8, 1, 2\}$
512	128	44.0	$\{37.2, 44.2, 47.3, 46.9\}$	$\{9, 1, 9, 7\}$

E5 processor.[7] The performance could probably be significantly boosted, e.g., by carefully optimizing the code and the implementation. Yet, even with this unoptimized code, response times are already of the order of 1 s to 1 min.

4   Related Work

Dimensionality Reduction. Dimensionality reduction for exploratory data analysis has been studied for decades. Early research into visual exploration of data led to approaches such as multidimensional scaling [11,21] and projection pursuit [6,9]. Most recent research on this topic (also referred to as manifold learning) is still inspired by the aim of multi-dimensional scaling; find a low-dimensional embedding of points such that their distances in the high-dimensional space are well represented. In contrast to Principal Component

---

[7] The R implementation used to produce Table 4 is available on our online demo page (footnote 5).

Analysis [16], one usually does not treat all distances equal. Rather, the idea is to preserve small distances well, while large distances are irrelevant, as long as they remain large; examples are Local Linear and (t-)Stochastic Neighbor Embedding [8,14,19]. Typically, it is not even possible to achieve this perfectly, and a trade-off between precision and recall arises [22]. Recent works are mostly spectral methods along this line.

Iterative Data Mining and Machine Learning. There are two general frameworks for iterative data mining: FORSIED [3,4] is based on modeling the belief state of the user as an evolving probability distribution in order to formalize subjective interestingness of patterns. This distribution is chosen as the Maximum Entropy distribution subject to the user beliefs as constraints, at that moment in time. Given a pattern syntax, one then aims to find the pattern that provides the most information, quantified as the pattern's 'subjective information content'. The other framework, which we here named CORAND [7,13], is similar, but the evolving distribution does not need to have an explicit form. Instead, it relies on sampling (randomization) of the data, using the user beliefs as constraints.

Both these frameworks are *general* in the sense that it has been shown they can be applied in various data mining settings; local pattern mining, clustering, dimensionality reduction, etc. The main difference is that in FORSIED, the background model is expressed analytically, while in CORAND it may be defined implicitly. This leads to differences in how they are deployed and when they are effective. Randomization schemes are easier to propose, or at least they require little mathematical skills. Explicit models have the advantage that they often enable faster search of the best pattern, and the models may be more transparent. Also, randomization schemes are computationally demanding when many randomizations are required. Yet, in cases like the current paper, a single randomization suffices, and the approach scales well. For both frameworks, the pattern syntax ultimately determines their relative tractability.

Many special-purpose methods have been developed for active learning, a form of iterative mining/learning, in diverse settings: classification, ranking, etc., as well as explicit models for user preferences. However, since these approaches do not target data exploration, we do not review them here. Finally, several special-purpose methods have been developed for visual iterative data exploration in specific contexts, for example for itemset mining and subgroup discovery [1,5, 12,15], information retrieval [20], and network analysis [2].

Visually Controllable Data Mining. This work was motivated by and can be considered an instance of *visually controllable data mining* [17], where the objective is to implement advanced data analysis method so that they are understandable and efficiently controllable by the user. Our proposed method satisfies the properties of a visually controllable data mining method (see [17], Sect. II B): (VC1) the data and model space are presented visually, (VC2) there are intuitive visual interactions that allow the user to modify the model space, and (VC3) the method is fast enough to allow for visual interaction.

Information Visualization and Visual Analytics. Many new interactive visualization methods are presented yearly at the IEEE Conference on Visual Analytics Science and Technology (VAST). The focus in these communities is less on the use or development of advanced data mining or machine learning techniques, and more on efficient use of displays and human cognition, as well as efficient exploration via selection of data objects and features, but the need to merge with the data mining community has been long recognized [10].

5   Conclusions

There is a growing need for generic tools that integrate advanced visualization with data mining techniques to facilitate visual data analysis by a human user. Our aim with this paper was to present a proof of concept for how this need can be addressed: a tool that initially presents the user with an 'interesting' projection of the data and then employs data randomization with constraints to allow users to flexibly express their interests or beliefs. These constraints expressed by the user are then taken into account by a projection-finding algorithm to compute a new 'interesting' projection, a process that can be iterated until the user runs out of time or finds that constraints explain everything the user needs to know about the data.

In our example, the user can associate two types of constraints on a chosen subset of data points: the appearance of the points in the particular projection or the fact that the points can be nearby also in other projections. We also tested the tool on two data sets, one controlled experiment on synthetic data and another on real census data. We found that the tool performs according to our expectations; it manages to find interesting projections, although interesting can be case specific and relies on the definition of an appropriate interestingness measure, here L_1 norm. More research into that is warranted. Nonetheless, we think this approach is useful in constructing new tools and methods for visually controllable interactive data analysis in variety of settings. In further work we intend to investigate the use of the FORSIED approach to formalizing the background model [3,4], as well as its use for computing the most informative data projections. Additionally, alternative types of constraints will be investigated.

Acknowledgements. This work was supported by the European Union through the ERC Consolidator Grant FORSIED (project reference 615517), Academy of Finland (decision 288814), and Tekes (Revolution of Knowledge Work project).

References

1. Boley, M., Mampaey, M., Kang, B., Tokmakov, P., Wrobel, S.: One click mining–interactive local pattern discovery through implicit preference and performance learning. In: Proceedings of KDD IDEA, pp. 27–35 (2013)

2. Chau, D.H., Kittur, A., Hong, J.I., Faloutsos, C.: Apolo: making sense of large network data by combining rich user interaction and machine learning. In: Proceedings of CHI, pp. 167–176 (2011)
3. De Bie, T.: An information-theoretic framework for data mining. In: Proceedings of KDD, pp. 564–572 (2011)
4. Bie, T.: Subjective interestingness in exploratory data mining. In: Tucker, A., Höppner, F., Siebes, A., Swift, S. (eds.) IDA 2013. LNCS, vol. 8207, pp. 19–31. Springer, Heidelberg (2013). doi:10.1007/978-3-642-41398-8_3
5. Dzyuba, V., Leeuwen, M.: Interactive discovery of interesting subgroup sets. In: Tucker, A., Höppner, F., Siebes, A., Swift, S. (eds.) IDA 2013. LNCS, vol. 8207, pp. 150–161. Springer, Heidelberg (2013). doi:10.1007/978-3-642-41398-8_14
6. Friedman, J.H., Tukey, J.W.: A projection pursuit algorithm for exploratory data analysis. IEEE Trans. Comp. 100(23), 881–890 (1974)
7. Hanhijärvi, S., Ojala, M., Vuokko, N., Puolamäki, K., Tatti, N., Mannila, H.: Tell me something I don't know: randomization strategies for iterative data mining. In: Proceedings of KDD, pp. 379–388 (2009)
8. Hinton, G.E., Roweis, S.T.: Stochastic neighbor embedding. In: Proceedings of NIPS, pp. 857–864 (2003)
9. Huber, P.J.: Projection pursuit. Ann. Stat. 13(2), 435–475 (1985)
10. Keim, D., Kohlhammer, J., Ellis, G., Mansmann, F. (eds.): Mastering the Information Age: Solving Problems with Visual Analytics. Eurographics Association (2010)
11. Kruskal, J.B.: Nonmetric multidimensional scaling: a numerical method. Psychometrika 29(2), 115–129 (1964)
12. Leeuwen, M., Cardinaels, L.: VIPER – visual pattern explorer. In: Bifet, A., May, M., Zadrozny, B., Gavalda, R., Pedreschi, D., Bonchi, F., Cardoso, J., Spiliopoulou, M. (eds.) ECML PKDD 2015, Part III. LNCS (LNAI), vol. 9286, pp. 333–336. Springer, Heidelberg (2015). doi:10.1007/978-3-319-23461-8_42
13. Lijffijt, J., Papapetrou, P., Puolamäki, K.: A statistical significance testing approach to mining the most informative set of patterns. DMKD 28(1), 238–263 (2014)
14. van der Maaten, L., Hinton, G.: Visualizing data using t-SNE. JMLR 9, 2579–2605 (2008)
15. Paurat, D., Garnett, R., Gärtner, T.: Interactive exploration of larger pattern collections: a case study on a cocktail dataset. In: Proceedings of KDD IDEA, pp. 98–106 (2014)
16. Pearson, K.: On lines and planes of closest fit to systems of points in space. Philos. Mag. 2(11), 559–572 (1901)
17. Puolamäki, K., Papapetrou, P., Lijffijt, J.: Visually controllable data mining methods. In: Proceedings of ICDMW, pp. 409–417 (2010)
18. R Core Team: R: A Language and Environment for Statistical Computing. R Foundation for Statistical Computing, Vienna, Austria (2016). https://www.R-project.org/
19. Roweis, S.T., Saul, L.K.: Nonlinear dimensionality reduction by locally linear embedding. Science 290(5500), 2323–2326 (2000)
20. Ruotsalo, T., Jacucci, G., Myllymäki, P., Kaski, S.: Interactive intent modeling: information discovery beyond search. CACM 58(1), 86–92 (2015)
21. Torgerson, W.S.: Multidimensional scaling: I. Theory and method. Psychometrika 17(4), 401–419 (1952)
22. Venna, J., Peltonen, J., Nybo, K., Aidos, H., Kaski, S.: Information retrieval perspective to nonlinear dimensionality reduction for data visualization. JMLR 11, 451–490 (2010)

Coupled Hierarchical Dirichlet Process Mixtures for Simultaneous Clustering and Topic Modeling

Masamichi Shimosaka[1]([⊠]), Takeshi Tsukiji[2], Shoji Tominaga[2],
and Kota Tsubouchi[3]

[1] Tokyo Institute of Technology, Tokyo, Japan
simosaka@miubiq.cs.titech.ac.jp
[2] The University of Tokyo, Tokyo, Japan
tsukiji@miubiq.cs.titech.ac.jp, tominaga@ics.t.u-tokyo.ac.jp
[3] Yahoo Japan Corporation, Tokyo, Japan
ktsubouc@yahoo-corp.jp

Abstract. We propose a nonparametric Bayesian mixture model that simultaneously optimizes the topic extraction and group clustering while allowing all topics to be shared by all clusters for grouped data. In addition, in order to enhance the computational efficiency on par with today's large-scale data, we formulate our model so that it can use a closed-form variational Bayesian method to approximately calculate the posterior distribution. Experimental results with corpus data show that our model has a better performance than existing models, achieving a 22 % improvement against state-of-the-art model. Moreover, an experiment with location data from mobile phones shows that our model performs well in the field of big data analysis.

Keywords: Non-parametric Bayes · Clustering · Hierarchical model · Topic modeling

1 Introduction

In this paper, we focus on a nonparametric Bayesian model in which the complexity of data can be controlled by using a stochastic process such as the Dirichlet process (DP) [9] as a prior distribution. Because of its flexibility against large-scale, complex data, this framework is useful for cluster analysis and has been applied to a wide range of research fields such as natural language processing, image processing, and bioinformatics. As well as cluster analysis, topic analysis on grouped data, e.g., topic modeling with corpus data, has long been studied. The hierarchical Dirichlet process (HDP) [22] is an example of successful nonparametric Bayesian model for topic analysis. Used as a prior distribution of a mixture model, HDP extracts the mixture components (= topics) across groups and allows all topics to be shared by all groups, with mixture weights of topics inferred independently for each group. The following model discussion is based on document analysis. As such, words, documents, and topic, which are

© Springer International Publishing AG 2016
P. Frasconi et al. (Eds.): ECML PKDD 2016, Part II, LNAI 9852, pp. 230–246, 2016.
DOI: 10.1007/978-3-319-46227-1_15

the expressions in document analysis, correspond to observations, groups, and mixture components, which are generic technical expressions, respectively. The following model discussion can be applied to various fields (e.g., urban dynamics analysis [17]) in addition to the research fields mentioned above.

These two fields of study have developed independently, but considering that the cluster structure, or relationship among groups, enhances the performance of topic modeling described in [20], it is useful to treat these two analyses at the same time. The naive approach is to follow a sequential process. For example, first we extract topics using HDP and then cluster the document, or we cluster documents on the basis of tf-idf [12] and then extract topics for each document cluster. However, as shown in [24], the sequential process possibly suffers from inaccurate results because the optimization criteria of topic extraction and group clustering are different. Therefore, a nonparametric Bayesian model that simultaneously optimizes the topic extraction and group clustering as a unified framework is required.

As an alternative to such naive approaches, the nested Dirichlet process (nDP) [21] has been proposed. The nDP simultaneously extracts topics and clusters groups as a unified framework. In this model, groups (documents) of data are clustered into various clusters and topics are extracted for each cluster. Since the topics are not shared with groups in different clusters, there is a risk of over-fitting in the clusters to which few groups belong due to the lack of training data for the mixture components of such a cluster.

In order to solve this problem in nDP, Ma et al. [15] proposed a hybrid nested/hierarchical Dirichlet process (hNHDP). The hNHDP extracts global topics, which are shared by all clusters, and local topics, which are shared only by groups in the same cluster. Using the idea of [16], hNHDP clusters groups and allows partial topics (global topics) to be shared by all clusters. However, as with the nDP, this framework has the risk of over-fitting with regard to the cluster specific local topics of a cluster to which few groups belong due to the lack of training data for each topic. As mentioned in [15], enhancing the computational efficiency is also important, since the sampling method is used to infer the model parameters of hNHDP.

In light of this background, in this paper, we propose a coupled hierarchical Dirichlet process (cHDP) that archives the desired framework mentioned above in order to solve the problems that hNHDP is currently facing. The cHDP extracts topics and clusters groups as well as nDP and hNHDP and allows all mixture components to be shared by all clusters, as with HDP. In addition, in order to enhance the computational efficiency for handling large-scale data, we formulate cHDP so that it can use a variational Bayesian method in which analytical approximation is provided and convergence speed is improved compared to conventional sampling methods.

To evaluate our cHDP performance against the existing models, we conduct experiments with corpus data on topic modeling and document clustering. In addition, using large-scale mobility logs from smartphones, we apply the cHDP to big data analysis – in this case, urban dynamics analysis – in order to show that cHDP works well in the fields other than document modeling where the data take continuous values, in contrast to the corpus data represented by discrete values. We perform experiments in which two simultaneous analyses are tackled: the extraction of the pattern of a daily transition of population common in target regions [17] and the clustering of these regions [25]. These analyses correspond to topic analysis and group clustering, respectively. As well as document modeling, since these two analyses have developed independently, and because even recent research [25] has proposed a sequential approach to such analysis, it is assumed that cHDP is useful in this urban dynamics analysis.

In order to clarify the position of our proposed cHDP, we introduce two existing models, nested hierarchical Dirichlet process (nHDP) [18] and coupled Dirichlet process (cDP) [13], whose names or motivation are similar to cHDP, and describe the differences between them and cHDP. The nHDP was proposed to extract tree structured, hierarchical topics, so unlike cHDP, it cannot realize simultaneous topic extraction or group clustering. In the case of cDP, its generic formulation is motivated by the same purpose as cHDP, but no concrete inference process was proposed in [13]. In this paper, we formulate a specific model equivalent to cDP and propose a closed-form variational inference that is superior to one in [13].

Our contributions are as follows. We developed a new nonparametric Bayesian method that simultaneously extracts topics and clusters groups in a unified framework while allowing all topics to be shared by all clusters. This is achieved by stochastic cluster assignment for both clustering processes. In order to enhance the computational efficiency, we formulate our model so that it can use a closed-form variational Bayesian method to approximately calculate the posterior distribution. We apply our proposed model to document analysis and big data analysis, in this case, urban dynamics analysis. The results of experiments with real data show that our model performs better in both research fields compared with existing models.

2   Related Works

As discussed in Sect. 1, for grouped data, we propose a new framework that simultaneously extracts topics and clusters groups, which allows all mixture components (topics) to be shared by all clusters. In this section, we briefly describe the existing nonparametric Bayesian models for grouped data. First, we describe HDP as a basic model for grouped data that focuses on topic analysis and then we introduce nDP and hNHDP, which simultaneously do two analyses, as a baseline for comparison with our model. In the following explanation, we assume that we have D groups of data, and the nth observation of group d is denoted as $x_{d,n}$.

2.1   Model for Topic Analysis

HDP. The hierarchical Dirichlet process (HDP) [22] is a nonparametric Bayesian model for grouped data. The generative process for a mixture model for grouped data is written as

$$G_0^* \sim \mathrm{DP}(\beta, H), \ G_d \sim \mathrm{DP}(\alpha, G_0^*), \tag{1}$$

where $G_0^* \sim \mathrm{DP}(\beta, H)$ denotes the Dirichlet process (DP) [8], which draws discrete distribution G_0^*. β is a concentration parameter and H is a base measure of DP. This process is described by stick-breaking representation as

$$G_0^* = \sum_{k=1}^{\infty} \pi_k \delta_{\phi_k}, \ \phi_k \sim H, \ \pi_k \sim \mathrm{GEM}(\beta), \tag{2}$$

where $\delta.$ is the Dirac's delta function. The expression GEM (named after Griffiths, Engen, and McCloskey [19]) is used as $\{\pi\}_{k=1}^{\infty} \sim \mathrm{GEM}(\beta)$ if we have $\pi_k = \pi_k' \prod_{j=1}^{k-1} (1 - \pi_j'), \ \pi_k' \sim \mathrm{Beta}(1, \beta)$ for $k = 1, \cdots, \infty$.

The group specific distribution G_d is drawn independently from $\mathrm{DP}(\alpha, G_0^*)$ and G_0^* is shared by all groups, which is itself drawn from another DP. As a result, mixture components (topics) are shared by all groups while the weights are independent of each group. The HDP cannot consider the relationship between groups, and since the mixture weights of each group are inferred independently, there is a risk of over-fitting.

2.2   Models that Simultaneously Extract Topics and Cluster Groups

NDP. The nested Dirichlet process (nDP) [21] clusters groups and extracts topics in a unified framework. The nDP is written as the following process, in which the DP itself is used as the base measure of different DP:

$$Q \sim \mathrm{DP}(\alpha, \mathrm{DP}(\beta, H)), G_d \sim Q. \tag{3}$$

This generative process induces the clustering of groups. The mixture components and weights are shared only in the same cluster of groups. The stick-breaking representation of the nDP is written as

$$Q = \sum_{g=1}^{\infty} \eta_g \delta_{G_g^*}, \ G_d \sim Q, \ \eta_g \sim \mathrm{GEM}(\alpha), \tag{4}$$

$$G_g^* = \sum_{t}^{\infty} \pi_{g,t} \delta_{g,t}, \ \phi_{g,t} \sim H, \ \pi_{g,t} \sim \mathrm{GEM}(\beta). \tag{5}$$

Let G_g^* denote the cluster specific distribution and $\phi_{g,t}$ denote the tth parameter of cluster g. In the mixture model with the nDP, as the mixture components in a cluster are not shared by different clusters, the clusters to which few groups belong suffer from over-fitting due to the lack of training data.

HNHDP. Ma et al. [15] proposed the hNHDP model, in which the advantages of the HDP and nDP are integrated. In the hNHDP, the cluster specific distribution F_g is modeled as the combination of two components, $G_0 \sim \mathrm{DP}(\alpha, H_0)$ and $G_g \sim \mathrm{DP}(\beta, H_1)$, and written as

$$F_g = \epsilon_g G_0 + (1 - \epsilon_g) G_g, \ \epsilon_g \sim \mathrm{Beta}(\alpha, \beta). \tag{6}$$

G_0 is shared by all group clusters and G_g is cluster-specific. α, β are concentration parameters and H_0, H_1 are base measures. Therefore, we have global mixture components shared by all clusters and cluster-specific local mixture components. With this modeling, we can cluster the groups while some mixture components are shared by all clusters, which enhances the modeling performance. However, as well as the nDP, this framework still has the risk of over-fitting due to the cluster specific mixture components. To tackle this problem, we need a framework in which all mixture components are shared among all group clusters.

3   Coupled Hierarchical Dirichlet Process (cHDP)

As described in Sect. 2, the existing nonparametric Bayesian models are facing various issues. In this section, we propose a coupled hierarchical Dirichlet process (cHDP) in which the advantages of HDP and nDP are integrated. The cHDP simultaneously extracts topics and clusters groups while allowing all mixture components to be shared by all group clusters, which solves the problem in the hNHDP. In addition, in order to enhance the computational efficiency, we modeled the cHDP so that it can use a variational Bayesian method in closed form for inferring the model parameters.

In this paper, we assume that we have D groups of data and let $\boldsymbol{x}_d = \{x_{d,1}, \ldots, x_{d,N_d}\}$ be the observations of group d, where $\{x_{d,n}\}$ denotes the nth observation and N_d is the total number of observations in group d. We assume that each observation $x_{d,n}$ is drawn from the probabilistic distribution $p(\theta_{d,n})$ with parameter $\theta_{d,n}$. The figure (D) in Fig. 1 shows the generative process of cHDP.

3.1   Definition and Formulation

We define the generative process of our proposed cHDP as follows

$$G_0^* \sim \mathrm{DP}(\gamma, H), \ Q \sim \mathrm{DP}(\alpha, \mathrm{DP}(\beta, G_0^*)), \ G_d \sim Q. \tag{7}$$

The second equation of (7) indicates that the DP is used as the base measure of another DP as with the nDP described in (3). The base measure of the nested DP in (7) is drawn from another DP whose base measure G_0^* is shared with all groups as with HDP described in (1). Considering this description, we can say cHDP is the generative process that holds the characteristics of HDP and nDP.

Several representations such as the Chinese restaurant franchise and the stick-breaking process are candidates for implementing the cHDP. In this paper, we adopt the stick-breaking representation, which enables us to use variational Bayesian inference, a computationally efficient approximation method, because we consider using the cHDP to handle large-scale data. We formulate the stick-breaking representation of the cHDP as

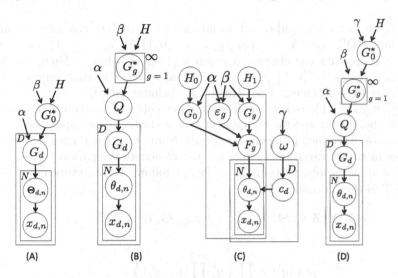

Fig. 1. Graphical model of (A) HDP, (B) nDP, (C) hNHDP, and (D) cHDP (proposed).

$$G_0^* = \sum_{k=1}^{\infty} \lambda_k \delta_{\phi_k^*}, \ \phi_k^* \sim H, \ \lambda_k \sim \text{GEM}(\gamma), \tag{8}$$

$$G_g^* = \sum_{t}^{\infty} \pi_{g,t} \delta_{\psi_{g,t}^*}, \ \psi_{g,t}^* \sim G_0^*, \ \pi_{g,t} \sim \text{GEM}(\beta), \tag{9}$$

$$Q = \sum_{g=1}^{\infty} \eta_g \delta_{G_g^*}, \ \eta_g \sim \text{GEM}(\alpha), \ G_d \sim Q, \tag{10}$$

where k is the index of mixture components shared by all groups and g is the index of the clusters of groups. Each group belongs to one of the clusters and cluster $g = 1 \cdots \infty$ has a cluster specific distribution G_g^* drawn as (9). Regarding the stick-breaking representation of the generative process of G_g^*, which is the same as the model structure of HDP in (7), there are different representations by Teh et al. [22] and Wang et al. [23]. The above representation is

$$G_g^* = \sum_{k=1}^{\infty} \pi_{g,k} \delta_{\phi_k}, \pi_{g,k} = \pi_{g,k}' \prod_{j=1}^{k-1} (1 - \pi_{g,j}'), \pi_{g,k}' \sim \text{Beta}\left(\alpha \lambda_k, \alpha \left(1 - \sum_{j=1}^{k} \lambda_j\right)\right). \tag{11}$$

With this representation, it is not possible to use a variational method in closed form in the inference of posterior distribution, so we formulate as (9) using the representation in the same way as [23], which enables us to use the variational method. This is achieved by introducing cluster specific parameter $\{\psi_{g,t}\}_{t=1}^{\infty}$ and a mapping variable that connects $\psi_{g,t}$ and mixture component ϕ_k, which is shared by all clusters.

Next, we introduce additional variables and formulate the mixture model using the cHDP. Let $\mathbf{Y} = \{y_{d,g}|y_{d,g} = \{0,1\}, \sum_g y_{d,g} = 1\}$ be a variable that represents the cluster to which a group d belongs. Then, we define $\mathbf{Z} = \{z_{d,n,t}|z_{d,n,t} = \{0,1\}, \sum_t z_{d,n,t} = 1\}$ as a variable that represents the cluster specific component t to which $x_{d,n}$ belongs and $\mathbf{C} = \{c_{g,t,k}|c_{g,t,k} = \{0,1\}, \sum_k c_{g,t,k} = 1\}$ as a variable that represents the mixture component k to which the cluster specific component t of a cluster g corresponds. As mentioned above, introducing the cluster specific component t and mapping variable c enables us to use variational inference. Let Θ denote the parameter set of distributions that the observations $\mathbf{X} = \{x_{d,n}\}$ follow. The mixture model using the cHDP is then formulated as

$$p(\mathbf{X}|\mathbf{Y}, \mathbf{Z}, \mathbf{C}, \Theta) = \prod_{d,g,n,t,k} p(x_{d,n}|\Theta_k)^{y_{d,g} z_{d,n,t} c_{g,t,k}}, \tag{12}$$

$$p(\mathbf{Y}|\boldsymbol{\eta}') = \prod_{d,g} \left\{ \eta'_g \prod_{f=1}^{g-1} (1 - \eta'_f) \right\}^{y_{d,g}}, \tag{13}$$

$$p(\mathbf{Z}|\mathbf{Y}, \boldsymbol{\pi}') = \prod_{d,g,n,t} \left\{ \pi'_{g,t} \prod_{s=1}^{t-1} (1 - \pi'_{g,s}) \right\}^{y_{d,g} z_{d,n,t}}, \tag{14}$$

$$p(\mathbf{C}|\boldsymbol{\lambda}') = \prod_{g,t,k} \left\{ \lambda'_k \prod_{j=1}^{k-1} (1 - \lambda'_j) \right\}^{c_{g,t,k}}, \tag{15}$$

$$p(\eta'_g) = \mathrm{Beta}(\eta'_g|1, \alpha), \tag{16}$$

$$p(\pi'_{g,t}) = \mathrm{Beta}(\pi'_{g,t}|1, \beta), \tag{17}$$

$$p(\lambda'_k) = \mathrm{Beta}(\lambda'_k|1, \gamma). \tag{18}$$

3.2   Variational Bayesian Inference with Closed Form Update

As with the general nonparametric Bayesian models, the posterior distribution of this cHDP mixture model cannot be calculated in closed form. We therefore need to apply an approximation method such as Gibbs sampling or variational Bayesian inference. In this paper, because we consider application to large-scale data, we opt to use variational Bayesian inference, which is characterized by its computational efficiency, to approximately calculate the posterior distribution and infer the model parameters. We approximate the posterior distribution as

$$q(\cdot) \equiv q(\mathbf{Y})q(\mathbf{Z})q(\mathbf{C})q(\boldsymbol{\eta}')q(\boldsymbol{\pi}')q(\boldsymbol{\lambda}')q(\Theta). \tag{19}$$

In variational inference, we update each parameter distribution q_i by $\ln q_i = \mathbb{E}_{q_{-i}}[\ln p(\mathbf{X}, \cdot)] + \mathrm{const}$.

Update $q(\mathbf{Y})$. We introduce $\xi_{d,g}$ that satisfies $\sum_g \xi_{d,g} = 1$ and

$$\ln \xi_{d,g} = \sum_{n,t} \mathbb{E}_q[z_{d,n,t}] \left(\sum_k \mathbb{E}_q[c_{g,t,k}] \mathbb{E}_q[\ln p(\boldsymbol{x}_{d,n}|\boldsymbol{\Theta}_k)] \right.$$

$$\left. + \mathbb{E}_q[\ln \pi_{g,t}] \right) + \mathbb{E}_q[\ln \eta_g] + \text{const}, \qquad (20)$$

then we have $q(\boldsymbol{y}_d) = \mathcal{M}(\boldsymbol{y}_d|\boldsymbol{\xi}_d)$ and $\mathbb{E}_q[y_{d,g}] = \xi_{d,g}$, where $\mathcal{M}(\cdot|\cdot)$ represents the multinomial distribution.

Update $q(\mathbf{Z})$, $q(\mathbf{C})$. As well as the update of $q(\mathbf{Y})$, both $q(\mathbf{Z})$ and $q(\mathbf{C})$ are represented as multinomial distribution by introducing variables.

Update $q(\boldsymbol{\eta}')$. We have $q(\eta_g') = \text{Beta}(\eta_g'|\alpha_{g,1}, \alpha_{g,2})$, where

$$\alpha_{g,1} = 1 + \sum_d \mathbb{E}_q[y_{d,g}], \qquad (21)$$

$$\alpha_{g,2} = \alpha_0 + \sum_{f=g+1}^{G} \sum_d \mathbb{E}_q[y_{d,f}]. \qquad (22)$$

G is a large truncation number for group clusters. We also have

$$\mathbb{E}_q[\ln \eta_g'] = \psi(\alpha_{g,1}) - \psi(\alpha_{g,1} + \alpha_{g,2}), \qquad (23)$$

$$\mathbb{E}_q[\ln (1 - \eta_g')] = \psi(\alpha_{g,2}) - \psi(\alpha_{g,1} + \alpha_{g,2}), \qquad (24)$$

$$\mathbb{E}_q[\ln \eta_g] = \mathbb{E}_q[\ln \eta_g'] \sum_{f=1}^{g-1} \mathbb{E}_q[\ln (1 - \eta_f')], \qquad (25)$$

where $\psi(\cdot)$ represents the digamma function $\psi(x) = \frac{d}{dx} \ln \Gamma(x)$.

Update $q(\boldsymbol{\pi}')$, $q(\boldsymbol{\lambda}')$. As well as the update of $q(\boldsymbol{\eta}')$, both $q(\boldsymbol{\pi}')$ and $q(\boldsymbol{\lambda}')$ are represented as the beta distribution.

3.3   Predictive Distribution for New Observation

By using the approximation $p(\mathbf{C}, \boldsymbol{\eta}, \boldsymbol{\pi}, \boldsymbol{\lambda}, \boldsymbol{\Theta}|\mathbf{X}) \simeq q(\mathbf{C})q(\boldsymbol{\eta})q(\boldsymbol{\pi})q(\boldsymbol{\lambda})q(\boldsymbol{\Theta})$ as with [23], the likelihood of new observation $\boldsymbol{x}^*$ of the cHDP model trained with data $\boldsymbol{X}$ is written as

$$p^*(\boldsymbol{x}^*|\mathbf{X}) \simeq \sum_g \mathbb{E}_q[\eta_g] \prod_n \sum_t \mathbb{E}_q[\pi_{g,t}] \sum_k \phi_{g,t,k} \mathbb{E}_q[p(\boldsymbol{x}_n^*|\boldsymbol{\Theta}_k)], \qquad (26)$$

where

$$\mathbb{E}_q[\eta_g] = \mathbb{E}_q[\eta_g'] \prod_{f=1}^{g-1} (1 - \mathbb{E}_q[\eta_f']), \ \mathbb{E}_q[\eta_g'] = \begin{cases} 1 & (g = G) \\ \frac{\alpha_{g,1}}{\alpha_{g,1}+\alpha_{g,2}} & (\text{o.w.}). \end{cases} \qquad (27)$$

$\mathbb{E}_q[\pi_{g,t}]$ is calculated in the same manner.

4   Experimental Results

4.1   Document Analysis with Corpus Data

We present the experiments with corpus data to evaluate our framework. We constructed a topic model, cHDP-LDA, in which our cHDP is applied to latent Dirichlet allocation (LDA) [6] as a prior distribution. In the experiment with corpus, the words, documents, and topic correspond to observations, groups, and mixture components. The cHDP-LDA simultaneously optimizes both words and document clustering, and topics are shared by all document clusters.

Suppose we have document $d \in \{1, \cdots, D\}$ whose number of words is N_d and the total number of words found in these documents is W. Let $\boldsymbol{x}_{d,n} = \{x_{d,n,w} | x_{d,n,w} = \{0,1\}, \sum_w x_{d,n,w} = 1\}$ be the nth words in document d. We assume that the word $\boldsymbol{x}_{d,n}$ is drawn from multinomial distribution $\mathcal{M}(\boldsymbol{x}_{d,n} | \boldsymbol{\mu}_k)$, where k is the topic index and $\boldsymbol{\mu}. \in \mathbb{R}^W$ is a parameter of the multinomial distribution. The Dirichlet distribution $\mathcal{D}(\boldsymbol{\mu} | \boldsymbol{\delta}) \propto \prod_i \mu_i^{\delta_i - 1}$, which is conjugate to multinomial distribution, is used as a prior distribution for $\boldsymbol{\mu}$, where $\boldsymbol{\delta} \in \mathbb{R}^W$ is the hyperparameter for the Dirichlet distribution. In this paper, we assume that $\{\delta_i\}_{i=1}^W = \delta$ and $\mathcal{D}(\boldsymbol{\mu} | \boldsymbol{\delta})$ is the symmetric Dirichlet distribution.

In the following experiments, we used three corpora: Reuters-21578 Corpus (Reuters corpus) [10], Nist Topic Detection and Tracking Corpus (TDT2 corpus) [2], and NIPS Conference Papers Vols. 012 Corpus (NIPS corpus) [4]. With these datasets, preprocessing (removal of stop words, etc.) has already been done. For the Reuters corpus, we chose the version used in [3] composed of uniquely labeled documents with a total of 65 categories. The TDT2 corpus was collected from six news services from January 4, 1998 to June 30, 1998, and we chose the version used in [3] composed of uniquely labeled documents with a total of 96 categories. The NIPS corpus [4] was made with the proceedings of the Neural Information Processing Systems (Advances in NIPS) [1] from Vols. 0 (1978) to 12 (1999).

Perplexity Evaluation. First, we evaluate the document modeling performance of our cHDP model and compare it to other existing topic models. All three corpora described above were used. As comparative models, we selected LDA models, each of whose prior distribution is an existing nonparametric Bayesian model, e.g., nested Chinese restaurant process (nCRP) [5], HDP [23], nDP [21], and hNHDP [15]. We refer to these models as hLDA, HDP-LDA, nDP-LDA, and hNHDP-LDA respectively. The hNHDP-LDA is a state-of-the-art framework that clusters both words and documents simultaneously. We set the hyperparameters of cHDP-LDA as $\alpha = \beta = \gamma = \delta = 1$, and those of nDP-LDA are also 1. As for hLDA, HDP-LDA and hNHDP-LDA, we followed the cited references.

We evaluate the models with the perplexity to test data. The perplexity indicates how well a trained model predicts new documents. Suppose we have D documents $\boldsymbol{X}^* = \{\boldsymbol{x}_d^*\}_{d=1}^D$ and the number of words in the dth document is N_d. In this case, the perplexity $\mathcal{P}(\boldsymbol{X}^*)$ is calculated as

Table 1. Test data perplexity (best score in boldface).

Corpus	Reuter		TDT2		NIPS	
Training → Test	A → B	B → A	A → B	B → A	A → B	B → A
cHDP-LDA	**1591**	**1529**	**4157**	**4200**	**2543**	**2463**
hLDA	1925	1864	6523	5600	2584	2560
HDP-LDA	2478	2390	6348	6406	3033	2998
nDP-LDA	4557	4460	10043	10189	3404	3374
hNHDP-LDA	2041	1939	5498	5350	2886	2817

$$\mathcal{P}(\boldsymbol{X}^*) = \exp\left(-\frac{\sum_d \ln p(\boldsymbol{x}_d^*)}{\sum_d N_d}\right). \tag{28}$$

The smaller the perplexity, the better the performance. In this experiment, we randomly divided each corpus into two groups, set A and set B, and then trained models with the one set and evaluated with the other.

For all corpora, the perplexities calculated with test sets A and B are shown in Table 1. The proposed cHDP-LDA performed best. The difference in performance between cHDP-LDA and HDP-LDA seems to be caused by the fact that cHDP can consider the relationship among documents. While the nCRP, which is the prior distribution of the hLDA, can indirectly consider the relationship of documents by partially sharing nodes (topics) in learning process, the hLDA performed worse than cHDP-LDA. We assume this is because the mixture weight to topics is independent of each document, resulting in over-fitting. HDP-LDA also suffers from this problem. Although the nDP-LDA can directly consider the relationship among documents, it exhibited a much worse performance than the others. This is because the topics in a document cluster to which few documents belong are inaccurate due to lack of training data, since topics in one document cluster are not shared by different clusters. The cHDP-LDA also outperformed the hNHDP-LDA, the state-of-the-art co-clustering model, in which partial topics are shared with different clusters. The same as the nDP-LDP, the hNHDP-LDA may suffer from over-fitting since hNHDP holds cluster specific topics (local topics). The above comparison clearly demonstrates that our cHDP-LDA, which clusters both words and documents while allowing all topics to be shared by all documents (or clusters), is suitable for topic modeling.

Document Clustering. We conducted an experiment to evaluate only the performance of document clustering against the existing methods, some of which do not extract topics. The datasets used here are the Reuters corpus and the TDT2 corpus, both of whose documents are categorically labeled. The evaluation criterion is the adjusted Rand index (ARI) [11], which indicates the accuracy of the clustering result against the true labeling. If the clustering result coincides with the true labeling, ARI takes 1 and if the result is from random clustering, ARI takes 0. The closer the ARI value to 1, the better the clustering accuracy.

As comparative models, we used spherical k-means (SPK) [7] and spectral clustering (SC) [14], which cluster documents without topic extraction. In addition, as nonparametric Bayesian models, we used nDP and hNHDP. For each model, we conducted 100 clustering trials and evaluated the ARI values. Figure 2 indicates the means and standard deviation of ARI at each number of document clusters and Table 2 shows the highest ARI value and the corresponding number of clusters. In the case of cHDP-LDA, nDP-LDA, and hNHDP-LDA, since the number of document clusters is not manually determined (inferred by model), we plotted the same value for each number of document clusters.

We firstly compare the cHDP-LDA with SPK and SC, which do not extract topics. For the Reuters corpus, the ARI value statistically exceeded that of SPK and SC at the most appropriate number of document clusters. Although the ARI of the cHDP-LDA was slightly lower than that of SPK with the TDT2 corpus, the difference was not statistically significant. Then, we argue the result against the nDP-LDA and the hNHDP-LDA, nonparametric Bayesian models that cluster documents with topic extraction. Against the nDP-LDA, the cHDP-LDA statistically outperformed with both corpora. In contrast, although the cHDP-LDA performed slightly worse than the hNHDP-LDA for the Reuters corpus, without statistically significant difference, it statistically outperformed for the TDT2 corpus. We found the cHDP is more robust against documents than the HNHDP-LDA. These results indicate that the document clustering performance of the cHDP-LDA is the same level or higher compared to the existing methods.

We summarize the results of both experiments. As for the perplexity evaluation for topic modeling, our cHDP-LDA outperformed all existing models with all corpora. Regarding the ARI evaluation for document clustering, although cHDP-LDA performed slightly worse than some combinations of model and corpus, no statistically significant difference was observed by t-testing. In other cases, cHDP-LDA performed best and the difference was statistically significant for each case. Therefore, we conclude our cHDP-LDA performs better and more stably than other models including the hNHDP-LDA, the state-of-the-art model.

Table 2. Results of document clustering.

Reuters	No. of clusters	ARI
cHDP-LDA	—	0.419 ± 0.045
nDP-LDA	—	0.195 ± 0.103
hNHDP-LDA	—	$\mathbf{0.424 \pm 0.050}$
SPK	5	0.391 ± 0.109
SC	4	0.385 ± 0.019

TDT2	No. of clusters	ARI
cHDP-LDA	—	0.640 ± 0.028
nDP-LDA	—	0.083 ± 0.042
hNHDP-LDA	—	0.520 ± 0.066
SPK	12	$\mathbf{0.646 \pm 0.065}$
SC	7	0.557 ± 0.008

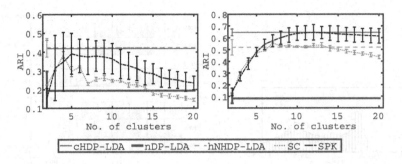

Fig. 2. Adjusted Rand indices with no. of clusters.

4.2 Big Data Analysis with Mobility Logs

In this section, using large-scale mobility logs from smartphones, we apply our cHDP to big data analysis, in this case, urban dynamics analysis. In this analysis, the following two analyses have been developed independently: extraction of patterns of the daily transition of population common in target regions [17], whose details are explained below, and clustering of regions [25]. Inspired by the success of cHDP in simultaneous topic modeling and document clustering, we apply cHDP to simultaneously tackle these analyses.

First, let us give an overview of this experiment. We set a square area (e.g., $300 \times 300\,\mathrm{m}$) as the target region and define this region as a point of interest (POI). In each POI, we divide a day into H time segments and describe the daily transition of population as a histogram, as shown in Fig. 3. Each bin in the histogram is the number of logs observed in a time segment in the POI. We define basic patterns in the transition of population as dynamics patterns and assume that a daily transition of population is generated from the mixture of dynamics patterns. Using an analogy from document modeling, POI, a daily transition, and dynamics pattern correspond to document, word, and topic, respectively. Figure 4 shows the framework of this big data analysis by cHDP. The left side of the figure shows the collections of the daily transition of population in each POI and the right side indicates the extracted dynamics pattern.

Let d, n, and h be the index of POI, day, and time segment, respectively. The transition of population in the nth day in POI d is described as $\boldsymbol{x}_{d,n} = \{x_{d,n,1}, \cdots, x_{d,n,H}\} \in \mathbb{R}^H$. We assume $x_{d,n,h}$ is drawn from the mixture of Gaussian distribution and the distribution of the kth dynamics pattern is written as $\mathcal{N}(x_{d,n,h}|\mu_{k,h}, \rho_{k,h}^{-1})$. $\mu_{.,.}, \rho_{.,.}$ are the mean and precision. We use the Gaussian distribution and gamma distribution as the prior distribution for $\mu_{k,h}$ and $\rho_{k,h}$.

The dataset and the problem settings in this experiment are as below. We use the large-scale GPS logs collected from the disaster alert mobile application released by Yahoo! JAPAN. The logs are anonymized and include no users' information. Each record has three components: timestamp, latitude, and longitude. We use data collected for 365 days, from 1 July 2013 to 30 June 2014, consisting

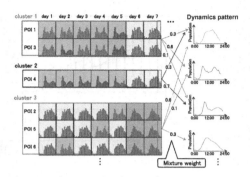

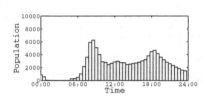

Fig. 3. Daily transition of population.   **Fig. 4.** Urban dynamics analysis by cHDP.

of 15 million logs per day in the Kanto region in Japan. We focus on the square area (approximately 8000×8000 m) indicated by the thick blue line in Fig. 6. We divide this focus area into 26×26 square pixels (each pixel is 300×300 m) and regard each pixel as a POI. A daily transition of population in each POI is characterized by its scale and shape (e.g., the population peak time). As in [17], to make the patterns depend only on shape, we use the log counts divided by the average number of logs per day for training and test data for each POI.

For quantitative evaluation of dynamics pattern modeling, we use mean log likelihood (MLL) for test data. The models are trained with data of 30, 60, 90, 120, 150, and 180 days and tested by 180 days of data. From the 365 days of the dataset, training data and test data are randomly selected without duplication. Five tests are conducted with each number of days and the average values of MLL are evaluated. As for the evaluation for POI clustering, we visualize the clustering result and argue the validity on the basis of the real geographical features. This is because numerical evaluation is difficult for POI clustering.

We use the HDP and nDP as comparative models. Parameters are inferred by variational method. As for the POI clustering of HDP, we used a DP Gaussian mixture model with the mixture weight to dynamics pattern for each POI. Due to the computational performance for large-scale data, we do not use the hNHDP model, which is trained by sampling. Note that neither SPK nor SC can be directly used for region clustering without pattern extraction because feature value must be ratio scale calculated from the set of discrete values such as words.

Results. As shown in Fig. 5, the cHDP model had the best performance for all the training data condition. We can see a big performance gap between the cHDP and the others in the test with a small amount of training data. This result indicates that the cHDP's framework, i.e., considering the POI's relationship and the sharing dynamics patterns among all POIs, enhances the modeling accuracy. The reason nDP exhibited a worse performance is that the dynamics patterns in a POI cluster where few POIs belong are inaccurate due to the lack of training data, since patterns are not shared among different clusters.

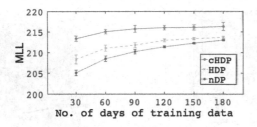

Fig. 5. Quantitative result of MLL.

Next, we evaluate the clustering performance. Since it is almost impossible to attach category labels by hand to such a small area, numerical evaluation like ARI is difficult. Therefore, we visualize the clustering result and qualitatively argue the validity. Figure 7 shows the POI clustering result by the cHDP model. POIs that belong to the same cluster are drawn in the same color, while similar colors do not indicate the similarity in dynamics pattern trends. As shown in Fig. 7, POIs distributed along railways are clustered into the same cluster (POIs around the Yamanote and Chuo lines are clustered in red and POIs around private railways are clustered in deep blue). In addition, yellow colored cluster corresponds to residential regions. Thus, it is shown that the cHDP model could cluster POIs corresponding to the actual geographical features.

The POI clustering by the HDP is shown in the left side of Fig. 8. We first extracted dynamics patterns by the HDP and then clustered POIs on the basis of the mixture weights by DP. The correlation between the result and the actual geographical feature such as railways is low compared to the cHDP. In addition, neighboring POIs tended to belong to different clusters. Since we mesh the focus area into small areas (300×300 m), we assumed that spatial continuity of POI clusters among neighboring POIs can be seen. Therefore, the result is not valid and we cannot say that this is a meaningful clustering result. The comparison between cHDP and HDP indicates the advantage of simultaneous extraction of patterns and POI clusters. In contrast, as shown in the right side of Fig. 8, the result of the nDP matches the geographical features to some extent. This is probably because the nDP simultaneously extracts patterns and clusters POIs as with cHDP. However, compared to the result of cHDP shown in Fig. 7, POIs along the Yamanote and Chuo liens are not clustered well. We assumed that this difference stems from over-fitting of the cluster specific dynamics patterns. Considering the above evaluation, we conclude that the cHDP is useful for big data analysis, i.e., dynamics pattern extraction and region clustering.

5   Conclusion

In this paper, we proposed cHDP, a new nonparametric Bayesian mixture model that simultaneously extracts topics and clusters groups while allowing all topics to be shared by all clusters. In order to achieve better computational efficiency,

244   M. Shimosaka et al.

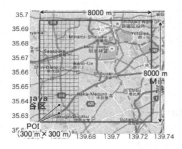

Fig. 6. Analysis area. (Color figure online)

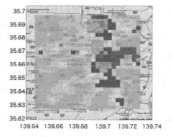

Fig. 7. Clustering by the cHDP model. (Color figure online)

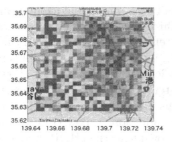

Fig. 8. Clustering by (left) HDP + DP clustering and (right) the nDP model.

we formulated our model in order to take variational Bayesian inference in closed form when inferring the model parameters.

We applied cHDP to document modeling and big data analysis, in this case, urban dynamics analysis. For the document modeling, we used cHDP as a prior distribution of LDA, which simultaneously conducts topic extraction and document clustering in a unified framework. Experiments with corpus data show that cHDP performs well in both tasks compared with existing models, achieving a 22 % improvement against the state-of-the-art model. For big data analysis, we simultaneously tackled dynamics patterns extraction and region clustering. Using the GPS logs from smartphones, we showed that the cHDP enhances performance in pattern modeling and obtains valid clustering results. The comparison with nDP indicates the superiority of cHDP's topic sharing among all clusters.

For future work, we will introduce an online approach in the learning process. This is necessary to handle the data that accumulate over time, such as GPS logs from smartphones, let alone much more large-scale data. One option for this is using the online variational Bayesian method proposed in [23].

Acknowledgement. We thank Tengfei Ma, Issei Sato, and Hiroshi Nakagawa for providing the hNHDP implementation. This work was partly supported by CREST, JST.

References

1. Advances in Neural Information Processing Systems. http://books.nips.cc/. Accessed 15 Jan 2013
2. Nist Topic Detection and Tracking Corpus. http://projects.ldc.upenn.edu/TDT2/. Accessed 15 Jan 2013
3. Popular Text Data Sets in Matlab Format. http://www.cad.zju.edu.cn/home/dengcai/Data/TextData.html. Accessed 15 Jan 2013
4. Roweis, S.: Data. http://www.cs.nyu.edu/~roweis/data.html. Accessed 15 Jan 2013
5. Blei, D.M., Griffiths, T.L., Jordan, M.I.: The nested Chinese restaurant process and Bayesian nonparametric inference of topic hierarchies. J. ACM **57**(2), 7:1–7:30 (2010)
6. Blei, D.M., Ng, A.Y., Jordan, M.I.: Latent Dirichlet allocation. J. Mach. Learn. Res. **3**, 993–1022 (2003)
7. Dhillon, I.S., Modha, D.S.: Concept decompositions for large sparse text data using clustering. Mach. Learn. **42**(1–2), 143–175 (2001)
8. Ferguson, T.S.: A Bayesian analysis of some nonparametric problems. Ann. Stat. **1**, 209–230 (1973)
9. Ghahramani, Z., Griffiths, T.L.: Infinite latent feature models and the Indian buffet process. In: Proceedings of NIPS, pp. 475–482 (2005)
10. Hayes, P.J., Weinstein, S.P.: CONSTRUE/TIS: a system for content-based indexing of a database of news stories. In: Proceedings of IAAI, pp. 49–64 (1991)
11. Hubert, L., Arabie, P.: Comparing partitions. J. Classif. **2**(1), 193–218 (1985)
12. Jones, K.S.: IDF term weighting and IR research lessons. J. Documentation **60**(5), 521–523 (2004)
13. Lin, D., Fisher, J.: Coupled Dirichlet processes: beyond HDP. In: Proceedings of NIPS Workshop (2012)
14. Luxburg, U.: A tutorial on spectral clustering. Stat. Comput. **17**(4), 395–416 (2007)
15. Ma, T., Sato, I., Nakagawa, H.: The hybrid nested/hierarchical Dirichlet process and its application to topic modeling with word differentiation. In: Proceedings of AAAI, pp. 2835–2841 (2015)
16. Muller, P., Quintana, F., Rosner, G.: A method for combining inference across related nonparametric Bayesian models. J. R. Stat. Soc. Ser. B (Stat. Method.) **66**(3), 735–749 (2004)
17. Nishi, K., Tsubouchi, K., Shimosaka, M.: Extracting land-use patterns using location data from smartphones. In: Proceedings of Urb-IoT, pp. 38–43 (2014)
18. Paisley, J., Wang, C., Blei, D.M., Jordan, M.I.: Nested hierarchical Dirichlet processes (2012). arXiv preprint: arXiv:1210.6738
19. Pitman, J.: Combinatorial stochastic processes. Technical report, Technical report 621, Dept. Statistics, UC Berkeley, 2002. Lecture notes for St. Flour course (2002)
20. Ramage, D., Hall, D., Nallapati, R., Manning, C.D.: Labeled lda: a supervised topic model for credit attribution in multi-labeled corpora. In: Proceedings of EMNLP, pp. 248–256 (2009)
21. Rodriguez, A., Dunson, D.B., Gelfand, A.E.: The nested Dirichlet process. J. Am. Stat. Assoc. **103**(483), 1131–1154 (2008)
22. Teh, Y.W., Jordan, M.I., Beal, M.J., Blei, D.M.: Hierarchical Dirichlet processes. J. Am. Stat. Assoc. **101**(476), 1566–1581 (2006)

23. Wang, C., Paisley, J.W., Blei, D.M.: Online variational inference for the hierarchical Dirichlet process. In: Proceedings of AISTATS, pp. 752–760 (2011)
24. Wang, X., Ma, X., Grimson, E.: Unsupervised activity perception by hierarchical Bayesian model. In: Proceedings of CVPR, pp. 1–8 (2007)
25. Yuan, J., Zheng, Y., Xie, X.: Discovering regions of different functions in a city using human mobility and pois. In: Proceedings of KDD, pp. 186–194 (2012)

Enhancing Traffic Congestion Estimation with Social Media by Coupled Hidden Markov Model

Senzhang Wang[1(✉)], Fengxiang Li[2], Leon Stenneth[3], and Philip S. Yu[4,5]

[1] Nanjing University of Aeronautics and Astronautics, Nanjing 211106, China
szwang@nuaa.edu.cn
[2] Peking University, Beijing 100871, China
1200012427@pku.edu.cn
[3] Nokia's HERE Connected Driving, Chicago, IL, USA
leon.stenneth@here.com
[4] University of Illinois at Chicago, Chicago, IL 60607, USA
psyu@uic.edu
[5] Institute for Data Science, Tsinghua University, Beijng 100084, China

Abstract. Estimating traffic conditions in arterial networks with GPS probe data is a practically important while substantially challenging problem. With the increasing availability of GPS equipments installed in various vehicles, GPS probe data is currently becoming a significant data source for traffic monitoring. However, limited by the lack of reliability and low sampling frequency of GPS probes, probe data are usually not sufficient for fully estimating traffic conditions of a large arterial network. For the first time this paper studies how to explore social media as an auxiliary data source and incorporate it with GPS probe data to enhance traffic congestion estimation. Motivated by the increasing amount of traffic information available in Twitter, we first extensively collect tweets that report various traffic events such as congestion, accident, and road construction. Next we propose an extended Coupled Hidden Markov Model which can effectively integrate GPS probe readings and traffic related tweets to more accurately estimate traffic conditions of an arterial network. To address the computational challenge, a sequential importance sampling based EM algorithm is also introduced. We evaluate the proposed model on the arterial network of downtown Chicago. The experimental results demonstrate the superior performance of the model by comparison with previous methods.

Keywords: Social media · Traffic estimation · CHMM

1 Introduction

Conventional traffic monitoring methods rely on road sensor data collected from various sensors such as loop detectors [14], surveillance cameras [4], and radars.

© Springer International Publishing AG 2016
P. Frasconi et al. (Eds.): ECML PKDD 2016, Part II, LNAI 9852, pp. 247–264, 2016.
DOI: 10.1007/978-3-319-46227-1_16

Due to the high cost of deploying and maintaining such devices, their spatialtemporal coverage is usually very limited. Recently, GPS based probe vehicle data have become a significant data source available for the arterials and highways not covered by dedicated sensing infrastructure. As such, there is considerable research interest in exploring GPS probes for conducting various traffic related applications [6,20]. However, the characteristics of probe data, including the lack of reliability, low sampling frequency, and the randomness of its spatiotemporal coverage, make it insufficient for fully estimating traffic states for large transportation networks [5].

Currently, it is a common practice for drivers and official transportation departments to release instant traffic information through social media [12,18]. By taking Twitter as an example, a large number of tweets that report traffic events like congestion and accident are posted instantly every day. Many such tweets, like *"Harrison St: accident at Kilbourn Ave, 2:04-4/2/2015"*, explicitly give the type of traffic event, time, and location information. Motivated by the rich traffic information available in social media, many recent efforts have been devoted to exploring social media data to facilitate traffic related applications, such as traffic event location identification [16,19], traffic event detection [1,21], as well as traffic congestion estimation [2,3,10]. Chen et al. made the first attempt to estimate urban traffic congestions by relying only on the traffic information collected from Twitter [10]. To improve long-term traffic prediction, He et al. tried to use rich semantic information in online social media [7]. Wang et al. proposed a coupled matrix and tensor factorization model to integrate social media data, road features, and other information to better estimate traffic congestions of a city [2]. However, existing works mainly focus on studying how to utilize social media as the major data source for traffic monitoring. How to use social media data and fuse it with GPS probe data to improve traffic congestion estimation is still not fully explored.

For the first time, this paper incorporates traffic information extracted from Twitter with the sparse and noisy GPS probe data to enhance urban traffic congestion estimation. The challenges of the studied problem are two-fold. Firstly, the traffic information extracted from Twitter can be associated to multi-typed traffic events including congestion, accident, road construction, etc. It is nontrivial to model the potential impacts of the diverse traffic events on traffic congestion. For example, given a tweet that reports a traffic accident, how can we quantitatively measure its impact on traffic congestion? Secondly, it is also difficult to combine the two types of data with totally different data formats seamlessly. A piece of GPS probe reading normally contains the time, speed, heading, and the exact location (longitude, latitude) information of a vehicle; while a tweet that reports a particular traffic event typically will mention the traffic event type, the time, and the road or road segment information. The differences of the two types of datasets on both traffic information and location granularity make the effective combination of them very challenging.

To address the above challenges, we first extensively collect traffic related tweets from both traffic authority Twitter accounts (explain later) and regular

Twitter user accounts, and extract the traffic event, time, and location information by data processing. Through data analysis, we discover that (1) there is a high occurrence correlation between traffic events like accident and traffic congestion, and (2) the data of traffic event related tweets is an important complementary to GPS probe data. Both discoveries indicate that the estimation performance could potentially improved if Twitter data are properly incorporated. To effectively fuse the two types of data, we propose an extended Coupled Hidden Markov Model (E_CHMM). Different from traditional models with the GPS probe observations only [6,8], in this model we consider the GPS probe data and traffic related tweets as two types of observations generated from two different distributions independently. As the exact solution of the E_CHMM model is infeasible for a large network due to the exponential space and time consumption, we utilize a sequential importance sampling method to more efficiently solve the E-step of the EM algorithm. In the M-step, we formulate the original optimization problem decomposable into smaller problems that can be independently optimized. We evaluate our model on the arterial network of downtown Chicago with 1,257 road links whose total length is nearly 700 miles. The result shows that by incorporating Twitter data, about 15 % GPS probes can be reduced to achieve the comparable performance to previous method with all the GPS probes. This research provides us with a promising way to reduce the cost and improve the performance of urban traffic congestion estimation.

2 Preliminary

In this section, we will start with some definitions, and introduce the framework of our method. Next we will make some basic assumptions in traffic congestion estimation to facilitate us model the studied problem.

Definition 1. *A tweet observation of traffic event $e_{t,l,i}$.* *We represent a tweet observation of traffic event occurring on the road link l at time t as such a tuple $e_{t,l,i} = (c, loc, t)$, where c is the traffic event category, loc represents the location or road segment of the event, and t denotes the time.*

Definition 2. *A GPS probe observation $y_{t,l,i}$.* *We represent a GPS probe observation on the road link l at time t as such a vector $y_{t,l,i} = (s, lat, lon, head, t)$, where s is the vehicle speed, lat is the latitude, lon is the longitude, $head$ is the heading of the probe, and t denotes the time.*

Definition 3. *A road link l.* *We use the intersections to partition an arterial road R into several road links $R = \{l_1, l_2, ...\}$. Each road link l can be represented as such a tuple $l = (Link\_ID, Start\_Inter, End\_Inter)$, where $link\_ID$ is the ID of the road link, $Start\_Inter$ is the start intersection, and End_Inter is the end intersection.*

Definition 4. *Neighbor road links.* *Two road links l_1 and l_2 are called neighbor road links if they connect to each other, namely they share an intersection. Particularly, the road link l is also considered as a neighbor road link of itself. We denote all the neighbor links of road link l as N_l.*

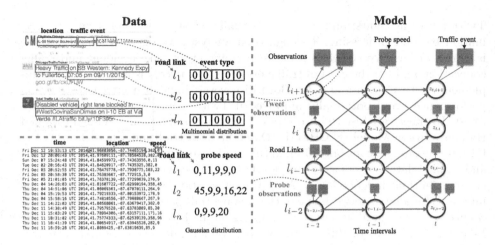

Fig. 1. Framework of the proposed model (Color figure online)

Figure 1 shows the framework of our method. It contains two major parts: data collection and processing part, and the model part. There are two types of data sources in our model, traffic related tweets and GPS probe readings. From each traffic related tweet, we first extract the traffic event type, location, and time information, and then map it to the corresponding road link by geocoding. Similarly, we extract the exact location and travel speed information from each GPS probe reading, and then map it to the corresponding road link. For each road link, we assume the occurrence of traffic events on it follows multinomial distribution, and the traveling speed of vehicles in a particular time interval follows Gaussian distribution [6].

We model the spatiotemporal conditional dependencies of arterial traffic using a probabilistic graphical model *Coupled Hidden Markov Model*. A CHMM models a system of multiple interaction processes which are assumed to be a Markov process with unobserved states. In our model, the multiple processes evolving over time are the discrete traffic states of each link in the road network (the circles in the model part of Fig. 1). Since we do not observe the state of each link for all times, we consider them as *hidden*. We can observe the vehicle speed and traffic events from GPS probe and tweets (the blue and red squares in the model part of Fig. 1), and the traffic speed and event on each link are conditioned on its hidden state. In addition, a coupled structure to the HMM specifies the local dependencies between adjacent links of the arterial network. As shown in the model part of Fig. 1, *the goal of this paper is to more accurately infer the hidden congestion states $z_{t,l}$ for each road link l in each time interval t by utilizing the traffic event observations $e_{t,l}$ and the probe observations $y_{t,l}$.*

Following the classical traffic congestion estimation models [6,8], we make the following assumptions for computational tractability.

- *Discrete traffic states:* For each time interval t, the traffic condition on link l is represented by a discrete value s_t^l, which indicates the level of congestion.
- *Conditional independence of link travel speed:* Conditioned on the state s_t^l of a link l, the travel speed distribution on l is independent from all other traffic variables.
- *Conditional independence of traffic events:* Conditioned on the state s_t^l of a link l, the probability of traffic event $e_{t,l,i}$ occurring on link l is independent from all other traffic variables.
- *Conditional independence of state transitions:* Conditioned on the states of link l and its neighbor links in time interval t, the state of link l at time $t+1$ is independent from all other current link states, all past link states, and all past observations.

The second and third assumptions show that the two types of observations are independent to each other and only determined by the current traffic state of the road link. The last assumption implies that the traffic state of each road link is only related to its neighbor links in the last time interval, but independent of the states of the rest road links in earlier time intervals.

3 Twitter Data Collection

In this section we introduce how we collect traffic event information from Twitter. This paper focuses on studying the traffic conditions in Chicago, and we collect traffic event tweets in Chicago from two types of accounts as in [2]: traffic authority Twitter accounts and regular Twitter user accounts.

Traffic Authority Twitter Accounts. Traffic authority Twitter accounts refer to the Twitter accounts that specialize in posting traffic related information. Such accounts are mostly operated by official transportation departments. Tweets posted by these accounts are formal and easy to process, and the exact location and time information are explicitly given such as the tweet "*Heavy Traffic on NB Western: Fullerton to Kennedy Expy. 06:15 pm 02/13/2015*". We identify 10 such Twitter accounts that report real-time traffic information of Chicago: *ChicagoDrives, ChiTraTracker, roadnowChicago, traffic_Chicago, IDOT_Illinois, WGNtraffic, TotalTrafficCHI, GeoTrafficChi, roadnowil,* and *rosalindrossi.*

Regular Twitter user accounts. We also crawl the tweets posted by the regular users registered in Chicago. In all we target on more than 100,000 such users and crawl more than 32.3 million corresponding tweets. Next, we preprocess the data as follows. (1) *Traffic Event Tweets Identification.* We select traffic event tweets from all the crawled tweets which match at least one term of the predefined vocabularies: "stuck", "congestion", "jam", "crowded", "pedestrian", "driver", "accident", "crash", "road blocked", "road construction", "slow traffic", "heavy traffic", and "disabled vehicle". Based on the keywords contained in the tweets, we can also identify the traffic event category. (2) *Tweet Geocoding.* We then geocode the tweets to the road links by matching their geo-tags and text content. By combing the geo-coordinates of tweets and the direction mentioned in the content, we can geocode the tweets to the road links. For most tweets without geo-tags, we first identify the streets, landmarks, and direction information from the content by using gazetteer, and then geocode them to the road segments.

Note that accurately identifying the locations of traffic events from tweets is itself a challenging task [1,16]. Traffic event location extraction from the short and noisy text is out of the scope of this work. In this paper we only keep the tweets that explicitly give the traffic event type and road segment information. For those with incomplete or obscure location information, we choose to omit them. In all we obtain 245,568 traffic event tweets from April 2014 to December 2014, around 80 % of which are collected from traffic authority accounts. Each tweet reports a traffic event. 163,742 of them are related to slow traffic, 77,454 are related to accident, and 4,372 report other traffic events such as road construction and road closure.

To investigate whether the traffic events reported by Twitter can reflect traffic conditions, we plot the probe speed observations on the road links with a traffic event reported by Twitter and on normal road links in Fig. 2. Each data point in the figure represents a probe speed observation on a road link. Blue data points represent the normal probe observations, while red data points represent probe observations on the road links where traffic congestions or accidents are reported by Twitter. One can see that the average probe speed on the road links with traffic events is lower than that on road links with normal traffic conditions. It implies that the traffic events reported by tweets usually indicate a slower traffic, and thus they can help us better estimate traffic conditions.

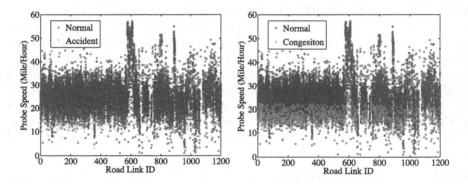

Fig. 2. Probe speed: Normal *vs* Accident (left figure), and Normal *vs* Congestion (right figure) (Color figure online)

4   Extended Coupled Hidden Markov Model: Incorporating Two Types of Observations

Before elaborating the method, we first give some notations and their meanings in Table 1. π_l^s denotes the initial probability of road link l in traffic state s.

A_l is the traffic state transition probability matrix for link l with respect to all its neighbor links. It is a matrix of size $S^{|N_l|} \times S$, where $S^{|N_l|}$ represents the number of all possible states of the neighbors N_l of link l. Based on our assumption, the state of link l in the time interval $t + 1$ is only related to the states of its neighbor links N_l in the last time interval t. Hence each element $A_l(R_i, s)$ represents the probability of link l to be in state s in the time interval $t + 1$ given that its neighbors N_l are in states $R_i = (r_{i1}, r_{i2}, ...r_{i|N_l|})$ in the time interval t. $g_l^s(\cdot)$ is the probability density function of vehicle speed for link l in state s. We assume it follows Gaussian distribution [6]. $f_l^s(\cdot)$ represents the distribution of traffic event number for link l in state s. We assume it follows Multinomial distribution. P_l^s contains all the parameters of the functions $f_l^s(\cdot)$ and $g_l^s(\cdot)$. $q_{t,l}^{R_i,s}$ is a variable to help estimate the transition probability matrix A_l. We use boldface capital letters to denote the observations or hidden state matrixes on all the road links in all the time intervals. For example, $\mathbf{Y}$ denotes all the GPS probe observations. We use capital letters with subscripts to denote the observations or hidden state vectors in a particular time interval. For example, Y_t denotes the GPS probe observations on all the road links from link l_1 to l_N in the time interval t.

Table 1. Notations and meanings

L	Number of road segment links
T	Number of time intervals
M	Number of traffic event types
S	Number of traffic states
N_l	The set of all the neighbor links of road link l
N_{li}	The i-th neighbor based on the lexicographical order of $link\_ID$, $link\_ID \in N_l$
$y_{t,l}$	The set of probe observations for link l in time slot t
$y_{t,l,i}$	One probe observation for link l in t, $y_{t,l,i} \in y_{t,l}$
$e_{t,l}$	The set of traffic event observations for link l in t
$e_{t,l,i}$	One traffic event observation for link l in t, $e_{t,l,i} \in e_{t,l}$
π_l^s	The initial probability that link l begins in state s
A_l	The state transition probability matrix for link l with respect to its neighbors N_l
$g_l^s(\cdot)$	The probability density function of travel speed for link l in traffic state s
$f_l^s(\cdot)$	The distribution function of traffic event for link l in traffic state s
P_l^s	The parameters of the probability density function $g_l^s(\cdot)$ and $f_l^s(\cdot)$
$z_{t,l}^s$	The probability of link l being in traffic state s in t
$q_{t,l}^{R_i,s}$	The probability of link l being in traffic state s for time period t given that its neighboring links N_l are in states $R_i = (r_{i1}, r_{i2}, ...r_{i|N_l|})$ in $t-1$

With above notations, we give the complete log likelihood of the observation data and hidden variables. Typically, the log likelihood of the hidden variables and observations of the CHMM can be written out as follows,

$$lnP(\mathbf{Y}, \mathbf{E}, \mathbf{Z}) = lnP(Z_1) + \sum_{t=2}^{T} lnP(Z_t|Z_{t-1}) + \sum_{t=1}^{T} lnP(Y_t, E_t|Z_t)$$

$$= lnP(Z_1) + \sum_{t=2}^{T} lnP(Z_t|Z_{t-1}) + \sum_{t=1}^{T} lnP(Y_t|Z_t) + \sum_{t=1}^{T} lnP(E_t|Z_t)$$

$$(1)$$

The first term of the formula (1) represents the initial probability of traffic states Z_1 for all the road links, the second term is the probability that traffic states Z_{t-1} in time interval $t-1$ transit to the states Z_t in the next time interval t, and the third term is the probability of observations Y_t, E_t conditioned on the traffic states Z_t. Since the GPS probe observations are independent from the traffic event observations, we can further decompose $\sum_{t=1}^{T} lnP(Y_t, E_t|Z_t)$ as shown in the second line of formula (1).

The initial probability of the congestion states in the first time interval is

$$lnP(Z_1) = \sum_{l=1}^{L} \sum_{s=1}^{S} z_{1,l}^{s} ln\pi_l^s \tag{2}$$

The log probability of congestion state transiting from time interval $t-1$ to t can be further represented as follows,

$$lnP(Z_t|Z_{t-1}) = \sum_{l=1}^{L} \sum_{s=1}^{S} \sum_{i=1}^{S^{|N_l|}} (\prod_{N_{lj} \in N_l} z_{t-1,N_{lj}}^{r_{ij}} z_{t,l}^{s} lnA_l(R_i, s)) \tag{3}$$

The third summation of formula (3) is over all the possible traffic states $S^{|N_l|}$ of the neighbors N_l, while the subsequent product is over terms on each of its individual neighbor state given the neighbor states $(r_{i1}, ..., r_{i|N_l|})$.

The probability of probe speed observations Y_t given the congestion states Z_t can be represented as

$$lnP(Y_t|Z_t) = \sum_{l=1}^{L} \sum_{s=1}^{S} z_{t,l}^{s} (\sum_{y_{t,l,i} \in y_{t,l}} ln(g_l^s(y_{t,l,i}))) \tag{4}$$

The probability of traffic event observations E_t given the congestion states Z_t can be represented as

$$lnP(E_t|Z_t) = \sum_{l=1}^{L} \sum_{s=1}^{S} z_{t,l}^{s} (\sum_{e_{t,l,i} \in e_{t,l}} ln(f_l^s(e_{t,l,i}))) \tag{5}$$

4.1   Solution of E_CHMM: EM Algorithm

Given the distribution function parameters P_l^s of observations and the state transition matrix A_l, it is possible to estimate the congestion states of the links based on the observations. Similarly, given the congestion states of the road links, we can estimate the parameters in the model. Motivated by this idea, EM algorithm can be applied to solve E_CHMM.

In the E-step, for road link l we compute the expected state probabilities $z_{t,l}^s$ and the transition probabilities $q_{t,l}^{R_i,s}$ given observations $(y_{t,l}, e_{t,l})$, distribution parameters P_l^s, and the state transition probability matrix A_l.

$$z_{t,l}^s \leftarrow E(z_{t,l}^s | y_{t,l}, e_{t,l}, P_l^s, A_l) \tag{6}$$

$$q_{t,l}^{R_i,s} \leftarrow E(q_{t,l}^{r,s} | y_{t,l}, e_{t,l}, P_l^s, A_l) \tag{7}$$

One can see that the traffic state $z_{t,l}^s$ is inferred based on both the GPS probe observation $y_{t,l}$ and the tweet observation $e_{t,l}$. To distinguish the importance of the two types of observations in estimating the traffic state $z_{t,l}^s$, we rewrite formula (7) as follows.

$$z_{t,l}^s \leftarrow \begin{cases} E(z_{t,l}^s | e_{t,l}, P_l^s, A_l) & \text{if } Cardinality(y_{t,l}) = 0 \\ w_{t,l} E(z_{t,l}^s | y_{t,l}, P_l^s, A_l) + (1 - w_{t,l}) E(z_{t,l}^s | e_{t,l}, P_l^s, A_l) & \text{otherwise} \end{cases} \tag{8}$$

If only the tweet observation $e_{t,l}$ is available on road link l in time interval t, the congestion state $z_{t,l}^s$ is estimated only based on $e_{t,l}$. Otherwise, $z_{t,l}^s$ is estimated by using both types of observations. $w_{t,l}$ is the confidence of the probe observations. The idea is that if sufficient probe observations are available, we trust more on the traffic state $z_{t,l}^s$ estimated by probe observations. If the probe data are very spare, we trust more on the estimation results with the tweet observations. Here we use a sigmoid function to estimate the importance of the coefficient $w_{t,l} = \frac{1}{1+e^{\theta - Cardinality(y_{t,l})}}$, where θ is a predefined threshold of the probe observation size. More probe observations result in a large $w_{t,l}$, and thus the final estimation result $z_{t,l}^s$ relies more on the probe observations. In this paper we set $\theta = 3$.

In the M-step, we maximize the expected complete log-likelihood, given the probabilities $z_{t,l}^s$ and the transition probabilities $q_{t,l}^{R_i,s}$.

$$(P_l^s, A_l, \pi_l^s) \leftarrow \underset{\mathbf{P,A,\Pi}}{\mathbf{argmax}} ln P(\mathbf{Y, Z, E, P, A, \Pi})$$

$$subject \ to \sum_{s=1}^{S} A_l(R_i, s) = 1, A_l(R_i, s) \in [0,1], \forall l, R_i, s; \tag{9}$$

$$\sum_{s=1}^{S} \pi_l^s = 1, \pi_l^s \in [0,1], \forall l, s.$$

5 Parameter Inference

On small networks, it is possible to do exact inference in the CHMM by converting the model to an HMM. However, it is intractable to do exact inference for any reasonable traffic network with the naive solution due to the following reasons. (1) Computation of the forward variable involves S^L additions and N multiplications at each of T time steps; (2) each forward variable requires $8S^L$ bytes of memory to store, and all T of them must be stored; (3) the transition matrix itself is $S^L \times S^L$. Next we will introduce a sequential importance sampling based approach to more efficiently address the computational challenge.

5.1 E-Step: Particle Filtering

As a popular sequential importance sampling method, particle filtering is widely used to approximately estimate the internal states in dynamical systems such as signal processing and Bayesian statistical inference. Due to the extremely high computational cost of the CHMM, particle filtering is introduced in previous works [9]. In our setting, each particle or sample represents an instantiation of the traffic state evolution on the traffic network. Given the observed probe data and traffic events from tweets, each particle or sample is assigned a weight proportional to the probability of the observations. Using a large number of sampled particles, we can estimate the probabilities of the traffic states of each link in each time interval, and the probabilities of traffic state transition among the neighbor road links in successive time intervals. Details of the algorithm is given in Algorithm 1.

Algorithm 1. Particle Filtering to Estimate Congestion States

---

Input: Number of samples K and time intervals T, the state transition matrix A_l, the parameters of the observation probability function P_l^s for each road link l.
Output: The state probability distribution matrix $\mathbf{Z}$, and the transition probability $q_{t,l}^{R_i,s}$.

1 Initialization: randomly sample K samples $\{x_k^0\}_{k=1}^K$;
2 **for** $t = 1 : T$ **do**
3   Generate K samples of the state x_k^t based on the sampled states x_k^{t-1} and state transition matrix A_l: $x_k^t \sim q(x_k|x_k^{t-1})$;
4   Compute the weights:
5       $w_k^t = p(Y_t, E_t|x_k^t) = p(Y_t|x_k^t)p(E_t|x_k^t)$;
6   Normalize the weights:
7       $\hat{w}_k^t = \frac{w_k^t}{\sum_{j=1}^K w_j^t}$;
8   Resample K random samples $\{\hat{x}_k^t\}_{k=1}^K$ from $\{x_k^t\}_{k=1}^K$ with replacement in proportion to the weights $\{\hat{w}_k^t\}_{k=1}^K$;
9   Replace the sample set with these new samples, *i.e.* $\{x_k^t\}_{k=1}^K \leftarrow \{\hat{x}_k^t\}_{k=1}^K$;
10   Set the weights to be equal: $\hat{w}_k^t = \frac{1}{N}, k = 1, ..., N$

11 Estimate the state probability matrix $\mathbf{Z}$ and transition probability $q_{t,l}^{R_i,s}$ with the K samples
     return $\mathbf{Z}, q_{t,l}^{R_i,s}$;

5.2 M-step: Road Network Decomposition

In the M-step, we update three groups of parameters: the initial congestion state probability π_l^s, the observation distribution function parameters P_l^s, and the transition probability matrix A_l. To update these parameters, the expected

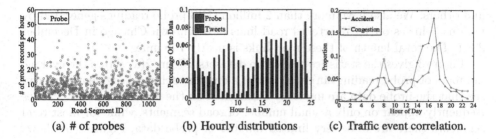

(a) # of probes   (b) Hourly distributions   (c) Traffic event correlation.

Fig. 3. Data Statistics: (a) Average # of probe readings for each road link in each hour. (b) Hourly distributions of probe readings and tweets on each road segment. (c) Hourly occurrence correlations between traffic accidents and congestions reported by tweets.

complete log-likelihood is maximized given the probability $z_{t,l}^s$ that each link l is in state s at time t and probability $q_{t,l}^{R_i,s}$ of link l to be in state s given that neighbors of link l are in states R_i at time $t-1$. Based on formulas (1)–(5), the expected complete log likelihood is as follows.

$$lnP(\mathbf{Y},\mathbf{E}|\mathbf{Z},\mathbf{Q},\mathbf{P},\mathbf{A},\mathbf{\Pi}) =$$

$$\sum_{l=1}^{L}\sum_{s=1}^{S}\sum_{t=1}^{T}z_{t,l}^s\Big(\sum_{y_{t,l,i}\in y_{t,l}}ln(g_l^s(y_{t,l,i})) + \sum_{e_{t,l,i}\in e_{t,l}}ln(f_l^s(e_{t,l,i}))\Big)$$

$$+\sum_{l=1}^{L}\sum_{t=2}^{T}\sum_{s=1}^{S}\sum_{i=1}^{S^{|N_l|}}q_{t,l}^{R_i,s}ln(A_l(R_i,s)) + \sum_{l=1}^{N}\sum_{s=1}^{S}z_{1,l}^s ln(\pi_{l,s}) \tag{10}$$

We can simplify the computation of formula (10) in the following two ways. (1) One can see that formula (10) is comprised of three parts. Different parameters appear in different parts, and thus the three parts can be solved separately. (2) The optimization problem on the entire road network can be further decomposed into $S \times L$ smaller optimization problems with each one associated to a particular congestion state and road link of the network. For example, for the road link l in state s the first part in the right-hand side of formula (10) can be decomposed to such an optimization problem.

$$\max_{P_{l,s}}\sum_{t=1}^{T}z_{t,l}^s\Big(\sum_{y_{t,l,i}\in y_{t,l}}ln(g_l^s(y_{t,l,i})) + \sum_{e_{t,l,i}\in e_{t,l}}ln(f_l^s(e_{t,l,i}))\Big) \tag{11}$$

6 Evaluation

6.1 Experiment Setup

Datasets and analysis. The Twitter data are described in Sect. 3. From each tweet, we extract the road segment, time and traffic event information. We categorize these tweets into three types by keywords matching: congestion, accident and others. We also have more than 2 million GPS probe readings generated by various vehicles on 1,257 arterial road links of downtown Chicago in December 2014. The total length of these road links is nearly 700 miles.

Figure 3 gives the statistics of the two datasets. Figure 2(a) shows the average numbers of probe readings in each hour of a day for each road segment. One can see that the probe data are unevenly distributed on the arterial network. Probes frequently appear on only a small number of road segments, while for most road segments there are only very limited number of probe data. Figure 3(b) shows the percentages of probe data and traffic related tweets in each hour of day. One can see that most probe data are distributed in the time interval from 14:00pm to 0:00am. Most traffic related tweets are posted in two time intervals from 5:00am to 10:00am and from 15:00pm to 22:00pm. The hourly distributions of the two datasets are not perfectly consistent, which implies the combination of them could provide us with more comprehensive information. Figure 3(c) shows the proportion curves of the traffic accident and congestion reported by tweets in each hour of a day. One can see that the two curves show very similar increasing and decreasing trends, which indicates a strong occurrence correlation. Traffic congestions can cause more accidents, and accidents in turn can make traffic even worse. The high occurrence correlation between accident and congestion implies that other types of traffic events captured from tweets may potentially help us better estimate traffic congestions.

Ground Truth. Obtaining the ground truth itself is a challenge problem. The manually annotated ground truth is very expensive, and thus is not feasible for a large transportation network. Previous studies show that the bus probe data in urban areas can provide a good approximation of the real traffic conditions [5,22]. Thus we use the traffic conditions reported by Chicago Transit Authority (CTA) as the ground truth. The traffic conditions are estimated based on more than 5 million GPS traces generated by more than 2,000 CTA public passenger buses from 11/25/2014 to 12/30/2014[1]. CTA defines 5-state traffic conditions in Chicago: heavy congestion, medium-heavy congestion, medium, light, and flow conditions, with the corresponding traffic speeds as 0–10, 10–15, 15–20, 20–25, and over 25 miles per hour. We assign the 5 congestion states with values 1.0, 0.8, 0.6, 0.4, and 0.2 respectively. As the real time GPS traces for some links are sparse, we also consider the historical average traffic speed for each road link in the last 3 years. Given a time interval t and a road link l, the traffic speed can be estimated as $speed_{t,l} = w \sum_{i=1}^{n} \frac{speed_{t,l,i}}{n} + (1 - w)speed_{t,l}^h$, where $speed_{t,l,i}$ is the ith real time probe speed record, $speed_{t,l}^h$ is the historical speed, and w is a weight. For simplicity, we consider a road segment is in congestion if the average speed is lower than 15 mph.

Competitive Methods. We compare E_CHMM with the following baselines.

– **CHMM with probe observations (P_CHMM)** [6]. Herring et al. proposed a CHMM model to estimate arterial traffic conditions with probe data.

---

[1] https://data.cityofchicago.org/Transportation/Chicago-Traffic-Tracker-Historical-Congestion-Esti/77hq-huss.

We use it as a baseline to evaluate whether incorporating the Twitter data can improve the performance.

- **CTCE model** [2]. CTCE is a recently proposed traffic congestion estimation model with social media as the primary data source. Instead of utilizing CHMM, CTCE models the traffic information on the road segments as matrices and tensors and apply matrix factorization technique to address the estimation task.
- **CHMM with tweet observations (T_CHMM).** In this model, only the tweet observations are available. We use this baseline to evaluate the performance of the CHMM model with the tweet observations only.
- **Linear combination of the two types of data (LC_CHMM).** We use two CHMMs with each one associated with one type of data to estimate the traffic conditions separately. Assuming the estimation results of the two models are $\mathbf{Z}_1$ and $\mathbf{Z}_2$, the final estimation is the linear combination of the two results, $\mathbf{Z} = \alpha\mathbf{Z}_1 + (1 - \alpha)\mathbf{Z}_2$.

Evaluation Metrics. We use the following metrics to evaluate the performance of the proposed model: *accuracy*, *precision@k*, and *Root Square Error (RMSE)*. We use *accuracy* to evaluate the estimation performance on all the road segments in all the time intervals. Normally, in a particular time interval only a small number of road segments are in congestion. Thus to better evaluate whether the proposed model can give good estimations on the road segments that are very likely to occur congestion, we also use *precision@k* as a metric. We first rank the congestion probabilities $z_{t,l}^s$ for all the road segments in all the time intervals. Then we only consider the road segments with the *top-k* congestion probabilities are in congestion. To further evaluate the performance of the model on the above mentioned 5-state traffic conditions, we use the *Root Mean Square Error (RMSE)* as the evaluation metric: $RMSE = \sqrt{\frac{\sum_{t,l}(z_{t,l}-\hat{z}_{t,l})^2}{L*T}}$, where $z_{t,l}$ is the estimated traffic state of link l in time interval t, and $\hat{z}_{t,l}$ is the ground truth.

6.2 Quantitive Evaluation Results

Evaluation with *precision@k*. Table 2 shows the average *precision@k* of different methods over various k. As the traffic conditions on weekdays and weekends can be quiet different, we present the results by weekday and weekend separately. We run the algorithm and calculate the *precision@k* on each day, and then average the results. The best results are highlighted in bold type. One can see that E_CHMM performs best among all the methods. LC_CHMM model is inferior to E_CHMM, but better than other methods. It is no surprise that T_CHMM presents the worst performance among all the methods. One can infer that the traffic event tweets are too sparse for the T_CHMM model to get an accurate estimation. P_CHMM can achieve comparable performance with CTCE, but both methods are inferior to LC_CHMM and E_CHMM. One can also see that in general the average *precision@k* on weekday is higher than

Table 2. Average Precision @k of different methods

Average Precision @k on weekday

	top-10	top-20	top-30	top-50	top-100	top-150	top-200	top-250	top-300
P_CHMM	0.870	0.850	0.845	0.832	0.812	0.792	0.773	0.744	0.732
T_CHMM	0.690	0.665	0.624	0.613	0.585	0.532	0.473	0.464	0.452
LC_CHMM	0.890	0.850	0.852	0.842	0.832	0.817	0.792	0.784	0.775
CTCE	0.870	0.860	0.853	0.840	0.824	0.816	0.718	0.705	0.712
E_CHMM	**0.920**	**0.900**	**0.894**	**0.887**	**0.864**	**0.826**	**0.810**	**0.795**	**0.786**

Average Precision @k on weekend

	top-10	top-20	top-30	top-50	top-100	top-150	top-200	top-250	top-300
P_CHMM	0.860	0.850	0.843	0.822	0.816	0.766	0.752	0.745	0.722
T_CHMM	0.660	0.650	0.612	0.625	0.570	0.464	0.453	0.415	0.425
LC_CHMM	0.870	0.850	0.845	0.825	0.820	0.812	0.805	0.785	0.768
CTCE	0.850	0.834	0.820	0.820	0.754	0.715	0.678	0.654	0.644
E_CHMM	**0.910**	**0.900**	**0.868**	**0.852**	**0.844**	**0.820**	**0.812**	**0.794**	**0.783**

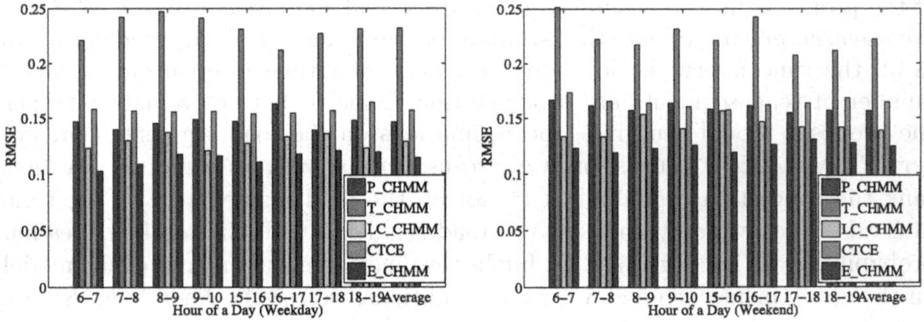

Fig. 4. RMSE of the four methods in rush hours

that on weekend. This is because most people travel on weekday more regularly than on weekend.

Performance evaluation in rush hours. People concern more on the traffic conditions in rush hours of a day. Thus we also evaluate the performance of different models in rush hours. Figure 4 shows the experiment results in the rush hours of 6:00–10:00 and 15:00–17:00 on weekday and on weekend, respectively. One can see that the RMSE of E_CHMM is mostly lower than all the baselines. The performance of T_CHMM is the worst among all the methods, which is consistent with the previous experiment results. LC_CHMM is consistently better than P_CHMM and CTCE, which means incorporating traffic event information from tweets does help us better estimate traffic conditions. However, LC_CHMM is inferior to the proposed E_CHMM. Thus we can conclude that E_CHMM is more efficient to fuse the two types of observations. By comparing the results on weekday and weekend, one can see that on average the RMSE of various methods on weekday is larger than that on weekend. This finding also verifies that traffic conditions on weekend is harder to estimate than on weekday.

Performance evaluation with various proportions of probe data. To examine how the probe data size affects the estimation performance, we display the estimation accuracy curves of the methods E_CHMM, LC_CHMM, and P_CHMM with different probe data sizes in Fig. 5. It shows that E_CHMM is consistently better than the two baselines. When the probe data are extremely sparse, say only 20 % probe data are available, the accuracy of P_CHMM is only 0.22 while E_CHMM is 0.42, which shows a significant improvement. However, with the increase of the probe data size, the difference between E_CHMM and the other two methods becomes smaller. This is probably because the information overlapping between the two datasets becomes larger when more probe date are available. When the probe data are sufficient, the traffic conditions inferred by traffic event tweets can also be captured by the probe readings. The LC_CHHM is better than P_CHHM but inferior to E_CHMM. One can see that E_CHMM only needs around 85 % probe data to achieve a comparable accuracy to P_CHMM with the whole probe data.

Scalability Analysis. As the optimization problem of the EM algorithm can be decomposed into many smaller optimization problems, we can easily solve it in parallel on multiple machines. Figure 6 shows the running time of solving the optimization problems by distributing them into multiple machines on the traffic data of a day on the studied road links. It shows a linearly decreasing trend of the running time with the increase of machine number. One can see that it needs more than 12 min for only one machine, but the time decreases to about 2 min if we distribute these independent smaller optimization problems on 5 machines. It demonstrates the proposed algorithm is very scalable to handle a large road network with thousands of road links.

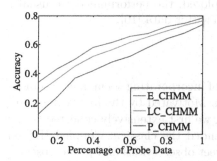

Fig. 5. Estimation accuracy *vs* probe data size

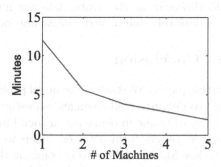

Fig. 6. Running time *vs* # of machines

7 Related Work

Traditionally, traffic monitoring and estimation mainly rely on various road sensors, and can be roughly categorized into traffic modeling on individual roads [11,13,14] and on a road network [20]. Helbing employed a Fundamental Diagram to learn the relations among vehicle speed, traffic density, and volume for a particular road to estimate traffic condition on an individual road [11]. Muoz et al. proposed a macroscopic traffic flow model SMM by utilizing the loop detector data to estimate the traffic density at unmonitored locations along a highway [14]. Porikli and Li proposed a Gaussian Mixture Hidden Markov Models to detect traffic condition with the MPEG video data [13]. Researches on traffic monitoring on a road network usually need to capture and model the correlations of the traffic conditions among the road segments connected to each other [6,15,20]. Such models mainly utilized the Floating Car Data (FCD) or probe data generated by the GPS sensors equipped in vehicles. Herring et al. proposed a coupled Hidden Markov Model which can effectively capture the traffic congestion correlations among the road segments [6]. Fabritiis et al. studied the problem of using FCD data based on traces of GPS positions to predict the traffic on Italian motorway network [15].

Recently, exploring traffic related information from social media like Twitter to detect traffic events or monitor traffic conditions has been a hot research topic [1,2,10,12]. Most previous works focused on investigating either how to extract and visualize the traffic event information from tweets [1,12] or how to locate the traffic events mentioned in the tweets [16,19]. As traffic event data are usually sparse and imbalanced, imbalanced learning techniques are usually explored [17]. The work in [10] is the first to estimate traffic congestion of an arterial network by collecting traffic related tweets from Twitter. Wang et al. further incorporated other information such as social events and road features with social media data to more effectively estimate citywide traffic congestions [2]. However, as the probe data are not explored, the performance are usually not desirable due to very sparse and noisy Twitter data [10].

8 Conclusion

In this paper, we studied the novel problem of incorporating social media semantics to enhance traffic congestion estimation. Motivated by the increasing availability of traffic information in social media, we first extensively collected traffic related tweets from Twitter. Then we extended the classical Coupled Hidden Markov Model to effectively combine the tweet observations and probe observations. To solve the proposed model, we also introduced an efficient EM algorithm to infer the parameters. Evaluation on the arterial network of Chicago showed the proposed model can both effectively combine the two types of observations and efficiently address the computational challenge.

Acknowledgments. This paper is supported by Beijing Advanced Innovation Center for Imaging Technology (No.BAICIT-2016001), the NSFC (Grant No.: 61272083,

61370126, 61202239), the NSF through grants III-1526499, the Natural Science Foundation of Hebei Province (No. F2014210068), and Collaborative Innovation Center of Novel Software Technology and Industrialization.

References

1. Liu, M.L., Fu, K.Q., Lu, C.T., Chen, G.S., Wang, H.Q.: A search and summary application for traffic events detection based on twitter data. In: ACM SIGSPATIAL GIS (2014)
2. Wang, S.Z., He, L.F., Stenneth, L., Yu, P.S., Li, Z.J.: Citywide traffic congestion estimation with social media. In: ACM SIGSPATIAL GIS (2015)
3. Wang, S.Z., He, L.F., Stenneth, L., Yu, P.S., Li, Z.J., Huang, Z.Q.: Estimating urban traffic congestions with multi-sourced data. In: IEEE MDM (2016)
4. Ozkurt, C., Camci, F.: Automatic traffic density estimation and vehicle classification for traffic surveillance systems using neural networks. Math. Comput. Appl. 14(3), 187–196 (2009)
5. Wang, Y., Zhu, Y.M., He, Z.C.: Challenges and Opportunities in Exploiting Large-Scale GPS Probe Data. Technical report, HPL-2011-109 (2011)
6. Herring, R., Hofleitner, A., Abbeel, P., Bayen, A.: Estimating arterial traffic conditions using sparse probe data. In: IEEE ITSC (2010)
7. He, J.R., Shen, W., Divakaruni, P., Wynter, L., Lawrence, R.: Improving traffic prediction with tweet semantics. In: IJCAI (2013)
8. Hofleitner, A., Herring, R., Abbeel, P., Bayen, A.: Learning the dynamics of arterial traffic from probe data using a dynamic Bayesian network. IEEE Trans. Intell. Transp. Syst. 13(4), 1679–1693 (2012)
9. Cheng, P., Qiu, Z.J., Ran, B.: Particle filter based traffic state estimation using cell phone network data. In: IEEE ITSC (2006)
10. Chen, P.T., Chen, F., Qian, Z.: Road traffic congestion monitoring in social media with hinge-loss Markov random fields. In: ICDM (2014)
11. Helbing, D.: Traffic and related self-driven many-particle systems. Rev. Mod. Phys. 73(4), 1067–1141 (2001)
12. Endarnoto, S.K., Pradipta, S., Nugroho, A.S., Purnama, J.: Traffic condition information extraction and visualization from social media twitter for android mobile application. In: ICEEI (2011)
13. Porikli, F., Li, X.K.: Traffic congestion estimation using HMM models without vehicle tracking. In: IEEE IV (2004)
14. Muoz L., Sun X.T., Horowitz R., Alvarez L.: Traffic density estimation with the cell transmission model. In: American Control Conference (2003)
15. Fabritiis, C.D., Ragona, R., Valenti, G.: Traffic estimation and prediction based on real time floating car data. In: ITCS (2008)
16. Ribeiro Jr., S.S., Davis Jr.,, C.A., Oliveira, D.R.R., Meira Jr., W., Gonalves, T.S., Pappa, G.L.: Traffic observatory: a system to detect and locate traffic events and conditions using twitter. In: ACM SIGSPATIAL LBSN (2012)
17. Wang, S.Z., Li, Z.J., Chao, W.H., Cao, Q.H.: Applying adaptive over-sampling technique based on data sparsity and cost-sensitive SVM to imbalanced learning. In: IJCNN (2012)
18. Bregman, S.: Uses of Social Media in Public Transportation. Transportation Research Board (2012)

19. Daly, E.M., Lecue, F., Bicer, V.: Westland row why so slow? Fusing social media and linked data sources for understanding real-time traffic conditions. In: IUI (2013)
20. Shang, J.B., Zheng, Y., Tong, W.Z., Chang, E., Yu, Y.: Inferring gas consumption and pollution emission of vehicles throughout a city. In: ACM SIGKDD (2014)
21. Sayyadi, H.: Event detection and tracking in social streams. In: ICWSM (2009)
22. Carli, R., Dotoli, M., Epicoco, N., Angelico, B., Vinciullo, A.: Automated evaluation of urban traffic congestion using bus as a probe. In: IEEE CASE (2015)

Learning Distributed Representations of Users for Source Detection in Online Social Networks

Simon Bourigault$^{(\boxtimes)}$, Sylvain Lamprier, and Patrick Gallinari

Sorbonne Universités, UPMC Univ Paris 06, UMR 7606, LIP6, 75005 Paris, France
{simon.bourigault,sylvain.lamprier,patrick.gallinari}@lip6.fr

Abstract. In this paper, we study the problem of source detection in the context of information diffusion through online social networks. We propose a representation learning approach that leads to a robust model able to deal with the sparsity of the data. From learned continuous projections of the users, our approach is able to efficiently predict the source of any newly observed diffusion episode. Our model does not rely neither on a known diffusion graph nor on a hypothetical probabilistic diffusion law, but directly infers the source from diffusion episodes. It is also less complex than alternative state of the art models. It showed good performances on artificial and real-world datasets, compared with various state of the art baselines.

1 Introduction

Today, hundreds of millions of people use online social networks to access, discuss, produce and share content. These social networks now have an important impact on the way information travels worldwide. This has motived a large amount of research on the topic of *information diffusion prediction*: how can we predict *which* users will be infected by a given piece of information in the future? This "word of mouth" phenomenon has been widely studied over the last decade (see [8] for a comprehensive survey).

More recently, the problem of *source detection* has emerged. This is the opposite task: the goal is to retrieve which user *started* some diffusion episode, given the set of *eventually* infected users. In an epidemiological context, this is also known as the *patient zero* problem. For social media, the main application of this problem is to retrieve the source of some rumor, leak or disinformation, either to remove it from the network or to take legal action against it.

While several works have already studied this problem (see Sect. 2), they are all based on the assumption that the social graph on which diffusion takes place is either known or can be inferred, and that information diffusion follows some known propagation model such as the SI model [16,20] or the NETRATE model [5]. These turn out to be strong assumptions in most applications.

In this paper, we drop the aforementioned assumptions by using a Representation Learning approach to embed users in a latent space and use their representations to directly retrieve infection sources. This method does not require

© Springer International Publishing AG 2016
P. Frasconi et al. (Eds.): ECML PKDD 2016, Part II, LNAI 9852, pp. 265–281, 2016.
DOI: 10.1007/978-3-319-46227-1_17

the influence graph to be known, and can be applied to partially observed diffusion episodes. Moreover, it allows us to easily consider the topic of the diffusion in concern, defining content-specific transformations of the representations of the users involved in the diffusion in concern. To the best of our knowledge, our approach is the first one to consider the content for source detection tasks. Our proposals are tested on *real* diffusion traces extracted from online social networks, something that is often missing in the literature of the field.

The rest of the paper is organized as follows. Section 2 reviews some related works and presents the motivations of our proposal. Section 3 introduces our model. Finally, Sect. 4 compares our model to various baselines.

2   Background and Motivations

While the topic of information diffusion prediction has been studied for a long time [8,9,18], source detection has been a subject of research for a few years.

As classically done in the field of diffusion modeling, existing approaches for source detection are based on the Susceptible-Infected framework defined on a given known graph of diffusion $G = (\mathcal{U}, E)$. When a user $u \in \mathcal{U}$ becomes infected at time t, each neighbor v in the graph becomes infected at time $t + d_{u,v}$, with $d_{u,v}$ being drawn from some delay distribution [4,5,16,19,20,22]. The various methods mainly differ in their way of reversing the process of diffusion to predict the most probable source when some infections are observed.

The work of [20] was the first one to introduce the key concept of *rumor centrality*, a measure rendering the likelihood, for any content emitted from a node $u \in \mathcal{U}$, to spread over a given subset of infected users $\mathcal{U}' \subseteq \mathcal{U}$, knowing some diffusion relationships between them $E' \subseteq E$. When a set $\mathcal{U}'$ of infected users is observed at some time T, the source user can be estimated with a maximum likelihood approach:

$$s^* = \arg\max_{s \in \mathcal{U}'} P(\mathcal{U}'|s) \propto \arg\max_{s \in \mathcal{U}'} R(s, \mathcal{U}')$$

where $R(s, V')$ stands for the *rumor centrality* measure, applied to the source candidate $s \in \mathcal{U}'$, which is computed by considering the number of possible sequences of infections of nodes from $\mathcal{U}'$, that start with s and are consistent with the precedence graph defined by $G' = (\mathcal{U}', E')$. This work was later extended in [4,22] to optimize the estimation of R on more complex graph structures. All these works assume that one observes a complete snapshot of the network (infected nodes and edges) at some time T, and that the source is among these infected nodes. The infection times of each node is left unknown. Later, [19] proposed to consider a framework in which only the states of a *subset* of all users (called "monitors") are observed, and compared various heuristics to select monitors and to retrieve rumor sources: reachability of infected nodes, distances to infected nodes in the graph, etc.

Some other works proposed to consider a framework in which we also observe *when* each node became infected. In this framework however, some infections (including the source one) remain unobserved (due for example to some API restrictions), and the goal is to retrieve the source node from the set of unobserved nodes $\mathcal{U} \setminus \mathcal{U}'$. A first model was developed in [16], which is based on the assumption that transmission delays in the network follow a Gaussian distribution. The predicted source then corresponds to an unobserved node that maximizes the likelihood of the observed infection times. They proposed a heuristic based on the extraction of trees from the graph, similar to the one described in [20, 22]. Recently, [5] proposed a more precise approach based on previous works on information diffusion and link prediction [7], where transmission delays follow a exponential distribution. The computation of the likelihood of a source being difficult, a method based on importance sampling is employed.

Finally, the problem of *multiple sources detection* has also been addressed. In [11], the authors define the *k-effectors* problem. They assume that diffusion follows an Independent Cascades Model (IC) and look for the set of k sources X that minimizes the cost:

$$C(X) = \sum_{u_i \in U} |a(i) - \alpha(i, X)|$$

where $a(i)$ indicates wether u_i is infected of not (1 or 0) and $\alpha(i, X)$ is the probability for a user i to become infected when the source set is X. In other words, they minimize an $\ell 1$ error. The minimization of C is shown to be NP-complete, so the authors study the problem on tree graphs, and propose a heuristic for this case. For general graphs, they suggest to extract a spanning tree and to apply the heuristic on that tree.

Another approach, NETSLEUTH, was proposed in [17]. It relies on the Minimum Description Length (MDL) principle. The authors propose an efficient method to describe the diffusion of an information (initial sources and list of all successive transmissions) in a minimum number of bits, assuming that the graph and the diffusion model are known. Given a set of infected users, they look for the set of sources and transmissions that minimizes the amount of bits required to be encoded. This approach is thus able to determine the number of sources as well as their identities.

While reasonable, considering the iterative diffusion process on a known graph faces the two following main limitations:

- The performances for source detection are strongly dependent on the quality of the diffusion graph that is considered. However, the information about the diffusion graph is often missing, incomplete or irrelevant. Various methods, such as those proposed in [10] or [7], can learn the graph from a training set of diffusion episodes, but their effectiveness greatly depend on the representativeness of the available training data.
- The estimation of the most probable source s^* usually requires to compute the shortest path between all user pairs in the graph, which is computationally expensive. For instance, the approach presented in [21] is #P-hard.

Finally, these models have usually been tested on synthetic datasets only, with episodes generated using the diffusion model used in prediction. While this definitely gives important insights on their performances, results on real diffusion episodes are necessary to assess the efficiency of these approaches. For instance, [5] performed experiments on the MEMETRACKER dataset, and the results are very low compared to those obtained on synthetic datasets.

In this paper, we propose to embed users in a latent space and use distance between them to retrieve the source. This is related to the work of [2], which applied representation learning to the information propagation prediction task. Recently, representation learning has been used in various domains like playlist prediction [3] or language models [15]. These methods aim at projecting some items like songs, users or words in a latent, euclidean space so that relationships between them can be modeled with the distances computed in that latent space. Representation Learning has at least the following main advantages in the context of source detection:

- Compression abilities offered by representation learning techniques enable the definition of more compact models, especially for dense social networks;
- Diffusion relationships, which are encoded in a shared representation space, are regularized naturally: users with similar behaviors are likely to be projected near each other, and then tend to share some similar transmission tendencies with other users, which improves the ability of the model to generalize from sparse data;
- A representation for the diffusion episode can be computed efficiently by combining individual representations of the infected users. This enables simple and fast source detection procedures;
- The diffused content or any other additional information can be taken into account, by considering specific transformations of the diffusion representation.

Rather than reversing a given diffusion model as classically done, we thus propose to consider the use of such techniques for source detection tasks, by directly learning projections of users that lead to an efficient retrieval of diffusion sources.

3   Diffusion Source Detection

Let $\mathcal{U} = \{u_1, \ldots, u_N\}$ be a population of N users who communicate and exchange information. When some piece of information *propagates* in that population, we observe a *diffusion episode*, which corresponds to a sequence of infected users associated with their timestamps of infection:

$$D = \{(u_i, t_i), (u_j, t_j)...\}$$

A diffusion episode can correspond, for instance, to a sequence of users who liked a specific video or retweeted a specific tweet. The first user of this sequence is the *source user*, and is denoted s_D. In the following, we note $\mathcal{U}_D$ the set of users infected in the episode D and $\hat{\mathcal{U}}_D$ the same set but without the source user of D (i.e., $\hat{\mathcal{U}}_D = \mathcal{U}_D \setminus \{s_D\}$).

3.1   A Representation Learning Model

Our ultimate goal is to be able to retrieve the source s_D of a given diffusion episode D from which that source is *missing* (i.e. observed infected users are those belonging to $\hat{\mathcal{U}}_D$).

Basic Idea. To this end, our idea is to embed all users of the network in a latent space, by defining a representation $z_i \in \mathbb{R}^d$ for every user $u_i \in \mathcal{U}$, such that it is possible to predict the source user of an episode by looking at the relative locations of users in this space. With z_D a representation of the episode D, constructed as a function of individual representations of users in $\hat{\mathcal{U}}_D$, we base our model on the following principle:

The representation of the source user s_D of any diffusion episode $D \in \mathcal{D}$ should be located at the point z_D, which corresponds to the synthetic representation of the diffusion episode.

Following this principle, the episode representation z_D corresponds to an initial diffusion point from which the emitted content can spread to reach all infected users of the episode D. In this context, building a source prediction model corresponds to making the representation of s_D coincide with this continuous initial diffusion point z_D. Like most representation techniques, we seek at defining a projection space where the similarities between user's representations render some interactions propensities between them, which allows the model to leverage the behavior correlations of the users. Following this, we wish to set the initial diffusion point z_D as the closest possible point to every individual representation of users in $\hat{\mathcal{U}}_D$, so that it can equivalently explains the set of all observed infections. Various functions $\phi : 2^N \rightarrow \mathbb{R}^d$ can hold this requirement to transform a set of individual representations of users in $\hat{\mathcal{U}}_D$ to the episode representation z_D. Another constraint is the low cost of computation. We therefore consider the following function ϕ, which corresponds to an averaged representation of infected users:

$$z_D = \phi(\hat{\mathcal{U}}_D) = \frac{1}{|\hat{\mathcal{U}}_D|} \sum_{u_i \in \hat{\mathcal{U}}_D} z_i \qquad (1)$$

Note that such a function ϕ also presents the advantage to be rather stable w.r.t. missing infected users (i.e., when $|\hat{\mathcal{U}}_D|$ is sufficiently large, $\forall u \in \mathcal{U} : \phi(\hat{\mathcal{U}}_D \cup \{u\}) \approx \phi(\hat{\mathcal{U}}_D)$), which allows the model to manipulate consistent representations in the case of incomplete observations of the episodes diffusions. An illustration of the targeted projection of a given observed diffusion episode is given in Fig. 1, where the source user of the episode D is projected at the center of the representations of infected user in $\hat{\mathcal{U}}_D$.

Source Prediction Model. Now that the basic idea of our proposal is presented, we can define our model for source prediction based on it. To begin with, lets us note that learning one representation per user would lead to a symetric model, where the tendency of a user to be an information source would be equal to its tendency to be infected in a diffusion episode. This setting is not realistic, as diffusion is an asymmetric process: while some users are opinion leaders, others only reproduce some collected content. To include this observation in our model, we therefore consider two representations for each user u_i: while the vector z_i embeds the behavior of u_i as a *receiver* of information, ω_i embeds his behavior as a *sender* of content (i.e., a source user in our context, since we only consider transmissions from the source to all eventually infected users). Defining these two embeddings per user allows us to model asymmetric relationships.

To retrieve the source of the diffusion D given $\hat{\mathcal{U}}_D$, the model then considers the user u_i whose sender embedding ω_i is the closest to the synthetic representation of the episode z_D:

$$s^* = \underset{u_i \in \mathcal{U} \setminus \hat{\mathcal{U}}_D}{\mathrm{argmin}} \ ||\omega_i - z_D||^2 \tag{2}$$

where z_D is computed by using formula 1 applied to users $\hat{\mathcal{U}}_D$. In order to learn both sets of embedding $\Omega = (\omega_i)_{\forall i \in \mathcal{U}}$ and $\mathcal{Z} = (z_i)_{\forall i \in \mathcal{U}}$ so that formula 2 returns accurate sources of diffusion, we consider the following pairwise loss on the learning set of diffusion episodes $\mathcal{D}$:

$$\mathcal{L}(\Omega, \mathcal{Z}) = \sum_{D \in \mathcal{D}} \sum_{u_i \notin \mathcal{U}_D} h\left(||\omega_i - z_D||^2 - ||\omega_{s_D} - z_D||^2\right) \tag{3}$$

where h corresponds to the hingeloss function: $h(x) = \max(1 - x, 0)$. This function is a pairwise ranking loss that follows the principle of the prediction function (formula 2). Basically, for our prediction function to be valid, we need the sender representation of the actual source to be closer to the representation of D (second term of the subtraction in h) than any other sender representation (first term of the subtraction), so that it is the one who would be predicted using formula 2.

This loss can easily be minimized by defining a stochastic gradient descent process, detailed in Algorithm 1. Intuitively, it can be summed up like this: first, we initialize all embeddings at random (lines 2 and 3). Then, at each iteration, we draw one episode D (line 6) and one "non-source" user u_j that is not in $\mathcal{U}_D$ (line 7). If the embedding ω_{s_D} of the actual source is not closer to the representation z_D than ω_j by at least 1 (line 11), all relevant embeddings are updated with one gradient step (lines 12, 13 and 15). This gradient step moves the representation z_D toward ω_{s_D} and away from ω_j. The learning goes on until convergence, which is tested by checking the variation of $\mathcal{L}$ every set number of iterations (100000 in our case).

Regularization of Embeddings. In the loss defined above, two representation vectors are learned for each user to account for the difference of its behavior as a sender and a receiver [1]. While these two representations can be quite different, it is reasonable to think that they are not uncorrelated: both behaviors are consequences of the centers of interests of that user. To account for this correlation,

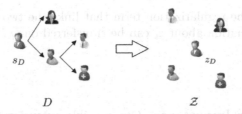

$$D \qquad\qquad\qquad \mathcal{Z}$$

Fig. 1. From a diffusion episode tree to its projection in our representation space.

Algorithm 1. Representation Learning for Source Detection

Data:
$\mathcal{U}$: Users set ;
$\mathcal{D}$: Learning set of diffusion episodes ;
d : Number of dimensions
ϵ : Gradient step size;
Result:
$Z = \{\forall u_i \in \mathcal{U} : z_i \in \mathbb{R}^d\}$; $\Omega = \{\forall u_i \in \mathcal{U} : \omega_i \in \mathbb{R}^d\}$;

1 **foreach** $u_i \in \mathcal{U}$ **do**
2   initialize z_i with random value in $[-1, 1]^d$
3   initialize ω_i with random value in $[-1, 1]^d$
4 **end**
5 **while** *non-convergence* **do**
6   Draw an episode $D \in \mathcal{D}$;
7   Draw $u_j \notin \mathcal{U}_D$;
8   Compute z_D with formula 1 ;
9      $d_s \leftarrow \|\omega_{s_D} - z_D\|^2$;
10      $d_j \leftarrow \|\omega_j - z_D\|^2$;
11   **if** $d_j - d_s < 1$ **then**
12          $\omega_{s_D} \leftarrow \omega_{s_D} - \epsilon \times 2\,(\omega_{s_D} - z_D)$;
13          $\omega_j \leftarrow \omega_j + \epsilon \times 2\,(\omega_j - z_D)$;
14   **forall** $u_x \in \hat{\mathcal{U}}_D$ **do**
15              $z_x \leftarrow z_x - \epsilon \times \frac{2}{|\hat{\mathcal{U}}_D|}\,(\omega_j - \omega_{s_D})$
16   **end**
17   **end**
18 **end**

---

we include a sender-receiver regularization term in the loss considered for the learning of the model:

$$\mathcal{L}(\Omega, \mathcal{Z}) + \lambda \sum_{u_i} \|\omega_i - z_i\|^2 \qquad\qquad (4)$$

where the second term corresponds to the desired regularization weighted by an hyper-parameter λ. This term favors embeddings such that ω_i and z_i are close, and improves the generalization ability of the model. For instance, without this term, no embedding ω_i for a user who never appears as a source in $\mathcal{D}$ could be learned. With the regularization term that links the two representations ω_i and z_i, some information about z_i can be transferred on ω_i. This also prevents over-fitting.

3.2 Extensions

Inclusion of User Importance. One possible extension of our model is to learn an additional *weight* $\alpha_i \in \mathbb{R}^+$ for each user in the training set and redefine z_D as:

$$z_D = \sum_{u_i \in \hat{\mathcal{U}}_D} \frac{e^{\alpha_i}}{\sum_{u_j \in \hat{\mathcal{U}}_D} e^{\alpha_j}} z_i$$

where the fraction corresponds to the softmax function that allows to map a vector of k real values to $[0;1]^k$. This formulation corresponds to the computation of the barycenter of the representations of users in $\hat{\mathcal{U}}_D$, with weights defined by relative values of the α parameters of these users. These parameters therefore model the relative importance of each user to predict the source of the diffusion. For instance, on Twitter, some user u_i could happen to actually be a spamming bot that just reuse all popular hashtags in order to gain visibility and post ads. In that case, the infection of this user gives little to no information about the source, and the system will learn a weight $\alpha_i \approx 0$. Beyond allowing the learning process to focus on more discriminant infections, and to discard users with very chaotic behaviors, it may also permit to select the most important users to select in a situation where only a subset of them can be simultaneously monitored [16].

Integration of Content. It is known that the content of a piece of information modifies the way it propagates [23]. For instance, two pieces of information shared by the same source, one about sports and the other about politics, will probably not spread to the same users. In this subsection, we propose a way to include it in the model. The content associated to an episode D is represented by some vector $w_D \in \mathbb{R}^a$. Depending on the application, this vector may for instance be a bag-of-words extracted from text, or some visual features extracted from an image. We learn content transformation parameters $\theta \in \mathbb{R}^{a \times d}$ that are used to map a given content to $\mathbb{R}^d$ by a linear application $< w_D, \theta >$. The resulting vector of this application is used to translate the episode representation z_D, which implies content-specific modifications of the prediction model:

$$z_D = \frac{1}{|\hat{\mathcal{U}}_D|} \sum_{u_i \in \hat{\mathcal{U}}_D} z_i + < w_D, \theta > \tag{5}$$

The parameters θ are learned at the same time as user's projection parameters, considering the optimization of the loss from formula 4 with this definition of translated representation z_D. Note that other content specific transformations have been investigated, but this simple translation of z_D allowed us to observe the best results on a validation set.

4 Experiments

4.1 Datasets

The following datasets have been used:

Artificial. Diffusion episodes generated using the IC model [18] on a scale-free network of 100 users.

Lastfm. Dataset extracted from a music streaming website. Each diffusion episode gathers the users who listened to a given song.[1]

Weibo. Retweet cascades extracted from the Weibo microbloging website using the procedure described in [12]. The dataset was collected by [6].

Twitter. Diffusion episodes of hashtags on Twitter, over a fixed population of about 5000 users during the US 2012 presidential campaign.

Each dataset was filtered to keep only a subset of about 5000 of its most active users. Table 1 gives some statistics on the datasets.

Table 1. Some statistics on the datasets: the number of users $|\mathcal{U}|$, of links $|\mathcal{E}|$ in the graph, of diffusion episodes in the training set, and the density of the graph.

	$|\mathcal{U}|$	$|\mathcal{E}|$	$|\mathcal{D}|$	Density
Artificial	100	262	10000	2 %
Lastfm	1984	235011	331829	5 %
Weibo	5000	20784	44345	0.08 %
Twitter	4107	128855	16824	1 %

4.2 Baselines

We compare our model to several graph-based baselines.

OutDeg: This simple baseline was used in [5]. First, we find all the "possible sources" i.e. all users who can reach every infected one through a series of hops in the graph. Then, we rank these possible sources by their out-degree, the higher one being the most likely source.

Jordan Center: The use of a Jordan Center as a source estimator was studied in [14]. Because our experimental context is not exactly the same as [14], we slightly adapt its formulation: the predicted source is the one with the minimum longest distance to any infected user.

Pinto's: The model described in [16], based on the assumption that infection delays follow a Gaussian law. It uses a heuristic based on the extraction of a tree subgraph.

---

[1] http://www.dtic.upf.edu/~ocelma/MusicRecommendationDataset/lastfm-1K.html.

For all these approaches, the diffusion graph used is obtained by using the Expectation-Maximization procedure described in [10] to learn the parameters of an Independent Cascades Model. This returns a probability of transmission $p_{i,j}$ for each pair of users. We then assume that a link (u_i, u_j) exists in E if and only if the learned probability of transmission $p_{i,j}$ is greater than S, where S is a threshold set empirically for each baseline to maximize its results on a validation set.

4.3   Experimental Contexts and Results

We now present the results obtained by all models on several experiments. We evaluate the ability of the models to retrieve the source on a testing set of diffusion episodes $\mathcal{D}'$ with a Top-K measure, for various values of K. The Top-K measure is computed by sorting users according to their "scores" (i.e. likelihood or distance to z_D, depending on the model). If the actual source is among the K best-ranked users, the Top-K value is 1, otherwise it is 0.

Choice of Latent Space Dimension. As a preliminary experiment, we study the effect of the number of dimensions of the latent space d. Figure 2 shows the time taken by our learning algorithm to converge for various values of d, and the performances obtained on the source detection task (described in the next subsection) on a validation set. Results are shown for the Weibo dataset, but we observed similar results on the other real datasets. We can see that while the time taken grows linearly with d, performances only grow a little for values of d beyond about 30. For these reasons, we use a value of $d = 30$ in all of our experiments.

Source Detection with Full Cascades. This is the regular experimental context: find s_D given $\hat{\mathcal{U}}_D$. Results for our model (RL) are given for a regularization parameter $\lambda = 10^{-4}$ that appeared to lead to the best results on a validation set. Results are presented in Fig. 3.

Firstly, we can see that on the artificial dataset, our model and the Jordan center model obtain better results that the other two baselines. Let us remember

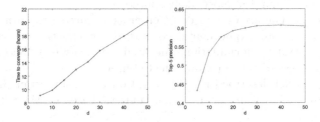

Fig. 2. Time to convergence and source detection performance (Top-5 measure) for various values of d, on the Weibo Dataset

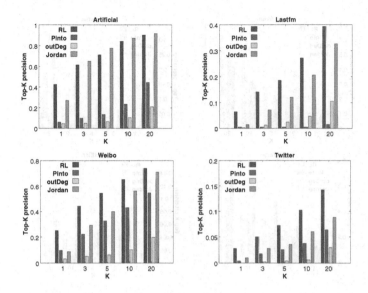

Fig. 3. Source detection with full cascades. Top-K precision

that on this dataset, diffusion episodes are generated using an Independent Cascade model (IC). Since the dataset is very small and since the data have been generated following this diffusion model, IC easily retrieves the actual transmission channels between users from the training set of diffusion episodes. In this context, the Jordan Center heuristic, which is based on an exhaustive computation of the number of hops between nodes, can achieve good performances, which our model is able to match *without the use of an external diffusion model*. Pinto's model, on the other hand, performs its calculation on a tree extracted from the graph with a Breath-First-Search, which ends up ignoring a lot of information and reducing its performances.

On the Weibo dataset, the IC algorithm cannot retrieve the real diffusion graph (to few data w.r.t. the complexity of the network and IC hypothesis not fully verified). Therefore, the performances of the Pinto model and the Jordan Center model are closer. Meanwhile, our approach outperforms every baseline, because it does not rely on any hypothesis about the graph structure or the diffusion model. The fact that Pinto's ends up slightly below Jordan can be explained by the fact that Pinto's makes the assumption that transmission delays follow a Gaussian distribution, which is unrealistic in real datasets [5].

Finally, Lastfm and Twitter are noiser datasets: the fact that two users listened to the same song or used the same hashtag does not always mean that one of them infected the other one, they might just happen to have similar centers of interest. In this context, infections may not be linked by causation but by correlation, which in turn could limit the relevance of the extracted graph. Since all the baselines are based on that graph, they exhibit poor performances on those datasets while our model outperforms them.

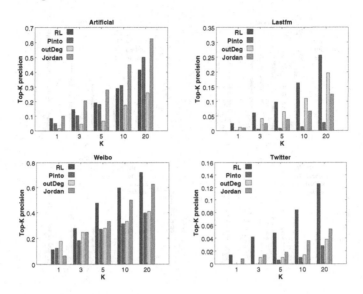

Fig. 4. Source detection on partial cascades (20 %). Top-K precision.

While the results of all approaches may appear to be rather low on Twitter, they can still be useful in some contexts like the one described in [13]: when the administrator of a network looking for the source of a rumor needs to decide which users to probe, any model that gives a non-trivial (which is the case here) result is important.

Source Detection on Partially Observed Cascades. As said in the introduction, diffusion episodes are often partially observed in real-world applications. We simulate this by removing random users from diffusion episodes in the testing set. For each diffusion episode D in $\mathcal{D}'$, we only keep a set percentage of $\hat{\mathcal{U}}_D$. Results are presented in Fig. 4.

On the Artificial dataset, all models see a large drop in performance. Our approach ends up below the Jordan Centers heuristic, and on par with Pinto's, which remains roughly at the same spot. Here, the superiority of the Jordan Center method can be explained by the fact that its shortest-path computation perfectly represents how information transits in an IC model. Also, because we use a scale-free random graph, a small number of observed users is enough to narrow down the number of possible sources in the graph.

On the other end, on the Weibo dataset, most models remain stable, and our approach stays superiors. Interestingly, results on Lastfm and Twitter are different. On the Lastfm dataset, outDeg ends up being better than the other baselines, while the Jordan center approach beats the other baselines on the Twitter dataset. On Lastfm long chains of diffusion are rather rare, as most of influences occur from one central user to a set of successors in the graph. This makes the out-degree of a user a good indicator of his tendency to be an "early adopter" which influences the whole set of infected users. On Twitter, longer chains of diffusion are observed, which results in better results for the Jordan centers heuristic, since it comes down to finding the source that minimizes the number hops required in the retweet graph to reach all infected users. Like on the Weibo dataset, Pinto's method performance is poor because it heavily relies on the modeling of transmission delays, which are very chaotic and hard to capture on such a noisy dataset [10]. In the end, in both cases, our approach exhibit better performances.

Overall, we can see that graph models can have very different results depending on the dataset considered: the best one on a given dataset can be the worse one on another. Meanwhile, our approach achieves consistent and better results on the real datasets, thanks to the use of a latent space that makes it more robust to noise and sparsity.

Learning on Partially Observed Cascades. In the previous experiments, we assumed that we had access to complete diffusion episodes during the learning step. However, it may not be the case in real applications. The same reasons that can make one unable to observe full episodes during the inference step may also prevent us from collecting full episodes for learning. To study this case, we used the filtering procedure described in the previous experiment, and kept only 20 % of the infections contained in each diffusion episode from $\mathcal{D}$ and $\mathcal{D}'$. Results are given in Fig. 5.

On most datasets, the relative performances of the models are similar to the ones obtained in the previous experiment, which is not surprising since the

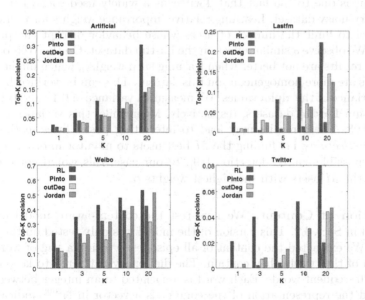

Fig. 5. Source detection with partially observed cascades (20 %) during learning *and* testing.

testing sets are the same. However, on the artificial dataset, our model clearly outperforms the Jordan baseline, contrary to the previous experiments. This is due to the fact that the influence graph inferred using $\mathcal{D}$ is not perfect at all, since the diffusion episodes in $\mathcal{D}$ are partial. This greatly reduces the quality of the Jordan model. Overall, our model also outperforms the baselines in such a setting.

Complexity. For the learning part, both our model and the graph-based approaches use a stochastic algorithm that roughly takes the same time to converge. However, our approach learns a fixed number of parameters for each user, which grows linearly with the number N of users, while the number of parameters for graph models is the number of links in the graph, which scales quadratically with N. Furthermore, during inference, our model is much faster: it usually takes less than a second to perform source detection for one episode, while baselines require minutes. We only need to compute the episode representation z_D and its distance to all possible sources, which has a linear complexity. The graph models usually need to compute shortest distances between all users in the graph, which is much more complex. This makes our approach more scalable, which is an important issue when dealing with large online social networks.

Inclusion of User Importance. In this subsection, we test the extension defined in Sect. 3.2. We compare the results of the model with weights to those of the base version, on the real datasets. Results are presented in Table 2. We can see that on the Twitter dataset, using weights improves our results by about 10 %. This is due to the fact that Twitter is a widely used social network and thus a very noisy dataset. Learning relative importance weights for users enables our model to limit the impact of users whose behavior disrupt the prediction process. We observe a similar effect on the Lastfm dataset. On the Weibo dataset, however, results are not better when learning such weights, which might indicate that users are more homogeneous on this dataset. This can be verified by looking at the variance of the alpha values. We measure a variance of 0.12 and 0.15 on the Twitter and Lastfm datasets, respectively. Meanwhile, this variance on Weibo is only 0.08. These results may lead to interesting possibilities in the task of *monitors selections*, i.e. finding the M best users to monitor in order to achieve the best possible source detection [19]. In our case, this would come down to selecting the M users with the highest weights α.

Integration of Content. We now test the content-aware model extension described in Sect. 3.2. This version of the model was only tested on the Twitter dataset. We extracted the content of all episodes by using a bag-of-word representation of the tweets they contain. The dictionary is filtered to keep only the most 2000 frequent words. Each word is associated to an integer between 0 and 1999, and the representation of the content is a vector in $\mathbb{N}^{2000}$ indicating the number of occurrences of each word in the tweets. The data collection was limited to tweets written in english, but the approach would remain valid for other

Table 2. Source Detection with users weights. Models are only tested on diffusion episodes of length 3 or higher (both models are equivalent on diffusions episodes of length 2)

Top-K	1	3	5	10	20
Twitter					
RL	0.020	0.042	0.058	0.099	0.141
RL w/weights	0.021	0.047	0.073	0.107	0.154
Gain	3%	10%	25%	8%	9%
Lastfm					
RL	0.052	0.12	0.166	0.2545	0.374
RL w/weights	0.065	0.1335	0.175	0.2605	0.378
Gain	25%	11%	5%	2%	1%
Weibo					
RL	0.31	0.51	0.59	0.72	0.82
RL w/weights	0.31	0.50	0.60	0.75	0.84
Gain	0%	−2.3%	+0%	+4%	+1%

Table 3. Source detection with content integration. Tested on the Twitter dataset

Top-K	1	3	5	10	20
RL	0.028	0.05	0.072	0.102	0.142
RL w/content	0.043	0.069	0.099	0.128	0.179
Gain	56%	38%	38%	26%	26%

languages. Results are presented in Table 3. We can see that the integration of the content greatly improves our prediction, especially in Top-1.

Since we use a bag-of-words representation of size 2000 and a linear projection of that content into a d-dimensional space, we learned a $2000 \times d$ projection matrix (the parameters θ) whose rows can be interpreted as representation of the words in $\mathbb{R}^d$. Table 4 lists the ten words with the largest representation norms. We can see that, apart from "new" and "retweet", all listed words are meaningful ones which greatly inform about the topic of the diffusion.

Furthermore, words with similar representations should tend to have similar effects on the diffusion. To verify this, we show in Table 5 the pairs of words with the highest cosine similarities. We can see that these pairs indeed either correspond to words with similar meanings (leisur/getawai and iran/iranian) or words used in similar contexts. OpESR (Operation Empire State Rebellion) and OccupyHQ are pages used by activists from the "Occupy Wall Street" movement. "Masen" and "Mapoli" stand for "Massachusetts Senate" and "Massachusetts Politics".

Table 4. Top 10 most important words according to our content-based model.

Word	Norm	Word	Norm
New	9.9646	Iran	7.9415
Obama2012	9.4358	NYC	7.2585
Music	9.2675	Game	7.223
2012	8.9344	Ohio	7.0147
President	8.1841	Retweet	6.8428

Table 5. Pair of words with the highest cosine similarities of their representations

opesr	leisur	music	iran	masen
occupyhq	getawai	hipster	iranian	mapoli

5 Conclusion

In this paper, we proposed a novel and efficient method to retrieve the sources of information from diffusion episodes. This method is based on the use of a latent space that embeds the influences and similarities between users. We tested this approach on artificial and real diffusion episodes, and found that it achieved better performances than state of the art approaches, while retaining a lower complexity. We also proposed a way to learn the importance of each user and to integrate the content of information in the model, and showed that both led to performance improvements. Ongoing works are focused on a unifying framework of *cascade completion*: how can we retrieve a whole diffusion episode given only a fraction of its users? Source detection and diffusion prediction are special cases of this task. An unifying model, fitted for this more general problem, would give important insights on the nature and dynamics of diffusion.

Acknowledgments. This work has been partially supported by the following projects: Xu Guangqi 2016 Deep learning for Large Scale Dynamic and Spatio-Temporal Data; REQUEST Investissement d'Avenir 2014 and LOCUST ANR 2015 (ANR-15-CE23-0027-01).

References

1. Barbieri, N., Bonchi, F., Manco, G.: Cascade-based community detection. In: Proceedings of the Sixth ACM International Conference on Web Search and Data Mining, WSDM 2013, NY, USA, pp. 33–42. ACM, New York (2013)
2. Bourigault, S., Lagnier, C., Lamprier, S., Denoyer, L., Gallinari, P.: Learning social network embeddings for predicting information diffusion. In: WSDM 2014, NY, USA, pp. 393–402. ACM, New York (2014)
3. Chen, S., Moore, J.L., Turnbull, D., Joachims, T.: Playlist prediction via metric embedding. In: KDD 2012, pp. 714–722. ACM (2012)
4. Dong, W., Zhang, W., Tan, C.W.: Rooting out the rumor culprit from suspects. In: ISIT 2013, pp. 2671–2675. IEEE (2013)

5. Farajtabar, M., Gomez-Rodriguez, M., Du, N., Zamani, M., Zha, H., Song, L.: Back to the past: source identification in diffusion networks from partially observed cascades. arXiv preprint (2015). arXiv:1501.06582
6. Fu, K., Chan, C.h., Chau, M.: Assessing censorship on microblogs in China: discriminatory keyword analysis and the real-name registration policy. IEEE Internet Comput. **17**(3), 42–50 (2013)
7. Gomez-Rodriguez, M., Balduzzi, D., Schölkopf, B.: Uncovering the temporal dynamics of diffusion networks. In: ICML 2011, pp. 561–568. ACM (2011)
8. Guille, A., Hacid, H., Favre, C., Zighed, D.A.: Information diffusion in online social networks: a survey. SIGMOD Rec. **42**(2), 17–28 (2013)
9. Kempe, D., Kleinberg, J., Tardos, E.: Maximizing the spread of influence through a social network. In: KDD 2003, pp. 137–146. ACM (2003)
10. Lamprier, S., Bourigault, S., Gallinari, P.: Extracting diffusion channels from real-world social data: a delay-agnostic learning of transmission probabilities. In: ASONAM 2015. IEEE Computer Society (2015)
11. Lappas, T., Terzi, E., Gunopulos, D., Mannila, H.: Finding effectors in social networks. In: KDD 2010, pp. 1059–1068. ACM (2010)
12. Leskovec, J., Backstrom, L., Kleinberg, J.: Meme-tracking and the dynamics of the news cycle. In: Proceedings of the 15th ACM SIGKDD International Conference on Knowledge Discovery and Data Mining, KDD 2009, NY, USA, pp. 497–506. ACM, New York (2009)
13. Luo, W., Tay, W.P., Leng, M.: Rumor spreading and source identification: a hide and seek game. arXiv preprint (2015). arXiv:1504.04796
14. Luo, W., Tay, W.P., Leng, M., Guevara, M.: On the universality of the Jordan center for estimating the rumor source in a social network. In: DSP 2015, pp. 760–764, July 2015
15. Mikolov, T., Chen, K., Corrado, G., Dean, J.: Efficient estimation of word representations in vector space. arXiv preprint (2013). arXiv:1301.3781
16. Pinto, P.C., Thiran, P., Vetterli, M.: Locating the source of diffusion in large-scale networks. Phys. Rev. Lett. **109**(6), 068702 (2012)
17. Prakash, B.A., Vreeken, J., Faloutsos, C.: Spotting culprits in epidemics: how many and which ones? In: ICDM 2012, pp. 11–20. IEEE (2012)
18. Saito, K., Nakano, R., Kimura, M.: Prediction of information diffusion probabilities for independent cascade model. In: Lovrek, I., Howlett, R.J., Jain, L.C. (eds.) KES 2008. Lecture Notes in Artificial Intelligence (LNAI), vol. 5179, pp. 67–75. Springer, Heidelberg (2008). doi:10.1007/978-3-540-85567-5_9
19. Seo, E., Mohapatra, P., Abdelzaher, T.: Identifying rumors and their sources in social networks. In: SPIE Defense, Security, and Sensing, pp. 83891I–83891I. International Society for Optics and Photonics (2012)
20. Shah, D., Zaman, T.: Detecting sources of computer viruses in networks: theory and experiment. In: ACM SIGMETRICS Performance Evaluation Review, vol. 38, pp. 203–214. ACM (2010)
21. Shah, D., Zaman, T.: Rumors in a network: who's the culprit? IEEE Trans. Inf. Theory **57**(8), 5163–5181 (2011)
22. Shah, D., Zaman, T.: Rumor centrality: a universal source detector. In: ACM SIGMETRICS Performance Evaluation Review, vol. 40, pp. 199–210. ACM (2012)
23. Tsur, O., Rappoport, A.: What's in a hashtag? Content based prediction of the spread of ideas in microblogging communities. In: Proceedings of the Fifth ACM International Conference on Web Search and Data Mining, pp. 643–652. ACM (2012)

Linear Bandits in Unknown Environments

Thibault Gisselbrecht[1,2(✉)], Sylvain Lamprier[2], and Patrick Gallinari[2]

[1] Technological Research Institute SystemX, 8 Avenue de la Vauve,
91120 Palaiseau, France
thibault.gisselbrecht@irt-systemx.fr
[2] Sorbonne Universités, UPMC Univ Paris 06, CNRS, LIP6 UMR 7606,
4 Place Jussieu, 75005 Paris, France
{thibault.gisselbrecht,sylvain.lamprier,patrick.gallinari}@lip6.fr

Abstract. In contextual bandit problems, an agent has to choose an action among a bigger set of available ones at each decision step, according to features observed on them. The goal is to define a decision strategy that maximizes the cumulative reward of actions over time. We focus on the specific case where the features of each action correspond to some kind of a constant profile, which can be used to determine its intrinsic utility for the task in concern. If there exists an unknown linear application that allows rewards to be mapped from profiles, this can be leveraged to greatly improve the exploitation-exploration trade-off of stationary stochastic methods like *UCB*. In this paper, we consider the case where action profiles are unknown beforehand. Instead, the agent only observes sample vectors, with mean equal to the true profiles, for a subset of actions at each decision step. We propose a new algorithm, called *SampLinUCB*, and derive a finite time high probability upper bound on its regret. We also provide numerical experiments on a task of focused data capture from online social networks.

1 Introduction

The multi-armed bandit is a learning problem aimed at tackling the trade off between exploration and exploitation in decision processes where, at each round, an agent chooses an action - or arm - among a finite set of size K and receives a reward which quantifies the quality of the chosen action. The goal for the agent is to maximize its cumulative reward trough time. Contextual bandit refers to an instance of multi-armed bandit problems where a context of decision is observed before selecting actions. Typically, this context corresponds to feature vectors observed for each possible action, which is used to drive the decision process. Contextual bandit strategies have been widely used recently, for example to design online personalized recommendation systems. Considering the current decision environment can indeed help to select the most fitted actions for the problem to solve. Those contexts vectors are used to better predict the rewards related to each action at each decision step, since rewards pertaining to each arm are connected by a common unknown parameter to be learned.

© Springer International Publishing AG 2016
P. Frasconi et al. (Eds.): ECML PKDD 2016, Part II, LNAI 9852, pp. 282–298, 2016.
DOI: 10.1007/978-3-319-46227-1_18

Here we assume that context vectors are constant through time as in [10]. They therefore stand as profile vectors, that can be leveraged to better explore the decision environment. Based on the assumption that there exists a linear application from these profiles to the utility of actions, we propose a new scenario where, for any possible reasons (technical, political, etc.), profile vectors are not directly available to the learner at each decision step. Instead, the agent only observes sample vectors, with mean equal to the true profiles, for a subset of actions at each decision step. This can happen in various situations where some restrictions might prevent us to observe the whole decision environment. For example in a technology intelligence scenario on a social media such as Twitter, where an agent is asked to capture relevant information for his need. Because of the extremely large number of users, the agent would have to focus only on a subset of relevant users to follow at each time step. However, given the strict restrictions of the media owners, current features of the users are only available for a small fraction of users. As we will see later, the mechanism that selects the subset of arms for which a sample vector is observed can be independent or part of the decision process. To the best of our knowledge this contextual bandit problem has not been studied in the literature. Moreover, existing bandit approaches are not adapted to tackle such problems for two reasons. Firstly, even if traditional algorithm such as *UCB* can be applied, the information provided by the sample profile vectors would be entirely lost. Secondly, existing contextual bandit policies cannot be applied since they do not take into account any potential uncertainty on context vectors. We face a bandit problem where uncertainty is not only to be considered on the regression parameters but also on the description of actions. In this context, we propose the following contributions:

- We propose a new instance of the contextual bandit problem where the representations of the actions are not known beforehand, but built from samples obtained at each iteration (3 cases are investigated for the sampling process);
- We propose the *SampLinUCB* algorithm to solve this problem; and derive the corresponding *sublinear* upper bound of our policy's regret;
- We perform experiments in a real world scenario of focused data capture on Twitter.

2   Related Work

The multi-armed bandit originally proposed in [16] in its stationary form has been widely studied in the literature. This learning problem is aimed at tackling the trade off between exploration and exploitation in decision processes where, at each round, an agent chooses an action - or arm - among a finite set of size K and receives a reward which quantifies the quality of the chosen action. The goal for the agent is to maximize its cumulative reward trough time. One of the simplest and most straightforward algorithms is the well-known ϵ-greedy [7], which exploits the action with best empirical mean with probabilty $1 - \epsilon$ and explore others with probability ϵ, where the parameter ϵ can possibly decrease with time.

Another class of algorithms known as Upper Confidence Bound use a cleaver way to balance exploitation and exploration. This type of strategies keeps an estimate of the confidence interval related to each reward distribution and plays the arm with highest upper confidence bound at each time step. Many extensions of the famous *UCB* algorithm proposed in [7] are known to converge (see *UCBV* in [5], *MOSS* in [4] or *KL-UCB* in [11]). Finally Thompson sampling algorithms, originally proposed in 1933 in [20] are based on a Bayesian approach and have also proven to be extremely efficient (see [2] and [15]). However, the performances of such approaches remain quite limited on instances of the problem where context features can be used to drive the decision process. The structured contextual bandit problem assumes the existence of a common unknown parameter (to be learned) linking the contexts of arms to their reward. For the linear case - in which we are interested in here - the first contextual upper confidence bound algorithm has been proposed in [6], while more recently the well-known *Lin-UCB* algorithm has been formalized. Many other UCB approaches have been developed since then to improve the performance the two last. In particular, algorithms such as *OFUL* or *ConfidenceBall* proposed in [1] and [9] have the advantage to enjoy a tighter regret upper bound (see also [14,19]). As in the tradional bandit case, Thompson sampling algorithms have also been designed to solve the contextual bandit problem. This approach has also proved to be powerful, first empirically in [8] and then theoretically in [3] and [17]. The problem we tackle in this paper is a variant of the linear case of the Generalized Linear Bandit described in [10], where each arm is associated with some constant profile vector. In our case, context vectors are **unknown** to the agent beforehand.

3   Model and Algorithm

3.1   Problem Setting

Notations: We use $||x||_p$ to denote the p-norm of a vector $x \in \mathbb{R}^d$. For a positive definite matrix $A \in \mathbb{R}^{d \times d}$, the weighted 2-norm of $x \in \mathbb{R}^d$ is defined by $||x||_A = \sqrt{x^T A x}$. Finally, $\lambda_{min}(A)$ denotes the smallest eigenvalue of the positive definite matrix A.

The contextual bandit problem we study proceeds in the following way: at each iteration $t \in \{1, .., n\}$, the learner receives a set $\{\mu_1, .., \mu_K\} \subset \mathbb{R}^d$ of K context vectors. He then selects an arm $a_t \in \{1, ..., K\}$ and observe the corresponding payoff $r_{a_t,t} \in [0..1]$. We assume a standard linear behavior of the reward distribution: $\forall t \in \{1, .., n\}, \forall a \in \{1, ..., K\} : r_{a,t} = \mu_a^T \theta_* + \eta_{a,t}$, where $\theta_* \in \mathbb{R}^d$ is an unknown parameter to be learned. As in [1], we consider $\eta_{a,t}$ as a zero-mean conditionally R sub-Gaussian random noise, with constant $R > 0$ i.e.: $\forall \lambda \in \mathbb{R} : \mathbb{E}[e^{\lambda \eta_{a,t}} | F_{t-1}] \leq e\left(\dfrac{\lambda^2 R^2}{2}\right)$, where $F_t = \{(a_t, r_{a_t}, \mu_{a_t})\}_{s=1..t}$.

As in [1], we consider the following instantaneous pseudo-regret reg_t:

$$reg_t = \theta_*^T \mu_{a_*} - \theta_*^T \mu_{a_t}$$

where $\mu_{a_*} = \arg\max_{a=1..K} \theta_*^T \mu_a$ represents the profile vector of the optimal action a_*. The goal of the algorithm is to bound with high probability the cumulative regret $R_n = \sum_{t=1}^{n} reg_t$ for the chosen sequence $\{a_1, ..., a_n\}$.

Our Problem: In our case, the set of context vectors $\{\mu_1, .., \mu_K\}$ cannot be observed directly. Instead at each time t the learner is given a subset of arms, denoted $\mathcal{O}_t$, such that for each arm $a \in \mathcal{O}_t$, a sample $x_{a,t}$ of a random variable with mean μ_a becomes available. Keeping the same hypothesis as before, the problem can be rewritten the following way:

$$\forall s \leq t : r_{a,s} = \theta_*^T \mu_a + \eta_{a,s} = \theta_*^T \hat{x}_{a,t} + (\mu_a - \hat{x}_{a,t})^T \theta_* + \eta_{a,s} = \theta_*^T \hat{x}_{a,t} + \theta_*^T \epsilon_{a,t} + \eta_{a,s} \tag{1}$$

where $\hat{x}_{a,t} = \dfrac{1}{n_{a,t}} \sum_{s \in T_{a,t}^{obs}} x_{a,s}$, with $T_{a,t}^{obs} = \{s \leq t, a \in O_s\}$ and $n_{a,t} = |T_{a,t}^{obs}|$.

Concretely, $n_{a,t}$ corresponds to the number of samples observed for arm a until time t, while $\hat{x}_{a,t}$ represents the empirical mean.

In the following we propose an algorithm to tackle this problem in the general case, i.e. when no specific assumption regarding the process that generates $\mathcal{O}_t \subset \{1, ..., K\}$ is made. We derive a general upper bound of the cumulative regret defined above. Then, we study three specific problems:

- **Case 1**: At each time t, every arm shows a profile sample i.e.: $\mathcal{O}_t = \{1, ..., K\}$.

- **Case 2**: At each time t, each arm has a probability p to deliver a profile sample. In this case, the content and the size of $\mathcal{O}_t$ changes over time.

- **Case 3**: At each time t, only the previously selected arm delivers a profile sample i.e.: $\mathcal{O}_t = a_{t-1}$.

3.2   Algorithm

3.2.1   Regression and Confidence Interval

Proposition 1. *Suppose that for any a, at every t, $x_{a,t} \in \mathbb{R}^d$ are iid drawn from an unknown distribution with mean $\mu_a \in \mathbb{R}^d$, and that $\|x_{a,t}\|_2 \leq L$ and $\|\theta_*\|_2 \leq S$: then, for any a, for every $s \leq t$, $\epsilon_{a,t}^T \theta_* + \eta_{a,s}$ is conditionally $R_{a,t}$ sub-Gaussian, with $R_{a,t} = \sqrt{R^2 + \dfrac{L^2 S^2}{n_{a,t}}}$.*

Proof. See Appendix A.1

Let use the following matrix notations (where we remove the t dependence to clarify): $\eta' = (\eta_{a_s,s} + \theta_*^T \epsilon_{a_s,t})_{s=1..t}^T$, $X = (\hat{x}_{a_s,t}^T)_{s=1..t}$, $Y = (r_{a_s,s})_{s=1..t}^T$, $A = diag(1/R_{a_s,t})_{s=1..t}$. Let $\hat{\theta}_t$ be the l^2-regularized least square estimator of θ_* with regularization parameter $\lambda > 0$, using the covariate X, where each learning example is weighted by its confidence factor $1/R_{a_s,t}$:[1]

$$\hat{\theta}_t = \arg\min_\theta \sum_{s=1}^{t} \frac{1}{R_{a_s,t}} (\theta^T \hat{x}_{a_s,t} - r_{a_s,s})^2 + \lambda||\theta||_2^2 = (X^T A X + \lambda I)^{-1} X^T A Y$$

(2)

Proposition 2. *Defining* $V_t = \lambda I + X^T A X$, *with the same hypothesis as in Proposition 1, for any* $\delta > 0$, *with probability at least* $1 - \delta$, *for all* $t \geq 0$:

$$||\hat{\theta}_t - \theta_*||_{V_t} \leq \sqrt{2\log\left(\frac{det(V_t)^{1/2}det(\lambda I)^{-1/2}}{\delta}\right)} + \sqrt{\lambda}S = \alpha_t$$

(3)

Proof. See Appendix A.2.

3.2.2 Algorithm

The principle behind any Upper Confidence Bound algorithm (UCB) is to maintain a confidence interval on the rewards and select *optimistically* the arm with highest UCB. As precised in [1], the smaller confidence sets on the rewards we are able to construct, the better regret bounds are, and, more importantly, the better the algorithm performs empirically. In the traditional contextual bandit problem, this is equivalent to build a confidence set for the unknown parameter θ_*, as done in the previous section. However in our case, we have an additional source of uncertainty due to the observed samples variability, that we need to deal with. Then, at each time step t, the *SampLinUCB* algorithm (described in Algorithm 1) selects the arm that maximizes the following UCB score[2]:

$$s_{a,t} = \hat{\theta}_t^T (\hat{x}_{a,t} + \bar{\epsilon}_{a,t}) + \alpha_t ||\hat{x}_{a,t} + \tilde{\epsilon}_{a,t}||_{V_{t-1}^{-1}}$$

(4)

where $\bar{\epsilon}_{a,t} = \frac{\rho_{a,t,\delta}\hat{\theta}_t}{||\hat{\theta}_t||_2}$, $\tilde{\epsilon}_{a,t} = \frac{\rho_{a,t,\delta}\hat{x}_{a,t}}{\sqrt{\lambda}||\hat{x}_{a,t}||_{V_{t-1}^{-1}}}$ and $\rho_{a,t,\delta} = Ld\sqrt{\frac{1}{2n_{a,t}}\log\left(\frac{2dt^2}{\delta}\right)}$.

$\bar{\epsilon}_{a,t}$ and $\tilde{\epsilon}_{a,t}$ aim at coping with uncertainty of the estimated profiles. Intuitively, they allow the algorithm to select actions that either have their estimated profile in the useful area or are enough uncertain to be assumed to belong to a useful area (despite their observed samples). The goal is to discard actions that lie in a useless area with a sufficiently high probability.

Note that to be selected, an arm must have been observed at least once, which is why we update the set of possible arms at line 5 of Algorithm 1. Moreover, instead of recalculating matrix V_t and vector b_t from scratch at every time step, their computation time can be reduced by only considering arms in O_t at time t, as done from line 13 to line 18.

---

[1] This weight increases as the number of observations of a specific arm becomes bigger.

[2] To simplify notations, $\hat{\theta}_t$ in the algorithm corresponds to $\hat{\theta}_{t-1}$ in the previous subsection.

Algorithm 1. SampLinUCB Algorithm

---

1 $V_0 \leftarrow \lambda I_{d \times d}$ (identity matrix of dimension d); $b_0 \leftarrow 0_d$ (zero vector of dimension d); $\mathcal{K} \leftarrow \emptyset$;

2 **for** $t \leftarrow 1$ **to** n **do**

3 $\quad \hat{\theta}_t \leftarrow V_{t-1}^{-1} b_{t-1}$;

4 $\quad$ Get $\mathcal{O}_t$;

5 $\quad \mathcal{K} \leftarrow \mathcal{O}_t \cup \mathcal{K}$;

6 $\quad$ **for** $a \in \mathcal{K}$ **do**

7 $\quad\quad$ **if** $a \in \mathcal{O}_t$ **then** Observe $x_{a,t}$, update $\hat{x}_{a,t}$ and $R_{a,t}$;

8 $\quad\quad$ Compute $s_{a,t}$ with formula 4 ;

9 $\quad$ **end**

10 $\quad a_t \leftarrow \underset{a \in \mathcal{K}}{\arg\max} \; s_{a,t}$;

11 $\quad$ Receive $r_{a_t,t}$;

12 $\quad V_t \leftarrow V_{t-1} + \hat{x}_{a_t,t} \hat{x}_{a_t,t}^T / R_{a_t,t}$; $\qquad b_t \leftarrow b_{t-1} + \hat{x}_{a_t,t} r_{a_t,t} / R_{a_t,t}$;

13 $\quad$ **for** $s \leftarrow 1$ **to** $t - 1$ **do**

14 $\quad\quad$ **if** $a_s \in \mathcal{O}_t$ **then**

15 $\quad\quad\quad V_t \leftarrow V_t - \hat{x}_{a_s,t-1} \hat{x}_{a_s,t-1}^T / R_{a_s,t-1} + \hat{x}_{a_s,t} \hat{x}_{a_s,t}^T / R_{a_s,t}$;

16 $\quad\quad\quad b_t \leftarrow b_t - \hat{x}_{a_s,t-1} r_{a_s,s} / R_{a_s,t-1} + \hat{x}_{a_s,t} r_{a_s,t} / R_{a_s,t}$;

17

18 $\quad$ **end**

19 **end**

3.3 Regret Analysis

Theorem 1. *In the general case, if $\lambda \geq \max(1, 2L^2)$, with probability at least $1 - 3\delta$, the cumulative regret of our algorithm $R_n = \sum\limits_{t=1}^{n} reg_t$ is bounded by:*

$$R_n \leq C + 4Ld \left(\sqrt{\frac{d}{\lambda} \log \left(\frac{1 + nL^2/\lambda}{\delta} \right)} + 2S \right) \sqrt{\log \left(\frac{2dn^2}{\delta} \right)} \sum_{t=2}^{n} \frac{1}{\sqrt{n_{a_t,t}}}$$

$$+ 2 \left(\sqrt{d \log \left(\frac{1 + nL^2/\lambda}{\delta} \right)} + \sqrt{\lambda} S \right)$$

$$\times \sqrt{nd \left(\sqrt{R^2 + L^2 S^2} \log \left(1 + \frac{nL^2}{\lambda d} \right) + \frac{L^2 d^2}{\lambda} \log \left(\frac{2dn}{\delta} \right) \sum_{t=1}^{n} \frac{1}{n_{a_t,t}} \right)} \quad (5)$$

Proof. See Appendix A.3.

Theorem 2. *Upper bounds of the regret in the three cases.*

- *For case 1, with probability at least $1 - 3\delta$:*

$$R_n = \mathcal{O} \left(d \sqrt{dn \log \left(\frac{n}{\delta} \right) \log \left(\frac{n^2}{\delta} \right)} \right) \quad (6)$$

- *For case 2, with probability at least $(1 - 3\delta)(1 - \delta)$, for $n \geq \lceil 2\log(1/\delta)/p^2 \rceil$:*

$$R_n = \mathcal{O}\left(d\sqrt{\frac{dn}{p}\log\left(\frac{n}{\delta}\right)\log\left(\frac{n^2}{\delta}\right)}\right) \tag{7}$$

- *For case 3, with probability at least $1 - 3\delta$:*

$$R_n = \mathcal{O}\left(d\sqrt{dnK\log\left(\frac{n}{\delta}\right)\log\left(\frac{n^2}{\delta}\right)}\right) \tag{8}$$

Proof. See Appendices A.4, A.5 and A.6 respectively for case 1, 2 and 3.

Note that in every case the regret is sublinear, and, generally $\mathcal{O}\left(\log(n)\sqrt{n}\right)$. As we could expect, in case 2, both the time from which the upper bound holds and the bound itself increases as the probability p decreases. Moreover, in case 3, the bound increases with the total number of arms, which seems normal since the algorithm only allows one observation at each iteration, and then naturally needs more time to converge when more actions are available.

In the following we consider an extended version of the Algorithm 1, where k arms are selected simultaneously. So, instead of a_t, the subset of chosen users is denoted $\mathcal{K}_t$. The algorithm works the same way, but chooses the top k arms with highest score at each time step. A similar theoretical upper bound of the regret can also be derived for every case, whose general idea is given in A.7.

4   Experiments

4.1   Task Definition

We apply our algorithm to the task of dynamic data capture on Twitter proposed in [12]. In this setting, we are given a social network where the whole set of K users cannot be monitored simultaneously. Given a time period divided in n steps of size T, the goal is to select, at each iteration $t \in \{1, ..., n\}$ of the process, a subset $\mathcal{K}_t$ of k user accounts to follow, among the whole set of possible users $\mathcal{K}$ ($\mathcal{K}_t \subseteq \mathcal{K}$), according to their likelihood of posting relevant tweets for the formulated information need. Given a relevance score $r_{a,t}$ assigned to the content posted by user $a \in \mathcal{K}_t$ during iteration t of the process (the set of tweets he posted during iteration t), the aim is then to select at each iteration t the set of user accounts that maximize the sum of collected relevance scores: $\max_{(\mathcal{K}_t)t=1..n} \sum_{t=1}^{n} \sum_{a \in \mathcal{K}_t} r_{a,t}$. On the one hand, listening to the subset of selected users $\mathcal{K}_t$ can be performed thanks to Twitter *Follow* Streaming API, which allows us to capture the messages produced by a restrained number of accounts - 5000 at most - during a certain time. The quality of messages, depending on the data need, is then measured through some utility function that gives a grade to a content (see below for some examples). This task can be treated as a bandit problem by representing each user as an arm. On the other hand, obtaining context samples can be performed through different ways. Firstly, with an independent random process where each arm has a probability p to belong to $\mathcal{O}_t$ (case 1 and 2)[3]. Note that in practice, the Twitter *Sample* streaming API furnishes real-time access to 1 % of all public tweets (i.e. $p = 0.01$). Secondly, as in case 3, it can be part of the decision process when $\mathcal{O}_t = \mathcal{K}_{t-1}$.

4.2 Dataset

While our approach behaves well in real-world settings, we report here offline experiments performed on a pre-collected dataset. This allows us to be able to easily compare different scenarios and policies. The used dataset is composed of the contents produced by $K = 5000$ users during ten days preceding the US presidential elections in in 2012. The 5000 chosen accounts are the first one who use either "Obama", "Romney" or "#USElections". At the end of the capture process the number of collected messages is equal to 2148651. We simulate data collection processes on this closed-world dataset to evaluate our approach.

4.3 Model Definition

4.3.1 Context Model

Each user a is associated to an unknown feature vector μ_a representing its profile. Considering this profile as the mean of the messages user a is likely to post, it can be estimated given a sample of messages we observed. Concretely, the content obtained by capturing data from a given user a at time step t is denoted $\omega_{a,t}$ (if we get several messages for a given author, these messages are concatenated). If user a belongs to $\mathcal{O}_t$, its profile sample $x_{a,t}$ for timestep t corresponds to the messages he posted during step $t - 1$. Given a dictionary of size m, messages could be represented as m dimensional bag of word vectors. However the size m might cause the algorithm to be computationally inefficient since it requires the inversion of a matrix of size m at each iteration. In order to reduce the dimension of those features, we use a *Latent Dirichlet Allocation* method for short messages [13] (tweets of a same user are aggregated in one document), which aims at modeling each tweet as a mixture of topics. We choose a number of $d = 30$ topics and learn the model on the whole corpus. Then, if we denote by $F : \mathbb{R}^m \longrightarrow \mathbb{R}^d$ the function that, given a message returns its representation in the topic space, the features sample of user a at time t is $x_{a,t} = F(\omega_{a,t-1})$.

4.3.2 Reward Model

We use a SVM classifier that we first trained on the *20 Newsgroups* dataset. For our experiments, we focus on 4 classes to test: *politics, religion, sport, science*. The reward obtained by a message equals the number of times it has been retweeted by other users if it belongs to the specified class according to our classifier, or 0 otherwise. Finally, if a user posted several messages during an iteration,

---

[3] Case 2 is equivalent to case 1 when $p = 1$.

its reward $r_{a,t}$ corresponds to the sum of the individual rewards obtained by the messages he posted during iteration t. This corresponds to the task of seeking to collect messages, related to some desired topic, that will have a strong impact on the network.

4.4 Setting

As done in [12], we set k, the number of listened users at each time step, to 100, and T the size of an iteration to 100 s. We implement the three cases for the process generating $\mathcal{O}_t$, and for case 2, we tested different values for p: 0.1, 0.05 and 0.01. Beyond a *Random* policy which uniformly chooses users in $\mathcal{K}_t$ at each time step, we compare our contextual algorithm to two existing bandit algorithms, which do not take features into account and are well fitted for the task of data capture: *CUCB* and *CUCBV* respectively proposed in [18] and [12]. Compared to *CUCB*, *CUCBV* adds the variance of reward distributions of users in the definition of the confidence intervals. This has been shown to well behave in cases such as our task of data capture [12], where a great variability can be observed in the content posted by users when they are selected by the process.

4.5 Results

Figure 1 represents the evolution of the cumulative reward through time for different policies and the four rewards defined above. First of all, as in [12], every algorithm perform better than the *Random* one. We also notice that *SampLin-UCB*, when every user delivers a sample vector at each time (i.e. case 1), always outperforms every other policies, which confirms the relevance of our approach: building a mean profile for each user allows us to better predict its reward. Generally speaking, in case 2, the performance of the algorithm increases with the probability of observation p, which fits well with the intuition. Moreover, *SampLinUCB* performs better than a classical *CUCB* in almost every cases. Except for the *sport* reward where the performances are almost equal for both approaches, *SampLinUCB* in case 3 provides better results than *CUCBV*. This result is very interesting since for our task, case 3 does not require any external process for the observation of users' sample vectors: what we observe is what we choose. Furthermore, it should be noticed that *SampLinUCB* in case 3 always performs significantly better than *SampLinUCB* in case 2 with $p = 0.01$ and usually better than case 2 with $p = 0.05$ (except for the *Politics* reward). This highlights the ability of our algorithm to be active for its learning, by selecting useful content: with less observed content samples (only 100 over 5000 each iteration, which would corresponds to the same observation rate than a *SampLinUCB* in case 2 with $p = 0.02$), it succeeds at capturing greatly more useful content in most cases. Considering the restrictions of Twitter (only 1 % of the messages observable at each iteration), this is a very interesting result for the task of data capture from streaming APIs.

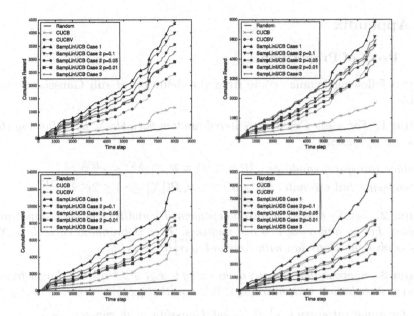

Fig. 1. Cumulative reward vs time for different policies on the *USElections* dataset. From upper left to lower right, rewards are: *politic, religion, science, sport.*

5 Conclusion

In this paper, we formalized a new contextual bandit problem, that differs from the traditional contextual bandit problem with constant feature vectors, in two major ways. Firstly, no context vectors is directly observable, but instead, the learner can only observe a sample of them. Secondly, those samples are available only for a subset of arms. We proposed an algorithm to deal with this problem and take advantage of the observed samples. Basically, the algorithm builds an upper confidence bound of the expected value of each arm's reward by taking into account the uncertainty on both the learned regression parameter and the sampled features. We studied three different cases for the process delivering profile samples of actions at each time, and proved a sublinear high-probability upper bound of our algorithm's regret for all of them. We finally conducted experiments on a focused data capture task on Twitter that shows the relevance of the proposed method. Ongoing works concern the extension of the proposal for non-stationary cases and for processes with non profile-centered samples.

Acknowledgments. This research work has been carried out in the framework of the Technological Research Institute SystemX, and therefore granted with public funds within the scope of the French Program "Investissements d'Avenir". Part of the work was supported by project Luxid'x financed by DGA on the Rapid program.

A   Appendix

A.1   Proof of Proposition 1

The two following lemmas come from the defintion of sub Gaussian random variable.

Lemma 1. *For a sub Gaussian centered random variable X, the following statements are equivalent:*

- *Laplace transform condition:* $\exists R > 0, \forall \lambda \in \mathbb{R}, \mathbb{E}[e^{\lambda X}] \leq e^{R^2 \lambda^2 / 2}$
- *Subgaussian tail estimate:* $\exists R > 0, \forall \gamma > 0, P(|X| \geq \gamma) \leq 2e^{-\gamma^2 / (2R^2)}$

Lemma 2. *Let X_1 and X_2 be two independent R_1 and R_2 sub-Gaussian random variables. Let α_1 and α_2 be two real numbers. Then the random variable $\alpha_1 X_1 + \alpha_2 X_2$ is also sub-Gaussian with constant $\sqrt{\alpha_1^2 R_1^2 + \alpha_2^2 R_2^2}$.*

Lemma 3. *Suppose that for any a, at every t, $x_{a,t} \in \mathbb{R}^d$ are iid drawn from an unknown distribution with mean $\mu_a \in \mathbb{R}^d$, and that $||x_{a,t}||_2 \leq L$ and $||\theta_*||_2 \leq S$. Then for any a, at every t, $\epsilon_{a,t}^T \theta_*$ is sub-Gaussian, with constant $\dfrac{LS}{\sqrt{n_{a,t}}}$.*

Proof. By Cauchy-Schwarz inequality, for any a and t, we have $|\theta_*^T \hat{x}_{a,t}| \leq ||\theta_*||_2 ||\hat{x}_{a,t}||_2 \leq LS$. Then given that for any a, $x_{a,t}$ are iid and $\mathbb{E}[\hat{x}_{a,t}] = \mu_a$, we can apply Hoeffding's Inequality:

$$\forall \gamma > 0, \mathbb{P}\left(|\theta_*^T \hat{x}_{a,t} - \theta_*^T \mu_a| > \gamma\right) = \mathbb{P}\left(|\theta_*^T \epsilon_{a,t}| > \gamma\right) \leq 2e^{-\frac{n_{a,t}\gamma^2}{2S^2 L^2}}$$

Applying Lemma 1 gives the desired result with $n_{a,t}\gamma^2 / (2S^2 L^2) = 1/(2R^2)$

Finally, using Lemma 2 with the sum of $\theta_*^T \epsilon_{a,t}$ and $\eta_{a,s}$ proves Proposition 1.

A.2   Proof of Proposition 2

$$\hat{\theta}_t = \arg\min_\theta \sum_{s=1}^t \frac{1}{R_{a_s,t}} (\theta^T \hat{x}_{a_s,t} - r_{a_s,s})^2 + \lambda ||\theta||_2^2$$

$$= (X^T A X + \lambda I)^{-1} X^T A Y$$

$$= (X^T A X + \lambda I)^{-1} X^T A (X\theta_* + \eta')$$

$$= (X^T A X + \lambda I)^{-1} X^T A \eta' + (X^T A X + \lambda I)^{-1} (X^T A X + \lambda I)\theta_*$$

$$\quad - (X^T A X + \lambda I)^{-1} \lambda I \theta_*$$

$$= (X^T A X + \lambda I)^{-1} X^T A \eta' + \theta_* - \lambda (X^T A X + \lambda I)^{-1} \theta_*$$

Using the same method than in the Proof of Theorem 2 of [1] we have:

$||\hat{\theta}_t - \theta_*||_{V_t} \leq ||X^T A \eta'||_{V_t^{-1}} + \lambda||\theta_*||_{V_t^{-1}}$ with $V_t = \lambda I + X^T A X$, which is positive definite because $\lambda > 0$. Given that $||\theta_*||_2 \leq S$ and $||\theta_*||_{V_t^{-1}}^2 \leq ||\theta_*||_2^2/\lambda_{min}(V_t) \leq ||\theta_*||_2^2/\lambda$ we have: $||\hat{\theta}_t - \theta_*||_{V_t} \leq ||X^T A \eta'||_{V_t^{-1}} + \sqrt{\lambda}S$

Using Theorem 1 of [1] and the fact that $\dfrac{\eta'_s}{R_{a_s,t}}$ is conditionally 1-sub Gaussian (Proposition 1), for any $\delta > 0$, with probability at least $1 - \delta$, for all $t \geq 0$:

$$||X^T A \eta'||_{V_t^{-1}} = ||\sum_{s=1}^{t} \frac{\eta'_s}{R_{a_s,t}}\hat{x}_{a_s,t}||_{V_t^{-1}} \leq \sqrt{2\log\left(\frac{det(V_t)^{1/2}det(\lambda I)^{-1/2}}{\delta}\right)}$$

A.3   Proof of Theorem 1

Lemma 4. *For any a and $t > 0$ with probability at least $1 - \delta/t^2 - \delta$:*
$$0 \leq s_{a,t} - \theta_*^T \mu_a \leq 2\alpha_t||\hat{x}_{a,t}||_{V_{t-1}^{-1}} + 4(\alpha_t/\sqrt{\lambda} + S)\rho_{a,t,\delta}$$

Proof. Applying Hoeffding to each dimension $i \in [1..d]$ gives:

$\forall \gamma > 0:$ $\mathbb{P}\left(|\hat{x}_{a,t}^i - \mu_a^i| > \gamma/d\right) \leq 2e^{-2\frac{n_{a,t}\gamma^2}{L^2 d^2}}$. Then, using $||\hat{x}_{a,t} - \mu_a||_2 \leq \sum_{i=1}^{d}|\hat{x}_{a,t}^i - \mu_a^i|$ and the union bound property gives: $\mathbb{P}\left(||\hat{x}_{a,t} - \mu_a||_2 \leq \gamma\right) \geq 1 - 2de^{-2\frac{n_{a,t}\gamma^2}{L^2 d^2}}$. Choosing the appropriate γ, for any a and $t > 0$ with probability at least $1 - \delta/t^2$:

$$||\hat{x}_{a,t} - \mu_a||_2 \leq Ld\sqrt{\frac{1}{2n_{a,t}}\log\left(\frac{2dt^2}{\delta}\right)} = \rho_{a,t,\delta}$$

Suppose that the previous inequality holds, then:

- $||\hat{x}_{a,t} - \mu_a||_{V_{t-1}^{-1}} \leq ||\hat{x}_{a,t} - \mu_a||_2/\sqrt{\lambda} \leq \rho_{a,t,\delta}/\sqrt{\lambda}$ so $||\mu_a||_{V_{t-1}^{-1}} \leq ||\hat{x}_{a,t}||_{V_{t-1}^{-1}} + \rho_{a,t,\delta}/\sqrt{\lambda} = ||\hat{x}_{a,t} + \tilde{\epsilon}_{a,t}||_{V_{t-1}^{-1}}$, with $\tilde{\epsilon}_{a,t} = \rho_{a,t,\delta}\hat{x}_{a,t}/(\sqrt{\lambda}||\hat{x}_{a,t}||_{V_{t-1}^{-1}})$.
- $|\hat{\theta}_t^T(\hat{x}_{a,t} - \mu_a)| \leq \hat{\theta}_t^T \bar{\epsilon}_{a,t}$, with $\bar{\epsilon}_{a,t} = \rho_{a,t,\delta}\hat{\theta}_t/||\hat{\theta}_t||_2$.

Using those two results, Proposition 2 and the union bound property, we are ready to prove the proposed result:
First part:

$$\hat{\theta}_t^T(\hat{x}_{a,t} + \bar{\epsilon}_{a,t}) + \alpha_t||\hat{x}_{a,t} + \tilde{\epsilon}_{a,t}||_{V_{t-1}^{-1}} - \theta_*^T \mu_a$$

$$= (\hat{\theta}_t - \theta_*)^T \mu_a + \alpha_t||\hat{x}_{a,t} + \tilde{\epsilon}_{a,t}||_{V_{t-1}^{-1}} - \hat{\theta}_t^T(\mu_a - \hat{x}_{a,t}) + \hat{\theta}_t^T \bar{\epsilon}_{a,t}$$

$$\geq - ||\hat{\theta}_t - \theta_*||_{V_t}||\mu_a||_{V_{t-1}^{-1}} + \alpha_t||\hat{x}_{a,t} + \tilde{\epsilon}_{a,t}||_{V_{t-1}^{-1}} + \hat{\theta}_t^T(\hat{x}_{a,t} - \mu_a + \bar{\epsilon}_{a,t})$$

$$\geq - \alpha_t||\mu_a||_{V_{t-1}^{-1}} + \alpha_t||\hat{x}_{a,t} + \tilde{\epsilon}_{a,t}||_{V_{t-1}^{-1}} + \hat{\theta}_t^T(\hat{x}_{a,t} - \mu_a + \bar{\epsilon}_{a,t})$$

$$\geq - \alpha_t||\hat{x}_{a,t} + \tilde{\epsilon}_{a,t}||_{V_{t-1}^{-1}} + \alpha_t||\hat{x}_{a,t} + \tilde{\epsilon}_{a,t}||_{V_{t-1}^{-1}} + \hat{\theta}_t^T(\hat{x}_{a,t} - \mu_a + \bar{\epsilon}_{a,t})$$

$$\geq 0$$

Second part: using the fact that $||\bar{\epsilon}_{a,t}||_{V_{t-1}^{-1}} \leq ||\bar{\epsilon}_{a,t}||_2/\sqrt{\lambda} = \rho_{a,t,\delta}/\sqrt{\lambda}$ and $||\tilde{\epsilon}_{a,t}||_{V_{t-1}^{-1}} = \rho_{a,t,\delta}/\sqrt{\lambda}$:

$$\hat{\theta}_t^T(\hat{x}_{a,t} + \bar{\epsilon}_{a,t}) + \alpha_t||\hat{x}_{a,t} + \tilde{\epsilon}_{a,t}||_{V_{t-1}^{-1}} - \theta_*^T \mu_a$$

$$=(\hat{\theta}_t - \theta_*)^T \mu_a + \alpha_t||\hat{x}_{a,t} + \tilde{\epsilon}_{a,t}||_{V_t^{-1}} - \hat{\theta}_t^T(\mu_a - \hat{x}_{a,t}) + \hat{\theta}_t^T \bar{\epsilon}_{a,t}$$

$$\leq ||\hat{\theta}_t - \theta_*||_{V_t}||\mu_a||_{V_{t-1}^{-1}} + \alpha_t||\hat{x}_{a,t} + \tilde{\epsilon}_{a,t}||_{V_{t-1}^{-1}} + \hat{\theta}_t^T(\hat{x}_{a,t} - \mu_a + \bar{\epsilon}_{a,t})$$

$$\leq 2\alpha_t||\hat{x}_{a,t} + \tilde{\epsilon}_{a,t}||_{V_{t-1}^{-1}} + 2||\hat{\theta}_t||_{V_{t-1}}||\bar{\epsilon}_{a,t}||_{V_{t-1}^{-1}}$$

$$\leq 2\alpha_t||\hat{x}_{a,t}||_{V_{t-1}^{-1}} + 2\alpha_t||\tilde{\epsilon}_{a,t}||_{V_{t-1}^{-1}} + 2(\alpha_t + S\sqrt{\lambda})||\bar{\epsilon}_{a,t}||_{V_{t-1}^{-1}}$$

$$\leq 2\alpha_t||\hat{x}_{a,t}||_{V_{t-1}^{-1}} + 4(\alpha_t/\sqrt{\lambda} + S)\rho_{a,t,\delta}$$

Proposition 3. *For every t, with probability at least $1 - \delta/t^2 - \delta$, the instantaneous regret $reg_t = \theta_*^T \mu_{a_*} - \theta_*^T \mu_{a_t}$ is bounded by:*

$$reg_t \leq 2\alpha_t||\hat{x}_{a_t,t}||_{V_{t-1}^{-1}} + 4(\alpha_t/\sqrt{\lambda} + S)\rho_{a_t,t,\delta} = reg_t^{(1)} + reg_t^{(2)}$$

Proof. Due to the selection policy and the first inequality of the previous lemma, at each time t: $s_{a_t,t} \geq s_{a_*,t} \geq \theta_*^T \mu_{a_*}$. On the other hand, due to the second inequality of the previous lemma, at each time t: $s_{a_t,t} \leq \theta_*^T \mu_{a_t} + 2\alpha_t||\hat{x}_{a,t}||_{V_{t-1}^{-1}} + 4(\alpha_t/\sqrt{\lambda} + S)\rho_{a,t,\delta}$, which concludes the proof.

Proof of the Main Theorem: On the one hand, using the fact that $\sum_{t=2}^{\infty} \delta_t = \delta(\pi^2/6 - 1) \leq \delta$, the union bound property, and the fact that in Proposition 2 the bound in uniform, with probability at least $1 - 2\delta$:

$$\sum_{t=1}^{n} reg_t^{(2)} \leq C + \sum_{t=2}^{n} 4(\alpha_t/\sqrt{\lambda} + S)\rho_{a_t,t,\delta}$$

$$\leq C + \sum_{t=2}^{n} 4(\alpha_t/\sqrt{\lambda} + S)Ld\sqrt{\frac{1}{2n_{a,t}}\log\left(\frac{2dt^2}{\delta}\right)}$$

$$\leq C + 4Ld(\alpha_n/\sqrt{\lambda} + S)\sqrt{\log\left(\frac{2dn^2}{\delta}\right)}\sum_{t=2}^{n}\frac{1}{\sqrt{n_{a_t,t}}}$$

On the other hand:

$$\sum_{t=1}^{n} reg_t^{(1)} \leq \sqrt{n\sum_{t=1}^{n} 4\alpha_t^2||\hat{x}_{a_t,t}||_{V_{t-1}^{-1}}^2} \leq 2\alpha_n\sqrt{n\sum_{t=1}^{n}||\hat{x}_{a_t,t}||_{V_{t-1}^{-1}}^2}$$

We need to upper bound the term $\sum_{t=1}^{n}||\hat{x}_{a_t,t}||_{V_{t-1}^{-1}}^2$. To do this, let us introduce $\nu_{a,t,\delta} = Ld\sqrt{1/(2n_{a_t,t})\log(2dn/\delta)}$. Still using Hoeffding, with probability at least $1 - \delta/n$ we have: $||\hat{x}_{a_s,t}||_2 \leq ||\mu_{a_s}||_2 + \nu_{a_s,t,\delta}$ for $s \leq t$.
Defining: $\check{\epsilon}_{a,t} = \nu_{a,t,\delta}\mu_a/||\mu_a||_2$ we have, for $s \leq t$:

$1/\sqrt{R_{a_s,s}}||\mu_{a_s} - \check{\epsilon}_{a_s,s}||_2 \leq 1/\sqrt{R_{a_s,t}}||\hat{x}_{a_s,t}||_2$. So the following holds:

$$V_t = \lambda I + \sum_{s=1}^{t} \frac{1}{R_{a_s,t}}\hat{x}_{a_s,t}\hat{x}_{a_s,t}^T \geq \lambda I + \sum_{s=1}^{t} \frac{1}{R_{a_s,s}}(\mu_{a_s} - \check{\epsilon}_{a_s,s})(\mu_{a_s} - \check{\epsilon}_{a_s,s})^T = W_t$$

Which means that for any vector x: $||x||_{V_t^{-1}} \leq ||x||_{W_t^{-1}}$.

We also define $\hat{\epsilon}_{a,t} = \nu_{a,t,\delta}\mu_a/(\sqrt{\lambda}||\mu_a||_{W_{t-1}^{-1}})$, such that for $s \leq t$:

$||\hat{x}_{a_s,t}||_{W_{t-1}^{-1}} \leq ||\mu_{a_s} + \hat{\epsilon}_{a_s,s}||_{W_{t-1}^{-1}}$ and $||\check{\epsilon}_{a_s,s}||_{W_{t-1}^{-1}} = \nu_{a_s,s,\delta}/\sqrt{\lambda}$.

So, using the union property and $\sum_{t=1}^{n} \frac{\delta}{n} = \delta$, with probability at least $1 - \delta$:

$$\sum_{t=1}^{n} ||\hat{x}_{a_t,t}||_{V_{t-1}^{-1}}^2 \leq \sum_{t=1}^{n} ||\hat{x}_{a_t,t}||_{W_{t-1}^{-1}}^2 \leq \sum_{t=1}^{n} ||\mu_{a_t} + \hat{\epsilon}_{a_t,t}||_{W_{t-1}^{-1}}^2 \leq \sum_{t=1}^{n} ||\mu_{a_t} + \hat{\epsilon}_{a_t,t} - \check{\epsilon}_{a_t,t}$$

$$+ \check{\epsilon}_{a_t,t}||_{W_{t-1}^{-1}}^2 \leq \sum_{t=1}^{n} ||\mu_{a_t} - \check{\epsilon}_{a_t,t}||_{W_{t-1}^{-1}}^2 + \sum_{t=1}^{n} ||\hat{\epsilon}_{a_t,t}||_{W_{t-1}^{-1}}^2 + \sum_{t=1}^{n} ||\check{\epsilon}_{a_t,t}||_{W_{t-1}^{-1}}^2 \leq \sum_{t=1}^{n} ||\mu_{a_t}$$

$$- \check{\epsilon}_{a_t,t}||_{W_{t-1}^{-1}}^2 + \frac{2}{\lambda}\sum_{t=1}^{n} \nu_{a_t,t,\delta}^2 \leq \sum_{t=1}^{n} ||\mu_{a_t} - \check{\epsilon}_{a_t,t}||_{W_{t-1}^{-1}}^2 + \frac{L^2 d^2}{\lambda}\log\left(\frac{2dn}{\delta}\right)\sum_{t=1}^{n} \frac{1}{n_{a_t,t}}$$

Secondly, let us notice that:

$$det(W_n) = det(W_{n-1} + \frac{1}{R_{a_n,n}}(\mu_{a_n} - \check{\epsilon}_{a_n,n})(\mu_{a_n} - \check{\epsilon}_{a_n,n})^T)$$

$$= det(W_{n-1})det(I + \frac{1}{R_{a_n,n}}W_{n-1}^{-1/2}(\mu_{a_n} - \check{\epsilon}_{a_n,n})(W_{n-1}^{-1/2}(\mu_{a_n} - \check{\epsilon}_{a_n,n}))^T)$$

$$= det(\lambda I)\prod_{t=1}^{n}(1 + \frac{1}{R_{a_t,t}}||\mu_{a_t} - \check{\epsilon}_{a_t,t}||_{W_{t-1}^{-1}}^2)$$

where we used that all the eigenvalues of a matrix of the form $I + xx^T$ are one except one eigenvalue, which is $1 + ||x||^2$ and which corresponds to the eigenvector x.

Given that $\lambda > \max(1, 2L^2)$: $||\mu_{a_t} - \check{\epsilon}_{a_t,t}||_{W_t^{-1}}^2 \leq ||x||_2^2/\lambda \leq 2L^2/\lambda \leq 1$

So using the fact $x \leq 2\log(1 + x)$ when $1 \leq x \leq 1$, we have:

$$2\log\left(\frac{det(W_n)}{det(\lambda I)}\right) \geq \sum_{t=1}^{n} \frac{1}{R_{a_t,t}}||\mu_{a_t} - \check{\epsilon}_{a_t,t}||_{W_{t-1}^{-1}}^2 \geq \frac{1}{\sqrt{R^2 + L^2 S^2}}\sum_{t=1}^{n} ||\mu_{a_t}$$

$$- \check{\epsilon}_{a_t,t}||_{W_{t-1}^{-1}}^2$$

And as in Lemma 11 of [1] we have that $\log\left(\frac{det(W_n)}{det(\lambda I)}\right) \leq d\log\left(1 + \frac{nL^2}{\lambda d}\right)$

Which gives: $\sum_{t=1}^{n} ||\mu_{a_t} - \check{\epsilon}_{a_t,t}||_{W_{t-1}^{-1}}^2 \leq \sqrt{R^2 + L^2 S^2}d\log\left(1 + \frac{nL^2}{\lambda d}\right)$.

Last, as in Lemma 10 of [1], the Determinant-Trace Inequality gives:

$$\alpha_n \leq \sqrt{d\log\left(\frac{1 + nL^2/\lambda}{\delta}\right)} + \sqrt{\lambda}S.$$

Putting all together, we get the proposed upper bound for the general case.

A.4 Proof for Case 1

On the one hand $\sum_{t=1}^{n} \frac{1}{n_{a_t,t}} = \sum_{t=1}^{n} \frac{1}{t} \leq 1 + \log(n)$. On the other hand $\sum_{t=1}^{n} \frac{1}{\sqrt{n_{a_t,t}}} =$

$\sum_{t=1}^{n} \frac{1}{\sqrt{t}} \leq \int_{0}^{n} \frac{1}{\sqrt{t}} \, dt \leq 2\sqrt{n}$. Using the bound of Theorem 1 gives the final result.

A.5 Proof for Case 2

Lemma 5. $\forall a, \forall t \geq \lceil 2\log(1/\delta)/p^2 \rceil$, with probability at least $1 - \delta$: $n_{a,t} \geq tp/2$

Proof. By Hoeffding's inequality, for every $\epsilon > 0$: $P(n_{a,t} \geq tp - \epsilon) \geq 1 - e^{-2\epsilon^2/t}$. Choosing $\epsilon = tp/2$ leads to $P(n_{a,t} \geq tp/2) \geq 1 - e^{-tp^2/2}$. If $t \geq 2\log(1/\delta)/p^2$, then $1 - e^{-tp^2/2} \geq 1 - \delta$, which proves the announced result.

In the following, we denote $u = \lceil 2\log(1/\delta)/p^2 \rceil$.

Main Proof: On the one hand, from Lemma 5, with probability at least $1 - \delta$:

$$\sum_{t=1}^{n} \frac{1}{n_{a_t,t}} \leq u + \frac{2}{p} \sum_{t=u+1}^{n} \frac{1}{t} \leq u + \frac{2}{p} \int_{u}^{n} \frac{1}{t} \, dt \leq u + \frac{2\log(n)}{p}$$

On the other hand, from Lemma 5, we have with probability at least $1 - \delta$:

$$\sum_{t=1}^{n} \frac{1}{\sqrt{n_{a_t,t}}} \leq u + \sqrt{\frac{2}{p}} \sum_{t=u+1}^{n} \frac{1}{\sqrt{t}} \leq u + \sqrt{\frac{2}{p}} \int_{u}^{n} \frac{1}{\sqrt{t}} \, dt \leq u + 2\sqrt{\frac{2n}{p}}$$

Then, using the general upper bound of Theorem 1 gives the final result.

A.6 Proof for Case 3

Let us decompose $n = \lfloor n/K \rfloor K + r$ with $r < K$.

Obviously, the sum $\sum_{t=1}^{n} \frac{1}{n_{a_t,t}}$ is maximized when each arm is observed exactly $\lfloor n/K \rfloor$ times during the first $\lfloor n/K \rfloor K$.

So: $\sum_{t=1}^{n} \frac{1}{n_{a_t,t}} \leq \sum_{a=1}^{K} \sum_{t=1}^{\lfloor n/K \rfloor + 1} \frac{1}{t} \leq K \sum_{t=1}^{\lceil n/K \rceil} \frac{1}{t} \leq 1 + \log(\lceil n/K \rceil)$. With the same argument: $\sum_{t=1}^{n} \frac{1}{\sqrt{n_{a_t,t}}} \leq K \sum_{t=1}^{\lceil n/K \rceil} \frac{1}{\sqrt{t}} \leq 2K\sqrt{\lceil n/K \rceil}$.

Finally, noting that $K\log(\lceil n/K \rceil) \sim K\log(n/K)$ and $K\sqrt{\lceil n/K \rceil} \sim \sqrt{Kn}$ and using the general upper bound of Theorem 1 gives the final result.

A.7 Proof of the Multiple Plays Extension

We follow a similar way than in [18] with the particular case of the summation: since we consider that the reward of a set of arms is the sum of its members, the instantaneous regret is defined as: $reg_t = \sum_{a_* \in \mathcal{K}_*} \theta_*^T \mu_{a_*} - \sum_{a_t \in \mathcal{K}_t} \theta_*^T \mu_{a_t}$, with $\mathcal{K}_* =$ $\underset{\hat{\mathcal{K}}, |\hat{\mathcal{K}}| = k}{\arg\max} \sum_{a \in \hat{\mathcal{K}}} \theta^T \mu_a$. We then use the fact that for all t: $\sum_{a_t \in \mathcal{K}_t} s_{a_t,t} \geq \sum_{a_* \in \mathcal{K}_*} s_{a_*,t}$, which leads to: $reg_t \leq \sum\limits_{a_t \in \mathcal{K}_t} 2\alpha_t ||\hat{x}_{a_t,t}||_{V_{t-1}^{-1}} + 4\sqrt{d}(\alpha_t/\sqrt{\lambda} + S)\rho_{a_t,t,\delta}$. Where V_t is defined considering that k learning examples come at each time step instead of 1: $V_t = \lambda I + \sum\limits_{s=1}^{t} \sum\limits_{a_s \in \mathcal{K}_s} \frac{1}{R_{a_s,t}} \hat{x}_{a_s,t} \hat{x}_{a_s,t}^T$.

Using the same methods than before in the three cases, the cumulative regret take the same form, expect that k appears explicitly in the regret because of the two terms: $\sum\limits_{t=1}^{n} \sum\limits_{a_t \in \mathcal{K}_t} \frac{1}{n_{a_t,t}}$ and $\sum\limits_{t=1}^{n} \sum\limits_{a_t \in \mathcal{K}_t} \frac{1}{\sqrt{n_{a_t,t}}}$.

References

1. Abbasi-yadkori, Y., Pál, D., Szepesvári, C.: Improved algorithms for linear stochastic bandits. In: NIPS (2011)
2. Agrawal, S., Goyal, N.: Analysis of thompson sampling for the multi-armed bandit problem. In: COLT (2012)
3. Agrawal, S., Goyal, N.: Thompson sampling for contextual bandits with linear payoffs. In: ICML (2013)
4. Audibert, J.Y., Bubeck, S.: Minimax policies for adversarial and stochastic bandits. In: COLT (2009)
5. Audibert, J.-Y., Munos, R., Szepesvári, C.: Tuning bandit algorithms in stochastic environments. In: Chaudhuri, K., Gentile, C., Zilles, S. (eds.) ALT 2015. Lecture Notes in Artificial Intelligence (LNAI), vol. 9355, pp. 150–165. Springer, Heidelberg (2007). doi:10.1007/978-3-540-75225-7_15
6. Auer, P.: Using confidence bounds for exploitation-exploration trade-offs. J. Mach. Learn. Res. **3**, 397–422 (2003)
7. Auer, P., Cesa-Bianchi, N., Fischer, P.: Finite-time analysis of the multiarmed bandit problem. Mach. Learn. **47**, 235–256 (2002)
8. Chapelle, O., Li, L.: An empirical evaluation of thompson sampling. In: NIPS, Curran Associates, Inc. (2011)
9. Dani, V., Hayes, T.P., Kakade, S.M.: Stochastic linear optimization under bandit feedback. In: COLT (2008)
10. Filippi, S., Cappe, O., Garivier, A., Szepesvári, C.: Parametric bandits: The generalized linear case. In: NIPS (2010)
11. Garivier, A.: The KL-UCB algorithm for bounded stochastic bandits and beyond. In: COLT (2011)
12. Gisselbrecht, T., Denoyer, L., Gallinari, P., Lamprier, S.: Whichstreams: a dynamic approach for focused data capture from large social media. In: ICWSM (2015)
13. Hong, L., Davison, B.D.: Empirical study of topic modeling in twitter. In: ECIR (2010)
14. Kaufmann, E., Cappe, O., Garivier, A.: On bayesian upper confidence bounds for bandit problems. In: AISTATS (2012)
15. Kaufmann, E., Korda, N., Munos, R.: Thompson sampling: an asymptotically optimal finite-time analysis. In: Chaudhuri, K., Gentile, C., Zilles, S. (eds.) ALT 2015. Lecture Notes in Artificial Intelligence (LNAI), vol. 9355, pp. 199–213. Springer, Heidelberg (2012). doi:10.1007/978-3-642-34106-9_18
16. Lai, T., Robbins, H.: Asymptotically efficient adaptive allocation rules. Adv. Appl. Math. **6**(1), 4–22 (1985)

17. May, B.C., Korda, N., Lee, A., Leslie, D.S.: Optimistic bayesian sampling in contextual-bandit problems. J. Mach. Learn. Res. **13**, 2069–2106 (2012)
18. Qin, L., Chen, S., Zhu, X.: Contextual combinatorial bandit and its application on diversified online recommendation. In: SIAM (2014)
19. Rusmevichientong, P., Tsitsiklis, J.N.: Linearly parameterized bandits. Math. Oper. Res. **35**, 395–411 (2010)
20. Thompson, W.: On the likelihood that one unknown probability exceeds another in view of the evidence of two samples. Bull. Am. Math. Soc. **25**, 285–294 (1933)

Ballpark Learning: Estimating Labels from Rough Group Comparisons

Tom Hope(✉) and Dafna Shahaf

The Hebrew University of Jerusalem, Jerusalem, Israel
tom.hope@mail.huji.ac.il, dshahaf@cs.huji.ac.il

Abstract. We are interested in estimating individual labels given only coarse, aggregated signal over the data points. In our setting, we receive sets ("bags") of unlabeled instances with constraints on label proportions. We relax the unrealistic assumption of known label proportions, made in previous work; instead, we assume only to have upper and lower bounds, and constraints on bag differences. We motivate the problem, propose an intuitive formulation and algorithm, and apply our methods to real-world scenarios. Across several domains, we show how using only proportion constraints and no labeled examples, we can achieve surprisingly high accuracy. In particular, we demonstrate how to predict income level using rough stereotypes and how to perform sentiment analysis using very little information. We also apply our method to guide exploratory analysis, recovering geographical differences in twitter dialect.

1 Introduction

In many classification problems, labeled instances are often difficult, expensive, or time-consuming to obtain. Unlabeled instances, on the other hand, are easier to obtain, but it is harder to use them for classification. Semi-supervised learning [6] addresses this problem, using unlabeled instances together with a small amount of labeled instances to improve performance.

We are interested in a learning setting where few, if any, labeled instances exist. Instead, we only know some coarse, aggregated signal over the data points. In particular, our instances are divided into sets (or *bags*), and we are given some aggregate information about the bags; for example, we might know that one bag has a higher percentage of positive-label instances than another.

There is recent interest in the task of estimating the labels of *individual* instances given aggregate information, due to the many real-world scenarios in which such information is available. In particular, aggregate information (e.g., summary statistics) is often published for sensitive data, when one cannot publish individual statistics. Being able to estimate individual labels from such data has important implications regarding privacy and data anonymization.

© Springer International Publishing AG 2016
P. Frasconi et al. (Eds.): ECML PKDD 2016, Part II, LNAI 9852, pp. 299–314, 2016.
DOI: 10.1007/978-3-319-46227-1_19

Constraining class proportions
of unlabeled data has been shown
to be useful for semi-supervised
learning [20, 25, 27]. Under this setting, we are given sets of unlabeled
instances with known label proportions (for example, one bag has
30 % positive instances and 70 %
negative instances).

Fig. 1. We are given bags of instances and
rough estimates about label proportions and
differences between bags. Here, the purple bag
has at least 50 % positive instances, more than
the red bag (but the magnitude of the difference is uncertain). (Color figure online)

We believe that the assumption
of known proportions is unrealistic,
and limits the applicability of such
methods. For example, suppose we want to classify Twitter users by political
orientation. We have some information about the users (for example, the text
of their tweets), but no explicit political affiliation to use as labels. We could,
however, use the commonly-known fact that political orientation is correlated
with geographic location. Thus, we can construct bags of users based on their
geographic location: bags would correspond to states whose residents predominantly vote for the Republican Party (red states) or Democratic Party (blue
states).

Estimating the proportion of Democrats on Twitter is hard, even using location information. Previous election data or polls are unlikely to accurately reflect
the behavior of Twitter users. Instead of assuming known proportions, we propose a setup where our input is much weaker: we only know some constraints on
bag proportions and on differences between bags. In other words, users from red
states do not necessarily vote for the Republicans, but it is safe to expect to see
more Republicans in the red-state bags. It is also reasonable to assume that, say,
at least 10 % of Blue-state users are Democrats. Using only this type of weak,
"ballpark" estimates, we would like to be able to classify individual users.

Figure 1 demonstrates this idea. Our input includes approximate information
on label proportions in some bags (left) and pairwise comparisons between bags
(middle) or sets of bags (right). Our contributions are as follows:

- We extend the Learning from Labeled Proportions setting by proposing a new,
  more realistic scenario in which label proportions in each bag are not assumed
  to be known, but rather some constraints on them. We suggest various domains
  that lend themselves to this setting.
- We propose a simple and intuitive bi-convex problem formulation and an efficient algorithm, including a novel form of cross-validation.
- We apply our algorithm to real data, perform sentiment analysis of movie
  reviews from a very coarse signal, and predict income using stereotypes.
- We demonstrate the use of our method for exploratory analysis. We find vernacular difference in geo-tagged tweets by incorporating expressive constraints
  such as "Alabama > Florida > New York".
- Our algorithm is designed to use when human labeling resources are scarce.
  Despite the simplicity of our methods, we achieve high accuracy with a very
  modest amount of input, and considerably loose (or misspecified) constraints.

2   Problem Formulation

We begin by formalizing our setting and problem. Consider a set of N training instances $\mathcal{X}_N = \{\mathbf{x}_1, \mathbf{x}_2, \ldots, \mathbf{x}_N\}$. Each $\mathbf{x}_i$ has a corresponding *unknown* label $y_i^* \in \{-1, 1\}$. In addition, we could be given a (possibly empty) set of L labeled training instances $\mathcal{X}_L = \{\mathbf{x}_{N+1}, \mathbf{x}_{N+2}, \ldots, \mathbf{x}_{N+L}\}$ with known binary labels y_i, where typically the vast majority of our instances are unlabeled: $N \gg L$. In addition, we are given a set of K subsets of $\mathcal{X}$, which we call *bags*:

$$\mathcal{B} = \{\mathcal{B}_1, \mathcal{B}_2, \ldots \mathcal{B}_K\}, \mathcal{B}_k \subseteq \mathcal{X}_N \cup \mathcal{X}_L.$$

Note that bags $\mathcal{B}$ may overlap, and do not have to cover all training instances $\mathcal{X}_N$. Let p_k be the proportion of positive-labeled instances in bag $\mathcal{B}_k$:

$$p_k = |\{i : i \in \mathcal{B}_k, y_i^* = 1\}| / |\mathcal{B}_k| \tag{1}$$

(where y_i^* is replaced with y_i for instances $\mathbf{x}_i \in \mathcal{X}_L$). Previous work [20] tackled the case of *known* label proportions, suggesting that precise proportions could be estimated using sampling. However, obtaining accurate estimates could be costly or impractical (e.g., for bags with high label skew). In this work we do not assume to know p_k. Rather, we are given weaker prior knowledge, in the form of constraints on proportions. We allow constraints of the following forms:

- **Lower and upper bounds** on bag proportions: $l_k \leq p_k \leq u_k$
- **Bag difference bounds**: $0 \leq l_{k_{12}} \leq p_{k_1} - p_{k_2} \leq u_{k_{12}}$

We are especially interested in the case where very little information is known: constraints are *loose*, and specified only for a small subset of the bags.

Our goal is to predict a label for each $\mathbf{x}_i$, using a function $f(\mathbf{x}) = \text{sign}(\mathbf{w}^T \varphi(\mathbf{x}))$, where $\mathbf{w}$ is a weight vector and $\varphi(\cdot)$ is a feature map (to simplify notation we drop a bias term $\mathbf{b}$ by assuming a vector $\mathbf{1}_{N+L}$ is appended to the features). To attain the classification goal, we use a maximum-margin approach. Let $\mathcal{R}$ be the subset of $\mathcal{B}$ for which we have upper and/or lower bounds. Let $\mathcal{D}$ be the set of tuples $(\mathcal{B}_{k_1}, \mathcal{B}_{k_2})$ for which we have difference bounds. To solve this problem we directly model the latent variable $\mathbf{y}^*$ – the vector of unknown labels $y_i^* \in \{-1, 1\}$, in an alternating optimization approach.

Noting that (1) can be written as $p_k = \frac{\sum_{i \in \mathcal{B}_k} y_i^*}{2|\mathcal{B}_k|} + \frac{1}{2}$, we formulate the following bi-convex optimization problem:

$$\operatorname*{argmin}_{\mathbf{y}, \mathbf{w}, \xi} \frac{1}{2} \mathbf{w}^T \mathbf{w} + \frac{C}{N} \sum_{i=1}^{N} \max(0, 1 - y_i \mathbf{w}^T \varphi(\mathbf{x}_i)) + \frac{C_L}{L} \sum_{j=N+1}^{N+L} \xi_j$$

$$s.t. -1 \leq y_i \leq 1 \quad \forall i \in 1, \ldots, N$$

$$y_j \mathbf{w}^T \varphi(\mathbf{x}_j) \geq 1 - \xi_j \quad \forall j \in \{N+1, \ldots, N+L\} \tag{2}$$

$$\xi_j \geq 0 \quad \forall j$$

$$l_k \leq \hat{p}_k \leq u_k \quad \forall\{k : \mathcal{B}_k \in \mathcal{R}\}$$

$$l_{k_{12}} \leq \hat{p}_{k_1} - \hat{p}_{k_2} \leq u_{k_{12}} \quad \forall\{k_1 \neq k_2 : (\mathcal{B}_{k_1}, \mathcal{B}_{k_2}) \in \mathcal{D}\},$$

where $\hat{p}_k = \frac{1}{2|\mathcal{B}_k|} \sum\limits_{i \in \mathcal{B}_k} y_i + \frac{1}{2}$ is the estimated positive label proportion in bag $\mathcal{B}_k$, l_k (or u_k) can be 0 (1) if not given as input, and analogously for difference bounds $l_{k_{12}}(u_{k_{12}})$. C and C_L are cost hyperparameters for unlabeled and labeled instances, respectively. Intuitively, the second term in the objective function helps find a weight vector $\mathbf{w}$ accurately predicting $\mathbf{y}$, and constraints ensure that we find an assignment to $\mathbf{y}$ that satisfies proportions constraints. C_L controls how much weight we give to our labeled instances versus our prior knowledge on $\mathcal{B}$. In our experiments we do not use any labeled instances, thus $C_L = 0$.

3   Algorithm

We have formalized our problem as a bi-convex optimization problem – holding either $\mathbf{w}$ or $\mathbf{y}$ fixed, we get a convex problem. We thus propose the following intuitive alternating algorithm to solve it.

– For a fixed $\mathbf{w}$, solve for $\mathbf{y}$:

$$\operatorname*{argmin}_{\mathbf{y}} \frac{1}{N} \sum_{i=1}^{N} \max(0, 1 - y_i \mathbf{w}^T \varphi(\mathbf{x}_i))$$

$$s.t. -1 \leq y_i \leq 1 \quad \forall i \in 1, \dots, N \qquad (3)$$
$$l_k \leq \hat{p}_k \leq u_k \quad \forall \{k : \mathcal{B}_k \in \mathcal{R}\}$$
$$l_{k_{12}} \leq \hat{p}_{k_1} - \hat{p}_{k_2} \leq u_{k_{12}} \quad \forall \{k_1 \neq k_2 : (\mathcal{B}_{k_1}, \mathcal{B}_{k_2}) \in \mathcal{D}\},$$

– For a fixed $\mathbf{y}$, solve w.r.t $\mathbf{w}$:

$$\operatorname*{argmin}_{\mathbf{w}} \frac{1}{2} \mathbf{w}^T \mathbf{w} + \frac{C}{N} \sum_{i=1}^{N} \max(0, 1 - y_i \mathbf{w}^T \varphi(\mathbf{x}_i)) + \frac{C_L}{L} \sum_{j=N+1}^{N+L} \xi_j$$

$$s.t. \ y_j \mathbf{w}^T \varphi(\mathbf{x}_j) \geq 1 - \xi_j \quad \forall j \in \{N+1, \dots, N+L\} \qquad (4)$$
$$\xi_j \geq 0 \quad \forall j$$

Intuitively, the first step finds an assignment to $\mathbf{y}$ that is "close" to predictions made by applying weights $\mathbf{w}$, and also satisfies proportions constraints. The second step re-adjusts $\mathbf{w}$. Our alternating algorithm for this bi-convex problem is thus guaranteed to descend, decreasing the objective in every iteration.

In practice, we replace $\mathbf{y}$ with $Sign(\mathbf{y})$ ($Sign(\cdot)$ applied elementwise) in order to use efficient off-the-shelf SVM solvers (See Fig. 2.). Empirically, in most cases we observed that $\mathbf{y}$ were very close to either 1 or -1.

To start off the alternation, we need to initialize $\mathbf{w}$. Specific label proportions constraints are handled by modeling the latent $\mathbf{y}^*$ directly, which is only possible in our alternating scheme once a vector $\mathbf{w}$ is fixed. Thus, we start the alternating optimization process by first solving the following simple convex program, which uses only the partial order between bags. Let the set of pairwise orderings $\mathcal{P}$ be the set of all tuples $(\mathcal{B}_{k_1}, \mathcal{B}_{k_2})$ such that $p_{k_1} \geq p_{k_2}$. To find our initial $\mathbf{w}$ we solve:

$$\operatorname*{argmin}_{\mathbf{w},\xi} \frac{1}{2}\mathbf{w}^T\mathbf{w} + \frac{1}{|\mathcal{P}|}\sum_{p=1}^{|\mathcal{P}|}\xi_p + \frac{C_L}{L}\sum_{j=N+1}^{N+L}\xi_j$$

$$s.t. \quad y_j\mathbf{w}^T\varphi(\mathbf{x}_j) \geq 1 - \xi_j \quad \forall j \in \{N+1,\ldots,N+L\} \quad\quad (5)$$

$$\mathbf{w}^T\frac{1}{|\mathcal{B}_{k_1}|}\sum_{i\in\mathcal{B}_{k_1}}\varphi(\mathbf{x}_i) \geq \mathbf{w}^T\frac{1}{|\mathcal{B}_{k_2}|}\sum_{i\in\mathcal{B}_{k_2}}\varphi(\mathbf{x}_i) - \xi_p$$

$$\forall\{k_1 \neq k_2 : (\mathcal{B}_{k_1},\mathcal{B}_{k_2}) \in \mathcal{P}\},$$

The second constraint in Problem 5 amounts to representing bags with their (**w**-weighted) mean in feature-space. Note that in order for a bag $\mathcal{B}_k$ to be well-approximated by its mean in feature-space, $\mathcal{B}_k$ should induce a low-variance distribution over bag instances. This is a strong assumption, but yields a simple quadratic program easy to solve quickly with standard solvers, and empirically leads to good starting points in parameter-space. We additionally note that when $C_L = 0$ (no labels), we recover as a special case the

---

Input: $\mathbf{x}, \mathcal{R}, \mathcal{D}, C$

1. **Init** $\mathbf{w}^0$: $\mathbf{w}^0 \leftarrow$ Solution to (5)

2. **Repeat**
   (a) Solve (3) for $\mathbf{y}^t$ w.r.t $\mathbf{w}^{t-1}$
   (b) Solve an SVM problem for $\mathbf{w}^t$
       w.r.t $Sign(\mathbf{y}^t)$ and cost parameter C
   until $\frac{\|\mathbf{w}^t - \mathbf{w}^{t-1}\|_2^2}{\|\mathbf{w}^{t-1}\|_2^2} \leq 10^{-5}$

Return w

Fig. 2. Alternating algorithm

Multiple-Instance (MI) ranking problem proposed in the image-retrieval framework of [13], albeit with a different objective (we are interested in classifying instances rather than learning to rank bags). We note that in Problem 3, we impose hard constraints on label proportions. Certain sets of constraints could, of course, be infeasible. In this case, a practitioner might adjust the constraints, or simply make them soft (by adding slack variables).

Optimizing C. In practice, we need to tune hyperparameter C. This is typically done with cross-validation (CV) grid search, measuring performance on held-out data. However, standard CV is impossible here, as we have no labeled examples.

We thus develop a novel variant of CV, suited for our setting. We run K-fold CV, splitting each bag $\mathcal{B}_k$ into training and held-out subsets. The intuition is that the label proportion in uniformly-sampled subsets of a bag is similar to the proportion p_k in the entire bag. For each split we run Algorithm 2 on training bags, and then compute by how much constraints are violated on *held-out* bags. More formally, we compute the average deviations from bounds, $max(\hat{p}_k - u_k, 0)$, $max(l_k - \hat{p}_k, 0)$ for $\hat{p}_k$ the estimated label proportion in the held-out subset of bag k. We do so over a grid, and select the C with lowest average violation.

4 Evaluation

In order to evaluate our algorithm, we prepared the following datasets:

- **Movie Reviews:** The Movie Reviews dataset [17] contains 1000 positive and 1000 negative movie reviews written before 2002. The task is to classify the sentiment of movie reviews as positive or negative.
- **Census:** The Adult dataset [1] (48842 instances) is from the Census bureau. The task is to predict whether a given adult makes more than $50,000 a year based on attributes such as education, hours of work per week, etc.

For each of the classification tasks described, we run 10-fold cross-validation and report average results (note that labels are used only for testing). For text classification tasks, feature map $\varphi(\cdot)$ is the standard TF-IDF features.

We formed bags corresponding to the different tasks (see below), demonstrating the wide applicability of the setting and our approach. In order to test our method's robustness we used approximate constraints, at times violating the true underlying proportions.

Baselines. To the best of our knowledge, no other method aims to solve the problem of Sect. 2. Thus, we compare ourselves to three natural baselines.

- **"High vs. low":** One reasonable approach in our setting is to create two sets of instances: The "high" set contains instances from bags with the highest label proportions, and the "low" set – from bags with the lowest proportions. The idea is to pretend all instances in the "high" set are positive, and in the "low" set – negative, and learn a classifier with the noisy labels. To make the baseline stronger, we use grid search to optimize hyper-parameter C (chosen from a commonly used grid for SVM C values, $[10^{-4}, 10^{-3}, \ldots, 10^{3}, 10^{4}]$, with 10-fold cross-validation and selecting C with best average). To counter the class-imbalance created, we apply a weighted SVM.
- **Supervised SVM:** Our method does not need labeled instances, but instead uses weaker, aggregate information. To show how many labels are needed to obtain comparable results to our method, we report SVM results over a labeled training set (note that this information is not available to our algorithm). We use grid search to optimize hyper-parameter C as above.
- **Learning from labeled proportions:** For the census data set, we compare our method's performance to results reported in [20] using known label proportions with various algorithms. Note that our method does not have access to the exact label proportions.

For our method, we select C using the constraint-violation approach described in the previous section.

We run the procedure for a maximum of 200 iterations, with convergence typically occurring long before. A typical iteration (for one value of C, one CV split) took at most a few seconds on a standard laptop. Our data is available on https://github.com/ttthhh/ballpark.git.

4.1   One-Word Classifier

Our first task is to classify sentiment of movie reviews. Our goal is not to compete with the host of previous sentiment-analysis algorithms [18] in terms of accuracy, but rather to provide a light-weight tool when very little information and resources are available: a "poor-man's" classifier. In this section, we show how we are able to obtain good results while assuming very scarce prior knowledge with simple, clean tools.

We envision a practitioner who knows a very simple fact – that reviews containing the word "great" are more likely to be positive than negative, but far from exclusively: many positive reviews do not use the word "great", and some negative reviews do use it ("horrific performance by a usually great actor").

We construct three bags: $\mathcal{B}_{\text{great}}, \mathcal{B}_{\text{good}}, \mathcal{B}_{\text{bad}}$, each containing reviews with the corresponding word in them (note the bags are not necessarily disjoint). For the three bags created on training set instances (10-fold CV) we find that $|\mathcal{B}_{\text{great}}| \approx 700$, $|\mathcal{B}_{\text{good}}| \approx 630$, $|\mathcal{B}_{\text{bad}}| \approx 160$, $p_{\text{great}} \approx 0.6$, $p_{\text{good}} \approx 0.45$, $p_{\text{bad}} \approx 0.25$.

For simplicity, we assume no labels are given, but the practitioner has rough estimates for proportions. This information could come from a sample or from domain knowledge. In our experiment, we assume an upper bound on the bag with the highest proportion and a lower bound for each bag. We used a weak bound for each bag, underestimating it by 50 %. We also assumed that $p_{\text{great}} > p_{\text{good}} > p_{\text{bad}}$. Again we use a weak bound, overestimating the real difference by 33 %. In Sect. 4.3 we explore how the tightness of the constraints affects accuracy, showing our method is robust to loose constraints.

For the *"high vs. low"* baseline, we take bag $\mathcal{B}_{\text{great}}, \mathcal{B}_{\text{good}}$ as the positive class, and $\mathcal{B}_{\text{bad}}$ as the other. As seen in Table 1, our method outperforms this naive baseline, and competes with supervised SVM trained on considerable amounts of labeled examples. Given fewer labels, supervised SVM is inferior to our label-free method: providing SVM with 25 labeled instances leads to accuracy of 0.51, 50 labels to accuracy of 0.63, and 75 labels increases accuracy to 0.69.

To test stability, we run the same experiment using different words to create the bags. The results are similar. Table 1 shows the results using "excellent", "nice", and "terrible". To make sure the classifier is not learning our input words, we test removing these words (e.g., "good") from the documents. In our experiment, the removal reduced accuracy by less than 1 %.

Table 1. Movie results for different sets of bags based on different choices of words. Our method outperforms the naive SVM baseline, and rivals a supervised SVM with a considerable number of labels.

Method	$\mathcal{B}_{\text{great}}, \mathcal{B}_{\text{good}}, \mathcal{B}_{\text{bad}}$	$\mathcal{B}_{\text{excellent}}, \mathcal{B}_{\text{nice}}, \mathcal{B}_{\text{terrible}}$
Bag constraints	**0.71**	**0.73**
"high vs. low" SVM	0.52	0.55
Supervised SVM	100 labels (0.71)	100 labels (0.71)

4.2 Learning from Stereotypes

In this section we simulate a scenario frequently occurring in practice. We have a large sample of individuals, and would like to predict their level of income using socio-demographic information. One variable that is known to be correlated with income is *education level*. This information is difficult to obtain (budgetary constraints, privacy issues, respondents' reluctance etc.) and is available only for a small sub-sample. In addition, we have no labels – individuals with known income. We do have ballpark-estimations on income proportions for different education levels, and the difference between them (based on an earlier census, expert assessments or other external sources).

In our first experiment we construct bags based on education level: $\mathcal{B}_{\text{Masters}}$, $\mathcal{B}_{\text{Bachelors}}$, $\mathcal{B}_{\text{Some-college}}$, $\mathcal{B}_{\text{High-School}}$. Over 20-fold CV (size of training set $\approx$ 1220) we find that $|\mathcal{B}_{\text{Masters}}| \approx 90$, $|\mathcal{B}_{\text{Bachelors}}| \approx 265$, $|\mathcal{B}_{\text{some-college}}| \approx 360$, $|\mathcal{B}_{\text{High-School}}| \approx 520$, $p_{\text{Masters}} \approx 0.55$, $p_{\text{Bachelors}} \approx 0.42$, $p_{\text{some-college}} \approx 0.19$, $p_{\text{High-School}} \approx 0.16$.

We use similar constraints to the previous section, but remove all lower bounds on bags, thus incorporating even less prior information than before. For the *SVM using "high vs. low"* baseline, we use $\mathcal{B}_{\text{Masters}}$, $\mathcal{B}_{Bachelors}$ as one class, and $\mathcal{B}_{\text{some-college}}$, $\mathcal{B}_{\text{High-School}}$ as the other.

We start with basic features: age, gender, race. After assigning individuals to education bags, we discard education features from the data – we assume not to have this information at test time (only for a small sub-sample available for training). We do retain those features for the *Supervised SVM* baseline. Our method achieves cross-validation accuracy of 0.74, while the baseline achieves 0.57. Supervised SVM, even with 1000 labeled examples, only reaches 0.71.

We also experiment with using less bags (removing "Masters"), and with an expanded feature set (age, race, gender, hours-per-week, capital-gain, capital-loss). See Table 2 for results. Here too, our method outperforms the baseline, and rivals supervised SVM with 900 labels.

Of course, we are not limited to using bags based on only education level. Another well-known correlation is between gender and income. Thus, we can also slice the data into bags based on education *and* gender. In another experiment we create 6 bags, $\mathcal{B}_{\text{Bachelors+Female}}$, $\mathcal{B}_{\text{Some-college+Female}}$, $\mathcal{B}_{\text{High-School+Female}}$, $\mathcal{B}_{\text{Bachelors+Male}}$, $\mathcal{B}_{\text{Some-college+Male}}$, $\mathcal{B}_{\text{High-School+Male}}$. There are stark differences in label proportions between the groups, notably in favor of males.

For the SVM using *"high vs. low"* baseline, we try two different class assignments. We start from Bachelors vs. everyone else. (It could seem more natural to take, for example, $\mathcal{B}_{\text{Bachelors+Male}}$ as the "high" bag and $\mathcal{B}_{\text{High-School+Female}}$ as "low", but this results in too small a sample). The baseline performed relatively well (Table 2) due to good class separation. However, when we tried females vs. males, performance of our method remained stable (with highest accuracy), but the baseline suffered a drastic drop (Table 2). This highlights the difficulty of using this baseline when using multiple bags based on richer information: it is not immediately clear how to create two well-separated classes. On the other hand, our method naturally compares groups based on given constraints.

More Baselines. Finally, we report classification accuracy on the same dataset, taken from [20]. The authors create two artificial bags, one retaining original label proportions and another containing only one class. With these bags, their method

Table 2. Census results for different sets of bags. Our method outperforms the naive SVM baseline, and rivals a supervised SVM with many labeled examples.

Method	Education bags	Edu + Gender
Bag constraints	0.75	0.77
"high vs. low" SVM (Bachelors vs. other)	0.52	0.6
"high vs. low" SVM (Female vs. Male)	-	0.38
Supervised SVM	0.75 (900 labels)	0.77 (900 labels)

(using known proportions) achieved 0.81 accuracy. They also report results for Kernel Density Estimation (0.75), Discriminative Sorting – a supervised method (0.77), MCMC sampling (0.81), and a baseline of predicting the major class (0.75). Our method achieves comparable performance despite having much less information on label proportions, fewer features, and using more realistic bags.

4.3   Sensitivity Analysis

In this section we give a short demonstration of how the tightness of constraints could affect model performance. We create artificial bags and vary the tightness of some constraints, reporting accuracy. This is a preliminary study, serving to illustrate some of the different factors that come into play.

We use the **20 Newsgroups** dataset [2] containing approximately 20,000 posts across 20 different newsgroups. Some of the newsgroups are closely related (e.g., comp.sys.ibm.pc.hardware and comp.sys.mac.hardware), while others are further apart (rec.sport.hockey and sci.space). The task is to classify messages according to the newsgroup to which they were posted.

We assume predefined bags and vary constraints on label proportions within and between bags. We do not use any labeled data at training time.

We examine three binary classification tasks, between different categories of posts: *space* vs. *medicine*, *ibm.pc* vs. *mac*, and *hockey* vs. *baseball*. For each of these binary classification tasks, we create six bags of training instances $\mathcal{B} = \{\mathcal{B}_1, \mathcal{B}_2, \ldots \mathcal{B}_6\}$. The sizes of each bag are $|\mathcal{B}_1| = |\mathcal{B}_2| = 200, |\mathcal{B}_3| = |\mathcal{B}_4| = 50, |\mathcal{B}_5| = |\mathcal{B}_6| = 100$. We thus use only 650 instances in this case – about half of the 1187 in the training set. The real label proportions within each bag are $p_1 = p_2 = 0.5, p_3 = p_4 = 0.3, p_5 = p_6 = 0.2$.

We test the effects of three different types of constraints, corresponding to common types of aggregate information:

- **Upper bounds on bag proportions:** Let k_{max} be the index of the bag with the highest proportion. We assume an upper bound multiplicative factor only on this bag: $p_{k_{max}} \times u_m$, where we control factor u_m.
- **Lower bounds:** For each true p_k, we take as a lower bound $l_p * p_k$.
- **Bag difference bounds:** For each true $p_{k_1} - p_{k_2}$ such that $p_{k_1} \geq p_{k_2}$, we lower-bound the difference with $l_d \times (p_{k_1} - p_{k_2})$.

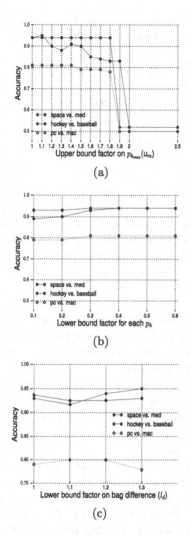

(a)

(b)

(c)

Fig. 3. Constraint effects. Accuracy results on a validation set: (a) Varying upper-bound factor on highest p_k (b) Varying individual lower-bound factor (c) Varying lower bound on bag differences. Results remain fairly robust (with fluctuations due to small-sample noise). The graph stops abruptly where constraints are no longer feasible.

Figure 3 shows the results of our experiments. In our initial setting, we take a fairly loose configuration of constraints to test our method's flexibility: $l_d = 1$ (no lower bound at all for bag differences), $l_p = 0.5$, and $u_m = 1$. In each experiment we vary one factor, keeping the others fixed: (a) upper bound on $p_{k_{max}}$, (b) individual lower bound, (c) lower bound on bag differences.

Notable in Fig. 3 is the overall robustness of the method to misspecified constraints. As u_m is gradually increased, performance remains overall stable for a long stretch (Fig. 3a). However, when u_m reaches extremely large values, the upper bound on $p_{k_{max}}$ becomes too loose (reaching 1) and robustness collapses. Increasing the lower bound on individual p_k slightly improves results, by tightening constraints (Fig. 3b). Results remain fairly robust to overestimating the lower bound on bag differences by increasing l_d, with fluctuations due to small-sample noise (Fig. 3c). The graph stops abruptly at $l_d = 1.3$ since beyond that point constraints are no longer feasible.

Finally, we compare results to the baselines of the previous section. For our method, we fix $u_m = 1, l_p = 0.5, l_d = 1.33$ (with no upper bound on bag differences, as in previous sections). For the *SVM using "high vs. low"* baseline, we take bags $\mathcal{B}_1, \mathcal{B}_2$ as one class, and $\mathcal{B}_5, \mathcal{B}_6$ as the other (adding $\mathcal{B}_3, \mathcal{B}_4$ led to inferior results). Our method outperforms this naive baseline, and also competes with supervised SVM trained on considerable amounts of labeled examples (Table 3). Given fewer labels, supervised SVM was inferior to our label-free method.

4.4 Simulation Study

To further test the behavior of our algorithm, we conduct simulation studies on synthetic data. We use the built-in simulation function *make_classification*

Table 3. 20 newsgroups results.

Method	med-space	pc-mac	baseball-hockey
Bag constraints	**0.94**	**0.81**	**0.94**
"high vs. low" SVM	0.82	0.62	0.64
Supervised SVM	110 labels (0.93)	95 labels (0.78)	140 labels (0.94)

provided in python package scikit-learn [19] to generate data for a binary classification problem. We create three equally-sized bags of instances $\mathcal{B}_1, \mathcal{B}_2, \mathcal{B}_3$ for our training set, with label proportions p_1, p_2, p_3, respectively. We vary bag sizes $|\mathcal{B}_k|$ and proportions p_k, as well as the number of features (*n_features*), number of informative features (*n_informative*), and class separation (*class_sep*).

We apply our cross-validation procedure to select $\mathcal{C}$, using 3 folds. We observe some typical behaviors, such as accuracy improvement with growing sample size. For instance, fixing *n_features*=20, *n_informative* = 1 and $p_1 = 0.4, p_2 = 0.3, p_3 = 0.2$, mean accuracy increases from 0.65 with $|\mathcal{B}_k| = 500$, to 0.77 with $|\mathcal{B}_k| = 1000$.

Accuracy suffered with smaller gaps between bag proportions p_k. However, with increasing sample size our algorithm got better at handling minuscule differences between p_k. For example, fixing $p_1 = 0.4, p_2 = 0.35, p_3 = 0.33$, mean accuracy increases from 0.6 with $|\mathcal{B}_k| = 500$ to 0.65 with $|\mathcal{B}_k| = 1000$, and further increases to 0.67 with $|\mathcal{B}_k| = 1500$.

Finally, we expect that labeled instances can improve performance, helping to counter bags that are very noisy or constraints that are not sufficiently tight. Preliminary experiments suggest that labeled instances can improve accuracy, but a comprehensive study of this effect is beyond the scope of this paper.

5   Exploratory Analysis

In previous sections we tackled classification problems with a clear objective. In this section our users have no specific classification in mind, but rather are interested in exploring the data. A sub-field within clustering allows users to guide the formation of clusters, usually in the form of pairwise constraints on instances (forcing data points to belong to the same cluster or to different clusters). A recent approach uses a maximum-margin framework [28], which extends the supervised large margin theory (such as SVMs) to an unsupervised setting. Similarly, we adapt our method to the exploratory setting. Rather than using instance-level constraints on cluster membership, we use *ranking* constraints based on prior knowledge – or on hypotheses we would like to explore.

We used the **Geo-tagged tweets** dataset, containing 377616 messages from 9475 geo-located microblog users over one week in March 2010 [9]. The user base is likely dominantly composed of teens and young adults (as some of the examples below will make clear). We combine all tweets for each user, and reverse-geocode the GPS coordinates to obtain the corresponding state.

Table 4. Geo-tagged tweets. For each set of geographic constraints, we show some of the top positive and negative words resulting from running our method.

Constraints	Positive terms	Negative terms
French > English: Quebec > Texas	je, est, et, le, pour	Houston, Texas, dallas, bro, tryna, boo
East Coast > West Coast: CA > NY, CA > PA, WA > NY, WA > PA	hella, coo, fasho, af, la, cali, san, washington	deadass, niggas, skool, wassup, dis, dat, philly, crib, lml, nah, dey, den
Ranking by religiosity: Alabama > Florida > New York	thank, easter, pray, road, trip, drove, loving, relationship, spring, folks, happy, dreams, laugh, friend	mad, bitches, neva, dis, dat, niggas,ova, spanish, girls, crazy, party, fun, high, dead

The dataset was used in [9] to analyze regional dialects. The authors used a cascading topic model to model geographic topic variation. The observed output of the generative process includes the texts and GPS coordinates of each user. We pursue this line of exploration too, but rather than positing a generative model of language, we investigate how various constraints on differences between geographic locations interact with dialect.

In Table 4, we show some of the constraints we explored and the resulting top positive and negative words. We start with a simple check with two bags $\mathcal{B}_{Quebec} \succeq \mathcal{B}_{Texas}$, combining tweets from Quebec and Texas, respectively. We discover obvious differences in language, with strong positive weights corresponding to French words and negative weights to English.

An ordinary classifier would likely recover similar results, as would standard unsupervised clustering algorithms. However, our method allows to pursue richer, more *expressive* constraints. First, we look into the difference between the East Coast and West Coast by imposing pairs of constraints such as $\mathcal{B}_{California} \succeq \mathcal{B}_{New York}$, $\mathcal{B}_{California} \succeq \mathcal{B}_{Pennsylvania}$. We recover various results previously highlighted by [9], such as the use of the slang terms "fasho" (for sure) "coo" (cool), "hella" in the West Coast, and "deadass", "wassup" and "niggas" in the East Coast. Our results agree with findings by [9,10], as well as suggest some potential new findings.

Finally, we look at a set of more expressive constraints, aiming to recover difference based on religiosity (or at least sociological confounders). We take states from the top, middle and bottom of a list of US states ranked by percentage of self-reported religiosity[1], and build sets of constraints that reflect this ordering. For instance, in Table 4, we show results for $\mathcal{B}_{Alabama} \succeq \mathcal{B}_{Florida} \succeq \mathcal{B}_{New York}$. Note that using such information in a standard classifier is unnatural. It is not

---

[1] https://en.wikipedia.org/wiki/List_of_U.S._states_by_religiosity.

clear how to construct the classes, and different splits could lead to very different results. Again, this artificial splitting is not required by our method.

We removed terms not in the wordnet [11] lexicon to mitigate the effects of local vernacular and highlight deeper differences. The differences in language are quite striking. As we traverse from Alabama to Florida to New York, discourse shifts from words such as "glad", "loving", "happy", "dreams", "easter" and "pray", to words including "mad", "bitches", "crazy", "party", "fun", "high" and other more profane content we spared from the reader. Similar results were obtained for other state tuples (e.g., Texas instead of Alabama).

Note that our method can be used for formulating new hypotheses. To test the hypotheses, more experiments (and often more data collection) are needed. We leave it up to sociologists to provide deeper interpretation of these results.

Our goal in this section was to use coarse prior information (in the form of relative rankings) for exploring a dataset. We note that the problem could be tackled with other approaches, such as topic models or classification. However, classification models assume a much stronger discriminative pattern or signal than taking a softer, weakly-supervised approach that seeks a direction (weight vector $\mathbf{w}$) along which one bag of instances is ranked higher than another. While clustering with pairwise memberships constraints is well-studied, we demonstrate clustering with expressive pairwise *ranking* constraints over sets. Many real-world settings naturally lend themselves to this formulation.

6 Discussion and Criticism

One clear practical issue with our method is the source of the constraints. We have illustrated several real-world cases where it is plausible to attain rough constraints on label proportions within and between groups of instances. In previous work [20], it is suggested that practitioners could sample from bags of instances to estimate label frequencies (e.g., in spam classification tasks). However, accurate estimations might require extensive sampling, exacting high costs. We thus relax this rather strong assumption, and propose that in many cases, it is possibly enough to get rough estimates. For example, after sampling 10 instances, we might observe 9 positives and only one negative, and rather conservatively declare "$\mathcal{B}$ should have more than 50 % positives". This sort of statement could of course be made more rigorous with probabilistic considerations (e.g., confidence intervals). We have demonstrated that even with considerably mis-specified constraints, we are still able to achieve good performance across various domains.

Furthermore, external sources of knowledge could be used to construct these constraints, such as previous surveys. In many cases taking exact figures from surveys (such as political polls) and expecting them to accurately reflect the distribution in new data is not realistic. This is the case, for instance, when looking at national political polls and wishing to extrapolate from them to new very different socio-demographic slices, such as Twitter users. Here too, we could use this external knowledge to *approximately* guide our model, rather than dictate precise hard proportions the model should match.

7   Related Work

There is a large body of work that is related to our problem.

Multiple Instance Learning. The field of *Multiple Instance Learning* (MIL) generally assumes instances come in "bags", each associated with a label modeled as a function of latent instance-level labels, which can be seen as a form of weak supervision. MIL methods vary by the assumptions made on this function. For a comprehensive review of assumptions and applications, see [7,12]. Most work in MIL focuses on making bag-level predictions rather than for individual instances. Recently, [15] used a convolutional neural network to predict labels for sentences given document-level labels.

Learning from Proportions. A niche within MIL which has seen growing interest and is closely-related to this paper, is concerned with predicting instance-level labels from known label proportions given for each bag. [20] assume to be given bags of unlabeled examples, each bag with *known* label proportions. Their method is based on estimating bag-means using given label proportions. The authors provide examples for scenarios in which such information could be available. In [21], the authors represent each bag with its mean, and model the known class proportions based on this representative "super-instance" with an SVM method, showing superior performance over [20]. In [27], instance-level labels are explicitly modeled to overcome issues the authors raise with representing bags with their means, such as when data distribution has high variance. The fundamental property these and other approaches share is that bag proportions are assumed be known or easily estimated, an assumption we relax.

Classification with Weak Signals. We applied our model to the problem of text classification when little or no labels are available but only a weaker signal. A vast amount of literature has tackled similar scenarios over the years, using tools from semi-supervised [6,14] active [16,22,24] and unsupervised [3] learning. Druck et al. [8] apply generalized expectation feature-labeling (GE-FL) approaches, using "labeled features" given by an oracle that encode knowledge such as "the word puck is a strong indicator of hockey". In practice, a Latent Dirichlet Allocation (LDA) [5] topic model is applied to the data to select top features per topic, for which a user provides labels. [23] propose a semi-supervised + active-learning method, with a human-in-the-loop who provides both feature-level and instance-level labels. We are also able to use labeled instances to refine the learning process, allowing for a trade-off between the user's trust in the (typically few) labeled instances available, and prior knowledge on bag proportions.

Similar to the above work on learning from labeled proportions, [25] considers a classification problem with no access to labels for individual training examples, but only average labels over subpopulations. They frame the problem as weakly-supervised clustering. When using our method for exploratory analysis, it can also be seen as a weakly-supervised clustering algorithm, using information on partial ordering between bags rather than assuming known proportions, within a max-margin framework (somewhat akin to clustering using maximum-margin as in [28]). The seminal work of [26] uses side-information for clustering in the form of pairwise constraints on cluster membership (pairwise similarity). Much work has since been done along these lines. We incorporate pairwise constraints in our maximum-margin approach, though with pairs representing bags of instances, and partial ordering with respect to relative label proportions.

Robust Optimization. Finally, robust optimization [4] research deals with uncertainty-affected optimization problems, by optimizing for the *worst-case* value of parameters. Because of its worst-case design, robust optimization can do poorly when the constraints are not tight. Our method, on the other hand, is designed to handle rough estimates and loose constraints.

8 Conclusions and Future Work

In this paper we proposed a new learning setting where we have bags of unlabeled instances with loose constraints on label proportions and difference between bags. Thus, we relax the unrealistic assumption of known bag proportions.

We formalized the problem as a bi-convex optimization problem and proposed an efficient algorithm. We showed how, surprisingly, our classifier performs well using very little input. We also demonstrated how the algorithm can guide exploratory classifications.

We have empirically studied the behavior of our algorithm under different types of constraints. One direction for future work is to analytically understand, for instance, how constraint tightness affects performance, obtain convergence guarantees, and provide generalization error bounds. This, in turn, could perhaps lead to better algorithms with theoretical justifications.

Finally, the relative-proportions setting is very natural, and can be found in various domains. We believe that this line of work will have interesting implications regarding privacy and anonymization of data – in particular, the amount of information one can recover using only weak, aggregated signals.

Acknowledgments. The authors thank the anonymous reviewers and Ami Wiesel for their helpful comments. Dafna Shahaf is a Harry&Abe Sherman assistant professor, and is supported by ISF grant 1764/15 and Alon grant.

References

1. https://archive.ics.uci.edu/ml/datasets/Adult/
2. http://qwone.com/~jason/20Newsgroups/
3. Aggarwal, C.C., Zhai, C.X.: A survey of text clustering algorithms. In: Aggarwal, C.C., Zhai, C.X. (eds.) Mining Text Data, pp. 77–128. Springer, New York (2012)
4. Ben-Tal, A., El Ghaoui, L., Nemirovski, A.: Robust Optimization. Princeton University Press, Princeton (2009)
5. Blei, D.M., Ng, A.Y., Jordan, M.I.: Latent Dirichlet allocation. J. Mach. Learn. Res. **3**, 993–1022 (2003)
6. Chapelle, O., Schölkopf, B., Zien, A., et al.: Semi-supervised Learning. MIT Press, Cambridge (2006)

7. Cheplygina, V., Tax, D., Loog, M.: On classification with bags, groups, sets. arXiv preprint arXiv:1406.0281 (2014)
8. Druck, G., Mann, G., McCallum, A.: Learning from labeled features using generalized expectation criteria. In: SIGIR 2008, pp. 595–602 (2008)
9. Eisenstein, J., Brendan, O., Smith, N., Xing, E.P.: A latent variable model for geographic lexical variation. In: Proceedings of the Conference on Empirical Methods in Natural Language Processing, Cambridge, MA (2010)
10. Eisenstein, J., Smith, N.A., Xing, E.P.: Discovering sociolinguistic associations with structured sparsity. In: Proceedings of the 49th Annual Meeting of the Association for Computational Linguistics: Human Language Technologies (2011)
11. Fellbaum, C.: WordNet: An Electronic Lexical Database. Bradford Books (1998)
12. Foulds, J., Frank, E.: A review of multi-instance learning assumptions. Knowl. Eng. Rev. **25**, 125 (2010)
13. Hu, Y., Li, M., Yu, N.: Multiple-instance ranking: learning to rank images for image retrieval. In: Proceedings of CVPR, p. 18 (2008)
14. Joachims, T.: Transductive inference for text classification using support vector machines. In: ICML 1999, pp. 200–209 (1999)
15. Kotzias, D., Denil, M., de Freitas, N., Smyth, P.: From group to individual labels using deep features. In: Proceedings of the 21th ACM SIGKDD International Conference on Knowledge Discovery and Data Mining, KDD 2015 (2015)
16. Li, L., Jin, X., Pan, S.J., Sun, J.-T.: Multi-domain active learning for text classification. In: Proceedings of the 18th ACM SIGKDD International Conference on Knowledge Discovery and Data Mining, pp. 1086–1094. ACM (2012)
17. Pang, B., Lee, L.: A sentimental education: sentiment analysis using subjectivity. In: Proceedings of ACL, pp. 271–278 (2004)
18. Pang, B., Lee, L.: Opinion mining and sentiment analysis. Found. Trends Inf. Retrieval **2**(1–2), 1–135 (2008)
19. Pedregosa, F., Varoquaux, G., Gramfort, A., Michel, V., Thirion, B., Grisel, O., Blondel, M., Prettenhofer, P., Weiss, R., Dubourg, V., Vanderplas, J., Passos, A., Cournapeau, D., Brucher, M., Perrot, M., Duchesnay, E.: Scikit-learn: Machine learning in python. J. Mach. Learn. Res. **12**, 2825–2830 (2011)
20. Quadrianto, N., Smola, A.J., Caetano, T.S., Le, Q.V.: Estimating labels from label proportions. J. Mach. Learn. Res. **10**, 2349–2374 (2009)
21. Rueping, S.: SVM classifier estimation from group probabilities. In: Proceedings of the 27th International Conference on Machine Learning (ICML 2010) (2010)
22. Settles, B.: Active learning literature survey. University of Wisconsin, Madison, 52(55-66):11
23. Settles, B.: Closing the loop: fast, interactive semi-supervised annotation with queries on features and instances. In: Proceedings of the Conference on Empirical Methods in Natural Language Processing (EMNLP), pp. 1467–1478 (2011)
24. Tong, S., Koller, D.: Support vector machine active learning with applications to text classification. J. Mach. Learn. Res. **2**, 45–66 (2002)
25. Wager, S., Blocker, A., Cardin, N.: Weakly supervised clustering: learning fine-grained signals from coarse labels. Ann. Appl. Stat. **9**(2), 801–820 (2015)
26. Xing, E.P., Jordan, M.I., Russell, S.J., Ng, A.Y.: Distance metric learning with application to clustering with side-information. In: NIPS 2003. MIT Press (2003)
27. Yu, F., Liu, D., Kumar, S., Jebara, T., Chang, S.: $\propto$-SVM for learning with label proportions. In: ICML 2013 (2013)
28. Zhou, G.-T., Lan, T., Vahdat, A., Mori, G.: Latent maximum margin clustering. In: Burges, C., Bottou, L., Welling, M., Ghahramani, Z., Weinberger, K. (eds.) Advances in Neural Information Processing Systems 26, pp. 28–36 (2013)

An Efficient Algorithm for Mining Frequent Sequence with Constraint Programming

John O.R. Aoga[1(✉)], Tias Guns[2], and Pierre Schaus[1]

[1] UCLouvain, ICTEAM, Louvain-la-Neuve, Belgium
{john.aoga,pierre.schaus}@uclouvain.be
[2] DTAI Research Group, KU Leuven, Leuven, Belgium
tias.guns@cs.kuleuven.be

Abstract. The main advantage of Constraint Programming (CP) approaches for sequential pattern mining (SPM) is their modularity, which includes the ability to add new constraints (regular expressions, length restrictions, etc.). The current best CP approach for SPM uses a global constraint (module) that computes the projected database and enforces the minimum frequency; it does this with a filtering algorithm similar to the PrefixSpan method. However, the resulting system is not as scalable as some of the most advanced mining systems like Zaki's cSPADE. We show how, using techniques from both data mining and CP, one can use a generic constraint solver and yet outperform existing specialized systems. This is mainly due to two improvements in the module that computes the projected frequencies: first, computing the projected database can be sped up by pre-computing the positions at which a symbol can become unsupported by a sequence, thereby avoiding to scan the full sequence each time; and second by taking inspiration from the *trailing* used in CP solvers to devise a backtracking-aware data structure that allows fast incremental storing and restoring of the projected database. Detailed experiments show how this approach outperforms existing CP as well as specialized systems for SPM, and that the gain in efficiency translates directly into increased efficiency for other settings such as mining with regular expressions. The data and software related to this paper are available at http://sites.uclouvain.be/cp4dm/spm/.

1 Introduction

Sequence mining is a widely studied problem concerned with discovering subsequences in a dataset of given sequences, where each (sub) sequence is an ordered list of symbols. It has applications ranging from web usage mining, text mining, biological sequence analysis and human mobility mining [7]. We focus on the problem of finding patterns in sequences of individual symbols, which is the most commonly used setting in those applications.

In recent years, constraint programming (CP) has been proposed as a general framework for pattern mining [3–5,8]. The main benefit of CP-based approaches over dedicated algorithms is that it is *modular*. In a CP framework, a problem is

© Springer International Publishing AG 2016
P. Frasconi et al. (Eds.): ECML PKDD 2016, Part II, LNAI 9852, pp. 315–330, 2016.
DOI: 10.1007/978-3-319-46227-1_20

expressed as a set of constraints that the solutions must satisfy. Each such a constraint can be seen as a *module*, and can range from being as simple as ensuring that a subsequence does not contain a certain symbol at a certain position, up to computing the frequency of a pattern in a database. This modularity allows for flexibility, in that certain constraints such as symbol restrictions, length, regular expressions etc. can easily be added and removed to existing problems. Another advantage is that improving the efficiency of one constraint will improve the efficiency of all problems involving this constraint.

However, this increased flexibility can come at a cost. Negrevergne et al. [8] have shown that a fine-grained modular approach to sequence mining can support any type of constraints, including gap and span constraints and any quality function beyond frequency, but that this is not competitive with state-of-the-art specialized methods. On the other hand, they showed that by using a global constraint (a module) that computes the pseudo-projection of the sequences in the database similar to PrefixSpan [10], this overhead can be reduced. Kemmar et al. [5,6] propose to use a single global constraint for pseudo-projection as well as frequency counting over all sequences. This approach is much more efficient than the one of [8] that uses many reified constraints. These CP-based methods obtain reasonable performance, especially for mining under regular expressions. While they improve scalability compared to each-other, they are not on par with some of the best specialized systems such as Zaki's cSpade [18]. In this work, we show for the first time that a generic CP system with a custom global constraint can outperform existing specialised systems including Zaki's.

The global constraint improves on earlier global constraints for sequence mining by combining ideas from both pattern mining and constraint programming as follows: first, we improve the efficiency of computing the projected database and the projected frequency using last-position lists, similar to the LAPIN algorithm [16] but within a PrefixSpan approach. Second, we take into account not just the efficiency of computing the projected database, but also that of storing and restoring it during depth-first search. For this we use the *trailing* mechanism from CP solvers to avoid unnecessary copying of the pseudo-projection data structure. Such an approach is in fact applicable to any depth-first algorithm in pattern mining and beyond.

By combining the right ingredients from both research communities in a novel way, we end up with an elegant algorithm for the projected frequency computation. When added as a module to a generic CP solver, the resulting system improves both on previous CP-based sequence miners as well as state-of-the-art specialized systems. Furthermore, we show that by improving this one module, these improvements directly translate to other problems using this module, such as regular-expression based sequence mining.

2   Related Works

We review specialized methods as well as CP-based approaches. A more thorough review of algorithmic developments is given in [7].

Specialized Methods. Introduced by Srikant and Agrawal [1], GSP was the first approach to extract sequential patterns from a sequential database. Many works have improved on this apriori-based method, typically employing depth-first search. A seminal work is that of PrefixSpan [10]. A prefix in this context is a sequential pattern that can only be extended by appending symbols to it. Given a prefix, one can compute the *projected database* of all suffixes of the sequences that have the prefix as a subsequence. This projected database can then be used to compute the frequency of the prefix and of all its 1-extensions (projected frequency). A main innovation in PrefixSpan is the use of a *pseudo-projected* database: instead of copying the entire (projected) database, one only has to maintain pointers to the position in each sequence where the prefix matched.

Alternative methods such as SPADE [18] and SPAM [2] use a vertical representation of the database, having for each symbol a list of sequence identifiers and positions at which that symbol appears.

Yang et al. have shown [17] that algorithms with either data representation can be improved by precomputing the last position of each symbol in a sequence. This can avoid scanning the projected database, as often the reason for scanning is to know whether a symbol still appears in the projected sequence.

The standard sequence mining settings have been extended in a number of directions, including user-defined constraints on length or on the gap or span of a sequence such as in the cSPADE algorithm [18], closed patterns [15] and algorithms that can handle regular expression constraints on the patterns such as SMA [14]. These constraints are typically hard-coded in the algorithms.

CP-Based Approaches for SPM. CP-based approaches for sequence mining are gaining interest in the CP community. Early work has focused on fixed-length sequences with wildcards [3]. More generally, [8] proposed two approaches: a full decomposition of the problem in terms of constraints and an approach using a global constraint to construct the pseudo-projected database similar to PrefixSpan. It uses one such constraint for each sequence. Kemmar et al. [6] propose to gather all these constraints into a unique global constraint to reduce the overhead of the multiple constraints. They further showed how the constraint can be modified to take a maximal gap constraint into account [5].

3   Sequential Pattern Mining Background

This section introduces the necessary concepts and definitions of sequence mining and constraint programming.

3.1   Sequence Mining Background

Let $I = \{s_1, \ldots, s_N\}$ be a set of N symbols. In the remaining of the paper when there is no ambiguity a symbol is simply denoted by its identifier i with $i \in \{1, \ldots, N\}$.

Table 1. A sequence database SDB_1 and list of last positions.

sid	Sequence	lastPosList	lastPosMap
sid_1	$\langle ABCBC \rangle$	[(C,5),(B,4),(A,1)]	{A→1, B→4, C→5, D→0}
sid_2	$\langle BABC \rangle$	[(C,4),(B,3),(A,2)]	{A→2, B→3, C→4, D→0}
sid_3	$\langle AB \rangle$	[(B,2),(A,1)]	{A→1, B→2, C→0, D→0}
sid_4	$\langle BCD \rangle$	[(D,3),(C,2),(B,1)]	{A→0, B→1, C→2, D→3}

(1) SDB, (2) lastPosList, (3) lastPosMap

Definition 1. *Sequence and sequence database.* *A sequence $s = \langle s_1 s_2 \ldots s_n \rangle$ over I is an ordered list of (potentially repeating) symbols s_j, $j \in [1, n]$ with $\#s = n$ the length of the sequence s. A set of tuples (sid,s) where sid is a sequence identifier and s a sequence, is called sequence database (SDB).*

Example 1. Table 1 shows an example SDB_1 over symbols $I = \{A, B, C, D\}$. For the sequence $s = \langle BABC \rangle$: $\#s = 4$ and $s_1 = B, s_2 = A, s_3 = B, s_4 = C$.

Definition 2. *Sub-sequence ($\preceq$),* super-sequence. *A sequence $\alpha = \langle \alpha_1 \ldots \alpha_m \rangle$ is called a sub-sequence of $s = \langle s_1 s_2 \ldots s_n \rangle$ and s is a super-sequence of α iff (i) $m \leq n$ and (ii) for all $i \in [1, m]$ there exist integers j_i s.t. $1 \leq j_1 \leq \cdots \leq j_m \leq n$, such that $\alpha_i = s_{j_i}$.*

Example 2. For instance $\langle BD \rangle$ is a sub-sequence of $\langle BCCD \rangle$, and inversely $\langle BCCD \rangle$ is the super-sequence of $\langle BD \rangle$: $\langle BD \rangle \preceq \langle BCCD \rangle$.

Definition 3. *Cover, Support, Pattern, Frequent Pattern.* *The cover of sequence p in SDB, denoted by $cover_{SDB}(p)$, is the subset of sequences in SDB that are a super-sequence of p, i.e. $cover_{SDB}(p) = \{(sid, s) \in SDB \mid p \preceq s\}$. The support of p in SDB, denoted by $sup_{SDB}(p)$, is the number of super-sequences of p in SDB: $sup_{SDB}(p) = \#cover_{SDB}(p)$. Any sequence p over symbols in I can be a pattern, and we call a pattern frequent iff $sup_{SDB}(p) \geq \theta$, where θ is a given minimum support threshold.*

Example 3. Assume that $p = \langle BC \rangle$ and $\theta = 2$, $cover_{SDB_1}(p) = \{(sid_1, \langle ABCBC \rangle), (sid_2, \langle BABC \rangle), (sid_4, \langle BCD \rangle)\}$ and hence $sup_{SDB_1}(p) = 3$. Hence, p is a frequent pattern for that given threshold.

The sequential pattern mining (SPM) problem, first introduced by Agrawal and Srikant [1], is the following:

Definition 4. *Sequential Pattern Mining (SPM).* *Given an minimum support threshold θ and a sequence database SDB, the SPM problem is to find all patterns p such that $sup_{SDB}(p) \geq \theta$.*

Our method uses the idea of a *prefix* and *prefix-projected* database for enumerating the frequent patterns. These concepts were first introduced in the seminal paper that presented the *PrefixSpan* algorithm [10].

Definition 5. Prefix, prefix-projected database. *Let* $\alpha = \langle \alpha_1 \ldots \alpha_m \rangle$ *be a pattern. If a sequence* $\beta = \langle \beta_1 \ldots \beta_n \rangle$ *is a super-sequence of* α: $\alpha \preceq \beta$, *then the prefix of* β *w.r.t.* α *is the smallest prefix of* β *that is still a super-sequence of* α: $\langle \beta_1 \ldots \beta_j \rangle$ *s.t.* $\alpha \preceq \langle \beta_1 \ldots \beta_j \rangle$ *and* $\nexists j' < j : \alpha \preceq \langle \beta_1 \ldots \beta_{j'} \rangle$. *The sequence* $\langle \beta_{j+1} \ldots \beta_n \rangle$ *is called the suffix and it represents the prefix-projection obtained by projecting the prefix away. A prefix-projected database of a pattern* α, *denoted by* $SDB|_\alpha$, *is the set of prefix-projections of all sequences in* SDB *that are super-sequences of* α.

Example 4. In SDB_1, assume $\alpha = \langle A \rangle$, then $SDB_1|_\alpha = \{(sid_1, \langle BCBC \rangle),$ $(sid_2, \langle BC \rangle), (sid_3, \langle B \rangle)\}$.

We say that the *prefix-projected frequency* of the symbols I in a prefix-projected database is the number of sequences in which these symbols appear. For $SDB_1|_{\langle A \rangle}$ the prefix-projected frequencies are $A : 0, B : 3, C : 2, D : 0$.

The PrefixSpan algorithm solves the SPM problem by starting from the empty pattern and extending this pattern using depth-first search. At each step it extends a pattern by a symbol and projects the database accordingly. The appended symbol is removed on backtrack. It hence grows the pattern incrementally, which is why it is called a pattern-growth method. A frequent pattern in the projected database is also frequent in the original database.

There are two important considerations for the efficiency of the method. The first is that one does not have to consider during search any symbol that is not frequent in the prefix-projected database. The second is that of *pseudo-projection*: to store the prefix-projected database during the depth-first search, it is not necessary to store (and later restore) an entire copy of the projected database. Instead, one only has to store for each sequence the pointer to the position j that marks the end of the prefix in that sequence (remember, the prefix of α in β is the smallest prefix $\langle \beta_1 \ldots \beta_j \rangle \succeq \alpha$).

Example 5. The projected database $SDB_1|_\alpha = \{(sid_1, \langle BCBC \rangle), (sid_2, \langle BC \rangle),$ $(sid_3, \langle B \rangle)\}$ can be represented as a pseudo-projected database as follows: $\{(sid_1, 2), (sid_2, 3), (sid_3, 2)\}$.

3.2 Constraint Programming Background

CP is a powerful declarative paradigm to solve combinatorial satisfaction and optimization problems (see, e.g., [12]). A CP problem (V, D, C) is defined by a set of variables V with their respective domains D (the values that can be assigned to a variable), and a set of constraints C on these variables. A solution of a CP problem is an assignment of the variables to a value from its domain, such that all constraints are satisfied.

At its core, CP solvers are depth-first search algorithms that iterate between *searching* over unassigned variables and *propagating* constraints. Propagation is the act of letting the constraints in C remove unfeasible values from the domains of its variables. This is repeated until *fixed-point*, that is, no more constraint can remove any unfeasible values. Then, a search exploration step is taken by choosing an unassigned variable and assigning it to a value from its current domain, after which propagation is executed again.

Example 6. Let there be 2 variables x, y with domains $D(x) = \{1, 2, 3\}, D(y) = \{3, 4, 5\}$. Then constraint $x + y \geq 5$ can derive during propagation that $1 \notin D(x)$ because the lowest value y can take is 3 and hence $x \geq 5 - \min(D(y)) \geq 5 - 3 \geq 2$.

Constraints and Global Constraints. Many different constraints and their propagation algorithms have been investigated in the CP community. This includes logical and arithmetic ones like the above, up to constraints for enforcing regular expressions or graph theoretic properties. A constraint that enforces some non-trivial or application-dependent property is often called a *global constraint*. For example, [8] introduced a global constraint for the pseudo-projection of a single sequence, and [5] for the entire projected frequency subproblem.

State Restoration in CP. In any depth-first solver, there must be some mechanism to store and restore some *state*, such that computations can be performed incrementally and intermediate values can be stored. In most of the CP solvers[1] a general mechanism, called *trailing* is used for storing and restoring the state (on backtrack) [13]. Externally, the CP solvers typically expose some "reversible" objects whose values are automatically stored and restored on the trail when they change. The most important example are the domains of CP variables. Hence, for a variable the domain modifications (*assign, removeValue*) are automatically reversible operations. A CP solver also exposes reversible versions of primitive types such as integers and sets for use within constraint propagators. They are typically used to store incremental computations. CP solvers consist of an efficient implementation of the DFS backtracking algorithm, as well as many constraints that can be called by the fix-point algorithm. The modularity of constraint solvers stems from this ability to add any set of constraints to the fix-point algorithm.

4   Global Constraints for Projected Frequency

We first introduce the basic CP model of frequent sequence mining introduced in [8] and extended in [6]. Then, we present how we improve the computation of the pseudo-projection, followed by the projected frequency counting and pruning.

4.1   Existing methods [6, 8]

As explained before, a constraint model consists of variables, domains and constraints. The CP model will be such that a single solution corresponds to a frequent sequence, meaning that all sequences can be extracted by enumerating all solutions.

---

[1] One notable exception is the Gecode copy-based solver.

Let L be an upper bound on the pattern length, e.g. the length of the longest sequence in the database. The variables used to represent the unknown pattern P is modeled as an array of L integer variables $P = [P_1, P_2, \ldots, P_L]$. Each variable has an initial domain $\{0, \ldots, N\}$, corresponding to all possible symbols identifiers and augmented with an additional identifier 0. The symbol with identifier 0 represents ϵ, the empty symbol. It will be used to denote the end of the sequence in P, using a trailing suffix of such 0's.

Definition 6. *A CP model over P represents the frequent sequence mining problem with threshold θ, iff the following three conditions are satisfied by every valid assignment to P:*

1. $P_1 \neq 0$
2. $\forall i \in \{2, \ldots, L-1\} : P_i = 0 \Rightarrow P_{i+1} = 0$
3. $\#\{(sid, s) \in SDB \ \langle P_1 \ldots P_j \rangle \preceq s\} \geq \theta, \ j = \max(\{i \in \{1 \ldots L\} | P_i \neq 0\})$.

The first requirement states that the sequence may not start with the empty symbol, e.g. no empty sequence. The second requirement enforces that the pattern is in a canonical form such that after the empty symbol, all other symbols are the empty symbol too. Hence, a sequence of length $l < L$ is represented by l non-zero symbols, followed by $L - l$ zero symbols. The last requirement states that the frequency of the non-zero part of the pattern must be above the threshold θ.

Prefix Projection Global Constraint. Initial work [8] proposed to decompose these three conditions into separate constraints, including a dedicated global constraint for the inclusion relation $\langle P_1 \ldots P_j \rangle \preceq s$ for each sequence separately. It used the pseudo-projection technique of PrefixSpan for this, with the projected frequency enforced on each symbol in separate constraints.

Kemmar et al. [6] extended this idea by encapsulating the filtering of all three conditions into one single (global) constraint called `PrefixProjection`. It also uses the pseudo-projection idea of PrefixSpan, but over the entire database. The propagation algorithm for this constraint, as executed when the next unassigned variable P_i is assigned during search, is given in Listing 1.1.

Listing 1.1. PrefixProjection(SDB,P,i,θ)

```
1  // pre: variables ⟨P₁,...,Pᵢ⟩ are bound, SDB is given
2  //        Pᵢ is the new instantiated variable since previous
       call.
3  if (Pᵢ == 0) {
4     foreach (j ∈ {i+1,...,L}) { Pⱼ.assign(0) }
5  } else if (i ≥ 2) {
6     projFreqs = ProjectAndGetFreqs(SDB,Pᵢ,θ)
7     foreach (j ∈ {i+1,...,L})
8        foreach (a ∈ D(Pⱼ))
9           if (a ≠ 0 and projFreqs[a] < θ) { Pⱼ.removeValue(a) }
10 }
```

An initial assumption is that the database SDB does not contain any infrequent symbols, which is a simple preprocessing step. The code is divided in three parts: (i) if P_i is assigned to 0 the remaining P_k with $k > i$ is assigned to 0; else (ii) from the second position onwards (remember that the first position can take any symbol and be guaranteed to be frequent as every symbol is known to be frequent), the projected database and the projected frequency of each symbol is computed; and (iii) all symbols that have a projected frequency below the threshold are removed from the domain of the subsequent pattern variables.

The algorithm for computing the (pseudo) projected database and the projected frequencies of the symbols is given in Listing 1.2. It operates as follows with a the new symbol appended to the prefix of assigned variables since previous call. The first loop at line 2 attempts to discover for each sequence s in the projected database if it can be a sub-sequence of the extended prefix. If yes, this sequence is added to the next projected database at line 5. The second loop at line 9 computes the frequency of each symbol occurring in the projected database but counting it at most once per sequence.

Listing 1.2. ProjectAndGetFreqs(SDB,a,θ)

```
1  PSDB_i = ∅
2  foreach (sid, start) ∈ PSDB_{i-1} {
3     s = SDB[sid]; pos = start
4     while (pos < #s and a ≠ s[pos]) { pos = pos + 1 }
5     if (pos < #s) { PSDB_i = PSDB_i ∪ {(sid, pos)} }
6  }
7  projFreqs[a]=0 ∀a ∈ {1,...,N}
8  if (#PSDB_i ≥ θ) {
9     foreach (sid, start) ∈ PSDB_i {
10        s = SDB[sid]; existsSymbol[b] = false ∀b ∈ {1,...,N}
11        foreach (i ∈ {start,...,#s}) {
12           if (!existsSymbol[s[i]]) {
13              projFreqs[s[i]] = projFreqs[s[i]]+1
14              existsSymbol[s[i]] = true
15           }
16  } } }
17 return projFreqs
```

4.2   Improving Propagation

Although being the state-of-art approach for solving SPM with CP, the filtering algorithm of Kemmar et al. [5] presents room for improvement. We identify four weaknesses and propose solutions to them.

Weakness 1. Databases with long sequences will have a large upper-bound L. For such databases, removing infrequent symbols from all remaining pattern variables P in the loop defined at line 7 of Listing 1.1 can take time. This is not only the case for doing the action, but also for restoring the domains on backtracking. On the other hand, only the next pattern variable P_{i+1} will be considered during search, and in most cases a pattern will never actually be of length L, so all subsequent domain changes are unnecessary. This weakness is a peculiarity of using a fixed-length array P to represent a variable-length sequence. Mining algorithms typically have a variable length representation of the pattern, and hence only look one position ahead. In our propagator we only remove values from the domain of P_{i+1}.

Weakness 2. When computing the projected frequencies of the symbols, one has to scan each sequence from its current pseudo-projection pointer *start* till the end of the sequence. This can be time consuming in case of many repetitions of only a few symbols for example. Thanks to the *lastPosList* defined next, it is possible to visit only the last position of each symbol occurring after *start*. This idea was first introduced in [17] and exploited in the LAPIN family of algorithms.

Definition 7 (Last position list). *For a current sequence s, lastPosList is a sequence of pairs $(symbol, pos)$ giving for each symbol that occurs in s its last position: $pos = \max\{p \leq \#s : s[p] = symbol\}$. The sequence is of length m, the number of distinct symbols in s. This sequence is decreasing according to positions: $lastPosList[i].pos > lastPosList[i+1].pos \; \forall i \in \{1, \ldots, m-1\}$.*

Example 7. Table 1 shows the *lastPosList* sequences for SDB_1. We consider the sequence with sid_1 and a prefix $\langle A \rangle$. The computation of the frequencies starts at position 2, remaining suffix is $\langle BCBC \rangle$. Instead of visiting all the 4 positions of this suffix, only the last two can be visited thanks to the information contained in $lastPosList[sid_1]$. Indeed according to $lastPosList[sid_1][1]$ the maximum last position is 5 (corresponding to the last C). Then according to $lastPosList[sid_1][2]$ the second maximum last position is 4 (corresponding to the last position of symbol B). The third maximum last position is 1 for symbol A. Since this position is smaller than 2 (our initial start), we can stop.

Weakness 3. Related to weakness 2, line 4 in Listing 1.2 finds the new position (pos_s) of a in $SDB[sid]$. This code is executed even if the new symbol no longer appears in that sequence. Currently, the code has to loop over the entire sequence until it reaches the end before discovering this.

Assume that the current position in the sequence s is already larger than the position of the last occurrence of a. Then we immediately know this sequence cannot be part of the projected database. To verify this in $O(1)$ time, we use a *lastPosMap* as follows:

Definition 8 (Last position map of symbols). *For a given sequence s with id sid, lastPosMap[sid] is a map such that lastPosMap[sid][i] is the last position of symbol i in the sequence s. In case the symbol i is not present: $lastPosMap[sid][i] = 0$ (positions are assumed to start at index 1).*

Example 8. Table 1 shows the *lastPosMap* arrays next to SDB_1. For instance for sid_2 the last position of symbol C is 4.

Weakness 4. Listing 1.2 creates a new set $PSDB_i$ to represent the projected database. This projected database is computed many times during the search, namely at least once in each node of the search tree (more if there are other constraints in the fixPoint set). This is a source of inefficiency for garbage collected languages such as Java but also for C since it induces many "slow" system calls such as free and malloc leading to fragmentation of the memory. We propose to store and restore the pseudo-projected databases with reversible vectors making use of CP trailing techniques. The idea is to use one and the same array throughout the search in the propagator, and only maintain the relevant start/stop position during search. Each call to propagate will read from the previous start to stop position, and write after the previous stop position plus store the new start/stop position. The projected databases are thus *stacked* in the array along a branch of the search tree. We implement the pseudo-projected database with two reversible vectors: *sids* and *poss* respectively for the sequence ids and the current position in the corresponding sequences. The position ϕ is the start entry (in *sids* and *poss*) of the current projected database, and φ is the size of the projected database. We thus have the current projected database contained in sub-arrays $sids[\phi, \ldots, \phi + \varphi - 1]$ and $poss[\phi, \ldots, \phi + \varphi - 1]$. In order to make the projected database reversible, ϕ and φ are reversible integers. That is on backtrack to an ancestor node those integers retrieve their previous value and entries of *sids* and *poss* starting from ϕ can be reused.

Example 9. Figure 1 is an example using SDB_1. Initially all the sequences are present $\varphi = 4$ and position is initialized $\phi = 0$. The A-projected database contains sequence $1, 2, 3$ at positions $1, 2, 1$ with $\phi = 4$ and $\varphi = 3$.

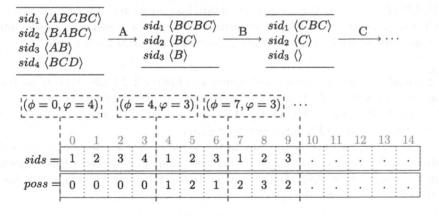

Fig. 1. Reversible vectors technique

Prefix Projection Incremental Counting Propagator (PPIC). Putting all the solutions to the identified weaknesses together, we list the code of the main function of our propagator's in Listing 1.3.

The main loop at line 3 iterates over the previous *(parent)* projected database. In case the sequence at index i in the projected database contains the new symbol at a subsequent position larger or equal to *start*, the matching position is searched and added to the new projected database (at index j of reversible vectors *sids* and *poss*) at line 9. Then the contribution of the sequence to the projected frequencies is computed in the loop at line 11. Only the entries in the lastPosList with position larger than current *pos* are considered (recall that his list is decreasing according to positions). Finally line 17 updates the reversible integers ϕ and φ to reflect the newly computed projected database. Based on these projected frequencies a filtering similar to the one of Listing 1.1 is achieved except that only the domain of the next variable $D(P_{i+1})$ is filtered according to the solution to Weakness 1.

Listing 1.3. ProjectAndGetFreqs(SDB,a,θ,$sids$,$poss$,ϕ,φ)

```
1  projFreqs[b]=0 ∀b ∈ {1,...,N}
2  i = φ;  j = φ + φ;  sup = 0
3  while (i < φ + φ) {
4      sid = sids[i];  pos = poss[i];  s = SDB[sid]
5      if (lastPosMap[sid][a] − 1 ≥ start) {
6          //find the next position of a in s
7          while (pos < #s and a ≠ s[pos]) { pos = pos + 1 }
8          // update projected database
9          sids[j] = sid;  poss[j] = pos + 1;  j = j + 1;  sup = sup + 1
10         // recompute projected frequencies
11         foreach ((symbol, pos_x) in lastPosList[sid]) {
12             if (pos_x ≤ pos) { break }
13             projFreqs[symbol] = projFreqs[symbol] + 1
14     } }
15     i = i + 1
16  }
17  φ = φ + φ;  φ = sup
18  return projFreqs
```

Prefix Projection Decreasing Counting Propagator (PPDC). The key idea of this approach is not to count the projected frequencies from scratch, but rather to *decrement* them. More specifically, when scanning the position of the current symbol at line 7, if *pos* happens to be the last position of a symbol (pos==lastPosMap[sid][s[pos]]) then projFreqs[s[pos]] is decremented. This requires projFreqs to be an array of reversible integers. With this strategy the loop at line 11 disappears, but in case the current sequence is not added to the projected database, the frequencies of all its last symbols occurring after *pos* must also be decremented. This can be done by adding an **else** block to the **if** defined at line 5 that will iterate over the lastPosList and decrement the symbol frequencies.

Example 10. Assume SDB_1. The initial projected frequency array is projFreqs= [A:3,B:4,C:3,D:1]. Consider now the A-projected database

illustrated on Fig. 1. The projected frequency array becomes
`projFreqs=[A:0,B:3,C:2,D:0]`. The entry at A is decremented three times
as *pos* moved beyond its *lastPos* for each of the sequences sid_1, sid_2 and sid_3.
Since sid_4 is removed from the projected database, the frequency of all its last
symbols occurring after *pos* is also decremented, that is for entries B, C and D.

PP-mixed. Both PPID and PPDC approaches can be of interest depending
on the number of removed sequences in the projected database. If the number
of sequences removed is large then PPIC is preferable. On the other hand is
only a few sequences are removed then PPDC can be more interesting. Inspired
from the *reset* idea of [11] the PP-mixed approach dynamically chooses the best
strategy: if $projFreqs_{SDB}(a) < \#PSDB_i/2$ (i.e., more than half of sequences
will be removed) then PPIC is used otherwise PPDC.

4.3 Constraints of SPM

We implemented common constraints such as minimum and maximum pattern
size, symbol inclusion/exclusion, and regular expression constraints. Time constraints (maxgap, mingap, maxspan, etc) are outside the scope of this work: they
change the definition of what a valid prefix is, and hence require changing the
propagator (as in [5]).

5 Experiments

In this section, we report our experimental results on the performance of our
approaches with six real-life datasets[2] and one synthetic (data200k [14]) with
various characteristics shown in Table 2. Sparsity, representing the average of
the number of symbols that appear in each sequence, is a good indicator of how
sparse or dense a dataset is.

Our work is implemented in Scala in OscaR solver [9] and run under JVM
with maximum memory set to 8GB. All our software, datasets and results are

Table 2. Dataset features. Sparsity is equal to $(\frac{1}{\#SDB} \times \sum \frac{\#s}{\#I_{/s}})$

SDB	#SDB	N	avg(#s)	avg(#I/s)	max(#s)	Sparsity	Description
BIBLE	36369	13905	21.64	17.85	100	1.2	Text
FIFA	20450	2990	36.24	34.74	100	1.2	Web click stream
Kosarak	69999	21144	7.98	7.98	796	1.0	Web click stream
Leviathan	5834	9025	33.81	26.34	100	1.3	Text
PubMed	17237	19931	29.56	24.82	198	1.2	Bio-medical text
Data200k	200000	26	50.25	18.25	86	2.8	Synthetic data
Protein	103120	25	482.25	19.93	600	24.2	Protein sequences

---

[2] http://www.philippe-fournier-viger.com/spmf/.

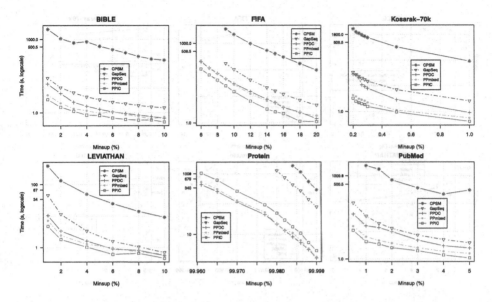

Fig. 2. CPU times for PPIC, PPDC, PPMIXED and GAP-SEQ for several minsup (missing points indicate a timeout)

available online as open source in order to make this research reproducible (http://sites.uclouvain.be/cp4dm/spm/).

We used a machine with a 2.7 Hz Intel core i5 processor and 8GB of RAM with Linux 3.19.0-32-generic 64 bits distribution Mint 17.3. Execution time limit is set to 3600 seconds (1 h). Our proposals are compared, first, with CPSM[3] [8] and GAP-SEQ[4] [5], the recently CP-based approaches including Gap constraint and the previous version of GAP-SEQ, PP[5] [6] without Gap but with regular expression constraint. Second, we made comparison with cSPADE[6] [18], PrefixSpan [10][7] and SPMF[8].

PPIC vs PPDC vs PPmixed. The CPU time of PPIC, PPDC and PPMIXED models are shown in Fig. 2. PPIC is more efficient than PPDC in 80 % of datasets. This is essentially because in many cases at the beginning of mining, there are many unsupported sequences for which the symbol counters must be decremented (compared to not having to increase the counters in PPIC). For instance with BIBLE SDB and $minsup = 10\%$ PPDC need to see 21,979,585 symbols to be complete while only 15,916,652 is needed for PPIC. Unsurprisingly, PPMIXED is between these approaches.

Our proposals vs* Gap-Seq *(CP method). Figure 2 confirms CPSM is outperformed by Gap-Seq which itself improves PP (without gap). We can clearly notice our approaches outperform GAP-SEQ (and hence PP) in all cases. In the case of FIFA

---

[3] https://dtai.cs.kuleuven.be/CP4IM/cpsm/.
[4] https://sites.google.com/site/cp4spm/.
[5] https://sites.google.com/site/prefixprojection4cp/.
[6] http://www.cs.rpi.edu/~zaki/www-new/pmwiki.php/Software.
[7] http://illimine.cs.uiuc.edu/software/.
[8] http://www.philippe-fournier-viger.com/spmf/index.php?link=download.php.

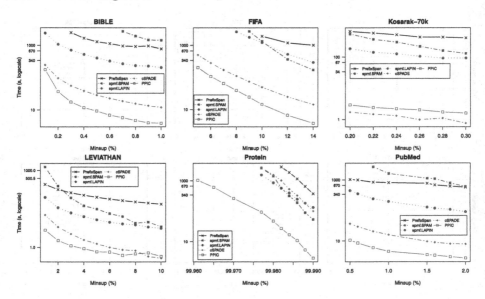

Fig. 3. CPU times for PPIC, PPDC, PPmixed and cSPADE for several *minsup*

SDB, GAP-SEQ reach time limit when *minsup* ≤ 9 %. PPIC is very effective in large and dense datasets regarding of CPU-times.

Comparison with Specialized Algorithms. Our third experience is the comparison with specialized algorithms. As we can see in the Fig. 3, we perform better on 84 % of the datasets. However, cSPADE is still the most efficient for Kosarak. In fact, Kosarak doesn't contain any symbol repetition in its sequences. So it is a bad case for prefix-projection-based algorithms which need to scan all the positions. On the contrary, with protein dataset (the sparse one) cSPADE requires much more CPU time. The SPMF implementation of SPAM, PrefixSpan and LAPIN appears to be consistently slower than cSPADE but there is no clear domination among these.

Impact of the Improvements. Figure 4 shows the incremental impact of our proposed solutions to the weaknesses defined in Sect. 4.2, starting from reversible vectors (fix of

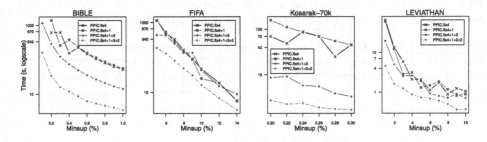

Fig. 4. Incremental impact of our solutions to the different weaknesses

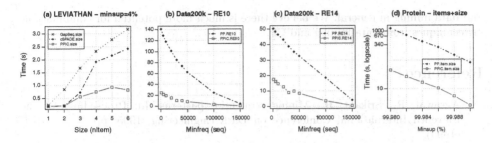

Fig. 5. Handling of different additional constraints

weakness 4) up to all our proposed modifications. Fix 1 has limited effect, while adding fix 3 is data dependent but adding fix 2 always improves further.

Handling Different Additional Constraints. In order to illustrate the modularity of our approach we compare with a number of user-defined constraints that can be added as additional modules without changing the main propagator (Fig. 5). (a) We compared PPIC and PP (unfortunately the GAP-SEQ tool does not support a regular expression command-line argument) under various size constraints on the protein dataset with *minsup* = 99.984. (b, c) We also selected data200k adding a regular expression constraint $RE10 = A * B(B|C)D * EF * (G|H)I*$ and $RE14 = A*(Q|BS*(B|C))D*E(I|S)*(F|H)G*R$ [14]. The last experiment reported on Fig. 5d consists in combining size and symbols constraints on the protein dataset: only sequential patterns that contain VALINE and GLYCINE twice and ASPARATE and SERINE once are valid. PPIC under constraints still dominates PP.

6   Conclusion

This work improved the existing CP-based sequential pattern mining approaches [5,8] up to the point that it also surpasses specialized mining systems in terms of efficiency. To do so, we combined and adapted a number of ideas from both the sequence mining literature and the constraint programming literature; correspondingly last-position information [16] and reversible data-structures for storing and restoring state during backtracking search. We introduced the PREFIXPROJECTION-INC (PPIC) global constraint and two variants proposing different strategies to compute the projected frequencies: from scratch, by decreasing the counters, or a mix of both. These can be plugged in as modules in a CP solver. These constraints are implemented in Scala and made available in the generic OscaR solver. Furthermore, the approach is compatible with a number of constraints including size and regular expression constraints. There are other constraints which change the subsequence relation and which would hence require hardcoding changes in the propagator (gap [5], span, etc.). We think many of our improvements can be applied to such settings as well.

Our work shows that generic CP solvers can indeed be used as framework to build scalable mining algorithms, not just for generic yet less scalable systems as was done for itemset mining [4]. Furthermore, advanced data-structures for backtracking search, such as trailing and reversible vectors, can also be used in non-CP algorithms. This appears to be an understudied aspect of backtracking algorithms in pattern mining and data mining in general. We believe there is much more potential for combinations of techniques from data mining and CP.

References

1. Agrawal, R., Srikant, R.: Mining sequential patterns. In: Proceedings of the Eleventh International Conference on Data Engineering, 1995, pp. 3–14. IEEE (1995)
2. Ayres, J., Flannick, J., Gehrke, J., Yiu, T.: Sequential pattern mining using a bitmap representation. In: ACM SIGKDD, pp. 429–435 (2002)
3. Coquery, E., Jabbour, S., Saïs, L., Salhi, Y.: A SAT-based approach for discovering frequent, closed and maximal patterns in a sequence. In: ECAI (2012)
4. Guns, T., Nijssen, S., De Raedt, L.: Itemset mining: a constraint programming perspective. Artif. Intell. **175**(12), 1951–1983 (2011)
5. Kemmar, A., Loudni, S., Lebbah, Y., Boizumault, P., Charnois, T.: A global constraint for mining sequential patterns with gap constraint. In: CPAIOR16 (2015)
6. Kemmar, A., Loudni, S., Lebbah, Y., Boizumault, P., Charnois, T.: PREFIX-PROJECTION global constraint for sequential pattern mining. In: Pesant, G. (ed.) CP 2015. LNCS, vol. 9255, pp. 226–243. Springer, Heidelberg (2015). doi:10.1007/978-3-319-23219-5_17
7. Mabroukeh, N.R., Ezeife, C.I.: A taxonomy of sequential pattern mining algorithms. ACM Comput. Surv. **43**(1), 3:1–3:41 (2010)
8. Negrevergne, B., Guns, T.: Constraint-based sequence mining using constraint programming. In: Michel, L. (ed.) CPAIOR 2015. LNCS, vol. 9075, pp. 288–305. Springer, Heidelberg (2015). doi:10.1007/978-3-319-18008-3_20
9. OscaR Team: OscaR: Scala in OR (2012). https://bitbucket.org/oscarlib/oscar
10. Pei, J., Han, J., Mortazavi-Asl, B., Pinto, H., Chen, Q., Dayal, U., Hsu, M.C.: Prefixspan: mining sequential patterns efficiently by prefix-projected pattern growth. In: ICCCN, p. 0215. IEEE (2001)
11. Perez, G., Régin, J.-C.: Improving GAC-4 for table and MDD constraints. In: O'Sullivan, B. (ed.) CP 2014. LNCS, vol. 8656, pp. 606–621. Springer, Heidelberg (2014). doi:10.1007/978-3-319-10428-7_44
12. Rossi, F., Van Beek, P., Walsh, T.: Handbook of CP. Elsevier (2006)
13. Schulte, C., Carlsson, M.: Finite domain constraint programming systems. In: Handbook of Constraint Programming, pp. 495–526 (2006)
14. Trasarti, R., Bonchi, F., Goethals, B.: Sequence mining automata: a new technique for mining frequent sequences under regular expressions. In: Eighth IEEE International Conference on Data Mining, 2008, ICDM 2008, pp. 1061–1066. IEEE (2008)
15. Yan, X., Han, J., Afshar, R.: Clospan: mining closed sequential patterns in large datasets. In: SDM, pp. 166–177. SIAM (2003)
16. Yang, Z., Kitsuregawa, M.: LAPIN-SPAM: an improved algorithm for mining sequential pattern. In: International Conference on Data Engineering (2005)
17. Yang, Z., Wang, Y., Kitsuregawa, M.: LAPIN: effective sequential pattern mining algorithms by last position induction for dense databases. In: DAFSAA, pp. 1020–1023 (2007)
18. Zaki, M.J.: Sequence mining in categorical domains: incorporating constraints. In: Proceedings of the Ninth International Conference on Information and Knowledge Management, pp. 422–429. ACM (2000)

Detecting Public Influence on News Using Topic-Aware Dynamic Granger Test

Lei Hou[1(✉)], Juanzi Li[1], Xiao-Li Li[2], and Jianbin Jin[3]

[1] Department of Computer Science and Technology,
Tsinghua University, Beijing 100084, China
houl10@mails.tsinghua.edu.cn, lijuanzi@tsinghua.edu.cn
[2] Institute for Infocomm Research, A*STAR, Singapore 138632, Singapore
xlli@i2r.a-star.edu.sg
[3] School of Journalism and Communication,
Tsinghua University, Beijing 100084, China
jinjb@mail.tsinghua.edu.cn

Abstract. With the rapid proliferation of Web 2.0, *user-generated content (UGC)*, which is formed by the public to reflect their views and voice, presents rich and timely feedback on news events. Existing research either studies the common and private features between news and UGC, or describes the ability of news media to influence the public opinion. However, in the current highly media-user interactive environment, investigating the public influence on news is of great significance to risk and credible management for government and enterprises. In this paper, we propose a novel topic-aware dynamic Granger test framework to quantify and characterize the public influence on news. In particular, we represent words and documents as distributed low-dimensional vectors which facilitates the subsequent topic extraction. Then, a topic-aware dynamic strategy is proposed to transfer news and UGC streams into topic series, and finally we apply Granger causality test to investigate the public influence on news. Extensive experiments on 45 diverse real-world events demonstrate the effectiveness of the proposed method, and the results show promising prospects on predicting whether an event will be properly handled at its early stage.

Keywords: News · User-generated content · Influence · Distributed representation · Granger causality

1 Introduction

Social media presents rich and timely feedback on news events that take place around the world. According to the report from *Pew Research Center*, 63 % of social users from Twitter and Facebook accessed news online, and roughly a quarter of them actively expressed their opinions on daily news through these social applications [2]. The various user-generated content not only fuels the news with different events from different perspectives, but also spurs additional

© Springer International Publishing AG 2016
P. Frasconi et al. (Eds.): ECML PKDD 2016, Part II, LNAI 9852, pp. 331–346, 2016.
DOI: 10.1007/978-3-319-46227-1_21

news coverage in the event. On the other hand, reading the social media, understanding and responding to public voice timely and objectively, can help news media promote its influence on the public.

Example. In the event *Asia-Pacific Economic Cooperation* (APEC) 2014 in China, *region cooperation, global economic* were the topics supposed to be reported by news media. However, in fact, social media users posted significant amount of comments on *APEC blue* – rare blue sky in Beijing during the summit due to emission reduction campaign directed by Chinese government. Following that, news media quickly followed and paid great attention on this topic which was beyond the original news agenda. We found that 38 of the 176 news articles on Sina were related to the *APEC blue*. Furthermore, how news responds to the public voice has a significant impact on the government credibility. For example, two severe earthquakes struck China in 2014, i.e., *Yunnan* and *Sichuan*. Both reports covered the major topics, but in *Yunnan* earthquake, news media responded the public timely and pictured a comprehensive image of event progress from the perspectives of the public, and thus harvested better support from the public. Therefore, investigating the public influence on news is of great benefit to public opinion management and government credibility improvement.

Related researches on mutually news and UGC stream analysis mainly follows three lines. The first studies event evolution within *individual* news stream [1,8], e.g., Mei and Zhai adapt PLSA to extract topics in news stream, then identify coherent topics over time, and finally analyze their lifecycle [17]. The second focuses on simultaneously modeling *multiple* news streams, e.g. identifying the characteristics of social media and news media [26] or their common and private features [10,23]. But both of them pay little attention on the interactions between two streams which could inspire their co-evolution. The last comes from the journalism communication. It applies agenda setting theory [16] to analyze the interactions between different news agenda, and it is often completed via questionnaire survey or manual work on limited events. However, in the era of social media, agenda setting is not a one-way pattern from news to the public, but rather a complex and dynamic interaction.

Detecting the public influence on news poses unique technical challenges: (i) most researches use latent topic to model news and UGC, but the traditional word distribution representation [5,17] suffers from the sparsity problem due to the UGC's short and fragmented characteristics, making it difficult to track topic changes; (ii) how to detect the cross-media influence links remains another problem, since the commonly-used measures (e.g., KL-divergence [12,17]) often leads to heuristic results without statistical explanation.

In this paper, we propose a novel topic-aware dynamic Granger test framework to automatically study the public influence on news media. To address the sparsity problem, we first represent word as low-dimensional word vectors through skip-gram model [19], and further reform word representation via sparse coding to capture the latent semantic of each dimension. Then we employ Granger causality test [9] to theoretically detect the public influence on news.

Particularly, for a pair of topics extracted from UGC and news respectively, we propose a topic-aware dynamic strategy to chronologically split those topic-related documents into disjoint bins with dynamic time intervals, calculate the topic representations based on the documents falling into each bin, and apply the multivariate Granger test to judge if the UGC-to-news influence exists. Finally, we quantify the influence [12] based on the discovered influence links, and validate the influence measures by calculating their correlations with the professional, manual results provided by *China Youth Online*.

The main contributions can be summarized as follows:

- We address problem of analyzing public influence analysis on news through a unified Granger-based framework. Extensive experiments are conducted on 45 real-world events to demonstrate its effectiveness, and results could provide useful guidance on handling public hot topics in event reporting.
- We propose a novel textual feature extraction method. Instead of directly using the popular word2vec, it further maps word and document into a low-dimensional space with each dimension denoting a more compact semantic thus facilitates topic extraction and representation.
- To track the temporal changes of topic pair from news and UGC respectively, we propose a novel topic-aware dynamic binning strategy, splitting both streams into chronological bins to achieve smooth topic representations of each bin.

The rest of the paper is organized as follows. In Sect. 2, we first define the related concepts and the problem of influence analysis from UGC to news. Section 3 presents our proposed textual feature extraction method and Granger-based influence analysis. Our experimental results are reported in Sect. 4. Section 5 reviews the related literatures, and finally Sect. 6 concludes this paper with future research directions.

2   Problem Definition

A particular event often brings forth two correlated streams, namely news articles from newsroom form a news stream and the public voice from different social applications converge into a UGC stream. Both news stream NS and UGC stream US are text streams, which are defined as follows:

Definition 1 *Text Stream. A text stream $TS = \langle s_1, s_2, \ldots, s_n \rangle$ is a sequence of documents, where s_i ($i = 1, 2, \ldots, n$) is associated with a pair (d_i, t_i), where d_i is a document comprising a list of words and t_i is the publish time in non-descending order, i.e. $t_i \leq t_{i+1}$.*

It has been shown that news and UGC streams are mutually dependent [24]. *Topic*, which bridges these two different streams, plays an important role. In order to study the cross-stream interactions, we first define topic as follows:

Definition 2 *Topic*. *Conceptually, topic z expresses an event related subject/theme within a time period. Mathematically, topic* **z** *is characterized as a vector with each dimension denoting a word feature or a latent aspect. Topic z covers multiple documents (news articles or users comments).*

The interaction between media, public and government has been theoretically studied in journalism communication, e.g., the agenda setting theory[1] evaluated the ability of mass media to influence the salience of topics on the public [15]. Nowadays, the proliferation of social media is changing the way of news diffusion, i.e., the public may inversely affect or even drive the news media. It is useful to explore to what extent the traditional news depends on social media and how long the public influence lasts, thus we condense the following research problem.

Definition 3 *Analyzing Public Influence on News*. *Given a news stream NS and a UGC stream US, influence analysis from UGC to news aims to discover a set of influence links $\{(\mathbf{z}_u, \mathbf{z}_n, \zeta)\}$, where $\mathbf{z}_u \in Z_u$ and $\mathbf{z}_n \in Z_n$ are topics extracted from US and NS respectively, and $\zeta \in \{0,1\}$ indicates whether $\mathbf{z}_u$ influences (or contributes to) $\mathbf{z}_n$.*

From the definition above, topic representation and extraction, influence detection are two major steps to complete the novel task. As mentioned in the introduction, existing methods suffer from various technical deficiencies, i.e., sparse representation and lack of theoretical foundation. To tackle these issues, we put effort on the following two problems: (i) given news and UGC streams, properly represent the documents and extract latent topics from both streams; (ii) given a topic pair $(\mathbf{z}_u, \mathbf{z}_n)$, determine if there exists a causality link and provide a statistical evaluation on how $\mathbf{z}_u$ contributes to $\mathbf{z}_n$.

3   Our Approach

In this paper, we propose a topic-aware dynamic Granger-based method to automatically detect the influence from UGC to news. Specifically, we develop a text representation method to better represent news and UGC in a low-dimensional space and extract their corresponding topics (Sect. 3.1). We incorporate temporal information to transform news and UGC topics into serialized representations and apply Granger causality test to detect the public influence on news (Sect. 3.2).

3.1   Text Representation and Topic Extraction

Text representation and topic extraction aims to properly represent the documents in *NS* and *US* and extract topics Z_n and Z_u. However, traditional TF-IDF representations suffer problems of the curse of dimensionality and feature independence assumption in dealing with the short and fragmented UGC.

---

[1] https://en.wikipedia.org/wiki/Agenda-setting_theory.

These methods often ignore the semantic relationships among word features which leads to document sparse representation with many zero features values.

To alleviate the sparse representation, many methods have been proposed to unveil the hidden semantics of words, such as topic models (e.g., LDA [3]) and external knowledge enrichment (e.g., ESA [7]). However, topic models rely much on the word co-occurrence that cannot be accurately computed with short texts, while ESA requires plenty of high-quality knowledge, which is often not available in practice. In this paper, we propose a novel textual feature extraction pipeline, which gradually maps word and document into a low dimensional space where each dimension represents a unique semantic meaning. It consists of the following three steps:

Word Vectorization. Word is the basic element in text, so we first transform words into continuous *low-dimensional* vectors. Let V denote the vocabulary in NS and US, we employ skip-gram model [19] to learn a mapping function: $V \rightarrow \mathbb{R}^M$, where $\mathbb{R}^M$ is a M-dimensional vector. Specifically, given a document $s \in NS \cup US$ associated with word sequence $\langle w_1, w_2, \ldots, w_W \rangle$, skip-gram model maximizes the co-occurrence probability among words that appear within a contextual window k:

$$\max_{\mathbf{w}} \frac{1}{W} \sum_{i=1}^{W} \sum_{j=i-k, j\neq 0}^{j=i+k} \log p(w_j | w_i) \tag{1}$$

The probability $p(w_j|w_i)$ is formulated as:

$$p(w_j|w_i) = \frac{\exp(\mathbf{w}_j^\mathrm{T} \mathbf{w}_i)}{\sum_{l=1}^{V} \exp(\mathbf{w}_l^\mathrm{T} \mathbf{w}_i)} \tag{2}$$

where $\mathbf{w}_i \in \mathbb{R}^M$ is the M-dimensional representation of w_i.

Mid-level Feature Learning. Intuitively, the document representation can be achieved via word vector composition. However, each dimension in word vector represents a latent meaning and word semantic scatters over almost all dimensions, simple composition of individual word vectors ignores the potential correlation between dimensions [20]. To prevent possible information loss by simple composition, we reconstruct each word vector into a mid-level feature [4], where each dimension represents a unique dense semantic. In other words, we learn a $\mathbb{R}^M \rightarrow \mathbb{R}^N$ mapping, and it is typically a sparse coding problem, whose objective is:

$$\min_{\mathbf{W}^*, \mathbf{D}} \sum_{i=1}^{V} \|\mathbf{w}_i - \mathbf{D}\mathbf{w}_i^*\|_2^2 + \lambda \|\mathbf{w}_i^*\|_1 \tag{3}$$

where $\mathbf{w}_i \in \mathbb{R}^M$ is the vector obtained in word vectorization; $\mathbf{w}_i^* \in W^* \subseteq \mathbb{R}^N$ is the N-dimensional sparse representation ($N > M$); $\mathbf{D}$ is an $M \times N$ matrix with each column denoting a dense sematic; $\| \cdot \|_1$ denotes the ℓ_1-norm of input vector; $\lambda > 0$ is a hyperparameter controlling the sparsity of the result representation, i.e., larger (or smaller) λ induces more (or less) sparseness of $\mathbf{w}_i^*$. Because the vocabulary V usually contains tens of thousands of words, optimization of the non-convex problem would be very time consuming.

To efficiently solve the problem, we apply a two-step approximation method. Firstly, we learn the matrix $\mathbf{D}$ offline. We cluster the learned word vectors into N clusters through K-means where each cluster denotes a compact semantic, and use the cluster centers as the columns of D. Secondly, based on the assumption that locality is more essential than sparsity [22], we select the K-nearest neighbors in D for each word $\mathbf{w}_i$ based on Euclidean distance, and then adopt the Locality-constraint Linear Coding (LLC) to learn its transformation $\mathbf{w}_i^*$:

$$\min_{\mathbf{W}^*} \sum_{i=1}^{V} \|\mathbf{w}_i - \mathbf{B}_i \mathbf{w}_i^*\|_2^2 + \lambda \|\mathbf{w}_i^*\|_2^2 \tag{4}$$
$$s.t. \mathbf{1}^{\mathrm{T}} \mathbf{w}_i^* = 1, \forall i$$

where $\mathbf{B}_i$ is the K-nearest neighbors to $\mathbf{w}_i$ in $\mathbf{D}$. The problem could be solved analytically by:

$$\widehat{\mathbf{w}_i^*} = (\mathbf{V}_i + \lambda \mathbf{I}) \backslash \mathbf{1}$$
$$\mathbf{w}_i^* = \widehat{\mathbf{w}_i^*} / \mathbf{1}^{\mathrm{T}} \widehat{\mathbf{w}_i^*} \tag{5}$$

where $\mathbf{V}_i = (\mathbf{B}_i - \mathbf{1}\mathbf{w}_i^{\mathrm{T}})^{\mathrm{T}} (\mathbf{B}_i - \mathbf{1}\mathbf{w}_i^{\mathrm{T}})$ denotes the covariance matrix.

Document Representation and Topic Extraction. We employ spatial pooling to represent each document as a N-dimensional vector $\mathbb{R}^N$ based on the learned sparse word vectors. Given a document s_i consisting W words with vector representations $\mathbf{w}_i^*, i = 1, 2, \ldots, W$, we try two different pooling functions to obtain the document representation $\mathbf{s}_i$:

$$s_{ij} = \underbrace{\frac{1}{W} \sum_{k=1}^{W} |\mathbf{w}_{kj}^*|}_{average} \quad or \quad s_{ij} = \underbrace{\sqrt{\frac{1}{W} \sum_{k=1}^{W} \mathbf{w}_{kj}^{*2}}}_{square\ root} \tag{6}$$

where $\mathbf{s}_i$ denotes the final representation of s_i and $s_{ij}|_{j=1}^{N}$ is the j-th entry. Note that different pooling functions assume the underlying distributions differently. Once completing the document representation, we feed the news and comment vectors into K-means algorithm separately to obtain topic sets Z_n and Z_u. The achieved topics have more compact distributed representations than TF-IDF, which is convenient to further computation and analysis.

3.2   Topic Influence Detection

Topic influence detection analyzes the relationship between news and UGC topics, which behaves as inter-stream *influence*. Normally, KL-divergence is employed to evaluate topic transition within news stream [13, 17] or topic interaction across streams [12], but the idea is heuristic and results are often restricted within a too short time period to track the topic evolution.

Therefore, we perform the influence detection in a more theoretical way through Granger causality test[2]. Its basic idea is that a *cause* should be helpful in predicting the future values of a time series, beyond what can be predicted solely based on its own historical values [9]. That is to say, a time series x is to Granger cause another time series y, if and only if regressing for y in terms of both past values of y and x is statistically significantly more accurate than regressing for y in terms of past values of y only.

Granger-Based Influence Detection. In this paper, Granger causality analysis is performed on two topics $z_u \in Z_u$ and $z_n \in Z_n$ to test whether z_u is the Granger cause of z_n.

In the previous subsection, we achieve the news and UGC topic sets and their associated documents, but the Granger causality test requires two time series. So we need to turn topics in Z_n and Z_u into time-varying topic series. For each $\mathbf{z} \in Z_n \cup Z_u$, we need to represent it as $\langle \mathbf{z}^t \rangle_{t=1}^T$ where $\mathbf{z}^t$ is the status of topic z at the t-th interval and T is the size of time intervals. A straightforward way is to partition both streams into disjoint slices with *fixed* time intervals (e.g., one day), i.e., equal-size binning. An alternative is equal-depth binning, i.e., evenly partitioning all documents into T bins. For an obtained partition $\langle S^t \rangle_{t=1}^T$, the representation of topic z at the t-th bin $\mathbf{z}^t$ could be simply computed via averaging the related document vectors within that bin:

$$\mathbf{z}^t = \frac{1}{|S_z^t|} \sum_{s_z \in S^t} \mathbf{s} \qquad (7)$$

where S_z^t denotes the documents within t-th bin that are related to topic z.

Once we get the time-varying representations of two target topics $\langle \mathbf{z}_n^t \rangle_{t=1}^T$ and $\langle \mathbf{z}_u^t \rangle_{t=1}^T$, we first fit two vector autoregressive models (VAR) over these two series:

$$\mathbf{z}_n^t = a_0 + \sum_{i=1}^q a_i \mathbf{z}_n^{t-i} + \mathbf{r} \qquad (8)$$

$$\mathbf{z}_n^t = a_0 + \sum_{i=1}^q (a_i \mathbf{z}_n^{t-i} + b_i \mathbf{z}_u^{t-i}) + \mathbf{r}_u \qquad (9)$$

where (8) predicts a news topic $\mathbf{z}_n^t$ at time stamp t purely based on its historical values, i.e., $\mathbf{z}_n^{t-i}$, while (9) considers the historical values from both news and UGC streams, i.e., $\mathbf{z}_n^{t-i}$ and $\mathbf{z}_u^{t-i}$, for prediction; q is a predefined maximum lag to measure how long the influence lasts; $\mathbf{r}_u$ and $\mathbf{r}$ denote the residuals with/without considering the topic z_u.

Then, to test whether or not (9) results in a better regression than (8) with statistical significance, we apply an F-test (some other similar tests could also be chosen). More specifically, we calculate the residual sum of squares RSS and RSS_u, based on which we obtain the F-statistic:

$$F = \frac{(RSS - RSS_u)/q}{RSS_u/(n - 2q - 1)} \sim F(q, n - 2q - 1) \qquad (10)$$

---

[2] http://en.wikipedia.org/wiki/Granger_causality.

Given a confidence coefficient α, we say $\mathbf{z}_u$ Granger causes $\mathbf{z}_n$ if F is greater than a predefined F_α, i.e. $\zeta = 1$ as defined in Sect. 2, and otherwise $\zeta = 0$.

However, both streams, especially news articles, are often generated nonuniformly. The equal-size binning performs poorly on such streams since it produces many empty intervals without any news or comments, and the equal-depth binning often leads to extremely unbalanced time spans. Either empty interval or unbalanced spans has side effect on Granger test, making it failed or meaningless.

Topic-aware Dynamic Granger Test. To address the uneven distribution problem, we propose a topic-aware dynamic binning strategy to partition both streams into several disjoint intervals. The motivation for *topic-aware* is that: different topics follow their unique patterns and show various distributions along timeline and the Granger causality test actually processes a topic pair rather than the whole streams at one time, thus one partition only need to deal with documents within target topics from news and UGC respectively. And the *dynamic binning* aims to alleviate problem of the uneven distribution. Let S_z denote the streaming documents associated with topic z, $\langle S_z^t \rangle_{t=1}^T$ is a partition result, we define the following two types of *dispersion*:

- dis_{amount}: the difference between the largest and the smallest bin size with bin size is defined as the number of contained documents;
- dis_{span}: the difference between the largest and the smallest time span.

Our objective is to balance these two dispersions, namely,

$$\min_{\langle S_z^t \rangle_{t=1}^T} dis_{amount} + dis_{span} \quad s.t. |S_z^t| > 0, \forall t \tag{11}$$

Due to the extremely unbalanced volume of news and comments, we perform the optimization on news stream and the comments just follow. The problem could be solved efficiently using dynamic programming (where *dynamic* comes) and the best solution is always available [13].

4   Experiments

In this section, we first briefly introduce our datasets, and then present the detailed experimental results on topic extraction, topic influence detection and further analysis.

4.1   Dataset Description

To evaluate the effectiveness of the proposed methods, we prepare the following two kinds of datasets:

Datasets from Hou's paper [12]. They are composed of five datasets containing four international events: *the Federal Government Shutdown* in both Chinese and English (cFGS/eFGS), *Jang Sung-taek's* (Jang), *The Boston Marathon Bombing* (Boston) and *India Election* (India). They are collected from

Table 1. Datasets: duration, numbers of comments and news articles, max and average number of comments per news

Dataset	Days	#Comments	#News	Comments/News	
				max	avg
cFGS	35	12,995	97	7,818	134
Jang	43	3,291	84	467	39.2
eFGS	53	17,295	136	1,112	127
Boston	46	7,521	211	518	29.4
India	66	4,723	88	113	53.7

influential news portals and social media platforms (i.e., Sina, New York Times, Twitter), and the detailed statistics are summarized in Table 1. These datasets are used to evaluate the effectiveness of our topic extraction and influence detection.

Datasets from China Youth Online (CYOL[3]). CYOL is one of the biggest and leading public opinion analysis website in China. It monthly publishes opinion index based on questionnaire surveys from experts and scholars, civil servants, media people, opinion leaders and ordinary Internet users. The index includes five well-defined metrics: *information coverage*, *activeness*, *response arrival rate*, *response recognition rate* and *satisfactory*. For each event reported by CYOL in 2014, we crawled the news articles and comments from Sina[4] if there existed a corresponding special issue. Finally, we collected 40 events, and for each event, there are 140 news articles and 12,849 comments on average. Due to the space limit, the detailed statistics and data will be published later. We incorporated these datasets and published opinion index to evaluate the influence measures that are automatically calculated based on our approach.

4.2 Results for Topic Extraction

In this section, we report the evaluation on text representation and topic extraction, including the experiment setup (settings, baselines and metrics), comparison results and the parameter analysis.

Settings. We use the gensim[5] implementation of word2vec to learn word vectors with $M = 200$, and K-means to generate the transform matrix D with $N = \{128, 256, 512, 1024\}$. For mid-level feature learning, we apply LLC with various K-nearest neighbors, with $K \in \{1, 5, 10, 50\}$. The parameter λ is set to be $1e^{-4}$ as the author suggested.

Baselines. We use DeepDoc to denote our proposed text representation method, and compare it with TF-IDF based method (TF-IDF) and state-of-the-art topic

---

[3] http://yuqing.cyol.com/.

[4] http://search.sina.com.cn/?t=zt.

[5] http://radimrehurek.com/gensim/models/word2vec.html.

models on news and UGC, i.e., Document Comment Topic Model (DCT) [11] and Cross Dependence Temporal Topic Model (CDTTM) [12].

Metrics. As for the evaluation metrics, we calculate the inner/inter-cluster distance for all topics. The inner-cluster distance (*inner*) is defined as the average distance between documents within topic, and a smaller value indicates a compact cluster. The inter-cluster distance (*inter*) is the average distance from one topic to all the other topics, and a larger value indicates a better result. We also calculate their relative ratios (*ratio*) where a bigger value shows better performance.

Comparison Results. Table 2 presents the comparison results, from which we can see: (i) macroscopically, our proposed DeepDoc outperforms three baselines consistently, and DCT is more steady than other methods while the TF-IDF representation obtains the worst performance. (ii) under this measurement, CDTTM is not so sensitive to the stream distribution as described in [12], and DeepDoc does not have this problem as we do not include temporal information in clustering.

Table 2. Results for topic extraction: *inner* and *inter* stand for the average inner/inter-cluster distances, and they are related to the dimension of document representation; *ratio* is calculated through dividing the *inter* by *inner* to measure the clustering performance, and a larger *ratio* indicates a better clustering result

Dataset	TF-IDF			DCT			CDTTM			DeepDoc		
	inner	*inter*	*ratio*	*inner*	*inter*	*ratio*	*inner*	*inter*	*ratio*	*inner*	*inter*	*ratio*
cFGS	35.86	92.30	2.574	.5211	1.517	2.911	.5406	1.631	3.017	4.091	15.70	3.838
Jang	17.94	40.28	2.245	.5412	1.596	2.949	.5333	1.697	3.182	4.024	15.19	3.774
eFGS	41.13	120.8	2.938	.5285	1.503	2.844	.5083	1.532	3.014	4.973	19.48	3.917
Boston	25.37	53.48	2.108	.5271	1.529	2.901	.5084	1.509	2.968	3.966	16.60	4.185
India	19.91	37.69	1.893	.5986	1.433	2.394	.6095	1.439	2.361	2.808	8.806	3.136

Parameter Analysis. Then the Boston dataset is chosen to investigate the effects of the number of neighbours, pooling function and the number of matrix columns, and the results are presented in Table 3. We have the following conclusions:

- **Number of neighbours (K).** Regardless of various other settings, generally small number of neighbors leads to better clustering results. This is a promising finding, because the smaller the number of neighbors used (i.e., the more sparse the codes are), the faster the computation will be run, and the less the memory will be consumed.
- **Pooling function.** Different choices of pooling functions lead to very different clustering results. The root mean square pooling achieves better performance under almost every settings than average pooling, and the smaller the code sparseness (larger K and smaller matrix), the gap between these two pooling functions is more significant.

– **Number of matrix columns.** It actually denotes the dimensions of transformed space. Intuitively, if the number of dimension is too small, the mid-level
representation will lose discriminative power, but words from the same category of documents will be less similar if the size is too big. Here, we mainly
focus on the trade-off between reasonably smaller and bigger size. As can be
seen from the results, larger size leads to better results when $K > 10$, while
it is likely that smaller matrix is sufficient under higher level of sparseness.

Table 3. Clustering results on *Boston* dataset with various number of neighbors (K),
pooling functions and number of matrix columns (N).

#Neighbors	Pooling func.	N			
		128	256	512	1,024
K = 1	Sqrt	4.1608	4.1625	3.9000	3.8275
	Avg	2.9550	2.9325	2.2900	1.9258
K = 5	Sqrt	4.0108	4.0925	4.1850	4.0825
	Avg	3.2475	3.2400	2.8442	2.5242
K = 10	Sqrt	4.0108	3.7325	3.7583	4.0033
	Avg	3.2475	3.5092	3.4850	3.5658
K = 50	Sqrt	2.6892	3.0642	3.2858	3.6292
	Avg	2.6267	2.8575	3.1892	3.3625

4.3 Results for Topic Influence Detection

To evaluate our proposed topic-aware dynamic Granger test method (TDG),
we perform three series of experiments, namely, (1) the overall comparison with
KL-divergence based method in [12], (2) the comparison of different binning
methods, and (3) the effect of the maximum lag.

TDG – KL Divergence. Hou *et al.* evaluated their method on manually
labeled data, and it achieved comparable results to the human annotation. To
make the comparison fair, we compare the Granger results with $\alpha = 0.9/0.8$ with
their top 10 %/20 % links (Hou *et al.* included links with distance less than the
median value). Through manual evaluation, the Granger test achieves 94 % precision while KL gets only 82 %, indicating our method significantly outperform
theirs. This comes as no surprise because their KL-divergence based method only
finds similar patterns in the other stream (it assumes *similar* topics share similar
patterns along timeline which may not hold) while our proposed Granger based
method discovers the most useful topics in UGC that contribute to predicting
the target news topic and thus are more likely to influence the news.

Dynamic Binning – Equal Size Binning. Table 4 shows the number of
detected Granger causal links when different time split methods are applied.

We can find that, (i) *equal-size*: the equal-size binning gets the worst performance because the streams (especially news stream) distribute nonuniformly and it often leads to zero vectors for bins with on documents. Though mean linear interpolation is employed to deal with the zeros, the results are still not so satisfactory. (ii) *dynamic*: dynamic binning optimize (11) over whole news stream without distinguish topics. It can handle the uneven distributed streams to some extent, thus finds more influence links. (iii) *topic*: since our proposed method tests a pair of topics every time and different topics may follow different patterns, while the dynamic binning is applied on the whole streams, thus it might not perform well on different topic pairs. Therefore, the topic-aware binning further improves the performance.

Table 4. Granger causality links with different time split methods (0.8 and 0.9 are confidence coefficients)

Dataset	equal-size		dynamic		topic	
	0.8	0.9	0.8	0.9	0.8	0.9
cFGS	1	0	4	2	6	4
Jang	1	0	5	2	6	3
eFGS	2	0	5	2	7	4
Boston	1	0	4	2	8	5
India	1	0	3	2	7	4

How Long the Influence Lasts. To choose a proper maximum lag q (i.e., how many historical values are included in the regression), we select five topic pairs to conduct Granger causality test with maximum lag ranging from 1 to 10, and determine the proper value that achieves the best F statistics (divided by $F_{0.8}$ due to the different time spans). Table 5 shows the results from 3 to 7, we observe that the $F/F_{0.8}$ increases initially until $q = 5$ to reach stable status. We therefore execute all the Granger test with q set as 5. Note that here $q = 5$ does not mean 5 days since *topic aware binning* is used for stream split, and actually the average time difference is about 3.2 days, which tells us that news and UGC in the previous 3 days have much more influence on the current news report.

Table 5. F-statistic with maximum lag (q): $F/F_{0.8}$ denotes the average values of the 5 selected topic pairs.

q	3	4	5	6	7
$F/F_{0.8}$	1.704	2.115	2.493	2.487	2.435

4.4   Influence Usage Analysis

This experiment exploits whether our automatically obtained results are consistent with the objective CYOL public opinion index. Specifically, with our achieved influence links $\{(z_u, z_n)\}$, we quantify the public influence on news through news response $rate(NRR)$, $promptness(NRP)$ and $effect(NRE)$ as defined in [12]. Their comparable measures in *CYOL Public Opinion Index* are *information coverage(IC)*, *response activity(RA)* and *satisfactory(SA)*. We compute three correlation coefficients for the 3 pairs of measures NRR-IC, NRP-RA, and NRE-SA respectively, and higher correlations indicates better results. For comparison, we use LDA+KL-divergence, Hou's methods(CDTTM+KL) as our baselines. We further try to only use first half of the event data for analysis (Ours$^{\frac{1}{2}}$) to test whether it is helpful in predicting the future influence. Table 6 shows the comparison results.

Table 6. Influence usage results.

Methods	Correlation Coefficient		
	NRR-IC	*NRP-RA*	*NRE-SA*
LDA+KL	0.6573	−0.5832	0.6029
Hou(CDTTM+KL)	0.6814	−0.5933	0.6157
Ours(DeepDoc+TDG)	0.7232	−0.6419	0.6483
Ours$^{\frac{1}{2}}$	0.7092	−0.6254	0.6085

As shown in Table 6, our method achieves higher correlations with the CYOL measures than other two methods. Furthermore, we notice that only using the first half of event data, our method can achieve comparable results with those on all data. This implies that it can be used on predicting whether an event could be handled properly at early stage.

Case Review. Now we review the events mentioned in the introduction. The APEC 2014 summit shows a good example that the social media can influence news media. Besides *APEC blue*, we identify another topic beyond the scheduled ones, i.e., *tourist*. It actually covered the part-time activities of the dignitaries Mrs, especially their clothing. The news media started to report the part-time activities causally. However, the public was very enthusiastic about the Mrs' tourist and discussed a lot about their clothing. To satisfy people's curiosity, news presented systematic introduction of the first lady's activities and dress. Then, we compare the news response in the two earthquakes: both reports covered the major topics — both NRRs are pretty high (*Yunnan* 84 % and *Sichuan* 82 %); but in *Yunnan* earthquake, news media responded the public more timely — the NRP in *Yunnan* is much smaller than that in *Sichuan*, roughly 0.8 day v.s. 1.4 days. The final satisfactory shows that it is very important to properly handle the heatedly-discussed topics. Our analysis could summarize about which topic that news should response at what time, thus benefits the public opinion management.

5   Related Work

Our work is related to three lines of research as follows:

5.1   Distributed Text Representation

Representing words in continuous vector space has been an appealing pursuit since 1986 [25]. Recently, Mikolov *et al.* developed efficient method to learn high quality word vectors [19], and a host of follow-up achievements have been made on phrase or document representation, such as paragraph-to-vector [20]. Different from these attempts, we are inspired to borrow the state-of-the-art feature extraction pipeline in computer vision [4] to represent word and document in a new space where each dimension denotes a more compact semantic than directly using word2vec.

5.2   Social News Analysis and Topic Evolution

The proliferation of social media encourages researchers to study its relationship between traditional news media, e.g., Zhao *et al.* employed Twitter-LDA to analyze Twitter and New York Times and found Twitter actively helped spread news of important world events although it showed low interests in them [26]. Petrovic *et al.* examined the relation between Twitter and Newsfeeds and concluded that neither streams consistently lead the other to major events [21]. Besides the common and specific characteristics of news and social media, we pay more attention on the cross-stream interaction.

As for the topic evolution, Mei *et al.* solved the problem of discovering evolutionary theme patterns from single text stream [17], Hu *et al.* modeled the topic variations and identified the topic breakpoints in news stream [13]. Wang *et al.* aimed at finding the burst topics from coordinated text streams based on their proposed mixture model [24]. Lin *et al.* formalized the evolution of a topic and its latent diffusion paths in social community as a joint inference problem, and solved it through a mixture model (for text generation) and a Gaussian Markov Random Field (for user-level social influence) [14]. In this paper, we study the interplay of news and UGC within specific events, trying to analyze the cross-media influence and figure out how they co-evolve over time.

5.3   Agenda Setting and Granger Causality

Agenda-setting is the creation of public awareness and concern of salient issues by the news media. Mccombs and Shaw discussed the function of mass media in agenda setting [16] in 1972. Many researchers studied the interactions between public agenda and news agenda, e.g., Meraz employed time series analysis to measure the influence in political blog and news media [18]. Our work falls into the second-level agenda-setting (also called attribute agenda-setting), and the major advantage of our framework is that, the attributes are predefined and we extract the latent topics automatically.

The Granger causality test [9] is a statistical hypothesis test for determining whether one time series is useful in forecasting the other one. It has been utilized in many areas for causality analysis or prediction, e.g., [6] adapted it to model the temporal dependence from large-scale time series data [6]; Chang *et al.* used it in Twitter user influence modeling. In this paper, we apply the agenda-setting theory and multivariate Granger test to automatically analyze how the social media influence traditional news.

6   Conclusion

In this paper, we analyze the public influence on news through a Granger-based framework: first represent words and documents in distributed low-dimensional space and extract topics from news and UGC streams, then dynamically split streams to achieve changing topic representations on which we employ Granger causality test to detect influence links. Experiments on real-world events demonstrate the effectiveness of the proposed methods and the results show good prospects on predicting whether an event could be properly handled.

It should be note that Granger test attempts to capture an interesting aspect of causality, but certainly is not meant to capture all, e.g., it has little to say about situations in which there is a hidden common cause for the two streams. In the future work, we will try to address the important but challenging issue.

Acknowledgement. The work is supported by 973 Program (No. 2014CB340504), NSFC-ANR (No. 61261130588), and NSFC key project (No. 61533018), Tsinghua University Initiative Scientific Research Program (No. 20131089256) and THU-NUS NExT Co-Lab. Thank Prof. Chua Tat-Seng, Dr. Hanwang Zhang and Xindi Shang from National University of Singapore for discussion on text representation.

References

1. Ahmed, A., Xing, E.: Timeline: A dynamic hierarchical dirichlet process model for recovering birth/death and evolution of topics in text stream. In: Proceedings of the 26th Conference on Uncertainty in Artificial Intelligence, pp. 20–29 (2010)
2. Barthel, M., Shearer, E., Gottfried, J., Mitchell, A.: The evolving role of news on twitter and facebook. Technical report, Pew Research Center, July 2015
3. Blei, D.M., Ng, A.Y., Jordan, M.I.: Latent dirichlet allocation. J. Mach. Learn. Res. **3**, 993–1022 (2003)
4. Boureau, Y.L., Bach, F., LeCun, Y., Ponce, J.: Learning mid-level features for recognition. In: Proceedings of the IEEE Conference on Computer Vision and Pattern Recognition, pp. 2559–2566 (2010)
5. Chang, Y., Wang, X., Mei, Q.Z., Liu, Y.: Towards twitter context summarization with user influence models. In: Proceedings of the 6th ACM International Conference on Web Search and Data Mining, pp. 527–536 (2013)
6. Cheng, D., Bahadori, M.T., Liu, Y.: FBLG: a simple and effective approach for temporal dependence discovery from time series data. In: Proceedings of the 20th ACM SIGKDD International Conference on Knowledge Discovery and Data Mining, pp. 382–391 (2014)
7. Gabrilovich, E., Markovitch, S.: Overcoming the brittleness bottleneck using wikipedia: enhancing text categorization with encyclopedic knowledge. In: AAAI, vol. 6, pp. 1301–1306 (2006)

8. Gohr, A., Hinneburg, A., Schult, R., Spiliopoulou, M.: Topic evolution in a stream of documents. In: the 9th SIAM International Conference on Data Mining, pp. 859–872 (2009)

9. Granger, C.W.: Investigating causal relations by econometric models and cross-spectral methods. Econometrica J. Econometric Soc. **37**, 424–438 (1969)

10. Hong, L., Dom, B., Gurumurthy, S., Tsioutsiouliklis, K.: A time-dependent topic model for multiple text streams. In: Proceedings of the 17th ACM International Conference on Knowledge Discovery in Data Mining, pp. 832–840 (2011)

11. Hou, L., Li, J., Li, X.L., Qu, J., Guo, X., Hui, O., Tang, J.: What users care about: a framework for social content alignment. In: Proceedings of the 23rd International Joint Conference on Artificial Intelligence, pp. 1401–1407 (2013)

12. Hou, L., Li, J., Li, X.L., Su, Y.: Measuring the influence from user-generated content to news via cross-dependence topic modeling. In: Proceedings of the 20th International Conference on Database Systems for Advanced Applications, pp. 125–141 (2015)

13. Hu, P., Huang, M., Xu, P., Li, W., Usadi, A.K., Zhu, X.: Generating breakpoint-based timeline overview for news topic retrospection. In: Proceedings of the 11th IEEE International Conference on Data Mining, pp. 260–269 (2011)

14. Lin, C.X., Mei, Q., Han, J., Jiang, Y., Danilevsky, M.: The joint inference of topic diffusion and evolution in social communities. In: Proceedings of the 11th IEEE International Conference on Data Mining, pp. 378–387 (2011)

15. Lippmann, W.: Public Opinion. Transaction Publishers, New Jersey (1946)

16. McCombs, M., Shaw, D.: The agenda-setting function of mass media. Public Opinion Quart. **36**, 176–187 (1972)

17. Mei, Q., Zhai, C.: Discovering evolutionary theme patterns from text: an exploration of temporal text mining. In: Proceedings of the 11th ACM International Conference on Knowledge Discovery in Data Mining, pp. 198–207 (2005)

18. Meraz, S.: Is there an elite hold? traditional media to social media agenda setting influence in blog networks. J. Comput. Mediated Commun. **14**(3), 682–707 (2009)

19. Mikolov, T., Chen, K., Corrado, G., Dean, J.: Efficient estimation of word representations in vector space. In: Proceedings of the ICLR (2013)

20. Mikolov, T., Sutskever, I., Chen, K., Corrado, G.S., Dean, J.: Distributed representations of words and phrases and their compositionality. In: Advances in Neural Information Processing Systems, pp. 3111–3119 (2013)

21. Petrovic, S., Osborne, M., McCreadie, R., Macdonald, C., Ounis, I., Shrimpton, L.: Can twitter replace newswire for breaking news? In: Proceedings of the 7th International AAAI Conference on Weblogs and Social Media (2013)

22. Wang, J., Yang, J., Yu, K., Lv, F., Huang, T., Gong, Y.: Locality-constrained linear coding for image classification. In: Proceedings of the 2010 IEEE Conference on Computer Vision and Pattern Recognition, pp. 3360–3367 (2010)

23. Wang, X., Zhang, K., Jin, X., Shen, D.: Mining common topics from multiple asynchronous text streams. In: Proceedings of the 2nd ACM International Conference on Web Search and Data Mining, pp. 192–201 (2009)

24. Wang, X., Zhai, C., Hu, X., Sproat, R.: Mining correlated bursty topic patterns from coordinated text streams. In: Proceedings of the 13th ACM International Conference on Knowledge Discovery in Data Mining, pp. 784–793 (2007)

25. Williams, D., Hinton, G.: Learning representations by back-propagating errors. Nature **323**, 533 (1986)

26. Zhao, W.X., Jiang, J., Weng, J., He, J., Lim, E.P., Yan, H., Li, X.: Comparing twitter and traditional media using topic models. In: Proceedings of the 33rd European Conference on Information Retrieval, pp. 338–349 (2011)

Exploring a Mixed Representation for Encoding Temporal Coherence

Jon Parkinson[✉], Ubai Sandouk, and Ke Chen

School of Computer Science, University of Manchester,
Oxford Road, Manchester M13 9PL, UK
{jon.parkinson,ubai.sandouk,ke.chen}@manchester.ac.uk

Abstract. Guiding representation learning towards temporally stable features improves object identity encoding from video. Existing models have applied temporal coherence uniformly over all features based on the assumption that optimal object identity encoding only requires temporally stable components. We explore the effects of mixing temporally coherent invariant features alongside variable features in a single representation. Applying temporal coherence to different proportions of available features, we introduce a mixed representation autoencoder. Trained on several datasets, model outputs were passed to an object classification task to compare performance. Whilst the inclusion of temporal coherence improved object identity recognition in all cases, the majority of tests favoured a mixed representation.

Keywords: Representation learning · Temporal coherence · Object recognition

1 Introduction

Real world objects likely to appear in video exhibit a natural permanence over time, a property known as temporal coherence. If an object is present in a given frame, whilst it might undergo small changes in position and pose, it is likely the same object will also appear in neighbouring frames [1]. By guiding representation learning from video towards the discovery of temporally coherent structure present in the raw image data, the capture of variance associated with the identity of individual objects is improved [2–4].

Existing models applying temporal coherence regularization in unsupervised representation learning have tended to apply the rule uniformly across all available features. This is based on the assumption that as the identity of objects remains temporally coherent, encoding this information in an abstract representation only requires the discovery of input structure exhibiting similar properties.

To investigate whether this approach improves object identity encoding, we explore the effects of discovering a mixture of temporally coherent and variable features, all contained in a single representation. Section 3 introduces a mixed autoencoder, based on a commonly used method for discovering the important

© Springer International Publishing AG 2016
P. Frasconi et al. (Eds.): ECML PKDD 2016, Part II, LNAI 9852, pp. 347–360, 2016.
DOI: 10.1007/978-3-319-46227-1_22

variance underlying visual data, the sparse autoencoder [5]. Sparse autoencoders have previously been adapted to encode temporal coherence [6], with the extra regularization enforced across every feature. We trained a range of models on three video datasets, applying temporal coherence regularisation over different sized subsets of the total available features. We describe our experimental process in Sect. 4.

Once trained, labelled examples were passed through each of the various models, providing the input to a classifier trained to recognise object identity. Temporally coherent representations have generally been used for encoding object identity, so comparing the performance boost each model provides a supervised classifier is an ideal test bed to evaluate performance. Our results are presented in Sect. 4.3.

The best classification accuracy from the supervised task came from a representation encoding temporal coherence in every test. Interestingly however, the majority of cases favoured our mixed representation over the all-invariant alternatives. A discussion of our results is provided in Sect. 5. This section is followed by our final conclusions.

2   Background

There are generally two methods for discovering temporally coherent structure from video. The first discovers variance changing as smoothly [7], also described as slowly [8], as possible over time. This approach is used in Slow Feature Analysis (SFA). The whole video dataset is analysed, with the slowest changing variance extracted on the assumption this encodes important properties of an input, whilst making sure the constant trivial solution is avoided. This approach has had considerable impact, producing behaviour similar to cells present in early areas of the visual cortex [9], and discovering features encoding object position and pose alongside identity [10]. SFA does suffer from drawbacks however. For complex problems, a non-linear expansion of an input is required, the details of which are not known in advance. The two instances of SFA on possibly expanded data can also render the algorithm computationally prohibitive.

The alternative approach manipulates the likelihood that an object present in a single frame is likely to appear in neighbouring frames. Temporally coherent structure is discovered by constraining network activations for each pair of neighbouring frames to be as similar as possible on the assumption of their likely semantic similarity. This method has improved object identity performance on the benchmark COIL100 dataset [11], applying the temporal coherence to the output of a Convolutional Neural Network [12], alongside sparsity in a deep invariant architecture [13] and during pre-training and network output regularization [14]. An architecture relatively similar in nature to ours is that used by Goroshin et al. [6]. Temporal coherence is applied to sparse autoencoders, as part of a convolutional architecture. Unlike our approach, regularization is applied across every feature.

To be the best of our knowledge, there is only one implementation of temporal coherence where regularization is not applied uniformly across all features.

The Temporal Product Network (TPN) learns two sets of features simultaneously, one encoding temporal coherence, the other discovering variable features [15]. This work differs from ours as the two sets of features are seeking different aspects from the input. The invariant features are designed to encode object identity, with the variable set discovering object position. We are learning the two types of features for the single purpose of object identity encoding.

3   Model Description

This section provides an overview of the models we have tested, with a full description of the network cost function to be minimized during training. For clarity, as features encoded with temporal coherence respond uniformly to small changes in an input, they are referred to as invariant. Their temporally unconstrained counterparts are denoted as variable features. A model with temporal coherence applied across all features will be described as all-invariant, with its opposite, a plain sparse autoencoder known as invariant-free. Versions with a mixture of invariant and variable features are called mixed-representation.

3.1   Overview

To investigate the effects of mixing invariant and variable features together in a single representation, we required a model capable of extracting good quality features from a visual dataset, in an unsupervised manner. The sparse autoencoder [5] is a well studied model fitting these requirements. An input layer is passed to a fully connected hidden layer, which is in turn passed to a fully connected output layer of the same dimensionality as the input. During training, network weights are adjusted via gradient descent to reconstruct the input as closely as possible at the output layer. Reconstruction of each input during training guides the representation to retain as much information present in the input as possible. This has the added benefit of avoiding the trivial constant solution, which is essential for temporal coherence to be applied successfully. Sparsity regularization forces a dictionary of distinct commonly occurring features to form in the hidden layer. Feature learning by this method is known to work well in unsupervised vision problems [16]. Once training is complete, the output layer is removed, with the new representation formed by the activations of all units in the hidden layer.

Whilst sparse autoencoders have previously been adapted to encode temporal coherence, our investigation mixes invariant and variable features in a single representation. Similar to existing methods, temporal coherence is applied by minimizing the difference in feature vectors between consecutive frames for the neurons encoding invariance. Instead of applying temporal coherence across all hidden layer neurons, we only apply the regularization to a variable size subset of the total available units. Remaining neurons are free to discover variable information components. Sparsity is applied across the entire hidden layer, making no distinction between invariant and variable neurons. To compare the mixed representation with existing architectures, we also trained all-invariant and invariant-free versions of our model. A diagram of the network is shown in Fig. 1.

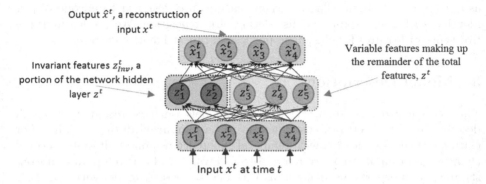

Fig. 1. Diagram of autoencoder architecture. An input x^t of dimensionality M is passed to a hidden layer, with dimensionality N (in this diagram, $M = 4$ and $N = 5$). The hidden layer z^t_{Inv} connects to an output of the same dimension as the input. Temporal coherence regularization is applied across the invariant portion of the hidden layer z^t_{Inv} of dimensionality $P \leq N$. In this case, $P = 2$. If $P = 0$, the network does not encode temporal coherence, when $P = N$ the representation is all-invariant.

3.2   Cost Function

An autoencoder is a fully connected network with two components, the encoder and decoder modules. For time-series data, an input at time t is given as $x^t \in \mathbb{R}^M$, out of a total T frames. Inputs are passed to the autoencoder's hidden layer of neurons via the encoder section. Hidden layer neuron activations $z^t \in \mathbb{R}^N$ are calculated using an affine transformation via the encoder weights $W^E \in \mathbb{R}^{N \times M}$ and bias $b^E \in \mathbb{R}^N$, followed by a suitable non-linearity $f()$ to give $z^t = f(W^E x^t + b^E)$. Similarly, the decoder section output activation values $\hat{x}^t \in \mathbb{R}^M$ are calculated using a similar process via the decoder weights $W^D \in \mathbb{R}^{M \times N}$ and bias $b^D \in \mathbb{R}^M$ to produce $\hat{x}^t = f(W^D z^t + b^D)$. For the non-linearity, we chose the commonly used sigmoidal activation function [17].

The first term in the model cost function is the reconstruction term. By constraining the difference between network input and output to be as small as possible, information loss is avoided. As mentioned previously, this ensures the trivial constant solution is avoided. The reconstruction term L_{Rec} is given as follows, with θ representing the encoder and decoder weights and biases:

$$L_{Rec}(x^t, \theta) = \sum_{t=1}^{T} \|x^t - \hat{x}^t\|^2 \tag{1}$$

We have imposed sparsity in the autoencoder using Kullback-Leibler (KL) Divergence across the hidden layer activations. KL-Divergence is an approximation to L_1-regularization known to perform well on vision problems [18].

During training, the average activation of each hidden layer neuron is calculated over all training examples as $\hat{\rho}_n = \frac{1}{T}\sum_t^T z_n^t$. The sigmoidal activation function constrains active and inactive neurons to generally have values very close to one and zero respectively. This characteristic enables the KL-Divergence term to enforce sparsity by constraining the average activation of each hidden layer neuron to be as close to a desired value ρ. The sparsity term L_{Spar} is calculated as:

$$L_{Spar}(x^t, \theta) = \sum_{n=1}^{N} KL(\rho||\hat{\rho}_n) = \sum_{n=1}^{N} \rho \log \frac{\rho}{\hat{\rho}_n} + (1 - \rho)\log\frac{1-\rho}{1-\hat{\rho}_n} \quad (2)$$

The invariant term L_{Inv} regularises hidden layer neurons with temporal coherence. The term is applied to the invariant subset of the total hidden layer neurons $z_{Inv} \in \mathbb{R}^P$, where $P \leq N$ is an adjustable parameter setting the number of invariant features. When $P = 0$ the network is an invariance-free sparse autoencoder with no additional temporal cost. When $P = N$, temporal coherence is encoded in every hidden layer neuron, similar to the architecture described in [6]. With values inbetween, a mixed representation of invariant and variable features forms. Remaining neurons in the hidden layer are not regularized with temporal coherence, and make up the variable features. It is important to note that the invariant and variable neurons combine to make a single representation. Temporal coherence is enforced in the invariant neurons by minimizing the difference in hidden layer activations for every pair of adjacent frames in the training set:

$$L_{Inv}(x^t, \theta) = \sum_{t=1}^{T-1} \|z_{Inv}^t - z_{Inv}^{t+1}\|^2 \quad (3)$$

Putting these components together, the complete cost function to be minimized during training is given as:

$$L(x^t, \theta) = \sum_{t=1}^{T} \|x^t - \hat{x}^t\|^2 + \alpha \sum_{t=1}^{T-1} \|z_{Inv}^t - z_{Inv}^{t+1}\|^2 + \beta \sum_{n=1}^{N} KL(\rho||\hat{\rho}_n) \quad (4)$$

where the α and β values are hyperparameters used to control the influence of the temporal coherence and sparsity terms respectively. Similar to when $P = 0$, when the α parameter is set to zero, the network is a sparse autoencoder, with no temporal coherence regularization.

4   Evaluation of Our Models

4.1   Datasets

To evaluate the representation learning capacity of our different models, we required datasets consisting of distinct objects moving through a visual field. To keep things simple, we restricted our tests to videos of single objects moving across a uniform blank background. Whilst the unsupervised architecture does not require labelled data, it was necessary that datasets contained object identity labels to allow the training of a supervised classifier acting as a performance metric.

Toy-Data Shapes Dataset. Initial proof-of-concept work was carried out on a toy dataset. The video consists of a sequence of five simple objects moving through a 12×12 pixel visual field (see Fig. 2). For each sequence, a shape is positioned at an edge of the visual field, moving in a perpendicular direction, without changing direction or speed until it hits the opposite side. At this point, the object moves one pixel in a perpendicular direction to its previous motion before making the journey back across the visual field. This process is repeated until an object has traversed the visual field in each of the four cardinal directions. Once an object has exhausted its motion, the next shape in the dataset appears and the sequence starts again until all five shapes have appeared. Object speed remains constant, and there is no rotation. Three sequences from this dataset are shown in Fig. 3.

Fig. 2. The five simple objects used in the toy-data shapes dataset

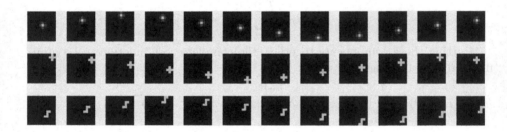

Fig. 3. Three sequences from the toy-data shapes dataset

COIL20-variant Dataset. COIL20 is a well known benchmark dataset for visual problems [11]. The dataset is comprised of 20 different labelled objects, in sequences of 72 images each. Each sequence shows an object rotating on its axis through 360°, in 5° increments. Whilst not strictly a video, the images can be shown in sequence to give a good approximation. As we wanted objects to be able to move around the visual field, COIL20 in its existing form was not appropriate, so we manipulated the dataset for our purposes. COIL20 was reduced from its original 128×128 pixel size down to 24×24 images. For every shape, each 72 image sequence was placed in turn over a blank 48×48 pixel background, with the positioning and movement for each object directed by a random generator. There is an 85 % chance an object carries on moving in the same direction as the previous frame, with reduced probability direction might change by either $\pm 22.5°, 45°$ or $67.5°$. Once an image reached an edge of the visual field, it was sent back in the direction it arrived. This process was repeated 20 times for each object producing 1440 examples for each different shape. Three sequences from our COIL20-variant dataset are shown in Fig. 4.

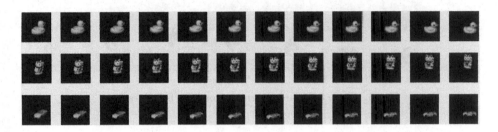

Fig. 4. Three 12 frame example sequences from the COIL20-variant dataset

Fish Dataset. For our final test, we wanted a dataset with increased image, motion and rotation complexity. In their 2011 paper describing object recognition and pose detection using SFA, Franzius, Wilbert and Wiskott used a dataset consisting of 3D models of fish. The fish rotate and move with variable speed and direction [10]. After getting in touch with the authors, they kindly provided us the code with which to create their fish dataset. The parameters used to control the motion of the artificial fish are user provided. We kept to the same settings described in the SFA paper, with a couple of exceptions. Firstly, we removed the chance a fish might randomly switch to a different type of fish between frames. This was done to ensure the dataset contained the same number of examples for each fish. Video sequences were created with 2000 frames of each type of fish. Secondly, to reduce the time taken to train our networks, we only used the first 15 out of the 25 available types of fish in our dataset. Images were resized from 155×155 pixels to 48×48 pixels. Three example sequences of frames from this dataset can be seen in Fig. 5.

4.2 Experimental Setting

Unsupervised Architecture. Our experiments proceeded as follows. For each dataset, the number of hidden layer neurons in the unsupervised architecture was fixed by training networks with an increasing number of neurons. We set the number of hidden layer neurons to be the network size at which the reconstruction term ceased reducing. The number of hidden layer units trained on the shapes dataset was fixed at 120 neurons, COIL20-variant at 400, and the fish dataset at 600. For each experiment, a range of invariant to variable feature splits for the mixed representations were picked. For the shapes dataset, five

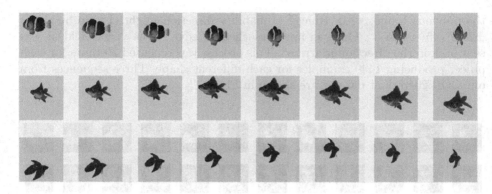

Fig. 5. Three 8 frame sequences from the fish dataset.

different models were chosen, with the number of invariant neurons increasing from 0 to 120 in increments of 30. All remaining neurons were trained without the invariant term applied. For the COIL20-variant and fish dataset, increments were set at 100 neurons, giving 5 and 7 different models respectively.

Due to the relatively small number of examples available in the shapes dataset, 2200 in total, the unsupervised architecture was trained using all available examples. As the COIL20-variant and fish dataset contain significantly more examples, we reduced the training set to half the available examples. To preserve the temporal integrity of the datasets, examples were split up into runs of 50 frames, each run containing image sequences of a single object. For each test, half the available runs for each object were picked at random for the training set, with remaining examples providing a validation set.

Unsupervised training of each model was carried by minimizing the cost function by gradient descent. Two measures were applied to determine training stopping conditions. At increments of 100 epochs, training was halted and the overall cost function and its individual terms were evaluated on the validation set. During each pause, a classifier was also trained to predict object identity using a labelled training set, recording classification accuracy, also on the validation set. Whilst we used a supervised measure for helping determine when to halt training, we did not apply any supervised fine-tuning to guide network weights.

Suitable values for the α and β cost function hyperparameters were chosen by performing a grid search and observing the evolution of the invariant and sparsity terms in the cost function as training progressed.

Supervised Classifier. Temporal coherence is generally used as a method for encoding input variance associated with object identity. For this reason, we passed labelled outputs from each of our various models to a supervised classifier trained to recognise the individual shapes present in the videos. Although we are using a supervised metric to test the performance of our unsupervised architecture, the labelled data has no effect on the representations learned by the various autoencoders.

For the supervised part of our experiments, we also wanted to observe whether the quantity of labelled data made available to the object identity classification task influenced which of the mixed representations produced the best results. For the shapes dataset, we trained a classifier to recognise object identity using one in every two, five and twenty labelled examples from the total available data. For the COIL20-variant and fish datasets, six different classifiers were trained, picking either every second, third, fifth, tenth, twentieth or fiftieth examples. All remaining examples provided the testing set.

4.3   Results

The classification accuracy results from each dataset are presented in Figs. 6, 7 and 8. In each figure, the groups of bars indicate results conducted with the same number of classifier training examples together, with the number indicated on the x-axis. The bars making up each group indicate an increasing number of invariant neurons as a percentage of total neurons available. The results displayed in each figure depict the highest classification accuracy recorded for that particular setting, over any of the cost function hyperparameter values tested.

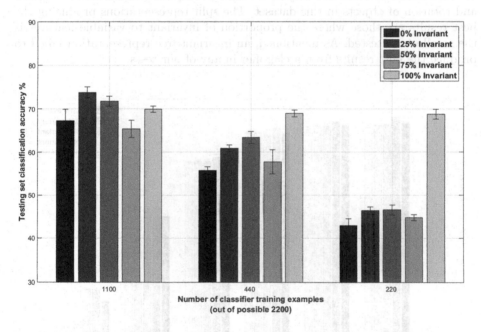

Fig. 6. Classification accuracies for the shapes dataset. Groups of bars display average classification accuracy, along with one standard deviation, from classifiers trained with a set number of examples, the number of which is given on the x-axis. Each bar in a group refers to a variant of the unsupervised architecture. Proportion of total features that are invariant for each bar is provided in the legend. The same notation is used in Figs. 7 and 8

Each test was run four times. The results displayed in each graph represent average classification accuracy on a testing set, with error bars corresponding to one standard deviation.

The most obvious thing standing out from each figure is the reduction in classification accuracy as the amount of labelled data is reduced. As the quantity of training data is reduced, a classifier has less information with which to make its predictions, which correspondingly suffer. More significantly, it is a representation encoding temporal coherence is some shape or form that produces the highest classification accuracy every time. Across every test, regularizing for temporal coherence has improved the capture of information related to object identity.

From the point of view of our experiments, the most important result is the performance observed from the mixed representations. For each dataset, when the quantity of labelled data passed to the classifier is at its largest, a mixed representation always produced the highest classification accuracy. In the case of our shapes and COIL20-variant datasets, optimal performance swings to the invariant only representation when the amount of classifier data is at its lowest. A split representation produces the best result every time for the fish dataset. We have attributed this anomaly to the higher complexity of the images, motion and rotation of objects in this dataset. The split representations producing the best results are those where the proportion of invariant to variable features is the smallest we tested. As mentioned, an invariant-free representation failed to produce the best results from a classifier in any of our tests.

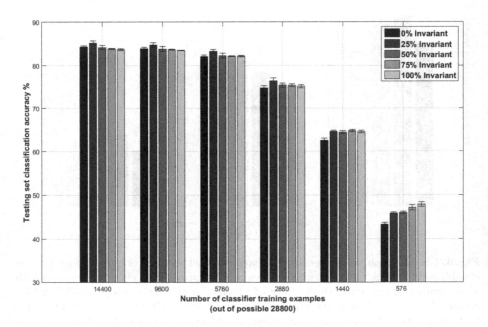

Fig. 7. Classification accuracies for the COIL20-Variant dataset.

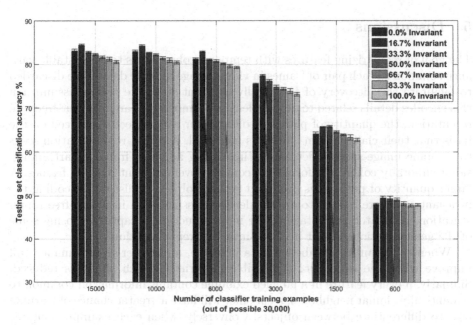

Fig. 8. Classification accuracies for the Fish dataset.

Table 1 gives the results gained when the classifiers are trained using half the available data. Classification accuracy is recorded for the all-invariant and invariant free models, and the best performance from a mixed representation, with the proportion of invariant features indicated. As a final comparison, classification accuracy is also provided from when a classifier is trained using the raw image data. In each of these cases, a mixed representation produced the best performance.

Table 1. Average object classification accuracy from each model when 50 % of data is passed to the classifier. For the mixed representation, values are given for the model producing the highest classification accuracy, with the percentage of invariant neurons that produced this result given. Classification accuracy from each dataset was also recorded for the situation where the classifier is trained on the raw image data, with no representation learning applied.

Dataset	Raw data	Best mixed (% invariant)	Sparse only	All invariant
Shapes	30.16 %	**73.82 %** (25 %)	69.38 %	66.77 %
COIL20-variant	34.06 %	**85.27 %** (25 %)	84.38 %	83.76 %
Fish	57.29 %	**84.51 %** (16.67 %)	83.21 %	80.59 %

5   Discussions

The goal of regularizing features with temporal coherence is to learn stable variance common to each pair of frames in video. Image-specific details are discarded to facilitate the discovery of temporally coherent structure on the assumption this encodes details related to object identity. Compared to an invariant-free representation, the quantity of patterns collected for each object is reduced on the basis that their classification quality is increased. A mixed-representation seeks to combine image-specific details, distinguishing adjacent frames apart, alongside temporally coherent information common over adjacent pairs of frames. A larger quantity of patterns is produced for each object, whilst still encoding the invariant structure related to object identity missing in an invariant-free representation. A greater percentage of the input structure is captured, losing some of the generalization present in an all-invariant representation.

When the quantity of labelled data is scarce, a classifier's performance will improve when the total pool of possible patterns for each object is reduced. Similarly, if every feature in a labelled example contains information common to semantically similar neighbours, a classifier will have a greater chance of learning how to differentiate between objects. Conversely, when each example contains image-specific details not applicable to other examples of the same object, this will confuse a classifier when there is not enough labelled data available to build a full picture of each object. When data is scarce, the generalized features learned by an all-invariant representation improves a classifier's chances of learning the details required to distinguish between objects.

When the quantity of labelled data is increased, classification performance is observed across every model tested, but the benefits of having an all-invariant representation start to dampen. Encoding the structure common to neighbouring examples of distinct objects, alongside a greater amount of image-specific detail, enables a classifier to build a more complete picture of each object from a mixed representation. Information discarded by an all-invariant representation, confusing to a classifier when data is scarce, starts to become useful. When this point is reached, object identity encoding benefits from having as much image-specific data as possible, so long as a portion of features remain invariant. This can clearly be seen from our tests, as a mixed representation predominantly composed of variable features performs optimally over every test where labelled data is abundant.

Optimal performance was observed when a classifier was passed a representation encoding temporal coherence in some manner in all of our tests. Consistent with previous work, extracting temporally stable aspects from video improves the encoding of information related to object identity [12]. When labelled data is plentiful, and a classifier is able to build a more complete picture of the structure underlying each distinct object, applying a mixed representation improves object identity encoding.

5.1   Future Work

The findings of this work lead to quite a wide range of possibilities. Firstly, we have applied our mixed-representation to a relatively simple single layered sparse autoencoder. Temporal coherence has been successfully applied to a wide range of models, including architecturally more complex deep networks. It would be interesting to investigate whether or not applying a mixed-representation to these existing models will boost object identity performance further.

Another route of study would be to consider more complex datasets. The videos in our study were all artificially generated, and only contain a single object at a time moving across a uniform background. Tests could be conducted to observe how a mixed representation copes with more complex images, including more than one object, and natural images.

Finally, whilst we have been capturing a mixture of invariant and variable features, this study has only been concerned with encoding object identity. As discussed in Sect. 2, there has been a small body of work applying temporal coherence to discover the what and the where of objects [10, 15]. As position and motion of objects generally change over quicker time scales than object identity, it is possible that our mixed representation also contains features associated with these properties. This is the primary direction we are going to be looking for our own future studies with this work.

6   Conclusions

We trained a range of sparse autoencoders, encoding temporal coherence over different portions of the available hidden layer units. We can summarize our findings as follows:

- Optimal performance across all tests involved an architecture encoding temporal coherence across a portion of the available features.
- For the majority of tests, the greatest classification performance was achieved when the classifier input received a representation mixing temporally coherent features alongside 'variable' counterparts.
- When a large amount of labelled data was available, a mixed representation always produced the best encoding.

We believe our work demonstrates there are situations where the previously accepted method of applying temporal coherence uniformly across all features is non-optimal. By discovering a mixed representation, consisting of both invariant and variable neurons, object identity encoding can be improved.

Acknowledgements. We'd like to pass on our gratitude to Laurenz Wiskott and Niko Wilbert for providing us with access to their fish dataset, and to Jon Shapiro for the invaluable contribution he has made to this project. This project has been funded by the EPSRC.

References

1. Becker, S.: Learning to categorize objects using temporal coherence. In: Giles, C.L., Hanson, S.J., Cowan, J.D. (eds.) Advances in Neural Information Processing Systems, vol. 5, pp. 361–368 (1993)
2. Hinton, G.E.: Connectionist learning procedures. Artif. Intell. **40**, 185–234 (1989)
3. Foldiak, P.: Learning invariance from transformation sequences. Neural Comput. **3**, 194–200 (1991)
4. Mitchison, G.: Removing time variation with the anti-hebbian differential synapse. Neural Comput. **3**, 312–320 (1991)
5. Olshausen, B., Field, D.J.: Emergence of simple-cell receptive field properties by learning a sparse code for natural images. Nature **381**, 607–609 (1996)
6. Goroshin, R., Bruna, J., Tompson, J., Eigen, D., LeCun, Y.: Unsupervised learning of spatiotemporally coherent metrics. In: Proceedings of the IEEE International Conference on Computer Vision, pp. 4086–4093 (2015)
7. Stone, J., Bray, A.: A learning rule for extracting spatio-temporal invariances. Netw. Comput. Neural Syst. **6**(3), 429–436 (1995)
8. Wiskott, L., Sijnowski, J.: Slow feature analysis: unsupervised learning of invariances. Neural Comput. **14**(4), 715–770 (2002)
9. Wiskott, L., Berkes, P.: Is slowness a learning principle of the visual cortex? Zoology **106**, 373–382 (2003)
10. Franzius, M., Wilbert, N., Wiskott, L.: Invariant object recognition and pose estimation with slow feature analysis. Neural Comput. **23**, 2289–2323 (2011)
11. Nene, S.A., Nayar, S.K., Murase, H.: Columbia Object Image Library (COIL-20), Technical report, CUCS-005-96 (1996)
12. Mobahi, H., Collobert, R., Weston, J.: Deep learning from temporal coherence in video. In: International Conference of Machine Learning 2009, pp. 737–744 (2009)
13. Zou, W.J., Zhu, S., Ng, A., Yu, K.: Deep learning of invariant features via simulated fixations in video. In: Advances in Neural Information Processing Systems, pp. 3212–3220 (2012)
14. Weston, J., Ratle, F., Mobahi, H., Collobert, R.: Deep learning via semi-supervised embedding. In: Montavon, G., Orr, G.B., Müller, K.-R. (eds.) NN: Tricks of the Trade. LNCS, vol. 7700, pp. 639–655. Springer, Heidelberg (2012). doi:10.1007/978-3-642-35289-8_34
15. Gregor, K., LeCun, Y.: Emergence of complex-like cells in a temporal product network with local receptive fields. Technical Report arXiv:1006.0448 (2010)
16. Bengio, Y.: Deep learning of representations: looking forward. In: Dediu, A.-H., Martín-Vide, C., Mitkov, R., Truthe, B. (eds.) SLSP 2013. LNCS (LNAI), vol. 7978, pp. 1–37. Springer, Heidelberg (2013). doi:10.1007/978-3-642-39593-2_1
17. Duch, W., Jankowski, N.: Survey of neural transfer functions. Comput. Surv. **2**, 163–213 (1999)
18. Bradley, D.M., Bagnell, J.A.: Differentiable sparse coding. Adv. Neural Inf. Process. Syst. **21**, 113–120 (2009)

Local Roots: A Tree-Based Subgoal Discovery Method to Accelerate Reinforcement Learning

Alper Demir[✉], Erkin Çilden, and Faruk Polat

Department of Computer Engineering,
Middle East Technical University, 06310 Ankara, Turkey
{ademir,ecilden,polat}@ceng.metu.edu.tr

Abstract. Subgoal discovery in reinforcement learning is an effective way of partitioning a problem domain with large state space. Recent research mainly focuses on automatic identification of such subgoals during learning, making use of state transition information gathered during exploration. Mostly based on the *options framework*, an identified subgoal leads the learning agent to an intermediate region which is known to be useful on the way to goal. In this paper, we propose a novel automatic subgoal discovery method which is based on analysis of predicted shortcut history segments derived from experience, which are then used to generate useful options to speed up learning. Compared to similar existing methods, it performs significantly better in terms of time complexity and usefulness of the subgoals identified, without sacrificing solution quality. The effectiveness of the method is empirically shown via experimentation on various benchmark problems compared to well known subgoal identification methods.

Keywords: Abstraction in reinforcement learning · Subgoal discovery · Options framework

1 Introduction

Subgoal discovery is a prominent way of coping with the scalability problem in reinforcement learning (RL). A subgoal in the problem is a natural hint to partition it into subproblems, so that the agent can focus on learning of smaller tasks, giving rise to opportunities like the transfer of the learned behaviour into a similar problem, and more importantly, increase in learning performance.

Subgoal discovery is almost always coupled with a temporal abstraction mechanism, by which the identified state acts as an artificial target for the problem partition that the agent is trying to solve. A widely accepted temporal abstraction formalism is the *options framework* [23]. An option –which is essentially an abstract action made up of consequent primitive actions through states– defines how to guide the learning agent by making it follow a route to a useful intermediate state, assuming it has the potential to improve learning performance. However, the formalism does not deal with how to decide that an

© Springer International Publishing AG 2016
P. Frasconi et al. (Eds.): ECML PKDD 2016, Part II, LNAI 9852, pp. 361–376, 2016.
DOI: 10.1007/978-3-319-46227-1_23

intermediate state is useful on the way to the ultimate goal. This requirement can effectively be fulfilled by subgoal discovery techniques.

There are a number of different approaches that attack the subgoal discovery problem in RL. Some of the methods are based on graph theory [8,15,19,24], some use statistical methods [3,13,18,20], while others invoke data mining approach [9,11].

Obviously, since the intrinsic focus of RL is on *on-line* performance, it is quite reasonable to expect that the identification of subgoals should better be confluent with the underlying learning procedure. While some methods natively support this paradigm [6,18,19], some others may require additional setup.

In this paper, we propose a subgoal discovery method based on sequence tree based episode history analysis. After each episode, the method first tries to generate a number of successful shortcut policies for every visited state, construct a tree of transitions from generated shortcut policies, and then analyze the tree to extract subgoal states. The method works concurrently with the underlying RL algorithm, and it performs no worse than existing similar methods in terms of solution quality. On the other hand, the method uses less CPU time, and does not depend on any external problem specific variables other than statistical decision parameters. The time complexity of the algorithm is $O((log_b(n))^2)$ on the average, where n is the number of nodes in the generated tree and b is the branching factor. The worst case scenario happens when the agent follows a path through which it visits each state only once causing a tree with branching factor of 1 (which is very unlikely to occur at the initial stages of learning), for which the time complexity is $O(n^2)$.

The paper is organized as follows: A compact summary of the related literature is given in Sect. 2. Section 3 contains the proposed method for subgoal discovery. Experimental evaluation of the proposed algorithm is given in Sect. 4, together with descriptions of problem domains used, parameter settings and a discussion of results. Section 5 includes concluding remarks and possible future research directions.

2   Background

Reinforcement learning (RL) has proven itself to be an effective on-line learning technique [22]. Basically, RL is about self improvements for decisions of a learning agent using environmental feedback. One of the recent advances in RL tries to diminish the diverse effects of increasing state space size, which is known to cause dramatic slow-downs in learning speed. A candidate solution is to partition the problem into manageable pieces and try to solve each first, then ensemble the solutions to obtain the overall result. Lately, subgoal discovery methods have taken attention for this purpose, and are usually coupled with *options framework*.

2.1   Reinforcement Learning

Generally, RL algorithms are constructed on top of a special form of decision process model, called *Markov decision process* (MDP), which possesses *Markov property*, meaning that future states of the process depends solely on the current state. With this restricted model, RL algorithms provide a convergence guarantee to the optimal solution under certain conditions, if one exists.

Formally, MDP is a tuple $\langle S, A, T, R \rangle$, consisting of a finite set of *states* S, a finite set of *actions* A, a transition function $T : S \times A \times S \rightarrow [0, 1]$ where $\forall s \in S$, $\forall a \in A$, $\sum_{s' \in S} T(s, a, s') = 1$, and a reward function $R : S \times A \rightarrow \Re$. $T(s, a, s')$ is the probability of being in state s' if action a is performed in state s. $R(s, a)$ gives the immediate reward from the environment after taking action a in state s. A policy $\pi : S \times A \rightarrow [0, 1]$ is a mapping defining the probability of selecting an action in a state. The aim is to find the optimal policy π^* which maximizes the total expected reward received by the agent. If reward and transition functions were known, the optimal policy could easily be found using classical dynamic programming techniques. Otherwise, π^* can effectively be found by estimating the *value function* (i.e. function giving the value of being in a state on the way to goal) incrementally. *Incremental estimation* approach makes use of the average cumulative rewards over different trajectories obtained by following a policy to calculate the value function and gives rise to the central idea of most RL algorithms, called the *temporal difference* (TD) [21].

A famous TD algorithm using action-values (i.e. Q-values) instead of state-values is named Q-Learning [25], and is widely respected due to its simplicity and ease of use. The update rule for Q-Learning is

$$Q(s, a) \leftarrow (1 - \alpha)Q(s, a) + \alpha[r + \gamma \max_{a' \in A} Q(s', a')] \tag{1}$$

where $\alpha \in [0, 1)$ is the *learning rate* and $\gamma \in [0, 1)$ is the *discount factor*. Q-Learning has been shown to converge to the optimal action-value function denoted by Q^*, under standard stochastic approximation assumptions.

2.2 Options Framework and Macro-Q Learning

An implicit assumption for the MDP model is that an action lasts for a single time step. However, there are acceptable rationales to relax this assumption. An obvious one would be the convenience in the reuse of a behaviour pattern (i.e. skill) in different situations within the problem space. This *abstraction* idea took attention by various researchers, and a few different mainstream approaches emerged [4, 16, 23].

A *Semi-Markov Decision Process* (SMDP) extends the MDP model with transitions of stochastic time duration. An SMDP is a tuple $\langle S, A, T, R, F \rangle$, where S, A, T and R define an MDP, and $F(t|s, a)$ is the probability that starting at s, action a completes within time t. MDP is clearly a specialization of SMDP, where a step function has a jump at 1. In the SMDP model, a policy is still a mapping from states to actions, thus the Bellman equations [1] still hold for an optimal policy [2].

As a prominent abstraction formalism based on the SMDP model, *options framework* [23] devices a way to define and invoke timed actions via incorporation of composite actions on top of an MDP model. It allows to create and use abstract actions (*options*) by using primitive actions, lasting for a finite number of discrete time steps. Briefly, an option is defined by three components: (1) a set of states that the option can be initiated at, called the *initiation set*, (2) option's local policy, and (3) a probability distribution induced by the *termination condition*.

A natural extension of Q-Learning to include options is Macro-Q Learning [14], where the value of each primitive action is again updated according to regular Q-Learning (as given in the update rule 1), while the value of an option is updated according to the following rule:

$$Q(s_t, o_t) \leftarrow Q(s_t, o_t) + \alpha \times (\gamma^n \times \max_{o'} Q(s_{t+n}, o') - Q(s_t, o_t)$$
$$+ r_{t+1} + \gamma r_{t+2} + ... + \gamma^{n-1} r_{t+n})$$

$$(2)$$

where s_t is the starting state of the option o_t, n is the number of steps taken while the option is employed, s_{t+n} is the state that the option terminates at, o' is the option from s_{t+n} that has the maximal value and r_{t+i} is the reward received at time $t + i$. The reward is discounted by the time it is received.

However, the options framework by itself does not guide or help the designer to grasp some useful abstractions. Thus, automatic generation of those abstractions is another interesting research topic, which has its own variety. A widely used approach is subgoal discovery, where the method seeks bottleneck states or regions in the problem space to derive artificial subgoals to be used as terminating points of the options to be generated.

2.3   Automatic Subgoal Discovery

Automatic discovery of subgoals deals with the problem of identifying a set of intermediate points or regions within an MDP, that are "subgoals" or "bottlenecks", naturally partitioning the problem in hand. Due to the vagueness of the concept, a number of different approaches had been developed for subgoal discovery in RL context.

Some of the methods transform the experience history to a transition graph and analyze it to find most suitable bottleneck regions that partitions the problem [8,15,19,24]. Some other methods rely on state visitation statistics to find frequently used states, based on the observation that frequently visited states are more likely to be a bottleneck on the way to goal [3,7,13,18,20]. A yet different approach interprets the same matter as a clustering problem, trying to find separate regions in state space using experiences and then identify access points between regions as subgoals [9,11]. Although not explicitly subgoal-based, a related family of methods focuses on the sequence analysis on episode histories, under the assumption that the subgoals are signaled via reward peaks [6,12].

However, it is not straightforward to determine whether a state is a subgoal or not. In the ideal case, one needs the complete transition function T in order to make an accurate decision, which is practically not possible. Nevertheless, majority of subgoal discovery methods rely on the assumption that an approximate T can be gathered throughout the RL experience. A drawback of this approach is that it may not be possible to decide that approximation of T is accurate enough to be used for subgoal discovery. Alternatively, few other methods use a hybrid approach that brings together locally collected transition information and a means to statistically test its sufficiency for subgoal discovery [17]. This paper focuses on the latter category of subgoal discovery algorithms.

[17] defines an *access state* as a state that connects two or more connected regions having few transitions in between. The key idea is that, a method searching for an access state is allowed to possess only local statistics throughout the experience, to classify a state as either a target state (an access state) or not. Observations that are collected for a state by this way are then used in the following decision rule:

$$\frac{n_+}{n} > \frac{\ln \frac{1-q}{1-p}}{\ln \frac{p(1-q)}{q(1-p)}} + \frac{1}{n} \frac{\ln(\frac{\lambda_{fa}}{\lambda_{miss}} \frac{p(N)}{p(T)})}{\ln \frac{p(1-q)}{q(1-p)}} \tag{3}$$

where n is the total number of observations for a state, n_+ is the total number of positive observations for a state, p is the probability of a positive observation given a target state (an access state), q is the probability of a positive observation given a non-target state, λ_{fa} is the cost of a false alarm, λ_{miss} is the cost of a miss, $p(N)$ is the prior probability of non-target states and $p(T)$ is the prior probability of target states. If the inequality holds, then the state is classified as an access state. This decision rule pinpoints the time step when the collected observations are enough to make a decision about the label of a state. This two-level mechanism enables the methods to avoid the time cost of traversing the whole problem domain.

In the same study, three access state identification methods are proposed. *Relative Novelty (RN)* is a frequency based subgoal identification method, based on the intuition that an *access state* allows the agent to pass from a highly visited region to a new region on the state space. *Local Cuts (L-Cut)* is a graph based method that aims to find a good cut, partitioning the local interaction graph into blocks with a low between-blocks transition probability. *Local Betweenness (LoBet)* is also a graph based algorithm which employs a betweenness measure [5] which is a centrality metric used in graph theory.

Discovered subgoals are of no use unless they are effectively used to diminish the adverse effects of the large state space. Usually, options framework is used to achieve this purpose. For each subgoal identified, an option towards that state is generated. The initiation set for the option is formed by adding the states observed before the subgoal in each trajectory.

One common way of generating the policy of an option to reach a subgoal is the *Experience Replay (ER)* mechanism [10]. ER reuses the past experiences of the agent to find a policy to reach the identified subgoal by providing artificial rewards rather than the actual reward yielded by the environment. A general convention is to provide a positive reward upon reaching the subgoal state, and a negative reward for any other transition.

Algorithm 1. *LOCAL_ROOTS*

Require: p, q, $\frac{\lambda_{fa}}{\lambda_{miss}}$, $\frac{p(N)}{p(T)}$

1: $o_i \leftarrow 0$, $o_i^+ \leftarrow 0$
2: **for** each episode **do**
3:     $h \leftarrow$ Interact with the environment   ▷ record episode history
4:     **if** h ended with a peak reward **then**
5:         $nt_{avg} \leftarrow$ calculate average number of distinct transitions in h
6:         $T \leftarrow CREATE\_TREE(h)$
7:         $CALCULATE\_ROOTING\_FACTORS(T, nt_{avg})$ ▷ for each vertex
8:         **for** $s \in V_T$ **do**
9:             $o_s \leftarrow o_s + 1$
10:            **if** s is a local maximum on T **then**
11:                $o_s^+ \leftarrow o_s^+ + 1$
12:            **end if**
13:            **if** the decision rule is satisfied **then**   ▷ use Decision Rule 3
14:                Classify s as a subgoal
15:            **end if**
16:        **end for**
17:    **end if**
18: **end for**

---

3 Local Roots Method for Subgoal Discovery

We define a subgoal as a state that serves as a junction point or a region of the known shortcut paths from each state to the goal state, which is in fact a likely bottleneck candidate. Following this intuitive definition, our approach depends on the notion of a successful trajectory, that is, a trajectory ending with a distinctive reward peak, which is usually the goal state of the problem domain.

Our method named *Local Roots* generates positive and negative observations for visited states and feeds them to the Decision Rule 3. This is a common pattern in local approaches in subgoal discovery, since the local information gathered from the episode history can be highly dependent on the particular way of state visitations, and thus, may give rise to noisy results without a high level decision filter (especially false positives). Decision Rule 3 is calculated for each visited state, aiming to distinguish the subgoals more accurately.

Local Roots method records the transition history for each episode (Algorithm 1, line 3). Upon completion of an episode, it first checks whether the last transition yields the maximum reward for that episode, or not. If so, it calculates the average number of distinct transitions made through a state (nt_{avg}) and creates a tree using shortcut paths derived using state equivalences, to serve as a collection of the best "memorized" trajectories starting from every visited state up to the goal state (Algorithm 1, line 6). Best trajectories are calculated by traversing from a leaf of the tree to the root, iteratively updating the transitions with the best values. The path from a vertex to the root in the tree forms the shortest path from the corresponding state to the last state in the episode. The tree generation procedure is given in Algorithm 2. The resulting

Algorithm 2. $CREATE\_TREE$

Require: a successful episode trajectory h
Ensure: a tree T representing shortcut histories to goal from each state
1: $t \leftarrow length(h) - 2$
2: $V_{s_{t+1}} \leftarrow 0$, $best_{s_{t+1}} \leftarrow null$
3: **while** $t \geq 0$ **do**
4:     **if** V_{s_t} is undefined $\vee$ $(r_{t+1} + \gamma * V_{s_{t+1}}) > V_{s_t}$ **then**
5:         $V_{s_t} \leftarrow r_{t+1} + (\gamma * V_{s_{t+1}})$
6:         $best_{s_t} \leftarrow s_{t+1}$
7:     **end if**
8:     $t \leftarrow t - 1$
9: **end while**
10: $V \leftarrow \{s_l\}$, $E \leftarrow \emptyset$
11: **for** each state $s \neq s_l$ **do**
12:     $V \leftarrow V \cup \{s\}$
13:     $E \leftarrow E \cup (s, best_s)$
14: **end for**
15: **return** (V, E)

---

tree is a collection of shortcut paths (i.e. free of loops) from every visited state to the goal state, based on the local transition graph derived from experiences.

The core idea of the proposed method lies in a state metric, what we call the *rooting factor*, due to the visual resemblance to a root structure of a tree in the nature fringing underground. To clarify the idea, Fig. 1a illustrates a sample grid world domain made up of three rooms with passageways between adjacent rooms, and Fig. 1b is a tree generated for the problem by using an episode history. The goal state is s_{76} which is located in the south-east corner of the room and the agent starts from the north-west corner, namely s_0. The agent can move to any one of the four compass directions at each time step, except that after a move attempt to the walls and the boundaries of the room, it stays still.

The grid world and the tree instance given in Fig. 1 are colored according to the rooting factor values of states scaled from black to white, where brighter color means a higher value. Note that the states in the doorways have high rooting factor values. In the case of state s_{48} located at level 9, the rooting factor metric focuses on the sub-tree having state s_{48} as the root. To calculate the rooting factor of state s_{48}, one should first find the *widest level* for the sub-tree rooted by state s_{48}, which is the level possessing the first peak in sub-tree width. In this sub-tree, the widths of each level are 1, 3, 5, 5, 5, 5, 3 consecutively, and the first peak value in terms of width is 5. The method considers level 11 as the widest level and ignores the second peak with value 5 starting in level 17 since level 11 has the first peak. This way, even if there is a wider level below, it is not taken into account for state s_{48}. Thus, each bottleneck state in the tree possesses its relative sub-tree, as is the case for another subgoal s_{22} in level 15 where the widest level of its relative sub-tree is level 17.

Upon identification of the widest level for state s_{48}, one can think of an imaginary triangle (i.e. the dotted triangle in Fig. 1b) where the vertices in the

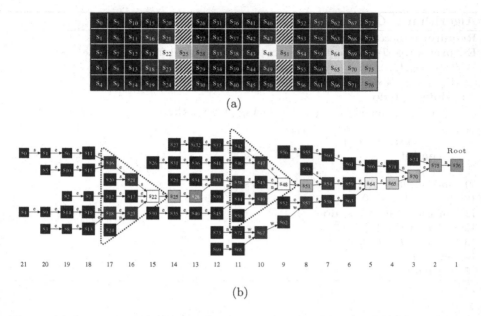

Fig. 1. (a) A sample grid world with two consecutive subgoals, colored according to rooting factor values of the states. Shaded cells represent walls. (b) The generated tree, using the same coloring scheme. Actions are noted on the edges. The numbers at the bottom are corresponding levels of the tree.

widest level compose its base, and its topmost corner is s_{48} (w.r.t. a portrait orientation of the tree, where root is at the top). The shape of this triangle is an indication of the "importance" of state s_{48} in the tree. A wider triangle suggests that, for relatively more states, the agent should pass through state s_{48} in order to reach the root state (i.e. goal). The height of the triangle, on the other hand, pinpoints a state which is the "first" junction point of the merging paths. That is why, the rooting factor of state s_{48} is higher than its parent's, state s_{51}.

As a mathematical interpretation of the above characteristics, the rooting factor of s can be defined as follows:

$$r_s = \frac{(n_{widest})^{nt_{avg}}}{d_{widest} - d_s} \tag{4}$$

where d_s is the depth of s in the tree, nt_{avg} is the average number of distinct transitions of states in the tree, n_{widest} is the number of vertices in the widest level and d_{widest} is the depth of the states in that level. In order to strengthen the effect of possible connections that a vertex can have, number of vertices in the widest level is powered by the average number of distinct transitions (nt_{avg}).

After the tree is constructed by using Algorithm 2, the rooting factor calculation takes place for every vertex (Algorithm 1, line 7). Algorithm 3 is employed for this purpose, where a tree traversal (via breadth first search, BFS) is employed first, to find the depth of each vertex in the tree. An addi-

---

Algorithm 3. $CALCULATE\_ROOTING\_FACTORS$

Require: a successful history tree T, nt_{avg}
1: calculate the depth of each state in T $\triangleright$ use BFS
2: $d_{max} \leftarrow \max_{s \in V_T}(\text{depth}(s))$ $\triangleright$ find maximum depth
3: **for** every $s \in V_T$ **do**
4:     $n_i(s) \leftarrow$ number of nodes at depth $i \geq \text{depth}(s)$ in the subtree rooted at s
5: **end for**
6: **for** each state s in V_T **do** $\triangleright$ rooting factor calculation for every vertex
7:     $d_s \leftarrow \text{depth}(s)$
8:     $i \leftarrow d_s, n_{widest} \leftarrow 1, d_{widest} \leftarrow d_s$
9:     **while** $i \leq d_{max}$ and $n_i(s) \geq n_{widest}$ **do**
10:        **if** $n_i(s) > n_{widest}$ **then**
11:           $n_{widest} \leftarrow n_i(s)$
12:           $d_{widest} \leftarrow i$
13:        **end if**
14:        $i \leftarrow i + 1$
15:     **end while**
16:     $r_s \leftarrow$ calculate the rooting factor of s using Eq. 4
17: **end for**

---

tional traversal is run afterwards, from the level with the deepest state(s) to the root, to find the number of vertices below each vertex classified by their depths (Algorithm 3, lines 3–7). Using this information, the rooting factor of each state can be calculated by traversing from the state under consideration to the deeper levels.

Having calculated the rooting factor values for every visited state, each state is checked whether it is a local maximum or not, in terms of rooting factors, among its children and parent in the tree. The state gets a positive observation if that is the case, or a negative observation otherwise. The root of the entire tree does not get a positive observation since it is a possible goal state and is obviously not a subgoal. The observations made for each state are fed to the Decision Rule 3 for a further classification.

The most time consuming portion of the Local Roots algorithm is the part where the rooting factor value is calculated for each state which has $O(n^2)$ worst time complexity. However, the worst case happens when the tree is linear (although it is usually unlikely at the initial states of learning due to the exploration component, the agent might execute a policy that visits each state only once) and the branch factor (b) is 1, which means, by our definition, there is no new subgoal to identify. Thus, the worst case can be avoided by a heuristic check on the shape of the generated tree. On the other hand, the algorithm performs $O((\log_b(n))^2)$ on the average, where n is the number of nodes in the tree. Since the algorithm makes use of local episode trajectories, the number of nodes in the tree (n) does not directly relate to the number of states in the whole domain.

4   Experiments

We tested the algorithms on four grid world navigation domains (Fig. 2), three of
the which are well known benchmark problems in the related literature. State and
action set sizes of problems and corresponding references are given in Table 1. A
new 3 rooms grid world problem (Fig. 2b) is designed to investigate the subgoal
identification behaviour of the methods in a 3-way junction situation. Local
Roots, just like the other similar methods, transforms the problem to a transition
graph. Thus, the method is essentially independent of domain specific structure.
Our motivation for experimenting on grid world domains is to better visualize
the bottleneck idea for the reader.

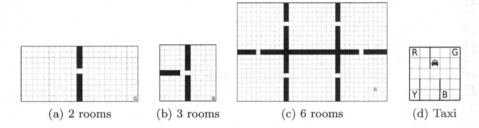

(a) 2 rooms   (b) 3 rooms   (c) 6 rooms   (d) Taxi

Fig. 2. Problem domains

In 2, 3 and 6 rooms problems, the agent can perform four movement actions,
which are *north, east, south* and *west*. The environment is non-deterministic,
since the agent moves to the intended direction with probability of 0.9 and moves
randomly in any of the movement directions with 0.1 probability. The reward
for reaching the goal state G is 1.0 while the reward for any other transition is 0.
In the 2 and 3 rooms problems, the agent starts from any cell in the left room(s)
while it starts from any cell in the upper left room in the 6 rooms domain.

The last problem is the famous Taxi domain (Fig. 2d, [4]), in which a taxi
tries to pick a passenger from its location and transfer it to a destination location
in a 5 × 5 grid world with designated locations. The taxi agent can perform 6
actions: movement actions *north, east, south, west*; a *pickup* action to get the
passenger, and a *putdown* action to drop the passenger. The passenger is initially
located in 4 different cells marked as R, Y, G and B, and the destination of the
passenger is one of these four designated cells. The actions are noisy, leading the
agent to its intended direction with probability of 0.8 and randomly moving it
to the left or the right of the intended direction with 0.1 probability each. The
agent is punished for wrong pickups and putdowns with −10 and it is rewarded
with +20 when it puts down its passenger in the desired location. Any other
transition is given the reward of −1.

4.1   Settings

We compared Local Roots method (LoRoots) with RN, LoBet and L-Cut, since
they use the same decision rule and can be employed on-line. The decision rule

Table 1. Problem sizes and parameter values used

Problem	Size		Parameters used							Ref.
	$|S|$	$|A|$	Method	p	q	t_c	t_{RN}	k	l_n	
2 rooms	201	4	RN	0.06	0.01	-	2.0	2	7	[19]
			L-Cut	0.3	0.01	0.05	-	-	-	
			LoBet	0.7	0.07	-	-	-	-	
			LoRoots	0.6	0.06	-	-	-	-	
3 rooms	106	4	RN	0.05	0.01	-	2.0	2	7	-
			L-Cut	0.1	0.01	0.05	-	-	-	
			LoBet	0.6	0.06	-	-	-	-	
			LoRoots	0.6	0.06	-	-	-	-	
6 rooms	605	4	RN	0.5	0.008	-	2.0	2	7	[15]
			L-Cut	0.2	0.01	0.05	-	-	-	
			LoBet	0.5	0.05	-	-	-	-	
			LoRoots	0.75	0.05	-	-	-	-	
Taxi	500	6	RN	0.712	0.01	-	2.0	2	7	[4]
			L-Cut	0.04	0.002	0.05	-	-	-	
			LoBet	0.3	0.03	-	-	-	-	
			LoRoots	0.24	0.03	-	-	-	-	

parameters were optimized separately for each method and problem so that they find subgoals in the early stages of learning and they eliminate noise properly. The cost ratio ($\lambda_{fa}/\lambda_{miss}$) and the prior ratio ($p(N)/p(T)$) parameters of Decision Rule 3 were set to 100 for all experiments. The visitation counts used by RN were reset at the end of each episode. The remaining parameters used by the subgoal identification methods are given in Table 1. Unfortunately, there is no practical way to find the correct values other than a number of trial-and-error experimentation sessions. Specifically, a heuristic we used to set p and q values is, to examine the outputs of the used subgoal discovery methods and calibrate them according to the subgoals that we manually identified. Other parameters are mostly inherited from [17] where a further analysis can be found.

When a subgoal is found, the agent generates an option to reach that subgoal. The initiation set of the new option contained the states before the first occurrence of the subgoal in each previous episode. *Option lag* (l_o), the number of time steps to look for states to add the initiation set, was 10. Termination probability for each state in the initiation set was set to 0.0, while 1.0 was used for the subgoal. The policy of the option was formed through ER by giving 100 reward upon reaching the subgoal, -10 punishment for leaving the initiation set and -1 punishment for any other transition. For the policy learning part of ER, $\alpha = 0.125$ and $\gamma = 0.9$ were used as the learning parameters. The replay is repeated 10 times for fast convergence.

The agent incorporated Macro-Q learning algorithm, where Q values of an option were updated according to Macro-Q learning while Q values of primitive actions were updated according to regular Q learning. ϵ-greedy was used as option selection strategy, with $\epsilon = 0.1$, and $\alpha = 0.05$ and $\gamma = 0.9$ were set as learning parameters. The same γ value is used in Algorithm 2. All of the results are averaged over 200 experiments.

4.2   Results and Discussion

Average number of steps to reach the goal state are compared among methods, and the results are sketched in Fig. 3. Plots are smoothed for visual clarity. All of the subgoal identification methods, including Local Roots, improve the learning speed of the agent by leading it to the goal state earlier and our proposed method matches the performance of the other methods. In general, subgoals discovered by Local Roots seem to be as useful as the ones found by L-Cut, LoBet and RN. We can conclude that the solution quality of Local Roots method is not worse than others on the average.

LoBet algorithm has $O(n \cdot m)$ and $O(n \cdot m + n^2 \cdot log n)$ time complexities on unweighted and weighted graphs respectively, while L-Cut algorithm requires $O(n^3)$ time when the local interaction graph has n vertices and m edges. On the other hand, RN algorithm is $O(1)$. The time complexity of our proposed method, Local Roots, depends on the maximum depth of a state in the tree it creates.

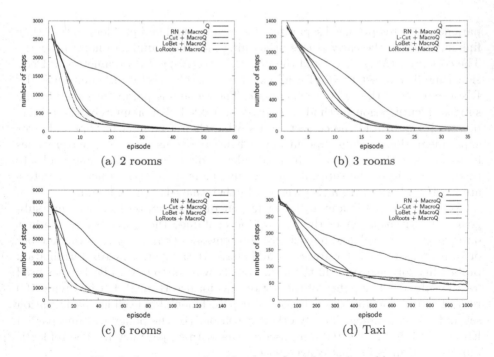

(a) 2 rooms

(b) 3 rooms

(c) 6 rooms

(d) Taxi

Fig. 3. Average number of steps to goal for each problem

It requires $O((\log_b(n))^2)$ time where b is the average branching factor and n is the number of vertices in the tree.

CPU time measurements, indicating the CPU times used by each subgoal discovery method excluding the underlying Macro-Q algorithm, are given in Table 2. The results shows that both LoBet and L-Cut require much more time than Local Roots because of their higher time complexities. The only exception is LoBet for Taxi problem, whose time consumption seems to be nearly the same as in Local Roots. On the other hand, although RN is $O(1)$, its time complexity is in fact associated with the number of steps taken, since it is invoked at every time step, unlike the other methods waiting for the episode end. Longer episodes in the earlier stages of an experiment causes RN to generally take more time than Local Roots. Table 2 implies that Local Roots algorithm shows a significant advantage in terms of CPU time compared to other methods.

Table 2. Average CPU time overhead per episode (msec)

Problem	RN	L-Cut	LoBet	LoRoots
2 rooms	0.63	5.18	0.65	0.45
3 rooms	0.33	1.78	0.32	0.24
6 rooms	11.00	289.30	7.10	4.08
Taxi	0.85	1.68	0.79	0.81

Figure 6 shows the subgoals discovered by all four methods for 2 rooms problem, marked with brighter color showing high frequency of identification. L-Cut, LoBet and RN finds more than one subgoals including the doorway and states one step near to it. On the other hand, our proposed method finds only the state before the doorway as it is the first merging point of the shortest paths of the states in the left room to the goal state in the right room. This characteristic causes Local Roots to find less number of subgoals than the other algorithms, especially in 2, 3 and 6 rooms domains. However, as seen Fig. 4, effectiveness of subgoals discovered are usually higher in Local Roots method compared to the others. We define the *effectiveness of a subgoal* as the $(100 \times n_{steps(option)}/n_{steps(episode)})/n_{subgoals}$, where $n_{steps(option)}$ is the total number steps passed within option sequences, $n_{steps(episode)}$ is the total number of steps taken during the episode, and $n_{subgoals}$ is the number of subgoals identified at the end of an episode. Subgoal effectiveness can be interpreted as the ability of a subgoal to trigger a useful option. Contribution of some of the additional subgoals found by the other three methods are not as significant as that are found by Local Roots in general.

Finally, average memory usage of Local Roots does not exceed the graph based methods (i.e. LoBet and L-Cut) in general, since it uses a tree instead of a graph, having less number of edges than the graphs used by the other methods. It is worth noting that, memory usage metrics also include ER repositories (Fig. 5).

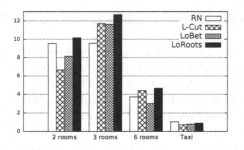

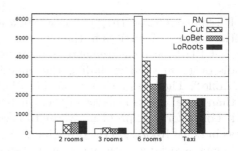

Fig. 4. Average subgoal effectiveness (average option trace % per subgoal)

Fig. 5. Average memory usage per episode in kilobytes

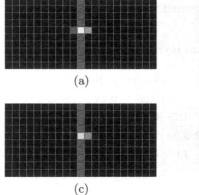

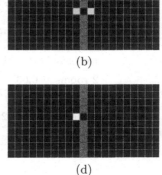

(a)   (b)

(c)   (d)

Fig. 6. Subgoals found in 2 rooms domain by (a) L-Cut (b) LoBet (c) RN (d) Local Roots.

In addition to the parameters of the Decision Rule 3, L-Cut requires one and RN requires four more parameters while LoBet and Local Roots require none. These extra parameters determine the quality of the subgoals found by L-Cut and RN and make them dependent on the structure of the domain. In that sense, Local Roots, like LoBet, is less dependent on the problem characteristics compared to L-Cut and RN. Moreover, as seen in Fig. 4, Local Roots outperforms LoBet in terms of subgoal quality.

5   Conclusion

In this paper, we propose a tree based automatic subgoal discovery method called Local Roots that helps the learning agent to identify important states on the way to the goal state in the early stages of learning. Local Roots method can be employed upon reward peaks, which are usually goal states. Using the options framework, the learning agent can devise abstractions to reach the identified subgoals. The method utilizes a tree based metric to locally identify the junction points of the shortcuts directed from each visited state towards the goal state.

In terms of learning speed, Local Roots outperforms the regular Q-Learning for all problem domains experimented. It also keeps up with the performance of the other local methods on the average, showing that subgoals identified by Local Roots are no worse than the ones found by other algorithms.

Compared to other graph based methods tested, Local Roots has lower time complexity. On the other hand, when average CPU times per episode are compared, Local Roots outperforms all other methods on the average, including Relative Novelty which has the lowest theoretical time complexity, but should be invoked at every time step. Local Roots is also shown to identify less number of subgoals with higher effectiveness in general. Moreover, it requires no additional parameters unlike Relative Novely and Local Cuts.

A possible future research direction is to find an alternative way of discriminating noise from local subgoal information with less domain specific parameters. Also, automatic detection of these parameters can be an important improvement for all the online methods presented here.

Acknowledgments. This work is partially supported by the Scientific and Technological Research Council of Turkey under Grant No. 215E250.

References

1. Bellman, R.E.: Dynamic Programming. Princeton University Press, Princeton (1957)
2. Bradtke, S.J., Duff, M.O.: Reinforcement learning methods for continuous-time markov decision problems. In: Tesauro, G., Touretzky, D., Leen, T. (eds.) Advances in Neural Information Processing Systems, NIPS 1994, vol. 7, pp. 393–400. MIT Press, Cambridge (1994)
3. Chen, F., Chen, S., Gao, Y., Ma, Z.: Connect-based subgoal discovery for options in hierarchical reinforcement learning. In: Lei, J., Yao, J., Zhang, Q. (eds.) Proceedings of the Third International Conference on Natural Computation, ICNC 2007, vol. 4, pp. 698–702. IEEE (2007)
4. Dietterich, T.G.: Hierarchical reinforcement learning with the MAXQ value function decomposition. J. Artif. Intell. Res. **13**(1), 227–303 (2000)
5. Freeman, L.C.: A set of measures of centrality based on betweenness. Sociometry **40**(1), 35–41 (1977)
6. Girgin, S., Polat, F., Alhajj, R.: Improving reinforcement learning by using sequence trees. Mach. Learn. **81**(3), 283–331 (2010)
7. Goel, S., Huber, M.: Subgoal discovery for hierarchical reinforcement learning using learned policies. In: Russell, I., Haller, S.M. (eds.) Proceedings of the 16th International FLAIRS Conference, pp. 346–350. AAAI Press (2003)
8. Kazemitabar, S.J., Beigy, H.: Automatic discovery of subgoals in reinforcement learning using strongly connected components. In: Köppen, M., Kasabov, N., Coghill, G. (eds.) ICONIP 2008, Part I. LNCS, vol. 5506, pp. 829–834. Springer, Heidelberg (2009). doi:10.1007/978-3-642-02490-0_101
9. Kheradmandian, G., Rahmati, M.: Automatic abstraction in reinforcement learning using data mining techniques. Robot. Auton. Syst. **57**(11), 1119–1128 (2009)
10. Lin, L.J.: Self-improving reactive agents based on reinforcement learning, planning and teaching. Mach. Learn. **8**(3), 293–321 (1992)

11. Mannor, S., Menache, I., Hoze, A., Klein, U.: Dynamic abstraction in reinforcement learning via clustering. In: Proceedings of the Twenty-first International Conference on Machine Learning, ICML 2004, pp. 71–78. ACM (2004)

12. McGovern, A.: acQuire-macros: an algorithm for automatically learning macro-actions. In: The Neural Information Processing Systems Conference Workshop on Abstraction and Hierarchy in Reinforcement Learning, NIPS 1998 (1998)

13. McGovern, A., Barto, A.G.: Automatic discovery of subgoals in reinforcement learning using diverse density. In: Proceedings of the Eighteenth International Conference on Machine Learning, ICML 2001, pp. 361–368. Morgan Kaufmann Publishers Inc. (2001)

14. McGovern, A., Sutton, R.S., Fagg, A.H.: Roles of macro-actions in accelerating reinforcement learning. In: Proceedings of the 1997 Grace Hopper Celebration of Women in Computing, pp. 13–18 (1997)

15. Menache, I., Mannor, S., Shimkin, N.: Q-Cut–Dynamic discovery of sub-goals in reinforcement learning. In: Elomaa, T., Mannila, H., Toivonen, H. (eds.) ECML 2002. LNCS, vol. 2430, pp. 295–306. Springer, Heidelberg (2002)

16. Parr, R., Russell, S.: Reinforcement learning with hierarchies of machines. In: Jordan, M., Kearns, M., Solla, S. (eds.) Advances in Neural Information Processing Systems, NIPS 1997, vol. 10, pp. 1043–1049. MIT Press (1998)

17. Simsek, O.: Behavioral building blocks for autonomous agents: description, identification, and learning. Ph.d. thesis, University of Massachusetts Amherst (2008)

18. Simsek, O., Barto, A.G.: Using relative novelty to identify useful temporal abstractions in reinforcement learning. In: Proceedings of the Twenty-First International Conference on Machine Learning, ICML 2004, pp. 95–102. ACM (2004)

19. Simsek, O., Wolfe, A.P., Barto, A.G.: Identifying useful subgoals in reinforcement learning by local graph partitioning. In: Proceedings of the Twenty-second International Conference on Machine Learning, ICML 2005, pp. 816–823. ACM (2005)

20. Stolle, M., Precup, D.: Learning options in reinforcement learning. In: Koenig, S., Holte, R.C. (eds.) SARA 2002. LNCS, vol. 2371, pp. 212–223. Springer, Heidelberg (2002)

21. Sutton, R.S.: Learning to predict by the methods of temporal differences. Mach. Learn. 3(1), 9–44 (1988)

22. Sutton, R.S., Barto, A.G.: Reinforcement Learning: An Introduction. Adaptive Computation and Machine Learning. MIT Press, Cambridge (1998)

23. Sutton, R.S., Precup, D., Singh, S.: Between MDPs and semi-MDPs: a framework for temporal abstraction in reinforcement learning. Artif. Intell. 112(1–2), 181–211 (1999)

24. Taghizadeh, N., Beigy, H.: A novel graphical approach to automatic abstraction in reinforcement learning. Robot. Auton. Syst. 61(8), 821–835 (2013)

25. Watkins, C.: Learning from delayed rewards. Ph.d. thesis, Cambridge University (1989)

Copula PC Algorithm for Causal Discovery from Mixed Data

Ruifei Cui[✉], Perry Groot, and Tom Heskes

Institute for Computing and Information Sciences, Radboud University,
Nijmegen, The Netherlands
{R.Cui,Perry.Groot,T.Heskes}@science.ru.nl

Abstract. We propose the 'Copula PC' algorithm for causal discovery from a combination of continuous and discrete data, assumed to be drawn from a Gaussian copula model. It is based on a two-step approach. The first step applies Gibbs sampling on rank-based data to obtain samples of correlation matrices. These are then translated into an average correlation matrix and an effective number of data points, which in the second step are input to the standard PC algorithm for causal discovery. A stable version naturally arises when rerunning the PC algorithm on different Gibbs samples. Our 'Copula PC' algorithm extends the 'Rank PC' algorithm, which has been designed for Gaussian copula models for purely continuous data. In simulations, 'Copula PC' indeed outperforms 'Rank PC' in cases with mixed variables, in particular for larger numbers of data points, at the expense of a slight increase in computation time.

Keywords: Causal discovery · Gaussian copula · Mixed data · Extended rank likelihood

1 Introduction

Causal discovery, or causal structure learning [23], aims to find an underlying directed acyclic graph (DAG), which represents direct causal relations between variables. It is a very popular approach for multivariate data analysis and therefore is widely studied in the past few years, resulting in lots of algorithms. The PC [27,28] algorithm can be considered the reference causal discovery algorithm. It makes use of conditional independence tests to build the underlying DAG from observations. Starting from a complete undirected graph, the PC algorithm removes edges recursively according to the outcome of the conditional independence tests. This procedure yields an undirected graph, also called the skeleton. After applying various edge orientation rules, it finally gives back a partially directed graph to represent the underlying DAGs.

One advantage of the PC algorithm is that it is computationally feasible for sparse graphs even with thousands of variables. Therefore, it is widely used in high-dimensional settings, generating a variety of applications [20,29]. Also, open-source software is available like *pcalg* [17] and the Tetrad project [25].

© Springer International Publishing AG 2016
P. Frasconi et al. (Eds.): ECML PKDD 2016, Part II, LNAI 9852, pp. 377–392, 2016.
DOI: 10.1007/978-3-319-46227-1_24

When applied to Gaussian models, the PC algorithm tests conditional independence using partial correlation based on Pearson correlations between variables: when the joint distribution is a multivariate Gaussian, pairwise conditional independence is equivalent to the vanishing of the corresponding partial correlation [18]. Following [12], we will refer to the PC algorithm for Gaussian models as the 'Pearson PC' algorithm. As input it takes the correlation matrix of the observed data and the number of data points. The number of data points is needed for the conditional independence tests: the higher the number of data points, the more reliable the observed correlation matrix as an estimate of the (unknown) true correlation matrix, and the more easily the null hypothesis of conditional independence (given the same value for the partial correlation and the significance level) gets rejected. Under relatively mild assumptions regarding the sparseness of the true underlying DAG, the 'Pearson PC' algorithm shows uniform consistency [16].

Harris and Drton [12] extend the PC algorithm to non-parametric Gaussian (nonparanormal) models, i.e., continuous data assumed to be generated from a Gaussian copula model. They propose to apply the standard PC algorithm, but then replacing the Pearson correlation matrix with rank-based measures of correlation. The so-called 'Rank PC' (RPC) algorithm works as well as the 'Pearson PC' algorithm on normal data and much better on non-normal data, and is shown to be uniformly consistent in high-dimensional settings.

In this paper, we aim to generalize the 'Pearson PC' and 'Rank PC' algorithm to Gaussian copula models that can also handle binary and ordinal variables. The 'Rank PC' algorithm is explicitly limited to the continuous situation, where ties appear with probability zero, making ranks well-defined. In the presence of binary and ordinal variables, ties make the rank correlations between observed variables different from those between the corresponding latent variables in the Gaussian copula setting. Ignorance of this difference typically leads to underestimates of the (absolute) correlations [13].

It is tempting to follow a similar two-step approach as for 'Rank PC': first estimate the correlation matrix in the latent space and then use this as input to the standard PC algorithm. This, however, is not as straightforward as it may seem, for two reasons. First, because of the ties, estimating the correlation matrix of Gaussian copula models for mixed data is considerably more complicated. Second, the ties imply a loss of information, which makes that our estimate of the correlation matrix will tend to be less reliable than in the fully continuous case, which should be accounted for when applying the conditional independence tests in the PC algorithm.

To solve both issues, we propose to make use of a Gibbs sampling procedure, specifically the one derived by Hoff [13] based on the so-called extended rank likelihood. This procedure is relatively straightforward and easy to implement (see the code in the Appendix of [13]). For purely Gaussian data, the correlation matrix samples follow a specific kind of inverse-Wishart distribution [3], which we refer to as the projected inverse-Wishart distribution. Projected inverse-Wishart distributions are characterized by two parameters: the scale matrix and the degrees of freedom; the former relates to the average correlation matrix and the latter to the number of data points. As we will show, under the projected inverse-Wishart, the variance of each off-diagonal element of the correlation matrix is an approximate function of its expectation and the degrees of freedom: the more degrees of freedom, the smaller the variance. The idea is now to estimate the scale matrix and degrees of freedom from the Gibbs samples of more general Gaussian copula models on mixed data, as if they were also drawn from a projected inverse-Wishart distribution. The scale matrix is translated into a correlation matrix and the degrees of freedom into a so-called 'effective number of data points', to take into account the reliability of our estimate of the correlation matrix. These are then input to the standard PC algorithm for causal discovery.

We refer to our two-step procedure as the 'Copula PC' (CoPC) algorithm. We also derive a stable version, referred to as 'Stable Copula PC' (SCPC), which runs PC repeatedly on a number of Gibbs samples. Experimental results show that both CoPC and SCPC outperform the current 'Rank PC' algorithm in mixed databases with discrete and continuous variables.

The rest of this paper is organized as follows. Section 2 reviews some relevant background information and analyzes issues of existing algorithms in more detail. Section 3 proposes an approximate inference method for the correlation matrix and the effective number of data points based on the projected inverse-Wishart distribution, and then derives the resulting algorithms CoPC and SCPC. Section 4 compares CoPC and SCPC with the 'Rank PC' algorithm on simulated data and provides an illustration on real-world data of ADHD patients. Section 5 gives conclusions and future work.

2 Preliminaries and Problem Analysis

In this section, we first review some necessary background information on causal discovery, then briefly introduce the PC and 'Rank PC' algorithm, and finally analyze the challenges current PC algorithms face for mixed data.

2.1 Causal Structure Learning

A graph $G = (V, E)$ consists of a set of vertices $V = \{X_1, \ldots, X_p\}$, representing random variables, and a set of edges E, representing relations between pairs of variables. A graph is *directed* if it only contains directed edges while it is *undirected* if it only contains undirected edges. Graphs containing both directed and undirected edges are called *partially directed* graphs. A graph with no directed cycles, e.g., $X_i \to X_j \to X_i$ is *acyclic*. A graph which is both directed and acyclic is a *Directed Acyclic Graph* (DAG). If there is an edge from X_i to X_j, X_i is the parent of X_j. The set of parents of X_j in graph G is denoted by $pa(G, X_j)$.

A multivariate probability distribution P over variables $V = \{X_1, \ldots, X_p\}$ is said to *factorize* according to a DAG $G = (V, E)$, if the joint probability density function of P can be written as the product of the conditional densities of each variable given its parents in G, i.e., $f(X_1, \ldots, X_p) = \prod_{i=1}^{p} f(X_i | pa(G, X_i))$.

If this condition holds, we can read off conditional independence relationships in distribution P from the DAG via a graphical criterion called *d-separation* [24]. *D-separation* implies that each variable is independent of its non-descendants given its parents.

Several DAGs may, via *d-separation*, correspond to the same set of conditional independencies. Such DAGs form a Markov equivalence class, which can be uniquely represented by a completed partially directed acyclic graph (CPDAG) [6]. Arcs in a CPDAG indicate a cause-effect relation between variables since the same arc occurs in all members of the CPDAG. Undirected edges $X_i - X_j$ in a CPDAG indicate that some of its members contain an arc $X_i \rightarrow X_j$ whereas other members contain an arc $X_j \rightarrow X_i$. The aim of causal discovery is to learn the Markov equivalence class of a DAG $G = (V, E)$ from n i.i.d. observations of V.

2.2   PC Algorithm and Rank PC Algorithm

The PC algorithm starts from a complete undirected graph, and then removes edges recursively according to conditional independencies yielding a partially connected undirected graph called the skeleton, after which some orientation rules are applied to direct as many edges as possible, resulting in a completed partially directed acyclic graph, i.e. the underlying CPDAG.

During the process, testing conditional independence plays the most important role. The PC algorithm uses partial correlation, denoted by $\rho_{uv|S}$, to do it. The correlation matrix from independent observations of a random vector Z can be used to estimate $\rho_{uv|S}$ [2]. Then, classical decision theory is applied to judge conditional independencies using significance level α,

$$Z_u \perp\!\!\!\perp Z_v | Z_S \Leftrightarrow \sqrt{n - |S| - 3} \left| \frac{1}{2} \log \left(\frac{1 + \hat{\rho}_{uv|S}}{1 - \hat{\rho}_{uv|S}} \right) \right| \leq \Phi^{-1}(1 - \alpha/2), \quad (1)$$

where $u \neq v$ and $S \subseteq \{1, \ldots, p\} \backslash \{u, v\}$. Thus in order to run the PC algorithm we need the correlation matrix corresponding to the data to estimate $\hat{\rho}_{uv|S}$ and the number of observations n.

The PC algorithm has been extended to a broader class of Gaussian copula by using rank correlations to replace Pearson correlations, resulting in the 'Rank PC' algorithm [12]. Rank correlations, typically Spearman's ρ and kendall's τ, only consider the ranks among observations, ignoring the actual variables.

Definition 1 (Gaussian Copula Model). *Consider two random vectors $Z = (Z_1, \ldots, Z_p)$ and $Y = (Y_1, \ldots, Y_p)$, satisfying the conditions $Z \sim \mathcal{N}(0, C)$ and $Y_i = F_i^{-1}(\Phi(Z_i))$ for $i = 1, \ldots, p$ where C denotes the correlation matrix of Z and $F_i^{-1}(t) = \inf\{y : F_i(y) \geq t\}$ is the pseudo-inverse of a cumulative distribution function F_i. Then this model is called Gaussian copula model with correlation matrix C and univariate margins F_i.*

In the Gaussian copula model, when all margins are continuous, ties occur with zero probability making ranks well-defined. For such so-called nonparanormal models, the sample correlations among ranks can naturally be used as an estimator for the Pearson correlation in the latent space. In this nonparanormal setting, RPC has been shown to perform well [12,19].

2.3 Challenges for Mixed Data

RPC works well on continuous data, because tied observations occur with probability zero. In the presence of discrete margins, however, the estimator used in RPC is no longer consistent because of the tied observations. In this case, standard rank-based correlation will be different from the true correlation in latent space [13], typically underestimating it. Hence, our first challenge is to estimate the underlying C efficiently and consistently from mixed data.

A second challenge concerns the information loss incurred by discrete variables. Specifically, simply setting n in Eq. (1) to the number of data points can lead to an underestimate of the p-values provided by the conditional independence tests. To solve this problem, we introduce the notion of an effective number of data points.

3   Approximate Inference and Copula PC Algorithm

In this section, we introduce an approximate inference approach for the underlying correlation matrix and the effective number of data points from mixed data. Subsection 3.1 introduces the projected inverse-Wishart distribution and its application to Gaussian models. Subsection 3.2 discusses how to obtain correlation matrix samples from mixed data using a Gibbs sampling procedure. Subsection 3.3 shows how to use these samples to estimate the two parameters of the projected inverse-Wishart distribution: the scale matrix (as the underlying correlation matrix) and the degrees of freedom (as the effective number of data points). Subsection 3.4 gives the resulting Copula PC algorithm and the Stable Copula PC algorithm.

3.1   Projected Inverse-Wishart Distribution

Priors on correlation matrices are typically derived by choosing the inverse-Wishart distribution, denoted by $\mathcal{W}^{-1}(\Sigma; \Psi_0, \nu)$, as a prior on covariance matrices and then turning the covariance matrices into a correlation matrix to end up with an implied distribution on the correlation matrix. We choose Σ from $\mathcal{W}^{-1}(\Sigma; \Psi_0, \nu)$ and write

$$P(C) = \mathcal{P}\mathcal{W}^{-1}(C; \Psi_0, \nu) \tag{2}$$

where $C_{ij} = \frac{\Sigma_{ij}}{\sqrt{\Sigma_{ii}\Sigma_{jj}}}$ for $\forall\, i, j$. Since many covariance matrices possibly correspond to the same correlation matrix, the above process can be considered as a projection from covariance matrices to a correlation matrix. Therefore, we refer to this distribution on correlation matrix C as a projected inverse-Wishart distribution.

For Gaussian models, the projected inverse-Wishart distribution gives exact inference [21]. Specifically, given data $\boldsymbol{Z} = (\boldsymbol{z_1}, \ldots, \boldsymbol{z_n})$, the posterior reads

$$P(\Sigma|\boldsymbol{Z}) = \mathcal{W}^{-1}(\Sigma; \Psi_0 + \Psi, \nu + n) \text{ and } P(C|\boldsymbol{Z}) = \mathcal{P}\mathcal{W}^{-1}(C; \Psi_0 + \Psi, \nu + n),$$

with $\Psi = \boldsymbol{Z}^T \boldsymbol{Z}$. Also, the projected inverse-Wishart is scale invariant [3,14], in the sense that we can make the posterior distribution on correlation matrices independent of the scale of the data by choosing $\Psi_0 = 0$, or perhaps better, $\Psi_0 = \epsilon \mathbb{1}$ in the limit $\epsilon \downarrow 0$.

Summarizing, we consider the prior distribution

$$P(\Sigma) = \mathcal{W}^{-1}(\Sigma; \epsilon \mathbb{1}, p + 1) \text{ in the limit } \epsilon \downarrow 0,$$

which in fact boils down to the well-known improper Jeffreys prior [32]:

$$P(\Sigma) \propto \|\Sigma\|^{-(p+1)}.$$

For Gaussian copula models, although there is no analytical expression, we still expect that the posterior $P(C|\boldsymbol{Y})$ can be approximated through a projected inverse-Wishart distribution, i.e., $P(C|\boldsymbol{Y}) \approx \mathcal{P}\mathcal{W}^{-1}(C; \Psi, \nu)$ for some Ψ and ν.

3.2   Gibbs Sampler Based on Extended Rank Likelihood

Hoff [13] describes an elegant procedure to obtain samples from $P(C|\boldsymbol{Y})$ for a Gaussian copula model. The essence is that we only consider the ranks among observations, hence the name extended rank likelihood, ignoring the actual variables. Since the cumulative distribution functions $F_i(Y_i)$ are non-decreasing, observing $y_{i_1,j} < y_{i_2,j}$ implies that $z_{i_1,j} < z_{i_2,j}$, where $y_{i_1,j}$ denotes the i_1^{th} observation of the jth component of random vector $\boldsymbol{Y}$, To be precise, observing $\boldsymbol{Y} = (\boldsymbol{y_1}, \ldots, \boldsymbol{y_n})$ tells us that $\boldsymbol{Z} = (\boldsymbol{z_1}, \ldots, \boldsymbol{z_n})$ must lie in the set

$$\left\{ \boldsymbol{Z} \in \mathbb{R}^{n \times p} : \max\left\{ z_{k,j} : y_{k,j} < y_{i,j} \right\} < z_{i,j} < \min\left\{ z_{k,j} : y_{i,j} < y_{k,j} \right\} \right\}.$$

Strong posterior consistency for C under the extended rank likelihood has been proved in the situation with both discrete and continuous marginal distribution functions [22].

An off-the-shelf sampling algorithm based on the extended rank likelihood is full Gibbs sampling [13]. The code of this sampling algorithm is provided in the Appendix of [13]. In this algorithm, each component of $\boldsymbol{Z}$ is initialized according to the rank information of the corresponding component of $\boldsymbol{Y}$, after which each component is resampled alternatively. Here we propose a slight modification by just resampling the discrete components instead of all of them. Experimental tests reveal that the results of this faster sampling approach are indistinguishable from Hoff's original Gibbs sampler. Although this modification is quite straightforward, it significantly reduces computation time because sampling continuous variables is far more time-consuming than sampling discrete ones in Hoff's Gibbs sampler. We will refer to this modified sampling algorithm as *SamplingAlgo*.

So, given the observed data $\boldsymbol{Y}$, samples on the underlying correlation matrix, denoted by $\{C^{(1)}, \ldots, C^{(m)}\}$, can be obtained using *SamplingAlgo*.

3.3 Estimation of the Correlation Matrix and the Effective Number of Data Points

This subsection aims to estimate the underlying correlation matrix and the effective number of data points from the obtained samples.

Theorem 1 suggests a procedure to estimate the parameters Ψ and ν from samples of a projected inverse-Wishart distribution $\mathcal{PW}^{-1}(C;\Psi,\nu)$.

Theorem 1. *If the correlation matrix C follows a projected inverse-Wishart distribution with parameters Ψ $(\Psi_{ii} = 1)$ and ν, i.e.,*

$$P(C) = \mathcal{PW}^{-1}(C;\Psi,\nu),$$

then for each off-diagonal element $C_{ij}(i \neq j)$ and large ν, we have

$$\mathrm{E}\left[C_{ij}\right] \approx \Psi_{ij} \; and \, \mathrm{Var}\left[C_{ij}\right] \approx \frac{(1-(\Psi_{ij})^2)^2}{\nu}.$$

The proof is given in the Appendix.

According to Theorem 1, the mean over samples of C is an excellent approximation of Ψ. As for ν, we have,

$$\nu \approx \frac{(1-(\mathrm{E}\left[C_{ij}\right])^2)^2}{\mathrm{Var}\left[C_{ij}\right]}. \tag{3}$$

The idea now is to apply the same estimates, as if the samples obtained by Gibbs sampling the Gaussian copula model on mixed data also (approximately) follow a projected inverse-Wishart distribution. Specifically, for the effective number of data points $\hat{n}$, we propose to take the average over all $p(p-1)/2$ estimates on ν that can be computed by applying (3) to each upper triangular element of a p-dimensional correlation matrix C.

3.4 Copula PC Algorithm and Stable Copula PC Algorithm

Now, we turn the previous results into a working algorithm. The two key input arguments of the 'Pearson PC' algorithm are the correlation matrix and the number of data points. In the general Gaussian copula model, we take the mean over $\{C^{(1)}, \ldots, C^{(m)}\}$ and the mean over $p(p-1)/2$ estimates on ν as the two arguments respectively, resulting in the Copula PC algorithm.

Next, we introduce a stable version of the Copula PC algorithm. We take l instances from all the m samples. For each instance, a corresponding graph can be obtained via the 'Pearson PC' algorithm using the earlier estimated effective number of data points, by which a collection of l graphs can be generated, denoted by $\{\widetilde{G}_1, \ldots, \widetilde{G}_l\}$. We keep those edge marks that emerge with a probability higher than a pre-defined threshold β and remove the others, leading to a resulting graph. Since this resulting graph seemingly contains only 'stable' edge

Algorithm 1. Copula PC algorithm and its stable version

1: **Input**: Observations $\boldsymbol{Y}$, Initialized parameters m, l, β
2: **Output**: Causal graph G_c by CoPC, G_s by SCPC
3: $C^{(1)}, \ldots, C^{(m)} = SamplingAlgo(\boldsymbol{Y})$
4: **for** all C_{ij} with $i < j$ (upper triangular elements) **do**
5:     Compute and store $\nu_k = \frac{(1-(\mathrm{E}\,[C_{ij}])^2)^2}{\mathrm{Var}\,[C_{ij}]}$ ▷ Equation (3)
6: **end for**
7: $\hat{n}$ = the average over $\{\nu_1, \ldots, \nu_{p(p-1)/2}\}$
8: **if** CoPC **then**   ▷ procedures for CoPC
9:     $\hat{C} = \frac{1}{m} \sum_{j=1}^{m} C^{(j)}$
10:     $G_c = pc(\hat{C}, \hat{n})$ ▷ the 'Pearson PC' algorithm
11: **else**   ▷ procedures for SCPC
12:     Choose l $(l < m)$ instances from $C^{(1)}, \ldots, C^{(m)}$
13:     **for** $i = 1 : l$ **do**
14:         Compute and store $\widetilde{G}_i = pc(C^{(l_i)}, \hat{n})$
15:     **end for**
16:     **for** all edge marks **do**
17:         e = the number of graphs containing the current edge mark
18:         **if** $e/l > \beta$ **then**
19:             keep the edge mark
20:         **end if**
21:         G_s = all kept edge marks among $\{\widetilde{G}_1, \ldots, \widetilde{G}_l\}$.
22:     **end for**
23: **end if**

---

marks, we call this method stable Copula PC algorithm (SCPC). The size of l has a linear influence on running time because choosing l means the 'Pearson PC' algorithm would run l times. As for β, a small value means keeping more edge marks and vice versa. The Copula PC algorithm and its stable version are summarized in Algorithm 1.

4   Experiments

In this section, we first verify the property of the projected inverse-Wishart distribution described by Eq. (3) and check whether it still holds in the presence of discrete variables. Then, we compare the proposed CoPC and SCPC with the 'Rank PC' algorithm on simulated data and give an illustration on real-world data of ADHD patients.

Following Kalisch and Bühlmann [16], we simulate random DAGs and draw samples from the distributions faithful to them. Firstly, we generate an adjacency matrix A, whose entries are zero or in the interval $[0.1, 1]$. There exists a directed edge from i to j in the corresponding DAG, if $i < j$ and $A_{ji} \neq 0$. The DAGs generated in this way have the property $\mathrm{E}\,(N_i) = s(p-1)$, where N_i is the number of neighbors of node i, and s is the probability that there is an edge between any two nodes, called the sparseness parameter. Then, the samples of a random vector $\boldsymbol{Z}$ are drawn through

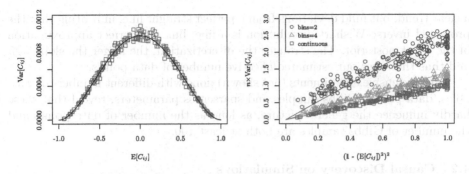

Fig. 1. The relationship between the expectation and the variance of the elements of sampled correlation matrices. Left panel: The samples are drawn from a given projected inverse-Wishart distribution. Right panel: The samples are drawn via *SamplingAlgo*, with circles for binary cases, triangles for ordinal cases with 4 levels, and squares for continuous cases.

$$Z = AZ + \epsilon, \tag{4}$$

where $\epsilon = (\epsilon_1, \ldots, \epsilon_p)$ is a vector of independent standard normal random variables. The data generated in this way follow a multivariate Gaussian distribution.

4.1 Estimation for the Effective Number of Data Points

As argued in Subsect. 3.3, the expectation and variance of the elements of correlation matrices drawn from a projected inverse-Wishart distribution are strongly related. To check this relationship, we proceed as follows: (1) we generate a random p-dimensional correlation matrix Ψ; (2) we draw 500 samples from a projected inverse-Wishart distribution with parameters Ψ and ν; (3) for each upper triangular element, we plot its variance against its expectation.

The left panel in Fig. 1 shows a typical result for $p = 20$ and $\nu = 1000$. We see that almost all pairs are distributed around the theoretical curve (solid line) especially when the expectation is far from zero, which indicates that it is indeed possible to infer ν of a projected inverse-Wishart distribution via the expectation and variance of off-diagonal elements.

Next, we study how our inference method works for estimating $\hat{n}$ in different cases. We first generate n samples of Z using Eq. (4) and discretize some of the variables to obtain the simulated samples of the observed random vector Y. Then, we run *SamplingAlgo* to get samples of the underlying C. The results for $p = 20$ and $n = 1000$ for different cases are shown in Fig. 1 (right panel), where 'bins $= 2$' means that all variables are binary, 'bins $= 4$' means that all variables are ordinal with 4 levels and 'continuous' means that all variables are kept continuous. We take $(1 - (\mathrm{E}\,[C_{ij}])^2)^2$ for the x-axis and $n \times \mathrm{Var}\,[C_{ij}]$ for the y-axis, so that all data points are expected to be distributed around a straight line with slope $n/\hat{n}$. For purely continuous variables, a straight line with slope 1 gives an almost perfect fit, as expected. For ordinal and binary variables, we still find a clear trend, but mild deviations from a perfect straight line, indicating that the projected inverse-Wishart distribution is a fine, but not perfect approximation of the exact posterior. The stronger the discretization, the larger the slope $n/\hat{n}$ and thus the lower our estimated effective number of data points.

More extensive experiments (not shown) done with different numbers of variables, data points, Gibbs samples and sparseness parameters, reveal that these hardly influence the general picture, as long as the number of data points and the number of Gibbs samples are both at least 100.

4.2   Causal Discovery on Simulations

In this subsection, we compare CoPC and SCPC with the 'Rank PC' [12] algorithm. All computations are implemented in the R-package *pcalg*.

We first generate multivariate normal data (p variables) via Eq. (4). After that, 25 % of all p variables are discretized into binary variables, and another 25 % of them are discretized into ordinal variables with 5 levels. In this way, we simulate the observations of Y which are generated from a Gaussian copula model with both discrete and continuous margins.

Three measures are used to test the performance: (1) percentage of correct edges in the resulting skeleton, usually called true positive rate (TPR); (2) percentage of spurious edges, usually called false positive rate (FPR); (3) Structural Hamming Distance (SHD), counting the number of edge insertions, deletions, and flips in order to transfer the estimated CPDAG into the correct CPDAG [30]. The first two measures are for the skeleton while SHD is for the CPDAG. A smaller SHD indicates better performance.

Next, we compare the performance of three versions of the PC algorithm, RPC, CoPC, and SCPC in terms of TPR, FPR, and SHD. We restrict the significance level to $\alpha = 0.01$, which has been shown to yield the best overall SHD [16]. For CoPC, we drop the first 20 Gibbs samples and save the next 100 samples ($m = 100$). For SCPC, we take $l = 20$ equidistant samples, so $\{C^{(1)}, C^{(6)}, \ldots, C^{(96)}\}$, and choose β such that the TPR for SCPC is more or less equal to that of RPC, which amounts to $\beta = 0.4$ for sparse graphs with 10 nodes, $\beta = 0.45$ for sparse graphs with 50 nodes, and $\beta = 0.3$ for dense graphs. The remaining parameters are set as follows: $p \in \{10, 50\}$, $n \in \{500, 1000, 2000, 5000\}$, and $E[N] \in \{2 \ (Sparse), 5 \ (Dense)\}$.

The comparative results in Fig. 2 (10 nodes) and Fig. 3 (50 nodes) provide the mean over 100 repeated experiments and errorbars representing 95 % confidence intervals. First, for sparse graphs (both small and large graphs), the three algorithms get nearly the same results w.r.t. TPR, but CoPC and SCPC show a large advantage over RPC w.r.t. FPR and SHD except SCPC with large graphs, which becomes more prominent with increasing sample size. Second, for dense graphs, the advantage of CoPC and SCPC over RPC still exists w.r.t. FPR, although seemingly CoPC performs a little worse than SCPC and RPC w.r.t. TPR. Third, we note that the performance of RPC deteriorates seriously w.r.t. FPR with the increase in sample size, while CoPC and SCPC are very stable. Apparently, using sample size as the effective number of data points, RPC incurs

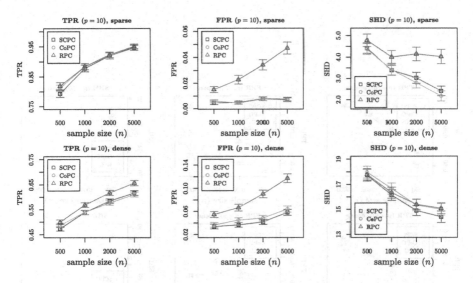

Fig. 2. Performance of Rank PC, Copula PC, and Stable Copula PC for 10 nodes, showing the mean of TPR, FPR, and SHD over 100 experiments together with 95 % confidence intervals. The first row represents the results with sparse graphs ($E[N] = 2$) while the second row represents those with dense graphs ($E[N] = 5$).

more false positives especially for larger sample sizes. Overall, CoPC and SCPC clearly outperform RPC, especially in the sparse cases with larger sample sizes.

4.3   Application to Real-World Data

In this subsection, we give an illustration on a real-world dataset on phenotypic information about children with Attention Deficit Hyperactivity Disorder (ADHD) [5]. It contains 23 variables for 245 subjects. We focus on nine variables as in [26], but keep all subjects with missing values since these are easily handled by the Gibbs sampler. The nine variables considered are: gender (G), attention deficit level (AD), hyperactivity/impulsivity level (HI), verbal IQ (VIQ), performance IQ (PIQ), full IQ (FIQ), aggressive behavior (Agg), medication status (Med), handedness (HN), where four of them (G, Agg, Med, HN) are binary.

We run CoPC and SCPC ($l = 30$, $\beta = 0.4$) on the dataset and consider prior knowledge that no variable can cause gender. The resulting graphs are shown in Figure 4. The graphs suggest that gender has an effect on attention deficit level, which then causes hyperactivity/impulsivity level. This point has been confirmed by many studies [4], [31]. It is common that AD and Agg cause patients to take medicine. Also, VIQ, PIQ, and FIQ are connected to each other by bi-directed edges. This indicates that the causal sufficiency assumption is violated, i.e., that there should be a latent common cause related to IQ, as also suggested in [26].

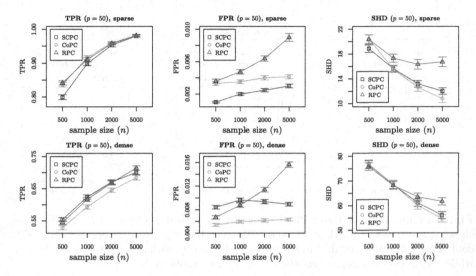

Fig. 3. Performance of Rank PC, Copula PC, and Stable Copula PC for 50 nodes, showing the mean of TPR, FPR, and SHD over 100 experiments together with 95 % confidence intervals.

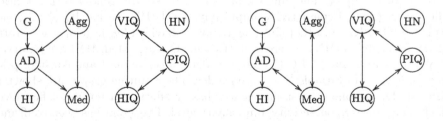

Fig. 4. The resulting graphs by CoPC (left panel) and SCPC (right panel) on ADHD dataset.

5   Conclusions and Future Work

In this paper we introduced a novel two-step approach for estimating the causal structure underlying a Gaussian copula model on mixed data. The essence is to estimate the correlation matrix in the latent space, which can then be given to any causal discovery algorithm to search for its underlying structure. Ties between the discretized observations incur information loss, making the estimate of correlation matrix less reliable than in fully continuous cases. For this, we introduced the notion of 'effective number of data points' that can be estimated from the expectation and variance of the correlation matrix elements. Our approach, based on ranks and correlation matrices, is fully scale invariant and has a natural uninformative setting when choosing a uniform distribution over pairwise correlations, which can be adjusted to account for different assumptions.

We like to think of our two-step approach as a general principle, where for each of the two steps one could plug in one's favorite choice: e.g., a different MCMC method [15] or a MAP approach along the lines of [1] for estimating the correlation matrix and its reliability, and another method, like FCI [28] or BCCD [7], for causal structure learning. Having generated samples, running the PC algorithm several times to gain an insight into the reliability of structure estimates is an obvious thing to do. Similar procedures have been proposed, e.g., by bootstrapping the original dataset [8,10]. In our simulations, the Gibbs sampler appears to converge quite fast, which makes Gibbs sampling cheap compared to running the PC algorithm, in particular for models with many variables. Our choice to only resample the discrete random variables and not the continuous ones, here also helps. Being fully Bayesian about structure learning as well may be very nice in theory [11], but is computationally infeasible in practice for any reasonable number of variables. Altogether, our Bayesian approach to sample correlation matrices in combination with a more frequentist approach towards structure learning attempts to combine the best of both worlds.

Our methods require the setting of just a few parameters: the significance level α to be used in the PC algorithm (typically 0.01 or 0.05), the number of Gibbs samples and burn-in (the more, the better), and for SCPC, the number of instances l in the ensemble (the more, the better), and the threshold β (the higher, the more conservative).

Our estimate of the 'effective number of data points' appears to work nicely in practice, but can and perhaps should be further improved. Instead of considering the variance of the elements of the correlation matrix, one may come up with another, more direct estimate, for example the entropy of the distribution and translate that into an effect number of data points. Preliminary attempts in that direction failed by being typically much less robust than the one described in this paper. Our current estimate gives a single, global value for the effective number of data points. Future work may consider estimating a different value for each conditional independence test, since each test only relies on a local structure, involving only part of the variables. Such estimates then can be integrated into the causal discovery algorithm itself. Another line of future research concerns the theoretical analysis of CoPC and SCPC, where it can be studied to what extent and under which conditions consistency can be proven. Our conjecture here is that consistency of our two-step procedure follows from the proven consistency of the two separate steps: Gibbs sampling to estimate the correct correlation matrix C [22] and the PC algorithm to arrive at the correct causal structure [16].

Appendix: Proof of Theorem 1

Consider partitioning the matrix Σ and Ψ as

$$\Sigma = \begin{bmatrix} \Sigma_{aa} & \Sigma_{ab} \\ \Sigma_{ba} & \Sigma_{bb} \end{bmatrix} \text{ and } \Psi = \begin{bmatrix} \Psi_{aa} & \Psi_{ab} \\ \Psi_{ba} & \Psi_{bb} \end{bmatrix}.$$

Then, if $P(\Sigma) = \mathcal{W}^{-1}(\Sigma; \Psi, \nu)$, we have

$$\begin{aligned} P(\Sigma_{aa}) &= \mathcal{W}^{-1}(\Sigma_{aa}; \Psi_{aa}, \nu - \dim(b)), \\ P(\Sigma_{bb|a}) &= \mathcal{W}^{-1}(\Sigma_{bb|a}; \Psi_{bb|a}, \nu), \\ P(\Sigma_{aa}^{-1}\Sigma_{ab}|\Sigma_{bb|a}) &= \mathcal{N}(\Sigma_{aa}^{-1}\Sigma_{ab}; \Psi_{aa}^{-1}\Psi_{ab}, \Sigma_{bb|a} \otimes \Psi_{aa}^{-1}), \end{aligned} \tag{5}$$

where $\dim(b)$ is the dimension of Σ_{bb} and $\Sigma_{bb|a} = \Sigma_{bb} - \Sigma_{ba}\Sigma_{aa}^{-1}\Sigma_{ab}$ [9].

Without loss of generality, we restrict our analysis to a two-dimensional system and suppose that we draw

$$\Sigma \sim \mathcal{W}^{-1}\left(\begin{pmatrix} 1 & \rho \\ \rho & 1 \end{pmatrix}, \nu\right).$$

Then, according to (5), we have

$$\Sigma_{11} \sim \mathcal{W}^{-1}(1, \nu - 1), \quad \Sigma_{22|1} \sim \mathcal{W}^{-1}(1 - \rho^2, \nu), \quad \Sigma_{11}^{-1}\Sigma_{12}|\Sigma_{22|1} \sim \mathcal{N}(\rho, \Sigma_{22|1}).$$

Rewriting the resulting $\hat{\rho}$ in terms of these variables, we obtain

$$\hat{\rho} = \frac{\Sigma_{12}}{\sqrt{\Sigma_{11}}\sqrt{\Sigma_{22}}} = \frac{(\Sigma_{11}^{-1}\Sigma_{12})\sqrt{\Sigma_{11}}}{\sqrt{\Sigma_{22|1} + \Sigma_{11}(\Sigma_{11}^{-1}\Sigma_{12})^2}}. \tag{6}$$

Since for large ν,

$$\mathrm{E}[\Sigma_{11}] = \frac{1}{\nu - 3} \approx \frac{1}{\nu}, \quad \mathrm{E}[\Sigma_{22|1}] = \frac{1 - \rho^2}{(\nu - 2)} \approx \frac{1 - \rho^2}{\nu},$$

$$\mathrm{Var}[\Sigma_{11}] = \frac{2}{(\nu - 3)^2(\nu - 5)} \approx \frac{2}{\nu^3}, \mathrm{Var}[\Sigma_{22|1}] = \frac{2(1 - \rho^2)^2}{(\nu - 2)^2(\nu - 4)} \approx \frac{2(1 - \rho^2)^2}{\nu^3}.$$

we can approximate,

$$\Sigma_{11} \approx \frac{1}{\nu}\left(1 + \sqrt{\frac{2}{\nu}}x\right), \Sigma_{22|1} \approx \frac{1 - \rho^2}{\nu}\left(1 + \sqrt{\frac{2}{\nu}}y\right)$$

$$\Sigma_{11}^{-1}\Sigma_{12} \approx \rho + \sqrt{\frac{1 - \rho^2}{\nu}}z, \tag{7}$$

where x, y, and z are independent random variables, all with mean zero and unit variance. Indeed, for large ν, all noise terms scale with $\sqrt{1/\nu}$ relative to the mean, and can hence be ignored when computing the expectation, to yield, as expected,

$$E\left[\hat{\rho}\right] \approx \rho. \tag{8}$$

To estimate the variance, we substitute (7) into (6), and compute (in leading order, and evaluated for $x = y = z = 0$),

$$\frac{\partial \hat{\rho}}{\partial x} \approx \rho(1-\rho^2)\sqrt{\frac{1}{2\nu}}, \quad \frac{\partial \hat{\rho}}{\partial y} \approx \rho(1-\rho^2)\sqrt{\frac{1}{2\nu}}, \quad \frac{\partial \hat{\rho}}{\partial z} \approx (1-\rho^2)^{3/2}\sqrt{\frac{1}{\nu}},$$

yielding the variance

$$\mathrm{Var}\left[\hat{\rho}\right] = \left(\frac{\partial \hat{\rho}}{\partial x}\right)^2 + \left(\frac{\partial \hat{\rho}}{\partial y}\right)^2 + \left(\frac{\partial \hat{\rho}}{\partial z}\right)^2 \approx \frac{(1-\rho^2)^2}{\nu}. \tag{9}$$

References

1. Abegaz, F., Wit, E.: Penalized EM algorithm and copula skeptic graphical models for inferring networks for mixed variables, arXiv preprint (2014). arXiv:1401.5264
2. Anderson, T.W.: An Introduction to Multivariate Statistical Analysis. Wiley, Hoboken (2003)
3. Barnard, J., McCulloch, R., Meng, X.L.: Modeling covariance matrices in terms of standard deviations and correlations, with application to shrinkage. Statistica Sinica, pp. 1281–1311 (2000)
4. Bauermeister, J.J., Shrout, P.E., Chávez, L., Rubio-Stipec, M., Ramírez, R., Padilla, L., Anderson, A., García, P., Canino, G.: ADHD and gender: are risks and sequela of ADHD the same for boys and girls? J. Child Psychol. Psychiatry 48(8), 831–839 (2007)
5. Cao, Q., Zang, Y., Sun, L., Sui, M., Long, X., Zou, Q., Wang, Y.: Abnormal neural activity in children with attention deficit hyperactivity disorder: a resting-state functional magnetic resonance imaging study. Neuroreport 17(10), 1033–1036 (2006)
6. Chickering, D.M.: Learning equivalence classes of Bayesian-network structures. J. Mach. Learn. Res. 2, 445–498 (2002)
7. Claassen, T., Heskes, T.: A Bayesian approach to constraint based causal inference, arXiv preprint (2012). arXiv:1210.4866
8. Dai, H., Li, G., Zhou, Z.: Ensembling MML causal discovery. In: Dai, H., Srikant, R., Zhang, C. (eds.) PAKDD 2004. LNCS (LNAI), vol. 3056. Springer, Heidelberg (2004)
9. Eaton, M.L.: The Wishart distribution. In: Multivariate Statistics. Lecture Notes-Monograph Series, vol. 53, pp. 302–333. Institute of Mathematical Statistics, Beachwood, Ohio, USA (2007)
10. Entner, D., Hoyer, P.O.: On causal discovery from time series data using FCI. Probabilistic graphical models, pp. 121–128 (2010)
11. Friedman, N., Koller, D.: Being Bayesian about network structure. A Bayesian approach to structure discovery in Bayesian networks. Mach. Learn. 50(1–2), 95–125 (2003)

12. Harris, N., Drton, M.: PC algorithm for nonparanormal graphical models. J. Mach. Learn. Res. **14**(1), 3365–3383 (2013)
13. Hoff, P.D.: Extending the rank likelihood for semiparametric copula estimation. Ann. Appl. Stat. **1**(1), 265–283 (2007)
14. Huang, A., Wand, M.P., et al.: Simple marginally noninformative prior distributions for covariance matrices. Bayesian Anal. **8**(2), 439–452 (2013)
15. Kalaitzis, A., Silva, R.: Flexible sampling of discrete data correlations without the marginal distributions. In: Advances in Neural Information Processing Systems, pp. 2517–2525 (2013)
16. Kalisch, M., Bühlmann, P.: Estimating high-dimensional directed acyclic graphs with the PC-algorithm. J. Mach. Learn. Res. **8**, 613–636 (2007)
17. Kalisch, M., Mächler, M., Colombo, D., Maathuis, M.H., Bühlmann, P.: Causal inference using graphical models with the R package pcalg. J. Stat. Softw. **47**(11), 1–26 (2012)
18. Lauritzen, S.L.: Graphical Models. Clarendon Press, Oxford (1996)
19. Liu, H., Han, F., Yuan, M., Lafferty, J., Wasserman, L.: High-dimensional semiparametric Gaussian copula graphical models. Ann. Stat. **40**(4), 2293–2326 (2012)
20. Maathuis, M.H., Colombo, D., Kalisch, M., Bühlmann, P.: Predicting causal effects in large-scale systems from observational data. Nat. Methods **7**(4), 247–248 (2010)
21. Murphy, K.P.: Conjugate Bayesian analysis of the Gaussian distribution. Def **1**(2), 16 (2007)
22. Murray, J.S., Dunson, D.B., Carin, L., Lucas, J.E.: Bayesian Gaussian copula factor models for mixed data. J. Am. Stat. Assoc. **108**(502), 656–665 (2013)
23. Pearl, J.: Causality. Cambridge University Press, New York (2009)
24. Pearl, J., et al.: Causal inference in statistics: an overview. Stat. Surv. **3**, 96–146 (2009)
25. Scheines, R., Spirtes, P., Glymour, C., Meek, C., Richardson, T.: The TETRAD project: constraint based aids to causal model specification. Multivar. Behav. Res. **33**(1), 65–117 (1998)
26. Sokolova, E., Groot, P., Claassen, T., Heskes, T.: Causal discovery from databases with discrete and continuous variables. In: Gaag, L.C., Feelders, A.J. (eds.) PGM 2014. LNCS (LNAI), vol. 8754, pp. 442–457. Springer, Heidelberg (2014). doi:10.1007/978-3-319-11433-0_29
27. Spirtes, P., Glymour, C.N., Scheines, R.: Causation, Prediction and Search. Lecture Notes in Statist, vol. 81. Springer, Heidelberg (1993)
28. Spirtes, P., Glymour, C.N., Scheines, R.: Causation, Prediction, and Search. MIT press, Cambridge (2000)
29. Stekhoven, D.J., Moraes, I., Sveinbjörnsson, G., Hennig, L., Maathuis, M.H., Bühlmann, P.: Causal stability ranking. Bioinformatics **28**(21), 2819–2823 (2012)
30. Tsamardinos, I., Brown, L.E., Aliferis, C.F.: The max-min hill-climbing Bayesian network structure learning algorithm. Mach. Learn. **65**(1), 31–78 (2006)
31. Willcutt, E.G., Pennington, B.F., DeFries, J.C.: Etiology of inattention and hyperactivity/impulsivity in a community sample of twins with learning difficulties. J. Abnorm. Child Psychol. **28**(2), 149–159 (2000)
32. Yang, R., Berger, J.O.: Estimation of a covariance matrix using the reference prior. Ann. Stat. **22**(3), 1195–1211 (1994)

Selecting Collaborative Filtering Algorithms Using Metalearning

Tiago Cunha[1]($\boxtimes$), Carlos Soares[1], and André C.P.L.F. de Carvalho[2]

[1] INESC-TEC/Faculdade de Engenharia da Universidade do Porto, Porto, Portugal
{tiagodscunha,csoares}@fe.up.pt
[2] ICMC - Universidade de São Paulo, São Paulo, Brasil
andre@icmc.usp.br

Abstract. Recommender Systems are an important tool in e-business, for both companies and customers. Several algorithms are available to developers, however, there is little guidance concerning which is the best algorithm for a specific recommendation problem. In this study, a metalearning approach is proposed to address this issue. It consists of relating the characteristics of problems (metafeatures) to the performance of recommendation algorithms. We propose a set of metafeatures based on the application of systematic procedure to develop metafeatures and by extending and generalizing the state of the art metafeatures for recommender systems. The approach is tested on a set of Matrix Factorization algorithms and a collection of real-world Collaborative Filtering datasets. The performance of these algorithms in these datasets is evaluated using several standard metrics. The algorithm selection problem is formulated as classification tasks, where the target attribute is the best Matrix Factorization algorithm, according to each metric. The results show that the approach is viable and that the metafeatures used contain information that is useful to predict the best algorithm for a dataset.

Keywords: Recommender system · Collaborative filtering · Model selection · Metalearning

1 Introduction

The digital economy enabled an important source of revenue for companies, by increasing the number of customers and markets available. However, e-commerce websites usually have an overwhelming amount of products in their catalog, which can easily result in the loss of purchase interest. This problem, known as information overload, has been reduced with the use of Recommender Systems (RSs), which recommend potentially interesting items [1]. Specifically in Collaborative Filtering (CF) algorithms, which is the focus of this work, these systems gather data from customers, products and relationships established between elements from these two groups (e.g. a customer visualizes the page of a product or buys that product) to extract patterns. These patterns can be used to recommend possibly interesting items in future sessions.

© Springer International Publishing AG 2016
P. Frasconi et al. (Eds.): ECML PKDD 2016, Part II, LNAI 9852, pp. 393–409, 2016.
DOI: 10.1007/978-3-319-46227-1_25

There are several recommendation methodologies, each one with a large variety of algorithms [1]. This makes it difficult to select the best algorithm for a new problem. The most common strategy is trial and error. However, it has a high computational cost. In fact, when the data size is large, it becomes virtually impossible to pursue this approach. The Metalearning (MtL) approach, which has proved successful in other Data Mining tasks, can provide a good solution to this problem. Besides, it allows the extraction of knowledge able to explain why a suggested algorithm is better suited for a specific dataset.

MtL studies how machine learning (ML) can be employed to understand the learning process and, improve the use of machine learning in future applications [6]. In MtL, learning occurs at two levels: base-level and meta-level [2]. At the base-level, base-learners (in this work, they are the CF algorithms) accumulate experience on a specific learning task (i.e., a single dataset). At the meta-level, meta-learners accumulate experience on the behavior of multiple base-learners on multiple learning tasks (i.e., multiple datasets). This experience is represented as a metamodel, which can be used to suggest the best base-learner for a specific dataset.

One of the main challenges in MtL is to define informative metafeatures, i.e. characteristics that effectively describe the area of competence of each algorithm [2]. In this study, the focus is on rating-based CF datasets and the metafeatures proposed here are based on three different perspectives on their distribution: in terms of user, item and global. These distributions are aggregated using simple, standard summary statistical functions [18]. These metafeatures are expected to contain some useful information about the (relative) performance of the algorithms. The experimental approach used in this work can be summarized as:

1. base-level experimental work to estimate the performance of the selected CF algorithms on the selected datasets;
2. extraction of metafeatures from the datasets;
3. meta-level learning to relate the metafeatures with the base-level algorithm performance;
4. extraction and presentation of metaknowledge extracted.

This work extends existing studies [7,14,28] by (1) proposing an approach with algorithm-independent metafeatures and (2) by performing the experimental work on a significantly larger number of datasets and base-level algorithms. The goal is to generalize the knowledge extracted from this process, rather than focus on specific application niches, unlike the related work studies.

This document is organized as follows: Sect. 2 presents the main aspects of CF and MtL, with emphasis on related work of model selection for RSs. Section 3 holds the explanation of the metafeature process to extract CF data characteristics. Section 4 describes the experimental procedure at the base and meta-levels, while Sect. 5 contains the results from both evaluation experiments. It also shows the knowledge extracted from the experiments performed. Section 6 presents the main conclusions and points out possible future works.

2   Related Work

2.1   Collaborative Filtering

RSs are inspired by human social behavior, where it is common to take into account the tastes and opinions of acquaintances when making decisions [1]. In this work, the application scope is limited to CF. Extensive surveys discussing other recommendation strategies can be found elsewhere [1,26].

CF recommendations are based on the premise that a user must like the items favored by a similar user. Thus, it uses the feedback from each individual user to recommended items to similar users [26]. There are two types of recommendation tasks in CF. In rating prediction, the goal is to train models to accurately estimate the ratings users would give to items. Alternatively, item recommendation aims to recommend ordered lists of items, according to the preference of the users. These are fundamentally different problems and CF algorithms have been designed for each task. In this study, we will address both tasks.

Data. Traditionally, the data used in CF approaches are numerical (implicit or explicit) feedback from the user, related with user preferences concerning some of the items [1]. Explicit feedback, also known as user ratings, is a numerical value, within a pre-defined scale, proportional to how the user likes the item. Probably, the most well known scale ranges from 1 to 5, based on the metaphor of 1 to 5 stars. Implicit feedback, on the other hand, derives a numerical value from the user interactions with the items on the website (e.g. clickstream data, click-through data from the search engine, the time users spends on the pages). Collecting user feedback through explicit and implicit methods present advantages and disadvantages: implicit methods are considered unobtrusive, but explicitly acquired data are more accurate in expressing the true preferences.

The data structure used in CF is known as the rating matrix R. It is described as $R = U \times I$, representing a set of users U, where $u \in \{1...N\}$ and a set of items I, where $i \in \{1...M\}$. Each element of this matrix is the numerical feedback provided by a user u relative to an item i, represented by r_{ui}.

Algorithms. CF algorithms can be divided into two major classes: memory-based and model-based [1,13,26]. Memory-based algorithms apply heuristics on a rating matrix to compute recommendations, whereas model-based algorithms induce a model from this matrix and use it to recommend items. Memory-based algorithms are usually based on Nearest Neighbor (NN) approaches, while model-based algorithms are mostly based on Matrix Factorization (MF). For reasons explained below (Sect. 4.1), this work focuses solely on MF algorithms.

MF is one of the most efficient and robust algorithms for CF [12]. It assumes that the original rating matrix values can be approximated by the multiplication of at least two matrices with latent features that capture the underlying data patterns. The computation is iterative and optimizes a performance measure, usually RMSE. In the simplest formulation of MF, the rating matrix R is approximated by the product of two matrices: $R \approx PQ$, where P is an $N \times K$ matrix and Q is a $K \times M$ matrix. P is the user feature matrix, Q is the item feature matrix and K is the number of features in the given factorization.

There are three characteristics to be analyzed in each algorithm: the factorization process, the learning strategy and user/item bias. The factorization process is usually the one explained previously. However, there are algorithms using other approaches, such as Singular Value Decomposition (SVD).

The most commonly used learning strategies are Alternating Least Squares (ALS) and Stochastic Gradient Descent (SGD). These strategies are used in an iterative fashion. In each iteration, a specific formula is optimized until a threshold value is reached. ALS alternates between two steps: the P-step, which fixes Q and recomputes P, and the Q-step, where P is fixed and Q is recomputed. The re-computation on the P-step employs a regression model for each user, whose input is the vector q_i and the output is the original user rating vector. In the Q-step, the input is the q_u vector and the output is the item rating vector. For SGD, the original rating r_{ui} is compared with the predicted value, giving an error measure: $e_{ui} = r_{ui} - q_i^T q_u$. Afterwards, user and item factors are modified to minimize this error and a new iteration starts.

Next, the user/item bias is introduced in MF as a regularization measure. This bias (either for users, items or both), tries to compensate the specific user/item difference against the average values of either users/items. The purpose is to take into account the fact that users have different rating habits. Note that the user/item bias is different from the model bias: while the first is used to compensate the specific user/item difference against the average values in the CF problem, the second refers to the ML model preference for choosing one hypothesis explaining the data over other (equally acceptable) hypothesis.

There are multiple frameworks with implementations of MF algorithms available (e.g. Apache Spark[1], Recommenderlab[2], Prediction.io[3]). In this work, we focus on the *MyMediaLite* framework of MF algorithms [4].

Rating Prediction. (**MF**) is the most basic algorithm for this task. It uses a standard factorization strategy, SGD, to perform the learning step and introduces no user/item bias. Another algorithm, *BiasedMatrixFactorization* (**BMF**) uses explicit user/item bias, but still learns through SGD and still uses the standard factorization approach [20]. SGD is also used as the learning technique in *LatentFeatureLogLinearModel* (**LFLLM**) [16]. However, it is inspired on logistic regression, instead of the standard MF. Besides, it has no user/item bias, since the authors state that the algorithm is insensitive to it. *SVDPlusPlus* (**SVD++**) is a MF strategy that extends the basic SVD strategy to include the items rated by the users in the optimization formula [11]. It is a combination of neighborhood algorithms with MF, which also includes user/item bias. Three asymmetric algorithms, which are variations of SVD, are also used. The asymmetric changes

---

[1] http://spark.apache.org/.
[2] https://cran.r-project.org/web/packages/recommenderlab/index.html.
[3] https://prediction.io/.

refer to the fact that the user (or item) factors are modeled by which items were rated by the users (or by which users rated the items). The algorithms focus on asymmetric changes on *item* (**SIAFM**), *user* (**SUAFM**) and *both user and item* (**SCAFM**) [17]. These algorithms assume that by modeling the problem in an asymmetric fashion, the prediction formula in SVD can be linearly combined with these factors to obtain more accurate results. All **these** algorithms have user/item bias and the learning stage is conducted with SGD. A MF-based algorithm was adopted as baseline: *UserItemBaseline* (**UIB**) [12]. It uses the average rating value plus a regularized user/item bias for prediction. The optimization problem is solved with ALS. Three average-based algorithms were also included: *GlobalAverage* (**GA**), *ItemAverage* (**IA**) and *UserAverage* (**UA**). These algorithms make the predictions based on the average rating value of all ratings of all users, all ratings of an item and all ratings of an user, respectively.

Item Recommendation. A different set of MF algorithms can be used to recommend rankings of items. **BPRMF** optimizes a criterion based on Bayesian logic [19]. It reduces the ranking problem to a pairwise classification task, optimizing the Area under the Curve (AUC) metric. It uses SGD as the learning strategy and no user/item bias. *MultiCoreBPRMF* (**MBPRMF**) is a parallel implementation of the previous algorithm. The algorithm *WeightedBPRMF* (**WBPRMF**) is a variation of *BPRMF* that includes a sampling mechanism that promotes low scored items and use/item bias. *SoftMarginRankingMF* (**SMRMF**) is another variation of *BPRMF*, but it replaces the optimization formula in SGD by a soft margin ranking loss inspired by SVM classifiers [24]. Another MF algorithm used is (**WRMF**) [10]. This algorithm uses ALS as the learning technique and introduces user/item bias to regularize the process. The only baseline algorithm available in this scope is *MostPopular* (**MP**). Here, items are ranked by how often they have been seen in the past.

Evaluation. Due to the experimental nature of this work, the CF algorithms are evaluated using an offline approach. This evaluation involves a data split strategy (usually k-fold cross-validation, although others can be used) and the application of suitable metrics, depending on the application scope. In the case of Rating Prediction, the metrics are error based and evaluate the rating accuracy. Examples of these metrics are the Mean Average Error (MAE), the normalized version of MAE (NMAE) and the Root Mean Squared Error (RMSE) [9]. The evaluation for the Item Recommendation task is based on predicted rankings, using metrics like Mean Average Precision (MAP), Normalized Discount Cumulative Gain (NDCG), Mean Reciprocal Rank (MRR) and AUC [9].

2.2 Metalearning

MtL looks for an hypothesis or function associating the characteristics of a dataset and the behavior of learning techniques, when applied to this dataset. Its use helps understand algorithm behavior on different conformations of data [22].

There are two model induction levels in this methodology: the base-level and the meta-level. In the problem investigated in this paper, the base-level refers to the application of CF algorithms on CF datasets, while the meta-level studies the effect of the characteristics of CF datasets on the performance of CF algorithms. The MtL process addresses the algorithm selection problem in two phases: training and prediction. In the training phase, datasets are characterized by a set of measurable characteristics and CF algorithms have their performance evaluated on these datasets. Next, a learning algorithm is trained on the metadata to induce a metamodel able to associate the characteristics of the dataset with the best base-level algorithm to analyze it. In the second phase, this metamodel is used to predict the best algorithm for a given dataset [21].

Metafeatures are dataset descriptors that are expected to correlate well with the performance of the models learned by different techniques [2]. The literature describes two main types of meta-features: (1) Statistical and/or information-theoretical measures and (2) Landmarkers. This study adopts the first type of meta-features for CF. More information on metafeatures can be found elsewhere [22].

The metatarget determines the type of prediction to be made by the MtL model for a dataset. Common metatargets are (1) the algorithm with the best performance on the dataset (2) a non-ordered subset of algorithms that performed well on the dataset, (3) a ranking of algorithms according to their performance on the dataset and (4) the performance of a set of techniques for the dataset [2]. This study will follow the first approach, namely addressing MtL as a classification task.

2.3   Model Selection for Recommender Systems

This section presents related work on model selection for RSs using MtL. Firstly, it is important to notice that, despite sharing the same nature, the problems have different goals: to predict the performance of CF algorithms at user level [7], to predict the performance of CF algorithms at dataset level [14] and to predict the best algorithm for group-oriented recommendations [28]. The studies diverge between using public [7,14] and private datasets [28], although none has the appropriate number of datasets required: the maximum found is 4. This is important since the generalization of the metalearning process requires a large and diverse collection of datasets. The base level algorithms are mostly based on NN, which despite being an important technique, have several drawbacks with larger datasets and are somehow outdated. The main exception is on the group-aware recommendations, since the algorithms are simply heuristics. The metafeatures used are of several types:

1. rating distribution analysis: the number of ratings per user, the average rating per user, the standard rating deviation per user [7], the ratings entropy, the ratings Gini index and ratings sparsity [14];
2. neighbourhood analysis: the number of neighbors, the average similarity to the top closest 30 neighbors, the clustering coefficient of a group of users, the

average Jaccard coefficient per user [7], group size, social contact level and dissimilarity level [28];

3. general user analysis: the user influence [7], experience level and activity level [28];

4. general item analysis: the item popularity, the item preference, the user influence and the average item entropy [7].

The techniques used in the meta-level are divided into 2 types: regression [7,14] and classification [28]. While the regression is evaluated with MAE measure, the classification problem uses error and rankings measures: RMSE and MRR.

3   Metafeatures for Recommender Systems Problems

One of the most important factors in the success of a metalearning approach is the definition of a set of metafeatures that contain information about the (relative) performance of the base-level algorithms [2]. Given that there is little work on MtL for recommender systems and that the nature of the data in these problems is quite different from traditional MtL problems (e.g. classification or regression), there is not much work we can build upon. The set of metafeatures proposed here is based on (1) the application of systematic procedure to develop metafeatures [18] and (2) extend and generalize the state of the art metafeatures for recommender systems [7,14,28].

The framework requires three main elements: the object that the metafeatures characterize, the function that analyzes the object and provides the result as a data distribution, and the post-processing functions that are applied on these distributions to extract their characteristics.

In the proposed approach, the objects can be of three types: dataset, row and column. As previously seen, row and column refer to user and item, respectively. On the dataset we analyze only the original rating distribution. However, for each row and column, we use three distinct functions: count the number of elements, mean value and sum of values. The post-processing functions used provide the following values: maximum, minimum, mean, standard deviation, median, mode, entropy, Gini index, skewness and kurtosis. The notation used to represent metafeatures follows the format: *object.function.post function*.

For each rating matrix $R = U \times I$, the set of meta-features, M, is extracted in two steps: (1) application of a function f to the ratings r_{ui} in each row ($f(U)$), column ($f(I)$) and the entire dataset ($f(R)$) to obtain three different ratings distributions and (2) post-process the outcome of each function f (in the shape of distribution) with the so-called post-functions pf by extracting statistics that can be used as meta-features. Therefore, the set of meta-features is described as $M = pf[f(U)] \cup pf[f(I)] \cup pf[f(R)]$. Four simple statistics were also included and presented in Table 1.

These combinations enable the exploration of the rating distribution analysis metafeatures commonly used in selection of CF algorithms and, more importantly, extend them in a systematic way.

4   Experimental Setup

4.1   Base-Level

The robustness of experimental results in MtL depends on the number of datasets available as each dataset represents a meta-example [2]. In most MtL studies, however, only a few dozen datasets are available. This is also true for CF tasks, as there are not many public datasets. Furthermore, very often these datasets are very large, which makes it hard to use them for MtL experiments, as it implies running all the base-level algorithms on the datasets. Thus, we selected 32 datasets for this study. Table 1 lists these datasets, providing their names, reference and a few simple statistics with approximate values for readability. To the best of our knowledge, this is the largest experimental study in terms of number of CF rating based datasets.

These datasets present different numbers of users, items and ratings. As expected, in most cases the sparsity is greater than 0.9 [26]. To ensure that the values of the metafeatures and the performance measures are comparable across datasets, it is necessary to normalize the rating scales. We decided to normalize all ratings to the scale $[1, 5]$, since it is the most common.

The performance of all selected CF algorithms on each dataset was estimated, using, as explained earlier, the *MyMediaLite* framework (Sect. 2.1). Some algorithms, namely the NN algorithms, were not able to obtain results on the largest datasets. Therefore, we decided to limit the algorithms to those able to process all the available datasets: the MF algorithms and the baselines. However, no tuning of these parameters was performed, as this is common practice in MtL.

We evaluated the algorithms using seven commonly employed recommendation metrics (Sect. 2.1). Each of those metrics evaluates the recommendation problem accordingly to a specific perspective. Thus, the best CF algorithm for a given dataset may vary for different evaluation metrics. Thus, we generate a different meta-target variable for each evaluation metric, yielding seven different classification meta-level tasks. The evaluation process uses 10-fold cross-validation and there is no parameter tuning at this stage. We decided to use the default parameters, since this is the usual approach in MtL experiments.

4.2   Meta Level

The meta-features defined in this work were implemented using the *recommenderlab* package,[4] which is based on the *Matrix* package.[5] These packages provide a flexible interface for CF data through a sparse matrix data structure. The implementations from these packages not only allow the application of functions to each row, column and entire dataset, but also worked efficiently.

Since all meta-features are somehow related to the original ratings distribution, it is necessary to ensure that the correlated features are removed. Therefore, a Correlation Feature Selection strategy (CFS) was applied to them, using

---

[4] https://cran.r-project.org/web/packages/recommenderlab/index.html.
[5] https://cran.r-project.org/web/packages/Matrix/index.html.

Table 1. Datasets used in the base-level experiments

dataset	#users	#items	#ratings	sparsity	ratings scale	ref.
amazon-apps	1.3 M	61 k	2.6 M	0.999	[1,5]	[15]
amazon-automotive	851 k	320 k	1.3 M	0.999	[1,5]	[15]
amazon-baby	531 k	64 k	915 k	0.999	[1,5]	[15]
amazon-beauty	1.2 M	249 k	2 M	0.999	[1,5]	[15]
amazon-cd	1.5 M	486 k	3.7 M	0.999	[1,5]	[15]
amazon-digital-music	478 k	266 k	836 k	0.999	[1,5]	[15]
amazon-food	768 k	166 k	1.2 M	0.999	[1,5]	[15]
amazon-games	826 k	50 k	1.3 M	0.999	[1,5]	[15]
amazon-garden	714 k	105 k	993 k	0.999	[1,5]	[15]
amazon-home	2.5 M	410 k	4.2 M	0.999	[1,5]	[15]
amazon-instant-video	426 k	24 k	584 k	0.999	[1,5]	[15]
amazon-instruments	339 k	83 k	500 k	0.999	[1,5]	[15]
amazon-movies	73 k	4 k	111 k	0.999	[1,5]	[15]
amazon-office	909 k	130 k	1.2 M	0.999	[1,5]	[15]
amazon-pet-supplies	741 k	103 k	1.2 M	0.999	[1,5]	[15]
amazon-phones	2.2 M	320 k	3.4 M	0.999	[1,5]	[15]
amazon-sports	1.9 M	479 k	3.3 M	0.999	[1,5]	[15]
amazon-tools	1.2 M	260 k	1.9 M	0.999	[1,5]	[15]
amazon-toys	1.3 M	328 k	2.3 M	0.999	[1,5]	[15]
flixter	148 k	49 k	8.2 M	0.998	[0,5]	[27]
jester1	25 k	100	1.8 M	0.275	[-10,10]	[5]
jester2	24 k	100	1.7 M	0.273	[-10,10]	[5]
jester3	25 k	100	617 k	0.753	[-10,10]	[5]
movielens100 k	1 k	2 k	100 k	0.937	[0,5]	[8]
movielens10m	70 k	11 k	10 M	0.987	[0,5]	[8]
movielens1m	6 k	4 k	1 M	0.955	[0,5]	[8]
movielens20m	138 k	27 k	20 M	0.995	[0,5]	[8]
movielens_latest	229 k	27 k	21 M	0.997	[0,5]	[8]
movietweetings_latest	37 k	21 k	389 k	0.999	[0,10]	[3]
movietweetings_recsys2014	25 k	15 k	211 k	0.999	[0,10]	[3]
tripadvisor	778 k	13 k	1.5 M	0.999	[1,5]	[23]
yahoo-music	6 k	10 k	364 k	0.994	[1,5]	[25]

a threshold $t \in [0.6, 0.9]$ with increments of 0.5. This decreased the number of features from 74 to the interval $[11, 28]$, depending on the threshold used.

Each set of meta-features originated seven meta-level datasets, one per each CFS threshold. Each metadataset is associated with 1 of the 7 recommendation targets, creating 49 metadatasets. As the model selection problem is approached here as a classification task, 11 classification algorithms representing several biases were chosen: ctree, C4.5, C5.0, kNN, LDA, Naive Bayes, SVM (linear, polynomial and radial kernels), random forest and a baseline algorithm: majority vote. Since the metadatasets have a reduced number of examples, the algorithms were evaluated for accuracy in a leave one out strategy and no tuning was performed. The goal is to reduce the potential overfitting of the meta-models.

5   Experimental Results

5.1   Base-Level Results

The results at the base-level are presented in Table 2. This table presents the best algorithm for each dataset and metric. Each metric is applicable only to a suitable type of recommendation algorithm: rating prediction or item recommendation. These results are used as the target attributes in the meta-datasets.

Regarding the rating prediction experiments, it can be observed that most datasets have for best algorithm either a baseline or BMF. In fact, only 6 datasets do not follow this process. Furthermore, the results show that, for the metrics MAE and NMAE, the best algorithms are almost always the same. Since the metrics are very similar, this behavior is expected.

In the item recommendation experiments, the distribution of best algorithm for each dataset is fairly distributed, although it is noticeable that these algorithms have the tendency to not change according to the different metrics. This is also expected since all of them evaluate the ranking accuracy of the algorithms. However, since AUC values are more concerned with accuracy assessment regardless of the ranking, it produces different results.

It is important to observe that the baseline algorithms often perform best on the largest datasets, regardless of the recommendation scope: IA, UA and GA in rating prediction and MP in item recommendation. This relates to the sparsity problem in CF and how difficult it is to make predictions in a cold start environment.

Another important observation is that there are few algorithms that are never chosen as the best in any pair dataset/metric. This may be a consequence of the lack of tuning on the base level methods. These are the cases of SUAFM and UIB in rating prediction and SMRMF in item recommendation. This means that it is not possible to extract useful knowledge from these algorithms in the meta-level. This can change if we can increase the number and diversify the nature of the datasets in order to expand the search space.

5.2   Meta-Level Results: Rating Prediction

Figures 1 and 2 show the meta-models performance across several CFS thresholds for the MAE and RMSE metrics, respectively. Each threshold was used to understand the effect of correlation in our metafeature framework. The NMAE

Table 2. Best models on multiple evaluation metrics for each dataset

dataset	Rating prediction			Item recommendation			
	MAE	NMAE	RMSE	MAP	MRR	NDCG	AUC
amazon-apps	BMF	BMF	BMF	MP	MP	MP	MP
amazon-automotive	IA	IA	BMF	MP	MP	MP	MP
amazon-baby	IA	IA	BMF	MP	MP	MP	MP
amazon-beauty	UA	UA	BMF	MP	MP	MP	MP
amazon-cd	UA	UA	BMF	MBPRMF	MBPRMF	MBPRMF	MBPRMF
amazon-digital-music	UA	UA	BMF	BPRMF	MP	MP	MP
amazon-food	IA	IA	BMF	MP	MP	MP	MP
amazon-games	BMF	BMF	BMF	MP	MP	MP	MP
amazon-garden	IA	IA	BMF	MP	MP	MP	MP
amazon-home	IA	IA	BMF	MBPRMF	MBPRMF	MBPRMF	MBPRMF
amazon-instant-video	IA	IA	BMF	MP	MP	MP	MP
amazon-instruments	IA	IA	BMF	MP	MP	MP	MP
amazon-movies	BMF	BMF	BMF	WBPRMF	WBPRMF	WBPRMF	MBPRMF
amazon-office	IA	IA	BMF	MP	MP	MP	MP
amazon-pet-supplies	IA	IA	BMF	MP	MP	MP	MP
amazon-phones	BMF	BMF	BMF	BPRMF	BPRMF	BPRMF	MBPRMF
amazon-sports	IA	IA	BMF	BPRMF	MBPRMF	MBPRMF	MBPRMF
amazon-tools	IA	IA	BMF	MP	MP	MP	MP
amazon-toys	IA	IA	BMF	MP	MP	MP	MP
flixter	BMF	BMF	BMF	MP	MBPRMF	MP	MBPRMF
jester1	SVD++	SVD++	SVD++	MP	MP	MP	MP
jester2	SVD++	SVD++	SVD++	MP	MP	MP	MP
jester3	SIAFM	SIAFM	SIAFM	MP	MP	MP	MP
movielens_latest	BMF	BMF	BMF	WRMF	WRMF	WRMF	MBPRMF
movielens100k	BMF	BMF	BMF	WRMF	WRMF	WRMF	WRMF
movielens10m	MF	MF	BMF	WRMF	WRMF	WRMF	WRMF
movielens1m	MF	MF	MF	WRMF	WRMF	WRMF	MBPRMF
movielens20m	BMF	BMF	BMF	WRMF	WRMF	WRMF	MBPRMF
movietweetings_latest	SCAFM	SCAFM	SCAFM	WRMF	WRMF	WRMF	MBPRMF
movietweetings_recsys2014	UA	GA	GA	MP	MP	MP	MBPRMF
tripadvisor	SIAFM	SIAFM	SIAFM	WBPRMF	WBPRMF	WBPRMF	MBPRMF
yahoo-music	SVD++	SVD++	LFLLM	WRMF	WRMF	WRMF	WRMF

analysis was discarded in the paper due to space restrictions. However, the performance is similar to the MAE metric.

The accuracy values are clearly different: while most algorithms, concerning MAE, performed always above the baseline, on the RMSE meta-level problem, only 2 of them achieve this goal. This is a consequence of the bias in the meta-dataset towards the BMF algorithm. Since this algorithm wins most of the times, the metalearning strategy becomes obsolete for this scope. Hopefully, using more and diversified datasets will enable to study this specific problem in further detail. This experiment shows that the meta-models created with our metafeature framework are useful for solving the algorithm selection problem for CF.

One important point lies in the fact that the performances are fairly constant across the thresholds. This was not expected beforehand and points to the fact that the metafeatures used are very different in nature, despite having for basis

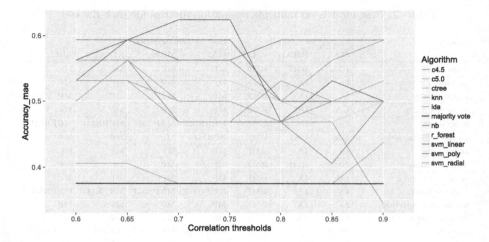

Fig. 1. Results of MAE meta-dataset on CFS thresholds

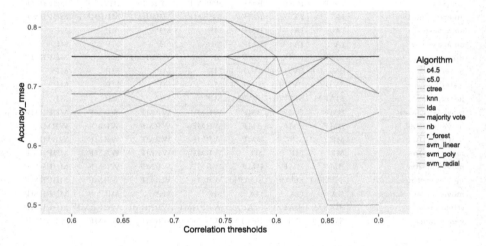

Fig. 2. Results of RMSE meta-dataset on CFS thresholds

the same rating distribution. Therefore, the CFS analysis does not have sufficient impact on selecting the best meta-models. This means that, in this experimental setup, if a meta-model beats the baseline, it is of low importance which is the CFS threshold used to build it.

The strategy to select the best algorithms follows the principle that the average accuracy across thresholds must be always better than the baseline. To ensure this principle, the algorithms whose average accuracy for all thresholds minus the standard deviation is above the performance of the baseline algorithm (majority voting) were selected as the best ones. Thus, the best algorithms for the MAE metric are all except the ctree. For the RMSE target, only the SVM with polynomial kernel satisfies this principle.

5.3 Meta-Level Results: Item Recommendation

Figures 3 and 4 present the accuracy results for the NDCG and AUC metrics, respectively. MRR and MAP were discarded due to space restrictions. However, the performances on these targets are also fairly similar to performance obtained with the NDCG metric. First of all, one notices that there are several algorithms whose performance was better than the baseline and remained stable across the CFS thresholds. This behavior is similar to the one found in the rating prediction problem. The only exception found shows that in both metrics, the Naive Bayes algorithm presents a poor predictive performance, scoring always below the baseline accuracy. The only explanation available is that the class

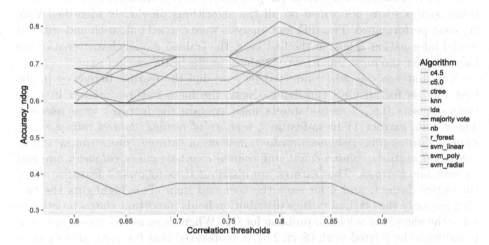

Fig. 3. Results of NDCG meta-dataset on CFS thresholds

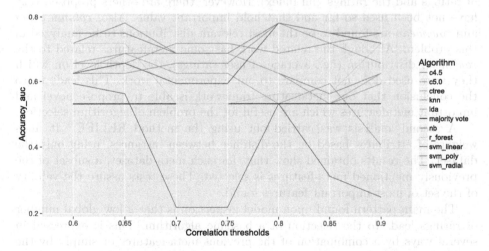

Fig. 4. Results of AUC meta-dataset on CFS thresholds

distribution is more balanced than the first problem and this affects the Naive Bayes algorithm specially. However, on the overall analysis, these meta-models perform better than in the rating prediction problem.

Following the previous strategy to select the best models, for the NDCG target, the algorithms with the best predictive performance were SVM (linear, polynomial and radial kernels), random forest, kNN and C4.5. For the AUC metric, the best algorithms are almost the same, with the difference that the random forest algorithm was replaced with the C5.0 algorithm.

5.4   Meta-Knowledge

To extract meta-knowledge from the previous MtL experiments, variable importance analysis was performed on all the algorithms previously identified with the best performance. Two different analysis were carried out: with and without model information. Thus, in the first case, the trained model characteristics are used to infer the most important variables, unlike in the second case.

The first analysis was conducted by assessing the feature frequency in the best models for all CFS thresholds. Next, the meta-features present in most meta-datasets (i.e., 5 meta-datasets must contain the feature) were selected. The results extract 11 meta-features: *number of ratings, dataset.ratings.mode, dataset.ratings.gini, row.mean.median, row.mean.entropy, row.mean.skewness, row.count.kurtosis, column.count.gini, column.count.skewness, col.mean.min* and *column.sum.kurtosis*. The features are distributed as follows: 3 features about the entire dataset, and 4 for each the user and item. This highlights the fact that not only the original ratings distribution holds important characteristics to solve the algorithm selection problem for CF. When these metafeatures are compared with the related work (Sect. 2.3), it is observed that few have already been used and their importance is confirmed in this study (for instance the number of ratings and the ratings gini index). However, there are others proposed that have not been used so far and that hold important value. Also, *column.count* and *row.mean* are found to be the most relevant distributions to be analyzed in this problem. Although the related work has some metafeatures related to the *row.mean* distribution (i.e., average of user ratings), the depth level on which they were used does not compare to our experimental work. This leads us to the conclusion that our metafeature framework is able to propose novel and important metafeatures which are useful for the problem of algorithm selection.

A second analysis was carried out using the method RELIEF.[6] It finds weights of attributes based on the distance between instances, using only the dataset. The results obtained show that, for each meta-dataset, a subset of the previously mentioned meta-features is selected. These tests assure the validity of the set of most important features found.

The main pattern found upon model inspection is that a low global number of ratings leads to the selection of a baseline algorithm. This is expressed in several ways by a combination of the previous meta-features or simply by the

---

[6] https://cran.r-project.org/web/packages/FSelector/index.html.

number of ratings. Other meta-features, despite being important in discerning the algorithms, are difficult to interpret. This has more impact if the functions or post-functions are themselves not easily understandable.

One important consideration lies in the fact that the meta-dataset has very few instances, which prevents a more detailed analysis of the meta-knowledge. Still, the fact remains that the meta-features proposed are informative and that help tackling the problem of algorithm selection for CF.

6   Conclusions

In this study, we have proposed a Metalearning approach to select Matrix Factorization algorithms on two scopes of the CF problem: rating prediction and item recommendation. The meta-features proposed follow a thorough analysis of the feature space and are based on combinations of the original rating distribution and generalize the meta-features used in recent studies. Each base-learner is trained on a collection of real-world datasets and evaluated on a range of suitable metrics, which serve as different targets in the meta-level. The meta-models induced have performed well above the baseline algorithm, even when the meta-dataset has very few examples. Furthermore, variable importance analysis has shown that the proposed meta-features provide added knowledge when compared with the usage of characteristics of only the original rating distribution. Future work may focus on increasing the number of datasets, perform dimensionality reduction to expand the range of algorithms available, proposal of meta-features related to the models characteristics, the extension of the meta-targets to label ranking problems and tuning of both the base and meta level algorithms.

Acknowledgments. This work is financed by the ERDF Fund through the Operational Programme for Competitiveness and Internationalization - COMPETE 2020 of Portugal 2020 through the National Innovation Agency (ANI) as part of the project 3506 and also through project «POCI-01-0145-FEDER-006961» via National Funds through the FCT – Fundação para a Ciência e a Tecnologia as part of project UID/EEA/50014/2013. The research was also funded from the ECSEL Joint Undertaking, the framework programme for research and innovation horizon 2020 (2014-2020) under grant agreement 662189-MANTIS-2014-1.

References

1. Bobadilla, J., Ortega, F., Hernando, A., Gutiérrez, A.: Recommender systems survey. Knowl.-Based Syst. **46**, 109–132 (2013)
2. Brazdil, P., Giraud-Carrier, C., Soares, C., Vilalta, R.: Metalearning: Applications to Data Mining, 1st edn. Springer Publishing Company, Incorporated, Heidelberg (2009)
3. Dooms, S., De Pessemier, T., Martens, L.: MovieTweetings: a movie rating dataset collected from twitter. In: Workshop on Crowdsourcing and Human Computation for Recommender Systems, CrowdRec at RecSys 2013 (2013)

4. Gantner, Z., Rendle, S., Freudenthaler, C., Schmidt-Thieme, L.: MyMediaLite: a free recommender system library. In: ACM Conference on Recommender Systems, pp. 305–308 (2011)
5. Goldberg, K., Roeder, T., Gupta, D., Perkins, C.: Eigentaste: a constant time collaborative filtering algorithm. Inf. Retrieval 4(2), 133–151 (2001)
6. Gomes, T.A., Prudêncio, R.B., Soares, C., Rossi, A.L., Carvalho, A.: Combining meta-learning and search techniques to select parameters for support vector machines. Neurocomputing 75(1), 3–13 (2012)
7. Griffith, J., O'Riordan, C., Sorensen, H.: Investigations into user rating information and predictive accuracy in a collaborative filtering domain. In: ACM Symposium on Applied Computing, pp. 937–942 (2012)
8. GroupLens: MovieLens datasets (2016). http://grouplens.org/datasets/movielens/
9. Herlocker, J.L., Konstan, J.A., Terveen, L.G., Riedl, J.T.: Evaluating collaborative filtering recommender systems. ACM Trans. Inf. Syst. 22(1), 5–53 (2004)
10. Hu, Y., Koren, Y., Volinsky, C.: Collaborative filtering for implicit feedback datasets. In: IEEE International Conference on Data Mining, pp. 263–272 (2008)
11. Koren, Y.: Factorization meets the neighborhood: a multifaceted collaborative filtering model. In: Proceeding of the 14th ACM SIGKDD International Conference on Knowledge Discovery and Data Mining, pp. 426–434 (2008)
12. Koren, Y.: Factor in the neighbors: scalable and accurate collaborative filtering. ACM Trans. Knowl. Discov. Data 4(1), 1–24 (2010)
13. Lü, L., Medo, M., Yeung, C.H., Zhang, Y.C., Zhang, Z.K., Zhou, T.: Recommender systems. Phys. Rep. 519(1), 1–49 (2012)
14. Matuszyk, P., Spiliopoulou, M.: Predicting the performance of collaborative filtering algorithms. In: International Conference on Web Intelligence, Mining and Semantics, pp. 38:1–38:6 (2014)
15. McAuley, J., Leskovec, J.: Hidden factors and hidden topics: understanding rating dimensions with review text. In: ACM Conference on Recommender Systems, pp. 165–172 (2013)
16. Menon, A.K., Elkan, C.: A log-linear model with latent features for dyadic prediction. In: Proceedings of the IEEE International Conference on Data Mining, ICDM, pp. 364–373 (2010)
17. Paterek, A.: Improving regularized singular value decomposition for collaborative filtering. In: Proceedings of KDD Cup and Workshop, pp. 2–5 (2007)
18. Pinto, F., Soares, C., Mendes-Moreira, J.: Towards automatic generation of metafeatures. In: Bailey, J., Khan, L., Washio, T., Dobbie, G., Huang, J.Z., Wang, R. (eds.) PAKDD 2016. LNCS (LNAI), vol. 9651, pp. 215–226. Springer, Heidelberg (2016). doi:10.1007/978-3-319-31753-3_18
19. Rendle, S., Freudenthaler, C., Gantner, Z., Schmidt-thieme, L.: BPR: bayesian personalized ranking from implicit feedback. In: Proceedings of the Twenty-Fifth Conference on Uncertainty in Artificial Intelligence, pp. 452–461 (2009)
20. Salakhutdinov, R., Mnih, A.: Probabilistic matrix factorization. In: Advances in Neural Information Processing Systems (NIPS 2008), pp. 1257–1264 (2008)
21. Serban, F., Vanschoren, J., Bernstein, A.: A survey of intelligent assistants for data analysis. ACM Comput. Surv. V(212), 1–35 (2013)
22. Vanschoren, J.: Understanding machine learning performance with experiment databases. Ph.D. thesis, Katholieke Universiteit Leuven (2010)
23. Wang, H., Lu, Y., Zhai, C.: Latent aspect rating analysis without aspect keyword supervision. In: Proceedings of the 17th ACM SIGKDD International Conference on Knowledge Discovery and Data Mining, pp. 618–626. ACM (2011)

24. Weimer, M., Karatzoglou, A., Smola, A.: Improving maximum margin matrix factorization. Mach. Learn. **72**(3), 263–276 (2008)
25. Yahoo!: Webscope datasets (2016). https://webscope.sandbox.yahoo.com/
26. Yang, X., Guo, Y., Liu, Y., Steck, H.: A survey of collaborative filtering based social recommender systems. Comput. Commun. **41**, 1–10 (2014)
27. Social Computing Data Repository at ASU (2009). http://socialcomputing.asu.edu
28. Zapata, A., Menéndez, V.H., Prieto, M.E., Romero, C.: Evaluation and selection of group recommendation strategies for collaborative searching of learning objects. Int. J. Hum.-Comput. Stud. **76**, 22–39 (2015)

A Bayesian Network Model
for Interesting Itemsets

Jaroslav Fowkes$^{(\boxtimes)}$ and Charles Sutton

School of Informatics, University of Edinburgh, Edinburgh EH8 9AB, UK
{jfowkes,csutton}@inf.ed.ac.uk

Abstract. Mining itemsets that are the most interesting under a statistical model of the underlying data is a commonly used and well-studied technique for exploratory data analysis, with the most recent interestingness models exhibiting state of the art performance. Continuing this highly promising line of work, we propose the first, to the best of our knowledge, generative model over itemsets, in the form of a Bayesian network, and an associated novel measure of interestingness. Our model is able to efficiently infer interesting itemsets directly from the transaction database using structural EM, in which the E-step employs the greedy approximation to weighted set cover. Our approach is theoretically simple, straightforward to implement, trivially parallelizable and retrieves itemsets whose quality is comparable to, if not better than, existing state of the art algorithms as we demonstrate on several real-world datasets.

1 Introduction

Itemset mining is one of the most important problems in data mining, with applications including market basket analysis, mining data streams and mining bugs in source code [1]. Early work on itemset mining focused on algorithms that identify all itemsets which meet a given criterion for pattern quality, such as all *frequent itemsets* whose support is above a user-specified threshold. Although appealing algorithmically, the list of frequent itemsets suffers from *pattern explosion*, i.e., is typically long, highly redundant and difficult to understand [1]. In an attempt to address this problem, more recent work focuses on mining *interesting itemsets*, smaller sets of high-quality, non-redundant itemsets that can be examined by a data analyst to get an overview of the data. Several different approaches have been proposed for this problem. Some of the most successful recent approaches, such as MTV [19], KRIMP [28] and SLIM [26] are based on the *minimum description length* (MDL) principle, meaning that they define an encoding scheme for compressing the database based on a set of itemsets, and search for the itemsets that best compress the data. These methods have been shown to lead to much less redundant pattern sets than frequent itemset mining.

In this paper, we introduce an alternative, but closely related, viewpoint on interesting itemset mining methods, by starting with a probabilistic model of the data rather than a compression scheme. We define a *generative model* of the data, that is, a probability distribution over the database, in the form

© Springer International Publishing AG 2016
P. Frasconi et al. (Eds.): ECML PKDD 2016, Part II, LNAI 9852, pp. 410–425, 2016.
DOI: 10.1007/978-3-319-46227-1_26

of a Bayesian network model, based on the interesting itemsets. To infer the interesting items, we use a probabilistic learning approach that directly infers the itemsets that best explain the underlying data. Our method, which we call the *Interesting Itemset Miner* (IIM)[1], is to the best of our knowledge, the first generative model for interesting itemset mining.

Interestingly, our viewpoint has a close connection to MDL-based approaches for mining itemsets that best compress the data (Sect. 3.9). Every probability distribution implicitly defines an optimal compression algorithm, and conversely every compression scheme implicitly corresponds to a probabilistic model. Explicitly taking the probabilistic modelling perspective rather than an MDL perspective has two advantages. First, focusing on the probability distribution relieves us from specifying the many book-keeping details required by a lossless code. Second, the probabilistic modelling perspective allows us to exploit powerful methods for probabilistic inference, learning, and optimization, such as submodular optimization and structural expectation maximization (EM).

The collection of interesting itemsets under IIM can be inferred efficiently using a structural EM framework [9]. One can think of our model as a probabilistic relative of some of the early work on itemset mining that formulates the task of finding interesting patterns as a covering problem [11,28], except that in our work, the set cover problem is used to identify itemsets that cover a transaction *with maximum probability*. The set cover problem arises naturally within the E step of the EM algorithm. On real-world datasets we find that the interesting itemsets seem to capture meaningful domain structure, e.g. representing phrases such as *anomaly detection* in a corpus of research papers, or regions such as *western US states* in geographical data. Notably, we find that IIM returns a much more diverse list of itemsets than current state of the art algorithms (Table 2), which seem to be of similar quality. Overall, our results suggest that the interesting itemsets found by IIM are suitable for manual examination during exploratory data analysis.

2   Related Work

Itemset mining was first introduced by Agrawal and Srikant [2], along with the Apriori algorithm, in the context of market basket analysis which led to a number of other algorithms for frequent itemset mining including Eclat and FPGrowth. Frequent itemset mining suffers from *pattern explosion*: a huge number of highly redundant frequent itemsets are retrieved if the given minimum support threshold is too low. One way to address this is to mine *compact representations* of frequent itemsets such as maximal frequent, closed frequent and non-derivable itemsets with efficient algorithms such as CHARM [31]. However, even mining such compact representations does not fully resolve the problem of pattern explosion (see Chap. 2 of [1] for a survey of frequent itemset mining algorithms).

An orthogonal research direction has been to mine *tiles* instead of itemsets, i.e., subsets of rows *and columns* of the database viewed as binary transaction

---

[1] https://github.com/mast-group/itemset-mining.

by item matrices. The analogous approach is then to mine *large tiles*, i.e., sub-matrices with only 1s whose area is greater than a given minimum area threshold. The Tiling algorithm [11] is an example of an efficient implementation that uses the greedy algorithm for set cover. Note that there is a correspondence between tiles and itemsets: every large tile is a closed frequent itemset and thus algorithms for large tile mining also suffer from pattern explosion to some extent.

In an attempt to tackle this problem, modern approaches to itemset mining have used the *minimum description length* (MDL) principle to find the set of itemsets that best summarize the database. MTV [20] uses MDL coupled with a *maximum entropy* (MaxEnt) model to mine the most informative itemsets. MTV mines the set of top itemsets with the highest likelihood under the model via an efficient convex bound that allows many candidate itemsets to be pruned and employs a method for more efficiently inferring the model itself. Due to the partitioning constraints necessary to keep computation feasible, MTV typically only finds in the order of tens of itemsets, whereas IIM has no such restriction.

KRIMP [28] employs MDL to find the subset of frequent itemsets that yields the best lossless compression of the database. While in principle this could be formulated as a set cover problem, the authors employ a fast heuristic that does not allow the itemsets to overlap (unlike IIM) even though one might expect that doing so could lead to better compression. In contrast, IIM employs a set cover framework to identify a set of itemsets that cover a transaction with highest probability. The main drawback of KRIMP is the need to mine a set of frequent itemsets in the first instance, which is addressed by the SLIM algorithm [26], an extension of KRIMP that mines itemsets directly from the database, iteratively joining co-occurring itemsets such that compression is maximised.

The MaxEnt model can also be extended to tiles, here known as the *Rasch* model, and, unlike in the itemset case, inference takes polynomial time. Kontonasios and De Bie [16] use the Rasch model to find the most surprising set of *noisy tiles* (i.e., sub-matrices with predominantly 1s but some 0s) by computing the likelihood of tile entries covered by the set. The inference problem then takes the form of weighted budgeted maximum set cover, which can again be efficiently solved using the greedy algorithm. The problem of Boolean matrix factorization can be viewed as finding a set of frequent noisy tiles which form a low-rank approximation to the data [22].

The MINI algorithm [10] finds the itemsets with the highest surprisal under statistical independence models of items and transactions from a precomputed set of closed frequent itemsets. OPUS Miner [29] is a branch and bound algorithm for mining the top *self-sufficient* itemsets, i.e., those whose frequency cannot be explained solely by the frequency of either their subsets or of their supersets.

In contrast to previous work, IIM maintains a generative model, in the form of a Bayesian network, *directly* over itemsets as opposed to indirectly over items. Existing Bayesian network models for itemset mining [14,15] have had limited success as modelling dependencies between the items makes inference for larger datasets prohibitive. In IIM inference takes the form of a weighted set cover problem, which can be solved efficiently using the greedy algorithm (Sect. 3.3).

The structure of IIM's statistical model is similar to existing models in the literature such as Rephil ([24], Sect. 26.5.4) for topic modelling and QMR-DT [25] for medical diagnosis. Rephil is a multi-level graphical model used in Google's AdSense system. QMR-DT is a bi-partite graphical model used for inferring significant diseases based on medical findings. However, the main contribution of our paper is to show that a binary latent variable model can be useful for selecting itemsets for exploratory data analysis.

3   Interesting Itemset Mining

In this section we will formulate the problem of identifying a set of interesting itemsets that are useful for explaining a database of transactions. First we will define some preliminary concepts and notation. An *item* i is an element of the universe $U = \{1, 2, \ldots, n\}$ that indexes database attributes. A *transaction* X is a subset of the universe U and an *itemset* S is simply a set of items i. The set of interesting itemsets $\mathcal{I}$ we wish to determine is therefore a subset of the power set (set of all possible subsets) of the universe. Further, we say that an itemset S is *supported* by a transaction X if $S \subseteq X$.

3.1   Problem Formulation

Our aim in this work is to infer a set of interesting itemsets $\mathcal{I}$ from a database of transactions. By *interesting*, we mean a set of itemsets that will best help a human analyst to understand the important properties of the database, that is, interesting itemsets should reflect the important probabilistic dependencies among items, while being sufficiently concise and non-redundant that they can be examined manually. These criteria are inherently qualitative, reflecting the fact that the goal of data mining is to build human insight and understanding. In this work, we formalize interestingness as those itemsets that best explain the transaction database under a *statistical model* of itemsets. Specifically we will use a *generative* model, i.e., a model that starts with a set of interesting itemsets $\mathcal{I}$ and from this set generates the transaction database. Our goal is then to infer the most likely generating set $\mathcal{I}$ under our chosen generative model. We want the model to be as simple as possible yet powerful enough to capture correlations between transaction items. A simple such model is to iteratively sample itemsets S from $\mathcal{I}$ and let their union form a transaction X. Sampling S from $\mathcal{I}$ uniformly would be uninformative, but if we associate each interesting itemset $S \in \mathcal{I}$ with a probability π_S, we can sample the indicator variable $z_S \sim \text{Bernoulli}(\pi_S)$ and include S in X if $z_S = 1$. We formally define this generative model next.

3.2   Bayesian Network Model

We propose a simple directed graphical model for generating a database of transactions $X^{(1)}, \ldots, X^{(m)}$ from a set $\mathcal{I}$ of interesting itemsets. The parameters of our model are Bernoulli probabilities π_S for each interesting itemset $S \in \mathcal{I}$. The generative story for our model is, independently for each transaction X:

1. For each itemset $S \in \mathcal{I}$, decide independently whether to include S in the transaction, i.e., sample

$$z_S \sim \text{Bernoulli}(\pi_S).$$

2. Set the transaction to be the set of items in all the itemsets selected above:

$$X = \bigcup_{S|z_S=1} S.$$

Note that the model allows individual items to be generated multiple times from different itemsets, e.g. *eggs* could be generated both as part of a breakfast itemset {*bacon, eggs*} and as part of a cake itemset {*flour, sugar, eggs*}.

Now given a set of itemsets $\mathcal{I}$, let $\mathbf{z}, \boldsymbol{\pi}$ denote the vectors of z_S, π_S for all $S \in \mathcal{I}$. Assuming $\mathbf{z}, \boldsymbol{\pi}$ are fully determined, it is evident from the generative model that the probability of generating a transaction X is

$$p(X, \mathbf{z}|\boldsymbol{\pi}) = \begin{cases} \prod_{S \in \mathcal{I}} \pi_S^{z_S} (1 - \pi_S)^{1-z_S} & \text{if } X = \bigcup_{z_S=1} S, \\ 0 & \text{otherwise} \end{cases} \tag{1}$$

3.3   Inference

Assuming the parameters $\boldsymbol{\pi}$ in the model are known, we can infer $\mathbf{z}$ for a specific transaction X by maximizing the posterior distribution $p(\mathbf{z}|X, \boldsymbol{\pi})$ over $\mathbf{z}$:

$$\max_{\mathbf{z}} \prod_{S \in \mathcal{I}} \pi_S^{z_S} (1 - \pi_S)^{1-z_S} \qquad \text{s.t. } X = \bigcup_{S|z_S=1} S. \tag{2}$$

Taking logs and rewriting (2) in a more standard form we obtain

$$\min_{\mathbf{z}} \sum_{S \in \mathcal{I}} z_S \left(-\ln(\pi_S) \right) + (1 - z_S) \left(-\ln(1 - \pi_S) \right)$$

$$\text{s.t. } \sum_{S|i \in S} z_S \geq 1 \quad \forall i \in X, \quad z_S \in \{0, 1\} \quad \forall S \in \mathcal{I} \tag{3}$$

which is (up to a penalty term) the weighted set-cover problem (see e.g. [17], Sect. 16.1) with weights $w_S \in \mathbb{R}^+$ given by $w_S := -\ln(\pi_S)$. This is an NP-hard problem in general and so impractical to solve directly in practice. It is important to note that the weighted set cover problem is a special case of minimizing a linear function subject to a submodular constraint,[2] which we formulate as follows (cf. [30]). Given the set of interesting itemsets $\mathcal{T} := \{S \in \mathcal{I} \mid S \subseteq X\}$ that support the transaction, a real-valued weight w_S for each itemset $S \in \mathcal{T}$ and a non-decreasing submodular function $f : 2^{\mathcal{T}} \to \mathbb{R}$, the aim is to find a covering $\mathcal{C} \subset \mathcal{T}$ of minimum total weight, i.e., such that $f(\mathcal{C}) = f(\mathcal{T})$ and $\sum_{S \in \mathcal{C}} w_S$ is minimized.

---

[2] Note that the posterior $p(z|X)$ would not be submodular if we were to use a noisy-OR model for the conditional probabilities.

Algorithm 1. HARD-EM

Input: Set of itemsets $\mathcal{I}$ and initial probability estimates $\boldsymbol{\pi}^{(0)}$

$\quad k \leftarrow 0$

$\quad$ **do**

$\qquad k \leftarrow k + 1$

$\qquad$ E-STEP: $\forall X^{(j)}$ solve (3) to get $z_S^{(j)} \ \forall S \in \mathcal{T}_j$

$\qquad$ M-STEP: $\pi_S^{(k)} \leftarrow \frac{1}{m} \sum_{j=1}^{m} z_S^{(j)} \ \forall S \in \mathcal{I}$

$\quad$ **while** $\|\boldsymbol{\pi}^{(k-1)} - \boldsymbol{\pi}^{(k)}\| > \varepsilon$

$\quad$ Remove from $\mathcal{I}$ itemsets S with $\pi_S = 0$

$\quad$ **return** $\mathcal{I}, \boldsymbol{\pi}^{(k)}$

For weighted set cover we simply define $f(\mathcal{C})$ to be the number of items in $\mathcal{C}$, i.e., $f(\mathcal{C}) := |\cup_{S \in \mathcal{C}} S|$. Note that $f(\mathcal{T}) = |X|$ by construction.

We can then approximately solve the weighted set cover problem (3) using the greedy approximation algorithm for submodular functions. The greedy algorithm builds a covering $\mathcal{C}$ by repeatedly choosing an itemset S that minimizes the weight w_S divided by the number of items in S not yet covered by the covering. In order to minimize CPU time spent solving the weighted set cover problem, we cache the itemsets and coverings for each transaction as needed.

It has been shown [4] that the greedy algorithm achieves a $\ln|X| + 1$ approximation ratio to the weighted set cover problem and moreover the following inapproximability theorem shows that this ratio is essentially the best possible.

Theorem 1 (Feige [7]). *There is no $(1 - o(1)) \ln|X|$-approximation algorithm to the weighted set cover problem unless* NP $\subseteq$ DTIME$(|X|^{O(\log \log |X|)})$*, i.e., unless* NP *has slightly superpolynomial time algorithms.*

The runtime complexity of the greedy algorithm is $O(|X||\mathcal{T}|)$, however by maintaining a priority queue this can be improved to $O(|X| \log|\mathcal{T}|)$ (see e.g. [5]). Note that there is also an $O(|X||\mathcal{T}|)$-runtime primal-dual approximation algorithm [3], however this has an approximation order of $f = \max_i |\{S \mid i \in S\}|$, i.e., the frequency of the most frequent element, which would be worse in our case.

3.4 Learning

Given a set of itemsets $\mathcal{I}$, consider now the case where both variables $\mathbf{z}, \boldsymbol{\pi}$ in the model are unknown. In this case we can use the hard EM algorithm [6] for parameter estimation with latent variables. The hard EM algorithm in our case is merely a simple layer on top of the inference algorithm (3). Suppose there are m transactions $X^{(1)}, \dots, X^{(m)}$ with supporting sets of itemsets $\mathcal{T}^{(1)}, \dots, \mathcal{T}^{(m)}$, then the hard EM algorithm is given in Algorithm 1. To initialize $\boldsymbol{\pi}$, a natural choice is simply the support (i.e., relative frequency) of each itemset in $\mathcal{I}$.

3.5 Inferring New Itemsets

We infer new itemsets using structural EM [9], i.e., we add a candidate itemset S' to $\mathcal{I}$ if doing so improves the optimal value $\overline{p}$ of the problem (3) averaged across

---

Algorithm 2. STRUCTURAL-EM (one iteration)

Input: Itemsets $\mathcal{I}$, probabilities $\boldsymbol{\pi}$, optima $p^{(j)}$ of (3) $\forall X^{(j)}$

    Set profit $\bar{p} \leftarrow \frac{1}{m} \sum_{j=1}^{m} p^{(j)}$

    do

        Generate candidate S' using CANDIDATE-GEN

        $\mathcal{I} \leftarrow \mathcal{I} \cup \{S'\}$, $\pi_{S'} \leftarrow 1$

        E-STEP: $\forall X^{(j)}$ solve (3) to get $z_S^{(j)}$ $\forall S \in \mathcal{T}_j$

        M-STEP: $\pi'_S \leftarrow \frac{1}{m} \sum_{j=1}^{m} z_S^{(j)}$ $\forall S \in \mathcal{I}$

        $\forall X^{(j)}$, solve (3) using $\pi'_S, z_S^{(j)}$ $\forall S \in \mathcal{T}_j$ to get the optimum $p^{(j)}$

        Set new profit $\bar{p}' \leftarrow \frac{1}{m} \sum_{j=1}^{m} p^{(j)}$

        $\mathcal{I} \leftarrow \mathcal{I} \setminus \{S'\}$

    while $\bar{p}' \leq \bar{p}$ {until one good candidate found}

    $\mathcal{I} \leftarrow \mathcal{I} \cup \{S'\}$

    return $\mathcal{I}, \boldsymbol{\pi}'$

---

transactions. Interestingly, there is an implicit regularization effect here. Observe from (3) that when a new candidate S' is added to the model, a corresponding term $\ln(1 - \pi_{S'})$ is added to the log-likelihood of all transactions that S' does not support. For large databases, this amounts to a significant penalty on candidates.

To get an estimate of maximum benefit to including candidate S', we must carefully choose an initial value of $\pi_{S'}$ that is not too low, to avoid getting stuck in a local optimum. To infer a good $\pi_{S'}$, we force the candidate S' to explain all transactions it supports by initializing $\pi_{S'} = 1$ and update $\pi_{S'}$ with the probability corresponding to its actual usage once we have inferred all the coverings. Given a set of itemsets $\mathcal{I}$ and corresponding probabilities $\boldsymbol{\pi}$ along with transactions $X^{(1)}, \ldots, X^{(m)}$, each iteration of the structural EM algorithm is given in Algorithm 2 above.

In practice, we cache the set of candidates that have been rejected by the STRUCTURAL-EM function to avoid reconsidering them.

3.6   Candidate Generation

The STRUCTURAL-EM algorithm (Algorithm 2) requires a method to generate new candidate itemsets S' that are to be considered for inclusion in the set of interesting itemsets $\mathcal{I}$. One possibility would be to use the Apriori algorithm to recursively suggest larger itemsets starting from singletons, however preliminary experiments found this was not the most efficient method. For this reason we take a slightly different approach and recursively combine the interesting itemsets in $\mathcal{I}$ with the *highest support first* (Algorithm 3). In this way our candidate generation algorithm is more likely to propose viable candidate itemsets earlier and in practice we find that this heuristic works well. We did try pruning potential itemset pairs to join using a χ^2-test, however this substantially slowed down the algorithm and barely improved the model likelihood.

In order to determine the supports of the itemsets to be combined, we store the transaction database in a Memory-Efficient Itemset Tree (MEI-TREE) [8]

Algorithm 3. CANDIDATE-GEN

Input: Itemsets $\mathcal{I}$, cached supports σ, queue length q
  if $\nexists$ priority queue $\mathcal{Q}$ for $\mathcal{I}$ **then**
    Initialize σ-ordered priority queue $\mathcal{Q}$
    Sort $\mathcal{I}$ by decreasing itemset support using σ
    for all distinct pairs $S_1, S_2 \in \mathcal{I}$, highest ranked first **do**
      Generate candidate $S' = S_1 \cup S_2$
      Cache support of S' in σ and add S' to $\mathcal{Q}$
      if $|\mathcal{Q}| = q$ **break**
    end for
  end if
  Pull highest-ranked candidate S' from $\mathcal{Q}$
  return S'

Algorithm 4. IIM (Interesting Itemset Miner)

Input: Database of transactions $X^{(1)}, \ldots, X^{(m)}$
  Initialize $\mathcal{I}$ with singletons, π with their supports
  Build MEI-TREE from transaction database
  while not converged **do**
    Add itemsets to $\mathcal{I}, \pi$ using STRUCTURAL-EM
    Optimize parameters for $\mathcal{I}, \pi$ using HARD-EM
  end while
  return $\mathcal{I}, \pi$

---

and query the tree for the support of a given itemset. A MEI-TREE stores itemsets in a tree structure according to their prefixes in a memory efficient manner. To minimize the memory usage of the MEI-TREE further, we first sort the items in order of decreasing support (as in the FPGrowth algorithm) as this often results in a sparser tree [13]. Note that a MEI-TREE is essentially an FP-tree [13] with node-compression and without node-links for nodes containing the same item. An itemset support query on the MEI-TREE efficiently searches the tree for all occurrences of the given itemset and adds up their supports (see Fig. 4 in [8] for the actual algorithm). With the wide availability of 100 GB+shared memory systems, it is reasonable to expect the MEI-TREE to fit into memory for all but the largest of datasets. The queue length parameter in the CANDIDATE-GEN algorithm effectively imposes a limit on the number of iterations the algorithm can spend suggesting candidate itemsets.

3.7 Mining Interesting Itemsets

Our complete interesting itemset mining (IIM) algorithm is given in Algorithm 4. Note that the HARD-EM parameter optimization step need not be performed at every iteration, in fact it is more efficient to suggest several candidate itemsets before optimizing the parameters. As all operations on transactions in our algorithm are trivially parallelizable, we perform the E and M-steps in both the hard and structural EM algorithms in parallel.

3.8   Interestingness Measure

Now that we have inferred the model variables $\mathbf{z}, \boldsymbol{\pi}$, we are able to use them to rank the retrieved itemsets in $\mathcal{I}$. There are two natural rankings one can employ, and both have their strengths and weaknesses. The obvious approach is to rank each itemset $S \in \mathcal{I}$ according to its probability under the model π_S, however this has the disadvantage of strongly favouring frequent itemsets over rare ones, an issue we would like to avoid. Instead, we prefer to rank the retrieved itemsets according to their *interestingness* under the model, that is the ratio of transactions they explain to transactions they support. One can think of interestingness as a measure of how necessary the itemset is to the model: the higher the interestingness, the more supported transactions the itemset explains. Thus interestingness provides a more balanced measure than probability, at the expense of missing some frequent itemsets that only explain some of the transactions they support. We define interestingness formally as follows.

Definition 1. The *interestingness* of an itemset $S \in \mathcal{I}$ retrieved by IIM (Algorithm 4) is defined as

$$int(S) = \frac{\sum_{j=1}^{m} z_S^{(j)}}{supp(S)}$$

and ranges from 0 (least interesting) to 1 (most interesting).

Any ties in the ranking can be broken using the itemset probability π_S.

3.9   Correspondence to Existing Models

There is a close connection between probabilistic models and the MDL principle [18]. Given a probabilistic model $p(X|\boldsymbol{\pi}, \mathcal{I})$ of a single transaction, by Shannon's theorem the optimal code for the model will encode X using approximately $-\log_2 p(X|\boldsymbol{\pi}, \mathcal{I})$ bits. So by finding a set of itemsets that maximizes the probability of the data, we are also finding itemsets that minimize description length. Conversely, any encoding scheme implicitly defines a probabilistic model: given an encoding scheme E that assigns each transaction X to a string of $L(X)$ bits, we can define $p(X|E) \propto 2^{-L(X)}$, and then E is an optimal code for $p(X|E)$. Interpreting previous MDL-based itemset mining methods in terms of their implicit probabilistic models provides interesting insights into these methods.

MTV uses a MaxEnt distribution over itemsets $S \in \mathcal{I}$, which for a transaction X can be written (cf. [20]):

$$p(X) = \pi_0 \prod_{S \in \mathcal{I}} \pi_S^{\mathbf{1}_X(S)}$$

where the indicator function $\mathbf{1}_X(S) = 1$ if X supports S and 0 otherwise. Thus if an itemset is present in the MaxEnt model *it must be used* to explain a supported transaction, contrast this with IIM (1) where there is a latent variable $z_S^{(j)}$ for each transaction $X^{(j)}$ that *infers if an itemset is used* to explain the transaction.

KRIMP by contrast, uses an itemset independence model, which for an itemset $S \in \mathcal{I}$ is given by (cf. [28]):

$$p(S) = \sum_{j=1}^{m} z_S^{(j)} \Big/ \sum_{I \in \mathcal{I}} \sum_{k=1}^{m} z_I^{(k)}$$

where the $z_S^{(j)}$, and therefore itemset coverings for $X^{(j)}$, are determined using a *heuristic approximation*. That is, unlike IIM, the itemset coverings are not chosen to maximise the probability under the statistical model. Instead, for each transaction X, frequent itemsets $S \in \mathcal{I}$ are chosen in order of *decreasing size and support* and added to the covering if they improve the compression, until all elements of X are covered. Additionally, itemsets in the covering are not allowed to overlap, in contrast to IIM which does allow overlap if it is deemed necessary.

SLIM uses the same approach as KRIMP but iteratively finds the candidate itemsets S directly from the dataset. It employs a greedy heuristic to do this: starting with a set of singleton itemsets $\mathcal{I}$, pairwise combinations of itemsets in $\mathcal{I}$ are considered as candidate itemsets S in order of highest estimated compression gain. IIM uses a very similar heuristic that iteratively extends itemsets by the most frequent itemset in its candidate generation step (Sect. 3.6).

However, IIM is different from these methods in that they all contain an explicit penalty term for the description length of the itemset database, which corresponds to a prior distribution $p(\mathcal{I})$ over itemsets. We did not find in practice that an explicit prior distribution was necessary but it would be possible to trivially incorporate it. Also, if we view IIM as an MDL-type method, not only the presence of an itemset, but also its absence is explicitly encoded (in the form of $(1 - \pi_S)^{1-z_S^{(j)}}$ in (1)). As a result, there is an implicit penalty for adding too many patterns to the model and one does not need to use a code table which would serve as an explicit penalty for greater model complexity.

One can also think of IIM as a probabilistic tiling method: each interesting itemset $S \in \mathcal{I}$ can be thought of as a binary submatrix of transactions for which $z_S = 1$ by items in S, where the choice of items and transactions in the tile are *inferred directly* from IIM's statistical model. That is, IIM formulates the inference problem (3) as a *weighted set cover* for *each transaction* where the weights correspond to *itemset probabilities*. This is in contrast to existing tiling methods: Geerts et al. [11] find k tiles covering the largest number of database entries and is thus an instance of *maximum coverage*. Kontonasios and De Bie [16] extend this to inferring a covering of noisy tiles using *budgeted maximum coverage*, that is, finding a covering that maximizes the sum of the *surprisal* of each tile, under a MaxEnt model constrained by expected row and column margins, subject to the sum of the *description lengths* of each tile being smaller than a given budget.

4   Numerical Experiments

In this section we perform a comprehensive qualitative and quantitative evaluation of IIM. On synthetic datasets we show that IIM returns a list of itemsets that

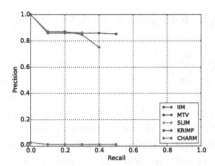
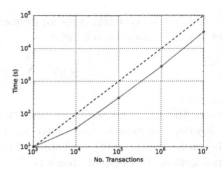

Fig. 1. Precision against recall for each algorithm on our synthetic database, using the top-k itemsets as a threshold.(Each curve is the 11-point interpolated precision i.e., the interpolated precision at 11 equally spaced recall points between 0 and 1 (inclusive), see [21], Sect. 8.4 for details.)

Fig. 2. IIM scaling as the number of transactions in our synthetic database increases.

is largely non-redundant, contains few spurious correlations and scales linearly with the number of transactions. On a set of real-world datasets we show that IIM finds itemsets that are much less redundant than state of the art methods, while being of similar quality.

Datasets. We use five real-world datasets in our numerical evaluation (Table 1). The plants dataset [27] is a list of plant species and the U.S. or Canadian states where they occur. The mammals dataset [23] consists of presence records of European mammals in 50×50 km geographical areas. The retail dataset consists of anonymized market basket data from a Belgian retail store [12]. The ICDM dataset [16] is a list of ICDM paper abstracts where each item is a stemmed word, excluding stop-words. The Uganda dataset consists of Facebook messages taken from a set of public Uganda-based pages with substantial topical discussion over a period of three months. Each transaction in the dataset is an English language message and each item is a stemmed English word from the message.

IIM Results. We ran IIM on each dataset for 1,000 iterations with a priority queue size of 100,000 candidates. The runtime and number of non-singleton itemsets returned is given in Table 1 (right). We also investigated the scaling of IIM as the number of transactions in the database increases, using the model trained on the plants dataset from Sect. 4.1 to generate synthetic transaction databases of various sizes. We then ran IIM for 100 iterations on these databases and one can see in Fig. 2 that the scaling is linear as expected. Our prototype implementation can process one million transactions in 30 s on 64 cores each iteration, so there is reason to hope that a more highly tuned implementation could scale to even larger datasets. All experiments were performed on a machine with 64 AMD Opteron 6376 CPUs and 256 GB of RAM.

Table 1. Summary of the real datasets used and IIM results after 1,000 iterations.

Dataset	Items	Trans.	$|\mathcal{I}|$†	Runtime
ICDM	4,976	859	798	163 min
Mammals	194	2,670	359	22 min
Plants	70	34,781	259	27 min
Retail	16,470	88,162	957	941 min
Uganda	33,278	124,566	928	1086 min

† excluding singleton itemsets.

Table 2. IID for the top 50 non-singleton itemsets returned by the algorithms.

	ICDM	Mam.	Plant	Retail	Ugan.
IIM	**4.00**	**7.42**	**4.80**	**3.26**	**3.78**
MTV	3.14	*5.50	*5.00	2.52	*1.60
SLIM	2.12	*1.76	*1.77	1.44	2.08
KRIMP	2.56	1.94	1.88	1.34	2.26
CHARM	1.42	1.44	1.50	1.32	1.72

*returned less than 50 non-singleton itemsets.

Evaluation Criteria. We will evaluate IIM along with MTV, SLIM, KRIMP and CHARM with χ^2-test ranking according to the following criteria:

1. *Spuriousness* – to assess the degree of spurious correlation in the mined set of itemsets.
2. *Redundancy* – to measure how redundant the mined set of itemsets is.
3. *Interpretability* – to informally assess how meaningful and relevant the mined itemsets actually are.

Note that we chose not to compare to the tiling methods from [11,16] as they have been shown to underperform on the ICDM dataset [20].

4.1 Itemset Spuriousness

The set-cover formulation of the IIM algorithm (3) naturally favours adding itemsets to the model whose items co-occur in the transaction database. One would therefore expect IIM to largely avoid suggesting itemsets of uncorrelated items and so generate more meaningful itemsets. To verify this is the case and validate our inference procedure, we check if IIM is able to recover the itemsets it used generate a synthetic database. To obtain a realistic synthetic database, we sampled 10,000 transactions from the IIM generative model trained on the plants dataset. We were then able to measure the precision and recall for each algorithm, i.e., the fraction of mined itemsets that are generating and the fraction of generating itemsets that are mined, respectively. We used a minimum support of 0.0575 for all algorithms (except IIM) as used in [20] for the plants dataset. Figure 1 shows the precision-recall curve for each algorithm using the top-k mined itemsets (according to each algorithm's ranking) as a threshold. One can clearly see that IIM was able to mine about 50 % of the generating itemsets and almost all the itemsets mined were generating. This not only provides a good validation of IIM's inference procedure and underlying generative model but also demonstrates that IIM returns few spurious itemsets. For comparison, SLIM and KRIMP exhibited very similar behaviour to IIM whereas MTV returned a very small set of generating itemsets. The set of top itemsets mined by CHARM contained many itemsets that were not generating. It is not our intention to draw conclusions about the performance of the other algorithms as this experimental setup naturally favours IIM. Instead, we compare the itemsets from IIM with those from MTV, SLIM and KRIMP on real-world data in the next sections.

4.2 Itemset Redundancy

We now turn our attention to evaluating whether IIM returns a less redundant list of itemsets than the other algorithms on real-world datasets. A suitable measure of redundancy for a single itemset is the minimum symmetric difference between it and the other itemsets in the list. Averaging this across all itemsets in the list, we obtain the *average inter-itemset distance* (IID). We therefore ran all the algorithms on the datasets in Table 1. This enabled us to calculate, for each dataset, the IID of the top 50 non-singleton itemsets, which we report in Table 2. For CHARM, we took the top 50 non-singleton itemsets ranked according to χ^2 from the top $100,000$ frequent itemsets it returned (as the χ^2 calculation would be prohibitively slow otherwise). One can clearly see that the top IIM itemsets have a larger IID on average, and are therefore less redundant, than the KRIMP, SLIM or CHARM itemsets. The top CHARM χ^2-ranked itemsets are the most redundant as expected. On all datasets, the IIM itemsets are less redundant than those mined by the other methods, with only one exception. On the Plants dataset, MTV is slightly less redundant than IIM, but this is because MTV is unable to return 50 items on this dataset, instead returning only 21.

4.3 Itemset Interpretability

For the datasets in Table 1 we can directly interpret the mined itemsets and informally assess how meaningful and relevant they are.

ICDM Dataset. We compare the top ten non-singleton itemsets mined by the algorithms in Table 3 (excluding KRIMP whose itemsets are similar for space reasons). The mined patterns are all very informative, containing technical concepts such as *support vector machine* and common phrases such as *pattern discovery*. The IIM itemsets suggest the stemmer used to process the dataset could be improved, as we retrieve {*parameter, parameters*} and {*sequenc, sequential*}.

Plants and Mammals Datasets. For both datasets, all algorithms find itemsets that are spatially coherent, but as we showed in Table 2, those returned by IIM are far less redundant. Our novel interestingness measure enables IIM to rank correlated itemsets above singletons and rare itemsets above frequent ones, in contrast to the other algorithms. For example, for the plants dataset, the top itemset retrieved by IIM is {*Puerto Rico, Virgin Islands*} whereas MTV returns {*Puerto Rico*}, not associating it with the *Virgin Islands* (which are adjacent) until the 20th ranked itemset. For the mammals dataset, the top two non-singleton IIM itemsets are a group of four mammals that coexist in Scotland and Ireland and a group of ten mammals that coexist on Sweden's border with Norway. By contrast, the top four SLIM and KRIMP itemsets list some of the most common mammals in Europe (see the supplementary material for details).

Table 3. Top ten non-singleton ICDM itemsets as found by IIM, MTV and SLIM.

IIM	MTV	SLIM
associ rule	experiment result	inform model
local global	synthetic real	cluster algorithm
support vector machin svm	real datasets	larg effici
parameter parameters	pattern discov	perform set
anomali detect	associ rule mine	propos problem
sequenc sequential	frequent pattern mine algorithm	method set
linear discriminant analysi	train classifi	associ rule
synthetic real life	address problem	problem result
background knowledg	classifi class	approach base method
semi supervised	machin learn	base method set

Table 4. Top six non-singleton Uganda itemsets for each algorithm.

IIM	MTV	SLIM	KRIMP
soul, rest, peace	heal, jesus, amen	!, ?	whi, ?
chris, brown	god, amen	2, 4	?, !
bebe, cool	2, 4	whi, ?	2, 4
airtel, red	whi, ?	god, amen	wat, ?
everi, thing	god, heal	da, dat	time, !
time, wast	2, !	heal, jesus, amen	soul, rest, peace

Uganda Dataset. The top six non-singleton itemsets found by the algorithms are shown in Table 4; the IIM itemsets provide much more information about the topics of the messages than those from the other algorithms. Figure 3 (left) plots the mentions of each of the top IIM itemsets per day. As one can see, usage of the top itemsets displays temporal structure (and exhibits spikes of popularity), even though our model does not explicitly capture this. Of particular interest are the large spikes of {*soul, rest, peac*} corresponding to notable deaths: wealthy businessman James Mulwana on the 15th January, President Museveni's father on the 22nd February and six school students in a traffic accident on the 29th March. Also of interest are the 285 mentions of {*airtel, red*} on New Year's Eve corresponding to mobile provider Airtel's Red Christmas competition for 10 K worth of airtime. The spike of {*bebe, cool*} on the 15th January corresponds to the Ugandan musician's wedding announcement and the spike on the 24th January of {*chris, brown*} refers to many enthusiastic mentions of the popular American singer that day. The last two itemsets capture common phrases.

In comparison, the top-six MTV itemsets are plotted in Fig. 3 (right). One can see that the itemsets {*heal, jesus, amen*};{*god, amen*} and {*god, heal*} substantially overlap and are strongly correlated with each other, sharing a large spike on the 8th February and a smaller spike on the 11th March. The remaining

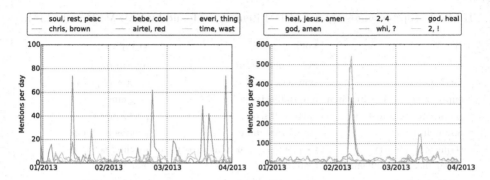

Fig. 3. Mentions per day of the top six non-singleton IIM (left) and MTV (right) itemsets from the Uganda messages dataset over three months.

itemsets exhibit no interesting spikes as one would expect. The top six SLIM and KRIMP itemsets in Table 4 all displayed random time evolution, as one would expect, except for the religious ones we have already encountered.

5   Conclusions

We presented a generative model that directly infers itemsets that best explain a transaction database along with a novel model-derived measure of interestingness and demonstrated the efficacy of our approach on both synthetic and real-world databases. In future we would like to extend our approach to directly inferring the association rules implied by the itemsets and parallelize our approach to large clusters so that we can efficiently scale to much larger databases.

Acknowledgements. This work was supported by the Engineering and Physical Sciences Research Council (grant number EP/K024043/1). We thank John Quinn for sharing the Uganda data.

References

1. Aggarwal, C., Han, J.: Frequent Pattern Mining. Springer (2014)
2. Agrawal, R., Srikant, R.: Fast algorithms for mining association rules. VLDB **1215**, 487–499 (1994)
3. Bar-Yehuda, R., Even, S.: A linear-time approximation algorithm for the weighted vertex cover problem. J. Algorithms **2**(2), 198–203 (1981)
4. Chvátal, V.: A greedy heuristic for the set-covering problem. Math. Oper. Res. **4**(3), 233–235 (1979)
5. Cormen, T., Leiserson, C., Rivest, R., Stein, C.: Introduction to Algorithms. MIT Press, Cambridge (2001)
6. Dempster, A., Laird, N., Rubin, D.: Maximum likelihood from incomplete data via the EM algorithm. J. R. Stat. Soc. Ser. B **39**, 1–38 (1977)
7. Feige, U.: A threshold of ln n for approximating set cover. J. ACM **45**(4), 634–652 (1998)

8. Fournier-Viger, P., Mwamikazi, E., Gueniche, T., Faghihi, U.: MEIT: memory efficient itemset tree for targeted association rule mining. In: Motoda, H., Wu, Z., Cao, L., Zaiane, O., Yao, M., Wang, W. (eds.) ADMA 2013. LNCS (LNAI), vol. 8347, pp. 95–106. Springer, Heidelberg (2013). doi:10.1007/978-3-642-53917-6_9
9. Friedman, N.: The Bayesian structural EM algorithm. In: UAI, pp. 129–138 (1998)
10. Gallo, A., Bie, T., Cristianini, N.: MINI: mining informative non-redundant itemsets. In: Kok, J.N., Koronacki, J., Lopez de Mantaras, R., Matwin, S., Mladenič, D., Skowron, A. (eds.) PKDD 2007. Lecture Notes in Artificial Intelligence (LNAI), vol. 4702, pp. 438–445. Springer, Heidelberg (2007). doi:10.1007/978-3-540-74976-9_44
11. Geerts, F., Goethals, B., Mielikäinen, T.: Tiling databases. In: Discovery science, pp. 278–289 (2004)
12. Goethals, B., Zaki, M.: FIMI repository (2004). http://fimi.ua.ac.be/
13. Han, J., Pei, J., Yin, Y.: Mining frequent patterns without candidate generation. SIGMOD Record 29, 1–12 (2000)
14. He, R., Shapiro, J.: Bayesian mixture models for frequent itemset discovery. arXiv preprint (2012). arXiv:1209.6001
15. Jaroszewicz, S., Simovici, D.A.: Interestingness of frequent itemsets using Bayesian networks as background knowledge. In: SIGKDD, pp. 178–186 (2004)
16. Kontonasios, K.N., De Bie, T.: An information-theoretic approach to finding informative noisy tiles in binary databases. In: SDM, pp. 153–164 (2010)
17. Korte, B., Vygen, J.: Combinatorial optimization: theory and algorithms. In: Korte, B., Vygen, J. (eds.) Algorithms and Combinatorics. Springer, Heidelberg (2012)
18. MacKay, D.J.C.: Information Theory, Inference, and Learning Algorithms. Cambridge University Press, Cambridge (2003)
19. Mampaey, M., Tatti, N., Vreeken, J.: Tell me what i need to know: succinctly summarizing data with itemsets. In: SIGKDD, pp. 573–581 (2011)
20. Mampaey, M., Vreeken, J., Tatti, N.: Summarizing data succinctly with the most informative itemsets. TKDD 6(4), 16 (2012)
21. Manning, C., Raghavan, P., Schütze, H.: Introduction to Information Retrieval. Cambridge University Press, Cambridge (2008)
22. Miettinen, P., Mielikainen, T., Gionis, A., Das, G., Mannila, H.: The discrete basis problem. IEEE TKDE 20(10), 1348–1362 (2008)
23. Mitchell-Jones, A., Amori, G., Bogdanowicz, W., Kryštufek, B., Reijnders, P., Spitzenberger, F., Stubbe, M., Thissen, J., Vohralík, V., Zima, J.: The Atlas of European Mammals. T & AD Poyser (1999)
24. Murphy, K.: Machine Learning: A Probabilistic Perspective. MIT Press, Cambridge (2012)
25. Shwe, M.A., Middleton, B., Heckerman, D., Henrion, M., Horvitz, E., Lehmann, H., Cooper, G.: Probabilistic diagnosis using a reformulation of the INTERNIST-1/QMR knowledge base. Methods Inf. Med. 30(4), 241–255 (1991)
26. Smets, K., Vreeken, J.: SLIM: Directly mining descriptive patterns. In: SDM, pp. 236–247 (2012)
27. USDA: The PLANTS Database (2008). http://plants.usda.gov/
28. Vreeken, J., Van Leeuwen, M., Siebes, A.: KRIMP: mining itemsets that compress. Data Min. Knowl. Discovery 23(1), 169–214 (2011)
29. Webb, G.I., Vreeken, J.: Efficient discovery of the most interesting associations. TKDD 8(3), 15 (2014)
30. Young, N.: Greedy set-cover algorithms (1974–1979, Chvátal, Johnson, Lovász, Stein). In: Kao, M. (ed.) Encyclopedia of Algorithms, pp. 379–381 (2008)
31. Zaki, M.J., Hsiao, C.J.: CHARM: an efficient algorithm for closed itemset mining. SDM 2, 457–473 (2002)

Top-N Recommendation via Joint Cross-Domain User Clustering and Similarity Learning

Dimitrios Rafailidis[1]([⊠]) and Fabio Crestani[2]

[1] Department of Informatics, Aristotle University of Thessaloniki,
Thessaloniki, Greece
draf@csd.auth.gr
[2] Faculty of Informatics, Università della Svizzera Italiana (USI),
Lugano, Switzerland
fabio.crestani@usi.ch

Abstract. A cross-domain recommendation algorithm exploits user preferences from multiple domains to solve the data sparsity and cold-start problems, in order to improve the recommendation accuracy. In this study, we propose an efficient Joint cross-domain user Clustering and Similarity Learning recommendation algorithm, namely JCSL. We formulate a joint objective function to perform adaptive user clustering in each domain, when calculating the user-based and cluster-based similarities across the multiple domains. In addition, the objective function uses an $L_{2,1}$ regularization term, to consider the sparsity that occurs in the user-based and cluster-based similarities between multiple domains. The joint problem is solved via an efficient alternating optimization algorithm, which adapts the clustering solutions in each iteration so as to jointly compute the user-based and cluster-based similarities. Our experiments on ten cross-domain recommendation tasks show that JCSL outperforms other state-of-the-art cross-domain strategies.

Keywords: Collaborative filtering · Cross-domain recommendation · Alternating optimization

1 Introduction

The collaborative filtering strategy has been widely followed in recommendation systems, where users with similar preferences tend to get similar recommendations [13]. User preferences are expressed explicitly in the form of ratings or implicitly in the form of number of views, clicks, purchases, and so on. Representative collaborative filtering strategies are matrix factorization techniques, which factorize the data matrix with user preferences in a single domain (e.g., music or video), to reveal the latent associations between users and items [14]. However, data sparsity and cold-start problems degrade the recommendation accuracy, as there are only a few preferences on which to base the recommendations in a single domain [5,13].

© Springer International Publishing AG 2016
P. Frasconi et al. (Eds.): ECML PKDD 2016, Part II, LNAI 9852, pp. 426–441, 2016.
DOI: 10.1007/978-3-319-46227-1_27

With the advent of social media platforms and e-commerce systems, such as Amazon and Epinions, users express their preferences in multiple domains. For example, in Amazon users can rate items from different domains, such as books and DVDs, or users express their opinion on different social media platforms, such as Facebook and Twitter. In the effort to overcome the data sparsity and cold-start problems, several cross-domain recommendation strategies have been proposed, which exploit the additional information of user preferences in multiple auxiliary domains to leverage the recommendation accuracy in a target domain [15]. However, generating cross-domain recommendations is a challenging task [5,23]; for example, if the auxiliary domains are richer than the target domain, algorithms learn how to recommend items in the auxiliary domains and consider the target domain as noise. Moreover, the auxiliary domains might be a potential source of noise, for example, if user preferences differ in the multiple domains, the auxiliary domains introduce noise in the learning of the target domain. Therefore, a pressing challenge resides on how to transfer the knowledge of user preferences from different domains.

In cross-domain recommendation, the auxiliary domains can be categorized based on users and items overlap, that is, full-overlap, and partial or non user/item overlap between the domains [5]. In this study, we focus on partial user overlap between the target and the auxiliary domains, as it reflects on the real-world setting [8]. Relevant methods, such as [8,20], form user and item clusters to capture the relationships between multiple domains at a cluster level, thus tackling the sparsity problem; and then weigh the cluster-based and user-based preferences to generate the top-N recommendations in the target domain. However, existing cluster-based cross-domain strategies have the following limitations, they form *non-adaptive* user and item clusters in a common latent space, when computing the cluster-based associations [8]; or they *linearly combine* the cluster-based and user-based relationships in the target domain [20].

1.1   Contribution

In this study, we overcome the aforementioned limitations in a novel approach for joint cross-domain recommendation based on user adaptive clustering and similarity learning. Our main contribution is summarized as follows, (i) we formulate an objective function to *jointly learn the user-based and cluster-based similarities across multiple domains, while adapting the user clusters in each domain at the same time.* To account for the fact that the user-based and cluster-based similarities across multiple domains are sparse, we use an $L_{2,1}$-norm regularization to force the similarities to be sparse. (ii) We propose *an efficient alternating optimization algorithm to minimize the joint objective function*, thus computing the user-based similarities across the multiple domains. The user latent factors are weighted based on the calculated user-based similarities, to generate the final top-N recommendations. Our experiments on ten cross-domain recommendation tasks demonstrate the superiority of the proposed approach over competitive cross-domain strategies.

The remainder of the paper is organized as follows, Sect. 2 reviews the related study and in Sect. 3 we formally define the cross-domain recommendation problem. Section 4 formulates the proposed joint objective function, Sect. 5 presents our alternating optimization algorithm, and in Sect. 6 we elaborate on how to generate the top-N cross-domain recommendations. Finally, Sect. 7 presents the experimental results and Sect. 8 concludes the study.

2   Related Work

Cross-domain recommendation algorithms differ in how the knowledge of user preferences from the auxiliary domains is exploited, when generating the recommendations in the target domain [15,23]. For example, various cross-domain approaches aggregate user preferences into a unified matrix, on which weighted single-domain techniques are applied, such as user-based kNN [2]. The graph-based method of [6] models the similarity relationships as a direct graph and explore all possible paths connecting users or items to capture the cross-domain relationships. Other methods exploit side information when linking multiple domains, on condition that the domains are linked by common knowledge, such as overlap of user/item attributes [4], social tags [1], and semantic networks [12]. However, such side information is not always available [20].

Other cross-domain techniques assume that the auxiliary and target domains are related by means of shared latent features. Representative methods are tri-matrix co-factorization, where user and item latent factors are shared between domains with different user preferences patterns. For example, Pan et al. [22] transform the knowledge of user preferences from different domains with heterogenous forms of user feedback, that is, explicit or implicit feedback, to compute the shared latent features. Hu et al. [10] model a cubic user-item-domain matrix (tensor), and by applying factorization the respective latent space is constructed, based on which the cross-domain recommendations are generated.

More closely related to our approach, cross-domain strategies transfer patterns of user preferences between domains at a cluster level. Li et al. [16] calculate user and item clusters for each domain, and then encode the cluster-based patterns in a shared codebook; finally, the knowledge of user preferences is transferred across domains through the shared codebook. Gao et al. [8] compute the latent factors of user-clusters and item-clusters to construct a common latent space, which represents the preference patterns e.g., rating patterns, of user clusters on the item clusters. Then, the common cluster-based preference pattern that is shared across domains is learnt following a subspace strategy, so as to control the optimal level of sharing among multiple domains. Mirbakhsh and Ling [20] factorize a cluster-based coarse matrix, to capture the shared interests among user and item clusters. By factorizing the coarse matrix the preferences of users on items are computed at a cluster-level. By linearly combining the factorized cluster-based preferences with the individual user preferences, the recommendation accuracy is improved. However, both [8] and [20] use non-adaptive clustering strategies, when computing the cluster-based similarities.

Table 1. Notation.

Symbol	Description
d	Number of domains
n_p	Number of users in the p-th domain, $p = 1, \ldots, d$
m_p	Number of items in the p-th domain
$\mathbf{R}_p \in \Re^{n_p \times m_p}$	User-item interaction (rating) matrix in the p-th domain
$\mathbf{A}_p \in \Re^{n_p \times n_p}$	Adjacency matrix of the users' graph in the p-th domain
c_p	Number of user clusters in the p-th domain
$\mathbf{C}_p \in \Re^{n_p \times c_p}$	Cluster assignment matrix in the p-th domain
$\mathbf{Y}_{pk} \in \Re^{c_p \times c_k}$	Cluster-based cross domain matrix between domains p and k
$\mathbf{X}_{pk} \in \Re^{n_p \times n_k}$	Cross domain matrix between the users in domains p and k

Meanwhile, there are several graph-based algorithms that perform clustering on multiple domains such as the studies reported in [3,7], However, these techniques focus on grouping instances e.g., users from different domains, and do not generate cross-domain recommendations.

3 Problem Formulation

3.1 Notation

Our notation is presented in Table 1. We assume that we have d different domains, where n_p and m_p are the numbers of users and items in the p-th domain, respectively. In matrix $\mathbf{R}_p$, we store the user preferences on items, in the form of explicit feedback e.g., ratings or in the form of implicit feedback e.g., number of views, clicks, and so on. Based on the matrix $\mathbf{R}_p$, we capture the user-based similarities in the p-th domain. If users i and j have interacted with at least a common item q, then users i and j are connected. The connections/similarities are stored in an adjacency matrix $\mathbf{A}_p$, whose ij-th entries are calculated as follows [24]:

$$
(\mathbf{A}_p)_{ij} = \begin{cases} \dfrac{\sum\limits_{q=1}^{m_p} (\mathbf{R}_p)_{iq}(\mathbf{R}_p)_{jq}}{\sqrt{\sum\limits_{q=1}^{m_p} (\mathbf{R}_p)_{iq}^2}\sqrt{\sum\limits_{q=1}^{m_p} (\mathbf{R}_p)_{jq}^2}} & \text{, if users } i \text{ and } j \text{ are connected} \\ 0 & \text{, otherwise} \end{cases} \tag{1}
$$

with $i, j = 1, \ldots, n_p$.

3.2 Cross-Domain Similarities

In our approach, we consider two types of cross-domain similarities, that is, the *cluster-based* and the *user-based* cross-domain similarities, defined as follows:

Definition 1 *(Cluster-based cross-domain similarities).* *For the p-th domain, we consider a cluster assignment matrix $\mathbf{C}_p$, with $(\mathbf{C}_p)_{ic}$ expressing the probability that user i belongs to cluster c. We define a cluster-based cross domain matrix $\mathbf{Y}_{pk}\Re^{c_p \times c_k}$, where c_p and c_k are the numbers of user clusters in the p-th and k-th domains respectively. The entry $(\mathbf{Y}_{pk})_{ij}$ expresses the similarity between clusters i and j in the p-th and k-th domains, accordingly.*

Definition 2 *(User-based cross-domain similarities).* *We define a cross-domain matrix $\mathbf{X}_{pk}$ between users in domains p and k. The entry $(\mathbf{X}_{pk})_{ij}$ expresses the similarity between users i and j in domains p and k, respectively.*

3.3 Problem Definition

In the cross-domain recommendation task, we assume that we have a target domain p and $d-1$ auxiliary domains. The goal is to predict the missing user preferences on items (recommendations) in the target domain p, while considering the user-based similarities in the rest of $d-1$ auxiliary domains. Following the notation of matrix factorization, let $\mathbf{U}_p \in \Re^{n_p \times l}$ and $\mathbf{V}_p \in \Re^{m_p \times l}$ be the user and item latent factor matrices, with the factorized matrix $\hat{\mathbf{R}}_p = \mathbf{U}_p \mathbf{V}_p^T \in \Re^{n_p \times m_p}$ containing the missing user preferences on items. As the i-th row of matrix $\mathbf{U}_p$, denoted by $(\mathbf{U}_p)_{i*}$, contains the l-dimensional user latent factor of user i, we can use a social regularization term $\Omega(\mathbf{U}_p)$, when learning the factorized matrix $\hat{\mathbf{R}}_p$ as follows [19]:

$$\min_{\mathbf{U}_p, \mathbf{V}_p} ||\mathbf{R}_p - \mathbf{U}_p \mathbf{V}_p^T||_F^2 + \gamma(||\mathbf{U}_p||_F^2 + ||\mathbf{V}_p||_F^2) + \Omega(\mathbf{U}_p) \qquad (2)$$

where the first term is the approximation error between the factorized matrix $\hat{\mathbf{R}}_p$ and the initial data matrix $\mathbf{R}_p$; the second one is the regularization term to avoid model overfitting with the parameter $\gamma > 0$; and the third term corresponds to the social regularization term based on the $d-1$ auxiliary domains between the partial user overlaps. In the social regularization term $\Omega(\mathbf{U}_p)$, we have to weigh the influence of the user latent factors based on the user-based cross-domain similarities in $\mathbf{X}_{pk}$ (Definition 2) as follows:

$$\Omega(\mathbf{U}_p) = \sum_{ij}^{n_p} \frac{1}{d-1} \sum_{k=1}^{d-1} (\mathbf{X}_{pk})_{ij} ||(\mathbf{U}_p)_{i*} - (\mathbf{U}_p)_{j*}||^2, \quad \text{with } p \neq k \qquad (3)$$

The term in the sum expresses the *approximation error between the user latent factors*, weighted by the user-based cross-domain similarities in $\mathbf{X}_{pk}$. The goal of the proposed approach is formally defined as follows:

Definition 3 *(Problem).* *The goal of the proposed approach is to calculate the weights in $\mathbf{X}_{pk}$ based on the preferences that users have in the different domains, in order to weigh the approximation error between the user latent factors $(\mathbf{U}_p)_{i*}$ and $(\mathbf{U}_p)_{j*}$ in (3).*

4   Joint Cross-Domain Objective Function

User clustering. To simplify the presentation, from now on we assume that we have a target domain p and an auxiliary domain k. Given the adjacency matrix $\mathbf{A}_p$ (computed in (1)), first we have to define the objective function for performing user clustering on the p-th domain, that is, to calculate the cluster assignment matrix $\mathbf{C}_p$, which corresponds to the following minimization problem:

$$\min_{\mathbf{C}_p} \sum_{ij} ||(\mathbf{C}_p)_{i*} - (\mathbf{C}_p)_{j*}||^2, \quad \text{with } i,j = 1,\ldots,n_p$$
$$\text{subject to } \mathbf{C}_p^T \mathbf{C}_p = \mathbf{I}, \, \mathbf{C}_p \geq 0 \tag{4}$$

with orthogonality constraints on the cluster matrix $\mathbf{C}_p$, and the user assignments to clusters being 0 or positive. According to the Laplacian method of [9], the minimization problem of (4) is equivalent to:

$$\min_{\mathbf{C}_p} \sum_{ij} ||(\mathbf{C}_p)_{i*} - (\mathbf{C}_p)_{j*}||^2 = \min_{\mathbf{C}_p} Tr(\mathbf{C}_p^T \mathbf{L}_p \mathbf{C}_p)$$
$$\text{subject to } \mathbf{C}_p^T \mathbf{C}_p = \mathbf{I}, \mathbf{C}_p \geq 0 \tag{5}$$

where $Tr(\cdot)$ is the trace operator. Matrix $\mathbf{L}_p \in \Re^{n_p \times n_p}$ is the Laplacian of the adjacency matrix $\mathbf{A}_p$, which is computed as follows: $\mathbf{L}_p = \mathbf{D}_p - \mathbf{A}_p$, where $\mathbf{D} \in \Re^{n_p \times n_p}$ is a diagonal matrix, whose entries are calculated as $(\mathbf{D}_p)_{ii} = \sum_{ij}(\mathbf{A}_p)_{ij}$. Similarly, we define the respective objective function in (5), for performing user clustering on the auxiliary domain k, denoted by matrix $\mathbf{C}_k \in \Re^{n_k \times c_k}$.

Cluster-based and User-based Similarities. To compute the cluster-based and user-based similarities between domains p and k, we follow a co-clustering strategy [7], where we have to minimize the following objective function:

$$\min_{\mathbf{Y}_{pk}, \mathbf{X}_{pk}} ||\mathbf{X}_{pk} - \mathbf{C}_p \mathbf{Y}_{pk} \mathbf{C}_k^T||_F^2 + \lambda_x ||\mathbf{X}_{pk}||_{2,1} + \lambda_y ||\mathbf{Y}_{pk}||_{2,1}$$
$$\text{subject to } \mathbf{Y}_{pk}^T \mathbf{Y}_{pk} = \mathbf{I}, \mathbf{Y}_{pk} \geq 0, \mathbf{X}_{pk} \geq 0 \tag{6}$$

with orthogonality constraints on the cluster-based matrix $\mathbf{Y}_{pk}$, and the user-based and cluster-based (cross-domain) similarities being 0 or positive. Symbol $||\cdot||_{2,1}$ denotes the $L_{2,1}$ norm of a matrix which is calculated as follows [21]:

$$||\mathbf{X}_{pk}||_{2,1} = \sum_{i=1}^{n_p} \sqrt{\sum_{j=1}^{n_k} (\mathbf{X}_{pk})_{ij}^2} = \sum_{i=1}^{n_p} ||(\mathbf{X}_{pk})_{i*}||_2 \tag{7}$$

The $L_{2,1}$ regularization terms in (6) force the solutions of matrices $\mathbf{X}_{pk}$ and $\mathbf{Y}_{pk}$ to be sparse, reflecting on the real-world scenario, where the user-based and cluster-based cross-domain similarities are usually sparse [5]. Parameters $\lambda_x, \lambda_y > 0$ control the respective $L_{2,1}$ regularization terms in (6).

Joint Objective Function. By combining (i) the objective function in (6) with (ii) the two clustering objective functions in (5) for domains p and k, we have to minimize the following *joint objective function*:

$$\min_{\mathbf{C}_k,\mathbf{C}_p,\mathbf{Y}_{pk},\mathbf{X}_{pk}} \mathcal{F} = ||\mathbf{X}_{pk} - \mathbf{C}_p\mathbf{Y}_{pk}\mathbf{C}_k^T||_F^2 + \lambda_x||\mathbf{X}_{pk}||_{2,1} + \lambda_y||\mathbf{Y}_{pk}||_{2,1}$$
$$+ \beta_p Tr(\mathbf{C}_p^T\mathbf{L}_p\mathbf{C}_p) + \beta_k Tr(\mathbf{C}_k^T\mathbf{L}_k\mathbf{C}_k) \tag{8}$$

subject to $\mathbf{C}_k^T\mathbf{C}_k = \mathbf{I}$, $\mathbf{C}_p^T\mathbf{C}_p = \mathbf{I}$, $\mathbf{Y}_{pk}^T\mathbf{Y}_{pk} = \mathbf{I}$, $\mathbf{C}_k, \mathbf{C}_p, \mathbf{Y}_{pk}, \mathbf{X}_{pk} \geq 0$

where $\beta_p, \beta_k > 0$ are the regularization parameters for the clusterings in domains p and k, respectively.

5   Alternating Optimization

As the joint objective function $\mathcal{F}(\mathbf{C}_k, \mathbf{C}_p, \mathbf{Y}_{pk}, \mathbf{X}_{pk})$ in (8) is not convex with respect to the four variables/matrices, we propose an alternating optimization algorithm, where we update one variable, while keeping the remaining three fixed. The cluster assignment matrices $\mathbf{C}_k, \mathbf{C}_p$ are initialized by performing k-means to the respective adjacency matrices $\mathbf{A}_k$ and $\mathbf{A}_p$, while $\mathbf{Y}_{pk}$ and $\mathbf{X}_{pk}$ are initialized by random (sparse) matrices. Next, we present the updating steps for each variable.

Step 1, fix $\mathbf{C}_p$, $\mathbf{Y}_{pk}$, $\mathbf{X}_{pk}$ and update $\mathbf{C}_k$. By considering the optimality condition $\partial F/\partial \mathbf{C}_k = 0$, we calculate the partial derivative of $\mathcal{F}$ with respect to $\mathbf{C}_k$:

$$\frac{\partial \mathcal{F}}{\partial \mathbf{C}_k} = -2\mathbf{X}_{pk}^T\mathbf{C}_p\mathbf{Y}_{pk} + 2\mathbf{C}_k\mathbf{Y}_{pk}^T\mathbf{C}_p^T\mathbf{C}_p\mathbf{Y}_{pk} + 2\beta_k\mathbf{L}_k\mathbf{C}_k \tag{9}$$

As the joint objective function $\mathcal{F}$ in (8) is subject to the orthogonality constraints $\mathbf{C}_p^T\mathbf{C}_p = \mathbf{I}$, $\mathbf{Y}_{pk}^T\mathbf{Y}_{pk} = \mathbf{I}$, the second term of (9) equals $2\mathbf{C}_k$. By setting the partial derivative equal to zero we have to solve the following equation with respect to $\mathbf{C}_k$:

$$- \mathbf{X}_{pk}^T\mathbf{C}_p\mathbf{Y}_{pk} + \mathbf{C}_k + \beta_k\mathbf{L}_k\mathbf{C}_k = 0 \tag{10}$$

As $(\mathbf{I} + \beta_k\mathbf{L}_k)$ is a positive definite matrix, we can obtain the following closed-form solution (*updating rule*) of $\mathbf{C}_k$:

$$\mathbf{C}_k = (\mathbf{I} + \beta_k\mathbf{L}_k)^{-1}\mathbf{X}_{pk}^T\mathbf{C}_p\mathbf{Y}_{pk} \tag{11}$$

Step 2, fix $\mathbf{C}_k$, $\mathbf{Y}_{pk}$, $\mathbf{X}_{pk}$ and update $\mathbf{C}_p$. The partial derivative of $\mathcal{F}$ with respect to $\mathbf{C}_p$ is equivalent to:

$$\frac{\partial \mathcal{F}}{\partial \mathbf{C}_p} = -2\mathbf{X}_{pk}\mathbf{C}_k\mathbf{Y}_{pk}^T + 2\mathbf{C}_p\mathbf{Y}_{pk}\mathbf{C}_k^T\mathbf{C}_k\mathbf{Y}_{pk}^T + 2\beta_p\mathbf{L}_p\mathbf{C}_p \tag{12}$$

Similarly, provided that $\mathcal{F}$ is subject to $\mathbf{C}_k^T\mathbf{C}_k = \mathbf{I}$, $\mathbf{Y}_{pk}^T\mathbf{Y}_{pk} = \mathbf{I}$, we have the optimality condition by setting the partial derivative equal to zero:

$$- \mathbf{X}_{pk}\mathbf{C}_k\mathbf{Y}_{pk}^T + \mathbf{C}_p + \beta_p\mathbf{L}_p\mathbf{C}_p = 0 \tag{13}$$

Given that $(\mathbf{I} + \beta_p \mathbf{L}_p)$ is positive definite, we have the following *updating rule* for $\mathbf{C}_p$:

$$\mathbf{C}_p = (\mathbf{I} + \beta_p \mathbf{L}_p)^{-1} \mathbf{X}_{pk} \mathbf{C}_k \mathbf{Y}_{pk}^T \qquad (14)$$

Step 3, fix $\mathbf{C}_p$, $\mathbf{C}_k$, $\mathbf{X}_{pk}$ and update $\mathbf{Y}_{pk}$. The presence of the $L_{2,1}$-norm regularization in the objective function $\mathcal{F}$ of (8) makes the model difficult to optimize, as the algorithm cannot be guaranteed to convergence based on the analysis at [21]. To overcome this issue, we define a diagonal matrix $\mathbf{Q}_y \in \Re^{c_p \times c_k}$ (with the same size as $\mathbf{Y}_{pk}$), whose entries are calculated as follows:

$$(\mathbf{Q}_y)_{ii} = \frac{1}{2\|(\mathbf{Y}_{pk})_{i*}\|_2} \qquad (15)$$

thus, we can calculate the partial derivative of $\mathcal{F}$ with respect to $\mathbf{Y}_{pk}$ as follows:

$$\frac{\partial \mathcal{F}}{\partial \mathbf{Y}_{pk}} = -2\mathbf{C}_p^T \mathbf{X}_{pk} \mathbf{C}_k + 2\mathbf{C}_p^T \mathbf{C}_p \mathbf{Y}_{pk} \mathbf{C}_k^T \mathbf{C}_k + 2\lambda_y \mathbf{Q}_y \mathbf{Y}_{pk} \qquad (16)$$

where the last term corresponds to the partial derivative of the $L_{2,1}$ regularization term of $\mathbf{Y}_{pk}$ in (8), with convergence guarantees [21]. As the joint objective function $\mathcal{F}$ is subject to $\mathbf{C}_k^T \mathbf{C}_k = \mathbf{I}$, $\mathbf{C}_p^T \mathbf{C}_p = \mathbf{I}$, by setting the partial derivative of (16) equal to zero, we have:

$$- \mathbf{C}_p^T \mathbf{X}_{pk} \mathbf{C}_k + \mathbf{Y}_{pk} + \lambda_y \mathbf{Q}_y \mathbf{Y}_{pk} = 0 \qquad (17)$$

which results in the following *update rule* for $\mathbf{Y}_{pk}$:

$$\mathbf{Y}_{pk} = (\mathbf{I} + \lambda_y \mathbf{Q}_y)^{-1} \mathbf{C}_p^T \mathbf{X}_{pk} \mathbf{C}_k \qquad (18)$$

where $(\mathbf{I} + \lambda_y \mathbf{Q}_y)$ is a positive definite matrix.

Step 4, fix $\mathbf{C}_k$, $\mathbf{C}_p$, $\mathbf{Y}_{pk}$ and update $\mathbf{X}_{pk}$. Similarly, given the $L_{2,1}$ regularization term of $\mathbf{X}_{pk}$ in the joint objective function $\mathcal{F}$, we define the diagonal matrix $\mathbf{Q}_x \in \Re^{n_p \times n_k}$, whose entries are computed as follows:

$$(\mathbf{Q}_x)_{ii} = \frac{1}{2\|(\mathbf{X}_{pk})_{i*}\|_2} \qquad (19)$$

Then, we take the partial derivative of $\mathcal{F}$ with respect to $\mathbf{X}_{pk}$:

$$\frac{\partial \mathcal{F}}{\partial \mathbf{X}_{pk}} = 2\mathbf{X}_{pk} - 2\mathbf{C}_p \mathbf{Y}_{pk} \mathbf{C}_k^T + \beta_x \mathbf{Q}_x \mathbf{X}_{pk} \qquad (20)$$

By setting the partial derivative of (20) equal to zero, we obtain the following *update rule* for $\mathbf{X}_{pk}$:

$$\mathbf{X}_{pk} = (\mathbf{I} + \beta_x \mathbf{Q}_x)^{-1} \mathbf{C}_p \mathbf{Y}_{pk} \mathbf{C}_k^T \qquad (21)$$

Analysis. The alternating optimization is performed iteratively, where in each iteration matrices $\mathbf{C}_k$, $\mathbf{C}_p$, $\mathbf{Y}_{pk}$ and $\mathbf{X}_{pk}$ are updated based on (11), (14), (18) and (21), respectively. More precisely, at each iteration each variable/matrix is recalculated based on the rest three matrices, which means that *each matrix is adapted to the values that the rest matrices have taken at the previous iteration, in order to reach a consensus solution for all four matrices over the iterations.* The alternating optimization algorithm is repeated, until the algorithm converges. The optimization algorithm converges on condition that the joint objection function $\mathcal{F}$ in (8) monotonically decreases after each iteration. Based on [21] the $L_{2,1}$-norm regularization terms of $\mathcal{F}$ are differentiable at zero, by using the diagonal matrices $\mathbf{Q}_y$ and $\mathbf{Q}_x$ in (15) and (19) when updating $\mathbf{Y}_{pk}$ and $\mathbf{X}_{pk}$ in (18) and (21), respectively[1]. By considering the optimality condition in each step, that is, setting the partial derivative of $\mathcal{F}$ with respect to each variable equal to zero when updating the four variables, the proof that the algorithm converges is similar to the convergence analysis of [11].

6   Generating Top-N Recommendations

The joint objective function $\mathcal{F}$ for $k = 1, \ldots, d - 1$ auxiliary domains can be extended to:

$$\min_{\mathbf{C}_k, \mathbf{C}_p, \mathbf{Y}_{pk}, \mathbf{X}_{pk}} \mathcal{F} = \sum_{k=1}^{d-1} \left[||\mathbf{X}_{pk} - \mathbf{C}_p \mathbf{Y}_{pk} \mathbf{C}_k^T||_F^2 + \lambda_x ||\mathbf{X}_{pk}||_{2,1} + \lambda_y ||\mathbf{Y}_{pk}||_{2,1} \right]$$
$$+ \beta_p Tr(\mathbf{C}_p^T \mathbf{L}_p \mathbf{C}_p) + \sum_{k=1}^{d-1} \beta_k Tr(\mathbf{C}_k^T \mathbf{L}_k \mathbf{C}_k)$$

$$\text{subject to } \mathbf{C}_k^T \mathbf{C}_k = \mathbf{I}, \ \mathbf{C}_p^T \mathbf{C}_p = \mathbf{I}, \mathbf{Y}_{pk}^T \mathbf{Y}_{pk} = \mathbf{I}, \ \mathbf{C}_k, \mathbf{C}_p, \mathbf{Y}_{pk}, \mathbf{X}_{pk} \geq 0$$
$$\tag{22}$$

where the variables of $\mathcal{F}$ are (i) the $d-1$ clustering matrices $\mathbf{C}_k$ of the auxiliary domains; (ii) the cluster matrix $\mathbf{C}_p$ of the target domain p; (iii) the $d-1$ matrices $\mathbf{Y}_{pk}$, and $\mathbf{X}_{pk}$. An overview of our approach is presented in Algorithm 1.

7   Experimental Evaluation

7.1   Settings

Cross-domain recommendation tasks. Our experiments were performed on ten cross-domain tasks from the Rich Epinions Dataset (RED)[2], which contains 131,228 users, 317,775 items and 1,127,673 user preferences, in the form of ratings. The items are grouped in categories/domains, and we evaluate the performance of cross-domain recommendation on the 10 largest domains. The evaluation data were provided by the authors of [20]. The main characteristics of the ten cross-domain recommendation tasks are presented in Table 2. The evaluation tasks are challenging, as the domains do not have item overlaps, but

---

[1] At the first iteration matrices $\mathbf{Q}_y$ and $\mathbf{Q}_x$ are initialized by using the identity matrix [21].

[2] http://liris.cnrs.fr/red/.

ALGORITHM 1. Joint Cross-Domain User Clustering and Similarity Learning

```
Input: R_p, A_p, R_k, A_k, with k = 1,..., d − 1
Output: factorized matrix R̂_p
1   Initialize C_p, C_k by performing k-means on A_p and A_k, ∀k = 1,..., d − 1
2   Initialize Y_pk and X_pk, ∀k = 1,..., d − 1
3   convergence = False
4   while convergence = False do
5       Update C_k based on (11), ∀k = 1,..., d − 1
6       Update C_p based on (14)
7       Update Y_pk based on (18), ∀k = 1,..., d − 1
8       Update X_pk based on (21), ∀k = 1,..., d − 1
9       Calculate F in (22) using the updated C_k, C_p, Y_pk, X_pk, ∀k = 1,..., d − 1
10      if F converges
11          convergence = true
12      end if
13  end while
14  Calculate Ω(U_p) based on the updated X_pk and (3), ∀k = 1,..., d − 1
15  Calculate U_p and V_p based on Ω(U_p) and (2)
16  R̂_p = U_p V_p^T
```

Table 2. The ten cross-domain recommendation tasks.

Domain	Users	Items	Ratings	Density (%)
Books	15,507	59,346	108,887	0.011
Baby Care	5,422	3,165	21,340	0.124
Destinations	9,290	3,615	31,418	0.093
Music	16,002	35,807	96,226	0.016
Online Stores & Services	28,643	5,518	54,734	0.034
Personal Care	6,214	10,786	28,945	0.043
Sport & Outdoor	6,750	9,597	19,181	0.029
Toys	9,040	18,681	51,152	0.030
Used Cars	17,041	4,174	28,598	0.040
Video & DVD	25,218	28,972	175,665	0.024

only user overlaps. In each cross-domain recommendation task, we consider one target domain and the remaining nine serve as auxiliary domains. For each task we preserve all the ratings of the auxiliary domains, and we randomly select 25 %, 50 % and 75 % of the target domain as training set [20]. For each split, the remaining ratings of the target domain are considered as test set. We repeated our experiment five times, and we report mean values and standard deviations over the runs.

Compared methods. In our experiments, we evaluate the performance of the following methods:

- *NMF* [14]: a *single-domain* baseline Nonnegative Matrix Factorization method, which generates recommendations based only on the ratings of the target domain, ignoring the ratings in the auxiliary domains.

- *CLFM* [8]: a *cross-domain* Cluster-based Latent Factor Model which uses joint nonnegative tri-factorization to construct a latent space to represent the rating patterns of user clusters on the item clusters from each domain, and then generates the cross-domain recommendations based on a subspace learning strategy.
- *CBMF* [20]: a *cross-domain* Cluster-based Matrix Factorization model, which defines a coarse cross-domain matrix to capture the shared preferences between user and item clusters in the multiple domains, and then reveals the latent associations at a cluster level by factorizing the coarse cross-domain matrix. The final recommendations are generated by linearly combining the cluster-based latent associations and the user-based latent associations in the target domain. CBMF controls the influence of the cluster-based relationships on the personalized recommendations based on a parameter α.
- *JCSL*: the proposed Joint *cross-domain* user Clustering and Similarity Learning model.

In all models we varied the number of latent factors from [10,100] by a step of 10. In the cross-domain methods of CLMF, CBMF and JCSL we fixed the number of clusters to 100, as suggested in [20]. The predefined clusters in both the CLMF and CBMF methods are computed by performing k-means, also used in [8,20]. Similarly, in the proposed JCSL method, the clusters are initialized by the k-means algorithm (Sect. 5). Following [8], in CLFM we tuned the dimensionality of the subspace up to the minimum number of latent factors of the multiple domains. In CBFM, we varied the α parameter in [0,1], where lower values of α consider to a fewer extent the cluster-based relationships, when computing the top-N recommendations. In JCSL the maximum number of iterations[3] of the alternating optimization algorithm is fixed to 50, and the regularization parameters of the objective function in (22) were varied in [0.0001,0.1]. In all examined models, the parameters were determined via cross validation and in our experiments we report the best results.

Evaluation protocol. Popular commercial systems make top-N recommendations to users, and relevant studies showed that rating error metrics, such as RMSE (Root Mean Squared Error) and MAE (Mean Absolute Error) do not necessarily reflect on the top-N recommendation performance [5]. Therefore, in our experiments we used the ranking-based metrics Recall and Normalized Discounted Cumulative Gain to evaluate the top-N performance of the examined models directly [20]. Recall ($R@N$) is defined as the ratio of the relevant items in the top-N ranked list over all the relevant items for each user. The Normalized Discounted Cumulative Gain ($NDCG@N$) metric considers the ranking of the relevant items in the top-N list. For each user the Discounted Cumulative Gain ($DCG@N$) is defined as:

---

[3] The algorithm terminates (converges) in less iterations, if $(\mathcal{F}^{(t+1)} - \mathcal{F}^{(t)})/\mathcal{F}^{(t)} \leq 1e - 03$, where $\mathcal{F}^{(t)}$ is the value of the objective function $\mathcal{F}$ after the t-th iteration.

$$DCG@N = \sum_{j=1}^{N} \frac{2^{rel_j} - 1}{\log_2 j + 1} \qquad (23)$$

where rel_j represents the relevance score of the item j to the user. $NDCG@N$ is the ratio of $DCG@N$ over the ideal $iDCG@N$ value for each user, that is, the $DCG@N$ value given the ratings in the test set. In our experiments we averaged $R@N$ and $NDCG@N$ over all users.

7.2 Results

In the first set of experiments, we use 75 % of the target domain as training set, while the remaining is considered as test set. Table 3 presents the experimental results in terms of $NDCG@10$. The cross-domain methods of CLFM, CBMF and JCSL significantly outperform the single-domain NMF method, by exploiting the auxiliary domains when generating the recommendations. This happens because the cross-domain methods incorporate the additional information of user preferences on items from the auxiliary domains, thus reducing the data sparsity in the target domain. The proposed JCSL method achieves an 8.95 % improvement on average when comparing with the second best method. Using the paired t-test we found that JCSL is superior over the rest approaches for $p < 0.05$. JCSL beats the competitive strategies, as it exploits the cluster-based similarities more efficiently than the competitive cluster-based models. The joint learning strategy of the adaptive user clustering while computing the user-based and cluster-based similarities, makes JCSL to efficiently incorporate the additional information of user preferences in the auxiliary domains. On the other hand, CLFM uses a subspace learning strategy on non-adaptive clusters in a common latent space. Finally, CBMF linearly combines the cluster-based and the individual latent associations by capturing the user preferences in the auxiliary domains based on predefined clusters. In this set of experiments there is the exceptional case of the "Baby Care" cross-domain task, where the proposed method has similar performance with CBMF. This happens because "Baby Care" is the less sparse domain, as presented in Table 2. Figure 1 compares the examined models in terms of recall $(R@N)$, by varying the number of the top-N recommendations. Similarly, JCSL achieves a 12.17 % improvement on average, for all the cross-domain recommendation tasks.

To evaluate the performance of the examined methods when sparsity increases, the training set is reduced to 25 % of the target domain, while keeping all the ratings of the auxiliary domains. Table 4 reports the experimental results in terms of recall $(R@10)$ based on the reduced training sets. In relation to the experimental results of Fig. 1, recall drops for all methods, due to the increased sparsity. As we can observe, the proposed JCSL method achieves relatively high recall. In all cross-domain recommendation tasks, JCSL is superior to the competitive cross-domain strategies (for $p < 0.05$), by achieving on average relative improvement of 14.49 %, when comparing with the second best method.

Figure 2 shows the effect on $NDCG@10$ of the cross-domain recommendation models, by varying the training sizes of three representative target domains,

Table 3. Effect on *NDCG@10* for the ten cross-domain recommendation tasks, using 75% of the target domain as training set. Bold values denote the best scores, for *$p < 0.05$ in paired t-test. The last column denotes the relative improvement (%), when comparing JCSL with the second best method (CBMF).

Target domain	NMF	CLFM	CBMF	JCSL	Improv. (%)
Books	.0997 ± .0119	.1502 ± 0308	.1780 ± .0619	**.1919 ± .0472***	**7.80***
Baby Care	.2054 ± .0506	.3144 ± .0345	.3875 ± .0153	.3749 ± .0460	−3.25
Destinations	.2991 ± .0810	.3648 ± .0961	.4271 ± .0903	**.4587 ± .0805***	**7.39***
Music	.1235 ± .0263	.1631 ± .0417	.1991 ± .0399	**.2109 ± .0420***	**7.43***
Online Stores	.1718 ± .0909	.2550 ± .0183	.3222 ± .0320	**.3665 ± .0509***	**13.74***
Personal Care	.1385 ± .0173	.1677 ± .0687	.2155 ± .0311	**.2491 ± .0460***	**15.59***
Sport & Outdoor	.1167 ± .0290	.1469 ± .0308	.1493 ± .0917	**.1665 ± .0922***	**11.52***
Toys	.1718 ± .0649	.2269 ± .0406	.2757 ± .0453	**.3017 ± .0434***	**9.43***
Used Cars	.0981 ± .0841	.1301 ± .0167	.1750 ± .0471	**.1897 ± .0209***	**8.15***
Video & DVD	.2413 ± .0381	.3225 ± .0350	.3984 ± .0428	**.4451 ± .0428***	**11.72***

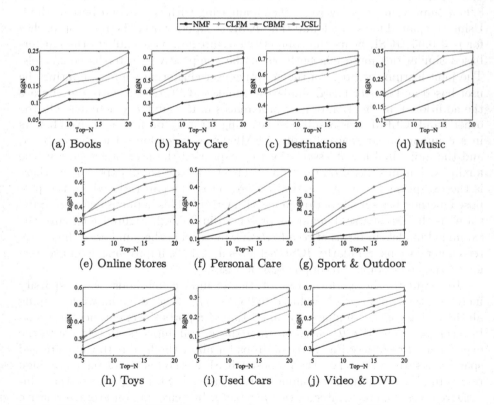

Fig. 1. Effect on *Recall (R@N)* by varying the number of the top-N recommendations. In the ten cross-domain tasks, 75% of the target domain is considered as training set.

Table 4. Effect on *Recall (R@10)* for the ten cross-domain recommendation tasks, using 25 % of the target domain as training set. Bold values denote the best scores, for $^*p < 0.05$ in paired t-test. The last column denotes the relative improvement (%), when comparing JCSL with the second best method (CBMF).

Target domain	NMF	CLFM	CBMF	JCSL	Improv. (%)
Books	.0856 ± .0348	.1108 ± .0744	.1468 ± .0390	**.1779 ± .0173***	**21.18***
Baby Care	.1796 ± .0947	.3411 ± .0872	.4474 ± .0752	**.4968 ± .1212***	**11.04***
Destinations	.2604 ± .0520	.3748 ± .0869	.4365 ± .0803	**.5013 ± .0720***	**14.84***
Music	.1485 ± .0431	.1805 ± .0450	.2073 ± .0869	**.2304 ± .0330***	**11.14***
Online Stores	.2099 ± .0112	.3151 ± .0860	.3689 ± .0401	**.4230 ± .0622***	**14.74***
Personal Care	.1345 ± .0152	.1690 ± .0622	.1961 ± .0560	**.2232 ± .0648***	**13.81***
Sport & Outdoor	.0800 ± .0853	.1239 ± .0751	.1622 ± .0306	**.1820 ± .0294***	**12.63***
Toys	.2448 ± .0679	.3137 ± .0585	.3497 ± .0751	**.3921 ± .0655***	**12.12***
Used Cars	.0730 ± .0250	.1098 ± .0959	.1160 ± .0699	**.1439 ± .0427***	**24.05***
Video & DVD	.2164 ± .0162	.3685 ± .0285	.4421 ± .0892	**.4834 ± .0733***	**9.34***

which are at different scale (Table 2). For presentation purposes, in this set of experiments the baseline NMF method is omitted, due to its low performance. We observe that all cross-domain methods increase the *NDCG* metric, when a larger training set is used. Figure 2 shows that the proposed JCSL method keeps *NDCG* relatively high in all settings, while outperforming CLFM and CBMF. The three cross-domain models differ in how the knowledge of user preferences is transferred between the domains when generating the recommendations, which explains their different performance when decreasing the training set size. JCSL adapts the clustering in each domain separately, while computing the cross-domain cluster-based similarities; whereas CLFM and CBMF compute the similarities between predefined/non-adaptive clusters, when generating the recommendations.

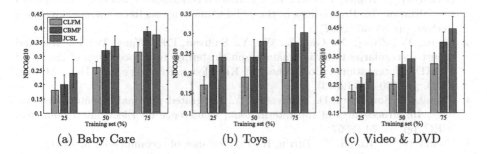

(a) Baby Care   (b) Toys   (c) Video & DVD

Fig. 2. Effect on *NDCG@10* by varying the size of the training set.

8   Conclusion

In this paper, we presented an efficient cross-domain recommendation algorithm based on a joint strategy to adapt the user clusters, while calculating the user-based and cluster-based similarities across multiple domains. The joint optimization problem is solved via an efficient alternating optimization algorithm. Our experiments on ten cross-domain tasks confirmed the superiority of the proposed approach over competitive cross-domain strategies at different levels of sparsity. The main advantages of our approach are that JCSL adapts the clusters in each domain separately, while computing the cross-domain cluster-based similarities, whereas the competitors compute the similarities between predefined/non-adaptive clusters when generating the recommendations. Instead of linearly combining the cluster-based and user-based similarities, as for example CBMF does, JCSL jointly learns both types of similarities. In this study we considered partial user overlaps, with the mapping of users being known between the different domains. An interesting future direction is to extend our proposed approach for unknown user-matching across multiple domains [17]. In addition, an interesting future direction is to evaluate the performance of different clustering algorithms, such as spherical k-means [25] or power iteration [18], to initialize the clusters in the different domains.

Acknowledgments. We would like to thank Nima Mirbakhsh and Charles Ling for providing us with the evaluation data of the ten cross-domain tasks.

References

1. Abel, F., Araújo, S., Gao, Q., Houben, G.-J.: Analyzing cross-system user modeling on the social web. In: Auer, S., Díaz, O., Papadopoulos, G.A. (eds.) ICWE 2011. LNCS, vol. 6757, pp. 28–43. Springer, Heidelberg (2011). doi:10.1007/978-3-642-22233-7_3
2. Berkovsky, S., Kuflik, T., Ricci, F.: Distributed collaborative filtering with domain specialization. In: Proceedings of the ACM Conference on Recommender Systems, RecSys, pp. 33–40 (2007)
3. Cheng, W., Zhang, X., Guo, Z., Wu, Y., Sullivan, P.F., Wang, W.: Flexible and robust co-regularized multi-domain graph clustering. In: Proceedings of the ACM SIGKDD International Conference on Knowledge Discovery and Data Mining, KDD, pp. 320–328 (2013)
4. Chung, R., Sundaram, D., Srinivasan, A.: Integrated personal recommender systems. In: Proceedings of the International Conference on Electronic Commerce, ICEC, pp. 65–74 (2007)
5. Cremonesi, P., Koren, Y., Turrin, R.: Performance of recommender algorithms on top-n recommendation tasks. In: Proceedings of the ACM Conference on Recommender Systems, RecSys, pp. 39–46 (2010)
6. Cremonesi, P., Tripodi, A., Turrin, R.: Cross-domain recommender systems. In: Proceedings of the IEEE International Conference on Data Mining Workshops, ICDMW, pp. 496–503 (2011)
7. Du, L., Shen, Y.: Towards robust co-clustering. In: Proceedings of the International Joint Conference on Artificial Intelligence, IJCAI (2013)

8. Gao, S., Luo, H., Chen, D., Li, S., Gallinari, P., Guo, J.: Cross-domain recommendation via cluster-level latent factor model. In: Blockeel, H., Kersting, K., Nijssen, S., Železný, F. (eds.) ECML PKDD 2013. LNCS, vol. 8189, pp. 161–176. Springer, Heidelberg (2013). doi:10.1007/978-3-642-40991-2_11

9. Gao, S., Tsang, I.W., Chia, L.: Laplacian sparse coding, hypergraph Laplacian sparse coding, and applications. IEEE Trans. Pattern Anal. Mach. Intell. **35**(1), 92–104 (2013)

10. Hu, L., Cao, J., Xu, G., Cao, L., Gu, Z., Zhu, C.: Personalized recommendation via cross-domain triadic factorization. In: Proceedings of the International World Wide Web Conference, WWW, pp. 595–606 (2013)

11. Jain, P., Netrapalli, P., Sanghavi, S.: Low-rank matrix completion using alternating minimization. In: Symposium on Theory of Computing Conference, STOC, pp. 665–674 (2013)

12. Kaminskas, M., Ricci, F., Schedl, M.: Location-aware music recommendation using auto-tagging and hybrid matching. In: Proceedings of the ACM Conference on Recommender Systems, RecSys, pp. 17–24 (2013)

13. Koren, Y., Bell, R.M., Volinsky, C.: Matrix factorization techniques for recommender systems. IEEE Comput. **42**(8), 30–37 (2009)

14. Lee, D.D., Seung, H.S.: Algorithms for non-negative matrix factorization. In: Advances in Neural Information Processing Systems 13, Papers from Neural Information Processing Systems, NIPS, pp. 556–562 (2000)

15. Li, B.: Cross-domain collaborative filtering: a brief survey. In: Proceedings of the IEEE International Conference on Tools with Artificial Intelligence, ICTAI, pp. 1085–1086 (2011)

16. Li, B., Yang, Q., Xue, X.: Can movies and books collaborate? Cross-domain collaborative filtering for sparsity reduction. In: Proceedings of the International Joint Conference on Artificial Intelligence, IJCAI, pp. 2052–2057 (2009)

17. Li, C., Lin, S.: Matching users and items across domains to improve the recommendation quality. In: Proceedings of the ACM SIGKDD International Conference on Knowledge Discovery and Data Mining, KDD, pp. 801–810 (2014)

18. Lin, F., Cohen, W.W.: Power iteration clustering. In: Proceedings of the 27th International Conference on Machine Learning, pp. 655–662 (2010)

19. Ma, H., Zhou, D., Liu, C., Lyu, M.R., King, I.: Recommender systems with social regularization. In: Proceedings of the International Conference on Web Search and Web Data Mining, WSDM, pp. 287–296 (2011)

20. Mirbakhsh, N., Ling, C.X.: Improving top-n recommendation for cold-start users via cross-domain information. TKDD **9**(4), 33 (2015)

21. Nie, F., Huang, H., Cai, X., Ding, C.H.Q.: Efficient and robust feature selection via joint l2, 1-norms minimization. In: 24th Annual Conference on Neural Information Processing Systems, NIPS. Advances in Neural Information Processing Systems 23, pp. 1813–1821 (2010)

22. Pan, W., Xiang, E.W., Liu, N.N., Yang, Q.: Transfer learning in collaborative filtering for sparsity reduction. In: Proceedings of the AAAI Conference on Artificial Intelligence, AAAI (2010)

23. Shi, Y., Larson, M., Hanjalic, A.: Collaborative filtering beyond the user-item matrix: a survey of the state of the art and future challenges. ACM Comput. Surv. **47**(1), 3:1–3:45 (2014)

24. Tang, J., Hu, X., Liu, H.: Social recommendation: a review. Social Netw. Analys. Mining **3**(4), 1113–1133 (2013)

25. Zhong, S.: Efficient online spherical k-means clustering. In: Proceedings of the IEEE Joint Conference on Neural Networks, pp. 3180–3185 (2005)

Measuring the Stability of Feature Selection

Sarah Nogueira$^{(\boxtimes)}$ and Gavin Brown

School of Computer Science, University of Manchester, Manchester M13 9PL, UK
{sarah.nogueira,gavin.brown}@manchester.ac.uk

Abstract. In feature selection algorithms, "stability" is the sensitivity of the chosen feature set to variations in the supplied training data. As such it can be seen as an analogous concept to the statistical variance of a predictor. However unlike variance, there is no unique definition of stability, with numerous proposed measures over 15 years of literature. In this paper, instead of defining a new measure, we start from an axiomatic point of view and identify what properties would be desirable. Somewhat surprisingly, we find that the simple Pearson's correlation coefficient has all necessary properties, yet has somehow been overlooked in favour of more complex alternatives. Finally, we illustrate how the use of this measure in practice can provide better interpretability and more confidence in the model selection process. The data and software related to this paper are available at https://github.com/nogueirs/ECML2016.

Keywords: Stability · Feature selection

1 Introduction

High-dimensional datasets can be very expensive in terms of computational resources and of data collection. Predictive models in this situation often suffer from the *curse of dimensionality* and tend to overfit the data. For these reasons, feature selection (FS) has become an ubiquitous challenge that aims at selecting a "useful" set of features [8].

Stability of FS is defined as the sensitivity of the FS procedure to small perturbations in the training set. This issue is of course extremely relevant with small training samples, e.g. in bioinformatics applications - if the alteration/exclusion of just one training example results in a very different choice of biomarkers, we cannot justifiably say the FS is doing a reliable job. In early cancer detection, stability of the identified markers is a strong indicator of reproducible research [6,12] and therefore selecting a stable set of markers is said to be equally important as their predictive power [7].

The study of stability poses several problems such as: What impacts stability? How can we make FS procedures more stable? How can we quantify it? A large

Electronic supplementary material The online version of this chapter (doi:10.1007/978-3-319-46227-1_28) contains supplementary material, which is available to authorized users.

© Springer International Publishing AG 2016
P. Frasconi et al. (Eds.): ECML PKDD 2016, Part II, LNAI 9852, pp. 442–457, 2016.
DOI: 10.1007/978-3-319-46227-1_28

part of the literature is dedicated to the later, which is the focus of this paper. Indeed, at a literature search conducted at the time of writing, we identified at least 10 different measures used to quantify stability [4,8,10,11,13,14,16,17,19, 21]. The existence of so many different measures without known properties may lead to an incorrect interpretation of the stability values obtained.

As described by [8], FS procedures can have 3 types of outputs: *a weighting* on the features also called scoring (e.g. ReliefF), a *ranking* on the features (e.g. ranking by mutual information of the features with the target class) or a *feature set* (e.g. any wrapper approach). A weighting can be mapped into a ranking, and by applying a threshold on a ranking, a ranking can be mapped into a feature set; but the reverse is clearly not possible. For this reason, there exist stability measures for each type of output. In this paper, we focus on FS procedures that return a feature set.

An Example. Imagine we have $d = 5$ features to choose from. We can model the output feature set of the FS procedure by a binary vector $\mathbf{s}$ of length 5, where a 1 at the fth position means the fth feature has been selected and a 0 means it has not been selected. For instance, the vector $[1\ 1\ 1\ 0\ 0]$ means that features 1–3 have been selected and features 4–5 have not been selected. Now imagine we apply two distinct FS procedures P_1 and P_2 to $M = 3$ different samples of the data and that we get the following output:

$$\mathcal{A}_1 = \begin{bmatrix} \mathbf{s}_1 \\ \mathbf{s}_2 \\ \mathbf{s}_3 \end{bmatrix} = \begin{bmatrix} 1\ 1\ 1\ 0\ 0 \\ 1\ 1\ 1\ 0\ 0 \\ 1\ 1\ 1\ 0\ 0 \end{bmatrix} \quad \mathcal{A}_2 = \begin{bmatrix} 1\ 1\ 1\ 0\ 0 \\ 1\ 0\ 1\ 1\ 0 \\ 1\ 0\ 1\ 0\ 0 \end{bmatrix} \Bigg\} M = 3 \text{ feature sets} \quad (1)$$

where the rows of $\mathcal{A}_1$ and $\mathcal{A}_2$ represent the feature sets respectively returned by P_1 and P_2. All the feature sets in $\mathcal{A}_1$ are identical, therefore there is no variation in the output of the procedure. Each column of the matrix $\mathcal{A}_1$ represents the selection of each one of the 5 features. The observed frequency of the first three features is equal to 1 while the one of the two last features is equal to 0. This situation corresponds to a fully stable selection. Now let us look at $\mathcal{A}_2$. In that situation, we can see that there is some variation in the output of the FS procedure since the feature sets in $\mathcal{A}_2$ are different. If we look at the second and fourth columns of $\mathcal{A}_2$ corresponding to the selection of the second and fourth feature over the 3 feature sets, we can see that they are selected with a frequency equal to $\hat{p}_2 = \hat{p}_4 = \frac{1}{3}$, which shows some instability in the FS.

Quantifying the stability of FS consists in defining a function $\hat{\Phi}$ that takes the output $\mathcal{A}$ of the FS procedure as an input and returns a stability value. It is important to note that this is an *estimate* of a quantity, as the true stability is a random variable. We present the general framework to quantify stability in Sect. 2. Coming from an axiomatic point of view, we derive a set of properties that we argue necessary for a stability measure and show that none of the existing measures have all desired properties in Sect. 3. In Sect. 4, we propose the use of the sample Pearson's correlation coefficient showing that it has all required properties and we provide an interpretation of the quantity estimated using this measure. Finally, we illustrate the use of stability in the context of FS by a $L1$-regularized logistic regression and show how when coupled with the error of the model, it can help select a regularizing parameter.

2   Background

2.1   General Framework

To quantify the stability of FS, the following steps are carried out [1]:

1. Take M perturbed versions of the original dataset $\mathcal{D}$ (e.g. by using a resampling technique [3] such as bootstrap or noise injection [2]).
2. Apply the FS procedure to each one of the M samples obtained. This gives a sequence $\mathcal{A} = [\mathbf{s}_1, ..., \mathbf{s}_M]^T$ of M feature sets.
3. Define a function $\hat{\Phi} : \{0, 1\}^{M \times d} \to \mathbb{R}$ taking the sequence of feature sets $\mathcal{A}$ as an input and measuring the stability of the feature sets in $\mathcal{A}$.

The main challenge here lies on the definition of an appropriate function $\hat{\Phi}$ that measures the stability in the choice of features in $\mathcal{A}$. Before looking into the approaches taken in the literature to define such a function $\hat{\Phi}$, we first establish the following notations that will be used in the remainder of the paper. We can denote the elements of the binary matrix $\mathcal{A}$ representing the M feature sets as follows:

$$
\mathcal{A} = \begin{bmatrix} \mathbf{s}_1 \\ \vdots \\ \mathbf{s}_M \end{bmatrix} = \begin{pmatrix} x_{1,1} & x_{1,2} & \cdots & x_{1,d} \\ x_{2,1} & x_{2,2} & \cdots & x_{2,d} \\ \vdots & \vdots & \ddots & \vdots \\ x_{M,1} & x_{M,2} & \cdots & x_{M,d} \end{pmatrix}
$$
$$
\begin{matrix} \uparrow & \uparrow & & \uparrow \\ X_1 & X_2 & & X_d \end{matrix}
$$

- For all $f \in \{1, ..., d\}$, the selection of the fth feature is modelled by a Bernoulli variable[1] X_f with unknown parameter p_f. Therefore, each column of the matrix $\mathcal{A}$ can be seen as a realisation of the variable X_f, from which we will assume they are random samples.
- For all f in $\{1, ..., d\}$, $\hat{p}_f = \frac{1}{M} \sum_{i=1}^{M} x_{i,f}$ is the observed frequency of the f_{th} feature and is the *maximum likelihood estimator* of p_f.
- For all i in $\{1, ..., M\}$, $k_i = |\mathbf{s}_i|$ is the cardinality of feature set $\mathbf{s}_i$ (i.e. the number of features in $\mathbf{s}_i$). When all feature sets in $\mathcal{A}$ are of identical cardinality, we will simply denote the cardinality of the sets by k.
- For all (i, j) in $\{1, ..., M\}^2$, $r_{i,j}$ denotes the size of the intersection between feature sets $\mathbf{s}_i$ and $\mathbf{s}_j$ (i.e. the number of features they have in common).

---

[1] We therefore have a set of d correlated Bernoulli variables $(X_1, ..., X_d)$.

2.2   Quantifying Stability

The main approach that can be found in the literature is the *similarity-based approach*. It consists in defining the stability as **the average pairwise similarities** between the feature sets in $\mathcal{A}$ [8]. Let $\phi : \{0,1\}^d \times \{0,1\}^d \to \mathbb{R}$ be a function that takes as an input two feature sets $\mathbf{s}_i$ and $\mathbf{s}_j$ and returns a similarity value between these two sets. Then the stability $\hat{\Phi}(\mathcal{A})$ is defined as[2]:

$$\hat{\Phi}(\mathcal{A}) = \frac{1}{M(M-1)} \sum_{i=1}^{M} \sum_{\substack{j=1 \\ j \neq i}}^{M} \phi(\mathbf{s}_i, \mathbf{s}_j).$$

This approach has been very popular in the literature and many similarity measures ϕ have been proposed to that end. Popular examples of similarity measures are the *Jaccard index* [8] defined as follows:

$$\phi_{Jaccard}(\mathbf{s}_i, \mathbf{s}_j) = \frac{|\mathbf{s}_i \cap \mathbf{s}_j|}{|\mathbf{s}_i \cup \mathbf{s}_j|} = \frac{r_{i,j}}{k_i + k_j - r_{i,j}}.$$

For instance, if we take back the examples given in Eq. 1, using the Jaccard index we get the stability values of:

$$\hat{\Phi}_{Jaccard}(\mathcal{A}_1) = \frac{1}{3} \left(\phi_{Jaccard}(\mathbf{s}_1, \mathbf{s}_2) + \phi_{Jaccard}(\mathbf{s}_1, \mathbf{s}_3) + \phi_{Jaccard}(\mathbf{s}_2, \mathbf{s}_3) \right) = 1$$

$$\hat{\Phi}_{Jaccard}(\mathcal{A}_2) = \frac{1}{3} \left(\frac{2}{4} + \frac{2}{3} + \frac{2}{3} \right) = \frac{11}{18} \simeq 0.61.$$

As expected, we get a smaller stability value in the second case.

Nevertheless, as we further discuss in Sect. 3, this similarity measure has been shown to provide stability estimates $\hat{\Phi}$ that are biased by the cardinality of the feature sets [11]. Based on this observation, Kuncheva [11] identifies a set of desirable properties and introduces a new similarity measure $\phi_{Kuncheva}$ between two feature sets $\mathbf{s}_i$ and $\mathbf{s}_j$ of identical cardinality as follows:

$$\phi_{Kuncheva}(\mathbf{s}_i, \mathbf{s}_j) = \frac{r_{i,j} - \mathbb{E}_{\nabla}[r_{i,j}]}{max(r_{i,j}) - \mathbb{E}_{\nabla}[r_{i,j}]} = \frac{r_{i,j} - \frac{k^2}{d}}{k - \frac{k^2}{d}},$$

where $\mathbb{E}_{\nabla}[r_{i,j}]$ is a correcting term equal to the expected value of $r_{i,j}$ when the FS procedure randomly selects k_i and k_j features from the d available features. As the random intersection of two sets of k_i and k_j objects follows a hypergeometric distribution, this term is known to be equal to $\frac{k_i k_j}{d}$ which is equal to $\frac{k^2}{d}$ here since $k_i = k_j = k$. This measure has been very popular in the literature because of its known properties. Nevertheless, because it is only defined for feature sets $\mathbf{s}_i$ and $\mathbf{s}_j$ of identical cardinality, it can only be used to measure the stability of FS algorithms that are guaranteed to select a constant number of features. As

---

[2] ϕ is not necessarily symmetric.

we have illustrated in example (1), the output of an FS procedure is not always guaranteed to be of constant cardinality. Examples of such FS procedures are in feature selection by hypothesis testing [15]. For this reason, several attempts at extending this measure to feature sets of varying cardinality have been made in the literature, somehow losing some of the important properties. Even though most similarity measures used to measure stability are increasing functions of the size of the intersection between the feature sets, they have shown to lack of some other required properties.

Other approaches have been taken in the literature to define a function $\hat{\Phi}$, without going through the definition of a similarity measure. A popular measure in this category is Somol's measure CW_{rel} [16] (also called Relative Weighted Consistency Measure). Its definition is a direct function of the observed frequencies of selection of each feature $\hat{p}_f$. This is the only measure in this category that is not biased by the cardinality of the feature sets in $\mathcal{A}$ and holds the property of *correction for chance*.

Due to the multitude of stability measures, it is necessary to discriminate between them with principled reasons which is the purpose of the next section.

3   Required Properties of a FS Stability Measure

In this section, we identify and argue for 4 properties which all stability measures should possess. These properties we will argue are necessary for a sensible measure of stability and if missing even one, a measure will behave nonsensically in certain situations. We will later demonstrate that from 10 stability measures published and widely used in the literature, none of them possesses all these properties.

Property 1: Fully Defined

Imagine we have an FS procedure: Procedure P. Procedure P sometimes returns 4 features, but sometimes 5, so the returned set size *varies*. It would seem sensible to have a stability measure which accounts for this. Unfortunately not all do - Krízek's and Kuncheva's measures [10,11] are *undefined* in this scenario.

Property 2: Upper/Lower Bounds

For useful *interpretation* of a stability measure and comparison across problems, the range of values of a stability measure should be finite. Imagine we wanted to evaluate the stability of an FS procedure and that we got a value of 0.9. How can we interpret this value? If we know that the stability values can take values in $[0, 1]$, then this corresponds to a fairly high stability value as it is close to its maximum 1. Let us imagine now that we have a stability value that can take values in $(-\infty, +\infty)$. A value of 0.9 is not meaningful any more.

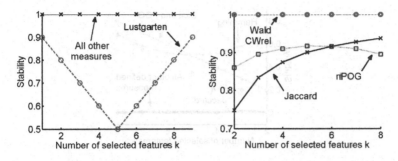

Fig. 1. Illustration of Property 3. Demonstration that Lustgarten's measure violates Property 3a [LEFT] by giving the stability when all feature sets in $\mathcal{A}$ are identical against k for $d = 10$. Demonstration that Wald's measure and CW_{rel} violate Property 3b [RIGHT]. Features $[1, ..., k]$ are selected half of the time and feature $[1, ..., k-1]$ are selected the other half of the time. Stability values against k for $d = 10$ and $M = 100$.

Property 3:

(a) **Deterministic Selection $\rightarrow$ Maximum Stability** Imagine that Procedure P selects the same k features every time, regardless of the supplied data. This is a completely *stable* method, so it would seem sensible that any stability *measure* should reflect this, returning its maximum value. Surprisingly, this is not always the case. Figure 1 [LEFT] shows the stability value using Lustgarten's measure [13] when for different values of k. The result clearly *varies* with k. That is, if Procedure P_1 were to repeatedly select features 1–4 and Procedure P_2 then repeatedly selects features 1–5: this measure judges P_1 and P_2 to have *different degrees of stability*, even though they are both completely deterministic procedures.

(b) **Maximum Stability $\rightarrow$ Deterministic Selection** The converse to the above should also hold. If a measure has a maximum possible value C, it should *only* return that value when Procedure P is deterministic. For example, imagine Procedure P selects features 1–4 half the time, and 1–5 the rest of the time. Wald's measure and CW_{rel} return a value of 1 in this scenario – their maximum possible value, even though clearly there is some variation in the feature sets. Figure 1 [RIGHT] illustrates this.

Property 4: Correction for Chance

This was first noted by Kuncheva [11]. This ensures that when the FS is **random**, the expected value of the stability estimate is constant, which we have set here to 0 by convention. Imagine that a procedure P_1 **randomly** selects 5 features and that a procedure P_2 randomly selects 6 features, the stability value should be the same. As illustrated by Fig. 2, this is not the case for all measures.

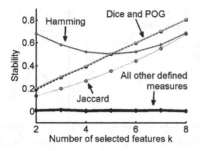

Fig. 2. Demonstration that Hamming, Jaccard, POG and Dice violate Property 4. **Random** selection of k features with probability 50 % and of $k-1$ and $k+1$ features with probability 25 % each. Stability against k for $d=10$ and $M=100$.

Table 1. Properties of Stability Measures

	Fully defined	Bounds	Maximum	Correction for chance	
Jaccard [8]	✓	✓	✓		⎫
Hamming [4]	✓	✓	✓		⎪
Dice [19]	✓	✓	✓		⎪
POG [14]	✓	✓	✓		⎬ Similarity-based
Kuncheva [11]		✓	✓	✓	⎪
nPOG [21]	✓		✓	✓	⎪
Lustgarten [13]	✓	✓		✓	⎭
Wald [17]	✓			✓	
Krízek [10]			✓		
CW_{rel} [16]	✓	✓		✓	

Summary

We provide a formal description of the required properties and sum up the properties of the different existing stability measures[3] in Table 1. We can observe that none of the measures satisfy all four desired properties.

1. **Fully defined.** $\hat{\Phi}$ is defined for any sequence $\mathcal{A}$ of feature sets.
2. **Bounds.** $\hat{\Phi}$ is bounded by constants.
3. **Maximum.** $\hat{\Phi}$ reaches its maximum $\iff$ All feature sets in $\mathcal{A}$ are identical.
4. **Correction for chance.** $\mathbb{E}_{\nabla}[\hat{\Phi}(\mathcal{A})] = 0$ when the selection is random.

---

[3] Sketches of proofs are given in the supplementary material available online at www. cs.man.ac.uk/~nogueirs/files/supplementary-material-ECML-2016.pdf.

4   The Sample Pearson's Correlation Coefficient

In this section, we first demonstrate that the stability measure using the sample Pearson's correlation coefficient[4] as a similarity measure satisfies all 4 properties. The sample Pearson's correlation coefficient between two feature sets $\mathbf{s}_i$ and $\mathbf{s}_j$ is by definition:

$$\phi_{Pearson}(\mathbf{s}_i, \mathbf{s}_j) = \frac{\frac{1}{d}\sum_{f=1}^{d}(x_{i,f} - \bar{x}_{i,.})(x_{j,f} - \bar{x}_{j,.})}{\sqrt{\frac{1}{d}\sum_{f=1}^{d}(x_{i,f} - \bar{x}_{i,.})^2}\sqrt{\frac{1}{d}\sum_{f=1}^{d}(x_{j,f} - \bar{x}_{j,.})^2}},$$

where $\forall i \in \{1, ..., M\}, \bar{x}_{i,.} = \frac{1}{d}\sum_{f=1}^{d}x_{i,f} = \frac{k_i}{d}$.

As other similarity measures, we can point out that $\phi_{Pearson}(\mathbf{s}_i, \mathbf{s}_j)$ is an increasing function of the size of the intersection of the selected features $r_{i,j}$ between the feature sets $\mathbf{s}_i$ and $\mathbf{s}_j$. Moreover, the sample Pearson correlation coefficient is already the similarity measure used when the FS outputs a scoring on the features [8], even though it has never been used or studied in the context of feature sets. The use of Pearson's correlation coefficient is therefore going towards a unification of the assessment of stability of FS.

The sample Pearson's correlation also subsumes other measures when the cardinality of the feature sets is constant, as stated by Theorem 2. This result is quite surprising, knowing that coming from an axiomatic point of view on a set of desirable properties, Kuncheva defined a measure that is indeed a specific case of the well-known sample Pearson's correlation coefficient $\phi_{Pearson}$.

Theorem 1. *For all $(i, j) \in \{1, ..., M\}^2$, the sample Pearson's correlation coefficient can be re-written:*

$$\phi_{Pearson}(\mathbf{s}_i, \mathbf{s}_j) = \frac{r_{i,j} - \mathbb{E}_\nabla[r_{i,j}]}{d\, v_i v_j} = \frac{r_{i,j} - \frac{k_i k_j}{d}}{d\, v_i v_j}, \tag{2}$$

where $\forall i \in \{1, ..., M\}, v_i = \sqrt{\frac{k_i}{d}(1 - \frac{k_i}{d})}$. Therefore it possesses the property of correction for chance.

Proof. *The proof is provided in the supplementary material.*

Theorem 2. *When k is constant, the stability using Pearson's correlation is equal to some other measures, that is:*

$$\hat{\Phi}_{Pearson} = \hat{\Phi}_{Kuncheva} = \hat{\Phi}_{Wald} = \hat{\Phi}_{nPOG}.$$

Proof. *Straightforward using Theorem 1 and the definition of the other similarity measures given in the supplementary material.*

---

[4] Also called the *Phi coefficient* in this case since we are dealing with binary vectors.

4.1   Required Properties

Property 1: Fully Defined. As most of the other similarity measures, we can see in Eq. 2 that the given expression presents indeterminate forms for $k_i = 0$, $k_j = 0$, $k_i = d$ and $k_j = d$. Because these indeterminate forms correspond to situations in which either all features or none of them are selected, these indeterminate forms are not critical in the context of feature selection since the main aim of FS is to identify a non-empty strict subset of relevant features taken from the available features. Nevertheless, for completeness, following the works on the correlation coefficient in [5], we set $\phi_{Pearson}$ to 0 when:

- $k_i = 0$ and $k_j \neq 0$ or vice-versa;
- $k_i = d$ and $k_j \neq d$ or vice-versa.

When $k_i = k_j = 0$ or $k_i = k_j = d$, then the feature sets are identical (either empty set $\emptyset$ or full set Ω) and in that case, we set $\phi_{Pearson}$ to be equal to 1 so it meets the property of maximum. Therefore, the resulting stability $\hat{\Phi}_{Pearson}$ has the property of being fully defined.

Property 2: Bounds. $\phi_{Pearson}$ is known to take values between -1 and 1: the similarity between two sets is minimal (i.e. equal to -1) when the two sets are fully anti-correlated (i.e. when $\mathbf{s}_i$ and $\mathbf{s}_j$ are complementary sets) and maximal (equal to 1) when the two sets are fully correlated (i.e. identical). Since $\hat{\Phi}_{Pearson}$ is the average value of $\phi_{Pearson}$ over all the possible pairs in $\mathcal{A}$, $\hat{\Phi}_{Pearson}$ will also be in the interval -1 and 1 and is therefore bounded by constants.

Theorem 3. *The stability estimate $\hat{\Phi}_{Pearson}$ is asymptotically in the interval $[0, 1]$ as M approaches infinity.*

Proof. *The proof is provided in the supplementary material.*

The asymptotic bounds on the stability estimates make the stability values obtained more interpretable. Indeed, knowing how the stability values behave as M increases allows us to understand better how to interpret these values. Theorem 3 tackles the misconception according to which negative stability values correspond to FS algorithms worse than random: asymptotically, any FS procedure will have a positive estimated stability.

Property 3: Maximum. When $\mathbf{s}_i = \mathbf{s}_j$, we have $\phi_{Pearson}(\mathbf{s}_i, \mathbf{s}_j) = 1$ and therefore $\hat{\Phi}_{Pearson} = 1$ when all the feature sets in $\mathcal{A}$ are identical. Conversely, $\phi_{Pearson}(\mathbf{s}_i, \mathbf{s}_j) = 1$ implies $\mathbf{s}_i = \mathbf{s}_j$, which gives us that $\hat{\Phi}_{Pearson} = 1$ implies all sets in $\mathcal{A}$ are identical.

Property 4: Correction for Chance. This property is given by Theorem 1.

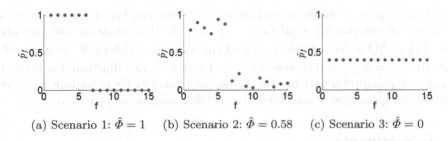

(a) Scenario 1: $\hat{\Phi} = 1$ (b) Scenario 2: $\hat{\Phi} = 0.58$ (c) Scenario 3: $\hat{\Phi} = 0$

Fig. 3. The parameters $\hat{p}_f$ of the random variables X_f in 3 scenarios for $d = 15$

4.2   Interpreting Stability

In this section, we aim at providing an interpretation of the stability value when using the sample Pearson's correlation. For simplicity, we focus on the case where the FS selects a constant number of features k. Hereafter, $\hat{\Phi}$ will denote $\hat{\Phi}_{Pearson}$. By phrasing the concept of stability in this way, it highlights an important point - that we are *estimating* a quantity. The stability is a random variable, from which we have a sample of size M.

Let $\widehat{Var}(X_f) = \frac{M}{M-1}\hat{p}_f(1 - \hat{p}_f)$ be the unbiased sample variance of the variable X_f. When the cardinality of the feature sets is constant, we can re-write the stability using the sample Pearson's correlation coefficient as follows:

$$\hat{\Phi}_{Pearson} = 1 - \frac{S}{S_{max}}, \tag{3}$$

where the average total variance $S = \frac{1}{d}\sum_{f=1}^{d} \widehat{Var}(X_f)$ is a measure of the variability in the choice of features and where $S_{max} = \frac{k}{d}\left(1 - \frac{k}{d}\right)$ the maximal value of S given that the FS procedure is selecting k features per feature set. In this situation, Eq. 3 shows that the stability decreases monotonically with the average variance of X_f.

Because $\widehat{Var}(X_f) = 0$ whenever $\hat{p}_f = 0$ or $\hat{p}_f = 1$, the maximum stability is achieved when all features are selected with an observed frequency equal to 0 or 1. Figure 3 illustrates how to interpret the value $\hat{\Phi}$ in 3 scenarios. Let us assume we have an FS procedure selecting $k = 6$ features out of $d = 15$ features. Scenario 1 illustrates the situation in which the FS algorithm always returns the same feature set made of the first k features. In that situation, the probability of selection of the k first features is equal to 1 and the one of the remaining features is equal to 0, which gives $S = 0$ and therefore a stability $\hat{\Phi}$ equal to its maximal value 1. Scenario 2 illustrates the case where the FS is not completely stable, even though we can still distinguish two group of features. In that scenario, the stability is equal to $\hat{\Phi} = 0.58$. Scenario 3 is the limit case scenario in which the selection of the k features is random. In that scenario, the d features have a frequency of selection all equal to $\hat{p}_f = \frac{k}{d} = \frac{6}{15}$. In that situation, the variance $Var(X_f) = \frac{k}{d}(1 - \frac{k}{d}) = 0.24$ of each of the random variables X_f is maximal. This gives $S = S_{max}$ and therefore $\hat{\Phi} = 0$. These scenarios illustrate the need to rescale the mean total variance by the one of a random FS procedure and give a useful interpretation of the estimated stability using Pearson's correlation.

5   Experiments

In the previous section we argued for an axiomatic treatment of stability measures—and demonstrated that the simple solution of using Pearson's correlation coefficient allows for all desirable properties.

In this section, we illustrate how stability can be used in practice to select a regularizing parameter in the context of feature selection by a $L1$-regularized regression. We show how without sacrificing a significant amount in terms of error, a regularizing parameter corresponding to a higher stability can be chosen. On the artificial dataset considered, we show how an increase in stability can help discarding the use of irrelevant features in the final model.

5.1   Description of Dataset

We use a synthetic dataset [9] – a binary classification problem, with 2000 instances and $d = 100$ features, where only the first 50 features are relevant to the target class. Instances of the positive class are i.i.d. drawn from a normal distribution with mean $\mu_+ = (\underbrace{1, ..., 1}_{50}, \underbrace{0, ..., 0}_{50})$ and covariance matrix:

$$\Sigma = \begin{bmatrix} \Sigma^*_{50\times50} & \mathbf{0}_{50\times50} \\ \mathbf{0}_{50\times50} & \mathbf{I}_{50\times50} \end{bmatrix}$$

where $\Sigma^*_{50\times50}$ is the matrix with ones on the diagonal and ρ, a parameter taken in $[0, 1]$ controlling the degree of redundancy everywhere else. The mean for the negative class is taken equal to $\mu_- = (\underbrace{-1, ..., -1}_{50}, \underbrace{0, ..., 0}_{50})$. The larger the value of ρ, the more the 50 relevant features will be correlated to each other.

5.2   Experimental Procedure and Results

We use $L1$-regularized logistic regression with 100 different regularizing parameters on the synthetic dataset for different degrees of redundancy ρ. The $L1$-regularization results in some coefficients being forced to zero – any coefficients left as non-zero after fitting the model are regarded as "selected" by the model.[5]

---

[5] You can reproduce these experiments in Matlab with the code given at https://github.com/nogueirs/ECML2016.

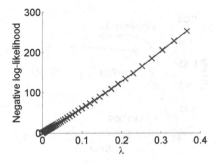
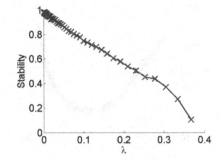

Fig. 4. Results for $\rho = 0$. Each point on the line corresponds to a different regularizing parameter λ. We can see that both high stability and low error are reached for $\lambda = 4.12 \times 10^{-4}$.

Our experimental procedure is as follows. We take the 2000 samples and divide into 1000 for model selection (the regularizing parameter λ) and 1000 for selection of the final set of features. The model selection set can be used simply to optimize error, or to optimize error/stability simultaneously – the experiments will demonstrate that the latter provides a lower false positive rate in the final selection of features.

For each regularizing parameter λ, we take $M = 100$ bootstrap samples to train our models. We then compute the stability $\hat{\Phi}$ and the out-of-bag (OOB) estimate of the error[6] using the coefficients returned.

Figure 4 shows the OOB error [LEFT] and the stability [RIGHT] versus the regularization parameter λ for a degree of redundancy $\rho = 0$ (i.e. the relevant features are independent from each other). On this case, picking up a value of λ that minimizes the OOB error is also the value of λ that maximizes the stability. Indeed for $\lambda = 4.12 \times 10^{-4}$, we get an error of 0.30 and a stability of 0.98, which means the same features are picked up on nearly all bootstrap samples.

Let us now take a degree of redundancy $\rho = 0.3$. In a normal situation, we would choose the regularizing parameter that minimizes the error which is $\lambda = 0.009$, shown in the left of Fig. 5. The right of the same figure shows the pareto optimal front, the trade-off of the two objectives – if we sacrifice some error, we can drastically increase stability.

Figure 6 gives the observed frequencies of selection $\hat{p}_f$ of each feature over the $M = 100$ bootstraps for $\lambda = 0.009$ [LEFT] and $\lambda = 0.023$ [RIGHT]. We can see on the right figure that nearly all irrelevant features have a frequency of selection of 0. Only two irrelevant features have a frequency of selection different from 0 with $\hat{p}_f = 0.01$, which means they have been selected on one of the 100 bootstrap samples only. From looking at the values of $\hat{p}_f$ for the value of λ minimizing the error on the left, we cannot discriminate the set of relevant features from the set of irrelevant ones by looking at the frequencies of selection.

---

[6] Here, the error is taken to be the negative log-likelihood, a measure of goodness-of-fit of the model. The lower the value, the better the model.

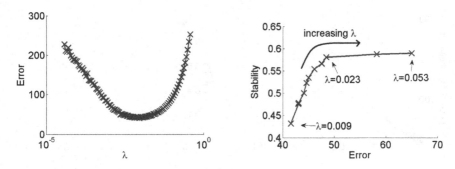

Fig. 5. If we optimize just OOB error [LEFT] we obtain $\lambda = 0.009$, but if we optimize a trade-off [RIGHT] of error/stability, sacrificing a small amount of error we get $\lambda = 0.023$, and can significantly increase feature selection stability.

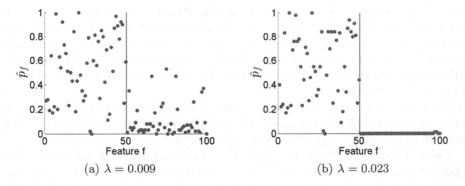

Fig. 6. The observed frequencies of selection $\hat{p}_f$ for each feature for two values of λ in the pareto front for $\rho = 0.3$. The Features on the left of the red vertical line correspond to relevant features and the ones on the right to irrelevant ones. (Color figure online)

Even though $\lambda = 0.023$ does not provide a *high* stability value, we can see how we can benefit from taking $\lambda = 0.023$ instead of $\lambda = 0.009$. The features used in the model (the ones with a non-zero coefficient) are indeed relevant to the target class. As explained in Sect. 4.2, the closer the observed frequencies are to 0 or 1, the higher the stability value will be.

Final feature set chosen: The model selection procedure on the first 1000 examples has suggested $\lambda = 0.009$ and $\lambda = 0.023$. We can now use these on the final 1000 holdout set to select a set of features, again with $L1$ logistic regression, and compare the 2 feature sets returned. Table 2 shows the false positives (irrelevant features that were falsely identified as relevant) and the false negatives (relevant features that were missed), for three different degrees of

Table 2. False positives and false negatives for different degrees of redundancy ρ

Redundancy	λ_{error}	λ_ϕ
Low	$FP = 4$, $FN = 17$	$FP = 0$, $FN = 17$
Medium	$FP = 7$, $FN = 24$	$FP = 0$, $FN = 25$
High	$FP = 5$, $FN = 35$	$FP = 0$, $FN = 33$

increasing redundancy. In all cases, the methodology involving stability reduces the FP rate to zero, with no significant effect on FN rate.

This case study also shows that feature redundancy is a source of instability of FS, as hypothesized by [8,9,18]. Similar results have been obtained for $\rho = 0.5$ and $\rho = 0.8$, with smaller stability values for the data points in the pareto front as we increased the degree of redundancy ρ.

6 Conclusions and Future Work

There are many different measures to quantify stability in the literature – we have argued for a set of properties that should be present in any measure, and found that several existing measures are lacking in this respect. Instead, we suggest the use of Pearson's correlation as a similarity measure, in the process showing that it is a generalization of the widely used Kuncheva index. We provide an interpretation of the quantity estimated through the typical procedure and illustrate its use in practice. We illustrate on synthetic datasets how stability can be beneficial and provides more confidence in the feature set returned.

Depending on the type of application, we might want the stability measure to take into account feature redundancy. Such measures attempt to evaluate the stability of the *information* in the feature sets returned by the FS procedure rather than the stability of the feature sets themselves [20,21]. These measures are generalizations of POG, $nPOG$ (called $POGR$ and $nPOGR$ [21]) and of the Dice coefficient [19] and reduce to these when there is no redundancy between the features. Because their simpler versions do not have the set of desired properties as shown in Table 1, we leave this type of measures to future work.

Acknowledgments. This work was supported by the EPSRC grant [EP/I028099/1].

References

1. Alelyani, S., Zhao, Z., Liu, H.: A dilemma in assessing stability of feature selection algorithms. In: HPCC (2011)
2. Altidor, W., Khoshgoftaar, T.M., Napolitano, A.: A noise-based stability evaluation of threshold-based feature selection techniques. In: IRI 2011 (2011)

3. Boulesteix, A.L., Slawski, M.: Stability and aggregation of ranked gene lists. Briefings Bioinform. **10**(5), 556–568 (2009)
4. Dunne, K., Cunningham, P., Azuaje, F.: Solutions to instability problems with sequential wrapper-based approaches to feature selection. Technical report, Journal of Machine Learning Research (2002)
5. Edmundson, H.P.: A correlation coefficient for attributes or events. In: Proceedings Statistical Association Methods for Mechanized Documentation (1966)
6. He, Z., Yu, W.: Review article: stable feature selection for biomarker discovery. Comput. Biol. Chem. **34**, 215–225 (2010)
7. Jurman, G., Merler, S., Barla, A., Paoli, S., Galea, A., Furlanello, C.: Algebraic stability indicators for ranked lists in molecular profiling. Bioinform. **24**(2), 258–264 (2008)
8. Kalousis, A., Prados, J., Hilario, M.: Stability of feature selection algorithms: a study on high-dimensional spaces. Knowl. Inf. Syst. **12**(1), 95–116 (2007)
9. Kamkar, I., Gupta, S.K., Phung, D., Venkatesh, S.: Stable feature selection with support vector machines. In: Pfahringer, B., Renz, J. (eds.) AI 2015. LNCS (LNAI), vol. 9457, pp. 298–308. Springer, Heidelberg (2015). doi:10.1007/978-3-319-26350-2_26
10. Křížek, P., Kittler, J., Hlaváč, V.: Improving stability of feature selection methods. In: Kropatsch, W.G., Kampel, M., Hanbury, A. (eds.) CAIP 2007. LNCS, vol. 4673, pp. 929–936. Springer, Heidelberg (2007). doi:10.1007/978-3-540-74272-2_115
11. Kuncheva, L.I.: A stability index for feature selection. In: Artificial Intelligence and Applications (2007)
12. Lee, H.W., Lawton, C., Na, Y.J., Yoon, S.: Robustness of chemometrics-based feature selection methods in early cancer detection and biomarker discovery. Stat. Appl. Genet. Mol. Biol. **12**(2), 207–223 (2012)
13. Lustgarten, J.L., Gopalakrishnan, V., Visweswaran, S.: Measuring stability of feature selection in biomedical datasets. In: AMIA Annual Symposium Proceedings, vol. 2009, p. 406 (2009)
14. MAQC consortium: The MicroArray quality control project shows inter- and intraplatform reproducibility of gene expression measurements. Nat. Biotech. **24**, 1151–1161 (2006)
15. Sechidis, K., Brown, G.: Markov blanket discovery in positive-unlabelled and semi-supervised data. In: ECML (2015)
16. Somol, P., Novovičová, J.: Evaluating stability and comparing output of feature selectors that optimize feature subset cardinality. IEEE Trans. Pattern Anal. Mach. Intell. **32**(11), 1921–1939 (2010)
17. Wald, R., Khoshgoftaar, T.M., Napolitano, A.: Stability of filter- and wrapper-based feature subset selection. In: International Conference on Tools with Artificial Intelligence. IEEE Computer Society (2013)
18. Woznica, A., Nguyen, P., Kalousis, A.: Model mining for robust feature selection. In: KDD (2012)
19. Yu, L., Ding, C.H.Q., Loscalzo, S.: Stable feature selection via dense feature groups. In: KDD (2008)

20. Yu, L., Han, Y., Berens, M.E.: Stable gene selection from microarray data via sample weighting. IEEE/ACM Trans. Comput. Biol. Bioinform. **9**(1), 262–272 (2012)
21. Zhang, M., Zhang, L., Zou, J., Yao, C., Xiao, H., Liu, Q., Wang, J., Wang, D., Wang, C., Guo, Z.: Evaluating reproducibility of differential expression discoveries in microarray studies by considering correlated molecular changes. Bioinformatics **25**(13), 1662–1668 (2009)

Aspect Mining with Rating Bias

Yitong Li[1], Chuan Shi[1(✉)], Huidong Zhao[1], Fuzhen Zhuang[2], and Bin Wu[1]

[1] Beijing Key Lab of Intelligent Telecommunications Software and Multimedia,
Beijing University of Posts and Telecommunications, Beijing, China
{liyitong,shichuan,wubin}@bupt.edu.cn, zhaohuidong1121@foxmail.com
[2] Key Lab of Intelligent Information Processing of
Chinese Academy of Sciences (CAS), Institute of Computing Technology,
CAS, Beijing, China
zhuangfz@ics.ict.ac.cn

Abstract. Due to the personalized needs for specific aspect evaluation on product quality, these years have witnessed a boom of researches on aspect rating prediction, whose goal is to extract ad hoc aspects from online reviews and predict rating or opinion on each aspect. Most of the existing works on aspect rating prediction have a basic assumption that the overall rating is the average score of aspect ratings or the overall rating is very close to aspect ratings. However, after analyzing real datasets, we have an insightful observation: there is an obvious rating bias between overall rating and aspect ratings. Motivated by this observation, we study the problem of aspect mining with rating bias, and design a novel RAting-center model with BIas (RABI). Different from the widely used review-center models, RABI adopts the overall rating as the center of the probabilistic model, which generates reviews and topics. In addition, a novel aspect rating variable in RABI is designed to effectively integrate the rating bias priori information. Experiments on two real datasets (Dianping and TripAdvisor) validate that RABI significantly improves the prediction accuracy over existing state-of-the-art methods.

Keywords: Aspect rating · Rating prediction · Rating bias · Topic model

1 Introduction

With the rapid development of the Internet, the information which people can gain from the Internet grows exponentially. Nowadays, people are used to viewing online reviews before making decisions. For example, if a user wants to go out for dinner, he or she may look at the reviews of restaurants around on the Internet and choose one according to his or her taste. These reviews contain mainly overall ratings which evaluate restaurants from a general view. However, people may expect more subtle aspect ratings, such as the taste, environment, service, and so on. This problem has inspired the research on aspect-level opinion mining. The goal of the aspect-level opinion mining (i.e., aspect identification

© Springer International Publishing AG 2016
P. Frasconi et al. (Eds.): ECML PKDD 2016, Part II, LNAI 9852, pp. 458–474, 2016.
DOI: 10.1007/978-3-319-46227-1_29

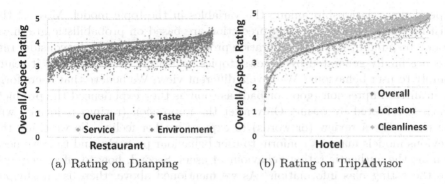

(a) Rating on Dianping   (b) Rating on TripAdvisor

Fig. 1. Distributions of ratings on Dianping and TripAdvisor

and aspect rating prediction) is to extract ad hoc aspects from online reviews and predict rating or opinion on each aspect.

Because of its great practical significance, there is a surge of researches on aspect identification and aspect rating prediction in recent years. Some works generate ratable aspects for reviews with whole overall ratings [7] or scarce overall ratings [6], and some works consider to integrate external knowledge [9]. Most of the existing works predict aspect ratings with the help of overall ratings, and they all have a basic assumption. That is, the overall rating is the average score of aspect ratings or the overall rating is close to aspect ratings.

However, the analysis on real datasets shows an insightful phenomenon: there is an obvious and systemic rating bias between overall ratings and aspect ratings. Figure 1 illustrates the rating distributions on two real datasets: Dianping[1] (a well-known social media platform in China, which contains the information and reviews of restaurant, hotel, entertainment, movie, etc.) and TripAdvisor[2] (a widely used dataset in this field, which is a social media platform about travel, hotel, scenic spot, etc.). The datasets we use are the restaurant data in Dianping and the hotel data in TripAdvisor. Note that the overall ratings of restaurants/hotels are sorted in an ascending order in Fig. 1. We can find that the overall ratings in TripAdvisor are obviously lower than two aspect ratings, while the overall ratings in Dianping are significantly larger than aspect ratings. The interesting observation implies that the previous aspect rating prediction approaches may achieve poor performance, if ignoring the rating bias between overall ratings and aspect ratings.

Motivated by the observed rating bias, we try to study the problem of aspect mining with rating bias. That is, the goal is to decompose the reviews into different aspects and predict the rating of different aspects on each entity, with the help of the overall rating and the rating bias priori information. However, aspect mining with rating bias may face two challenges. First, the rating process of users may conform to some behaviour patterns, which determine the

---

[1] http://www.dianping.com/.

[2] http://www.tripadvisor.com/.

dependency relationship among the variables in the topic model. Most of the existing works on aspect rating prediction are based on probabilistic graphical model. Inspired by the word generation process, these works usually consider ratings are finally generated by reviews, topics or aspects. However, does it really comply to user behaviour? We have a different view. We believe that users form an intuitive impression (good or bad) as soon as they experienced the product, which is reflected by rating. Only after the impression (rating) is formed will the user write a review (or words) to express his/her feeling. So we think the previous models may not conform to user behaviour properly, and thus we need to mine the authentic rating behaviour of users. Second, how to effectively utilize the rating bias information? As we mentioned above, there is an obvious bias between overall rating and aspect ratings. The rating bias may cause the inaccuracy of aspect rating prediction, and influence the results tremendously. Luo et al. [6] have discovered the rating bias, but nobody has considered it in the model until now. So how to use the rating bias priori information properly to improve the prediction accuracy is also a challenge.

To solve the challenges mentioned above, we design a novel RAting-center model with BIas (RABI). Different from traditional rating generating process [6,7,9], RABI considers rating as the center of the model, which generates the reviews and topics. This idea stems from users' real experiences. When users decide to write a review, they usually have intuitional opinions (i.e., overall ratings) on the products, and then they will use proper phrases to represent their opinions. In addition, RABI introduces a novel latent aspect rating variable which can effectively learn the correlation of the overall rating, aspects, and rating bias. Experimental results on two real datasets (i.e., Dianping and TripAdvisor) validate the effectiveness of RABI on both Chinese and English reviews, compared to existing state-of-the-art methods. The results also show that RABI can accurately decompose the reviews into different aspects.

Our contributions are summarized as follows:

- We first analyze the rating bias between overall rating and aspect ratings in real data, and put forward the problem of aspect mining with rating bias.
- We propose a novel RABI model for aspect mining with rating bias. Different from existing models, RABI considers rating as the center of the model, which simulates the generation of the review better. In addition, an aspect rating variable is proposed to effectively utilize the rating bias information.
- Experiments on real datasets have shown the effectiveness of our algorithm over existing state-of-the-art methods.

2   Data Analysis

In order to show the rating bias phenomenon, we analyze two real datasets. The first dataset is crawled from Dianping website, a well-known social media platform in China, which provides a review platform for businesses and entertainments. In Dianping website, a user can give a review to a business after

Table 1. Statistics of the datasets

Datasets	#Products	#Reviews	#Phrases	Avg. overall rating
Dianping	1,097	216,291	696,608	3.97
TripAdvisor	1,850	197,970	2,571,902	3.81

Table 2. Rating bias on each aspect on both datasets

Dataset	Category	Avg. rating	Rating bias
Dianping	*Overall*	3.97	
	Taste	3.69	+0.28
	Service	3.48	+0.48
	Environment	3.43	+0.54
TripAdvisor	*Overall*	3.81	
	Value	3.80	+0.01
	Room	3.82	−0.01
	Location	4.14	−0.33
	Cleanliness	4.07	−0.26
	Front Desk/Staff	3.96	−0.15
	Service	3.92	−0.11
	Business	3.59	+0.22

enjoying a service in this business. Besides an overall rating, the review information includes Chinese comments and three aspect ratings on Taste, Service, and Environment, respectively. In addition, we also employ the widely used TripAdvisor dataset [10]. Accompanying with English comments, reviews in this dataset are not only associated with overall ratings, but also with ground truth aspect ratings on 7 aspects: Value, Room, Location, Cleanliness, Front desk/staff, Service, and Business. All the ratings in the datasets are in the range from 1 to 5. The statistic information of these datasets is shown in Table 1.

We first intuitively show the distributions of overall and aspect ratings on these two datasets in Fig. 1. Note that, we only show the distributions of some aspect ratings due to the space limitation. Moreover, we sort products according to their overall ratings for clarity. From Fig. 1, we can find that there are obvious rating biases between overall rating and aspect rating on both datasets. In Dianping dataset, the overall rating is far above the aspect ratings in all three aspects, while the overall rating is smaller than two aspect ratings in TripAdvisor.

Furthermore, we calculate the rating bias on each aspect on both datasets. The calculating process can be seen in Eq. (1) and the results are listed in Table 2. The rating biases in Dianping are huge on most aspects, especially +0.48 for Service and +0.54 for Environment, which are pretty huge values. So the rating biases in Dianping should be well considered. The rating biases in TripAdvisor are small on some aspects (e.g., +0.01 for Value and −0.01 for Room), but huge on other aspects (e.g., −0.33 for Location and −0.26 for Cleanliness). Although the rating biases in TripAdvisor are not as much as those in Dianping, they all truly exist. The interesting observation implies that the previous aspect rating prediction approaches may achieve poor performance, if ignoring the rating bias. As shown in Table 2, the rating biases are different in different datasets and aspects, which can influence the results to varying degrees and cause the inaccuracy of aspect rating prediction. So the proper consideration of the rating bias can improve the prediction accuracy.

3   Preliminary Notations and Problem Definition

In this section, we first introduce the notations and concepts used in this paper, and then formally propose the problem of aspect mining with rating bias.

Entity: An entity e indicates a product which belongs to the product set E (e.g., a restaurant in Dianping dataset or a hotel in TripAdvisor dataset). N_e indicates the number of entities in E.

Review: A review d is the user's opinion about the entity e. An entity e can have many reviews from different users. A review consists of the text content, the overall rating and many aspect ratings. There are N_d reviews in total.

Phrase: A phrase $f = (h, m)$ consists of a pair of words, which are extracted from the review's text content. h denotes the head term, and m is the modifier term which modifies h. A review d contains several phrases f.

Head term: The head term h is used to describe the aspect information. It decides which aspect the phrase f is expressing. For instance, "attitude" is a head term, and it belongs to the aspect "Service".

Modifier term: The modifier term m is used to describe the sentiment information. It is used to describe the aspect, which is decided by h, is good or bad. For instance, for the head term "attitude", "cold" or "passionate" may be used as the modifier term.

Overall rating: An overall rating r of a review d is a numerical rating, which indicates the user's overall sentiment tendency on the entity e. The number of the values of rating is N_r and it is usually 5, which means the values of rating r are from 1 to 5.

Aspect: An aspect A_i is a specific side of the entity e, e.g., the taste of the restaurant. It is a set of many similar characteristic of the entity e. N_A indicates the number of aspects.

Aspect rating: An aspect rating r_{A_i} is a numerical rating, which indicates the user's sentiment tendency on the aspect A_i of the entity e, and is also from 1 to 5. And a review d has N_A aspect ratings, which corresponds to N_A aspect.

Rating bias: The rating bias is the gap between the average of overall ratings and the average of aspect ratings. There are N_A biases on N_A aspects, and they are in connection with the current aspect A_i. The rating bias b_{A_i} on aspect A_i can be calculated as follows:

$$b_{A_i} = \frac{\sum_d r}{N_d} - \frac{\sum_d r_{A_i}}{N_d}. \tag{1}$$

Aspect mining with rating bias: The problem of *aspect mining with rating bias* is to predict the rating on each aspect with the rating bias prior information. Specifically, given a set of reviews $D = \{d_1, d_2, \cdots, d_{N_d}\}$ about entities $E = \{e_1, e_2, \cdots, e_{N_e}\}$, we know that each review $d_i \in D$ contains text content (Chinese or English) and overall rating r on an entity $e_j \in E$, as well as the rating bias b_{A_i} between the overall rating and the aspect rating on N_A aspects for all reviews. The goal is to decompose the phrases f, which are extracted from texts in D, into N_A aspects $\{A_1, A_2, \cdots, A_{N_A}\}$, and rate the aspects of each entity e with $\{r_{A_1}, r_{A_2}, \cdots, r_{A_{N_A}}\}$.

In fact, our goal includes two sub-tasks. (1) The first sub-task is aspect identification, which is to correctly identify the aspect label A_i given phrase f. (2) The second sub-task is aspect rating prediction, which is to predict the aspect rating r_{A_i} given the entity e and aspect A_i.

The problem of aspect mining with rating bias is very important in real applications. The problem is also the base of many tasks, such as overall rating prediction and aspect-level product recommendation. Compared to overall ratings, the aspect ratings are always missing and more unreliable. The aspect rating prediction is an effective way to repair the missing ratings and correct the unreliable ratings. However, the existent rating bias may make current methods on aspect rating prediction not effective anymore, so it is desired to consider rating bias for aspect rating prediction. Please note that the rating bias is known in our problem setting. Moreover, the rating bias can be easily obtained through limited reliable aspect ratings or a small quantity of manual labeling in real applications. So we can use the information of rating bias to correct the aspect rating prediction.

4   Rating-Center Model with Bias

The simplest way to handle rating bias is to subtract rating bias from the rating prediction results of existing models. However, it does not consider the correlations of ratings, aspects, and rating bias, so it may result in poor performances. In this section, we propose a novel RABI to handle the problem of the existent rating bias. Furthermore, we derive an iterative optimization solution with the EM algorithm.

4.1   Model Description

Existing models on aspect rating prediction usually consider reviews as the center to generate ratings and topics [6,7,9]. However, it does not conform to the

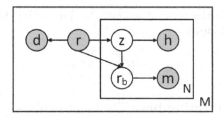

Fig. 2. Graphical model of RABI

authentic rating behaviour of users. In daily life, we form an intuitive impression as soon as we experienced a product. Only after we form an intuitive opinion (like or dislike, quantitatively represented by a rating) on a product, will we write a review to express our opinion. In addition, our opinion may involve multiple aspects of the product, such as taste, service and environment. So in the generative process of a product review, we will choose proper head terms to represent the aspect we want to express, and proper modifier terms to express sentiments on corresponding head terms. Finally, we organize these terms and other words to form a review. Therefore, we believe it is more reasonable to consider rating (overall rating) as the center to generate topics and reviews, which conforms to the authentic rating behaviour of users. Following this idea, we design the probabilistic model of RABI, shown in Fig. 2.

In Fig. 2, d indicates the reviews, r indicates the overall rating, h indicates the head term and m indicates the modifier term. These four variables are represented as the shaded circles, which means these four variables are observable. z indicates the aspect A_i. In order to keep consistent with the topic model, the aspect A_i is expressed as the topic z. And r_b indicates aspect rating, which will be introduced in the following. These two variables are represented as the open circles, which means these two variables are latent variables. Furthermore, N indicates the number of phrases in a review. And M indicates the number of reviews, which is equal to N_d.

To utilize the rating bias information effectively, we bring in a new latent aspect rating variable r_b. The modifier term m is used to modify the head term h to express the opinion (like or dislike) on aspect A_i (represented with z in the model), so m is actually influenced by the corresponding aspect rating r_{A_i}. As we mentioned above, there is an obvious rating bias between overall rating and aspect ratings. This observation causes that we cannot use the overall rating r to influence the modifier term m directly. So we bring in a new variable r_b between r and m to eliminate the influence of rating bias. r_b indicates an unknown aspect rating, so it is a latent variable. For a certain aspect A_i, the value of r_b is set as the overall rating r minus the rating bias b_{A_i}. Note that r_b can take N_r values in A_i, since the variable r can take N_r values. By bringing in the latent variable r_b, the association between r and m is modeled more reasonably in RABI.

According to the RABI model shown in Fig. 2, as the origin of the model, the overall rating r generates the review d and the latent topic z. The latent aspect rating r_b depends on the topic z and the overall rating r. And the head term h and the modifier term m are influenced by the topic z and the aspect rating r_b, respectively. So the joint probability over all variables is as follows:

$$p(h, m, r, d, z, r_b) = p(m|r_b)p(r_b|r, z)p(h|z)p(z|r)p(d|r)p(r). \tag{2}$$

All the parameters can be iteratively calculated using the EM algorithm [4], which is a common method to solve the problem with latent variable. The detail derivation is given in next section.

4.2 EM Solution

In the E-step, we need to maximize the lower bound function $\mathcal{L}_0$ (i.e., Jensens inequality [2]),

$$\mathcal{L}_0 = \sum_{z,r_b} q(z, r_b) \log\{\frac{p(h, m, r, d, z, r_b|\Lambda)}{q(z, r_b)}\}. \tag{3}$$

Here, as usual, $q(z, r_b)$ is set as follows:

$$q(z, r_b) = p(z, r_b|h, m, r, d; \Lambda^{old}). \tag{4}$$

Then we simplify Eq. (3), we can get

$$
\begin{aligned}
\mathcal{L}_0 &= \sum_{z,r_b} q(z, r_b) \log\{\frac{p(h, m, r, d, z, r_b|\Lambda)}{q(z, r_b)}\} \\
&= \underbrace{\sum_{z,r_b} q(z, r_b) \log p(h, m, r, d, z, r_b|\Lambda)}_{\mathcal{L}} - \underbrace{\sum_{z,r_b} q(z, r_b) \log q(z, r_b)}_{const} \\
&= \mathcal{L} - const.
\end{aligned}
\tag{5}
$$

So the second part is a *const*, which can be ignored. Then we ignore the *const*, and only consider the $\mathcal{L}$.

The function for the posterior probabilities of the latent variables is as follows:

$$\mathcal{L} = \sum_{h,m,r,d,z,r_b} n(h, m, r, d)q(z, r_b) \log p(h, m, r, d, z, r_b|\Lambda), \tag{6}$$

where Λ includes all parameters, i.e., $p(m|r_b)$, $p(r_b|r, z)$, $p(h|z)$, $p(z|r)$, $p(d|r)$ and $p(r)$, which are mentioned in Eq. (2). Besides, $n(h, m, r, d)$ is the number of co-occurrences of h, m, r and d.

The function $q(z, r_b)$ and $p(h, m, r, d, z, r_b|\Lambda)$ in Eq. (3) are expanded as follows:

$$q(z, r_b) = p(z, r_b|h, m, r, d; \Lambda^{old}) = \frac{p(m|r_b)p(r_b|r, z)p(h|z)p(z|r)p(d|r)p(r)}{\sum_{z,r_b} p(m|r_b)p(r_b|r, z)p(h|z)p(z|r)p(d|r)p(r)}, \tag{7}$$

$$p(h, m, r, d, z, r_b | \Lambda) = p(m|r_b)p(r_b|r, z)p(h|z)p(z|r)p(d|r)p(r). \qquad (8)$$

In the M-step, the Lagrangian Multiplier method is used to maximize $\mathcal{L}$ and calculate the parameters.

For $p(m|r_b)$, there is a basic constraint as follows:

$$\sum_m p(m|r_b) = 1. \qquad (9)$$

Applying the Lagrangian Multiplier method, we can get a function for $p(m|r_b)$ as follows:

$$\frac{\partial[\mathcal{L}_{[p(m|r_b)]} + \lambda(\sum_m p(m|r_b) - 1)]}{\partial p(m|r_b)} = 0. \qquad (10)$$

After calculation, we have

$$p(m|r_b) \propto n(h, m, r, d)p(z, r_b|h, m, r, d; \Lambda^{old}). \qquad (11)$$

Then the update function for $p(m|r_b)$ is as follows:

$$p(m|r_b) = \frac{\sum\limits_{h,r,d,z} n(h, m, r, d)p(z, r_b|h, m, r, d; \Lambda^{old})}{\sum\limits_{h,m',r,d,z} n(h, m', r, d)p(z, r_b|h, m', r, d; \Lambda^{old})}. \qquad (12)$$

Similarly, the update functions for other parameters are as follows:

$$p(r_b|r, z) = \frac{\sum\limits_{h,m,d} n(h, m, r, d)p(z, r_b|h, m, r, d; \Lambda^{old})}{\sum\limits_{h,m,d,r_b'} n(h, m, r, d)p(z, r_b'|h, m, r, d; \Lambda^{old})}, \qquad (13)$$

$$p(h|z) = \frac{\sum\limits_{m,r,d,r_b} n(h, m, r, d)p(z, r_b|h, m, r, d; \Lambda^{old})}{\sum\limits_{h',m,r,d,r_b} n(h', m, r, d)p(z, r_b|h', m, r, d; \Lambda^{old})}, \qquad (14)$$

$$p(z|r) = \frac{\sum\limits_{h,m,d,r_b} n(h, m, r, d)p(z, r_b|h, m, r, d; \Lambda^{old})}{\sum\limits_{h,m,d,z',r_b} n(h, m, r, d)p(z', r_b|h, m, r, d; \Lambda^{old})}, \qquad (15)$$

$$p(d|r) = \frac{\sum\limits_{h,m,z,r_b} n(h, m, r, d)p(z, r_b|h, m, r, d; \Lambda^{old})}{\sum\limits_{h,m,d',z,r_b} n(h, m, r, d')p(z, r_b|h, m, r, d'; \Lambda^{old})}, \qquad (16)$$

$$p(r) = \frac{\sum\limits_{h,m,d,z,r_b} n(h, m, r, d)p(z, r_b|h, m, r, d; \Lambda^{old})}{\sum\limits_{h,m,r',d,z,r_b} n(h, m, r', d)p(z, r_b|h, m, r', d; \Lambda^{old})}. \qquad (17)$$

Through these functions above, we can iteratively calculate the parameters until the model has converged.

4.3   Aspect Rating Prior

To verify our model's effectiveness, we need to compare the predicted aspect ratings with the real aspect ratings. So the aspects should correspond to the real aspects which are set by the e-commerce review sites. To make the predicted aspects similar to the real aspects, we need to assign some seed words to each aspect. For instance, the aspect "Taste" may include a few prior words, such as "taste" and "flavor".

In our model, we inject the prior knowledge for the aspect z. The function is as follows:

$$p(h|z) = \frac{\sum\limits_{m,r,d,r_b} n(h,m,r,d)p(z,r_b|h,m,r,d;\Lambda^{old}) + \tau(h,z)}{\sum\limits_{h',m,r,d,r_b} n(h',m,r,d)p(z,r_b|h',m,r,d;\Lambda^{old}) + \sum\limits_{h'} \tau(h',z)}, \quad (18)$$

where $\tau(h,z)$ indicates the prior knowledge of the prior words. Only when there is a relationship between the head term h and the topic z, in other words, h belongs to z, does $\tau(h,z)$ have a value δ, otherwise 0.

Note that, in the real applications, we can set aspects manually or generate aspects by the model directly. Moreover, manual aspect setting usually has better performances.

4.4   Aspect Identification and Aspect Rating Prediction

We can get $p(z,r_b|h,m)$ from the model by the following function,

$$rlp(z,r_b|h,m) = \frac{\sum_{r,d} p(h,m,r,d,z,r_b)}{\sum_{r,d,z,r_b} p(h,m,r,d,z,r_b)}$$
$$= \frac{\sum_{r,d} p(m|r_b)p(r_b|r,z)p(h|z)p(z|r)p(d|r)p(r)}{\sum_{r,d,z,r_b} p(m|r_b)p(r_b|r,z)p(h|z)p(z|r)p(d|r)p(r)}. \quad (19)$$

The goal of aspect identification is to find the mapping function $\mathcal{G}$ that correctly assigns the aspect label for given phrase f.

$$\mathcal{G}(f = (h,m)) = \arg\max_z \sum_{r_b} p(z,r_b|h,m). \quad (20)$$

The goal of aspect rating prediction is to predict the aspect rating r_{A_i} of the entity e given all the phrases f from all reviews and aspect $A_i(z)$. The aspect rating function is as follows:

$$r_{e,A_i} = \frac{\sum_{(h,m)\in\text{all reviews of } e} \sum_{r_b} r_b \cdot p(z,r_b|h,m)}{\sum_{(h,m)\in\text{all reviews of } e} \sum_{r_b} p(z,r_b|h,m)}, \quad (21)$$

where r_{e,A_i} indicates the aspect rating on the aspect A_i of the entity e.

In this way, RABI learns the joint probability distribution of phrases, aspects and ratings, and predicts aspect ratings with bias.

5   Evaluation

In this section, we introduce experimental preparation, evaluation metric and baselines. Then we conduct extensive experiments to evaluate the effectiveness of RABI on two real datasets.

Table 3. Prior words for aspect prior

Dataset	Category	Prior words
Dianping	Taste	taste, flavor, dish, dishes
	Service	serving, attitude, waitress, service
	Environment	environment, location, room, decoration
TripAdvisor	Value	value, price, quality, worth
	Room	room, suite, view, bed
	Location	location, traffic, place, area
	Cleanliness	clean, dirty, maintain, smell
	Front Desk/Staff	staff, check, help, reservation
	Service	service, food, breakfast, buffet
	Business	business, center, computer, internet

5.1   Experimental Preparation

Experiments are conducted on two real datasets (i.e., Dianping and TripAdvisor), which are introduced in Sect. 2. The preprocessing of TripAdvisor is similar to that in [6]. But the preprocessing of Dianping is slightly different. Since Dianping is a Chinese website, the Word Segmenter[3] and the rules from [8] are adopted for preprocessing. To inject the prior knowledge for the aspect, we select some words as prior for each aspect, and Table 3 lists some of the prior words (not all of the prior words due to the space limitation). For better understanding, we translate the Chinese words in Dianping into English.

Besides, all of the initial parameters ($p(m|r_b)$, $p(r_b|r, z)$, $p(h|z)$, $p(z|r)$, $p(d|r)$ and $p(r)$ in Eq. (2)) are assigned uniformly and randomly. δ in the Sect. 4.3 is set as 1 after some preliminary tests. The number of aspects or topics K is set as 3 for Dianping and 7 for TripAdvisor. The experiments are done on different-size of datasets (i.e., 25 %, 50 %, 75 %, and 100 % of review data) from Dianping and TripAdvisor, respectively. The maximum number of iterations is set as 500.

5.2   Evaluation Metric

RMSE (Root Mean Square Error) is one of the most common metrics for rating prediction. RMSE can measure the difference between the real values and the

---

[3] http://nlp.stanford.edu/software/segmenter.shtml.

predicted values. For every entity e, we have the real aspect rating vector r_{e,A_i} and the predicted aspect rating vector $\hat{r}_{e,A_i}$. The function of RMSE is as follows:

$$RMSE = \sqrt{\frac{\sum_{e=0}^{N_e} \sum_{A_i=0}^{N_A} (\hat{r}_{e,A_i} - r_{e,A_i})^2}{N_e * N_A}} \tag{22}$$

Smaller value of RMSE indicates a stronger predictor, which means the real values and the predicted values are nearer.

Besides, we use Pearson Correlation Coefficient ρ [10] to measure the relative ordering of products based on the predicted aspect rating and the real aspect rating. The correlation is stronger when the absolute value of ρ is closer to 1, and weaker when the absolute value of ρ is closer to 0. The function is as follows:

$$\rho = \frac{N \sum \hat{r}_{e,A_i} r_{e,A_i} - \sum \hat{r}_{e,A_i} \sum r_{e,A_i}}{\sqrt{N \sum (\hat{r}_{e,A_i})^2 - (\sum \hat{r}_{e,A_i})^2} \sqrt{N \sum (r_{e,A_i})^2 - (\sum r_{e,A_i})^2}}, \tag{23}$$

where N indicates the total amount, which is $N_e * N_A$.

5.3 Baseline Methods

We compare the proposed model with three representative methods and one variation of RABI. Since all of these baselines do not consider the rating bias, we adjust the results of these baselines through subtracting the rating bias for fair comparison. The adjusted method is marked with "*" to distinguish from the original method.

- QPLSA/QPLSA* [7] uses quad-tuples information to build a model based on PLSA framework. The model not only can generate fine-granularity aspects of products, but also capture the relationship between words and ratings.
- GRAOS/GRAOS* [6] is a semi-supervised model based on LDA framework. It also uses the quad-tuples information to capture the relationship between words and ratings. The model considers the rating distribution as a Gaussian distribution.
- SATM/SATM* [9] is a sentiment-aligned model based on LDA framework. The model uses two kinds of external knowledge: productlevel overall rating distribution and wordlevel sentiment lexicon.
- RA/RA* is a simplified model which removes the latent aspect rating variable r_b from our model RABI. It only considers the rating-center assumption. Through comparing RA* and RABI, we can testify the importance of the good mechanism to utilize rating bias information.

5.4 Results Evaluation

We firstly validate the effectiveness of aspect identification of RABI through a case study, and then compare the results of different methods on the accuracy of aspect rating prediction with two criteria mentioned above.

Table 4. Representative phrases for different aspects on two datasets

Datasets	Aspects	Representative phrases (ratings)
Dianping	Taste	amazing mouthfeel (4.71), first-rate taste (4.58), common taste (2.75), so-so flavor (1.77)
	Service	smart waiter (4.51), passive service (3.51), slow serving (2.51), cold attitude (1.67)
	Environment	great location (4.45), sumptuous fitment (4.26), common environment (2.88), small room (2.45)
TripAdvisor	Value	perfect price (4.81), standard charge (4.65), delightful priceline (4.05), astronomical deal (1.59)
	Room	greatest setting (4.81), cool room (4.19), beautiful decor (4.18), worst setting (1.57)
	Location	wonderful location (4.90), central location (4.63), nice place (4.13), remote area (1.27)
	Cleanliness	normal maintained (4.61), standard cleanliness (4.38), well homey (4.34), dirty housekeeping (1.26)
	Front Desk/Staff	hospitable staff (4.95), great staff (4.71), friendly hotel (4.65), so-so staff (1.82)
	Service	super singer (4.71), great wine (4.50), valuable amenities (4.27), worst experience (1.23)
	Business	best wifi (4.63), common websites (4.22), nice desktop (4.17), standard business (3.52)

Aspect Identification. RABI extracts a set of rated phrases to describe the product for each aspect. We list the top 20 automatically mined phrases for each aspect, from which we select several meaningful phrases to be shown in Table 4. The phrases are ranked by their ratings for every aspect.

Generally, the extracted phrases properly describe the corresponding aspects and accurately embody the opinion in both English and Chinese reviews. On one hand, the head terms can indicate the aspects well, such as "attitude" for service, "fitment" for environment, "setting" for room, and "area" for location. When a user sees the head term, he can understand which aspect is talked about. On the other hand, a positive modifier term indicates a positive attitude and is likely to obtain a higher rating, and a negative modifier term indicates a negative attitude and is likely to obtain a lower rating. For example, in the Service aspect of Dianning, the phrase "cold attitude" is rated as 1.67 because "cold" is a negative modifier term, while the phrase "smart waiter" has a score of 4.51 because "smart" is a positive modifier term. In addition, the phrases and their ratings are also able to reflect the different rating styles in Chinese and English. That is, users tend to give relatively lower ratings in Chinese reviews. The distribution of the predicted ratings on phrases also conforms to that of aspect-level ratings

Table 5. RMSE performances of different methods on two datasets

	Dianping				TripAdvisor			
	25 %	50 %	75 %	100 %	25 %	50 %	75 %	100 %
QPLSA	0.5816	0.5799	0.5714	0.5635	0.6374	0.6248	0.6129	0.6119
QPLSA*	0.3656	0.3584	0.3554	0.3435	0.6262	0.6180	0.6125	0.6071
GRAOS	0.4751	0.4668	0.4624	0.4500	0.6056	0.6072	0.6011	0.5968
GRAOS*	0.4228	0.4152	0.4155	0.4136	0.5790	0.5724	0.5691	0.5668
SATM	0.5804	0.5767	0.5639	0.5594	0.5587	0.5502	0.5406	0.5300
SATM*	0.3979	0.3890	0.3816	0.3738	0.5419	0.5322	0.5310	0.5188
RA	0.5789	0.5601	0.5511	0.5451	0.6081	0.5935	0.5826	0.5784
RA*	0.3471	0.3404	0.3312	0.3267	0.5785	0.5612	0.5599	0.5471
RABI	**0.3248**	**0.3162**	**0.3047**	**0.2919**	**0.5346**	**0.5267**	**0.5204**	**0.5089**

Table 6. Pearson correlation coefficient of different methods on two datasets

	Dianping				TripAdvisor			
	25 %	50 %	75 %	100 %	25 %	50 %	75 %	100 %
QPLSA	0.5792	0.5809	0.5836	0.5985	0.3167	0.3451	0.3508	0.3827
GRAOS	0.1281	0.1280	0.1328	0.1376	0.3238	0.3407	0.3463	0.3569
SATM	0.3522	0.3605	0.3742	0.3906	0.3315	0.3521	0.3621	0.3679
RA	0.5248	0.5330	0.5430	0.5494	0.4065	0.4167	0.4291	0.4377
RABI	**0.6059**	**0.6137**	**0.6174**	**0.6211**	**0.5328**	**0.5522**	**0.5597**	**0.5657**

on these two datasets in Table 2. It also confirms the effectiveness of RABI on Chinese and English datasets.

Accuracy Experiment. Then we validate the performances of different methods through comparing predicted aspect ratings with real aspect ratings using the RMSE criterion by Eq. (22).

From the results shown in Table 5, we can clearly find that the integration of the rating bias information can significantly improve the prediction accuracy for all methods (e.g., QPLSA* has better performances than QPLSA), and RABI always performs best on both datasets. The improvement is particularly obvious for Dianping, because this dataset has large rating biases. Although the rating bias is small in TripAdvisor, the methods considering rating bias all achieve better performances than original methods. It illustrates that it is necessary to consider the rating bias for aspect rating prediction.

Besides, the rating-center model (i.e., RA) also achieves good performances among four baselines, which confirms the correctness of the rating-center assumption. Compared to simply subtracting the rating bias in four baselines, the best performances of RABI imply that the good mechanism to utilize rating bias information is also necessary. We think the rating-center and the latent aspect rating variable contribute to the good performances of RABI.

In addition, with the increment of review data, the accuracy of RABI increases steadily and slowly, which reflects that RABI is a steady method.

Relative Order Experiment. Furthermore, we verify the ability of different methods to maintain the relative order among products with the Pearson Correlation Coefficient ρ. The results are shown in Table 6. Note that the rating bias has slight effect on the order of products, so we only display the results of original methods and ignore the adjusted methods. We can see that RABI obtains much higher ρ than other methods in all datasets. It once again shows that RABI is more effective to model the correlations between aspects and ratings, and thus better maintains aspect ranking orders compared to other methods. The results also imply that RABI is very promising for aspect-level recommender system, since it can generate very similar product order to the real order.

6   Related Work

In recent years, sentiment analysis on reviews becomes a research hotspot. Reviews focus on the products in each aspect, so sentiment analysis on reviews usually involves aspect. This situation leads to the aspect rating prediction. Aspect rating prediction usually contains two subtasks, aspect identification and aspect rating prediction.

Topic model is widely used to solve aspect identification. It mainly contains LSI [3], PLSA [5] and LDA [1]. Xu et al. [12] centered on implicit feature identification in Chinese product reviews via LDA and SVM. An AEP-based Latent Dirichlet Allocation (AEP-LDA) [13] model was also proposed to extract product and service aspect words automatically from reviews. Fu et al. [11] proposed an approach to automatically discover the aspects discussed in Chinese social reviews and classified the polarity of the associated sentiment by HowNet lexicon. Our model RABI is designed based on the PLSA framework.

To solve aspect identification and aspect rating prediction simultaneously, many researches adopted the topic-sentiment mixture models. QPLSA [7] adopted the quad-tuples, which consist of head, modifier, rating and entity. It can generate fine-granularity aspects and capture the correlations between words and ratings. SATM [9] used external knowledge, product-level overall rating distribution and word-level sentiment lexicon, to extract the product aspects and predict aspect ratings simultaneously. Luo et al. [6] proposed a model based on LDA to predict aspect ratings and overall ratings for unrated reviews and made two assumptions for the rating distribution. However, all of these works did not consider the existing rating bias, which is firstly studied in this paper.

7   Conclusion

Aspect rating prediction for reviews is a hot research issue nowadays. Most of researches base on such a basic assumption, the overall rating is the average score of aspect ratings or the overall rating is close to aspect ratings. However in the real world, there may be rating biases between overall rating and aspect ratings, and existing works did not consider these rating biases.

In this paper, we study the problem of aspect mining with rating bias and propose a novel probabilistic model RABI based on PLSA framework. The RABI model makes rating as the center to generate ratings and topics, and introduces a latent aspect rating variable to integrate the rating bias information. Experiments on two real datasets validate the effectiveness of RABI. In the future, we can import the Dirichlet prior and redesign our model based on LDA framework. The effectiveness will be enhanced further.

Acknowledgements. This work is supported in part by the National Key Basic Research and Department (973) Program of China (No. 2013CB329606), and the National Natural Science Foundation of China (No. 61375058, 61473273), and the Co-construction Project of Beijing Municipal Commission of Education, and the CCF-Tencent Open Fund, and 2015 Microsoft Research Asia Collaborative Research Program.

References

1. Blei, D.M., Ng, A.Y., Jordan, M.I.: Latent Dirichlet allocation. J. Mach. Learn. Res. **3**, 993–1022 (2003)
2. Chandler, D.: Introduction to modern statistical mechanics. Phys. Today **1**, 288 (1987)
3. Deerwester, S.C., Dumais, S.T., Landauer, T.K., Furnas, G.W., Harshman, R.A.: Indexing by latent semantic analysis. J. Am. Soc. Inform. Sci. **41**(6), 391–407 (1990)
4. Dempster, A.P., Laird, N.M., Rubin, D.B.: Maximum likelihood from incomplete data via the EM algorithm. J. Roy. Stat. Soc. Ser. B (Methodol.) **39**, 1–38 (1977)
5. Hofmann, T.: Probabilistic latent semantic indexing. In: Proceedings of the 22nd Annual International ACM SIGIR Conference on Research and Development in Information Retrieval, pp. 50–57. Association for Computing Machinery, Berkeley, August 1999
6. Luo, W., Zhuang, F., Cheng, X., He, Q., Shi, Z.: Ratable aspects over sentiments: predicting ratings for unrated reviews. In: 2014 IEEE International Conference on Data Mining (ICDM), pp. 380–389. Institute of Electrical and Electronics Engineers, Shenzhen, December 2014
7. Luo, W., Zhuang, F., He, Q., Shi, Z.: Quad-tuple PLSA: incorporating entity and its rating in aspect identification. In: Tan, P.-N., Chawla, S., Ho, C.K., Bailey, J. (eds.) PAKDD 2012. LNCS (LNAI), pp. 392–404. Springer, Heidelberg (2012). doi:10.1007/978-3-642-30217-6_33
8. Moghaddam, S., Ester, M.: On the design of LDA models for aspect-based opinion mining. In: Proceedings of the 21st ACM International Conference on Information and Knowledge Management, pp. 803–812. Association for Computing Machinery, Sheraton, October 2012
9. Wang, H., Ester, M.: A sentiment-aligned topic model for product aspect rating prediction. In: Proceedings of the 2014 Conference on Empirical Methods in Natural Language Processing, pp. 1192–1202. The Association for Computational Linguistics, Doha, October 2014

10. Wang, H., Lu, Y., Zhai, C.: Latent aspect rating analysis on review text data: a rating regression approach. In: Proceedings of the 16th ACM SIGKDD International Conference on Knowledge Discovery and Data Mining, pp. 783–792. Association for Computing Machinery, Washington DC, July 2010

11. Fu, X., Liu, G., Guo, Y., Wang, Z.: Multi-aspect sentiment analysis for Chinese online social reviews based on topic modeling and HowNet lexicon. Knowl.-Based Syst. **37**, 186–195 (2013)

12. Xu, H., Zhang, F., Wang, W.: Implicit feature identification in Chinese reviews using explicit topic mining model. Knowl.-Based Syst. **76**, 166–175 (2015)

13. Zheng, X., Lin, Z., Wang, X., Lin, K.J., Song, M.: Incorporating appraisal expression patterns into topic modeling for aspect and sentiment word identification. Knowl.-Based Syst. **61**, 29–47 (2014)

Planning with Information-Processing Constraints and Model Uncertainty in Markov Decision Processes

Jordi Grau-Moya[1,2,3]([✉]), Felix Leibfried[1,2,3], Tim Genewein[1,2,3], and Daniel A. Braun[1,2]

[1] Max Planck Institute for Intelligent Systems, Tübingen, Germany
jordi.grau@tuebingen.mpg.de
[2] Max Planck Institute for Biological Cybernetics, Tübingen, Germany
[3] Graduate Training Centre for Neuroscience, Tübingen, Germany

Abstract. Information-theoretic principles for learning and acting have been proposed to solve particular classes of Markov Decision Problems. Mathematically, such approaches are governed by a variational free energy principle and allow solving MDP planning problems with information-processing constraints expressed in terms of a Kullback-Leibler divergence with respect to a reference distribution. Here we consider a generalization of such MDP planners by taking model uncertainty into account. As model uncertainty can also be formalized as an information-processing constraint, we can derive a unified solution from a single generalized variational principle. We provide a generalized value iteration scheme together with a convergence proof. As limit cases, this generalized scheme includes standard value iteration with a known model, Bayesian MDP planning, and robust planning. We demonstrate the benefits of this approach in a grid world simulation.

Keywords: Bounded rationality · Model uncertainty · Robustness · Planning · Markov decision processes

1 Introduction

The problem of planning in Markov Decision Processes was famously addressed by Bellman who developed the eponymous principle in 1957 [2]. Since then numerous variants of this principle have flourished in the literature. Here we are particularly interested in a generalization of the Bellman principle that takes information-theoretic constraints into account. In the recent past there has been a special interest in the Kullback-Leibler divergence as a constraint to limit deviations of the action policy from a prior. This can be interesting in a number of ways. Todorov [31,32], for example, has transformed the general MDP problem into a restricted problem class without explicit action variables, where control directly changes the dynamics of the environment and control costs are measured by the Kullback-Leibler divergence between controlled and uncontrolled

© Springer International Publishing AG 2016
P. Frasconi et al. (Eds.): ECML PKDD 2016, Part II, LNAI 9852, pp. 475–491, 2016.
DOI: 10.1007/978-3-319-46227-1_30

dynamics. This simplification allows mapping the Bellman recursion to a linear algebra problem. This approach can also be generalized to continuous state spaces leading to path integral control [4,5]. The same equations can also be interpreted in terms of *bounded rational* decision-making where the decision-maker has limited computational resources that allow only limited deviations from a prior decision strategy (measured by the Kullback-Leiber divergence in bits) [19]. Such a decision-maker can also be instantiated by a sampling process that has restrictions in the number of samples it can afford [20]. Disregarding the possibility of a sampling-based interpretation, the Kullback-Leibler divergence introduces a control information cost that is interesting in its own right when formalizing the perception action cycle [30].

While the above frameworks have led to interesting computational advances, so far they have neglected the possibility of model misspecification in the MDP setting. Model misspecification or model uncertainty does not refer to the uncertainty arising due to the stochastic nature of the environment (usually called risk-uncertainty in the economic literature), but refers to the uncertainty with respect to the latent variables that specify the MDP. In Bayes-Adaptive MDPs [7], for example, the uncertainty over the latent parameters of the MDP is explicitly represented, such that new information can be incorporated with Bayesian inference. However, Bayes-Adaptive MDPs are not robust with respect to model misspecification and have no performance guarantees when planning with wrong models [15]. Accordingly, there has been substantial interest in developing robust MDP planners [13,16,33]. One way to take model uncertainty into account is to bias an agent's belief model from a reference Bayesian model towards worst-case scenarios; thus avoiding disastrous outcomes by not visiting states where the transition probabilities are not known. Conversely, the belief model can also be biased towards best-case scenarios as a measure to drive exploration—also referred in the literature as *optimism in face of uncertainty* [28,29].

When comparing the literature on information-theoretic control and model uncertainty, it is interesting to see that some notions of model uncertainty follow exactly the same mathematical principles as the principles of relative entropy control [32]. In this paper we therefore formulate a unified and combined optimization problem for MDP planning that takes *both*, model uncertainty and bounded rationality into account. This new optimization problem can be solved by a generalized value iteration algorithm. We provide a theoretical analysis of its convergence properties and simulations in a grid world.

2   Background and Notation

In the MDP setting the agent at time t interacts with the environment by taking action $a_t \in \mathcal{A}$ while in state $s_t \in \mathcal{S}$. Then the environment updates the state of the agent to $s_{t+1} \in \mathcal{S}$ according to the transition probabilities $T(s_{t+1}|a_t, s_t)$. After each transition the agent receives a reward $R_{s_t,a_t}^{s_{t+1}} \in \mathcal{R}$ that is bounded. For our purposes we will consider $\mathcal{A}$ and $\mathcal{S}$ to be finite. The aim of the agent is to choose its policy $\pi(a|s)$ in order to maximize the total discounted expected reward or value function for any $s \in \mathcal{S}$

$$V^*(s) = \max_\pi \lim_{T \to \infty} \mathbb{E} \left[\sum_{t=0}^{T-1} \gamma^t R_{s_t,a_t}^{s_{t+1}} \right]$$

with discount factor $0 \leq \gamma < 1$. The expectation is over all possible trajectories $\xi = s_0, a_0, s_1 \ldots$ of state and action pairs distributed according to $p(\xi) = \prod_{t=0}^{T-1} \pi(a_t|s_t) \, T(s_{t+1}|a_t, s_t)$. It can be shown that the optimal value function satisfies the following recursion

$$V^*(s) = \max_\pi \sum_{a,s'} \pi(a|s) T(s'|a, s) \left[R_{s,a}^{s'} + \gamma V^*(s') \right]. \tag{1}$$

At this point there are two important implicit assumptions. The first is that the policy π can be chosen arbitrarily without any constraints which, for example, might not be true for a bounded rational agent with limited information-processing capabilities. The second is that the agent needs to know the transition-model $T(s'|a, s)$, but this model is in practice unknown or even misspecified with respect to the environment's true transition-probabilities, specially at initial stages of learning. In the following, we explain how to incorporate both bounded rationality and model uncertainty into agents.

2.1   Information-Theoretic Constraints for Acting

Consider a one-step decision-making problem where the agent is in state s and has to choose a single action a from the set $\mathcal{A}$ to maximize the reward $R_{s,a}^{s'}$, where s' is the next state. A perfectly rational agent selects the optimal action that is $a^*(s) = \text{argmax}_a \sum_{s'} T(s'|a, s) R_{s,a}^{s'}$. However, a bounded rational agent has only limited resources to find the maximum of the function $\sum_{s'} T(s'|a, s) R_{s,a}^{s'}$. One way to model such an agent is to assume that the agent has a prior choice strategy $\rho(a|s)$ in state s *before* a deliberation process sets in that refines the choice strategy to a posterior distribution $\pi(a|s)$ that reflects the strategy *after* deliberation. Intuitively, because the deliberation resources are limited, the agent can only afford to deviate from the prior strategy by a certain amount of information bits. This can be quantified by the relative entropy $D_{\text{KL}}(\pi||\rho) = \sum_a \pi(a|s) \log \frac{\pi(a|s)}{\rho(a|s)}$ that measures the average information cost of the policy $\pi(a|s)$ using the source distribution $\rho(a|s)$. For a bounded rational agent this relative entropy is bounded by some upper limit K. Thus, a bounded rational agent has to solve a constrained optimization problem that can be written as

$$\max_\pi \sum_a \pi(a|s) \sum_{s'} T(s'|a, s) R_{s,a}^{s'} \qquad \text{s.t.} \quad D_{\text{KL}}(\pi||\rho) \leq K$$

This problem can be rewritten as an unconstrained optimization problem

$$F^*(s) = \max_\pi \sum_a \pi(a|s) \sum_{s'} T(s'|a, s) R_{s,a}^{s'} - \frac{1}{\alpha} D_{\text{KL}}(\pi||\rho) \tag{2}$$

$$= \frac{1}{\alpha} \log \sum_a \rho(a|s) e^{\alpha \sum_{s'} T(s'|a,s) R_{s,a}^{s'}}. \tag{3}$$

where F^* is a free energy that quantifies the value of the policy π by trading off the average reward against the information cost. The optimal strategy can be expressed analytically in closed-form as

$$\pi^*(a|s) = \frac{\rho(a|s)e^{\alpha \sum_{s'} T(s'|a,s)R^{s'}_{s,a}}}{Z_\alpha(s)}$$

with partition sum $Z_\alpha(s) = \sum_a \rho(a|s) \exp\left(\alpha \sum_{s'} T(s'|a,s)R^{s'}_{s,a}\right)$. Therefore, the maximum operator in (2) can be eliminated and the free energy can be rewritten as in (3). The Lagrange multiplier α quantifies the boundedness of the agent. By setting $\alpha \to \infty$ we recover a perfectly rational agent with optimal policy $\pi^*(a|s) = \delta(a - a^*(s))$. For $\alpha = 0$ the agent has no computational resources and the agent's optimal policy is to act according to the prior $\pi^*(a|s) = \rho(a|s)$. Intermediate values of α lead to a spectrum of bounded rational agents.

2.2 Information-Theoretic Constraints for Model Uncertainty

In the following we assume that the agent has a model of the environment $T_\theta(s'|a, s)$ that depends on some latent variables $\theta \in \Theta$. In the MDP setting, the agent holds a belief $\mu(\theta|a, s)$ regarding the environmental dynamics where θ is a unit vector of transition probabilities into all possible states s'. While interacting with the environment the agent can incorporate new data by forming the Bayesian posterior $\mu(\theta|a, s, D)$, where D is the observed data. When the agent has observed an infinite amount of data (and assuming $\theta^*(a, s) \in \Theta$) the belief will converge to the delta distribution $\mu(\theta|s, a, D) = \delta(\theta - \theta^*(a, s))$ and the agent will act optimally according to the true transition probabilities, exactly as in ordinary optimal choice strategies with known models. When acting under a limited amount of data the agent cannot determine the value of an action a with the true transition model according to $\sum_{s'} T(s'|a, s)R^{s'}_{s,a}$, but it can only determine an expected value according to its beliefs $\int_\theta \mu(\theta|a, s) \sum_{s'} T_\theta(s'|a, s)R^{s'}_{s,a}$.

The Bayesian model μ can be subject to model misspecification (e.g. by having a wrong likelihood or a bad prior) and thus the agent might want to allow deviations from its model towards best-case (optimistic agent) or worst-case (pessimistic agent) scenarios up to a certain extent, in order to act more robustly or to enhance its performance in a friendly environment [12]. Such deviations can be measured by the relative entropy $D_{KL}(\psi|\mu)$ between the Bayesian posterior μ and a new biased model ψ. Effectively, this allows for mathematically formalizing model uncertainty, by not only considering the specified model but all models within a neighborhood of the specified model that deviate no more than a restricted number of bits. Then, the effective expected value of an action a while having limited trust in the Bayesian posterior μ can be determined for the case of optimistic deviations as

$$F^*(a, s) = \max_\psi \int_\theta \psi(\theta|a, s) \sum_{s'} T_\theta(s'|a, s)R^{s'}_{s,a} - \frac{1}{\beta}D_{KL}(\psi||\mu) \qquad (4)$$

for $\beta > 0$, and for the case of pessimistic deviations as

$$F^*(a,s) = \min_{\psi} \int_{\theta} \psi(\theta|a,s) \sum_{s'} T_{\theta}(s'|a,s) R^{s'}_{s,a} - \frac{1}{\beta} D_{\mathrm{KL}}(\psi\|\mu) \qquad (5)$$

for $\beta < 0$. Conveniently, both equations can be expressed as a single equation

$$F^*(a,s) = \frac{1}{\beta} \log Z_{\beta}(a,s)$$

with $\beta \in \mathbb{R}$ and $Z_{\beta}(s,a) = \int_{\theta} \mu(\theta|a,s) \exp\left(\beta \sum_{s'} T_{\theta}(s'|a,s) R^{s'}_{s,a}\right)$ when inserting the optimal biased belief

$$\psi^*(\theta|a,s) = \frac{1}{Z_{\beta}(a,s)} \mu(\theta|a,s) \exp\left(\beta \sum_{s'} T_{\theta}(s'|a,s) R^{s'}_{s,a}\right)$$

into either Eq. (4) or (5). By adopting this formulation we can model any degree of trust in the belief μ allowing deviation towards worst-case or best-case with $-\infty \leq \beta \leq \infty$. For the case of $\beta \to -\infty$ we recover an infinitely pessimistic agent that considers only worst-case scenarios, for $\beta \to \infty$ an agent that is infinitely optimistic and for $\beta \to 0$ an agent that fully trusts its model.

3   Model Uncertainty and Bounded Rationality in MDPs

In this section, we consider a bounded rational agent with model uncertainty in the infinite horizon setting of an MDP. In this case the agent must take into account all future rewards and information costs, thereby optimizing the following free energy objective

$$F^*(s) = \max_{\pi} \operatorname*{ext}_{\psi} \lim_{T \to \infty} \mathbb{E} \sum_{t=0}^{T-1} \gamma^t \left(R^{s_{t+1}}_{s_t,a_t} - \frac{1}{\beta} \log \frac{\psi(\theta_t|a_t,s_t)}{\mu(\theta_t|a_t,s_t)} - \frac{1}{\alpha} \log \frac{\pi(a_t|s_t)}{\rho(a_t|s_t)} \right) \tag{6}$$

where the extremum operator ext can be either max for $\beta > 0$ or min for $\beta < 0$, $0 \leq \gamma < 1$ is the discount factor and the expectation $\mathbb{E}$ is over all trajectories $\xi = s_0, a_0, \theta_0, s_1, a_1, \ldots a_{T-1}, \theta_{T-1}, s_T$ with distribution $p(\xi) = \prod_{t=0}^{T-1} \pi(a_t|s_t)$ $\psi(\theta_t|a_t,s_t) T_{\theta_t}(s_{t+1}|a_t,s_t)$. Importantly, this free energy objective satisfies a recursive relation and thereby generalizes Bellman's optimality principle to the case of model uncertainty and bounded rationality. In particular, Eq. (6) fulfills the recursion

$$F^*(s) = \max_{\pi} \operatorname*{ext}_{\psi} \mathbb{E}_{\pi(a|s)} \left[-\frac{1}{\alpha} \log \frac{\pi(a|s)}{\rho(a|s)} + \right.$$

$$\mathbb{E}_{\psi(\theta|a,s)} \left[-\frac{1}{\beta} \log \frac{\psi(\theta|a,s)}{\mu(\theta|a,s)} + \right.$$

$$\left. \left. \mathbb{E}_{T_{\theta}(s'|a,s)} \left[R^{s'}_{s,a} + \gamma F^*(s') \right] \right] \right]. \tag{7}$$

Applying variational calculus and following the same rationale as in the previous sections [19], the extremum operators can be eliminated and Eq. (7) can be re-expressed as

$$F^*(s) = \frac{1}{\alpha} \log \mathbb{E}_{\rho(a|s)} \left[\mathbb{E}_{\mu(\theta|a,s)} \left[\exp \left(\beta \mathbb{E}_{T_\theta(s'|a,s)} \left[R^{s'}_{s,a} + \gamma F^*(s') \right] \right) \right]^{\frac{\alpha}{\beta}} \right] \quad (8)$$

because

$$F^*(s) = \max_\pi \mathbb{E}_{\pi(a|s)} \left[\frac{1}{\beta} \log Z_\beta(a, s) - \frac{1}{\alpha} \log \frac{\pi(a|s)}{\rho(a|s)} \right] \quad (9)$$

$$= \frac{1}{\alpha} \log \mathbb{E}_{\rho(a|s)} \left[\exp \left(\frac{\alpha}{\beta} \log Z_\beta(a, s) \right) \right], \quad (10)$$

where

$$Z_\beta(a, s) = \underset{\psi}{\text{ext}}\, \mathbb{E}_{\psi(\theta|a,s)} \left[\mathbb{E}_{T_\theta(s'|a,s)} \left[R^{s'}_{s,a} + \gamma F^*(s') \right] - \frac{1}{\beta} \log \frac{\psi(\theta|a,s)}{\mu(\theta|a,s)} \right] \quad (11)$$

$$= \mathbb{E}_{\mu(\theta|a,s)} \exp \left(\beta \mathbb{E}_{T_\theta(s'|a,s)} \left[R^{s'}_{s,a} + \gamma F^*(s') \right] \right)$$

with the optimizing arguments

$$\psi^*(\theta|a, s) = \frac{1}{Z_\beta(a, s)} \mu(\theta|a, s) \exp \left(\beta \mathbb{E}_{T_\theta(s'|a,s)} \left[R^{s'}_{s,a} + \gamma F(s') \right] \right)$$

$$\pi^*(a|s) = \frac{1}{Z_\alpha(s)} \rho(a|s) \exp \left(\frac{\alpha}{\beta} \log Z_\beta(a, s) \right) \quad (12)$$

and partition sum

$$Z_\alpha(s) = \mathbb{E}_{\rho(a|s)} \left[\exp \left(\frac{\alpha}{\beta} \log Z_\beta(a, s) \right) \right].$$

With this free energy we can model a range of different agents for different α and β. For example, by setting $\alpha \to \infty$ and $\beta \to 0$ we can recover a Bayesian MDP planner and by setting $\alpha \to \infty$ and $\beta \to -\infty$ we recover a robust planner. Additionally, for $\alpha \to \infty$ and when $\mu(\theta|a, s) = \delta(\theta - \theta^*(a, s))$ we recover an agent with standard value function with known state transition model from Eq. (1).

3.1   Free Energy Iteration Algorithm

Solving the self-consistency Eq. (8) can be achieved by a generalized version of value iteration. Accordingly, the optimal solution can be obtained by initializing the free energy at some arbitrary value F and applying a value iteration scheme $B^{i+1}F = BB^i F$ where we define the operator

$$BF(s) = \max_{\pi} \text{ext}_{\psi} \, \mathbb{E}_{\pi(a|s)} \left[-\frac{1}{\alpha} \log \frac{\pi(a|s)}{\rho(a|s)} + \right.$$

$$\mathbb{E}_{\psi(\theta|a,s)} \left[-\frac{1}{\beta} \log \frac{\psi(\theta|a,s)}{\mu(\theta|a,s)} + \right.$$

$$\left. \left. \mathbb{E}_{T_\theta(s'|a,s)} \left[R_{s,a}^{s'} + \gamma F(s') \right] \right] \right] \quad (13)$$

with $B^1 F = BF$, which can be simplified to

$$BF(s) = \frac{1}{\alpha} \log \mathbb{E}_{\rho(a|s)} \left[\mathbb{E}_{\mu(\theta|a,s)} \left[\exp \left(\beta \mathbb{E}_{T_\theta(s'|a,s)} \left[R_{s,a}^{s'} + \gamma F(s') \right] \right) \right]^{\frac{\alpha}{\beta}} \right]$$

In Algorithm (1) we show the pseudo-code of this generalized value iteration scheme. Given state-dependent prior policies $\rho(a|s)$ and the Bayesian posterior beliefs $\mu(\theta|a,s)$ and the values of α and β, the algorithm outputs the equilibrium distributions for the action probabilities $\pi(a|s)$, the biased beliefs $\psi(\theta|a,s)$ and estimates of the free energy value function $F^*(s)$. The iteration is run until a convergence criterion is met. Assuming dimensionality A for the action space, S for the state space, and B for the (discretized) belief space we have a complexity of $O(AS^2B)$ per iteration, similar to other value iteration algorithms running on the belief space. The convergence proof of the algorithm is shown in the next section.

Algorithm 1. Iterative algorithm solving the self-consistency Eq. (8)

Input: $\rho(a|s), \mu(\theta|a,s), \alpha, \beta$
Initialize: $F \leftarrow 0, F_{old} \leftarrow 0$
while *not converged* **do**
  forall $s \in \mathcal{S}$ **do**
    $F(s) \leftarrow \frac{1}{\alpha} \log \mathbb{E}_{\rho(a|s)} \left[\mathbb{E}_{\mu(\theta|a,s)} \left[\exp \left(\beta \mathbb{E}_{T_\theta(s'|a,s)} \left[R_{s,a}^{s'} + \gamma F_{old}(s') \right] \right) \right]^{\frac{\alpha}{\beta}} \right]$
  end
  $F_{old} \leftarrow F$
end
$\pi(a|s) \leftarrow \frac{1}{Z_\alpha(s)} \rho(a|s) \exp \left(\frac{\alpha}{\beta} \log Z_\beta(a,s) \right)$
$\psi(\theta|a,s) \leftarrow \frac{1}{Z_\beta(a,s)} \mu(\theta|a,s) \exp \left(\beta \mathbb{E}_{T_\theta(s'|a,s)} \left[R_{s,a}^{s'} + \gamma F(s') \right] \right)$
return $\pi(a|s), \psi(\theta|a,s), F(s)$

---

4   Convergence

Here, we show that the value iteration scheme described through Algorithm 1 converges to a unique fixed point satisfying Eq. (8). To this end, we first prove the existence of a unique fixed point (Theorem 1) following [3,25], and subsequently prove the convergence of the value iteration scheme presupposing that a unique fixed point exists (Theorem 2) following [27].

Theorem 1. *Assuming a bounded reward function $R_{s,a}^{s'}$, the optimal free-energy vector $F^*(s)$ is a unique fixed point of Bellman's equation $F^* = BF^*$, where the mapping $B : \mathbb{R}^{|S|} \to \mathbb{R}^{|S|}$ is defined as in Eq. (13)*

Proof. Theorem 1 is proven through Propositions 1 and 2 in the following.

Proposition 1. *The mapping $T_{\pi,\psi} : \mathbb{R}^{|S|} \to \mathbb{R}^{|S|}$*

$$T_{\pi,\psi}F(s) = \mathbb{E}_{\pi(a|s)}\left[-\frac{1}{\alpha}\log\frac{\pi(a|s)}{\rho(a|s)} + \right.$$

$$\mathbb{E}_{\psi(\theta|a,s)}\left[-\frac{1}{\beta}\log\frac{\psi(\theta|a,s)}{\mu(\theta|a,s)} + \right.$$

$$\left.\left.\mathbb{E}_{T_\theta(s'|a,s)}\left[R_{s,a}^{s'} + \gamma F(s') \right]\right]\right]. \quad (14)$$

converges to a unique solution for every policy-belief-pair (π, ψ) independent of the initial free-energy vector $F(s)$.

Proof. By introducing the matrix $P_{\pi,\psi}(s, s')$ and the vector $g_{\pi,\psi}(s)$ as

$$P_{\pi,\psi}(s, s') := \mathbb{E}_{\pi(a|s)}\left[\mathbb{E}_{\psi(\theta|a,s)}\left[T_\theta(s'|a, s) \right] \right],$$

$$g_{\pi,\psi}(s) := \mathbb{E}_{\pi(a|s)}\left[\mathbb{E}_{\psi(\theta|a,s)}\left[\mathbb{E}_{T_\theta(s'|a,s)}\left[R_{s,a}^{s'} \right] - \frac{1}{\beta}\log\frac{\psi(\theta|a,s)}{\mu(\theta|a,s)} \right] - \frac{1}{\alpha}\log\frac{\pi(a|s)}{\rho(a|s)} \right],$$

Equation (14) may be expressed in compact form: $T_{\pi,\psi}F = g_{\pi,\psi} + \gamma P_{\pi,\psi}F$. By applying the mapping $T_{\pi,\psi}$ an infinite number of times on an initial free-energy vector F, the free-energy vector $F_{\pi,\psi}$ of the policy-belief-pair (π, ψ) is obtained:

$$F_{\pi,\psi} := \lim_{i\to\infty} T_{\pi,\psi}^i F = \lim_{i\to\infty}\sum_{t=0}^{i-1}\gamma^t P_{\pi,\psi}^t g_{\pi,\psi} + \underbrace{\lim_{i\to\infty}\gamma^i P_{\pi,\psi}^i F}_{\to 0},$$

which does no longer depend on the initial F. It is straightforward to show that the quantity $F_{\pi,\psi}$ is a fixed point of the operator $T_{\pi,\psi}$:

$$T_{\pi,\psi}F_{\pi,\psi} = g_{\pi,\psi} + \gamma P_{\pi,\psi}\lim_{i\to\infty}\sum_{t=0}^{i-1}\gamma^t P_{\pi,\psi}^t g_{\pi,\psi}$$

$$= \gamma^0 P_{\pi,\psi}^0 g_{\pi,\psi} + \lim_{i\to\infty}\sum_{t=1}^{i}\gamma^t P_{\pi,\psi}^t g_{\pi,\psi}$$

$$= \lim_{i\to\infty}\sum_{t=0}^{i-1}\gamma^t P_{\pi,\psi}^t g_{\pi,\psi} + \underbrace{\lim_{i\to\infty}\gamma^i P_{\pi,\psi}^i g_{\pi,\psi}}_{\to 0} = F_{\pi,\psi}.$$

Furthermore, $F_{\pi,\psi}$ is unique. Assume for this purpose an arbitrary fixed point F' such that $T_{\pi,\psi}F' = F'$, then $F' = \lim_{i\to\infty} T_{\pi,\psi}^i F' = F_{\pi,\psi}$.

Proposition 2. *The optimal free-energy vector* $F^* = \max_\pi \text{ext}_\psi F_{\pi,\psi}$ *is a unique fixed point of Bellman's equation* $F^* = BF^*$.

Proof. The proof consists of two parts where we assume ext = max in the first part and ext = min in the second part respectively. Let ext = max and $F^* = F_{\pi^*,\psi^*}$, where (π^*, ψ^*) denotes the optimal policy-belief-pair. Then

$$F^* = T_{\pi^*,\psi^*}F^* \leq \underbrace{\max_\pi \max_\psi T_{\pi,\psi}F^*}_{=BF^*} =: T_{\pi',\psi'}F^* \overset{\text{Induction}}{\leq} F_{\pi',\psi'},$$

where the last inequality can be straightforwardly proven by induction[1] and exploiting the fact that $P_{\pi,\psi}(s,s') \in [0;1]$. But by definition $F^* = \max_\pi \max_\psi F_{\pi,\psi} \geq F_{\pi',\psi'}$, hence $F^* = F_{\pi',\psi'}$ and therefore $F^* = BF^*$. Furthermore, F^* is unique. Assume for this purpose an arbitrary fixed point $F' = F_{\pi',\psi'}$ such that $F' = BF'$ with the corresponding policy-belief-pair (π', ψ'). Then

$$F^* = T_{\pi^*,\psi^*}F^* \geq T_{\pi',\psi'}F^* \overset{\text{Induction}}{\geq} F_{\pi',\psi'} = F',$$

and similarly $F' \geq F^*$, hence $F' = F^*$.

Let ext = min and $F^* = F_{\pi^*,\psi^*}$. By taking a closer look at Eq. (13), it can be seen that the optimization over ψ does not depend on π. Then

$$F^* = T_{\pi^*,\psi^*}F^* \geq \min_\psi T_{\pi^*,\psi}F^* =: T_{\pi^*,\psi'}F^* \overset{\text{Induction}}{\geq} F_{\pi^*,\psi'}.$$

But by definition $F^* = \min_\psi F_{\pi^*,\psi} \leq F_{\pi^*,\psi'}$, hence $F^* = F_{\pi^*,\psi'}$. Therefore it holds that $BF^* = \max_\pi \min_\psi T_{\pi,\psi}F^* = \max_\pi T_{\pi,\psi^*}F^*$ and similar to the first part of the proof we obtain

$$F^* = T_{\pi^*,\psi^*}F^* \leq \underbrace{\max_\pi T_{\pi,\psi^*}F^*}_{=BF^*} =: T_{\pi',\psi^*}F^* \overset{\text{Induction}}{\leq} F_{\pi',\psi^*}.$$

But by definition $F^* = \max_\pi F_{\pi,\psi^*} \geq F_{\pi',\psi^*}$, hence $F^* = F_{\pi',\psi^*}$ and therefore $F^* = BF^*$. Furthermore, F_{π^*,ψ^*} is unique. Assume for this purpose an arbitrary fixed point $F' = F_{\pi',\psi'}$ such that $F' = BF'$. Then

$$F' = T_{\pi',\psi'}F' \leq T_{\pi',\psi^*}F' \overset{\text{Induction}}{\leq} F_{\pi',\psi^*} \overset{\text{Induction}}{\leq} T_{\pi',\psi^*}F^* \leq T_{\pi^*,\psi^*}F^* = F^*,$$

and similarly $F^* \leq F'$, hence $F^* = F'$.

Theorem 2. *Let ϵ be a positive number satisfying $\epsilon < \frac{\eta}{1-\gamma}$ where $\gamma \in [0;1)$ is the discount factor and where u and l are the bounds of the reward function $R_{s,a}^{s'}$ such*

---

[1] Base case: $T_{\pi,\psi}F \leq F$. Inductive step: assume $T_{\pi,\psi}^i F \leq T_{\pi,\psi}^{i-1}F$ then $T_{\pi,\psi}^{i+1}F = g_{\pi,\psi} + \gamma P_{\pi,\psi}T_{\pi,\psi}^i F \leq g_{\pi,\psi} + \gamma P_{\pi,\psi}T_{\pi,\psi}^{i-1}F = T_{\pi,\psi}^i F$ and similarly for the base case $T_{\pi,\psi}F \geq F$ □.

that $l \leq R^{s'}_{s,a} \leq u$ and $\eta = \max\{|u|, |l|\}$. *Suppose that the value iteration scheme from Algorithm 1 is run for* $i = \lceil \log_\gamma \frac{\epsilon(1-\gamma)}{\eta} \rceil$ *iterations with an initial free-energy vector* $F(s) = 0$ *for all* s. *Then, it holds that* $\max_s |F^*(s) - B^i F(s)| \leq \epsilon$, *where* F^* *refers to the unique fixed point from Theorem 1.*

Proof. We start the proof by showing that the L_∞-norm of the difference vector between the optimal free-energy F^* and $B^i F$ exponentially decreases with the number of iterations i:

$$\max_s |F^*(s) - B^i F(s)| =: |F^*(s^*) - B^i F(s^*)|$$

$$\overset{\text{Eq. (9)}}{=} \left| \max_\pi \mathbb{E}_{\pi(a|s^*)} \left[\frac{1}{\beta} \log Z_\beta(a, s^*) - \frac{1}{\alpha} \log \frac{\pi(a|s^*)}{\rho(a|s^*)} \right] \right.$$
$$\left. - \max_\pi \mathbb{E}_{\pi(a|s^*)} \left[\frac{1}{\beta} \log Z^i_\beta(a, s^*) - \frac{1}{\alpha} \log \frac{\pi(a|s^*)}{\rho(a|s^*)} \right] \right|$$

$$\leq \max_\pi \left| \mathbb{E}_{\pi(a|s^*)} \left[\frac{1}{\beta} \log Z_\beta(a, s^*) - \frac{1}{\beta} \log Z^i_\beta(a, s^*) \right] \right|$$

$$\leq \max_a \left| \frac{1}{\beta} \log Z_\beta(a, s^*) - \frac{1}{\beta} \log Z^i_\beta(a, s^*) \right|$$

$$=: \left| \frac{1}{\beta} \log Z_\beta(a^*, s^*) - \frac{1}{\beta} \log Z^i_\beta(a^*, s^*) \right|$$

$$\overset{\text{Eq. (11)}}{=} \left| \underset{\psi}{\text{ext}} \, \mathbb{E}_{\psi(\theta|a^*,s^*)} \left[\mathbb{E}_{T_\theta(s'|a^*,s^*)} \left[R^{s'}_{s,a} + \gamma F^*(s') \right] - \frac{1}{\beta} \log \frac{\psi(\theta|a^*,s^*)}{\mu(\theta|a^*,s^*)} \right] \right.$$
$$\left. - \underset{\psi}{\text{ext}} \, \mathbb{E}_{\psi(\theta|a^*,s^*)} \left[\mathbb{E}_{T_\theta(s'|a^*,s^*)} \left[R^{s'}_{s,a} + \gamma B^{i-1} F(s') \right] - \frac{1}{\beta} \log \frac{\psi(\theta|a^*,s^*)}{\mu(\theta|a^*,s^*)} \right] \right|$$

$$\leq \max_\psi \left| \mathbb{E}_{\psi(\theta|a^*,s^*)} \left[\mathbb{E}_{T_\theta(s'|a^*,s^*)} \left[\gamma F^*(s') - \gamma B^{i-1} F(s') \right] \right] \right|$$

$$\leq \gamma \max_s |F^*(s) - B^{i-1} F(s)| \overset{\text{Recur.}}{\leq} \gamma^i \max_s |F^*(s) - F(s)| \leq \gamma^i \frac{\eta}{1-\gamma},$$

where we exploit the fact that $|\text{ext}_x f(x) - \text{ext}_x g(x)| \leq \max_x |f(x) - g(x)|$ and that the free-energy is bounded through the reward bounds l and u with $\eta = \max\{|u|, |l|\}$. For a convergence criterion $\epsilon > 0$ such that $\epsilon \geq \gamma^i \frac{\eta}{1-\gamma}$, it then holds that $i \geq \log_\gamma \frac{\epsilon(1-\gamma)}{\eta}$ presupposing that $\epsilon < \frac{\eta}{1-\gamma}$.

5   Experiments: Grid World

This section illustrates the proposed value iteration scheme with an intuitive example where an agent has to navigate through a grid-world. The agent starts at position $\mathbf{S} \in \mathcal{S}$ with the objective to reach the goal state $\mathbf{G} \in \mathcal{S}$ and can choose one out of maximally four possible actions $a \in \{\uparrow, \rightarrow, \downarrow, \leftarrow\}$ in each time-step. Along the way, the agent can encounter regular tiles (actions move the agent deterministically one step in the desired direction), walls that are represented as *gray tiles* (actions that move the agent towards the wall are not possible), holes that are represented as *black tiles* (moving into the hole causes a negative reward) and *chance tiles* that are illustrated as white tiles with a question mark (the transition probabilities of the chance tiles are unknown to the agent). Reaching the goal **G** yields a reward $R = +1$ whereas stepping into a hole results in a negative reward $R = -1$. In both cases the agent is subsequently teleported back to the starting position **S**. Transitions to regular tiles have a small negative reward of $R = -0.01$. When stepping onto a chance tile, the agent is pushed stochastically to an adjacent tile giving a reward as mentioned above. The true state-transition probabilities of the chance tiles are not known by the agent, but the agent holds the Bayesian belief

$$\mu(\boldsymbol{\theta}_{s,a}|a, s) = \mathsf{Dirichlet}\big(\Phi^{s_1'}_{s,a}, \ldots, \Phi^{s_{N(s)}'}_{s,a}\big) = \prod_{i=1}^{N(s)} (\theta^{s_i'}_{s,a})^{\Phi^{s_i'}_{s,a}-1}$$

where the transition model is denoted as $T_{\boldsymbol{\theta}_{s,a}}(s'|s, a) = \theta^{s'}_{s,a}$ and $\boldsymbol{\theta}_{s,a} = (\theta^{s_1'}_{s,a} \ldots \theta^{s_{N(s)}'}_{s,a})$ and $N(s)$ is the number of possible actions in state s. The data are incorporated into the model as a count vector $(\Phi^{s_1'}_{s,a}, \ldots, \Phi^{s_{N(s)}'}_{s,a})$ where $\Phi^{s'}_{s,a}$ represents the number of times that transition (s, a, s') occurred. The prior $\rho(a|s)$ for the actions at every state is set to be uniform. An important aspect of the model is that in the case of unlimited observational data, the agent will plan with the correct transition probabilities.

We conducted two experiments with discount factor $\gamma = 0.9$ and uniform priors $\rho(a|s)$ for the action variables. In the first experiment, we explore and illustrate the agent's planning behavior under different degrees of computational limitations (by varying α) and under different model uncertainty attitudes (by varying β) with fixed uniform beliefs $\mu(\theta|a, s)$. In the second experiment, the agent is allowed to update its beliefs $\mu(\theta|a, s)$ and use the updated model to re-plan its strategy.

5.1   The Role of the Parameters α and β on Planning

Figure 1 shows the solution to the variational free energy problem that is obtained by iteration until convergence according to Algorithm 1 under different values of α and β. In particular, the first row shows the free energy function $F^*(s)$ (Eq. (8)). The second, third and fourth row show heat maps of the position of an agent that follows the optimal policy (Eq. (12)) according to the agent's biased beliefs (plan) and to the actual transition probabilities in a friendly and unfriendly environment, respectively. In chance tiles, the most likely transitions in these two environments are indicated by arrows where the agent is teleported with a probability of 0.999 into the tile indicated by the arrow and with a probability of 0.001 to a random other adjacent tile.

In the first column of Fig. 1 it can be seen that a stochastic agent ($\alpha = 3.0$) with high model uncertainty and optimistic attitude ($\beta = 400$) has a strong

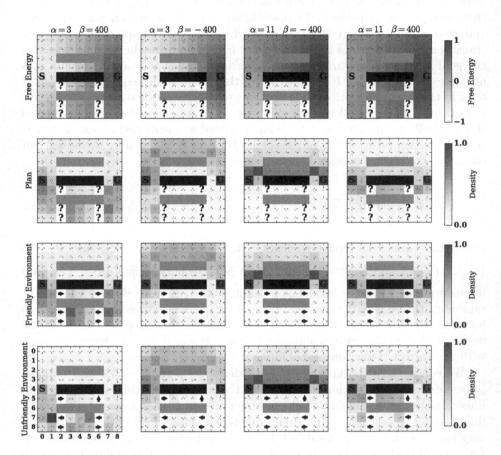

Fig. 1. The four different rows show free energy values and heat-maps of planned trajectories according to the agent's beliefs over state-transitions in chance tiles, heat-maps of real trajectories in a friendly environment and in an unfriendly environment respectively. The Start-position is indicated by **S** and the goal state is indicated by **G**. Black tiles represent holes with negative reward, gray tiles represent walls and chance tiles with a question mark have transition probabilities unknown to the agent. The white tiles with an arrow represent the most probable state-transition in chance tiles (as specified by the environment). Very small arrows in each cell encode the policy $\pi(a|s)$ (the length of each arrow encodes the probability of the corresponding action under the policy, highest probability action is indicated as a red arrow). The heat map is constructed by normalizing the number of visits for each state over 20000 steps, where actions are sampled from the agent's policy and state-transitions are sampled according to one of three ways: in the second row according to the agent's belief over state-transitions $\psi(\theta|a,s)$, in the third and fourth row according to the actual transition probabilities of a friendly and an unfriendly environment respectively. Different columns show different α and β cases. (Color figure online)

preference for the broad corridor in the bottom by assuming favorable transitions for the unknown chance tiles. This way the agent also avoids the narrow corridors that are unsafe due to the stochasticity of the low-α policy. In the second column of Fig. 1 with low $\alpha = 3$ and high model uncertainty with pessimistic attitude $\beta = -400$, the agent strongly prefers the upper broad corridor because unfavorable transitions are assumed for the chance tiles. The third column of Fig. 1 shows a very pessimistic agent ($\beta = -400$) with high precision ($\alpha = 11$) that allows the agent to safely choose the shortest distance by selecting the upper narrow corridor without risking any tiles with unknown transitions. The fourth column of Fig. 1 shows a very optimistic agent ($\beta = 400$) with high precision. In this case the agent chooses the shortest distance by selecting the bottom narrow corridor that includes two chance tiles with unknown transition.

5.2 Updating the Bayesian Posterior μ with Observations from the Environment

Similar to model identification adaptive controllers that perform system identification while the system is running [1], we can use the proposed planning algorithm also in a reinforcement learning setup by updating the Bayesian beliefs about the MDP while executing always the first action and replanning in the next time step. During the learning phase, the exploration is governed by both factors α and β, but each factor has a different influence. In particular, lower α-values will cause more exploration due to the inherent stochasticity in the agent's action selection, similar to an ϵ-greedy policy. If α is kept fixed through time, this will of course also imply a "suboptimal" (i.e. bounded optimal) policy in the long run. In contrast, the parameter β governs exploration of states with unknown transition-probabilities more directly and will not have an impact on the agent's performance in the limit, where data has eliminated model uncertainty. We illustrate this with simulations in a grid-world environment where the agent is allowed to update its beliefs $\mu(\theta|a, s)$ over the state-transitions every time it enters a chance tile and receives observation data acquired through interaction with the environment—compare left panels in Fig. 2. In each step, the agent can then use the updated belief-models for planning the next action.

Figure 2 (right panels) shows the number of data points acquired (each time a chance tile is visited) and the average reward depending on the number of steps that the agent has interacted with the environment. The panels show several different cases: while keeping $\alpha = 12.0$ fixed we test $\beta = (0.2, 5.0, 20.0)$ and while keeping $\beta = 0.2$ fixed we test $\alpha = (5.0, 8.0, 12.0)$. It can be seen that lower α leads to better exploration, but it can also lead to lower performance in the long run—see for example the rightmost bottom panel. In contrast, optimistic β values can also induce high levels of exploration with the added advantage that in the limit no performance detriment is introduced. However, high β values can in general also lead to a detrimental persistence with bad policies, as can be seen for example in the superiority of the low-β agent at the very beginning of the learning process.

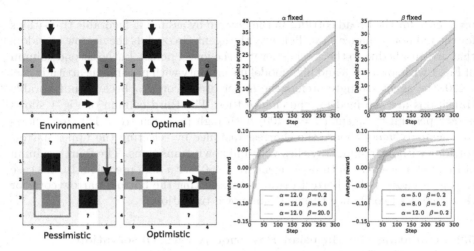

Fig. 2. The effect of α and β when updating beliefs over 300 interaction steps with the environment. The four panels on the left show the grid-world environment and the pertaining optimal policy if the environment is known. The lower left panels show paths that the agent could take depending on its attitude towards model uncertainty. The panels on the right show the number of acquired data points, that is the number of times a chance tile is entered, and the average reward (bottom panels) for fixed α (varying β) or fixed β (varying α). The average reward at each step is computed as follows. Each time the agent observes a state-transition in a chance tile and updates its belief model, 10 runs of length 2000 steps are sampled (using the agent's current belief model). The average reward (bold lines) and standard-deviation (shaded areas) across these 10 runs are shown in the figure.

6   Discussion and Conclusions

In this paper we are bringing two strands of research together, namely research on information-theoretic principles of control and decision-making and robustness principles for planning under model uncertainty. We have devised a unified recursion principle that extends previous generalizations of Bellman's optimality equation and we have shown how to solve this recursion with an iterative scheme that is guaranteed to converge to a unique optimum. In simulations we could demonstrate how such a combination of information-theoretic policy and belief constraints that reflect model uncertainty can be beneficial for agents that act in partially unknown environments.

Most of the research on robust MDPs does not consider information-processing constraints on the policy, but only considers the uncertainty in the transition probabilities by specifying a set of permissible models such that worst-case scenarios can be computed in order to obtain a robust policy [13,16]. Recent extensions of these approaches include more general assumptions regarding the set properties of the permissible models and assumptions regarding the data generation process [33]. Our approach falls inside this class of robustness methods that use a restricted set of permissible models, because we extremize the biased belief $\psi(\theta|a, s)$ under the constraint that it has to be within some information bounds measured by the Kullback-Leibler divergence from a reference Bayesian posterior. Contrary to these previous methods, our approach additionally considers robustness arising from the stochasticity in the policy.

Information-processing constraints on the policy in MDPs have been previously considered in a number of studies [14, 23, 25, 32], however not in the context of model uncertainty. In these studies a free energy value recursion is derived when restricting the class of policies through the Kullback-Leibler divergence and when disregarding separate information-processing constraints on observations. However, a small number of studies has considered information-processing constraints both for actions and observations. For example, Polani and Tishby [30] and Ortega and Braun [19] combine both kinds of information costs. The first cost formalizes an information-processing cost in the policy and the second cost constrains uncertainty arising from the state transitions directly (but crucially not the uncertainty in the latent variables). In both information-processing constraints the cost is determined as a Kullback-Leibler divergence with respect to a reference distribution. Specifically, the reference distribution in [30] is given by the marginal distributions (which is equivalent to a rate distortion problem) and in [19] is given by fixed priors. The Kullback-Leibler divergence costs for the observations in these cases essentially correspond to a risk-sensitive objective. While there is a relation between risk-sensitive and robust MDPs [6, 22, 26], the innovation in our approach is at least twofold. First, it allows combining information-processing constraints on the policy with model uncertainty (as formalized by a latent variable). Second, it provides a natural setup to study learning.

The algorithm presented here and Bayesian models in general [7] are computationally expensive as they have to compute possibly high-dimensional integrals depending on the number of allowed transitions for action-state pairs. Nevertheless, there have been tremendous efforts in solving unknown MDPs efficiently, especially by sampling methods [10, 11, 24]. An interesting future direction to extend our methodology would therefore be to develop a sampling-based version of Algorithm 1 to increase the range of applicability and scalability [21]. Moreover, such sampling methods might allow for reinforcement learning applications, for example by estimating free energies through TD-learning [8], or by Thompson sampling approaches [17, 18] or other stochastic methods for adaptive control [1].

Acknowledgments. This study was supported by the DFG, Emmy Noether grant BR4164/1-1. The code was developed on top of the RLPy library [9].

References

1. Åström, K.J., Wittenmark, B.: Adaptive control. Courier Corporation, Mineola (2013)
2. Bellman, R.: Dynamic Programming, 1st edn. Princeton University Press, Princeton (1957). http://books.google.com/books?id=fyVtp3EMxasC&pg=PR5& dq=dynamic+programming+richard+e+bellman&client=firefox-a#v=onepage& q=dynamic%20programming%20richard%20e%20bellman&f=false

3. Bertsekas, D., Tsitsiklis, J.: Neuro-Dynamic Programming. Athena Scientific, Belmont (1996)
4. Braun, D.A., Ortega, P.A., Theodorou, E., Schaal, S.: Path integral control and bounded rationality. In: 2011 IEEE Symposium on Adaptive Dynamic Programming And Reinforcement Learning (ADPRL), pp. 202–209. IEEE (2011)
5. van den Broek, B., Wiegerinck, W., Kappen, H.J.: Risk sensitive path integral control. In: UAI (2010)
6. Chow, Y., Tamar, A., Mannor, S., Pavone, M.: Risk-sensitive and robust decision-making: a CVaR optimization approach. In: Advances in Neural Information Processing Systems, pp. 1522–1530 (2015)
7. Duff, M.O.: Optimal learning: computational procedures for Bayes-adaptive Markov decision processes. Ph.d. thesis, University of Massachusetts Amherst (2002)
8. Fox, R., Pakman, A., Tishby, N.: G-learning: Taming the noise in reinforcement learning via soft updates. arXiv preprint (2015). arXiv:1512.08562
9. Geramifard, A., Dann, C., Klein, R.H., Dabney, W., How, J.P.: Rlpy: a value-function-based reinforcement learning framework for education and research. J. Mach. Learn. Res. **16**, 1573–1578 (2015)
10. Guez, A., Silver, D., Dayan, P.: Efficient Bayes-adaptive reinforcement learning using sample-based search. In: Advances in Neural Information Processing Systems, pp. 1025–1033 (2012)
11. Guez, A., Silver, D., Dayan, P.: Scalable and efficient Bayes-adaptive reinforcement learning based on Monte-Carlo tree search. J. Artif. Intell. Res. **48**, 841–883 (2013)
12. Hansen, L.P., Sargent, T.J.: Robustness. Princeton University Press, Princeton (2008)
13. Iyengar, G.N.: Robust dynamic programming. Math. Oper. Res. **30**(2), 257–280 (2005)
14. Kappen, H.J.: Linear theory for control of nonlinear stochastic systems. Phys. Rev. Lett. **95**(20), 200201 (2005)
15. Mannor, S., Simester, D., Sun, P., Tsitsiklis, J.N.: Bias and variance approximation in value function estimates. Manag. Sci. **53**(2), 308–322 (2007)
16. Nilim, A., El Ghaoui, L.: Robust control of Markov decision processes with uncertain transition matrices. Oper. Res. **53**(5), 780–798 (2005)
17. Ortega, P.A., Braun, D.A.: A Bayesian rule for adaptive control based on causal interventions. In: 3rd Conference on Artificial General Intelligence (AGI-2010), Atlantis Press (2010)
18. Ortega, P.A., Braun, D.A.: A minimum relative entropy principle for learning and acting. J. Artif. Intell. Res. **38**(11), 475–511 (2010)
19. Ortega, P.A., Braun, D.A.: Thermodynamics as a theory of decision-making with information-processing costs. Proc. R. Soc. A. **469**, 20120683 (2013). The Royal Society
20. Ortega, P.A., Braun, D.A.: Generalized Thompson sampling for sequential decision-making and causal inference. Complex Adapt. Syst. Model. **2**(1), 2 (2014)
21. Ortega, P.A., Braun, D.A., Tishby, N.: Monte Carlo methods for exact & efficient solution of the generalized optimality equations. In: 2014 IEEE International Conference on Robotics and Automation (ICRA), pp. 4322–4327. IEEE (2014)
22. Osogami, T.: Robustness and risk-sensitivity in Markov decision processes. In: Advances in Neural Information Processing Systems, pp. 233–241 (2012)
23. Peters, J., Mülling, K., Altun, Y., Poole, F.D., et al.: Relative entropy policy search. In: Twenty-Fourth National Conference on Artificial Intelligence (AAAI-10), pp. 1607–1612. AAAI Press (2010)

24. Ross, S., Pineau, J., Chaib-draa, B., Kreitmann, P.: A Bayesian approach for learning and planning in partially observable Markov decision processes. J. Mach. Learn. Res. **12**, 1729–1770 (2011)
25. Rubin, J., Shamir, O., Tishby, N.: Trading value and information in MDPs. In: Guy, T.V., Kárný, M., Wolpert, D.H. (eds.) Decision Making with Imperfect Decision Makers. Intelligent Systems Reference Library, vol. 28, pp. 57–74. Springer, Heidelberg (2012)
26. Shen, Y., Tobia, M.J., Sommer, T., Obermayer, K.: Risk-sensitive reinforcement learning. Neural Comput. **26**(7), 1298–1328 (2014)
27. Strehl, A.L., Li, L., Littman, M.L.: Reinforcement learning in finite MDPs: Pac analysis. J. Mach. Learn. Res. **10**, 2413–2444 (2009)
28. Szita, I., Lőrincz, A.: The many faces of optimism: a unifying approach. In: Proceedings of the 25th International Conference on Machine Learning, pp. 1048–1055. ACM (2008)
29. Szita, I., Szepesvári, C.: Model-based reinforcement learning with nearly tight exploration complexity bounds. In: Proceedings of the 27th International Conference on Machine Learning (ICML-10), pp. 1031–1038 (2010)
30. Tishby, N., Polani, D.: Information theory of decisions and actions. In: Cutsuridis, V., Hussain, A., Taylor, J.G. (eds.) Perception-Action Cycle. Springer Series in Cognitive and Neural Systems, pp. 601–636. Springer, New York (2011)
31. Todorov, E.: Linearly-solvable Markov decision problems. In: Advances in Neural Information Processing Systems, pp. 1369–1376 (2006)
32. Todorov, E.: Efficient computation of optimal actions. Proc. Nat. Acad. Sci. **106**(28), 11478–11483 (2009)
33. Wiesemann, W., Kuhn, D., Rustem, B.: Robust Markov decision processes. Math. Oper. Res. **38**(1), 153–183 (2013)

Subgroup Discovery with Proper Scoring Rules

Hao Song[1]($\boxtimes$), Meelis Kull[1], Peter Flach[1], and Georgios Kalogridis[2]

[1] Intelligent Systems Laboratory, University of Bristol, Bristol, UK
{Hao.Song,Meelis.Kull,Peter.Flach}@bristol.ac.uk
[2] Toshiba Research Europe Ltd.,
Telecommunications Research Laboratory, Bristol, UK
George@toshiba-trel.com

Abstract. Subgroup Discovery is the process of finding and describing sufficiently large subsets of a given population that have unusual distributional characteristics with regard to some target attribute. Such subgroups can be used as a statistical summary which improves on the default summary of stating the overall distribution in the population. A natural way to evaluate such summaries is to quantify the difference between predicted and empirical distribution of the target. In this paper we propose to use proper scoring rules, a well-known family of evaluation measures for assessing the goodness of probability estimators, to obtain theoretically well-founded evaluation measures for subgroup discovery. From this perspective, one subgroup is better than another if it has lower divergence of target probability estimates from the actual labels on average. We demonstrate empirically on both synthetic and real-world data that this leads to higher quality statistical summaries than the existing methods based on measures such as Weighted Relative Accuracy.

1 Introduction

Statistical models intend to capture the distributional information in a domain of interest. While a global statistical model is useful, it is often also of interest to capture local variations exhibited in a subset of the data. Recognising such subsets can provide valuable knowledge and opportunities to improve performance at tasks relying on the statistical model. In the area of machine learning and data mining, the problem of obtaining such statistically different subsets is known as Subgroup Discovery (SD) [6,7,10,17], loosely defined as the process of finding and describing sufficiently large subsets of a given population that have unusual distributional characteristics with regard to some target attribute.

Consider a synthetic toy data set relating to someone's dietary habits. It contains two (discretised) features: the time of the day, denoted as $X_1 \in \{Morning, Afternoon, Evening\}$ and the calorie consumption in the diet, denoted as $X_2 \in \{Low, Medium, High\}$. The target variable is $Y \in \{Weekday, Weekend\}$. Figure 1 visualises the data, with two potentially interesting subgroups (shaded areas). The subgroup on the right concentrates on the area of maximum statistical deviation (high calorie intake in the evening is more common during weekend), while the one on the left covers both medium and high calorie intake

© Springer International Publishing AG 2016
P. Frasconi et al. (Eds.): ECML PKDD 2016, Part II, LNAI 9852, pp. 492–510, 2016.
DOI: 10.1007/978-3-319-46227-1_31

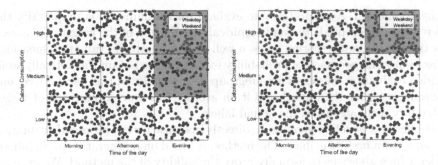

Fig. 1. An example bivariate data set with two subgroups (shaded areas) defined on the discretised features, both capturing an area of statistical deviation in comparison to the overall population. The subgroup on the left is preferred by a commonly used evaluation measure (WRAcc) while the right subgroup is preferred by the one of the measures we propose in this paper.

in the evening. In this paper we study reasons why one of these subgroups might be preferred over the other.

Clearly, if a subgroup is small, distributional differences may arise purely because of random chance in sampling, so a trade-off between subgroup size and distributional deviation needs to be made. Statistical tests such as χ^2 can be used, but are usually over-emphasising size: a very large subgroup with small deviation is more likely to be picked up than a medium-sized subgroup with considerable deviation. p-values as reported in rule-based approaches [10] tend to suffer from the same issue.

Historically, SD developed as a variation on rule-learning and other logic-based approaches, and hence it is not surprising that many existing quality measures have been adapted from decision trees and rule-based classifiers. For instance, [1] explored the use of Gini-split (among several others) as quality measure for subgroups, which hypothesises that a good binary split in a decision tree also establishes a good subgroup. One of the most commonly used measures is Weighted Relative Accuracy (WRAcc), which can be seen as an adaptation of precision, a measure that is used as a search heuristic in rule learners such as CN2 [3]. Many other subgroup quality measures have been introduced in the literature, see [6] for an overview.

Evaluation methods for SD depend on the task for which subgroups need to be found. In [10], the subgroups are used to construct a ranking model, and the area under the corresponding ROC curve is used as an evaluation measure. In [1] the obtained subgroups are used as features for a decision tree and hence they can be evaluated according to the classification performance of the trees. However, the predictive task used in evaluation (ranking or classification) is then different from the descriptive Subgroup Discovery (SD) task, and it is unclear how the predictive task affects the choice of subgroup quality measure.

In this paper we propose a novel approach to evaluate subgroups as summaries which improve on the default summary of stating the overall distribution in the population. A natural way to evaluate such summaries is to quantify the difference between predicted and empirical distribution of the target. This obviates the use of proper scoring rules, a well-known family of evaluation measures for assessing the goodness of probability estimators, to obtain theoretically well-founded evaluation measures for subgroup discovery. From this perspective, one subgroup is better than another if it on average has lower divergence of target probability estimates from the actual labels.

We derive a novel SD method to directly optimise for the proposed evaluation measure, from first principles. The method is based on a generative probabilistic model, which allows us to formally prove the validity of the method. We perform experiments on a synthetic data set where the theoretically optimal subgroup is known, and demonstrate that our method outperforms alternative methods in the sense that it finds subgroups that are closer to the theoretically optimal one. Additionally, we perform experiments on 20 UCI data sets which demonstrate that the proposed method is superior in summarising the statistical properties of the data.

The structure of this paper is as follows. Section 2 introduces the notations and concepts for SD. Section 3 provides an overview of Proper Scoring Rules (PSRs) and describes related quality measures. In Sect. 4 we propose a novel generative modelling approach to address the summarisation problem, and derive the corresponding measures. Section 5 evaluates the proposed quality measures against existing measures and Sect. 6 presents related work. Section 7 concludes this paper and discusses possible future research directions.

2   Subgroup Discovery

We start by introducing some notation. Consider a dataset (X_i, Y_i), $i = 1, \ldots, n$ in the instance space $(\mathbb{X}, \mathbb{Y})$. We assume a multi-class target variable, representing the k classes in $\mathbb{Y}$ by unit vectors, i.e. class j is represented by the vector with 1 at position j and 0 everywhere else. The set of all considered subgroups is indicated by $\mathbb{G} \subset 2^{\mathbb{X}}$. This set is typically generated by a *subgroup language* (e.g., the set of all conjunctions over some fixed set of literals) but here it suffices to deal with subgroups extensionally. A subgroup $g \in \mathbb{G}$ can then be identified with its characteristic function $g : \mathbb{X} \to \{0, 1\}$ determining whether an instance X_i is in the subgroup $(g(X_i) = 1)$ or not $(g(X_i) = 0)$. A subgroup quality measure is a function $\phi : \mathbb{G} \to \mathbb{R}$ such that better subgroups g get a higher $\phi(g)$. The task of SD is then to find the subgroup g^* with the highest value of ϕ, i.e. $g^* = \arg\max_{g \in \mathbb{G}} \phi(g)$.

A wide range of proposed quality measures can be found in the literature. The common way of defining a quality measure is to separate them into two factors: the deviation factor and the size factor. The deviation factor is in charge of comparing the local statistic to the global statistic. In the case of a discrete target variable, the deviation factor can be seen as a function that takes two estimates of class probabilities as input and outputs a single number to indicate how different these two estimates are. The size factor is normally treated as a correction term to encourage the method to find larger subgroups, as small subgroups tend to be less valuable.

One of the most widely adopted quality measures is the Weighted Relative Accuracy (WRAcc) family [1,2,9,10]. For a binary target this essentially is the covariance between the target variable and subgroup membership: since these are both Bernoulli variables this takes values in the interval $[-0.25, 0.25]$. For a multi-class target we take the average of all one-against-rest binary WRAcc values, taking the absolute value of the latter to avoid positive and negative covariances cancelling out [1]. For our purposes we derive a related but unnormalised quantity, as follows.

Denote the overall class distribution in the data set by $\pi = (\sum_{i=1}^{n} Y_i)/n$ (note that Y_i and π are vectors of length k). Let m denote the number of training set instances belonging to the subgroup g, i.e. $m = \sum_{i=1}^{n} g(X_i)$. Denote the class distribution in the subgroup by $\rho^{(g)}$, i.e., $\rho^{(g)} = (\sum_{i=1}^{n} g(X_i) \cdot Y_i)/m$. Then an unnormalised version of Multi-class Weighted Relative Accuracy (MWRAcc) can be calculated as:

$$\phi_{MWRAcc}(g) = m \cdot \sum_{j=1}^{k} |\rho_j^{(g)} - \pi_j| \tag{1}$$

The definition of [1] is obtained from this by normalising with $n \cdot k$, where n is the number of training instances and k is the number of classes (both constant). Our version can be interpreted as absolute differences between observed and expected counts.

3   Proper Scoring Rules

The class distribution π is a very simple way to summarise the target variable across the whole training dataset. That is, we summarise the labels vectors $Y_1, \ldots, Y_n$ with the summary S^π where we define $S_i^\pi = \pi$ for $i = 1, \ldots, n$. Another possibility is to separately summarise a particular subgroup g with its class distribution $\rho^{(g)}$ while its complement is summarised with π. We denote this summary by $S^{g, \rho^{(g)}, \pi}$, and for an instance i this summary predicts $S_i^{g, \rho^{(g)}, \pi} = \rho^{(g)}$ if $g(X_i) = 1$ and $S_i^{g, \rho^{(g)}, \pi} = \pi$ if $g(X_i) = 0$, which can be jointly written as $S_i^{g, \rho^{(g)}, \pi} = \rho^{(g)} g(X_i) + \pi(1 - g(X_i))$. One could then ask which of the subgroups gives the best summary, and whether the summary is better than the default summary S^π. In order to assess this, we need a way to calculate the extent to which the probability estimates within the summary deviate from the actual labels.

Proper Scoring Rules (PSRs) have been widely adopted in the area of machine learning and statistics to assess the goodness of probability estimates [16]. A scoring rule is a function $\psi : \mathbb{S} \times \mathbb{Y} \to \mathbb{R}$ that assigns a real-valued loss to the estimate S_i within the summary S with respect to the actual label Y_i of instance i. Two of the most commonly adopted scoring rules are the Brier Score (BS) and Log-loss (LL), which are defined as:

$$\psi_{BS}(S_i, Y_i) = \sum_{j=1}^{k}(S_{i,j} - Y_{i,j})^2 \tag{2}$$

$$\psi_{LL}(S_i, Y_i) = -\log(S_{i,*}) \tag{3}$$

where $Y_{i,j} = 1$ if the i-th instance is of the j-th class and 0 otherwise, $S_{i,j}$ is the probability estimate of class j for the i-th instance, and $S_{i,*}$ denotes the probability estimate of the i-th instance for the true class as determined by Y_i.

The distance from a whole summary S to the actual labels can then be calculated as follows:

$$\psi'(S, Y) = \sum_{i=1}^{n} \psi(S_i, Y_i) \tag{4}$$

The scoring rule ψ is *proper* if $\arg\min_p \psi'(S^p, Y) = \pi$ for any Y, i.e., if the actual class distribution is the minimiser of the scoring rule. In particular, both BS and LL are proper.

For every proper scoring rule ψ there is a corresponding divergence measure d which quantifies how much a class probability distribution diverges from another class distribution. Formally, the divergence $d(p, q)$ is the expected value of the difference $\psi(p, Y) - \psi(q, Y)$ where Y is drawn from the distribution q. The divergences corresponding to BS and LL are the squared error and Kullback-Leibler (KL) divergence, respectively.

$$d_{BS}(p, q) = \sum_{j=1}^{k}(p_j - q_j)^2 \tag{5}$$

$$d_{LL}(p, q) = \sum_{j=1}^{k} q_j \cdot \log\frac{q_j}{p_j} \tag{6}$$

For more details see [8].

3.1   Information Gain

Suppose we now want to decide whether to summarise the whole dataset by S^π or by $S^{g,\rho^{(g)},\pi}$ for some g. For this let us take a proper scoring rule ψ' to quantify the loss of a summary with respect to actual labels. We can now define the quality of a subgroup g as the gain in ψ' of the summary $S^{g,\rho^{(g)},\pi}$ over the default summary S^π, that is:

$$\phi_{IG}(g) = \psi'(S^\pi, Y) - \psi'(S^{g,\rho^{(g)},\pi}, Y) \tag{7}$$

In principle, we could consider summaries $S^{g,\rho,\pi}$ for any other class distribution ρ. However, the summary with $\rho^{(g)}$ is special among these, as it is maximising the gain over the summary S^π due to properness of the scoring rule. This is stated in the following theorem:

Theorem 1. *Let ψ, ψ', d be a proper scoring rule, its sum across the dataset, and its corresponding divergence measure, respectively. Then for any given subgroup g the following holds:*

$$\arg\max_{\rho} \psi'(S^\pi, Y) - \psi'(S^{g, \rho, \pi}, Y) = \rho^{(g)} \tag{8}$$

where $\rho^{(g)}$ denotes the class distribution within the subgroup g. The maximum value achieved is $m \cdot d(\pi, \rho^{(g)})$ where m is the size of the subgroup g.

Proofs of all theorems are provided in Appendix A.

The theorem implies that Eq. (7) can be rewritten as follows:

$$\phi_{IG}(g) = m \cdot d(\pi, \rho^{(g)}) \tag{9}$$

In words, this quality measure multiplies the size of the subgroup by the divergence of the overall class distribution from the distribution within the subgroup[1].

If we consider Log-loss as the proper scoring rule, then the corresponding information gain measure is:

$$\phi_{IG\text{-}LL}(g) = m \cdot KL(\pi, \rho^{(g)}) \tag{10}$$

where KL is the KL-divergence. For Brier Score the corresponding measure is quadratic error:

$$\phi_{IG\text{-}BS}(g) = m \cdot \sum_{j=1}^{k} (\pi_j - \rho_j^{(g)})^2 \tag{11}$$

where $\rho_j^{(g)}$ is the proportion of the j-th class in the subgroup g.

These information gain measures have a long history in machine learning, for example in decision tree learning where they measure the decrease in impurity when splitting a parent node into two children nodes. If we measure impurity by Shannon entropy this leads to Quinlan's information gain splitting criterion; and if we measure impurity by the Gini index we obtain Gini-split. We have shown how they can be unified from the perspective of Proper Scoring Rules; we now proceed to improve them.

4   Generative Modelling

The general context in which SD is applied is where one observes a set of data points that belongs to a particular domain and the task is to extract information from the data. As mentioned in the introduction, such information can then be adopted to improve the performance of corresponding applications. Therefore, it is desirable that the subgroups as the representation of obtained knowledge would generalise to future data observed in the same domain.

---

[1] In general, divergence measures are not symmetric, so $d(\pi, \rho^{(g)})$ is different from $d(\rho^{(g)}, \pi)$.

Two problems need addressing when generalising to future data. First, the class distribution $\rho^{(g)}$ is calculated on a (small) sample and can therefore be a poor estimate of the actual distribution in the future. Second, it is not certain whether the actual distribution of the subgroup is different from the overall distribution π. In order to capture these aspects we employ a generative model to generate a new *test* instance Y of the subgroup g. We assume that the observed (training) instances of subgroup g were generated according to the same model, which is defined as follows.

4.1   The Generative Model

First, we fix the default k-class distribution π. We then decide whether the distribution of the subgroup g is different from the default ($Z = 1$) or the same as default ($Z = 0$):

$$Z \sim Bernoulli[\gamma] \tag{12}$$

where γ is our prior belief that $Z = 1$. If $Z = 1$ then we generate the class distribution Q for the subgroup g:

$$Q \sim Dir[\beta] \tag{13}$$

where $Dir[\beta]$ is the k-dimensional Dirichlet distribution with parameter vector β. Finally, we assume that the test instance Y and the training instances of the subgroup g are all independent and identically distributed (iid). For simplicity of notation, let us assume that the training instances within g are the first m instances $Y_1, \ldots, Y_m$. The distribution of $Y_1, \ldots, Y_m$ and the test label Y is as follows:

$$Y, Y_1, \ldots, Y_m \sim Cat[ZQ + (1 - Z)\pi] \tag{14}$$

where Cat is the categorical distribution with the given class probabilities. In the experiments reported later we used non-informative priors for Z and Q ($\gamma = 0.5$ and $\beta = (1, \ldots, 1)$, respectively).

4.2   Proposed Quality Measures

The above model can be used to generate instances for a subgroup g. We will now exploit this model to derive two subgroup quality measures, the first one of which takes into account the uncertainty about the true class distribution in the subgroup, while the second one also models our uncertainty whether it is different from the background distribution. Therefore, we consider the task of choosing ρ which would maximise the expected gain in ψ' on the *test* instances. The following theorem solves this task, conditioning on the observed class distribution within the subgroup and on the assumption that this subgroup is different from background ($Z = 1$).

Theorem 2. *Consider a subgroup as generated with the model above. Denote the counts of each class in the training set of this subgroup by $C = \sum_{i=1}^{m} Y_i$. Then*

$$\arg\max_{\rho} \mathbb{E}[\psi'(\pi, Y) - \psi'(\rho, Y)|C = c, Z = 1] = \frac{c + \beta}{\sum_{j=1}^{k} c_j + \beta_j} \qquad (15)$$

Denoting this quantity by $\hat{\rho}$, the achieved maximum is $d(\pi, \hat{\rho})$, where d is the divergence measure corresponding to ψ.

In the experiments we use $\beta = (1, \ldots, 1)$ and hence the gain is maximised when predicting the Laplace-corrected probabilities, i.e., adding 1 to all counts and then normalising. According to this theorem we propose a novel quality measure which takes into account the uncertainty about the class distribution:

$$\phi_d(g) = m \cdot d(\pi, \hat{\rho}) \qquad (16)$$

where m is the size of the subgroup.

The following theorem differs from the previous theorem by not conditioning on $Z = 1$. Hence, it additionally takes into account the uncertainty about whether the distribution of the subgroup is different from the background.

Theorem 3. *Consider a subgroup as generated with the model above and denote C as above. Then*

$$\arg\max_{\rho} \mathbb{E}[\psi'(\pi, Y) - \psi'(\rho, Y)|C = c] = a\frac{c + \beta}{\sum_{j=1}^{k} c_j + \beta_j} + (1 - a)\pi \qquad (17)$$

where $a = \mathbb{P}[Z = 1|C = c]$. Denote this quantity by $\hat{\hat{\rho}}$. Then the achieved maximum value is $d(\pi, \hat{\hat{\rho}})$, where d is the divergence measure corresponding to ψ.

Following this theorem we propose another novel quality measure, which takes into account both the uncertainty about the class distribution and about whether it is different from the background distribution:

$$\phi_{PSR}(g) = m \cdot d(\pi, \hat{\hat{\rho}}) \qquad (18)$$

where m is the size of the subgroup. In order to calculate the value of $a = \mathbb{P}[Z = 1|C = c]$ we have the following theorem:

Theorem 4. *Consider a subgroup as generated with the model above and denote C as above. Then the following equalities hold:*

$$\mathbb{P}[Z = 1|C = c] = \frac{\gamma \cdot \mathbb{P}[C = c \mid Z = 1]}{\gamma \cdot \mathbb{P}[C = c \mid Z = 1] + (1 - \gamma) \cdot \mathbb{P}[C = c \mid Z = 0]}$$

$$\mathbb{P}[C = c \mid Z = 1] = \binom{m}{c} \cdot \frac{\Gamma(\sum_{j=1}^{k} \beta_j)}{\prod_{j=1}^{k} \Gamma(\beta_j)} \cdot \frac{\prod_{j=1}^{k} \Gamma(c_j + \beta_j)}{\Gamma(m + \beta_0)} \qquad (19)$$

$$\mathbb{P}[C = c \mid Z = 0] = \binom{m}{c} \cdot \prod_{j=1}^{k} \pi_j^{c_j}$$

where $\beta_0 = \sum_{j=1}^{k} \beta_j$.

Referring back to Fig. 1 in the introduction, the subgroup on the left was discovered with ϕ_{WRAcc} as quality measure and the right one by ϕ_{PSR} with Brier Score. While WRAcc provides a larger coverage, it can be seen that the PSR measure captures a more distinct statistical deviation of the class distribution in the subgroup.

5   Experiments

In this section we experimentally investigate the performance of our proposed measures. The experiments are separated into two parts. For the first part we generated synthetic data, such that we know the true subgroup. In the second part we applied our methods to UCI data to investigate summarisation performance.

For our proposed measures, we adopt the generalised divergences of BS and LL as given in Sect. 3, Eqs. (5 and 6). Plugging these into Eqs. (16) and (18) we obtain four novel measures d-BS, d-LL, PSR-BS and PSR-LL. We compare these proposals against a range of subgroup evaluation measures used in the literature: Weighted Relative Accuracy (WRAcc), IG-LL (Eq. (10)), IG-BS (Eq. (11)), as well as the χ^2 statistic, which is defined as follows:

$$\phi_{Chi2} = C \cdot \sum_{j=1}^{K} \frac{(\rho_j - \pi_j)^2}{\pi_j} \tag{20}$$

5.1   Synthetic Data

In the experiments on the synthetic data we evaluate how good the methods are in revealing the true subgroup used in generating the data, as well as in producing good summaries of the data.

To provide a more intuitive illustration, we construct our data set according to a real-life scenario. Suppose one has been using a wearable device to record whether daily exercises were performed or not, for a whole year. As it turned out, there were 146 out of 365 days when the exercises were performed, which gives a probability about 2/5 that the exercises were performed on a random day. According to the website of the wearable device, the same statistics are about 1/3 for the general population. It is possible that the overall exercise frequency was different, but perhaps a more plausible explanation might be that more exercises were performed during a particular period only. SD can hence be applied to recognise the period of more intensive exercise and summarise the corresponding exercise frequency.

Following this scenario, the feature space consists of the 52 weeks of the year, hence $\mathbb{X} = \{1, ..., 52\}$. We define the subgroup language as the set of all intervals of weeks of length from 2 to 8 weeks. The data set is assumed to contain a single year from January to December. This setting allows us to perform exhaustive search on the subgroup language. As here our aim is to compare the performance among different quality measures, applying exhaustive search can avoid the bias introduced by other greedy search algorithms.

The way to generate the data is then as described in the previous section. Given the default class distribution π, the subgroup class distribution Q is sampled from a Dirichlet prior and a true subgroup is selected uniformly within the language. Therefore, all the 7 days within each week can be distributed either according to π or according to Q.

We evaluate each subgroup quality measure by comparing the obtained subgroup against the true subgroup. This is done by measuring similarity of the respective indicator functions Z and $\hat{Z}$. For similarity we use the F-score as we are not really interested in the 'true negatives' (instances in the complements of both true and discovered subgroups). The F-score for this case can be computed as (Z_i and $\hat{Z}_i$ are used to represent whether an instance belongs to the true subgroup and the obtained subgroup respectively):

$$F_1 = \frac{2 \cdot \sum_{i=1}^{N} \mathbb{I}(Z_i = 1, \hat{Z}_i = 1)}{\sum_{i=1}^{N}(2 \cdot \mathbb{I}(Z_i = 1, \hat{Z}_i = 1) + \mathbb{I}(Z_i = 1, \hat{Z}_i = 0) + \mathbb{I}(Z_i = 0, \hat{Z}_i = 1))} \tag{21}$$

The results are given in Table 1 as the micro-averaged F-scores from 5 000 synthetic sequences, for different values of π_1 (the first component of the class distribution vector). We can see that the PSR-based approaches generally outperform existing measures, with a slight advantage for Log-loss over Brier score. The information gain-based methods perform particularly poorly, as they have a preference for pure subgroups, whereas for skewed π it would be advantageous to look for subgroups with a more uniform class distribution. As π becomes more uniform, the 'true' subgroup becomes more random and harder to identify, which is why all methods are expected to perform poorly for $\pi_1 \approx 0.5$. The variance is quite high across all methods, probably because the data set is quite small ($52 \cdot 7 = 364$ instances).

Since a better statistical summary is essentially our aim, the results are also evaluated according to their overall loss on a test set (also of length 1 year) drawn from the same distribution. For each quality measure, a subgroup is obtained from the training fold together with the local statistical summary ($\hat{\rho}$ for ϕ_{PSR}, $\hat{\rho}$ for other quality measures). The loss for the obtained summarisation can then be

Table 1. Micro-averaged F-scores on the artificial data, for different class distributions (π_1). The best results for each row are shown in bold.

π_1	PSR-BS	PSR-LL	WRAcc	Chi2	IG-BS	IG-LL	d-BS	d-LL
.1	**.744**	.736	.597	.526	.030	.029	.742	.716
.2	.636	**.638**	.510	.436	.089	.091	.628	.631
.3	.587	**.589**	.480	.403	.218	.223	.581	.585
.4	.558	**.564**	.454	.390	.372	.379	.550	.559
.5	.567	**.569**	.458	.410	.561	.565	.561	.565

Table 2. Average Brier scores on the artificial data. The best results are shown in bold

π_1	PSR-BS	PSR-LL	WRAcc	Chi2	IG-BS	IG-LL	d-BS	d-LL
.1	**.195 ± .03**	**.195 ± .03**	.207 ± .03	.212 ± .03	.231 ± .04	.231 ± .04	**.195 ± .03**	**.195 ± .03**
.2	**.326 ± .03**	**.326 ± .03**	.334 ± .03	.337 ± .03	.350 ± .04	.350 ± .04	**.326 ± .03**	**.326 ± .03**
.3	**.419 ± .02**	**.419 ± .02**	.424 ± .02	.426 ± .02	.430 ± .03	.430 ± .03	.420 ± .02	.420 ± .02
.4	**.475 ± .02**	**.475 ± .02**	.479 ± .02	.480 ± .01	.478 ± .02	.478 ± .02	.476 ± .02	.476 ± .02
.5	**.494 ± .02**	**.494 ± .02**	.497 ± .01	.498 ± .01	**.494 ± .02**	.495 ± .02	**.494 ± .02**	**.494 ± .02**

Table 3. Average Log-loss on the artificial data. The best results are shown in bold.

π_1	PSR-BS	PSR-LL	WRAcc	Chi2	IG-BS	IG-LL	d-BS	d-LL
.1	**.344 ± .04**	**.344 ± .04**	.359 ± .04	.368 ± .04	.406 ± .06	.407 ± .06	**.344 ± .04**	.347 ± .04
.2	**.507 ± .03**	**.507 ± .03**	.517 ± .03	.520 ± .03	.539 ± .05	.540 ± .05	.508 ± .03	.509 ± .03
.3	**.610 ± .03**	**.610 ± .03**	.616 ± .02	.618 ± .02	.624 ± .03	.624 ± .03	.611 ± .03	.611 ± .03
.4	**.668 ± .02**	**.668 ± .02**	.673 ± .02	.674 ± .02	.671 ± .02	.671 ± .02	.670 ± .02	.669 ± .02
.5	.687 ± .02	**.686 ± .02**	.690 ± .01	.691 ± .01	.688 ± .02	.687 ± .02	.688 ± .02	.687 ± .02

calculated as in Eq. (4). The corresponding results are given in Tables 2 and 3 for both Brier score and Log-loss. We see a similar pattern as with the F-score results.

5.2   UCI Data

We proceed to compare our method with existing approaches on UCI data sets [13]. We selected the same 20 UCI datasets as described in [1]. The information regarding the number of attributes and instances are provided in the appendix.

The subgroup language we used here is conjunctive normal form, with disjunctions (only) between values of the same feature, and conjunctions among disjunctions involving different features. All features are treated as nominal. If the original feature is numeric and contains more than 100 values, it is discretised into 16 bins.

Since for most data sets in this experiment exhaustive search is intractable we perform beam search instead. The beam width is set to be 32 (i.e., 32 candidate subgroups are kept to be refined in the next round). The number of refinement rounds is set to 8.

The resulting average Brier scores and Log-loss are given in Tables 4 and 5. All the results are obtained by 10-fold cross-validation. As in the previous experiment, a subgroup is learned on the training folds and the class distribution estimated on the test fold is then used to compute the corresponding loss.

Given these results, it can be seen that our proposed measures generally outperform WRAcc, $Chi2$ and both versions of information gain. The PSR measures (first two columns) are never outperformed by the generalised divergence (last two columns) so we recommend using the former unless simplicity of implementation is an issue (as the latter don't need estimation of a). Regarding the choice between (BS, LL), this is still an ongoing debate in the community. Here we

Table 4. Average Brier scores for the UCI data sets. The best results are shown in bold.

Data set	PSR-BS	PSR-LL	WRAcc	Chi2	IG-BS	IG-LL	d-BS	d-LL
Abalone	**.872** ± **.005**	.874 ± .005	.879 ± .006	.897 ± .004	.878 ± .01	.884 ± .006	**.872** ± **.005**	.874 ± .005
Balance-scale	.539 ± .043	.572 ± .027	**.527** ± **.047**	.578 ± .024	.561 ± .032	.562 ± .032	.539 ± .043	.572 ± .027
Car	**.379** ± **.023**	.380 ± .032	.381 ± .030	.466 ± .031	.406 ± .036	.406 ± .036	**.379** ± **.024**	.380 ± .032
Contraceptive	**.618** ± **.019**	.647 ± .013	.638 ± .015	.650 ± .012	.619 ± .021	.616 ± .021	**.618** ± **.019**	.647 ± .013
Contact-lens	.624 ± .283	.651 ± .285	.579 ± .226	.611 ± .151	.461 ± .438	.461 ± .438	.627 ± .284	.655 ± .287
Credit	**.351** ± **.047**	**.351** ± **.047**	**.351** ± **.047**	.500 ± .012	**.351** ± **.047**	**.351** ± **.047**	**.351** ± **.047**	**.351** ± **.047**
Dermatology	**.633** ± **.073**	.708 ± .027	.721 ± .026	.806 ± .026	**.633** ± **.073**	.635 ± .077	**.633** ± **.073**	.708 ± .027
Glass	**.698** ± **.050**	**.698** ± **.051**	.725 ± .065	.745 ± .046	.716 ± .068	.719 ± .048	**.698** ± **.050**	**.698** ± **.051**
Haberman	.427 ± .083	**.387** ± **.092**	.391 ± .096	.398 ± .068	.394 ± .094	.394 ± .094	.430 ± .082	**.387** ± **.092**
Hayes-roth	.634 ± .029	.625 ± .040	.632 ± .046	.659 ± .028	.608 ± .048	**.602** ± **.044**	.634 ± .029	.625 ± .040
House-votes	**.269** ± **.041**	.271 ± .037	.309 ± .061	.482 ± .027	.306 ± .055	.306 ± .055	**.269** ± **.041**	.271 ± .037
Ionosphere	**.389** ± **.061**	**.389** ± **.062**	.411 ± .115	.470 ± .054	.401 ± .114	.398 ± .112	**.389** ± **.061**	**.389** ± **.062**
Iris	**.460** ± **.077**	**.460** ± **.077**	**.460** ± **.077**	.675 ± .005	**.460** ± **.077**	**.460** ± **.077**	**.460** ± **.077**	**.460** ± **.077**
Labor	.478 ± .237	.466 ± .249	.500 ± .338	.491 ± .152	.397 ± .328	.397 ± .328	.478 ± .237	.467 ± .249
Mushroom	**.253** ± **.010**	**.253** ± **.010**	.279 ± .012	.505 ± .001	.279 ± .012	**.253** ± **.010**	**.253** ± **.010**	**.253** ± **.010**
Pima-indians	**.416** ± **.029**	.458 ± .044	.422 ± .062	.462 ± .035	.425 ± .058	.427 ± .060	**.416** ± **.029**	.458 ± .044
Soybean	**.826** ± **.046**	.882 ± .019	.882 ± .018	.920 ± .011	**.826** ± **.046**	.861 ± .026	**.826** ± **.046**	.882 ± .019
Tic-Tac-Toe	**.395** ± **.019**	.455 ± .039	.434 ± .053	.460 ± .034	.424 ± .051	.403 ± .046	**.395** ± **.019**	.455 ± .039
Breast Cancer	**.274** ± **.035**	.306 ± .053	.325 ± .051	.459 ± .030	.318 ± .050	.306 ± .053	**.274** ± **.035**	.306 ± .053
Zoo	**.582** ± **.135**	.684 ± .052	.675 ± .058	.781 ± .077	**.582** ± **.135**	**.582** ± **.135**	**.582** ± **.135**	.684 ± .052

Table 5. Average Log-loss for the UCI data sets. The best results are shown in bold.

Data set	PSR-BS	PSR-LL	WRAcc	Chi2	IG-BS	IG − LL	d-BS	d-LL
Abalone	**2.430** ± **.055**	2.436 ± .057	2.450 ± .062	2.608 ± .051	2.504 ± .061	2.511 ± .061	2.430 ± .055	2.436 ± .057
Balance-scale	.958 ± .077	**.918** ± **.064**	.918 ± .084	1.026 ± .064	.986 ± .067	.993 ± .067	.958 ± .077	**.918** ± **.064**
Car	.766 ± .037	**.764** ± **.047**	.766 ± .052	.946 ± .056	.797 ± .066	.797 ± .066	.766 ± .037	**.764** ± **.047**
Contraceptive	1.119 ± .031	**1.068** ± **.021**	1.089 ± .022	1.173 ± .021	1.122 ± .035	1.115 ± .036	1.119 ± .031	**1.068** ± **.021**
Contact-lens	**1.166** ± **.483**	1.212 ± .485	1.042 ± .336	1.076 ± .239	.884 ± .735	.884 ± .735	1.175 ± .488	1.223 ± .492
Credit	**.563** ± **.069**	**.563** ± **.069**	**.563** ± **.069**	.794 ± .014	**.563** ± **.069**	**.563** ± **.069**	**.563** ± **.069**	**.563** ± **.069**
Dermatology	1.459 ± .178	**1.424** ± **.075**	1.443 ± .077	1.807 ± .084	1.459 ± .178	1.464 ± .185	1.459 ± .178	**1.424** ± **.075**
Glass	1.479 ± .130	**1.477** ± **.131**	1.478 ± .211	1.635 ± .154	1.552 ± .188	1.493 ± .192	1.479 ± .130	1.478 ± .131
Haberman	.695 ± .104	**.601** ± **.111**	.617 ± .121	.686 ± .083	.623 ± .117	.622 ± .117	.693 ± .105	**.601** ± **.111**
Hayes-roth	1.142 ± .050	1.054 ± .116	**1.045** ± **.103**	1.180 ± .051	.968 ± .116	.953 ± .108	1.142 ± .050	1.054 ± .116
House-votes	.491 ± .074	.476 ± .071	.476 ± .101	.774 ± .029	.467 ± .088	**.467** ± **.088**	.491 ± .074	.476 ± .071
Ionosphere	.667 ± .098	.670 ± .102	.629 ± .139	.763 ± .062	.620 ± .147	**.616** ± **.145**	.667 ± .098	.670 ± .102
Iris	**.836** ± **.132**	**.836** ± **.132**	**.836** ± **.132**	1.210 ± .008	**.836** ± **.132**	**.836** ± **.132**	**.836** ± **.132**	**.836** ± **.132**
Labor	.775 ± .332	**.747** ± **.359**	.787 ± .482	.785 ± .176	.622 ± .470	.622 ± .470	.775 ± .333	**.747** ± **.359**
Mushroom	**.408** ± **.016**	**.408** ± **.016**	.455 ± .019	.798 ± .001	.455 ± .019	**.408** ± **.016**	**.408** ± **.016**	**.408** ± **.016**
Pima-indians	.688 ± .034	.659 ± .060	**.655** ± **.077**	.754 ± .041	.669 ± .076	.669 ± .076	.688 ± .034	.659 ± .060
Soybean	2.579 ± .157	**2.447** ± **.079**	2.452 ± .083	2.810 ± .103	2.579 ± .157	2.455 ± .172	2.579 ± .157	**2.447** ± **.079**
Tic-Tac-Toe	.660 ± .022	.647 ± .040	.663 ± .061	.752 ± .040	.669 ± .067	**.641** ± **.061**	.660 ± .022	.647 ± .040
Breast Cancer	.507 ± .048	**.455** ± **.087**	.508 ± .078	.751 ± .035	.491 ± .077	.456 ± .086	.507 ± .048	**.455** ± **.087**
Zoo	**1.435** ± **.329**	1.439 ± .118	1.447 ± .139	1.825 ± .228	**1.435** ± **.329**	**1.435** ± **.329**	**1.435** ± **.329**	1.439 ± .118

used both to demonstrate that our novel measure can apply either as the two most well-known Proper Scoring Rules.

6   Related Work

As is the case for supervised rule learning in general, SD comprises three major components: description language, quality measure and search algorithm. A detailed comparison with rule learning can be found in [15]. While early work in SD has been surveyed in [6], we briefly describe some recent progress in the area.

Regarding the subgroup description language, most existing work defines it through logical operations on attribute values. In [14] the authors present an approach to construct more informative descriptions on numeric and nominal attributes in linear time. The proposed algorithm is able to find the optimal interval for numeric attributes and optimal set of values for nominal attributes. The results show improvements on the quality of obtained subgroups comparing to traditional descriptions.

In terms of quality measures, recent work has focused on the extension of traditional measures with improved statistical modelling. In [4,11] Exceptional Model Mining (EMM) was introduced as a framework to support improved target concepts with different model classes. For example, if linear regression models are trained on the whole data set and different candidate subgroups, the quality of subgroups can be evaluated by comparing the regression coefficient between the global model and the local subgroup model. In [5] the authors extend the framework to support predictive statistical information. This further allows subgroups to be found where a scoring classifier's performance deviates from its overall performance.

With respect to the search algorithm, while greedy search algorithms have been widely adopted in existing implementations, recent work in [12] presents a fast exhaustive search strategy for numerical target concepts. The authors propose and illustrate novel bounds on different types of quality measures. The exhaustive search can then be performed efficiently via additional pruning techniques.

7   Conclusion

In this paper we investigated how to discover subgroups that are optimal in the sense of maximally improving the global statistical summary of a given data set. By assuming that the (discrete) statistical summary is to be evaluated by the Proper Scoring Rule, we derived the corresponding quality measures from first principles. We also proposed a generative model to consider the optimal statistical summary for any candidate subgroup. By performing experiments on both synthetic data and UCI data, we showed that our measures provide better summaries in comparison with existing methods.

The major advantage of adopting our generative model is that it prevents finding small subgroups with extreme distributions. This can be seen as applying a regularisation on the class distribution, similar to performing Laplace smoothing in decision tree learning. Given the experiments, we can observe that the novel measures tend to perform better on small data sets (e.g. Contact-lenses, Labor).

Since in this paper we assume that only the subgroup with the highest gain will be discovered, one major direction for further work is to investigate multiple subgroups that can together improve the overall statistical summary. Previous Subgroup Discovery algorithms have extended the covering algorithm to weighted covering in order to promote the discovery of overlapping subgroups [10]. We expect that the PSR approach will be able to derive appropriate weight updates in a principled fashion.

Another direction would be to generalise our approach to numeric target variables. Although in general PSRs are designed to work with discrete random variables, Log-loss has been widely adopted in Bayesian analysis, which provides an interface to extend our approach to a general form of statistical modelling.

Acknowledgements. This work was supported by the SPHERE Interdisciplinary Research Collaboration, funded by the UK Engineering and Physical Sciences Research Council under grant EP/K031910/1; and the REFRAME project granted by the European Coordinated Research on Long-Term Challenges in Information and Communication Sciences &Technologies ERA-Net (CHIST-ERA), and funded by the Engineering and Physical Sciences Research Council in the UK under grant EP/K018728/1. Hao Song would like to thank Toshiba Research Europe Ltd, Telecommunications Research Laboratory, for funding his doctoral research within SPHERE.

Appendix A: Proofs

Lemma 1. *Let ψ be a proper scoring rule and d its respective divergence measure. If S, S' are random vectors representing two sets of class probability estimates for a random variable T representing the actual class, then*

$$\mathbb{E}[\psi(S,T) - \psi(S',T)] = \mathbb{E}[d(S,T) - d(S',T)] = \mathbb{E}[d(S,\mathbb{E}[T]) - d(S',\mathbb{E}[T])] \quad (22)$$

Proof. By using Lemma 1 from the supplementary of [8] we get the decomposition $\mathbb{E}[\psi(S,T)] = \mathbb{E}[d(S,T)] = \mathbb{E}[d(S,\mathbb{E}[T])] + \mathbb{E}[d(\mathbb{E}[T],T)]$ and the analogous decomposition for S'. The second term is shared and hence when subtracting it cancels, yielding the required result.

Theorem 1. *Let ψ, ψ', d be a proper scoring rule, its sum across the dataset, and its corresponding divergence measure, respectively. Then for any given subgroup g the following holds:*

$$\arg\max_{\rho} \psi'(S^{\pi}, Y) - \psi'(S^{g,\rho,\pi}, Y) = \rho^{(g)} \quad (23)$$

where $\rho^{(g)}$ denotes the class distribution within the subgroup g. The value of achieved maximum is $m \cdot d(\pi, \rho^{(g)})$ where m is the size of the subgroup g.

Proof. For simplicity of notation, let us assume that the training instances within g are $Y_1, \ldots, Y_m$ (the first m instances). Consider a random variable T obtaining its value by uniformly choosing one Y_i that belongs to g among $Y_1, \ldots, Y_m$. The summaries S^π and $S^{g, \rho^{(g)}, \pi}$ are equal for instances $m + 1, \ldots, n$, hence $\psi'(S^\pi, Y) - \psi'(S^{g, \rho^{(g)}, \pi}, Y) = m \cdot \mathbb{E}[\psi(\pi, T) - \psi(\rho^{(g)}, T)]$. Using Lemma 1 this is in turn equal to $m \cdot \mathbb{E}[d(\pi, \mathbb{E}[T]) - m \cdot \mathbb{E}[d(\rho^{(g)}, \mathbb{E}[T])]$. However, since $\mathbb{E}[T] = \rho^{(g)}$ then the second term is zero and the first is $m \cdot d(\pi, \rho^{(g)})$, which is exactly the required result.

Theorem 2. *Consider a subgroup as generated with the model above. Denote the counts of each class in the training set of this subgroup by $C = \sum_{i=1}^m Y_i$. Then*

$$\arg \max_\rho \mathbb{E}[\psi'(\pi, Y) - \psi'(\rho, Y) | C = c, Z = 1] = \frac{c + \beta}{\sum_{j=1}^k c_j + \beta_j} \tag{24}$$

Denoting this quantity by $\hat{\rho}$, the achieved maximum is $d(\pi, \hat{\rho})$, where d is the divergence measure corresponding to ψ.

Proof. Consider a random variable T obtaining its value by uniformly choosing one Y_i that belongs to g among $Y_1, \ldots, Y_m$. Then $\mathbb{E}[\psi'(\pi, Y) - \psi'(\rho, Y) | C = c, Z = 1] = \mathbb{E}[\psi(\pi, T) - \psi(\rho, T) | C = c, Z = 1]$. Using Lemma 1 this is in turn equal to $d(\pi, \mathbb{E}[T | C = c, Z = 1]) - d(\rho, \mathbb{E}[T | C = c, Z = 1])$. Since the first term does not depend on ρ this quantity is maximised by minimising the second divergence. As with any divergence, the minimal value is zero and it is obtained if the two terms are equal, i.e., $\rho = \mathbb{E}[T | C = c, Z = 1]$. It remains to prove that $\mathbb{E}[T | C = c, Z = 1] = \frac{c + \beta}{\sum_{j=1}^k c_j + \beta_j}$. This holds because it is a posterior distribution under the Dirichlet prior $Dir(\beta)$ after observing $c_1, \ldots, c_k$ of the classes $1, \ldots, k$, respectively.

Theorem 3. *Consider a subgroup as generated with the model above and denote C as above. Then*

$$\arg \max_\rho \mathbb{E}[\psi'(\pi, Y) - \psi'(\rho, Y) | C = c] = a \frac{c + \beta}{\sum_{j=1}^k c_j + \beta_j} + (1 - a)\pi \tag{25}$$

where $a = \mathbb{P}[Z = 1 | C = c]$. Denote this quantity by $\hat{\rho}$. Then the achieved maximum value is $d(\pi, \hat{\rho})$, where d is the divergence measure corresponding to ψ.

Proof. Consider a random variable T obtaining its value by uniformly choosing one Y_i that belongs to g among $Y_1, \ldots, Y_m$. Then $\mathbb{E}[\psi'(\pi, Y) - \psi'(\rho, Y) | C = c] = \mathbb{E}[\psi(\pi, T) - \psi(\rho, T) | C = c]$. Using Lemma 1 this is in turn equal to $d(\pi, \mathbb{E}[T | C = c]) - d(\rho, \mathbb{E}[T | C = c])$. Since the first term does not depend on ρ this quantity is maximised by minimising the second divergence. As with any divergence, the minimal value is zero and it is obtained if the two terms are equal, i.e., $\rho = \mathbb{E}[T | C = c]$. It remains to prove that $\mathbb{E}[T | C = c] = a\hat{\rho} + (1 - a)\hat{\rho}$ where $\hat{\rho}$ is defined in the previous Theorem 2. Indeed, $\mathbb{E}[T | C = c] = \mathbb{P}(Z = 1 | C =$

$c)\mathbb{E}[T|C = c, Z = 1] + \mathbb{P}(Z = 0|C = c)\mathbb{E}[T|C = c, Z = 0] = a\hat{\rho} + (1 - a)\pi$, where $\mathbb{E}[T|C = c, Z = 0] = \pi$ due to Y (and therefore T) drawn from Bernoulli with the mean $ZQ + (1 - Z)\pi$. The achieved maximum is $d(\pi, \hat{\rho})$.

Theorem 4. *Consider a subgroup as generated with the model above and denote C as above. Then the following equalities hold:*

$$\mathbb{P}[Z = 1|C = c] = \frac{\gamma\mathbb{P}[C = c \mid Z = 1]}{\gamma\mathbb{P}[C = c \mid Z = 1] + (1 - \gamma)\mathbb{P}[C = c \mid Z = 0]}$$

$$\mathbb{P}[C = c \mid Z = 1] = \frac{\Gamma(\sum_{j=1}^{k}\beta_j)}{\prod_{j=1}^{k}\Gamma(\beta_j)} \cdot \frac{\prod_{j=1}^{k}\Gamma(c_j + \beta_j)}{\Gamma(m + \beta_0)} \cdot \binom{m}{c} \quad (26)$$

$$\mathbb{P}[C = c \mid Z = 0] = \binom{m}{c} \cdot \prod_{j=1}^{k}\pi_j^{c_j}$$

where $\beta_0 = \sum_{j=1}^{k}\beta_j$.

Proof. Due to $\mathbb{P}[Z = 1] = \gamma$, we can obtain the first result from the Bayes formula with $\mathbb{P}[Z = 1|C = c] = \frac{\mathbb{P}[C=c|Z=1]\mathbb{P}[Z=1]}{\mathbb{P}[C=c]}$. To obtain the second result we note that in the subgroup $Z = 1$ the class distribution is drawn from $Dir(\beta)$, therefore the distribution of C follows the Dirichlet-Multinomial distribution. The stated result represents simply the probability distribution function of the Dirichlet-Multinomial with $Dir(\beta)$ and multinomial of size m. The third result is simply the probability distribution function of the Multinomial Distribution.

Appendix B: Information for the UCI Data

See Table. 6

Table 6. The 20 UCI data sets used in the experiments.

Name	# instances	# features	# classes
Abalone	4 176	9	3
Balance-scale	624	5	3
Car	1 727	7	4
Contraceptive	1 472	10	3
Contact-lenses	24	5	3
Credit	589	16	2
Dermatology	365	35	6
Glass	213	11	6

(Continued)

Table 6. *(Continued)*

Name	# instances	# features	# classes
Haberman	305	4	2
Hayes-roth	131	5	3
House-votes	434	17	2
Ionosphere	350	34	2
Iris	150	5	3
Labor	57	17	2
Mushroom	8 123	23	2
Pima-indians	767	9	2
Soybean	683	36	19
Tic-Tac-Toe	957	10	2
Breast Cancer	197	34	2
Zoo	100	18	7

References

1. Abudawood, T., Flach, P.: Evaluation measures for multi-class subgroup discovery. In: Buntine, W., Grobelnik, M., Mladenić, D., Shawe-Taylor, J. (eds.) ECML PKDD 2009. LNCS (LNAI), vol. 5781, pp. 35–50. Springer, Heidelberg (2009). doi:10.1007/978-3-642-04180-8_20
2. Atzmueller, M., Lemmerich, F.: Fast subgroup discovery for continuous target concepts. In: Rauch, J., Raś, Z.W., Berka, P., Elomaa, T. (eds.) ISMIS 2009. LNCS (LNAI), vol. 5722, pp. 35–44. Springer, Heidelberg (2009). doi:10.1007/978-3-642-04125-9_7
3. Clark, P., Boswell, R.: Rule induction with CN2: some recent improvements. In: Kodratoff, Y. (ed.) EWSL 1991. LNCS, vol. 482, pp. 151–163. Springer, Heidelberg (1991). doi:10.1007/BFb0017011
4. Duivesteijn, W., Feelders, A.J., Knobbe, A.: Exceptional model mining. Data Min. Knowl. Discovery **30**(1), 47–98 (2016)
5. Duivesteijn, W., Thaele, J.: Understanding where your classifier does (not) work-the SCaPE model class for EMM. In: 2014 IEEE International Conference on Data Mining (ICDM), pp. 809–814. IEEE (2014)
6. Herrera, F., Carmona, C.J., González, P., del Jesus, M.J.: An overview on subgroup discovery: foundations and applications. Knowl. Inf. Syst. **29**(3), 495–525 (2011)
7. Klösgen, W.: Explora: a multipattern and multistrategy discovery assistant. In: Fayyad, U.M., Piatetsky-Shapiro, G., Smyth, P., Uthurusamy, R. (eds.) Advances in Knowledge Discovery and Data Mining, pp. 249–271. American Association for Artificial Intelligence, Menlo Park (1996)
8. Kull, M., Flach, P.: Novel decompositions of proper scoring rules for classification: score adjustment as precursor to calibration. In: Appice, A., Rodrigues, P.P., Santos Costa, V., Soares, C., Gama, J., Jorge, A. (eds.) ECML PKDD 2015. LNCS (LNAI), vol. 9284, pp. 68–85. Springer, Heidelberg (2015). doi:10.1007/978-3-319-23528-8_5

9. Lavrač, N., Flach, P., Zupan, B.: Rule evaluation measures: a unifying view. In: Džeroski, S., Flach, P. (eds.) ILP 1999. LNCS (LNAI), vol. 1634, pp. 174–185. Springer, Heidelberg (1999). doi:10.1007/3-540-48751-4_17

10. Lavrač, N., Kavšek, B., Flach, P., Todorovski, L.: Subgroup discovery with CN2-SD. J. Mach. Learn. Res. **5**, 153–188 (2004)

11. Leman, D., Feelders, A., Knobbe, A.: Exceptional model mining. In: Daelemans, W., Goethals, B., Morik, K. (eds.) ECML PKDD 2008. LNCS (LNAI), vol. 5212, pp. 1–16. Springer, Heidelberg (2008). doi:10.1007/978-3-540-87481-2_1

12. Lemmerich, F., Atzmueller, M., Puppe, F.: Fast exhaustive subgroup discovery with numerical target concepts. Data Min. Knowl. Disc. **30**(3), 711–762 (2016)

13. Lichman, M.: UCI machine learning repository (2013). http://archive.ics.uci.edu/ml

14. Mampaey, M., Nijssen, S., Feelders, A., Knobbe, A.: Efficient algorithms for finding richer subgroup descriptions in numeric and nominal data. In: IEEE International Conference on Data Mining, pp. 499–508 (2012)

15. Novak, P.K., Lavrač, N., Webb, G.I.: Supervised descriptive rule discovery: a unifying survey of contrast set, emerging pattern and subgroup mining. J. Mach. Learn. Res. **10**, 377–403 (2009)

16. Winkler, R.L.: Scoring rules and the evaluation of probability assessors. J. Am. Stat. Assoc. **64**(327), 1073–1078 (1969)

17. Wrobel, S.: An algorithm for multi-relational discovery of subgroups. In: Komorowski, J., Zytkow, J. (eds.) PKDD 1997. LNCS, pp. 78–87. Springer, Heidelberg (1997). doi:10.1007/3-540-63223-9_108

Consistency of Probabilistic Classifier Trees

Krzysztof Dembczyński[1]([⊠]), Wojciech Kotłowski[1], Willem Waegeman[2],
Róbert Busa-Fekete[3], and Eyke Hüllermeier[3]

[1] Poznan University of Technology, Poznań, Poland
{kdembczynski,wkotlowski}@cs.put.poznan.pl
[2] Ghent University, Ghent, Belgium
willem.waegeman@ugent.be
[3] Paderborn University, Paderborn, Germany
{busarobi,eyke}@upb.de

Abstract. Label tree classifiers are commonly used for efficient multi-class and multi-label classification. They represent a predictive model in the form of a tree-like hierarchy of (internal) classifiers, each of which is trained on a simpler (often binary) subproblem, and predictions are made by (greedily) following these classifiers' decisions from the root to a leaf of the tree. Unfortunately, this approach does normally not assure consistency for different losses on the original prediction task, even if the internal classifiers are consistent for their subtask. In this paper, we thoroughly analyze a class of methods referred to as probabilistic classifier trees (PCTs). Thanks to training probabilistic classifiers at internal nodes of the hierarchy, these methods allow for searching the tree-structure in a more sophisticated manner, thereby producing predictions of a less greedy nature. Our main result is a regret bound for 0/1 loss, which can easily be extended to ranking-based losses. In this regard, PCTs nicely complement a related approach called filter trees (FTs), and can indeed be seen as a natural alternative thereof. We compare the two approaches both theoretically and empirically.

1 Introduction

Multi-class and multi-label classification problems are nowadays characterized not only by large sample sizes and feature spaces, but also by a large number of labels. In application fields like image classification [12], text classification [8], online advertising [3], and video recommendation [23], it is not uncommon to deal with tens or hundreds of thousands [11], or even millions of labels [20].

Label tree classifiers belong to the most efficient approaches for problems at this scale [2]. In this approach, a solution to the original problem is represented in the form of a hierarchy of classifiers, each of which is trained on a simpler subproblem. A prediction for a new example is then derived from the predictions of these (internal) classifiers, each of which corresponds to a node in the tree-like hierarchical structure; typically, each label in the original classification problem is uniquely represented by a path from the root to a leaf of that tree.

© Springer International Publishing AG 2016
P. Frasconi et al. (Eds.): ECML PKDD 2016, Part II, LNAI 9852, pp. 511–526, 2016.
DOI: 10.1007/978-3-319-46227-1_32

However, combining conventional training of the internal classifiers with greedy inference, namely, following a single root-to-leaf path in the tree, does not guarantee consistency of this approach [4, 10]. Thus, even perfect (zero regret) classifiers in each node of the tree do not imply a perfect (global) classification of new examples. There are two ways to remedy this problem: adjusting training and adjusting inference. The first idea is to modify the training of the internal classifiers so as to assure the consistency of greedy inference later on. The second approach, while training more conventionally, guarantees consistency by searching the tree-structure for an optimal prediction in a less greedy way.

The first idea is realized by the *filter tree* (FT) approach [4]. By constructing label trees in a bottom-up manner, an internal classifier can anticipate the decisions of its successor classifiers, and exploit this information to properly condition its own behavior to these classifiers. In the case of 0/1 loss, this is accomplished thanks to a specific filter technique, which removes examples from the training data on which successor classifiers made incorrect predictions. For this training procedure, a regret bound connecting the global performance with the average performance of node classifiers can be proved [4]. This bound can be generalized from 0/1 loss to any cost-based loss function, albeit at the price of a more expensive training procedure; ranking-based losses, which require the ordering of labels, cannot be tackled by FTs. Since inference can be done in a greedy way, the complexity of prediction is only logarithmic in the number of labels. More recently, the training of FTs has been further improved in the context of multi-label classification [17].

The second approach ensures consistency thanks to more sophisticated search of label trees in the inference phase [10, 16, 18]. To this end, probabilistic classifiers in each node of the tree are required, which allow for assessing the usefulness of different search directions. Label trees with probabilistic classifiers have already been considered in multi-class classification under the name of conditional probability trees [3] and nested dichotomies [14]. In multi-label classification, a similar approach has been referred to as probabilistic classifier chains [9]. The same concept also appears in neural networks and natural language processing under the name of hierarchical softmax [19]. In the following we unify all these approaches and jointly refer to them as *probabilistic classifier trees* (PCTs).

We restrict to binary label trees, which are especially natural for multi-label classification; here, each level of the binary tree directly corresponds to one label. Higher order trees (including nodes with more than two children) are often used in multi-class classification. This usually improves the predictive performance at the cost of an increase in prediction time. We also assume the tree structure to be given beforehand, or to have been induced using any of the methods developed for this purpose [2, 3, 23], and focus on the (orthogonal) problem of how training and prediction should be performed to ensure consistency (given the tree structure).

The main contribution of the paper is a regret bound for PCT in the case of 0/1 loss, which is expressed in terms of the search error and the Kullback-Leibler (KL) divergence (i.e., log-loss regret) of the internal classifiers. The regret bound implies the consistency of the method, a good "sanity check" for any learning algorithm. Its form quantifies a trade-off between the computational complexity and the statistical accuracy. Moreover, we show that under log-loss we do not theoretically pay any price in terms of performance for representing the joint distribution over classes by a tree structure. Our regret analysis significantly extends and improves the results of [3] for the estimation error of conditional probability trees expressed in terms of squared error loss. We also point out that the bound can be further generalized to ranking-based losses, e.g., recall at k. We also generalize the tree search algorithms of [10,18] to get an anytime A^*-like algorithm and study its theoretical guarantees, extending the previous results given in [10]. Our theoretical contributions are complemented by a comparison of PCTs with filter trees, both conceptually and experimentally.

The paper is organized as follows. We formally state the problem in Sect. 2. Section 3 describes PCTs and gives a theoretical analysis of the generalized tree search algorithm. In Sect. 4, we prove the regret bound for 0/1 loss. Section 5 compares PCTs with other label tree approaches, particularly with conditional probability and filter trees. Section 6 discusses the use of PCTs for predicting top-k labels and its extension to multi-label classification. Section 7 presents experimental results, prior to concluding the paper in Sect. 8.

2   Problem Statement

We formalize our problem in the setting of multi-class classification. Let (x, y) be an example coming from a probability distribution $P(X = x, Y = y)$ (later denoted $P(x, y)$) on $\mathcal{X} \times \mathcal{Y}$, where $x \in \mathcal{X} = \mathbb{R}^d$ and $y \in \mathcal{Y} = \{1, \ldots, m\}$. A classifier h predicts a label $\hat{y} = h(x) \in \mathcal{Y}$ for each $x \in \mathcal{X}$. The prediction accuracy of h can be measured in terms of 0/1 loss:[1]

$$\ell_{0/1}(y, h(x)) = [\![y \neq h(x)]\!]$$

We are interested in minimizing the expected loss, also referred to as the *risk*:

$$L_{0/1}(h) = \mathbb{E}_{(x,y) \sim P} \left[\ell_{0/1}(y, h(x)) \right] = \int_{\mathcal{X} \times \mathcal{Y}} [\![y \neq h(x)]\!] \, dP(x, y)$$

The *Bayes classifier*

$$h^* = \arg \min_h L_{0/1}(h)$$

minimizes the risk among all possible classifiers. While h^* may not be unique in general, the risk of h^*, denoted $L_{0/1}^*$, is unique, and is called the *Bayes risk*. Decomposing the risk over classes, i.e., writing $L_{0/1}(h)$ in the form

$$L_{0/1}(h) = \int_{\mathcal{X}} \left(\underbrace{\sum_{y \in \mathcal{Y}} [\![y \neq h(x)]\!] P(y|x)}_{=1 - P(h(x)|x)} \right) dP(x) ,$$

---

[1] We use $[\![P]\!]$ to denote a number that is 1 if condition P is satisfied, and 0 otherwise.

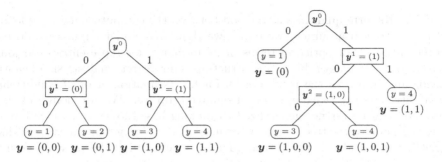

Fig. 1. Different binary codes in multi-class classification.

reveals that h^* minimizes risk in a pointwise manner, i.e., for every $\boldsymbol{x}$,

$$h^*(\boldsymbol{x}) = \arg\min_{y \in \mathcal{Y}} \{1 - P(y|\boldsymbol{x})\} = \arg\max_{y \in \mathcal{Y}} P(y \mid \boldsymbol{x}).$$

Given a classifier h, the *regret* of h is defined as

$$\operatorname{reg}_{0/1}(h) = L_{0/1}(h) - L^*_{0/1} = \int_{\mathcal{X}} \Big(P(h^*(\boldsymbol{x})|\boldsymbol{x}) - P(h(\boldsymbol{x})|\boldsymbol{x}) \Big) dP(\boldsymbol{x}). \quad (1)$$

The regret quantifies the suboptimality of h compared to the optimal classifier h^*. The goal is to train a classifier h with a small regret, ideally equal to zero.

In the following, we assume h to be represented as a label tree classifier. To this end, we encode the labels $\{1, \ldots, m\}$ using a prefix code. Any such code can be represented by a tree with 0/1 splits. Each path from the root to a leaf node then corresponds to a code word. Recall that codes of fixed length are also prefix codes. Figure 1 shows two examples of coding trees for multi-class classification with 4 classes. Under the coding, we represent each label y by a binary vector $\boldsymbol{y} = (y_1, \ldots, y_l)$, where l is the maximum length of the code. The set of all code words we denote by $\mathcal{C}$. As another special case, consider the problem of multi-label (instead of multi-class) classification, where the goal is to predict the set of labels assigned to a given instance $\boldsymbol{x}$. Such a set can be represented by a binary vector $\boldsymbol{y} = (y_1, \ldots, y_m)$, which in turn can be used as a prefix code.

In the label tree approach, we put a binary classifier in each non-leaf node of the tree. An internal node can be uniquely identified by the partial code word $\boldsymbol{y}^i = (y_1, \ldots, y_i)$. We denote the root node by $\boldsymbol{y}^0$, which is an empty vector (without any elements). The final prediction is determined by a sequence of decisions of internal classifiers. In the next section, we present a specific instance of the label tree approach that uses probabilistic classifiers in internal nodes of the tree.

3   Probabilistic Classifier Trees

Probabilistic classifier trees (PCTs) are designed to estimate probabilities $P(y \mid \boldsymbol{x})$ by following a path from the root to a leaf node, which corresponds to a code word $\boldsymbol{y} = (y_1, \ldots, y_l)$ assigned to label $y \in \mathcal{Y}$. Recalling the chain rule of probability, the process corresponds to computing

$$P(y \mid \boldsymbol{x}) = P(\boldsymbol{y} \mid \boldsymbol{x}) = \prod_{i=1}^{l} P(y_i \mid \boldsymbol{y}^{i-1}, \boldsymbol{x}), \tag{2}$$

where $P(y_i \mid \boldsymbol{y}^{i-1}, \boldsymbol{x})$ are probabilities of $y_i \in \{0, 1\}$, estimated in non-leaf nodes $\boldsymbol{y}^{i-1}$. In the next two subsections, training and inference (classification of new examples) for PCT will be discussed in more detail.

3.1 Training

Training of PCT naturally decomposes into learning problems over non-leaf nodes of the tree. In each node $\boldsymbol{y}^{i-1}$, the task is to train a probabilistic classifier (e.g., logistic regression) to estimate $P(y_i \mid \boldsymbol{y}^{i-1}, \boldsymbol{x})$.

Looking at PCTs as a reduction technique, it is worth mentioning that its training complexity could be much lower than that of the 1-vs-all approach, since each example $(\boldsymbol{x}, y)$ is used in only l instead of m binary problems, where l is the height of the tree (i.e., $l = \lceil \log_2 m \rceil$ if the tree is balanced). To further improve the training time complexity, one can use online learning methods, such as stochastic gradient descent [5]. Moreover, internal classifiers in PCT can be trained independently of each other, thereby allowing for a massive parallelization of the training procedure. Let us also remark that the learning process can be defined as a single task; this is the so-called one-classifier trick [4], in which a node indicator is used as an additional feature. Alternatively, one can use a separate task for each level of the tree. This approach is used in multi-label classification, as will be discussed in Sect. 6.

3.2 Inference

The classification procedure in PCTs is more involved. To begin with, note that a probability estimate $Q(y \mid \boldsymbol{x})$ for any label y (given instance $\boldsymbol{x}$) is obtained quite easily, simply by following the corresponding path in the tree and applying the chain rule:

$$Q(y \mid \boldsymbol{x}) = Q(\boldsymbol{y} \mid \boldsymbol{x}) = \prod_{i=1}^{l} Q(y_i \mid \boldsymbol{y}^{i-1}, \boldsymbol{x})$$

However, being interested in minimization of 0/1 loss, we actually seek to find

$$\hat{\boldsymbol{y}}^* = \arg\max_{\boldsymbol{y} \in \mathcal{C}} Q(\boldsymbol{y} \mid \boldsymbol{x}), \tag{3}$$

preferably without computing the probability of each label first. A simple idea is to follow a single path in the tree, starting in the root and always choosing the branch $y_i \in \{0, 1\}$ for which $Q(y_i \mid \boldsymbol{y}^{i-1}, \boldsymbol{x}) > 0.5$. However, while being efficient, this approach is not guaranteed to find the optimal solution [4,10].

Algorithm 1. Inference with ϵ-approximate A^*

1: **input:** x (test example)
2: priority list $\mathcal{Q} \leftarrow \{y_0\}$ (contains root node initially)
3: priority list $\mathcal{K} \leftarrow \{\}$ (contains nodes whose both children were not inserted to $\mathcal{Q}$)
4: $\epsilon \leftarrow 2^{-c}$ with $1 \leq c \leq m$
5: **while** $\mathcal{Q} \neq \emptyset$ **do**
6:     $v \leftarrow$ pop first element in $\mathcal{Q}$
7:     **if** v is a leaf **then** delete all elements in $\mathcal{K}$ and **break the while loop**
8:     $v_1 \leftarrow (v,1)$ (left child of v) and $v_0 \leftarrow (v,0)$ (right child of v)
9:     compute $E(v_1 \mid x)$ and $E(v_0 \mid x)$ recursively from $E(v \mid x)$ using Eq. (4)
10:     **if** $E(v_1 \mid x) \geq \epsilon$ **then** add $(v_1, E(v_1 \mid x))$ to $\mathcal{Q}$ sorted in descending order of E
11:     **if** $E(v_0 \mid x) \geq \epsilon$ **then** add $(v_0, E(v_0 \mid x))$ to $\mathcal{Q}$ sorted in descending order of E
12:     **if** v_1 and v_0 are not inserted to $\mathcal{Q}$ **then** add v to $\mathcal{K}$ in descending order of E
13: $\theta \leftarrow 0$
14: **while** $\mathcal{K} \neq \emptyset$ **do**
15:     $v' \leftarrow$ pop first element in $\mathcal{K}$
16:     $v' \leftarrow$ apply greedy search downward on v'
17:     **if** $Q(v' \mid x) \geq \theta$ **then** $v \leftarrow v'$ and $\theta \leftarrow Q(v' \mid x)$
18: **return** $h_\epsilon(x) = \hat{y}_\epsilon = v$

Better inference methods have been presented in recent years, based on search algorithms such as uniform-cost search [10], beam search [16], and A^* [18].

All three approaches allow for trading complexity against optimality, and hence for using PCTs in an anytime fashion, thanks to a hyper-parameter ϵ. This parameter controls the degree of optimality, i.e., of finding the true loss minimizer (3), as a function of the runtime (it finds a solution $\hat{y}_\epsilon$ the conditional probability $Q(\hat{y}_\epsilon \mid x)$ of which is not much worse than the probability of the optimal solution $\hat{y}^*$ defined in Eq. 3). In the analysis that follows, we will use this property to give a formal bound on the error made by such inference algorithms, with a particular focus on uniform-cost and A^* search. An extension of the analysis to beam search is straightforward and omitted due to lack of space. The pseudo code in Algorithm 1 unifies the approaches of [10,18]. This general algorithm, which we denote $h_\epsilon(x)$, is a variant of A^*. It fulfills the anytime property, i.e., the search can be stopped at any time and the algorithm will deliver a valid though possibly suboptimal solution.

Recall that each node in the tree is uniquely defined by a path from the root to this node, i.e., by the partial code word y^i. We use v to denote the node currently visited by the algorithm, and associate with this node the following value:

$$E(v \mid x) = E(y^i \mid x) = Q(y^i \mid x) \times H(y^i \mid x)$$

This value can be interpreted as an approximation of the maximal value of $Q(y \mid x)$, in which $Q(y^i \mid x)$ is the part of the path that can be computed when moving from the root to node v, and $H(y^i \mid x)$ is a heuristic that optimistically guesses the part of the path that has not yet been computed (in the considered case, $E(y^i \mid x)$ has to overestimate or to be the same as the maximal value of

$Q(\boldsymbol{y}\,|\,\boldsymbol{x}))$. $Q(\boldsymbol{y}^i\,|\,\boldsymbol{x})$ can be computed recursively as follows: $Q(\boldsymbol{y}^0\,|\,\boldsymbol{x}) = 1$ and

$$Q(\boldsymbol{y}^i\,|\,\boldsymbol{x}) = Q(y_i = 1|\boldsymbol{y}^{i-1},\boldsymbol{x}) \times Q(\boldsymbol{y}^{i-1}\,|\,\boldsymbol{x})\,. \tag{4}$$

In [18], a procedure for computing $H(\boldsymbol{y}^i\,|\,\boldsymbol{x})$ is proposed for the specific case of logistic regression as a base learner, whereas the heuristic is simply $H(\boldsymbol{y}^i\,|\,\boldsymbol{x}) = 1$ in uniform-cost search used in [10]. The former approach has the advantage of providing a more accurate estimation of maximal $Q(\boldsymbol{y}\,|\,\boldsymbol{x})$, albeit with an additional computing cost, while the latter approach makes a more rough estimation without any additional cost. Interestingly, as shown in experiments in [18], the former approach is still more expensive in terms of the total search cost than the latter.

In a nutshell, Algorithm 1 starts from the root of the label tree, which is the single element of priority list $\mathcal{Q}$, sorted in descending order of E. In every iteration, the top element of the list is popped and the children $\boldsymbol{v}_0$ and $\boldsymbol{v}_1$ of the corresponding node $\boldsymbol{v}$ are visited. $E(\boldsymbol{y}^i\,|\,\boldsymbol{x})$ is then recursively computed for the children of node $\boldsymbol{v}$, which are added to the list if this quantity exceeds the threshold $\epsilon = 2^{-c}$ with $1 \le c \le l$, where l is the maximal length of the path in the tree. Basically, they are inserted into the list at the appropriate position, so that the order imposed by $E(\boldsymbol{y}^i\,|\,\boldsymbol{x})$ is respected. The first while-loop of the algorithm stops in two situations: (i) when the element popped from the list $\mathcal{Q}$ corresponds to a leaf of the tree, or (ii) when the list $\mathcal{Q}$ is empty. The label corresponding to the leaf is then returned in the former case, while in the latter case, inference by greedy search is applied to define a path from all nodes from the list $\mathcal{K}$. This list, also sorted in descending order of E, contains nodes for which none of their children has been added to $\mathcal{Q}$. The use of list $\mathcal{K}$ ensures that by decreasing the value of ϵ, the algorithm will always find a solution that is not worse than a solution that would be found with greater ϵ.

Algorithm 1 enjoys strong theoretical guarantees. Assuming the cost for computing $H(\boldsymbol{y}^i\,|\,\boldsymbol{x})$ to be constant, the following result immediately follows from a theorem proved in [10].

Theorem 1. *Let $1 \le c \le l$. Algorithm 1 with $\epsilon = 2^{-c}$ needs at most $\mathcal{O}(l\epsilon^{-1})$ iterations to find a prediction $h_\epsilon(\boldsymbol{x}) = \hat{\boldsymbol{y}}_\epsilon$ such that*

$$Q(\hat{\boldsymbol{y}}^*\,|\,\boldsymbol{x}) - Q(\hat{\boldsymbol{y}}_\epsilon\,|\,\boldsymbol{x}) \le \epsilon - 2^{-l}\,.$$

From the theorem, we see that the quality of the solution found by the algorithm improves with the length of the running time. Consequently, the algorithm will always find the optimal solution, provided its probability mass is greater than ϵ. Reformulating the above, we can say that the algorithm finds the solution in time linear in $1/q_{max}$, where q_{max} is the probability mass of the best solution in the estimated distribution Q. For problems with low noise (high values of q_{max}), this method should work very fast.

The theorem also implies that the greedy search, which corresponds to the algorithm with $\epsilon = 0.5$, has very poor guarantees that approach the bound of 0.5 with $m \to \infty$.

4   Regret Bounds for PCT

In this section, we are concerned with the generalization ability of the PCT classifier, measured by means of the regret (1). Assume for a moment that $Q(\cdot|\boldsymbol{x})$, the label distribution produced by PCT, coincides with the true conditional distribution $P(\cdot|\boldsymbol{x})$ for every $\boldsymbol{x}$. Then, if the ϵ-approximate inference algorithm is used for classification, Theorem 1 implies the regret of the PCT classifier is at most ϵ, i.e., the expected classification error of PCT is at most ϵ larger than the expected classification error of the Bayes classifier.

It is, however, unrealistic to assume that PCT is able to perfectly match the true data distribution, hence $Q(\cdot|\boldsymbol{x})$ and $P(\cdot|\boldsymbol{x})$ will differ in general. Thus, the question arises whether the expected classification error of PCT is still not much worse than the expected classification error of the Bayes classifier if $Q(\cdot|\boldsymbol{x})$ and $P(\cdot|\boldsymbol{x})$ do not coincide, but are *close* to each other in some sense. This section presents an affirmative answer to this question, delivering a regret bound on the classification error that takes into account the predictive performance of the internal classifiers. More precisely, we bound the PCT regret for 0/1 loss in terms of the difference between Q and P, quantified in terms of *log-loss regret*.

We start with a general definition of the log-loss. Consider a problem of estimating a probability distribution on some outcome space $\mathcal{S}$. The log-loss of probability estimate $Q(\cdot)$ on $\mathcal{S}$ when the observed outcome is $y \in \mathcal{S}$ is given by

$$\ell_{\log}(y, Q) = -\log Q(y)\,.$$

The log-loss is by far the most popular measure for quantifying the accuracy of probabilistic predictions, and plays an important role in information theory, data compression, and statistics [7] (we briefly analyze the other loss function, squared loss, in Sect. 5). The *log-loss risk* is the expected log-loss of $Q(\cdot)$:

$$L_{\log}(Q) = \mathbb{E}_{y\sim P}[\ell_{\log}(y, Q)]\,,$$

where $P(\cdot)$ is the true distribution of y. The log-loss is a *strictly proper loss*, which means that the unique minimizer of the risk is achieved at $Q(\cdot) \equiv P(\cdot)$ (see, e.g., [21]). We thus define the *log-loss regret* as:

$$\mathrm{reg}_{\log}(Q) = L_{\log}(Q) - L_{\log}(P) = \mathbb{E}_{y\sim P}\left[\log \frac{P(y)}{Q(y)}\right] = D(P\|Q),$$

where $D(\cdot\|\cdot)$ is the Kullback-Leibler (KL) divergence.

We now turn back to PCTs. Let us first fix an instance $\boldsymbol{x} \in \mathcal{X}$ and consider the distribution over code words $\boldsymbol{y} \in \mathcal{C}$. There are two ways in which log-loss can be used in this setting:

- To measure the quality of the estimate of the joint distribution of labels given $\boldsymbol{x}$, $Q(\boldsymbol{y}|\boldsymbol{x})$, i.e., the outcome space is $\mathcal{S} = \mathcal{C}$, and the log-loss is $\ell_{\log}(\boldsymbol{y}, Q(\cdot|\boldsymbol{x})) = -\log Q(\boldsymbol{y}|\boldsymbol{x})$. The log-loss regret is then the KL divergence between true joint conditional distribution $P(\boldsymbol{y}|\boldsymbol{x})$ and its estimate $Q(\boldsymbol{y}|\boldsymbol{x})$, $\mathrm{reg}_{\log}(Q(\cdot|\boldsymbol{x})) = D(P(\cdot|\boldsymbol{x})\|Q(\cdot|\boldsymbol{x}))$.

– To measure the quality of individual classifiers in each node of the tree. Given a node $y^{i-1} = (y_1, \ldots, y_{i-1})$, the probability estimate for label $y_i \in \{0,1\}$ at this node is $Q(\cdot|y^{i-1}, x)$. Thus, the outcome space is $S = \{0,1\}$, and $\ell_{\log}(y_i, Q(\cdot|y^{i-1}, x)) = -\log Q(y_i|y^{i-1}, x)$. The log-loss regret is then $\mathrm{reg}_{\log}(Q(\cdot|y^{i-1}, x)) = D(P(\cdot|y^{i-1}, x)\|Q(\cdot|y^{i-1}, x))$.

Both ways described above turn out to be equivalent. Indeed, we have

$$\ell_{\log}(y, Q(\cdot|x)) = -\log Q(y|x) = \sum_{i=1}^{l} -\log Q(y_i|y^{i-1}, x)$$

$$= \sum_{i=1}^{l} \ell_{\log}(y_i, Q(\cdot|y^{i-1}, x)),$$

so that the log-loss of the joint distribution is equal to the sum of log-losses of individual node classifiers along the path from the root to leaf y. Similarly,

$$\mathrm{reg}_{\log}(Q(\cdot|x)) = \mathbb{E}_{y\sim P(\cdot|x)}\left[\log \frac{P(y|x)}{Q(y|x)}\right] = \mathbb{E}_{y\sim P(\cdot|x)}\left[\sum_{i=1}^{l} \log \frac{P(y_i|y^{i-1}, x)}{Q(y_i|y^{i-1}, x)}\right]$$

$$= \mathbb{E}_{y\sim P(\cdot|x)}\left[\sum_{i=1}^{l} \mathrm{reg}_{\log}(Q(\cdot|y^{i-1}, x))\right], \tag{5}$$

i.e., the log-loss regret of the joint distribution is equal to the sum of the regrets of node classifiers along the random path from the root to leaf y, where y is drawn from $P(\cdot|x)$. This basically expresses the chain rule for KL divergence [7]. The consequence of the above is that under log-loss we theoretically do not pay any price in terms of performance for representing the joint distribution by a tree structure.

We are now ready to present the main result of this section, which states that the 0/1-regret of the PCT classifier is bounded by means of the sum of log-loss regrets along a random path from the root to the leaf (or, equivalently, by the log-loss regret of the joint distribution) and the search error ϵ of the inference procedure.

Theorem 2. *Consider PCT, which estimates the probability $Q(\cdot|y^{i-1}, x)$ in each non-leaf node y^{i-1}, and let h_ϵ be the classifier which for any x, outputs $\hat{y}_\epsilon$ found by the ϵ-approximate inference procedure (Algorithm 1). Then, for any distribution P,*

$$\mathrm{reg}_{0/1}(h_\epsilon) \leq \sqrt{2\mathrm{reg}_{\log}(Q)} + \epsilon - 2^{-l},$$

where $\mathrm{reg}_{\log}(Q) = \mathbb{E}_{(x,y)\sim P}\left[\sum_{i=1}^{l} \mathrm{reg}_{\log}(Q(\cdot|y^i, x))\right]$ is the expected sum of regrets at internal classifiers along a path from the root to the leaf.

Proof. We first condition everything on a fixed x. Let $y^* = \arg\max_y P(y|x)$ be the mode of $P(\cdot|x)$, and let $\hat{y}_\epsilon = h_\epsilon(x)$ be the output of Algorithm 1 for input x.

Moreover, we let $\hat{\boldsymbol{y}}^* = \arg\max_{\boldsymbol{y}} Q(\boldsymbol{y}|\boldsymbol{x})$ denote the mode of $Q(\cdot|\boldsymbol{x})$, and note that from Theorem 1,

$$Q(\hat{\boldsymbol{y}}^*|\boldsymbol{x}) - Q(\hat{\boldsymbol{y}}_\epsilon|\boldsymbol{x}) \leq \epsilon - 2^{-l}. \tag{6}$$

According to (1), the 0/1-regret of $\hat{\boldsymbol{y}}_\epsilon$ conditioned at $\boldsymbol{x}$ is given by

$$\text{reg}_{0/1}(\hat{\boldsymbol{y}}_\epsilon) = P(\boldsymbol{y}^*|\boldsymbol{x}) - P(\hat{\boldsymbol{y}}_\epsilon|\boldsymbol{x}).$$

Note that the regret is 0 if $\boldsymbol{y}^* = \hat{\boldsymbol{y}}_\epsilon$, hence we assume $\boldsymbol{y}^* \neq \hat{\boldsymbol{y}}_\epsilon$ in what follows. From the definition of $\hat{\boldsymbol{y}}^*$, $Q(\hat{\boldsymbol{y}}^*|\boldsymbol{x}) - Q(\boldsymbol{y}^*|\boldsymbol{x}) \geq 0$, which together with (6) gives $Q(\hat{\boldsymbol{y}}_\epsilon|\boldsymbol{x}) - Q(\boldsymbol{y}^*|\boldsymbol{x}) + \epsilon - 2^{-l} \geq 0$. Hence, we obtain the upper bound

$$\begin{aligned}
\text{reg}_{0/1}(\hat{\boldsymbol{y}}_\epsilon) &\leq \left(P(\boldsymbol{y}^*|\boldsymbol{x}) - Q(\boldsymbol{y}^*|\boldsymbol{x})\right) + \left(Q(\hat{\boldsymbol{y}}_\epsilon|\boldsymbol{x}) - P(\hat{\boldsymbol{y}}_\epsilon|\boldsymbol{x})\right) + \epsilon - 2^{-l} \\
&\leq |P(\boldsymbol{y}^*|\boldsymbol{x}) - Q(\boldsymbol{y}^*|\boldsymbol{x})| + |Q(\hat{\boldsymbol{y}}_\epsilon|\boldsymbol{x}) - P(\hat{\boldsymbol{y}}_\epsilon|\boldsymbol{x})| + \epsilon - 2^{-l} \\
&\leq \sum_{\boldsymbol{y} \in C} |P(\boldsymbol{y}|\boldsymbol{x}) - Q(\boldsymbol{y}|\boldsymbol{x})| + \epsilon - 2^{-l},
\end{aligned}$$

where the last inequality is from $\boldsymbol{y}^* \neq \hat{\boldsymbol{y}}_\epsilon$. We now make use of Pinsker's inequality

$$\frac{1}{2}\sum_{\boldsymbol{y} \in C} |P(\boldsymbol{y}\,|\,\boldsymbol{x}) - Q(\boldsymbol{y}\,|\,\boldsymbol{x})| \leq \sqrt{\frac{1}{2}D(P(\cdot\,|\,\boldsymbol{x})\|Q(\cdot\,|\,\boldsymbol{x}))},$$

which together with (5) implies

$$\text{reg}_{0/1}(\hat{\boldsymbol{y}}_\epsilon) \leq \sqrt{2\mathbb{E}_{\boldsymbol{y}\sim P(\cdot|\boldsymbol{x})}\left[\sum_{i=1}^{l} \text{reg}_{\log}(Q(\cdot|\boldsymbol{y}^{i-1},\boldsymbol{x}))\right]} + \epsilon - 2^{-l}. \tag{7}$$

Note that the 0/1-regret of h_ϵ, $\text{reg}_{0/1}(h_\epsilon)$, is just the expectation of the left-hand side of (7) with respect to $\boldsymbol{x}$. Thus, taking expectation on both sides of (7), and using $\mathbb{E}\left[\sqrt{\cdot}\right] \leq \sqrt{\mathbb{E}[\cdot]}$ on the right-hand side (which is Jensen's inequality applied to the concave function $x \mapsto \sqrt{x}$) gives

$$\begin{aligned}
\text{reg}_{0/1}(h_\epsilon) &\leq \sqrt{2\mathbb{E}_{(\boldsymbol{x},\boldsymbol{y})\sim P}\left[\sum_{i=1}^{l} \text{reg}_{\log}(Q(\cdot|\boldsymbol{y}^{i-1},\boldsymbol{x}))\right]} + \epsilon - 2^{-l} \\
&= \sqrt{2\text{reg}_{\log}(Q)} + \epsilon - 2^{-l}.
\end{aligned}$$

$\square$

Theorem 2 states that if the log-loss regret of node classifiers is small, the resulting ϵ-approximate classifier will be close to the Bayes classifier in terms of 0/1 loss. This suggests to use node classifiers which minimize log-loss on the training sample, examples of which include logistic regression, Gradient Boosting Machines, deep neural networks,[2] and many others. One can show that the square-root dependence in the bound of Theorem 2 cannot be improved in general, since when the tree consists only of the root node, our bound essentially specializes to the bound in [1], which also exhibits square-root dependence.

---

[2] In this case, the log-loss if often referred to as "soft-max" function.

5   Relation to Other Label Tree Approaches

5.1   Conditional Probability Trees

Conditional probability trees (CPTs) [4] estimate a conditional probability distribution $P(y|\boldsymbol{x})$ in the multiclass setting and have the same structure as PCTs. What distinguishes this approach from ours is that CPTs are used for probability estimation, with squared loss $\ell_{\mathrm{sq}}(y_i, Q(\cdot|\boldsymbol{y}^{i-1}, \boldsymbol{x})) = \left(y_i - Q(y_i|\boldsymbol{y}^{i-1}, \boldsymbol{x})\right)^2$ as a performance measure, whence there is no inference phase to determine the mode of the conditional distribution. The main result in [4] relates the squared loss regret on the joint distribution to the expected squared loss over the nodes of the tree. This result is analogous to the identity (5), except that an additional $O(\sqrt{l})$ factor appears in the squared loss bound. Moreover, no result analogous to Theorem 2 is given, which would relate expected squared loss regret to the 0/1 classification regret.

In fact, we can show a *lower bound* on the 0/1 regret in terms of expected squared loss, which is at least a factor of $\Omega(\sqrt{l})$ worse than our bound. To be more precise, one can show that for any $l > 2$, there exists a true distribution P and an estimate Q with the following property: even when assuming that the inference algorithm can identify the mode of the distribution exactly, it holds that $\mathrm{reg}_{0/1}(h_\epsilon) > \sqrt{l\,\mathrm{reg}_{\mathrm{sq}}(Q)}$, where $\mathrm{reg}_{\mathrm{sq}}(Q)$ is the corresponding regret with log-loss replaced by squared loss.[3] In other words, using squared loss yields a bound for classification error that is at least a factor $\Omega(\sqrt{l})$ worse than the bound we obtained for log-loss.

5.2   Filter Trees

The filter tree (FT) approach [3] is the first label tree algorithm for which a regret bound for the classification error has been proved. Interestingly, the specific training procedure used in FTs ensures that the greedy classification procedure is sufficient for obtaining consistent predictions.

FT uses the same tree structure as PCT, but with binary classifiers instead of class probability estimators in the non-leaf nodes of the tree. The method follows a bottom-up strategy, which can be interpreted as a single elimination tournament on the set of labels. A classifier in node $\boldsymbol{y}^{i-1}$ is trained to predict y_i, but FT implicitly transforms the underlying distribution of examples in the node. The transformation for 0/1 loss relies on filtering out all training examples that have been misclassified by successor classifiers on a path to a leaf. The learning algorithm starts with classifiers on the lowest non-leaf level of the tree. The correctly classified examples are then moved upward to nodes one level above. This process is repeated until the root node is reached.

In [3], a regret bound for 0/1 loss has been proved that is conceptually similar to the one given in Theorem 2. The difference is that the right side of the

---

[3] We skip the details of the construction of P and Q due to the space limit.

bound is expressed in terms of 0/1 loss of the binary classifiers in non-leaf nodes. Therefore, these two bounds are not directly comparable.

Another advantage of FTs is that they can be used with any cost-based loss function. An appropriate bound has also been proved in [3]. The classification procedure still follows a greedy search, but training is more demanding. It requires weighting of examples, the use of cost-sensitive learners, and each training example generally occurs in each internal classifier.

6   Extensions of PCTs

Since PCT estimates the entire conditional distribution over labels, it can be used with any loss function. This comes with no additional cost during training, but may lead to very costly inference. Actually, inference can be performed efficiently only for certain losses, such as 0/1 loss as discussed in Sect. 3.2, but also some ranking-based loss functions. As an example, consider recall at kth position defined as

$$R_{@k}(\boldsymbol{y}, \boldsymbol{x}, \mathcal{Y}_k) = [\![\boldsymbol{y} \in \mathcal{Y}_k]\!],$$

where $\mathcal{Y}_k$ is a set of k labels predicted for $\boldsymbol{x}$. One can easily verify that an optimal $\mathcal{Y}_k$ should contain k top-labels with largest $P(y \mid \boldsymbol{x})$. This can be approximated by k top-labels with largest $Q(y \mid \boldsymbol{x})$, which are easily obtained by PCT and a small extension of the ϵ-approximate algorithm: it is enough to continue the search procedure until k leaves are visited. Moreover, the bound in Theorem 2 can be easily extended to this case.

As already mentioned, PCTs can also be used in multi-label classification. In this case, the tree is of height m and is fully balanced. Each path from the root to a leaf corresponds to one of possible label combinations. In principle, PCT contains a single classifier in each non-leaf node. In multi-label case, storing $2^m - 1$ classifiers for large m is not feasible. One can, however, follow a trick used in probabilistic classifier chains [9] and condensed filter trees [17], which relies on using one binary classifier per tree level. In other words, prediction of the ith label corresponds to the prediction made by the classifier on level i with additional features that indicate a given node of the tree.

7   Experimental Results

We empirically evaluate PCTs and FTs in two scenarios: multi-label classification (MLC) and multi-class classification (MCC). We test the algorithms in terms of 0/1 loss and the computational costs of their training and testing procedures. For PCTs, we additionally report $R_{@k}$.

We conduct experiments on 3 multi-class and 3 multi-label datasets.[4] Table 1 provides a summary of basic statistics of the datasets. Notice that the number of

---

[4] Taken from the libsvm repository https://www.csie.ntu.edu.tw/~cjlin/libsvmtools/ datasets and the image net competition webpage http://www.image-net.org/ challenges/LSVRC/2010.

Table 1. Multi-class (MCC) and multi-label (MLC) datasets and their properties: the number of training (#train) and test (#test) examples, the number of labels (m) and features (d).

MCC					MLC				
Dataset	#train	#test	m	d	Dataset	#train	#test	m	d
Sector	6412	3207	105	55197	Yeast	1500	917	14	103
Aloi	97200	10800	1000	128	TMC	21519	7077	22	30438
ILSVR2010	1261406	150000	1000	1000	Mediamill	30993	12914	101	120

leaf nodes is equal to m (the number of labels) in case of multi-class problems, and 2^m (the number of all possible label combinations) in case of multi-label problems. We therefore use multi-label datasets up to around 100 labels. For datasets with a greater number of labels, the 0/1 loss is usually very close to 1. We use the original split into a training and test set if available; otherwise, we use 90/10 train/test splits. For the ILSVR2010 dataset, we use the visual code words (sbow) vectors provided by the organizers of the challenge. Features were generated on the basis of the guidance contained in the ILSVR development kit.

7.1 Implementation

We carefully implemented PCTs and FTs in Java. As internal classifiers, we use L_2 linear logistic regression trained by a variant of stochastic gradient descent (SGD) introduced in [13]. To deal with a large number of weights, we use feature hashing [22] shared over all tree nodes using hashes up to size of 2^{24}. We use a random complete binary tree to code class labels in the MCC scenarios and train a classifier in each node of the tree. For MLC problems we take the original order of the labels to obtain the code words. We use one classifier per tree level. We tune the hyper-parameters of SGD in a 80/20 simple validation on the training set. We applied an off-the-shelf hyper-parameter optimizer [15] with a wide range of parameters. We tune PCTs to optimize the log-loss as suggested by our theoretical analysis. FTs are tuned to perform well on 0/1 loss.

We use PCTs with the ϵ-approximate inference algorithm with different values of $\epsilon \in \{0, 0.25, 0.5\}$. The variant with $\epsilon = 0.5$ corresponds to greedy search, while the algorithm with $\epsilon = 0$ will always find the optimal solution, but may visit all nodes of the tree in the worst case (in fact, ϵ should be set to 2^{-l} instead to 0 to be concordant with the description of the algorithm; to keep the notation simple, we use 0 to indicate the smallest possible value of ϵ for a given dataset).

7.2 Results

The results are given in Table 2. We can observe that PCTs improve with decreasing value of ϵ. PCT with $\epsilon = 0.5$ gets worse results than FT, which confirms the theoretical results, i.e., filtering of misclassified examples during training in FT

Table 2. Experimental results for 0/1 loss and 1-$R_{@5}$ (both in %), train (t_{trn}) and test (t_{test}) running times (in seconds), and the average (A) number of inner products per a test example. The *Top 1* column indicates the results for top-1 prediction, while column *Top 5* the results for top-5 prediction (only for PCT with $\epsilon < 0.5$). The best results are indicated in bold (except for wall-clock times which can be affected by many factors). The value in subscript of PCT corresponds to the value of ϵ.

	MCC							MLC						
	t_{trn}	Top 1			Top 5			t_{trn}	Top 1			Top 5		
		0/1	t_{test}	A	1-$R_{@5}$	t_{test}	A		0/1	t_{test}	A	1-$R_{@5}$	t_{test}	A
	Sector, $m = 105$							Yeast, $m = 14$						
FT	11.75	13.43	0.144	6.81	–			2.49	78.73	0.07	**14**	–		
PCT$_{.5}$	11.56	17.18	0.154	6.81	–			3.12	80.15	0.04	**14**	–		
PCT$_{.25}$	11.56	13.68	0.16	7.04	12.61	**0.24**	7.5	3.12	79.28	0.05	17.15	76.22	**0.12**	21.3
PCT$_0$	11.56	**13.28**	0.198	7.13	**7.23**	0.48	18.2	3.12	**78.62**	0.09	23.82	**58.77**	0.17	64.6
	Aloi, $m = 105$							TMC, $m = 22$						
FT	15.11	88.98	0.14	**9.97**	–			30.7	77.06	0.47	**22**	–		
PCT$_{.5}$	13.43	88.99	0.14	**9.97**	–			34.3	75.06	0.39	**22**	–		
PCT$_{.25}$	13.43	**88.95**	0.15	9.98	88.64	**0.21**	10.2	34.3	73.74	0.45	27.97	68.50	**0.57**	**34.0**
PCT$_0$	13.43	**88.95**	0.21	9.98	**76.19**	0.55	26.1	34.3	**73.18**	0.73	33.50	**41.18**	1.29	87.9
	ILSVR2010, $m = 1000$							Mediamill, $m = 101$						
FT	1710	95.10	10.12	**8.39**	–			220	90.79	2.24	**101**	–		
PCT$_{.5}$	1825	99.96	10.13	**8.39**	–			274	90.78	2.22	**101**	–		
PCT$_{.25}$	1825	95.30	13.23	10.03	95.30	**20.10**	14.4	274	90.06	2.79	107	89.14	**3.02**	129
PCT$_0$	1825	**94.76**	15.20	10.57	**92.33**	44.31	34.3	274	**89.65**	5.23	274	**74.22**	9.50	529

improves for the greedy inference. For $\epsilon = 0.25$, the results are already very competitive to FT. For $\epsilon = 0$, PCT consistently outperforms FT, but the difference is not always large.

From a computational perspective, FTs achieve better performance. The training time of both approaches is very similar, but the testing time is in favor of FTs (and PCTs with $\epsilon = 0.5$). To give a deeper insight into the time costs we also report the average number of inner products computed by internal classifiers per test example. Interestingly, PCT with $\epsilon = 0$ always finds the solution in a reasonable time. Its testing time is never longer than three times that of FT. Similarly, the number of inner products is only up to three times greater than that of FT or PCT with $\epsilon = 0.5$.

Recall at kth position ($R_{@k}$) can be measured only for PCTs. There is no way to deliver top-k predictions in FTs, since this algorithm uses binary decisions in non-leaf nodes, so the search process results only in a single path from the root to a leaf node. From the results we observe that PCT efficiently finds topmost results. The positive label appears more often in the top-5 predictions than in the top-1. Similarly as for 0/1 loss, $R_{@5}$ improves with decreasing value of ϵ. Unfortunately, predicting top-k labels increases test time. Therefore, the label tree search for $\epsilon = 0$ requires about 2–3 times more steps to find top-5 labels.

8   Conclusions

In this paper, we analyzed probabilistic classifier trees for efficient multi-class and multi-label classification. In particular, we proved a regret bound for 0/1 loss, which provides a strong theoretical foundation of PCTs, and which can also be extended to ranking-based losses. Moreover, we compared PCTs with the closely related filter tree method. We conclude the paper by summarizing the main theoretical and empirical results of FTs and PCTs, pointing out advantages and disadvantages of both approaches.

An unquestionable advantage of FTs is their prediction time, which is logarithmic in the number of classes or possible label combinations. FT can be used with any type of binary classifier as base learner and relies on simple 0/1 predictions. However, to guarantee the consistency of greedy inference, it requires more demanding training. In the naïve implementation, classifiers are trained sequentially in a bottom-up manner. The most important disadvantage is a significant reduction of the number of training examples in the top levels of the tree, which is caused by filtering examples in each level from bottom to top. This sparsity of training data may deteriorate predictive performance. However, thanks to filtering, an internal classifier is aware of errors of the successor classifiers. FT can be used with any cost-based loss function, but it is not able to predict top-k labels.

Prediction with PCTs requires search techniques, whence it is usually more demanding than FTs (yet significantly faster than 1-vs-all). Moreover, anytime algorithms can be used for searching the tree. The time complexity of PCT strongly depends on the noise contained in the data. If the signal-to-noise ratio is high, we can expect prediction time to be small. However, learning is much simpler for PCT than for FT, and can be easily parallelized. There is no filtering of training examples, so all examples are used for training on each level of the tree. The probabilistic nature of PCTs allows for delivering a list of top-labels and to work efficiently for $R_{@k}$.

The results we obtained for FTs are comparable with those reported in [6]. We stress that better results can be obtained by other algorithms, for example LomTrees introduced in the same paper. This is mainly because LomTrees train the tree structure online, along with the internal classifiers, whereas PCTs and FTs use random trees/coding. Interestingly, LomTrees are not consistent. Thus, an important challenge for future research is to find an algorithm that is able to train the tree structure online while ensuring consistency.

Acknowledgments. The work of Krzysztof Dembczyński and Wojciech Kotłowski has been supported by the Polish National Science Centre under grant no. 2013/09/D/ST6/03917 and 2013/11/D/ST6/03050, respectively.

References

1. Bartlett, P.L., Jordan, M.I., McAuliffe, J.D.: Convexity, classification, and risk bounds. J. Am. Stat. Assoc. **101**(473), 138–156 (2006)
2. Bengio, S., Weston, J., Grangier, D.: Label embedding trees for large multi-class tasks. In: NIPS, vol. 23, pp. 163–171. Curran Associates, Inc. (2010)
3. Beygelzimer, A., Langford, J., Lifshits, Y., Sorkin, G.B., Strehl, A.L.: Conditional probability tree estimation analysis and algorithms. In: UAI, pp. 51–58 (2009)
4. Beygelzimer, A., Langford, J., Ravikumar, P.: Error-correcting tournaments. In: Chaudhuri, K., Gentile, C., Zilles, S. (eds.) ALT 2015. LNCS (LNAI), vol. 9355, pp. 247–262. Springer, Heidelberg (2009). doi:10.1007/978-3-642-04414-4_22
5. Bottou, L.: Large-scale machine learning with stochastic gradient descent. In: Lechevallier, Y., Saporta, G. (eds.) Proceedings of COMPSTAT 2010, pp. 177–187. Springer, Heidelberg (2010)
6. Choromanska, A., Langford, J.: Logarithmic time online multiclass prediction. In: NIPS, vol. 29 (2015)
7. Cover, T., Thomas, J.: Elements of Information Theory. Wiley, New York (1991)
8. Dekel, O., Shamir, O.: Multiclass-multilabel learning when the label set grows with the number of examples. In: AISTATS (2010)
9. Dembczyński, K., Cheng, W., Hüllermeier, E.: Bayes optimal multilabel classification via probabilistic classifier chains. In: ICML, pp. 279–286. Omnipress (2010)
10. Dembczyński, K., Waegeman, W., Cheng, W., Hüllermeier, E.: An analysis of chaining in multi-label classification. In: ECAI (2012)
11. Deng, J., Dong, W., Socher, R., Li, L.J., Li, K., Li, F.F.: ImageNet: a large-scale hierarchical image database. In: CVPR, pp. 248–255 (2009)
12. Deng, J., Satheesh, S., Berg, A.C., Li, F.F.: Fast and balanced: efficient label tree learning for large scale object recognition. In: NIPS, vol. 24, pp. 567–575 (2011)
13. Duchi, J., Singer, Y.: Efficient online and batch learning using forward backward splitting. JMLR **10**, 2899–2934 (2009)
14. Fox, J.: Applied Regression Analysis, Linear Models, and Related Methods. Sage, Thousand Oaks (1997)
15. Hutter, F., Hoos, H.H., Leyton-Brown, K.: Sequential model-based optimization for general algorithm configuration. In: Coello, C.A.C. (ed.) LION 2011. LNCS, vol. 6683, pp. 507–523. Springer, Heidelberg (2011). doi:10.1007/978-3-642-25566-3_40
16. Kumar, A., Vembu, S., Menon, A.K., Elkan, C.: Beam search algorithms for multilabel learning. Mach. Learn. **92**(1), 65–89 (2013)
17. Li, C.L., Lin, H.T.: Condensed filter tree for cost-sensitive multi-label classification. In: ICML, pp. 423–431 (2014)
18. Mena, D., Montañés, E., Quevedo, J.R., del Coz, J.J.: Using A* for inference in probabilistic classifier chains. In: IJCAI, pp. 3707–3713 (2015)
19. Morin, F., Bengio, Y.: Hierarchical probabilistic neural network language model. In: AISTATS, pp. 246–252 (2005)
20. Prabhu, Y., Varma, M.: Fastxml: A fast, accurate and stable tree-classifier for extreme multi-label learning. In: KDD, pp. 263–272. ACM (2014)
21. Reid, M.D., Williamson, R.C.: Composite binary losses. JMLR **11**, 2387–2422 (2010)
22. Weinberger, K., Dasgupta, A., Langford, J., Smola, A., Attenberg, J.: Feature hashing for large scale multitask learning. In: ICML, pp. 1113–1120. ACM (2009)
23. Weston, J., Makadia, A., Yee, H.: Label partitioning for sublinear ranking. In: ICML, pp. 181–189 (2013)

Actively Interacting with Experts: A Probabilistic Logic Approach

Phillip Odom[(✉)] and Sriraam Natarajan

Indiana University, Bloomington, IN, USA
{phodom,natarasr}@indiana.edu

Abstract. Machine learning approaches that utilize human experts combine domain experience with data to generate novel knowledge. Unfortunately, most methods either provide only a limited form of communication with the human expert and/or are overly reliant on the human expert to specify their knowledge upfront. Thus, the expert is unable to understand what the system could learn without their involvement. Allowing the learning algorithm to query the human expert in the most useful areas of the feature space takes full advantage of the data as well as the expert. We introduce *active advice-seeking* for relational domains. Relational logic allows for compact, but expressive interaction between the human expert and the learning algorithm. We demonstrate our algorithm empirically on several standard relational datasets.

1 Introduction

Probabilistic logic models (PLMs) [3,8] combine the expressive power of first-order logic and the ability of probability theory to model noise and uncertainty. They have been inspired by databases [6,9] and by logic [4,5]. Given their expressivity, several powerful learning algorithms have been developed that allow for learning from interpretations [5,18] and learning from entailment [4,23]. While efficient algorithms have been developed to learn the parameters of these models (either weights or probabilities), full model-learning (also called *structure learning* to denote learning of the logical structure) remains a challenging task. Recently, methods based on ensemble learning have been proposed that allow for efficient structure learning for PLMs [16].

These methods essentially rely only on data. Given that the primary assumption is that data can be noisy, restricting humans to be mere labelers of the data, as is done in many popular approaches, is inefficient. Recently, a formulation for incorporating prior knowledge as *preferences* over labels for the ensemble learning method was proposed [19]. The key idea was to explicitly trade-off between the label preferences suggested by the human expert and the posterior label distributions obtained from the data. It was demonstrated that advice was particularly useful where there was targeted noise. For example, missing certain regions in a segmentation task, or missing stop signs when creating driving demonstrations.

While the framework of Odom et al. [19] does not merely treat the given advice as "prior" knowledge, it assumes that all the advice is provided up-front

© Springer International Publishing AG 2016
P. Frasconi et al. (Eds.): ECML PKDD 2016, Part II, LNAI 9852, pp. 527–542, 2016.
DOI: 10.1007/978-3-319-46227-1_33

before the learning takes place. Not only is this a potentially time consuming task for the experts, but it is also highly likely that they, not being experts in machine learning or probabilistic logic, would find it difficult to identify the domain knowledge that might be optimal for the learning algorithm. Hence, inspired by active learning [24], we propose *active advice-seeking* that aims to determine the regions of (relational/logical) feature space that is ideal for obtaining advice. For instance, will the accuracy of a model learned to predict heart attacks be higher if advice is given about the population who is overweight and has high blood pressure or about the population which smokes but exercises regularly? The answer is not clear but this is where active advice-seeking should be helpful. The goal of active advice-seeking is to lessen the responsibility of the expert both in terms of the effort that must be spent in specifying the advice, as well as the necessity that the expert understands the intricacies of the algorithm. The algorithm will automatically identify the regions of the feature space where the advice will be useful.

More precisely, the proposed algorithm presents a set of conjunctions of predicates as queries to the expert. The size of the set is pre-determined by a budget given by the expert (i.e., the algorithm and the expert agree in advance for the number of allowable queries). In order to compute the clause that should be queried, the algorithm learns a model from only the data to compute a score for each example, then it uses a regression clause learner to fit the scores. The best clause is presented to the expert who provides a preference over the labels. For instance, in a university domain, the clause could be of the form $prof(X) \land student(Y) \land paper(P,X) \land paper(P,Y)$. The expert could then prefer the label to be $advisedBy(X,Y)$. Essentially the system is asking the expert, what is your choice of label if a student and a professor are co-authors? The expert replies saying, I prefer the student to be advised by the professor. Note that this is a "soft" preference in that this preference may not always hold. This preference is then explicitly weighed against the data while learning the model.

We make the following key contributions: first, we introduce the notion of advice-seeking to the probabilistic logic model (PLM) community (and the general AI community). Second, we adapt a recent successful knowledge-based probabilistic logic learning algorithm to seek advice from the human expert. Third, we present the first relational algorithm that can go beyond data and interactively solicit input from the expert. Finally, we demonstrate using experiments, that such an approach is robust in learning from noisy data.

The rest of the paper is organized as follows: we first introduce the required background on PLMs and active learning. Then, we present our learning approach before presenting empirical evaluations. Finally, we conclude the paper by outlining areas for future research.

2   Background

Techniques for incorporating expert knowledge into learning are a key precondition for any active advice-seeking approach to be successful. We aim to introduce a broad learning paradigm that can use any method that incorporates prior knowledge. To that effect, we cover one advice-based framework which we will use to empirically validate our approach.

2.1   Advice-Based PLMs

While there have been many knowledge-based systems developed for propositional models [7, 11, 14, 26, 27], work on probabilistic logic models (PLMs) has not progressed as far. In PLMs, the expert is typically used to define some prior structure that can either be used as the complete structure or locally refined.

Recently, Odom et al. [19] introduced a knowledge-based PLM method that learns seamlessly from data and any expert knowledge. While making use of Relational Functional Gradient-Boosting (RFGB) to learn the structure and parameters of the model simultaneously [17], they incorporate expert preferences which guide the structure and parameters to more robust models.

Extending previous work that considered knowledge as propositional Horn clauses [7, 12, 27], they considered their advice as first-order logic Horn clauses. Thereby, allowing experts to give advice over different granularities of examples. The body of the clause specifies the examples over which the expert would like to give advice, while the head of the clause gives the preferred and avoided labels. For example, a cardiologist might suggest that patients whose close relatives had heart problems are more likely to have a heart problem.

Odom et al. [19] incorporate this expert knowledge into RFGB [17] which learns a series of relational regression trees [2]. These relational regression trees have first-order logic literals in the nodes and regression values at the leaves. Functional gradient-boosting aims to capture the error in the current model in a regression tree and then adds this regression tree to the model. The final model is a sum over all of the learned trees.

The gradients used by Odom et al. [19] incorporate an additional term in the optimization function that pushes the model in the direction of the expert advice (represented by n_t and n_f, the number of advice which say that example x_i should be preferred/avoided)[1]

$$\Delta(x_i) = \alpha \cdot (I(y_i) - P(y_i; \psi)) + (1 - \alpha) \cdot [n_t(x_i) - n_f(x_i)]$$

While this approach has shown positive results in several difficult tasks, it still requires the expert to specify *all of the advice in advance*. Given a particular dataset, deciding the most useful advice is not a trivial problem. This problem is exacerbated by the fact that the expert could potentially have no expertise in machine learning. *Active advice-seeking* aims to alleviate this issue by querying the expert directly, using the training data as a guide to select the most useful queries. Previous work on *active advice-seeking* is limited to propositional queries in sequential decision making problems [20]. Grouping ground states into queries allowed the proposition algorithm to maximize the impact of the human expert. However, lifting advice to be relational as we do in this work is a more powerful and principled approach.

---

[1] Note the difference to standard (only data) RFGB which optimizes $(I(y_i) - P(y_i; \psi))$.

2.2 Active Learning

Active Learning is a related research problem where the goal is to make use of an expert that can provide the labels of examples [24]. Pool-based active learning approaches assume a pool of unlabeled examples from which the learning algorithm should choose. In *active advice-seeking*, this pool of examples is the training set. While there are labels in the training set, it is assumed that either there is not sufficient training data (and thus there is missing knowledge) or the training data is noisy and so the labels should not be fully trusted. So while active learning aims for finding the labels of the examples, we are soliciting advice.

Most active learning methods repeat the following general steps:

1. Learn a model from training data
2. Compute uncertainty over unlabeled data
3. Select examples based on uncertainty and solicit label
4. Add labeled examples to training set

The process begins by learning a model with the current set of labeled data. This model is then used to compute some measure of uncertainty (this could be entropy, KL-divergence or other measures) that suggests how likely the model would correctly predict the unlabeled examples. Consider a simple, linear classifier with two possible unlabeled examples, one located close to the decision boundary with the other located far from the boundary. The example close to the decision boundary is more likely to effect the decision boundary and would be selected for labeling.

This cycle accumulates the best examples to label at each step and has been shown to be effective especially in domains where there is a dearth of data available. However, labeling individual examples is not an effective use of human experts availability. Allowing expert's to give advice results in the expert being able to select the ideal granularity of advice (over a single example or many examples). *Active advice-seeking* aims to effectively use human experts by providing clauses instead of ground examples. Not only does this allow for automatically selecting the granularity of advice, but it also provides a compact description of the most uncertain examples.

A particular active learning paradigm that is closer to our work is the work of Rashidi and Cook [21]. In their work, they cluster informative examples and run a rule induction algorithm (such as C4.5) to generate a rule based query to which the expert can provide a label. The similarity to our approach lies in the use of a rule to ask the query. The two key differences are that, first, ours is a relational learning algorithm that goes beyond flat feature vectors. Second, the rule was used to obtain a label that was used for all the examples that satisfy that rule. In our case, we go beyond labels and solicit human advice as preferences over logical rules.

Active learning has been considered for relational data particularly, with the focus of querying for node labels based on the structure of the network [1,13,15,22] which have been studied under the broad area of active inference in relational domains. Particularly relevant to our paper are three of the most recent works - ALFNET [1], the RAL algorithm [13] and FLIP [25]. ALFNET employed uncertainty sampling to generate committee-based network clusters (which consisted of three classifiers) in order to query the expert. A related work in this direction is the RAL algorithm that used a utility metric with network variance as the criteria. This variance was used since the RAL algorithm is interested in across-network classifications. While we do not employ this heuristic, our algorithm can handle across-network classifications due to the underlying logic-based ensemble learner. Finally, the FLIP algorithm by Saha et al., extends the notion of active inference by considering several query selection methods and evaluates them on single and multi-labeled networks. Our algorithm is similar in spirit to ALFNET in that we employ uncertainty sampling as well but our query is generated using clauses learned through logic programming. An important difference to the RAL, FLIP and ALFNET algorithms is that we query for preferences over the relations instead of the actual labels.

3 Relational Active Advice-Seeking

The aim of relational active advice-seeking is to offload the task of selecting areas of the feature space to give targeted advice from the human expert to the learning algorithm. In relational models, experts are often asked to define the logical structure of the model with the parameters learned from data. However, it is important to be able to learn the full model (structure and parameters) especially in complex, real world domains. Experts can still provide valuable input about targeted areas of the feature space. The wide variety of potential expert advice complicates the advice-giving process and can lead the expert to give correct, but not relevant advice.

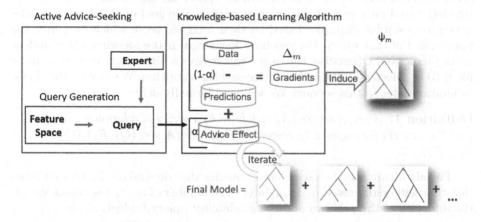

Fig. 1. An overview of our framework for actively interacting with human experts. The learner is responsible for selecting where to query the expert.

Previous work on advice-giving requires significant effort on the part of the expert to determine the relevant advice [12,19]. If the expert provides exhaustive advice, the learning algorithm will be able to learn an accurate classifier. However, the experts time is often limited and only a few queries can be answered. These queries should not be redundant, focusing on areas that are well covered by the data. Instead, they should focus on areas where the learning algorithm cannot distinguish the correct label or behavior. Thus, we extend relational advice-taking methods to active advice-seeking. Each part of our formulation is shown in Fig. 1. It consist of the active advice-seeking component that is capable of generating queries and interacting with the human expert as well as the knowledge-based learning algorithm which learns from the expert provided knowledge and any available training data.

4   Problem Formulation

The overall goal of our algorithm is to identify regions of the feature space that the agent is most uncertain about and query the expert for advice on these regions. In the propositional case, this was handled by simply clustering examples based on the distribution over the labels and querying the expert over this cluster [20]. However, this heuristic may not suffice for relational tasks since there are typically more negative examples than there are positives. Fortunately, the use of a rich representation such as first-order logic naturally allows us to query over the most uncertain regions of the feature space.

We represent the regions of feature space as conjunctions of predicates. Intuitively, this corresponds to grouping examples such that a particular condition is satisfied. More precisely, the goal of our algorithm is to select a set of conjunctions of first-order logic atoms about which to query the expert. These queries concisely describe the set of training examples to which the advice will apply. In order to select relevant areas of the feature space, the algorithm learns a clause (model) based on scores of the given examples. The goal of this learned model is to group similar examples based on their assigned score which measures the importance of that query. Queries have low scores if the algorithm is confident in its prediction, Otherwise, the query will receive a high score, making it more likely to be selected by the active advice-seeking algorithm. We explain the clause generation later in this section. We will now formally define advice:

Definition 1. *A set of advice (A) is defined as a series of relational queries (Q_i) and the experts corresponding response (R_i), i.e. (A =< $(Q_1, R_1), (Q_2, R_2), ...,$ (Q_n, R_n) >).*

The algorithm solicits a sequence of queries that depend on the scoring function that will be discussed in detail later. The number of queries is dependent on the difficulty of the problem and the availability (query budget) of the expert.

Definition 2. *A Relational Query (Q) is defined as a conjunction of literals ($\wedge f_i$), which defines the set of examples to which the advice will be applied. Q will be shown to the human expert.*

Definition 3. *An Expert Response (R) is defined as a set of preferred labels (l+), and a set of avoided label (l−) given with respect to a relational query. Note that both l + /l− could be empty if the expert does not understand Q or if the query does not separate different classes.*

If the expert is not satisfied with the query - possibly because the query does not properly delineate between labels - then the expert can provide no preferred or avoided labels. Such a query is not useful to the learning algorithm and squanders the time of the expert. The relational query and its accompanied response represent a single piece of advice that can be utilized by the knowledge-based learning algorithm. We now present an illustrative example before discussing the algorithm in detail.

4.1   Illustrative Example

Consider the example of heart attack prediction given clinical information about the patients such as their blood pressure. The training set (e.g. one particular county in Wisconsin) might show all patients having a lower risk of heart attack, with patients having high blood pressure having an especially low incidence of heart attacks. This systematic difference could be attributed to local factors. The local county data (the training set) could be shown in Fig. 2 in blue, while the true distribution for the entire nation could be shown in red.

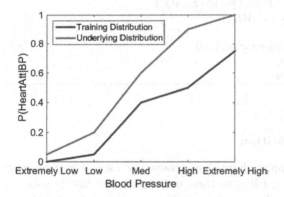

Fig. 2. Example showing the distribution of heart attacks given blood pressure for an observed and underlying distribution. The difference in these distributions could cause an expert to give advice that is not customized with respect to the training distribution. (Color figure online)

Now consider soliciting advice about heart attacks and blood pressure from a cardiologist in California. Being unfamiliar with Wisconsin, the cardiologist might give broad, straight-forward advice. However, such knowledge might already follow from the training data. Examples of such advice include "extremely high blood pressure leads to heart attacks" and "heart attacks are not likely with low blood pressure". While these pieces of advice are valid, they are not the most relevant advice for this particular learning problem.

If the algorithm had the ability to solicit advice, then it could direct the expert to give the most relevant advice at any point. Our proposed algorithm will identify areas in the data that are unclear and will instead query the expert automatically with "How likely are heart attacks when the blood pressure is high, but not extreme". This is likely the most useful advice given the data. This approach not only benefits the learning algorithm, but reduces the burden on the expert who is only required to answer specific questions.

Algorithm 1. Actively Seeking Advice for PLMs (ASAPlm)

    function ASAPLM(D,E,$MaxQuery$)
2:    $A = \emptyset$
      M=RFGB(D)   ▷ Model from Noisy Data
4:    **for** $x_i \in D$ **do**   ▷ Compute Uncertainty per Example
          $R(x_i) = H(x_i)$
6:    **end for**
      AQ=LRC(D, R)   ▷ Learn Regression Clauses
8:    **for** $i = 1$ to $MaxQuery$ **do**   ▷ Query Expert
          AQ_q =MAXSCORE(AQ)
10:       $AQ = AQ - AQ_q$
          $< AQ_q, R >$=QUERY(E, AQ_q)
12:       $A = A\cup < AQ_q, R >$
        end for
14:    M_F=ADVLEARNER(A, D)   ▷ Learn with Advice
      return M_F
16: **end function**

4.2 The Algorithm

Our proposed approach involves generating a set of queries, scoring those queries to rank them according to their usefulness, and finally soliciting the most useful queries to the human expert. The number of queries that can be requested depends on the problem (more difficult domains require more knowledge) and the availability of the human expert. The complete active advice-seeking algorithm (ASAPlm) is shown in Algorithm 1. We will address each of these vital components in turn.

Generating and Scoring Queries. Recall that in standard active learning, a model is learned from labeled data and using this model, some uncertainty measure is calculated to identify the most uncertain unlabeled example to query the expert. We take a similar approach with an important change. We learn an ensemble of relational regression trees using RFGB on the noisy data (line 3 of the algorithm) and compute the entropy over the examples given this model (lines 4–6). Following active learning, we define the score of an example as the entropy of the model's prediction (line 5 of Algorithm 1), i.e.,

$$H(x_i) = \sum_{l \in Labels} P_l(y_i|x_i) log(P_l(y_i|x_i))$$

where $P(y_i|x_i)$ is learned using RFGB. Such uncertainty measures have performed extremely well in many active learning methods and similar results can be shown over relational data. The key difference is that the uncertainty is based on all of the training examples that satisfy the query. In our empirical evaluation, we focus on entropy as our uncertainty measure. However, the framework is broad and allows for the selection of the most appropriate uncertainty function for the problem at hand.

Then these scores are used as regression values for the corresponding example and a set of weighted first-order-logic clauses are learned that can potentially group these examples (line 7, function LRC). We learn relational regression trees using RFGB as our implementation. These clauses are presented to the expert according to the learned weights. We learn these weighted clauses through an adaptation of RFGB where instead of learning $P(y_i|x_i)$, we want to learn a model for the uncertainty values of x_i (by fitting regression trees). The key intuition is that the regression trees find clauses that apply to examples with similar uncertainties. Note that unlike in discriminative learning where there are positive and negative examples, regression does not treat positive and negative examples differently. Every example has a uncertainty value and regression is just trying to fit those values. The learned clauses represent a set of possible queries from which the algorithm can select.

Querying the Expert. After the queries have been generated and ranked, they can be used to solicit advice from the human expert. For a given relational query, the expert should supply the suggested preferred labels (should be considered more likely) and the avoided labels (should be considered less likely). Alternatively, the expert could decline to answer if the query is too general or incomprehensible. Declining is an indication that the active advice-seeking algorithm is not selecting appropriate queries.

Advice-Based Learner. Given the advice, the final step is to utilize the advice-based learner to learn from both the training data as well as the expert advice. An ideal algorithm should trade-off between the sources of knowledge when they offer contradictory information. For the purposes of empirical validation, we utilize KBPLL [19] as our advice-based learner. It combines the target distribution of the training data and the distribution suggested by the advice to find a robust model (refer to Sect. 2).

Overall, the proposed approach to active advice-seeking aims to effectively utilize the human expert by generating queries. These queries are targeted based on the perceived weaknesses in the training data. We now thoroughly investigate the active advice-seeking algorithm.

5 Experiments

Through our experiments, we aim to answer the following questions:

Q1: Does active advice-seeking result in more effective learning?
Q2: Is our algorithm robust to both random and systematic noise?
Q3: Is advice an efficient form of communication between algorithm and expert?

5.1 Methods

We compare our method against two baselines. To evaluate our query generation method, we compare against learning with randomly generated queries (*Random Queries*). Note that the expert still gives the correct answer for the particular query generated. To evaluate the effectiveness of active advice-seeking, we compare against learning with no advice (*No Advice*). This represents the effectiveness of traditional machine learning systems that do not make use of expert knowledge. We also discuss the quality of the advice that is generated in each domain. Given our experience with the domains, we take the role of expert to answer the queries.

In all the experiments, we compare the accuracy of learned model. To show that our algorithm is capable of correcting noisy data, we added noise equal to 25 % of the positive examples. Note that in the relational space the number of negative examples typically greatly outnumbers the positive examples. This means that the impact of the noise is much less than 25 %. To show that our algorithm is capable of correcting systematic noise, we label examples incorrectly in a targeted region of the feature space. The synthetic heart attack dataset and driving domain are domains where systematic noise is natural. Heart problems effect different regions or ethnic groups in different ways and many drivers consistently drive over the speed limit and roll through stop signs. For the remaining datasets, imdb, webkb and uw, we have experimented with both systematic and noisy data. Each randomly noisy experimental domain has either 4 or 5 folds and we randomly add noise 5 times for each fold. Each systematically noisy experiment generated data for each fold or was repeated 5 times. For our relational advice-based learning algorithm, we use KBPLL [19] with $\alpha = 0.25$.

5.2 Domains

We have a variety of standard relational datasets as well as an imitation learning dataset focused on driving. An overview of each domain and the corresponding typed of noise (the datasets are either systematically noisy or randomly noisy) used in conjunction with that domain is shown in Table 1.

Table 1. Describes the prediction task of each of the experimental domains as well as the kind of noise used in the experiments.

Domain	Prediction task (possible labels)	Type of noise
Driving	moveLeft,moveRight,stayInLane	Systematic
Synthetic	heartAttack	Systematic
IMDB	workedUnder	Systematic/Noisy
WEBKB	faculty	Systematic/Noisy
UW	advisedBy	Systematic/Noisy

IMDB: This dataset is a movie database that consists of movies, actors, directors and their various genres. Our goal is to predict the workedunder relationship (i.e. which actors worked on movies under a particular director). This dataset consist of 5 folds.

WEBKB: This dataset is a university dataset that consists of webpages and their hyperlinks. Our goal in this domain is to predict which webpage belongs to a faculty member based on the webpages and their linking structure. This dataset has 5 folds.

UW: This dataset is a university dataset that consists of professors, students, courses, and publications each having various relationships and features. Our goal is to predict the advisedby relationship. This dataset has 4 folds.

Synthetic: The goal of the synthetic dataset, from the illustrative example, is to predict heart attacks given the blood pressure. There is a systematic difference (see Fig. 2) between the training set and the testing set. This dataset was generated 5 independent times.

Driving: The driving domain focuses on navigating down a 5-lane highway, avoiding the other cars on the road [10]. The possible actions are to stay in the current lane or change lanes to the left or right. The size of the training set and testing set are 100 trajectories consisting of 10000 total training examples.

5.3 Systematic Noise

The results with systematic noise (Fig. 3) are shown for the synthetic and driving domains as well as each of the standard relational datasets. Together they show the power of our proposed approach when dealing with systematic noise. In most datasets, the algorithm is capable of selecting useful queries immediately, providing significant impact. Random queries demonstrate gradual performance gains in the synthetic and webkb domains, but fail to have a positive effect on

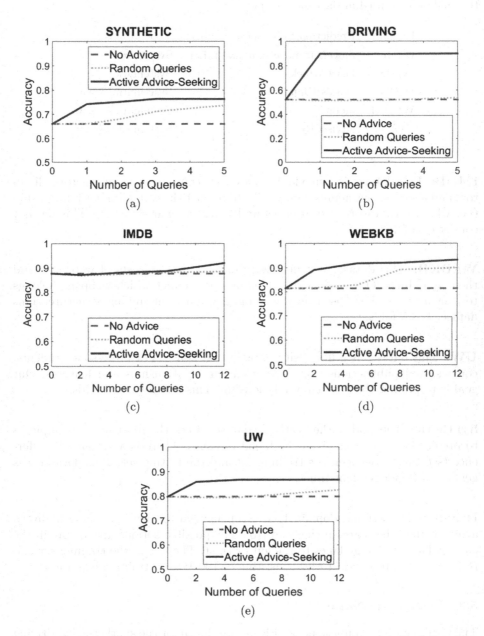

Fig. 3. The learning curves for the experiments with systematic noise. Each learning curve shows accuracy as the number of queries to the expert increases. We compare Active Advice-Seeking to Random Queries and No Advice.

the other domains. While random queries do not cause performance to degrade, they have an extremely difficult time isolating systematic noise especially when there are more features. A key reason there is very little change in these domains is that the queries generated were ambiguous and useful only for a few examples. For instance, a common query in the driving domain is "What action should I take if there is a car both to my left, right, AND in front". While this is a possible scenario, it is not likely in this dataset and there is no obvious advice to give for these states. Alternatively, the queries generated from the active advice-seeking algorithm select more relevant and overall useful queries. Thus, **Q1** is answered affirmatively in that our proposed approach is able to learn effectively in the presence of systematic noise.

Table 2. The top queries generated in each domain for the systematically noisy datasets. Experts respond to these queries by providing $l + /l-$ from Table 1.

Domain	Query generated
Driving	What if there is a car in the left lane?
Synthetic	What if a person has medium to high blood pressure?
IMDB	Do female actors work under people in crime movies?
WEBKB	What is the title of students working on projects?
UW	What is the relationship between students and TA's?

5.4 Random Noise

The standard relational domains (Fig. 4) are used to show that even when noise is random, our proposed method can still generate high-quality queries to the expert. Random noise should be more difficult for our algorithm, as there may not be specific regions of the feature space that need attention. However, across all three domains, our proposed approach achieves consistent success, generating performance gains with each query. In contrast, randomly generated queries can yield positive performance (as in imdb or uw), or actually result in a model that is worse than relying on the data (as in webkb). It may seem counter intuitive for advice to be harmful. However, consider the query "Is a student advised by a professor". While it may seem that the advice should be that students are advised by professors, there are many student and many professors. Therefore, such an advice could result in many false positives as a student is not advised by most professors. Thus, our proposed approach is robust to random noise as well as systematic noise (**Q2**).

5.5 Quality of Advice

The preceding empirical results show that our proposed approach is able to generate relevant queries that yield significantly higher accuracy in nearly all of the

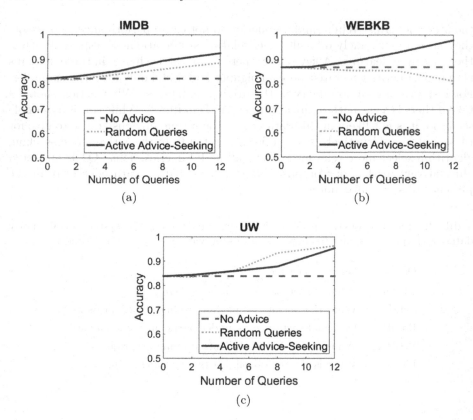

Fig. 4. The learning curves for the experiments with random noise. Each learning curve shows accuracy as the number of queries to the expert increases. We compare Active Advice Seeking to Random Queries and No Advice. As previously, randomness (for Random Queries) does not come from incorrect answer by the experts, but rather from randomly generated queries.

domains for both systematic and noisy experiments. However, the interpretability of the queries is vital as the experts need to easily comprehend the queries in order to give the proper advice. Table 2 shows the top query generated for each domain (systematic noise). In the driving domain, the query asks what action to take when there is a car in the left lane. The expert response would be to stay in the current lane. As another example, in the uw domain, the query asks about the relationship between students and TAs. While TAs might help teach students, the advice would say that TAs cannot advise students. The best queries are heavily influenced by the noise in the training set. Overall, the queries are concise (as shown in Table 2) and effective (as shown in the empirical validation). Thus, advice is an efficient form of communication (**Q3**).

6   Conclusion

We presented the first advice seeking framework for PLMs. Our method, inspired by active learning, queries the expert with sub-spaces of the feature space where advice can be provided as preferences over labels. The key insight is that the learning algorithm can better query the expert based on the uncertainty in the data as compared to the expert providing all advice pieces in advance. Our experimental results across standard data sets proved that such a method is indeed effective in soliciting useful advice. It must be mentioned our work is inspired by and bridges three promising areas of research inside machine learning - knowledge elicitation, active learning and PLMs. It extends knowledge elicitation to PLMs for the first time. It builds upon the success of active learning in relational tasks by soliciting advice (as preferences) instead of simple labels as done in previous research. Finally, it contributes to PLMs by making the learning algorithm go beyond merely using data by providing a natural way of interacting with the human expert.

Evaluating on larger data sets such as electronic health records is an important future direction. EHRs in particular can provide the opportunity to interact with domain experts who could provide advice potentially as qualitative statements - increase in one risk factor can increase the risk of a disease. Another interesting direction is exploring the different measures of uncertainty for grouping the different examples. A third direction could be to consider more types of advice that have been previously employed in machine learning. Learning from multiple experts by weighing them explicitly is another direction that we will explore. Finally, performing user studies on more sophisticated test beds is an interesting research direction.

Acknowledgments. The authors thank the Army Research Office (ARO) grant number W911NF-13-1-0432 under the Young Investigator Program.

References

1. Bilgic, M., Mihalkova, L., Getoor, L.: Active learning for networked data. In: ICML (2010)
2. Blockeel, H.: Top-down induction of first order logical decision trees. AI Commun. **12**(1–2), 119–120 (1999)
3. De Raedt, L., Kersting, K.: Probabilistic inductive logic programming. In: De Raedt, L., Frasconi, P., Kersting, K., Muggleton, S. (eds.) Probabilistic ILP 2007. LNCS(LNAI), vol. 4911, pp. 1–27. Springer, Heidelberg (2008)
4. De Raedt, L., Kimmig, A., Toivonen, H.: Problog: a probabilistic prolog and tis application in link discovery. In: IJCAI (2007)
5. Domingos, P., Lowd, D.: Markov Logic: An Interface Layer for Artificial Intelligence. Morgan & Claypool, San Rafael (2009)
6. Friedman, N., Getoor, L., Koller, D., Pfeffer, A.: Learning probabilistic relational models. In: IJCAI (1999)
7. Fung, G., Mangasarian, O.L., Shavlik, J.W.: Knowledge-based support vector machine classifiers. In: NIPS, pp. 1–9 (2002)

8. Getoor, L., Taskar, B.: Introduction to Statistical Relational Learning. MIT Press, Cambridge (2007)
9. Heckerman, D., Meek, C., Koller, D.: Probabilistic entity-relationship models, prms, and plate models. In: ICML (2004)
10. Judah, K., Fern, A., Tadepalli, P., Goetschalckx, R.: Imitation learning with demonstrations and shaping rewards. In: AAAI (2014)
11. Kunapuli, G., Bennett, K.P., Shabbeer, A., Maclin, R., Shavlik, J.: Online knowledge-based support vector machines. In: Balcázar, J.L., Bonchi, F., Gionis, A., Sebag, M. (eds.) ECML PKDD 2010. LNCS (LNAI), vol. 6322, pp. 145–161. Springer, Heidelberg (2010). doi:10.1007/978-3-642-15883-4_10
12. Kunapuli, G., Odom, P., Shavlik, J., Natarajan, S.: Guiding autonomous agents to better behaviors through human advice. In: ICDM (2013)
13. Kuwadekar, A., Neville, J.: Relational active learning for joint collective classification models. In: ICML (2011)
14. Le, Q.V., Smola, A.J., Gärtner, T.: Simpler knowledge-based support vector machines. In: ICML, pp. 521–528 (2006)
15. Macskassy, S.: Using graph-based metrics with empirical risk minimization to speed up active learning on networked data. In: KDD (2009)
16. Natarajan, S., Kersting, K., Khot, T., Shavlik, J.: Boosted Statistical Relational Learners: From Benchmarks to Data-Driven Medicine. Springer, Heidelberg (2015)
17. Natarajan, S., Khot, T., Kersting, K., Gutmann, B., Shavlik, J.: Gradient-based boosting for statistical relational learning: the relational dependency network case. Mach. Learn. **86**(1), 25–56 (2012)
18. Natarajan, S., Tadepalli, P., Dietterich, T., Fern, A.: Learning first-order probabilistic models with combining rules. Ann. Math. AI **54**(1), 223–256 (2008)
19. Odom, P., Khot, T., Porter, R., Natarajan, S.: Knowledge-based probabilistic logic learning. In: AAAI (2015)
20. Odom, P., Natarajan, S.: Active advice seeking for inverse reinforcement learning. In: AAMAS (2016)
21. Rashidi, P., Cook, D.: Ask me better questions: active learning queries based on rule induction. In: KDD (2011)
22. Rattigan, M., Maier, M., Jensen, D.: Exploiting network structure for active inference in collective classification. In: IDM (2007)
23. Sato, T., Kameya, Y.: Prism: A symbolic statistical modeling language. In: IJCAI (1997)
24. Settles, B.: Active Learning. Morgan & Claypool, San Rafael (2012)
25. Saha, T., Rangwala, H., Domeniconi, C.: FLIP: active learning for relational network classification. In: Calders, T., Esposito, F., Hüllermeier, E., Meo, R. (eds.) ECML PKDD 2014, Part III. LNCS (LNAI), vol. 8726, pp. 1–18. Springer, Heidelberg (2014). doi:10.1007/978-3-662-44845-8_1
26. Torrey, L., Walker, T., Shavlik, J., Maclin, R.: Using advice to transfer knowledge acquired in one reinforcement learning task to another. In: Gama, J., Camacho, R., Brazdil, P.B., Jorge, A.M., Torgo, L. (eds.) ECML 2005. LNCS(LNAI), vol. 3720, pp. 412–424. Springer, Heidelberg (2005). doi:10.1007/11564096_40
27. Towell, G., Shavlik, J.: Knowledge-based artificial neural networks. Artif. Intell. **69**, 119–165 (1994)

Augmented Leverage Score Sampling
with Bounds

Daniel J. Perry[✉] and Ross T. Whitaker

SCI Institute, University of Utah, 72 S. Central Campus Drive,
Salt Lake City, UT 84112, USA
{dperry,whitaker}@cs.utah.edu

Abstract. We introduce a modification to the well studied *leverage score* sampling algorithm which takes into account data scale, called the *augmented leverage score*, and introduce an initial error bound proof for the case of deterministic sampling – which to our knowledge is the first bound for this augmented leverage score. We discuss the implications of the error bounds proof and present an empirical evaluation of the proposed augmented leverage score performance on the column subsample selection problem (CSSP) as compared to the traditional leverage score and other methods in both a deterministic and probabilistic sampling paradigm. We show that the augmentation of the leverage score improves the empirical performance on CSSP significantly for many kinds of data.

Keywords: Subset selection · Low-rank matrix approximation · Leverage scores · Spectral analysis

1 Introduction

In many situations within machine learning and data analysis it is desirable to represent a data set using either a subset of the features, to ease interpretation, or a subset of the data, to identify a representative coreset of the data, which can be thought of as column (or row) subset selection of the data matrix. For example, many kernelized machine learning methods can be more quickly approximated using a Nyström method which requires the selection of a coreset [4]. There has been substantial interest in selecting an optimal subset of columns for a data matrix [1,5,15,16].

To be precise, we are interested in the *Column Subset Selection Problem* (CSSP) recently examined in [15]:

Definition 1. *Column Subset Selection Problem.* Let $\mathbf{A} \in \mathbb{R}^{d \times n}$ and let $c < n$ be a sampling parameter. Find c columns for A - denoted as $\mathbf{C} \in \mathbb{R}^{d \times c}$ that minimize

$$\|\mathbf{A} - \mathbf{C}\mathbf{C}^\dagger \mathbf{A}\|_\eta, \tag{1}$$

for $\eta \in \{F, 2\}$, and where $\mathbf{C}^\dagger$ denotes the Moore-Penrose pseudo-inverse.

© Springer International Publishing AG 2016
P. Frasconi et al. (Eds.): ECML PKDD 2016, Part II, LNAI 9852, pp. 543–558, 2016.
DOI: 10.1007/978-3-319-46227-1_34

Previous work has investigated the use of leverage scores, defined below, for use in selecting the column samples using both deterministic and random methods [1]. We use the definition provided in [15],

Definition 2. *Leverage scores. Let $\mathbf{V_k} \in \mathbb{R}^{n \times x}$ contain the top k right singular vectors of a $d \times n$ matrix A with rank $\rho = \mathbf{rank}(\mathbf{A}) \geq k$. Then the (rank-$k$) leverage score of the i-th column of $\mathbf{A}$ is defined as*

$$l_i^{(k)} = \|[\mathbf{V_k}]_{i,:}\|_2^2, \ i = 1, 2, \ldots, n. \tag{2}$$

Here, $[\mathbf{V_k}]_{i,:}$ denotes the i-th row of $\mathbf{V_k}$.

Leverage scores have a long tradition of being very useful in both deterministic and random solutions. However the SVD produces other useful information and one could ask whether taking advantage of the other components in the SVD could improve the subsample selection?

Consider this alternative formulation of the CSSP objective, where we make the assumption that the matrix $\mathbf{CC^T}$ is invertible (only for this specific discussion):

$$\|\mathbf{A} - \mathbf{CC^\dagger A}\|_\eta \tag{3}$$
$$\|\mathbf{A} - \mathbf{C(C^T(CC^T)^{-1})A}\|_\eta \tag{4}$$
$$\|\mathbf{U\Sigma V^T} - \mathbf{W_d W_d^T U\Sigma V^T}\|_\eta \tag{5}$$
$$\|(\mathbf{U} - \mathbf{W_d W_d^T U})\mathbf{\Sigma V^T}\|_\eta \tag{6}$$

where $\mathbf{A} = \mathbf{U\Sigma V^T}$ and $\mathbf{C} = \mathbf{W\Psi H^T}$ are the respective SVDs. In other words, this objective encodes the idea of selecting a $\mathbf{C}$ such that the columns space $\mathbf{W}$ aligns with the column space of $\mathbf{U}$ (corresponding to full $\mathbf{A}$), and the projection of the data points $\mathbf{\Sigma V^T}$ onto that basis.

In this form it becomes more apparent why the $\mathbf{V^T}$ is important in selecting a subset of columns. The standard leverage scores are drawn from the rows of $\mathbf{V} = \mathbf{A^T U\Sigma^\dagger}$ which are the data projected onto the principal components and rescaled ("whitened") according to the singular values ("variances") so that the covariance is identity. The leverage score is essentially the distance from the origin in the k-subspace spanned by $\mathbf{V_k}$. However the above objective function dictates the "unscaled" or "unwhitened" data points $\mathbf{\Sigma V^T}$, not $\mathbf{V^T}$ are projected. This indicates an "unscaled" leverage score would be more appropriate for CSSP.

This intuition informs the core premise of this paper – to propose an *augmented leverage score*, specifically

Definition 3. *Augmented leverage scores. Let $\mathbf{Y_k} = \mathbf{V_k \Sigma_k} \in \mathbb{R}^{n \times k}$ contain the top k singular values multiplied with the right singular vectors of a $d \times n$ matrix $\mathbf{A}$ with rank $\rho = \mathbf{rank}(\mathbf{A}) \geq k$. Then the (rank-$k$) augmented leverage score of the i–th column of $\mathbf{A}$ is defined as*

$$\hat{l}_i^{(k)} = \|[\mathbf{Y_k}]_{i,:}\|_2^2, \ i = 1, 2, \ldots, n. \tag{7}$$

Here we prove error bounds with respect to the optimal rank-k approximation for the corresponding deterministic sampling algorithm, and demonstrate through various examples how this augmented leverage score performs with respect to previous work as both a deterministic and random column sampling method. We show that by using this augmentation there is substantial improvement in the empirical performance for both deterministic and probabilistic sampling strategies.

2   Related Work

As described in Sect. 1, leverage score sampling has a long tradition and recently there as been substantial interest in analyzing and extending it [13,14]. Specifically, [1] presents a very thorough examination of CSSP in general, presenting several proofs to help in understanding and analyzing various approaches to CSSP. [15] presented a deterministic error bound for the traditional leverage score sampling with a bound very similar to ours,

$$\|\mathbf{A} - \mathbf{C}\mathbf{C}^\dagger \mathbf{A}\|_\zeta^2 < \frac{1}{1-\epsilon} \cdot \|\mathbf{A} - \mathbf{A_k}\|_\zeta^2. \tag{8}$$

for $\zeta \in \{2, F\}$, where $\mathbf{A_k}$ is the best rank-k approximation to $\mathbf{A}$.

Here instead we propose making a very simple modification to leverage scores which *substantially* improves the quality of the column subsample found, with regards to the CSSP error shown in (1). This change is fundamental enough that existing error bounds in (8) no longer hold for the proposed approach. We further compare empirically the proposed augmented leverage score sampling to traditional leverage score sampling in both a deterministic and probabilistic setting.

Related work also includes CUR and Nyström approximation techniques which aren't specifically targeted at solving CSSP, but are performing related subsampling of datasets with error bound guarantees. CUR [5] is an approximation technique which first computes a column subsample $\mathbf{C}$ of $\mathbf{A}$, then a row subsample $\mathbf{R}$ of $\mathbf{A}$, then an interpolative matrix $\mathbf{U}$ which minimizes the approximation error $\|\mathbf{A} - \mathbf{C}\mathbf{U}\mathbf{R}\|_{2,F}$. However, as described in [5], the column subsample step uses a leverage sample, and so by comparing to leverage sampling directly, we gain some idea of how the proposed method compares to CUR. The Nyström method [4,8] is a matrix approximation technique where a column subsample $\mathbf{C}$ of $\mathbf{A}$ is found such that $\|\mathbf{A} - \mathbf{C}\mathbf{U}\mathbf{C}^\mathbf{T}\|_{2,F}$ is minimized. [7] obtains a direct relationship between Nyström and CSSP by formulating the Nyström approximation as a CSSP by setting $\mathbf{C} = \mathbf{A}^{1/2}\mathbf{S}$. However all of the subsampling approaches discussed use either a naive subsampling (uniformly random) or a leverage score sampling, so a comparison of our method to leverage score and uniform random sampling is sufficient.

Note that our new bound adds a multiplicative factor to previous bounds, $\hat{\sigma}_1(\mathbf{\Sigma_k}) = \frac{\sigma_1(\mathbf{\Sigma_k})}{\sigma_k(\mathbf{\Sigma_k})}$, where $\sigma_i(\mathbf{M})$ is the i-th singular value of the matrix $\mathbf{M}$, which is based on the data scaling. One central premise of the proposed augmentation is that we are taking into consideration the data scale, and so it is logical that the bound now includes singular value terms from the data (for example consider the bounds in [8] which also include data scale terms). The new factor has some implications on our error bound – specifically if k is chosen such that $\sigma_k(\boldsymbol{\Sigma_k})$ is very small compared to $\sigma_1(\boldsymbol{\Sigma_k})$, the factor becomes large making the bound very loose. On the other hand if they are similar in magnitude the bound has similar tightness to the previous bound in (8). Although this scaling ratio has implications on the theoretical bound, we observed very good performance in the empirical results for problems with a large gap in the spectrum, indicating this is primarily a theoretical problem we hope to improve in future work.

3   Deterministic Column Sampling

In the following we make use of the singular value decomposition (SVD) of $\mathbf{A} \in \mathbb{R}^{d \times n}$, $\mathbf{A} = \mathbf{U \Sigma V^T}$, with the left singular vectors $\mathbf{U^T U} = \mathbf{U U^T} = \mathbf{I}, \mathbf{U} \in \mathbb{R}^{d \times d}$, the right singular vectors $\mathbf{V^T V} = \mathbf{V V^T} = \mathbf{I}, \mathbf{V} \in \mathbb{R}^{n \times n}$, and the singular value matrix $\boldsymbol{\Sigma} \in \mathbb{R}^{d \times n}$ a diagonal matrix with entries $\boldsymbol{\Sigma}_{ii} = \sigma_i(\mathbf{A})$ being the singular values of $\mathbf{A}$, sorted so that $\sigma_1(\mathbf{A}) \geq \sigma_2(\mathbf{A}) \geq \ldots$. The rank-$k$ approximation $\mathbf{A_k} = \mathbf{A V_k V_k^T} = \mathbf{U_k \Sigma_k V_k^T} \in \mathbb{R}^{d \times n}$, with $\mathbf{U_k} \in \mathbb{R}^{d \times k}, \boldsymbol{\Sigma_k} \in \mathbb{R}^{k \times k}, V \in \mathbb{R}^{n \times k}$ is the optimal rank-k matrix approximation under the Frobenius norm of the difference, $\|\mathbf{A} - \mathbf{A_k}\|_F^2 = \|\boldsymbol{\Delta_k}\|_F^2 = \sum_{i=1}^k \sigma_i(\boldsymbol{\Delta_k})$ and under the spectral norm $\|\boldsymbol{\Delta_k}\|_2^2 = \sigma_1(\boldsymbol{\Delta_k})$. We will use the shorthand, $\hat{\sigma}_i^2(\mathbf{A_k}) = \frac{\sigma_i^2(\mathbf{A_k})}{\sigma_k^2(\mathbf{A_k})}$, which is a scaled (unitless) singular value. Note also that for a symmetric matrix $\mathbf{M} = \mathbf{A^T A} = \mathbf{U \Sigma^2 U^T}$, the eigenvalues $\lambda_i(\mathbf{M})$ correspond to the singular values, so that $\lambda_i(\mathbf{M}) = \sigma_i^2(\mathbf{M})$.

The corresponding *probabilistic* algorithm for augmented leverage scores samples points randomly according to the probability $p_i = \frac{\hat{l}_i}{\sum_j \hat{l}_i}$, instead of sorting.

We prove the following convergence theorem with respect to Algorithm 1.

Theorem 1. *Let $\theta = k \cdot \hat{\sigma}_1^2(\boldsymbol{\Sigma_k}) - \epsilon$ for some $\epsilon \in (0,1)$, and let $\mathbf{S} \in \mathbb{R}^{n \times c}$ be the sampling matrix from Algorithm 1, then, for $\mathbf{C} = \mathbf{AS}$ and $\zeta = \{2, F\}$*

$$\|\mathbf{A} - \mathbf{C C}^\dagger \mathbf{A}\|_\zeta^2 < \frac{\hat{\sigma}_1^2(\boldsymbol{\Sigma_k})}{1 - \epsilon} \cdot \|\mathbf{A} - \mathbf{A_k}\|_\zeta^2. \tag{11}$$

We can rewrite the bound as

$$\|\mathbf{A} - \mathbf{C C}^\dagger \mathbf{A}\|_\zeta^2 < (1 + 2\epsilon) \cdot \hat{\sigma}_1^2(\boldsymbol{\Sigma_k}) \cdot \|\mathbf{A} - \mathbf{A_k}\|_\zeta^2 \tag{12}$$

if $\epsilon < \frac{1}{2}$ [15].

We note that the parameter θ in Algorithm 1 provides the primary connection between the column subset size and the spectrum of the data, and that while the algorithm can be run with any choice of θ the provided bounds only hold for the given choice of θ in Theorem 1.

Algorithm 1. AugmentedLeverageScoreSampler($\mathbf{A}, k, \theta$)

Input $\mathbf{A} \in \mathbb{R}^{d \times n}, k, \theta$

Compute $\mathbf{Y_k} = \mathbf{V_k}\hat{\mathbf{\Sigma}}_\mathbf{k} \in \mathbb{R}^{n \times k}$ (top k singular values normalized so that $\mathbf{\Sigma}_{ii} = \hat{\sigma}_i(\mathbf{\Sigma}_k)$ and right singular vectors of $\mathbf{A}$ multiplied together)

for $i = 1, 2, \ldots, n$

$\hat{l}_i^{(k)} = \|[\mathbf{Y_k}]_{i,:}\|_2^2$

end for

Let $\hat{l}_i^{(k)}$'s **be sorted:**

$$\hat{l}_1^{(k)} \geq \cdots \geq \hat{l}_i^{(k)} \geq \hat{l}_{i+1}^{(k)} \geq \cdots \geq \hat{l}_n^{(k)} \tag{9}$$

Find index $c \in \{1, \ldots, n\}$ **such that:**

$$c = \arg\min_c \left(\sum_{i=1}^{c} \hat{l}_i^{(k)} > \theta \right). \tag{10}$$

If $c < k$, **set** $c = k$.

Output $\mathbf{S} \in \mathbb{R}^{n \times c}$, **s.t. AS has the top c columns of A.**

---

4 Experiments

4.1 Algorithms

For the deterministic subsampling problem we compare the proposed deterministic augmented leverage score sampling method (aug-det) from Algorithm 1 with the traditional deterministic leverage score sampling method (lev-det) from [15] and the QR-pivot based sampling method (qr) from [9]. QR-pivot based sampling is done by computing the QR decomposition with pivoting and then using the resulting pivot matrix for column selection as described in [3]. The leverage score methods are described in more detail in Sect. 2.

In the probabilistic case, we compare the proposed probabilistic augmented leverage score sampling method (aug-prob) (similar to Algorithm 1 as described in Sect. 3) with the traditional probabilistic leverage score sampling method (lev-prob) from [15], and the ubiquitous uniformly random sampling (unif).

The algorithms were compared on the basis of the CSSP projection space error, $\|\mathbf{A} - \mathbf{CC}^\dagger\mathbf{A}\|_\eta$ for both the Frobenius norm ($\eta = F$) and spectral norm ($\eta = 2$), as well as the projection error sum over all the data points, which is $\sum_{i=1}^{n} \|\mathbf{A}_i - \mathbf{CC}^\dagger\mathbf{A}_i\|_2$, where $\mathbf{C}$ is selected using the corresponding algorithm.

4.2 Datasets

To examine how each of the methods performed we compared them using synthetic datasets, with specific spectral properties, and several real datasets.

We use two small synthetic datasets to examine how the spectral makeup of the dataset effects the results. The first dataset we sampled from a multidimensional normal distribution with a spectrum, $s_L(i)$ that decays linearly from

$d = 10$, $s_L(i) = d - i$, which was then rotated to take it off axis. The second synthetic dataset was chosen to see the effect of a power law decay on the spectrum. As described in [15], this model is of interest because many datasets will have a spectrum that decays quickly. This was constructed similarly to the first, but a power-law decay was used, $s_P(i) = \frac{d}{i^{1+\eta}}$, where for this dataset we used $\eta = 0.5$. The size of both datasets are summarized in Table 1, and the spectrum of each is shown in Fig. 1, in the top left and bottom left plots. As both leverage score sampling techniques rely on the selection of a rank-k subspace, we selected the SVD subspace of size k that captures 90 % of the spectral energy and report the specific k chosen in Table 1. We chose a diverse set of real datasets to provide insight into how the algorithms compare on real data. We summarize the size and dimensionality of each dataset used in Table 1, and show the corresponding spectrum plots along the left column of Fig. 4. The first dataset, *diabetes*, is a small dataset of various health measurements, with the goal of finding some relationship to diabetes used in [6], which has some spectral properties somewhat similar to the power law decaying synthetic dataset. The second dataset, *enron*, is an email social graph derived from the popular Enron public email dataset from [10,11]. This dataset exhibits a mostly linear decay in the spectrum. The forest cover, adult, and census dataset are all taken from the UCI data repository [12], and were chosen because of their use in other related machine learning and data analysis task, which helps in comparison to other (future) algorithms.

Table 1. Real and synthetic datasets examined with number of points n, dimension d, and subspace dimension k which captures 90 % of the spectral energy.

Name	n	d	k
Synthetic linear decay	1000	10	9
Synthetic power law decay	1000	10	6
Diabetes	442	9	6
Enron	3000	3000	2108
Forest cover	58102	54	5
Adult	16281	123	73
Census	2273	119	8

4.3 Synthetic

The leverage and augmented leverage scores for the synthetic datasets are shown in the middle column of Fig. 1, and a visualization of the resulting samples selected using the deterministic approach are shown in the far right column. In the center column plots the leverage scores have been normalized so that the largest value is 1.0 and they have been sorted in descending order to make the visual comparison easier. This view of the scores makes the effect of the augmentation on the scores more obvious – for any point whose "magnitude energy" is in the directions with dominant spectral values the value is preserved or increased, while for any point with "magnitude energy" in the directions corresponding to smaller spectral values the score is decreased. This contrast in point selection is illustrated in the 2D principal component analysis (PCA) visualization in the right column of Fig. 1. While the traditional leverage score selects points that are important in several directions, the augmented leverage score selects points that are primarily important in the dominant PC modes – note how the augmented leverage score points are mainly on the outer edges of this data projection.

Now consider how the augmented score performs on these synthetic datasets as shown in Figs. 2 and 3. The first column of both figures shows results measured under the Frobenius norm, the center column under the spectral norm, and the right column under projection error. The rank-k approximation error is included in each plot for reference – it is constant because it is independent of the column subsample size. Consider the deterministic results in Fig. 2, specifically the first row showing the linear decay dataset. In each error measurement, the augmented leverage score performs better at column selection, and the QR-pivot based selection performs the best. Now consider the power law decay dataset in the second row, in this case, both the augmented leverage score and the QR-pivot have significant advantage over the traditional leverage score for the first several samples. This indicates that both QR and augmented leverage are taking advantage of the fast drop in spectral error, while traditional leverage sampling does not – this provides a nice insight into not only why augmented leverage scores perform better, but also why QR based sampling performs so well and confirms that taking into account the spectrum of the dataset can be important to the CSSP, for certain kinds of data.

Observe the probabilistic results in Fig. 3 and note that for this case we compare the two leverage score approaches to a uniformly random sampling of the dataset. For these synthetic results each algorithm was run 100 times and the solid plot line shows the mean result, while the vertical error bars indicate the standard deviation at each sample point. The first row shows the probabilistic results for the linear decaying spectrum dataset, and for this specific case the three algorithms actually perform very similarly. In contrast, in the bottom row, when considering the power law spectrum decay data, the augmented leverage score performs significantly better than both the uniformly random and the traditional leverage score sampling. Again these results reinforce the idea that, especially for datasets with a fast decay, taking into account the spectral magnitudes is important to the probabilistic CSSP.

4.4   Real

A similar comparison was run on the real datasets. In Fig. 4, in the right column, the leverage scores are shown – again these are normalized so that the maximum leverage score is 1.0, and sorted in descending order for visual comparison. Note how the varying spectra strongly influence the resulting scores. The *enron* dataset has one of the most dramatic changes - it shows a very gradual decrease in leverage scores over the samples, while after augmentation the leverage score

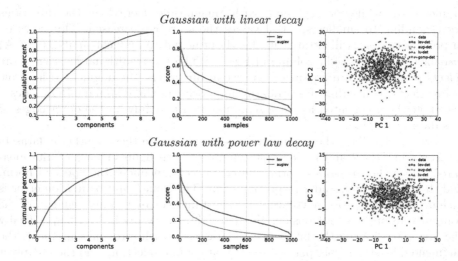

Fig. 1. Synthetic Gaussian datasets. The first row shows the spectrum (top left), the leverage and augmented leverage scores (top center), and a PCA-based visualization of the leverage score samples for the deterministic algorithm (top right), all for the linear decay dataset. The second row shows the same results but for the power law decay dataset.

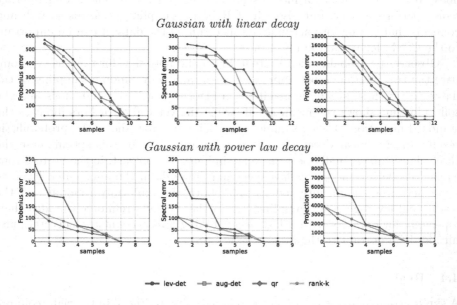

Fig. 2. Deterministic results on the synthetic datasets. The first row is a synthetic Gaussian dataset with linear spectral decay. The second row is a synthetic Gaussian dataset with a power law decay. Results are shown for each data set measured under a Frobenius norm error (left column), a spectral norm error (center column), and a projection error (right column).

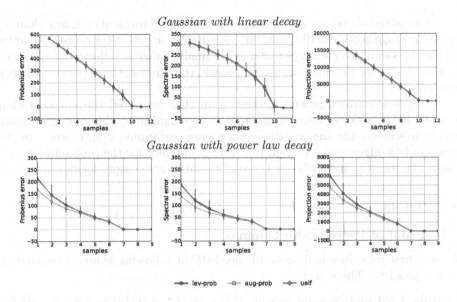

Fig. 3. Probabilistic results on the synthetic datasets. The first row is a synthetic Gaussian dataset with linear spectral decay. The second row is a synthetic Gaussian dataset with a power law decay. Results are shown for each data set measured under a Frobenius norm error, a spectral norm error, and a projection error. For these synthetic results the algorithm was run 100 times and the line shows the mean result, while the vertical error bar indicates the standard deviation.

values drop off very quickly. The *adult* dataset also has a dramatic decrease in leverage score value (and increased decay rate), while the *forest* dataset actually decreases the decay rate of the leverage scores.

Figure 5 shows the spatial layout of the deterministic leverage samples in 2D PCA space for the *diabetes* and *enron* datasets, the other datasets have so many points the scatter plot visualization becomes very difficult to read. The left plot in Fig. 5 for the *diabetes* dataset shows similar results to the synthetic datasets - the leverage score appears to select points independent of the dominant directions, while the augmented and QR based samples select points that are clearly important in the dominant directions. The right side plot shows the resulting sample for the *enron* graph dataset. In this case there is actually a lot of overlap between the methods – they all pick very similar columns.

The deterministic method performance comparison is shown in Fig. 6, while the probabilistic results are in Fig. 7. For these probabilistic results, the algorithms were each run 100 times (except *enron* which was only run 5 times) and the average is plotted with the standard deviation shown by the error bars. Consider the *diabetes* dataset in the top row, where we noted a moderate change in leverage score distribution, and significant differences in actual leverage sampling in Fig. 5. These modifications resulted in a significant improvement in the errors reported - in almost every case the augmented leverage scores performed significantly better than the traditional leverage score.

As we noted above, the *enron* dataset in the second row had dramatic changes in the leverage score distribution, but viewing the actual deterministic samples taken, the samples were very similar. This small change is reflected in the error scoring where, for the deterministic approach, the three methods performed very similarly.

The *forest* dataset proved interesting in that we noted the augmentation actually led to a slower decay in leverage score magnitude. This change actually results in some of the augmented leverage score performing slightly worse in the deterministic algorithm – though the story is different for the probabilistic case where the mean is always below the traditional leverage score mean error.

5   Proofs

5.1   Proof of Prerequisite Lemmas

We will first introduce and prove (as needed) the following lemmas, concluding with a proof for Theorem 1.

Lemma 1 *(A more general version of Lemma 3.1 in [1]). Consider the decomposition* $\mathbf{A} = \mathbf{A}\mathbf{Y}\mathbf{Y}^\dagger \frac{1}{\sigma_k^2(\mathbf{L})} + \mathbf{E}$, *with* $\mathbf{Y} = \mathbf{Z}\mathbf{L}$, $\mathbf{Z} \in \mathbb{R}^{n\times k}$, $\mathbf{Z}^\mathbf{T}\mathbf{Z} = \mathbf{I_k}$, *and* $\mathbf{L} \in \mathbb{R}^{k\times k}$ *a diagonal matrix where* $\mathbf{L}\mathbf{L}^\dagger = \mathbf{I_k}$. *Let* $\mathbf{S} \in \mathbb{R}^{n\times c}$ *be any matrix such that* $\mathbf{rank}(\mathbf{Y}^\mathbf{T}\mathbf{S}) = k$. *Let* $\mathbf{C} = \mathbf{A}\mathbf{S} \in \mathbb{R}^{d\times c}$. *Then for* $\zeta = \{2, F\}$

$$\|\mathbf{A} - \mathbf{C}\mathbf{C}^\dagger \mathbf{A}\|_\zeta^2 \leq \|\mathbf{E}\|_\zeta^2 \|\mathbf{S}(\mathbf{Y}^\mathbf{T}\mathbf{S})^\dagger \mathbf{L}\|_2^2 \tag{13}$$

Proof. Remember that a projection operator $\mathbf{P}$ is equal to the square $\mathbf{P}^2 = \mathbf{P}$. We will use the following two Lemmas,

Lemma 2 *(Lemma 7 in [1]). Let* $\mathbf{P}$ *be a non-null projection, then* $\|\mathbf{I} - \mathbf{P}\|_2 \leq \|\mathbf{P}\|_2$.

Lemma 3. *For all matrices* $\mathbf{X} \in \mathbb{R}^{d\times n}$ *of rank at most* k *in the column space of* $\mathbf{C}$,

$$\|\mathbf{A} - \mathbf{C}\mathbf{C}^\dagger \mathbf{A}\|_\zeta^2 \leq \|\mathbf{A} - \mathbf{X}\|_\zeta^2 \tag{14}$$

See Lemma 9 in [1].

Now consider the matrix $\mathbf{X} = \mathbf{C}(\mathbf{Y}^\mathbf{T}\mathbf{S})^\dagger \mathbf{Y}^\mathbf{T}$, where $\mathbf{X}$ is clearly in the column space of $\mathbf{C}$ and $\mathbf{rank}(\mathbf{X}) = k$, because $\mathbf{rank}(\mathbf{Y}) = k$. Then using Lemma 3,

$$\|\mathbf{A} - \mathbf{C}(\mathbf{Y}^\mathbf{T}\mathbf{S})^\dagger \mathbf{Y}^\mathbf{T}\|_\zeta^2 \tag{15}$$

$$= \|\mathbf{A}\mathbf{Y}\mathbf{Y}^\dagger + \mathbf{E} - (\mathbf{A}\mathbf{Y}\mathbf{Y}^\dagger + \mathbf{E})\mathbf{S}(\mathbf{Y}^\mathbf{T}\mathbf{S})^\dagger \mathbf{Y}^\mathbf{T}\|_\zeta^2 \tag{16}$$

$$= \|\mathbf{A}\mathbf{Y}\mathbf{Y}^\dagger - \mathbf{A}\mathbf{Y}\mathbf{Y}^\dagger \mathbf{S}(\mathbf{Y}^\mathbf{T}\mathbf{S})^\dagger \mathbf{Y}^\mathbf{T} + \mathbf{E} - \mathbf{E}\mathbf{S}(\mathbf{Y}^\mathbf{T}\mathbf{S})^\dagger \mathbf{Y}^\mathbf{T}\|_\zeta^2 \tag{17}$$

$$= \|\mathbf{E} - \mathbf{E}\mathbf{S}(\mathbf{Y}^\mathbf{T}\mathbf{S})^\dagger \mathbf{Y}^\mathbf{T}\|_\zeta^2 \tag{18}$$

$$\leq \|\mathbf{E}\|_\zeta^2 \|\mathbf{I} - \mathbf{S}(\mathbf{Y}^\mathbf{T}\mathbf{S})^\dagger \mathbf{Y}^\mathbf{T}\|_2^2 \tag{19}$$

$$\leq \|\mathbf{E}\|_\zeta^2 \|\mathbf{S}(\mathbf{Y}^\mathbf{T}\mathbf{S})^\dagger \mathbf{Y}^\mathbf{T}\|_2^2 \tag{20}$$

$$= \|\mathbf{E}\|_\zeta^2 \|\mathbf{S}(\mathbf{Y}^\mathbf{T}\mathbf{S})^\dagger \mathbf{L}\|_2^2. \tag{21}$$

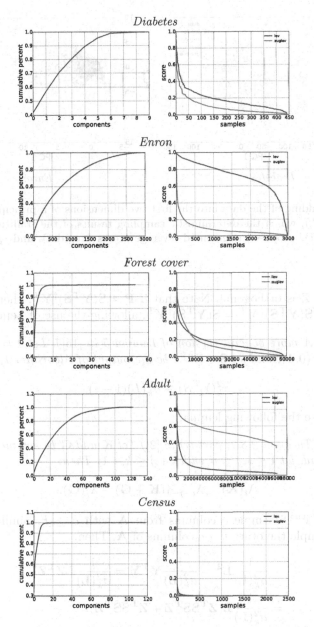

Fig. 4. A plot of the spectrum, and a plot comparing the leverage and augmented leverage scores for each of the real datasets evaluated.

The first 4 lines are simple substitution and the definition of the pseudo-inverse, the 5th line bound is due to strong submultiplicativity, the 6th line is bound using Lemma 2, and the 7th line uses the definition of $\mathbf{Y} = \mathbf{ZL}$

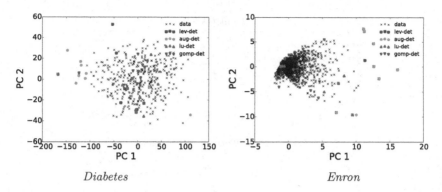

Diabetes   *Enron*

Fig. 5. Embedding of the data into the first two dimensions of a principal component analysis (PCA), which the deterministic sampling results of the traditional leverages scores (lev-det), proposed augmented leverage scores (aug-det), and QR-pivot samples (qr).

and because $\mathbf{Z}$ is orthogonal. Note that if $\mathbf{P} = \mathbf{S}(\mathbf{Y^T S})^\dagger \mathbf{Y^T}$, then $\mathbf{P^2} = \mathbf{P} = \mathbf{S}(\mathbf{Y^T S})^\dagger \mathbf{Y^T S}(\mathbf{Y^T S})^\dagger \mathbf{Y^T} = \mathbf{S}(\mathbf{Y^T S})^\dagger \mathbf{Y^T}$, allowing the use of Lemma 2 above.

Lemma 4 *(A more general version of Lemma 5 in [15]). Let $\theta = k \cdot \hat{\sigma}_1^2(L) - \epsilon$ for some $\epsilon \in (0,1)$, and let $\mathbf{S} \in \mathbb{R}^{n \times c}$ be the output of Algorithm 1, then,*

$$\sigma_k^2(Y_k^T S) > \sigma_k^2(L)(1 - \epsilon). \tag{22}$$

Proof. We use the following lemma,

Lemma 5 *(Theorem 2.8.1; part (i) in [2]) Let $\mathbf{K}$ and $\mathbf{G}$ be symmetric matrices of order k and, let $1 \le i, j \le n$ with $i + j \le k + 1$. Then,*

$$\lambda_i(\mathbf{K}) \ge \lambda_{i+j-1}(\mathbf{K} + \mathbf{G}) - \lambda_j(\mathbf{G}). \tag{23}$$

Let $\mathbf{S} \in \mathbb{R}^{n \times c}$ sample c columns from $\mathbf{A}$ with $c \ge k$. Similarly let $\hat{\mathbf{S}} \in \mathbb{R}^{n \times (n-c)}$ sample the other $n - c$ columns of $\mathbf{A}$. Then,

$$\frac{1}{\sigma_k^2(\mathbf{L})}\mathbf{L^2} = \frac{1}{\sigma_k^2(\mathbf{L})}\mathbf{Y^T Y} = \frac{1}{\sigma_k^2(\mathbf{L})}\mathbf{L Z^T Z L} \tag{24}$$

$$= \frac{1}{\sigma_k^2(\mathbf{L})}\mathbf{L}(\mathbf{Z^T S S^T Z} + \mathbf{Z^T \hat{S} \hat{S}^T Z})\mathbf{L} \tag{25}$$

$$= \frac{1}{\sigma_k^2(\mathbf{L})}(\mathbf{L Z^T S S^T Z L} + \mathbf{L Z^T \hat{S} \hat{S}^T Z L}) \tag{26}$$

$$= \frac{1}{\sigma_k^2(\mathbf{L})}(\mathbf{Y^T S S^T Y} + \mathbf{Y^T \hat{S} \hat{S}^T Y}) \tag{27}$$

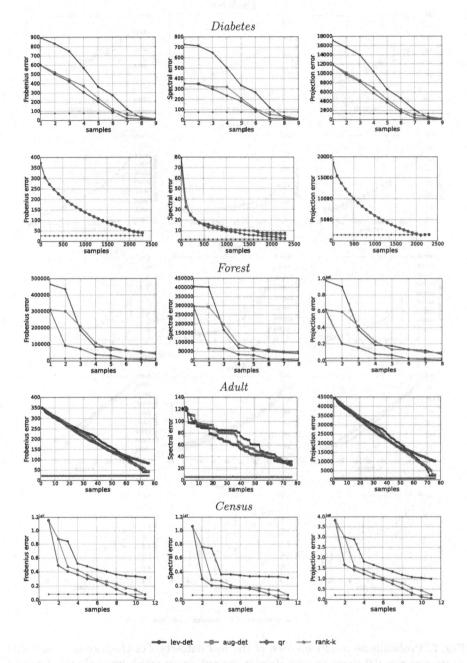

Fig. 6. Deterministic results on each of the real datasets. Results are shown for each data set measured under a Frobenius norm error (left column), a spectral norm error (center column), and a projection error (right column).

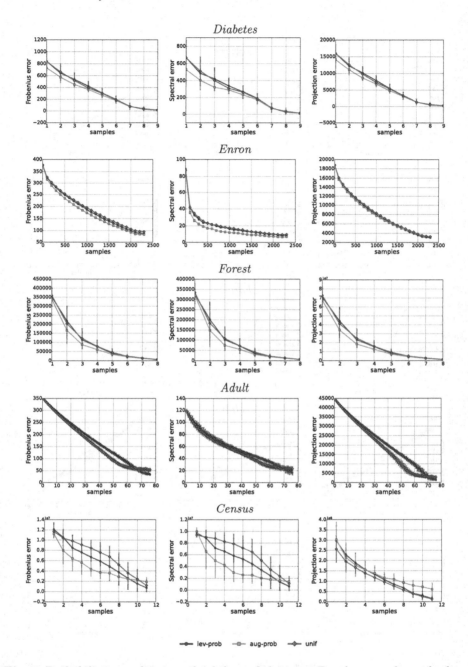

Fig. 7. Probabilistic results on each of the real datasets. For these results each algorithm was run 100 times on each dataset (except *enron* which was only run 5 times), the mean is then plotted with a line, and the standard deviation shown with error bars. Results are shown for each data set measured under a Frobenius norm error (left column), a spectral norm error (center column), and a projection error (right column).

Using Lemma 5 where $\mathbf{K} = \frac{1}{\sigma_k^2(\mathbf{L})}\mathbf{Y}^T\mathbf{S}\mathbf{S}^T\mathbf{Y}, \mathbf{G} = \frac{1}{\sigma_k^2(\mathbf{L})}\mathbf{Y}^T\hat{\mathbf{S}}\hat{\mathbf{S}}^T\mathbf{Y},$

$$\frac{1}{\sigma_k^2(\mathbf{L})}\lambda_k(\mathbf{Y}^T\mathbf{S}\mathbf{S}^T\mathbf{Y}) \geq \frac{1}{\sigma_k^2(\mathbf{L})}\lambda_k(\mathbf{L}^2) - \frac{1}{\sigma_k^2(\mathbf{L})}\lambda_1(\mathbf{Y}^T\hat{\mathbf{S}}\hat{\mathbf{S}}^T\mathbf{Y}) \tag{28}$$

$$= 1 - \frac{1}{\sigma_k^2(\mathbf{L})}\|\mathbf{Y}^T\hat{\mathbf{S}}\|_2^2 \tag{29}$$

$$\geq 1 - \frac{1}{\sigma_k^2(\mathbf{L})}\|\mathbf{Y}^T\hat{\mathbf{S}}\|_F^2 \tag{30}$$

$$> 1 - (k \cdot \hat{\sigma}_1^2(\mathbf{L}) - \theta). \tag{31}$$

The last step is true because $\|\mathbf{Y}^T\hat{\mathbf{S}}\|_F^2 \leq (k \cdot \hat{\sigma}_1^2(\mathbf{L}))$, which can be shown by considering that $\|\mathbf{Y}^T\|_F^2 = \|\mathbf{Y}^T\mathbf{S}\|_F^2 + \|\mathbf{Y}^T\hat{\mathbf{S}}\|_F^2$, $\|\mathbf{Y}^T\|_F^2 \leq \mathbf{rank}(\mathbf{Y}^T)\|Y_k^T\|_2^2$, and $\|Y^T S\| > \theta$, so that $\|Y^T S\|_F^2 = \|Y^T\|_F^2 - \|Y^T S\|_F^2 \leq k\hat{\sigma}_1(L) - \|Y^T S\|_F^2 < k\hat{\sigma}_1(L) - \theta$.

This implies that,

$$\sigma_k^2(\mathbf{Y}^T\mathbf{S}) > (1 - \epsilon)\sigma_k^2(\mathbf{L}). \tag{32}$$

5.2   Proof of Theorem 1

The proof of Theorem 1 is a combination of Lemmas 1 and 4.

Proof. We start with Lemma 1 where $\mathbf{Z} = \mathbf{V_k}$ and $\mathbf{L} = \mathbf{\Sigma_k}$ so that

$$\|\mathbf{A} - \mathbf{C}\mathbf{C}^\dagger\mathbf{A}\|_\zeta^2 \leq \|\mathbf{A} - \mathbf{A_k}\|_\zeta^2\|\mathbf{S}(\mathbf{\Sigma_k}\mathbf{V_k}^T\mathbf{S})^\dagger\mathbf{\Sigma_k}\|_2^2 \tag{33}$$

$$\leq \|\mathbf{A} - \mathbf{A_k}\|_\zeta^2\|\mathbf{S}\|_2^2\|\mathbf{\Sigma_k}\|_2^2\|(\mathbf{\Sigma_k}\mathbf{V_k}^T\mathbf{S})^\dagger\|_2^2 \tag{34}$$

$$= \|\mathbf{A} - \mathbf{A_k}\|_\zeta^2\sigma_1^2(\mathbf{\Sigma_k})\|(\mathbf{\Sigma_k}\mathbf{V_k}^T\mathbf{S})^\dagger\|_2^2 \tag{35}$$

$$= \frac{\|\mathbf{A} - \mathbf{A_k}\|_\zeta^2\sigma_1^2(\mathbf{\Sigma_k})}{\sigma_k^2(\mathbf{\Sigma_k}\mathbf{V_k}^T\mathbf{S})} \tag{36}$$

$$< \frac{\|\mathbf{A} - \mathbf{A_k}\|_\zeta^2\sigma_1^2(\mathbf{\Sigma_k})}{(1 - \epsilon)\sigma_k^2(\mathbf{L})} \tag{37}$$

$$= \frac{\|\mathbf{A} - \mathbf{A_k}\|_\zeta^2\hat{\sigma}_1^2(\mathbf{\Sigma_k})}{1 - \epsilon}. \tag{38}$$

6   Conclusion

In conclusion, we have shown motivation for taking into account the scale when computing leverage scores, which we call an *augmented leverage score*. We have shown the first error bound for the deterministic case of an augmented leverage score. This proof highlights some of the effect of the spectral scales on the bound – specifically that a steep dropoff in spectral scale makes the bound very loose, while a very gradual decay makes the bound much more tight.

We have shown empirically the method benefits both the probabilistic and the deterministic leverage score sampling algorithms on a variety of datasets, as well as a few cases where the improvement is less noticeable or even made the result slightly worse. In contrast to the theoretical results, we have shown that the augmented leverage score performs exceptionally well, as compared to the traditional leverage score, on datasets with a steep dropoff in spectral scale. This indicates there is some room for improvement in the error bounds, something we hope to obtain in future work.

References

1. Boutsidis, C., Drineas, P., Magdon-Ismail, M.: Near-optimal column-based matrix reconstruction. SIAM J. Comput. **10598**(i), 1–27 (2014). http://epubs.siam.org/doi/abs/10.1137/12086755X
2. Brouwer, A.E., Haemers, W.H.: Spectra of Graphs. Springer, New York (2011). https://dx.doi.org/10.1007/978-1-4614-1939-6
3. Chan, T.F., Hansen, P.C.: Some applications of the rank revealing QR factorization. SIAM J. Sci. Comput. **13**(3), 727–741 (1992)
4. Drineas, P., Mahoney, M.W.: On the Nystrom method for approximating a gram matrix for improved kernel-based learning (Extended abstract). In: Proceedings of Learning Theory, vol. 3559, pp. 323–337 (2005). <Go to ISI>://WOS:000230769100022
5. Drineas, P., Mahoney, M., Muthukrishnan, S.: Relative-error CUR matrix decompositions. SIAM J. Matrix Anal. Appl. **30**(2), 844–881 (2008)
6. Efron, B., Hastie, T., Johnstone, I., Tibshirani, R.: Least angle regression. Ann. Stat. **32**(2), 407–499 (2004)
7. Gittens, A.: The spectral norm error of the naive Nystrom extension. arXiv preprint, pp. 1–9 (2011). http://arxiv.org/pdf/1110.5305, http://arxiv.org/abs/1110.5305
8. Gittens, A., Mahoney, M.W.: Revisiting the Nystrom method for improved large-scale machine learning. ICML **28**, 1–45 (2013)
9. Golub, G.: Numerical methods for solving linear least squares problems. Numer. Math. **7**(1), 206–216 (1965)
10. Klimt, B., Yang, Y.: Introducing the Enron corpus. In: CEAS (2004)
11. Leskovec, J., Lang, K.J., Dasgupta, A., Mahoney, M.W.: Community structure in large networks: natural cluster sizes and the absence of large well-defined clusters. Internet Math. **6**(1), 29–123 (2009)
12. Lichman, M.: UCI machine learning repository (2013). http://archive.ics.uci.edu/ml
13. Mahoney, M.W.: Algorithmic and Statistical Perspectives on Large-Scale Data Analysis, p. 33 (2010). http://arxiv.org/abs/1010.1609
14. Mahoney, M.M.W.: Randomized algorithms for matrices and data. arXiv preprint arXiv:1104.5557, p. 49 (2011)
15. Papailiopoulos, D., Kyrillidis, A., Boutsidis, C.: Provable deterministic leverage score sampling. arXiv preprint, pp. 997–1006 (2014). http://dl.acm.org/citation.cfm?doid=2623330.2623698
16. Paul, S., Magdon-Ismail, M., Drineas, P.: Column Selection via Adaptive Sampling, pp. 1–9 (2015). http://arxiv.org/abs/1510.04149

Anti Imitation-Based Policy Learning

Michèle Sebag$^{(\boxtimes)}$, Riad Akrour, Basile Mayeur, and Marc Schoenauer

TAO, CNRS – Inria – UPSud, Université Paris-Saclay, Orsay, France
michele.sebag@lri.fr

Abstract. The *Anti Imitation-based Policy Learning* (AIPoL) approach, taking inspiration from the Energy-based learning framework (LeCun et al. 2006), aims at a pseudo-value function such that it induces the same order on the state space as a (nearly optimal) value function. By construction, the greedification of such a pseudo-value induces the same policy as the value function itself. The approach assumes that, thanks to prior knowledge, not-to-be-imitated demonstrations can easily be generated. For instance, applying a random policy on a good initial state (e.g., a bicycle in equilibrium) will on average lead to visit states with decreasing values (the bicycle ultimately falls down). Such a demonstration, that is, a sequence of states with decreasing values, is used along a standard learning-to-rank approach to define a pseudo-value function. If the model of the environment is known, this pseudo-value directly induces a policy by greedification. Otherwise, the bad demonstrations are exploited together with off-policy learning to learn a pseudo-Q-value function and likewise thence derive a policy by greedification. To our best knowledge the use of bad demonstrations to achieve policy learning is original. The theoretical analysis shows that the loss of optimality of the pseudo value-based policy is bounded under mild assumptions, and the empirical validation of AIPoL on the mountain car, the bicycle and the swing-up pendulum problems demonstrates the simplicity and the merits of the approach.

1 Introduction

Reinforcement learning aims at building optimal policies by letting the agent interact with its environment [32,33]. Among the signature challenges of RL are the facts that the agent must sufficiently explore its environment in order to ensure the optimality of its decisions, and that the consequences of its actions are delayed. Both facts raise severe scalability issues in large search spaces, which have been addressed in two ways in the last decade (Sect. 2). One way relies on the human expert's support to speed up the discovery of relevant behaviors, ranging from inverse reinforcement learning [1,27] and learning by imitation [5, 19,29] to learning from the expert's feedback [2,15,25,35]. Another way relies on the extensive interaction of the agent with its environment; it mostly operates in simulated environments, where the agent can interact with the environment and tirelessly evaluate and improve its policy without suffering exploration hazards;

© Springer International Publishing AG 2016
P. Frasconi et al. (Eds.): ECML PKDD 2016, Part II, LNAI 9852, pp. 559–575, 2016.
DOI: 10.1007/978-3-319-46227-1_35

following the pioneering TD-Gammon [34], are the Monte-Carlo Tree Search approaches [3,11,13,18] and Deep Reinforcement Learning [26].

Yet another approach is investigated in this paper, taking some inspiration from the Inverse Reinforcement Learning setting, although it almost entirely relaxes the expertise requirement on the human teacher. Specifically, the proposed approach referred to as *Anti Imitation-based Policy Learning* (AIPoL) is based on a weak prior knowledge: *when in a good state, some trivial (random or constant) policies will on average tend to deteriorate the state value, and lead to a sequence of states with decreasing value*. For instance, starting from the state where the bicycle is in equilibrium, a random policy will lead the bicycle to sooner or later fall down. This knowledge provides an operational methodology to tackle RL with very limited support from the human expert: the human expert is only asked to set the agent in a target state (e.g., the car on the top of the mountain or the bicycle in equilibrium); from this initial state, the agent applies a random policy, defining a trajectory. Contrasting with the IRL setting, this demonstration is a *bad demonstration*, showing something that should *not be done*. One merit of the approach is that it is usually much easier, and requires significantly less expertise, to generate a bad demonstration than a good one. However, such bad demonstrations provide an operational methodology to derive a good value function, as follows. Assuming that the sequence of states visited by the demonstration is such that the state value likely decreases along time (the bicycle falls down and the car arrives at the bottom of the slope), a value function can thus be derived on the state space along a learning-to-rank framework [16]. If the model of the environment is known, this value function directly defines a policy, enabling the agent to reach the target state. Otherwise, the bad demonstrations are exploited together with off-policy learning to build a Q-value function. The optimality loss of the resulting policy is bounded under mild assumptions (Sect. 3).

The empirical validation of the approach is conducted on three benchmark problems – the mountain car, the bicycle balancing and the swing-up pendulum problems – and the performances are compared to the state of the art (Sect. 4). The paper concludes with a discussion about the limitations of the AIPoL approach, and some research perspectives.

Notations. In the rest of the paper the standard Markov decision process notations $(\mathcal{S}, \mathcal{A}, p, r)$ are used: $\mathcal{S}$ and $\mathcal{A}$ respectively stand for the state and action spaces, $p : \mathcal{S} \times \mathcal{A} \times \mathcal{S} \mapsto \mathbb{R}$ is the transition model (when known), with $p(s, a, s')$ the probability of reaching state s' after selecting action a in state s, and $r : \mathcal{S} \mapsto \mathbb{R}$ is the deterministic, bounded reward function. At the core of mainstream RL approaches are the value functions associated to every policy π. $V^\pi : \mathcal{S} \mapsto \mathbb{R}$, yields for each state the expected discounted cumulative reward gathered by following π from this state, with discount factor γ in $[0, 1]$. Likewise, Q-value $Q^\pi : \mathcal{S} \times \mathcal{A} \mapsto \mathbb{R}$ associates to each state-action pair s, a the expected discounted cumulative reward $Q^\pi(s, a)$ gathered by selecting action a in state s and following policy π ever after.

$$V^\pi(s) = r(s) + \mathbb{E}_{s_t \sim p(s_{t-1}, \pi(s_{t-1}), \cdot)} \left[\sum_t \gamma^t r(s_t) | s_0 = s \right]$$

$$Q^\pi(s, a) = r(s) + \gamma \sum_{s'} p(s, a, s') V^\pi(s') \tag{1}$$

2   State of the Art

While RL traditionally relies on learning value functions [32], their learning (using dynamic programming and approximate dynamic approaches [4]) faces scalability issues w.r.t. the size of state and action spaces.[1] In the meanwhile, there is some debate about the relevance of learning value functions to achieve reinforcement learning, on the ground that solving an RL problem and defining a policy only requires to associate an action with each state. Associating a value with each state or each state-action pair thus requires more effort than needed to solve the problem. Along this line, direct policy search (DPS) (see [9] for a comprehensive presentation) directly tackles the optimization of the policy. Furthermore, DPS does not need to rely on the Markovian assumption, thus making it possible to deal with a more agile description of the search space. DPS faces two main difficulties: (i) the choice of the parametric representation, granted that the optimization landscape involves many local optima; (ii) the optimization criterion, the policy return expectation, is approximated by an empirical estimate thereof, thus defining a noisy and expensive optimization problem.

Other RL trends addressing the limitations of learning either value functions or policies are based on the expert's help. In early RL, the human expert stayed behind the stage, providing a precious and hidden help through the design of the representation space and the reward function. In inverse reinforcement learning (IRL), the expert explicitly sets the learning dynamics through demonstrating a few expert trajectories, narrowing down the exploration in the vicinity of these trajectories [1,5,19,28]. In preference-based reinforcement learning, the expert is on the stage, interacting with the learning agent. Contrasting with IRL, preference-based RL does not require the human expert to be able to demonstrate a competent behavior; the expert is only assumed to be able to rank state-action pairs [12,17], fragments of behaviors [35], or full-length trajectories [2,15] while the RL agent achieves active ranking, focusing on the generation of most informative pairs of state-actions, behaviors or trajectories. In summary, RL increasingly puts the human expert in the learning loop, and relaxes the expertise requirement; in counterpart, the RL agent becomes more and more autonomous, striving to ask more informative preference queries to the expert and to best exploit her input [25]. Supervised learning-based policy learning, pioneered by [20], also increasingly relies on expert knowledge. In [21], a sequence of reward-sensitive classification problems is built for each time step, assuming that the optimal actions will be executed in the remaining steps (akin structured

---

[1] A most appealing approach sidestepping these scalability issues, Deep Reinforcement Learning (see, e.g., [26]) requires intensive interactions between the learning agent and the environment. It is outside the scope of this paper.

learning [6]). In [22], Direct Policy Iteration (DPI) handles cost-sensitive classification problems where the loss function is defined from the Q-regret of the
current policy. Further work on DPI [7] use expert demonstrations. In [10], the
Classification-based Approximate Policy Iteration also relies on the estimation
of the Q-value based on the current policy.

As these classification-based RL approaches involve loss functions related to
the value regret, they rely on the estimation of the value function, which is computationally or expertise-wise demanding. Another limitation of classification-
based RL approaches is related to the ties, that is, the fact that there might be
several optimal actions in a given state. A supervised learning approach addressing the tie issue is energy-based learning (EBL) [23]: When aiming at finding
a classifier $h : \mathcal{X} \mapsto \mathcal{Y}$, mapping an instance space $\mathcal{X}$ onto a (possibly structured) output space $\mathcal{Y}$, the EBL claim is that in some cases, learning an energy
function $g : \mathcal{X} \times \mathcal{Y} \mapsto \mathbb{R}$, and defining $h(x)$ as $\left(\arg\max_{y\in\mathcal{Y}} g(x,y)\right)$ leads to significantly more robust results than learning directly h. An appealing argument
for the EBL approach is that g is only defined up to a monotonous transformation.[2] Along these lines, RL might be content with learning an energy-like value
function $U(s,a)$, such that policy $\pi_U(s) = \arg\max_{a\in\mathcal{A}} U(s,a)$ is a (nearly) optimal policy, regardless of whether U satisfies the Bellman optimality equation.
Next section will present a methodology for learning such an energy-based value
function, however with limited help from the human expert.

3   Overview of AIPoL

3.1   Rationale

AIPoL is based on the assumption that, while quite some expertise is required
to perform an expert demonstration, it is usually very easy to generate terrible
demonstrations, in a sense defined below. Let V^* be the (unknown) optimal
value function, satisfying the Bellman equation for some $0 < \gamma \le 1$:

$$V^*(s) = r(s) + \gamma \arg\max_{a\in\mathcal{A}} \sum_{s'\in\mathcal{S}} p(s,a,s')V^*(s') \tag{2}$$

Definition 1. *A counter-demonstration (CD), is a sequence of states*
$(s_1, \dots s_T)$ *with decreasing V^* values, i.e., s.t.*

$$\forall 1 \le i < j \le T,\ V^*(s_i) > V^*(s_j)$$

Definition 2. *Let $\mathcal{E} = \{CD_1, \dots CD_n\}$ be a set of n CDs, with $CD_i = (s_{i,t}, t = 1\dots T_i)$. The learning-to-rank problem associated to $\mathcal{E}$ is defined from the set of constraints $s_{i,t} \prec s_{i,t'}\ \forall i$ in $[1..n]$ and $\forall 1 \le t < t' \le T_i$.*

$$Find\ \widehat{U} = \arg\min_{U:\mathcal{S}\mapsto\mathbf{R}} \{\mathcal{L}(U,\mathcal{E}) + \mathcal{R}(U)\} \tag{3}$$

with $\mathcal{L}$ a ranking loss function and $\mathcal{R}(U)$ a regularization term.

---

[2] For any non-decreasing scalar function f, g and $f \circ g$ define the same classifier.

Following the learning-to-rank setting [16] and denoting $(A)_+ = \max(0, A)$, the ranking loss function used in AIPoL is the sum of the hinge loss between $U(s_{i,t})$ and $U(s_{i,t'})$ over all pairs (t, t') such that $1 \le t < t' < T_i$, with $i = 1 \dots n$:

$$\mathcal{L}(U, \mathcal{E}) = \sum_{i=1}^{n} \sum_{1 \le t < t' \le T_i} (U(s_{i,t'}) + 1 - U(s_{i,t}))_+ \tag{4}$$

A solution $\widehat{U}$ of Pb (3) (solved using e.g., [16, 24]) is thereafter referred to as *pseudo-value function*. Of course, it is unlikely that pseudo-value $\widehat{U}$ satisfies the Bellman optimality equation (2). Nevertheless, it will be seen that the optimality of the policy based on $\widehat{U}$ can be assessed in some cases, when V^* and $\widehat{U}$ define sufficiently similar orderings on the state space.

AIPoL relies on the assumption that CDs can be easily generated without requiring any strong expertise. CDs are additionally required to sufficiently visit the state space, or the interesting regions of the state space. How much does sufficient mean will be empirically assessed in Sect. 4.

3.2   AIPoL with a Known Transition Model

In this section, the transition model is assumed to be known. In the case of a **deterministic transition** model, let $s'_{s,a}$ denote the arrival state when selecting action a in state s.

Definition 3. *In the deterministic transition model case, the greedy policy based on $\widehat{U}$ selects the action leading to the arrival state with maximal $\widehat{U}$ value:*

$$\pi_{\widehat{U}}(s) = \arg\max_{a \in \mathcal{A}} \left\{ \widehat{U}(s'_{s,a}) \right\} \tag{5}$$

It immediately follows from this definition that:

Proposition 1. *In the deterministic transition setting, if $\widehat{U}$ derives the same order on the state space as the optimal value function V^*, i.e.,*

$$\forall (s, s') \in \mathcal{S},\ (\widehat{U}(s) > \widehat{U}(s')) \Leftrightarrow (V^*(s) > V^*(s'))$$

then greedy policy $\pi_{\widehat{U}}$ is an optimal policy.

In the case of a **stochastic transition** model, by slight abuse of notation let $s'_{s,a}$ denote a state drawn after distribution $p(s, a, \cdot)$.

Definition 4. *In the stochastic transition model case, the greedy policy based on $\widehat{U}$ selects the action leading to the maximal $\widehat{U}$ value expectation:*

$$\pi_{\widehat{U}}(s) = \arg\max_{a \in \mathcal{A}} \left\{ \mathbb{E}_{p(s,a,\cdot)} \widehat{U}(s'_{s,a}) \right\} \tag{6}$$

Some assumptions on the regularity of the transition model, on V^* and $\widehat{U}$ are required to establish a result analogous to Proposition 1 in the stochastic case. The main assumption regards the sub-Gaussianity of the transition model. Note that this assumption does hold, of course, for Gaussian transition models, and also in robotic settings, where the distance between two consecutive states is bounded due to physical and mechanical constraints.

A guarantee on the $\pi_{\widehat{U}}$ optimality can be obtained under the following assumptions:

Proposition 2. *Assuming (1) a continuous optimal value function V^* on $\mathcal{S}$; (2) a pseudo-value $\widehat{U}$ deriving the same order on the state space as V^*; (3) a β-sub-Gaussian transition model, that is,*

$$\forall t \in \mathbb{R}^+, \mathbb{P}(\|\mathbb{E}s'_{s,a} - s'_{s,a}\|_2 > t) < 2e^{-\beta t^2}$$

(4) $\widehat{U}$ being Lipschitz with constant M, that is,

$$\forall s, s' \in \mathcal{S}, |\widehat{U}(s) - \widehat{U}(s')| < L\|s - s'\|_2$$

(5) for all every s, there exists a margin between the best and the second best action after $\widehat{U}$, such that:

$$\forall a' \neq a = \pi_{\widehat{U}}(s), \mathbb{E}\widehat{U}(s'_{s,a}) > \mathbb{E}\widehat{U}(s'_{s,a'}) + M \tag{7}$$

Then, if $2L < M\beta$, $\pi_{\widehat{U}}$ is an optimal policy.

Proof. The idea of the proof is the following: Consider the average value $\mathbb{E}V^*(s'_{s,a})$, where the expectation is taken over $p(s, a, \cdot)$. By continuity of V^* there exists a state noted $s'_{s,a,V}$ in the neighborhood of $\mathbb{E}s'_{s,a}$ such that $V^*(s'_{s,a,V}) = \mathbb{E}V^*(s'_{s,a})$. Likewise there exists a state $s'_{s,a',V}$ such that $V^*(s'_{s,a',V}) = \mathbb{E}V^*(s'_{s,a'})$.

Let us assume by contradiction that optimal policy π^* is such that $\pi^*(s) = a' \neq a$. It follows that

$$V^*(s'_{s,a',V}) > V^*(s'_{s,a,V})$$

and therefore, as $\widehat{U}$ and V^* define same orderings on $\mathcal{S}$,

$$\widehat{U}(s'_{s,a',V}) > \widehat{U}(s'_{s,a,V}) \tag{8}$$

Let us denote $K(u) = \|s'_{s,a',V} - s'_{s,a',u}\|$. For any $\varepsilon > 0$, we have

$$
\begin{aligned}
|\widehat{U}(s'_{s,a',V}) - \mathbb{E}_u\widehat{U}(s'_{s,a'})| &= \mathbb{E}_u|\widehat{U}(s'_{s,a',V}) - \widehat{U}(s'_{s,a',u})| \\
&= \mathbb{E}_{K(u)<\varepsilon}|\widehat{U}(s'_{s,a',V}) - \widehat{U}(s'_{s,a',u})| \\
&\quad + \mathbb{E}_{K(u)>\varepsilon}|\widehat{U}(s'_{s,a',V}) - \widehat{U}(s'_{s,a',u})| \\
&\leq L\varepsilon + L \int_\varepsilon^\infty t \cdot 2e^{-\beta t^2} dt \quad (*) \\
&\leq L\varepsilon + \tfrac{L}{\beta} e^{-\beta \varepsilon^2} < L(\varepsilon + \tfrac{1}{\beta})
\end{aligned}
$$

where (*) is derived using the Lipschitz property of $\widehat{U}$ and the sub-Gaussian property of the transition model.

If $L(\varepsilon + \frac{1}{\beta}) < M/2$, then from Eq. (8) it comes:

$$\mathbb{E}\widehat{U}(s'_{s,a'}) + M/2 \geq \widehat{U}(s'_{s,a',V}) > \widehat{U}(s'_{s,a,V}) \geq \mathbb{E}\widehat{U}(s'_{s,a}) - M/2$$

which contradicts the margin assumption (Eq. 7), hence the result.   □

Overall, in the known transition model case, AIPoL proceeds by generating the CDs, defining the associated ranking problem, finding a solution $\widehat{U}$ thereof and building policy $\pi_{\widehat{U}}$ by greedification of $\widehat{U}$ (Algorithm 1).

Algorithm 1. Model-based AIPoL

Input: $\mathcal{E} = \{CD_1, \ldots CD_n\}$
$\widehat{U} = \arg\min \{\mathcal{L}(U, \mathcal{E}) + \mathcal{R}(U)\}$ (Pb (3))
      with $\mathcal{L}(U, \mathcal{E})$ (from Eq. 4) and $\mathcal{R}(U)$ an L_2 regularization.
Return: $\pi_{\widehat{U}}$ (Eq. (5))

3.3 AIPoL with Unknown Transition Model

When the transition model is unknown, a pseudo Q-value $\widehat{Q}$ is built from the pseudo-value $\widehat{U}$ learned from the CDs using off-policy learning. The intuition is that, given $\widehat{U}$ and triplets (s_1, a_1, s'_1) and (s_2, a_2, s'_2), the pseudo Q-value of state-action pair (s_1, a_1) is lower than for state action pair (s_2, a_2) if state s'_1 has a lower pseudo-value than s'_2 $(\widehat{U}(s'_1) < \widehat{U}(s'_2))$.

Definition 5. *With the same notations as above, let $\widehat{U}$ be a pseudo value function solution of Pb (3), and let*

$$\mathcal{G} = \{(s_i, a_i, s'_i), i = 1 \ldots m\}$$

be a set of state-action-next-state triplets. The learning-to-rank problem associated to $\mathcal{G}$ is defined from the set of ranking constraints $(s_i, a_i) \prec (s_j, a_j)$ for all i, j such that $\widehat{U}(s'_i) < \widehat{U}(s'_j)$.

$$Find \; \widehat{Q} = \underset{Q:\mathcal{S}\times\mathcal{A}\mapsto\mathbf{R}}{\arg\min} \; \{\mathcal{L}(Q, \mathcal{G}) + \mathcal{R}(Q)\} \tag{9}$$

with $\mathcal{L}$ a loss function and $\mathcal{R}(Q)$ a regularization term.

In AIPoL, the ranking loss function is set to:

$$\mathcal{L}(Q, \mathcal{G}) = \sum_{i=1}^{m} \sum_{j=1}^{m} y_{ij} \left(Q((s_i, a_i) + 1 - Q(s_j, a_j) \right)_+ \tag{10}$$

with $y_{ij} = 1$ iff $\widehat{U}(s'_i) < \widehat{U}(s'_j)$ and 0 otherwise.

Definition 6. *Letting* $\widehat{Q}$ *be a pseudo Q-value function learned from Pb (9), policy* $\pi_{\widehat{Q}}$ *is defined as:*

$$\pi_{\widehat{Q}}(s) = \arg\max_{a \in \mathcal{A}} \left\{ \widehat{Q}(s, a) \right\} \tag{11}$$

Some more care must however be exercized in order to learn accurate pseudo Q-value functions. Notably, comparing two triplets (s_1, a_1, s_1') and (s_2, a_2, s_2') when s_1 and s_2 are too different does not yield any useful information. Typically, when $\widehat{U}(s_1) \gg \widehat{U}(s_2)$, it is likely that $\widehat{U}(s_1') > \widehat{U}(s_2')$ and therefore the impact of actions a_1 and a_2 is very limited: in other words, the learned $\widehat{Q}$ does not deliver any extra information compared to $\widehat{U}$. This drawback is addressed by filtering the constraints in Pb (9) and requiring that the triplets used to learn $\widehat{Q}$ be such that:

$$||s_1 - s_2||_2 < \eta \tag{12}$$

with η a hyper-parameter of the AIPoL algorithm (set to 10 % or 1 % of the state space diameter in the experiments). Empirically, another filter is used, based on the relative improvement brought by action a_1 in s_1 compared to action a_2 in s_2. Specifically, the constraint $(s_1, a_1) \succ (s_2, a_2)$ is generated only if selecting action a_1 in s_1 and going to s_1' results in a higher value improvement than selecting action a_2 in s_2 and going to s_2':

$$\widehat{U}(s_1) - \widehat{U}(s_1') > \widehat{U}(s_2) - \widehat{U}(s_2') \tag{13}$$

Overall, the model-free AIPoL (Algorithm 2) proceeds by solving the model-based problem (Algorithm 1), using traces (s, a, s') to build the learning-to-rank problem (9), finding a solution $\widehat{Q}$ thereof and building policy $\pi_{\widehat{Q}}$ by greedification of $\widehat{Q}$.

Algorithm 2. Model-free AIPoL

Input: $\mathcal{E} = \{CD_1, \ldots CD_n\}$
Input: $\mathcal{G} = \{(s_i, a_i, s_i'), i = 1 \ldots m\}$
$\widehat{U} = \arg\min \{\mathcal{L}(U, \mathcal{E}) + \mathcal{R}(U)\}$ (Pb (3))
                         with $\mathcal{L}(U, \mathcal{E})$ (Eq. 4) and $\mathcal{R}(U)$ an L_2 regularization.
$\widehat{Q} = \arg\min \{\mathcal{L}(Q, \mathcal{G}) + \mathcal{R}(Q)\}$ (Pb (9))
                         with $\mathcal{L}(Q, \mathcal{G})$ (Eq. 10) and $\mathcal{R}(Q)$ an L_2 regularization.
Return: $\pi_{\widehat{Q}}(s)$ (Eq. 11)

3.4 Discussion

In the model-based setting, the quality of the AIPoL policy essentially depends on sufficiently many CDs to be generated with limited expertise, and on the coverage of the state space enforced by these CDs. In many benchmark problems, the goal is to reach a target state (the car on the mountain or the bicycle in equilibrium). In such cases, CDs can be generated by simply setting the starting state to the target state, and following a random or constant policy ever after. Such a trivial policy is likely to deviate from the good state region, and visit states with lower and lower values, thus producing a CD. Additionally, the CDs will sample the neighborhood of the target state; the pseudo value function $\widehat{U}$ learned from these CDs will then provide a useful guidance toward the (usually narrow) good region. The intuition behind AIPoL is similar to that of TD-gammon [34]: the value function should steadily increase when reaching the desirable states, regardless of satisfying the Bellman equation.

In the known transition model case, under the assumption that the pseudo value $\widehat{U}$ induces the same ordering on the state space as V^*, policy $\pi_{\widehat{U}}$ is optimal in the deterministic case (Proposition 1). Under additional assumptions on the regularity of the optimal value function V^*, of $\widehat{U}$ and on the transition noise, the optimality still holds in the stochastic transition case (Proposition 2). This is true even though no reward is involved in the definition of $\widehat{U}$.

In the unknown transition model case, the constraints used to learn $\widehat{Q}$ introduce a systematic bias, except in the case where the reward function r is equal to 0 almost everywhere. Let us consider the deterministic case for simplicity. By definition,

$$Q^*(s_1, a_1) = r(s_1) + \gamma V^*(s_1')$$

If the reward function is equal to 0 almost everywhere, then with high probability:

$$(V^*(s_1') > V^*(s_2')) \Leftrightarrow (Q^*(s_1, a_1) > Q^*(s_2, a_2))$$

Then, if $\widehat{U}$ induces the same ordering on the state space as V^*, the constraints on $\widehat{Q}$ derived from the traces are satisfied by Q^*, and the learning-to-rank problem (9) is noiseless.

Otherwise, the difference between the instantaneous rewards $r(s_1)$ and $r(s_2)$ can potentially offset the difference of values between s_1' and s_2', thus leading to generate noisy constraints. The generation of such noisy constraints is alleviated by the additional requirement on the constraints (Eq. 13), requiring the $\widehat{U}$ value gap between s_1 and s_1' be larger than between s_2 and s_2'.

Overall, the main claim of the AIPoL approach is that generating CDs, though requiring much less expertise than that required to generate quasi expert behavior or asking the expert to repair or compare behaviors, can still yield reasonably competent policies. This claim will be examined experimentally in next section.

4   Experimental Validation

This section presents the experimental setting used for the empirical validation of AIPoL, before reporting and discussing the comparative results.

4.1   Experimental Setting

The AIPoL performance is assessed on three standard benchmark problems: The *mountain car* problem, using SARSA as baseline [31]; The *bicycle balancing* problem, using preference-based reinforcement learning as baseline [2,35]; the *under-actuated swing-up pendulum* problem, using [14] as baseline. In all experiments, the pseudo-value $\widehat{U}$ and $\widehat{Q}$ functions are learned using Ranking-SVM with Gaussian kernel [16]. The hyper-parameters used for all three benchmark problems are summarized in Table 1.

The first goal of the experiments is to investigate how much knowledge is required to generate sufficiently informative CD, enabling AIPoL to yield state-of-art performances. This issue regards (i) the starting state of an CD, (ii) the controller used to generate an CD, (iii) the number and length of the CDs. A second goal is to examine whether and to which extent the performances obtained in the model-free setting (transition model unknown) are degraded compared to the model-based setting. A third goal is to investigate the sensitivity of the AIPoL performance w.r.t. the algorithm hyper-parameters (Table 1), including: (i) the number and length of the CDs; (ii) the Ranking-SVM hyper-parameters C (weight of the data fitting term) and σ (Gaussian kernel width); (iii) the AIPoL parameter η used to filter the ordering constraints used to learn $\widehat{Q}$ (Eq. 12).

Table 1. AIPoL hyper parameters on the three benchmark problems: number and length of CDs, starting state and controllers used to generate the CDs; hyper parameters used to learn pseudo-value $\widehat{U}$ (parameters C_1 and $1/\sigma_1^2$; hyper-parameters used to learn pseudo value $\widehat{Q}$ (parameters C_2 and $1/\sigma_2^2$; # of constraints).

		Mountain car	Bicycle	Pendulum
CD	number	1	20	1
	length	1,000	5	1,000
	starting state	target st	random	target st
	controller	neutral	random	neutral
$\widehat{U}$	C_1	10^3	10^3	10^{-5}
	$1/\sigma_1^2$	10^{-3}	10^{-3}	.5
$\widehat{Q}$	nb const	500	5,000	–
	C_2	10^3	10^3	–
	$1/\sigma_2^2$	10^{-3}	10^{-3}	–

4.2   The Mountain Car

Following [31], the mountain car problem involves a 2D state space (position, speed), and a discrete action space (backward, neutral position, forward). The friction coefficient ranges in $[0, .02]$. AIPoL is compared to the baseline SARSA with $\lambda = .9$, $\alpha = .05/10m$, $\varepsilon = 0$, with 9×9 tile coding, with 100 episodes for

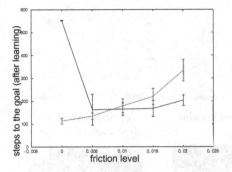

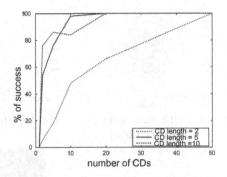

Fig. 1. Mountain car: Number of time steps to reach the goal for AIPoL (solid blue line) and SARSA (dashed red line) *vs* friction value (average and standard deviation over 20 runs) (Color figure online)

Fig. 2. Bicycle (model-based setting): Fraction of success of AIPoL w.r.t. number and length of CDs.

learning, stopping an episode whenever a terminal state is reached, or after 1000 steps. AIPoL uses 1 CD of length 1000. In the unknown transition model case, the 500 constraints are generated from random trajectories where pairs (s, a) and $(s'a')$ were selected subject to constraints (Eqs. 12 and 13) with proximity threshold $\eta = 10\%$.

The performances are excellent for both known (1 CD with length 1,000) and unknown transition model cases, with negligible runtime. Figure 3a depicts the CD in the 2D (position, speed) space, starting from the target state and selecting the neutral action for 1,000 time steps. The pseudo-value $\widehat{U}$ function learned by AIPoL and the approximation of V^* learned by SARSA in two representative runs are respectively displayed in Fig. 3b and c, showing that $\widehat{U}$ is very smooth compared to that approximate value.

Figure 3e and f display the policy based on $\widehat{Q}$ in the model-free case, and the optimal policy learned by SARSA, suggesting that the policy learned by AIPoL is much simpler than for SARSA. Figure 3d shows a typical trajectory based on the AIPoL policy in the model-free case.

The sensitivity analysis shows that the main parameter governing the AIPoL performance on the mountain car problem is the friction (Fig. 1). For low friction values, the dynamics is quasi reversible as there is no loss of energy; accordingly, letting the car fall down from the target state does not generate a sequence of states with decreasing value (the value of the state intuitively increases with its energy). In the low friction region (friction in [0, .05]), AIPoL is dominated by SARSA. For high friction values (> .02), the car engine lacks the required power to climb the hill and both approaches fail. For moderate friction values (in [.01, .02]), AIPoL significantly outperforms SARSA.

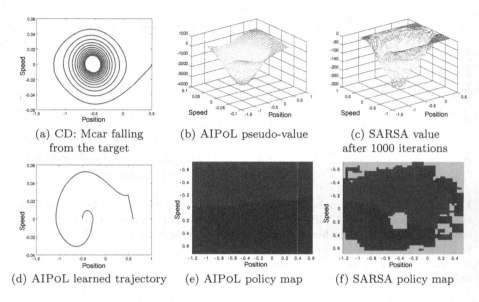

(a) CD: Mcar falling
from the target

(b) AIPoL pseudo-value

(c) SARSA value
after 1000 iterations

(d) AIPoL learned trajectory

(e) AIPoL policy map

(f) SARSA policy map

Fig. 3. The mountain car problem: Comparative evaluation of AIPoL and SARSA in the model-free setting, on two representative runs (friction = .01). The policy map visually displays the selected action for each state in the 2D (position, speed) space (best seen in color: red= forward, blue= backward, green= neutral). (Color figure online)

4.3   The Bicycle Balancing

Following [20], the bicycle balancing problem involves a 4-dimensional state space (the angles of the handlebar and of the bicycle and the angular velocities), and a 3-action action space (do nothing, turn the handlebar left or right, lean the rider left or right). The goal is to maintain the bicycle in equilibrium for 30,000 time steps; note that a random controller starting from the equilibrium state $(0, 0, 0, 0)$ leads the bicycle to fall after 200 steps on average.

Known Transition Model Case. To assess the sensitivity of the approach w.r.t. the starting state, the CDs are generated using a random starting state and a random controller. The definition of the policy $\pi_{\hat{U}}$ (Eq. 5) is adapted to account for the fact that, due to the temporal discretization of the transition model [20], the effect of action a_t on the angle values is only visible in state s_{t+2}. Some look-ahead is thus required to define the greedy policy $\pi_{\hat{U}}$. Formally, the selected action is obtained by maximization of the value obtained after two time steps:

$$\pi_{\hat{U}}(s) = \underset{a \in \mathcal{A}}{argmax} \left\{ \max_{a' \in \mathcal{A}} \mathbb{E} \widehat{U}(s''_{\bar{s}'_{s,a}, a'}) \right\}$$

Given this definition, AIPoL only requires 20 CD of length 5 to learn a competent policy, keeping the bicycle in equilibrium for over 30,000 time steps with high probability (in all of the 100 runs, Fig. 2). In comparison, the state of the art requires a few dozen trajectories to be ranked by the expert (15 for [2] and 20 for [35]), the starting point of which is close to the equilibrium. With same starting point, AIPoL reaches the goal (keeping the bicycle in equilibrium 100 times out of 100 runs) with a single CD of length 5.

Unknown Transition Model Case. The ordering constraints on the state-action pairs likewise take into account the temporal discretization and the delayed impact of the actions. Formally, from sequences $(s_1, a_1, s_1', a_1', s_1'')$ and $(s_2, a_2, s_2', a_2', s_2'')$, constraint $(s_1, a_1) \succ (s_2, a_2)$ is generated if

$$\widehat{U}(s_1'') - \widehat{U}(s_1) > \widehat{U}(s_2'') - \widehat{U}(s_2)$$

The proximity threshold is set to $\eta = 1\%$. 5,000 constraints are required to achieve the same performance as in the model-based setting.

4.4   The Under-Actuated Swing-Up Pendulum

Following [14], the swing-up pendulum involves a 2-dimensional state space $(s = (\theta, \dot{\theta}))$ and a 3 action space. The pendulum has two equilibrium states, a stable one and an unstable one. The goal, starting from the stable state (bottom position) is to reach the unstable one (top position). The task is under-actuated since the agent has a limited torque and must gain some momentum before achieving its swing-up. The task is stopped after 20 s or when the agent successfully maintains the pendulum in an up-state ($\theta < \pi/4$) for 3 consecutive time steps. Only the model-based setting has been considered for the pendulum problem, with a computational cost of 3 s.

On the pendulum problem, the sensitivity of the approach w.r.t. the Ranking-SVM hyper-parameters is displayed in Fig. 4. Two failure regions appear when learning the pseudo-value $\widehat{U}$ from a single CD of length 1,000: if the kernel width is too small, there is no generalization and the pendulum does not reach the top.

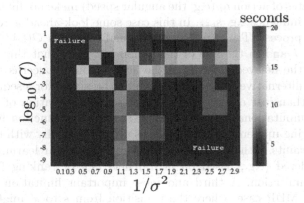

Fig. 4. The pendulum problem: Sensitivity of AIPoL performance (average over 10 runs) w.r.t Ranking-SVM hyper-parameters C and $1/\sigma^2$ (see text for details).

If the kernel width is too large, the accuracy is insufficient and the pendulum does not decrease its speed sufficiently early: it reaches the top and falls down on the other side. For good hyper-parameter settings ($C = 1$ and $1/\sigma^2$ ranging in $[1.7, 2.7]$; or C and $1/\sigma^2$ very small), the pendulum reaches the target state in 3 s and stays there. The AIPoL performance matches the state of the art [14], which relies on a continuous variant of the Bayes-adaptive planning, and achieves the goal (staying in an up-state for 3 s) after on average 10 s of interaction.

5   Discussion and Perspectives

The AIPoL approach to reinforcement learning has been presented together with an analytic and empirical study of its performances. Its main novelty is twofold compared to the state of the art. On the one hand, AIPoL learns a pseudo-value function and derives a policy by greedification; computationally-wise, it tackles a much less complex problem than e.g., inverse reinforcement learning [1,19] (learning a reward function and solving a complete RL problem) or preference-based RL [2] (learning a return value and solving a difficult optimization problem). In addition, AIPoL significantly relaxes the requirements on the human teacher. She is not required to perform (nearly) optimal demonstrations as in IRL, or to compare, and possibly repair, trajectories as in preference-based learning: she is only required to know *what will go wrong*.

In the mountain car and the pendulum problems, AIPoL uses informed CDs (starting in the target state). In the bicycle problem however, the CD sequences start in a random state. In this latter case, the pseudo value function coarsely leads to get away from state regions with low value: the inadequacy of the pseudo value in low value regions is (almost) harmless should the learning agent spend little or no time in these regions.

A first limitation of the AIPoL approach, illustrated on the bicycle problem, is when the effect of the selected actions is fully visible after a few time steps, that is, when the transition dynamics involves some latency. This latency occurs when some coordinates of action a_t (e.g. the angular speed) make no difference on state s_{t+1} and only influence e.g. $s_{t+\ell}$. In this case some look-ahead is required in the greedification process. The extra computational cost is in $\mathcal{O}(|\mathcal{A}|^\ell)$, exponential in the latency ℓ and in the size of the action set. Note that this phenomenon, distinct from the delayed rewards of the actions, only concerns the transition dynamics: An alternative could be to commit to an action for a sequence of time-steps, rather than just a single step [26]. A second limitation of the approach is that the computational cost of building the Q-value function might be high (e.g. on the swing-up pendulum) as it scales up quadratically with the number of ranking constraints. Other ranking approaches with linear learning complexity will be considered (e.g., based on neural nets [30] or ranking forests [8]) to address this limitation. A third and most important limitation concerns the non-reversible MDP case, where the transition from s to s' might take much longer than from s' to s. Further work is on-going to address the non reversible case. A main theoretical perspective is to investigate the quality of the AIPoL policy in the unknown transition model case, depending on the structure of the MDP dynamics and the sparsity of the reward function.

Acknowledgments. We wish to thank the anonymous reviewers for their insightful comments, which helped to greatly improve the paper.

References

1. Abbeel, P.: Apprenticeship learning and reinforcement learning with application to robotic control. Ph.d. thesis, Stanford, CA, USA (2008). aAI3332983
2. Akrour, R., Schoenauer, M., Sebag, M., Souplet, J.C.: Programming by feedback. In: Proceedings of the International Conference on Machine Learning (ICML). JMLR Proceedings, vol. 32, pp. 1503–1511. JMLR.org (2014)
3. Auer, P., Ortner, R.: Logarithmic online regret bounds for undiscounted reinforcement learning. In: Schölkopf, B., et al. (eds.) NIPS 19, pp. 49–56. MIT Press (2007)
4. Bertsekas, D.P.: Dynamic Programming and Optimal Control, 2nd edn. Athena Scientific (2000)
5. Calinon, S., Billard, A.: Active teaching in robot programming by demonstration. In: Proceedings of the 16th IEEE RO-MAN, pp. 702–707. IEEE (2007)
6. Chang, K., Krishnamurthy, A., Agarwal III, A., H.D., Langford, J.: Learning to search better than your teacher. In: Bach, F.R., Blei, D.M. (eds.) Proceedings of the 32nd ICML. JMLR Proceedings, vol. 37, pp. 2058–2066. JMLR.org (2015)
7. Chemali, J., Lazaric, A.: Direct policy iteration with demonstrations. In: Yang, Q., Wooldridge, M. (eds.) Proceedings of the 24th IJCAI, pp. 3380–3386. AAAI Press (2015)
8. Clémençon, S., Depecker, M., Vayatis, N.: Ranking forests. J. Mach. Learn. Res. **14**(1), 39–73 (2013)
9. Deisenroth, M.P., Neumann, G., Peters, J.: A survey on policy search for robotics. Found. Trends Robot. **2**(1–2), 1–142 (2013)
10. Farahmand, A.M., Precup, D., Barreto, A.M.S., Ghavamzadeh, M.: Classification-based approximate policy iteration. IEEE Trans. Autom. Control **60**(11), 2989–2993 (2015)
11. Filippi, S., Cappé, O., Garivier, A.: Optimism in reinforcement learning and kullback-leibler divergence. In: Proceedings of the Allerton Conference on Communication, Control, and Computing, pp. 115–122 (2010). https://hal.archives-ouvertes.fr/hal-00476116
12. Fürnkranz, J., Hüllermeier, E., Cheng, W., Park, S.: Preference-based reinforcement learning: a formal framework and a policy iteration algorithm. Mach. Learn. **89**(1–2), 123–156 (2012)
13. Gelly, S., Silver, D.: Combining online and offline knowledge in UCT. In: Ghahramani, Z. (ed.) Proceedings of the 24th ICML, pp. 273–280. ACM (2007)
14. Guez, A., Heess, N., Silver, D., Dayan, P.: Bayes-adaptive simulation-based search with value function approximation. In: Ghahramani, Z., et al. (eds.) NIPS 27, pp. 451–459. Curran Associates, Inc. (2014)
15. Jain, A., Wojcik, B., Joachims, T., Saxena, A.: Learning trajectory preferences for manipulators via iterative improvement. In: Burges, C.C., et al. (eds.) NIPS 26 (2013)
16. Joachims, T.: A support vector method for multivariate performance measures. In: Raedt, L.D., Wrobel, S. (eds.) Proceedings of the 22nd ICML, pp. 377–384. ACM (2005)

17. Knox, W.B., Stone, P., Breazeal, C.: Training a robot via human feedback: a case study. In: Herrmann, G., et al. (eds.) Proceedings of the 5th International Conference on Social Robotics, pp. 460–470 (2013)

18. Kocsis, L., Szepesvári, C.: Bandit based Monte-Carlo planning. In: Fürnkranz, J., Scheffer, T., Spiliopoulou, M. (eds.) ECML 2006. LNCS(LNAI), vol. 4212, pp. 282–293. Springer, Heidelberg (2006). doi:10.1007/11871842_29

19. Konidaris, G., Kuindersma, S., Barto, A., Grupen, R.: Constructing skill trees for reinforcement learning agents from demonstration trajectories. In: Lafferty, J.D., et al. (eds.) NIPS 23, pp. 1162–1170. MIT Press (2010)

20. Lagoudakis, M.G., Parr, R., Bartlett, L.: Least-squares policy iteration. JMLR **4**, 2003 (2003)

21. Langford, J., Zadrozny, B.: Relating reinforcement learning performance to classification performance. In: Proceedings of the 22nd ICML, pp. 473–480. ACM (2005)

22. Lazaric, A., Ghavamzadeh, M., Munos, R.: Analysis of a classification-based policy iteration algorithm. In: Fürnkranz, J., Joachims, T. (eds.) Proceedings of the 27th ICML, pp. 607–614. Omnipress (2010)

23. LeCun, Y., Chopra, S., Hadsell, R., Ranzato, M., Huang, F.: A tutorial on energy-based learning. In: Bakir, G., Hofman, T., Schölkopf, B., Smola, A., Taskar, B. (eds.) Predicting Structured Data. MIT Press (2006)

24. Li, P., Burges, C.J.C., Wu, Q.: Mcrank: Learning to rank using multiple classification and gradient boosting. In: Advances in Neural Information Processing Systems, vol. 20, pp. 897–904 (2007)

25. Loftin, R.T., Peng, B., MacGlashan, J., Littman, M.L., Taylor, M.E., Huang, J., Roberts, D.L.: Learning behaviors via human-delivered discrete feedback: modeling implicit feedback strategies to speed up learning. Auton. Agent. Multi-Agent Syst. **30**(1), 30–59 (2016)

26. Mnih, V., Kavukcuoglu, K., Silver, D., Rusu, A.A., Veness, J., Bellemare, M.G., Graves, A., Riedmiller, M., Fidjeland, A.K., Ostrovski, G., Petersen, S., Beattie, C., Sadik, A., Antonoglou, I., King, H., Kumaran, D., Wierstra, D., Legg, S., Hassabis, D.: Human-level control through deep reinforcement learning. Nature **518**, 529–533 (2015)

27. Ng, A.Y., Russell, S.: Algorithms for inverse reinforcement learning. In: Langley, P. (ed.) Proceedings of the International Conference on Machine Learning (ICML), pp. 663–670. Morgan Kaufmann (2000)

28. Ross, S., Bagnell, J.A.: Reinforcement and imitation learning via interactive no-regret learning. CoRR abs/1406.5979 (2014)

29. Schaal, S., Ijspeert, A., Billard, A.: Computational approaches to motor learning by imitation. Philos. Trans. R Soc. Lond. B Biol. Sci. **358**(1431), 537–547 (2003)

30. Severyn, A., Moschitti, A.: Learning to rank short text pairs with convolutional deep neural networks. In: Baeza-Yates, R.A., Lalmas, M., Moffat, A., Ribeiro-Neto, B.A. (eds.) SIGIR, Research and Development in Information Retrieval, pp. 373–382. ACM (2015)

31. Sutton, R.S.: Generalization in reinforcement learning: successful examples using sparse coarse coding. In: Touretzky, D.S., et al. (eds.) NIPS 8, pp. 1038–1044. MIT Press (1995)

32. Sutton, R.S., Barto, A.G.: Introduction to Reinforcement Learning, 1st edn. MIT Press, Cambridge (1998)

33. Szepesvári, C.: Algorithms for Reinforcement Learning. Synthesis Lectures on Artificial Intelligence and Machine Learning. Morgan & Claypool Publishers, San Rafael (2010)

34. Tesauro, G., Sejnowski, T.J.: A parallel network that learns to play backgammon. Artif. Intell. **39**(3), 357–390 (1989)
35. Wilson, A., Fern, A., Tadepalli, P.: A bayesian approach for policy learning from trajectory preference queries. In: Bartlett, P.L., et al. (eds.) NIPS 25, pp. 1142–1150 (2012)

Cancer: Another Algorithm for Subtropical Matrix Factorization

Sanjar Karaev(✉) and Pauli Miettinen

Max-Planck-Institut für Informatik, Saarbrücken, Germany
{skaraev,pmiettin}@mpi-inf.mpg.de

Abstract. Subtropical algebra is a semi-ring over the nonnegative real numbers with standard multiplication and the addition defined as the maximum operator. Factorizing a matrix over the subtropical algebra gives us a representation of the original matrix with element-wise maximum over a collection of nonnegative rank-1 matrices. Such structure can be compared to the well-known Nonnegative Matrix Factorization (NMF) that gives an element-wise sum over a collection of nonnegative rank-1 matrices. Using the maximum instead of sum changes the 'parts-of-whole' interpretation of NMF to 'winner-takes-it-all' interpretation. We recently introduced an algorithm for subtropical matrix factorization, called Capricorn, that was designed to work on discrete-valued data with discrete noise [Karaev & Miettinen, SDM '16]. In this paper we present another algorithm, called Cancer, that is designed to work over continuous-valued data with continuous noise – arguably, the more common case. We show that Cancer is capable of finding sparse factors with excellent reconstruction error, being better than either Capricorn, NMF, or SVD in continuous subtropical data. We also show that the winner-takes-it-all interpretation is usable in many real-world scenarios and lets us find structure that is different, and often easier to interpret, than what is found by NMF.

1 Introduction

Matrix factorizations such as Singular Value Decomposition (SVD) or Nonnegative Matrix Factorization (NMF) are among the most-used methods in data analysis. One way to interpret the factorization is the so-called 'components view' that considers the factorization as a sum of rank-1 matrices. The rank-1 matrices can be considered as patterns found from the data, and different constraints on the factorizations yield different types of patterns. The non-negativity constraint in NMF, for example, yields patterns that are 'parts-of-whole'.

Instead of – or in addition to – constraining the rank-1 matrices, we can also change how we aggregate them. For factorizations made under the standard algebra, the aggregation is always the standard sum, but if we change the algebra, we can have different kinds of aggregations. One possible algebra is the so-called *subtropical algebra*: a semi-ring over the non-negative real numbers with the standard multiplication but with the addition defined as the maximum operation.

© Springer International Publishing AG 2016
P. Frasconi et al. (Eds.): ECML PKDD 2016, Part II, LNAI 9852, pp. 576–592, 2016.
DOI: 10.1007/978-3-319-46227-1_36

A subtropical factorization gives us non-negative rank-1 matrices, just as NMF, but unlike NMF's parts-of-whole interpretation, the subtropical factors are best interpreted using the 'winner-takes-it-all' interpretation: for each element of the matrix, only the largest value in any of the rank-1 components matter.

The winner-takes-it-all interpretation means that each rank-1 component tries to present a dominant pattern: the elements should be as close to the original matrix's elements as possible (but without being much larger) to have any effect to the final outcome of the factorization. Consequently, the values of a component that do not contribute to the final result (i.e. are not the largest ones) can be made as small as possible without any adverse effects; often, many of them can simply be set to 0.

Recently, we introduced an algorithm for subtropical matrix factorization called Capricorn [9]. Capricorn aims at finding subtropical factorizations from discrete-valued (e.g. integer) data and consequently, it also assumes a discrete noise model where only some of the entries are perturbed. We also empirically validated that Capricorn is capable of finding the exact subtropical decomposition if it exists. Many real-world data, however, are better modelled using Gaussian noise, where every element is slightly perturbed, but Capricorn often fails finding good factorizations from such data sets. In this paper we present Cancer, another algorithm for subtropical factorizations. Cancer is complementary to Capricorn as it is designed to work well on data perturbed with Gaussian noise; conversely, Cancer does not do well if the noise follows the model Capricorn was designed for. One could say that if Capricorn is the south, Cancer is the north.

2   Related Work

Our recent work on the Capricorn algorithm [9] is, to the best of our knowledge, the only existing work using tropical or subtropical algebra in data analysis. It also provides a number of theoretical results regarding the subtropical algebra and its close cousin, the tropical algebra (see below). Another application of subtropical algebra is [12], where it is used as a part of a recommender system.

In general, though, matrix factorization methods are ubiquitous in data analysis. A popular example is the nonnegative matrix factorization (NMF) (see, e.g. [5]), where the factorization is restricted to the semi-ring of the nonnegative real numbers. Another example of a matrix factorization over a non-standard algebra is the Boolean matrix factorization (see [11]), where the factorization is restricted to binary matrices and the algebra is the Boolean one (i.e. the summation is defined as $1 + 1 = 1$).

The tropical, or max-plus, algebra [1] is another semi-ring over the extended set of reals $\mathbb{R} \cup \{-\infty\}$ with addition defined as the maximum operator and the multiplication defined as the standard plus operator. Tropical and subtropical algebras are isomorphic (take the logarithm of the latter to obtain the former), and as such, many results obtained for max-plus automatically hold for max-times, although this is not directly true in the case of approximate matrix factorizations (see [9]). Despite the theory of max-plus algebra being relatively young, it has been thoroughly studied in recent years. The reason for this is an explosion of interest in so called discrete event systems (DES) [4], where max-plus algebra has become ubiquitously used for modeling (see e.g. [2,6]).

Yet another approach of computing the matrix factorization over non-standard algebras involves using the Łukasiewicz algebra. They have been recently applied to decompose matrices with grade values [3].

3   Notation and Definitions

Throughout this paper, we will denote a matrix by upper-case boldface letters (A), and vectors by lower-case boldface letters (a). The ith row of matrix A is denoted by A_i and the jth column by A^j. The matrix A with the ith column removed is denoted by A^{-i}, and A_{-i} is the respective notation for A with a removed row. Most matrices and vectors in this paper are restricted to the nonnegative real numbers $\mathbb{R}_+$.

In this paper we consider matrix factorization over so called max-times algebra. It differs from the standard algebra of real numbers in that addition is replaced with the operation of taking the maximum. Also the domain is restricted to the set of nonnegative real numbers.

Definition 1. *The* max-times *(or* subtropical*) algebra is a set* $\mathbb{R}_+$ *of nonnegative real numbers together with operations* $a \boxplus b = \max\{a, b\}$ *(addition) and* $a \boxdot b = ab$ *(multiplication) defined for any* $a, b \in \mathbb{R}_+$. *The identity element for addition is* 0 *and for multiplication it is* 1.

In the future we will use the notation $a \boxplus b$ and $\max\{a, b\}$ and the names *max-times* and *subtropical* interchangeably. It is straightforward to see that the max-times algebra is a *dioid*, that is, a semiring with idempotent addition ($a \boxplus a = a$). It is important to note that subtropical algebra is anti-negative, that is, there is no subtraction operation.

The subtropical matrix algebra follows naturally:

Definition 2. *The* max-times matrix product *of two matrices* $B \in \mathbb{R}_+^{n \times k}$ *and* $C \in \mathbb{R}_+^{k \times m}$ *is defined as*

$$(B \boxtimes C)_{ij} = \max_{s=1}^{k} B_{is} C_{sj}. \tag{1}$$

The definition of a *rank-1* matrix over the max-times algebra is the same as over the standard algebra, i.e. a matrix that can be expressed as an outer product of two vectors. We will use the term *block* to mean a rank-1 matrix. The general *rank* of a matrix over the max-times algebra is defined analogously to the standard Schein rank:

Definition 3. *The* max-times rank *of a matrix* $A \in \mathbb{R}_+^{n \times m}$ *is the least integer* k *such that* A *can be expressed as a max of* k *rank-1 matrices,* $A = F_1 \boxplus F_2 \boxplus \cdots \boxplus F_k$, *where all* F_i *are rank-1.*

Now that we have sufficient notation, we can formally introduce the main problem considered in the paper.

Problem 1. Given a matrix $A \in \mathbb{R}_+^{n \times m}$ and an integer $k > 0$, find factor matrices $B \in \mathbb{R}_+^{n \times k}$ and $C \in \mathbb{R}_+^{k \times m}$ minimizing

$$E(A, B, C) = \|A - B \boxtimes C\|_F^2 = \sum_{i,j}(A_{ij} - (B \boxtimes C)_{ij})^2. \qquad (2)$$

4 Algorithm

As we work over the max-times algebra, the common approaches for finding matrix factorizations under normal algebra do not work as such. The main problem is the non-linear behavior of the maximum function, and our algorithm tries to alleviate the problems caused by it. The two main ideas we employ are updating the rank-1 factors one-by-one in an iterative fashion, and approximating the max-times reconstruction error with a low-degree polynomial. The first idea is similar to what we used in [9], except that here we only update parts of the rank-1 factors. The motivation behind this is to avoid building few factors that try to explain the whole data (badly), but instead build many factors that explain small parts of the data well.

4.1 The Main Algorithm

Our proposed algorithm, `Cancer`, is outlined in Algorithm 1. It accepts as input the data to be decomposed A, the required rank k, and three additional scalar parameters M, t, and f. Integer M is the number of cycles that the algorithm will make, that is each one of k blocks will be visited M times. A reasonable value for M would be 15 since further rounds provide only marginal improvement, although to make sure that the algorithm has converged, a value as high as 40 might be required. The next parameter, $t \in \mathbb{N}$, represents the maximum allowed degree of polynomials: after each cycle the degree of polynomials used for approximation is incremented, but when it reaches t, it is reset to the value of 2. Typically, we can set $t = 16$. Finally, $f \in (0, 1)$ controls how much of each block (rank-1 matrix) is revealed on each iteration. Namely, each block $bc \in \mathbb{R}_+^{n \times m}$ consists a total of $n + m$ variables, and the maximum number of variables we can change when a block is visited is $\lfloor f(n + m)/2 \rfloor$. The algorithm outputs two factor matrices $B \in \mathbb{R}_+^{n \times k}$ and $C \in \mathbb{R}_+^{k \times m}$ whose product is a rank-k max-times approximation of A.

`Cancer` starts with empty blocks (line 2) and updates them iteratively (lines 6–14) using the `UpdateBlock` routine (line 9). `UpdateBlock` updates one block while keeping all others fixed. We then compare the current decomposition to the best one seen so far, and if it provides an improvement, then the best solution is replaced with the current one (lines 10–12). The final step of the loop is to increment the degree of polynomials used for approximation (lines 13–14). Intuitively, lower degrees polynomials give more latitude for varying the variables,

---

Algorithm 1. Cancer

Input: $A \in \mathbb{R}_+^{n \times m}$, $k > 0$, $M > 0$, $t > 2$, $0 < f < 1$
Output: $B^* \in \mathbb{R}_+^{n \times k}$, $C^* \in \mathbb{R}_+^{k \times m}$
1: **function** Cancer (A, k, M, t, f)
2:     $B \leftarrow 0^{n \times k}$, $C \leftarrow 0^{k \times m}$
3:     $B^* \leftarrow B, C^* \leftarrow C$
4:     $bestError \leftarrow E(A, B, C)$
5:     $deg \leftarrow 2$
6:     **for** $count \leftarrow 0$ **to** $k \times M - 1$ **do**
7:         $i \leftarrow count \pmod{k} + 1$ ▷ Index of the current block.
8:         $N \leftarrow B^{-i} \boxtimes C_{-i}$ ▷ Reconstructed matrix without the i-th block.
9:         $[B^i, C_i] \leftarrow$ UpdateBlock(A, N, B^i, C_i, deg, f)
10:        **if** $E(A, B, C) < bestError$ **then**
11:            $B^* \leftarrow B, C^* \leftarrow C$
12:            $bestError \leftarrow E(A, B, C)$
13:        **if** $count > k$ **and** $count \pmod{k} = 0$ **then**
14:
$$deg \leftarrow \begin{cases} deg + 1 & \text{if } deg < t \\ 2 & \text{otherwise} \end{cases}$$

15:    **return** B^*, C^*

---

whereas polynomials of higher degrees are better suited for finalizing results since they provide better approximations. This is similar to an execution of a simulated annealing algorithm, where high temperatures are used to make big steps and get out of local minima, and lower temperatures are better suited for converging to a particular minimum. In our case low degrees correspond to high temperatures and vice versa.

Most of the time the reconstruction error decreases gradually with increased iterations of Cancer. There are however rare cases where it would remain almost constant for some time or even increase slightly, and then start dropping again. For this reason the algorithm is run until all cycles are complete and is not stopped using any sort of convergence criteria.

4.2 Updating a Block

The UpdateBlock procedure (Algorithm 2) performs the work of updating a single block $bc \in \mathbb{R}_+^{n \times m}$ on one iteration of Cancer. It takes a block bc, where $b \in \mathbb{R}_+^{n \times 1}$ and $c \in \mathbb{R}_+^{1 \times m}$, and performs alternating updates of b and c one element at a time using the AdjustOneElement function. That function is called $\lfloor f(n + m)/2 \rfloor$ times to update only a part of the block, as explained above.

The AdjustOneElement function (Algorithm 3) updates a single entry in either a column vector b or a row vector c. Let us consider the case when b is fixed and c varies. In order to decide which element of c to change, we need to compare the best changes to all m entries and then choose the one that yields the most improvement to the objective. A single element c_l only has an effect on

---

Algorithm 2. UpdateBlock

Input: $A \in \mathbb{R}_+^{n \times m}$, $N \in \mathbb{R}_+^{n \times m}$, $b \in \mathbb{R}_+^{n \times 1}$, $c \in \mathbb{R}_+^{1 \times m}$, $deg \geq 2$, $0 < f < 1$
Output: $b \in \mathbb{R}_+^{n \times 1}$, $c \in \mathbb{R}_+^{1 \times m}$
1: **function** UpdateBlock (A, N, b, c, deg, f)
2:     $niters \leftarrow \lfloor f(n+m)/2 \rfloor$
3:     **for** $count \leftarrow 1$ **to** $niters$ **do**
4:         $c = $ AdjustOneElement(A, N, b, c, deg)
5:         $b = $ AdjustOneElement$(A^T, N^T, c^T, b^T, deg)^T$
6:     **return** b, c

Algorithm 3. AdjustOneElement

Input: $A \in \mathbb{R}_+^{n \times m}$, $N \in \mathbb{R}_+^{n \times m}$, $b \in \mathbb{R}_+^{n \times 1}$, $c \in \mathbb{R}_+^{1 \times m}$, $deg \geq 2$
Output: $c \in \mathbb{R}_+^{1 \times m}$
1: **function** AdjustOneElement (A, N, b, c, deg)
2:     **for** $j \leftarrow 1$ **to** m **do**
3:         $baseError \leftarrow \sum_{i=1}^{n} (A_{ij} - \max\{N_{ij}, b_i c_j\})^2$
4:         $[err, x_i] \leftarrow$ PolyMin(A^j, N^j, b, deg)
5:         $u_i \leftarrow baseError - err$
6:     $i \leftarrow$ the index i of largest value of u
7:     $c_i \leftarrow x_i$
8:     **return** c

---

the error along the column l. Assume that we are currently updating block with index q and let N denote the reconstruction matrix without this block, that is $N = B^{-q} \boxtimes C_{-q}$. Minimizing $E(A, B, C)$ with respect to c_l is then equivalent to minimizing

$$\gamma(A_l, N_l, b, c_l) = \sum_{i=1}^{n} (A_{il} - \max\{N_{il}, b_i c_l\})^2. \qquad (3)$$

Instead of minimizing (3) directly, we use polynomial approximation in the PolyMin routine (line 4). The routine returns the (approximate) error err and the value x achieving that. Since we are only interested in the improvement of the objective achieved by updating a single entry of c, we compute the improvement of the objective after the change (line 5). After trying every column of c, we update only the column that yield the largest improvement.

4.3   The PolyMin Procedure

The function γ that we need to minimize in order to find the best change to the vector c in AdjustOneElement is hard to work with directly since it is not convex, and also not smooth because of the presence of the maximum operator. To alleviate this, we approximate the error function γ with a polynomial g of degree deg. Notice that when updating c_l, other variables of γ are fixed and we only need to consider function $\gamma'(x) = \gamma(A_l, N_l, b, x)$. To build g we sample

$deg + 1$ points from $(0, 1)$ and fit g to the values of γ' at these points. We then find the $x \in \mathbb{R}_+$ that minimizes $g(x)$ and return $g(x)$ (the approximate error) and x (the optimal value).

4.4 Computational Complexity

We will express the complexity of the algorithm asymptotically in terms of the dimensions of the input data n and m and the rank k of the factorization. Most of the work in `Cancer` is performed in the `UpdateBlock` routine, which is called Mk times. `UpdateBlock` is in turn just a loop that calls `AdjustOneElement` $\lfloor f(n + m) \rfloor$ times. In `AdjustOneElement` the contributors to the complexity are computing the base error (line 3) and a call to `PolyMin` (line 4). Both of them are performed n or m times depending on whether we supplied the column vector b or the row vector c to `AdjustOneElement`. Finding the base error takes time $O(m)$ for b and $O(n)$ for c. The complexity of `PolyMin` boils down to that of evaluating the max-times objective at $deg + 1$ points and then minimizing a degree deg polynomial. Hence, `PolyMin` runs in time $O(m)$ or $O(n)$ depending on whether we are optimizing b or c, and the complexity of `AdjustOneElement` is $O(nm)$. Since the parameters f and M are fixed, this gives the complexity $O((n + m)nm)$ for `UpdateBlock` and $O((n + m)nmk) = O(\max\{n, m\}nmk)$ for `Cancer`.

5 Experiments

In this section we evaluate the performance of `Cancer` against other algorithms on various synthetic and real-world datasets. The purpose of the synthetic experiments is to verify that `Cancer` is capable of finding subtropical structure from data where we know it is present, and to evaluate its performance under different data characteristics in a controlled manner. Our tests demonstrate that `Cancer` not only provides better approximations than other methods, but also produces much sparser factors. The main purpose of experiments with real-world datasets is to see if they possess the max-times structure and whether `Cancer` is capable of extracting it.

Setting the parameters for `Cancer`*.* For all synthetic experiments we used $M = 14$, $t = 16$, and $f = 0.1$. For the real world experiments we set $t = 16$, $f = 0.1$, and $M = 40$ (except for `Eigenfaces` for which we used $M = 50$).

5.1 Other Methods

We compared `Cancer` against `Capricorn`, which is our previous tropical matrix factorization algorithm designed for discrete data [9],[1] `SVD`, and four different

---

[1] The source code for `Cancer` and `Capricorn` and the scripts to repeat the experiments are available at http://people.mpi-inf.mpg.de/~pmiettin/tropical/.

versions of NMF. The first form of NMF is a sparse NMF algorithm by Hoyer [8],[2] which we call SNMF. Hoyer's algorithm [8] defines the sparsity of a vector $x \in \mathbb{R}^n_+$ as

$$\text{sparsity}(x) = \frac{\sqrt{n} - \left(\sum_i |x_i|\right) / \sqrt{\sum_i x_i^2}}{\sqrt{n} - 1}, \tag{4}$$

and returns factorization where the sparsity of the factor matrices is user-controllable. In all of our experiments, we used the sparsity of Cancer's factors as the sparsity parameter of SNMF.

The second form of NMF is a standard alternating least squares algorithm called ALS [5]. The remaining two versions of NMF are essentially the same as ALS, but they use L_1 regularization for increased sparsity [5], that is, they aim at minimizing

$$\|A - BC\|_F + \alpha \|B\|_1 + \beta \|C\|_1.$$

The first method is called ALSR and uses regularizer coefficient $\alpha = \beta = 1$, and the other, called ALSR 5, has regularizer coefficient $\alpha = \beta = 5$. All NMF algorithms were restarted 10 times, and the best result was selected.

5.2 Synthetic Experiments

The general setup of synthetic experiments is as follows. First we create data that is guaranteed to have the subtropical structure by generating random factors of some density with nonzero elements drawn from a uniform distribution on the $[0, 1]$ interval and then multiplying them using the max-times matrix product. Then we add noise and feed the obtained noisy matrices into algorithms to see how well they can approximate the original data. We distinguish two types of noise. One is the normal, or Gaussian, noise with 0 mean, for which we define the level of noise to be its standard deviation. Since adding this noise to the data might result in negative entries, we truncate all values in a resulting matrix that are below zero. We use two noise levels, 0.01 and 0.08, called low and high noise levels, respectively.

The other type of noise is a discrete (tropical) noise, which is introduced in the following way. Assume that we are given an input matrix A of size n-by-m. We first generate an n-by-m noise matrix N with elements drawn from a uniform distribution on the $[0, 1]$ interval. Given a level of noise l, we then turn $\lfloor (1-l)nm \rfloor$ random elements of N to 0, so that its resulting density is l. Finally, the noise is applied by taking elementwise maximum between the original data and the noise matrix $F = \max\{A, N\}$.

All synthetic experiments were performed on 1000-by-800 matrices. In all tests, except those with varying rank, the true max-times rank of the data was 10. For all experiments we report errors, which are measured as relative Frobenius errors between original and reconstructed matrices, that is, $E(A, B, C)/\|A\|_F^2$.

---

[2] https://github.com/aludnam/MATLAB/tree/master/nmfpack.

We also report the sparsity s of factor matrices obtained by algorithms, which is defined as a fraction of zero elements in the factor matrices,

$$s(\boldsymbol{A}) = |\{(i,j) : \boldsymbol{A}_{ij} = 0\}|/(nm), \tag{5}$$

for an n-by-m matrix $\boldsymbol{A}$. The results were averaged over 10 repetitions. The reconstruction errors are reported in Fig. 1 and the sparsities in Fig. 2.

Varying Gaussian noise. Here we investigate how the algorithms respond to different levels of Gaussian noise, which was varied from 0 to 0.14 with increments of 0.01. A level of noise is a standard deviation of the noise matrix as described earlier. The factor density was kept at 50 %. The results are given on Fig. 1(a) (reconstruction error) and Fig. 2(a) (sparsity of factors).

Here, Cancer is generally the best method in reconstruction error, and second in sparsity only to Capricorn. The sole exception to reconstruction error is the no-noise case, where Capricorn – as designed – obtains essentially a perfect decomposition, though its results deteriorate rapidly with increased noise levels.

Varying density. In this experiment we studied what effects the density of factor matrices used in data generation has on the algorithms' performance. For this purpose we varied the density from 10 % to 100 % with increments of 10 % while keeping the other parameters fixed. There are two versions of this experiment, one with low noise level of 0.01 (Figs. 1(b) and 2(b)), and a more noisy case at 0.08 (Figs. 1(c) and 2(c)).

Cancer provides the least reconstruction error in this experiment, being clearly the best until the density is 0.7, from which point on it is tied with SVD and the NMF-based methods (the only exception being the least-dense high-noise case, where ALSR obtains slightly better reconstruction error). Capricorn is the worst by a wide margin, but this is not surprising, as the data does not follow its assumptions. On the other hand, Capricorn does produce generally the sparsest factorization, but these are of little use given its bad reconstruction error. Cancer produces the sparsest factors from the remaining methods, except in the first few cases where ALSR 5 is sparser (and worse in reconstruction error), meaning that Cancer produces factors that are both the most accurate and very sparse.

Varying rank. The purpose of this test is to study the performance of algorithms on data of different max-times ranks. We varied the true rank of the data from 2 to 20 with increments of 2. The factor density was fixed at 50 % and Gaussian noise at 0.01. The results are shown on Fig. 1(d) (reconstruction error) and Fig. 2(d) (sparsity of factors). The results are similar to the two above ones, with Cancer returning the most accurate and second-most sparsest factorizations.

Varying tropical noise. In this setup we used the tropical noise instead of the Gaussian one. The level of noise represents the density of the noise matrix with which the original data is 'maxed'. We varied the noise from 0 % to 14 % with increments of 1%. There are two forms of this experiment, one with density 50% (Fig. 1(e) shows the reconstruction error and Fig. 2(e) shows the sparsity of factors) and with density 90% (Figs. 1(f) and 2(f), respectively).

As Capricorn was designed for tropical noise, unlike Cancer that was designed for standard 'white' noise, it obtains the least reconstruction error of all the methods (albeit with high deviation when the noise density is higher). Cancer is generally the second-best method, although with the high-density noise, it is mostly tied with SVD, ALS and ALSR. In the sparsity of the factors, Cancer and Capricorn are quite similar, with Capricorn having slightly sparser factors in the low-density noise case, but Cancer having an edge in the high-density noise case. In the latter case, ALSR 5 is also comparable on sparsity, but clearly the worst in reconstruction error.

Discussion. The synthetic experiments verify that Cancer can find the max-times structure from the data when it is present and potentially perturbed with Gaussian noise. It also shows strong invariance over the level of noise, rank, or density of the factors. The experiments also highlight the design differences between Cancer and Capricorn: the former is superior in Gaussian noise situation, while the latter excels with tropical noise. If the type of noise cannot be predetermined, it seems it is best to try both methods.

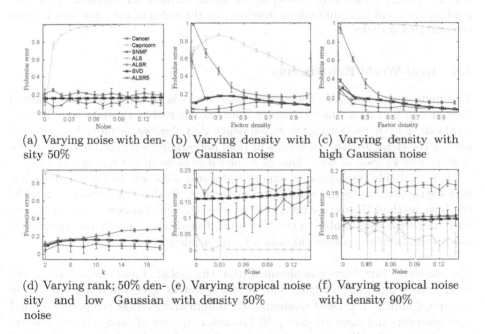

(a) Varying noise with density 50%

(b) Varying density with low Gaussian noise

(c) Varying density with high Gaussian noise

(d) Varying rank; 50% density and low Gaussian noise

(e) Varying tropical noise with density 50%

(f) Varying tropical noise with density 90%

Fig. 1. Reconstruction error (Frobenius norm) for synthetic data. The markers are averages of 10 random matrices and the width of the error bars is twice the standard deviation.

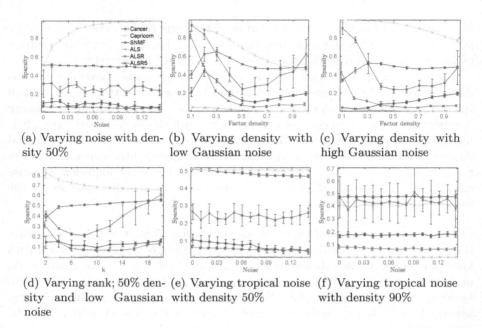

(a) Varying noise with density 50%

(b) Varying density with low Gaussian noise

(c) Varying density with high Gaussian noise

(d) Varying rank; 50% density and low Gaussian noise

(e) Varying tropical noise with density 50%

(f) Varying tropical noise with density 90%

Fig. 2. Sparsity (fraction of zeroes) of the factor matrices for synthetic data. The markers are averages of 10 random matrices and the width of the error bars is twice the standard deviation.

5.3   Real-World Experiments

The main purpose of the real-world experiments is to study to which extend Cancer can find max-times structure from various real-world data sets. Having established with the synthetic experiments that Cancer is indeed capable of finding the structure when it is present (and potentially perturbed with Gaussian noise), here we look at what kind of results it obtains in the real-world data.

It is probably unrealistic to expect real-world data sets to have 'pure' max-times structure, as in the synthetic experiments. Rather, we expect SVD to be the best method (in reconstruction error's sense), and Cancer to obtain reconstruction error comparable to the NMF-based methods. We will also verify that the results from the real-world data sets are intuitive.

The datasets. Worldclim was obtained from the global climate data repository[3] It describes historical climate data across different geographical locations in Europe. Columns represent minimum, maximum and average temperatures and precipitation, and rows are 50-by-50 kilometer squares of land where measurements were made. We preprocessed every column of the data by first subtracting its mean, dividing by the standard deviation, and then subtracting its minimum value, so that the smallest value becomes 0.

---

[3] The raw data is available at http://www.worldclim.org/.

NPAS is a nerdy personality test that uses different attributes to determine the level of nerdiness of a person.[4] It contains answers by 1418 respondents to a set of 36 questions that asked them to self assess various statements about themselves on a scale of 1 to 7. We preprocessed NPAS analogously to Worldclim.

Eigenfaces is a subset of the Extended Yale Face collection of face images [7]. It consists of 32-by-32 images under different lighting conditions. We used a preprocessed data by Xiaofei He et al.[5] We selected a subset of pictures with lighting from the left and then preprocessed the input matrix by first subtracting from every column its mean and then dividing it by its standard deviation.

4News is a subset of the 20Newsgroups dataset,[6] containing the usage of 800 words over 400 posts for 4 newsgroups.[7] Before running the algorithms we represented the dataset as a TF-IDF matrix, and then scaled it by dividing each entry by the greatest entry in the matrix.

HPI is a land registry house price index.[8] Rows represent months, columns are locations, and entries are residential property price indices. We preprocessed the data by first dividing each column by its standard deviation and then subtracting its minimum, so that each column has minimum 0.

The basic properties of these data sets are listed in Table 1.

Table 1. Real world datasets specs.

Algorithm	Rows	Columns	Density
Worldclim	2575	48	99.9 %
NPAS	1418	36	99.6 %
Eigenfaces	1024	222	97.0 %
4News	400	800	3.5 %
HPI	253	177	99.5 %

Reconstruction error, sparsity, and convergence. Table 2 provides the relative Frobenius reconstruction errors for the real-world data sets. We omitted ALSR 5 from these experiments due to its bad performance with the synthetic data. SVD is, as expected, consistently the best method. Somewhat surprisingly, Hoyer's SNMF is usually the second-best method, even though in the synthetic experiments it usually was the second-worst of the NMF-based methods. Cancer is usually the third-best method (with the exception of 4News and NPAS), and often very close to SNMF in reconstruction error. Overall, it seems Cancer is

[4] Tha dataset can be obtained on the online personality website http://personality-testing.info/_rawdata/NPAS-data.zip.

[5] http://www.cad.zju.edu.cn/home/dengcai/Data/FaceData.html.

[6] http://qwone.com/~jason/20Newsgroups/.

[7] The authors are grateful to Ata Kabán for pre-processing the data, see [10].

[8] Available at https://data.gov.uk/dataset/land-registry-house-price-index-backg round-tables/.

capable of finding max-times structure that is comparable to what NMF-based methods provide. Consequently, we can study the max-times structure found by Cancer, knowing that it is (relatively) accurate.

Table 2. Reconstruction error for various real-world datasets.

	Worldclim	NPAS	Eigenfaces	4News	HPI
$k =$	10	10	40	20	15
Cancer	0.071	0.240	0.204	0.556	0.027
Capricorn	0.392	0.395	0.972	0.987	0.217
SNMF	0.046	0.225	0.178	0.546	0.023
ALS	0.087	0.227	0.313	0.538	0.074
ALSR	0.122	0.226	0.294	1.000	0.045
SVD	0.025	0.209	0.140	0.533	0.015

The sparsity of the factors for real-world data is presented in Table 3, except for SVD. Here, Cancer often returns the second-sparsest factors (being second only to Capricorn), but with 4News and HPI, ALSR obtains sparser decompositions.

Table 3. Factor sparsity for various real-world datasets.

	Worldclim	NPAS	Eigenfaces	4News	HPI
$k =$	10	10	40	20	15
Cancer	0.645	0.528	0.571	0.812	0.422
Capricorn	0.795	0.733	0.949	0.991	0.685
SNMF	0.383	0.330	0.403	0.499	0.226
ALS	0.226	0.120	0.434	0.513	0.331
ALSR	0.275	0.117	0.480	1.000	0.729

We also studied the convergence behavior of our algorithm using some of the real-world data sets. The results can be seen in Fig. 3, where we plot the relative error with respect to the iterations over the main for-loop in Cancer. As we can see, in both cases Cancer has obtained a good reconstruction error already after few full cycles, with the remaining runs only providing minor improvements. We can deduce that Cancer reaches quickly an acceptable solution.

Interpretability of the results. The crux of using max-times factorizations instead of standard (nonnegative) ones is that the factors (are supposed to) exhibit the 'winner-takes-it-all' structure instead of the 'parts-of-whole' structure. To demonstrate this, we plotted the left factor matrices for the Eigenfaces

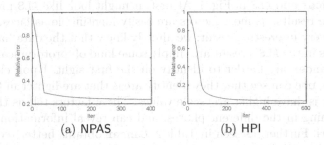

(a) NPAS   (b) HPI

Fig. 3. Convergence rate of Cancer for two real-world datasets. Each iteration is a single run of UpdateBlock, that is if a factorization has rank k, then one full cycle would correspond to k iterations.

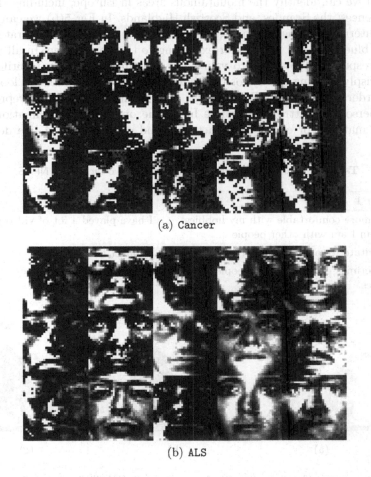

(a) Cancer

(b) ALS

Fig. 4. Cancer finds the dominant patterns from the Eigenfaces data. Pictured are the left factor matrices for the Eigenfaces data.

data for Cancer and ALS in Fig. 4. At first, it might look like ALS provides more interpretable results, as most factors are easily identifiable as faces. This, however, is not very interesting result: we already knew that the data has faces, and many factors in the ALS's result are simply some kind of 'prototypical' faces. The results of Cancer are harder to identify on the first sight. Upon closer inspection, though, one can see that they identify areas that are lighter in the different images, that is, have higher grayscale values. These factors tell us the variances in the lightning in the different photos, and can reveal information we did not know a priori. Further, as seen in Table 2, Cancer obtains better reconstruction error than ALS with this data, confirming that these factors are indeed useful to recreate the data.

In Fig. 5, we show some factors from Cancer when applied to the Worldclim data. These factors clearly identify different bioclimatic areas from Europe: In Fig. 5(a) we can identify the mountainous areas in Europe, including the Alps, the Pyrenees, the Scandes, and Scottish Highlands. In Fig. 5(b) we can identify the mediterranean coastal regions, while in Fig. 5(c) we see the temperate climate zone in blue, with the green color extending to the boreal zone. In all pictures, red corresponds to (near) zero values. As we can see, Cancer identifies these areas crisply, making it easy for the analyst to know which areas to look at.

In order to interpret NPAS we first observe that each column represents a single personality attribute. Denote by A the obtained approximation of the original matrix. For each rank-1 factor X and each column A_i we define the

Table 4. Top three attributes for the first two factors of NPAS.

Factor 1	Factor 2
I am more comfortable with my hobbies than I am with other people	I have played a lot of video games
I gravitate towards introspection	I collect books
I sometimes prefer fictional people to real ones	I care about super heroes

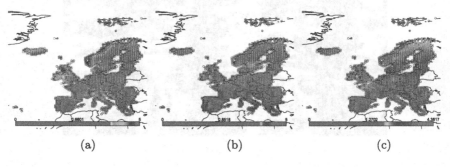

(a)   (b)   (c)

Fig. 5. Cancer can find interpretable factors from the Worldclim data. Shown are the values for three columns in the left-hand factor matrix B on a map. Red is zero.

score $\sigma(i)$ as the number of elements in A_i that are determined by X. By sorting attributes in descending order of $\sigma(i)$ we obtain relative rankings of the attributes for a given factor. The results are shown in Table 4. The first factor clearly shows introvert tendencies, while the second one can be summarized as having interests in fiction and games.

6 Conclusions

Using max-times algebra instead of the standard (nonnegative) algebra, we can find factors that adhere to the 'winner-takes-it-all' interpretation instead of the 'parts-of-whole' interpretation of NMF. The winner-takes-it-all factors give us the most dominant features, building a sharper contrast between what is and is not important for that factor, making the factors potentially easier to interpret. As we saw in our experimental evaluation, the factors are also sparse, emphasizing the winner-takes-it-all interpretation.

Finding a good max-times factorization, unfortunately, seems harder than – or at least as hard as – finding a good nonnegative factorization. Our earlier algorithm, Capricorn, was designed to work with discrete-valued data and what we call 'tropical' noise; in this paper we presented Cancer that is designed to work with Gaussian noise and matrices with continuous values. It seems that this latter case is more applicable to real-world data, as witnessed by Cancer's good results with real-world data.

There are still questions that need to be addressed by future research. Could these two approaches be merged? That is, is it possible to design an algorithm that works well for both tropical and Gaussian noise? Can one achieve provable approximation ratios for max-times factorizations? In addition to data analysis, can max-times factorizations be used in other data mining and machine learning tasks (e.g. to do matrix completion or latent topic models)? We hope our initial work in this paper (and its predecessor [9]) helps to increase data mining and machine learning community's interest to max-times algebras so that the above question could be answered.

References

1. Akian, M., Bapat, R., Gaubert, S.: Max-plus algebra. In: Hogben, L. (ed.) Handbook of Linear Algebra. Chapman & Hall/CRC, London (2007)
2. Baccelli, F., Cohen, G., Olsder, G.J., Quadrat, J.P.: Synchronization and Linearity: An Algebra for Discrete Event Systems. John Wiley & Sons, New York (1992)
3. Bělohlávek, R., Krmelova, M.: Factor analysis of ordinal data via decomposition of matrices with grades. Ann. Math. Artif. Intell. **72**(1–2), 23–44 (2014)
4. Cassandras, C.G., Lafortune, S.: Introduction to Discrete Event Systems, 2nd edn. Springer, Secaucus (2008)
5. Cichocki, A., Zdunek, R., Phan, A.H., Amari, S.I.: Nonnegative Matrix and Tensor Factorizations: Applications to Exploratory Multi-way Data Analysis and Blind Source Separation. John Wiley & Sons, Chichester (2009)

6. Cohen, G., Gaubert, S., Quadrat, J.P.: Max-plus algebra and system theory: where we are and where to go now. Ann. Rev. Control **23**, 207–219 (1999)
7. Georghiades, A.S., Belhumeur, P.N., Kriegman, D.J.: From few to many: generative models for recognition under variable pose and illumination. In: IEEE AFGR, pp. 277–284 (2000)
8. Hoyer, P.O.: Non-negative matrix factorization with sparseness constraints. J. Mach. Learn. Res. **5**, 1457–1469 (2004)
9. Karaev, S., Miettinen, P.: Capricorn: an algorithm for subtropical matrix factorization. In: SDM, pp. 702–710 (2016)
10. Miettinen, P.: Matrix decomposition methods for data mining: computational complexity and algorithms. Ph.d. thesis, University of Helsinki (2009)
11. Miettinen, P., Mielikäinen, T., Gionis, A., Das, G., Mannila, H.: The discrete basis problem. IEEE Trans. Knowl. Data Eng. **20**(10), 1348–1362 (2008)
12. Weston, J., Weiss, R.J., Yee, H.: Nonlinear latent factorization by embedding multiple user interests. In: RecSys, pp. 65–68 (2013)

Robust Principal Component Analysis by Reverse Iterative Linear Programming

Andrea Visentin[1]($\boxtimes$), Steven Prestwich[1], and S. Armagan Tarim[2]

[1] Insight Centre for Data Analytics, Department of Computer Science,
University College Cork, Cork, Ireland
andrea.visentin@insight-centre.org
[2] Department of Management, Cankaya University, Ankara, Turkey

Abstract. Principal Components Analysis (PCA) is a data analysis technique widely used in dimensionality reduction. It extracts a small number of orthonormal vectors that explain most of the variation in a dataset, which are called the Principal Components. Conventional PCA is sensitive to outliers because it is based on the L_2-norm, so to improve robustness several algorithms based on the L_1-norm have been introduced in the literature. We present a new algorithm for robust L_1-norm PCA that computes components iteratively in reverse, using a new heuristic based on Linear Programming. This solution is focused on finding the projection that minimizes the variance of the projected points. It has only one parameter to tune, making it simple to use. On common benchmarks it performs competitively compared to other methods. The data and software related to this paper are available at https://github.com/visentin-insight/L1-PCAhp.

Keywords: Principal components analysis · Linear programming · L1-norm · Robust

1 Introduction

Principal Components Analysis (PCA) is a data analysis technique to find orthonormal vectors that explain the variance structure of the data [12]. These are the vectors in which lie most of the variance and they are called *principal components* (PCs). Since the number of PCs is generally low compared to the total dimension of the dataset, PCA is often use for dimensional reduction or data analysis. Its applications include quality control [2], image reconstruction [20], wave reconstruction [25], and outlier detection [11].

PCs can be computed *forward* or *backward*. Classical PCA uses the forward approach: in each iteration it aims to find the direction that yields the maximum information. The backward approach aims to find the direction with least information, so that when it is eliminated the projection of the data into the remaining subspace retains the maximum amount of variation. If only the few PCs with most variation in them are needed then the forward approach is more

© Springer International Publishing AG 2016
P. Frasconi et al. (Eds.): ECML PKDD 2016, Part II, LNAI 9852, pp. 593–605, 2016.
DOI: 10.1007/978-3-319-46227-1_37

suitable, while if the aim is to eliminate only the few least useful components then the backward approach is appropriate [1]. The two approaches are equivalent under the L_2-norm (Euclidean space) but under other norms they can give different results.

The quality of PCA algorithms can be evaluated in two ways: by reconstruction error or variance of the projected points. Under the L_2-norm these two measures are equivalent. However, L_2-norm based methods are rather sensitive to outliers so they are not well-suited to noisy datasets. To improve robustness different methods have been applied to PCA. The algorithms in [5–7] are based on the *projection pursuit* method for which the L_1-norm is widely used. [16] describes a greedy method that aims to maximize the L_1-norm distance of the projected points. In [15] a heuristic estimate for the general L_1 reconstruction error is presented, which assumes that the projected and lifted data are the product of two matrices. [3] proposes a backward L_1-norm method. [22] introduces two algorithms to compute PCs by minimizing different objectives: an iterative weighted approach that minimizes the reconstruction error, and one that uses linear programming (LP) to maximize the L_1-norm of the projected points. The LP-based method aims to compute directly the orthonormal matrix of the PCs. It uses LP to perturb a projection matrix, and if this is an improvement it forces the new one to be orthonormal. Technically this can not be done in LP because maximizing a sum of absolute values is NP-hard (see [13,14,23,24,26] for a discussion of this issue). It can be done by adding binary decision variables to obtain a mixed integer programming (MIP) model but this is far less scalable than a pure LP approach. To resolve this problem [22] relax the MIP model to an LP and restrict moves to small distances. They show that their search algorithm is locally convergent.

Our approach is also based on an LP model, but instead of computing the PCs simultaneously it iteratively computes them in backward order. This enables us to use a direct LP model, so we do not need to restrict it to small steps. The rest of the paper is organized as follows. In Sect. 2 the problem is formulated. In Sect. 3 we introduce a new algorithm for the L_1-PCA. In Sect. 4 the algorithm is compared with others for robustness and the competitiveness. Section 5 concludes the paper.

2   Problem Formulation

Let $A \in \Re^{n \times m}$ be the data set, where m denotes the number of attributes (dimension of the original input space) and n the number of instances. The goal of PCA is to find a projection matrix $X \in \Re^{m \times k}$ whose rows are the bases of a k-dimensional linear subspace (so $X^T X = I_k$). This subspace will be referred as the *feature space*. This matrix must minimize the reconstruction error:

$$E_p(X, Y) = \left\| A - Y X^T \right\|_p^p \tag{1}$$

where $\|\cdot\|_p^p$ is the L_p-norm, and $Y \in \Re^{n \times k}$ is the coefficient matrix whose rows correspond to the coordinate of each instance of A projected into the feature space spanned by X. If X is fixed then the Y minimizing the error function is uniquely determined by $Y = AX$, by the projection theorem [19]. So the error function becomes:

$$E_p(X) = \|A - AXX^T\|_p^p \tag{2}$$

Another possible approach is to maximize the variance of the projected point in the feature space:

$$\arg\max_X \|AX\|_p^p \tag{3}$$

Classical PCA is based on L_2-norm ($p = 2$) in which case minimizing (1) and maximizing (3) are equivalent because of singular value decomposition.

Consider the case $p = 1$ so that we use the L_1-norm. We minimize the error as in Eq. (2) by solving the problem

$$\min_{X \in \Re^{m \times k}} \sum_{i \in I} \sum_{j \in J} |e_{ij}| \tag{4a}$$

$$\text{s.t.} \quad X^T X = I_k, \; E = A - AXX^T \tag{4b}$$

where $I = \{1, \ldots, n\}$ and $J = \{1, \ldots, m\}$. Even if (4) is more robust compared to the L_2-norm version it is not invariant under rotation of the data, and the shape of equidistance surfaces becomes very skewed [9].

Similarly, setting $p = 1$ in (3):

$$\max_{X \in \Re^{m \times k}} \sum_{i \in I} \sum_{h \in K} |y_{ih}| \tag{5a}$$

$$\text{s.t.} \quad X^T X = I_k, \; Y = AX \tag{5b}$$

where $K = \{1, \ldots, k\}$. The solution of this problem is also invariant under data rotation because the maximization is done in the feature space [9].

3   Rotation-Invariant L_1-PCA by LP

In this section we present an iterative LP-based algorithm that aims to solve problem (5). In a forward approach we would iteratively look for the projection that maximizes variance in the feature space. Denoting by t the projection we are looking for, the problem would be:

$$\max \sum_{i \in I} |y_i| \tag{6a}$$

$$\text{s.t.} \quad y_i = \sum_{j \in [1,m]} a_{ij} t_j \qquad i \in I \tag{6b}$$

$$\sum_{j \in J} t_j^2 = 1 \tag{6c}$$

Having found a solution $\boldsymbol{proj}_0$ we would then look for the next most interesting projection $\boldsymbol{proj}_1$ under the added constraint $\boldsymbol{proj}_1 \cdot \boldsymbol{proj}_0 = 0$; and so on, adding a new orthogonality constraint for each new projection.

Unfortunately (6) can not be solved by LP. As pointed out above, maximizing a sum of absolute values is NP-hard. Even under the L_2-norm we can not solve it: quadratic objectives are allowed in quadratically constrained quadratic programs, but only if the resulting matrix is positive semidefinite, and in this case it is not because of constraint (6c). Instead we use a backward approach which does not require the solution of any NP-hard problems.

Like many other methods we start by centering and normalizing the dataset by subtracting the centroid and dividing by standard deviations. Then at each dimension iteration we choose a starting vector p and iteratively improve it via an LP-based heuristic. To find such an improvement we find a vector transformation that minimizes the LP without the (6c) constraint, but forcing the vector to lie in the hyperplane tangent to the unit hypersphere at the point defined by t. This idea is illustrated in Fig. 1. Solving the LP yields a vector that is likely to be quite close to the direction we want, but the result is not a unit vector because it lies in the hyperplane and not on the unit hypersphere. We therefore project the solution onto the hypersphere to obtain a unit vector, which is the new starting vector t. We iterate this procedure until the improvement in the objective function of the LP is smaller than a fixed threshold, which is the only parameter needed by this solution. In practice we found that the procedure converges in a small number of iterations, and that a value such as 0.01 is reasonable.

The LP to solve at every iteration is:

$$\min \sum_{i \in I} l_i \tag{7a}$$

$$\text{s.t.} \quad l_i \geq -\sum_{j \in J} a_{ij} t_j \qquad i \in I \tag{7b}$$

$$l_i \geq \sum_{j \in J} a_{ij} t_j \qquad i \in I \tag{7c}$$

$$\sum_{j \in J} t_j proj_{qj} = 0 \qquad proj_q \in X \tag{7d}$$

$$\sum_{j \in J} t_j p_j = 1 \tag{7e}$$

where X is the matrix that contains all the PCs already computed, (7d) guarantees that t is orthogonal to them. (7e) assures that t satisfy the hyperplane equation. p is the starting vector or the solution of the previous iteration, so the coordinates of the point in which the hyperplane is tangent to the unit hypersphere. We introduced real auxiliary variables l_i ($i \in I$). The inequality constraints enforce $l_i \geq |t \cdot A_i|$, and because we minimize the sum of the l_i this forces the objective to be equal to the sum of the absolute values of the projected points. This approach exploits that fact that *minimizing* a sum of absolute values is trivial in LP.

We call this algorithm L_1-PCAhp, where *hp* stands for *hyperplane*: it is summarised in Algorithm 1. L_1-PCAhp is not globally convergent as it can become trapped in local minima, but it is locally convergent. We can improve the probability of finding a global optimum restarting from different starting vectors,

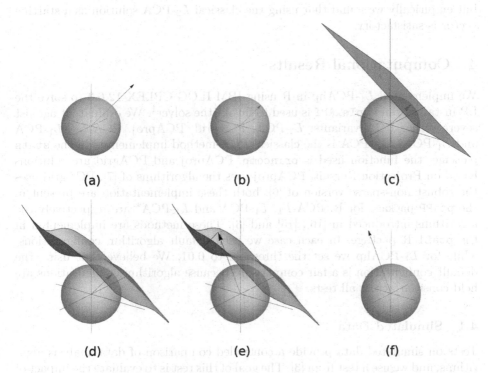

Fig. 1. Hyperplane iterative heuristic. (a) the starting vector p. (b) p is projected on the unit hypersphere. (c) the hyperplane tangent to the hypersphere in p. (d) the objective function of (7) projected on the hyperplane. (e) t the solution of (7). (f) t will be the new starting vector of the new iteration

---

Algorithm 1. L_1-PCAhp

1: P ▷ Contains the starting vectors
2: Initialize P randomly or with L_2-PCA
3: $X \leftarrow \emptyset$ ▷ Contains the PCs
4: **for** $j \in J$ **do**
5:     $p \leftarrow P_j$
6:     $Obj_{LP} \leftarrow \infty$
7:     **do**
8:         $Prev_{Obj} \leftarrow Obj_{LP}$
9:         $t \leftarrow SolutionLP(7)$
10:        $Obj_{LP} \leftarrow objectiveValueLP(7)$
11:        **if** $Obj_{LP} \leq Prev_{Obj}$ **then**
12:            $p \leftarrow t/|t|$
13:        **while** $Prev_{Obj} - Obj_{LP} \geq threshold$
14:     $X \leftarrow X \parallel p$
15: **return** X

---

but empirically we found that using the classical L_2-PCA solution as a starting vector is satisfactory.

4 Computational Results

We implemented L_1-PCAhp in R using IBM ILOG CPLEX 12.6.3 to solve the LP instances, and `cplexAPI` is used to invoke the solver[1]. We evaluate it against several other PCA variants: L_2-PCA, PCAgrid, PCAproj, PCA-L_1, L_1-PCA and L_1-PCA*. L_2-PCA is the classical PCA method implemented in the `stats` package, the function used is `princomp`. PCAproj and PCAgrid are solutions based on Projection Pursuit. PCAproj uses the algorithms of [7], PCAgrid uses the robust non-sparse version of [6]; both these implementation are present in the `pcaPP` package for R. PCA-L_1, L_1-PCA and L_1-PCA* are respectively the algorithms introduced in [16], [15] and [3]. These methods are implemented in the `pcaL1` R package. In each case we used default algorithm configurations, while for L_1-PCAhp we set the threshold to 0.01. We believe that using the default configuration is a fair comparison, because algorithm configurations are held constant across all tests.

4.1 Simulated Data

Tests on simulated data provide a controlled comparison of data analysis algorithms, and we use a test from [3]. The goal of this test is to evaluate the impact of outliers while varying their magnitude. Each dataset contains a "true" subspace and a subset of attributes that contain the information; the other attributes contain only noise. A fixed percentage of the instances are outliers with high norm. If a solution is robust to these outliers the reconstruction of the true subspace should be accurate. We also take into account the variance of the projected points.

Each dataset has $n = 1000$ instances and $m = 10$ attributes. The first q attributes define the true subspace, and 10 % of the observations have outliers. These observations have the first p of the $m - q$ noise attributes affected by the outliers, where p is the number of outlier-contaminated attributes. The true subspace attributes, the noise attributes and the outliers are sampled respectively from Laplace(0,10), Laplace(0,1) and Laplace(μ,0.01) distributions, where μ is the outlier magnitude. We take averages over 100 runs for every possible combination of $q \in \{2, 5\}$, $p \in \{1, 2, 3\}$ and $\mu \in \{25, 50, 75\}$. We repeat the experiment with Gaussian distributions while maintaining the same parameters.

Figure 2 contains the average performance (variance in feature space and reconstruction error) versus outlier magnitude (μ). Plots in which the x-axis is the number of outlier-contaminated attributes (p) are similar, but we omit them for space reasons. L_1-PCA, PCA-L_1 and L_1-PCAhp perform better than the other algorithms with respect to variance in feature space. Regarding reconstruction error L_1-PCA* is the best method, and its breakdown point is higher:

---

[1] The code is available at https://github.com/visentin-insight/L1-PCAhp.

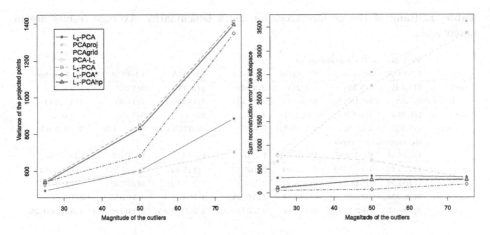

Fig. 2. Performance on the simulated datasets. Variance of the projected points against magnitude of the outliers on the left. Reconstruction error of the "true" subspace against magnitude of the outliers on the right.

the *breakdown point* is reached where the methods begin to fit the outlier observations better than the non-contaminated data. This confirms the robustness of L_1-PCAhp as it ranks joint second with L_1-PCA.

4.2   Datasets with Known Outliers

We also applied the various PCA methods to datasets with known outliers. This experiment has been used to prove the robustness of PCA algorithms to outliers [3]. The *artificial* dataset generated by [10] has 75 instances of which the first 14 are outliers. The others dataset are real-world data. The *milk* and the *pollution* dataset were introduced respectively by [8] and [21]. In the *milk* dataset the outliers are instances 70, 45, 31, 12, 14 and 15 according to [5]. In the *pollution* dataset instances 29 and 48 are identified as outliers by [7]. We centered and normalized the datasets as in Sect. 3.

To evaluate the robustness of the solutions the PCs are computed as usual with the full datasets. Then we measured the L_1-norm reconstruction error using Formula (4), and the variance of the projected points. Both were calculated only on the non-outlier instances. If the quality of the solution is good and robust to the outliers then the total reconstruction error should be small and the variance should be large.

Table 1 shows the results. Regarding the variance of the projected points, the best method depends on the dataset. Overall L_1-PCA has the best average rank, but L_1-PCAhp has the better average variance (slightly better than L_1-PCA*). Regarding the reconstruction error L_1-PCAhp clearly outperforms the others, even L_1-PCA* which was the previous winner [3]: this is quite surprising because L_1-PCAhp is based on the maximization of the projected variance and not on the reconstruction error.

Table 1. Rank of the performances for the 2 benchmarks. Average results are in parentheses.

	Variance on the feature space						
	L2-PCA	PCAproj	PCAgrid	PCA-L1	L1-PCA	L1-PCA*	L1-PCAhp
milk	7(4.43)	5(5.52)	6(5.21)	2(5.59)	**1(5.71)**	3(5.59)	4(5.57)
pollution	7(7.31)	5(11.01)	6(10.35)	3(11.17)	4(11.05)	2(11.2)	**1(11.24)**
artificial	7(0.09)	2(0.12)	**1(0.12)**	4(0.1)	3(0.1)	6(0.09)	5(0.09)
average	7(3.94)	4(5.55)	4.33(5.22)	3(5.62)	**2.67(5.62)**	3.67(5.63)	**3.33(5.63)**
	Reconstruction error						
	L2-PCA	PCAproj	PCAgrid	PCA-L1	L1-PCA	L1-PCA*	L1-PCAhp
milk	7(175.78)	4(112.76)	6(131.04)	3(105.17)	5(115.91)	2(103.54)	**1(100.62)**
pollution	7(400.9)	5(214.18)	6(248.3)	3(209.11)	4(213.54)	2(207.33)	**1(201.69)**
artificial	7(48.95)	5(36.95)	6(43.56)	**1(16.34)**	4(16.61)	3(16.55)	2(16.43)
average	7(208.55)	4.67(121.3)	6(140.96)	2.33(110.21)	4.33(115.35)	2.33(109.14)	**1.33(106.25)**

Generally the methods based on the L_1-norm without projection pursuit outperform the others. As expected, the worst performance is L_2-PCA, as it is well-known to be sensitive to outliers. Scaling the dataset by the standard deviation mitigates the effects of the outliers with high norm, but it is not enough to impart robustness. We removed it from the table to make the results more readable.

4.3   UCI Datasets

This test is used in several PCA papers [9,16,17]. For data analysis problems with a large number of input variables, dimensionality reduction methods are typically used to reduce the number of input variables to simplify the problems without degrading performance. The test is based on the idea that if the dimensionality reduction of the PCA preserves the information then the performance of a classifier should not degrade with a reduced dataset. Moreover, the projection in the feature space can eliminate some information that is not useful for the classifier, thus improving its performance.

The algorithms were applied to several datasets from the UCI machine learning repository [18]. All datasets were centered and normalized as in Sect. 3. For each dataset we extracted all possible features from 1 to 10 or from 1 to m (total number of attributes if smaller, $k \in [1, \ldots, \min(10, m)]$. We choose to limit to 10 the number of maximum extracted features because all the datasets can be reduced to 10 or fewer dimensions without affecting the classifier performances. Adding more PCs only makes the performances fluctuate almost randomly, as shown in Fig. 3. So, as in other work that uses this test [9,16,17], we limit the number of extracted features.

We used the one-nearest neighbourhood (1-NN) coded in the `class` package as a classifier, trained and tested with a 10-fold cross validation. As the results depends on the instances order the tests were conducted 10 times with different shuffles, and we report the average results.

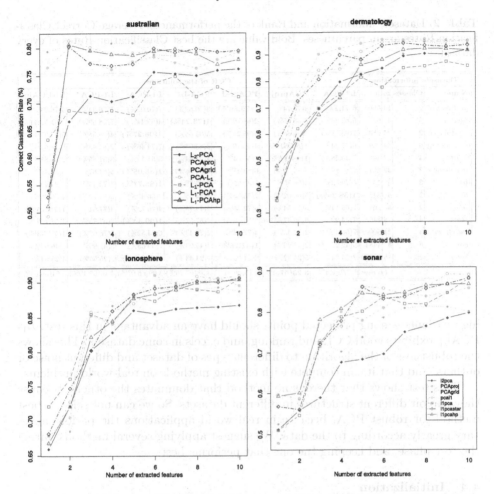

Fig. 3. Average Correct Classification Rate versus dimension of the projected space of 4 UCI datasets. The performances of the methods get almost stable after some features are extracted. Adding tests on the remaining dimensions will only dilute the ones relative to the dimensional reduction. We omitted error bars since the variances are quite high and make the graphs hard to read.

Table 2 shows the dataset information and solution rank, with the average *Correct Classification Rate* (CCR) in parentheses. Algorithm performances vary significantly across datasets: even L_2-PCA, which usually performs poorly in this test, wins in one case. This is caused by the heterogeneity of the datasets. The algorithms with best average CCRs are PCA-L_1 and PCAgrid. The winner in the highest number of datasets is L_1-PCA, but its average ranking and CCR are worse than those of the other algorithms. We believe that the poor performance of L_1-PCA* is explained by the fact that it is designed to minimize reconstruction error: classification is applied to projected points, so algorithms that preserve a

Table 2. Datasets information and Rank of the performances. Average Correct Classification Rates are in parentheses. Bold value are the best Classification Rates of every dataset

Datasets information			CCR of the solutions						
Name	Classes	m n	L2-PCA	PCAproj	PCAgrid	PCA-L1	L1-PCA	L1-PCA*	L1-PCAhp
australian	2	14 690	6(71.2%)	2(77.74%)	**1(79.59%)**	3(76.62%)	7(70.09%)	5(76.24%)	4(76.24%)
balance	3	4 625	5(62.84%)	3(63.65%)	2(68.79%)	**1(73.24%)**	4(63.44%)	7(58.69%)	6(61.52%)
brestcancer	2	9 683	7(93.77%)	2(95.88%)	6(95.21%)	3(95.66%)	**1(96.27%)**	4(95.58%)	5(95.55%)
dermatology	6	34 358	6(77.24%)	**1(87.51%)**	4(80.19%)	3(84.03%)	7(74.46%)	2(85.43%)	5(78.91%)
heart	4	13 303	7(46.58%)	**1(54.11%)**	6(50.64%)	4(51.44%)	5(51.12%)	3(51.85%)	2(51.85%)
ionoosphere	2	32 351	7(81.87%)	4(84.62%)	6(83.47%)	2(84.86%)	**1(85.08%)**	3(84.67%)	5(84.1%)
iris	3	4 150	6(79.53%)	4(90.25%)	5(89.2%)	3(90.7%)	**1(93.77%)**	7(72.45%)	2(92%)
liver	2	6 345	**1(58.34%)**	5(56.02%)	2(56.55%)	7(54.5%)	3(56.43%)	6(54.77%)	4(56.09%)
sonar	2	60 208	7(68.14%)	5(75.54%)	6(72.75%)	2(78.78%)	4(75.72%)	3(77.61%)	**1(79.15%)**
vehicle	4	18 846	7(54.9%)	6(56.47%)	5(56.56%)	3(57.57%)	**1(60.34%)**	2(57.75%)	4(57.17%)
waveform	3	21 5000	7(55.36%)	2(77.43%)	4(77.42%)	2(77.43%)	6(74.32%)	5(76.62%)	**1(77.53%)**
yeast	8	8 1484	6(42.22%)	2(47.23%)	**1(47.28%)**	4(45.93%)	7(39.65%)	5(42.99%)	3(46.04%)
letter	26	16 20000	3(26.34%)	5(24.82%)	6(24.65%)	7(24.43%)	**1(26.93%)**	2(26.69%)	4(25.54%)
average			5.77(62.47%)	**3.23(67.4%)**	4.15(66.81%)	3.38(67.75%)	3.69(66.07%)	4.15(65.19%)	3.54(66.84%)

high variance among projected points should have an advantage in this test. L_1-PCAhp exhibits good CCR and ranking, and excels in some datasets. This shows the robustness of the algorithm to different types of dataset and different noise or outliers, and that it can compete with existing methods on real-world problems.

This test shows that there is no method that dominates the others, because they exploit different structures in different datasets. So we can not select a best method for robust PCA, because in real world applications the performances vary greatly according to the data. We suggest applying several methods during the test phase, and keeping the one that performs best.

4.4   Initialization

We wanted to analyze the effect different initialization have on our solution. We repeated the test in Sect. 4.3 applying every time the random and the L_2-PCA initialization. The Correct Classification Rate obtained is almost identical, with very small fluctuation. The only difference is the computational time, since the L_2-PCA initialization is 5 % faster on average.

We think that the performance does not vary significantly because it generally converges to similar projections under different initializations, so hardly it get stuck in a completely different minimum. The improvement in the computational times are due to the small number of hyperplanes required for convergence. We believe this is because the local minimum to which our solution converges is closer to the classical PCA solution than to a random projection.

4.5   Computational Times

Making a correct time comparison is not an easy task in this situation, because the different methods are implemented in different programming languages and

use different MIP solvers. Some can improve the performances with parallelization, others can not. Backwards methods compute all components every time,
while the others only compute the required components. Some depend more on
the number of original dimensions than on the number of instances.

We coded our method using the same language code and the same MIP solver
as pcaL1 R package. We decided to use this setting to have a direct comparison
with the others LP-based methods that are implemented in that package. An R
wrap function call C code that uses Coin Clp to solve the LP instances.

We used the same dataset generator used in Sect. 4.1, because it allows us to
manage directly the size of the problem. The algorithms configuration is the kept
the same. All the computations are done in a single core, so no parallelization is
involved.

We kept constant the number of attributes. We varied the number of instances
from 100 to 50000. We computed each time a number of components equal to
half the number of attributes.

Figure 4 shows the comparison with $m = 10$, both the computational time
and the instances number use logarithmic scale. As expected the fastest method
is always L_2-PCA. The only method with comparable execution time on large
instances is PCA-L_1, which is approximately 10 times slower on average. The
two methods that use projection pursuit, PCAproj and PCAgrid, are approximately 100 times slower than L_2-PCA. We omitted the error bars since both the
standard deviation and the mean error were too small to be plotted.

All the methods that use LP are generally slower than the others. The L_1-
PCA* and L_1-PCAhp computational times are not dependent on the number of
components required, because they always compute all components. Our solution

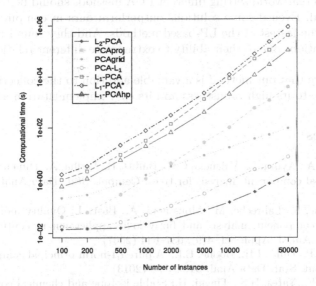

Fig. 4. Computational time comparison. The number of attributes is 10.

is always faster than the other LP-based methods, and this gap increases with the dimension of the dataset. This may seems odd, since our solution needs to solve numerous LP models for every component, even if the dimension of our LP are smaller. The main reason is that we do not have to solve a new model everytime, but simply edit the hyperplane constraint (7e) in every iteration and add a new orthogonality constraint (7d) every time it finds a components.

The execution time of our solution does not depend on the number of components we are searching. In contrast, the performance of L_1-PCA strongly varies according to the number of required components. If we need only a few components then it has good performance.

The performances of our solution can be improved by using a more improved MIP solver. Also the parallelization will strongly improve the performances.

5   Conclusion

In this paper we introduced a new algorithm for PCA called L_1-PCAhp which is intuitively straightforward, easy to implement and has only one runtime parameter to tune. Its novelty lies in the way it uses LP, and in its hyperplane-based iterative improvement approach. Moreover it is the only L_1-PCA LP-based methods with a backward approach that tries to maximize the variance of the projected points in the projection space. In experiments its performance was consistently good compared to other algorithms in the literature. We showed its versatility and good performance both in variance maximization and in reconstruction error minimization, where it excels on real datasets with outliers. In classification tests the different algorithms exploit different information in each dataset, so we suggest that in a real-world setting different PCA methods should be tested and the best one used. Even if some solutions outperform ours in computational time, our PCA is the fastest of the LP-based methods. And they found plenty of real world applications due to their ability to extrapolate different relationship in the data [4,27].

We believe that our method is a valuable addition to the collection of known methods, due to its high robustness and its easy implementation and tuning.

References

1. Alfaro, C.A., Aydın, B., Valencia, C.E., Bullitt, E., Ladha, A.: Dimension reduction in principal component analysis for trees. Comput. Stat. Data Anal. **74**, 157–179 (2014)
2. Bouhouche, S., Lahreche, M., Moussaoui, A., Bast, J.: Quality monitoring using principal component analysis and fuzzy logic application in continuous casting process 1. Am. J. Appl. Sci. **4**(9), 637–644 (2007)
3. Brooks, J.P., Dulá, J.H., Boone, E.L.: A pure L1-norm principal component analysis. Comput. Stat. Data Anal. **61**, 83–98 (2013)
4. Carter, J.F., Yates, H.S., Tinggi, U.: Stable isotope and chemical compositions of European and Australasian ciders as a guide to authenticity. J. Agric. Food Chem. **63**(3), 975–982 (2015)

5. Choulakian, V.: L1-norm projection pursuit principal component analysis. Comput. Stat. Data Anal. **50**(6), 1441–1451 (2006)
6. Croux, C., Filzmoser, P., Fritz, H.: Robust sparse principal component analysis. Technometrics **55**(2), 202–214 (2013)
7. Croux, C., Ruiz-Gazen, A.: High breakdown estimators for principal components: the projection-pursuit approach revisited. J. Multivar. Anal. **95**(1), 206–226 (2005)
8. Daudin, J.J., Duby, C., Trecourt, P.: Stability of principal component analysis studied by the bootstrap method. Statistics: J. Theoret. Appl. Stat. **19**(2), 241–258 (1988)
9. Ding, C., Zhou, D., He, X., Zha, H.: R1-PCA: rotational invariant L1-norm principal component analysis for robust subspace factorization. In: Proceedings of the 23rd International Conference on Machine Learning, pp. 281–288. ACM (2006)
10. Hawkins, D.M., Bradu, D., Kass, G.V.: Location of several outliers in multiple-regression data using elemental sets. Technometrics **26**(3), 197–208 (1984)
11. Hodge, V.J., Austin, J.: A survey of outlier detection methodologies. Artif. Intell. Rev. **22**(2), 85–126 (2004)
12. Jolliffe, I.: Principal Component Analysis. Wiley Online Library, New York (2002)
13. Hill Jr., T.W., Ravindran, A.: On programming with absolute-value functions. J. Optim. Theory Appl. **17**(1–2), 181–183 (1975)
14. Kaplan, S.: Comment on a precis by Shanno and Weil. Manag. Sci. **17**(11), 778–780 (1971)
15. Ke, Q., Kanade, T.: Robust L1-norm factorization in the presence of outliers and missing data by alternative convex programming. In: IEEE Computer Society Conference on Computer Vision and Pattern Recognition, vol. 1, pp. 739–746. IEEE (2005)
16. Kwak, N.: Principal component analysis based on L1-norm maximization. IEEE Trans. Pattern Anal. Mach. Intell. **30**(9), 1672–1680 (2008)
17. Kwak, N.: Principal component analysis by-norm maximization. IEEE Trans. Cybern. **44**(5), 594–609 (2014)
18. Lichman, M.: UCI machine learning repository (2013). http://archive.ics.uci.edu/ml
19. Luenberger, D.G.: Optimization by Vector Space Methods. Wiley, New York (1997)
20. Malagón-Borja, L., Fuentes, O.: Object detection using image reconstruction with PCA. Image Vis. Comput. **27**(1), 2–9 (2009)
21. McDonald, G.C., Schwing, R.C.: Instabilities of regression estimates relating air pollution to mortality. Technometrics **15**(3), 463–481 (1973)
22. Park, Y.W., Klabjan, D.: Algorithms for L1-norm principal component analysis (2014)
23. Rao, M.R.: Technical note - some comments on 'linear' programming with absolute-value functionals. Oper. Res. **21**(1), 373–374 (1973)
24. Ravindran, A., Hill Jr., W.H.: Note - a comment on the use of simplex method forabsolute value problems. Manag. Sci. **19**(5), 581–582 (1973)
25. Röver, C., Bizouard, M.A., Christensen, N., Dimmelmeier, H., Heng, I.S., Meyer, R.: Bayesian reconstruction of gravitational wave burst signals from simulations of rotating stellar core collapse and bounce. Phys. Rev. D **80**(10), 102004 (2009)
26. Shanno, D.F., Weil, R.L.: Technical note - 'linear' programming with absolute-value functionals. Oper. Res. **19**(1), 120–124 (1971)
27. Zhuo, S., Guo, D., Sim, T.: Robust flash deblurring. In: 2010 IEEE Conference on Computer Vision and Pattern Recognition (CVPR), pp. 2440–2447. IEEE (2010)

Multilabel Classification on Heterogeneous Graphs with Gaussian Embeddings

Ludovic Dos Santos[✉], Benjamin Piwowarski, and Patrick Gallinari

Sorbonne Universities, UPMC Univ Paris 06, CNRS, LIP6 UMR 7606,
4 Place Jussieu, 75005 Paris, France
{ludovic.dossantos,benjamin.piwowarski,patrick.gallinari}@lip6.fr

Abstract. We consider the problem of node classification in heterogeneous graphs, where both nodes and relations may be of different types, and different sets of categories are associated to each node type. While graph node classification has mainly been tackled for homogeneous graphs, heterogeneous classification is a recent problem which has been motivated by applications in fields such as social networks, where graphs are intrinsically heterogeneous. We propose a transductive approach to this problem based on learning graph embeddings, and model the uncertainty associated to the node representations using Gaussian embeddings. A comparison with representative baselines is provided on three heterogeneous datasets.

Keywords: Node graph classification · Representation learning · Gaussian embeddings

1 Introduction

Classification of nodes in graphs is a relational classification problem where the labels of each node depend on its neighbors. Many problems in domains like image, biology, text or social data labeling can be formulated as graph node classification and this problem has been tackled with different approaches like collective classification [21], random walks [1], and transductive regularized models [10]. Most approaches consider homogeneous graphs, where all the nodes share the same set of labels, propagating labels from seed nodes to their neighbors. Many problems in domains like biology or social data analysis involve heterogeneous networks where the nodes and the relations between nodes are of different types, each node type being associated to a specific set of labels. For example, the LastFM social network, one of the datasets used in our experiments, links users, tracks, artists and albums via seven different types of relations such as *friendship*, *most listened tracks*, and *authorship*. In such a network, nodes of different types influence each other and their labels are interdependent. The dependency is, however, more complex than with homogeneous networks and depends both on the nodes type and on their specific relation. Classical methods for homogeneous graphs based for example on label propagation, usually relies on a simple

P. Frasconi et al. (Eds.): ECML PKDD 2016, Part II, LNAI 9852, pp. 606–622, 2016.
DOI: 10.1007/978-3-319-46227-1_38

relational hypothesis like homophily in social networks. They cannot be easily extended to heterogeneous networks, and new methods have to be developed for dealing with this relational classification problem.

In this paper, we consider the problem of node classification in heterogeneous graphs. We propose a transductive approach based on graph embeddings where the node embeddings are learned so as to reflect both the classification objective for the different types of nodes and the relational structure of the graph. When most embedding techniques consider deterministic embeddings where each node is represented as a point in a representation space, we focus here on density-based embeddings which capture some form of uncertainty about the learned representations. Uncertainty can have various causes related to the lack of information (isolated nodes in the graph) or because of the contradiction between neighboring nodes (different labels). Our hypothesis is that, because of these different factors, training will result in learned representations with different confidence, and that this uncertainty is important for this classification problem. For that, we will use Gaussian embeddings which have been recently proposed for learning word [23] and knowledge graph [7] embeddings in an unsupervised setting. More precisely, each graph node representation corresponds to a Gaussian distribution where the mean and the variance are learned. The variance term is a measure of uncertainty associated to the node representation. The objective function is composed of two terms, one reflecting the classification task and the other one reflecting the relations between the nodes. Both mono and multi-label classification can be handled by the model. For the experiments, we focus on classification in social network data. This type of data offers a variety of situations which allows us to illustrate the behavior and the performance of the model for different types of heterogeneous classification problems.

To summarize, our contributions are as follows: (i) We propose a new method for learning to classify nodes adapted to heterogeneous graph data; (ii) We model the uncertainty associated with the nodes representation; (iii) We provide a comparison with state of the art baselines on a series of social data classification problems representative of different situations.

2 Related Work

2.1 Graph Node Classification

Several different models have been proposed to solve the graph node classification task. We discuss below three main families [4] (i) collective classification, (ii) random walk type methods, and (iii) semi-supervised/transductive graph regularized models.

Random Walks. This family gathers methods where labels are iteratively propagated from seed nodes to all the other nodes in a network. Propagation follows a random walk or a similar iterative mechanism. [8,28] are among the early ML models using random walks for classification in homogeneous graphs. [27] propose an extension of these models for heterogeneous graphs. It relies on hand-defined projections of the graph onto homogeneous graphs, the approach being difficult to adapt automatically to new datasets. The Graffiti random surfer model [1] is a state of the art random walk classifier for heterogeneous graphs. It is based on two intertwined random walks. Both are between nodes of the same type, but allowing either one hop (standard) or two-hop (extended) steps in the graph. It models up to a certain extent the influence among nodes of different types. In our preliminary tests on different datasets, this model was among the best ones.

Collective Classification. Collective classification algorithms are extensions of classical inductive classification to relational data. They take as input a fixed size vector composed of node features and of statistics on the node neighbors current labels. Sen et al. [21] provide an introduction and a comparison of some of these models. They distinguish between two families: local and global models. The former make use of local classifiers. In [14,15] for example, naive Bayes classifiers are used iteratively, dynamically updating the attributes of nodes as inferences are made about their neighbors. Along these lines, [18] recently introduced an iterative model for sparsely labeled network which forces the label predictions to map the distribution of the observed data with a maximum entropy constraint. Global classifiers optimize a global loss function using graphical models, like e.g. Markov Random Fields. Iterative methods suppose features associated with nodes to learn the classifier, which is not the case in our work.

Random Walk Type Methods. This family gathers methods where labels are iteratively propagated from seed nodes to all the other nodes in a network. Propagation follows a random walk or a similar iterative mechanism. [8,28] are among the early Machine Learning (ML) models using random walks for classification in homogeneous graphs. [27] propose an extension of these models for heterogeneous graphs. It relies on hand-defined projections of the graph onto homogeneous graphs, the approach being difficult to adapt automatically to new datasets. The Graffiti random surfer model [1] is a state of the art random walk classifier for heterogeneous graphs. It is based on two intertwined random walks. Both are between nodes of the same type, but allowing either one hop (standard) or two-hop (extended) steps in the graph. It models up to a certain extent the influence among nodes of different types. In our preliminary tests on different datasets, this model was among the best ones.

Semi-Supervised Transductive Learning. The third family has been developed for exploiting the manifold assumption in semi-supervised learning. The loss function is composed of two main terms, one is for classification on the labeled nodes, the other one is a propagation equation which encourages neighbor nodes to share similar labels. Seminal works in this direction include [2,19,24,26]. All these models have been developed for homogeneous graphs and perform some form of label propagation similar to random walks. The difference with the latter is that the problem is formulated as a loss minimization one, which is more general than simply formulating a propagation rule. Relations between random walks and loss-based models are discussed more at length in [4,29]. Extensions have been proposed over the years to handle more general situations. Multi-relational graphs where nodes are all of the same type, but can be linked by different relations are considered in [9,12]. This also allows them to extend the transductive models to inductive formulations. Some authors have attempted to extend homogeneous formulations to the heterogeneous setting. All follow more or less the idea of projecting the heterogeneous graph onto a series of homogeneous ones, thus creating a series of homogeneous classification problems. Work in this direction includes [11] which is a direct extension of the homogeneous formulation in [25]. Graph projections have to be defined for each new problem and none of these models is able to directly exploit the correlation between nodes of different types. The work closer to ours is [10] who was among the first to propose an embedding model for transductive heterogeneous graph classification. This has been the starting point of our work, but they only consider deterministic representations while we use a more general transductive formulation with probabilistic embeddings.

To summarize heterogeneous graph classification approaches, very few allow modeling the influences between nodes of different types. In the experimental section, we will compare our model to [1,10] which have been designed specifically for heterogeneous classification, as well as an unsupervised graph embedding model [22] and a homogeneous graph model [28].

2.2   Learning Representations for Graphs and Relational Data

In the last years, there has been a growing interest in learning latent representations. This has led to breakthroughs in domains like image recognition, speech or natural language processing [3,13]. Graph node embeddings have been proposed for unsupervised learning where the goal is to learn node representations that preserves the graph structure and that can be exploited latter for different purposes like visualization, clustering or classification. [17] learns node representations by performing truncated walks on the graph – and supposing that nodes along the path should be close together in the representation space. [22] propose an algorithm designed for very large graphs, which can be used for different types of graphs (undirected, directed, weighted or not) – we use their method as our unsupervised baseline that embeds all data points, and then train a classifier on labeled ones. Somewhat related to this topic is the learning of embeddings for graphs where a unique representation of the whole graph is learned [20] and the learning of triplets in knowledge graphs where both relations and nodes representations are learned for ranking positive triplets over negative ones [5–7]. The setting is, however, quite different from the one considered here. Finally, modeling uncertainty via Gaussian embeddings has been proposed recently for unsupervised learning in [7,23]. Based on sentences in the former and knowledge graph in the latter, they propose energy-based models to learn Gaussian embeddings. In this paper, we borrow their formalization and graph regularization cost in a transductive setting.

3   Model

In this section we present our model, namely *Heterogeneous Classification with Gaussian Embeddings (HCGE)*.

We first introduce the notations used throughout this paper. A heterogeneous network is modeled as a directed weighted graph $\mathcal{G} = (\mathcal{N}, \mathcal{E}, \mathcal{W})$ where $\mathcal{N}$ is the set of nodes, $\mathcal{E}$ the set of edges and $\mathcal{W}$ the weights associated to the edges. Each node $x_i \in \mathcal{N}$ of the graph has a type $t_i \in \mathcal{T}$, where $\mathcal{T} = 1, 2, \ldots, T$. We denote N_i the neighbors of x_i.

Regarding the classification task, let $\mathcal{Y}^t$ denotes the set of categories associated with nodes of type t, and $\#\mathcal{Y}^t$ the cardinality of $\mathcal{Y}^t$. $\mathcal{L} \subset \mathcal{N}$ is the set of indices of labeled nodes. For $i \in \mathcal{L}$, y_i is the class vector associated to x_i: node x_i belongs to category c if $y_i^c = 1$ and does not belong if $y_i^c = -1$.

In our model, each node x_i is mapped onto a representation which is a Gaussian distribution over the space $z_i \sim \mathcal{N}(\mu_i, \Sigma_i)$ in $\mathbb{R}^Z$. The latent space is common to all nodes. In this paper, we compare two different parameterizations of Σ. We experimented with a spherical ($\Sigma_i = \sigma_i Id$) and a diagonal ($\Sigma_i = diag(\sigma_i^p)_p$) covariance matrix. We use a weight w_r for each type of relation. To simplify we use w_{ij} for the weight $w_{r_{ij}}$ of the edge (i, j) linking node i to node j with a relation r_{ij}.

Loss Function. We learn the representations of nodes and classifiers parameters by minimizing an objective loss function. It takes the general form of transductive regularized loss [11,25], with a classification (Δ_C) and a regularization term (Δ_G), both being detailed later:

$$L(z, \theta) = \sum_{i \in \mathcal{L}} \Delta_C(f_{\theta^{t_i}}(z_i), y_i) + \lambda \sum_{i \in \mathcal{N}} \sum_{j \in N_i} w_{ij} \Delta_G(z_i, z_j) \tag{1}$$

As for classical transductive graph losses, the minimization in (1) aims at finding a trade-off between the difference between observed and predicted labels in $\mathcal{Y}^t$, and the amount of information shared between two connected nodes. There are however major differences, since here z is not a label as in classical formulations, but a node embedding. Finally, the function $f_{\theta^t}(.)$ is a parametric classifier for a node of type t – there is one such classifier for each node type. Since we are using Gaussian embeddings, the zs are random variables and the regularization term is a dissimilarity measure between distributions.

To avoid overfitting, following [23], we regularize the mean and the covariance matrix associated to each node. We add two constraints to prevent means and covariances to be too large and to keep the covariance matrices positive definite (this also prevents degenerate solutions):

$$\|\mu_i\| \leq C \text{ and } \forall p, \, m \leq \sigma_i^p \leq M \tag{2}$$

where the different parameters C, m and M have been set manually after some trials on a subset of the DBLP training set to respectively 10, 0.01 and 10 (and not changed after that), but any other reasonable value will do.

The two following paragraphs refer to the respective parts of (1).

Classifier. The mapping onto the latent space is learned so that the labels of each type of node can be predicted from the (Gaussian) embedding. For that, we use a parametric classification function f_{θ^t} depending on the type t of the node. This multivariate function takes as input a node representation and outputs a vector of scores for each label corresponding to the node type. The parameters θ^t of the classifier are learned by minimizing the following loss on labeled data:

$$L_{Classification} = \sum_{i \in \mathcal{L}} \Delta_C(f_{\theta^{t_i}}(z_i), y_i) \tag{3}$$

where $\Delta_C(f_{\theta^{t_i}}(z_i), y_i)$ is the loss associated with predicting labels $f_{\theta^{t_i}}(z_i)$ given the observed labels y_i. We recall that in this equation $f_{\theta^{t_i}}(z_i)$ and y_i have values in $\mathbb{R}^{\#\mathcal{Y}^t}$.

In our experiments, we used different losses for Δ_C. We first considered the case where a class decision is simply the expectation of the classifier score together with a hinge loss, adapting the loss proposed in [10]. For a given node x of type t with an embedding z, this gives:

$$\Delta_C(f_{\theta^t}(z), y) = \Delta_{EV}(f_{\theta^t}(z), y) \stackrel{\text{def}}{=} \sum_{k=1}^{\#\mathcal{Y}^t} \max\left(0; 1 - y^k \mathbb{E}_z[f_{\theta^t}^k(z)]\right) \tag{4}$$

where y^k is 1 if x belongs to category k and -1 otherwise, and $f_{\theta^t}^k(z)$ is a random variable for category k.

Alternatively, the density based formulation allows us to leverage the density-based representation through a probabilistic criterion, even in the case of linear classifiers. We used here for Δ_C the log-probability that $y^k f_{\theta^t}(z)$ take a positive value. In this case, the variance will be influenced by the two loss terms: if the two terms act in opposite directions, one solution will be to increase variance. As we will see, this is confirmed by the experiments.

$$\Delta_C(f_{\theta^t}(z), y) = \Delta_{Pr}(f_{\theta^t}(z), y) \stackrel{\text{def}}{=} -\sum_{k=1}^{\#\mathcal{Y}^t} \log \mathbb{P}\left(y^k f_{\theta^t}^k(z) > 0\right) \tag{5}$$

In our experiments and for both costs, we used a linear classifier for $f_{\theta^t}^k$, which allows to easily compute the different costs and gradients, since the random variable $f_{\theta^t}^k(z)$, being a linear combination of Gaussian variables, is Gaussian too. A basic derivation shows that:

$$\mathbb{P}\left(y^k f_{\theta^t}^k(z) > 0\right) = \frac{1}{2}\left(1 + \text{erf}\left(\frac{\mu \cdot \theta^t}{\sqrt{2 \sum_p (\theta_p^t \sigma^p)^2}}\right)\right) \tag{6}$$

where erf is the Gauss error function.

There are some notable differences between the two classification losses during learning. In the case of a linear classifier f_{θ^t}, $\mathbb{E}_z[f_{\theta^t}^k(z)] = \mu \cdot \theta_k^t$. Thus, minimizing Δ_{EV} only updates the mean of the Gaussian embedding: the covariance matrix of the embedding does not interfere with the classification term, and is only present in the second term of (1).

For the Δ_{Pr} loss, the probability is proportional to erf $\left(\frac{\mu \cdot \theta^t}{\sqrt{2 \sum_p (\theta_p^t \sigma^P)^2}}\right)$ where the variance is present. When the graph regularization and classification cost pull the representation mean in opposite directions (opposite gradients), the model will respond by increasing the variance for the spherical variance model[1]: this behavior is interesting since it transforms an opposition between regularization and classification costs into increased uncertainty.

Graph Embedding. We make the hypothesis that two nodes connected in the graph should have similar representations, whatever their type is. Intuitively, this will force nodes of the same type which are close in the graph to be close in the representation space. The strength of this attraction between nodes of the same class will be proportional to their closeness in the graph and to the weight of the path(s) linking them. We use the asymmetric loss proposed in [7,23]:

$$L_{Graph} = \sum_i \sum_{j \in N_i} w_{ij} D_{KL}(z_j \| z_i) \tag{7}$$

where $\Delta_G(z_i, z_j) = D_{KL}(z_j \| z_i)$ is the Kullback-Leibler divergence between the distributions of z_i from z_j:

$$D_{KL}(z_j \| z_i) = \int_{x \in \mathbb{R}} \mathcal{N}(x; \mu_j, \Sigma_j) \log \frac{\mathcal{N}(x; \mu_j, \Sigma_j)}{\mathcal{N}(x; \mu_i, \Sigma_i)} dx$$
$$= \frac{1}{2} \left(\mathrm{tr}(\Sigma_i^{-1} \Sigma_j) + (\mu_i - \mu_j)^T \Sigma_i^{-1} (\mu_i - \mu_j) - d - \log \frac{\det(\Sigma_j)}{\det(\Sigma_i)} \right) \tag{8}$$

The loss L_{Graph} is a sum over the neighbors N_i of i, where w_{ij} is the weight of the edge between x_i and x_j. Other similarity measures between distributions could be used as well, the Kullback-Leibler divergence having the advantage of being asymmetric, which fits well the social network datasets used in the experiments.

Algorithm. Learning the Gaussian embeddings $z \sim \mathcal{N}(\mu, \Sigma)$ and the classifiers parameters θ consists in minimizing loss function in (1). We used here a Stochastic Gradient Descent Method to learn the latent representations, i.e. the μ_i, Σ_i as well as the parameters θ of the classifiers.

Our algorithm samples a pair of connected nodes and then makes a gradient update of the nodes parameters. For each sampled node z_i that is part of the labeled training set $\mathcal{L}$, the algorithm performs an update according to the first term of (3). This update consists in successively modifying the parameters of the classification functions θ^{t_i} and of the latent representations μ_i and Σ_i so as

---

[1] the increase will be in the direction of the normal to the classifier hyperplane for the diagonal variance model.

to minimize the classification loss term. Then, the model updates its parameters with respect to the smoothness term of (7). Note that, while we use a stochastic gradient descent, other methods like mini-batch gradients or batch algorithms could be used as well.

4 Experiments

4.1 Datasets

Experiments have been performed on three datasets respectively extracted from DBLP, Flickr and LastFM. For all but the first dataset (DBLP), each node can have multiple labels. The three datasets are described below and summarized in Table 1.

Table 1. Datasets

DBLP	Nodes	Type	Nb. Nodes	Nb. Labeled Nodes	Nb. Labels
		Paper	14,376	14,376	20
		Author	14,475	4,057	4
	Edges	Type	Nb. Edges		
		Author↔Paper	41,794		
Flickr	Nodes	Photos	46,926	8,766	21
		User	4,760	3,476	42
	Edges	User↔User	175,779		
		User↔Photo	46,926		
LastFM	Nodes	Users	1,013	321	59
		Tracks	35,181	24,562	28
		Albums	32,118	15,966	47
		Artists	17,138	11,564	47
	Edges	User↔User	1,109		
		User↔Album	47,541		
		User↔Artist	47,812		
		User↔Track	47,807		
		Track↔Album	29,647		
		Track↔Artist	35,181		
		Album↔Artist	32,118		

The **DBLP** dataset is a bibliographical network composed of authors and papers. Authors are labeled with their research domain (4 different domains) while papers are labeled with the conference name they were published in (20 labels). Authors and papers are connected through an *authorship* relation.

The graph is thus composed of two types of nodes and is bipartite with only one relation type. Classification is monolabel on papers and authors.

The **Flickr** corpus is a dataset composed of photos and users. The photo labels correspond to different possible tags while the user labels correspond to their subscribed groups. The classification problem is multi-label: images and users may belong to more than one category. Photos are related to users through an *authorship* relation, while users are related to others through a *following* relation. We have kept the image tags that appear in at least 500 images, and user categories that also appear at least 500 times in the dataset resulting in 21 possible labels for photos and 42 for authors.

The **LastFM** dataset is a social network composed of users, tracks, albums and artists. This dataset was extracted using the LastFM API[2]. The task is multi-label, and all node types have their specific set of labels. Users are labeled with the type of music they like (59 labels), tracks with the kind of music they belong to (28 labels), albums with their type (47 labels) and artists with the kind of music they play the most (47 labels). Users are related to users (*friendship*), tracks (*favorite tracks*), albums (*favorite albums*) and artists (*favorite artists*). Tracks are related to albums (*belong to*) and artists (*singer*). Finally, albums are related to artists (*sing in*). Note that one track can be related to several artists, and an album can be related to several artists. This dataset contains tracks labeled by their genre (*rock, indie, ...*), users by the type of music they like (*female vocalists, ambient, ...*), albums by their type (*various artists, live, ...*) and artists by the kind of music they make (*folk, singer songwriter, ...*). Some labels may be the same string-wise for different types of nodes, but we consider that labels of different types of nodes are distinct, e.g. *pop* is not the same for an artist or a track.

We compare our approach with four state-of-the-art models (see Sect. 2):

- **LINE** [22], which is representative of unsupervised learning of graph embeddings suitable for various tasks such as classification. We performed a logistic regression with the learned representations as inputs.
- **HLP** [28], which is representative of transductive graph algorithms developed for semi-supervised learning. As HLP is designed for homogeneous graphs, we perform as many random walks as the number of node types, considering each time that all the nodes are of a same given type.
- **Graffiti** [1], which is a state of the art model for the task of classification with random walk in heterogeneous graph.
- **LSHM** [10], which is another state of the art model for the task of classification with deterministic vector representations in heterogeneous graph.

Evaluation Measures and Protocol. For the evaluation, we have considered two different evaluation measures. The **Precision at 1 (P@1)** measures the percentage of nodes for which the category with the highest score is among the observed labels. The **Precision at k (P@k)** is the proportion of correct labels

---

[2] To access the API go to http://www.lastfm.fr/api.

in the set of k labels with the highest predicted scores. Here micro P@k is an average on all the node types, with k set to the number of relevant categories. This is a measure of the capacity of a model to correctly pick the k relevant categories of any node. In the case of DBLP (mono-label dataset), we consider that the predicted category is the category with the highest score. We make use of the **Precision at 1 (P@1)** measure as there is at most one label per node. We optimize and compare the different models with regard to micro-average, and also report macro-average.

Regarding the experimental protocol, we partition a dataset into two different subsets, namely a training set and a testing set. As all the models have hyperparameters, one subset of the training set is used as a validation set to optimize by grid search the hyperparameters. The optimization is done with respect to the **Micro P@k** measure, which corresponds to the mean of P@k over all nodes. The other part of the training set is used to learn the parameters of the different models. We then compare the different models based on the results on the testing set, by using the model for which the performance over the validation set was the best.

Experiments are performed with different training set sizes: 10 %, 30 %, 50 %. Within our transductive setting, the training set size refers to **the proportion of labeled nodes** used in the training set[3]. The training nodes are selected at random. The proportion of nodes used during the parameters training phase and used for the hyperparameters selection depends on the size of the training set. We use 50–50 for a training set size of 10 % and 80–20 (train/validation) for the others. Experiments are performed with 5 random splits. The hyper-parameters are selected for each split using the validation set. We then average 5 runs over each split.

4.2   Results

In this section we present the results of four variants of our Gaussian embedding model, and compare to LINE [22], Graffiti [1], HLP [28] and LSHM [10]. The experiments are performed on the three datasets described in Table 1 and the results are described in Tables 2 (DBLP), 3 (FlickR) and 4 (LastFM). The best performing classifier (on the test set) is presented in bold.

Concerning the four variants of our model, HCGE($\Delta_\bullet$, X) refers to the HCGE model with the classification loss $\Delta_\bullet$ (Δ_{EV} or Δ_{Pr}) and a spherical (X=S) or diagonal (X=D) covariance matrix.

For micro P@k, our model generally outperforms the others on all the datasets. Supervised models (HLP, Graffiti, LSHM and HCGE) using the class information outperform unsupervised representation learning, which matches the results reported in [10]. On all datasets, the performances of HLP are below the performances of Graffiti, LSHM and HCGE. This clearly shows that modeling the heterogeneity of the graph brings noteworthy improvements. Comparing the heterogeneous models, both LSHM and HCGE outperform Graffiti on all

---

[3] We did not prune the graph.

Table 2. P@1 DBLP

Train size	Model	Train	Val	Test			
		Micro		Micro	Macro	Author	Paper
10 %	LINE	25.1	18.9	19.5	23.0	29.1	16.8
	HLP	100	24.7	24.1	27.2	32.6	21.8
	Graffiti	100	32.4	30.9	38.1	50.8	25.3
	LSHM	99.8	33.8	**32.1**	**40.0**	**53.9**	**26.0**
	HCGE(Δ_{EV}, S)	99.7	33.1	30.9	38.5	52.1	24.9
	HCGE(Δ_{EV}, D)	95.6	31.4	30.4	37.4	49.9	24.9
	HCGE(Δ_{Pr}, S)	83.8	29.0	27.9	34.3	45.6	22.9
	HCGE(Δ_{Pr}, D)	92.9	29.0	28.3	34.3	45.1	23.6
30 %	LINE	24.0	21.5	21.9	24.8	30.1	19.5
	HLP	100	35.8	36.0	41.9	52.4	31.4
	Graffiti	100	39.6	38.5	46.6	61.1	32.1
	LSHM	99.7	43.0	41.2	52.9	73.8	31.9
	HCGE(Δ_{EV}, S)	98.5	44.4	**42.3**	52.6	71.0	**34.3**
	HCGE(Δ_{EV}, D)	98.8	42.9	41.2	50.8	68.0	33.6
	HCGE(Δ_{Pr}, S)	97.5	41.8	41.3	52.1	71.4	32.8
	HCGE(Δ_{Pr}, D)	97.4	43.8	**42.3**	**54.1**	**75.0**	33.1
50 %	LINE	24.2	21.1	22.3	25.0	29.8	20.2
	HLP	100	39.7	39.4	46.5	59.3	33.7
	Graffiti	100	41.5	41.2	49.4	64.1	34.8
	LSHM	99.9	45.5	44.4	56.8	**79.2**	34.5
	HCGE(Δ_{EV}, S)	99.3	45.6	44.6	55.2	74.1	36.3
	HCGE(Δ_{EV}, D)	98.1	44.7	43.9	53.7	71.0	36.3
	HCGE(Δ_{Pr}, S)	99.4	45.8	45.5	57.1	77.8	**36.4**
	HCGE(Δ_{Pr}, D)	97.6	45.9	**45.7**	**57.7**	**79.2**	36.2

datasets. On average, compared to Graffiti, LSHM is 2.4 better on DBLP, 2.1 better on FlickR and 2.5 better on LastFM. We observed the same behavior for HCGE, with +2.8 on DBLP, +4.4 on FlickR and +6.0 on LastFM. We can note that the more complex the dataset, the higher the gap compared to the baselines. This also shows that the use of representations can clearly improve the performances.

On each dataset, our model outperforms LSHM (and the other competitors) 8 times over 9, with on average +1.0 points for DBLP, +2.3 for FlickR, and +3.8 for LastFM over the second ranked model. According to the results, introducing uncertainty in representations clearly improves results when compared to LSHM. Let us also point out that, according to our initial intuition, the effect of using uncertainty has more impact when the amount of training data is lower: the

Table 3. P@k FlickR

| Train size | Model | Train | Val | Test | | | |
		Micro		Micro	Macro	User	Photo
10%	LINE	24.4	19.4	20.7	23.2	29.1	17.3
	HLP	100	26.0	26.3	27.8	31.3	24.3
	Graffiti	100	24.3	24.5	27.0	32.7	21.2
	LSHM	99.3	29.6	29.3	29.1	28.6	29.5
	HCGE(Δ_{EV}, S)	98.9	33.5	**32.7**	**32.6**	32.4	**32.8**
	HCGE(Δ_{EV}, D)	99.1	33.4	32.6	**32.6**	32.7	32.5
	HCGE(Δ_{Pr}, S)	96.0	30.4	29.7	29.2	28.1	30.3
	HCGE(Δ_{Pr}, D)	98.7	31.7	31.9	32.2	**33.0**	31.5
30%	LINE	23.0	21.6	21.5	24.2	30.6	17.9
	HLP	100	47.6	47.7	43.7	34.5	53.0
	Graffiti	100	47.5	47.0	43.7	**36.1**	51.3
	LSHM	100	49.2	48.4	43.6	32.5	54.7
	HCGE(Δ_{EV}, S)	99.1	51.5	50.0	45.6	35.4	55.8
	HCGE(Δ_{EV}, D)	98.7	51.6	**50.1**	45.7	35.3	**56.0**
	HCGE(Δ_{Pr}, S)	98.3	50.1	49.0	44.4	33.8	55.1
	HCGE(Δ_{Pr}, D)	98.5	50.6	50.0	**45.8**	**36.1**	55.5
50%	LINE	23.2	21.8	21.8	24.6	31.0	18.2
	HLP	100	54.2	54.1	48.6	35.8	61.4
	Graffiti	100	54.4	54.0	48.8	36.9	60.8
	LSHM	99.9	55.1	54.0	47.9	33.7	62.0
	HCGE(Δ_{EV}, S)	97.9	56.7	55.8	50.0	36.5	**63.4**
	HCGE(Δ_{EV}, D)	97.3	56.6	55.8	50.0	36.5	**63.4**
	HCGE(Δ_{Pr}, S)	98.8	55.7	54.8	49.0	35.5	62.5
	HCGE(Δ_{Pr}, D)	98.4	56.4	**55.9**	**50.3**	**37.2**	63.3

difference between LSHM and HCGE decreases in general when more training data is available (except for DBLP).

Let us compare the performance of the variants Δ_{EV} and Δ_{Pr}. Globally, Δ_{Pr} seems to be disadvantaged by a low number of training examples, when Δ_{EV} seems to be more stable in comparison to other baselines. However, the more training data, the closer the Δ_{Pr} variant is to Δ_{EV}. For example, on the DBLP dataset, moving from 10% to 30% improves on average Δ_{Pr} results by +13.7 but only by +11.1 for Δ_{EV}. For a training set size of 50%, the difference between Δ_{Pr} and Δ_{EV} is +1.1 on DBLP, and +0.1 on FlickR. For LastFM, the difference is resp. −14.6 for 10%, −6.5 for 30% and −1.5 for 50% of the dataset used for training. On the three datasets, the lower the training set size, the better Δ_{EV} seems to be compared to Δ_{Pr}. We could not explain this difference

Table 4. P@k LastFM

Train size	Model	Train	Val	Test					
		Micro		Micro	Macro	User	Track	Album	Artist
10 %	LINE	20.8	20.6	20.4	15.9	5.6	26.0	14.5	17.4
	HLP	98.7	38.1	38.4	30.0	9.9	47.8	27.2	35.1
	Graffiti	100	40.1	40.0	31.4	**10.6**	49.0	28.1	38.1
	LSHM	99.9	36.4	36.3	27.2	9.0	48.4	26.2	25.3
	HCGE(Δ_{EV}, S)	99.8	44.4	**44.0**	**34.1**	9.6	**52.3**	**35.0**	**39.7**
	HCGE(Δ_{EV}, D)	99.3	44.0	43.6	34.0	10.5	52.2	34.4	38.7
	HCGE(Δ_{Pr}, S)	97.6	27.7	27.8	20.7	4.1	34.9	21.0	23.0
	HCGE(Δ_{Pr}, D)	96.0	30.3	29.4	21.9	6.7	38.7	22.1	20.2
30 %	LINE	20.5	20.9	20.5	17.0	10.1	25.9	14.4	17.5
	HLP	98.9	50.2	49.7	40.0	**17.2**	60.5	37.7	44.8
	Graffiti	100	50.8	50.3	40.4	**17.2**	61.7	36.2	46.5
	LSHM	99.8	54.2	53.3	40.3	9.7	65.8	42.7	42.9
	HCGE(Δ_{EV}, S)	99.6	58.2	**57.3**	45.0	14.8	**68.2**	**45.9**	51.2
	HCGE(Δ_{EV}, D)	99.5	57.9	57.0	**45.3**	16.8	67.5	45.7	**51.3**
	HCGE(Δ_{Pr}, S)	97.5	50.5	50.4	37.7	9.9	66.4	32.6	42.0
	HCGE(Δ_{Pr}, D)	96.9	51.5	50.8	38.5	13.2	65.0	41.4	34.4
50 %	LINE	20.5	20.5	20.5	17.0	10.3	26.0	14.4	17.5
	HLP	98.8	51.9	52.1	42.3	19.4	63.1	40.2	46.4
	Graffiti	100	53.2	53.5	43.2	19.1	65.4	39.5	48.7
	LSHM	99.7	56.6	56.7	43.2	11.0	68.8	45.6	47.6
	HCGE(Δ_{EV}, S)	99.4	60.3	**60.4**	**48.7**	**20.4**	**71.2**	**48.8**	**54.4**
	HCGE(Δ_{EV}, D)	99.9	60.1	60.3	48.6	20.1	71.1	48.7	54.3
	HCGE(Δ_{Pr}, S)	99.2	58.6	58.5	45.0	11.8	69.8	47.4	51.0
	HCGE(Δ_{Pr}, D)	99.9	58.9	58.9	47.2	18.9	70.2	46.4	53.4

in the behavior between Δ_{EV} and Δ_{Pr}, but believe that this is due to the fact that the covariance matrix is only optimized in the graph regularization term in the case of Δ_{EV}. Let us now compare the use of a spherical and a diagonal covariance matrix. For the Δ_{EV} variant, it looks like moving from a spherical covariance matrix to a diagonal one brings no improvement. It even decreases the performance on DBLP. Concerning the Δ_{Pr} variant, for which the covariance matrix plays a role in the classification cost, conclusions are reversed and using diagonal covariance matrices improves the results. On the FlickR dataset, using a diagonal variance improves the results by 1.4 on average. However, it looks like the more training data, the less the improvement, with +2.2 improvement for a training set size of 10 %, +1.0 for 30 % and +1.1 for 50 %.

4.3   Qualitative Discussion

In this section, we focus on studying qualitatively the representations found by HCGE. We consider the most robust variant of our model (Δ_{EV}, S), and the most challenging dataset, LastFM (similar observations were made on the other datasets). We will examine the respective role of regularization and classification costs on labeled training nodes, and the relationship between the learned variance of a node and the local node properties (like its number of neighbors).

We first examined the respective role of classification and regularization costs. In (1), the max-margin classification cost implies that the gradient of a node x is 0 if $y^k \mathbb{E}_z[f^k_{\theta^t}(z)]$ is above 1. In this case, the only constraints on the node are due to the graph regularization cost. We can see how many of the nodes are used by the classification cost by looking at the number of cases for which $y^k \mathbb{E}_z[f^k_{\theta^t}(z)]$ is below or equal to 1. In Fig. 1a, we have shown a histogram of $y^k \mathbb{E}_z[f^k_{\theta^t}(z)]$ for labeled nodes in the training set (after convergence). For around one third of the nodes, the value of the classifier is above 1.1 – they could be removed from the labeled set without harming the solution (however, these could have been useful in early stages of optimization). This is clearly in agreement with the experiments where we have shown that representation-based models were performing better than the others, and suggests that it would be interesting to use these statistics to predict the performance of the model on held-out data.

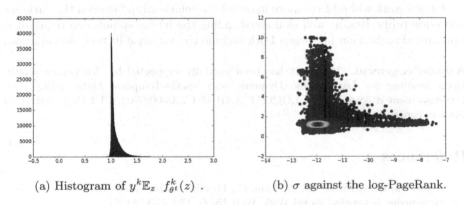

(a) Histogram of $y^k \mathbb{E}_z \; f^k_{\theta^t}(z)$.   (b) σ against the log-PageRank.

Fig. 1. Qualitative results for the model HCGE(Δ_{EV}, S) on the LastFM dataset with 50 % of the dataset used for training. In Fig. 1b, we computed Gaussian kernel density to show high density regions in the plot.

Regarding the relationship between the learned variance and the local properties of each node, we looked at the relationship between the PageRank[4] [16] of a node and its variance. Figure 1b shows that high PageRank implies a small variance. Which means that for central nodes, representations are less uncertain. However, the reverse implication is not true.

---

[4] Using a standard damping factor of 0.15.

5 Conclusion

We have explored the use of uncertainty for learning to represent nodes in the challenging task of heterogeneous graph node classification. The proposed model, Heterogeneous Classification with Gaussian Embeddings (HCGE), learns for each node a Gaussian distribution over the representation space, parameterized by its mean and covariance matrix, by optimizing a loss function that includes a classification loss and graph regularization loss. We have examined four variants of this model, by using either spherical and diagonal covariance matrices, and by using two different loss functions for classification. Our model can easily be extended to inductive learning by defining the Gaussian representation z as a parameterized function of the input features.

Based on the experimental results obtained on datasets representative of different situations, our main findings are that (i) integrating uncertainty in representations improved classification (ii) according to our initial intuition, the effect of using uncertainty has generally more impact when the amount of training data is lower and (iii) according to our expectation, highly central nodes seem to have less variance associated to their representation.

Future work will address more in detail the relationship between the variance and node properties, as well as understanding the interplay between regularization and classification loss when both include the variance in their formulation.

Acknowledgement. This work has been partially supported by: Xu Guangqi 2016 Deep learning for Large Scale Dynamic and Spatio-Temporal Data; REQUEST Investissement d'Avenir 2014; LOCUST ANR-15-CE23-0027-01; FUI PULSAR (BPI France, Rgion Ile de France).

References

1. Angelova, R., Kasneci, G., Weikum, G.: Graffiti: graph-based classification in heterogeneous networks. World Wide Web **15**(2), 139–170 (2012)
2. Belkin, M., Niyogi, P., Sindhwani, V.: Manifold regularization: a geometric framework for learning from labeled and unlabeled examples. J. Mach. Learn. Res. **7**, 2399–2434 (2006). http://portal.acm.org/citation.cfm?id=1248547.1248632
3. Bengio, Y., Courville, A., Vincent, P.: Representation learning: a review and new perspectives. IEEE Trans. Pattern Anal. Mach. Intell. **35**(8), 1798–1828 (2013)
4. Bengio, Y., Delalleau, O., Le Roux, N.: Label propagation and quadratic criterion. Semi-supervised learning 10 (2006)
5. Bordes, A., Usunier, N., Garcia-Duran, A., Weston, J., Yakhnenko, O.: Translating embeddings for modeling multi-relational data. In: Advances in Neural Information Processing Systems, pp. 2787–2795 (2013)
6. Bordes, A., Weston, J., Collobert, R., Bengio, Y.: Learning structured embeddings of knowledge bases. In: Conference on Artificial Intelligence, No. EPFL-CONF-192344 (2011)
7. He, S., Liu, K., Ji, G., Zhao, J.: Learning to represent knowledge graphs with gaussian embedding. In: Proceedings of the 24th ACM CIKM, pp. 623–632. ACM (2015)

8. Jaakkola, M.S.T., Szummer, M.: Partially labeled classification with Markov Random walks. Adv. Neural Inform. Process. Syst. (NIPS) **14**, 945–952 (2002)
9. Jacob, Y., Denoyer, L., Gallinari, P.: Classification and annotation in social corpora using multiple relations. In: Proceedings of the 20th ACM international CIKM, pp. 1215–1220. ACM Press (2011)
10. Jacob, Y., Denoyer, L., Gallinari, P.: Learning latent representations of nodes for classifying in heterogeneous social networks. In: Proceedings of the 7th ACM International Conference on Web Search and Data Mining, pp. 373–382. ACM (2014)
11. Ji, M., Sun, Y., Danilevsky, M., Han, J., Gao, J.: Graph regularized transductive classification on heterogeneous information networks. In: Balcázar, J.L., Bonchi, F., Gionis, A., Sebag, M. (eds.) ECML PKDD 2010. LNCS (LNAI), vol. 6321, pp. 570–586. Springer, Heidelberg (2010). doi:10.1007/978-3-642-15880-3_42
12. Kato, T., Kashima, H., Sugiyama, M.: Integration of multiple networks for robust label propagation. In: SIAM Conference on Data Mining, pp. 716–726. Citeseer (2008)
13. LeCun, Y., Bengio, Y., Hinton, G.: Deep learning. Nature **521**(7553), 436–444 (2015)
14. Neville, J., Jensen, D.: Iterative classification in relational data. In: Proceedings of the AAAI-2000 Workshop on Learning Statistical Models from Relational Data, pp. 13–20 (2000)
15. Oh, H.J., Myaeng, S.H., Lee, M.H.: A practical hypertext catergorization method using links and incrementally available class information. In: Proceedings of the 23rd Annual International ACM SIGIR Conference on Research and Development in Information Retrieval, pp. 264–271. ACM (2000)
16. Page, L., Brin, S., Motwani, R., Winograd, T.: The pagerank citation ranking: Bringing order to the web. Technical Report 1999–66, Stanford InfoLab, November 1999. http://ilpubs.stanford.edu:8090/422/, previous number = SIDL-WP-1999-0120
17. Perozzi, B., Al-Rfou, R., Skiena, S.: Deepwalk: Online learning of social representations. In: Proceedings of the 20th ACM SIGKDD International Conference on Knowledge Discovery and Data Mining, pp. 701–710. ACM (2014)
18. Pfeiffer III., J.J., Neville, J., Bennett, P.N.: Overcoming relational learning biases to accurately predict preferences in large scale networks. In: Proceedings of the 24th International Conference on World Wide Web, pp. 853–863. International World Wide Web Conferences Steering Committee (2015)
19. Pimplikar, R., Garg, D., Bharani, D., Parija, G.: Learning to propagate rare labels. In: Proceedings of the 23rd ACM International Conference on Conference on Information and Knowledge Management, pp. 201–210. ACM (2014)
20. Scarselli, F., Gori, M., Tsoi, A.C., Hagenbuchner, M., Monfardini, G.: The graph neural network model. IEEE Trans. Neural Netw. **20**(1), 61–80 (2009)
21. Sen, P., Namata, G., Bilgic, M., Getoor, L., Gallagher, B., Eliassi-Rad, T.: Collective classification in network data. AI Mag. **29**(3), 93–106 (2008)
22. Tang, J., Qu, M., Wang, M., Zhang, M., Yan, J., Mei, Q.: Line: Large-scale information network embedding. In: Proceedings of the 24th International Conference on World Wide Web, pp. 1067–1077. International World Wide Web Conferences Steering Committee (2015)
23. Vilnis, L., McCallum, A.: Word representations via gaussian embedding. arXiv preprint, arXiv:1412.6623 (2014)
24. Zhou, D., Bousquet, O., Lal, T.N., Weston, J., Schölkopf, B.: Learning with local and global consistency. In: Thrun, S., Saul, L., Schölkopf, B. (eds.) Advances in Neural Information Processing Systems 16. MIT Press, Cambridge (2004)

25. Zhou, D., Bousquet, O., Lal, T.N., Weston, J., Schölkopf, B.: Learning with local and global consistency. Adv. Neural Inform. Process. Syst. **16**(16), 321–328 (2004)
26. Zhou, D., Huang, J., Schölkopf, B.: Learning from labeled and unlabeled data on a directed graph. In: Proceedings of the 22nd International Conference on Machine Learning, ICML 2005, NY, USA, pp. 1036–1043 (2005). http://doi.acm.org/10.1145/1102351.1102482
27. Zhou, Y., Liu, L.: Activity-edge centric multi-label classification for mining heterogeneous information networks. In: Proceedings of the 20th ACM SIGKDD International Conference on Knowledge Discovery and Data Mining, pp. 1276–1285. ACM (2014)
28. Zhu, X., Ghahramani, Z.: Learning from labeled and unlabeled data with label propagation. Technical report, Citeseer (2002)
29. Zhu, X., Ghahramani, Z., Lafferty, J., et al.: Semi-supervised learning using gaussian fields and harmonic functions. ICML **3**, 912–919 (2003)

M-Flash: Fast Billion-Scale Graph Computation Using a Bimodal Block Processing Model

Hugo Gualdron[1]([⊠]), Robson Cordeiro[1]([⊠]), Jose Rodrigues Jr.[1]([⊠]),
Duen Horng (Polo) Chau[2]([⊠]), Minsuk Kahng[2]([⊠]), and U. Kang[3]([⊠])

[1] University of Sao Paulo, Sao Carlos, SP, Brazil
{gualdron,robson,junio}@icmc.usp.br
[2] Georgia Institute of Technology, Atlanta, USA
{polo,kahng}@gatech.edu
[3] Seoul National University, Seoul, Republic of Korea
ukang@snu.ac.kr

Abstract. Recent graph computation approaches have demonstrated that a single PC can perform efficiently on billion-scale graphs. While these approaches achieve scalability by optimizing I/O operations, they do not fully exploit the capabilities of modern hard drives and processors. To overcome their performance, in this work, we introduce the Bimodal Block Processing (*BBP*), an innovation that is able to boost the graph computation by minimizing the I/O cost even further. With this strategy, we achieved the following contributions: (1) M-Flash, the fastest graph computation framework to date; (2) a flexible and simple programming model to easily implement popular and essential graph algorithms, including the *first* single-machine billion-scale eigensolver; and (3) extensive experiments on real graphs with up to 6.6 billion edges, demonstrating M-Flash's consistent and significant speedup. The software related to this paper is available at https://github.com/M-Flash.

Keywords: Graph algorithms · Graph processing · Graph mining · Complex networks

1 Introduction

Large graphs with *billions* of nodes and edges are increasingly common in many domains and applications, such as in studies of social networks, transportation route networks, citation networks, and many others. Distributed frameworks (find a thorough review in the work of Lu *et al.* [13]) have become popular choices for analyzing these large graphs. However, distributed approaches may not always be the best option, because they can be expensive to build [11], and hard to maintain and optimize.

These potential challenges prompted researchers to create single-machine, billion-scale graph computation frameworks that are well-suited to essential graph algorithms, such as eigensolver, PageRank, connected components and many others. Examples are GraphChi [11] and TurboGraph [5]. Frameworks

© Springer International Publishing AG 2016
P. Frasconi et al. (Eds.): ECML PKDD 2016, Part II, LNAI 9852, pp. 623–640, 2016.
DOI: 10.1007/978-3-319-46227-1_39

in this category define sophisticated processing schemes to overcome challenges induced by limited main memory and poor locality of memory access observed in many graph algorithms. However, when studying previous works, we noticed that despite their sophisticated schemes and novel programming models, they do not optimize for disk operations and data locality, which are the core of performance in graph processing frameworks.

In the context of *single-node, billion-scale*, graph processing frameworks, we present **M-Flash**, a novel scalable framework that overcomes critical issues found in existing works. The innovation of M-Flash confers it a performance many times faster than the state of the art. More specifically, our contributions include:

1. **M-FlashFramework & Methodology:** we propose a novel framework named M-Flash that achieves fast and scalable graph computation. M-Flash (https://github.com/M-Flash) introduces the Bimodal Block Processing, which significantly boosts computation speed and reduces disk accesses by dividing a graph and its node data into blocks (dense and sparse) to minimize the cost of I/O.
2. **Programming Model:** M-Flash provides a flexible and simple programming model, which supports popular and essential graph algorithms, e.g., PageRank, connected components, and the *first* single-machine eigensolver over billion-node graphs, to name a few.
3. **Extensive Experimental Evaluation:** we compared M-Flash with state-of-the-art frameworks using large graphs, the largest one having 6.6 billion edges (YahooWeb https://webscope.sandbox.yahoo.com). M-Flash was consistently and significantly faster than GraphChi [11], X-Stream [15], TurboGraph [5], MMap [12], and GridGraph [19] across all graph sizes. Furthermore, it sustained high speed even when memory was severely constrained, e.g., 6.4X faster than X-Stream, when using 4 GB of Random Access Memory (RAM).

2 Related Works

A typical approach to scalable graph processing is to develop a distributed framework. This is the case of Gbase [7], Powergraph, Pregel, and others [13]. Among these approaches, Gbase is the most similar to M-Flash. Despite the fact that Gbase and M-Flash use a block model, Gbase is distributed and lacks an adaptive edge processing scheme to optimize its performance. Such scheme is the greatest innovation of M-Flash, conferring to it the highest performance among existing approaches, as demonstrated in Sect. 4.

Among the existing works designed for single-node processing, some of them are restricted to SSDs. These works rely on the remarkable low-latency and improved I/O of SSDs compared to magnetic disks. This is the case of Turbo-Graph [5], which relies on random accesses to the edges — not well supported over magnetic disks. Our proposal, M-Flash, avoids this drawback by avoiding random accesses.

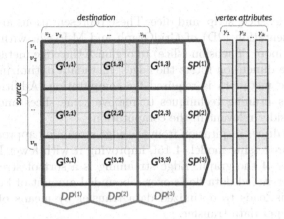

Fig. 1. Organization of edges and vertices in M-Flash. **Edges (left):** example of a graph's adjacency matrix (in light blue color) using 3 logical intervals ($\beta = 3$); $G^{(p,q)}$ is an edge block with source vertices in interval $I^{(p)}$ and destination vertices in interval $I^{(q)}$; $SP^{(p)}$ is a *source-partition* containing all blocks with source vertices in interval $I^{(p)}$; $DP^{(q)}$ is a *destination-partition* containing all blocks with destination vertices in interval $I^{(q)}$. **Vertices (right):** the data of the vertices as k vectors ($\gamma_1 \ldots \gamma_k$), each one divided into β logical segments. (Color figure online)

GraphChi [11] was one of the first single-node approaches to avoid random disk/edge accesses, improving the performance over mechanical disks. GraphChi partitions the graph on disk into units called *shards*, requiring a preprocessing step to sort the data by source vertex. GraphChi uses a vertex-centric approach that requires a shard to fit entirely in memory, including both the vertices in the shard and all their edges (in and out). As we demonstrate, this fact makes GraphChi less efficient when compared to our work. M-Flash requires only a subset of the vertex data to be stored in memory.

MMap [12] introduced an interesting approach based on OS-supported mapping of disk data into memory (virtual memory). It allows graph data to be accessed as if they were stored in unlimited memory, avoiding the need to manage data buffering. Our framework uses memory mapping when processing edge blocks but, with an improved engineering, M-Flash consistently outperforms MMap, as we demonstrate.

GridGraph [19] divides the graphs into blocks and processes the edges reusing the vertices' values loaded in main memory (in-vertices and out-vertices). Furthermore, it uses a two-level hierarchical partitioning to increase the performance, dividing the blocks into small regions that fit in cache. When comparing GridGraph with M-Flash, both divide the graph using a similar approach with a two-level hierarchical optimization to boost computation. However, M-Flash adds a bimodal partition model over the block scheme to optimize even more the computation for sparse blocks in the graph.

GraphTwist [18] introduces a 3D cube representation of the graph to add support for multigraphs. The cube representation divides the edges using three partitioning levels: slice, strip, and dice. These representations are equivalent to the block representation (2D) of GridGraph and M-Flash, with the difference that it adds one more dimension (slice) to organize the edge metadata for multi-graphs. The slice dimension filters the edges' metadata optimizing performance when not all the metadata is required for computation. Additionally, Graph-Twist introduces pruning techniques to remove some slices and vertices that they do not consider relevant in the computation.

M-Flash also draws inspiration from the edge streaming approach introduced by X-Stream's processing model [4,15], improving it with fewer I/O operations for dense regions of the graph. Edge streaming is a sort of stream processing referring to unrestricted data flows over a bounded amount of buffering. As we demonstrate, this leads to optimized data transfer by means of less I/O and more processing per data transfer.

3   M-Flash

The design of M-Flash considers the fact that real graphs have a varying density of edges; that is, a given graph contains dense regions with many more edges than other regions that are sparse. In the development of M-Flash, and through experimentation with existing works, we noticed that these dense and sparse regions could not be processed in the same way. We also noticed that this was the reason why existing works failed to achieve superior performance. To cope with this issue, we designed M-Flash to work according to two distinct processing schemes: Dense Block Processing (DBP) and Streaming Partition Processing (SPP). For full performance, M-Flash uses a theoretical I/O cost-based scheme to decide the kind of processing to use in face of a given block, which can be dense or sparse. The final approach, which combines DBP and SPP, was named Bimodal Block Processing (BBP).

3.1   Graph Representation in M-Flash

A graph in M-Flash is a directed graph $G = (V, E)$ with vertices $v \in V$ labeled with integers from 1 to $|V|$, and edges $e = (source, destination)$, $e \in E$. Each vertex has a set of attributes $\gamma = \{\gamma_1, \gamma_2, \ldots, \gamma_K\}$; edges also might have attributes for specific processings.

Blocks **in M-Flash:** Given a graph G, we divide its vertices V into β intervals denoted by $I^{(p)}$, where $1 \leq p \leq \beta$. Note that $I^{(p)} \cap I^{(p')} = \varnothing$ for $p \neq p'$, and $\bigcup_p I^{(p)} = V$. Consequently, as shown in Fig. 1, the edges are divided into β^2 *blocks*. Each block $G^{(p,q)}$ has a *source node interval* p and a *destination node interval* q, where $1 \leq p, q \leq \beta$. In Fig. 1, for example, $G^{(2,1)}$ is the block that contains edges with source vertices in the interval $I^{(2)}$ and destination vertices in the interval $I^{(1)}$. We call this on-disk organization as *partitioning*. Since M-Flash works by alternating one entire block in memory for each running thread, the value of β is automatically determined by the following equation:

$$\beta = \left\lceil \frac{\phi(T+1)\,|V|}{M} \right\rceil \tag{1}$$

where the constant 1 refers to the need of one buffer to store the input vertex values that are shared between threads (read-only), ϕ is the amount of data to represent each vertex, T is the number of threads, $|V|$ is the number of vertices, and M is the available RAM. For example, 4 bytes of data per node, 2 threads, a graph with 2 billion nodes, and for 1 GB RAM, $\beta = \lceil (4 \times (2+1) \times (2 \times 10^9))/(2^{30}) \rceil = 23$, thus requiring $23^2 = 529$ blocks. The number of threads enters the equation because all the threads access the same block to avoid multiple seeks on disk, and they use an exclusive memory buffer to store the vertex data processed (one buffer per thread), so to prevent "race" conditions.

3.2 The M-Flash Processing Model

This section presents our proposed processing model. We first describe the two strategies targeted at processing dense or sparse blocks. Next, we present the novel cost-based optimization used to determine the best processing strategy.

Dense Block Processing (DBP): Figure 2 illustrates the DBP; notice that vertex intervals are represented by vertical (Source I) and horizontal (Destination I) vectors. After partitioning the graph into *blocks*, we process them in a vertical zigzag order, as illustrated. There are three reasons for this order: (1) we store the computation results in the destination vertices; so, we can "pin" a

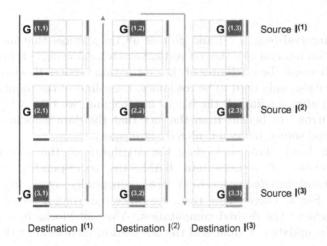

Fig. 2. M-Flash's computation schedule for a graph with 3 intervals. Vertex intervals are represented by vertical (Source I) and horizontal (Destination I) vectors. Blocks are loaded into memory, and processed in a vertical zigzag manner, indicated by the sequence of red, orange and yellow arrows. This enables the reuse of input, as when going from $G^{(3,1)}$ to $G^{(3,2)}$, M-Flash reuses source node interval $I^{(3)}$), which reduces data transfer from disk to memory. (Color figure online)

Fig. 3. Example of DBP I/O operations to process the *dense* block $G^{(2,1)}$.

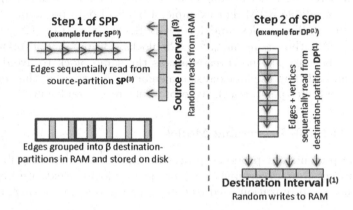

Fig. 4. I/O operations for SPP taking $SP^{(3)}$ and $DP^{(1)}$ as illustrative examples. Step 1: the edges of *source-partition* $SP^{(3)}$ are sequentially read and combined with the values of their source vertices from $I_{(3)}$. Next, edges are grouped by destination, and written to β files, one for each *destination partition*. Step 2: the files corresponding to *destination-partition* $DP^{(1)}$ are sequentially processed according to a given desired computation, with results written to destination vertices in $I_{(1)}$.

destination interval (e.g., $I^{(1)}$) and process all the vertices that are sources to this destination interval (see the red vertical arrow); (2) using this order leads to fewer reads because the attributes of the destination vertices (horizontal vectors in the illustration) only need to be read once, regardless of the number of source intervals. (3) after reading all the blocks in a column, we take a "U turn" (see the orange arrow) to benefit from the fact that the data associated with the previously-read source interval is already in memory.

Within a block, besides loading the attributes of the source and destination intervals of vertices into RAM, the corresponding edges $e = \langle source, destination, edge\ properties \rangle$ are sequentially read from disk, as explained in Fig. 3. These edges, then, are processed using a user-defined function so to achieve the desired computation. After all blocks in a column are processed, the updated attributes of the destination vertices are written to disk.

Streaming Partition Processing (SPP): The performance of DBP decreases for graphs with sparse blocks; this is because, for a given block, we have to read more data from the source intervals of vertices than from the very blocks of edges. In such cases, SPP processes the graph using partitions instead of blocks. A graph *partition* is a set of *blocks* sharing the same *source node interval* – a line in the logical partitioning, or, similarly, a set of *blocks* sharing the same *destination node interval* – a column in the logical partitioning. Formally, a *source-partition* $SP^{(p)} = \bigcup_q G^{(p,q)}$ contains all the blocks with edges having source vertices in the interval $I^{(p)}$; a *destination-partition* $DP^{(q)} = \bigcup_p G^{(p,q)}$ contains all the blocks with edges having destination vertices in the interval $I^{(q)}$. For example, in Fig. 1, $SP^{(1)}$ is the union of blocks $G^{(1,1)}$, $G^{(1,2)}$, and $G^{(1,3)}$; meanwhile, $DP^{(3)}$ is the union of blocks $G^{(1,3)}$, $G^{(2,3)}$, and $G^{(3,3)}$. In a graph, hence, there are β *source-partitions* and β *destination-partitions*.

Considering the graph organized into partitions, SPP takes two steps (see Fig. 4). In the first step, for a given *source-partition* $SP^{(p)}$, it loads the values of the vertices of the corresponding interval $I^{(p)}$; next, it reads the edges of the partition $SP^{(p)}$ sequentially from disk, storing them in a buffer together with their source-vertex values. At this point, it sorts the buffer in memory, grouping the edges by destination. Finally, it stores the edges on disk into β files, one for each of the β *destination-partitions*. This processing is performed for each *source-partition* $SP^{(p)}$, $1 \leq p \leq \beta$, so to iteratively build the β *destination-partitions*.

In the second step, after processing the β *source-partitions* (each with β blocks), it is possible to read the β files according to their destinations, so to logically "build" β *destination-partitions* $DP^{(q)}$, $1 \leq q \leq \beta$, each one containing edges together with their source-vertex values. For each *destination-partition* $DP^{(q)}$, we read the vertices of interval $I^{(q)}$; next, we sequentially read the edges, processing their values through a user-defined function. This function uses the properties of the vertices and of the edges to perform specific computations whose results will update the vertices. Finally, SPP stores the updated vertices of interval $I^{(q)}$ back on disk.

Bimodal Block Processing (BBP): Schemes *DBP* and *SPP* improve the performance in complementary circumstances. But, *How can we decide which processing scheme to use when we are given a graph block to process?* To answer this question, we join DBP and SPP into a single scheme – the Bimodal Block Processing (BBP). The combined scheme uses the theoretical I/O cost model proposed by Aggarwal and Vitter [1] to decide for *SBP* or *SPP*. In this model, the I/O cost of an algorithm is equal to the number of *disk blocks* with size B transferred between disk and memory plus the number of non-sequential reads (seeks). Since we use this model to choose the scheme with the smaller cost, we need to algebraically determine the cost of each scheme, as follows.

For processing a graph $G = \{V, E\}$, *DBP* performs the following operations: it reads the $|V|$ vertices β times and it writes the $|V|$ vertices once; it also reads the $|E|$ edges once – *disk blocks* of size B, vertices and edges with constant sizes omitted from the equation for simplification. β^2 seeks are necessary because the edges are read sequentially. Hence, the I/O cost for *DBP* is given by:

$$\mathcal{O}\left(\mathrm{DBP}\left(G\right)\right) = \mathcal{O}\left(\frac{(\beta+1)\,|V| + |E|}{B} + \beta^2\right) \tag{2}$$

In turn, *SPP* initially reads the $|V|$ source vertices and the $|E|$ edges; then, still in its first step, it sorts (simple shuffling) the $|E|$ edges grouping them by destination into a set of edges and vertices $|\hat{E}|$, writing them to disk. In its second step, it reads the $\hat{E}$ edges/vertices to perform the update operation, writing $|V|$ destination vertices back to disk. The I/O cost for *SPP* comes to be:

$$\mathscr{O}\left(\text{SPP}\left(G\right)\right) = \mathscr{O}\left(\frac{2\,|V| + |E| + 2\,\left|\hat{E}\right|}{B} + \beta\right) \tag{3}$$

Equations 2 and 3 define the I/O cost for one processing iteration over the whole graph G. However, in order to decide in relation to the graph blocks, we are interested in the costs of Eqs. 2 and 3 divided by the number of graph blocks β^2. The result, after the appropriate algebra, reduces to Eqs. 4 and 5.

$$\mathscr{O}\left(\text{DBP}\left(G^{(p,q)}\right)\right) = \mathscr{O}\left(\frac{\vartheta\phi\left(1 + 1/\beta\right) + \xi\psi}{B}\right) \tag{4}$$

$$\mathscr{O}\left(\text{SPP}\left(G^{(p,q)}\right)\right) = \mathscr{O}\left(\frac{2\vartheta\phi/\beta + 2\xi(\phi + \psi) + \xi\psi}{B}\right) \tag{5}$$

where ξ is the number of edges in $G^{(p,q)}$, ϑ is the number of vertices in the interval, and ϕ and ψ are, respectively, the number of bytes to represent a vertex and an edge e. Once we have the costs per graph block of DBP and SPP, we can decide between one and the other by simply analyzing the ratio SPP/DBP:

$$\mathscr{O}\left(\frac{\text{SPP}}{\text{DBP}}\right) = \mathscr{O}\left(\frac{1}{\beta} + \frac{2\xi}{\vartheta}\left[1 + \frac{\psi}{\phi}\right]\right) \tag{6}$$

This ratio leads to the final decision equation:

$$\text{BlockType}\left(G^{(p,q)}\right) = \begin{cases} \text{sparse,} & \text{if } \mathscr{O}\left(\frac{\text{SPP}}{\text{DBP}}\right) < 1 \\ \text{dense,} & \text{otherwise} \end{cases} \tag{7}$$

We apply Eq. 6 to select the best option according to Eq. 7. With this scheme, BBP is able to select the best processing scheme for each graph block. In Sect. 4, we demonstrate that this procedure yields a performance superior than the current state-of-the-art frameworks.

Algorithm 1. *MAlgorithm*: Algorithm Interface for coding in M-Flash

---

   initialize (Vertex v)
   process (Vertex u, Vertex v, EdgeData data)
   gather (Accum v_1, Accum v_2, Accum v_out)
   apply (Vertex v)

Algorithm 2. PageRank in M-Flash

---

   degree(v): = out degree for Vertex v
   initialize(v): v.value = 0
   process(u, v, data): v.value += u.value/ *degree(u)*
   gather(v_1, v_2, v_out): v_out = v_1 + v_2
   apply(v): v.value = 0.15 + 0.85 * v.value

Algorithm 3. Algorithm M-Flash

Input: Graph $G(V, E)$ and vertex attributes γ
Input: user-defined *MAlgorithm* program
Input: memory size M and number of iterations *iter*
Output: vector v with vertex results

1: set ϕ from γ attributes, and β using Eq. 1. $\vartheta = |V| / \beta$
2: execute graph preprocessing and *partitioning*
3: **for** $i = 1$ to *iter* **do**
4:     execute the first step of *SPP* (Fig. 4) to process the sparse source-partitions
5:     **for** $q = 1$ to β **do**
6:         load vertex values of destination interval $I^{(q)}$
7:         initialize $I^{(q)}$ of v using *MAlgorithm*.initialize
8:         **if** there is a sparse destination-partition associated with $I^{(q)}$ **then**
9:             **for each** edge
10:                 invoke *MAlgorithm*.process storing results in vector v
11:         **if** q is odd **then**
12:             partition-order $= \{1$ to $\beta\}$
13:         **else**
14:             partition-order $= \{\beta$ to $1\}$
15:         **for** $p = \{$partition-order$\}$ **do**
16:             **if** $G^{(p,q)}$ is dense **then**
17:                 load vertex values of source interval $I^{(p)}$
18:                 **for each** edge in $G^{(p,q)}$
19:                     invoke *MAlgorithm*.process storing results in vector v
20:             invoke *MAlgorithm*.gather for $I^{(q)}$ of v
21:             invoke *MAlgorithm*.apply for $I^{(q)}$ of v
22:             store interval $I^{(q)}$ of vector v

3.3   Programming Model in M-Flash

The M-Flash's computational model, which we named *MAlgorithm* (short for *Matrix Algorithm Interface*) is shown in Algorithm 1. Since *MAlgorithm* is a vertex-centric model, it stores computation results in the destination vertices, allowing for a vast set of iterative computations, such as PageRank, Random Walk with Restart, Weakly Connected Components, Sparse Matrix Vector Multiplication, Eigensolver, and Diameter Estimation.

The *MAlgorithm* interface has four operations: **initialize**, **process**, **gather**, and **apply**. The *initialize* operation, optionally, loads the initial value of each destination vertex; the *process* operation receives and processes the data from incoming edges (neighbors) – this is where the desired processing occurs; the *gather* operation joins the results from the multiple threads so to consolidate a single result; finally, the *apply* operation is able to perform finalizing operations, such as normalization – apply is optional.

3.4   System Design and Implementation

M-Flash starts by preprocessing an input graph dividing the edges into β partitions and counting the number of edges per logical block (β^2 blocks), at the same time that the blocks are classified as sparse or dense using Eq. 7. Note that M-Flash does *not* sort the edges during preprocessing, it simply divides them into β^2 blocks, $\beta^2 \ll |V|$. In a second preprocessing, M-Flash processes the graph according to the organization given by the concept of *source-partition* as seen in Sect. 3.2. At this point, blocks are only a logical organization, while partitions are physical. The *source-partitions* are read and, whenever a dense block is found, the corresponding edges are extracted from the partition and a file is created for this block in preparation for DBP; the remaining edges in the *source-partition* will be ready for processing using SPP. Notice that, after the second preprocessing, the logical blocks classified as dense, are materialized into physical files. The total I/O cost for preprocessing is $\frac{4|E|}{B}$, where B is the size of each block transferred between disk and memory. Algorithm 3 shows the pseudo-code of M-Flash.

4   Evaluation

We compare M-Flash (https://github.com/M-Flash) with multiple state-of-the-art approaches: GraphChi, TurboGraph, X-Stream, MMap, and GridGraph. For a fair comparison, we used the experimental setups recommended by the respective authors. GridGraph did not publish nor share its code, so the comparison is based on the results reported in its publication. We omit the comparison with GraphTwist because it is not accessible and its published results are based on a hardware that is less powerful than ours. We use four graphs at different scales (See Table 1), and we compare the runtimes of all approaches for two well-known essential algorithms PageRank (Subsect. 4.2) and Weakly Connected Components (Subsect. 4.3). To demonstrate how M-Flash generalizes to more algorithms, we implemented the Lanczos algorithm (with *selective orthogonalization*), which is one of the most computationally efficient approaches to computing eigenvalues and eigenvectors [8] (Subsect. 4.4). To the best of our knowledge, M-Flash provides the **first design and implementation** of Lanczos that can handle graphs with more than one billion nodes. Next, in Subsect. 4.5, we show that M-Flash maintains its high speed even when the machine has little RAM (including extreme cases, like 4 GB), in contrast to the other methods. Finally, through a theoretical analysis of I/O, we show the reasons for the performance increase using the BBP strategy (Subsect. 4.6).

4.1   Experimental Setup

All experiments ran on a standard personal computer equipped with a four-core Intel i7-4500U CPU (3 GHz), 16 GB RAM, and 1 TB 540-MB/s (max) SSD disk. Note that M-Flash does *not* require an SSD to run, which is not the case for

Table 1. Graph datasets used in our experiments.

Graph	Nodes	Edges	Size
LiveJournal [2]	4,847,571	68,993,773	Small
Twitter [10]	41,652,230	1,468,365,182	Medium
YahooWeb	1,413,511,391	6,636,600,779	Large
R-Mat (Synthetic graph)	4,000,000,000	12,000,000,000	Large

all frameworks, like TurboGraph. We used an SSD, nevertheless, to make sure that all methods can perform at their best. Table 1 shows the datasets used in our experiments. GraphChi, X-Stream, MMap, and M-Flash ran on Linux Ubuntu 14.04 (x64). TurboGraph ran on Windows (x64). All the reported times correspond to the average time of three **cold** runs, that is, with all caches and buffers purged between runs to avoid any potential advantage due to caching or buffering.

Table 2. Runtime (in seconds) with 8 GB of RAM. The symbol "-" indicates that the corresponding system failed to process the graph or the information is not available in the respective papers.

	GraphChi	X-Stream	TurboGraph	MMap	GridGraph	M-Flash
PageRank						
LiveJournal (10 iter.)	33.1	10.5	7.9	18.2	6.4	**5.3**
Twitter (10 iter.)	1,199	962	241	186	269	**173**
YahooWeb (1 iter.)	642	668	628	1,245	235.95	**195**
R-Mat (1 iter.)	2,145	1,360	-	-	-	**745**
Connected Components						
LiveJournal (Union Find)	3.2	5.7	4.4	10.7	4.4	**1.3**
Twitter (Union Find)	70	1,038	128	45	287	**25**
YahooWeb (WCC - 1 iter.)	668	889	-	-	-	**125**
R-Mat (WCC - 1 iter.)	3,334	2,167.63	-	-	-	**663.17**

4.2   PageRank

Table 2 presents the PageRank runtime of all the methods, as discussed next.

LiveJournal (small graph): Since the whole graph fits in RAM, all approaches finish in seconds. Still, M-Flash was the fastest, up to 6X faster than GraphChi, 3X than MMap, and 2X than X-Stream.

Twitter (medium graph): The edges of this graph do not fit in RAM (it requires 11.3 GB) but its node vectors do. M-Flash had a similar performance if compared to MMap, however, MMap is not a generic framework, rather it is based on dedicated implementations, one for each algorithm. Still, M-Flash was faster. In comparison to GraphChi and X-Stream, the related works that offer generic programming models, M-Flash was the fastest, 5.5X and 7X faster, respectively.

YahooWeb (large graph): For this billion-node graph, neither its edges nor its node vectors fit in RAM; this challenging situation is where M-Flash has notably outperformed the other methods. The results of Table 2 confirm this claim, showing that M-Flash provides a speed that is 3X to 6.3X faster that those of the other approaches.

R-Mat (Synthetic large graph): For our big graph, we compared only GraphChi, X-Stream, and M-Flash because TurboGraph and MMap require indexes or auxiliary files that exceed our current disk capacity. GridGraph was not considered in the comparison because its paper does not provide information about R-Mat graphs with a similar scale. Table 2 shows that M-Flash is 2X and 3X faster that X-Stream and GraphChi respectively.

4.3   Weakly Connected Components (WCC)

When there is enough memory to store all the vertex data, the *Union Find* algorithm [16] is the best option to find all the WCCs in one single iteration. Otherwise, with memory limitations, an iterative algorithm produces identical solutions. Hence, in this round of experiments, we use Algorithm *Union Find* to solve WCC for the small and medium graphs, whose vertices fit in memory; and we use an iterative algorithm for the YahooWeb graph.

Table 2 shows the runtimes for the LiveJournal and Twitter graphs with 8 GB RAM; all approaches use Union Find, except X-Stream. This is because of the way that X-Stream is implemented, which handles only iterative algorithms. In the WCC problem, M-Flash is again the fastest method with respect to the entire experiment: for the LiveJournal graph, M-Flash is 3x faster than GraphChi, 4.3X than X-Stream, 3.3X than TurboGraph, and 8.2X than MMap. For the Twitter graph, M-Flash's speed is 2.8X faster than GraphChi, 41X than X-Stream, 5X than TurboGraph, 2X than MMap, and 11.5X than GridGraph.

In the results of the YahooWeb graph, one can see that M-Flash was significantly faster than GraphChi, and X-Stream. Similarly to the PageRank results, M-Flash is pronouncedly faster: 5.3X faster than GraphChi, and 7.1X than X-Stream.

4.4   Spectral Analysis Using the Lanczos Algorithm

Eigenvalues and eigenvectors are at the heart of numerous algorithms, such as singular value decomposition (SVD) [3], spectral clustering, triangle counting [17], and tensor decomposition [9]. Hence, due to its importance, we demonstrate M-Flash over the *Lanczos algorithm*, a state-of-the-art method for eigen computation. We implemented it using method *Selective Orthogonalization (LSO)*. To the best of our knowledge, M-Flash provides the **first design and implementation** that can handle Lanczos for graphs with more than one billion nodes. Different from the competing works, M-Flash provides functions for basic vector operations using secondary memory. Therefore, for the YahooWeb graph, we are not able to compare it with the other frameworks using only 8 GB of memory.

To compute the top 20 eigenvectors and eigenvalues of the YahooWeb graph, one iteration of *LSO* over M-Flash takes 737 s when using 8 GB of RAM. For a comparative panorama, to the best of our knowledge, the closest comparable result of this computation comes from the HEigen system [6], at 150 s for one iteration; note however that, it was for a much smaller graph with 282 million edges (23X fewer edges), using a 70-machine Hadoop cluster, while our experiment with M-Flash used a single personal computer and a much larger graph.

4.5   Effect of Memory Size

Since the amount of memory strongly affects the computation speed of single-node graph processing frameworks, here, we study the effect of memory size. Figure 5 summarizes how all approaches perform under 4 GB, 8 GB, and 16 GB of RAM when running one iteration of PageRank over the YahooWeb graph. M-Flash continues to run at the highest speed even when the machine has very little RAM, 4 GB in this case. Other methods tend to slow down. In special, MMap does not perform well due to *thrashing*, a situation when the machine spends a lot of time on mapping disk-resident data to RAM or unmapping data from RAM, slowing down the overall computation. For 8 GB and 16 GB, respectively, M-Flash outperforms all the competitors for the most challenging graph, the YahooWeb. Notice that all the methods, but for M-Flash and X-Stream, are strongly influenced by restrictions in memory size; according to our analyses, this is due to the higher number of data transfers needed by the other methods when not all the data fit in the memory. Despite that X-Stream worked efficiently for any memory setting, it still has worse performance if compared to M-Flashbecause it demands three full disk scans in every case – actually, the innovations of M-Flash, as presented in Sect. 3, were designed to overcome such problem.

4.6   Theoretical (I/O) Analysis

Following, we show the theoretical scalability of M-Flash when we reduce the available memory at the same time that we demonstrate why the performance of M-Flash improves when we combine DBP and SPP into BBP, instead of using

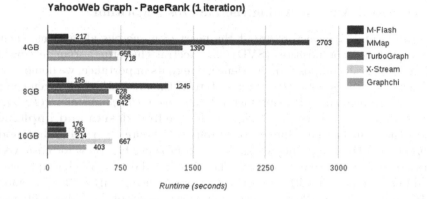

Fig. 5. Runtime comparison for PageRank over the YahooWeb graph. M-Flash is significantly faster than all the state-of-the-art competitors for three different memory settings, 4 GB, 8 GB, and 16 GB.

DBP or SPP alone. Here, we use a measure that we named *t-cost*; 1 unit of t-cost corresponds to three operations, one reading of the vertices, one writing of the vertices, and one reading of the edges. In terms of computational complexity, t-cost is defined as follows:

$$\text{t-cost}(G(E, V)) = 2\,|V| + |E| \tag{8}$$

Notice that this cost considers that reading and writing the vertices have the same cost; this is because the evaluation is given in terms of computational complexity. For more details, please refer to the work of McSherry *et al.* [14], who draws the basis of this kind of analysis.

We measure the t-cost metric to analyze the theoretical scalability for processing schemes *DBP* only, *SPP* only, and *BBP*. We perform these analyses using MatLab simulations that were validated empirically. We considered the characteristics of the three datasets used so far, LiveJournal, Twitter, and YahooWeb. For each case, we calculated the t-cost (y-axis) as a function of the available memory (x-axis), which, as we have seen, is the main constraint for graph processing frameworks.

Figure 6 shows that, for all the graphs, DBP-only processing is the least efficient when memory is reduced; however, when we combine DBP (for dense region processing) and SPP (for sparse region processing) into BBP, we benefit from the best of both worlds. The result corresponds to the best performance, as seen in the charts. Figure 7 shows the same simulated analysis – t-cost (y-axis) in function of the available memory (x-axis), but now with an extra variable: the density of hypothetical graphs, which is assumed to be uniform in each analysis. Each plot, from (a) to (d) considers a different density in terms of average vertex degree, respectively, 3, 5, 10, and 30. In each plot, there are two curves, one corresponding to DBP-only, and one for SPP-only; and, in dark blue, we depict the behavior of M-Flash according to combination BBP. Notice that, as

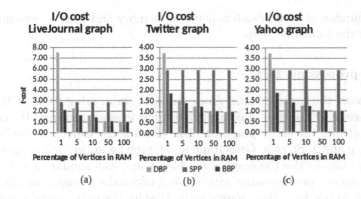

Fig. 6. I/O cost using *DBP*, *SPP*, and *BBP* for LiveJournal, Twitter and YahooWeb Graphs using different memory sizes. *BBP* model always performs fewer I/O operations on disk for all memory configurations.

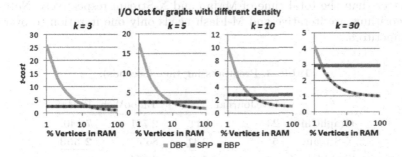

Fig. 7. I/O cost using DBP, SPP, and BBP for a graph with average degree (density) $k = \{3, 5, 10, 30\}$, where $|E| \approx k|V|$, and varying amount of memory (Color figure online)

the amount of memory increases, so does the performance of DBP, which takes less and less time to process the whole graph (decreasing curve). SPP, in turn, has a steady performance, as it is not affected by the amount of memory (light blue line). In dark blue, one can see the performance of BBP; that is, which kind of processing will be chosen by Eq. 7 at each circumstance. For sparse graphs, Figs. 7(a) and (b), SPP answers for the greater amount of processing; while the opposite is observed in denser graphs, Figs. 7(c) and (d), when DBP defines almost the entire dark blue line of the plot.

These results show that the graph processing must take into account the density of the graph at each moment (block) so to choose the best strategy. It also explains why M-Flash improves the state of the art. It is *important* to note that no former algorithm considered the fact that most graphs present varying density of edges (dense regions with many more edges than other regions that are sparse). Ignoring this fact leads to a decreased performance in the form of a higher number of data transfers between memory and disk, as we empirically verified in the former sections.

4.7   Preprocessing Time

Table 3 shows the preprocessing times for each graph using 8 GB of RAM. As one can see, M-Flash has a competitive preprocessing runtime. It reads and writes two times the entire graph on disk, which is the third best performance, after MMap and X-Stream. GridGraph and GraphTwist, in turn, demand a preprocessing that divides the graph using blocks in a way similar to M-Flash. We did not compare preprocessing with these frameworks because, as already discussed, we do not have their source code. Despite the extra preprocessing time required by M-Flash – if compared to MMap and X-Stream, the total processing time (preprocessing + *processing with only one iteration*) for algorithms PageRank and WCC over the YahooWeb graph, is of 1, 460s and 1, 390s, still, 29 % and 4 % better than the total time of MMap and X-Stream respectively. Note that the algorithms are iterative and M-Flash needs only one iteration to overcome its competitors.

Table 3. Preprocessing time (seconds).

	LiveJournal	Twitter	YahooWeb	R-Mat
GraphChi	23	511	2,781	7,440
X-Stream	**5**	**131**	865	**2,553**
TurboGraph	18	582	4,694	-
MMap	17	372	**636**	-
M-Flash	10	206	1,265	4,837

5   Conclusions

We proposed M-Flash, a *single-machine, billion-scale* graph computation framework that uses a block partition model to optimize the disk I/O. M-Flash uses an innovative design that takes into account the variable density of edges observed in the different blocks of a graph. Its design uses Dense Block Processing (DBP) when the block is dense, and Streaming Partition Processing (SPP) when the block is sparse. In order to take advantage of both worlds, it uses the combination of DBP and SPP according to the Bimodal Block Processing (BBP) scheme, which is able to analytically determine whether a block is dense or sparse, so to trigger the appropriate processing. To date, our proposal is the first framework that considers a bimodal approach for I/O minimization, a fact that, as we demonstrated, granted M-Flash the best performance compared to the state of the art (GraphChi, X-Stream, TurboGraph, MMap, and GridGraph); notably, even when memory is severely limited.

The findings observed in the design of M-Flash are a step further in determining an ultimate graph processing paradigm. We expect the research in this field to consider the criterion of block density as a mandatory feature in any such framework, consistently advancing the research on high-performance processing.

Acknowledgments. This work received support from Brazilian agencies CNPq (grant 444985/2014-0), Fapesp (grants 2016/02557-0, 2014/21483-2), and Capes; from USA agencies NSF (grants IIS-1563816, TWC-1526254, IIS-1217559), and GRFP (grant DGE-1148903); and Korean (MSIP) agency IITP (grant R0190-15-2012).

References

1. Aggarwal, A., Vitter, J.: The input/output complexity of sorting and related problems. Commun. ACM **31**, 1116–1127 (1988)
2. Backstrom, L., Huttenlocher, D., Kleinberg, J., Lan, X.: Group formation in large social networks: membership, growth, and evolution. In: KDD, pp. 44–54 (2006)
3. Berry, M.: Large-scale sparse singular value computations. Int. J. High Perform. Comput. Appl. **6**(1), 13–49 (1992)
4. Cheng, J., Liu, Q., Li, Z., Fan, W., Lui, J., He, C.: Venus: vertex-centric streamlined graph computation on a single PC. In: IEEE International Conference on Data Engineering, pp. 1131–1142 (2015)
5. Han, W.S., Lee, S., Park, K., Lee, J.H., Kim, M.S., Kim, J., Yu, H.: Turbograph: a fast parallel graph engine handling billion-scale graphs in a single PC. In: KDD, pp. 77–85 (2013)
6. Kang, U., Meeder, B., Papalexakis, E., Faloutsos, C.: Heigen: spectral analysis for billion-scale graphs. IEEE TKDE **26**(2), 350–362 (2014)
7. Kang, U., Tong, H., Sun, J., Lin, C.Y., Faloutsos, C.: Gbase: an efficient analysis platform for large graphs. VLDB J. **21**(5), 637–650 (2012)
8. Kang, U., Tsourakakis, C., Faloutsos, C.: Pegasus: a peta-scale graph mining system implementation and observations. In: ICDM, pp. 229–238. IEEE (2009)
9. Kolda, T., Bader, B.: Tensor decompositions and applications. SIAM Rev. **51**(3), 455–500 (2009)
10. Kwak, H., Lee, C., Park, H., Moon, S.: What is twitter, a social network or a news media? In: WWW, pp. 591–600. ACM (2010)
11. Kyrola, A., Blelloch, G., Guestrin, C.: Graphchi: large-scale graph computation on just a PC. In: OSDI, pp. 31–46. USENIX Association (2012)
12. Lin, Z., Kahng, M., Sabrin, K., Chau, D.H., Lee, H., Kang, U.: Mmap: fast billion-scale graph computation on a PC via memory mapping. In: BigData (2014)
13. Lu, Y., Cheng, J., Yan, D., Wu, H.: Large-scale distributed graph computing systems: an experimental evaluation. VLDB **8**(3), 281–292 (2014)
14. McSherry, F., Isard, M., Murray, D.G.: Scalability! but at what cost. In: HotOS (2015)
15. Roy, A., Mihailovic, I., Zwaenepoel, W.: X-stream: edge-centric graph processing using streaming partitions. In: SOSP, pp. 472–488. ACM (2013)
16. Tarjan, R.E., van Leeuwen, J.: Worst-case analysis of set union algorithms. J. ACM **31**(2), 245–281 (1984)
17. Tsourakakis, C.: Fast counting of triangles in large real networks without counting: algorithms and laws. In: ICDM, pp. 608–617. IEEE (2008)

18. Zhou, Y., Liu, L., Lee, K., Zhang, Q.: Graphtwist: fast iterative graph computation with two-tier optimizations. Proc. VLDB Endowment **8**(11), 1262–1273 (2015)
19. Zhu, X., Han, W., Chen, W.: Gridgraph: large-scale graph processing on a single machine using 2-level hierarchical partitioning. In: USENIX ATC 2015, pp. 375–386. USENIX Association (2015)

Improving Locality Sensitive Hashing Based Similarity Search and Estimation for Kernels

Aniket Chakrabarti[✉], Bortik Bandyopadhyay, and Srinivasan Parthasarathy

Department of Computer Science and Engineering,
The Ohio State University, Columbus, Ohio, USA
{chakrabarti.14,bandyopadhyay.14}@osu.edu, srini@cse.ohio-state.edu

Abstract. We present a novel data embedding that significantly reduces the estimation error of locality sensitive hashing (LSH) technique when used in reproducing kernel Hilbert space (RKHS). Efficient and accurate kernel approximation techniques either involve the kernel principal component analysis (KPCA) approach or the Nyström approximation method. In this work we show that extant LSH methods in this space suffer from a bias problem, that moreover is difficult to estimate apriori. Consequently, the LSH estimate of a kernel is different from that of the KPCA/Nyström approximation. We provide theoretical rationale for this bias, which is also confirmed empirically. We propose an LSH algorithm that can reduce this bias and consequently our approach can match the KPCA or the Nyström methods' estimation accuracy while retaining the traditional benefits of LSH. We evaluate our algorithm on a wide range of realworld image datasets (for which kernels are known to perform well) and show the efficacy of our algorithm using a variety of principled evaluations including mean estimation error, KL divergence and the Kolmogorov-Smirnov test.

Keywords: Locality sensitive hashing · Kernel similarity measure · Similarity estimation · Nyström method

1 Introduction

In recent past, Locality Sensitive Hashing (LSH) [1] has gained widespread importance in the area of large scale machine learning. Given a high dimensional dataset and a distance/similarity metric, LSH can create a small sketch (low dimensional embedding) of the data points such that the distance/similarity is preserved. LSH is known to provide approximate and efficient solution for estimating the pairwise similarity among data points, which is critical in solving applications for many domains ranging from image retrieval to text analytics and from protein sequence clustering to pharmacogenomics. Recently kernel-based similarity measures [22] have found increased use in such scenarios in part because the data becomes easily separable in the kernel induced feature space. The challenges of working with kernels are two fold – (1) explicit embedding of data points in the kernel induced feature space (RKHS) may be unknown

© Springer International Publishing AG 2016
P. Frasconi et al. (Eds.): ECML PKDD 2016, Part II, LNAI 9852, pp. 641–656, 2016.
DOI: 10.1007/978-3-319-46227-1_40

or infinite dimensional and (2) generally the kernel function is computationally expensive. The first problem prohibits building of a smart index structure such as kdtrees [3] that can allow efficient querying, while the second problem makes constructing the full kernel matrix infeasible.

LSH has been used in the context of kernels to address both of the aforementioned problems. Existing LSH methods for kernels [12,13] leverage the KPCA or Nyström techniques to estimate the kernel. The two methods differ only in the form of covariance operator that they use in the eigenvector computation step to approximately embed the data in RKHS. While KPCA uses the centered covariance operator, Nyström method uses the uncentered one (second moment operator). Without loss of generality, for the rest of the paper, we will use the Nyström method and hence by covariance operator we will mean the uncentered one. The LSH estimates for kernel differ significantly from the Nyström approximation. This is due to the fact that the projection onto the subspace (spanned by the eigenvectors of covariance operator) results in reduction of norms of the data points. This reduction depends on the eigenvalue decay rate of the covariance operator. Therefore, this norm reduction is difficult to estimate apriori. Assume that the original kernel was normalized with norm of the data points (self inner product) equaling 1. As a consequence of this norm reduction, in the resulting subspace the Nyström approximated kernel is not normalized (self inner product less than 1). Now, it is shown in [6] that LSH can only estimate normalized kernels. Thus in the current setting, instead of the Nyström approximated kernel, it estimates the *re-normalized version of it*. The bias arising out of this re-normalization depends on the eigenvalue decay rate of the covariance operator, and is unknown to the user apriori. This is particularly problematic, since for the LSH applications (index building and estimation) in the context of similarity (not distance), accurate estimation is paramount. For instance, the *All Pairs Similarity Search* (APSS) [2,4,5] problem finds all pairs of data points whose similarity is above a user defined threshold. Therefore, APSS quality will degrade in case of high estimation error. In APSS using LSH [5], it is clearly noticeable that the quality for non-kernel similarity measures is better than their kernel counterparts.

We propose a novel embedding of data points that is amenable to LSH sketch generation, while still estimating the Nyström approximated kernel matrix instead of the re-normalized version (which is the shortcoming of existing work). Specifically the contributions of this paper are as follows:

1. We show that Nyström embedding based LSH generates the LSH embedding for a slightly different kernel rather than the Nyström approximated one. This bias becomes particularly important during the LSH index construction where similarity threshold (or distance radius) is a mandatory parameter. Since this radius parameter is given in terms of the original similarity (kernel) measure, if the LSH embedding results in a bias (estimating a slightly different kernel), then the resulting index generated will be incorrect.

2. We propose an LSH scheme to estimate the Nyström approximation of the original input kernel and develop an algorithm for efficiently generating the LSH embedding.

3. Finally we empirically evaluate our methods against state-of-the-art KLSH [12,13] and show that our method is substantially better in estimating the original kernel values. We additionally run statistical tests to prove that the statistical distribution of pairwise similarity in the dataset is better preserved by our method. Preserving the similarity distribution correctly is particularly important in applications such as clustering.

Our results indicate upto 9.7x improvement in the kernel estimation error and the KL divergence and Kolmogorov-Smirnov tests [15] show that the estimates from our method fit the pairwise similarity distribution of the ground truth substantially better than the state-of-the-art KLSH method.

2   Background and Related Works

2.1   LSH for Cosine Similarity

A family of hash functions F is said to be locality sensitive with respect to some similarity measure, if it satisfies the following property [6]:

$$P_{h \in F}(h(x) = h(y)) = sim(x, y) \tag{1}$$

Here x, y is a pair of data points, h is a hash function and sim is a similarity measure of interest. LSH for similarity measures can be used in two ways:

1. **Similarity Estimation:** If we have k i.i.d. hash functions $\{h_i\}_{i=1}^{k}$, then a maximum likelihood estimator (MLE) for the similarity is:

$$\widehat{sim(x, y)} = \frac{1}{k} \sum_{i=1}^{k} I(h_i(x) = h_i(y)) \tag{2}$$

2. **LSH Index Search:** The concatenation of the aforementioned k hash functions form a signature and suppose l such signatures are generated for each data point. Then for a query data point q, to find the nearest neighbor, only those points that have at least one signature in common with q need to be searched. This leads to an index construction algorithm that results in a sublinear time search. It is worth noting that a similarity threshold is a mandatory parameter for an LSH index construction. Consequently, a bias in its estimation may lead to a different index than the one intended based on input similarity measure (Table 1).

Charikar [6] introduced a hash family based on the rounding hyperplane algorithm that can very closely approximate the cosine similarity. Let $h_i(x) = sign(r_i x^T)$, where $r_i, x \in R^d$ and each element of r_i is drawn from i.i.d. $N(0, 1)$.

Table 1. Key symbols

n, d	Number of data points, Dimensionality of data
p, c	Parameters: Number of eigenvectors to use, Number of extra dimensions to use
$\kappa(x, y)$	Kernel function over data points x, y ($= < \Phi(x), \Phi(y) >$)
$\Phi(x)$	Kernel induced feature map for data point x
X_i	i^{th} data point (i^{th} row of matrix $X_{n \times d}$)
$K_{i,j}$	$(i, j)^{th}$ value of true kernel matrix $K_{n \times n}$ ($= \kappa(X_i, X_j)$)
Y_i	p dimensional Nyström embedding of $\Phi(X_i)$ (i^{th} row of $Y_{n \times p}$ matrix)
$\widehat{K_{i,j}}$	Approximation of $K_{i,j}$ due to Nystroöm embedding ($= < Y_i, Y_j >$)
Z_i	Our $(p + n)$ dimensional Augmented Nyström embedding of $\Phi(X_i)$
$\widehat{K_{(Z)i,j}}$	Approximation of $K_{i,j}$ due to Augmented Nyström embedding ($= < Z_i, Z_j >$)
Z_i'	Our $(p + c)$ dimensional Remapped Augmented Nyström embedding of $\Phi(X_i)$

Essentially the hash functions are signed random projections (SRP). It can be shown that in this case,

$$P(h_i(x) = h_i(y)) = 1 - \frac{\theta(x, y)}{\pi} = sim(x, y)$$
$$\implies cos(\theta(x, y)) = cos(\pi(1 - sim(x, y)))$$

where $\theta(x, y)$ is the angle between x, y. The goal of this work is to find a locality sensitive hash family for the Nyström approximation $\hat{\kappa}$ of any arbitrary kernel κ that will satisfy the following property:

$$P(h_i(x) = h_i(y)) = 1 - \frac{cos^{-1}(\widehat{\kappa(x, y)})}{\pi} \tag{3}$$

2.2 Existence of LSH for Arbitrary Kernels

Kernel similarity measures are essentially the inner product in some transformed feature space. The transformation of the original data into the kernel induced feature space is usually non-linear and often explicit embedding in the kernel induced space are unknown, only the kernel function can be computed. Shrivastava *et al.* [23] recently proved the non-existence of LSH functions for general inner product measures. In spite of the non-existence of LSH for kernels in the general case, LSH can still exist for a special case, where the kernel is normalized – in other words the inner product is equal to the cosine similarity measure.

As mentioned in previous section, Charikar [6] showed that using signed random projections, cosine similarity can be well approximated using LSH. To summarize, LSH in kernel context is meaningful in the following two cases:

1. The case where the kernel is normalized with each data object in the kernel induced feature space having unit norm.

$$\|\Phi(x)\|^2 = \kappa(x,x) = 1 \tag{4}$$

Here $\kappa(.,.)$ is the kernel function and $\Phi(.)$ is the (possibly unknown) kernel induced feature map in RKHS.

2. In the case Eq. 4 does not hold, LSH does not exist for $\kappa(.,.)$. But it exists for a normalized version of κ, say $\kappa_N(.,.)$, where:

$$\kappa_N(x,y) = \frac{\kappa(x,y)}{\sqrt{\kappa(x,x)}\sqrt{\kappa(y,y)}} \tag{5}$$

2.3 Kernelized Locality Sensitive Hashing

KLSH [13] is an early attempt to build an LSH index for any arbitrary kernel similarity measure. Later work by Xia et al. [26] tries to provide bounds on kernel estimation error using Nyström approximation [25]. This work also provides an evaluation of applying LSH directly on explicit embedding generated by KPCA [21]. A follow up [12] to KLSH provided further theoretical insights into KLSH retrieval performance and proved equivalence of KLSH and KPCA+LSH.

KLSH computes the dot product of a data point and a random Gaussian in the approximate RKHS spanned by the first p principal components of the empirical centered covariance operator. It uses an approach similar to KPCA to find out a data point's projection onto the eigenvectors in the kernel induced feature space and it approximates the random Gaussian in the same space by virtue of the central limit theorem (CLT) of Hilbert spaces by using a sample of columns of the input kernel matrix. Let $X_{n \times d}$ denote the dataset of n points, each having d dimensions. We denote the ith row/data point by X_i and i,j^{th} element of X by $X_{i,j}$. Let $K_{n \times n}$ be the full kernel matrix ($K_{i,j} = \kappa(X_i, X_j)$). KLSH takes as input p randomly selected columns from kernel matrix - $K_{n \times p}$. The algorithm to compute the hash bits is as follows:

1. *Extract $K_{p \times p}$ from input $K_{n \times p}$. $K_{p \times p}$ is a submatrix of $K_{n \times n}$ created by sampling the same p rows and columns.*
2. *Center the matrix $K_{p \times p}$.*
3. *Compute a hash function h by forming a binary vector e by selecting t indices at random from $1,...,p$, then form $w = K_{p \times p}^{-1/2} e$ and assign bits according to the hash function*

$$h(\Phi(X_a)) = sign(\sum_i w(i)\kappa(X_i, X_a))$$

One thing worth noting here is, unlike vanilla LSH, where an LSH estimator tries to estimate the similarity measure of interest directly, in case of KLSH, the estimator tries to estimate the kernel similarity that is approximated by the KPCA embedding. The idea is that the KPCA embedding should lead to good approximations of the original kernel and hence KLSH should be able to approximate the original kernel as well. Alternatively, instead of directly computing the dot product in RKHS, one may first explicitly compute the KPCA/Nyström $p-$dimensional embedding of the input data and generate a $p-$dimensional multivariate Gaussian, and then compute the dot product. The two methods are equivalent [12]. Next, we discuss why approximation error due to applying LSH on kernels may be significant.

3   Estimation Error of LSH for Kernels

According to Mercer's theorem [16], the kernel induced feature map $\Phi(x)$ can be written as $\Phi(x) = [\phi_i(x)]_{i=1}^{\infty}$ where $\phi_i(x) = \sqrt{\sigma_i}\psi_i(x)$ and σ_i and ψ_i are the eigenvalues and eigenfunctions of the covariance operator whose kernel is κ. The aforementioned infinite dimensional kernel induced feature map can be approximated explicitly in finite dimensions by using Nyström style projection [25] as described next. This can be written as $\widehat{\Phi(x)} = [\widehat{\phi_i(x)}]_{i=1}^{p}$ where $\widehat{\phi_i(x)} = \frac{1}{\sqrt{\lambda_i}} < K(x,.), u_i >$. Here $K(x,.)$ is a vector containing the kernel values of data point x to the p chosen points, λ_i and u_i are the ith eigenvalue and eigenvector of the sampled $p \times p$ kernel matrix $K_{p \times p}$. Note that, both the KPCA and Nyström projections are equivalent other than the fact that in case of KPCA, $K_{p \times p}$ is centered, whereas in case of Nyström, it is uncentered. Essentially, $\widehat{\Phi(x)} = P_{\hat{S}}\Phi(x)$, where $P_{\hat{S}}$ is the projection operator that projects $\Phi(x)$ onto the subspace spanned by first p eigenvectors of the empirical covariance operator. Let $Y_{n \times p}$ represent this explicit embedding of the data points.

In the next lemma, we show that the above approach results in a bias for kernel similarity approximation from LSH.

Lemma 1. *If $\widehat{K_{(LSH)i,j}}$ is the quantity estimated by using LSH on Nyström embedding, then $\widehat{K_{(LSH)i,j}} \geq \widehat{K_{i,j}}$.*

Proof. Since $\widehat{K_{(LSH)}}$ is the quantity estimated by the LSH estimator for cosine similarity on embedding $Y_{n \times p}$, then by Eq. 5

$$\widehat{K_{(LSH)i,j}} = \frac{Y_i Y_j^T}{||Y_i|| ||Y_j||} = \frac{\widehat{K_{i,j}}}{\sqrt{\widehat{K_{i,i}}}\sqrt{\widehat{K_{j,j}}}} \tag{6}$$

where Y_i is the ith row of Y.

By assumption, $||\Phi(X_i)|| = 1$, $\forall i$. Hence

$$\widehat{K_{i,i}} = < P_{\hat{S}}\Phi(X_i), P_{\hat{S}}\Phi(X_i) > = ||P_{\hat{S}}\Phi(X_i)||^2 \leq 1, \ \forall i$$

(since $P_{\hat{S}}$ is a projection operator onto a subspace). Specifically if $i \in p$, then $\widehat{K_{i,i}} = K_{i,i}$. Putting $\widehat{K_{i,i}} \leq 1$ in Eq. 6, we get the following.

$$K_{\widehat{(LSH)i,j}} \geq \widehat{K_{i,j}}$$

Thus, applying LSH to the Nyström embedding results in an overestimation of the kernel similarity when compared to the Nyström approximation to the kernel similarity. In terms of our goal, Eq. 3 will have $\widehat{K_{(LSH)}}$ instead of $\hat{K}$ (Nyström approximated kernel). Unlike $\hat{K}$, $\widehat{K_{(LSH)}}$ does not approximate K (true kernel) well, unless p is extremely large. This is not feasible since eigendecomposition is $O(p^3)$. Interestingly, the above bias $\|\Phi(x) - P_{\hat{S}}\Phi(x)\|$ depends on the eigenvalue decay rate [28], that in turn depends on the data distribution and the kernel function. Hence this error in estimation is hard to predict beforehand.

Additionally, another cause of estimation error, specifically for KLSH is due to the fact that KLSH relies on the CLT in Hilbert space to generate the random Gaussians in the kernel induced feature space. Unlike the single dimensional CLT, Hilbert space's CLT's convergence rate could be much worse [20], implying that the sample size requirement may be quite high. However, the number of available samples is limited by p (number of sampled columns). Typically p is set very small for performance consideration (in fact we found that $p = 128$ performs extremely well for dataset size upto one million).

We next propose a transformation over the Nyström embedding on which the SRP technique can be effectively used to create LSH that approximates the input kernel $\kappa(.,.)$ (K) well. Our methods apply to centered KPCA case as well.

4   Augmented Nyström LSH Method (ANyLSH)

In this section we propose a data embedding that along with the SRP technique forms an LSH family for the RKHS. Given n data points and p columns of the kernel matrix, we first propose a $p+n$ dimensional embedding for which the bias is 0 (LSH estimator is an unbiased one for the Nyström approximated kernel). Since $p + n$ dimensional embedding is infeasible in practice due to large n, we propose a $p + c$ dimensional embedding, where c is a constant much smaller than n. In this case the estimator is biased, but that bias can be bounded by setting c and this bound hence is independent of the eigenvalue decay rate of the covariance operator. We provide theoretical analysis regarding the preservation of the LSH property and we also give the runtime and memory cost analysis.

4.1   Locality Sensitive Hash Family

We identify that the major problem with using Nyström embedding for LSH is the underestimation bias of the norms ($\widehat{K_{i,i}}$) of these embedding. Hence, though the estimates of the numerator of Eq. 6 are very good, the denominator causes estimation bias. We propose a new embedding of the data points such that the numerator will remain the same, but the norms of the embedding will become 1.

Definition 1. *We define the augmented Nyström embedding as the feature map $Z_{n \times (p+n)}$ such that $Z_{n \times (p+n)} = [Y_{n \times p} \ V_{n \times n}]$, where $V_{n \times n}$ is an $n \times n$ diagonal matrix with the diagonal elements as $\{\sqrt{1 - \sum_{j=1}^{p} Y_{i,j}^2}\}_{i=1}^{n}$.*

Lemma 2. *For $Z_{n \times (p+n)}$, if $\widehat{K_{(Z)}}_{n \times n}$ is the inner product matrix, then for (i) $i = j$, $\widehat{K_{(Z)i,j}} = 1$ and (ii) for $i \neq j$, $\widehat{K_{(Z)i,j}} = \widehat{K_{i,j}}$*

Proof. Case (i):

$$
\begin{aligned}
\widehat{K_{(Z)i,j}} &= Z_i Z_j^T \\
&= \sum_{k=1}^{p} Y_{i,k}^2 + \sum_{l=1}^{n} V_{i,l}^2 \\
&= \sum_{k=1}^{p} Y_{i,k}^2 + \left(\sqrt{1 - \sum_{j=1}^{p} Y_{i,j}^2} \right)^2 \\
&= 1
\end{aligned}
$$

Case (ii):

$$
\begin{aligned}
\widehat{K_{(Z)i,j}} &= Z_i Z_j^T \\
&= \sum_{k=1}^{p} Y_{i,k} Y_{j,k} + \sum_{l=1}^{n} V_{i,l} V_{j,l} \\
&= \sum_{k=1}^{p} Y_{i,k} Y_{j,k} + 0 \ (V \text{ is a diagonal matrix}) \\
&= Y_i Y_j^T \\
&= \widehat{K_{i,j}}
\end{aligned}
$$

Hence Z_i gives us a $p + n$ dimensional embedding of the data point X_i where Z_i approximates $\Phi(X_i)$. The inner product between two data points using this embedding gives the cosine similarity as the embedding are unit norm and the inner products are exactly same as that of Nyström approximation. Hence we can use SRP hash family on $Z_{n \times (p+n)}$ to compute the LSH embedding related to cosine similarity. Essentially we have:

$$
P(h(Z_i) = h(Z_j)) = 1 - \frac{\cos^{-1}(\widehat{K_{i,j}})}{\pi} \tag{7}
$$

Hence we are able to achieve the LSH property of the goal Eq. 3.

4.2   Quality Implications

The quality of an LSH estimator depends on (i) similarity and (ii) number of hash functions. It is independent of the original data dimensionality. From Eq. 1, it is easy to see that each hash match is a i.i.d. Bernoulli trial with success probability $sim(x, y)$ (s). For k such hashes, the number of matches follow a binomial distribution. Hence the LSH estimator $\hat{s}$ of Eq. 2 is an MLE for the binomial proportion parameter. The variance of this estimator is known to be $\frac{s(1-s)}{k}$. Therefore, even with the increased dimensionality of $p+n$, the estimator variance remains the same.

4.3   Performance Implications

The dot product required for a single signed random projection for Z_i can be computed as follows:

$$
\begin{aligned}
Z_i r_j^T &= \sum_{l=1}^{p+n} Z_{i,l} R_{j,l} \\
&= \sum_{l=1}^{p} Y_{i,l} R_{j,l} + \sum_{k=1}^{n} V_{i,k} R_{j,p+k} \\
&= \sum_{l=1}^{p} Y_{i,l} R_{j,l} + V_{i,i} R_{j,p+i}
\end{aligned}
$$

Hence there are (p+1) sum operations ($O(p)$). Though $Z_i \in R^{p+n}$, the dot product for SRP ($Z_i r_j^T$) can be computed in $O(p)$ (which is the case for vanilla LSH). Since $V_{n \times n}$ is a diagonal matrix, the embedding storage requirement is increased only by n (still $O(np)$). However, the number of $N(0,1)$ Gaussian samples required is $O(k(p+n))$, where as in case of vanilla LSH it was only $O(kp)$ (k is the number of hash functions). In the next section, we develop an algorithm with probabilistic guaranty that can substantially reduce the number of hashes required for the augmented Nyström embedding.

4.4   Two Layered Hashing Scheme

Next we define a $p + c$ dimensional embedding of a point X_i to approximate $\Phi(X_i)$. The first p dimensions contain projections onto p eigenvectors (same as first p dimensions of Z_i). In the second step, the norm residual (to make the norm of this embedding 1.0) will be randomly projected to 1 of c remaining dimensions, other remaining dimensions will be set zero.

Definition 2. *Remapped augmented Nyström embedding is an embedding $Z'_{n \times (p+c)}$ ($\forall i, Z'_i \in R^{p+c}$) obtained from $Z_{n \times (p+n)}$ ($\forall i, Z_i \in R^{p+n}$) such that, (i) $\forall j \leq p$, $Z'_{i,j} = Z_{i,j}$ and (ii) $Z'_{i,p+a_i} = Z_{i,p+i}$, where $a_i \sim unif\{1, c\}$.*

Definition 3. *$C(i, j)$ is a random event of collision that is said to occur when for two vectors Z'_i, $Z'_j \in Z$, $a_i = a_j$.*

Since this embedding is in R^{p+c} rather than R^{p+n}, the number of $N(0,1)$ samples required will be $O(k(p+c))$, rather than $O(k(p+n))$. Next we show that using SRP on $Z'_{n \times (p+c)}$ yields LSH embedding, where the estimator converges to $\widehat{K_{n \times n}}$ with $c \to n$.

Lemma 3. *For $Z'_{n \times (p+c)}$, the LSH property that will be satisfied is*

$$P(h(Z'_i) = h(Z'_j)) =$$

$$\frac{1}{c}\left[1 - \frac{cos^{-1}(\widehat{K_{i,j}} + \sqrt{1 - \sum_{l=1}^{p} Y_{i,l}^2}\sqrt{1 - \sum_{l=1}^{p} Y_{j,l}^2})}{\pi}\right] + \frac{c-1}{c}\left[1 - \frac{cos^{-1}(\widehat{K_{i,j}})}{\pi}\right]$$

Proof. For the remap we used, collision probability is given by,

$$P(C(i,j)) = \frac{1}{c} \tag{8}$$

If there is a collision, then the norm correcting components will increase the dot product value.

$$P(h(Z'_i) = h(Z'_j)|C(i,j)) =$$
$$1 - \frac{cos^{-1}(\widehat{K_{i,j}} + \sqrt{1 - \sum_{l=1}^{p} Y_{i,l}^2}\sqrt{1 - \sum_{l=1}^{p} Y_{j,l}^2})}{\pi} \tag{9}$$

If there is no collision, LSH will be able to approximate the Nyström method.

$$P(h(Z'_i) = h(Z'_j)|\neg\, C(i,j)) = 1 - \frac{cos^{-1}(\widehat{K_{i,j}})}{\pi} \tag{10}$$

We can compute the marginal distribution as follows:

$$P(h(Z'_i) = h(Z'_j)) = P(h(Z'_i) = h(Z'_j)|C(i,j))P(C(i,j))$$
$$+ P(h(Z'_i) = h(Z'_j)|\neg\, C(i,j))P(\neg C(i,j))$$

Applying Eqs. 8, 9 and 10 above, we get the result.

There are two aspects to note about the aforementioned lemma:

1. According to Nyström approximation [25], as we increase p (higher rank approx.), the quantity $\sqrt{1 - \sum_{l=1}^{p} Y_{i,l}^2}$ tends to 0 and the lemma leads to the desired goal of Eq. 3, but at a computational cost of $O(p^3)$ for the eigen-decomposition operation. Of course increasing p improves the overall quality of Nyström approximation itself, however in practice small values of p suffice.
2. Interestingly, instead of p, if we increase c, then also we converge to the goal of Eq. 3 as the first term of the lemma converges to 0. The computational cost is $O(k(p+c))$ which usually is much less than $O(p^3)$. This is the strategy we adopt and as we will show shortly, small values of c are sufficient even for large scale datasets. Hence c can be used to bound the bias (difference from the probability of Eq. 3).

5 Evaluation

5.1 Datasets and Kernels

We evaluate our methodologies on five real world image datasets varying from 3030 data points to 1 million and three popular kernels known to work well on them. Summary of the datasets can be found in Table 2.

Caltech101: This is a popular image categorization dataset [9]. We use 3030 image from this data. Following KLSH [12,13] we use this dataset with the CORR kernel [27].

PASCAL VOC: This is also an image categorization dataset [8]. We use 5011 images from this data. Following [7] we use the additive χ^2 kernel for this data.

Notre Dame image patches: This dataset contains 468159 small image patches of Notre Dame [10] and the image patch descriptors used are as per [24]. We use the Gaussian RBF kernel on this data.

INRIA holidays: To test at large scale, we use 1 million SIFT as well as 1 million GIST descriptors from the INRIA holidays dataset [11]. Following KLSH [12,13] we use the additive χ^2 kernel with this data.

Table 2. Dataset and kernel details

Dataset	Size	Kernel
Caltech101	3030	CORR
PASCAL VOC 2007	5011	Additive χ^2
Notre Dame image patches	468159	Gaussian RBF
INRIA holidays SIFT-1M	1000000	Additive χ^2
INRIA holidays GIST-1M	1000000	Additive χ^2

5.2 Evaluation Methodology

The focus of this work is accurate estimation of the input kernel similarity measure through LSH. For evaluating the quality of similarity estimation, we use two approaches - (i) we take a sample of pairs from each dataset, and compute the average estimation error directly and (ii) we use a sample of pairs from each dataset, compute the similarity of the pairs, both accurately (ground truth) and approximately (ANyLSH) and then compare the statistical distribution of the pairwise similarity of ground truth with ANyLSH. The former gives a direct measure of estimation accuracy, while the latter gives us insights on how well the pairwise similarity distribution is preserved. In terms of execution times, our algorithm performs the same as the baseline we compare against.

We use state-of-the-art KLSH as our baseline. We randomly sample 1000 pairs of data points from each dataset for our experiments. We use the values

64 and 128 for p, and vary h from 1024 to 4096 in steps of 1024. For ANyLSH, we set $c = 1000$. In our evaluation, we see that c generalizes well to varying data sizes. For KLSH, we set $q = 16, 32$ for $p = 64, 128$ respectively as per the guideline in the source code [13].

5.3   Results

Similarity Estimation Comparisons. Figures 1(a), (c), (e), (g) and (i) report the results on estimation error. We clearly see that our ANyLSH method outperforms KLSH in every single case by a large margin. The improvement of estimation error varies from a minimum of 2.4x (Fig. 1(e), $p = 128, h = 1024$) to a maximum of 9.7x (Fig. 1(e), $p = 64, h = 4096$), with average reduction in error of 5.9x across all datasets. With fixed p, the estimation error of our method decreases consistently across all datasets with the increase of hashes, as should be the case per Eq. 2. Interestingly, for KLSH, there are multiple cases when with the increase in hashes, the estimation error also increased. For instance, in Fig. 1(i), at $p = 64$, by increasing h from 2048 to 4096, KLSH's error increased from 0.076 to 0.078. This provides empirical evidence as well that not only the estimates are off, but in case of KLSH, they are converging towards a biased value as described in Lemma 1. Additionally note that our average absolute error varies between $0.011 - 0.038$ across all datasets and there is no trend that the error increases with larger datasets. This provides strong empirical evidence to the theoretical insight that at fixed c (1000 in our case), the average estimation error generalizes extremely well to different datasets of varied sizes and different kernels. Though the error is a function of the eigenvalue decay rate, it is upper bounded by ANyLSH.

Similarity Distribution Comparisons. As second part of our qualitative evaluation, in this section, we investigate how well the pairwise similarity distribution of the data is preserved. This is particularly important in applications that rely heavily on similarity distribution such as clustering. Our goal is to compare the two distributions in a non-parametric fashion as we do not have any prior knowledge of these distribution. Our first approach is to compare normalized histograms (probabilities). We choose the popular KL divergence measure to compare probability distributions represented by histograms. We discretized both our data and the ground truth by splitting the similarity range 0–1 into fixed length bins of length 0.1. Figures 1(b), (d), (f), (h) and (j) report the KL divergence numbers. The improvement in terms of KL divergence is even better, with upto two orders of magnitude improvement over KLSH. This improvement can be partly attributed to the discretization process – since we used length 0.1 bins and our estimation errors are significantly less than 0.1, most of our errors get absorbed in the discretization process. With KLSH's error being substantially higher than 0.1, it's KL divergence becomes very high.

To account for the binning issue, we additionally run the non-parametric Kolmogorov-Smirnov two sample test that is more suitable for comparing empirical distributions of continuous data. This test is particularly challenging in our

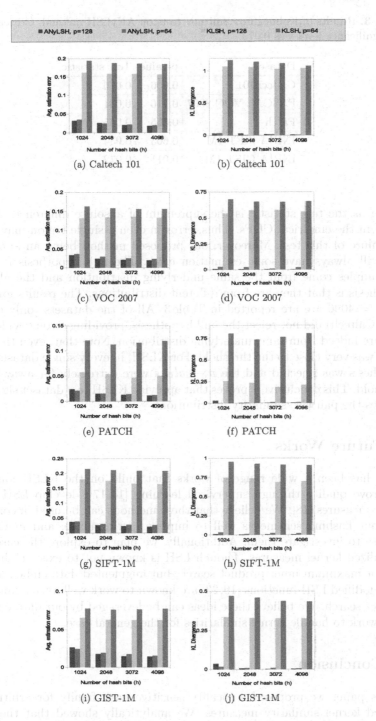

Fig. 1. Estimation error and KL divergence are reported in the first and second columns respectively for all datasets.

Table 3. Results of Kolmogorov Smirnov tests on ANyLSH method. Critical value at 1 % significance level was 0.073.

Dataset	p-value	Test statistic
Caltech101	0.006	0.076
PASCAL VOC 2007	0.716	0.031
Patch	0.565	0.035
INRIA SIFT-1M	0.603	0.034
INRIA GIST-1M	0.011	0.072

setting as the test statistic is the supremum of absolute differences across all values in the empirical CDFs. Thus, error in even a single region may result in the failure of this test. Moreover, our proposed method being an approximate one, will always have some estimation error. The null hypothesis is that the two samples come from the same underlying distribution and the alternative hypothesis is that they are from different distributions. The results for $p = 128$ and $h = 4096$ are are reported in Table 3. All of the datasets (only exception being Caltech) did not reject the null hypothesis, providing strong evidence that they are indeed from same underlying distribution. Note that, even the Caltech result was very close to the threshold. For KLSH, in every single dataset the null hypothesis was rejected and the $p - values$ were extremely far away from the threshold. This conclusively proves that applying KLSH to a dataset significantly changes the pairwise similarity distribution.

6   Future Works

There has been a wide range of works that build on the KLSH foundations - improve quality through supervised learning [14,17]; develop LSH for non-metric measures [18]; We believe that these methods can be used in conjunction with our hashing scheme as well to improve performance, and in future, we propose to investigate them. Additionally, we plan to explore the case of non-normalized kernel measures. Though LSH is known not to exist in the general case for maximum inner product search, but augmented data embedding along with modified LSH functions [19,23] are known to work well for maximum inner product search. We believe these ideas can be leveraged by our data embedding framework to handle kernel similarities for the general case.

7   Conclusion

In this paper we proposed a locality sensitive hash family for arbitrary normalized kernel similarity measures. We analytically showed that the existing methods of LSH for kernel similarity measures based on KPCA/Nyström projections suffer from an estimation bias, specific to the LSH estimation technique.

In other words, these LSH estimates differ from the KPCA/Nyström estimates of the kernel. This bias depends on the eigenvalue decay rate of the covariance operator and as such unknown apriori. Our method, ANyLSH, can directly estimate the KPCA/Nyström approximated input kernel efficiently and accurately in a principled manner. Key to our method are novel data embedding strategies. We showed that, given p columns of the input kernel matrix, the bias can be completely removed by using a $p + n$-dimensional embedding. Since n can be rather large and also not fixed, we additionally propose a $p + c$-dimensional embedding where c is fixed and much smaller than n. In our analysis we showed that in this case the worst case bias can be controlled by the user by setting c. Consequently, we overcame the short coming that resulted from the bias term being unknown to the user apriori. Our methods, when compared to the state-of-the-art KLSH improves the kernel similarity estimation error by upto 9.7x. Further evaluations based on the KL divergence and Kolmogorov-Smirnov tests provide strong evidence that pairwise similarity distribution is well preserved by ANyLSH.

Acknowledgments. We thank the anonymous reviewers for their feedback. This work is supported in part by NSF grants CCF-1217353 and DMS-1418265.

References

1. Andoni, A., Indyk, P.: Near-optimal hashing algorithms for approximate nearest neighbor in high dimensions. Commun. ACM **51**, 117–122 (2008)
2. Bayardo, R., Ma, Y., Srikant, R.: Scaling up all pairs similarity search. In: WWW (2007)
3. Bentley, J.L.: Multidimensional binary search trees used for associative searching. Commun. ACM **18**(9), 509–517 (1975)
4. Chakrabarti, A., Parthasarathy, S.: Sequential hypothesis tests for adaptive locality sensitive hashing. In: Proceedings of the 24th International Conference on World Wide Web, pp. 162–172. ACM (2015)
5. Chakrabarti, A., Satuluri, V., Srivathsan, A., Parthasarathy, S.: A bayesian perspective on locality sensitive hashing with extensions for kernel methods. ACM Trans. Knowl. Discov. Data (TKDD) **10**(2), 19 (2015)
6. Charikar, M.S.: Similarity estimation techniques from rounding algorithms. In: STOC 2002 (2002). http://doi.acm.org/10.1145/509907.509965
7. Chatfield, K., Lempitsky, V., Vedaldi, A., Zisserman, A.: The devil is in the details: an evaluation of recent feature encoding methods. In: British Machine Vision Conference (2011)
8. Everingham, M., Van Gool, L., Williams, C.K.I., Winn, J., Zisserman, A.: The PASCAL Visual Object Classes Challenge 2007 (VOC 2007) Results (2007). http://www.pascal-network.org/challenges/VOC/voc2007/workshop/index.html
9. Fei-Fei, L., Fergus, R., Perona, P.: Learning generative visual models from few training examples: an incremental bayesian approach tested on 101 object categories. In: Conference on Computer Vision and Pattern Recognition Workshop, CVPRW 2004, pp. 178–178. IEEE (2004)
10. Goesele, M., Snavely, N., Curless, B., Hoppe, H., Seitz, S.M.: Multi-view stereo for community photo collections. In: IEEE 11th International Conference on Computer Vision, ICCV 2007, pp. 1–8. IEEE (2007)

11. Jegou, H., Douze, M., Schmid, C.: Hamming embedding and weak geometric consistency for large scale image search. In: Forsyth, D., Torr, P., Zisserman, A. (eds.) ECCV 2008. LNCS, vol. 5305, pp. 304–317. Springer, Heidelberg (2008). doi:10. 1007/978-3-540-88682-2_24
12. Jiang, K., Que, Q., Kulis, B.: Revisiting kernelized locality-sensitive hashing for improved large-scale image retrieval. arXiv preprint arXiv:1411.4199 (2014)
13. Kulis, B., Grauman, K.: Kernelized locality-sensitive hashing. IEEE Trans. Pattern Anal. Mach. Intell. **34**(6), 1092–1104 (2012)
14. Liu, W., Wang, J., Ji, R., Jiang, Y.G., Chang, S.F.: Supervised hashing with kernels. In: 2012 IEEE Conference on Computer Vision and Pattern Recognition (CVPR), pp. 2074–2081. IEEE (2012)
15. Massey Jr., F.J.: The kolmogorov-smirnov test for goodness of fit. J. Am. Stat. Assoc. **46**(253), 68–78 (1951)
16. Mercer, J.: Functions of positive and negative type, and their connection with the theory of integral equations. Philos. Trans. R. Soc. Lond. Ser. A, Containing Papers of a Mathematical or Physical Character **209**, 415–446 (1909)
17. Mu, Y., Shen, J., Yan, S.: Weakly-supervised hashing in kernel space. In: 2010 IEEE Conference on Computer Vision and Pattern Recognition (CVPR), pp. 3344–3351. IEEE (2010)
18. Mu, Y., Yan, S.: Non-metric locality-sensitive hashing. In: AAAI (2010)
19. Neyshabur, B., Srebro, N.: On symmetric and asymmetric LSHs for inner product search. In: Proceedings of the 32nd International Conference on Machine Learning, pp. 1926–1934 (2015)
20. Paulauskas, V.: On the rate of convergence in the central limit theorem in certain Banach spaces. Theory Probab. Appl. **21**(4), 754–769 (1977)
21. Schölkopf, B., Smola, A., Müller, K.R.: Nonlinear component analysis as a kernel eigenvalue problem. Neural Comput. **10**(5), 1299–1319 (1998)
22. Shawe-Taylor, J., Cristianini, N.: Kernel Methods for Pattern Analysis. Cambridge University Press, Cambridge (2004)
23. Shrivastava, A., Li, P.: Asymmetric LSH (ALSH) for sublinear time maximum inner product search (MIPS). In: Advances in Neural Information Processing Systems, pp. 2321–2329 (2014)
24. Simonyan, K., Vedaldi, A., Zisserman, A.: Descriptor learning using convex optimisation. In: Fitzgibbon, A., Lazebnik, S., Perona, P., Sato, Y., Schmid, C. (eds.) ECCV 2012. LNCS, vol. 7578, pp. 243–256. Springer, Heidelberg (2012). doi:10. 1007/978-3-642-33718-5_18
25. Williams, C., Seeger, M.: Using the nyström method to speed up kernel machines. In: Proceedings of the 14th Annual Conference on Neural Information Processing Systems, pp. 682–688 (2001). No. EPFL-CONF-161322
26. Xia, H., Wu, P., Hoi, S.C., Jin, R.: Boosting multi-kernel locality-sensitive hashing for scalable image retrieval. In: Proceedings of the 35th International ACM SIGIR Conference on Research and Development in Information Retrieval, pp. 55–64. ACM (2012)
27. Zhang, H., Berg, A.C., Maire, M., Malik, J.: SVM-KNN: discriminative nearest neighbor classification for visual category recognition. In: 2006 IEEE Computer Society Conference on Computer Vision and Pattern Recognition, vol. 2, pp. 2126–2136. IEEE (2006)
28. Zwald, L., Blanchard, G.: On the convergence of eigenspaces in kernel principal component analysis. In: NIPS (2005)

Continuous Kernel Learning

John Moeller[(⊠)], Vivek Srikumar[(⊠)], Sarathkrishna Swaminathan[(⊠)],
Suresh Venkatasubramanian[(⊠)], and Dustin Webb[(⊠)]

School of Computing, University of Utah, Salt Lake City, UT 84112, USA
{moeller,svivek,sarath,suresh,dustin}@cs.utah.edu

Abstract. Kernel learning is the problem of determining the best kernel (either from a dictionary of fixed kernels, or from a smooth space of kernel representations) for a given task. In this paper, we describe a new approach to kernel learning that establishes connections between the Fourier-analytic representation of kernels arising out of Bochner's theorem and a specific kind of feed-forward network using cosine activations. We analyze the complexity of this space of hypotheses and demonstrate empirically that our approach provides scalable kernel learning superior in quality to prior approaches.

1 Introduction

Kernel methods have been a powerful tool in machine learning for decades and *kernel learning* is the problem of learning the "right" or "best" kernel for a given task. Broadly speaking, we can divide kernel learning methods into two categories. Multiple kernel learning (MKL) methods largely assume that the desired kernel can be represented as a combination of a dictionary of *fixed* kernels, and seeks to learn their mixing weights. The other approach is based on a Fourier-analytic representation of shift-invariant kernels via Bochner's theorem: roughly speaking, a kernel can be represented (in dual form) as a probability distribution, and so the search for a kernel becomes a search over distributions.

In both approaches, training the model is challenging with many thousands of training points and hundreds of dimensions. Standard training approaches either employ some form of convex or alternating optimization (for MKL) or parameterize the space of distributions in terms of known distributions and try to optimize their parameters.

In this paper, we describe *continuous kernel learning (CKL)*, a new way of tackling this problem by establishing and exploiting a connection to feed-forward networks. Working within the Fourier-analytic framework for kernel learning, we propose to search directly over the space of shift-invariant kernels instead of optimizing the parameters of a known family of distributions. In doing so, though we lose the ability to isolate parameters of a single learned kernel, we

This research was partially supported by the NSF under grants CCF-0953066, IIS-1251049 and CNS-1302688.

P. Frasconi et al. (Eds.): ECML PKDD 2016, Part II, LNAI 9852, pp. 657–673, 2016.
DOI: 10.1007/978-3-319-46227-1_41

gain representability in terms of a nonlinear basis of cosines that can be naturally interpreted as activations for a feed-forward network. This interpretation allows us to deploy the power of backpropagation on this network to learn the desired kernel representation. In addition, the generalization power of the cosine representation can be established formally using machinery from learning theory: this also helps guide the regularization that we use to learn the resulting kernel. We support these arguments with a suite of experiments on relatively large data sets (tens of thousands of points, hundreds of dimensions) that demonstrate that our learned kernels are more accurate than the state-of-the-art MKL methods.

In summary, our main contributions are:

- We develop the *continuous kernel learning* (CKL) framework, a kernel learning method that learns an implicit representation of a kernel. We show that we can interpret the learning task as a feed-forward network. This allows us to utilize recent advances in optimization technology from deep learning to train a classifier.
- We prove VC-dimension and generalization bounds for a single Fourier embedding, which yields natural regularization techniques for CKL.
- We show via experiments that CKL outperforms existing scalable MKL methods.

1.1 Technical Overview

The starting point for our work is the representation of any shift-invariant kernel[1] as an infinite linear combination of cosine basis elements via Bochner's theorem [9], as first demonstrated by Rahimi and Recht [41]. This representation is typically used to generate a *random* low-dimensional embedding of the associated Hilbert space.

If we move away from a random low-dimensional embedding and embrace the entire distribution that we sample from, we reach infinite-width embeddings. Dealing with infinite-width embeddings simply means that we consider the expectation of the embedding over the distribution. Neal [36] linked infinite-width networks to Gaussian processes when the distribution is Gaussian. Much later, Cho and Saul [11] applied the technique to infinite-width *rectified linear units* (ReLUs), and showed a correspondence to a kernel they called the *arc-cosine kernel*. Hazan and Jaakkola [21] extended this result further, and analyzed the kernel corresponding to two infinite layers stacked in series. In all of this, a *specific distribution* is chosen in order to obtain a kernel.

In our work, we return to the infinite representation provided by Bochner's theorem. Rather than picking a specific distribution over weights, we *learn* a distribution based on our training data. This effectively means we learn a *representation* of a kernel. While we cannot learn an infinite-width embedding directly, since the space of functions is itself infinite, we are able to construct approximate representations from a finite number of Fourier embeddings. Since the learned

---

[1] A kernel $\kappa(x, y)$ expressible as $\kappa(x, y) = k(x - y)$.

kernel representations are a form of kernel learning, we dub our technique *continuous kernel learning* (CKL).

2   Prior Kernel Learning Work

2.1   Multiple Kernel Learning (MKL)

Multiple Kernel Learning, or MKL, is an extension to kernelized support vector machines (SVMs) that employs a combination of kernels to extend the space of possible kernel functions. MKL algorithms learn not only the parameters of the SVM, but also the parameters of the kernel combination. In this sense, MKL algorithms seek to find the correct kernel function for the training data instead of relying on a predefined kernel function.

Lanckriet et al. [28] describe several convex optimization problems that learn the coefficients of a linear combination of kernel functions $\kappa_\gamma(\cdot, \cdot) = \sum_i \gamma_i \kappa_i(\cdot, \cdot)$. There are several algorithms to solve the MKL problem, including [1,3,17,18,42]. In addition to solving the MKL problem, MWUMKL [34] and SPG-GMKL [22] also work at scale.

2.2   Approaches Utilizing Bochner's Theorem

The key mathematical tool that drives much of kernel learning work is Bochner's theorem:

Theorem 1 (Bochner [9]). *A continuous function $k : \mathbb{R}^d \to \mathbb{R}$ is positive-definite iff $k(\cdot)$ is the Fourier transform of a non-negative measure.*

Several papers have been published that explore the connection between Bochner's theorem and learning a kernel[2]. A Bayesian view produces an interpretation of this optimization as learning the kernel of a Gaussian process (GP). Wilson and Adams [46] equate stationary (shift-invariant) kernels to the spectral density function of a GP. They observe that linear combinations of squared-exponential kernels are dense in the space of stationary kernels. The resulting kernel has few parameters and is relatively easy to interpret.

Yang et al. [51] extend the ideas in [46] and combine it with the principles from Fastfood [29]. The authors also discuss variants of their algorithms such as computing a piecewise linear kernel. Similarly, the BaNK method by Oliva et al. [37] learns a kernel using the GP technique and trains the kernel using MCMC. Finally in the GP vein, Wilson et al. [47] integrate a deep network as input to the GP, treating the GP as an "infinite-dimensional" layer of the network, and optimize the parameters of the GP simultaneously with the parameters of the network using backpropagation.

---

[2] Note that Yang et al. [50] are not producing a kernel learning method, but an effective way to sparsify CNNs. No comparison to other kernel learning methods is made in [50].

Băzăvan et al. [10], in contrast, optimize Fourier embeddings, but decompose each ω_i into a parameter σ_i multiplied by a nonlinear function of a uniform random variable to represent the sample. The uniform variable is resampled during optimization as the parameter is learned.

3 Continuous Kernel Learning

3.1 Bochner's Theorem

A couple observations must be made in order for Theorem 1 to be relevant to our setting. First, we observe that (for the purposes of this paper) a positive definite *function* $k(\cdot)$ is a positive definite *kernel* $\kappa(\cdot,\cdot)$ when $\kappa(\mathbf{x},\mathbf{x}') = k(\mathbf{x}-\mathbf{x}')$. A kernel of this type is a *shift-invariant* kernel. Examples include the Gaussian or RBF kernel ($e^{-\|\mathbf{x}-\mathbf{x}'\|^2/\sigma^2}$) and the Laplacian kernel ($e^{-\lambda\|\mathbf{x}-\mathbf{x}'\|}$).

Next, any non-negative measure $\mu : \mathbb{R}^d \to \mathbb{R}^+$ can be converted to a probability distribution if we normalize by $Z = \int_{\mathbb{R}^d} d\mu$. Since Fourier transforms are linear, we can normalize the kernel by the same factor Z and maintain the equivalence. So without loss of generality, we can assume that the measure μ is a probability measure. This equivalence between shift-invariant kernel and distribution is important in the rest of this paper.

3.2 Fourier Embeddings

Rahimi and Recht [41] built on Bochner's theorem by observing that the Fourier transform of μ is also an expectation:

$$k(\mathbf{x}-\mathbf{x}') = \int_{\mathbb{R}^d} e^{i\boldsymbol{\omega}^\top(\mathbf{x}-\mathbf{x}')} f_\mu(\boldsymbol{\omega}) \, d\boldsymbol{\omega} = E_{\boldsymbol{\omega}}[\zeta_{\boldsymbol{\omega}}(\mathbf{x})\overline{\zeta_{\boldsymbol{\omega}}(\mathbf{x}')}],$$

if $\zeta_{\boldsymbol{\omega}}(\mathbf{x}) = e^{i\boldsymbol{\omega}^\top\mathbf{x}}$ and $\boldsymbol{\omega} \sim \mathcal{D}_\mu$, where $\mathcal{D}_\mu$ is the probability distribution over Borel sets on $\mathbb{R}^d$ with measure μ. This shows that $\zeta_{\boldsymbol{\omega}}(\mathbf{x})\overline{\zeta_{\boldsymbol{\omega}}(\mathbf{x}')}$ is an unbiased estimate of $k(\mathbf{x}-\mathbf{x}')$. Because $k(\mathbf{x}-\mathbf{x}')$ is real, we know that $E_{\boldsymbol{\omega}}[\zeta_{\boldsymbol{\omega}}(\mathbf{x})\overline{\zeta_{\boldsymbol{\omega}}(\mathbf{x}')}]$ has no imaginary component. A straightforward Chernoff-type argument [35, see Ch.4] shows that averaging $\zeta_{\boldsymbol{\omega}}(\mathbf{x})\overline{\zeta_{\boldsymbol{\omega}}(\mathbf{x}')}$ over D samples of $\boldsymbol{\omega}$ produces a bound on the error of the estimate that diminishes exponentially in D. The lifting map then becomes $\Phi(\mathbf{x}) = \sqrt{1/D}(\zeta_{\boldsymbol{\omega}_1}(\mathbf{x}),\ldots,\zeta_{\boldsymbol{\omega}_D}(\mathbf{x}))$. The inner product $\langle\Phi(\mathbf{x}),\overline{\Phi(\mathbf{x}')}\rangle$ is obviously the desired average.

We can avoid complex numbers by using $z_{\boldsymbol{\omega},b}(\mathbf{x}) = \sqrt{2}\cos(\boldsymbol{\omega}^\top\mathbf{x} + b)$ with $\boldsymbol{\omega} \sim \mathcal{D}_\mu$ and $b \sim U[0,2\pi)$, which offers the same unbiased estimate (see [41]). The lifting map in this case is $\Phi(\mathbf{x}) = \sqrt{2/D}(z_{\boldsymbol{\omega}_1,b_1}(\mathbf{x}),\ldots,z_{\boldsymbol{\omega}_D,b_D}(\mathbf{x}))$. In this work we will refer to these maps (of the real or complex type) as *Fourier embeddings*. In [41] these embeddings are called *random Fourier features*, because they are selected at random from the distribution that is Fourier-dual to the approximated kernel. We will demonstrate that Fourier embeddings of this type need not be selected at random, and can in fact be optimized.

Our Approach. Our approach is most similar to that in Băzăvan et al. [10]. Like the authors of [10], we recognize that we can optimize the parameters $\{\boldsymbol{\omega}_i\}$ of a Fourier embedding. Băzăvan et al. [10] decompose $\boldsymbol{\omega}_i$ as follows:

$$\boldsymbol{\omega}_i = \boldsymbol{\sigma}_i \odot h(\mathbf{u}_i),$$

where $\boldsymbol{\sigma}_i$ is the parameter of a shift-invariant kernel, $\odot$ is the Hadamard (element-wise) product of two vectors, h is an element-wise nonlinear function (essentially an inverse quantile function), and $\mathbf{u}_i$ is a sample drawn from a multivariate uniform distribution (cube). The procedure is to optimize $\boldsymbol{\sigma}_i$ and periodically resample $\mathbf{u}_i$. This has the advantage of being able to represent the kernel with its parameter $\boldsymbol{\sigma}_i$, which adds to clarity, but the kernel must be one of a particular class of shift-invariant kernels that decomposes into this form. A Gaussian kernel, however, does decompose this way.

In contrast, we sample the vectors $\boldsymbol{\omega}_i$ from the distribution $\mathcal{D}_\mu$, and then optimize them directly. The weights $\{\boldsymbol{\omega}_i\}$ become different vectors $\{\boldsymbol{\omega}'_i\} \subset \mathbb{R}^d$ – and are now very unlikely to be drawn i.i.d. from the distribution $\mathcal{D}_\mu$ anymore. As in prior approaches, by learning the embeddings, we learn the kernel, because the Bochner equivalence between distributions and kernels guarantees this. We use backpropagation to learn the weights, avoiding the need to resample at every step, and allowing us to take advantage of recent neural network technology to perform scalable optimization. While other approaches focus on decomposing the representation of the kernels into individual kernel components and learn their parameters, we avoid this and focus only on producing the final weights $\boldsymbol{\omega}'_i$. We lose the clarity and sparsity of individual kernel parameters but gain the flexibility of learning a representation of a shift-invariant kernel free of individual base kernels, and recent technology allows us to do this training quickly.

For brevity, we refer to the $d \times D$ matrices $\mathbf{W}$ (for the $\{\boldsymbol{\omega}_i\}$) and $\mathbf{W}'$ (for the $\{\boldsymbol{\omega}'_i\}$), since there are D samples from $\mathbb{R}^d$.

3.3 Generalization Bounds in Fourier Embeddings

We now examine the capacity of this class of kernels by analyzing its VC-dimension. Note that the cosine function complicates this analysis since it has nontrivial gradient almost everywhere.

Fortunately we can exploit an observation already well-known in kernel learning that a narrow kernel function, for example, a Gaussian kernel with a small variance, is more likely to overfit (and therefore have higher capacity). This is because a narrow kernel function only allows the model to examine a very small range around each point, so a new point is unlikely to be affected by the model at all. Because the kernel is the Fourier transform of a distribution, a narrow kernel function corresponds to a distribution with high variance – using the same example, a Gaussian kernel with variance parameter σ^2 is the Fourier transform of a Gaussian distribution with variance $1/\sigma^2$. So a small variance in the kernel corresponds to a high variance in the distribution, and vice-versa. In fact, we can demonstrate that if the norm of the embedding parameter ω is high, then this translates to higher capacity.

Let $z(x) = e^{2\pi i x}$, $\mathsf{Re}(z)$ and $\mathsf{Im}(z)$ be the real and imaginary components of z, respectively, let $[a..b]$ refer to the set of integers between a and b, inclusive (i.o.w., $\{n \in \mathbb{Z} \mid a \le n \le b\}$), and let $\mathbf{1}_P(x)$ be the indicator (or characteristic) function of $P : \mathbb{R} \to \{0, 1\}$.

Definition 1. *An $(\boldsymbol{\omega}, \beta, d)$-range is the set $\{\mathbf{x} \in \mathbb{R}^d \mid \mathsf{Im}(z(\boldsymbol{\omega} \cdot \mathbf{x} + \beta)) \ge 0, \|\mathbf{x}\| < 1\}$ where $d \ge 1$ is an integer, $\boldsymbol{\omega} \in \mathbb{R}^d$, and $\beta \in [0, 1)$.*

Definition 2. *Let $\mathcal{G}_d(R)$ be the set of all $(\boldsymbol{\omega}, \beta, d)$-ranges such that $\|\boldsymbol{\omega}\|_2 \le R$.*

Lemma 1. *The decision function $\mathbf{1}_{\mathsf{Im}(z(wx+\beta)) \ge 0}$ induces a unique binary labeling for the set $x \in \{1/2^i\}_{i=1}^n$ for every integer value of $w \in [1..2^n]$, and any $\beta \in (0, 2^{-(n+1)})$.*

Proof. For any integer $w \in [1..2^n]$ and $i \in [1..n]$, choose the binary label as 0 if $z(w/2^i + \beta)$ lands in the upper half-plane of $\mathbb{C}$, and 1 if the lower half-plane. The label can be read as the most significant fractional digit of the binary representation of $w/2^i$, as long as $\beta \in (0, 2^{-(n+1)})^3$. The labeling is then unique for integer values of w up to 2^n. □

Clearly, every $(\boldsymbol{\omega}, \beta, d)$-range corresponds to a binary classifier and the range space $(\mathbb{R}^d, \mathcal{G}_d(R))$ is the hypothesis space of interest. We denote the unbounded range space $\cup_R \mathcal{G}_d(R)$ by $\mathcal{G}_d(\infty)$.

Theorem 2. *The VC-dimension of the range space $(\mathbb{R}^d, \mathcal{G}_d(R))$ is $\Theta(\max\{d \log R, d + 1\})$.*

We prove this theorem in two parts.

Lemma 2. *The VC-dimension of $(\mathbb{R}^d, \mathcal{G}_d(R))$ is at least $d \max\{\lfloor \log_2 R \rfloor, 1\} + 1$.*

Proof. Let $n = \lfloor \log_2 R \rfloor$, for $R \ge 2$. We now construct a set of dn points. Along each axis of $\mathbb{R}^d$, place n points with corresponding coordinate from the set $\{1/2^i\}_{i=1}^n$. From Lemma 1, we know that we can induce a binary labeling on every axis-restricted set, using integers $[1..2^n]$. Given $\boldsymbol{\omega} \in [1..2^n]^d$, each $\omega_j \in [1..2^n]$ will give a unique labeling to the points on axis $j \in [1..d]$, independent of any other axis j. Therefore we can uniquely label the whole set of dn points, for all possible labelings.

To add one more point to the set, we select a point $\mathbf{c}$, the d-dimensional vector with all coordinates equal to a constant c, and make sure that we can find values β_+ and β_- so that $\langle \mathbf{c}, \boldsymbol{\omega} \rangle + \beta_+ \ge 0$ and $\langle \mathbf{c}, \boldsymbol{\omega} \rangle + \beta_- < 0$, independently of $\boldsymbol{\omega}$. Observe that $\langle \mathbf{c}, \boldsymbol{\omega} \rangle = c \sum_j \omega_j$, and that $d \le \sum_j \omega_j \le d2^n$. For $\langle \mathbf{c}, \boldsymbol{\omega} \rangle + \beta_- < 0$ we need that $\beta_+ < -\langle \mathbf{c}, \boldsymbol{\omega} \rangle$ for all $\boldsymbol{\omega}$, since the choice of β must be independent of $\boldsymbol{\omega}$. This means that first, $c < 0$ since $\beta_- > 0$ and $\sum_j \omega_j > 0$. Then $-cd \le -\langle \mathbf{c}, \boldsymbol{\omega} \rangle \le -cd2^n$, so we need to pick $\beta_+ < -cd$. Similarly, we require $\beta_+ \ge -cd2^n$, and since $\beta_+ < 2^{-(n+1)}$, we need $-c < 1/d2^{-(2n+1)}$. Set $c = -1/d2^{2n+2}$, $\beta_+ = 2^{-(n+2)}$,

---

^{3} To avoid ambiguity, we require $\beta > 0$, to prevent $z(w/2^i)$ from landing on the real axis when 2^i divides w.

and $\beta_- = 2^{-(2n+3)}$. We can now uniquely label $dn + 1$ points for all possible labelings, when $R > 2$.

Regardless of the value of R, there is always a unique labeling of $d+1$ points induced by the range space, since we can restrict to a ball small enough that $\text{Im}(z(\omega x + \beta)) = \sin(2\pi(\omega x + \beta))$ is monotonic for appropriate values of β. Within the ball, the range space is effectively the range of half-spaces, which has VC-dimension $d + 1$. □

Corollary 1. *The VC-dimension of the range space* $(\mathbb{R}^d, \mathcal{G}_d(\infty))$ *is unbounded.*

To prove the corresponding upper bound, we use the notion of the *shatter function* of $(\mathbb{R}^d, \mathcal{G}_d(R))$ [20]. For a positive integer n, the shatter function of a range space is the maximum highest number of subsets induced by the range space on any set of n points X_n. That is, any range $\mathcal{R}$ induces a subset of X_n simply by the intersection $\mathcal{R} \cap X_n$, and the shatter function counts all unique subsets of this type.

Lemma 3. *The shatter function of* $(\mathbb{R}^d, \mathcal{G}_d(R))$ *is* $O(R^d n^{d+1})$.

Proof. We can first observe that $\|\omega\|_2 \le R$ implies that $\|\omega\|_\infty \le R$. This implies that $|\omega_j| \le R$ for every $j \in [1..d]$. Treating each coordinate separately this way, each term in $\langle \omega, \mathbf{x} \rangle + \beta$ contributes a factor in the growth function.

For a fixed ω, the number of subsets of a set of n points selected by (ω, β, d)-ranges is $O(n)$, because as β changes, at most one point exits or leaves the upper half-plane (because the points all travel at the same speed around the unit circle).

For fixed β, and fixed ω save for some coordinate ω_j, on the other hand, how often a point enters or leaves the upper half-plane as ω_j varies in $(0, R]$ depends upon the value of x_j. For higher values of x_j, the mapped point travels more rapidly. In fact, for $x = 1$, z takes R revolutions around the circle, so enters and exits the upper half-plane $2R$ times. The number of subsets is bounded by $\sum_{i=1}^n 2R|x_i| = 2R \sum_{i=1}^n |x_i| \le 2Rn$. We take the absolute value because a negative x_i simply changes the direction of travel of $z(\omega_j x_i + \beta)$. Everything else remains the same. For ω and β varying independently, we now have the bound stated in the lemma. □

Lemma 4. *The VC-dimension of* $(\mathbb{R}^d, \mathcal{G}_d(R))$ *is* $O(d \log R)$.

Proof. Follows directly from the relationship between the shatter function and VC-dimension [20]. □

With Lemmas 2 and 4, we have proven Theorem 2. The VC dimension also gives us a generalization bound, due to Bartlett and Mendelson [4]:

Theorem 3. *Let F be a class of ± 1-valued functions defined on a set $\mathcal{X}$. Let P be a probability distribution on $\mathcal{X} \times \{\pm 1\}$, and suppose that $(X_1, Y_1), \ldots, (X_n, Y_n)$ and (X, Y) are chosen independently according to P. Then for any positive integer n, w.p. $(1 - \delta)$ over samples of length n, every $f \in F$ satisfies*

$$P(Y \ne f(X)) \le \frac{1}{n} \sum_{i=1}^n \mathbf{1}_{Y_i \ne f(X_i)} + O\left(\sqrt{\frac{\max\{d \log R, d+1\}}{n}} + \sqrt{\frac{\ln 1/\delta}{n}} \right)$$

Regularization. Theorems 2 and 3 immediately suggest three regularization techniques: First, we limit the norm of the Fourier weights with weight decay (a.k.a. L_2 regularization). Alternatively, we simply cap the norm of each Fourier weight vector to some constant at each round of the training. We can further control the initial capacity by setting the variance of the initializing distribution.

4   From an Embedding to a Feed-Forward Network

We now return to the single Fourier embedding

$$z_{\omega,b} = \sqrt{2}\cos(\omega^\top \mathbf{x} + b)$$

If we fix an input $\mathbf{x}$, then we can view the mapping $z_{\omega,b}$ as a neuron with a cosine activation function and biases of the form $b \in [0, 2\pi)$. We call this type of neuron a *cosine neuron*. Such a neuron, with a cutoff to ensure zero support outside an interval, was introduced in [15]. We impose no such cutoff in this work.

Consider a layer of cosine neurons, each with associated weight vector ω_j. Each of these weights can be viewed as a sample from *some* distribution, and therefore the entire ensemble is a (dual) representation of some shift-invariant kernel (by Bochner's theorem). We can then write the associated classifier for such a combination. Denoting the bias vector by $\mathbf{b}$ and the collection of all the weight vectors ω_j by W, the resulting classifier (with a `softmax` layer to combine the individual activations and logarithmic loss), can be written as

$$\ell_{\log}(\mathrm{softmax}(\cos(\mathbf{W} \cdot \mathbf{x_i} + \beta)), y_i),$$

where $\mathrm{softmax}(\mathbf{r})_j = e^{r_j} / \sum_k e^{r_k}$, and $\ell_{\log}$ is the log loss.

What we now have is a standard (shallow) 2-layer network that we can train using backpropagation and stochastic gradient descent.

5   Experiments

We have designed our experiments to answer the following questions: (1) Does allowing the learning algorithm to pick an arbitrary kernel improve performance over standard MKL techniques that are only allowed to select from a fixed library of kernels? (2) How does the learning algorithm for CKL adapt to large datasets and higher dimensions?

5.1   MKL Vs. CKL on Small Datasets

Since CKL is proposed as an alternative to MKL, we compare CKL to two scalable MKL algorithms, namely SPG-GMKL [22] and MWUMKL [34].

Data Sets. All of the datasets used for the experiments are taken from the `libsvm` repository[4]. See Table 1 for details of the datasets.

---

[4] https://www.csie.ntu.edu.tw/~cjlin/libsvmtools/datasets/binary.html.

Table 1. Summary of datasets

LibSVM datasets	Features	Examples	
Liver	6	345	
Diabetes	8	768	
Cod-RNA	8	59535	
Breast cancer	10	683	
German-Numeric	24	1000	
Mushroom	112	8192	
Adult	123	32561	
Gisette	5000	6000	
Million Song Datasets (MSD)	Features	Examples	Notes
Genre 1	182	37,037	*"Classic pop and rock"* vs. *"folk"*
Genre 2	182	59,485	*"Classic pop and rock"* vs. everything else
Year pred	90	515,345	Prior to year 2000 vs. after year 2000

Experimental Procedure. The data for Adult and Mushroom datasets consist of binary features (one-hot representations of categorical features), so no scaling was applied. Features were scaled to the range $[-1, 1]$ for other datasets.

For MKL experiments, we used the *Scikit-Learn* Python package [40] for much of the testing infrastructure. For testing with MKL methods, the training data is split randomly into 75 % training and 25 % validation data. The random splits were repeated 100 times for all sets except Mushroom, Gisette, and Adult, which received 20 splits for considerations of time. The C parameter was selected through cross validation and for MWUMKL, the ϵ parameter was chosen to be 0.005, to achieve high accuracy while allowing all of the experiments to complete (the number of iterations of the algorithm in [34] is proportional to $1/\epsilon$). We use two kernels: a linear kernel and a Gaussian kernel. For the Gaussian kernel, a wide range of γ are tried and the the best accuracy observed is used in the results.

For CKL experiments, the same test/train split was applied, and additionally, the training portion was split further into 75 % training and 25 % validation. We apply early stopping and momentum, and random searches for: the width (h_0) of the hidden layer, a parameter (σ) used for initializing the weights of the hidden layer, and the learning rate (ℓ) hyperparameters. Training was stopped if the validation objective did not decrease within 100 epochs and was otherwise permitted to run for up to 10,000 epochs. Momentum was applied from the first epoch with a value of 0.5 that was increased to 0.99 over the course of 10 epochs.

Values for h_0 were selected from $\{2^i\}$ with i sampled uniformly from $[0..9]$, except for Gisette, where i was sampled uniformly from $[0..14]$. The weights of the hidden layer were sampled from a Gaussian distribution with variance σ selected from $\{2^i\}$ where i was sampled uniformly from $[-6..0]$. The weights of the softmax layer were selected from $U[-0.1, 0.1]$. Finally ℓ was sampled from $LU[10^{-5}, 0.2]$.[5] 100 models with random hyperparameters were trained, and

---

[5] A random variable X is drawn from $LU[a, b]$ if $X = e^Y$, where $Y \sim U[\ln(a), \ln(b)]$.

then the one with the highest performance was chosen and validated with 100 random splits (as described in the previous paragraph).

Results. The results are shown in Table 2. CKL is not different in any significant capacity from either GMKL or MWUMKL on very small datasets. Letting the learning algorithm pick an arbitrary kernel improves performance over standard MKL techniques that only choose a mixture of kernels. Additionally, we see that CKL adapts to large datasets and higher dimensions better than MKL.

Table 2. Mean accuracies (standard deviations) for various datasets on MKL and CKL. If a mean, minus the standard deviation, is greater than all other means plus standard deviations in the row, then the mean is bold. Note that for all MSD tests, the difference is more than three standard deviations.

Small datasets	GMKL	MWUMKL	CKL
Liver	67.78 % (4.78 %)	59.34 % (6.04 %)	66.45 % (6.19 %)
Diabetes	77.06 % (2.66 %)	75.59 % (2.92 %)	76.08 % (2.95 %)
Cod-RNA	**87.31 %** (0.13 %)	72.42 % (7.30 %)	85.7 % (1.14 %)
Breast cancer	97.14 % (1.20 %)	91.89 % (2.22 %)	96.87 % (1.22 %)
German-Numeric	73.05 % (3.25 %)	74.40 % (3.01 %)	76.14 % (2.57 %)
Mushroom	99.80 % (0.08 %)	99.93 % (0.04 %)	**100 %** (0.0042 %)
Adult income	83.94 % (0.28 %)	76.90 % (0.82 %)	**84.80 %** (0.35 %)
Gisette	95.15 % (0.53 %)	93.50 % (0.72 %)	**96.90 %** (0.52 %)
Million Song Dataset	GMKL	MWUMKL	CKL
Genre 1	77.62 % (0.36 %)	68.14 % (1.06 %)	**81.68 %** (0.39 %)
Genre 2	69.12 % (0.33 %)	53.02 % (0.55 %)	**74.16 %** (0.36 %)
Year pred	75.38 % (0.1 %)	57.72 % (1.64 %)	**77.57 %** (0.11 %)

5.2   MKL Vs. CKL on Million Song Datasets

In this section, we compare MKL methods with CKL on the Million Song Dataset [6]. The Million Song Dataset consists of audio features and metadata of one million contemporary popular music tracks. For the experiments, we utilized three different subsets of the Million Song Dataset, all binary. The features are the average and covariance of the pitch and timbre vectors for each track:

- **Genre 1:** The two most common genres in Million Song Dataset - *"classic pop and rock"* and *"folk."* The tracks which have both genres as tags are removed to avoid confusion.
- **Genre 2:** The ten most common genres in the Million Song Dataset. Since the *"classic pop and rock"* genre has significantly more tracks than any other genre, *"classic pop and rock"* is considered as one class and everything else together as another class.

– **Year Prediction:** Taken from the UCI Machine Learning Repository. All tracks prior to the year 2000 are considered as one class and all tracks after and including the year 2000 are considered as the other class. The dimensions of the dataset are summarized in Table 1.

Results. The results are shown in Table 2. CKL is clearly superior to the scalable MKL methods that we tested against, adding to the evidence that higher-dimensional and larger datasets can benefit from our technique.

5.3 MKL Vs. CKL on Images

We compare MKL and CKL on CIFAR10. CIFAR10 [27] is a labeled image dataset containing 60,000 1,024-dimensional (32×32) images and 10 classes used extensively for testing image classification algorithms. While image classification is an important benchmark for neural networks, we wish to point out that our objective is *not* to classify the CIFAR10 dataset better than all other previous techniques. Instead, we wish to provide comparisons between the methods described in this paper on a large and very challenging task using a simple convolutional neural architecture.

Preprocessing. We first centered the CIFAR10 training set by mean, and then used Pylearn2 [19] to apply two transformations: global contrast normalization [12] and ZCA whitening [5][6]. We applied the same transformations computed for the training set to the testing set.

Feature extraction. For MKL, we used a convolutional neural network (CNN) [30] to learn a representation from the data. In total, we trained 100 models and we extracted the features from the model with the best performance. All of the models had the form $conv_{ReLU} \rightarrow pool_{max} \rightarrow fc_{ReLU} \rightarrow softmax$ where $conv_{ReLU}$ is a convolutional layer using ReLU non-linearities, $pool_{max}$ is a max-pool layer, fc_{ReLU} was a fully-connected layer using ReLU non-linearities, and *softmax* was a *softmax* layer.

We trained the models with (1) momentum, initialized to 0.5 and increased to 0.99 over the first 100 epochs, and (2) early stopping: we set aside the last 10,000 samples of the training set as a validation set for early stopping, and trained the models for at most 5,000 epochs. We initialized the weights of all layers by selecting values uniformly at random from the range $[-0.01, 0.01]$. The parameters of best performing model were as follows: (1) the convolutional layer (with ReLU activations): a 5×5 kernel with 1×1 stride, 32 channels, a max

---

[6] PCA whitening attempts to decorrelate features and normalize singular values ("whitening") of the original data by rotating the data by singular vectors, and then normalizing singular values. ZCA whitening, in contrast, attempts to do the same, but make the resulting data as close to the original as possible, in a least-squares sense. The ZCA transformation is simply to multiply by the inverse square root of the covariance matrix of the data.

kernel norm of 1.8, and cross channel normalization with $\alpha = 3.2 \times 10^{-4}$ and $\beta = 0.75$, (2) the max pooling layer: a 3×3 kernel with 2×2 stride, (3) the fully connected layer: $1{,}000$ rectified linear units, and (4) the softmax layer: one output for each CIFAR10 class. Each sample of CIFAR10 was passed through the CNN and the activations of the fully connected layer were recorded as the new representation.

CIFAR10 with MKL. For MKL experiments, the testing infrastructure and the experimental procedures are similar to the experimental procedure of Sect. 5.1 except for the following details: (1) One-vs-one multiclass strategy is used for the classification task, (2) Random 75% of the training data is used for training and tested on the standard test data. The runs were repeated 20 times, and (3) We used two Gaussian kernels, one with $\gamma = 1$ and the other with a range of γ from 2^{-7} to 2^7. The best accuracy observed is used in Table 3.

CIFAR10 with CKL. For comparison with MKL, we trained a network of the form $conv_{ReLU} \rightarrow pool_{max} \rightarrow fc_{ReLU} \rightarrow fc_{cos} \rightarrow$ softmax. A CKL model of this form uses the same structure as the CNN used for the MKL/CKL experiments (defined in the paragraph "Feature Extraction"), up to and including the fully connected layer of rectified linear units. Instead of a softmax layer, the units of the fully connected layer were connected to a CKL model with $1{,}000$ hidden units (untuned).

The primary difference between this model and MKL trained on features extracted from a CNN (see Sect. 5.3) is that this model is trained all at once, while in the MKL experiments the CNN used for feature learning and the MKL model were trained separately. This end-to-end learning allows the features of each layer to adapt to the features that appear later in the network. It is also important to note that the MKL experiments were trained on a one-vs.-one basis, while the CKL model uses multinomial (softmax) regression with log loss.

Experimental procedure. The models in these experiments were trained using stochastic gradient descent for a maximum of $1{,}000$ epochs with early stopping and momentum. The initial momentum rate was 0.5 and was adjusted from the first epoch to 0.99 over the first 500 epochs of the training.

Results. The CKL model outstrips the MKL methods by a wide margin. We conjecture that this is due to two effects: (1) the end-to-end training allows for better adaptation in the training process and (2) the search space of kernels is much larger. The first effect demonstrates that CKL is more adaptable than MKL in these settings. It is also important to note that training is a crucial component for CKL models when operating on large datasets. For CIFAR10, evaluating any random model upon initialization yielded an accuracy of only 10.1% with standard deviation of 0.235%. In contrast, evaluating random models on smaller datasets frequently yields accuracies that are better than chance.

Table 3. Accuracy for CIFAR10 on MKL and CKL with CNN.

GMKL	MWUMKL	CKL+CNN
44.43 % (0.57 %)	48.2 % (0.41 %)	**67.77 %** (0.61 %)

CIFAR10 with Two Layer Convnets. One might ask whether stacking two cosine layers has any beneficial effect, since stacking two cosine layers is similar to composing two lifting maps, which if defined, yields a kernel. Zhuang et al. [53] construct an algorithm specifically for the composition of two kernels – essentially layering the kernels. Lu et al. [31] discuss extensions to [41] that cover products, sums, and compositions of kernels. Since these are based on the sampling methodology of [41], there is a direct analogy to composing two cosine layers (fixed, in this case). We did not observe significant improvement in accuracy when we employed combinations of two cosine layers. One possible explanation is that since the composition of a kernel is itself a kernel, it can be argued that optimizing a network that contains two consecutive cosine layers accomplishes no more than doing so with one individual cosine layer.

6 Related Work

Multiple kernel learning. The general area of kernel learning was initiated by Lanckriet et al. [28] who proposed to simultaneously train an SVM as well as learn a convex combination of kernel functions. The key contribution was to frame the learning problem as an optimization over positive semi-definite kernel matrices which in turn reduces to a QCQP. Soon after, Bach et al. [3] proposed a block-norm regularization method based on *second order cone programming* (SOCP).

For efficiency, researchers started using optimization methods that alternate between updating the classifier parameters and the kernel weights. Many authors then explored the MKL landscape, including Rakotomamonjy et al. [42], Sonnenburg et al. [43], Xu et al. [48,49]. However, as pointed out by Cortes [13], most of these methods do not compare favorably (both in accuracy as well as speed) even with the simple *uniform* heuristic. More recently, Moeller et al. [34] developed a multiplicative-weight-update based approach that has a much smaller memory footprint and scales far more effectively. Other kernel learning methods include [14,33,38,39,44] and notably methods using the ℓ_p-norm [25,26,45].

Infinite-width networks. Early work on infinite-width networks was done by Neal [36], who tied infinite networks to Gaussian processes, assuming that the distribution is Gaussian. Cho and Saul [11] analyzed the case where the network is either a step network (the output is 1 if the input is positive, 0 otherwise) or a rectified linear unit (ReLU), a type of network used frequently in deep networks (the input z is passed through the function $\max\{0, z\}$). They showed that if the distribution is Gaussian in these settings, the function $\phi_{\mathbf{x}}$ output by the network is a lifting map corresponding to a kernel they dub the *arc-cosine kernel*. Hazan and Jaakkola [21] extended this result further, and analyzed the kernel corresponding to two infinite layers stacked in series. They showed that such a network, when the distribution of the first layer is Gaussian, and the second layer is treated as a Gaussian process, (a process is a distribution of distributions), corresponds to a kernel that can be computed explicitly. Globerson and Livni [16] produce an online algorithm for infinite-layer networks that *avoids* the kernel trick. They demonstrate a sample complexity equal to methods that use the kernel trick, demonstrating that sampling can be as effective as methods that have access to kernel values.

Layered kernels. Zhuang et al. [53] develop a multiple kernel learning technique where they use a layered kernel to combine the output of several other kernels. Their algorithm alternates the use of standard SVM and stochastic gradient descent. Lu et al. [31] scale up [41] by making some interesting mathematical observations about kernels and distributions. Their work relies heavily on the correspondence between distributions and kernels, a theme that we explore as well. Yu et al. [52] also seek to optimize a kernel, using alternating optimization and also based on Bochner's theorem. Jiu and Sahbi [23,24] exploit *kernel map networks* and Laplacians of nearest-neighbor graphs [24] to produce "deep" kernels for use in SVMs.

Neural networks as kernels. Yang et al. [50] exploit the correspondence between ReLUs and arc-cosine kernels [11], and the sparsity of the Fastfood transform [29] to reduce the complexity of a convolutional neural net.

Aslan et al. [2] seek to make the optimization of neural networks convex through kernels and matrix techniques. Mairal et al. [32] extend hierarchical kernel descriptors [7,8] to act as convolutional layers. Very recently, Wilson et al. [47] combine neural networks with Gaussian processes, drawing on the infinite-width network setting, to produce "deep" kernels.

References

1. Aiolli, F., Donini, M.: EasyMKL: a scalable multiple kernel learning algorithm. Neurocomputing **169**, 215–224 (2015)
2. Aslan, O., Zhang, X., Schuurmans, D.: Convex deep learning via normalized kernels. In: Ghahramani, Z., Welling, M., Cortes, C., Lawrence, N.D., Weinberger, K.Q. (eds.) NIPS, pp. 3275–3283. Curran Associates, Inc. (2014)
3. Bach, F.R., Lanckriet, G.R.G., Jordan, M.I.: Multiple kernel learning, conic duality, and the SMO algorithm. In: ICML, Banff, Canada (2004)
4. Bartlett, P.L., Mendelson, S.: Rademacher and Gaussian complexities: risk bounds and structural results. JMLR **3**, 463–482 (2003)
5. Bell, A.J., Sejnowski, T.J.: Edges are the 'Independent Components' of natural scenes. In: Mozer, M.C., Jordan, M.I., Petsche, T. (eds.) NIPS, pp. 831–837. MIT Press (1997)
6. Bertin-Mahieux, T., Ellis, D.P.W., Whitman, B., Lamere, P.: The million song dataset. In: ISMIR (2011)

7. Bo, L., Lai, K., Ren, X., Fox, D.: Object recognition with hierarchical kernel descriptors. In: CVPR, pp. 1729–1736, June 2011
8. Bo, L., Ren, X., Fox, D.: Kernel descriptors for visual recognition. In: Lafferty, J.D., Williams, C.K.I., Shawe-Taylor, J., Zemel, R.S., Culotta, A. (eds.) NIPS, pp. 244–252. Curran Associates, Inc. (2010)
9. Bochner, S.: Lectures on Fourier Integrals. Annals of Mathematics Studies, vol. 42. Princeton University Press, Princeton (1959)
10. Băzăvan, E.G., Li, F., Sminchisescu, C.: Fourier kernel learning. In: Fitzgibbon, A., Lazebnik, S., Perona, P., Sato, Y., Schmid, C. (eds.) ECCV 2012, Part II. LNCS, vol. 7573, pp. 459–473. Springer, Heidelberg (2012)
11. Cho, Y., Saul, L.K.: Kernel methods for deep learning. In: Bengio, Y., Schuurmans, D., Lafferty, J.D., Williams, C.K.I., Culotta, A. (eds.) NIPS, pp. 342–350. Curran Associates, Inc. (2009)
12. Coates, A., Ng, A.Y., Lee, H.: An analysis of single-layer networks in unsupervised feature learning. In: AIStats, pp. 215–223 (2011)
13. Cortes, C.: Invited talk: can learning kernels help performance? In: ICML, Montreal, Canada (2009)
14. Cortes, C., Mohri, M., Rostamizadeh, A.: Learning non-linear combinations of kernels. In: NIPS, Vancouver, Canada (2009)
15. Gallant, A., White, H.: There exists a neural network that does not make avoidable mistakes. In: ICNN, vol. 1, pp. 657–664, July 1988
16. Globerson, A., Livni, R.: Learning infinite-layer networks: beyond the kernel trick. arXiv:1606.05316 [cs], June 2016
17. Gönen, M., Alpaydın, E.: Localized multiple kernel learning. In: ICML, Helsinki, Finland (2008)
18. Gönen, M., Alpaydın, E.: Localized algorithms for multiple kernel learning. Pattern Recogn. 46(3), 795–807 (2013)
19. Goodfellow, I.J., Warde-Farley, D., Lamblin, P., Dumoulin, V., Mirza, M., Pascanu, R., Bergstra, J., Bastien, F., Bengio, Y.: Pylearn2: a machine learning research library. arXiv:1308.4214 [cs, stat], August 2013
20. Har-Peled, S.: Geometric Approximation Algorithms. American Mathematical Society, Boston (2011)
21. Hazan, T., Jaakkola, T.: Steps toward deep kernel methods from infinite neural networks. arXiv:1508.05133 [cs], August 2015
22. Jain, A., Vishwanathan, S.V.N., Varma, M.: SPG-GMKL: generalized multiple kernel learning with a million kernels. In: KDD, pp. 750–758 (2012)
23. Jiu, M., Sahbi, H.: Deep kernel map networks for image annotation. In: ICASSP, pp. 1571–1575, March 2016
24. Jiu, M., Sahbi, H.: Laplacian deep kernel learning for image annotation. In: ICASSP, pp. 1551–1555, March 2016
25. Kloft, M., Brefeld, U., Sonnenburg, S., Laskov, P., Müller, K.R., Zien, A.: Efficient and accurate Lp-norm multiple kernel learning. In: NIPS, Vancouver, Canada (2009)
26. Kloft, M., Brefeld, U., Sonnenburg, S., Zien, A.: Lp-norm multiple kernel learning. JMLR 12, 953–997 (2011)
27. Krizhevsky, A.: Learning Multiple Layers of Features from Tiny Images. Citeseer (2009)
28. Lanckriet, G.R.G., Cristianini, N., Bartlett, P., Ghaoui, L.E., Jordan, M.I.: Learning the kernel matrix with semidefinite programming. JMLR 5, 27–72 (2004)

29. Le, Q., Sarlos, T., Smola, A.: Fastfood - computing Hilbert space expansions in loglinear time. In: ICML, pp. 244–252 (2013)
30. LeCun, Y.: Generalization and network design strategies. In: Pfeifer, R., Schreter, Z., Fogelman, F., Steels, L. (eds.) Connectionism in Perspective. Elsevier, Zurich (1989). An extended version was published as a technical report of the University of Toronto
31. Lu, Z., May, A., Liu, K., Garakani, A.B., Guo, D., Bellet, A., Fan, L., Collins, M., Kingsbury, B., Picheny, M., Sha, F.: How to scale up kernel methods to be as good as deep neural nets. arXiv:1411.4000 [cs, stat], November 2014
32. Mairal, J., Koniusz, P., Harchaoui, Z., Schmid, C.: Convolutional kernel networks. In: NIPS, pp. 2627–2635 (2014)
33. Micchelli, C.A., Pontil, M.: Learning the kernel function via regularization. JMLR **6**, 1099–1125 (2005)
34. Moeller, J., Raman, P., Venkatasubramanian, S., Saha, A.: A geometric algorithm for scalable multiple kernel learning. In: AIStats, pp. 633–642 (2014)
35. Motwani, R., Raghavan, P.: Randomized Algorithms. Cambridge University Press, Cambridge (1995)
36. Neal, R.M.: Priors for infinite networks. Bayesian Learning for Neural Networks. Lecture Notes in Statistics, vol. 118, pp. 29–53. Springer, New York (1996)
37. Oliva, J., Dubey, A., Poczos, B., Schneider, J., Xing, E.P.: Bayesian Nonparametric Kernel-learning. arXiv:1506.08776 [stat], June 2015
38. Ong, C.S., Smola, A.J., Williamson, R.C.: Learning the kernel with hyperkernels. JMLR **6**, 1043–1071 (2005)
39. Orabona, F., Luo, J.: Ultra-fast optimization algorithm for sparse multi kernel learning. In: ICML, Bellevue, USA (2011)
40. Pedregosa, F., Varoquaux, G., Gramfort, A., Michel, V., Thirion, B., Grisel, O., Blondel, M., Prettenhofer, P., Weiss, R., Dubourg, V., Vanderplas, J., Passos, A., Cournapeau, D., Brucher, M., Perrot, M., Duchesnay, E.: Scikit-learn: machine learning in python. JMLR **12**, 2825–2830 (2011)
41. Rahimi, A., Recht, B.: Random features for large-scale kernel machines. In: NIPS, pp. 1177–1184 (2007)
42. Rakotomamonjy, A., Bach, F., Canu, S., Grandvalet, Y.: More efficiency in multiple kernel learning. In: ICML, Corvalis, USA (2007)
43. Sonnenburg, S., Rätsch, G., Schäfer, C., Schölkopf, B.: Large scale multiple kernel learning. JMLR **7**, 1531–1565 (2006)
44. Varma, M., Babu, B.R.: More generality in efficient multiple kernel learning. In: ICML, Montreal, Canada (2009)
45. Vishwanathan, S.V.N., Sun, Z., Ampornpunt, N., Varma, M.: Multiple kernel learning and the SMO algorithm. In: NIPS, Vancouver, Canada (2010)
46. Wilson, A., Adams, R.: Gaussian process kernels for pattern discovery and extrapolation. In: ICML, pp. 1067–1075 (2013)
47. Wilson, A.G., Hu, Z., Salakhutdinov, R., Xing, E.P.: Deep Kernel Learning. arXiv:1511.02222 [cs, stat], November 2015
48. Xu, Z., Jin, R., King, I., Lyu, M.R.: An extended level method for efficient multiple kernel learning. In: NIPS, Vancouver, Canada (2008)
49. Xu, Z., Jin, R., Yang, H., King, I., Lyu, M.R.: Simple and efficient multiple kernel learning by group lasso. In: ICML, Haifa, Israel (2010)
50. Yang, Z., Moczulski, M., Denil, M., de Freitas, N., Smola, A., Song, L., Wang, Z.: Deep Fried Convnets. arXiv: 1412.7149, December 2014
51. Yang, Z., Wilson, A., Smola, A., Song, L.: À la Carte - learning fast kernels. In: AIStats, pp. 1098–1106 (2015)

52. Yu, F.X., Kumar, S., Rowley, H., Chang, S.F.: Compact Nonlinear Maps and Circulant Extensions. arXiv:1503.03893 [cs, stat], March 2015
53. Zhuang, J., Tsang, I.W., Hoi, S.: Two-layer multiple kernel learning. In: AIStats, pp. 909–917 (2011)

Temporal PageRank

Polina Rozenshtein[(✉)] and Aristides Gionis

Department of Computer Science, Helsinki Institute for Information Technology,
Aalto University, Espoo, Finland
{polina.rozenshtein,aristides.gionis}@aalto.fi

Abstract. PageRank is one of the most popular measures for ranking the nodes of a network according to their importance. However, PageRank is defined as a steady state of a random walk, which implies that the underlying network needs to be fixed and static. Thus, to extend PageRank to networks with a temporal dimension, the available temporal information has to be judiciously incorporated into the model.

Although numerous recent works study the problem of computing PageRank on dynamic graphs, most of them consider the case of updating static PageRank under node/edge insertions/deletions. In other words, PageRank is always defined as the static PageRank of the current instance of the graph.

In this paper we introduce *temporal PageRank*, a generalization of PageRank for temporal networks, where activity is represented as a sequence of time-stamped edges. Our model uses the random-walk interpretation of static PageRank, generalized by the concept of *temporal random walk*. By highlighting the actual information flow in the network, temporal PageRank captures more accurately the network dynamics.

A main feature of temporal PageRank is that it adapts to concept drifts: the importance of nodes may change during the lifetime of the network, according to changes in the distribution of edges. On the other hand, if the distribution of edges remains constant, temporal PageRank is equivalent to static PageRank.

We present temporal PageRank along with an efficient algorithm, suitable for online streaming scenarios. We conduct experiments on various real and semi-real datasets, and provide empirical evidence that temporal PageRank is a flexible measure that adjusts to changes in the network dynamics. The data and software related to this paper are available at https://github.com/polinapolina/temporal-pagerank.

Keywords: PageRank · Graph mining · Social-network analysis · Dynamic graphs · Time-evolving networks · Interaction networks

1 Introduction

PageRank is a classic algorithm for estimating the importance of nodes in a network. It has been considered a success story on applying link analysis information seeking and ranking, and has been listed as one of the ten most influential

© Springer International Publishing AG 2016
P. Frasconi et al. (Eds.): ECML PKDD 2016, Part II, LNAI 9852, pp. 674–689, 2016.
DOI: 10.1007/978-3-319-46227-1_42

data-mining algorithms [24]. PageRank has been applied to numerous settings and it has inspired a family of fixed-point computation algorithms, such as, TopicRank [6], TrustRank [8], SimRank [11], and more.

PageRank is defined to be the steady-state distribution of a random walk. As such, it is implied that the underlying network structure is fixed and does not change over time. Even though numerous works have studied the problem of computing PageRank on dynamic graphs, the emphasis has been given on maintaining PageRank efficiently under network updates [12,19], or on computing PageRank efficiently in streaming settings [22]. Instead there has not been much work on how to incorporate temporal information and network dynamicity in the PageRank definition.

To make the previous claim more clear imagine that starting from an initial network G we observe k elementary updates in the network structure $e_1, \ldots, e_k$ (such as edge additions or deletions), resulting on a modified network G'. A typical question is how to compute the PageRank of G' efficiently, possibly by taking into consideration the PageRank of G, and the incremental updates. Nevertheless, the PageRank of G' is defined as a steady-state distribution and as the network G' would "freeze" at that time instance.

Our goal in this paper is to extend PageRank so as to incorporate temporal information and network dynamics in the definition of node importance. The proposed measure, called temporal PageRank, is designed to provide estimates of the importance of a node u *at any given time* t. If the network dynamics and the importance of nodes change over time, so does temporal PageRank, and it duly adapts to reflect these changes.

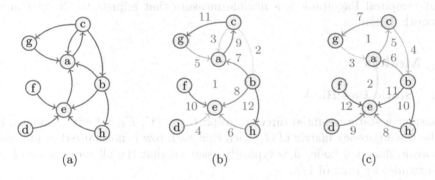

Fig. 1. (a) A static graph, in which hubs a and e have the highest static PageRank score; (b) and (c) represent two different temporal networks: in (b) the temporal PageRank score of nodes a and e are expected to be stable over time; in (c) node e becomes more important than a as the time goes by, and the temporal PageRank scores of a and e are expected to change accordingly.

An example illustrating the concept of temporal PageRank, and presenting the main difference with classic PageRank, is shown in Fig. 1. First, a static (directed) graph is shown in Fig. 1(a). Vertices a and e are the hubs of the graph, and thus, the nodes with the highest static PageRank score. Figures 1(b) and (c) show two *temporal networks*; the number next to each edge denotes the time-stamp that the edge arrives. In Fig. 1(b) the in-coming edges of nodes a and e are arriving in an interleaving manner, so we expect that the importance of a and e will be stable over time, and that their temporal PageRank scores will be approximately equal to their static PageRank scores. On the other hand, in Fig. 1(c) we are witnessing a *concept drift*: node a receives its in-coming edges in the initial phase, while node e receives its in-coming edges later on. Due to this change, node e becomes more important than a as time goes by. Accordingly the scores of temporal PageRank for a and e are changing over time reflecting the change in the network dynamics.

Note also that a dynamic algorithm for computing PageRank is required to report the same output (the static PageRank of the graph in Fig. 1(a)) independently of whether it receives its input as in Fig. 1(b) or (c).

As illustrated in the previous example, temporal PageRank is defined for *temporal networks* [9,18], i.e., networks with time-stamped edges. We generalize the random-walk interpretation of static PageRank by using *temporal random walks*, i.e., *time-respecting* random walks on the temporal network.

We provide a simple update algorithm for computing temporal PageRank. Our algorithm processes the graph edges in order of arrival and it is proven to converge to the correct temporal PageRank scores. We also prove that if the edge distribution remains constant, temporal PageRank converges to the static PageRank of the underlying graph that the edge distribution is drawn.

We conduct extensive experimental evaluation on various real and semi-real datasets, which support our theoretical results and provide empirical evidence that temporal PageRank is a flexible measure that adjusts to changes in the network dynamics.

2 Models

2.1 Static PageRank

Consider a static weighted directed graph $G_s = (V, E_s, w)$ with n nodes. Let P be the adjacency matrix of G_s, such that each row is normalized to unit sum. To avoid dangling nodes it is typically assumed that the all-zero rows of P are substituted by rows of $1/n$.

Given adjacency matrix $P \in \mathbb{R}^{n \times n}$ and a unit-normalized *personalization* row vector $h \in \mathbb{R}^n$, we consider a random walk that visits the nodes of the graph G_s at discrete steps $i = 1, 2, \ldots$. At step $i = 1$ the random walk starts at a node $u \in V$ with probability $h(u)$. Given that at step i the random walk has visited a node u, at step $i + 1$ it visits a node v selected as follows: with probability $1 - \alpha$ the node v is chosen according to the distribution h, while with probability α the node v is chosen according to the distribution specified by the u-th row of P.

Consider now a *Markov chain* with nodes V as its *state space* and *transition matrix*

$$P' = \alpha P + (1 - \alpha)\mathbf{1}h,$$

where $\mathbf{1}$ is a unit column vector. This Markov chain models the random walk defined above. Assuming that the matrix P' is stochastic, aperiodic, and irreducible, by the Perron–Frobenius theorem there exists a unique row vector π, such that $\pi P' = \pi$ and $\pi \mathbf{1} = 1$. The vector π is the stationary distribution of the Markov chain, and it is also known as the *PageRank vector*. The u-th coordinate of π is the *PageRank score* of node u.

A closed-form expression for π can be derived as

$$\pi = (1 - \alpha)h(I - \alpha P)^{-1} = (1 - \alpha)h \sum_{k=0}^{\infty} \alpha^k P^k,$$

and the PageRank score of a node u can be written as

$$
\begin{aligned}
\pi(u) &= \sum_{v \in V} h(v) \sum_{k=0}^{\infty} (1 - \alpha)\alpha^k \sum_{\substack{z \in Z(v,u)\,(i,j) \in z \\ |z|=k}} \prod_{(i,j) \in z} P(i,j) \\
&= \sum_{v \in V} \sum_{k=0}^{\infty} (1 - \alpha)\alpha^k \sum_{\substack{z \in Z(v,u) \\ |z|=k}} h(v)\mathrm{Pr}\,[z \mid v] \\
&= \sum_{v \in V} \sum_{k=0}^{\infty} (1 - \alpha)\alpha^k \sum_{\substack{z \in Z(v,u) \\ |z|=k}} \mathrm{Pr}\,[z]\,,
\end{aligned}
\tag{1}
$$

where $Z(v, u)$ is a set of all walks from v to u, and (i, j) is used to denote two consecutive nodes of a certain walk $z \in Z(v, u)$. The product $\prod_{(i,j) \in z} P(i,j) = \mathrm{Pr}\,[z \mid v] = \mathrm{Pr}\,[z]\,/h(v)$ expresses the probability that a random walk reaches node u, provided that it starts at node v and it follows only graph edges.

In the definition of PageRank, it is assumed that the transition probability matrix P' is given in advance, and it does not change. A number of works address the problem of computing PageRank *incrementally*, when nodes and edges are added or removed. However, PageRank is still defined by its static version, as the stationary distribution of the graph that contains all nodes and edges that are currently *active* [4,5,12,19]. Here we propose another view of PageRank, where temporal information and network dynamics are explicitly incorporated in the underlying random walk that defines the PageRank distribution.

2.2 Temporal PageRank

Temporal PageRank extends static PageRank by incorporating temporal information into the random-walk model. Our model uses *temporal networks* [9,13,

18,21]. A temporal network $G = (V, E)$ consists of a set of n nodes V and a set of m timestamped edges (or interactions) E between pairs of nodes

$$E = \{(u_i, v_i, t_i)\}, \text{ with } i = 1, \ldots, m, \text{ such that } u_i, v_i \in V \text{ and } t_i \in \mathbb{R},$$

where t_i represents the timestamp when an interaction between u_i and v_i is taking place. For generality we assume that the edges of the temporal graph are *directed*. We also assume more than one different edge may exist between a given pair of nodes, with different timestamps, representing multiple interactions in time between a pair of nodes.

Following previous studies on temporal networks [9,18], given a temporal network G, we define a *temporal walk* on G, or a *time-respecting walk*, to be a sequence of edges $(u_1, u_2, t_1), (u_2, u_3, t_2), \ldots, (u_j, u_{j+1}, t_j)$, such that $t_i \leq t_{i+1}$ for all $1 \leq i \leq j - 1$.

Our extension of static PageRank to temporal PageRank is based on modifying the PageRank definition of Eq. (1) so that only temporal walks are considered instead of all possible walks.

The intuition behind the idea can be illustrated by the example shown in Fig. 1(c). Node a initially receives many in-links and it should be considered important. After time $t = 8$, however, it does not receive any more in-links and thus, its importance should diminish. By using time-respecting walks one can accurately model the fact that the probability of the random walk being at node a decreases as time increases beyond time $t = 8$. Essentially, the probability that a random walk being at node a after time $t = 8$ corresponds to the probability that the random walk has arrived at node a *before* time $t = 8$ and it has not left yet. Clearly this probablity decreases as time increases beyond $t = 8$.

We now define temporal PageRank more formally. Let $Z^T(v, u \mid t)$ be a set of all possible *temporal walks* that start at node v and reach node u before time t. We can compute the probability of a particular walk $z \in Z^T(v, u \mid t)$ as the number $c(z \mid t)$ of all such walks (starting at v and reaching u before time t) normalized by a number of all temporal walks that start at node v and have the same length

$$\Pr{}' \left[z \in Z^T(v, u \mid t) \right] = \frac{c(z \mid t)}{\sum_{\substack{z' \in Z^T(v,x\mid t) \\ x \in V,\, |z'|=|z|} } c(z' \mid t)}. \tag{2}$$

To compute the number $c(z \mid t)$ of temporal walks that start at v and reach u before time t one can consider the unweighted count of all possible temporal walks. Such a count implies that once reaching u at time t_1 the random walk selects uniformly at random one of the future interactions (u, x, t_2), with $t_2 > t_1$, to move out of u. This model is not very intuitive as it assumes that the random walk has knowledge of the future interations. Instead, once reaching u by an interaction (v, u, t_1) it is more likely to move out of u in one of the next interactions (u, x, t_2). Thus, we assume that the probability of taking (v, u, t_1) followed by (u, x, t_2) increases as the time difference $(t_2 - t_1)$ decreases.

To model this decreasing probability we consider an exponential distribution. Our motivation for this definition is the exponential-decay model in data-stream processing, which is commonly used. We define the probability that interaction (v, u, t_1) is followed by (u, x, t_2):

$$\Pr\left[(v, u, t_1), (u, x, t_2)\right] = \beta^{|(u,y,t')|t' \in [t_1, t_2], \, y \in V|}.$$

We will refer to β as transition probability. The weighted number of temporal walks is then defined as

$$c(z \mid t) = (1 - \beta) \prod_{((u_{i-1}, u_i, t_i), (u_i, u_{i+1}, t_{i+1})) \in z} \beta^{|(u_i, y, t')|t' \in [t_i, t_{i+1}], \, y \in V|},$$

where $(1 - \beta)$ is a normalization term. Note that $\beta = 1$ with omitted normalization corresponds to the unweighted case. In this case we view temporal network as a sequence of samples from some unknown and changing distribution P'.

By combining Eqs. (1) and (2), the *temporal PageRank* score of a node u at time t is defined as

$$\mathbf{r}(u, t) = \sum_{v \in V} \sum_{k=0}^{t} (1 - \alpha) \alpha^k \sum_{\substack{z \in Z^T(v, u | t) \\ |z| = k}} \Pr'\left[z \mid t\right]. \tag{3}$$

Note that according to this definition, the temporal PageRank score of a node u is a function of time. Thus, although our definition is an adaptation of the path-counting formulation of static PageRank (Eq. (1)), the temporal PageRank is not a limiting distribution as static PageRank.

Also note that the definition of temporal PageRank (Eq. (3)) does not incorporate explicitly a personalization vector h. Instead, in the temporal PageRank model presented above, the probability of starting a temporal walk at a node u is proportional to the number of temporal edges that start in u. The vector that contains the starting probabilities for all nodes is referred to as *walk starting probability vector* and it is denoted by h'. The vector h' is learned from the data, in particular, for each node u, it is $h'(u) = \frac{|(u,v,t) \in E: \, \forall v \in V|}{|E|}$.

On the other hand, given a personalization vector h^*, the personalized temporal PageRank is defined as

$$\mathbf{r}(u, t) = \sum_{v \in V} \sum_{k=0}^{t} (1 - \alpha) \alpha^k \frac{h^*(v)}{h'(v)} \sum_{\substack{z \in Z^T(v, u | t) \\ |z| = k}} \Pr'\left[z \mid t\right] \tag{4}$$

Equation (4) assumes that the walk starting probability vector h' is known. In practice, h' can be learned by one scan of the edges of the temporal network.

3 Algorithms

3.1 Computing Temporal PageRank

In order to compute temporal PageRank we need to process the sequence of interactions E and calculate the weighted number of temporal walks. When a

---

Algorithm 1. stream processing

---
input : E, transition probability $\beta \in (0, 1]$, jumping probability α

1 $r = 0, \, s = 0$;
2 **foreach** $(u, v, t) \in E$ **do**
3 $\quad$ $r(u) = r(u) + (1 - \alpha)$;
4 $\quad$ $s(u) = s(u) + (1 - \alpha)$;
5 $\quad$ $r(v) = r(v) + s(u)\alpha$;
6 $\quad$ **if** $\beta \in (0, 1)$ **then**
7 $\quad\quad$ $s(v) = s(v) + s(u)(1 - \beta)\alpha$;
8 $\quad\quad$ $s(u) = s(u)\beta$;
9 $\quad$ **else if** $\beta = 1$ **then**
10 $\quad\quad$ $s(v) = s(v) + s(u)\alpha$;
11 $\quad\quad$ $s(u) = 0$;

12 normalize r;
13 **return** r;

---

new interaction (u, v, t) arrives it can be used to advance any of the temporal walks that end in u, or it can be the start of a new walk. To keep count of the number of walks ending at each node we use an *active mass* vector $s(t) \in \mathbb{R}^{|V|}$, with $s(u, t)$ being equal to the weighted count of walks ending at node u at time t. We also use a vector $r(t) \in \mathbb{R}^{|V|}$ to keep temporal PageRank estimates, where $r(u, t)$ stores the value of temporal PageRank (t-PR) of node u at time t. Algorithm 1 processes a sequence of interactions E, updates the counts $s(t)$ and $r(t)$ for each new interaction (u, v, t), and outputs r as a t-PR estimate.

Proposition 1. *Algorithm 1 computes temporal PageRank defined in Eq. (3).*

Proof. Algorithm 1 counts explicitly the weighted number of temporal walks. Lines 3 and 4 correspond to initiating a new walk with probability $1 - \alpha$. With probability α the last interaction is chosen to continue active walks that wait in node u (line 5). Line 7 (or 10, depending on transition probability β) increments the active walks (active mass) count in the node v with appropriate normalization $1 - \beta$. Line 8 (or 11) decrements the active mass count in node u. If the transition probability is $\beta = 1$, then the random walk chooses the first suitable arrived interaction to continue the walk. $\square$

Algorithm 1 processes all interactions E in *one* pass and $\mathcal{O}(n)$ space. We need $\mathcal{O}(1)$ space per node, leading to total $\mathcal{O}(n)$ space, while every interaction initiates a constant number of updates, leading to $\mathcal{O}(1)$ update time per interaction.

To compute personalized temporal PageRank for a given personalization h^* we perform normalization, defined by Eq. (4), and multiply terms $(1 - \alpha)$ in lines 3 and 4 by $\frac{h^*(u)}{h'(u)}$. Unless we know the distribution of temporal edges in advance, we need to learn h'. Thus, we obtain a 2-pass algorithm to calculate personalized temporal PageRank for a given personalization vector h^*.

3.2   Temporal Vs. Static PageRank

Temporal PageRank is defined to handle network dynamics and concept drifts. An intuitive property that one may expect is that if the edge distribution of the temporal edges remains constant, then temporal PageRank approximates static PageRank. In this section we show that indeed this is the case.

Consider a weighted directed graph $G_s = (V, E_s, w)$ and a time period $T = [1, .., T]$. Without loss of generality assume $\sum_{e \in E_s} w(e) = 1$ and let $N_{out}(u)$ be the out-link neighbors of u. Let edges $e \in E_s$ be associated with a sampling distribution $S_E : p[e = (u, v)] = w(e)$. A temporal graph $G = (V, E)$ is constructed by sampling T edges from G_s using S_E (probability to pick an edge into E is proportional to the weight of this edge in the static graph). We will consider a simple case of transition probability $\beta = 1$: a random walk takes the first suitable interaction to continue.

In the setting described above we can prove the following statement.

Proposition 2. *The expected values of temporal PageRank on graph $G = (V, E)$ converge to the values of static PageRank on graph $G_s = (V, E_s, w)$, with personalization vector $h(u) = \sum_{v \in N_{out}(u)} w(e = (u, v))$ (weighted out-degree).*

Proof. At any time moment t every vertex $u \in V$ has PageRank score $r(u, t)$ and active mass (number of walkers that wait to continue) equal to $s(u, t)$.

The expected value $\mathbb{E}(r(v, T))$ of the PageRank count of node v at time T is a sum over expected increments of $r(v)$ over time:

$$\mathbb{E}(r(v, T)) = \sum_{t=1}^{T} \mathbb{E}(\Delta r(v, t)).$$

At time t the increment of $r(v)$ can be caused by selecting an edge $e(t) = (v, q)$ with starting point in v and $q \in V$. In this case $r(v)$ is incremented by $(1 - \alpha)$. Another possibility to increment $r(v)$ is to select an edge $e(t) = (q, v)$ with u as an end point and $q \in V$. In this case $r(v)$ is incremented by $\alpha s(q, t)$, where $s(q, t)$ is a value of active mass in node q at time t. Let $p(u)$ be a probability that sampled interaction has u as its start point. Note, that

$$p(u) = \frac{\sum_{j \in V} w(e = (u, j))}{\sum_{i \in V} \sum_{j \in V} w(e = (i, j))},$$

that is, the normalized out-degree of u. Thus, $\mathbb{E}(\Delta r(v, t))$ can be written as

$$\mathbb{E}(\Delta r(v, t)) = (1 - \alpha)p(v) + \alpha \sum_{u \in V} p(u)p(v|u)\mathbb{E}(s(u, t)).$$

To calculate expected amount of active mass in $s(u, t)$, notice that $s(u, t)$ equals to total increments of $r(u)$ happened between the time moment, when edge with starting point in u was selected to update, and t:

$$\mathbb{E}(s(u, t)) = \Delta r(u, t)p(u) + (\Delta r(u, t) + \Delta r(u, t - 1))p(u)(1 - p(u)) + \cdots$$

$$\cdots + p(u)(1 - p(u))^{t-1} \sum_{t'=0}^{t-1} \Delta r(u, t - t') = \sum_{t'=0}^{t-1} \mathbb{E}(\Delta r(u, t - t'))p(u) \sum_{k=t'}^{t-1} (1 - p(u))^k$$

The inner sum is a geometric progression:

$$\mathbb{E}(s(u,t)) = \sum_{t'=0}^{t-1} \mathbb{E}(\Delta r(u,t-t'))p(u)\frac{1}{p(u)}\left[(1-p(u))^{t'} - (1-p(u))^{t}\right].$$

We sum $\mathbb{E}(s(u,t))$ over time and consider the two summations separately:

$$\sum_{t=1}^{T}\mathbb{E}(s(u,t)) = \sum_{t=1}^{T}\sum_{t'=0}^{t-1}\mathbb{E}(\Delta r(u,t-t'))(1-p(u))^{t'}$$

$$- \sum_{t=1}^{T}\sum_{t'=0}^{t-1}\mathbb{E}(\Delta r(u,t-t'))(1-p(u))^{t}.$$

The first summation term can be written as:

$$\sum_{t=1}^{T}\sum_{t'=0}^{t-1}\mathbb{E}(\Delta r(u,t-t'))(1-p(u))^{t'} = \sum_{t=1}^{T}\mathbb{E}(\Delta r(u,t))\sum_{t'=0}^{T-t}(1-p(u))^{t}.$$

The second summation term is:

$$\sum_{t=1}^{T}\sum_{t'=0}^{t-1}\mathbb{E}(\Delta r(u,t-t'))(1-p(u))^{t} = \sum_{t=1}^{T}\mathbb{E}(\Delta r(u,t))(1-p(u))^{t}\sum_{t'=0}^{T-t}(1-p(u))^{t}.$$

Putting the parts together:

$$\sum_{t=1}^{T}\mathbb{E}(s(u,t)) = \sum_{t=1}^{T}\mathbb{E}(\Delta r(u,t))(1-(1-p(u))^{t})\sum_{t'=0}^{T-t}(1-p(u))^{t}$$

$$= \sum_{t=1}^{T}\mathbb{E}(\Delta r(u,t))(1-(1-p(u))^{t})\frac{1}{p(u)}(1-(1-p(u))^{T-t+1}).$$

Now the expected total increment $\mathbb{E}(r(v,T))$ can be expressed as:

$$\mathbb{E}(r(v,T)) = (1-\alpha)\sum_{t=1}^{T}p(v)$$

$$+ \alpha\sum_{u\in V}p(v|u)\sum_{t=1}^{T}\mathbb{E}(\Delta r(u,t))(1-(1-p(u))^{t})(1-(1-p(u))^{T-t+1}).$$

We need to show that

$$\lim_{T\to\infty}\frac{\mathbb{E}(r(v,T))}{T} = (1-\alpha)p(v) + \alpha\lim_{T\to\infty}\sum_{u\in V}p(v|u)\frac{\mathbb{E}(r(u,T))}{T}. \qquad (5)$$

Let us upper-bound $\mathbb{E}(\Delta r(v,t))$. Consider a time moment $t' \leq t$. A value of mass introduced to the system at t' is $(1-\alpha)$. This mass can arrive to the node

v at time moment t through a sequence of $t - t'$ steps of transmission (when a node u, which currently holds this mass, was chosen for action) or retainment (a node u was not chosen for action and the mass remains in u). Transmission happens with probability $p(u)\alpha$; the probability of retainment is $1 - p(u)$. Define $p = \max_{v \in V}\{1 - p(v), \alpha p(v)\}$. Then the expected value remained from this mass is upper-bounded by $(1 - \alpha)p^{(t-t')}$. The sum of all introduced bits of mass is an upper-bound for the active mass expected to enter node v at time t:

$$\mathbb{E}(\Delta r(v,t)) \le \sum_{t'=1}^{T}(1-\alpha)p^{t'} \le (1-\alpha)\frac{p(1-p^t)}{1-p} \le \frac{1}{1-p}$$

Now we need to show that the following limit goes to 0:

$$\lim_{T\to\infty}\frac{1}{T}\sum_{t=1}^{T}\mathbb{E}(\Delta r(u,t))((1-p(u))^{T+1}-(1-p(u))^t-(1-p(u))^{T-t+1})$$

$$= \lim_{T\to\infty}\frac{1}{T}\sum_{t=1}^{T}\mathbb{E}(\Delta r(u,t))(1-p(u))^{T+1} - \lim_{T\to\infty}\frac{1}{T}\sum_{t=1}^{T}\mathbb{E}(\Delta r(u,t))(1-p(u))^t$$

$$- \lim_{T\to\infty}\frac{1}{T}\sum_{t=1}^{T}\mathbb{E}(\Delta r(u,t))(1-p(u))^{T-t+1}$$

Consider three limits separately. The first one:

$$\lim_{T\to\infty}\frac{1}{T}\sum_{t=1}^{T}\mathbb{E}(\Delta r(u,t))(1-p(u))^{T+1} = \lim_{T\to\infty}\frac{(1-p(u))^{T+1}}{T}\sum_{t=1}^{T}\mathbb{E}(\Delta r(u,t))$$

$$\le \lim_{T\to\infty}\frac{p^{T+1}}{T}\sum_{t=1}^{T}\frac{1}{1-p} = 0$$

The second one:

$$\lim_{T\to\infty}\frac{1}{T}\sum_{t=1}^{T}\mathbb{E}(\Delta r(u,t))(1-p(u))^t \le \lim_{T\to\infty}\frac{\sum_{t=1}^{T}p^t}{T(1-p)} = \lim_{T\to\infty}\frac{p-p^{T+1}}{T(1-p)^2} = 0$$

The third one:

$$\lim_{T\to\infty}\frac{1}{T}\sum_{t=1}^{T}\mathbb{E}(\Delta r(u,t))(1-p(u))^{T-t+1} \le \lim_{T\to\infty}\frac{p^{T+1}}{T(1-p)}\sum_{t=1}^{T}p^{-t}$$

$$= \lim_{T\to\infty}\frac{p^{T+1}}{T(1-p)}\frac{p^{-T}-1}{1-p} \le \lim_{T\to\infty}\frac{p}{T(1-p)^2} = 0$$

It follows that Expression (5) is true. Now, if we define $pr(v) = \lim_{T\to\infty}\frac{1}{T}\mathbb{E}(r(v,T))$, then Expression (5) can be written as personalized PageRank in a steady state:

$$pr(v) = (1-\alpha)p(v) + \alpha\sum_{u\in V}p(v|u)pr(u) \qquad \square$$

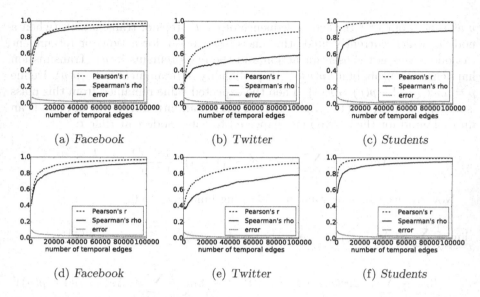

Fig. 2. Convergence of temporal PageRank to static PageRank. The first row (a, b, c) corresponds to degree personalization, the second row (d, e, f) corresponds to random personalization, given a priori.

4   Experimental Evaluation

To further support our theoretical analysis, we provide an empirical evaluation of temporal PageRank. The implementation of all algorithms and scripts are publicly available.[1] We first describe our experimental setup.

Datasets. We consider semi-real temporal networks, constructed by using real-world directed networks with edge weights equal to the frequency of corresponding interaction. In particular, we consider the following networks: *Facebook*, *Twitter* and *Students*. For each such network we extract static subgraphs $G_s = (V, E_s, w)$ with $n = 100$ nodes, obtained by BFS from a random node. We normalize edge weights w to sum to 1. Then we sample a sequence of temporal edges E, such that each edge $e \in E_s$ is sampled with probability proportional to its weight $w(e)$; the distribution of sampled edges is denoted by $\mathcal{S}_{E(w)}$. The number of temporal edges E is set to $m = 100\,\mathrm{K}$.

The *Facebook* dataset is a 3-month subset of Facebook activity in a New Orleans regional community [23]. The dataset contains an anonymized list of wall posts (interactions). The *Twitter* dataset tracks activity of Twitter users in Helsinki during 08.2010–10.2010. As interactions we consider tweets that contain mentions of other users. The *Students* dataset[2] is an activity log of a student online community at the University of California, Irvine. Nodes represent students and edges represent messages.

---

[1] https://github.com/polinapolina/temporal-pagerank.
[2] https://toreopsahl.com/datasets/online_social_network.

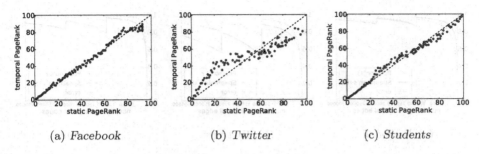

(a) *Facebook*   (b) *Twitter*   (c) *Students*

Fig. 3. Comparison of temporal PageRank ranking with static PageRank ranking, degree personalization is used.

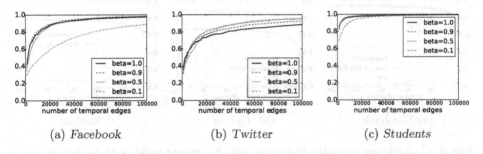

(a) *Facebook*   (b) *Twitter*   (c) *Students*

Fig. 4. Rank quality (Pearson corr. coeff.) and transition probability β.

Measures. To evaluate the settings in which temporal PageRank is expected to converge to the static PageRank of a corresponding graph, we compare temporal and static PageRank using three different measures: we use (i) Spearman's ρ to compare the induced rankings, we also use (ii) Pearson's correlation coefficient r, and (iii) Euclidean distance ϵ on the PageRank vectors.

All the reported experimental results are averaged over 100 runs. Damping parameter is set of $\alpha = 0.85$. Waiting probability β for temporal PageRank is set to 0 unless specified otherwise.

4.1 Results

Convergence. In the first set of experiments we test how fast the temporal PageRank algorithm converges to corresponding static PageRank. In this setting we process datasets with m temporal edges and compare the temporal PageRank ranking with the corresponding static PageRank ranking. In the plots of Fig. 2 we report Pearson's r, Spearman's ρ and Euclidean error ϵ. The first column corresponds to the calculation of temporal PageRank without any a priori knowledge of personalization vector. Thus, the resulting temporal PageRank corresponds to the static PageRank with out-degree personalization: $h(u) = \sum_{v \in N_{\text{out}}(u)} w(u, v)$, where $N_{\text{out}}(u)$ are out-link neighbors of u. The second column shows convergence in the case when the personalization vector h^* is given and appropriate renormalization of t-PR counts is taking place.

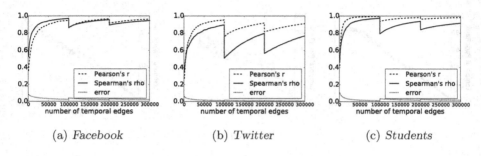

(a) *Facebook*   (b) *Twitter*   (c) *Students*

Fig. 5. Adaptation for the change of sampling distribution.

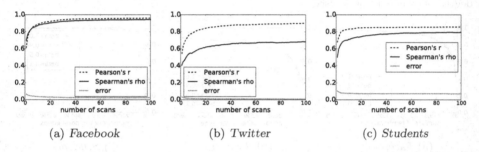

(a) *Facebook*   (b) *Twitter*   (c) *Students*

Fig. 6. Convergence to static PageRank with increasing number of random scans of edges.

The plots in Fig. 2 show that in both variants of personalization the behavior is similar: in most cases the correlation of the PageRank counts reaches high values already after 20 K temporal edges. Pearson's r is remarkably high, while Spearman's ρ is typically lower. This can be explained by the large number of discordant pairs in the tail of ranking — due to producing a power-law distribution PageRank is known to give robust rankings only at the top of the ranking list. The Euclidean error ϵ also decreases to near-zero values fast.

In Fig. 3 we show direct comparison between rankings, obtained by static and temporal PageRanks after processing all temporal edges. We observe that the rank correlation is high for top-ranked nodes and decreases towards the tail of ranking.

Transition probability β. In this experiment we evaluate the dependence of the resulting ranking and the speed of convergence on the transition probability β. The plots in Fig. 4 show that lower transition probability β corresponds to slower convergence rate. On the other hand, smaller values of β produce better correlated rankings. This behavior is intuitive, as a lower value for β implies accumulation of more information regarding the possible walks, which in turn implies a slower convergence rate.

Adaptation to concept drifts. In this experiment we test whether temporal PageRank is adaptive to concept drifts. We start with a temporal network sampled from some static network $G_s^1 = (V, E_s, w_1)$. After sampling m temporal edges E_1, we change the weights of the static graph and sample another m temporal edges E_2 from $G_s^2 = (V, E_s, w_2)$. A final sequence of m edges E_3 is sampled from $G_s^3 = (V, E_s, w_3)$. We run our algorithm on the concatenated sequence $E = \langle E_1, E_2, E_3 \rangle$, without a priori personalization. On Figure 5 we report correlation with the corresponding ground-truth static PageRank. The transition probability β is set to 0.5. In all cases, temporal PageRank is able to adapt to the changing distribution quite fast. Note however, that the previous history is not completely eliminated and for each change of the distribution an increasing number of edges is required to reach a certain correlation level.

Random scans. In the last experiment, given a static graph $G_s = (V, E_s, w)$ we generate a sequence of temporal E by scanning the edges E_s in random order several times. Figure 6 shows that as the number of scans increasing, our estimate for temporal PageRank converges to the static PageRank of the graph. We see that the correlation obtains high values even after a few (around 10) scans. This experiments suggests a very simple and efficient algorithm to compute the static PageRank of a graph, by running our algorithm on a small number of linear scans (randomly ordered) on the graph edges.

5   Related Work

PageRank is one of the most popular measures for ranking the nodes of a network according to their importance. The original idea was introduced by Page and Brin [20] for application to web search, and since then it is widely used as a graph-mining tool. As the size of typical networks has increased significantly over the last years, and as networks tend to grow and evolve fast, research on designing scalable algorithms for computing PageRank is still active [16].

A different line of research is dedicated to efficient approaches for updating PageRank in dynamic and/or online scenarios [4,5,12,19,22]. The term "dynamic" is typically used to refer to the model of edge additions and deletions. However, we discussed in the introduction, even in these dynamic settings PageRank is defined as a stationary distribution over a static graph (the current graph). Another research direction uses temporal information to calculate weights of edges of a static graph [10,17].

On the contrary, temporal PageRank intends to capture the continuous interaction between individuals. Temporal PageRank is defined over temporal networks [9,18], where each edge has an associated time-stamp recording an interaction at that point. To our knowledge there is no published work, which considers temporal generalization of PageRank. The closest work is dedicated to Bonacich's centrality [15]. It focuses on empirical study of a citation network with coarse snapshots, aggregated over a year. In contrast, we are interested in theoretical relation between temporal and static PageRanks and test our methods on several networks with fine granularity.

The static Pagerank definition has multiple interpretations, extensively discussed in a survey by Langville et al. [14]. Our definition of temporal PageRank has a random walk-based interpretation inspired by the one given for static

PageRank [3]. Methodologically, the closest papers to our work, are Monte-Carlo simulation algorithms [2] and PageRank calculation by local updates [1, 7].

6   Concluding Remarks

We proposed a generalization of static PageRank for the case of temporal networks. The novelty of our approach relies on the fact that we explicitly take into account the exact time that nodes interact, which leads to more accurate ranking. The main feature of the generalization is that it takes into account structural network changes, and models the fact that the importance of nodes may change during the lifetime of the network, according to changes in the distribution of edges. Additionally, we showed that if the distribution of edges remains stable, the temporal PageRank converges to the static PageRank. We provided an efficient algorithm to calculate temporal PageRank and demonstrated its quality and convergence rate through multiple experiments on diverse datasets.

Acknowledgements. This work is partially supported by the Academy of Finland project "Nestor" (286211) and the EC H2020 RIA project "SoBigData" (654024).

References

1. Andersen, R., Chung, F., Lang, K.: Local graph partitioning using PageRank vectors. In: 2006 47th Annual IEEE Symposium on Foundations of Computer Science, FOCS 2006, pp. 475–486. IEEE (2006)
2. Avrachenkov, K., Litvak, N., Nemirovsky, D., Osipova, N.: Monte Carlo methods in PageRank computation: when one iteration is sufficient. SIAM J. Numer. Anal. **45**(2), 890–904 (2007)
3. Baeza-Yates, R., Boldi, P., Castillo, C.: Generalizing PageRank: damping functions for link-based ranking algorithms. In: Proceedings of the 29th Annual International ACM SIGIR Conference on Research and Development in Information Retrieval, pp. 308–315. ACM (2006)
4. Bahmani, B., Chowdhury, A., Goel, A.: Fast incremental and personalized PageRank. Proc. VLDB Endowment **4**(3), 173–184 (2010)
5. Bahmani, B., Kumar, R., Mahdian, M., Upfal, E.: PageRank on an evolving graph. In: Proceedings of the 18th ACM SIGKDD International Conference on Knowledge Discovery and Data Mining, pp. 24–32. ACM (2012)
6. Berlocher, I., Lee, K.i., Kim, K.: TopicRank: bringing insight to users. In: SIGIR (2008)
7. Bianchini, M., Gori, M., Scarselli, F.: Inside PageRank. ACM Trans. Internet Technol. (TOIT) **5**(1), 92–128 (2005)
8. Gyöngyi, Z., Garcia-Molina, H., Pedersen, J.: Combating web spam with TrustRank. In: VLDB (2004)
9. Holme, P., Saramäki, J.: Temporal networks. Phy. Rep. **519**(3), 97–125 (2012)
10. Hu, W., Zou, H., Gong, Z.: Temporal PageRank on social networks. In: Wang, J., Cellary, W., Wang, D., Wang, H., Chen, S.-C., Li, T., Zhang, Y. (eds.) WISE 2015. LNCS, pp. 262–276. Springer, Heidelberg (2015). doi:10.1007/978-3-319-26190-4_18

11. Jeh, G., Widom, J.: SimRank: a measure of structural-context similarity. In: KDD (2002)
12. Kim, K.S., Choi, Y.S.: Incremental iteration method for fast PageRank computation. In: Proceedings of the 9th International Conference on Ubiquitous Information Management and Communication, p. 80. ACM (2015)
13. Kumar, R., Calders, T., Gionis, A., Tatti, N.: Maintaining sliding-window neighborhood profiles in interaction networks. In: Appice, A., Rodrigues, P.P., Santos Costa, V., Gama, J., Jorge, A., Soares, C. (eds.) ECML PKDD 2015. LNCS (LNAI), pp. 719–735. Springer, Heidelberg (2015). doi:10.1007/978-3-319-23525-7_44
14. Langville, A.N., Meyer, C.D.: Deeper inside PageRank. Internet Math. 1(3), 335–380 (2004)
15. Lerman, K., Ghosh, R., Kang, J.H.: Centrality metric for dynamic networks. In: Proceedings of the Eighth Workshop on Mining and Learning with Graphs, pp. 70–77. ACM (2010)
16. Lofgren, P.A., Banerjee, S., Goel, A., Seshadhri, C.: FAST-PPR: scaling personalized PageRank estimation for large graphs. In: Proceedings of the 20th ACM SIGKDD International Conference on Knowledge Discovery and Data Mining, pp. 1436–1445. ACM (2014)
17. Manaskasemsak, B., Teerasetmanakul, P., Tongtip, K., Surarerks, A., Rungsawang, A.: Computing personalized PageRank based on temporal-biased proximity. In: Park, J.J.J.H., Barolli, L., Xhafa, F., Jeong, H.Y. (eds.) Information Technology Convergence. LNEE, pp. 375–385. Springer, Heidelberg (2013). doi:10.1007/978-94-007-6996-0_39
18. Michail, O.: An introduction to temporal graphs: an algorithmic perspective. In: Zaroliagis, C., Pantziou, G., Kontogiannis, S. (eds.) Algorithms, Probability, Networks, and Games. LNCS, vol. 9295, pp. 308–343. Springer, Heidelberg (2015). doi:10.1007/978-3-319-24024-4_18
19. Ohsaka, N., Maehara, T., Kawarabayashi, K.i.: Efficient PageRank tracking in evolving networks. In: Proceedings of the 21th ACM SIGKDD International Conference on Knowledge Discovery and Data Mining, pp. 875–884. ACM (2015)
20. Page, L., Brin, S., Motwani, R., Winograd, T.: The PageRank citation ranking: bringing order to the web (1999)
21. Rozenshtein, P., Tatti, N., Gionis, A.: Discovering dynamic communities in interaction networks. In: Machine Learning and Knowledge Discovery in Databases, pp. 678–693 (2014)
22. Sarma, A.D., Gollapudi, S., Panigrahy, R.: Estimating PageRank on graph streams. J. ACM (JACM) 58(3), 13 (2011)
23. Viswanath, B., Mislove, A., Cha, M., Gummadi, K.P.: On the evolution of user interaction in facebook. In: Proceedings of the 2nd ACM Workshop on Online Social Networks, pp. 37–42. ACM (2009)
24. Wu, X., et al.: Top 10 algorithms in data mining. Knowl. Inf. Syst. 14(1), 1–37 (2008)

Discovering Topically- and Temporally-Coherent Events in Interaction Networks

Han Xiao[1]([⊠]), Polina Rozenshtein[2], and Aristides Gionis[2]

[1] Department of Computer Science, Helsinki Institute for Information Technology,
University of Helsinki, Helsinki, Finland
`hxiao@cs.helsinki.fi`

[2] Department of Computer Science, Helsinki Institute for Information Technology,
Aalto University, Espoo, Finland
`{polina.rozenshtein,aristides.gionis}@aalto.fi`

Abstract. With the increasing use of online communication platforms, such as email, Twitter, and messaging applications, we are faced with a growing amount of data that combine *content* (what is said), *time* (when), and *user* (by whom) information. Discovering meaningful patterns and understand what is happening in this data is an important challenge. We consider the problem of mining online communication data and finding top-k *temporal events*. A temporal event is a coherent topic that is discussed frequently in a relatively short time span, while its information flow respects the underlying network.

Our method consists of two steps. We first introduce the notion of *interaction meta-graph*, which connects associated interactions. Using this notion, we define a *temporal event* to be a subset of interactions that (*i*) are topically and temporally close and (*ii*) correspond to a tree that captures the information flow. Finding the best temporal event leads to a budget version of the prize-collecting Steiner-tree (PCST) problem, which we solve using three different methods: a greedy approach, a dynamic-programming algorithm, and an adaptation to an existing approximation algorithm. Finding the top-k events maps to a maximum set-cover problem, and thus, solved by greedy algorithm. We compare and analyze our algorithms in both synthetic and real datasets, such as Twitter and email communication. The results show that our methods are able to detect meaningful temporal events. The software related to this paper are available at https://github.com/xiaohan2012/lst.

Keywords: Social-network analysis · Temporal networks · Event detection

1 Introduction

Event detection is a fundamental data-mining problem in many different domains, such as, time series and data streams [10], point clouds and vector

Electronic supplementary material The online version of this chapter (doi:10.1007/978-3-319-46227-1_43) contains supplementary material, which is available to authorized users.

© Springer International Publishing AG 2016
P. Frasconi et al. (Eds.): ECML PKDD 2016, Part II, LNAI 9852, pp. 690–705, 2016.
DOI: 10.1007/978-3-319-46227-1_43

spaces [4], and networks [3]. In this paper we focus on the problem of detecting events in networks, in particular, networks that contain both *content* and *time* information. An *interaction* (u, v, α, t) occurs whenever a piece of information α is exchanged between two network entities u and v at time t. Examples of interaction networks include data communication networks, such as email, Twitter, or online messaging systems.

Our goal is to summarize the network activity by finding the top-k events. We consider an interaction network $H = (N, \mathcal{I})$, where interactions $\mathcal{I}$ take place among a set of network entities N. The interactions in $\mathcal{I}$ are directed, annotated with content information, and time-stamped. We define an *event* in the interaction graph H to be a subset of interactions, $\mathcal{I}' \subseteq \mathcal{I}$ that are (*i*) temporally close, (*ii*) topically similar, and (*iii*) correspond to a tree that captures the information flow in the network. The intuition behind representing events as trees is similar to the work by Yang [19].

We convert the interaction network $H = (N, \mathcal{I})$ into a weighted *interaction meta-graph* $G = (\mathcal{I}, E)$, that is, a graph whose vertices are the interactions $\mathcal{I}$. Two interactions $i, j \in \mathcal{I}$ are connected in G if it is possible to explain the information flow between i and j. In particular, we consider three types of flow: *broadcast*, *relay* and *reply*. The edge weights of the interaction meta-graph G measure the topic dissimilarity between connected interactions. Our transformation from the interaction network to the interaction meta-graph has the interesting property that an *event* in the interaction graph H corresponds to a tree T in the interaction meta-graph G. The root of the tree T is interpreted as the source of the event. Downstream interactions (interactions that are reachable from the root) are due to information propagation.

Motivated by the previous discussion, we formalize the task of interaction-network summarization as the problem of finding top-k trees in the transformed interaction meta-graph $G = (\mathcal{I}, E)$. We decompose this task into two sub-problems. First, we find a set of independent candidate events that are temporally and topically coherent. Since our goal is to summarize the interaction network we aim to find large events. We show that this problem is the budget version of prize-collecting Steiner-tree problem in directed acyclic graphs. We provide three algorithms, among which a greedy approach performs the best.

The second sub-problem is to select k events that maximize the overall node coverage. This task maps to the maximum set-cover problem, and it can be approximated using a standard greedy algorithm. To speed up further our algorithm, we also propose a search strategy that avoids evaluating candidate events at all possible tree roots, but heuristically selects the most promising ones.

Example 1. Consider the email communication network of a company, such as the one shown in Fig. 1. The interaction network is shown in Fig. 1(a) and the corresponding interaction meta-graph in Fig. 1(b). The edges between interactions $(2, 4)$, $(1, 2)$, $(2, 3)$ in Fig. 1(b) are examples of edge types broadcast, relay, and reply, respectively. In this toy example there are two main events. (*i*) *progress*: The CEO asks a project manager (PM) about progress on a project, and the PM forwards the request to team members 1 (TM1) and 2 (TM2). Later, TM1 reports

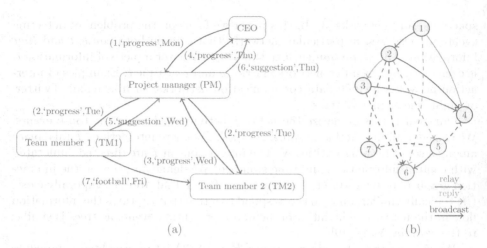

Fig. 1. A toy example showing the email communication network within a company. (a) The interaction network. Each edge corresponds to one interaction/email, labeled as (`interaction id`, `message topic`, `timestamp`). (b) The corresponding interaction meta-graph. Topics in both graphs are depicted using different colors. Edges in the interaction meta-graph are depicted by a different color according to their type (relay, reply, or broadcast). Edges are solid if they have small weight (topic dissimilarity). Otherwise, they are dashed.

back to PM, who in turn reports back to CEO. The information flow of this event follows the interactions $1 \rightarrow 2 \rightarrow 3 \rightarrow 4$. (*ii*) *suggestion*: Motivated by the first event, TM2 comes up with some suggestion, which she sends to PM. The PM finds the suggestion useful and forwards it to CEO. The information flow of this second event is $5 \rightarrow 6$. A third event, *football*, is smaller in size, and it is not included in the top-2 events. Note that due to time ordering of the interactions, the interaction meta-graph G is a directed acyclic graph.

The problem considered in this paper has many applications in different domains. In our experimental evaluation, we focus on analyzing textual data in social media. We experiment with one email dataset (Enron) and three Twitter datasets. We provide a comparison of the different approaches, as well as many examples in which our methods discover meaningful events.

The contributions of this paper are summarized as follows[1]:

- We propose a novel formulation for the problem of discovering events that are temporally and topically coherent in interaction networks, such as, online communication networks.
- We present a transformation of the interaction network to an interaction meta-graph, which captures temporal and topical association of interactions as well as the information flow in the network. This transformation helps to provide a cleaner abstraction to the event-detection problem.

---

[1] All scripts are available at https://github.com/xiaohan2012/lst.

- For the problem of finding high-volume events while satisfying constraints of temporal and topical coherence we present and we evaluate three different algorithms: a greedy approach, a dynamic-programming algorithm, and an adaptation to an existing approximation algorithm.
- We address the problem of finding the top-k events that summarize the network activity. The classic greedy algorithm is the standard way to approach this problem, but here, to speed-up the computations, we also propose and evaluate a search strategy that avoids construction of candidate events at all possible tree roots, but adaptively selects the most promising ones.
- We compare and analyze our algorithms on both synthetic and real datasets, such as Twitter and email communication. We show that our methods are able to detect meaningful temporal events.

2   Related Work

Phrase-based event detection. The problem of detecting events in social media has attracted significant attention. Leskovec et al. [14] and Yang et al. [20] treat events as short, distinctive phrases that propagate relatively intact through in a network. Their work offers a graph formulation for clustering variants of phrases based on string edit distance. Although their objective is similar to ours, there are significant differences. First, our methods focus on interaction networks, aiming to capture information flows in communication networks, rather than action networks. Second, we explicitly impose topic-coherence constraints, where the edit distance is insufficient for this goal. Third, instead of representing events by phrases, we derive higher-level representation using topic terms.

Text summarization. Text summarization techniques attempt to select a subset of sentences [6] or tweets [11] to summarize textual content. Similarly, we select a subset of interactions under a topic-coherence constraint. However, we also impose temporal coherence constraint, whereas they take a static view.

Statistical methods. Statistical and machine learning approaches for event detection are gaining increasing attention in recent years. Mathioudakis et al. [15] develop an interactive system for identifying trends (events). The system first identifies "bursty" keywords, then clusters them based on co-occurrence and later performs trend analysis using dimension-reduction methods. Becker et al. [1] focus on online event identification. Their approach relies on online clustering techniques in order to discover topically-related tweets as an event and feature-based modeling in order to distinguish events from non-events. The difference of this approach with our work is that we offer a graph-theoretic formulation.

Graph-based methods. Other event-detection methods are based on constructing a word graph [5,16,18]. Weng et al. [18] combines wavelet analysis and graph-partitioning techniques to cluster the words into events. Meladianos et al. [16] construct a word graph to represent a sequence of tweets, however, they focus on identifying key sub-events inside the sequence. Cataldi et al. [5] detect events by locating strongly connected components. Compared to those approaches, in this paper we explicitly model interactions, and take into account temporal constraints and topical-coherence constraints.

3   Model

An *interaction network* $H = (N, \mathcal{I})$ consists of a set of n nodes N and a set of m time-stamped interactions $\mathcal{I}$ between pairs of nodes. $\mathcal{I}$ is represented as:

$$\mathcal{I} = \{(u_i, v_i, \alpha_i, t_i)\}, \text{ with } i = 1, \dots, m, \text{ such that } u_i, v_i \in N, \ t_i \in \mathbb{R}, \ \alpha_i \in \mathbb{R}^L,$$

indicating that nodes u_i and v_i interacted at time t_i. Each interaction is annotated with textual content represented by α_i. The representation is independent to our main methodology. We can use various text modeling techniques such as bag-of-words representation or latent Dirichlet allocation (LDA) [2].

For generality we consider that interactions are directed. More than one interaction may take place between a pair of nodes, with different timestamps. Conversely, more than one interaction may take place at the same time, between different nodes. Online communication networks, such as email networks, are examples of interaction networks.

Given an interaction network H we construct a directed weighted *interaction meta-graph* $G = (\mathcal{I}, E, c)$. The vertices $\mathcal{I}$ in G correspond to the interactions $\mathcal{I}$ in H. There is an edge from vertex $i = (u_i, v_i, \alpha_i, t_i) \in \mathcal{I}$ to a vertex $j = (u_j, v_j, \alpha_j, t_j) \in \mathcal{I}$ if the following holds:

1. Interaction i takes place before interaction j (time comprehension): $t_i \leq t_j$.
2. Information comprehension takes place in one of the following ways:
   (a) interactions i and j share the same start node in N: $u_i = u_j$ (broadcast);
   (b) the end node of interaction i is the start node of interaction j and the end node of j is not the start node of i: $v_i = u_j$ and $v_j \neq u_i$ (relay);
   (c) the end node of an interaction i is the start node of an interaction j and the end node of j is the start node of i: $v_i = u_j$ and $v_j = u_i$ (reply).

Note, that due time comprehension the G is a directed acyclic graph (DAG).

For the edges of the interaction meta-graph G we use weights to measure the topical (dis)similarity between interactions. Thus, given two interactions $(u_i, v_i, \alpha_i, t_i)$ and $(u_j, v_j, \alpha_j, t_j)$ connected by an edge in G, our edge-weighting function $c : E \to \mathbb{R}$ is a distance function between topic vectors α_i and α_j.

Finally, given a meta-graph $G = (\mathcal{I}, E, c)$ and a time interval $[s, f]$ we define the *time-induced meta-graph* $G([s, f]) = (\mathcal{I}([s, f]), E, c)$, where $\mathcal{I}([s, f])$ are the interactions that occur in $[s, f]$: $\mathcal{I}([s, f]) = \{(u, v, \alpha, t) \in \mathcal{I} \mid s \leq t \leq f\}$.

4   Problem Formulation

We aim at summarizing the top-k events in an interaction network. We define an *event* to be a rooted subtree T of the interaction meta-graph G. An event naturally has a source vertex (or interaction) and is spread in the network. We are interested in events of high volume, which translates into a large number of iterations included into the tree T. We are also interested in events with temporally close and topically coherent interactions.

These aspects can be incorporated into the optimization cost function in different ways. Our primary objective is to obtain k events that have high enough coverage to represent the whole network, and thus, we aim to maximize the number of interactions that are included in the event. To incorporate temporal and topical coherence we set constraints on the time interval spanned by the event tree (temporal coherence), and on total weight of its edges (topical coherence).

To simplify the problem of finding the best k events, we decompose the main task into two subproblems: (1) finding a set of independent candidate events that satisfy the constraints and maximize volume of interactions, and (2) selecting the top-k events to maximize total coverage. The first problem is defined as follows.

Problem 1. **Time-constrained maximum tree** (TMAXTREE): Given an interaction meta-graph $G = (\mathcal{I}, E, c)$, a root vertex $r \in \mathcal{I}$, time budget I, and dissimilarity budget B, find a directed subtree $T = (V_e, E_e) \subseteq G$, rooted at r, which satisfies the constraints

$$\sum_{e \in E_e} c(e) \leq B \quad \text{and} \quad (\max_{i \in V_e} t_i - \min_{j \in V_e} t_j) \leq I,$$

while maximizing the number of vertices $|V_e|$.

Note that the time constraint can be omitted, if we restrict the input graph to be induced by the time interval $[t_r, t_r + I]$, where t_r is the root timestamp. By omitting the time constraint, our problem can be written as follows.

Problem 2. **Maximum tree** (MAXTREE): Given a weighted directed acyclic graph $G([s, f]) = (\mathcal{I}([s, f]), E, c)$, a root vertex r, and cost budget B, find a subtree $T = (V_e, E_e) \subseteq G([s, f])$, rooted at r, that satisfies $\sum_{e \in E_e} c(e) \leq B$ while maximizing the number of vertices $|V_e|$.

We observe that MAXTREE is directly related to *budget* version of the prize-collecting Steiner-tree problem (PCST) [12]. However, we are dealing with a special case of the budget PCST, as vertex prize is uniform and our input graph is a DAG. Despite so, this special case is still **NP**-hard.

Proposition 1. MAXTREE *is **NP**-hard.*[2]

As the interaction network is likely to contain more than one event, we are interested in finding k events that describe different aspects of the whole network while covering as much activity as possible. This is captured in the following problem formulation.

---

[2] Proof can be found at appendix.

Problem 3. **Maximum k trees** (k-MAXTREES): We are given an interaction meta-graph $G = (\mathcal{I}, E, c)$ and $k \in \mathbb{N}$. Find a set of k vertex-disjoint trees $\mathcal{T} = \{T_1, \dots T_k\}$, with each event tree $T = (V_e, E_e) \in \mathcal{T}$ to be a subgraph of G rooted in some $r_i \in \mathcal{I}$, such that the total number of spanned interactions $|\cup_{T=(V_e,E_e)\in\mathcal{T}} V_e|$ is maximized.

It is easy to observe that this problem is equivalent to maximum k-coverage problem and thus is **NP**-hard. To solve k-MAXTREES efficiently, we consider the question of sampling as few root vertices as possible so that the major events can still be captured. Real-world networks consist of millions of interactions so it is impractical to calculate candidate event trees rooted at each vertex.

5 Algorithms

5.1 Approximating MAXTREE

For finding the best tree, as defined by MAXTREE, we consider three algorithms. Recall that for MAXTREE we are working with the interaction meta-graph G, and that a root vertex is fixed.

Greedy tree growing: The greedy algorithm starts from the root and builds the event tree by adding one vertex (interaction) at a time. At each step the algorithm selects the edge with the minimum cost (topic dissimilarity) from the cutset of the current tree. This choice aims to maximize the topical coherent of the event discovered. The running time is $\mathcal{O}(|\mathcal{I}|^2)$.

Directed Steiner tree algorithm (DST): Recall that MAXTREE corresponds to the *budget* PCST problem. Our second algorithm is inspired by an approach proposed by Johnson et al. [12], where the the *budget* PCST problem can be solved by the *quota* PCST problem using binary search. In our case, the prizes of all vertices are uniform, thus the *quota* PCST problem is equivalent to *k-minimum spanning tree*. The latter problem can be solved by an algorithm for finding directed Steiner trees (DST), such as the algorithm proposed by Charikar et al. [7]. Thus, our second algorithm uses the DST algorithm, within a binary search to find an event that satisfies the budget constraint. The DST algorithm takes four arguments, G, r, X, and ℓ, where X is a set of terminal nodes and ℓ is a parameter that provides a quality-of-approximation vs. efficiency trade-off. The running time of the algorithm is $\mathcal{O}(|\mathcal{I}|^\ell |X|^{2\ell})$. In our case, $X = \mathcal{I}$, thus the running time is $\mathcal{O}(|\mathcal{I}|^{3\ell})$. We use $\ell = 1$ but still the algorithm is mainly of theoretical interest and not practical for large datasets.

Dynamic programming algorithm (DP): The third algorithm we present is inspired by the idea that when the input DAG is a tree, the problem can be solved optimally using a simple dynamic programming approach. We investigate two approaches to adapt this algorithm for general (non-tree) DAGs.

In the first approach, we slightly modify the dynamic programming algorithm to make sure the result is a tree. Specifically, when attempting to connect the current node with the subtrees of its children, we enforce the condition that the subtrees cannot have any common nodes. In the second approach, we transform the input DAG into a tree and then apply the original dynamic programming algorithm. Specifically, we first calculate single-source shortest paths from r to all vertices of G using Dijkstra's algorithm and then apply the dynamic programming algorithm. For integer edge weights and a tree input, the running time is $\mathcal{O}(|\mathcal{I}|B^2)$. In our case, edge weights are real numbers, so we discretize the weights to some decimal digits.

5.2 Approximating k-MaxTrees

Once we have computed a set of candidate event trees using any algorithm for MaxTree, we need to select k event trees from the candidate set so that vertex coverage is maximized. This is essentially the *maximum coverage* problem. A standard greedy algorithm gives approximation ratio $(1 - \frac{1}{e})$ in time $\mathcal{O}(|\mathcal{I}|^2)$ [17].

5.3 Root Sampling Strategy

One issue with the greedy max-cover algorithm discussed above, is that all candidate root vertices need to be tested before selecting the one that greedily optimizes the coverage. This is an expensive computational task. To speed up the algorithm for finding top-k trees, we propose a simple root-sampling strategy that ranks roots according to their potential of maximizing MaxTree.

For every sampled root r we construct a candidate tree D and evaluate *event size upper bound* $U(D, B)$ of DAG D with budget B, defined as:

$$U(D, B) = \max_{F' \in F(D)} \left\{ |F'.\mathcal{I}| \text{ such that } \sum_{e \in F'.E} c(e) \leq B \right\}$$

where $F(D)$ is a set of all forests containing $D.r$ (the root of DAG D).

It is easy to see the optimal tree $T(D, r, B)$ cannot have size greater than $U(D, B)$, thus $U(D, B)$ is indeed an upper bound.

Table 1. Network statistics on real datasets. Singleton interactions in the interaction meta-graph are removed.

Datasets	Interaction networks		Interaction meta-graphs		
	#nodes	#edges	#nodes	#edges	Period
Enron	1144	2106	812	21297	1998-10-30 - 2002-02-13
#beefban	11895	33584	26317	75870	2015-03-03 - 2015-03-05
#ukraine	16218	59096	46540	142746	2015-02-27 - 2015-03-03
#baltimore	38541	102139	61501	132012	2015-04-26 - 2015-04-28

Define the *minimum in-edge* of a vertex u as

$$e^*(G, u) = \underset{e' \in \delta^+(G,u)}{\operatorname{argmin}} \; c(e'),$$

where $\delta^+(G, u) = \{e \in G.E \mid e.i = u\}$. U can be computed efficiently as follows. Consider only nodes, which belong to $[t_r, t_r + I]$ time interval, where t_r is the root timestamp. Start constructing an event D by adding root r and its child with the lightest edge. Now sort all other nodes by cost of their minimum in-edge cost in increasing order; greedily add nodes with their minimum in-edge to the event D and stop when budget constraint B is reached. U is a number of nodes in the event D. Note that D is a forest, as we do not care about connectivity during construction.

Our root sampling strategy first ranks all the vertices by U. Then it sequentially selects vertices from the ranked list.

6 Experimental Evaluation

As no datasets with ground-truth events are available to us, we validate our approach by using synthetic datasets and by case studies. For the experiments with synthetic datasets: (1) we plant events (considered as ground truth) within random interaction networks; (2) we then apply our algorithm to find events in those synthetic data; (3) we measure the precision and recall of the discovered events with respect to ground-truth. For the case studies we apply our algorithm on *Enron* and on Twitter data, then examine the events we discover, and map them on real known historical events based on textual content and time period. As means of exploratory data analysis, we also visualize the event trees in order to show the information flow within the event.

6.1 Datasets and Preprocessing

Synthetic data. We generate synthetic datasets in two steps: (1) we generate ground-truth event trees; (2) we inject noise interactions. Each event is generated independently using the model by Kumar et al. [13], which constructs a tree by iteratively adding random edges. We sample a sender, recipients, timestamp and a topic vector randomly for each node.

Real-world data. We use two real-world datasets: email (*Enron*) and Twitter. Dataset statistics are given in Table 1. ***Enron:*** we use a preprocessed version of the original Enron dataset [8]. ***Twitter:*** we use Twitter datasets extracted for three hashtags, each one containing a specific hashtag. The hashtags are *#beefban*, *#baltimore* and *#ukraine*. There is a interaction from a user u to a user v, if the tweet of user v contains username of u. The Twitter datasets are provided by Garimella et al. [9].

Preprocessing. We observe the phenomenon that the same person sends the same (or very similar) messages multiple times, especially on Twitter. Our methods are easily misled by the sheer amount of redundant messages. To avoid this problem, we merge similar messages from the same sender into one. We consider two messages *similar* if (1) they are sent by the same user, (2) their Levenshtein edit distance ratio is below 10 %, (3) their time distance is relatively small (e.g., one day). In the newly-merged message, the text content, timestamps are copied from the earliest message. Recipients are the union of all recipients.

We take different approaches for representing interaction content in *Enron* and Twitter. For *Enron*, we train a topic model using *gensim*[3]. We assign each interaction a topic vector and use cosine distance to compute edge weight.

Measuring tweet similarity is an open challenge due to its short length and conciseness. We took an ensemble approach where vector representation comes from several models. Besides topic vectors, we use also (*i*) bag-of-word (BoW) with tf-idf re-weighting and (*ii*) hashtags included in each tweet. For BoW and hashtag representations, we use cosine and Jarccard distance for weight assignment, respectively. Last, we sum up the three distances. For topic modeling, for both *Enron* and Twitter datasets, we use 10 topics, batch size 100 and run it for 10 iterations.

6.2   Results on Synthetic Datasets

We evaluate five different algorithms for finding the best event: (1) greedy tree growing (*greedy*), (2) binary search using Charikar's DSP algorithm (*binary_search*), (3) dynamic programming without preprocessing (*DP*), (4) dynamic programming with Dijkstra preprocessing (*DP+dij*), and (5) random tree growing (*random*) as a baseline. The *random* algorithm mimics the *greedy*, but it selects a random edge to grow at each step. We compare quality of solutions obtained on datasets with various noise level. We define *noise level* as a number of noise interactions divided by the total number of interactions of all events. For the DSP algorithm we set level parameter $\ell = 1$, as we have insufficient memory for experiments with larger values.

Different noise levels. To compare the capability of the algorithms to find *one* best event, we generate a sequence of datasets with increasing noise levels and only one event of size 20 (containing 20 nodes). We set ground-truth values of I, B, r for parameters in MAXTREE. We consider three types of measurements: (1) precision, recall, and F1, (2) the value of objective function, and (3) the running time. Log scale is applied in the case of running time as difference between algorithms is of magnitudes order.

In Fig. 2(a), we see that all our algorithms outperform the trivial *random* baseline. Although *greedy* is a simple heuristic, its performance is among the top. Dijkstra preprocessing for *DP* improves both F1 and computational time. In the contrary, *binary_search* consumes much time, even though it is among the best in other measurements. Notice that *random* achieves high precision because it can select a wrong edge that violates the budget constraint at the first few steps and terminate.

---

[3] https://radimrehurek.com/gensim/models/ldamodel.html.

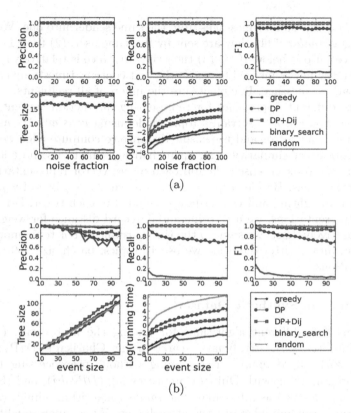

Fig. 2. (a) Performance of the algorithms under noise levels from 0 to 100 with step size 0.5. Results are averaged over 50 repetitions. (b) Performance of the algorithms on synthetic dataset with noise level 20 and varying event size from 10 to 100 at step size 10. Measurement values are averaged from 50 rounds.

Different event sizes. We also study how the algorithms perform in extracting events of different sizes. The experiment setting is similar to the above, but the noise level is fixed to 20, while the event size varies. In Fig. 2(b), *greedy*, *binary_search* are among the best in terms of precision, recall, F1 and set cover objective, whereas $DP+dij$ is slightly worse due to needed edge weight discretization. Again, preprocessing for DP improves performance. Running time comparison is consistent with the previous case.

6.3   Parameter Effects on Real Datasets

Effect of B. We evaluate the effect of topic dissimilarity budget B on the tree size objective in MAXTREE. We randomly sample 100 roots for each dataset. B varies from 0 to 100 at a step size of 5.0. For Twitter and *Enron* dataset, I is set to 1 day and 4 weeks respectively. We take the median of all trees returned by each algorithm (Fig. 3).

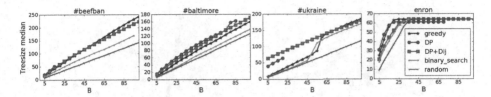

Fig. 3. Effect of B on the median of tree sizes for different datasets. Note that for #ukraine and $B > 25$, the DP algorithm fails to complete the experiments as it consumes excessive amount of memory.

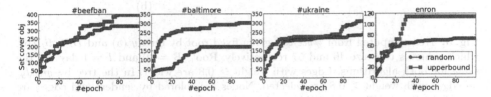

Fig. 4. Performance of Different sampling schemes on real datasets: $k = 10$. For Twitter, $B = 15.0, I = 1$ day. For *Enron*, $B = 10.0, I = 4$ weeks. 100 unique roots are selected based on the sampling scheme.

In *Enron*, we observe a converging effect on both objectives as the dataset is relatively small, while this is not the case in all Twitter datasets. In practice, *greedy* is the best performing algorithm, as it is both competitive in maximizing the objective function and it is computationally efficient.

Sampling scheme comparison. We compare two sampling schemes in real data setting: (1) random root sampling (*random*) as the baseline, (2) ranking roots by event size upperbound (*upperbound*) Sect. 5. For each scheme, the set cover objective is recorded whenever a new candidate is added. As we can see in Fig. 4, the event size upper-bound heuristic helps to discover better solutions, especially for *#baltimore* and *Enron*.

Event trees by different algorithms. We compare the behaviours of the algorithms for MAXTREE in real-world datasets. In Fig. 5, the trees are produced by *greedy*, and *DP+dij* are given the same root and budget. The *greedy* algorithm avoids to select heavy edges with weights larger than 0.8 due to its local search strategy whereas *DP+dij* achieves larger tree by selecting a few heavy edges. Therefore we expect *greedy* to produce more topically-coherent events as the pairwise dissimilarity between nodes tend to be smaller.

6.4 Case Study in *Enron* Dataset

We sample 50 nodes using *upperbound* scheme and applied *greedy* algorithm with $B = 10, I = 28$ days. First, we observed that the events can be grouped into two

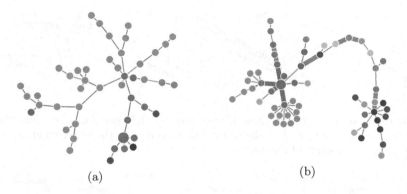

(a)   (b)

Fig. 5. Tree computed from *#beefban* given fixed root by *greedy* (a) and *DP+dij* (b), which achieves tree size 46 and 57 respectively. Root, $B = 30$ and $I = 1$ day are the same for both algorithms. Edges with weight ≥ 0.8 are wider. In the tree by *greedy*, no edges with weight ≥ 0.8 are selected. Nodes are colored by senders and edges are colored by its type (broadcast: blue, relay: green, reply: orange) (Color figure online)

types: (1) California Energy Crisis,[4] (2) investigation into Enron's scandal.[5] In Fig. 6(a), we annotated the real world events about the crisis happening during the timespan of the dataset. We found shortly after each major blackout, there is at least one extracted events about it. And before Enron filed bankruptcy, Federal Energy Regulatory Commission (FERC) investigated Enron. Second, in Fig. 7(a), extracted events tend to occur at the peak of the volume plot.

6.5   Case Study in Twitter Datasets

We use the same parameters for all three datasets as they have similar size and timespan. Events are extracted by selecting 100 roots using *upperbound* and using *greedy* algorithm with $B = 50$ and $I = 1$ day.

#ukraine. Ukraine crisis arouses media war on Ukraine and Russia.[6] We observe some of the detected events align well topically and temporally with the actual events in Fig. 6(b). However, topics are mixed inside some other events. For example, topics on both #nemstsov and #freesavchenko are detected in event 2. This is expected due to the local similarity measurement in MAXTREE.

#beefban. For the controversial "beef ban"[7] law in India, results demonstrate clear separation of opinions among events. In Fig. 8, the 1st and 2nd event represents opinions opposing and supporting the law. However, we are not able to interpret any temporal pattern in the events due to the short timespan (3 days). We also observe the following. First, certain event (Fig. 8(a)) display evidence of

---

[4] https://en.wikipedia.org/wiki/California_electricity_crisis.
[5] https://en.wikipedia.org/wiki/Enron_scandal.
[6] https://en.wikipedia.org/wiki/Ukrainian_crisis.
[7] http://indianexpress.com/article/explained/explained-no-beef-nation/.

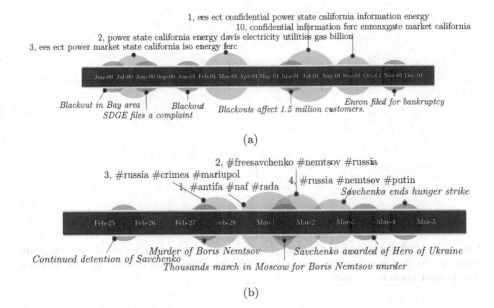

(a)

(b)

Fig. 6. Timeline with extracted events (larger red circle) and publicly recognised events (smaller black circle and italic text) for Enron (a) and *#ukraine* (b). (a) highlights events on Enron's energy scandal and bankruptcy. Event 3, 2, 1 and 10 are displayed. The larger the circle, the larger the event size is. For each event, top topic terms are displayed. In (b), top-4 events are displayed with the top hashtags. Event 2 and 4 maps to the murder of Boris Nemstsov (#nemstsov), while event 4 also contains tweets on freeing Savchenko (#freesavchenko). Event 1, 3 is about other related issues. (Color figure online)

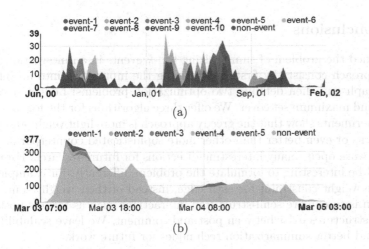

(a)

(b)

Fig. 7. Stacked area graph of interaction frequency against time. *Enron* (a) contains top-10 events. *#beefban* (b) contains top-5 events.

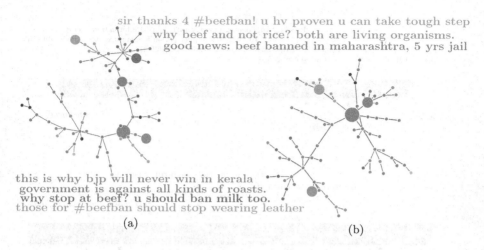

sir thanks 4 #beefban! u hv proven u can take tough step
why beef and not rice? both are living organisms.
good news: beef banned in maharashtra, 5 yrs jail

this is why bjp will never win in kerala
government is against all kinds of roasts.
why stop at beef? u should ban milk too.
those for #beefban should stop wearing leather

(a)   (b)

Fig. 8. Extracted events for *#beefban*. (a) 1st event demonstrates sign of opinion propagation. (b) 2nd event containing mixed opinions. Nodes are colored by the senders. The largest node is the root.

information propagation. For example, opposing opinions spreads along the user network and affected users also express their objection. Second, for some event (Fig. 8(b)), dominant user exists who sent more than half of the tweets. Third, we observe events with mixed opinions (Fig. 8(b)). Last, our method tends to discover events at the "peak" as the set cover objective is better than the "bottom" (Fig. 7(b)).

#baltimore. We discovered two types of events: (1) "emotional" events showing anger towards the riot, (2) "descriptive" events reporting current situation.

7   Conclusions

We defined the problem of summarizing top-k events in an interaction network. Our approach consists by first transforming the input data into an interaction meta-graph and then defining two optimization problems: budgeted version of PCST and maximum set cover. We offer three algorithms for the former problem. Our experiments show that the greedy approach is more lightweight and performs as good as or even better than other more sophisticated counterparts.

Our work opens many interesting directions for future research. For example, it would be interesting to formulate the problem differently, for example, imposing edge weight constraint for each edge, instead of their weight sum. Another direction is to explore semi-structured interactions such as forums posts, where nesting structures exist between post and comment. We leave scalability experiment and better summarization techniques for future work.

Acknowledgements. This work is partially supported by the Academy of Finland project "Nestor" (286211) and the EC H2020 RIA project "SoBigData" (654024).

References

1. Becker, H., Naaman, M., Gravano, L.: Beyond trending topics: real-world event identification on twitter. In: ICWSM (2011)
2. Blei, D., Ng, A., Jordan, M.: Latent Dirichlet allocation. J. Mach. Learn. Res. **3**, 993–1022 (2003)
3. Boden, B., Günnemann, S., Hoffmann, H., Seidl, T.: Mining coherent subgraphs in multi-layer graphs with edge labels. In: KDD (2012)
4. Breunig, M., Kriegel, H.P., Ng, R., Sander, J.: LOF: identifying density-based local outliers. SIGMOD Rec. **29**(2), 93–104 (2000)
5. Cataldi, M., Di Caro, L., Schifanella, C.: Emerging topic detection on twitter based on temporal and social terms evaluation. In: IWMDM (2010)
6. Celikyilmaz, A., Hakkani-Tür, D.: Discovery of topically coherent sentences for extractive summarization. In: ACL (2011)
7. Charikar, M., Chekuri, C., Cheung, T., Dai, Z., Goel, A., Guha, S., Li, M.: Approximation algorithms for directed Steiner problems. J. Algorithms **33**, 73–91 (1999)
8. Fiore, A., Heer, J.: UC Berkeley Enron email analysis (2004)
9. Garimella, K., De Francisci Morales, G., Gionis, A., Mathioudakis, M.: Quantifying controversy in social media. In: WSDM (2016)
10. Guralnik, V., Srivastava, J.: Event detection from time series data. In: KDD (1999)
11. Inouye, D., Kalita, J.K.: Comparing twitter summarization algorithms for multiple post summaries. In: PASSAT and SocialCom (2011)
12. Johnson, D., Minkoff, M., Phillips, S.: The prize collecting Steiner tree problem: theory and practice. In: SODA (2000)
13. Kumar, R., Mahdian, M., McGlohon, M.: Dynamics of conversations. In: KDD (2010)
14. Leskovec, J., Backstrom, L., Kleinberg, J.: Meme-tracking and the dynamics of the news cycle. In: KDD (2009)
15. Mathioudakis, M., Koudas, N.: Twittermonitor: trend detection over the twitter stream. In: SIGMOD (2010)
16. Meladianos, P., Nikolentzos, G., Rousseau, F., Stavrakas, Y., Vazirgiannis, M.: Degeneracy-based real-time sub-event detection in twitter stream. In: ICWSM (2015)
17. Vazirani, V.: Approximation Algorithms. Springer, Heidelberg (2013)
18. Weng, J.: Event detection in twitter. In: ICWSM (2011)
19. Yang, J., Leskovec, J.: Modeling information diffusion in implicit networks. In: ICDM (2010)
20. Yang, J., Leskovec, J.: Patterns of temporal variation in online media. In: WSDM (2011)

Proactive Transfer Learning for Heterogeneous Feature and Label Spaces

Seungwhan Moon[✉] and Jaime Carbonell

School of Computer Science, Language Technologies Institute,
Carnegie Mellon University, 5000 Forbes Avenue, Pittsburgh, PA 15213, USA
{seungwhm,jgc}@cs.cmu.edu

Abstract. We propose a framework for learning new target tasks by leveraging existing heterogeneous knowledge sources. Unlike the traditional transfer learning, we do not require explicit relations between source and target tasks, and instead let the learner actively *mine* transferable knowledge from a source dataset. To this end, we develop (1) a transfer learning method for source datasets with heterogeneous feature and label spaces, and (2) a proactive learning framework which progressively builds *bridges* between target and source domains in order to improve transfer accuracy. Experiments on a challenging transfer learning scenario (learning from *hetero-lingual* datasets with non-overlapping label spaces) show the efficacy of the proposed approach.

1 Introduction

The notion of enabling a machine to learn a new task by leveraging an auxiliary source of knowledge has long been the focus of transfer learning. While many different flavors of transfer learning approaches have been developed, most of these methods assume explicit relatedness between source and target tasks, such as the availability of source-target correspondent instances (*e.g.* multi-view/multimodal learning), or the class relations information for multiple datasets sharing the same feature space (*e.g.* zero-shot learning, domain adaptation), etc. These approaches have been effective in their respective scenarios, but very few limited studies have investigated learning from heterogeneous knowledge sources that lie in both different feature and label spaces. See Sect. 2 for the detailed literature review.

Given an unforeseen target task with limited label information, we seek to mine useful knowledge from a plethora of heterogeneous knowledge sources that have already been curated, albeit in different feature and label spaces. To address this challenging scenario we first need an algorithm to estimate *how* the source and the target datasets may be related. One common aspect of any dataset for a classification task is that each instance is eventually assigned to some abstract concept(s) represented by its category membership, which often has its own *name*. Inspired by the Deep Visual-Semantic Embedding (DeViSE) model [7] which assigns the unsupervised word embeddings to label terms, we propose to map heterogeneous source and target labels into the same word embedding

P. Frasconi et al. (Eds.): ECML PKDD 2016, Part II, LNAI 9852, pp. 706–721, 2016.
DOI: 10.1007/978-3-319-46227-1_44

space, from which we can obtain their semantic class relations. Using information from the class relations as an anchor, we first attempt to uncover a shared latent subspace where both source and target features can be mapped. Simultaneously, we learn a shared projection from this intermediate layer into the final embedded labels space, from which we can predict labels using the shared knowledge.

The quality of transfer essentially depends on how well we can uncover the *bridge* in the projected space where the two datasets are semantically linked. Intuitively, if the two datasets describe completely different concepts, very little information can be transferred from one to the other. We therefore also propose a proactive transfer learning framework which expands the labeled target data to actively *mine* transferable knowledge and to progressively improve the target task performance.

We evaluate the proposed combined approach on a unique learning problem of a *hetero-lingual* text classification task, where the objective is to classify a novel target text dataset given only a few labels along with a source dataset in a different language, describing different classes from the target categories. While this is a challenging task, the empirical results show that the proposed approach improves over the baselines.

The rest of the paper is organized as follows: we position our approach in relation to the previous work in Sect. 2, and formulate the heterogeneous transfer learning problem in Sect. 3. Section 4 describes in detail the proposed proactive transfer learning framework and presents the optimization problem. The empirical results are reported and analyzed in Sect. 5, and we give our concluding remarks and proposed future work in Sect. 6.

2 Related Work

Transfer Learning with Heterogeneous Feature Spaces: Multi-view representation learning aims at aggregating multiple heterogeneous "views" (feature sets) of an instance that describe the same concept to train a model. Most notably, [30] proposes Deep Canonically Correlated Autoencoders (DCCAE) which learn a representation that maximizes the mutual information between different views under an autoencoder regularization. While DCCAE is reported to be state of the art on multi-view representation learning using Canonical Correlation Analysis (CCA) [4], their approach (as well as other CCA-based methods) strictly require access to paired observations from two views belonging to the same class. [3] proposes translated learning which aims to learn a target task in the same label space as the source task, using source-correspondent instances such as image-text parallel captions as an anchor. [34] proposes Hybrid Heterogeneous Transfer Learning (HHTL) which extends the previous translated learning work with an added objective of learning an unbiased feature mapping through marginalized stacked denoising autoencoders (mSDA), given correspondent instances. [29] develops a similar approach in bilingual content classification tasks, and proposes to generate correspondent samples through an available machine translation system. [22] proposes a Transfer Deep Learning

(TDL) framework for fine-tuning intermediate layers of a target network with transferred source data, where the mapping between source and target layers is learned from corresponding instances. [5] propose the Heterogeneous Feature Augmentation (HFA) method for a shared homogeneous binary classification task, which relaxes the previous limitations that require correspondent instances, and instead aims to discover a common subspace that can map two heterogeneous features. Our approach generalizes all the previous work by allowing for heterogeneous label spaces between source and target, thus not requiring explicit source-target correspondent instances or classes.

Transfer Learning with a Heterogeneous Label Space: Zero-shot learning aims at building a robust classifier for unseen novel classes in the target task, often by relaxing categorical label space into a distributed vector space via transferred knowledge. For instance, [19] uses image co-occurrence statistics to describe a novel image class category, while [7,8,15,25,28,31,33] embed labels into semantic word vector space according to their label terms, where textual embeddings are learned from auxiliary text documents in an unsupervised manner. More recently, [13] proposes to learn domain-adapted projections to the embedded label space. While these approaches are reported to improve robustness and generalization on novel target classes, they assume that source datasets are in the same feature space as the target dataset (*e.g.* image). We extend the previous research by adding the joint objective of uncovering relatedness among datasets with heterogeneous feature spaces, via anchoring the semantic relations between the source and the target label embeddings.

Domain Adaptation approaches aim to minimize the marginal distribution difference between source and target datasets, assuming their class conditional distribution remains the same for homogeneous feature and label spaces. This is typically implemented via instance re-weighting [2,11,14], subspace mapping [32], or via identification of transferable features [16]. [24] provide an exhaustive survey on other traditional transfer learning approaches.

Active learning provides an alternative solution to the label scarcity problem, which aims at reducing sample complexity by iteratively querying the most informative samples with the highest utility given the labeled sampled thus far [21,26]. **Transfer active learning** approaches [2,6,12,27,36] aim to combine transfer learning with the active learning framework by conditioning transferred knowledge as priors for optimized selection of target instances. Specifically, [9] overcomes the common *cold-start* problem at the beginning phase of active learning with zero-shot class-relation priors. However, many of the previously proposed transfer active learning methods do not apply to our setting because they require source and target data to be in either homogeneous feature space or the same label space or both. Therefore, we propose a *proactive transfer learning* approach for heterogeneous source and target datasets, where the objective is to progressively find and query *bridge* instances that allow for more accurate transfer, given a sampling budget.

Our contributions are three-fold: we propose (1) a novel transfer learning method with both heterogeneous feature and label spaces, and (2) a proactive transfer learning approach for identifying and querying *bridge* instances between target and source tasks to improve transfer accuracy effectively. (3) We evaluate the proposed approach on a novel transfer learning problem, the *hetero-lingual* text classification task.

3   Problem Formulation

We formulate the proposed framework for learning a target multiclass classification task given a source dataset with heterogeneous feature and label spaces as follows: We first define a dataset for the target task $\mathbf{T} = \{\mathbf{X_T}, \mathbf{Y_T}, \mathbf{Z_T}\}$, with the target task features $\mathbf{X_T} = \{\mathbf{x_T}^{(i)}\}_{i=1}^{N_T}$ for $\mathbf{x_T} \in \mathbb{R}^{M_T}$, where N_T is the target sample size and M_T is the target feature dimension, the ground-truth labels $\mathbf{Z_T} = \{\mathbf{z_T}^{(i)}\}_{i=1}^{N_T}$, where $\mathbf{z_T} \in \mathcal{Z}_T$ for a categorical target label space $\mathcal{Z}_T$, and the corresponding high-dimensional label descriptors $\mathbf{Y_T} = \{\mathbf{y_T}^{(i)}\}_{i=1}^{N_T}$ for $\mathbf{y_T} \in \mathbb{R}^{M_E}$, where M_E is the dimension of the embedded labels, which can be obtained from *e.g.* unsupervised word embeddings, etc. We also denote L_T and UL_T as a set of indices of labeled and unlabeled target instances, respectively, where $|L_T| + |UL_T| = N_T$. For a novel target task, we assume that we are given zero or a very few labeled instances, thus $|L_T| = 0$ or $|L_T| \ll N_T$. Similarly, we define a heterogeneous source dataset $\mathbf{S} = \{\mathbf{X_S}, \mathbf{Y_S}, \mathbf{Z_S}\}$, with $\mathbf{X_S} = \{\mathbf{x_S}^{(i)}\}_{i=1}^{N_S}$ for $\mathbf{x_S} \in \mathbb{R}^{M_S}$, $\mathbf{Z_S} = \{\mathbf{z_S}^{(i)}\}_{i=1}^{N_S}$ for $\mathbf{z_S} \in \mathcal{Z}_S$, $\mathbf{Y_S} = \{\mathbf{y_S}^{(i)}\}_{i=1}^{N_S}$ for $\mathbf{y_S} \in \mathbb{R}^{M_E}$, and L_S, accordingly. For the source dataset we assume $|L_S| = N_S$. Note that in general, we assume $M_T \neq M_S$ (heterogeneous feature space) and $\mathcal{Z}_T \neq \mathcal{Z}_S$ (heterogeneous label space).

Our goal is then to build a robust classifier $\mathbf{f} : \mathcal{X}_T \rightarrow \mathcal{Z}_T$ for the target task, trained with $\{\mathbf{x_T}^{(i)}, \mathbf{y_T}^{(i)}, \mathbf{z_T}^{(i)}\}_{i \in L_T}$ as well as transferred knowledge from $\{\mathbf{x_S}^{(i)}, \mathbf{y_S}^{(i)}, \mathbf{z_S}^{(i)}\}_{i \in L_S}$.

4   Proposed Approach

Our approach aims to leverage a source data that lies in different feature and label spaces from a target task. Transferring knowledge directly from heterogeneous spaces is intractable, and thus we begin by obtaining a unified vector representation for different source and target categories. Specifically, we utilize a skip-gram based language model that learns semantically meaningful vector representations of words, and map our categorical source and target labels into the word embedding space (Sect. 4.1). In parallel, we learn compact representations for the source and the target features that encode abstract information of the raw features (Sect. 4.2), which allows for more tractable transfer through affine projections. Once the label terms for the source and the target datasets are anchored in the word embedding space, we first learn projections into a new latent common feature space from the source and the target feature spaces ($\mathbf{W_S}$ and $\mathbf{W_T}$), respectively, from which $\mathbf{W_f}$ maps the joint features into the

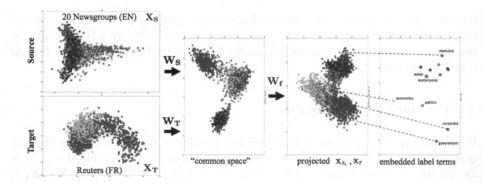

Fig. 1. An illustration of the proposed approach. Source (20 Newsgroups: English) and target (Reuters Multilingual: French) datasets lie in different feature spaces ($\mathbf{x_S} \in \mathbb{R}^{M_S}$, $\mathbf{x_T} \in \mathbb{R}^{M_T}$), and describe different categories ($\mathcal{Z}_S \neq \mathcal{Z}_T$). First, categorical labels are embedded into the dense continuous vector space (*e.g.* via text embeddings learned from unsupervised documents.) The objective is then to learn $\mathbf{W_f}$, $\mathbf{W_S}$, and $\mathbf{W_T}$ jointly such that $\mathbf{W_S}$ and $\mathbf{W_T}$ map the source and target data to the latent common feature space, from which $\mathbf{W_f}$ can project to the same space as the embedded label space. Note that the shared projection $\mathbf{W_f}$ is learned from both the source and the target datasets, thus we can more robustly predict a label for a projected instance by finding its nearest label term projection.

embedded label space (Sect. 4.3). Lastly, we actively query and expand the labeled set L_T to jointly improve the joint classifier $\mathbf{W_f}$ and the transfer accuracy (Sect. 4.4). Figure 1 shows the illustration of the proposed approach, visualized with the real datasets (20 Newsgroups and Reuters Multilingual Datasets).

4.1   Language Model Label Embeddings

The skip-gram based language model [20] has proven effective in encoding semantic information of words, which can be trained from unsupervised text. We use the obtained label term embeddings as *anchors* for source and target datasets, and drive the target model to learn indirectly from source instances that belong to semantically similar categories. In this work, we use 300-D word embeddings trained from the `Google News` dataset[1] (about 100 billion words).

4.2   Unsupervised Representation Learning for Features

In order to project source and target feature spaces into the joint latent space effectively, as a pre-processing step we first obtain abstract and compact representations of raw features to allow for more tractable transformation. Unlike the similar zero-shot learning approaches [7], we do not use the embeddings obtained from a fully supervised network (e.g. the activation embeddings at the

---

[1] word2vec: https://code.google.com/archive/p/word2vec/.

top of the trained visual model), because we assume the target task is scarce in labels. For our experiments with text features, we use the latent semantic analysis (LSA) method [10] to transform the raw tf-idf features into a 200-D low-rank approximation.

4.3 Transfer Learning for Heterogeneous Feature and Label Spaces

We define $\mathbf{W_S}$ and $\mathbf{W_T}$ to denote the sets of learnable parameters that project source and target features into a latent joint space, where the mappings can be learned with deep neural networks, kernel machines, etc. For simplicity, we treat $\mathbf{W_S}$ and $\mathbf{W_T}$ as linear transformation layers, thus $\mathbf{W_S} \in \mathbb{R}^{M_S \times M_C}$ and $\mathbf{W_T} \in \mathbb{R}^{M_T \times M_C}$ for projection into the M_C-dimension common space. Similarly, we define $\mathbf{W_f} \in \mathbb{R}^{M_C \times M_E}$ which maps from the common feature space into the embedded label space.

To learn these parameters simultaneously, we solve the following joint optimization problem with hinge rank losses (similar to [7]) for both source and target.

$$\min_{\mathbf{W_f}, \mathbf{W_S}, \mathbf{W_T}} \frac{1}{|L_S|} \sum_{i=1}^{|L_S|} l(\mathbf{S}^{(i)}) + \frac{1}{|L_T|} \sum_{j=1}^{|L_T|} l(\mathbf{T}^{(j)}) \tag{1}$$

where

$$l(\mathbf{S}^{(i)}) = \sum_{\tilde{y} \neq y_S^{(i)}} \max[0, \epsilon - \mathbf{x}_S^{(i)} \mathbf{W_S} \mathbf{W_f} \mathbf{y}_S^{\mathbf{T}(i)} + \mathbf{x}_S^{(i)} \mathbf{W_S} \mathbf{W_f} \tilde{\mathbf{y}}^{\mathbf{T}}]$$

$$l(\mathbf{T}^{(j)}) = \sum_{\tilde{y} \neq y_T^{(j)}} \max[0, \epsilon - \mathbf{x}_T^{(j)} \mathbf{W_T} \mathbf{W_f} \tilde{\mathbf{y}}_T^{\mathbf{T}(j)} + \mathbf{x}_T^{(j)} \mathbf{W_T} \mathbf{W_f} \tilde{\mathbf{y}}^{\mathbf{T}}]$$

where $l(\cdot)$ is a per-instance hinge loss, $\tilde{y}$ refers to the embeddings of other label terms in the source and the target label space except the ground truth label of the instance, and ϵ is a fixed margin. We use $\epsilon = 0.1$ for all of our experiments.

In essence, we train the weight parameters to produce a higher dot product similarity between the projected source or target instance and the word embedding representation of its correct label than between the projected instance and other incorrect label term embeddings. The intuition of the model is that the learned $\mathbf{W_f}$ is a shared and more generalized linear transformation capable of mapping the joint intermediate subspace into the embedded label space.

We solve Eq. 1 efficiently with stochastic gradient descent (SGD), where the gradient is estimated from a small minibatch of samples.

Once $\mathbf{W_S}$, $\mathbf{W_T}$, and $\mathbf{W_f}$ are learned, at test time we build a label-producing nearest neighbor (NN) classifier for the target task as follows:

$$\text{NN}(\mathbf{x_T}) = \underset{\mathbf{z} \in \mathcal{Z}_T}{\operatorname{argmax}} \ \mathbf{x_T} \mathbf{W_T} \mathbf{W_f} \mathbf{y}_{\mathbf{z}}^{\mathbf{T}} \tag{2}$$

where $\mathbf{y_z}$ maps a categorical label term $\mathbf{z}$ into its word embeddings space. Similarly, we can build a NN classifier for the source task as well, using the projection $\mathbf{W_S} \mathbf{W_f}$.

4.4 Proactive Transfer Learning

The quality of the learned parameters for the target task $\mathbf{W_T}$ and $\mathbf{W_f}$ depends on the available labeled target training samples (L_T). As such, we propose to expand L_T by querying a near-optimal subset of the unlabeled pool UL_T, which once labeled will improve the performance of the transfer accuracy and ultimately the target task, assuming the availability of unlabeled data and (limited) annotators. In particular, we relax this problem with a greedy pool-based active learning framework, where we iteratively select a small subset of unlabeled samples that maximizes the expected utility to the target model:

$$\hat{\mathbf{x}}_\mathbf{T} = \underset{\mathbf{x_T} \in \{\mathbf{x}_\mathbf{T}^{(i)}\}_{i \in UL_T}}{\text{argmax}} \quad U(\mathbf{x_T}) \tag{3}$$

where $U(\mathbf{x_T})$ is a utility function that measures the value of a sample $\mathbf{x_T}$ defined by a choice of the query sampling objective. In traditional active learning, the uncertainty-based sampling [17,26] and the density-weighted sampling strategies [23,35] are often used for the utility function $U(\mathbf{x_T})$ in the target domain only. However, the previous approaches in active learning disregard the knowledge that we have in the source domain, thus being prone to query samples of which the information can be potentially redundant to the transferable knowledge. In addition, these approaches only aim at improving the target classification performance, whereas querying *bridge* instances to maximally improve the transfer accuracy instead can be more effective by allowing more information to be transferred in bulk from the source domain. Therefore, we propose the following two proactive transfer learning objectives for sampling in the target domain that utilize the source knowledge in various ways:

Maximal Marginal Distribution Overlap (MD): We hypothesize that the overlapping projected region is where the heterogeneous source and target data are semantically related, thus a good candidate for a *bridge* that maximizes the information transferable from the source data. We therefore propose to select unlabeled target samples $(\mathbf{x_T})$ in regions where the marginal distributions of projected source and target samples have the highest overlap:

$$U_{\mathrm{MD}}(\mathbf{x_T}) = \min\left(\hat{P}_\mathbf{T}(\mathbf{x_T}|\mathbf{W_T}, \mathbf{W_f}), \hat{P}_\mathbf{S}(\mathbf{x_T}|\mathbf{W_S}\mathbf{W_f})\right) \tag{4}$$

where $\hat{P}_\mathbf{T}$ and $\hat{P}_\mathbf{S}$ are the estimated marginal probability of the projected target and source instances, respectively. Specifically, we estimate each density with the non-parametric kernel method:

$$\hat{P}_\mathbf{T}(\mathbf{x_T}|\mathbf{W_T}, \mathbf{W_f}) = \frac{1}{N_T} \sum_{i=1}^{N_T} K_h(\mathbf{x_T}\mathbf{W_T}\mathbf{W_f} - \mathbf{x}_\mathbf{T}^{(i)}\mathbf{W_T}\mathbf{W_f})$$

$$\hat{P}_\mathbf{S}(\mathbf{x_T}|\mathbf{W_S}, \mathbf{W_f}) = \frac{1}{N_S} \sum_{j=1}^{N_S} K_h(\mathbf{x_T}\mathbf{W_T}\mathbf{W_f} - \mathbf{x}_\mathbf{S}^{(j)}\mathbf{W_S}\mathbf{W_f}) \tag{5}$$

Algorithm 1. Proactive Transfer Learning

Input: source data **S**, target data **T**, active learning policy $U(\cdot)$, budget B, query size per iteration Q

Randomly initialize $\mathbf{W_f}, \mathbf{W_T}, \mathbf{W_S}$

for $iter = 1$ **to** B **do**

1. Learn $\mathbf{W_f}, \mathbf{W_T}, \mathbf{W_S}$ by solving

$$\min_{\mathbf{W_f},\mathbf{W_S},\mathbf{W_T}} \frac{1}{|L_S|} \sum_{i=1}^{|L_S|} l(\mathbf{S}^{(i)}) + \frac{1}{|L_T|} \sum_{j=1}^{|L_T|} l(\mathbf{T}^{(j)})$$

2. Query Q new samples

for $q = 1$ **to** Q **do**

$$\hat{\mathbf{i}} = \underset{i \in UL_T}{\operatorname{argmax}} \, U(\mathbf{x}_T^{(i)})$$
$$UL_T := UL_T \backslash \{\hat{\mathbf{i}}\}, L_T := L_T \cup \{\hat{\mathbf{i}}\}$$

end for

end for

Output: $\mathbf{W_f}, \mathbf{W_T}, \mathbf{W_S}$

---

where K_h is a scaled Gaussian kernel with a smoothing bandwidth h. Solving $\max_{\mathbf{x_T}} \min(\hat{P}_T(\mathbf{x_T}), \hat{P}_S(\mathbf{x_T}))$ finds such instance $\mathbf{x_T}$ whose projection lies in the highest density overlap between source and target instances.

Maximum Projection Entropy (PE) aims at selecting an unlabeled target sample that has the maximum entropy of dot product similarities between a *projected* instance and its possible label embeddings:

$$U_{\text{PE}}(\mathbf{x_T}) = - \sum_{z \in \mathcal{Z}_T} \log(\mathbf{x_T} \mathbf{W_T} \mathbf{W_f} \mathbf{y_z^T}) \mathbf{x_T} \mathbf{W_T} \mathbf{W_f} \mathbf{y_z^T} \tag{6}$$

The projection entropy utilizes the information transferred from the source domain (via $\mathbf{W_f}$), thus avoiding information redundancy between source and target. After samples are queried via the maximum projection entropy method and added to the labeled target data pool, we re-train the weights such that projections of the target samples have less uncertainty in label assignment.

To reduce the active learning training time at each iteration, we query a small fixed number of samples ($= Q$) that have the highest utilities. Once the samples are annotated, we re-train the model with Eq. 1, and select the next batch of samples to query with Eq. 3. The overall process is summarized in Algorithm 1.

5 Empirical Evaluation

We evaluate the proposed approach on a hetero-lingual text classification task (Sect. 5.2) with the baselines described in Sect. 5.1.

5.1 Baselines

In our experiments we use a source dataset within heterogeneous feature and label spaces from a target dataset. Most of the previous transfer learning

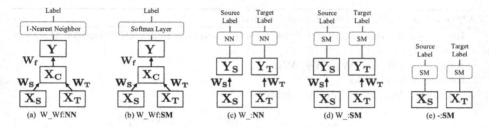

Fig. 2. The proposed method (a) and the baseline networks (b–e). At test time, the nearest neighbor-based models (a, c) return the nearest label in the embedding space ($\mathcal{Y}$) to the projection of a test sample, whereas the n-way softmax layer (SM) classifiers (b, d, e) are trained to produce categorical labels from their respective final projection. We use the notation $\mathbf{W}_-$ to refer to $\mathbf{W}t$ and $\mathbf{W}s$, as they share the same architecture.

approaches that allow only one of input or output spaces to be heterogeneous thus cannot be used as baselines (see Sect. 2 for the detailed comparison). We therefore compare the proposed heterogeneous transfer approach with the following baseline networks (illustrated in Fig. 2):

- **W_Wf:NN (proposed approach**; heterogeneous transfer learning network): We learn the projections $\mathbf{W_S}$, $\mathbf{W_T}$, and $\mathbf{W_f}$ by solving the joint optimization problem in Eq. 1. At test time, we use the 1-nearest neighbor classifier (NN) defined in Eq. 2 and look for a category embedding that is closest to the projected source ($\mathbf{x_S}\mathbf{W_S}\mathbf{W_f}$) or target instance ($\mathbf{x_T}\mathbf{W_T}\mathbf{W_f}$) at the final layer. We use the notation $\mathbf{W}_-$ to denote a placeholder for a source model (**WsWf:NN**) and a target model (**WtWf:NN**), as they share the same architecture.
- **W_Wf:SM**: We train the weights in the same way (Eq. 1), and we add a softmax layer (SM) at the top projection layer (word embedding space) in replacement of the NN classifier.
- **W_:NN** ([7]; zero-shot learning networks with distributed word embeddings): We learn the projections $\mathbf{W_S} \in \mathbb{R}^{M_S \times M_E}$ and $\mathbf{W_T} \in \mathbb{R}^{M_T \times M_E}$ by solving two separate optimization problems for source and target networks respectively:

$$\min_{\mathbf{W_S}} \frac{1}{|L_S|} \sum_{i=1}^{|L_S|} l(\mathbf{S}^{(i)}), \quad \min_{\mathbf{W_T}} \frac{1}{|L_T|} \sum_{j=1}^{|L_T|} l(\mathbf{T}^{(j)}) \tag{7}$$

where the loss functions are defined in a similar way as in Eq. 1:

$$l(\mathbf{S}^{(i)}) = \sum_{\tilde{y} \neq y_S^{(i)}} \max[0, \epsilon - \mathbf{x}_S^{(i)} \mathbf{W_S} \mathbf{y}_S^{T(i)} + \mathbf{x}_S^{(i)} \mathbf{W_S} \tilde{\mathbf{y}}^T]$$

$$l(\mathbf{T}^{(j)}) = \sum_{\tilde{y} \neq y_T^{(j)}} \max[0, \epsilon - \mathbf{x}_T^{(j)} \mathbf{W_T} \mathbf{y}_T^{T(j)} + \mathbf{x}_T^{(j)} \mathbf{W_T} \tilde{\mathbf{y}}^T] \tag{8}$$

At test time, we use the NN classifier with projected source ($\mathbf{x_S}\mathbf{W_S}$) and target ($\mathbf{x_T}\mathbf{W_T}$) instances. The target task thus does not use the transferred information from the source task, but only uses the semantic word embeddings transferred from a separate unannotated corpus. This baseline can be regarded as an application of DeViSE [7] on non-image classification tasks.

- **W_:SM**: We train the weights with Eq. 7, and we add a softmax layer.
- **-:SM**: We train two separate networks with logistic regression softmax layers for source and target tasks with $\mathbf{X_S}$ and $\mathbf{X_T}$, respectively.

5.2 Application: Hetero-Lingual Text Classification

We apply the proposed approach to learn a target text classification task given a source text dataset with both a heterogeneous feature space (*e.g.* a different language) and a label space (*e.g.* describing different categories).

The datasets we use are summarized in Table 1. Note that the 20 News-groups[2] (English: 18,846 documents), the Reuters Multilingual [1] (French: 26,648, Spanish: 12,342, German: 24,039, Italian:12,342 documents), the R8 of RCV-1[3] (English: 7,674 documents) datasets describe different categories with varying degrees of relatedness. The original categories of some of the datasets were not in the format compatible to our word embeddings dictionary. We manually replaced those label terms to the semantically close words that exist in the dictionary (*e.g.* sci.med → 'medicine', etc.).

Task 1: Transfer Learning for Scarce Target

Setup: We assume a scenario where only a small fraction of the target samples are labeled ($\%_{L_T} = 0.1\%$ or 1% depending on the size of the dataset) whereas the source dataset is fully labeled, and create various heterogeneous source-target pairs from the datasets summarized in Table 1. Table 2 reports the text classification results for both source and target tasks in this experimental setting. The results are averaged over 10-fold runs, and for each fold we randomly select $\%_{L_T}$ of the target train instances to be labeled as indicated in Table 2. Bold denotes the best performing model for each test, and * denotes the statistically significant improvement ($p < 0.05$) over other methods.

Main results: Table 2 shows that the proposed approach (**WtWf:NN**) improves upon the baselines on several source-target pairs on the target classification task. Specifically, **WtWf:NN** shows statistically significant improvement over the single-modal baseline (**Wt:NN**) on the source-target pairs 20NEWS→SP, 20NEWS→GR, R8→SP, R8→GR, and SP→R8. The performance boost demonstrates that the transferred knowledge from a source dataset (in the form of $\mathbf{W_f}$) does improve the projection pathway from the target feature space to the embedded label space. Note that the transfer learning (**WtWf:NN**) from Reuters Multilingual datasets (FR, SP, GR, IT) to 20 News-groups (20NEWS) dataset specifically does not improve over the single-modal baseline (**Wt:NN**). The 20 Newsgroups dataset is in general harder to discriminate and spans over a larger label space than the Reuters Multilingual datasets,

---

[2] http://qwone.com/~jason/20Newsgroups/.

[3] http://csmining.org/index.php/r52-and-r8-of-reuters-21578.html.

Table 1. Overview of datasets. $|\mathcal{Z}|$: the number of categories

Dataset	$|\mathcal{Z}|$	Label terms (*e.g.*)
20 Newsgroups (20NEWS)	20	'politics', 'religion', 'electronics', 'motorcycles', 'baseball', 'sale', $\cdots$
Reuters Multilingual (FR,SP,GR,IT)	6	'corporate', 'finance', 'economics', 'performance', 'government', 'equity'
Reuters R8 (R8)	8	'acquisition', 'interest', 'money', 'crude', 'trade', 'grain', $\cdots$

Table 2. Hetero-lingual text classification test accuracy (%) on (1) the target task and (2) the source task, given a fully labeled source dataset and a partially labeled target dataset, averaged over 10-fold runs ($M_C = 320$). $\%_{L_T}$: the percentage of target samples labeled. The baselines are described in Fig. 2.

Source	Target	$\%_{L_T}$	$W_T W_f$:NN	$W_T W_f$:SM	W_T:NN	W_T:SM	-:SM	$W_S W_f$:NN	$W_S W_f$:SM	W_S:NN	W_S:SM	-:SM
20NEWS	FR	0.1	**57.7**	46.1	55.7	44.4	39.4	**78.2**	77.1	78.0	77.3	77.6
	SP		**52.1***	43.0	46.6	42.7	43.8	77.8	77.3			
	GR		**56.2***	44.2	51.1	41.0	37.7	**78.5**	77.3			
	IT		**47.3**	39.8	46.2	35.2	31.8	77.2	77.4			
R8	FR	0.1	**56.5**	42.1	55.6	44.4	39.4	97.0	96.9	**97.2**	96.6	96.7
	SP		**50.6***	43.5	46.6	42.7	43.8	**97.2**	96.8			
	GR		**57.8***	45.1	51.1	41.0	37.7	97.0	96.8			
	IT		**49.7**	32.7	46.2	35.2	31.8	96.9	96.9			
FR	20NEWS	1	**44.7**	35.2	44.4	35.7	27.5	86.1	**86.0**	85.9	**86.0**	**86.0**
SP			44.2	36.0				**88.3**	88.1	88.2	88.2	88.1
GR			43.3	35.5				83.4	83.3	**83.5**	83.2	**83.5**
IT			**44.9**	34.1				**85.5**	85.3	85.3	85.1	85.1
FR	R8	0.1	61.8	52.1	62.8	52.3	48.1	86.0	**86.0**	85.9	**86.0**	**86.0**
SP			**67.3***	52.3				**88.3**	88.1	88.2	88.2	88.1
GR			**64.1**	50.9				83.3	83.1	**83.5**	83.2	**83.5**
IT			62.0	54.7				**85.4**	85.2	85.3	85.1	85.1

and thus this result indicates that the heterogeneous transfer is not as reliable if the target label space is more densely distributed than the source label space.

We observe that the nearest neighbor (**NN**) classifiers outperform the softmax (**SM**) classifiers in general. This is because the objectives in Eq. 1 aim at learning a mapping such that each instance is mapped close to its respective label term embedding (in terms of dot product similarity), thus making the nearest neighbor-finding approach a natural choice. The networks with a softmax layer perform poorly on our target classification task, possibly due to the small number of categorical training labels, making the task very challenging.

We also present the summary of cosine similarities in the embedded label space between the source and the target label terms in Table 3, which approximates the inherent *distance* between the source and the target tasks. While the

Table 3. Label terms (word embeddings) cosine similarities summary for heterogeneous dataset pairs.

Datasets	Cosine similarity		
	max	min	avg
20NEWS ↔ FR,SP,GR,IT	0.460	−0.085	0.090
R8 ↔ FR,SP,GR,IT	0.342	−0.039	0.114

R8 dataset tends to be more semantically related with the Reuters Multilingual datasets than the 20 Newsgroups dataset on average, we only observe marginal difference in their knowledge transfer performance, given the same respective source or target dataset.

Note also that both **WtWf:SM** and **Wt:SM** significantly outperform -:**SM**, a single softmax layer that does not use the auxiliary class relations information learned from word embeddings. This result demonstrates that the projection of samples into the embedded label space improves the discriminative quality of feature representation.

We observe that for a small portion of the target dataset neither helps nor hurts the source classification task, showing no statistically significant difference between the proposed approach (**WsWf:NN**) and other baselines. The learned W_f can thus be considered as a robust projection that maps the intermediate common subspace instances into the embedded label space which can describe both the source and the target categories.

Feature visualization: To visualize the projection quality of the proposed approach, we plot the t-SNE embeddings [18] of the source and the target instances (R8→GR; $\%_{L_T} = 0.1$), projected with **W_Wf:NN** and **W_:NN**, respectively (Fig. 3). We make the following observations: (1) The target instances are generally better discriminated with the projection learned from **WtWf:NN** which transfers knowledge from the source dataset, than the one learned from **Wt:NN**.

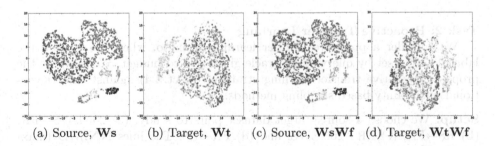

(a) Source, **Ws**   (b) Target, **Wt**   (c) Source, **WsWf**   (d) Target, **WtWf**

Fig. 3. t-SNE visualization of the projected source (R8) and target (GR) instances, where (a), (b) are learned without the transferred knowledge (**W_:NN**), and (c), (d) use the transferred knowledge (**W_Wf:NN**).

Table 4. Comparison of performance (**WtWf:NN**) with varying intermediate embedding dimensions, averaged over 10-fold runs.

Datasets		Test Accuracy (%) vs. M_C					
S	T	20	40	80	160	320	640
20NEWS	FR	54.6	56.8	55.3	56.4	**57.7**	57.1
R8	FR	55.9	54.3	55.1	**57.0**	56.5	56.7

(2) The projection quality of the source samples remains mostly the same. Both of these observations accord with the results in Table 2.

Sensitivity to the embedding dimension: Table 4 compares the performance of the proposed approach (**WtWf:NN**) with varying embedding dimensions (M_C) at the intermediate layer. We do not observe statistically significant improvement for any particular dimension, and thus we simply choose the embedding dimension that yields the highest average value on the two dataset pairs ($M_C = 320$) for all of the experiments.

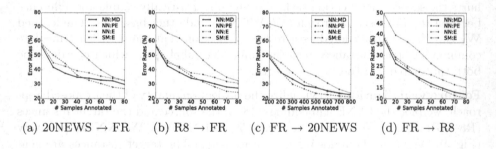

(a) 20NEWS → FR   (b) R8 → FR   (c) FR → 20NEWS   (d) FR → R8

Fig. 4. Proactive transfer learning results. X-axis: the number of queried samples, Y-axis: error rate. (Color figure Online)

Task 2: Proactive Transfer Learning

We consider a proactive transfer learning scenario, where we expand the labeled target set by querying an oracle given a fixed budget. We compare the proposed proactive transfer learning strategies (Sect. 4.4) against the conventional uncertainty-based sampling methods.

Setup: We choose 4 source-target dataset pairs to study: (a) 20NEWS→FR, (b) R8→FR, (c) FR→20NEWS, and (d) FR→R8. The lines **NN:MD** (maximal marginal distribution overlap; solid black) and **NN:PE** (maximum projection entropy; dashed red) refer to the proposed proactive learning strategies in Sect. 4.4, respectively, where the weights are learned with **WtWf:NN**. The baseline active learning strategies **NN:E** (entropy; dashdot green) and **SM:E**

(entropy; dotted blue) select target samples that have the maximum class-posterior entropy given the original target input features only, which quantifies the uncertainty of samples in multiclass classification. The uncertainty-based sampling strategies are widely used in conventional active learning [17,26], however these strategies do not utilize any information from the source domain. Once the samples are queried, **NN:E** learns the classifier **WtWf:NN**, whereas **SM:E** learns a 1-layer softmax classifier.

Main results: Figure 4 shows the target task performance improvement over iterations with various active learning strategies. We observe that both of the proposed active learning strategies (**NN:MD, NN:PE**) outperform the base-lines on all of the source-target dataset pairs. Specifically, **NN:PE** outperforms **NN:E** on most of the cases, which demonstrates that reducing entropy in the projected space is significantly more effective than reducing class-posterior entropy given the original features. Because we re-train the joint network after each query batch, avoiding information redundancy between source and target while reducing target entropy is critical. Note that **NN:MD** outperforms **NN:PE** generally at the beginning, while the performance of **NN:PE** improves faster as it gets more samples annotated. This result indicates that selecting samples with the maximal source and target density overlap (**MD**) helps in building a *bridge* for transfer of knowledge initially, while this information may eventually get redundant, thus the decreased efficacy. Note also that the all of the projection-based methods (**NN:MD, NN:PE, NN:E**) significantly outperform **SM:E**, which measures the entropy and learns the classifier at the original feature space. This result demonstrates that the learned projections $\mathbf{W_T W_f}$ effectively encode input target features, from which we can build a robust classifier efficiently even with a small number of labeled instances.

6 Conclusions

We summarize our contributions as follows: We address a unique challenge of mining and leveraging transferable knowledge in the heterogenous case, where labeled source data differs from target data in both feature and label spaces. To this end, (1) we propose a novel framework for heterogeneous transfer learning to discover the latent subspace to map the source into the target space, from which it simultaneously learns a shared final projection to the embedded label space. (2) In addition, we propose a proactive transfer learning framework which expands the labeled target data with the objective of actively improving transfer accuracy and thus enhancing the target task performance. (3) An extensive empirical evaluation on the hetero-lingual text classification task demonstrates the efficacy of each part of the proposed approach.

Future Work: While the empirical evaluation was conducted on the text domain, our formulation does not restrict the input domain to be textual. We thus believe the approach can be applied broadly, and as future work, we plan to investigate the transferability of knowledge with diverse heterogeneous settings, such as image-aided text classification tasks, etc., given suitable source and target data. In addition, extending the proposed approach for learning selectively from *multiple* heterogeneous source datasets also remains as a challenge.

References

1. Amini, M., Usunier, N., Goutte, C.: Learning from multiple partially observed views-an application to multilingual text categorization. In: NIPS, pp. 28–36 (2009)
2. Chattopadhyay, R., Fan, W., Davidson, I., Panchanathan, S., Ye, J.: Joint transfer and batch-mode active learning. In: ICML, pp. 253–261 (2013)
3. Dai, W., Chen, Y., Xue, G.R., Yang, Q., Yu, Y.: Translated learning: transfer learning across different feature spaces. In: NIPS, pp. 353–360 (2008)
4. Dhillon, P., Foster, D.P., Ungar, L.H.: Multi-view learning of word embeddings via CCA. In: NIPS, pp. 199–207 (2011)
5. Duan, L., Xu, D., Tsang, I.: Learning with augmented features for heterogeneous domain adaptation. In: ICML (2012)
6. Fang, M., Yin, J., Tao, D.: Active learning for crowdsourcing using knowledge transfer. In: AAAI (2014)
7. Frome, A., Corrado, G., Shlens, J., Bengio, S., Dean, J., Ranzato, M., Mikolov, T.: Devise: a deep visual-semantic embedding model. In: NIPS (2013)
8. Fu, Z., Xiang, T., Kodirov, E., Gong, S.: Zero-shot object recognition by semantic manifold distance. In: CVPR, pp. 2635–2644 (2015)
9. Gavves, E., Mensink, T.E.J., Tommasi, T., Snoek, C.G.M., Tuytelaars, T.: Active transfer learning with zero-shot priors: reusing past datasets for future tasks. In: ICCV (2015)
10. Halko, N., Martinsson, P.G., Tropp, J.A.: Finding structure with randomness: probabilistic algorithms for constructing approximate matrix decompositions. SIAM Rev. **53**(2), 217–288 (2011)
11. Huang, J., Gretton, A., Borgwardt, K.M., Schölkopf, B., Smola, A.J.: Correcting sample selection bias by unlabeled data. In: NIPS (2007)
12. Kale, D., Liu, Y.: Accelerating active learning with transfer learning. In: ICDM, pp. 1085–1090 (2013)
13. Kodirov, E., Xiang, T., Fu, Z., Gong, S.: Unsupervised domain adaptation for zero-shot learning. In: ICCV (2015)
14. Kshirsagar, M., Carbonell, J., Klein-Seetharaman, J.: Multisource transfer learning for host-pathogen protein interaction prediction in unlabeled tasks (2013)
15. Li, X., Guo, Y., Schuurmans, D.: Semi-supervised zero-shot classification with label representation learning. In: ICCV, pp. 4211–4219 (2015)
16. Long, M., Wang, J.: Learning transferable features with deep adaptation networks. In: ICML (2015)
17. Loy, C., Hospedales, T., Xiang, T., Gong, S.: Stream-based joint exploration-exploitation active learning. In: CVPR, pp. 1560–1567, June 2012
18. Van der Maaten, L., Hinton, G.: Visualizing data using t-SNE. J. Mach. Learn. Res. **9**(2579–2605), 85 (2008)
19. Mensink, T., Gavves, E., Snoek, C.G.: Costa: Co-occurrence statistics for zero-shot classification. In: CVPR, pp. 2441–2448 (2014)
20. Mikolov, T., Chen, K., Corrado, G., Dean, J.: Efficient estimation of word representations in vector space. In: ICLR (2013)

21. Moon, S., Carbonell, J.: Proactive learning with multiple class-sensitive labelers. In: International Conference on Data Science and Advanced Analytics (DSAA) (2014)
22. Moon, S., Kim, S., Wang, H.: Multimodal transfer deep learning with applications in audio-visual recognition. In: NIPS MMML Workshop (2015)
23. Nguyen, H., Smeulders, A.: Active learning using pre-clustering. In: International Conference on Machine Learning (ICML) (2004)
24. Pan, S.J., Yang, Q.: A survey on transfer learning. IEEE Trans. Knowl. Data Eng. **22**(10), 1345–1359 (2010)
25. Rohrbach, M., Ebert, S., Schiele, B.: Transfer learning in a transductive setting. In: NIPS. pp. 46–54 (2013)
26. Settles, B., Craven, M.: Training text classifiers by uncertainty sampling. In: EMNLP, pp. 1069–1078 (2008)
27. Shi, X., Fan, W., Ren, J.: Actively transfer domain knowledge. In: Daelemans, W., Goethals, B., Morik, K. (eds.) ECML PKDD 2008, Part II. LNCS (LNAI), vol. 5212, pp. 342–357. Springer, Heidelberg (2008)
28. Socher, R., Ganjoo, M., Manning, C.D., Ng, A.Y.: Zero shot learning through cross-modal transfer. In: NIPS (2013)
29. Sun, Q., Amin, M., Yan, B., Martell, C., Markman, V., Bhasin, A., Ye, J.: Transfer learning for bilingual content classification. In: KDD, pp. 2147–2156 (2015)
30. Wang, W., Arora, R., Livescu, K., Bilmes, J.: On deep multi-view representation learning. In: ICML (2015)
31. Weston, J., Bengio, S., Usunier, N.: Wsabie: Scaling up to large vocabulary image annotation. In: IJCAI 2011 (2011)
32. Xiao, M., Guo, Y.: Semi-supervised subspace co-projection for multi-class heterogeneous domain adaptation. In: Appice, A., Rodrigues, P.P., Santos Costa, V., Gama, J., Jorge, A., Soares, C. (eds.) ECML PKDD 2015. LNCS (LNAI), vol. 9285, pp. 525–540. Springer, Heidelberg (2015). doi:10.1007/978-3-319-23525-7_32
33. Zhang, Z., Saligrama, V.: Zero-shot learning via semantic similarity embedding. In: ICCV (2015)
34. Zhou, J.T., Pan, S.J., Tsang, I.W., Yan, Y.: Hybrid heterogeneous transfer learning through deep learning. In: AAAI (2014)
35. Zhu, J., Wang, H., Tsou, B., Ma, M.: Active learning with sampling by uncertainty and density for data annotations. IEEE Trans. Audio Speech Lang. Process. **18**, 1323–1331 (2010)
36. Zhu, Z., Zhu, X., Ye, Y., Guo, Y.F., Xue, X.: Transfer active learning. In: CIKM, pp. 2169–2172 (2011)

Asynchronous Distributed Incremental Computation on Evolving Graphs

Jiangtao Yin[(✉)] and Lixin Gao

University of Massachusetts Amherst, Amherst, USA
{jyin,lgao}@ecs.umass.edu

Abstract. Graph algorithms have become an essential component in many real-world applications. An essential property of graphs is that they are often dynamic. Many applications must update the computation result periodically on the new graph so as to keep it up-to-date. Incremental computation is a promising technique for this purpose. Traditionally, incremental computation is typically performed synchronously, since it is easy to implement. In this paper, we illustrate that incremental computation can be performed asynchronously as well. Asynchronous incremental computation can bypass synchronization barriers and always utilize the most recent values, and thus it is more efficient than its synchronous counterpart. Furthermore, we develop a distributed framework, GraphIn, to facilitate implementations of incremental computation on massive evolving graphs. We evaluate our asynchronous incremental computation approach via extensive experiments on a local cluster as well as the Amazon EC2 cloud. The evaluation results show that it can accelerate the convergence speed by as much as 14x when compared to recomputation from scratch.

1 Introduction

A large class of data routinely produced and collected by large corporations can be modeled as graphs, such as web pages crawled by Google (e.g., the web graph) and tweets collected by Twitter (e.g., the mention graph for users). Since graphs can capture complex dependencies and interactions, graph algorithms have become an essential component in many real-world applications [2,8,24], including business intelligence, social sciences, and data mining.

An essential property of graphs is that they are often dynamic. As new data and/or updates are being collected (or produced), the graph will evolve. For example, search engines will periodically crawl the web, and the web graph is evolving as web pages and hyper-links are created and/or deleted. Many applications must utilize the up-to-date graph in order to produce results that can reflect the current state. However, rerunning the computation over the entire graph is not efficient (considering the huge size of the graph), since it discards the work done in earlier runs no matter how little changes have been made.

The dynamic nature of graphs implies that performing incremental computation can improve efficiency dramatically. Incremental computation exploits the

© Springer International Publishing AG 2016
P. Frasconi et al. (Eds.): ECML PKDD 2016, Part II, LNAI 9852, pp. 722–738, 2016.
DOI: 10.1007/978-3-319-46227-1_45

fact that only a small portion of the graph has changed. It reuses the result of the prior computation and performs computation only on the part of the graph that is affected by the change. Although a number of distributed frameworks have been proposed to support incremental computation on massive graphs [3,6,15–17,23], most of them apply synchronous updates, which require expensive synchronization barriers. In order to avoid the high synchronization cost, asynchronous updates have been proposed. In the asynchronous update model, a vertex performs the update using the most recent values instead of the values from the previous iteration (and there is no waiting time). Intuitively, we can expect asynchronous updates outperform synchronous updates since more up-to-date values are used and the synchronization barriers are bypassed. However, asynchronous updates might require more communications and perform useless computations (e.g., when no new value available to a vertex), and thus result in limited performance gain over synchronous updates.

In this paper, we provide an approach to efficiently apply asynchronous updates to incremental computation. We first describe a broad class of graph algorithms targeted by this paper. We then present our incremental computation approach through illustrating how to apply asynchronous updates to incremental computation. In order to address the challenge that asynchronous updates might require more communication and computation, we present a scheduling scheme to coordinate updates. Furthermore, we develop a distributed system to support our proposed asynchronous incremental computation approach. We evaluate our approach on a local cluster of machines as well as the Amazon EC2 cloud. More specifically, our main contributions are as follows:

- We propose an approach to efficiently apply asynchronous updates to incremental computation on evolving graphs for a broad class of graph algorithms. In order to improve efficiency, a scheduling scheme is presented to coordinate asynchronous updates. The convergence of our proposed asynchronous incremental computation approach is proved.
- We develop an asynchronous distributed framework, GraphIn, to support incremental computation. GraphIn eases the process of implementing graph algorithms with incremental computation in a distributed environment and does not require users to have the distributed programming experience.
- We extensively evaluate our asynchronous incremental computation approach with several real-world graphs. The evaluation results show that our approach can accelerate the convergence speed by as much as 14x when compared to recomputation from scratch. Moreover, a scalability test on a 50-machine cluster demonstrates our approach works with massive graphs having tens of millions of vertices and a billion of edges.

2   Problem Setting

In this section, we first define the problem of performing algorithms on evolving graphs. We then describe a broad class of graph algorithms which we target.

2.1   Problem Formulation

Many graph algorithms leverage iterative updates to compute states (e.g., scores of importance, closenesses to a specified vertex) of the vertices until convergence points are reached. For example, PageRank iteratively refines the rank scores of the vertices (e.g., web pages) of a graph. Such a graph algorithm typically starts with some initial state and then iteratively refines it until convergence. We refer to this kind of graph algorithms as *iterative graph algorithms*.

We are interested in how to efficiently perform iterative graph algorithms on evolving graphs. More formally, if we use G to denote the original graph and G' to represent the new graph, the question we ask is: for an iterative graph algorithm, given G' and the convergence point on G, how to efficiently reach the convergence point on G'.

2.2   Iterative Graph Algorithms

We here describe the iterative graph algorithms targeted by this paper. Typically, the update function of an iterative graph algorithm has the following form:

$$x^{(k)} = f(x^{(k-1)}), \tag{1}$$

where the n-dimensional vector $x^{(k)}$ presents the state of the graph at iteration k, each of its elements is the state for one vertex (e.g., $x^{(k)}[i]$ for vertex i), and $x^{(0)}$ is the initial state. A convergence point is a fixed point of the update function. That is, if $x^{(*)}$ is a convergence point, we have $x^{(*)} = f(x^{(*)})$.

The update function usually can be decomposed into a series of individual functions. In other words, we can update a vertex's state (e.g., x_j) as follows:

$$x_j^{(k)} = c_j \star \sum_{i=1}^{n} \star f_{\{i,j\}}(x_i^{(k-1)}), \tag{2}$$

where '$\star$' is an abstract operator ($\sum_{i=1}^{n} \star$ represents an operation sequence of length n by '$\star$'), c_j is a constant, and $f_{\{i,j\}}(x_i^{k-1})$ is an individual function denoting the impact from vertex i to vertex j in the kth iteration. The operator '$\star$' typically has three candidates, '$+$', 'min', and 'max'. In this paper, we target the iterative graph algorithm that can compute the state in the form of Eq. (2).

2.3   Example Graph Algorithms

We next illustrate a series of well-known iterative graph algorithms, the update functions of which can be converted into the form of Eq. (2).

PageRank and Variants: PageRank is a well-known algorithm, which ranks vertices in a graph based on the stationary distribution of a random walk on the graph. Each element (e.g., r_j) of the score vector r can be computed iteratively as follows: $r_j^{(k)} = \sum_{\{i|\{i \rightarrow j\} \in E\}} \frac{dr_i^{(k-1)}}{|N(i)|} + (1-d)e_j$, where d $(0 < d < 1)$ is the damping factor, $\{i \to j\}$ represents the edge from vertex i to vertex j, E is the set of edges, $|N(i)|$ is the number of outgoing edges of vertex i, and e is a size-n vector with each entry being $\frac{1}{n}$. We can convert the update function of PageRank into the form of Eq. (2). If $\{i \to j\} \in E$, $f_{\{i,j\}}(x_i^{(k-1)}) = dx_i^{(k-1)}/|N(i)|$, otherwise $f_{\{i,j\}}(x_i^{(k-1)}) = 0$, $c_j = (1-d)e_j$, and '$\star$' is '$+$'.

The update function of Personalized PageRank [9] differs from that of PageRank only at vector e. Vector e of Personalized PageRank assigns non-zero values only to the entries indicating the personally preferred pages. Rooted PageRank [19] is a special case of Personalized PageRank. It captures the probability for two vertices to run into each other and uses this probability as the similarity score of those two vertices.

Shortest Paths: The shortest paths algorithm is a simple yet common graph algorithm which computes the shortest distances from a source vertex to all other vertices. Given a weighted graph, $G = (V, E, W)$, where V is the set of vertices, E is the set of edges, and W is the weight matrix of the graph (if there is no edge between i and j, $W[i,j] = \infty$). Then the shortest distance (i.e., d_j) from the source vertex s to a vertex j can be calculated by performing the iterative updates: $d_j^{(k)} = \min\{d_j^{(0)}, \min_i(d_i^{(k-1)} + W[i,j])\}$. For the initial state, we usually set $d_s^{(0)} = 0$ and $d_j^{(0)} = \infty$ for any vertex j other than s. We can map the update function of the shortest paths algorithm into the form of Eq. (2). If there is an edge from vertex i to vertex j, $f_{\{i,j\}}(x_i^{(k-1)}) = x_i^{(k-1)} + W[i,j]$, otherwise $f_{\{i,j\}}(x_i^{(k-1)}) = \infty$, $c_j = d_j^{(0)}$, and '$\star$' is 'min'.

Connected Components: The connected components algorithm is an important algorithm for understanding graphs. It aims to find the connected components in a graph. The main idea of the algorithm is to label each vertex with the maximum vertex id across all vertices in the component which it belongs to. Initially, a vertex j sets its component id $p_j^{(0)}$ as its own vertex id, i.e., $p_j^{(0)} = j$. Then the component id of vertex j can be iteratively updated by $p_j^{(k)} = \max\{p_j^{(0)}, \max_{i \in N(j)}(p_i^{(k-1)})\}$, where $N(j)$ denotes vertex j's neighbors. When no vertex in the graph changes its component id, the algorithm converges. As a result, the vertices having the same component id belong to the same component. We can map the update function of the connected components algorithm into the form of Eq. (2). If there is an edge from vertex i to vertex j, $f_{\{i,j\}}(x_i^{(k-1)}) = x_i^{(k-1)}$, otherwise $f_{\{i,j\}}(x_i^{(k-1)}) = -\infty$, $c_j = j$, and '$\star$' is 'max'.

Other Algorithms: There are many more iterative graph algorithms, update functions of which can be mapped into the form of Eq. (2). We name several ones here. Hitting time is a measure based on a random walk on the graph. Penalized hitting probability [8] and discounted hitting time [18] are variants of hitting time. The adsorption algorithm [2] is a graph-based label propagation algorithm proposed for personalized recommendation. HITS [10] utilizes a two-phase iterative update approach to rank web pages of a web graph. SALSA [13] is another link-based ranking algorithm for web graphs. Effective Importance [4] is a proximity measure to capture the local community structure of a vertex.

3   Asynchronous Incremental Computation

As the underlying graph evolves, the states of the vertices also change. Obviously, rerunning the computation from scratch over the new graph is not efficient, since it discards the work done in earlier runs. Intuitively, performing computations incrementally can improve efficiency. In this section, we present our asynchronous incremental computation approach. The convergence of our approach is proved.

3.1   Asynchronous Updates

In order to describe our asynchronous incremental computation approach, we define a time sequence $\{t_0, t_1, \ldots, t_\infty\}$. Let $\hat{x}^{(k)}$ denote the state vector at time t_k. Also, we introduce the delta state vector $\Delta\hat{x}^{(k)}$ to represent the difference between $\hat{x}^{(k+1)}$ and $\hat{x}^{(k)}$ in the operator '$\star$' manner, i.e., $\hat{x}^{(k+1)} = \hat{x}^{(k)} \star \Delta\hat{x}^{(k)}$. The goal of introducing $\Delta\hat{x}^{(k)}$ is to perform accumulative computations. When the operator '$\star$' has the commutative property and the associative property and the function $f_{\{i,j\}}(x_i)$ has the distributive property over '$\star$', the computation can be performed accumulatively. All the graph algorithms discussed in Sect. 2.3 satisfy these properties. It is straightforward to verify that accumulative computations are equivalent to normal computations. The benefit of performing accumulative computations is that only changes of the states (i.e., delta states) are used to compute new changes. If there is no change for the state of a vertex, no communication or computation is necessary. The general idea of separating fixed parts from changes and leveraging changes to compute new changes also shows efficiency in many other algorithms, such as Nonnegative Matrix Factorization [21] and Expectation-Maximization [22].

In our asynchronous incremental computation approach, each vertex i updates its $\Delta\hat{x}_i^{(k)}$ and $\hat{x}_i^{(k)}$ independently and asynchronously, starting from $\Delta\hat{x}_i^{(0)}$ and $\hat{x}_i^{(0)}$ (we will illustrate how to construct them soon). In other words, there are two separate operations for vertex j:

– *Accumulate* operation: whenever receiving a value (e.g., $f_{\{i,j\}}(\Delta\hat{x}_i)$) from a neighbor (e.g., i), perform $\Delta\hat{x}_j = \Delta\hat{x}_j \star f_{\{i,j\}}(\Delta\hat{x}_i)$;
– *Update* operation: perform $\hat{x}_j = \hat{x}_j \star \Delta\hat{x}_j$; for any neighbor l, if $f_{\{j,l\}}(\Delta x_j^{(1)}) \neq o$, send $f_{\{j,l\}}(\Delta\hat{x}_j)$ to l; and then reset $\Delta\hat{x}_j$ to o;

where o is the identity value of the operator '$\star$'. That is, for $\forall z \in R$, $z = z \star o$ (if '$\star$' is '$+$', $o = 0$; if '$\star$' is 'min', $o = \infty$; if '$\star$' is 'max', $o = -\infty$). Basically, the *accumulate* operation accumulates received values between two consecutive updates on $\hat{x}_j$. The *update* operation adjusts $\hat{x}_j$ by absorbing $\Delta\hat{x}_j$, sends useful values to other vertices, and resets $\Delta\hat{x}_j$.

We now illustrate how to construct $\hat{x}_i^{(0)}$ and $\Delta\hat{x}_i^{(0)}$ by leveraging the computation result on the previous graph, G. We need to make sure that the constructed

$\hat{x}_i^{(0)}$ and $\Delta\hat{x}_i^{(0)}$ can guarantee the correctness of the result on the new graph. Let $\bar{x}^{(*)}$ denote the convergence point on G. We next show how to construct $\hat{x}_i^{(0)}$ and $\Delta\hat{x}_i^{(0)}$ when the operator '$\star$' is '$+$' (for all the graph algorithms discussed in Sect. 2.3 except shortest paths and connected components) and when '$\star$' is 'min/max' (shortest paths and connected components), respectively.

For an iterative graph algorithm with the operator '$\star$' as '$+$', we first leverage $\bar{x}^{(*)}$ to construct $\hat{x}^{(0)}$ in the following way: for a kept vertex (e.g., i), we set $\hat{x}_i^{(0)} = \bar{x}_i^{(*)}$; for a newly added vertex (e.g., j), we set $\hat{x}_j^{(0)} = 0$. In contrast, *recomputation from scratch* typically utilizes $\mathbf{0}$ as $\hat{x}^{(0)}$ (where $\mathbf{0}$ is a vector with all its elements being zero). In order to construct $\Delta\hat{x}^{(0)}$, we compute $\hat{x}^{(1)}$ using $\hat{x}^{(1)} = f(\hat{x}^{(0)})$ and then construct $\Delta\hat{x}^{(0)}$ by making sure $\Delta\hat{x}^{(0)}$ satisfying $\hat{x}^{(1)} = \hat{x}^{(0)} \star \Delta\hat{x}^{(0)}$. Since '$\star$' is '$+$', we can calculate $\Delta\hat{x}^{(0)}$ by $\Delta\hat{x}^{(0)} = \hat{x}^{(1)} - \hat{x}^{(0)}$. It is important to note that here the deleted vertices and/or edges do not affect the way we construct $\hat{x}_i^{(0)}$ and $\Delta\hat{x}_i^{(0)}$. In other words, no matter whether there are deleted vertices and/or edges, the way we construct $\hat{x}_i^{(0)}$ and $\Delta\hat{x}_i^{(0)}$ can guarantee the correctness of the result on the new graph.

For an iterative graph algorithm with the operator '$\star$' as 'min/max', we construct $\hat{x}_i^{(0)}$ and $\Delta\hat{x}_i^{(0)}$ as follows. When the operator '$\star$' is 'min' (e.g., shortest paths), if any vertex's initial state is not smaller than its final converged state, the algorithm will converge. This is because of the following reason. When the algorithm has not converged, in each iteration there must be at least one vertex whose state is becoming smaller, and thus the overall state vector is becoming closer to the final converged state vector. When there is no vertex changing its state, the algorithm converges. Generally, it is hard to know the final converged state vector. Therefore, for the shortest paths algorithm, recomputation from scratch usually sets the initial state of a vertex (other than the source vertex) as ∞ to guarantee that it is not smaller than the final converged state. Fortunately, when the graph grows (vertices and/or edges are added and no vertices or edges are deleted), the previous converged state of a kept vertex must be not smaller than its converged state on the new graph. Therefore, for the graph growing scenario, we construct $\hat{x}_i^{(0)}$ in the following way: for a kept vertex (e.g., i), we set $\hat{x}_i^{(0)} = \bar{x}_i^{(*)}$; for a newly added vertex (e.g., j), we set $\hat{x}_j^{(0)} = \infty$. Similarly, for the connected component algorithm, whose operator '$\star$' is 'max', we can construct $\hat{x}_i^{(0)}$ (for the graph growing scenario) as follows: for a kept vertex (e.g., i), we set $\hat{x}_i^{(0)} = \bar{x}_i^{(*)}$; for a newly added vertex (e.g., j), we set $\hat{x}_j^{(0)} = j$. To construct $\Delta\hat{x}^{(0)}$, we also compute $\hat{x}^{(1)}$ using $\hat{x}^{(1)} = f(\hat{x}^{(0)})$ and then simply set $\Delta\hat{x}_j^{(0)} = \hat{x}_j^{(1)}$. It can satisfy $\hat{x}^{(1)} = \hat{x}^{(0)} \star \Delta\hat{x}^{(0)}$, no matter '$\star$' is 'min' or 'max'.

3.2   Selective Execution

One potential problem of basic asynchronous updates is that they might require more computation and communication when compared to their synchronous counterparts. This is because vertices are updated in a round-robin manner no matter how many new values available to a vertex. To solve this problem, instead of updating vertices in a round-robin manner, we update vertices selectively by identifying their importance. The motivation behind it is that not all vertices contributes the same to the convergence. We refer to this scheduling scheme as *selective execution*. The vertices are selected according to their importance (in terms of contribution to the convergence).

Our selective execution scheduling scheme selects a block of m vertices (instead of one) to update each round. The reason is that if one vertex is chosen to update at a time, the scheduling overhead (e.g., maintaining a priority queue to always choose the vertex with the highest importance) is high. Once the block of the selected vertices are updated, it selects another block to update. Every time our scheme selects the top-m vertices in terms of the importance value. The size of the block (i.e., m) balances the tradeoff between the gain from selective execution and the cost of selecting vertices. Setting m too small may incur considerable overhead, while setting m too large may degrade the effect of selective execution, e.g., if setting m as the number of total vertices, it degrades to the round-robin scheduling. We will discuss how to determine m in Sect. 4.1.

We then illustrate how to quantify a vertex's importance when '$\star$' is 'min/max' and when the operator '$\star$' is '+', respectively. Ideally, the vertex whose update decreases the distance to the fixed point (i.e., $||x^{(*)} - \hat{x}^{(k)}||_1$) most should have the highest importance. For an iterative graph algorithm with the operator '$\star$' as 'min/max', the iterative updates either monotonically decrease (e.g., shortest paths) or monotonically increase (e.g., connected components) any element of $\hat{x}^{(k)}$. For ease of exposition, we assume the monotonically decreasing case. In this case, $x_j^{(*)} \le \hat{x}_j^{(k)}$ for any j, and thus we have $||x^{(*)} - \hat{x}^{(k)}||_1 = ||\hat{x}^{(k)}||_1 - ||x^{(*)}||_1$. An update on vertex j decrease $||\hat{x}^{(k)}||_1$ by $|\hat{x}_j^{(k)} \star \Delta\hat{x}_j^{(k)} - \hat{x}_j^{(k)}|$. Therefore, we use $|\hat{x}_j^{(k)} \star \Delta\hat{x}_j^{(k)} - \hat{x}_j^{(k)}|$ to represent the importance of the vertex j (denoted as η_j), i.e. $\eta_j = |\hat{x}_j^{(k)} \star \Delta\hat{x}_j^{(k)} - \hat{x}_j^{(k)}|$.

For an iterative graph algorithm with the operator '$\star$' as '+', it is difficult to directly measure how the distance to the fixed point decreases. Update one single vertex may even increase the distance to the fixed point. Fortunately, for such an algorithm, its update function ($f()$) typically can be seen as a $||\cdot||$-*contraction mapping*. That is, there exists an α ($0 \le \alpha < 1$), such that $||f(x) - f(y)|| \le \alpha||x-y||, \forall x, y \in R^n$. Therefore, we can provide an upper bound on the distance, as stated in Theorem 1. The proof is omitted due to the space limitation. We then analyze how the upper bound decreases.

Theorem 1. $||x^{(*)} - \hat{x}^{(k+1)}||_1 \le \frac{||\Delta\hat{x}^{(k+1)}||_1}{1-\alpha}$.

Without loss of generality, assume that current time is t_k and that during interval $[t_k, t_{k+1}]$ we only update vertex j. When updating vertex j, we accumulate $\Delta\hat{x}_j^{(k)}$ to $\hat{x}_j$, send $f_{(j,l)}(\Delta\hat{x}_j^{(k)})$ to a vertex l (and the total sending out value is no larger than $\alpha|\Delta\hat{x}_j^{(k)}|$), and reset $\Delta\hat{x}_j^{(k)}$ to 0. Therefore, we have the following theorem.

Theorem 2. $||\Delta\hat{x}^{(k+1)}||_1 \leq ||\Delta\hat{x}^{(k)}||_1 - (1 - \alpha)|\Delta\hat{x}_j^{(k)}|.$

Theorem 2 implies that the upper bound monotonically decreases. When updating vertex j, we have $\frac{||\Delta\hat{x}^{(k+1)}||_1}{1-\alpha} \leq \frac{||\Delta\hat{x}^{(k)}||_1}{1-\alpha} - |\Delta\hat{x}_j^{(k)}|$. It shows that the reduction in the upper bound is at least $|\Delta\hat{x}_j^{(k)}|$. Given a graph, α is a constant. Hence, we define the importance of the vertex j to be $|\Delta\hat{x}_j^{(k)}|$, i.e., $\eta_j = \arg\max_j |\Delta\hat{x}_j^{(k)}|$.

3.3 Convergence

Our asynchronous incremental computation approach yields the same result as recomputation from scratch. To prove it, we first show that if synchronous updates (i.e., $x^{(k)} = f(x^{(k-1)})$) converge (and synchronous updates converge for all the graph algorithms discussed in Sect. 2.3), any asynchronous update scheme that can guarantee every vertex is updated infinitely often (until its state is fixed) will yield the same result as synchronous updates, as stated in Lemma 1.

Lemma 1. *If updates* $x^{(k)} = f(x^{(k-1)})$ *converge to* $x^{(*)}$, *any asynchronous update scheme that guarantees every vertex is updated infinitely often will converge to* $x^{(*)}$ *as well, i.e.,* $\hat{x}^{(\infty)} = x^{(*)}$.

We then show that our asynchronous incremental computation approach fulfills this requirement, as stated in Lemma 2. The proofs of both Lemmas 1 and 2 are omitted.

Lemma 2. *Our asynchronous incremental computation approach can guarantee that every vertex is updated infinitely often (until its state is fixed).*

We can also prove that recomputation from scratch converges to $x^{(*)}$ (no matter what type of updates it uses). As a result, we have the following theorem.

Theorem 3. *Our asynchronous incremental computation approach converges and yields the same result as recomputation from scratch.*

4 Distributed Framework

Oftentimes, iterative graph algorithms in real-world applications need to process massive graphs. Hence, it is desirable to leverage the parallelism of a cluster of machines to run these algorithms. Furthermore, it is troublesome to implement asynchronous incremental computation for each individual algorithm. Therefore, we propose GraphIn, an in-memory asynchronous distributed framework, for supporting iterative graph algorithms with incremental computation. GraphIn provides several high-level APIs to users for implementing asynchronous incremental computation and meanwhile hides the complexity of distributed computation. It leverages the proposed selective execution to accelerate convergence.

GraphIn consists of a number of workers and one master. Workers perform vertex updates, and the master controls the flow of computation. The new graph and the previous computed result are taken as the input of GraphIn. The input graph is split into partitions and each worker is responsible for one partition. Each worker leverages an in-memory table to store the vertices assigned to it. A worker has two main operations for its stored vertices: the accumulate operation and the update operation, as illustrated in Sect. 3.1. The accumulate operation utilizes a user-defined function to aggregate incoming messages for a vertex and also triggers another user-defined function to calculate the vertex's importance. The update operation uses a user-defined function to update the states of scheduled vertices and compute outgoing messages.

The prototype of GraphIn is built upon Maiter [26]. Maiter is designed for processing static graphs, and thus has inherent impediments to the execution of graph algorithms with incremental computation. First, it relies on the specific initial state to guarantee the convergence of a graph algorithm. However, incremental computation leverages the previous result as the initial state, which can be arbitrary. Second, although Maiter supports prioritized updates, its scheduling scheme assumes that Δx_i is always positive for any vertex i, which can be not true under incremental computation. Last, the termination check mechanism of Maiter assumes that $||x||_1$ varies monotonically, which can be not true as well under incremental computation. GraphIn removes all these impediments to efficiently support incremental computation.

4.1 Distributed Selective Execution

GraphIn leverages the proposed selective execution scheduling as its default scheduling scheme. Since a centralized approach of finding the top-m elements is inefficient in a distributed environment, GraphIn allows each worker to build its own selective execution scheduling. Round by round (except the first round in which all vertices are selected to derive $\hat{x}^{(0)}$ and $\Delta\hat{x}^{(0)}$), each worker selects its local top-m vertices in terms of the importance. The number m is crucial to the effect of selective execution.

For the iterative graph algorithm with the operator '$\star$' as '$+$', GraphIn learns m online. We use $\mu \cdot n$ to quantify the overhead of selecting such m vertices (where μ represents the amortized overhead), which is proportional to the total number (n) of vertices with an efficient selection algorithm (e.g., quick-select). Also, we assume that the average cost of updating one vertex is ν, and then the cost of updating those m vertices is $\nu \cdot m$. Let $c(m)$ be the total cost of updating those m vertices (including both selection and update), then $c(m) = \mu \cdot n + \nu \cdot m$. Let $g(m) = \sum_{j \in S} |\Delta\hat{x}_j|$ (recall that $|\Delta\hat{x}_j|$ represents the importance of vertex i), where S denotes the set of the m selected vertices. For each round, we aim to find the m that can achieve the largest efficiency, i.e., $m = \arg\max_m \frac{g(m)}{c(m)}$. It is computationally impossible to try every value (from 1 to n) to figure out the best m. Therefore, our practical approach chooses several values ($0.05n$, $0.1n$, $0.25n$, $0.5n$, n), which cover the entire range of possible m, as the candidates.

For each candidate m, we leverage quick-select to find the m-th $|\Delta\hat{x}_j|$, which is used as a threshold, and all $|\Delta\hat{x}_i|$ no less than the threshold are counted into $g(m)$. By testing each candidate (we set ν/μ as 4 by default), we can figure out the best m and the set S. The practical approach leverages quick-select to avoid the time-consuming sorting, and thus takes $O(n)$ time on extracting the top-m vertices instead of $O(n \log n)$ time. For the iterative graph algorithm with the operator '$\star$' as 'min/max', the importance of a vertex might be close to ∞. If we still use the above idea, $g(m)$ might easily be overflown. Therefore, in this case, we simply set m as $0.1n$, which shows good performance in experiments. Note that if there are only m' ($m' < m$) vertices with the importance being larger than 0, we only select these m' vertices to update.

4.2 Distributed Termination Check

We design termination check mechanisms for the iterative graph algorithm with the operator '$\star$' as 'min/max' and for that with the operator '$\star$' as '$+$', respectively. When '$\star$' is 'min/max', $||\hat{x}^{(k)}||_1$ monotonically decreases or increases. Therefore, we can utilize $||\hat{x}^{(k)}||_1$ to perform the termination check. If $||\hat{x}^{(k)}||_1 - ||\hat{x}^{(k-1)}||_1 = 0$, the algorithm has converged, and thus the computation can be terminated. When '$\star$' is '$+$', $||x^{(*)} - \hat{x}^{(k)}||_1$ is the choice for measuring convergence. However, it is difficult to directly quantify $||x^{(*)} - \hat{x}^{(k)}||_1$, since the fixed point $x^{(*)}$ is always unknown during the computation. Fortunately, we know $||x^{(*)} - \hat{x}^{(k)}||_1 \leq ||\Delta\hat{x}^{(k)}||_1/(1 - \alpha)$ from Theorem 1, and thus can leverage $||\Delta\hat{x}^{(k)}||_1$ to measure convergence. We use the convergence criterion, $||\Delta\hat{x}^{(k)}||_1 \leq \epsilon$, where the convergence tolerance ϵ is a pre-defined constant.

GraphIn adopts a passively monitoring model to perform the termination check, which works by periodically (and the period is configurable) measuring $||\hat{x}^{(k)}||_1$ if the operator '$\star$' is 'min/max' (or $||\Delta\hat{x}^{(k)}||_1$ if '$\star$' is '$+$'). To complete the measure, each worker computes the sum of $|\hat{x}_j^{(k)}|$ (or $|\Delta\hat{x}_j^{(k)}|$) of its local vertices and sends the local sum to the master. The master aggregates the local sums into a global sum. The challenge of performing such a distributed termination is to make sure that the local sum at each worker are calculated from the snapshot of the values at the same time (especially for $|\Delta\hat{x}_j^{(k)}|$). To address the challenge, GraphIn asks all the workers to pause vertex updates before starting to calculate the local sums. The procedure of the distributed termination check is as follows.

1. When it is the time to perform the termination check, the master broadcasts a chk_{pre} message to all the workers.
2. Upon receiving the chk_{pre} message, every worker pauses vertex updates and then replies a chk_{ready} message to the master.
3. The master gathers those chk_{ready} messages from all the workers, and then broadcasts a chk_{begin} message to them.
4. Upon receiving the chk_{begin} message, every worker calculates the local sum, $\sum_j |\hat{x}_j^{(k)}|$ (or $\sum_j |\Delta\hat{x}_j^{(k)}|$), and reports it to the master.

5. The master aggregates the local sums to the global sum $||\hat{x}^{(k)}||_1$ (or $||\Delta\hat{x}^{(k)}||_1$). If $||\hat{x}^{(k)}||_1 - ||\hat{x}^{(k-1)}||_1 \neq 0$ (or $||\Delta\hat{x}^{(k)}||_1 > \epsilon$), the master broadcasts a chk_{fin} message to all the workers. Otherwise, it broadcasts a $term$ message.
6. When a worker receives the chk_{fin} message, it resumes vertex updates. When a worker receives the $term$ message, it dumps the result to a local disk and then terminates the computation.

It is important to note that since calculating the local sums is inexpensive and it is done periodically, the overhead of the termination check is ignorable.

5   Evaluation

In this section, we evaluate the performance of our asynchronous incremental computation approach. We compare it with re-computation from scratch. Both approaches are supported by GraphIn. To show the performance of the selective execution scheduling, we compare it with the round-robin scheduling. The performance of other distributed frameworks that can support synchronous incremental computation are also evaluated.

5.1   Experiment Setup

The experiments are performed on both a local cluster and a large-scale cluster on Amazon EC2. The local cluster consists of 4 machines. The large-scale cluster consists of 50 EC2 medium instances.

Two graph algorithms are implemented on GraphIn, PageRank and the shortest paths algorithm. For PageRank, the damping factor is set to 0.8, and if not stated otherwise, the convergence tolerance ϵ (which is discussed in Sect. 4.2) is set to $10^{-2}/n$ (n is the number of vertices of the corresponding graph). The shortest paths algorithm stops running only when the convergence point is reached (i.e., all the vertices reach their shortest paths to the source vertex). The measurement of each experiment is averaged over 10 runs. Real-world graphs of various sizes are used in the experiments and are summarized in Table 1.

Table 1. Graph Dataset Summary

Dataset	Vertices	Edges
Amazon co-purchasing graph (Amz) [14]	403K	3.4M
Web graph from Google (Gog) [14]	876K	5.1M
LiveJournal social network (LJ) [14]	4.8M	69M
Web graph from UK (UK) [5]	39M	936M
Web graph from IT (IT) [5]	41M	1.2B

5.2 Overall Performance

We first show the convergence time of PageRank on the local cluster. The convergence time is measured as the wall-clock time that PageRank uses to reach the convergence criterion (i.e., $||\Delta \hat{x}^{(k)}||_1 \leq \epsilon$). We consider both the edge change case and the vertex change case. Under the edge change case, we randomly pick a number of vertices to change their edges. In the graph evolving process, there are usually more added edges than deleted edges. Therefore, for 80 % of the picked vertices, we add one outgoing edge to it with a randomly picked neighbor. For the rest 20 % vertices, we remove one randomly picked edge from it. Under the vertex change case, we pick a number (e.g., p, some percentage of the total number of vertices) for each experiment. We add $0.8p$ new vertices to the graph and delete $0.2p$ vertices. For each added vertex, we put two edges (one incoming edge and one outgoing edge) with randomly picked neighbors. For each deleted vertex, we also delete all its edges.

Figure 1 shows the performance on the Amz graph under the edge change case. We can see that incremental computation (denoted as "Incr") is much faster than re-computation from scratch (denoted as "Re") for different percentages of vertices with edge change. The selective execution scheduling (denoted as "Sel") is faster than the round-robin scheduling (denoted as "R-R") with either approach. The efficiency of incremental computation is more prominent when the change is smaller. For example, when the percentage of vertices with edge change is 0.01 %, incremental computation with the selective execution scheduling is about 10x faster than recomputation from scratch with the round-robin scheduling and 7x faster than recomputation from scratch with the selective exe-

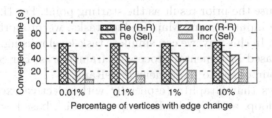

Fig. 1. PageRank on Amz graph (edge change).

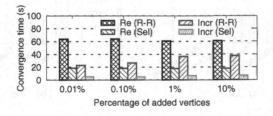

Fig. 2. Shortest paths on weighted Amz graph.

cution scheduling. Not surprisingly, incremental computation takes longer time as the percentage of vertices with edge change becomes larger, and the convergence time of the re-computation is almost the same since the change to the graph is relatively small. Similar trends are observed for the vertex change case.

We then present the result of the shortest paths algorithm, which runs on weighted graphs. Here the convergence time is measured as the wall-clock time that the shortest paths algorithm uses to reach the convergence point. All the graphs summarized in Table 1 are unweighed. We generate a weighted graph by assigning weights to the Amz graph. The weight of each edge is an integer, which is randomly drawn from the rang $[1, 100]$. Figure 2 plots the performance comparison under the vertex adding case. The percentage means the ratio between the number of added vertices to the number of original vertices. For each added vertex, we put two weighted edges (one incoming edge and one outgoing edge) with randomly picked neighbors. From the figure, we can see that incremental computation with the selective execution scheduling is about 14x faster than recomputation from scratch with the round-robin scheduling when the percentage of added vertices is 0.01 % and still 9x faster even when the percentage is 10 %. Similar results are observed for the edge adding case as well.

5.3   Comparison with Synchronous Incremental Computation

It is also possible to build a framework to support incremental computation upon other systems, such as Hadoop and Spark. To demonstrate the efficiency of GraphIn, we compare it with both Hadoop and Spark for the 1 % of vertices with edge change scenario. We restrict our performance comparison to PageRank, since it is a representative graph algorithm. For fair comparison, we instruct both systems to use the prior result as the starting point. For Hadoop, if there is no change in the input of some Map/Reduce tasks, we proportionally discount the running time. In this way, we can simulate task-level reusing, which is the key of MapReduce-based incremental processing frameworks. For Spark, we choose its Graphx [7] component to implement PageRank.

Figure 3 shows that GraphIn (especially with selective execution) is much faster than Hadoop and Spark. Hadoop is a disk-based system and uses

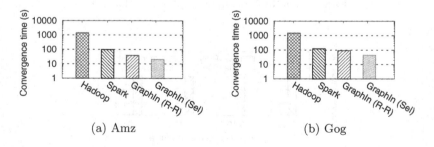

(a) Amz   (b) Gog

Fig. 3. PageRank on different frameworks.

synchronous updates. Even though Spark is a memory-based system, it also utilizes synchronous updates. Therefore, it is still slower than GraphIn.

5.4   Scaling Performance

We further evaluate incremental computation on the large-scale Amazon cluster to test its scalability. We consider the 1 % of vertices with edge change scenario, and concentrate on PageRank (and set the convergence tolerance ϵ to 10^{-4}). We first use the three large real-world graphs, LJ, UK, and IT (both UK and IT have tens of millions of vertices and a billion of edges), as input graphs when all the 50 instances are used. As shown in Fig. 4a (note that the y-axis is in log scale), on the large-scale cluster incremental computation is still much faster than re-computation from scratch, and both approaches can benefit from the selective execution scheduling.

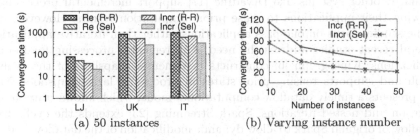

(a) 50 instances   (b) Varying instance number

Fig. 4. Performance on Amazon cluster.

We then show the performance of incremental computation when different numbers of instances are used. Figure 4b shows the convergence time on LJ as we increase the number of instances from 10 to 50. It can be seen that by increasing the number of instances, the convergence time is reduced, and that the selective execution scheduling is always faster than the round-robin scheduling.

6   Related Work

Due to the dynamic nature of graphs in real-world applications, incremental computation has been studied extensively. In terms of iterative graph algorithms, most of the studies [1,11,12] focus on PageRank. The basic idea behind approaches in [11,12] is that when a change happens in the graph, the effect of the change on the PageRank scores is mostly local. These approaches start with the exact PageRank scores of the original graph but provide approximate scores for the graph after the change, and the estimations may drift away from the exact scores. On the contrary, our approach can provide exact scores. The work in [1] utilizes the Monte Carlo method to approximate PageRank scores on evolving graphs. It precomputes a number of random walk segments for each vertex and stores them in distributed shared memory. Besides of the approximate result, it also incurs high memory overhead.

In recent years, the growing scale and importance of graph data have driven the development of a number of distributed graph systems. Graphx [7] is a graph system built on top of Spark. It stores graphs as tabular data and implements graph operations using distributed joins. PrIter [25], Maiter [26], and Prom [20], introduce prioritized updates to accelerate convergence. PrIter is a MapReduce-based framework, which requires synchronous iterations. Maiter and Prom utilize asynchronous iterative computation. All these graph systems aim at supporting graph computation on static graph structures.

There are several systems for supporting incremental parallel processing on massive datasets. Incoop [3] extends the MapReduce programming model to support incremental processing. It saves and reuses states at the granularity of individual Map or Reduce tasks. Continuous bulk processing (CBP) [15] provides a groupwise processing operator to reuse prior state for incremental analysis. Similarly, other systems like DryadInc [17] support incremental processing by allowing their applications to reuse prior computation results. However, most of the studies focus on one-pass applications rather than iterative applications. Several recent studies address the need of incremental processing for iterative applications. Kineograph [6] constructs incremental snapshots of the evolving graph and supports reusing prior states in processing later snapshots. Naiad [16] presents a timely dataflow computational mode, which allows stateful computation and nested iterations. Spark Streaming [23] extends the cyclic batch dataflow of original Spark to allow dynamic modification of the dataflow and thus supports iteration and incremental processing. However, most of these systems apply synchronous updates to incremental computation. Our work illustrates how to efficiently apply asynchronous updates to incremental computation.

7   Conclusion

In this paper, we propose an approach to efficiently apply asynchronous updates to incremental computation on evolving graphs. Our approach works for a family of iterative graph algorithms. We also present a scheduling scheme, selective execution, to coordinate asynchronous updates so as to accelerate convergence. Furthermore, to facilitate the implementation of iterative graph algorithms with incremental computation in a distributed environment, we design and implement an asynchronous distributed framework, GraphIn. Our evaluation results show that our asynchronous incremental computation approach can significantly boost the performance.

Acknowledgments. We would like to thank anonymous reviewers for their insightful comments. This work is partially supported by NSF grants CNS-1217284 and CCF-1018114.

References

1. Bahmani, B., Chowdhury, A., Goel, A.: Fast incremental and personalized pagerank. Proc. VLDB Endow. **4**(3), 173–184 (2010)
2. Baluja, S., Seth, R., Sivakumar, D., Jing, Y., Yagnik, J., Kumar, S., Ravichandran, D., Aly, M.: Video suggestion and discovery for youtube: taking random walks through the view graph. In: WWW 2008, pp. 895–904 (2008)
3. Bhatotia, P., Wieder, A., Rodrigues, R., Acar, U.A., Pasquin, R.: Incoop: Mapreduce for incremental computations. In: SoCC 2011, pp. 7:1–7:14 (2011)
4. Bogdanov, P., Singh, A.: Accurate and scalable nearest neighbors in large networks based on effective importance. In: CIKM 2013. pp. 1009–1018 (2013)
5. Boldi, P., Vigna, S.: The WebGraph framework I: compression techniques. In: WWW 2004, pp. 595–601 (2004)
6. Cheng, R., Hong, J., Kyrola, A., Miao, Y., Weng, X., Wu, M., Yang, F., Zhou, L., Zhao, F., Chen, E.: Kineograph: taking the pulse of a fast-changing and connected world. In: EuroSys 2012, pp. 85–98 (2012)
7. Gonzalez, J.E., Xin, R.S., Dave, A., Crankshaw, D., Franklin, M.J., Stoica, I.: Graphx: graph processing in a distributed dataflow framework. In: OSDI 2014, pp. 599–613 (2014)
8. Guan, Z., Wu, J., Zhang, Q., Singh, A., Yan, X.: Assessing and ranking structural correlations in graphs. In: SIGMOD 2011, pp. 937–948 (2011)
9. Jeh, G., Widom, J.: Scaling personalized web search. In: WWW 2003, pp. 271–279 (2003)
10. Kleinberg, J.M.: Authoritative sources in a hyperlinked environment. J. ACM **46**, 604–632 (1999)
11. Langville, A.N., Meyer, C.D.: Updating PageRank with iterative aggregation. In: WWW 2004, pp. 392–393 (2004)
12. Langville, A.N., Meyer, C.D.: Updating markov chains with an eye on Google's PageRank. SIAM J. Matrix Anal. Appl. **27**(4), 968–987 (2006)
13. Lempel, R., Moran, S.: Salsa: the stochastic approach for link-structure analysis. ACM Trans. Inf. Syst. **19**(2), 131–160 (2001)
14. Leskovec, J., Krevl, A.: SNAP datasets: stanford large network dataset collection, Jun 2014. http://snap.stanford.edu/data
15. Logothetis, D., Olston, C., Reed, B., Webb, K.C., Yocum, K.: Stateful bulk processing for incremental analytics. In: SoCC 2010, pp. 51–62 (2010)
16. Murray, D.G., McSherry, F., Isaacs, R., Isard, M., Barham, P., Abadi, M.: Naiad: a timely dataflow system. In: SOSP 2013, pp. 439–455 (2013)
17. Popa, L., Budiu, M., Yu, Y., Isard, M.: Dryadinc: reusing work in large-scale computations. In: HotCloud 2009 (2009)
18. Sarkar, P., Moore, A.W.: Fast nearest-neighbor search in disk-resident graphs. In: KDD 2010, pp. 513–522 (2010)
19. Song, H.H., Cho, T.W., Dave, V., Zhang, Y., Qiu, L.: Scalable proximity estimation and link prediction in online social networks. In: IMC 2009, pp. 322–335 (2009)
20. Yin, J., Gao, L.: Scalable distributed belief propagation with prioritized block updates. In: CIKM 2014, pp. 1209–1218 (2014)
21. Yin, J., Gao, L., Zhang, Z.M.: Scalable nonnegative matrix factorization with block-wise updates. In: Calders, T., Esposito, F., Hüllermeier, E., Meo, R. (eds.) ECML PKDD 2014, Part III. LNCS, vol. 8726, pp. 337–352. Springer, Heidelberg (2014)

22. Yin, J., Zhang, Y., Gao, L.: Accelerating expectation-maximization algorithms with frequent updates. In: CLUSTER 2012, pp. 275–283 (2012)
23. Zaharia, M., Das, T., Li, H., Hunter, T., Shenker, S., Stoica, I.: Discretized streams: fault-tolerant streaming computation at scale. In: SOSP 2013, pp. 423–438 (2013)
24. Zhang, C., Jiang, S., Chen, Y., Sun, Y., Han, J.: Fast inbound Top-K query for random walk with restart. In: Appice, A., Rodrigues, P.P., Santos Costa, V., Gama, J., Jorge, A., Soares, C. (eds.) ECML PKDD 2015. LNCS, vol. 9285, pp. 608–624. Springer, Heidelberg (2015)
25. Zhang, Y., Gao, Q., Gao, L., Wang, C.: PrIter: a distributed framework for prioritized iterative computations. In: SoCC 2011, pp. 13:1–13:14 (2011)
26. Zhang, Y., Gao, Q., Gao, L., Wang, C.: Maiter: an asynchronous graph processing framework for delta-based accumulative iterative computation. IEEE Trans. Parallel Distrib. Syst. **25**(8), 2091–2100 (2014)

Infection Hot Spot Mining
from Social Media Trajectories

Roberto C.S.N.P. Souza$^{(\boxtimes)}$, Renato M. Assunção, Derick M. de Oliveira,
Denise E.F. de Brito, and Wagner Meira Jr.

Department of Computer Science, Universidade Federal de Minas Gerais,
Belo Horizonte, Brazil
{nalon,assuncao,derickmath,denise.brit,meira}@dcc.ufmg.com

Abstract. Traditionally, in health surveillance, high risk zones are identified based only on the residence address or the working place of diseased
individuals. This provides little information about the places where people are infected, the truly important information for disease control. The
recent availability of spatial data generated by geotagged social media
posts offers a unique opportunity: by identifying and following diseased
individuals, we obtain a collection of sequential geo-located events, each
sequence being issued by a social media user. The sequence of map positions implicitly provides an estimation of the users' social trajectories as
they drift on the map. The existing data mining techniques for spatial
cluster detection fail to address this new setting as they require a single
location to each individual under analysis. In this paper we present two
stochastic models with their associated algorithms to mine this new type
of data. The *Visit Model* finds the most likely zones that a diseased person visits, while the *Infection Model* finds the most likely zones where a
person gets infected while visiting. We demonstrate the applicability and
effectiveness of our proposed models by applying them to more than 100
million geotagged tweets from Brazil in 2015. In particular, we target the
identification of infection hot spots associated with dengue, a mosquito-
transmitted disease that affects millions of people in Brazil annually,
and billions worldwide. We applied our algorithms to data from 11 large
cities in Brazil and found infection hot spots, showing the usefulness of
our methods for disease surveillance.

Keywords: Hot spots · Spatial cluster detection · Trajectories · Disease
surveillance · Social media

1   Introduction

There is an increasing availability of geolocated data generated by mobile phones,
connected vehicles and geotagged social media, among other sources. This is
enabling a broad spectrum of applications and services that exploit such data
and demand the development of novel data mining models and algorithms that
support those tasks. Building such models and algorithms require that we are

© Springer International Publishing AG 2016
P. Frasconi et al. (Eds.): ECML PKDD 2016, Part II, LNAI 9852, pp. 739–755, 2016.
DOI: 10.1007/978-3-319-46227-1_46

able to handle novel types of data, such as user's movement record as well as the noisy and the incomplete nature of the data. We may glance the problem complexity by checking the right plot of Fig. 1, which depicts the movement of Twitter users in 2015 in a Brazilian city, Rio de Janeiro (each line segment shows a user's movement – location change – between two consecutive messages).

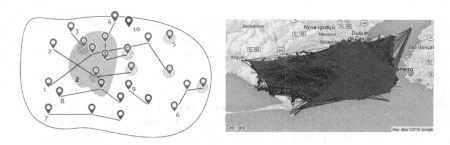

Fig. 1. *Left:* Schematic drawing of a potential infection hot spot (shaded area) and the individuals trajectories of cases (red) and controls (blue). *Right:* Individuals trajectories of cases (red) and controls (blue) in the city of Rio de Janeiro during the year of 2015. (Color figure online)

In this paper we tackle one of these disruptive application scenarios: determining infection hot spots, that is, the high risk zones where people got infected by a disease. Our proposal adopts the case-control framework, where, by contrasting the case and control individuals' characteristics, we learn about the disease dissemination process. The input is composed of trajectories, which are sequences of user locations that provide an estimate of the users' movements as they drift on the map. We depict this application scenario in the left plot of Fig. 1, where each polygonal line is a trajectory that represents either a case (red) or a control (blue) individual, and we want to determine whether there are regions (represented by the shaded area in the left plot of Fig. 1) where infection is more likely, manifested by a larger number of case trajectories than control trajectories, among other evidences. Such information may be key to surveillance and disease mitigation actions.

Although the main idea seems simple, there are a large number of challenging data mining issues that require the development of novel models and algorithms to the problem. At first, this task resembles spatial cluster detection, which aims at detecting localized spatial regions or zones, called *spatial clusters*, where the likelihood of some event occurrence is higher than in the rest of the map [8,11,17]. There have been several different proposals for detecting spatial clusters, but all of them are based on the same premise: each entity is associated with one or at most two locations. Thus, our proposal differs significantly from current spatial cluster detection strategies in the sense that there is no limit on the number of events, and then locations, associated with a person. Further, there are two other characteristics that make our problem more challenging: (i) the number of events per trajectory may vary significantly, and (ii) we do not know in advance which events represent the actual infection.

In this paper we propose two stochastic models with their associated algorithms to mine this new type of data. The Visit Model finds the most likely zones that a diseased person visits, while the Infection Model finds the most likely zones where a person gets infected while visiting. To the best of our knowledge this is the first work that goes beyond predicting disease incidence rate from social media data. Our approach leverages the geo-tagged social media messages in order to discover potentially high infection risk zones. Specifically our contributions are as follows:

- We describe the problem of detecting infection hot spots from trajectory data in a case-control framework (Sect. 2).
- We propose two novel models, and the respective algorithms, the *Visit Model* and the *Infection Model*, for the discovery of significant infection hot spots. Our algorithms address all three aforementioned issues (Sect. 3).
- We propose an extraction and modeling strategy of Twitter data to the hot spot detection problem in the context of dengue (Sect. 4).
- We present detailed experimental results to illustrate our approach in action by applying our algorithms to a set of 11 Brazilian municipalities analyzing more than 100 million tweets issued in 2015 (Sect. 5).

2 Problem Description

Social media data represent a rich and promising source of plenty, cheap, and timely data that has been only tapped in its usefulness. The excitement involving the use of social media as a social sensor could be felt by the countless number of research works using this kind of data [6,16,18]. In our case, we are interested in probing the usefulness of social data spontaneously generated by users as a way to identify the location, shape, and size of high risk zones and to determine its statistical significance. Depending on the application, we believe that these data may be more precise in the detection of such hot spots than other more standard data and, in many cases, they may be the only data available. Indeed, this latter observation is exactly the case of dengue surveillance (see Sect. 4.1), since there is scarce, if any, information about the place where people are being infected with dengue by the transmitter mosquito. As dengue usually is a debilitating disease that causes much pain, our assumption is that infected individuals will report what they are experiencing in social media [6,19].

In this work we use dengue and Twitter to instantiate our proposal, but it is obviously general and can be applied to a large range of other situations (see Sect. 7). Each user in the database is classified either as a case or a control individual. The separation of cases and controls is based on the content of tweets text: users mentioning personal experience with dengue are labelled as cases, otherwise, they are labelled as controls. In the left plot of Fig. 1, we have $N = 6$ cases and $M = 4$ controls identified by the red and blue polygonal lines, respectively. The vertices of the polygonal lines correspond to the locations of tweets issued by each individual. The tweets of a single individual are connected in chronological order and hence we refer to the polygonal lines as *trajectories*. For the cases, the specific dengue-labelled messages are marked by a hatched ground area. We also show in the same figure a candidate hot spot Z (shaded area), a spatial zone potentially riskier than other regions in the map.

The mining task is to scan the map varying the position, shape, and size of the candidate zones, looking for the zone $\hat{Z}$ that most likely is a higher risk area. After finding this most likely hot spot $\hat{Z}$, we calculate its probability of occurrence to evaluate whether there is enough evidence to call it a real cluster. The simple schematic illustration is put in due perspective when we look at the right-hand map in Fig. 1. The large amount of data and the impossibility to visually identify any meaningful pattern supports the demand for new data mining models and algorithms.

Usual approaches [8,11,17] for spatial cluster detection can not be used here. All spatial detection methods have a single location associated with each case or control individual, usually their residential addresses or working places. In our case, we have a completely different spatial data structure. First, each i-th individual is not associated with a single location, but with a series of n_i successive positions $\mathbf{x}_i$ in the map. There is no single unambiguous position to assign each case or control but rather a sequence of positions. Usual methods are not able to handle this scenario.

Second, the number n_i of positions of each individual is quite variable. For some individuals, n_i is small, with less than 10 positions. Others may contribute a large number of positions, reaching more than 100 tweets. Clearly, the locations can not be put on the map ignoring the variable contribution of each individual. To make this point clearer, imagine an extreme situation where 3 case individuals contribute each one with two positions, one in a risky zone, and another one outside. At the same time, an additional case individual has 200 tweets spread all over the map. This extreme individual would dominate the analysis if we do not take the sample size n_i into account. Again, this is not considered in the usual techniques, where each individual contributes with a single point.

Third, and more challenging, the positions of the dengue-labelled tweets are not necessarily those where the infection risk is higher. Indeed, our assumption is that the individual entire trajectory (and not a single position) will be informative of the risk areas. Someone affected by dengue could tweet about his condition days after recovery and at a location not associated with its infection place. This challenge is addressed through sampling the controls. We expect that contrasting between the spatial pattern of trajectories of the case and the control individuals, riskier zones should be pinpointed by our algorithms.

3   Mining for Hot Spots from Trajectories

As mentioned, we adopted a case-control framework, where the data consist of locations, within a specified geographical region, of all known cases of a particular disease, and of a random sample of controls drawn from the population at risk. Each individual carries a set of features corresponding to known or hypothesized risk-factors for the disease in question.

In our analysis, the key innovation is that the input is a series of locations rather than a single location for each individual. As in a standard case-control study, each sampled person is classified either as a dengue case or a non-dengue (control) individual. We labelled the individuals such that the first N of them are the cases and the last M are the controls. Let $\mathbf{x}_i = (x_{i,1}, \ldots, x_{i,n_i})$ be the point events associated with the n_i tweets issued by the i-th individual, $i = 1, \ldots, N + M$. Each $x_{i,k}$ represents the geographical tweet location such as a latitude-longitude coordinate pair. For the cases $i = 1, \ldots, N$, at least one tweet in $\mathbf{x}_i$ refers to a personal dengue experience and their specific locations will be denoted dengue-labelled tweets hereafter. Typically, there will be a small percentage of dengue-labelled tweets for each individual. None of the control individual tweets are dengue-labelled.

Let $\mathcal{Z}$ be a (large) set of geographical zones that are candidates to be infection hot spots. The left plot of Fig. 1 helps us to describe how our algorithm works. There are potentially infinite zones in $\mathcal{Z}$ and they cover the entire region under analysis. By varying $Z \in \mathcal{Z}$ we scan the map looking for the zone $\hat{Z}$ that most likely is a higher risk area. After finding this most likely hot spot $\hat{Z}$, we calculate its likelihood to evaluate whether there is enough evidence to identify it as a hot spot. Secondary clusters are also searched, as we explain later.

Our approach is to contrast the number of cases and controls visiting the potential zone. With a meaningful contrasting score, we should then scan the map to find the most likely zone. We considered two different probability scores, depending on how we calculate conditional probabilities of relevant events. In the first, we use the probability that someone visits the candidate zone Z given that she is either a case or a control individual. Intuitively, a risky zone Z should have this visit probability higher for cases than for controls. This first approach is called the *Visit Model*. In the second, we use the probability that someone gets infected given that it visits the candidate zone Z. Intuitively, we anticipate that cases visit Z more often than controls. This second approach is called the *Infection Model*. We present them formally next.

3.1   Visit Model

Let $V_{i,z}$ be the random number of tweets in Z among the n_i total number of tweets issued by the i-th individual. Use $\mathbb{1}[A]$ to represent the indicator random variable that the event A occurs. Hence $\mathbb{1}[V_{i,z} \geq 1]$ is the binary random variable indicating whether the i-th individual ever tweeted inside the candidate zone Z. These random variables can be assumed independent, but they are not identically distributed as the success probability depends on the number n_i of tweets issued by each individual. Denote by $p = p(Z)$ the probability that, giving that a case individual is tweeting, she does it from within Z. Let $\bar{p} = \bar{p}(Z)$ be the similar probability for a control individual. We are interested in zones where $p(Z) > \bar{p}(Z)$.

For a user who is a case, we have $\mathbb{P}(V_{i,z} \geq 1)$ equals to $1 - (1 - p)^{n_i}$ and, for a control user, it is equal to $1 - (1 - \bar{p})^{n_i}$. Considering a fixed zone Z, the visit model likelihood for the observed $N + M$ binary indicators $\mathbb{1}[V_{i,z} \geq 1]$ is given by

$$L_1(Z, p, \bar{p}) = \prod_{i=1}^{N} \left[(1 - (1 - p)^{n_i})^{\mathbb{1}[V_{i,z} \geq 1]} ((1 - p)^{n_i})^{\mathbb{1}[V_{i,z}=0]} \right]$$

$$\prod_{i=N+1}^{N+M} \left[(1 - (1 - \bar{p})^{n_i})^{\mathbb{1}[V_{i,z} \geq 1]} ((1 - \bar{p})^{n_i})^{\mathbb{1}[V_{i,z}=0]} \right]$$

$$= (1 - p)^{\sum_{i=1}^{N} n_i \mathbb{1}[V_{i,z}=0]} (1 - \bar{p})^{\sum_{i=N+1}^{N+M} n_i \mathbb{1}[V_{i,z}=0]}$$

$$\prod_{i=1}^{N} \left[(1 - (1 - p)^{n_i})^{\mathbb{1}[V_{i,z} \geq 1]} \right] \prod_{i=N+1}^{N+M} \left[(1 - (1 - \bar{p})^{n_i})^{\mathbb{1}[V_{i,z} \geq 1]} \right]$$

Let $N(\bar{Z}) = \sum_{i=1}^{N} n_i \mathbb{1}[V_{i,z} = 0]$ and $M(\bar{Z}) = \sum_{i=N+1}^{N+M} n_i \mathbb{1}[V_{i,z} = 0]$ be the total number of tweets from users (both cases and controls) who did not visit zone Z, respectively. Hence, the log-likelihood $\ell_1(Z, p, \bar{p}) = \log(L_1(Z, p, \bar{p}))$ for this first model can be written as

$$\ell_1(Z, p, \bar{p}) = \log(1 - p)N(\bar{Z}) + \log(1 - \bar{p})M(\bar{Z})$$

$$+ \sum_{i=1}^{N} \mathbb{1}[V_{i,z} \geq 1] \log(1 - (1 - p)^{n_i}) + \sum_{i=N+1}^{N+M} \mathbb{1}[V_{i,z} \geq 1] \log(1 - (1 - \bar{p})^{n_i}) \quad (1)$$

3.2   Infection Model

We will estimate the probability that someone issues a dengue-labelled tweet (and becomes a case) given that she visited k times the region Z. Let $r = r(Z)$ be the infection risk inside the candidate cluster and $\bar{r} = r(\bar{Z})$ the infection risk in $\bar{Z}$, the region outside Z. We are interested in zones Z where $r(Z) > r(\bar{Z})$.

Let I_i be the binary indicator that the individual i is a case. We assume that these binary random variables are independent. They are not identically distributed since their probability of $I_i = 1$ depends on the number of visits $V_{i,z}$ by the i-th individual to the zone Z. We have

$$\mathbb{P}(I_i = 1 | V_{i,z} = k_i) = 1 - \mathbb{P}(I_i = 0 | V_{i,z} = k) = 1 - (1 - r)^{k_i} (1 - \bar{r})^{n_i - k_i}$$

$$= \pi(k_i, r, \bar{r}) \quad (2)$$

Therefore, the likelihood of the pattern of cases and controls is given by

$$L_2(Z, r, \bar{r}) = \prod_{i=1}^{N+M} (\pi(k_i, r, \bar{r}))^{I_i} (1 - \pi(k_i, r, \bar{r}))^{1 - I_i}$$

and therefore the log-likelihood expression is given by

$$\ell_2(Z, r, \bar{r}) = \sum_{i=1}^{N+M} I_i \log\left(1 - (1 - r)^{k_i}(1 - \bar{r})^{n_i - k_i}\right)$$
$$+ (1 - I_i)(k_i \log(1 - r) + (n_i - k_i)\log(1 - \bar{r})) \tag{3}$$

3.3   Evaluating the Data Evidence

Recall that $\mathcal{Z}$ is the set of candidate zones to be scanned. The test statistic we adopt for the Visit Model is

$$T_1 = \ell_1(\hat{Z}, \hat{p}, \hat{\bar{p}}) = \sup_{\substack{Z \in \mathcal{Z} \\ \hat{p}(Z) > \hat{\bar{p}}(Z)}} \ell_1(\hat{Z}, \hat{p}(Z), \hat{\bar{p}}(Z)) \tag{4}$$

and an analogous formula defines T_2 for the Infection Model. In order to verify its statistical significance, we must use Monte Carlo simulation to obtain the null hypothesis distribution of T_1 and T_2 as the exact or asymptotic analytic calculation is not feasible. The null hypothesis is given by either $H_0 : p = \bar{p}$ or $H_0 : r = \bar{r}$ for all $Z \in \mathcal{Z}$ for the Visit Model and the Infection Model, respectively.

The Monte Carlo distribution is determined by randomly permuting the labels of cases and controls among all individuals. Using this pseudo dataset, we proceed the entire scan over all $Z \in \mathcal{Z}$ to obtain a pseudo value for T_1 and T_2. As this will be replicated several times, we call these values $T_1^{(1)}$ and $T_2^{(1)}$. We then select another random permutation of the labels, scan the zones and find $T_1^{(2)}$ and $T_2^{(2)}$. Independently, we repeat this procedure a large number $B - 1$ of times generating a set of pseudo values plus the values calculated with the actually observed dataset: $T_1, T_1^{(1)}, T_1^{(2)}, \ldots, T_1^{(B-1)}$ and $T_2, T_2^{(1)}, T_2^{(2)}, \ldots, T_2^{(B-1)}$. Under the null hypothesis, these values are independent and identically distributed. Therefore, the rank of the real observed statistics T_1 and T_2 are uniformly distributed on the integers $1, \ldots, B$. This implies that an exact p-value for the null hypothesis of each model is given by

$$p_1 = \frac{1}{B}(1 + \#\{T_1^{(k)} \geq T_1, k = 1, \ldots, B - 1\})$$

and

$$p_2 = \frac{1}{B}(1 + \#\{T_2^{(k)} \geq T_2, k = 1, \ldots, B - 1\})$$

The test is significant at the level $\alpha \in (0, 1)$ if $p_m < \alpha$. When either test is significant, the most likely zone is given by the corresponding maximizing argument $\hat{Z}$ in (4).

We also identify secondary clusters, zones with highly significant p-values, which do not intersect with the most likely zone $\hat{Z}$. The non-intersecting restriction is necessary because, if one zone $\hat{Z}$ is the most anomalous in $\mathcal{Z}$, many other sets in $\mathcal{Z}$ that are only slightly different from $\hat{Z}$ will produce very similar likelihood numbers. These zones should be ignored since the most anomalous among them has already been pinpointed. Among the non-intersection zones, we look for those whose p-value p_m is smaller than α where the p-values are calculated as described above.

3.4   Contrasting the Two Models

In this section, we discuss in more detail the two proposed models aiming at providing an understanding of the differences between them. In particular, we want to distinguish between the two approaches in an intuitive way and hence explain when and how we can have one of the models detecting a certain hot spot while the other model is insensitive to this same cluster presence.

Avoiding the rigorous mathematical notation, let us define two random events. The first one is denoted by C and represents the random selection of an individual from the database that is dengue-affected or simply a case. Its complementary event is $\bar{C}$ and represents the selection of a control individual. Given that a tweet is posted by a user, we denote by W_Z the event that it is issued from Z while $W_{\bar{Z}}$ means that it is from outside Z.

The visit model considers two conditional probabilities, $p = \mathbb{P}(W_Z|C)$ and $\bar{p} = \mathbb{P}(W_Z|\bar{C})$, while the infection model considers the corresponding inverse conditional probabilities, $r = \mathbb{P}(C|W_Z)$ and $\bar{r} = \mathbb{P}(C|W_{\bar{Z}})$. Intuitively, the visit model scans the map looking for a zone Z where p and $\bar{p}$ are quite different. The infection model searches for a zone where the difference between r and $\bar{r}$ is large. They can find distinct and separate zones in this process. The main reason is the usual large difference we find between conditional probabilities $\mathbb{P}(A|B)$ and $\mathbb{P}(B|A)$ of events A and B. The connection between the two is given by the Bayes rule: $\mathbb{P}(A|B) = \mathbb{P}(B|A)\mathbb{P}(A)/\mathbb{P}(B)$. Since the factor $\mathbb{P}(A)/\mathbb{P}(B)$ is the link between the two, when we have very different values for $\mathbb{P}(A)$ and $\mathbb{P}(B)$ we can expect large differences on the two directions for the two conditional probabilities, $\mathbb{P}(A|B)$ and $\mathbb{P}(B|A)$.

This is indeed what one can expect in our dengue application. The unconditional probabilities $\mathbb{P}(C)$ and $\mathbb{P}(W_Z)$ are typically very different. As we take about 3 times more controls than cases, we anticipate $\mathbb{P}(C) \approx 1/4$. For a localized zone Z, even if it is highly infectious, we should not expect $\mathbb{P}(W_Z) > 1/20$. Hence, zones detected by one of the models should not be predicted as the likely output by the other model.

An additional enlightening way to contrast the two models is to consider the extreme situation in which each user has issued a single tweet (that is, $n_i = 1$). As a consequence, k_i is equal to 0 or 1 and the likelihood for the two models may be considerably simplified. Remember that $N(\bar{Z})$ is the number of tweets from cases posted from outside Z (or $\bar{Z}$) while $M(\bar{Z})$ is the analogous count for the controls. The notation $N(Z)$ and $M(Z)$ has the obvious definition: counts of tweets from inside Z.

For the visit and infection models, their respective log-likelihood functions $\ell_1' = \ell_1(Z, p, \bar{p})$ and $\ell_2' = \ell_2(Z, r, \bar{r})$ are reduced to

$$\ell_1' = N(Z)\log(p) + N(\bar{Z})\log(1 - p) + M(Z)\log(\bar{p}) + M(\bar{Z})\log(1 - \bar{p}) \quad (5)$$
$$\ell_2' = N(Z)\log(r) + N(\bar{Z})\log(\bar{r}) + M(Z)\log(1 - r) + M(\bar{Z})\log(1 - \bar{r}) \quad (6)$$

These likelihood functions for this extreme situation show that the two models use the data differently to search for suspicious zones Z. They both point out to likely high infection risk areas but they use different approaches in the process and may spot different potential candidates. The two approaches are logically consistent and produce meaningful results. They are complementary to each other and should not be seen as opposites.

3.5 Spatial Scan Statistics as a Particular Case

Expression (6) shows that the usual spatial scan statistic [8,11,17] is a particular case of our infection model. Assuming that each sampled individual has a single spatial location (usually her residential address), the notation r represents now the probability that she is a disease case given that she is within Z. The probability $\bar{r}$ is the same probability for someone living outside Z. Then, (6) is the Bernoulli likelihood used by the original spatial scan statistic. That is, when there is a single location for each individual, we obtain the classic spatial scan statistic by applying our infection model.

4 Case Study: Dengue in Brazil

In this section we present the motivation behind our evaluation scenario, dengue disease surveillance in Brazil. Also, we describe the Twitter data collection process and how we properly filter the data in order to obtain the case-control individuals' trajectories.

4.1 Context

Despite all the progress achieved in the twenty-first century, diseases transmitted by insects are still challenging our health services and policy makers. The recent outbreak of Zika virus in Brazil and other Latin American countries, potentially associated with thousands of microcephalic birth cases, prompted The World Health Organization (WHO) to declare the Zika Infection a world health threat[1]. Other disease that is transmitted by the same mosquito, *Aedes aegypti*, is dengue.

With an estimated 50–100 million infections globally per year [3], dengue is currently regarded as the most important mosquito-borne viral disease. Dengue affects over 100 endemic countries in tropical and sub-tropical regions of the

---

[1] http://www.who.int/mediacentre/news/statements/2016/1st-emergency-committee -zika/en/.

world, mostly in Asia, the Pacific Region and the Americas. Presenting four distinct viral serotypes, dengue fever may range from severe flu-like illness up to a potentially lethal complication known as severe (or hemorrhagic) dengue. The World Health Organization estimates that 3.9 billion people are at risk of infection with dengue viruses. However, the true impact of the disease is, sometimes, difficult to assess due to misdiagnosis and underreporting [2]. Global dengue incidence still grows in number and severity of cases and also in the amount of new affected areas. This is most due to modern climate changes and to socioeconomic, and viral evolution [12]. However, the potential drivers of dengue are often difficult to detect and factor out. Since there is no current approved vaccine to protect the population against the virus [12], epidemiological surveillance and effective vector control are still the mainstay of dengue prevention.

Dengue is a serious concern in Brazil. In 2015, more than US$ 300 million were spent in surveillance and prevention actions[2]. This is a significant figure for Brazilian standards and, despite its magnitude, more than 1.6 million cases were recorded in 2015. This number represents a rate of 813 cases per 100 thousand inhabitants, well above the redline indicated by the WHO (300 cases).

Most studies for diseases such as dengue place the cases at individuals' residential addresses, which may quite often not be the infection location. The relatively easy to obtain residential address may be a poor indicator of the zones where humans and infected mosquitoes tend to meet each other. These zones are hard to determine, since the necessary information about them is scarcely available. Indeed, such information comprises data on the mosquito prevalence, its infection rate, and the human movement in each potential zone. Notwithstanding the task difficulty, identifying the most risky places would be invaluable because we could focus the expensive and diffuse preventive efforts undertaken until now.

4.2 Data Acquisition and Preprocessing

The data used in our experimental analysis were acquired through the Twitter streaming application programming interface (API) [1], using a geographic boundary box that covers the whole Brazilian territory. Consequently, all collected tweets are geo-tagged with lat/long GPS coordinates. The collecting period comprises from January 1st, 2015 to December 31th, 2015. During this time we were able to collect 106,784,441 tweets comprising a multitude of subjects. We want to use this data to search for zones that increase the likelihood that an initially control individual becomes a case.

Since the majority of users usually moves within the same city, we decided to perform our analysis at the city level. This granularity is also interesting because, in Brazil, the decision process regarding dengue surveillance actions is under the responsibility of each city hall. Thus, a fine geographic scale analysis would lead to focused preventive efforts. Since the messages are geocoded, to obtain the data from a specific city is straightforward. The Twitter API provides the location

---

[2] http://www.brasil.gov.br/saude/2015/04/orcamento-2015-para-acoes-de-combate-a-dengue-cresce-37.

Table 1. Data summary: #msg is the total number of tweets from the city; #unq_usr is the number of unique users; #case_usr and #ctrl_usr are the number of case and control individuals; #case_usr and #ctrl_msg are the number of tweets they issued.

City name	#msg	#unq_usr	#case_msg	#case_usr	#ctrl_msg	#ctrl_usr
Belém	1,049,433	19,611	8,134	23	18,416	65
B. Horizonte	3,134,497	50,360	60,968	104	168,820	302
Curitiba	1,694,301	35,775	3,028	18	9,066	54
Goiânia	566,114	16,849	15,933	54	33,750	147
Natal	522,689	16,689	3,847	15	8,748	42
R. de Janeiro	9,875,435	167,567	71,115	163	213,168	490
São Paulo	6,965,165	174,544	167,772	413	486,264	1229
Campinas	574,226	20,335	37,313	90	64,442	226
Limeira	91,454	2,991	11,614	47	16,830	108
SJ. Campos	407,143	9,697	19,883	58	40,251	148
Sorocaba	230,224	7,471	32,734	91	39,352	206

based on the lat/long coordinates. We use the assigned location by filtering the corresponding tweet field. We choose 11 municipalities (see Sect. 5 for the explanation) to analyze. Table 1 summarizes the data for each selected city.

For each city analyzed, we filtered the data indicating whether the user is a case individual. We defined the keywords *dengue* and *aedes*, and started a search throughout the data. Previous works showed a high correlation between official dengue reports and Twitter data collected with such keywords [6,19]. We also check for misspelling and ignore letter case. Since the vocabulary in text-based social media is very dynamic, the retrieved messages based on keywords may not be actually associated with people reporting personal experience with the disease. Hence, we classified the messages according to the sentiment expressed in the textual content. To classify the messages, we preprocessed texts by filtering out accents marks and URL's. Bi-grams were created by joining adjacent words with a separator, and stop-words were removed as well as bi-grams composed of two stop-words. The classification was performed in a supervised manner. We manually labelled a set of tweets from a different Twitter collection specifically about dengue disease. This collection is performed based on the same keywords. Similar to [6,19], the tweets were classified into one out of five categories: Personal Experience, Information, Opinion, Campaign and Irony/Sarcasm, using the the Lazy Associative Classification algorithm (LAC) [20]. Next, we separated the messages assigned to the Personal Experience category, since they may indicate a closer relationship between the user and the disease. These messages represent the dengue-labelled tweets for the case individuals.

4.3   Case-Control Trajectories

Recall that, each user in the database is classified as either a case or a control individual, and the separation of cases and controls is based on the content of tweets text, as described above. Then, for each city we build the case-control trajectories as follows.

Case-trajectories. In order to build the case individuals trajectories we started by separating all unique users who posted a dengue-labelled message. Then, we retrieved all other tweets sent by these users. For each case individual, her list of messages composes the trajectory. Such strategy is interesting because we are implicitly considering that the users must have been infected at some point in their daily movements and not exactly where the dengue-labelled messages were sent. After that, we excluded highly active users to avoid, for instance, bots. We adopted a 5-message-per-day threshold, which represents a maximum of 1825 messages per year. The users with total number of messages above this threshold are excluded from the dataset.

Control-trajectories. The control individuals group comprises all users who never posted a message containing any of the keywords used to define the case individuals group. Therefore, none of the control individuals tweets are dengue-labelled. We defined the same threshold to exclude highly active users. The number of control individuals is much larger than the number of case individuals. Thus, we sampled the control individuals. We stratified the case individuals according to the total number of messages in ranges of 10. Then, for each range we sampled the number of control users as 3 times the number of case users in that same range. When the number of control users in a given range was not enough to reach the amount required, we used the total available.

5   Experimental Analysis

After generating the dataset for each selected city as described in the previous section, we proceeded to the experimental analysis. For each one of the 11 selected cities (see Table 1) we applied the Visit Model and the Infection Model to search for infection hot spots. Among the selected cities we included 7 state capitals (Belém, Belo Horizonte, Curitiba, Goiânia, Natal, Rio de Janeiro and São Paulo) with at least one capital from a major Brazilian region. We also decided to assess our models using data from municipalities facing high epidemics bursts. Therefore, we included 4 other cities: Campinas, Limeira, São José dos Campos and Sorocaba. For instance, while in 2014 Sorocaba reported less than 400 dengue cases, in 2015 the same city reported more than 50 thousand cases.

In order to run the algorithms, the zones Z are defined by overlaying different grids on the map and each grid cell corresponds to a zone to be scanned. The size of the grid cells vary in order to accommodate risk zones that present different characteristics. We set the number of Monte Carlo replicas to $B - 1 = 999$ and define the significance level as $\alpha = 0.05$. Among the 11 selected cities, in 4 of

Table 2. Results obtained by the Visit and Infection models in the respective cities. We present the log-likelihood value of the zone (Log-Lik); the respective probabilities considered by the models ($r \mid p$ and $\bar{r} \mid \bar{p}$); the obtained p-value based on the Monte Carlo reference distribution; the number of case and control individuals inside the zone (#cases and #ctrl); and the amount of messages issued inside the zone by case and control individuals (#case$\_k_i$ and #ctrl$\_k_i$).

City	Log-Lik	$r \mid p$	$\bar{r} \mid \bar{p}$	p-value	#cases	#case$\_k_i$	#ctrl	#ctrl$\_k_i$
Visit Model								
Goiânia	−135.32151	0.04379	0.01	0.01	48	6352	115	14600
Limeira	−89.51999	0.04379	0.01	0.019	43	5655	80	7940
Infection Model								
Limeira	−198.51340	0.48310	0.01	0.014	5	11	1	1
	−200.16361	0.07759	0.01	0.02	4	8	3	10
	−200.35639	0.07759	0.01	0.02	3	97	7	9
SJ. Campos	−427.44342	0.14517	0.01	0.055	5	28	2	4
Sorocaba	−446.94606	0.04379	0.01	0.002	3	150	8	16

them at least one of the models was able to find one or more significant hot spots. Table 2 summarizes the results.

First of all, we point out that our models were able to find infection hot spots in 3 cities that faced the aforementioned strong surges. Despite the significance level being $\alpha = 0.05$, we considered the borderline region found in SJ. Campos as significant. In the context of disease surveillance, it would be also important to check such zones. We observe that, in Goiânia and Limeira, the zones pinpointed by the Visit Model were visited by most of the case individuals, since the Visit Model searches for the most likely zones where case individuals visit. On the other hand, the zones identified by the Infection Model comprise a lower number of case individuals seeking for more restricted areas. In fact, the size of the zones found by the models differ. The Visit Model usually finds larger regions whilst the Infection Model finds smaller regions. Figure 2 depicts the zones found by each model in the corresponding cities. Notice that in Limeira the models identified different regions within the same city. These results also point out the complementarity of the models, so that they may be used together towards establishing two different levels of surveillance.

After we find the significant zones, we may analyze them in detail to observe their characteristics. We show this more detailed analysis for Goiânia. Figure 3 displays a zoom in the zone identified by the Visit Model and the respective case-control trajectories. We point out that there are many places, such as, college campi, hospitals and parks inside the zone. Since those places are non-residential, current techniques would never consider them as potential infection hotspots, in the face of a rise in the number of cases. This is another interesting feature of our algorithms, they can point out places which represent a better approximation of where people might have been infected, being worthy to investigate those areas.

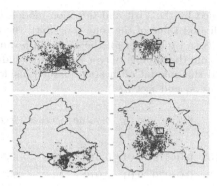

Fig. 2. Maps of the cities with the hot spots found by both models. The cities are Goiânia, Limeira, São José dos Campos and Sorocaba. The green and black squares depict the zones found by the Visit and Infection models respectively. We also display the case and control individuals trajectories as red and blue points, respectively. (Color figure online)

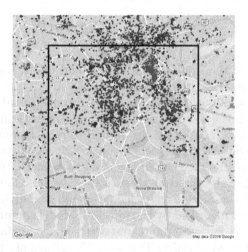

Fig. 3. Zoom in to the zone found by the Visit Model in Goiânia. Red and blue points represent the case-control trajectories respectively. (Color figure online)

6   Related Work

Spatial cluster detection is a special class of data mining problem within the more general anomalous pattern detection problem. The assumed structure of the input data is a spatial point location, such as latitude-longitude pair, besides the usual features associated with each of them. The seminal paper [7] originated a flow of work and its large impact may be explained by a breakthrough contribution. They developed a practical way, *the spatial scan statistics*, to take into account the multiple testing involved in the search of anomalous regions. They showed how a simple Monte Carlo reference distribution could be obtained from the data and how it controls the false positive level of the potentially infinite statistical tests involved. This idea opened the door to many additional developments [4,5,9,13–15,17,21,24]. While the recent availability of spatial data offers a unique opportunity, the existing data mining techniques for spatial cluster detection fail to address this new setting as they require a single location to each individual under analysis.

On the other hand, there has been fruitful research exploiting spatial data for a variety of purposes, such as, discovering the spatial dependency of objects [22], understanding mobility patterns [10] and clustering similar trajectories [23], to name a few. However, none of the strategies proposed so far focused on searching for hot spots by contrasting trajectory data of targeted populations with those from control populations as we have done here. In this sense, this paper has a two-fold contribution. First, it generalizes the spatial cluster detection approaches by considering the individual trajectory data instead of a single point. Second, it describes the aforementioned problem in the context of disease surveillance and proposes two algorithms to mine the data.

7   Concluding Remarks

Exploiting the large amount of available data for addressing relevant social problems has been one of the key challenges in data mining. In this paper we attempt to help on this task by proposing two stochastic models to search for infection hot spots using social media trajectories. Our application scenario is a major infectious disease in Brazil and other tropical countries, dengue. We applied our models to data from 11 Brazilian cities and were able to detect infection hot spots in 4 of them. This result shows the usefulness of our methods to disease surveillance. To identify the high risk regions would be invaluable to direct preventive efforts and mitigation actions. Currently, we are carrying out a validation procedure of our results with local health officials.

We see our proposal as a first step on the direction of a more general and comprehensive framework. In fact, future research directions abound, both from theoretical and practical perspectives. One direction is to incorporate a richer data structure allowing features to be included at the individual level. In this paper, we only considered a binary indicator (case or control). However, we could add other features such as age and sex of the individuals. Another possibility is to associate features to the events that constitute the trajectories. For instance, distinguishing whether the event occurred in the summer or winter is potentially useful. A third possible direction is to consider the social links between the individuals as a means to create a social network between the trajectories. Notwithstanding these further developments, our models are useful for the difficult task of infection hot spots detection.

Acknowledgements. This work was partially funded by Fapemig, CNPq, CAPES, and by the projects MASWeb (FAPEMIG-PRONEX APQ-01400-14), InWeb (MCT/CNPq 573871/2008-6) and EUBra-BIGSEA (H2020-EU.2.1.1 690116, Brazil/MCTI/RNP GA-000650/04).

References

1. Twitter: The Streaming API. https://dev.twitter.com/streaming/overview
2. World Health Organization. http://www.who.int/csr/disease/dengue/denguenet
3. Bhatt, S., et al.: The global distribution and burden of dengue. Nature **496**, 504–507 (2013)
4. Chen, F., Neill, D.B.: Non-parametric scan statistics for event detection and forecasting in heterogeneous social media graphs. In: Proceedings of the 20th ACM SIGKDD Conference, pp. 1166–1175 (2014)
5. Duczmal, L., Assunção, R.: A simulated annealing strategy for the detection of arbitrarily shaped spatial clusters. Comput. Stat. Data Anal. **45**(2), 269–286 (2004)
6. Gomide, J., Veloso, A., Meira Jr., W., Almeida, V., Benevenuto, F., Ferraz, F., Teixeira, M.: Dengue surveillance based on a computational model of spatio-temporal locality of twitter. In: Proceedings of the ACM WebSci Conference (2011)
7. Kulldorff, M., Nagarwalla, N.: Spatial disease clusters: detection and inference. Stat. Med. **14**(8), 799–810 (1995)
8. Kulldorff, M.: A spatial scan statistic. Comm. Stat. Theory Meth. **26**(6), 1481–1496 (1997)
9. Kulldorff, M., Heffernan, R., Hartman, J., Assunção, R., Mostashari, F.: A space time permutation scan statistic for disease outbreak detection. PLoS Med (2005)
10. Lima, A., Stanojevic, R., Papagiannaki, D., Rodriguez, P., González, M.C.: Understanding individual routing behaviour. Roy. Soc. Interface **13**(116) (2016)
11. McFowland III, E., Speakman, S., Neill, D.B.: Fast generalized subset scan for anomalous pattern detection. J. Mach. Learn. Res. **14**, 1533–1561 (2013)
12. Murray, N.E.A., Quam, M.B., Wilder-Smith, A.: Epidemiology of dengue: past, present and future prospects. Clin. Epidemiol. **5**, 299–309 (2013)
13. Neill, D.B., Cooper, G.F.: A multivariate Bayesian scan statistic for early event detection and characterization. Mach. Learn. **79**(3), 261–282 (2010)
14. Neill, D.B., Moore, A.W.: Rapid detection of significant spatial clusters. In: Proceedings of the 10th ACM SIGKDD, pp. 256–265 (2004)
15. Neill, D.B., Moore, A.W., Sabhnani, M., Daniel, K.: Detection of emerging space-time clusters. In: Proceedings of the 11th ACM SIGKDD Conference, pp. 218–227 (2005)
16. Sakaki, T., Okazaki, M., Matsuo, Y.: Earthquake shakes twitter users: real-time event detection by social sensors. In: Proceedings of WWW, pp. 851–860 (2010)
17. Shi, L., Janeja, V.P.: Anomalous window discovery through scan statistics for linear intersecting paths (SSLIP). In: Proceedings of the 15th SIGKDD, pp. 767–776 (2009)
18. Silva, T.H., de Melo, P.O.V., Almeida, J., Loureiro, A.A.: Large-scale study of city dynamics and urban social behavior using participatory sensing. IEEE Wirel. Commun. **21**(1), 42–51 (2014)
19. Souza, R.C.S.N.P., de Brito, D.E.F., Cardoso, R.L., de Oliveira, D.M., Meira Jr., W., Pappa, G.L.: An evolutionary methodology for handling data scarcity and noise in monitoring real events from social media data. In: Bazzan, A.L.C., Pichara, K. (eds.) IBERAMIA 2014. LNCS, vol. 8864, pp. 295–306. Springer, Heidelberg (2014)
20. Veloso, A., Meira Jr., W., Zaki, M.J.: Lazy associative classification. In: Proceedings of the International Conference on Data Mining, pp. 645–654 (2006)
21. Wu, M., Song, X., Jermaine, C., Ranka, S., Gums, J.: A LRT framework for fast spatial anomaly detection. In: 15th ACM SIGKDD, pp. 887–896 (2009)

22. Yoo, J.S., Bow, M.: Mining spatial colocation patterns: a different framework. Data Min. Knowl. Disc. **24**, 159–194 (2012)
23. Zheng, Y.: Trajectory data mining: an overview. ACM Trans. Intell. Syst. Technol. **6** (2015)
24. Zhou, R., Shu, L., Su, Y.: An adaptive minimum spanning tree test for detecting irregularly-shaped spatial clusters. Comput. Stat. Data Anal. **89**, 134–146 (2015)

Learning to Aggregate Using Uninorms

Vitalik Melnikov and Eyke Hüllermeier[✉]

Department of Computer Science, Paderborn University, Paderborn, Germany
melnikov@mail.upb.de, eyke@upb.de

Abstract. In this paper, we propose a framework for a class of learning problems that we refer to as "learning to aggregate". Roughly, learning-to-aggregate problems are supervised machine learning problems, in which instances are represented in the form of a composition of a (variable) number on constituents; such compositions are associated with an evaluation, score, or label, which is the target of the prediction task, and which can presumably be modeled in the form of a suitable aggregation of the properties of its constituents. Our learning-to-aggregate framework establishes a close connection between machine learning and a branch of mathematics devoted to the systematic study of aggregation functions. We specifically focus on a class of functions called uninorms, which combine conjunctive and disjunctive modes of aggregation. Experimental results for a corresponding model are presented for a review data set, for which the aggregation problem consists of combining different reviewer opinions about a paper into an overall decision of acceptance or rejection.

1 Introduction

In spite of certain generalizations that have been proposed in the recent past, the bulk of methods for supervised machine learning still proceeds from a formal setting in which data objects (instances) are represented in the form of feature vectors. Thus, an instance x is described in terms of a vector $(x_1, \ldots, x_d) \in \mathcal{X} = \mathcal{X}_1 \times \cdots \times \mathcal{X}_d$, where $\mathcal{X}_i$ is the domain of the ith attribute or feature. The corresponding view of instances as *points* in a *space* of fixed dimension d has largely influenced the way in which learning problems are studied and methods developed: Supervised learning is considered as *embedding* objects as data points in the space $\mathcal{X}$, and then *separating* these points (in the case of classification) or *fitting* them (in the case of regression) using models that have a natural geometric interpretation, such as hyperplanes or any other type of decision boundary or manifold in the space $\mathcal{X}$; a prediction $\hat{y}$ of the output $y \in \mathcal{Y}$ associated with an instance x is then obtained by means of a corresponding function $f : \mathcal{X} \longrightarrow \mathcal{Y}$. Alternatively, instead of modeling dependencies with a deterministic function, a model may correspond to a probability distribution on $\mathcal{X} \times \mathcal{Y}$.

While this approach to formalizing and tackling learning problems proved to be highly successful, there are problems for which the production of predictions $\hat{y}$ by means of a (single) function f defined on the space $\mathcal{X}$ is arguably

P. Frasconi et al. (Eds.): ECML PKDD 2016, Part II, LNAI 9852, pp. 756–771, 2016.
DOI: 10.1007/978-3-319-46227-1_47

less appropriate. This paper is devoted to one such class of problems that we refer to as *aggregation problems*. The view we promote is to consider data objects as *compositions* of individual *constituents*; moreover, we assume that the output associated with such a composition is obtained as an aggregation of the properties of the individual constituents, using a suitable type of aggregation function. Thus, the *learning-to-aggregate* framework we envision establishes a close connection between machine learning and a branch of mathematics devoted to the systematic study of aggregation functions [10].

Needless to say, the idea of aggregation is not new to machine learning. On the contrary, aggregation problems seem to abound in this field and appear in various guises; for example, combining the information of the neighbors in nearest neighbor estimation, the predictions of base learners in stacking, etc., can all be seen as specific types of aggregation problems. Yet, to the best of our knowledge, a common framework of learning-to-aggregate has not been proposed so far. We believe that such a framework, and the specific view on learning problems it comes along with, is useful for different reasons. In particular, it allows for looking at different learning problems as specific instances of the same problem class, thereby connecting and cross-fertilizing subfields that would otherwise remain separated. Moreover, it may of course motivate new learning problems and trigger the development of novel methods.

The remainder of the paper is organized as follows. In the next section, we outline our learning-to-aggregate framework. The description of this framework is completed in Sect. 3, which is devoted to a discussion of aggregation functions. In Sect. 4, we propose a specific instance of the framework, namely a model for learning to aggregate based on so-called uninorms. Related work is briefly reviewed in Sect. 5. Finally, to illustrate our approach, some experiments on a data set consisting of reviews on papers submitted the the ECML/PKDD 2014 conference are presented in Sect. 6, prior to concluding the paper in Sect. 7.

2   Learning to Aggregate

In this section, we introduce a formal framework of learning-to-aggregate and elaborate on some of its properties. Prior to doing so, we give a simple example that already highlights important aspects of aggregation problems as well as limitations of standard vectorial (feature-based) representations in this context.

2.1   A Simple Example

Suppose compositions c are multisets (*bags*) of real numbers from the unit interval, such as $\{0.8, 0.7\}$ or $\{0.2, 0.6, 0.3\}$. Moreover, suppose the output y associated with a composition $c \subset [0, 1]$ is an aggregation of the constituents; to be concrete, consider the product as an example. The goal of the learner is to induce the dependency between inputs c and outputs y based on corresponding training examples, such as $(c, y) = (\{0.8, 0.7\}, 0.56)$.

Although this toy example is actually very simple, tackling it with standard machine learning methods is non-trivial. As one important reason, note that, in contrast to a feature vector of fixed dimension, compositions are of variable length. In fact, the sought dependency is a mapping of the form $\mathcal{X} \longrightarrow \mathcal{Y}$, with the instance space

$$\mathcal{X} = \bigcup_{n \in \mathbb{N}} \mathcal{Y}^n \tag{1}$$

and $\mathcal{Y} = [0, 1]$ in our case. This instance space is a union of spaces of finite dimension but does not have a finite dimension itself; indeed, in our example, we allow compositions c of any size. It is thus neither clear how to define a suitable hypothesis space on $\mathcal{X}$, i.e., a set of functions with domain $\mathcal{X}$, nor how to learn in this space.

To make the problem amenable to standard methods, it is of course possible to map compositions c to feature vectors $x = (x_1, \ldots, x_d) = (f_1(c), \ldots, f_d(c))$ of finite length, on which a model of the form $y = f(x)$ could then be learned; in fact, this is a common approach to dealing with structured data objects, which are given as bags in our case but could also be sequences or graphs, for example. Like in learning on structured objects in general, the success of this approach strongly hinges on the definition of the right features. In our example, features would be needed that allow for reconstructing, for any bag of numbers, the product of these numbers. Making sure that such features are available arguably presumes that the dependency between c and y is already known.

2.2 Formal Setting and Notation

We proceed from a set of training data

$$\mathcal{D} = \{(c_1, y_1), \ldots, (c_N, y_N)\} \subset \mathcal{C} \times \mathcal{Y}, \tag{2}$$

where $\mathcal{C}$ is the space of *compositions* and $\mathcal{Y}$ a set of possible (output) values associated with a composition; since aggregation is often used for the purpose of evaluating a composition, we also refer to the values y_i as *scores*. A composition $c_i \in \mathcal{C}$ is a multiset (*bag*) of constituents

$$c_i = \{c_{i,1}, \ldots, c_{i,n_i}\},$$

where $n_i = |c_i|$ is the size of the composition; scores y_i are typically scalar values (real numbers or values from an ordinal scale, such as 1 to 5 star ratings in recommender systems). Constituents $c_{i,j}$ can be of different type. In particular, the description of a constituent may or may not contain the following information:

− A *label* specifying the role of the constituent in the composition. For example, suppose a composition is a menu consisting of constituents in the form of dishes; each dish could then be labeled with appetizer, main dish, or dessert, thereby providing information about the part of the menu it belongs to (and hence adding additional structure to the composition).

– A description of *properties* of the constituent. For example, each dish could be described in terms of certain nutritional values. Formally, we assume properties to be given in the form of a feature vector $\boldsymbol{v}_{i,j} \in \mathcal{V}$, where $\mathcal{V}$ is a corresponding feature space. We note, however, that more complex descriptions are conceivable; for example, the description could itself be a composition.
– A *quantity* $q_{i,j} \in \mathbb{R}_+$ representing the amount of the constituent in the composition (instead of simply informing about the presence or absence of the constituent).
– A *local evaluation* in the form of a score $y_{i,j} \in \mathbb{R}_+$.

Finally, a composition can also be equipped with an additional structure in the form of a (binary) relation on its constituents. In this case, a composition is not simply an unordered set (or bag) of constituents but a more structured object, such as a sequence or a graph.

Like in standard supervised learning, the goal in learning-to-aggregate is to induce a model $h : \mathcal{C} \longrightarrow \mathcal{Y}$ that predicts scores for compositions. More specifically, given a hypothesis space $\mathcal{H}$ and a loss function $L : \mathcal{Y}^2 \longrightarrow \mathbb{R}_+$, the goal is to find a risk-minimizing hypothesis

$$h^* \in \underset{h \in \mathcal{H}}{\operatorname{argmin}} \int_{\mathcal{C} \times \mathcal{Y}} L\big(y, h(\boldsymbol{c})\big) \, d\mathbf{P}(\boldsymbol{c}, y)$$

on the basis of the training data $\mathcal{D}$ (but without knowledge of the data-generating process, i.e., the joint probability distribution $\mathbf{P}$ generating composition/score tuples $(\boldsymbol{c}, y)$).

2.3 Learning Aggregation Functions

Our simple example in Sect. 2.1 already illustrates one of the key problems in learning-to-aggregate, namely the combination of a variable number of scores $y_{i,j}$, pertaining to evaluations of the constituents $c_{i,j}$ in a composition $\boldsymbol{c}$, into a single score y_i. In Fig. 1, which provides an overview of our setting, this step corresponds to the part marked by the dashed rectangle.

Now, suppose that we know, or can at least reasonably assume, that y_i is obtained from $y_{i,1}, \ldots, y_{i,n_i}$ through an aggregation process defined by a binary aggregation function $A : \mathcal{Y}^2 \longrightarrow \mathcal{Y}$:

$$y_i = A\Big(\ldots A\big(A(y_{i,1}, y_{i,2}), y_{i,3}\big), \ldots, y_{i,n_i} \Big)$$

In the simplest case, where the constituents do not have labels and hence cannot be distinguished, the aggregation should be invariant against permutation of the constituents in the bag. Thus, it is reasonable to assume A to be associative and symmetric. Besides, one may of course restrict an underlying class of candidate functions $\mathcal{A}$ by additional assumptions. In our example, for instance, we may know that the aggregation is monotone decreasing.

Starting from a class $\mathcal{A}$ of aggregation functions, instead of a hypothesis space $\mathcal{H}$ on the instance space (1) directly, has at least two important advantages.

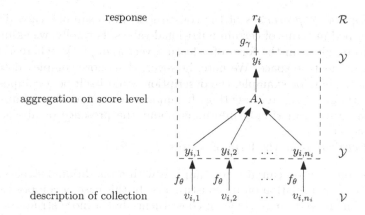

Fig. 1. Illustration of a basic version of the learning-to-aggregate model.

First, as just said, it allows for incorporating prior knowledge about the aggregation, which may serve as a suitable inductive bias of the learning process. Second, it naturally solves the problem that hypotheses $h \in \mathcal{H}$ must accept inputs of any size. Indeed, under the assumption of associativity and symmetry, a binary aggregation function A is naturally extended to any arity, and can hence be used as a "generator" of a hypothesis $h = h_A$:

$$h(y_1, \ldots, y_n) = A^{(n)}(y_1, \ldots, y_n) = A\big(A^{(n-1)}(y_1, \ldots, y_{n-1}), y_n\big)$$

for all $n \geq 1$, where $h(y_1) = A^{(1)}(y_1) = y_1$ by definition.

For these reasons, we consider the learning of (binary) aggregation functions, and related to this the specification of a suitable class $\mathcal{A}$ of candidates, as an integral part of learning-to-aggregate. In Sect. 3, such classes and different types of aggregation functions will be discussed in more detail. Before doing so, we elaborate on some extensions of our learning-to-aggregate setting.

2.4   Disaggregation

The aggregation we have been speaking about so far is an aggregation on the level of scores. Thus, we actually assume that local scores $y_{i,j}$ of the constituents $c_{i,j}$ are already given, and that we are interested in aggregating them into an overall score y_i of the composition c_i. This is indeed the genuine purpose of aggregation functions, which typically assume that all scores are elements of the same scale $\mathcal{Y}$. For example, we might be interested in how the scores on a conference paper (strong reject, reject, ..., strong accept) coming from a (variable) number of reviewers are aggregated into an overall rating by the program chairs.

Now, suppose that local scores $y_{i,j}$ are not part of the training data. Instead, the constituents $c_{i,j}$ are only described in terms of properties in the form of feature vectors $\boldsymbol{v}_{i,j} \in \mathcal{V}$ (and perhaps quantities $q_{i,j}$, which we subsequently ignore for simplicity). A natural way to tackle the learning problem, then, is to consider the local scores as latent variables, and to induce them as functions $f : \mathcal{V} \longrightarrow \mathcal{Y}$ of the properties.

In the following, we assume these functions to be parameterized by a parameter vector θ, and the aggregation function A by a parameter λ. The model is then of the form

$$y_i = A_\lambda(y_{i,1}, \ldots, y_{i,n_i}) = A_\lambda\big(f_\theta(\boldsymbol{v}_{i,1}), \ldots, f_\theta(\boldsymbol{v}_{i,n_i})\big),$$

and the problem consists of learning both the aggregation function A, i.e., the parameter λ, and the mapping from features to local scores, i.e., the parameter θ, simultaneously. Here, supervision only takes place on the level of the entire composition, namely in the form of scores y_i, whereas the "explanation" of these scores via induction of local scores is part of the learning problem.

The decomposition of global scores into several local scores is sometimes referred to as *disaggregation* (because it inverts the direction of aggregation, which is from local scores to global ones). For example, suppose we observe a user's ratings of different playlists, each one considered as a collections of songs, but not of the individual songs themselves. In order to predict the user's rating of new playlists, we could then try to learn how she rates individual songs and, simultaneously, how she aggregates several (local) ratings into a global rating.

Obviously, there is a strong interaction between the local ratings and their aggregation into a global score. For example, if we consistently observe low scores for different playlists, this could be either because the user dislikes (almost) all songs, or because she dislikes only a few but aggregates very strictly (i.e., a playlist gets a low score as soon as is contains a single or a few poor songs). An important question, therefore, concerns the *identifiability* of the model, i.e., the question whether different parameterizations imply different models (or, more formally, whether $(\lambda, \theta) \neq (\lambda', \theta')$ implies that the corresponding models assign different scores $y_i \neq y_i'$ for at least one composition).

2.5 Further Extensions

Sometimes, not even the (aggregate) scores y_i can be observed directly, but only certain response values $r_i \in \mathcal{R}$ related to these scores, i.e., training data is of the form

$$\mathcal{D} = \{(c_1, r_1), \ldots, (c_N, r_N)\} \subset \mathcal{C} \times \mathcal{R}, \tag{3}$$

For example, in the case of the playlist, direct feedback of the user might not be available. Instead, it might only be possible to observe a user's behavior, e.g., how long she listens to the playlist, or whether or not she decides to buy it. The response must then be modeled by another link function g (parameterized by γ), for example a discrete choice model like logit, which assumes $\mathcal{R} = \{0, 1\}$ and models the probability of a positive response according to $\mathbf{P}(r_i = 1) = (1 + \exp(-\gamma_1(y_i - \gamma_2)))^{-1}$. The model discussed so far, including indirect feedback in the form of a response, is summarized and illustrated graphically in Fig. 1.

Instead of absolute feedback in the form of a (binary) response, one may also assume relative feedback in the form of pairwise comparisons $c_i \succ c_j$ between compositions, suggesting that c_i is preferred to c_j (and hence that y_i is larger than y_j). This type of feedback and corresponding training data

$$\mathcal{D} = \left\{ c_{i(1)} \succ c_{j(1)}, \ldots, c_{i(N)} \succ c_{j(N)} \right\} \subset \mathcal{C} \times \mathcal{C}, \tag{4}$$

is especially interesting from the point of view of preference learning [9]. Model induction could then be based, for example, on discrete choice models like Bradley-Terrey [1].

Further extensions of the model are possible thanks to additional information provided about the constituents or structural information about the composition (cf. Sect. 2.2). In particular, the aggregation step can be generalized in the case where a label is assigned to the constituents. For example, we may assume that a user first rates the appetizer, main dish, and dessert (each of which may consist of several dishes) separately, and then aggregates the corresponding scores into an overall rating. Note that, since the intermediate scores are now associated with roles, the last aggregation step does not necessarily need to be invariant against permutation (for example, the user may give a higher weight to the main dish and a lower one to the starter), so that a larger class of aggregation functions could be used.

2.6   Learning Problems

Even in its basic form shown in Fig. 1, our learning-to-aggregate framework can be instantiated in various ways and gives rise to a number of different learning problems, in particular depending on the type of data that is observed and can be used for training. In the most general case, compositions are of different size, and training data consists of properties of constituents together with a corresponding response. Then, the learning problem essentially comes down to estimating the full set of parameters $(\gamma, \lambda, \theta)$.

Learning becomes simpler for various special cases. For example, if scores are observed directly (i.e., $r_i = y_i$), the link from scores to responses, specified by g_γ, does not need to be learned (or, stated differently, g can be taken as the identity). The case where individual scores $y_{i,j}$ are observed, too, is often considered in decision analysis and related fields [12, 26], typically even with the assumption that each individual score corresponds to a *criterion* (which, in our terminology, means that it has a unique label, and that n_i is given by the number of criteria and hence the same for each composition). The main question, then, is how the rating of an alternative on different criteria is aggregated into an overall rating. For example, one might be interested in how reviewers combine their ratings on criteria such as readability, novelty, etc. into an overall rating of a paper.

3   Aggregation Functions

Aggregation functions have been studied intensively as a branch of applied mathematics; we refer to the monograph [10] for a comprehensive treatment

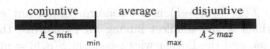

Fig. 2. Aggregation functions: conjunctive, disjunctive, and generalized averages.

of the topic. Roughly speaking, the purpose of an aggregation function operating on a scale $\mathcal{Y}$ is to combine values $y_1, \ldots, y_n \in \mathcal{Y}$ into another value y on the same scale. Typically, $\mathcal{Y}$ is taken as the unit interval $[0,1]$; this is not a strong restriction, since aggregation functions on other domains can be studied via suitable transformations in the form of monotone bijections [11].

The study of aggregation functions is of axiomatic nature and proceeds from specific properties such functions should obey. Natural requirements, for example, include properties like symmetry (the result of the aggregation should not depend on the order of the values) and monotonicity. Especially interesting are binary aggregation functions A in the form of associative and commutative $[0,1]^2 \longrightarrow [0,1]$ mappings, because, as already said, these can be extended to n-ary aggregation functions in a canonical way:

$$A^{(n)}(y_1, \ldots, y_n) = A\big(A^{(n-1)}(y_1, \ldots, y_{n-1}), y_n\big),$$

where $A^{(1)}(y_1) = y_1$. One can then simply identify A with the family of functions thus defined, and write $A(y_1, \ldots, y_n)$ for any number n of arguments.

A natural order on (binary) aggregation functions is defined as follows: $A \leq B$ if $A(y_1, y_2) \leq B(y_1, y_2)$ for all $y_1, y_2 \in [0,1]$. Based on this order relation, three important classes of aggregation functions are often distinguished: conjunctive, disjunctive, and generalized averaging operators. An aggregation A is called conjunctive if $A \leq \min$ and disjunctive if $A \geq \max$; all aggregations in-between min and max are called (generalized) averaging operators (see Fig. 2).

3.1 Conjunctive and Disjunctive Aggregation

In this paper, we are specifically interested in conjunctive and disjunctive aggregation, that is, aggregation functions that can be seen, respectively, as generalizations of the classical logical conjunction and disjunction. Important classes of such functions are given by the so-called t-norms and t-conorms [16].

Triangular norms (t-norms), which emerged in the context of probabilistic metric spaces [21], play a central role is many-valued and fuzzy logic, where they are used to generalize the logical conjunction [13]. A t-norm T is a monotone increasing, associative and commutative $[0,1]^2 \longrightarrow [0,1]$ mapping with neutral element 1 and absorbing element 0. Important examples include the minimum, which is the largest among all t-norms, the product $T(a,b) = ab$, and the Lukasiewicz t-norm $T(a,b) = \min(a + b, 1)$.

A t-conorm S is a monotone increasing, associative and commutative mapping $[0,1]^2 \longrightarrow [0,1]$ with neutral element 0 and absorbing element 1. These operators are dual to t-norms in the sense that, if T is a t-norm, then S defined by $S(a,b) = 1 - T(1-a, 1-b)$ is a t-conorm. Important examples include the maximum, which is the smallest among all t-conorms, the algebraic sum $S(a,b) = a + b - ab$, and the Lukasiewicz t-conorm $S(a,b) = \max(a+b-1, 0)$.

3.2 Uninorms

Generalized conjunctions and disjunctions share the properties of being monotone, associative and commutative, and actually only differ in their neutral element, which is 1 for the former and 0 for the latter. The location of the neutral element in the unit interval is also reflected by the characteristics of these two types of operators: For t-norms, the overall aggregation remains unchanged only when adding the highest value 1, i.e., $T(y_1, \ldots, y_n) = T(y_1, \ldots, y_n, y_{n+1})$ only if $y_{n+1} = 1$; otherwise, the overall aggregation can only decrease. Thus, t-norms aggregate very strictly and are fully non-compensatory: it is not possible to compensate for low evaluations by adding high ones. The dual class of t-conorms behaves in exactly the opposite way: aggregation via t-conorms is fully compensatory.

One may wonder whether a neutral behavior is only possible with respect to 0 and 1, or perhaps also some other value $e \in (0,1)$. Is there is a class of aggregation functions that shares the properties of t-norms and t-conorms, except for having an arbitrary value e as neutral element? This question is answered affirmatively by the class of so-called *uninorms* [27]. A uninorm U is a monotone increasing, associative and commutative $[0,1]^2 \longrightarrow [0,1]$ mapping with neutral element $e \in (0,1)$, i.e., such that $U(a,e) = U(e,a) = a$ for all $a \in [0,1]$.

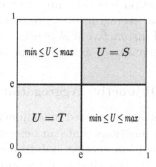

Fig. 3. Structure of a uninorm.

Uninorms U can be shown to have a specific structure: For arguments exceeding e, they behave like a t-conorm, i.e., there is a t-conorm S such that $U(a,b) = S(a,b)$ for all $a,b \in [e,1]^2$. Likewise, for arguments below e, they behave like a t-norm: $U(a,b) = T(a,b)$ for all $(a,b) \in [0,e]$. On the remaining part of the unit square $[0,1]^2$, U can be completed in different ways, though always remaining between the minimum and the maximum; see Fig. 3 for an illustration. A concrete family of uninorms called min-uninorms is constructed from a t-norm T and a t-conorm S as follows:

$$U_e(a,b) = \begin{cases} e\,T\left(\frac{a}{e}, \frac{b}{e}\right) & \text{if } a,b \in [0,e] \\ e + (1-e)S\left(\frac{a-e}{1-e}, \frac{b-e}{1-e}\right) & \text{if } a,b \in [e,1] \\ \min(a,b) & \text{otherwise} \end{cases} \tag{5}$$

3.3 Complex Aggregation

Basic aggregation functions like those discussed above can be combined into more complex ones, for example in a hierarchical way [22,23]. An example is shown graphically in Fig. 4: The output produced by one aggregation serves as an input of another one on a higher level. In the particular example shown, the aggregation function is of the form

$$A_{\lambda_t, \lambda_s} : [0,1]^4 \longrightarrow [0,1], \ (y_1, y_2, y_3, y_4) \mapsto T_{\lambda_t}\big(S_{\lambda_s}(y_1, y_2), S_{\lambda_s}(y_3, y_4)\big), \tag{6}$$

and thanks to the logical interpretation of t-norms and t-conorms, A itself can be interpreted as a degree of truth of a generalized logical expression. For example, if y_1, y_2, y_3, y_4 correspond, respectively, to the evaluation of a job candidate on skills in math (M), programming (P), French (F), and Spanish (S), then A evaluates the expression $(M \wedge P) \vee (F \wedge S)$. In other words, a good candidate needs to be strong in math or programming, and also have good language skills, either in French or Spanish. Thus, there is no compensation between language and analytical skills, but full compensation within each of the two categories.

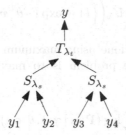

Fig. 4. Example of a complex (hierarchical) aggregation functions.

As shown by (6), complex aggregation functions typically assign different roles to different inputs. In our framework, this means that constituents must be identified by a label (such as M or P above). In principle, of course, structures more general than hierarchies (trees) could be used to design complex aggregation functions, for example directed acyclic graphs. Such structures appear to be especially useful in the case where the constituents $c_{i,j}$ in a composition c_i are equipped with a structure (i.e., c_i is not simply a bag). However, as extensions of this kind are beyond the scope of this paper, we refrain from a deeper discussion.

4 A Model Based on Uninorms

Suppose a component c_i is a multiset of constituents $c_{i,j}$ described in terms of feature vectors $v_{i,j}$. We assume scores $y_{i,j} \in [0,1]$ to be of the form

$$y_{i,j} = f_\theta(v_{i,j}) = \left(1 + \exp\left(-\theta^\top v_{i,j}\right)\right)^{-1},$$

i.e., θ is a vector that assigns weights for the different entries in $v_{i,j}$. The local scores $y_{i,j}$ are then aggregated using a uninorm U_λ parameterized by λ:

$$y_i = U_\lambda(\{y_{i,j}\}_{j=1}^{n_i}) = U_\lambda(y_{i,1}, \ldots, y_{i,n_i})$$

Finally, the response r_i is a binary decision, for which

$$\mathbf{P}(r_i = 1) = \left(1 + \exp\left(-\gamma_1(y_i - \gamma_2)\right)\right)^{-1}.$$

Thus, the higher the score, the higher is the probability of a positive decision. More specifically, the probability of a positive decision is controlled by two parameters $\gamma = (\gamma_1, \gamma_2)$. The second parameter, $\gamma_2 \in [0,1]$, is a kind of aspiration level or ambition threshold, since $\mathbf{P}(r_i = 1) > 1/2$ for $y_i > \gamma_2$ and $\mathbf{P}(r_i = 1) < 1/2$ for $y_i < \gamma_2$. Moreover, $\gamma_1 \geq 0$ is a scaling parameter that models the precision with which decisions are made: For $\gamma_1 \to \infty$, decisions become deterministic, whereas for $\gamma_1 = 0$, decisions are made completely at random (i.e., without actually taking the score y_i into account).

Overall, we thus end up with a probabilistic model of the following form:

$$\mathbf{P}(r_i = 1) = \left(1 + \exp\left(-\gamma_1\left(U_\lambda\left(\left\{\left(1 + \exp\left(-\theta^\top v_{i,j}\right)\right)^{-1}\right\}_{j=1}^{n_i}\right) - \gamma_2\right)\right)\right)^{-1} \quad (7)$$

Learning this model can be done using maximum likelihood estimation. Thus, given training data (4), the problem is to maximize the (regularized) log-likelihood function

$$L(\gamma, \lambda, \theta) = \sum_{i=1}^{N} \log\left(\mathbf{P}(r_i \mid \gamma, \lambda, \theta, c_i)\right) - \alpha R(\gamma, \lambda, \theta), \quad (8)$$

where $\mathbf{P}(r_i \mid \gamma, \lambda, \theta, c_i)$ is given by the expression on the right-hand side of (7) if $r_i = 1$ and by 1 minus this expression if $r_i = 0$, and $R(\gamma, \lambda, \theta)$ is a regularization term that is used to penalize large feature weights.

The above model simplifies in the case where the local scores $y_{i,j}$ are already given, i.e., training data is of the form (4):

$$\mathbf{P}(r_i = 1) = \left(1 + \exp\left(\gamma_1\left(U_\lambda\left(\{y_{i,j}\}_{j=1}^{n_i}\right) - \gamma_2\right)\right)\right)^{-1} \quad (9)$$

5   Related Work

As our framework is quite general, it has connections to various other branches of machine learning. These are either established by the non-standard representation of instances, or by the idea of using aggregation functions in one way or the other. This section is meant to point to some of the related fields, although space restrictions obviously prevent from a comprehensive discussion.

Compositions c_i can of course be seen as a specific types of structured objects, on which kernel functions can be defined; for example, kernel functions for "bags of feature vectors" have been studied in image processing and other fields [5]. Then, given such a kernel function, the large arsenal of kernel-based machine learning methods can be applied. Yet, an approach of that kind is not fully in line with our idea of learning to aggregate. First, kernel methods eventually produce a vectorial representation (in some feature space), which, for the reasons already mentioned, might not be fully appropriate. More importantly, they do not easily allow for incorporating knowledge about the process of aggregation, which is a key idea of our approach, nor do they lead to well interpretable models.

In the special case where compositions c_i are bags (i.e., multisets without additional structure) of feature vectors $v_{i,j}$, our framework is similar to *multi-instance learning* (MIL) [2], especially with regard to the representation of data objects. Yet, there are also some notable differences. In MIL, for example, a bag is normally not viewed as a composition of constituents that belong together and form a whole; in the simplest case, one proceeds from a binary setting with positive and negative instances, and assumes a bag to be labeled positive as soon as it contains at least one positive instance. Correspondingly, aggregation over predictions for individual instances is done, either explicitly or implicitly, via the maximum (or generalizations like the noisy OR [14]), whereas less attention has been payed to a systematic study of the aggregation process.

Specific types of aggregation functions have attracted attention in machine learning in recent years. For example, copulas can be seen as a specific type of conjunctive aggregation that allows for combining marginal into joint probability distributions [6]. In preference learning, the so-called Choquet integral has been used as a generalization of the weighted average that is able to capture interactions between different variables [24]. Yet, these approaches still proceed from a feature representation of data objects.

There are other generalizations of supervised learning in which aggregation plays an important role. For example, in learning from aggregate outputs [18], the assumption is that output values cannot be observed for each training instance individually; instead, only an aggregation of these values is observed for sets of instances. Here, however, the aggregation function is supposed to be known.

As already mentioned, the scores assigned to a composition can often be interpreted as a kind of evaluation. Thus, there is also an obvious connection to the field of preference learning [9]. From the point of view of preference learning, a composition can be seen as a *bundle of goods*, to which a user assigns a degree of utility [25]. In comparison to learning preferences on items represented in terms of feature vectors, work on preference learning on bundles is still very scarce.

6   Illustration

As an illustration of our framework, we consider the problem of aggregating reviewer recommendations into an overall decision about the acceptance or rejection of a conference submission. Or, stated differently, we adopt a data-driven approach to modeling the way in which the program chairs of a conference aggregate different reviews of a paper into an overall decision.

To this end, we collected data about the reviewing process of ECML/PKDD 2014. More concretely, our data set consists of 481 submitted papers with corresponding reviews. While most papers have three reviews, there are also papers with two or four reviews. Each review consists of a rating of the originality and quality of the paper, an overall recommendation, and a level of confidence of the reviewer. The underlying scale comprises five categories (strong reject, weak reject, weak accept, accept, strong accept), which we embedded in the unit interval by mapping them to $\{0, 0.25, 0.5, 0.75, 1\}$. Finally, the decision of acceptance or rejection is known for each paper (with an acceptance rate of 23,9 %).

As already said, the problem we consider consists of learning to aggregate reviewer recommendations into a final decision. Here, a paper is modeled as a composition c_i, the constituents of which consist of feature vectors $v_{i,j}$ with values for originality, quality, and overall recommendation given by a reviewer, as well as his or her confidence (and an intercept). Moreover, the final decision is treated as a response (0 for rejection and 1 for acceptance).

We applied the model (7) introduced in Sect. 4 with two uninorms: The so-called 3-Π uninorm [3], and the uninorm (5) with the product t-norm $T(a, b) = ab$ and the dual t-conorm $S(a, b) = a + b - ab$; in the latter case, the parameter λ of U_λ is thus given by the neutral element e in (5). Note that a uninorm is a quite plausible aggregation function for this application: The neutral element e can be seen as kind of "borderline" recommendation. A recommendation better than e expresses a positive reviewer option and can only increase the probability of acceptance, whereas a recommendation worse than e has the opposite effect. For comparison, we also present results for purely conjunctive ($U_\lambda = \min$) and purely disjunctive aggregation ($U_\lambda = \max$).

To learn the parameters, we maximize the likelihood function (8) with L_2 regularization ($\alpha = 0.01$) using the L-BFGS-B algorithm [4]. To avoid local optima, we did 10 random restarts, choosing initial parameters according to Latin hypercube sampling [17] with 10 samples.

The problem considered in this study can in principle also be formalized in the setting of multi-instance (MI) learning: papers are considered as bags and the reviews as instances, represented in the form of feature vectors. Therefore, we also compare our method with several state-of-the-art MI algorithms [2].

A standard approach based on a feature representation of submissions does not appear meaningful. In fact, even if the number of reviewers would be the same for each paper, the order of reviewers should not play any role, i.e., the aggregation should be invariant against permutation (renumbering of the reviewers). For example, it does not make sense to give a higher weight to the first reviewer and a lower weight to the second one. Since all features are discrete, it is still

Table 1. Mean ± standard deviation for classification rate, AUC and F-measure.

Approach	Algorithm/Aggregation	Accuracy	AUC	F_1
Aggregation	3-Π uninorm	.921 ± .035	.974 ± .025	.823 ± .091
	min-uninorm	.890 ± .029	.949 ± .021	.767 ± .059
	minimum	.885 ± .045	.923 ± .034	.756 ± .100
	maximum	.831 ± .064	.903 ± .064	.568 ± .179
MIL	MILR [20]	.916 ± .038	.973 ± .017	.811 ± .097
	MIBoost [8]	.911 ± .037	.960 ± .027	.807 ± .087
	MISMO-PolyKernel [19]	.906 ± .041	.858 ± .070	.791 ± .099
	MISMO-RBFKernel [19]	.909 ± .041	.870 ± .067	.804 ± .095
	MIWrapper [7]	.880 ± .040	.955 ± .031	.664 ± .146
Feature	AdaBoostM1-Dec.Table	.858 ± .045	.892 ± .052	.683 ± .112
Vector	AdaBoostM1-Dec.Stumps	.873 ± .043	.906 ± .048	.742 ± .091
	Decision Table	.855 ± .045	.900 ± .049	.687 ± .123
	C4.5 (J48)	.856 ± .038	.860 ± .073	.671 ± .101
	KNN	.862 ± .038	.904 ± .045	.667 ± .109
	LBR	.857 ± .041	.924 ± .038	.731 ± .079
	RandomForest	.838 ± .043	.891 ± .048	.633 ± .109
	Logistic Regression	.872 ± .042	.911 ± .046	.738 ± .097
	SVM (SMO)	.868 ± .042	.773 ± .073	.675 ± .122

possible to create a vector representation, simply by counting, for each feature value, its total number of occurrences in all reviews. Obviously, this transformation comes with a loss of information, since the reviews are merged and cannot be distinguished anymore. Nevertheless, we used it as another baseline (with several standard learning methods implemented in WEKA [15]).

All performance measures were estimated using 10-fold cross validation repeated 10 times. The mean values and standard deviations of classification rate, AUC, and F-measure are reported in Table 1. As can be seen, our approach compares quite favorably with the baselines. Moreover, the estimated model appears to be quite plausible. For example, the parameter γ_2, which plays the role of an acceptance threshold, equals (on average) 0.687; moreover, $\gamma_1 \approx 8$, which means that the reviewer recommendations determine decisions quite precisely. The vector θ has a plausible interpretation too: the overall recommendation has the highest influence, with a relative importance of about 0.78, followed by originality and quality with around 0.11 and 0.09, respectively.

7   Summary and Conclusion

The learning-to-aggregate framework introduced in this paper is meant to provide a basis for learning (predictive) models in which aggregation plays an integral role. We believe that, first, there are many applications of this kind of modeling, and second, that machine learning can strongly benefit from the large repertoire of existing work on aggregation functions in the mathematical literature. More specifically, we argue that this field offers interesting mathematical tools for constructing model classes, thereby helping to learn models that are not only accurate but also interpretable, as well as important theoretical insights about aggregation functions and their properties, thereby supporting the design of efficient learning algorithms.

We illustrated our framework by looking at one of its particular instances and applying that instance on a review data set, where the aggregation problem consists of combining a (variable) number of reviews of a paper submission into a final decision of acceptance or rejection. While this is only a specific example, we look forward to developing the learning-to-aggregate framework both more broadly and more deeply in future work. As explained in Sect. 2, various learning problems can be defined based on the representation of compositions, assumptions about the aggregation process, and the type of training data to learn from. Developing and analyzing learning-to-rank methods for concrete, practically relevant settings is a major goal of our future work.

Acknowledgments. We thank Pritha Gupta and Karlson Pfannschmidt for their helpful suggestions. This work is part of the Collaborative Research Center "On-the-Fly Computing", which is supported by the German Research Foundation (DFG).

References

1. Alvo, M., Yu, P.: Statistical Methods for Ranking Data. Springer, New York (2014)
2. Amores, J.: Multiple instance classification: review, taxonomy and comparative study. Artif. Intell. **201**, 81–105 (2013)
3. Beliakov, G., Calvo, T., James, S.: Aggregation of preferences in recommender systems. In: Recommender Systems Handbook, pp. 705–734. Springer, US (2011)
4. Byrd, R.H., Lu, P., Nocedal, J., Zhu, C.: A limited memory algorithm for bound constrained optimization. SIAM J. Sci. Comput. **16**(5), 1190–1208 (1995)
5. Csurka, G., Dance, C.R., Fan, L., Willamowski, J., Bray, C.: Visual categorization with bags of keypoints. In: Workshop on Statistical Learning in Computer Vision, ECCV (2004)
6. Elidan, G.: Copula bayesian networks. In: Proceedings of the NIPS, Advances in Neural Information Processing Systems 23, pp. 559–567 (2010)
7. Frank, E.T., Xu, X.: Applying propositional learning algorithms to multi-instance data. Technical report, University of Waikato, Department of Computer Science, University of Waikato, Hamilton, NZ, June 2003
8. Freund, Y., Schapire, R.E.: Experiments with a new boosting algorithm. In: Thirteenth International Conference on Machine Learning, pp. 148–156. Morgan Kaufmann, San Francisco (1996)
9. Fürnkranz, J., Hüllermeier, E. (eds.): Preference Learning. Springer, Heidelberg (2011)
10. Grabisch, M., Marichal, J., Mesiar, R., Pap, E.: Aggregation Functions. Cambridge University Press, Cambridge (2009)

11. Grabisch, M., Marichal, J., Mesiar, R., Pap, E.: Aggregation functions: construction methods, conjunctive, disjunctive and mixed classes. Inf. Sci. **181**, 23–43 (2011)
12. Greco, S., Mousseau, V., Slowinski, R.: Robust ordinal regression for value functions handling interacting criteria. Eur. J. Oper. Res. **239**(3), 711–730 (2014)
13. Hajek, P.: Metamathematics of Fuzzy Logic. Springer, Dordrecht (1998)
14. Hajimirsadeghi, H., Mori, G.: Multiple instance real boosting with aggregation functions. In: Proceedings of the ICPR, 21st International Conference on Pattern Recognition, pp. 2706–2710 (2012)
15. Hall, M., Frank, E., Holmes, G., Pfahringer, B., Reutemann, P., Witten, I.: The WEKA data mining software: an update. SIGKDD Explor. **11**(1), 10–18 (2009)
16. Klement, E., Mesiar, R., Pap, E.: Triangular Norms. Kluwer Academic Publishers, Dordrecht (2002)
17. McKay, M.D., Beckman, R.J., Conover, W.J.: A comparison of three methods for selecting values of input variables in the analysis of output from a computer code. Technometrics **21**(2), 239 (1979)
18. Musicant, D.R., Christensen, J.M., Olson, J.F.: Supervised learning by training on aggregate outputs. In: Proceedings of the ICDM, 7th IEEE International Conference on Data Mining, Omaha, Nebraska, USA, pp. 252–261 (2007)
19. Platt, J.: Machines using sequential minimal optimization. In: Schoelkopf, B., Burges, C., Smola, A. (eds.) Advances in Kernel Methods - Support Vector Learning. MIT Press (1998)
20. Ray, S., Page, D.: Multiple instance regression. In: ICML, vol. 1, pp. 425–432 (2001)
21. Schweizer, B., Sklar, A.: Probabilistic Metric Spaces. North-Holland, New York (1983)
22. Senge, R., Hüllermeier, E.: Top-down induction of fuzzy pattern trees. IEEE Trans. Fuzzy Syst. **19**(2), 241–252 (2011)
23. Senge, R., Hüllermeier, E.: Fast fuzzy pattern tree learning for classification. IEEE Trans. Fuzzy Syst. **23**(6), 2024–2033 (2015)
24. Tehrani, A.F., Cheng, W., Dembczynski, K., Hüllermeier, E.: Learning monotone nonlinear models using the Choquet integral. Mach. Learn. **89**(1), 183–211 (2012)
25. Tschiatschek, S., Djolonga, J., Krause, A.: Learning probabilistic submodular diversity models via noise contrastive estimation. In: Proceedings of the AISTATS, 19th International Conference on Artificial Intelligence and Statistics (2016)
26. Narukawa, Y., Torra, T.: Modeling Decisions: Information Fusion and Aggregation Operators. Springer, Berlin (2007)
27. Yager, R., Rybalov, A.: Uninorm aggregation operators. Fuzzy Sets Syst. **80**, 111–120 (1996)

Sequential Labeling with Online Deep Learning: Exploring Model Initialization

Gang Chen[✉], Ran Xu, and Sargur N. Srihari

Department of Computer Science and Engineering, SUNY at Buffalo,
Buffalo, NY 14260, USA
{gangchen,rxu2,srihari}@buffalo.edu

Abstract. In this paper, we leverage both deep learning and conditional random fields (CRFs) for sequential labeling. More specifically, we explore parameter initialization and randomization in deep CRFs and train the whole model in a simple but effective way. In particular, we pretrain the deep structure with greedy layer-wise restricted Boltzmann machines (RBMs), followed with an independent label learning step. Finally, we re-randomize the top layer weight and update the whole model with an online learning algorithm – a mixture of perceptron training and stochastic gradient descent to estimate model parameters. We test our model on different challenge tasks, and show that this simple learning algorithm yields the state of the art results. The data and software related to this paper are available at https://github.com/ganggit/deepCRFs.

Keywords: Sequential labeling · Deep learning · Online learning · Parameter initialization

1 Introduction

Recent advances in deep learning [1,16,43] have sparked great interest in dimension reduction [15,44] and classification [16,26]. In a sense, the success of deep learning lies on learned features, which are useful for supervised/unsupervised tasks [1,4,11]. For example, the binary hidden units in the discriminative Restricted Boltzmann Machines (RBMs) [12,25] and deep belief networks (DBN) [16] can model latent features of raw data to improve classification. Unfortunately, one major difficulty in deep learning [16] is structured output prediction [31], where output space typically may have an exponential number of possible configurations. As for sequential labeling, the joint classification of all the items is also difficult because observations are of an indeterminated dimensionality and the number of possible classes is exponentially growing in the length of the sequences.

To address the sequential prediction, the architecture of recurrent neural networks (RNNs) have cycles incorporating the activations from previous time steps as input to the network to make a decision for the current input, which

© Springer International Publishing AG 2016
P. Frasconi et al. (Eds.): ECML PKDD 2016, Part II, LNAI 9852, pp. 772–788, 2016.
DOI: 10.1007/978-3-319-46227-1_48

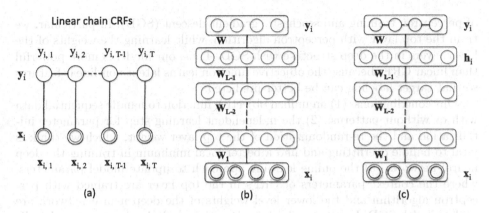

Fig. 1. (a) linear chain CRFs; (b) deep neural networks (for classification with 1 of K (encoding) vector representation); (c) our deep neural networks for sequential labeling. The two main differences between (b) and (c) are: (1) y_i in (c) is a label sequence, which has links between labels, while y_i in (b) is a single label without correlation; (2) the input of (c) is a sequence with multiple instances (or frames), while the input of (b) is an independent instance (or vector).

makes RNNs better suited for sequence labeling tasks. Long short term memory (LSTM), as an improved version of RNNs, shows good results on handwritten recognition [13]. And bi-directional LSTM trained on unsegmented sequence data has also outperformed the state of the art HMM-based system.

Another direction is to combine deep learning with CRFs for sequential labeling [9,28,33]. One of key advantages of linear CRFs can be attributed to its exploitation on context information and its structured output prediction. However, linear CRFs with the raw data as input strongly restricts its representation power for classification tasks. More recently, one trend is to generalize CRFs to learn discriminative and non-linear representations, such as kernel CRFs [23], hidden-unit CRFs [28,33] and CRFs with multilayer perceptrons [27,35]. As an alternative, some studies have trained CRFs on features learned by unsupervised deep learning [32]. Also the work in [9,33] has exploited to learn non-linear mappings by combing CRFs and neural networks. However, how to learn a better deep CRFs model is still a challenge, considering the overfitting issue with large parameter space and the non-convex objective function.

In this paper, we propose a deep model for sequential labeling, which inherits both advantages of linear chain CRFs and deep learning. Hence, our model can learn non-linear features and also handle structured output, refer to Fig. 1 for visual understanding about the model. Because the deep CRFs model is non-convex, it can be easily trapped into local minimum. Thus, how to learn a good model and generalize well in unseen dataset is still an challenge. Compared to the traditional deep CRFs [9,29,33], we take a different learning approach. We pre-train our model with stacked RBMs, followed with an independent learning step with backpropagation. Then, we re-randomize the top layer weight and optimize the whole deep model using an online learning algorithm, which is a mixture of perceptron training and stochastic gradient descent (SGD). In particular, we train the top layer with perceptron algorithm, while learning the weights of the lower layers in the deep structure with SGD. Thus, our model is more powerful than linear CRFs because the objective function learns latent non-linear features so that target labeling can be better predicted.

Our contributions: (1) an unified objective function to handle sequential data with or without patterns; (2) the independent learning step for parameter initialization and the re-randomization of the top layer weight, which we think is vital to handle overfitting and find a better local minimum in training the deep neural network; (3) the online learning approach to update model parameters, where the context parameters of CRFs in the top layer are trained with perceptron algorithm and the lower level weights of the deep neural network are updated with SGD. Lastly, we also introduce the regularization terms to handle overfitting in the deep neural network. We test our model over a range of tasks and show that it yields accuracy significantly better than the state of the art.

2   Sequential Labeling with Deep Learning

Let $D = \{\langle \mathbf{x}_i, \mathbf{y}_i \rangle\}_{i=1}^N$ be a set of N training examples. Each example is a pair of a time series $\langle \mathbf{x}_i, \mathbf{y}_i \rangle$, with $\mathbf{x}_i = \{\mathbf{x}_{i,1}, \mathbf{x}_{i,2}, ..., \mathbf{x}_{i,T_i}\}$ and $\mathbf{y}_i = \{y_{i,1}, y_{i,2}, ..., y_{i,T_i}\}$, where $\mathbf{x}_{i,t} \in \mathbb{R}^d$ is the i-th observation at time t and $y_{i,t}$ is the corresponding label (we indicate its encoded vector as $\mathbf{y}_{i,t}$ that uses a so-called 1-of-K encoding). Linear first-order CRFs [24] is a conditional discriminative model over the label sequence given the data

$$p(\mathbf{y}_i | \mathbf{x}_i) = \frac{\exp\{-E(\mathbf{x}_i, \mathbf{y}_i)\}}{Z(\mathbf{x}_i)} \tag{1}$$

where $Z(\mathbf{x}_i)$ is the partition function and $E(\mathbf{x}_i, \mathbf{y}_i)$ is the energy function given by

$$-E(\mathbf{x}_i, \mathbf{y}_i) = \mathbf{y}_{i,1}^T \boldsymbol{\pi} + \mathbf{y}_{i,T_i}^T \boldsymbol{\tau}$$
$$+ \sum_{t=1}^{T_i} (\mathbf{x}_{i,t}^T \mathbf{W} \mathbf{y}_{i,t} + \mathbf{b}^T \mathbf{y}_{i,t}) + \sum_{t=2}^{T_i} \mathbf{y}_{i,t-1}^T \mathbf{A} \mathbf{y}_{i,t} \tag{2}$$

where $\mathbf{y}_{i,1}^T \boldsymbol{\pi}$ and $\mathbf{y}_{i,T_i}^T \boldsymbol{\tau}$ are the initial-state and final-state factors respectively, $\mathbf{b}^T \mathbf{y}_{i,t}$ is the bias term for labels, $\mathbf{A} \in \mathbb{R}^{K \times K}$ represents the state transition parameters and $\mathbf{W} \in \mathbb{R}^{d \times K}$ represents the parameters of the data-dependent term. Compared to linear SVMs, the linear CRFs has an additional item $\mathbf{y}_{i,t-1}^T \mathbf{A} \mathbf{y}_{i,t}$ to model the label correlation. However, one of the main disadvantages of linear CRFs is the linear dependence on the raw input data. Thus, we introduce our sequential labeling model with deep feature learning, which leverages both context information, as well as the nonlinear representations from deep learning [15].

2.1 Objective Function

Although it is possible to leverage the deep neural networks for structured prediction, its output space is explosively growing because of non-determined length of sequential data. Thus, we consider a compromised model, which combine CRFs and deep learning in an unified framework, refer Fig. (1). On the one hand, we hope the independent label prediction is as accuracy as possible via the representation learning. On the other hand, we need to handle overfitting problem in the deep network. We propose an objective function with L layers neural network structure,

$$\mathcal{L}(D; \boldsymbol{\theta}, \boldsymbol{\omega}) = - \sum_{i=1}^{N} \log p(\mathbf{y}_{i,1}, \ldots, \mathbf{y}_{i,T_i} | \mathbf{h}_{i,1}, \ldots, \mathbf{h}_{i,T_i})$$

$$+ \frac{\lambda_1}{2} \sum_{i=1}^{N} \sum_{t=1}^{T_i} \| \underbrace{f_L \circ f_{L-1} \circ \cdots \circ f_1(\mathbf{x}_{i,t})}_{L \text{ times}} - \mathbf{y}_{i,t} \|^2$$

$$+ \lambda_2 \|\boldsymbol{\theta}\|^2 + \lambda_3 \|\boldsymbol{\omega}\| \tag{3}$$

where $\boldsymbol{\theta}$ and $\boldsymbol{\omega}$ are the top layer parameters and lower layer ($l = \{1, ..., L-1\}$) parameters respectively, which will be explained later. The first row on the right side of the equation is from the linear CRFs in Eq. (1), but with latent features. The conditional likelihood depends respectively on $\boldsymbol{\theta}$ and the latent non-linear features $\mathbf{h}_i = \{\mathbf{h}_{i,1}, .., \mathbf{h}_{i,T_i}\}$ in the coding space, with

$$\log p(\mathbf{y}_{i,1}, \ldots, \mathbf{y}_{i,T_i} | \mathbf{h}_{i,1}, \ldots, \mathbf{h}_{i,T_i})$$

$$= \sum_{t=2}^{T_i} \mathbf{y}_{i,t-1}^T \mathbf{A} \mathbf{y}_{i,t} + \sum_{t=1}^{T_i} \left(\mathbf{h}_{i,t}^T \mathbf{W} \mathbf{y}_{i,t} + \mathbf{b}^T \mathbf{y}_{i,t} \right)$$

$$+ \mathbf{y}_{i,1}^T \boldsymbol{\pi} + \mathbf{y}_{i,T_i}^T \boldsymbol{\tau} - \log(Z(\mathbf{h}_i)) \tag{4}$$

and non-linear mappings $\mathbf{h}_i$ is the output with $L-1$ layers neural network, s.t.

$$\mathbf{h}_i = \underbrace{f_{L-1} \circ f_{L-2} \circ \cdots \circ f_1(\mathbf{x}_i)}_{L-1 \text{ times}} \tag{5}$$

where $\circ$ indicates function composition, and f_i is logistic function with the weight parameter $\mathbf{W}_l$ respectively for $l = \{1, .., L-1\}$, refer more details in Sect. 2.2. With a bit abuse of notation, we denote $\mathbf{h}_{i,t} = f_{1 \to (L-1)}(\mathbf{x}_{i,t})$.

The least square (the second term) in the right hand side of Eq. (3) is for deep feature learning, with the top layer defined as

$$\underbrace{f_L \circ f_{L-1} \circ \cdots \circ f_1(\mathbf{x}_{i,t})}_{L \text{ times}}$$

$$= f_{1 \to L}(\mathbf{x}_{i,t}) = f_L(\mathbf{h}_{i,t}) = \mathbf{h}_{i,t}^T \mathbf{W} + \mathbf{c}^T \tag{6}$$

where $\mathbf{W}$ has been defined in Eq. (4), and $\mathbf{c}$ is the bias term. Note that $\mathbf{W}$ is the same in both Eqs. 4 and 6. Hence, the second term in Eq. (3) can be thought as the independent label prediction without considering context information. Note that other objective functions such as softmax (which has the same gradient as least square, in other words, the model updating with SGD is the same as here) can be applied here too. The weighing variable λ_1 can control the balance between the first term and the second one on the RHS in Eq. (3). If $\lambda_1 \to +\infty$, then Eq. (3) can be thought as deep learning [15] for classification without context information, and it can handle the cases where outputs are independent (no significant patterns in the label sequences). If $\lambda_1 \to 0$, then Eq. (3) is the CRFs with non-linear deep feature learning, which generalizes the linear CRFs to learn non-linear deep mappings. The main purpose we incorporate the second term in our model is to introduce this parameter initialization step via the independent learning (see further). Note that we can vary λ_1 to achieve this purpose.

The last two terms in Eq. (3) are for regularization on all parameters with $\theta = \{\mathbf{A}, \mathbf{W}, \boldsymbol{\pi}, \boldsymbol{\tau}, \boldsymbol{b}, \boldsymbol{c}\}$, and $\boldsymbol{\omega} = \{\mathbf{W}_l | l \in [1, .., L-1]\}$. We add the ℓ_2 regularization to θ as most linear CRFs does, while we have the ℓ_1-regularized term on weight parameters $\boldsymbol{\omega}$ in the deep neural network to avoid overfitting in the learning process.

The aim of our objective function in Eq. (3) is for sequential labeling, which explores both the advantages of Markov properties in CRFs and latent representations in deep learning. Our model is different from the common deep learning structure in Fig. 1(b). Firstly, the input to our model in Fig. 1(c) is the sequential data, such as sequences with non-determined length, while the input to Fig. 1(b) is generally an instance with fixed length. Secondly, our model can predict structured outputs or label sequences, while the output in Fig. 1(b) is just one label for each instance, which is independent from each other. Note that we use the first-order CRFs for clarity in Eq. 4, which can be easily extended to the second or high-order cases. Moreover, our model is also different from other deep CRFs [9,34]. Our mixture objective function can handle sequential data with or without patterns. And we have the independent label learning (a pretraining step in our model) to learn better representations. Lastly, we use an online algorithm in our deep learning model for parameter updating, which has the potential to handle large scale dataset.

2.2   Parameter Learning

We use RBMs to initialize the weights layer by layer greedily, with Contrastive Divergence [16] (we used CD-1 in our experiments). Then we compute the sub-gradients w.r.t. θ and $\boldsymbol{\omega}$ in the objective function, and initialize the whole deep CRFs with independent learning. Finally, we re-randomize the top layer weight and update the whole framework with online learning.

Initialization: The second term on the right hand side of Eq. (3) is from the deep belief network (DBN) for classification [16]. In our deep model, the weights from the layer 1 to $L-1$ are $\mathbf{W}_l$ respectively, for $l = \{1, .., L-1\}$, and the top layer L has weight $\mathbf{W}$. We first pre-train the L-layer deep structure with RBMs layer by layer greedily. Specifically, we think RBM is a 1-layer DBN, with weight $\mathbf{W}_1$. Thus, DBN can learn a parametric nonlinear mapping from input $\mathbf{x}$ to output $\mathbf{h}$, $f : \mathbf{x} \rightarrow \mathbf{h}$. For example, for 1-layer DBN, we have $\mathbf{h} = f_1(\mathbf{x}) = \text{logistic}(\mathbf{W}_1^T[\mathbf{x}, 1])$, where we extend $\mathbf{x} \in \mathbb{R}^d$ into $[\mathbf{x}, 1] \in \mathbb{R}^{(d+1)}$ in order to handle bias in the non-linear mapping. Note that we use the logistic function from layer 1 to $L - 1$, and the top layer is a linear mapping with weight $\mathbf{W}$ in our deep neural network.

After initializing all weights in the deep neural network, we use the independent label learning by minimizing $\lambda_1 \sum_{i=1}^{N} \sum_{t=1}^{T_i} \| \underbrace{f_L \circ f_{L-1} \circ \cdots \circ f_1}_{L \text{ times}}(\mathbf{x}_{i,t}) - \mathbf{y}_{i,t} \|$

with L-BFGS (backpropagation is used to compute sub-gradient w.r.t. weight in each layer) to fine-tune all the weights in the deep neural network. More specifically, to learn the initial weights in the deep network, we think each instance $\mathbf{x}_{i,t} \in \mathbf{x}_i$ has its corresponding label $\mathbf{y}_{i,t} \in \mathbf{y}_i$ independently. Then, the parameters can be finetuned with backpropagation [15]. Note that it does not leverage the context information in this stage, and we will show the independent label learning step is helpful to boost the recognition accuracy in the experiments. Finally, we will update the parameters θ and ω in an online fashion simultaneously, which will be introduced in the following parts.

Learning: After we initialize the deep CRFs with independent learning and re-randomization, we need to minimize the final objective function $\mathcal{L}(D; \theta, \omega)$ in Eq. (3). Because we introduce the deep neural network here for feature learning, the objective is not convex anymore. However, we can find a local minimum in Eq. (3). In our learning framework, we optimize the objective function with an online learning algorithm, by mixing perceptron training and stochastic gradient descent.

Firstly, we can calculate the (sub)gradients w.r.t. all parameters. Considering different regularization methods for θ and ω respectively, we can calculate gradients w.r.t. them separately. As for the parameters in the negative log likelihood in Eq. 3, we can compute the gradients w.r.t. θ as follows

$$\frac{\partial \mathcal{L}}{\partial \mathbf{A}} = \sum_{i=1}^{N} \sum_{t=2}^{T_i} \mathbf{y}_{i,t-1}(\mathbf{y}_{i,t})^T - \boldsymbol{\gamma}_{i,t-1}(\boldsymbol{\gamma}_{i,t})^T; \tag{7a}$$

$$\frac{\partial \mathcal{L}}{\partial \boldsymbol{\pi}} = \sum_{i=1}^{N} (\mathbf{y}_{i,1} - \boldsymbol{\gamma}_{i,1}); \tag{7b}$$

$$\frac{\partial \mathcal{L}}{\partial \boldsymbol{\tau}} = \sum_{i=1}^{N} (\mathbf{y}_{i,T_i} - \boldsymbol{\gamma}_{i,T_i}); \tag{7c}$$

$$\frac{\partial \mathcal{L}}{\partial \mathbf{b}} = \sum_{i=1}^{N} \left(\sum_{t=1}^{T_i} (\mathbf{y}_{i,t} - \boldsymbol{\gamma}_{i,t}) \right); \tag{7d}$$

$$\frac{\partial \mathcal{L}}{\partial \mathbf{W}} = \sum_{i=1}^{N} \sum_{t=1}^{T_i} \left(\mathbf{h}_{i,t}(\mathbf{y}_{i,t} - \boldsymbol{\gamma}_{i,t})^T \right.$$
$$\left. + \lambda_1 f_{1 \rightarrow (L-1)}(\mathbf{x}_{i,t})(\mathbf{y}_{i,t} - \hat{\mathbf{y}}_{i,t})^T \right) \tag{7e}$$

where $\boldsymbol{\gamma}_{i,t} \in \mathbb{R}^K$ is the vector of K dimensions, which can be thought as the posterior probability for labels in the sequence and will be introduced in Sect. 2.3, and $\hat{\mathbf{y}}_{i,t} = f_{1 \to L}(\mathbf{x}_{i,t})$ is the output from Eq. (6). Note that it is easy to derive the gradients of the ℓ_2 regularization term w.r.t. $\boldsymbol{\theta}$ in the objective in Eq. (3), which can be added to the gradients in Eq. (7).

As for the gradients of weights $\boldsymbol{\omega} = \{\mathbf{W}_l | l \in [1, .., L-1]\}$, we first use backpropagation to get the partial gradient in the neural network, refer to [15] for more details. Then the gradient of the ℓ_1 term in Eq. (3) can be attached to get the final gradients w.r.t. $\mathbf{W}_l$ for $l = \{1, .., L-1\}$.

Finally, we use a mixture of perceptron learning and stochastic gradient descent to optimize the objective function. There are various optimization methods, such as L-BFGS [3,37], stochastic gradient descent (SGD) and perceptron-based learning [28]. L-BFGS as a gradient descent method, has been widely used to optimize weights in the deep structure [16]. However, it can be slow, and there are no guarantees if there are multiple local minima in the error surface. SGD and perceptron training both are the online learning algorithms by updating the parameters using the gradient induced by a single time series, so they have significant computational advantages over L-BFGS. Furthermore, perceptron-based online learning can be viewed as a special case of SGD, but it is more flexible than SGD on parameter updating (i.e. parallelization). In our experiments, we tried L-BFGS, but it can be easily trapped into the bad local minimum, and performs worse than other optimization methods in almost all experiments. Thus, in this work, we use perceptron-based learning for the CRF related parameters and stochastic gradient descent for the weights in the deep neural structure.

If the perceptron incorrectly classifies a training example, each of the input weights is nudged a little bit in the right direction for that training example. In other words, we only need to update the CRF model only for frames that are misclassified in each training example. To update the CRF's parameters, we need to find the most violated constraints for each example. Basically, given a training example $\langle \mathbf{x}_i, \mathbf{y}_i \rangle$, we infer its most violated label $\mathbf{y}_i^*$. If the frame is misclassified, then it directly performs a type of stochastic gradient descent on the energy gap between the observed label sequence and the predicted label sequence. Otherwise, we do not need to update the model parameters. Thus, for the parameters $\boldsymbol{\theta}$ from the negative log likelihood in Eq. (3), we first project $\mathbf{x}_i$ into the code $\mathbf{h}_i$ according to Eq. (5). Then, the updating rule takes the form below

$$\boldsymbol{\theta} \leftarrow \boldsymbol{\theta} + \eta_{\boldsymbol{\theta}} \frac{\partial}{\partial \boldsymbol{\theta}} \left(E(\mathbf{h}_i, \mathbf{y}_i) - E(\mathbf{h}_i, \mathbf{y}_i^*) \right) \tag{8}$$

where $\mathbf{y}_i^*$ is the most violated constraint in the misclassificated case, and $\eta_{\boldsymbol{\theta}}$ is the step size. Note that the posterior probability $\boldsymbol{\gamma}_{i,t} \in \mathbb{R}^K$ in Eq. (7) should be changed into the hard label assignment $\mathbf{y}_{i,t}^*$ in the inference stage. Note that this is the key difference between perceptron training (using hard label) and SGD (using label likelihood) while updating parameters.

While for the weights $\boldsymbol{\omega} = \{\mathbf{W}_l | l \in [1, .., L-1]\}$ in the deep neural network, we first use backpropagation to compute the gradients, and then update it as follows

$$\omega \leftarrow \omega - \eta_\omega \frac{\partial \mathcal{L}}{\partial \omega} \tag{9}$$

where η_ω is the learning rate to update the parameters. Note that one vital step before our online learning over the parameters is that we re-randomize the top layer weight in the deep neural network. We note that this randomization step is very important for the model to generalize well in the testing data. Note that in the independent learning step, we already have trained a very good model. However, it may overfit to the data and trap into a bad local minimum. For example, the parameters in all layers have fitted the data well. Thus, if we only introduce additional CRF parameters (i.e. the transition matrix) into our whole framework, it may have no chance to update the low lever weights in the neural network (because of overfitting). On the contrary, if we re-randomize the top layer weight, we can update all parameters effectively.

2.3 Inference

Given the observation $\mathbf{x}_i = \{\mathbf{x}_{i,1}, ..., \mathbf{x}_{i,T_i}\}$, we first use Eq. (5) to compute the non-linear code $\mathbf{h}_i = \{\mathbf{h}_{i,1}, ..., \mathbf{h}_{i,T_i}\}$, and then we use Viterbi algorithm [36] to infer labels. To simplify the problem, we assume the first-order CRFs here. To estimate the parameters θ, there are two main inferential problems that need to be solved during learning: (1) the posterior probability (or the marginal distribution of a label given the codes) $\gamma_{i,t}(k) = p(y_{i,t} = k | \mathbf{h}_{i,1}, ..., \mathbf{h}_{i,T_i})$; (2) the distribution over a label edge $\xi_{i,t}(j, k) = p(y_{i,t} = j, y_{i,t+1} = k | \mathbf{h}_{i,1}, ..., \mathbf{h}_{i,T_i})$. Apparently, the inference problem can be solved efficiently with Viterbi algorithm [2,36].

For the given hidden sequence $\mathbf{h}_i = \{\mathbf{h}_{i,1}, ..., \mathbf{h}_{i,T_i}\}$, we assume the corresponding states $\{q_{i,1}, ..., q_{i,T_i}\}$. Furthermore, we define the forward messages $\alpha_{i,t}(k) \propto p(y_{i,1}, .., y_{i,t}, q_{i,t} = k | \mathbf{h}_{i,1}, ..., \mathbf{h}_{i,T_i})$, and the backward messages $\beta_{i,t}(k) \propto p(y_{i,t+1}, .., y_{i,T_i} | q_{i,t} = k, \mathbf{h}_{i,1}, ..., \mathbf{h}_{i,T_i})$

$$\alpha_{i,t+1}(j) = \left[\sum_{k=1}^{K} \alpha_{i,t}(k) A_{kj} \right] B(j, y_{i,t+1}); \tag{10}$$

$$\beta_{i,t}(j) = \sum_{k=1}^{K} A_{jk} B(k, y_{i,t+1}) \beta_{i,t+1}(k); \tag{11}$$

where $B(k, y_{i,t})$ is the probability to emit $y_{i,t}$ at the state k. We can compute it as follows

$$B(:, y_{i,t}) = exp\{\mathbf{h}_{i,t}^T \mathbf{W} + \mathbf{b}^T + \lambda_1 f_L(\mathbf{h}_{i,t})\} \tag{12}$$

After calculating $\alpha_{i,t+1}(j)$ and $\beta_{i,t}(j)$, we can compute the marginal probability for $\gamma_{i,t}$ and $\xi_{i,t}$ respectively

$$\gamma_{i,t}(k) \propto \alpha_{i,t}(k) \beta_{i,t}(k), \tag{13}$$

$$\xi_{i,t}(k, j) \propto \alpha_{i,t}(k) A_{kj} B(j, y_{i,t+1}) \beta_{i,t+1}(j); \tag{14}$$

Then, we can compute $\gamma_{i,t}$ in Eq. (7), which is the concatenation: $[\gamma_{i,t}(1), ..., \gamma_{i,t}(K)]$.

In the testing stage, the main inferential problem is to compute the most likely label sequence $\mathbf{y}^*_{1,...,T}$ given the data $\mathbf{x}_{1,...,T}$ by $\mathrm{argmax}_{\mathbf{y}'_{1,...,T}} p(\mathbf{y}'_{1,...,T}|\mathbf{x}_{1,...,T})$, which can be addressed similarly using the Viterbi algorithm mentioned above.

The learning algorithm is shown in Algorithm 1. The steps 2 and 3 are the pretraining processes (independent training and re-randomization) in our model, which is different from traditional deep CRFs [9,28,33]. There two steps initialize the model parameters, which offers the advantages over traditional deep CRFs approaches and makes our model to yield significantly better results, see further in the experimental parts. Also our online mixture learning strategy is different from traditional approaches.

Algorithm 1. Deep CRFs with online learning

Input: sequential training data $D = \{\langle \mathbf{x}_i, \mathbf{y}_i \rangle\}_{i=1}^N$, C, λ_1 and λ_2, η_ω, η_θ, iterations T
Output: ω and θ

1: Initialize the deep neural networks layer by layer with RBMs $\omega = \{\mathbf{W}_l | l \in [1, .., L-1]\}$ and the top layer weight $\mathbf{W}$;
2: Fine-tune the deep neural network parameters (ω and $\mathbf{W}$) without context information, just like to learn a deep SVM classifier;
3: Re-randomize the top layer weight $\mathbf{W}$;
4: Learn all parameters of our deep CRFs online, including ω, $\mathbf{W}$ and correlation matrix $\mathbf{A}$;
5: Return ω, $\mathbf{W}$ and $\mathbf{A}$;

---

3   Experiments

To test our method, we compared our method to the state of the art approaches and performed experiments on four sequential labeling tasks: (1) optical character recognition, (2) labeling questions and answers, (3) protein secondary structure prediction, and (4) part-of-speech tagging. Below, we described the datasets we used and also the parameter setting in the experiments.

3.1   Data Sets

1. The OCR dataset [42] contains data with 6, 877 handwritten words, in which there are total 55 unique words, and each word $\mathbf{x}_i$ is represented as a series of handwritten characters $\{\mathbf{x}_{i1}, ..., \mathbf{x}_{i,T_i}\}$. The data consists of a total of 52, 152 characters (i.e., frames), with 26 unique classes. Each character is a binary image of size 16×8 pixels, leading to a 128-dimensional binary feature vector.
2. The FAQ data set [30] contains data of 48 files on questions and answers, with a total of 55,480 sentences (i.e., frames). Each sentence is a 24- dimensional

binary feature that describes lexical characteristics of the sentence. Each sentence in the FAQ data set belongs to one of four labels: (1) question, (2) answer, (3) header, or (4) footer.

3. The CB513 contains amino acid structures of 513 proteins [6], and has been widely used for protein secondary structure prediction. For each of the proteins, it has 20-dimensional position-specific score matrix features. In the experiment, we concatenate the features from the surrounding 13 frames into the 260 dimensional vector [28]. As common in protein secondary structure prediction, we convert the eight-class labeling into a three-class labeling. The resulting data set has 513 sequences with total 74, 874 frames (260 dimensions), belonging to 3 classes.

4. The Penn Treebank corpus[1] has 74, 029 sentences with a total of 1, 637, 267 words. The whole data set contains 49, 115 unique words, and each word in each sentence is labeled according to its part of speech with total 43 different tags. To represent each word, all features are measured in a window with width 3 around the current word, which leads to a total of 212, 610 features. If we use 1000 hidden nodes, then we need to store 2×10^8 parameters in the one-layer neural network. Considering the high storage demanding for the personal computer, we calculated the frequency for each dimension in the total 212, 610 features, and selected the most frequent 5000 features as our codebook. Then we can represent each word with 5000 dimensions in our experiment. All the four data sets in [28] are available on the author's website[2].

3.2 Experimental Setup

In our experiments, we randomly initialized the weight $\mathbf{W}$ by sampling from the normal Gaussian distribution, and all other parameters in θ to be zero (i.e. biases b and c, and the transition matrix $\mathbf{A}$ all to be zero). As for $\omega = \{\mathbf{W}_l | l \in [1, .., L-1]\}$, we initialized them with DBN, which had been mentioned before. As for the number of layers and the number of hidden units in each layer, we set differently according to the dimensionality for different datasets. In all the experiments, we use the 3-layer deep neural networks on the four datasets. Considering the OCR dataset has 128 dimensional binary feature, while FAQ is the dataset with 24 dimensional vector, we set the number of hidden nodes [100 100 64] in each layer respectively on both the OCR dataset and FAQ dataset. For the CB513 dataset, we set the number of hidden units to be [400 200 100]. For the treebank dataset, the hidden units [1000 400 200] are used in the 3-layer network. We did not try other deep structure in the experiments.

After weight initialization with independent learning, we set $\lambda_1 = 0$ and then used perceptron training [12] to estimate the CRF related parameters and SGD to learn the weights in the deep neural network. We did not use regularization by setting $\lambda_2 = 0$ in perceptron learning, and set $\lambda_3 = 2 \times 10^{-4}$ for SGD weights in the deep network. From the base step size, we computed parameter-specific step sizes η_θ and η_ω as suggested by [12]. For each dataset, we divided

---

[1] www.cis.upenn.edu/~treebank.

[2] https://github.com/ganggit/deepCRFs.

it into 10 folds (9 folds as the training set, and the rest as the testing/validation set), and performed 100 full sweeps through the training data, to update the model parameters. We tuned the base step size based on the error on a small held-out validation set. Unless otherwise indicated, we use the average generalization error to measure all methods in 10-fold cross-validation experiments.

3.3 Results

We test our method on the four data sets mentioned above with the second-order CRFs with deep learning, and compare our method with linear CRFs, deep neural networks, traditional deep CRFs and LSTM. We also test whether the pretraining step: independent learning and re-randomization of the top layer weight is helpful or not in the sequential labeling tasks.

Table 1. The experimental comparisons on the OCR dataset. Our method (without pretraining) is from the result without steps 2 and 3 in Algorithm 1. The results reveal the merits of our method over other methods.

Hand-written recognition (Error rate %)	
Linear-chain CRF [9]	14.2
Max-margin Markov net [9]	13.4
Searn [8]	9.09
SVM + CRF [17]	5.76
Deep learning (DBN+Labeling) [16]	4.0
NeuroCRFs (Deep learning + CRFs)[9]	4.44
Cond. graphical models [34]	2.7
LSTM [19]	0.40
Bidirectional LSTM [19]	0.36
Hidden-unit CRF [28]	1.9
Our method (without pretraining)	1.56
Our method	**0.2**

In Table 1, we compared the performance of our method with the performance of competing models on the handwriting recognition task. It shows that the pretraining stage (independent learning and re-randomization) in our model is helpful to improve the recognition accuracy (boosting error rate from 1.56 % to 0.2 %). It also shows that the label correlation is helpful in this case. For example, the deep learning without label correlation yields accuracy 4.0 %, while is significantly lower than our model. Compared to previous deep CRFs and RNNs, our method with all steps in Algorithm 1 yields a generalization error of 0.2 %, while the best performance of other methods is 0.36 %. Note that we change the LSTM code a little bit in [19] to handle the handwritten images.

Table 2. The comparison (generalization errors) on the FAQ dataset using different methods. It shows that our method is significantly better than the Hidden-unit CRF.

Model	Error rate (%)
Linear SVM	9.87
Linear CRF [28]	6.54
NeuroCRFs (Deep learning + CRFs)[9]	6.05
Hidden-unit CRF [28]	4.43
Deep learning (DBN + Labeling) [16]	7.75
Our method (without pretraining)	7.44
Our method	**3.34**

It demonstrates that our learning approach is significantly better than other methods, and the deep structure is definitely helpful than the shallow models.

On the FAQ data set, the lowest generalization error of hidden-unit CRFs is 4.43 %, compared to 3.34 % for our method in Table 2. And again, our method outperforms other competitive baselines. It also shows that the CRF with deep feature learning (3 layers) in this case, is better than the one hidden layer CRF. Note that we just used the original 24 dimension features in the experiment, instead of extending the feature set into a $24 + 242 = 600$-dimensional feature representation in [28].

We also test our method on the protein secondary structure prediction task. The results of these experiments are presented in Table 3. In particular, our method achieves a generalization error of only 3.16 %, compared to 19.5 % error with the conditional neural field on the CB513 data set. The results presented

Table 3. The comparison on the CB513 dataset for protein secondary structure prediction task. It demonstrates that our method significantly outperforms other approaches. And the pretraining step with the independent label learning is very helpful to boost the performance.

Model	Error rate (%)
PSIRED [18]	24.0
SVM [14]	23.4
SPINE [10]	23.2
YASSP [20]	22.2
Cond. neural field (Deep learning + CRFs)[33]	19.5
NeuroCRFs (Deep learning + CRFs)[9]	28.4
Hidden-unit CRF [28]	20.2
Deep learning (DBN+Labeling) [16]	8.57
Our method (without pretraining)	27.1
Our method	**3.16**

Table 4. The experimental comparison on the treebank data set by varying the number of training data. It demonstrates that given few training data, our method is generalized well and more robust in the recognition task.

Model	generalization error rate (%)						
	1000	2000	4000	5000	8000	10000	20000
Linear SVM	12.22	12.0	10.6	9.33	8.96	8.71	8.27
Deep learning [15]	58.73	11.4	9.44	9.28	8.36	8.17	7.60
Hidden-unit CRF [28]	10.2	8.74	7.59	7.45	7.02	6.79	6.29
Our method	**9.79**	**7.97**	**6.66**	**6.5**	**6.26**	**6.24**	**6.01**

in the figure indicate that the CRFs with deep feature learning can significantly improve the performance, compared to hidden-unit CRFs.

Lastly, we also tested our method on part-of-speech tagging task. Note that we already take context information into consideration by using a window width 3 for feature representations. And the final representation is based on only 5,000 codebooks because of storage problem for model parameters. To test whether our method can tackle overfitting problem effectively, we randomly sampled a subset from the Penn Treebank corpus, and did the 10 fold cross validation. We show the experimental results in Table 4. It demonstrates that when there's a few data set available for training, deep learning with L-BFGS has overfitting problems. As the number of training data increasing, the performance of the deep learning also is increasing. While our method outperforms other baselines remarkably, and show stable and better performance with increasing training data. It also shows that our method can generalize well effectively, and it is more robust with few training data in the recognition task.

4   Related Work

Deep learning has significantly improvement on classification, such as object recognition, handwriting and natural language processing [5,19,22]. However, to predict structured output with deep learning is still a challenge in machine learning [31]. The difficulty of this problem is that the input and output data have non-determinated length, which may lead to an exponential number of possible configurations. Recently, a conditional RMBs is proposed for structured output prediction [31]. Unfortunately, the model is shallow with only one hidden layer, and also cannot deal with large output configurations well. Typically, it either considers a small output space or uses semantic hashing in order to efficiently compute a small set of possible outputs to predict. LSTM has attracted great attention for sequential labeling. For example, a 2-branch LSTM model has been proposed for machine translation [40]. Also LSTM is combined with CNN for image to text task [19].

On the other hand, graphic models, such as hidden Markov model (HMM) and CRFs have been an popular method for segmentation and labeling time series data [24]. And CRFs, as a discriminative probabilistic model for structured prediction [24], has been widely used in natural language processing [38], handwriting recognition [28,42] and scene parsing [39]. Over the last decade, many different approaches have been proposed to improve its performance on the sequential labeling problems. One trend is to extend the linear CRFs into the high-order graphical model, by exploiting more context information [7,21]. However, the main weakness of those approaches is the time-consuming inference in the high-order graphical model. Another trend in the CRFs is to discover discriminative features to improve classification performance. One related work is a multilayer CRF (ML-CRF) [35]. The system uses a multilayer perceptron (MLP), with one layer of hidden units, with a linear activation function for the output layer units and a sigmoid activation function for the hidden layer units. Similarly, hidden-unit CRFs [28,33] also assumes one-hidden layer for feature representation. The main idea of these two methods is similar to our approach here, in that we also transform the input to construct hidden features from the data so that these hidden units are discriminative in classification. But, unlike those systems, our model inherits the advantages of deep learning, and feature functions do not have any direct interpretation and are learned implicitly. Moreover, the deep features learned with large hidden units are powerful enough to represent the data, and generalize well in the classification tasks. As demonstrated by previous work, the performance of linear CRFs on a given task is strongly dependent on the feature representations [41]; while deep learning [16] can learn representations that are helpful for classification. Thus, it is possible to unify these two methods into one framework.

Our sequential labeling model with deep learning also bears some resemblance to approaches that train a deep network, and then train a linear CRF or Viterbi decoder on the output of the resulting network [9,29,32]. However, these methods differ from our approach in that (1) The initialization step in our approach with independent learning and weight re-randomization can significantly boost performance; (2) they do not learn all state-transition, data-dependent parameters and weights in the deep networks jointly. As a result, the top hidden units in these models may not discover latent distributed representations that are discriminative for classification. (3) Previous approaches [27,35] does not take an online learning strategy to estimate model parameters. But we consider to update the weights with an online algorithm in our deep learning model, which can learn more useful representations [1] to handle large scale dataset. Our work here inherits both advantages of CRFs and deep learning. Thus, our model can effectively handle structured prediction, and also learn discriminative features automatically for better sequential labeling under an unified framework.

5   Conclusions

In this paper, we introduce a model for sequential labeling with online deep CRFs. More specifically, we propose a mixture objective function, which learns the non-linear features with deep learning and predict labels with CRFs in the sequential data. Hence, our approach leverage both feature learning and context information for the classification and segmentation of time series. One vital issue arising while training the deep model is how to handle overfitting and local bad minimum. We take an effective initialization step (with dependent label learning and re-randomization) to address this problem. Finally, we use a simple but efficiently online learning method to update the whole model (end-to-end), which has the potential to handle large-scale learning problem. In the experiments, we show that our model outperforms the current state of the art remarkably on a wide range of tasks.

References

1. Bengio, Y., Courville, A., Vincent, P.: Representation learning: a review and new perspectives. PAMI **35**, 1798–1828 (2013)
2. Bishop, C.M.: Pattern Recognition and Machine Learning. Springer, Secaucus (2006)
3. Byrd, R.H., Lu, P., Nocedal, J., Zhu, C.: A limited memory algorithm for bound constrained optimization. SIAM J. Sci. Comput. **16**(5), 1190–1208 (1995)
4. Chen, G.: Deep transductive semi-supervised maximum margin clustering. CoRR abs/1501.06237 (2015)
5. Chen, G., Srihari, S.N.: Removing structural noise in handwriting images using deep learning. In: ICVGIP 2014, pp. 28:1–28:8. ACM, New York (2014)
6. Cuff, J.A., Barton, G.J.: Evaluation and improvement of multiple sequence methods for protein secondary structure prediction. PSFG **34**, 508–519 (1999)
7. Cuong, N.V., Ye, N., Lee, W.S., Chieu, H.L.: Conditional random field with high-order dependencies for sequence labeling and segmentation. JMLR **15**(1), 981–1009 (2014)
8. Daumé III, H., Langford, J., Marcu, D.: Search-based structured prediction. Mach. Learn. **75**(3), 297–325 (2009)
9. Do, T.M.T., Artires, T.: Neural conditional random fields. In: AISTATS, vol. 9, pp. 177–184 (2010)
10. Dor, O., Zhou, Y.: Achieving 80 % ten-fold cross-validated accuracy for secondary structure prediction by large-scale training (2007)
11. Erhan, D., Bengio, Y., Courville, A., Manzagol, P.A., Vincent, P., Bengio, S.: Why does unsupervised pre-training help deep learning? J. Mach. Learn. Res. **11**, 625–660 (2010)
12. Gelfand, A., Chen, Y., van der Maaten, L., Welling, M.: On herding and the perceptron cycling theorem. In: NIPS, pp. 694–702 (2010)
13. Graves, A., Schmidhuber, J.: Offline handwriting recognition with multidimensional recurrent neural networks. In: NIPS, pp. 545–552. Curran Associates, Inc. (2008)
14. Kim, H., Park H.: Protein secondary structure prediction based on an improved support vector machines approach (2003)
15. Hinton, G.E., Salakhutdinov, R.R.: Reducing the dimensionality of data with neural networks. Science **313**(5786), 504–507 (2006)
16. Hinton, G.E., Osindero, S., Teh, Y.W.: A fast learning algorithm for deep belief nets. Neural Comput. **18**(7), 1527–1554 (2006)
17. Hoefel, G., Elkan, C.: Learning a two-stage svm/crf sequence classifier. In: CIKM, pp. 271–278. ACM (2008)

18. Jones, D.T.: Protein secondary structure prediction based on position-specific scoring matrices. J. Mol. Biol. **292**, 195–202 (1999)
19. Karpathy, A., Li, F.F.: Deep visual-semantic alignments for generating image descriptions. In: CVPR, pp. 3128–3137. IEEE (2015)
20. Karypis, G.: Yasspp: Better kernels and coding schemes lead to improvements in protein secondary structure prediction (2006)
21. Krähenbühl, P., Koltun, V.: Efficient inference in fully connected CRFs with gaussian edge potentials. In: NIPS (2011)
22. Krizhevsky, A., Sutskever, I., Hinton, G.E.: Imagenet classification with deep convolutional neural networks. In: NIPS (2012)
23. Lafferty, J., Zhu, X., Liu, Y.: Kernel conditional random fields: representation and clique selection. In: ICML, p. 64. ACM (2004)
24. Lafferty, J.D., McCallum, A., Pereira, F.C.N.: Conditional random fields: probabilistic models for segmenting and labeling sequence data. In: ICML, pp. 282–289 (2001)
25. Larochelle, H., Bengio, Y.: Classification using discriminative restricted boltzmann machines. In: ICML, pp. 536–543. ACM, New York (2008)
26. Larochelle, H., Mandel, M., Pascanu, R., Bengio, Y.: Learning algorithms for the classification restricted Boltzmann machine. J. Mach. Learn. Res. **13**(1), 643–669 (2012)
27. Lecun, Y., Bottou, L., Bengio, Y., Haffner, P.: Gradient-based learning applied to document recognition. Proc. IEEE **86**(11), 2278–2324 (1998)
28. van der Maaten, L., Welling, M., Saul, L.K.: Hidden-unit conditional random fields. In: AISTATS, pp. 479–488 (2011)
29. Chen, G., Li, Y., Srihari, S.N.: Word recognition with deep conditional random fields. In: ICIP (2016)
30. McCallum, A., Freitag, D., Pereira, F.C.N.: Maximum entropy markov models for information extraction and segmentation. In: ICML, pp. 591–598 (2000)
31. Mnih, V., Larochelle, H., Hinton, G.E.: Conditional restricted boltzmann machines for structured output prediction. In: UAI, pp. 514–522 (2011)
32. Mohamed, A.-r., Dahl, G.E., Hinton, G.E.: Deep belief networks for phone recognition. In: NIPS Workshop on Deep Learning for Speech Recognition and Related Applications (2009)
33. Peng, J., Bo, L., Xu, J.: Conditional neural fields. In: NIPS, pp. 1419–1427 (2009)
34. Pérez-Cruz, F., Ghahramani, Z., Pontil, M.: Conditional Graphical Models (2007)
35. Prabhavalkar, R., Fosler-Lussier, E.: Backpropagation training for multilayer conditional random field based phone recognition. In: ICASSP 2010, pp. 5534–5537 (2010)
36. Rabiner, L.R.: A tutorial on hidden markov models and selected applications in speech recognition. Proc. IEEE **77**(2), 257–287 (1989)
37. Rasmussen, C.E., Williams, C.K.I.: Gaussian Processes for Machine Learning. The MIT Press, Cambridge (2005)
38. Sha, F., Pereira, F.: Shallow parsing with conditional random fields. In: NAACL, pp. 134–141 (2003)
39. Shotton, J., Winn, J., Rother, C., Criminisi, A.: *TextonBoost*: joint appearance, shape and context modeling for multi-class object recognition and segmentation. In: Leonardis, A., Bischof, H., Pinz, A. (eds.) ECCV 2006. LNCS, vol. 3951, pp. 1–15. Springer, Heidelberg (2006). doi:10.1007/11744023_1
40. Sutskever, I., Vinyals, O., Le, Q.V.: Sequence to sequence learning with neural networks. In: NIPS (2014)

41. Sutton, C., Mccallum, A.: Introduction to Conditional Random Fields for Relational Learning. MIT Press, Cambridge (2006)
42. Taskar, B., Guestrin, C., Koller, D.: Max-margin markov networks. In: NIPS. MIT Press (2003)
43. Vincent, P., Larochelle, H., Lajoie, I., Bengio, Y., Manzagol, P.A.: Stacked denoising autoencoders: Learning useful representations in a deep network with a local denoising criterion. J. Mach. Learn. Res. **11**, 3371–3408 (2010)
44. Weston, J., Ratle, F.: Deep learning via semi-supervised embedding. In: International Conference on Machine Learning (2008)

Warped Matrix Factorisation for Multi-view Data Integration

Naruemon Pratanwanich[1,2,3](✉), Pietro Lió[3], and Oliver Stegle[1,2](✉)

[1] European Molecular Biology Laboratory, European Bioinformatics Institute,
Wellcome Genome Campus, Hinxton, Cambridge CB10 1SD, UK
{np394,stegle}@ebi.ac.uk
[2] Open Targets, Wellcome Genome Campus, Hinxton, Cambridge CB10 1SD, UK
[3] Computer Laboratory, University of Cambridge, 15 JJ Thomson Avenue,
Cambridge CB3 0FD, UK

Abstract. Matrix factorisation is a widely used tool with applications in collaborative filtering, image analysis and in genomics. Several extensions of the classical model have been proposed, such as modelling of multiple related "data views" or accounting for side information on the latent factors. However, as the complexity of these models increases even subtle mismatches of the distributional assumptions on the input data can severely affect model performance. Here, we propose a simple yet effective solution to address this problem by modelling the observed data in a transformed or *warped* space. We derive a joint model of a multi-view matrix factorisation model that infers view-specific data transformations and provide a computationally efficient variational approximation for parameter inference. We first validate the model on synthetic data before applying it to a matrix completion problem in genomics. We show that our model improves the imputation of missing values in gene-disease association analysis and allows for discovering enhanced consensus structures across multiple data views The data and software related to this paper are available at https://github.com/PMBio/WarpedMF.

Keywords: Multi-view learning · Matrix factorisation · Data transformation · Side information

1 Introduction

Probabilistic matrix factorisation is a widely used tool to impute missing values in dyadic data [16,19,26]. Using these models, the unobserved entries in the data matrix can be recovered by the inner product of a (typically low-rank) representation of factors and loadings, which can be inferred from the observed entries in the data matrix. Several extensions of the classical matrix factorisation model (MF) have been considered, including multi-view approaches to combine multiple related matrix factorisation tasks as well as methods to integrate prior (side) information. Intuitively, multi-view models use a set of common latent variables to explain shared structure in multiple complementary views, thereby borrowing

© Springer International Publishing AG 2016
P. Frasconi et al. (Eds.): ECML PKDD 2016, Part II, LNAI 9852, pp. 789–804, 2016.
DOI: 10.1007/978-3-319-46227-1_49

statistical strength across datasets. A number of alternative implementations of multi-view models have been proposed, assuming different extents of sharing using a common loading matrix [3,7], or using a shared subset of the latent factors [27]. A second widely considered extension is modelling additional side information, either on the inferred factors and/or the loadings. The inclusion of such additional data can improve the recovery of the latent variables, in particular if the input matrices are spares or if the number of latent factors is large compared to the dimensionality of the observed data matrix. Existing methods use linear regression on the latent factors [12,15,21] or employ multivariate normal priors on the latent factors [1,28].

However, while in principle powerful, multi-view methods are challenging to apply in practice. This is because the underlying representation of the raw data frequently differs between views and in particular the assumption of marginal Gaussian residuals is hardly met.

To address this limitation, we here show that a simple parametric transformation of the observed data can substantially improve the performance of matrix factorisation models that span multiple views. We fit one parametric transformation for each view, assuming a common latent space representation, such that a common set of factors and loadings explain the observed data across all views. We derive an efficient variational inference scheme that scales to tens of views, each consisting of thousands of rows and columns, where view-specific transformations are estimated as part of the inference. Additionally, our model allows incorporating side information in the form of a covariance prior on either factors and/or loadings.

We first validate our model using synthetic data before applying it to a biomedical problem. We use our model to impute gene-disease associations that have been acquired from multiple complementary data sources. Our results show that learning warping functions within the matrix factorisation framework in conjunction with low-rank side information substantially outperforms previous methods.

2   Related Work

Multi-view formulations differ in the assumptions how specific latent variables are coupled between views [3,7,9,23]. In this work, we assume that all views are consistent and related to the same entities (e.g. diseases and genes), however reflect complementary sources of evidence. We require both latent factor matrices from MF to be shared, of which the inner product represents the consensus across all data sets. The ability to require such consensus structures is strongly dependent on appropriate data pre-processing steps. Several parametric and non-parametric transformations have been considered for this purpose. One objective is to decouple mean and variance relationships [8,13], for example using the BoxCox transformation [5]. Within the class of transformations, the BoxCox transformation can recover natural logarithmic, square root, and reciprocal functions. In the context of Gaussian processes (GP) regression, more general parametric transformations have been considered, for example a sum of (a small number of) step functions [25]. The parameters of these transformations can be learned jointly with the remaining GP hyper-parameters. Similar principles have also been considered for linear mixed models in statistical genetics [11], as well as for collective link prediction [6]. Moreover, there is some albeit limited work on using warping transformations in conjunction with GP-based function factorisation [22]. However, to the best of our knowledge, there are no methods that consider warping for multi-view matrix factorisation.

There are also a number of existing methods to incorporate side information within the matrix factorisation, where it is available. One approach is to place a regression-based prior that relates the side information in the form of covariates for rows and columns of the data matrix [2,12,15,21]. Scalable inference within the regression-based matrix factorisation models (RBMF-SI) can be achieved through variational approximations that assume a fully-factorised form [15]. Alternatively, side information can also be encoded as row and column covariance priors on the latent factors and loadings [28]. Inference in such models can be prohibitively expensive, mainly since naive implementations require the inversion of matrices with the same dimension as the number of rows or columns of the observed data matrix. We here show how this bottleneck can be addressed using low-rank approximations, which is similar to approaches that have been used for parameter inference in linear mixed models [17].

3 RBMF-SI

We start by briefly reviewing the standard matrix factorisation model that incorporates side information via linear regression [2,15]. In the RBMF-SI model, each entry (i,j) of the observed data matrix $\boldsymbol{Y} \in \mathbb{R}^{I \times J}$ is modelled as the inner product of two factor matrices of rank $K \ll I, J$ which are $\boldsymbol{U} \in \mathbb{R}^{I \times K}$ and $\boldsymbol{V} \in \mathbb{R}^{J \times K}$, with Gaussian distributed residuals with variance τ^{-1}. The corresponding likelihood is then:

$$p(\boldsymbol{Y}|\boldsymbol{U}, \boldsymbol{V}, \tau^{-1}) = \prod_{(i,j) \in \mathcal{O}} \mathcal{N}(Y_{ij}|\boldsymbol{U}_{i:}\boldsymbol{V}_{j:}^{\top}, \tau^{-1}), \qquad (1)$$

where $\mathcal{O}$ denotes the set of observed indices in $\boldsymbol{Y}$ and $\mathcal{N}(\cdot)$ denotes a normal distribution.

Side information $\boldsymbol{F} \in \mathbb{R}^{I \times N_F}$ and $\boldsymbol{G} \in \mathbb{R}^{J \times N_G}$ for the factors $\boldsymbol{U}$ and the loadings $\boldsymbol{V}$ respectively is incorporated as a multivariate normal prior on factors and loadings using a regression model in the prior mean:

$$p(\boldsymbol{U}|\boldsymbol{F}, \boldsymbol{A}, \sigma_{uk}^2) = \prod_{k=1}^{K} \mathcal{N}(\boldsymbol{U}_{:k}|\boldsymbol{F}\boldsymbol{A}_{:k}, \sigma_{uk}^2\boldsymbol{I}),$$

$$p(\boldsymbol{V}|\boldsymbol{G}, \boldsymbol{B}, \sigma_{vk}^2) = \prod_{k=1}^{K} \mathcal{N}(\boldsymbol{V}_{:k}|\boldsymbol{G}\boldsymbol{B}_{:k}, \sigma_{vk}^2\boldsymbol{I}), \qquad (2)$$

where I denotes the identity matrix. Here, the regression coefficient matrices $A \in \mathbb{R}^{N_F \times K}$ and $B \in \mathbb{R}^{N_G \times K}$ are shrunk using an L_2 prior with variances specific for each factor k:

$$p(A|\sigma_A^2) = \prod_{f=1}^{N_F} \prod_{k=1}^{K} \mathcal{N}(A_{fk}|0, \sigma_{Ak}^2), \; p(B|\sigma_B^2) = \prod_{g=1}^{N_G} \prod_{k=1}^{K} \mathcal{N}(B_{gk}|0, \sigma_{Bk}^2). \quad (3)$$

We will show later that by marginalising out the weights A and B, these regression-priors can be cast as linear covariance matrices derived from the side information F and G, which results in low rank covariances in case of $N_F < I$ and $N_G < J$ (see Sect. 4.1).

4   MV-WarpedMF-SI

In this section, we derive MV-WarpedMF-SI, a multi-view warped matrix factorisation model that accounts for side information (MV-WarpedMF-SI). The model unifies the inference of data transformations and matrix factorisation, performing joint inference for the model parameters of both components.

4.1   Model Description

Let $Y^n \in \mathbb{R}^{I \times J}$ be an observed data matrix for a data view n where $n = 1, \ldots, N$. An entry (i, j) from each view could for example represent an association score between a row i and a column j (e.g. gene-disease associations). Rather than modelling the observed data directly, we introduce a deterministic function that maps (warps) the observation space Y^n into a latent space $Z^n \in \mathbb{R}^{I \times J}$. In principle, any monotonic function could be used. Here, we follow [25] and consider a superposition of a (typically small) set of tanh functions (we used $T = 3$ in the experiments):

$$Z_{ij}^n = \phi_n(Y_{ij}^n) = Y_{ij}^n + \sum_{t=1}^{T} \alpha_t^n \tanh(\beta_t^n(Y_{ij}^n + \gamma_t^n)). \quad (4)$$

In this parametrization, $\alpha_t^n, \beta_t^n \geq 0$ adjust the step size and the steepness respectively, and γ_t^n adjusts the relative position of each tanh factor. We use distinct warping functions for each data view.

In the transformed data space, we assume that the data in all views can be explained by the same lower dimensional factor representation $U \in \mathbb{R}^{I \times K}$ and $V \in \mathbb{R}^{J \times K}$, where $K \ll I, J$ denotes the number of latent factors. Consequently, the latent variables capture common structure across views. Additionally, we incorporate individual row b^{r^n} and column b^{c^n} bias vectors for each view. Finally, residual variation in the latent space is modelled as multivariate normal $\varepsilon_{ij}^n \sim \mathcal{N}(0, 1/\tau^n)$, assuming view-specific residual variances $1/\tau^n$. The conditional likelihood of the transformed data Z follows as:

$$p(Z|\mu, \tau) = \prod_{n=1}^{N} \prod_{(i,j) \in \mathcal{O}^n} \mathcal{N}(Z_{ij}^n|\mu_{ij}^n, 1/\tau^n), \quad (5)$$

where $\mu_{ij}^n = U_{i:} V_{j:}^\top + b_i^{r^n} + b_j^{c^n}$ and $\mathcal{O}^n$ denotes the set of the observed indices (i, j) in view n.

Suppose that side information is available in the form of similarity matrices, $\Sigma^u \in \mathbb{R}^{I \times I}$ and $\Sigma^v \in \mathbb{R}^{J \times J}$, that indicate the relatedness over rows and columns of Y respectively. If the side information is given as a feature matrix, the similarity matrix can also be computed from these features using a suitable kernel function e.g. a linear kernel or a Gaussian kernel for real-valued features, or a Jaccard kernel for binary features.

We assume that the factor matrix and the loadings have multivariate normal priors whose covariance matrices correspond to Σ^u and Σ^v respectively:

$$p(U | \Sigma^u, \sigma_u^2) = \prod_{k=1}^{K} \mathcal{N}(U_{:k} | 0, \Sigma^u + \sigma_{uk}^2 I), \tag{6}$$

$$p(V | \Sigma^v, \sigma_v^2) = \prod_{k=1}^{K} \mathcal{N}(V_{:k} | 0, \Sigma^v + \sigma_{vk}^2 I). \tag{7}$$

The additional variance parameters σ_{uk}^2 and σ_{vk}^2 control the prior strength for each factor k of U and V respectively.

We note that there is a close relationship between employing a covariance matrix to encode side information and the use of a regression-based model on factors and their coefficients. In fact, the marginal likelihood of a regression model is a special case of our approach with a linear kernel:

$$p(U | F, \sigma_A^2, \sigma_u^2) = \prod_{k=1}^{K} \int p(U_{:k} | F A_{:k}, \sigma_{uk}^2 I) p(A_{:k} | 0, \sigma_{Ak}^2 I) \, dA_{:k}$$

$$= \prod_{k=1}^{K} \mathcal{N}(U_{:k} | 0, \sigma_{Ak}^2 F F^\top + \sigma_{uk}^2 I), \tag{8}$$

$$p(V | G, \sigma_B^2, \sigma_v^2) = \prod_{k=1}^{K} \int p(V_{:k} | G B_{:k}, \sigma_{vk}^2 I) p(B_{:k} | 0, \sigma_{Bk}^2 I) \, dB_{:k}$$

$$= \prod_{k=1}^{K} \mathcal{N}(V_{:k} | 0, \sigma_{Bk}^2 G G^\top + \sigma_{vk}^2 I). \tag{9}$$

Finally, in order to avoid overfitting, we regularise the bias parameters for row and column bias terms by a zero mean and a variance prior over each element:

$$p(b^r) = \prod_{n=1}^{N} \prod_{i=1}^{I} \mathcal{N}(b_i^{r^n} | 0, 1/\tau^{r^n}), \; p(b^c) = \prod_{n=1}^{N} \prod_{j=1}^{J} \mathcal{N}(b_j^{c^n} | 0, 1/\tau^{c^n}) \tag{10}$$

Figure 1 shows a graphical model of MV-WarpedMF-SI, representing the relationships of all variables in the model.

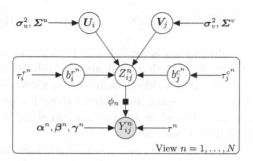

Fig. 1. Graphical model representation of MV-WarpedMF-SI. Nodes inside the rectangular plate correspond to view-specific variables. All remaining variables are shared across views. Observed variables are shaded in grey.

4.2   Training the MV-WarpedMF-SI

We need to make inference of the joint posterior distribution $p(\boldsymbol{U}, \boldsymbol{V}, \boldsymbol{b}^r, \boldsymbol{b}^c | \boldsymbol{Y}, \theta)$, where θ denotes the set of all model parameters $(\boldsymbol{\alpha}, \boldsymbol{\beta}, \boldsymbol{\gamma}, \boldsymbol{\tau}, \boldsymbol{\tau}^r, \boldsymbol{\tau}^c, \sigma_u^2, \sigma_v^2)$. Closed-form inference in this matrix factorization model is not tractable. For efficient parameter inference, we here revert to a variational approach to approximate the true posterior over the latent variables with a factorised form. An iterative inference scheme can then be derived by minimising the Kullback-Leibler (KL) divergence between the true posterior and the factorised approximation; see for example [4] for a comprehensive overview. The parameters of the warping functions $(\boldsymbol{\alpha}, \boldsymbol{\beta}, \boldsymbol{\gamma})$ and the variances $(1/\boldsymbol{\tau}, 1/\boldsymbol{\tau}^r, 1/\boldsymbol{\tau}^c, \sigma_u^2, \sigma_v^2)$ are inferred using maximum likelihood type II, i.e. by maximising the variational lower bound.

Using a standard change of variable, we first derive the marginal log-likelihood in the observation space. This results in an additional Jacobian term evaluated at each observed data point which appears additively in the marginal log-likelihood of the latent space, leading to:

$$\log p(\boldsymbol{Y}|\theta) = \log p(\boldsymbol{Z}|\theta) + \sum_{n=1}^{N} \sum_{(i,j)\in\mathcal{O}^n} \log \phi_n'(Y_{ij}^n) \tag{11}$$

where $\phi_n'(Y_{ij}^n) = \dfrac{\partial \phi_n(y)}{\partial y}\bigg|_{Y_{ij}^n}$ is a Jacobian term.

Equivalent to minimising the KL divergence, we maximise the variational lower bound of the marginal log-likelihood conditioned on the model parameters, which is:

$$\begin{aligned}
\log p(\boldsymbol{Z}|\theta) \geq &\mathbb{E}_q[\log p(\boldsymbol{Z}|\boldsymbol{U}, \boldsymbol{V}, \boldsymbol{b}^r, \boldsymbol{b}^c, \boldsymbol{\tau})] + \mathbb{E}_q[\log p(\boldsymbol{U}|\boldsymbol{\Sigma}^u, \sigma_u^2)] \\
&+ \mathbb{E}_q[\log p(\boldsymbol{V}|\boldsymbol{\Sigma}^v, \sigma_v^2)] + \mathbb{E}_q[\log p(\boldsymbol{b}^r|\boldsymbol{\tau}^r)] \\
&+ \mathbb{E}_q[\log p(\boldsymbol{b}^c|\boldsymbol{\tau}^c)] - \mathbb{E}_q[\log q(\boldsymbol{U}, \boldsymbol{V}, \boldsymbol{b}^r, \boldsymbol{b}^c)],
\end{aligned} \tag{12}$$

where $\mathbb{E}_q[\cdot]$ denotes the statistical expectation with respect to $q(\boldsymbol{U}, \boldsymbol{V}, \boldsymbol{b}^r, \boldsymbol{b}^c)$ as defined below.

To achieve scalable inference, we assume a fully factorise variational distribution q for all latent variables except the factors U and the loadings V, for which the prior factorisation is maintained. Thus, we choose a multivariate normal distribution parameterised by a mean and a covariance matrix for each latent factor, which enables automatic relevance determination, i.e. the number of effective factors within the model can be pruned by shrinking unused factors to zero [20]. The resulting variational distribution is:

$$q(U, V, b^r, b^c) = q(U)q(V)q(b^r)q(b^c) \tag{13}$$

$$\text{where} \quad q(U) = \prod_k^K \mathcal{N}(U_{:k}|\tilde{U}_{:k}, C_k^u), \qquad q(V) = \prod_k^K \mathcal{N}(V_{:k}|\tilde{V}_{:k}, C_k^v)$$

$$q(b^r) = \prod_n^N \prod_i^I \mathcal{N}(b_i^{r^n}|\tilde{b}_i^{r^n}, s_i^{r^n}), \qquad q(b^c) = \prod_n^N \prod_j^J \mathcal{N}(b_j^{c^n}|\tilde{b}_j^{c^n}, s_j^{c^n})$$

Training of the model is done by optimising the variation lower bound and the Jacobian term with respect to each of the unknown variables including the warping parameters in turn until convergence.

4.3 Efficient Inference of Low-Rank Side Information

The computational limitation for imposing a Gaussian process prior on each latent factor is inverting the covariance matrix. The naive update equations for the covariance matrices of the variational distributions $q(U)$ and $q(V)$ are given by:

$$C_k^u = \left(\sum_{n=1}^N \sum_{j=1}^J \text{diag}\left\{ \tau^n \left(\tilde{V}_{jk}^2 + C_{k\,jj}^v \right) \mathcal{O}_{:j}^n \right\} + (\Sigma^u + \sigma_{uk}^2 I)^{-1} \right)^{-1} \tag{14}$$

$$C_k^v = \left(\sum_{n=1}^N \sum_{i=1}^I \text{diag}\left\{ \tau^n \left(\tilde{U}_{ik}^2 + C_{k\,ii}^u \right) \mathcal{O}_{i:}^n \right\} + (\Sigma^v + \sigma_{vk}^2 I)^{-1} \right)^{-1} \tag{15}$$

The matrix inversions entail cubic time complexity per iteration in the variational EM algorithm, which renders applications to larger datasets intractable. If the side information is low rank, the matrix inversion lemma can be exploited to invert the matrix efficiently, reducing the complexity to cubical scaling in the rank of the prior matrix.

We start by exploiting a standard spectral decomposition of the full covariance matrix:

$$(\Sigma + \sigma^2 I)^{-1} \simeq (PXP^\top + \sigma^2 I)^{-1} = P(X + \sigma^2 I)^{-1}P^\top, \tag{16}$$

where $X = \begin{bmatrix} x & 0 \\ 0 & 0 \end{bmatrix}$ and $PP^\top = P^\top P = I$.

More specifically, we apply single value decomposition (SVD) on the covariance to obtain a rank H approximation by forcing all remaining eigenvalues to zero, resulting in $\boldsymbol{PXP}^\top$. Using the matrix inversion lemma, the updating rule is reformed to:

$$\{\boldsymbol{D} + \boldsymbol{P}(\boldsymbol{X} + \sigma^2 \boldsymbol{I})^{-1}\boldsymbol{P}^\top\}^{-1} = \boldsymbol{D}^{-1} - \boldsymbol{D}^{-1}\boldsymbol{P}\boldsymbol{W}^{-1}\boldsymbol{P}^\top\boldsymbol{D}^{-1}, \qquad (17)$$

where $\boldsymbol{W} = \boldsymbol{X} + \boldsymbol{P}^\top(\boldsymbol{D}^{-1} + \sigma^2 \boldsymbol{I})\boldsymbol{P}$ and $\boldsymbol{D}$ is the diagonal matrix from the first part of the updating rule in Eqs. (14) and (15).

Since the eigen decomposition needs only to be performed once at initialisation, the effective computational cost per iteration is therefore dominated by calculating the inverse $\boldsymbol{W} \in \mathbb{R}^{H \times H}$, which is cubic in $H \ll I, J$.

4.4   Missing-Value Imputation with the MV-WarpedMF-SI

The trained model can be used to make predictions of missing values in the transformed space. A consensus prediction using evidence across views can be obtained by calculating $\tilde{\boldsymbol{U}}\tilde{\boldsymbol{V}}^\top$, where $\tilde{\boldsymbol{U}}$ and $\tilde{\boldsymbol{V}}$ correspond to the expected latent factors and loadings under the variational posterior respectively. For each data view, the predictive distribution for any entry in the transformed space Z_{ij}^n is a univariate normal distribution with the learned mean and variance:

$$p(Z_{ij}^n | \mathcal{M}) = \mathcal{N}(Z_{ij}^n | \tilde{\mu}_{ij}^n, \tilde{\xi}_{ij}^n), \qquad (18)$$

where $\mathcal{M}$ is the set of learned variables, $\tilde{\mu}_{ij}^n = \tilde{U}_{i:}\tilde{V}_{j:}^\top + \tilde{b}_i^{r^n} + \tilde{b}_j^{c^n}$, and $\tilde{\xi}_{ij}^n = \sum_k^K (\tilde{U}_{ik}^2 C_{k\,jj}^v + \tilde{V}_{jk}^2 C_{k\,ii}^u + C_{k\,ii}^u C_{k\,jj}^v) + s_i^{r^n} + s_j^{c^n} + 1/\tilde{\tau}^n$.

The predictive distribution in the observation space can then be obtained by reversing the warping transformation. This is done by squashing the predictive normal distribution in the latent space through the learned warping function, parameterised by $\tilde{\alpha}, \tilde{\beta}, \tilde{\gamma}$, leading to:

$$p(Y_{ij}^n | \mathcal{M}) = \phi_n'(\phi_n^{-1}(Z_{ij}^n)) \cdot \mathcal{N}(Z_{ij}^n | \tilde{\mu}_{ij}^n, \tilde{\xi}_{ij}^n). \qquad (19)$$

To compute a point estimate of a missing value, we use the predictive expectation of the warped Gaussian distribution in Eq. (19). Effectively, this operation marginalises over the latent space, integrating over all possible values through the inverse warping function ϕ^{-1} under its predictive distribution:

$$Y_{ij}^n = \int \phi_n^{-1}(Z_{ij}^n) \cdot \mathcal{N}(Z_{ij}^n | \tilde{\mu}_{ij}^n, \tilde{\xi}_{ij}^n)\, dZ_{ij}^n. \qquad (20)$$

Since we parameterise the function in the observation space, its inverse $\phi_n^{-1}(Z_{ij}^n)$ cannot be analytically computed in a closed form. However, computing the inverse function $\phi_n^{-1}(Z_{ij}^n)$ is similar to finding the root of $\phi_n(Y_{ij}^n) - Z_{ij}^n = 0$. This problem can be solved using the Newton-Raphson method, which typically converges within a few iterations. Although convergence of this method in principle depends on the initialisation, we observed that a random initialisation yields robust results in practice. Finally, we estimate the integral in Eq. (20)

by reformulating the one dimensional Gaussian distribution into the form of a Hermite polynomial. This approach allows to approximate the integral using a Gauss-Hermite quadrature, estimating the integral with a weighted sum of a relatively small number of the function evaluated at appropriate points (we used ten evaluations in the experiments).

The implementation of MV-WarpedMF-SI is available at https://github. com/PMBio/WarpedMF.

5 Results

We first applied the MV-WarpedMF-SI model on synthetic datasets to investigate its transformation capability in a multi-view setting. Subsequently we used the model for a genomic imputation task to fill in missing values and recover the consensus structure in a gene-disease prioritisation study.

5.1 Simulation Studies

We simulated synthetic data drawn from the generative model of MV-WarpedMF-SI. Firstly, we simulated covariance matrices from an inverse Wishart distribution and used them to generate latent factors U and loadings V by assuming $K = 5$ hidden factors. We then created two $1,000 \times 1,000$ data matrices with 90% missing values from the inner product of the same latent factors, $UV^\top$, corrupted with Gaussian noise, resulting in Z^1 and Z^2. To investigate to what extent the model is able to recover a data transformation, we finally created Y^1 and Y^2 by using a linear superposition of the untransformed data and a non-linear transformation, $Y^n = (1 - \lambda) \cdot Z^n + \lambda \cdot \phi_n(Z^n)$, where the parameter λ determines the intensity of the transformation and ϕ denotes an exponential and a logarithmic data transformation for the view $n = 1$ and 2 respectively. In total, we generated six datasets with a variable degree of non-linear warping. We also simulated side information regarding row and column similarities using rank $H = 10$ approximations to the true simulated covariances of U and V.

The proposed models, MV-WarpedMF and MV-WarpedMF-SI were trained on each dataset. For comparison, we also considered a standard (non-warping) multi-view matrix factorisation model (MV-MF) applied to the same data. Both Y^1 and Y^2 were modelled simultaneously by each model. For each simulated dataset, we evaluated the model performance using five-fold cross validation, calculating the correlation coefficient (R^2) between observed and predicted matrix values on the hold-out test set.

The prediction results in Fig. 2(a) show that the warped models performed markedly better than the un-warped MV-MF, where the differences were largest for strong non-linearities and the best model was the combination of learning warping function and incorporating side information (MV-WarpedMF-SI). Figure 2(b) shows a comparison of the true transformations and the warping

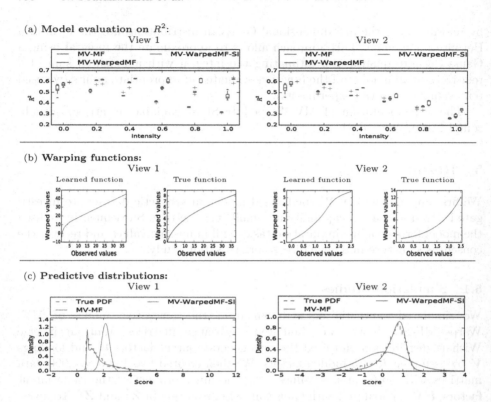

Fig. 2. Impact of the inference of warping functions in multi-view learning. Considered are the proposed MV-WarpedMF and MV-WarpedMF-SI as well as a standard multi-view matrix factorisation model (MV-MF) applied to the raw untransformed data. Box plots show the out of sample prediction accuracy (shown is variation in R^2 across the five folds in each of six datasets) for increasing degrees of non-linear distortion (a). The true generative warping functions and the parametric fits recovered by the model are shown in (b). The predictive distributions in the observed space for each view are depicted in (c).

functions inferred using MV-WarpedMF-SI. Representative examples of the predictive density for one entry of the data matrix are shown in Fig. 2(c). The warping model employed in MV-WarpedMF-SI can capture complex and asymmetric distributions, providing a substantially better approximation to the true density than a normal distribution as used in a standard MV-MF.

5.2   Analysis of Therapeutic Gene-Disease Associations

Data. Next, we applied the MV-WarpedMF-SI to a gene-disease association task. The dataset consisted of disease × gene matrices. We considered six evidence sources of therapeutic gene-disease relationships as well as the additional validation set of gene-disease associations derived from drugs in clinical trials. These data are freely available via the Open Targets platform[1]:

- ANIM, Y^1: drug effects on animal models where scores were calculated using the phenodigm similarity to human diseases [24].
- EXPR, Y^2: differential gene expression profiles of control-disease experiments from Expression Atlas[2] where scores were calculated from the p-value and $\log_2$ fold change.
- GEAS, Y^3: gene association studies in GWAS Catalog[3] which were scored by the p-value, sample size, and severity effect.
- LITR, Y^4: literature mining of scientific articles on Pubmed database[4], scoring gene-disease associations by the co-occurrence of the gene and disease terms in the same sentence.
- PATH, Y^5: evidences of pathway analysis from REACTOME[5].
- SOMU, Y^6: evidences of somatic mutation studies from COSMIC[6].
- An independent validation set of 22,138 known associations covering 372 diseases and 614 therapeutic genes, derived from ChEMBL[7], scored by drug development pipeline progression. This dataset was not included for training the models.

We also considered side information of a disease similarity matrix (Σ^u) derived from disease ontology trees [18] and a gene similarity matrix (Σ^v), which was estimated from gene expression networks [10]. To define the disease similarity covariance, we considered the inverse of the shortest path distance between diseases through the lowest common ancestor. For the gene similarity network we used the pre-computed 1,000 eigenvectors and eigenvalues of the gene-gene correlation matrix derived from 33,427 gene expression profiles [10].

In total, we constructed six matrices of 426 diseases and 10,721 gene targets, with an average of 95 % missing values. These datasets represent typical examples of evidences that differ in scale and distributional properties.

Considered Methods. We compared the following models in single-view learning, where each data view was trained and validated independently, as well as multi-view learning, where all the data views were considered simultaneously. We applied a standard matrix factorisation (MF) [14] and a regression-based MF model with side information (RBMF-SI) [15] to each view separately, both of which were trained on the raw (un-warped) data as baselines. As an additional comparison partner, we also considered preprocessing the raw data using the Box-Cox transformation before applying an MF and an RBMF-SI. We denote

---

[1] https://www.targetvalidation.org/
[2] https://www.ebi.ac.uk/arrayexpress.
[3] https://www.ebi.ac.uk/gwas.
[4] https://europepmc.org.
[5] https://www.reactome.org.
[6] https://cancer.sanger.ac.uk/cosmic.
[7] https://www.ebi.ac.uk/chembl.

Table 1. Summary of the considered methods in this work.

Model	Data transformation	Side information
Single-view learning		
MF[b] [14]	None	None
RBMF-SI[b] [15]	None	Regression-based
BoxCoxMF	Box-Cox preprocessing	None
BoxCoxRBMF-SI	Box-Cox preprocessing	Regression-based
WarpedMF[a]	Built-in warping functions	None
WarpedMF-SI[a]	Built-in warping functions	Covariance priors
Multi-view learning		
MV-MF	None	None
MV-RBMF-SI	None	Regression-based
MV-BoxCoxMF	Box-Cox preprocessing	None
MV-BoxCoxRBMF-SI	Box-Cox preprocessing	Regression-based
MV-WarpedMF[a]	Built-in warping functions	None
MV-WarpedMF-SI[a]	Built-in warping functions	Covariance priors

[a] Our proposed model variants
[b] We modified the original model by adding bias terms.

these methods as BoxCoxMF and BoxCoxRBMF-SI respectively. The Box-Cox transformation was fit for each data view independently.[8] Moreover, we applied all the models in multi-view learning, denoting them with the prefix 'MV'. Finally, the proposed model of learning warping functions during matrix factorisation was used either without (WarpedMF) or with the inclusion of side information (WarpedMF-SI), and in its multi-view form either without (MV-WarpedMF) or with side information (MV-WarpedMF-SI). Table 1 summarises the methods considered in this analysis.

Evaluation of Prediction Accuracy Using Cross Validations. We first assessed the predictive accuracy of the considered methods in terms of their ability to impute held-out values. We trained each model using a five-fold cross validation experiment, and compared the predicted scores to the true values in the hold-out test predictions using the Spearman rank correlation coefficients (Rs). Predictions from all models were assessed on the raw data scale.

While we assessed each method in terms of the imputation task by using both the latent factors and the bias terms ($\tilde{U}\tilde{V}^\top + \tilde{b}^r + \tilde{b}^c$), we also explored the alternative ability to impute gene-disease scores when considering only the inferred latent factors ($\tilde{U}\tilde{V}^\top$) without the learned bias terms. Table 2 shows the average test Rs under these two prediction schemes.

---

[8] This was done by using a SciPy library.

Table 2. Average test Rs from the five-fold cross validations using all learned variables $(\tilde{U}\tilde{V}^\top + \tilde{b}^r + \tilde{b}^c)$ are presented. In the parentheses are the average test Rs of the imputing gene-disease relationships using only the inner product of the shared latent factors $(\tilde{U}\tilde{V}^\top)$, which is considered the inferred consensus in multi-view learning.

Model	5-fold cross validation					
	ANIM	EXPR	GEAS	LITR	PATH	SOMU
Single-view learning						
MF	.76 (.28)	.71 (.22)	.89 (.27)	.60 (.26)	.94 (.26)	.84 (.51)
RBMF-SI	.60 (.08)	.59 (.20)	.84 (.26)	.39 (.02)	.76 (.22)	.81 (.20)
BoxCoxMF	.76 (.27)	.74 (.29)	**.92** (.38)	.62 (.23)	.94 (.04)	.84 (.25)
BoxCoxRBMF-SI	.77 (.38)	.75 (.46)	.87 (.37)	**.69** (.45)	**.95** (.02)	.78 (.22)
WarpedMF[a]	.77 (.32)	.75 (.49)	**.92** (**.55**)	.62 (.30)	**.95** (.38)	.84 (.60)
WarpedMF-SI[a]	**.81** (**.47**)	**.77** (**.67**)	**.92** (**.55**)	**.69** (**.52**)	**.95** (**.44**)	**.87** (**.76**)
Multi-view learning						
MV-MF	.68 (-.11)	.64 (.03)	.85 (.25)	.59 (.26)	.91 (-.13)	.80 (.18)
MV-RBMF-SI	.64 (.01)	.62 (.02)	.85 (.05)	.46 (.01)	.59 (-.06)	.79 (.08)
MV-BoxCoxMF	.72 (-.02)	.70 (.05)	.89 (.22)	.57 (.07)	.91 (**.13**)	.82 (.04)
MV-BoxCoxRBMF-SI	.70 (-.16)	.83 (.42)	.84 (.05)	.59 (.44)	.09 (.03)	.66 (.01)
MV-WarpedMF[a]	.60 (**.68**)	.69 (**.65**)	.37 (**.53**)	.52 (**.52**)	.19 (.08)	.71 (**.59**)
MV-WarpedMF-SI[a]	.75 (.36)	.72 (.43)	.89 (.22)	.61 (.50)	.90 (.09)	.81 (.38)

[a] Our proposed model variants.

For imputation performance, it is not surprising that modelling each view independently can yield better results, where the best performing model combined learning warping function within matrix factorisation with low-rank side information (WarpedMF-SI). The inclusion of side information via low-rank covariance priors (WarpedMF-SI) consistently increased prediction accuracy for all data views, whereas other methods, i.e. the linear regression based MF models (RBMF-SI and BoxCoxRBMF-SI) yielded variable performance.

When considering the inferred latent representations without the bias terms, the WarpedMF-SI model had the highest predictive performance. The proposed warped matrix factorisation models without side information (WarpedMF) was substantially more accurate than un-wapred factorisation models (MF) or the Box-Cox preprocessing models (BoxCoxMF). This is more evident in multi-view learning where the un-warped factorisation (MV-MF) and the Box-Cox preprocessing (MV-BoxCoxMF) failed to capture the consensus across views; very little structure was remained for the shared latent factors to discover. In contrast, learning warping functions in multi-view learning of the MV-WarpedMF model as well as the MV-WarpedMF-SI model maximised the mutual latent structures across views, promoting our confidence in true associations (see the next section).

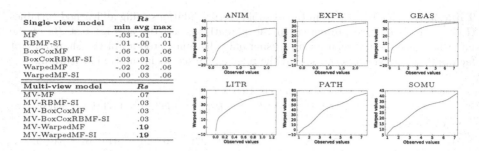

Single-view model	Rs		
	min	avg	max
MF	-.03	-.01	.01
RBMF-SI	-.01	-.00	.01
BoxCoxMF	-.06	-.00	.06
BoxCoxRBMF-SI	-.03	.01	.05
WarpedMF	-.02	.02	.06
WarpedMF-SI	.00	.03	.06
Multi-view model	Rs		
MV-MF	.07		
MV-RBMF-SI	.03		
MV-BoxCoxMF	.03		
MV-BoxCoxRBMF-SI	.03		
MV-WarpedMF	.19		
MV-WarpedMF-SI	.19		

Fig. 3. Test Rs are shown when validating with known association scores (left). Learned transformation functions inferred by MV-WarpedMF on each data set (right).

Evaluation of Consensus Discovery Using Known Associations. To further explore the benefit of the consensus discovery captured by the shared latent factors, we assessed each model using the independent out-of-sample association scores of 22,138 known gene-disease associations. Figure 3(left) shows the test correlation coefficient (Rs) obtained from each model, where the minimum, average and maximum of Rs across views are shown for single-view models. These results show that single-view learning did fail to identify true gene-disease associations, despite the strong predictive performance. Multi-view learning consistently resulted in improved performance, where the best models were the combination of warping and multi-view modelling with or without side information (MV-WarpedMF and MV-WarpedMF-SI), followed by the un-warped factorisation (MV-MF). This confirms that learning warping functions in conjunction with the parameters of matrix factorisation modelling rather than the Box-Cox preprocessing or the un-warped factorisation can capture complex transformations and in particular is an effective approach to adjust for differences in scale between views, leading to significantly improved imputation accuracies. Figure 3(right) depicts the six warping functions inferred by MV-WarpedMF-SI.

6   Conclusion

We have proposed a method to jointly infer a parametric data transformation function while performing inference in matrix factorisation models. Our approach unifies previous efforts, including models that combine data across views and the incorporation of side information. In experiments on real data, we demonstrate that learning warping functions within the matrix factorisation framework and incorporating low-rank side information yield increased accuracy for imputing missing values in single-view learning, and in multi-view learning where joint inference was made across all views. Flexible data transformations will be particularly useful if distant data types are integrated. Our experiments illustrate an example application of such a setting, where we consider gene-disease associations obtained using complementary sources of evidence. We show that learning warping functions in multi-view matrix factorisation can enhance the discovery of the shared latent structures (consensus) underlying across views.

The proposed variational inference scheme is computationally efficient and allows to incorporate side information in the form of multivariate normal (covariance) priors. Combined with suitable low-rank approximations, the proposed strategy is directly applicable to thousands of rows and columns with robust performance.

Acknowledgement. This work was supported by Open Targets. We are thankful to Dr. Ian Dunham, Dr. Gautier Koscielny, Dr. Samiul Hasan, and Dr. Andrea Pierleoni for their helpful discussion and contributions in curating and managing the data we have used for the described experiments. NP has received support from the Royal Thai Government Scholarship.

References

1. Adams, R., Dahl, G., Murray, I.: Incorporating side information in probabilistic matrix factorization with gaussian processes. In: Proceedings of the 26th Conference Annual Conference on Uncertainty in Artificial Intelligence (UAI), pp. 1–9. AUAI Press, Corvallis (2010)
2. Agarwal, D., Chen, B.C.: Regression-based latent factor models. In: Proceedings of the 15th ACM SIGKDD International Conference on Knowledge Discovery and Data Mining, pp. 19–28. ACM (2009)
3. Akata, Z., Thurau, C., Bauckhage, C.: Non-negative matrix factorization in multimodality data for segmentation and label prediction. In: 16th Computer Vision Winter Workshop (2011)
4. Bishop, C.M.: Pattern Recognition and Machine Learning. Springer, New York (2006)
5. Box, G.E., Cox, D.R.: An analysis of transformations. J. Roy. Stat. Soc. Ser. B (Methodol.) **26**(2), 211–252 (1964)
6. Cao, B., Liu, N.N., Yang, Q.: Transfer learning for collective link prediction in multiple heterogenous domains. In: Proceedings of the 27th International Conference on Machine Learning (2010)
7. Damianou, A., Ek, C., Titsias, M., Lawrence, N.: Manifold relevance determination. In: Proceedings of the 27th International Conference on Machine Learning (2012)
8. Durbin, B.P., Hardin, J.S., Hawkins, D.M., Rocke, D.M.: A variance-stabilizing transformation for gene-expression microarray data. Bioinformatics **18**(Suppl. 1), S105–S110 (2002)
9. Fang, Y., Si, L.: Matrix co-factorization for recommendation with rich side information and implicit feedback. In: Proceedings of the 2nd International Workshop on Information Heterogeneity and Fusion in Recommender Systems, pp. 65–69. ACM (2011)
10. Fehrmann, R.S., Karjalainen, J.M., Krajewska, M., Westra, H.J., Maloney, D., Simeonov, A., Pers, T.H., Hirschhorn, J.N., Jansen, R.C., Schultes, E.A., et al.: Gene expression analysis identifies global gene dosage sensitivity in Cancer. Nat. Genet. **47**(2), 115–125 (2015)
11. Fusi, N., Lippert, C., Lawrence, N.D., Stegle, O.: Warped linear mixed models for the genetic analysis of transformed phenotypes. Nat. Commun. **5** (2014)
12. Gonen, M., Kaski, S.: Kernelized bayesian matrix factorization. IEEE Trans. Pattern Anal. Mach. Intell. **36**(10), 2047–2060 (2014)

13. Kelmansky, D.M., Martínez, E.J., Leiva, V.: A new variance stabilizing transformation for gene expression data analysis. Stat. Appl. Genet. Mol. Biol. **12**(6), 653–666 (2013)
14. Kim, Y.D., Choi, S.: Scalable variational bayesian matrix factorization. In: Proceedings of the First Workshop on Large-Scale Recommender Systems (LSRS) (2013)
15. Kim, Y.D., Choi, S.: Scalable variational bayesian matrix factorization with side information. In: Proceedings of the International Conference on Artificial Intelligence and Statistics (AISTATS), Reykjavik, Iceland (2014)
16. Lim, Y.J., Teh, Y.W.: Variational Bayesian approach to movie rating prediction. In: Proceedings of KDD Cup and Workshop (2007)
17. Lippert, C., Listgarten, J., Liu, Y., Kadie, C.M., Davidson, R.I., Heckerman, D.: Fast linear mixed models for genome-wide association studies. Nat. Methods **8**(10), 833–835 (2011)
18. Malone, J., Holloway, E., Adamusiak, T., Kapushesky, M., Zheng, J., Kolesnikov, N., Zhukova, A., Brazma, A., Parkinson, H.: Modeling sample variables with an experimental factor ontology. Bioinformatics **26**(8), 1112–1118 (2010)
19. Mnih, A., Salakhutdinov, R.: Probabilistic matrix factorization. In: Advances in Neural Information Processing Systems, pp. 1257–1264 (2007)
20. Neal, R.M.: Bayesian Learning for Neural Networks, vol. 118. Springer Science & Business Media, New York (2012)
21. Porteous, I., Asuncion, A.U., Welling, M.: Bayesian matrix factorization with side information and dirichlet process mixtures. In: AAAI (2010)
22. Schmidt, M.N.: Function factorization using warped gaussian processes. In: Proceedings of the 26th Annual International Conference on Machine Learning, pp. 921–928. ACM (2009)
23. Singh, A.P., Gordon, G.J.: Relational learning via collective matrix factorization. In: Proceedings of the 14th ACM SIGKDD International Conference on Knowledge Discovery and Data Mining, pp. 650–658. ACM (2008)
24. Smedley, D., Oellrich, A., Köhler, S., Ruef, B., Westerfield, M., Robinson, P., Lewis, S., Mungall, C., et al.: Phenodigm: analyzing curated annotations to associate animal models with human diseases. Database **2013**, bat025 (2013)
25. Snelson, E., Rasmussen, C.E., Ghahramani, Z.: Warped gaussian processes. Adv. Neural Inf. Process. Syst. **16**, 337–344 (2004)
26. Takács, G., Pilászy, I., Németh, B., Tikk, D.: Scalable collaborative filtering approaches for large recommender systems. J. Mach. Learn. Res. **10**, 623–656 (2009)
27. Virtanen, S., Klami, A., Khan, S.A., Kaski, S.: Bayesian group factor analysis. In: Proceedings of the Fifteenth International Conference on Artificial Intelligence and Statistics (2012)
28. Zhou, T., Shan, H., Banerjee, A., Sapiro, G.: Kernelized probabilistic matrix factorization: exploiting graphs and side information. In: SDM, vol. 12, pp. 403–414. SIAM (2012)

Communication-Efficient Distributed Online Learning with Kernels

Michael Kamp[(✉)], Sebastian Bothe, Mario Boley, and Michael Mock

Fraunhofer IAIS, Sankt Augustin, Germany
{michael.kamp,sebastian.bothe,mario.boley,
michael.mock}@iais.fraunhofer.de

Abstract. We propose an efficient distributed online learning protocol for low-latency real-time services. It extends a previously presented
protocol to kernelized online learners that represent their models by a
support vector expansion. While such learners often achieve higher predictive performance than their linear counterparts, communicating the
support vector expansions becomes inefficient for large numbers of support vectors. The proposed extension allows for a larger class of online
learning algorithms—including those alleviating the problem above
through model compression. In addition, we characterize the quality of
the proposed protocol by introducing a novel criterion that requires the
communication to be bounded by the loss suffered.

1 Introduction

We consider the problem of distributed online learning for low-latency real-time
services [4,10]. In this scenario, a learning system of $m \in \mathbb{N}$ connected local learners provides a real-time prediction service on multiple dynamic data streams. In
particular, we are interested in generic distributed online learning protocols that
treat concrete learning algorithms as a black-box. The goal of such a protocol
is to provide, in a communication efficient way, a service quality similar to a
serial setting in which all examples are processed at a central location. While
such an optimal predictive performance can be trivially achieved by centralizing
all data, the required continuous communication usually exceeds practical limits
(e.g., bandwidth constraints [1], latency [8,21], or battery power [5,16]). Similarly, communication limits can be satisfied trivially by letting all local learners
work in isolation. However, this usually comes with a loss of service quality that
increases with the number of local learners.

In previous work, we presented a protocol that effectively reduces communication while providing strict loss bounds for a class of algorithms that perform
loss-proportional convex updates of linear models [10]. That is, algorithms that
update linear models in the direction of a convex set with a magnitude proportional to the instantaneous loss (e.g., Stochastic Gradient Descent [2], or Passive
Aggressive [3]). The protocol is able to cease communication as soon as no loss is
suffered anymore. However, for most realistic problems this cannot be achieved

© Springer International Publishing AG 2016
P. Frasconi et al. (Eds.): ECML PKDD 2016, Part II, LNAI 9852, pp. 805–819, 2016.
DOI: 10.1007/978-3-319-46227-1_50

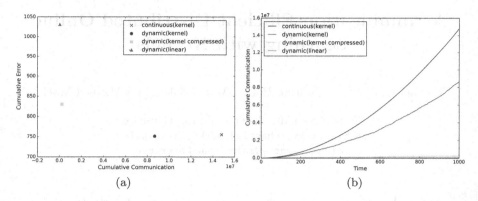

Fig. 1. (a) Trade-off between cumulative error and cumulative communication, and (b) cumulative communication over time of a distributed learning system using the proposed protocol. The learning task is classifying instances from the UCI SUSY dataset with 4 learners, each processing 1000 instances. Parameters of the learners are optimized on a separate set of 200 instances per learner.

by linear models. Thus, a more complex hypothesis class is desirable that enables the learners to achieve zero loss and thus reach quiescence.

Kernelized online learning algorithms can provide such an extended hypothesis class, but practical versions of these algorithms do not perform loss-proportional convex updates (e.g., [12,15,20]). Therefore, in this paper we extend the class of algorithms to *approximately* loss-proportional convex updates (Sect. 2). This relaxation is particularly crucial for kernelized online learners for streams that represent the model by its support vector expansion. These learners use this relaxation in order to reduce the number of support vectors, since otherwise a monotonically increasing model size would render them infeasible in streaming settings.

Also, for the first time we characterize the quality of the proposed protocol by introducing a novel criterion for efficient protocols that requires a strict loss bound and ties the loss to the allowed amount of communication. In particular, the criterion implies that the communication vanishes whenever the loss approaches zero. We bound the loss and communication of the proposed protocol and show for which class of learning algorithms it fulfills the efficiency criterion (Sect. 3). While the strict loss bound required in our criterion can be achieved by periodically communicating protocols [4,14], their communication never vanishes, independent of their loss, which is also required for efficiency. By communicating only when it significantly improves the service quality, our protocol achieves similar service quality as any periodically communicating protocol while communicating less by a factor depending on its in-place loss.

To further amplify this advantage, we apply methods from serial kernelized in-stream learning approaches. These approaches reduce the number of support vectors, e.g., by truncating individual support vectors with small weights [12], or by projecting a single support vector on the span of the remaining ones [15,20].

We illustrate the impact of the choice of the hypothesis class on the predictive performance and communication as well as the impact of model compression on an example dataset in Fig. 1. In this example, we predicted the class of instances drawn from the SUSY dataset from the UCI machine learning repository [13]. The learning systems using linear models continuously suffer loss resulting in a large cumulative error, but since the linear models are small compared to support vector expansions, the cumulative communication is small. A continuously synchronizing protocol using support vector expansions has a significantly smaller loss at the cost of very high communication, since each synchronization requires to send models with a growing number of support vectors. Using the proposed dynamic protocol, this amount of communication can be reduced without losing in prediction quality. In addition, when using model compression the communication can be further reduced to an amount similar to the linear model, but at the cost of prediction quality.

We further discuss the behavior of our protocol with respect to the trade-off between predictive performance and communication, and point out the strengths and weaknesses of the protocol in Sect. 4.

2   Distributed Online Learning with Kernels

In this section, we provide preliminaries and describe the protocol, extend it from linear function spaces to kernel Hilbert spaces, and provide an effectiveness criterion for distributed online learning. For that, we consider **distributed online learning protocols** $\Pi = (\mathcal{A}, \sigma)$ that run an online learning algorithm $\mathcal{A}$ on a distributed system of $m \in \mathbb{N}$ local learners and exchange information between these learners using a synchronization operator σ.

Preliminaries: The **online learning algorithm** $\mathcal{A} = (\mathcal{H}, \varphi, \ell)$ run at each **local learner** $i \in [m]$ maintains a **local model** $f^i \in \mathcal{H}$ from a function space $\mathcal{H}$ using an update rule φ and a loss function ℓ. That is, at each time point $t \in \mathbb{N}$, each learner i observes an individual input (x_t^i, y_t^i) drawn independently from a time-variant distribution $P_t \colon X \times Y \to [0,1]$ over an input space $X \times Y$. Based on this input and the local model, the local learner provides a service whose quality is measured by the **loss function** $\ell \colon \mathcal{H} \times X \times Y \to \mathbb{R}_+$. After providing the service, the local learner updates its local model using the **update rule** $\varphi \colon \mathcal{H} \times X \times Y \to \mathcal{H}$ in order to minimize the cumulative loss. The **synchronization operator** $\sigma \colon \mathcal{H}^m \to \mathcal{H}^m$ transfers the current **model configuration** $\mathbf{f} = (f^1, \dots, f^m)$ of m local models to the synchronized configuration $\sigma(\mathbf{f})$. In the following, we recapitulate the dynamic protocol presented in [10] as well as two baseline protocols, i.e., a continuously and a periodic protocol.

Given an online learning algorithm $\mathcal{A}$, the **periodic protocol** $\mathcal{P} = (\mathcal{A}, \sigma_b)$ synchronizes every $b \in \mathbb{N}$ time steps the current model configuration $\mathbf{f}$ by replacing all local models by their joint **average** $\bar{\mathbf{f}} = 1/m \sum_{i=1}^{m} f^i$. That is, the synchronization operator is given by

$$\sigma_b(\mathbf{f}_t) = \begin{cases} (\overline{\mathbf{f}}_t, \ldots, \overline{\mathbf{f}}_t), & \text{if } b \mid t \\ \mathbf{f}_t = (f_t^1, \ldots, f_t^m), & \text{otherwise} \end{cases}.$$

A special case of this is the **continuous protocol** $\mathcal{C} = (\mathcal{A}, \sigma_1)$ that continuously synchronizes every round, i.e., $\sigma_1(\mathbf{f}) = (\overline{\mathbf{f}}, \ldots, \overline{\mathbf{f}})$.

The **dynamic protocol** $\mathcal{D} = (\mathcal{A}, \sigma_\Delta)$ synchronizes the local learners using a **dynamic operator** σ_Δ [10]. This operator only communicates when the **model divergence**

$$\delta(\mathbf{f}) = \frac{1}{m} \sum_{i=1}^{m} \left\| f^i - \overline{\mathbf{f}} \right\|^2 \tag{1}$$

exceeds a **divergence threshold** Δ. That is, the dynamic averaging operator is defined as

$$\sigma_\Delta(\mathbf{f}_t) = \begin{cases} (\overline{\mathbf{f}}_t, \ldots, \overline{\mathbf{f}}_t), & \text{if } \delta(\mathbf{f}_t) > \Delta \\ \mathbf{f}_t, & \text{otherwise} \end{cases}.$$

In order to decide when to communicate, each local learner $i \in [m]$ monitors the **local condition** $\|f_t^i - r_t\|^2 \leq \Delta$ for a **reference model** $r_t \in \mathcal{H}$ that is common among all learners (see [6,7,11,19] for a more general description of this method). The local conditions guarantee that if none of them is violated, the divergence does not exceed the threshold Δ. The closer the reference model is to the true average of local models, the tighter are the local conditions. Generally, the first choice for the reference model is the average model from the last synchronization step. Note, however, that there are several refinements of this choice that can be used in practice to further reduce communication.

Efficiency Criterion: In the following, we introduce performance measures in order to analyze the dynamic protocol and compare it to the continuous and periodic protocols. We measure the predictive performance of a distributed online learning system until time $T \in \mathbb{N}$ by its cumulative loss

$$L(T, m) = \sum_{t=1}^{T} \sum_{i=1}^{m} \ell(f_t^i, (x_t^i, y_t^i)).$$

Performance guarantees are typically given by a **loss bound** $\mathbf{L}(T, m)$, i.e., for all possible input sequences it holds that $L(T, m) \leq \mathbf{L}(T, m)$. These bounds can be defined with respect to a sequence of reference models, in which case they are referred to as (shifting) **regret bounds**.

We measure its performance in terms of communication by its cumulative communication

$$C(T, m) = \sum_{t=1}^{T} c(\mathbf{f}_t),$$

where $c : \mathcal{H}^m \to \mathbb{N}$ measures the number of bytes required by the learning protocol to synchronize models $\mathbf{f}_t = (f_t^1, \ldots, f_t^m)$ at time t.

There is a natural trade-off between communication and loss of a distributed online learning system. On the one hand, a loss similar to a serial setting can be trivially achieved by continuous synchronization. On the other hand, communication can be entirely omitted. The trade-off for these two extreme protocols can be easily determined: if the cumulative loss of an online learning algorithm $\mathcal{A}$ is bounded by $\mathbf{L}_{\mathcal{A}}(T)$, the loss of a permanently centralizing system with m local learners running $\mathcal{A}$ is bounded by $\mathbf{L}_C(T, m) = \mathbf{L}_{\mathcal{A}}(mT)$, i.e., the loss bound of a serial online learning algorithm processing mT inputs. The protocol transmits $\mathcal{O}(m)$ messages of size up to $\mathcal{O}(T)$ in every of the T points in time. At the same time, the loss of a distributed system without any synchronization is bounded by $\mathbf{L}(T, m) = m\mathbf{L}_{\mathcal{A}}(T)$, whereas the communication is $C(T) = 0$.

The communication bound of an adaptive protocol should only depend on $\mathbf{L}_{\mathcal{A}}(T)$ and not on T, while at the same time retaining the loss bound of the serial setting. In the following definition we formalize this in order to provide a strong criterion for effectiveness of distributed online learning protocols.

Definition 1. *A distributed online learning protocol $\Pi = (\mathcal{A}, \sigma)$ processing mT inputs is **consistent** if it retains the loss bound of the serial online learning algorithm $\mathcal{A}$, i.e.,*

$$\mathbf{L}_\Pi(T, m) \in \mathcal{O}\left(\mathbf{L}_{\mathcal{A}}(mT)\right).$$

*The protocol is **adaptive** if its communication bound is linear in the number of local learners m and the loss bound $\mathbf{L}_{\mathcal{A}}(mT)$ of the serial online learning algorithm, i.e.,*

$$C_\Pi(T, m) \in \mathcal{O}\left(m\mathbf{L}_{\mathcal{A}}(mT)\right).$$

An **efficient** protocol is adaptive and consistent at the same time. In the following section we theoretically analyze the performance of the dynamic protocol with respect to this efficiency criterion.

Extension to Kernel Methods: The protocols presented above are defined for models from a Euclidean vector space. In this paper, we generalize $\mathcal{H}$ to be a **reproducing kernel Hilbert space** $\mathcal{H} = \{f : X \to \mathbb{R} | f(\cdot) = \sum_{j=1}^{\dim F} w_j \Phi_j(\cdot)\}$ with **kernel function** $k : X \times X \to \mathbb{R}$, **feature space** F, and a mapping $\Phi : X \to F$ into the feature space [18]. The kernel function corresponds to an inner product of input points mapped into feature space, i.e., $k(x, x') = \sum_{j=1}^{\dim F} \xi_j \Phi_j(x) \Phi_j(x')$ for constants $\xi_1, \xi_2, \cdots \in \mathbb{R}$. Thus, we can express the model in its **support vector expansion**, or dual representation, i.e., $f(\cdot) = \sum_{x \in S} \alpha_x k(x, \cdot)$ with a set of **support vectors** $S = \{x_1, \ldots, x_{|S|}\} \subset X$ and corresponding **coefficients** $\alpha_x \in \mathbb{R}$ for all $x \in S$. This implies that the linear weights $w = (w_1, w_2, \ldots) \in F$ defining f are given implicitly by $w_i = \sum_{x \in S} \xi_i \alpha_x \Phi_i(x)$. In order to apply the previously defined synchronization protocols to models from a reproducing kernel Hilbert space, we determine how to calculate the average of a model configuration and its divergence. For that, let $\mathbf{f} = (f^1, \ldots, f^m) \subset \mathcal{H}$ be a model configuration with corresponding weight vectors $(w^1, \ldots, w^m) \subset F$, where each model $i \in [m]$ has support vectors $S^i = \{x_1^i, \ldots, x_{|S^i|}^i\} \subset X$ and coefficients α_x^i for all $x \in S^i$. The average is given by

$$\bar{\mathbf{f}}(\cdot) = \frac{1}{m} \sum_{i=1}^{m} f^i(\cdot) = \frac{1}{m} \sum_{i=1}^{m} \sum_{j=1}^{\dim F} w_j^i \Phi_j(\cdot) = \frac{1}{m} \sum_{i=1}^{m} \sum_{j=1}^{\dim F} \sum_{x \in S^i} \xi_j \alpha_x^i \Phi_j(x) \Phi_j(\cdot).$$

We can simplify the above equation to $\bar{\mathbf{f}}(\cdot) = \frac{1}{m} \sum_{i=1}^{m} \sum_{x \in S^i} \alpha_x^i k(x, \cdot)$. By defining the union of support vectors $\overline{S} = \bigcup_{i \in [m]} S^i = \{s_1, \ldots, s_{|\overline{S}|}\}$ and augmented coefficients $\overline{\alpha}_s^i \in \mathbb{R}$, which are given by

$$\overline{\alpha}_s^i = \begin{cases} \alpha_x^i, & \text{if } x = s \\ 0, & \text{otherwise} \end{cases},$$

the dual representation of the average directly follows.

Proposition 2. *For a model configuration $\mathbf{f} = (f^1, \ldots, f^m) \subset \mathcal{H}$, where each model $i \in [m]$ has augmented coefficients $\overline{\alpha}_s^i$ for $s \in \overline{S}$, the average $\bar{\mathbf{f}} \in \mathcal{H}$ is given by*

$$\bar{\mathbf{f}}(\cdot) = \sum_{s \in \overline{S}} \left(\frac{1}{m} \sum_{i=1}^{m} \overline{\alpha}_s^i \right) k(s, \cdot),$$

with support vectors $\overline{S}$ and coefficients $\overline{\alpha}_s = 1/m \sum_{i=1}^{m} \overline{\alpha}_s^i$ for all $s \in \overline{S}$.

Using this definition of the average, we now define the distance between models in $\mathcal{H}$ and the divergence δ of a model configuration $\mathbf{f} \subset \mathcal{H}$. For an individual model f^i and the average $\bar{\mathbf{f}}$, the distance induced by the inner product of $\mathcal{H}$ is defined by $\|f^i - \bar{\mathbf{f}}\| = \langle f^i, f^i \rangle + \langle \bar{\mathbf{f}}, \bar{\mathbf{f}} \rangle - 2 \langle f^i, \bar{\mathbf{f}} \rangle$, i.e.,

$$\|f^i - \bar{\mathbf{f}}\| = \sum_{x \in S^i} \left(\alpha_x^i \right)^2 k(x, x) + \sum_{s \in \overline{S}} \left(\overline{\alpha}_s \right)^2 k(s, s) - 2 \sum_{x \in S^i} \sum_{s \in \overline{S}} \alpha_x^i \overline{\alpha}_s k(x, s).$$

Using this distance, we can compute the divergence (Eq. 1) for models from a reproducing kernel Hilbert space.

3   Performance Guarantees

In order to determine the performance of the dynamic protocol, we start by extending the definition of loss-proportional convex update rules. This allows us to bound the loss for kernelized online learning algorithms that reduce their model size using a compression step.

Let $\varphi \colon \mathcal{H} \times X \times Y \to \mathcal{H}$ be a loss-proportional convex update rule, then $\widetilde{\varphi}$ is an **approximately loss-proportional convex update rule** if for all $f \in \mathcal{H}$, $x \in X$, and $y \in Y$ it holds that $\|\widetilde{\varphi}(f, x, y) - \varphi(f, x, y)\| \leq \epsilon$. With this, we can bound the distance between two models after the approximate update step.

Lemma 3. *For two models $f, g \in \mathcal{H}$ and an approximately loss-proportional convex update rule $\widetilde{\varphi}$, with $\|\widetilde{\varphi}(f, x, y) - \varphi(f, x, y)\| \leq \epsilon$ for the corresponding loss-proportional convex update rule φ, it holds that*

$$\|\widetilde{\varphi}(f, x, y) - \widetilde{\varphi}(g, x, y)\|^2 \leq \|f - g\|^2 - \gamma^2 \left(\ell(f, x, y) - \ell(g, x, y)\right)^2 + 2\epsilon^2.$$

Proof. We abbreviate $\varphi(f, x, y)$ as $\varphi(f)$. Then $\|\widetilde{\varphi}(f) - \varphi(f)\| \leq \epsilon$ implies for $f, g \in \mathcal{H}$ that $\|\widetilde{\varphi}(f) - \widetilde{\varphi}(g)\|^2 \leq \|\varphi(f) - \varphi(g)\|^2 + 2\epsilon^2$. Together with the result from Lemma 4 in [10], i.e., $\|\varphi(f) - \varphi(g)\|^2 \leq \|f - g\|^2 - \gamma^2 \left(\ell(f) - \ell(g)\right)^2$, follows the result. □

Using Lemma 3, we can bound the loss of our protocol.

Theorem 4. *Let $\mathcal{A}$ be an online learning algorithm with γ-loss-proportional convex update rule φ. Let $\mathbf{d}_1, \dots \mathbf{d}_T$ and $\mathbf{p}_1, \dots, \mathbf{p}_T$ be two sequences of model configurations such that $\mathbf{d}_1 = \mathbf{p}_1$ and the first sequence is maintained by the dynamic protocol $\mathcal{D} = (\mathcal{A}, \sigma_\Delta)$ and the second by the periodic protocol $\mathcal{P} = (\mathcal{A}, \sigma_b)$. That is, for $t = 1, \dots, T$ the sequence is defined by $\mathbf{d}_{t+1} = \sigma_\Delta \left(\varphi(\mathbf{d}_t)\right)$, and $\mathbf{p}_{t+1} = \sigma_b \left(\varphi(\mathbf{p}_t)\right)$ respectively. Then it holds that*

$$L_\mathcal{D}(T, m) \leq L_\mathcal{P}(T, m) + \frac{T}{\gamma^2}(\Delta + 2\epsilon^2).$$

Proof. First note that for simplicity we abbreviate $\ell(f_t, x_t, y_t)$ by $\ell(f_t)$. We combine our Lemma 3 with Lemma 3 from [10] which states that

$$\frac{1}{m} \sum_{i=1}^{m} \|\sigma_\Delta(\mathbf{d})^i - \sigma_b(\mathbf{p})^i\|^2 \leq \frac{1}{m} \sum_{i=1}^{m} \|d^i - p^i\|^2 + \Delta.$$

This yields for all $t \in [T]$ that

$$\sum_{i=1}^{m} \left\|d_{t+1}^i - p_{t+1}^i\right\|^2 \leq \sum_{i=1}^{m} \left\|d_t^i - p_t^i\right\|^2 - \gamma^2 \sum_{i=1}^{m} \left(\ell(d_t^i) - \ell(p_t^i)\right)^2 + \Delta + 2\epsilon^2.$$

By applying this inequality recursively for $t = 1, \dots, T$ it follows that

$$\sum_{i=1}^{m} \left\|d_{t+1}^i - p_{t+1}^i\right\|^2 \leq \sum_{i=1}^{m} \left\|d_1^i - p_1^i\right\|^2 + T(\Delta + 2\epsilon^2) - \gamma^2 \sum_{t=1}^{T} \sum_{i=1}^{m} \left(\ell(d_t^i) - \ell(p_t^i)\right)^2.$$

Using $\mathbf{d}_1 = \mathbf{p}_1$, we conclude that

$$\sum_{t=1}^{T} \sum_{i=1}^{m} \left(\ell(d_t^i) - \ell(p_t^i)\right)^2 \leq \frac{1}{\gamma^2} \left(T(\Delta + 2\epsilon^2) - \sum_{i=1}^{m} \left\|d_{t+1}^i - p_{t+1}^i\right\|^2\right) \leq \frac{1}{\gamma^2} T\Delta$$

$$\Leftrightarrow L_\mathcal{D}(T)^m - L_\mathcal{P}(T)^m \leq \frac{1}{\gamma^2} T(\Delta + 2\epsilon^2) \qquad \square$$

By setting the communication period $b = 1$, this result also holds for the continuous protocol $\mathcal{C}$.

The result of Theorem 4 is similar to the original loss bound of the dynamic protocol but also accounts for the inaccuracy of the update rule, e.g., because of model compression. We can apply the original consistency result: if the continuous protocol is consistent, then the dynamic protocol is consistent as well. For Stochastic Gradient Descent it has been shown that the dynamic protocol is consistent for linear models [10]. From Theorem 4 follows that the dynamic protocol remains consistent for approximately loss-proportional update rules. Note that for static target distributions, consistency can be achieved by a decreasing divergence threshold and compression error, i.e., $\Delta_t = t^{-1/2}$ and $\epsilon = t^{-1/4}$.

We now provide communication bounds for the dynamic protocol. For that, assume that the m learners maintain models in their support vector expansion. Let $S_t^i \subset \mathbb{R}^d$ denote the set of support vectors of learner $i \in [m]$ at time t and α_t^i the corresponding coefficients. Let $B_x \in \mathcal{O}(d)$ be the number of bytes required to transmit one support vector and $B_\alpha \in \mathcal{O}(1)$ be the number of bytes required for the corresponding weight. Furthermore, let $I : \mathbb{N} \times [m] \to \{0,1\}$ be an indicator function that is 1 if for learner i at time t a new support vector has been added during the update.

We assume that a designated coordinator node performs the synchronizations, i.e., all local learners transmit their models to the coordinator which in turn sends the synchronized model back to each learner. Furthermore, we assume that all protocols apply the following trivial communication reduction strategy. Let t' be the time of last synchronization. Assume the coordinator stored the support vectors of the last average model $\overline{S}_{t'}$. Whenever a learner i has to send its model to the coordinator, it sends all support vector coefficients α but only the new support vectors, i.e., only $S_t^i \setminus S_{t'}^i$. This avoids redundant communication at the cost of higher memory usage at the coordinator side. In turn, after averaging the models, the coordinator sends to learner i all support vector coefficients, but only the support vectors $\overline{S}_t \setminus S_t^i$.

We start by bounding the communication of a continuous protocol $\mathcal{C}$, i.e., one that transmits all models from each learner in each round. The trivial communication reduction technique discussed above implies that in each round, a learner transmits its full set of support vector coefficients and potentially one support vector—depending on whether a new support vector was added in this round. Thus, at time t learner i submits

$$|S_t^i| B_\alpha + I(t, i) B_x \tag{2}$$

bytes to the coordinator. The coordinator transmits to learner $i \in m$ all support vector coefficients of the average model and all its support vectors, except the support vectors S_t^i of the local model at learner i. Thus, it transmits the following amount of bytes.

$$\left|\overline{S}_t\right| B_\alpha + \left|\overline{S}_t \setminus S_t^i\right| B_x = \left|\bigcup_{j=1}^m S_t^j\right| B_\alpha + \left|\bigcup_{j=1}^m S_t^j \setminus S_t^i\right| B_x. \tag{3}$$

With this we can derive the following communication bound.

Proposition 5. *The communication of the continuous protocol $\mathcal{C}$ on $m \in \mathbb{N}$ learners until time $T \in \mathbb{N}$ is bound by*

$$C_\mathcal{C}(T, m) \leq Tm2|\overline{S}_T|B_\alpha + m|\overline{S}_T|B_x \leq m^2T^2B_\alpha + m^2TB_x \in \mathcal{O}\left(m^2T^2\right).$$

Proof. The constantly synchronizing protocol transmits at each time step from each learner a set of support vector coefficients and potentially one support vector to the coordinator. The amount of bytes is given in Eq. 2. The coordinator transmits the averaged model back to each learner with an amount of bytes as given in Eq. 3. Summing up the communication over $T \in \mathbb{N}$ time points and m learners yields

$$C_\mathcal{C}(T, m) = \sum_{t=1}^{T} \sum_{i=1}^{m} \left(|S_t^i|B_\alpha + I(t, i)B_x + \left|\bigcup_{j=1}^{m} S_t^j\right| B_\alpha + \left|\bigcup_{j=1}^{m} S_t^j \setminus S_t^i\right| B_x \right)$$

$$= \sum_{t=1}^{T} \sum_{i=1}^{m} \left(|S_t^i|B_\alpha + |\overline{S}_t| B_\alpha + I(t, i)B_x + |\overline{S}_t \setminus S_t^i| B_x \right).$$

We analyze this sum separately in terms of bytes required for sending the support vectors and bytes for sending the coefficients. The amount of bytes for sending the support vectors is bounded by $m|S_T^i|B_x$, as we show in the following.

$$\sum_{t=1}^{T} \sum_{i=1}^{m} I(t, i)B_x + |\overline{S}_t \setminus S_t^i| B_x = \underbrace{\sum_{t=1}^{T} \sum_{i=1}^{m} I(t, i)B_x}_{=|\overline{S}_T|B_x} + \sum_{t=1}^{T} \sum_{i=1}^{m} |\overline{S}_t \setminus S_t^i| B_x$$

$$= |\overline{S}_T|B_x + \sum_{t=1}^{T} \sum_{i=1}^{m} \left| \left(\bigcup_{j=1}^{m} S_t^j \setminus \bigcup_{j=1}^{m} S_{t-1}^j \right) \setminus (S_t^i \setminus \overline{S}_{t-1}) \right| B_x$$

$$\leq |\overline{S}_T|B_x + \sum_{t=1}^{T} \sum_{i=1}^{m} \sum_{\substack{j=1 \\ j \neq i}}^{m} I(t, i)B_x \leq |\overline{S}_T|B_x + \sum_{t=1}^{T} \sum_{i=1}^{m} (m-1)I(t, i)B_x$$

$$\leq |\overline{S}_T|B_x + (m-1)|\overline{S}_T|B_x = m|\overline{S}_T|B_x.$$

We now bound the amount of bytes required for sending the support vector coefficients.

$$\sum_{t=1}^{T} \sum_{i=1}^{m} \underbrace{|S_t^i|}_{\leq|\overline{S}_T|} B_\alpha + \underbrace{|\overline{S}_t|}_{\leq|\overline{S}_T|} B_\alpha \leq \sum_{t=1}^{T} \sum_{i=1}^{m} 2|\overline{S}_T|B_\alpha = Tm2|\overline{S}_T|B_\alpha.$$

From $|\overline{S}_T| \leq mT$ and the fact that we regard $B_\alpha \in \mathcal{O}(1)$ and $B_x \in \mathcal{O}(d)$ as constants we can follow that

$$C_\mathcal{C}(T, m) \leq 2Tm|\overline{S}_T|B_\alpha + m|\overline{S}_T|B_x \leq m^2T^2B_\alpha + m^2TB_x \in \mathcal{O}\left(m^2T^2\right). \quad \square$$

Note that this communication bound implies that—unlike for linear models—synchronizing models in their support vector expansion requires even more communication than centralizing the input data. However, in real-time prediction applications, the latency induced by central computation can exceed the time constraints, rendering continuous synchronization a viable approach nonetheless.

Similarly, the communication of a periodic protocol $\mathcal{P}$ that communicates every $b \in \mathbb{N}$ steps (b is often referred to as mini-batch size) can be bounded by

$$C_{\mathcal{P}}(T,m) \leq \frac{T}{b} 2m|\overline{S}_T|B_\alpha + m|\overline{S}_T|B_x \leq \frac{T}{b} m^2 T B_\alpha + m^2 T B_x \in \mathcal{O}\left(\frac{1}{b} m^2 T^2\right).$$

We now for the first time provide a communication bound for the dynamic protocol $\mathcal{D}$. For that, we first bound the number of synchronization steps and then analyze the amount of communication per synchronization.

Proposition 6. *Let $\mathcal{A} = (\mathcal{H}, \widetilde{\varphi}, \ell)$ be an online learning algorithm with an approximately loss-proportional convex update rule $\widetilde{\varphi}$ for which holds that $\|f - \widetilde{\varphi}(f,x,y)\| \leq \eta \ell(f,x,y)$. The number of synchronizations $V_{\mathcal{D}}(T)$ of the dynamic protocol $\mathcal{D}$ running $\mathcal{A}$ in parallel on m nodes until time $T \in \mathbb{N}$ with divergence threshold Δ is bounded by*

$$V_{\mathcal{D}}(T) \leq \frac{\eta}{\sqrt{\Delta}} L_{\mathcal{D}}(T,m).$$

where $L_{\mathcal{D}}(T,m)$ denotes the cumulative loss of $\mathcal{D}$.

Proof. For this proof, we abbreviate $\ell(f_t^i, x_t^i, y_t^i)$ as $\ell(f_t^i)$ and $\widetilde{\varphi}(f_t^i, x_t^i, y_t^i)$ as $\widetilde{\varphi}(f_t^i)$. The dynamic protocol synchronizes if a local condition $\|f_t^i - r_t\|^2 \leq \Delta$ is violated. Now assume that at $t = 1$ all models are initialized with $f_1^1 = \cdots = f_1^m$ and $r_1 = \overline{f}_1$, i.e., for all local learners i it holds that $\|f_1^i - r_1\| = 0$. A violation, i.e., $\|f_t^i - r_t\| > \sqrt{\Delta}$, occurs if one local model drifts away from r_t by more than $\sqrt{\Delta}$. After a violation, a synchronization is performed and $r_t = \overline{f}_t$, hence $\|f_t^i - r_t\| = 0$ and the situation is again similar to the initial setup for $t = 1$. In the worst case, a local learner drifts continuously in one direction until a violation occurs. Hence, we can bound the number of violations $V_i(T)$ at a single learner i by the sum of its drifts divided by $\sqrt{\Delta}$:

$$V_i(T) \leq \frac{1}{\sqrt{\Delta}} \sum_{t=1}^{T} \|f_t^i - f_{t+1}^i\| = \frac{1}{\sqrt{\Delta}} \sum_{t=1}^{T} \underbrace{\|f_t^i - \widetilde{\varphi}(f_t^i)\|}_{\leq \eta \ell(f_t^i)} \leq \frac{1}{\sqrt{\Delta}} \sum_{t=1}^{T} \eta \ell(f_t^i).$$

With this, we can bound the amount of points in time $t \in [T]$ where at least one learner l has a violation, i.e., $V(T)$. In the worst case, all violations at all local learners occur at different time points, so that we can upper bound $V(T)$ by the sum of local violations $V_i(T)$ which is again upper bounded by the cumulative sum of drifts of all local models:

$$V(T) \leq \sum_{i=1}^{m} V_i(T) \leq \frac{1}{\sqrt{\Delta}} \sum_{t=1}^{T} \sum_{i=1}^{m} \eta \ell(f_t^i) = \frac{\eta}{\sqrt{\Delta}} L_{\mathcal{D}}(T,m). \qquad \square$$

In the following theorem we bound the overall communication by combining this bound on the number of synchronizations with an analysis of the amount of bytes transfered per synchronization.

Theorem 7. *Let $\mathcal{A} = (\mathcal{H}, \widetilde{\varphi}, \ell)$ be an online learning algorithm with approximately loss-proportional update rule $\widetilde{\varphi}$ and $\|f - \widetilde{\varphi}(f, x, y)\| \leq \eta \ell(f, x, y)$. The amount of communication $C_{\mathcal{D}}(T, m)$ of the dynamic protocol $\mathcal{D}$ running $\mathcal{A}$ in parallel on m nodes until time $T \in \mathbb{N}$ with divergence threshold Δ is bounded by*

$$C_{\mathcal{D}}(T, m) \leq \frac{\eta}{\sqrt{\Delta}} L_{\mathcal{D}}(T, m) \left(2m \left|\overline{S}_T\right| B_\alpha\right) + m \left|\overline{S}_T\right| B_x$$

Proof. Assume that at time T, the dynamic protocol performs a synchronization. Then, similar to the argument for the continuous protocol, the support vector set at time T is similar for all learners and independent of the number of synchronization steps before. In particular, it is the same if a synchronization was performed in every time step. Thus, again the amount of bytes required for sending the support vectors is bounded by $m \left|\overline{S}_T\right| B_x$. Let $\theta: \mathbb{N} \to \{0, 1\}$ be an indicator function such that $\theta(t) = 1$ if at time t the dynamic protocol performed a synchronization and $\theta(t) = 0$ otherwise. Then, the amount of bytes required to send all the support vector coefficients until time T is

$$\underbrace{\sum_{t=1}^{T} \theta(t) \sum_{i=1}^{m} \left(\left|S_t^i\right| + \left|\overline{S}_t\right|\right) B_\alpha}_{=V_{\mathcal{D}}(T)} \leq \sum_{t=1}^{T} \theta(t) \sum_{l=1}^{m} 2|\overline{S}_T| B_\alpha \leq \underbrace{\frac{\eta}{\sqrt{\Delta}} L_{\mathcal{D}}(T, m) \left(2m|\overline{S}_T| B_\alpha\right)}_{\text{Proposition 6}}$$

Together with the amount of bytes required for exchanging all support vectors this yields $C_{\mathcal{D}}(T, m) \leq \frac{\eta}{\sqrt{\Delta}} L_{\mathcal{D}}(T, m) \left(2m|\overline{S}_T| B_\alpha\right) + m \left|\overline{S}_T\right| B_x$. □

Note that the loss bounds for online learning algorithms are typically sublinear in T, e.g., optimal regret bounds for static target distributions are in $\mathcal{O}(\sqrt{T})$. In these cases, the dynamic protocol has an amount of communication in $\mathcal{O}(m^2 T \sqrt{T})$ which is smaller than $\mathcal{O}(m^2 T^2)$ of the continuously and periodic protocols by a factor of $\sqrt{T}$.

In the original case of linear models instead, the dynamic protocol only transmits m weight vectors of fixed size per synchronization. In this case the amount of communication per synchronization is bounded by a constant. If for an online learning algorithm $\mathcal{A}$ and the periodic protocol $\mathcal{P}$ it holds that $\mathbf{L}_{\mathcal{P}}(T, m) \leq \mathbf{L}_{\mathcal{A}}(mT)$, then by Theorem 4 it also holds that $\mathbf{L}_{\mathcal{D}}(T, m) \leq \mathbf{L}_{\mathcal{A}}(mT)$. This implies that the dynamic protocol is adaptive. In the following corollary, we show that for linear models, the dynamic protocol is adaptive when using the Stochastic Gradient Descent algorithm.

Corollary 8. *The dynamic protocol $\mathcal{D} = (SGD, \sigma_\Delta)$ using Stochastic Gradient Descent SGD with linear models is adaptive, i.e.,*

$$C_{\mathcal{D}}(T, m) \in \mathcal{O}\left(m\mathbf{L}_{SGD}(mT)\right)$$

Proof. The amount of synchronizations of the dynamic protocol is bounded by $V(T)$ (see Proposition 6). In each synchronization, each learner transmits one linear model, i.e., one weight vector of fixed size to the coordinator. The coordinator submits one averaged weight vector back to each learner. Thus, the amount of communication per synchronization is bounded by $\mathbf{c}_m \in \mathbb{N}$, where $\mathbf{c}_m \in \mathcal{O}(m)$. Then, the total communication is bounded by

$$C_{\mathcal{D}}(T, m) \leq \mathbf{c}_m \frac{\eta}{\sqrt{\Delta}} L_{\mathcal{D}}(T, m) \in \mathcal{O}\left(m L_{\mathcal{D}}(T, m)\right).$$

The dynamic protocol retains the loss bound of Stochastic Gradient Descent [10], i.e., $L_{\mathcal{D}}(T, m) \leq \mathbf{L}_{\text{SGD}}(mT)$. □

Unfortunately, from Theorem 4 also follows that the dynamic protocol applied to kernelized online learning algorithms that do not bound the size of their models does not comply to the strict notion of adaptivity as given in Definition 1. That is, because the model size and thus the size of each message to and from the coordinator can grow with T. Nonetheless, the theorem guarantees that if the learners do not suffer loss anymore, the dynamic protocol reaches quiescence.

In order to make the dynamic protocol adaptive in the strict sense of Definition 1, the model size has to be bounded. For kernelized online learning in streams, several **model compression** techniques have been proposed [12,15,20]. These techniques typically guarantee that the compression error is bounded, i.e., for the **compressed model** $\tilde{f}$ it holds that $\left\| f - \tilde{f} \right\| \leq \epsilon$. From this directly follows that if the base algorithm uses a loss-proportional convex update rule φ, the compressed version is an approximately loss-proportional convex update rule $\tilde{\varphi}$.

One approach to compressing the support vector expansion is to project a new support vector on the span of the remaining ones and thus avoid adding it to the support set. Another one is to truncate support vectors with small coefficients. For the projection approach (e.g., described in [15]) the error bound is independent of the learning algorithm. However, there is no bound on the number of support vectors. Thus, even though the model size is reduced in practice, there is no formal bound on the model size. For the truncation approach, however, [12] have shown that an error bound as well as a bound on the number of support vectors can be achieved when using Stochastic Gradient Descent. Specifically, for a fixed model size of τ support vectors, they have shown that the compression error is bound by $\left\| f - \tilde{f} \right\| \leq \epsilon \in \mathcal{O}\left(\frac{1}{\lambda}(1 - \lambda)^\tau\right)$, where $\lambda \in \mathbb{R}$ is the learning rate of the Stochastic Gradient Descent algorithm (SGD). Therefore, we can follow that the dynamic protocol with SGD using kernel models compressed by truncation is adaptive. Specifically for SGD, [4] have shown that periodic synchronizations retain the serial loss bound of SGD. It is consistent in this setting, because the dynamic protocol in turn retains the loss bounds of any periodic protocol. Since it is both consistent and adaptive, the dynamic protocol is efficient.

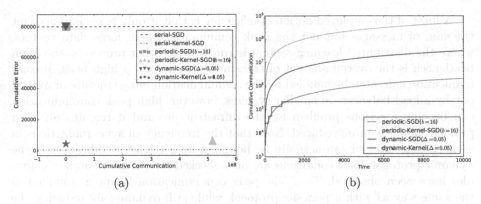

Fig. 2. (a) Trade-off between cumulative error and cumulative communication and (b) cumulative communication over time of the dynamic protocol versus a periodic protocol. 32 learners perform a stock price prediction task using SGD (learning rate η and regularization parameter λ optimized over 200 instances, with $\eta = 10^{-10}$, $\lambda = 1.0$ for the periodic protocol, and $\eta = 1.0$, $\lambda = 0.01$ for the dynamic protocol) updates, either with linear models or with non-linear models (Gaussian kernel with number of support vectors limited to 50 using the truncation approach of [12]).

4   Discussion

The dynamic protocol, extended to kernel methods, yields for the first time a theoretically efficient tool to learn non-linear models for distributed real-time services, in settings where communication is a major bottleneck. For that, it can employ online kernel methods together with model compression techniques, which reduce, or bound the number of support vectors. The efficiency of the protocol is characterized by a novel criterion that ties a tight loss bound to the required amount of communication—a criterion which is not satisfied by the state of the art of periodically communicating protocols.

While we provided a theoretical analysis, the advantage of the dynamic protocol in combination with kernel methods can also be shown in practice: Fig. 2 shows the results of an experiment on financial data [9], where 32 learners predicted the stock price of a target stock. We can see that for this difficult learning task linear models perform poorly compared to non-linear models using a Gaussian kernel function. Simultaneously, the communication required to periodically synchronize these non-linear models is larger than for linear models by more than two orders of magnitude. Using the dynamic protocol with kernel models we could reduce the error by an order of magnitude compared to using linear models (a reduction by a factor of 18). At the same time, the communication is reduced by more than three orders of magnitude compared to the static protocol (by a factor of 2433), which is yet an order of magnitude smaller than the communication when using linear models (by a factor of 10). Moreover, within less than 2000 rounds, the dynamic protocol reaches quiescence, as it is implied by the efficiency criterion.

A limit of the employed notion of efficiency is that it only takes into account the sum of messages but not the peak communication. In large data centers, where the distributed learning system is run next to other processes, the main bottleneck is the overall amount of transmitted bytes and a high peak in communication can often be handled by the communication infrastructure or evened out by a load balancer. In smaller systems, however, high peak communication can become a serious problem for the infrastructure and it remains an open problem how it can be reduced. Note that the frequency of synchronizations in a short time interval can actually be bounded by a trivial modification of the dynamic protocol: local conditions are only checked after a mini-batch of examples have been observed. Thus, the peak communication is upper bounded in the same way as with a periodic protocol, while still dynamically reducing the overall amount of communication.

When analyzing the reason for practical efficiency, model compression has proven to be a crucial factor, since storing and evaluating models with large numbers of support vectors can become infeasible—even in serial settings. In a distributed setting, transmitting large models furthermore induce high communication costs, which is aggravated by averaging local models, because the synchronized model consists of the union of all local support vectors. For the model truncation approach of [12], we have shown that the efficiency criterion is satisfied, but other model compression approaches might be favorable in certain scenarios. Thus, an interesting direction for future research is to study the relationship between loss and model size of those model compression techniques in order to extend the results on efficiency.

Also, alternative approaches to ensuring constant model size could be investigated, e.g., a finite dimensional approximation of the feature map $\Phi \colon X \to \mathcal{H}$ of a reproducing kernel Hilbert space $\mathcal{H}$, such as Random Fourier Features [17]. It remains an open problem how tight loss bounds combined with communication bounds can be derived in these settings.

Finding the right divergence threshold for the dynamic protocol, i.e., one that suits the desired trade-off between service quality and communication, is in practice a neither intuitive nor trivial task. The threshold can be selected using a small data sample, but the communication for a given threshold can vary over time and is also influenced by other parameters of the learner. Thus, another direction for future research is to investigate an adaptive divergence threshold. This could allow for a more direct selection of the desired trade-off between service quality (i.e., predictive performance) and communication.

Acknowledgments. This research has been supported by the EU FP7-ICT-2013-11 under grant 619491 (FERARI).

References

1. Barroso, L.A., Clidaras, J., Hölzle, U.: The datacenter as a computer: an introduction to the design of warehouse-scale machines. Synth. Lect. Comput. Archit. **8**(3), 1–154 (2013)

2. Boyd, S., Vandenberghe, L.: Convex Optimization. Cambridge University Press, Cambridge (2004)
3. Crammer, K., Dekel, O., Keshet, J., Shalev-Shwartz, S., Singer, Y.: Online passive-aggressive algorithms. J. Mach. Learn. Res. **7**, 551–585 (2006)
4. Dekel, O., Gilad-Bachrach, R., Shamir, O., Xiao, L.: Optimal distributed online prediction using mini-batches. J. Mach. Learn. Res. **13**, 165–202 (2012)
5. Deligiannakis, A., Kotidis, Y., Roussopoulos, N.: Bandwidth-constrained queries in sensor networks. VLDB J. **17**(3), 443–467 (2008)
6. Gabel, M., Keren, D., Schuster, A.: Communication-efficient distributed variance monitoring and outlier detection for multivariate time series. In: Proceedings of the 28th International Parallel and Distributed Processing Symposium (IPDPS), pp. 37–47. IEEE (2014)
7. Giatrakos, N., Deligiannakis, A., Garofalakis, M., Sharfman, I., Schuster, A.: Prediction-based geometric monitoring over distributed data streams. In: Proceedings of the 2012 ACM SIGMOD International Conference on Management of Data, pp. 265–276 (2012)
8. Heires, K.: Budgeting for latency: if i shave a microsecond, will i see a 10x profit. Securities Ind. News **22**(1), 4–5 (2010)
9. Kamp, M., Boley, M., Gärtner, T.: Beating human analysts in nowcasting corporate earnings by using publicly available stock price and correlation features. In: 2013 IEEE 13th International Conference on Data Mining Workshops, pp. 384–390 (2013)
10. Kamp, M., Boley, M., Keren, D., Schuster, A., Sharfman, I.: Communication-efficient distributed online prediction by dynamic model synchronization. In: Calders, T., Esposito, F., Hüllermeier, E., Meo, R. (eds.) ECML PKDD 2014, Part I. LNCS(LNAI), vol. 8724, pp. 623–639. Springer, Heidelberg (2014). doi:10.1007/978-3-662-44848-9_40
11. Keren, D., Sharfman, I., Schuster, A., Livne, A.: Shape sensitive geometric monitoring. IEEE Trans. Knowl. Data Eng. **24**(8), 1520–1535 (2012)
12. Kivinen, J., Smola, A.J., Williamson, R.C.: Online learning with kernels. IEEE Trans. Signal Process. **52**(8), 2165–2176 (2004)
13. Lichman, M.: UCI machine learning repository (2013). http://archive.ics.uci.edu/ml
14. Mcdonald, R., Mohri, M., Silberman, N., Walker, D., Mann, G.S.: Efficient large-scale distributed training of conditional maximum entropy models. In: Advances in Neural Information Processing Systems, pp. 1231–1239 (2009)
15. Orabona, F., Keshet, J., Caputo, B.: Bounded kernel-based online learning. J. Mach. Learn. Res. **10**, 2643–2666 (2009)
16. Predd, J.B., Kulkarni, S.R., Poor, H.V.: Distributed Learning in Wireless Sensor Networks. John Wiley & Sons, Chichester (2007)
17. Rahimi, A., Recht, B.: Random features for large-scale kernel machines. In: Advances in Neural Information Processing Systems (NIPS), pp. 1177–1184 (2007)
18. Schölkopf, B., Smola, A.J.: Learning with Kernels: Support Vector Machines, Regularization, Optimization, and Beyond. MIT Press, Cambridge (2001)
19. Sharfman, I., Schuster, A., Keren, D.: A geometric approach to monitoring threshold functions over distributed data streams. Trans. Database Syst. (TODS) **32**(4), 23 (2007)
20. Wang, Z., Vucetic, S.: Online passive-aggressive algorithms on a budget. In: International Conference on Artificial Intelligence and Statistics, pp. 908–915 (2010)
21. Yuan, S., Wang, J., Zhao, X.: Real-time bidding for online advertising: measurement and analysis. In: Proceedings of the Seventh International Workshop on Data Mining for Online Advertising, p. 3 (2013)

Author Index

Printed in the United States
By Bookmasters